Giuseppe Di Bucchianico
Editor

Advances in Design for Inclusion

Proceedings of the AHFE 2019 International Conference on Design for Inclusion and the AHFE 2019 International Conference on Human Factors for Apparel and Textile Engineering, July 24–28, 2019, Washington D.C., USA

Springer

Editor
Giuseppe Di Bucchianico
Department of Architecture
University of Chieti-Pescara
Pescara, Pescara, Italy

ISSN 2194-5357 ISSN 2194-5365 (electronic)
Advances in Intelligent Systems and Computing
ISBN 978-3-030-20443-3 ISBN 978-3-030-20444-0 (eBook)
https://doi.org/10.1007/978-3-030-20444-0

This Springer imprint is published by the registered company Springer Nature Switzerland AG
The registered company address is: Gewerbestrasse 11, 6330 Cham, Switzerland

Advances in Intelligent Systems and Computing

Volume 954

The series "Advances in Intelligent Systems and Computing" contains publications on theory, applications, and design methods of Intelligent Systems and Intelligent Computing. Virtually all disciplines such as engineering, natural sciences, computer and information science, ICT, economics, business, e-commerce, environment, healthcare, life science are covered. The list of topics spans all the areas of modern intelligent systems and computing such as: computational intelligence, soft computing including neural networks, fuzzy systems, evolutionary computing and the fusion of these paradigms, social intelligence, ambient intelligence, computational neuroscience, artificial life, virtual worlds and society, cognitive science and systems, Perception and Vision, DNA and immune based systems, self-organizing and adaptive systems, e-Learning and teaching, human-centered and human-centric computing, recommender systems, intelligent control, robotics and mechatronics including human-machine teaming, knowledge-based paradigms, learning paradigms, machine ethics, intelligent data analysis, knowledge management, intelligent agents, intelligent decision making and support, intelligent network security, trust management, interactive entertainment, Web intelligence and multimedia.

The publications within "Advances in Intelligent Systems and Computing" are primarily proceedings of important conferences, symposia and congresses. They cover significant recent developments in the field, both of a foundational and applicable character. An important characteristic feature of the series is the short publication time and world-wide distribution. This permits a rapid and broad dissemination of research results.

**** Indexing: The books of this series are submitted to ISI Proceedings, EI-Compendex, DBLP, SCOPUS, Google Scholar and Springerlink ****

More information about this series at http://www.springer.com/series/11156

Advances in Human Factors and Ergonomics 2019

AHFE
International
Applied
Human Factors
and Ergonomics
Conference
2019
www.ahfe.org

AHFE 2019 Series Editors

Tareq Ahram, Florida, USA
Waldemar Karwowski, Florida, USA

10th International Conference on Applied Human Factors and Ergonomics and the Affiliated Conferences

Proceedings of the AHFE 2019 International Conference on Design for Inclusion and the AHFE 2019 International Conference on Human Factors for Apparel and Textile Engineering, held on July 24–28, 2019, in Washington D.C., USA

Advances in Affective and Pleasurable Design	Shuichi Fukuda
Advances in Neuroergonomics and Cognitive Engineering	Hasan Ayaz
Advances in Design for Inclusion	Giuseppe Di Bucchianico
Advances in Ergonomics in Design	Francisco Rebelo and Marcelo M. Soares
Advances in Human Error, Reliability, Resilience, and Performance	Ronald L. Boring
Advances in Human Factors and Ergonomics in Healthcare and Medical Devices	Nancy J. Lightner and Jay Kalra
Advances in Human Factors and Simulation	Daniel N. Cassenti
Advances in Human Factors and Systems Interaction	Isabel L. Nunes
Advances in Human Factors in Cybersecurity	Tareq Ahram and Waldemar Karwowski
Advances in Human Factors, Business Management and Leadership	Jussi Ilari Kantola and Salman Nazir
Advances in Human Factors in Robots and Unmanned Systems	Jessie Chen
Advances in Human Factors in Training, Education, and Learning Sciences	Waldemar Karwowski, Tareq Ahram and Salman Nazir
Advances in Human Factors of Transportation	Neville Stanton

(continued)

(continued)

Advances in Artificial Intelligence, Software and Systems Engineering	Tareq Ahram
Advances in Human Factors in Architecture, Sustainable Urban Planning and Infrastructure	Jerzy Charytonowicz and Christianne Falcão
Advances in Physical Ergonomics and Human Factors	Ravindra S. Goonetilleke and Waldemar Karwowski
Advances in Interdisciplinary Practice in Industrial Design	Cliff Sungsoo Shin
Advances in Safety Management and Human Factors	Pedro M. Arezes
Advances in Social and Occupational Ergonomics	Richard H. M. Goossens and Atsuo Murata
Advances in Manufacturing, Production Management and Process Control	Waldemar Karwowski, Stefan Trzcielinski and Beata Mrugalska
Advances in Usability and User Experience	Tareq Ahram and Christianne Falcão
Advances in Human Factors in Wearable Technologies and Game Design	Tareq Ahram
Advances in Human Factors in Communication of Design	Amic G. Ho
Advances in Additive Manufacturing, Modeling Systems and 3D Prototyping	Massimo Di Nicolantonio, Emilio Rossi and Thomas Alexander

Preface

This book has two underlying messages, the emerging importance of the social issue of inclusion and human diversity in contemporary society and the increasing awareness that there is no such thing as a 'standard human being.' The first question has to do with inequality and social disparity as a necessary milestone toward economic revival, among other things, while a vital role in the strategies adopted by the European Union's Horizon2020 framework program is played by the ones that focus on strengthening equality, participation, and accessibility for all to goods, services and what Dahrendorf called 'life chances.' What this means is that the issue of social inclusion of diversity and for equality is firmly on political agendas all over the world, not least because of increasing awareness that new visions, new strategies, new tools and new approaches are needed, if we are to tackle the challenges arising from recent phenomena of economic and cultural globalization, demographic change; economic migration from poorer countries and an ageing population in wealthier countries, a phenomena that are destined to upset the entire planet's micro and macro-economic and social structures in years to come.

The second issue tackled in this book is more technical in nature, since the paradigm change from 'designing for standards' and 'inclusive products and service design' to the enlightened awareness that there are no such concepts to fit the standard human being, this has immediate, direct repercussions on the specialized dimension of designing. The realization is at last taking hold not only that those individuals are physically, psychologically and culturally 'diverse,' but they also have widely diversified skills, abilities, aspirations, and desires that make each one of us unique and not at all replicable. Since the diversity of individuals is the rule, not the exception, it makes sense to consider it as a resource, not as a limiting factor or a restriction on design, while equality between individuals, communities and peoples should be treated as fundamental strategic inputs to the sustainable development of contemporary society, where everybody should have the same opportunities to experience places, products and services. Numerous design approaches have been adopted to facilitate social and cultural inclusion in recent decades: Design for Disability, Universal Design, Inclusive Design, and Design for All. All of these philosophies, approaches, and methodologies aim to build value on all aspects of human diversity, from psychophysical to cultural issues, and to offer equal opportunity to everyone in order to experience places, products, services and systems. With this in mind, this book sets out to forge a climate conducive to discussion and comparison between these approaches, without any prejudice in favor or against any one of them, but attempting to identify the elements they hold in common and to build each one's heritage of originality, because we are convinced that the true resource of Design for Inclusion may well be found in this very diversity of opinions.

In particular, this book describes the state of the art of recent research conducted in a variety of fields that share the focus on Design for Inclusion and was presented in the fourth international conference on Design for Inclusion (AHFE 2019, Washington D.C.). On this occasion, the numerous research papers presented were collected together into two different thematic areas, articulated in six sections of this book:

Part 1	**Design for Inclusion**
Section 1	Designing for Inclusion in Learning Experiences
Section 2	Industrial Design for Inclusion
Section 3	Designing for Inclusion in the Information Society
Section 4	Public Spaces, Building Environment and Communities
Section 5	Global Perspectives on People-Centered Design and Cultural Heritage

Part 2	**Human Factors for Apparel and Textile Engineering**
Section 6	Design for Inclusion for Apparel and Textile Design

Special thanks to Gianni Montagna and Cristina Carvalho from the Lisbon School of Architecture, CIAUD for their valuable contribution and for co-chairing the conference track on Human Factors for Apparel and Textile Engineering.

Each section contains research paper that has been reviewed by members of the International Editorial Board. Our sincere thanks and appreciation to the board members as listed below:

Design for Inclusion

Tsai Lu Liu, USA
Giuseppe Mincolelli, Italy
Matteo Zallio, Ireland

Human Factors for Apparel and Textile Engineering

Luis Almeida, Portugal
Ana Cristina Broega, Portugal
Carlos Figueiredo, Portugal
Sandra Heffernan, New Zealand
Anne Marr, UK
Carla Morais, Portugal
Fernando Moreira da Silva, Portugal
Delfina Gabriela Garrido Ramos, Portugal
Maria Antonietta Sbordone, Italy

July 2019 Giuseppe Di Bucchianico

Contents

Designing for Inclusion in the Information Society

Public Spaces, Building Environment and Communities

Designing for Inclusion in Learning Experiences

Inclusion in Danish Architectural Education and Design Practice

Masashi Kajita(✉)

Institute of Architecture and Design, The Royal Danish Academy of Fine Arts, School of Architecture, Design and Conservation, Philip de Langes Allé 10, 1435 Copenhagen, Denmark
Masashi.Kajita@kadk.dk

Abstract. In wishing to make our physical environment inclusive, this paper investigates the complexity and challenge of design process, focussing on the Danish distinction between architect and *bygningskonstruktør* (constructing architect) in both qualifications and practice. The paper looks at the division of the roles in relation to the knowledge acquired during their education; and discusses the kind of knowledge on inclusion they gain, mobilise and share during the actual design process. The findings from nine qualitative interviews highlight two key areas. Firstly, there is a need to develop shared knowledge on inclusion between architects and *bygningskonstruktører*; secondly, team structure and communication strategy needs to be critically reviewed and appropriated to optimise knowledge allocation and dissemination within a team and practice at large.

Keywords: Accessibility · Architectural education · Design process · Knowledge sharing · Spatial inclusion

1 Introduction

There is a growing emphasis on providing equal opportunities for all people regardless of age, ability and many more possible limitations. In November 2010, the European Commission adopted a disability strategy which addresses this challenge supporting the rights of disabled persons to full and equal participation in society [1]. In October 2013, the Danish government presented a new disability policy action plan which sets visions for an inclusive society characterised by diversity, equality and respect for the difference [2]. These followed on from the United Nations *Convention on the Rights of Persons with Disability* [3]. Article 9 of the Convention obligates State Parties to take appropriate measures to ensure that disabled people have access to facilities and services on an equal basis with others. Accessibility in architecture provides an essential foothold, upon which the independence, participation and enjoyment of people can be realised. Yet it is a slow process to make the built environment accessible and inclusive. This is not only because architects struggle to recognise and engage with diverse bodily variations and their complex relations to the built environment, but also, the perpetual existence of disabling environments elucidates the complexity of the actual design process and its associated challenges.

G. Di Bucchianico (Ed.): AHFE 2019, AISC 954, pp. 3–14, 2020.
https://doi.org/10.1007/978-3-030-20444-0_1

In wishing to advance architecture as a means by which to improve future conditions, this paper addresses the challenges in the design process by focusing on the Danish distinction between architect and *bygningskonstruktør* (constructing architect) in both qualifications and practice. *Bygningskonstruktører* work with architects and other consultants in practice, focusing on construction details and technical specifications. The paper specifically looks at the division of roles in between the two professions, and questions how might this division relate to the *loss of accessibility* during the process of building and planning. How differently do they carry out their work in practice? And what are the consequences of the division of roles and responsibilities in practice? Furthermore, the paper investigates this separation of professions in relation to their knowledge acquired during their education. How differently have they been taught during their education? And how does the difference in education been embodied in their practices?

Empirically this paper is based on nine qualitative interviews conducted with Danish practicing architects and *bygningskonstruktører* who have experience of designing for inclusion. It was to get first-hand knowledge about the design process and their roles, as well as the type of knowledge on accessibility they use when designing. Their voices create a picture of the situation, which describes how projects were developed and executed in their everyday practice. To have a deeper understanding of how they gain knowledge on accessibility and inclusion, the curriculum of four leading schools in Denmark for architects and *bygningskonstruktører* were studied. The four institutions are: The Royal Danish Academy of Fine Arts, School of Architecture, Design and Conservation (KADK) and Aarhus School of Architecture (AAA) for architects, as well as Copenhagen school of Design and Technology (KEA) and Via University College (VIA) for *bygningskonstruktører*. The author represents KADK.

The analysis draws upon the theoretical distinction between context-dependent and context-independent knowledge of inclusion. This distinction is positioned within the critical discourse on the limited use of accessibility operated within the regulatory framework. The codification of the concept involves the generalisation of diverse human bodies and their relations to the built environment which further separates the lives of people from their contexts. The standardisation of architectural solutions with the potential of furthering stigma and social exclusion must be avoided. Accordingly, the paper discusses the knowledge on accessibility with a strong focus on the experience of users: how should designers – architects and *bygningskonstruktører* – gain, mobilise and share knowledge on disabled persons in order to produce inclusive architecture? The findings challenge this study in two perspectives: firstly, to develop a framework that internalises codified rules and standards, as well as knowledge on users into the creative process which can then be shared by both architects and *bygningskonstruktører*; and secondly, to continuously question the process and method of transferring and sharing knowledge between architects and *bygningskonstruktører*.

2 The Process of Building and Planning

The Danish construction industry is made up of a complex aggregation of professionals, specialists and suppliers. These different actors have different roles and involve themselves in different work stages. Amongst these actors involved, this paper focuses on architects and *bygningskonstruktører*; and looks at the separation of their tasks in relation to the process of developing building projects. In what follows, schematic overviews of work stages for building and planning are described, based on *Service Descriptions for Building and Planning 2018* prepared by the Danish Association of Consulting Engineers and the Danish Association of Architectural Firms [4], and with reference to RIBA Plan of Works 2013 [5]. A general model, which organises the process of briefing, designing, constructing, operating, maintaining and using building projects into four phases, is presented in Fig. 1. The structure of projects can vary substantially from one development to another. The choice of project structure or of procurement route depends on numerous factors including the duration of project, cost certainty, flexibility, price competition and quality of design and workmanship required. Accordingly, the content of stages may differ or overlap to suit specific project requirements.

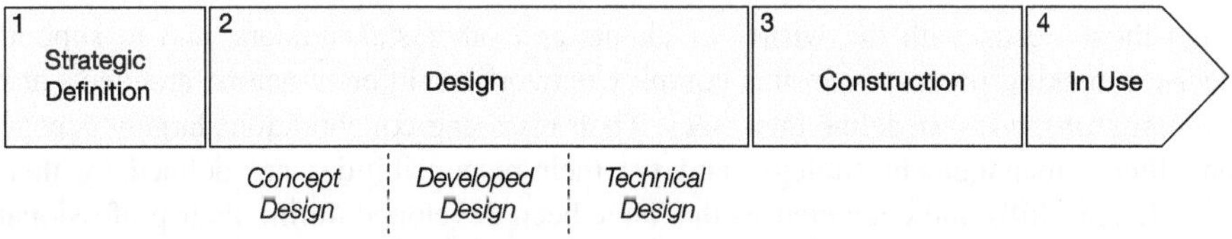

Fig. 1. An over view of generic work phases for building and planning

The first phase is strategic definition. This is the phase of appraisal and brief development. Consultancy typically starts from identifying client's business case, and later develops the project brief defining project objectives, project programme, project budget. Defined tasks in this phase also include clarifying the physical, social and legislative context around the project. The second phase is to develop design, where the project progresses from concept design stage, developed design stage to technical design stage. The concept design stage develops outline proposals for structural design, building services systems, outline specification and preliminary cost information in accordance with relevant project strategies and project programme. The developed design stage reaches to coordinated and updated proposals. It is the process of taking on and further developing the approved outline proposal to an extent that can provide sufficient information, describing all the main components of the building and how they fit together, for applying and obtaining statutory approval for building activities. Accordingly, planning applications are typically made using the output from this middle stage of the design phase. However, technical aspects may not yet have been fully developed and incorporated into the design at this stage of developed design.

The last stage of the design phase, the technical design stage, develops the final proposal that all architectural, structural and building service information, specialist subcontractor design and specifications are fully developed. This is typically the stage before the construction contract is tendered or construction begins. However increasingly, technical design may continue through the preparation of production information and tender documentation. It might even be extended to the construction phase, particularly when the technical design aspects are subcontracted to specialists. It should be noted that, during these stages of design development, the chosen procurement route defines when the planning application to be made, as well as specific tendering and procurement activities. The third phase is for construction, where onsite construction and offsite manufacturing take place in accordance with the construction programme. Typical activities include administration and conclusion of building contract, the handover of building and the provision of *as-built* information. The last and fourth phase covers in-use service provisions and resource building. It deals with the operation and maintenance of buildings. Typical activities include reviewing of project performance and outcomes, undertaking post-occupancy evaluation and updating project information.

The above described the schematic overview of generic phases for building work. Specific tendering and procurement activities occur at different stages in the design phase in relation to the chosen procurement route. Accordingly, roles of architects and *bygningskonstruktører,* and the duration of their involvement in the process can vary from one project to another. In addition, with regard to accessibility, there is an increase of disabled users' and experts' involvements during the process, in order to understand their needs, to align those needs with the wishes of clients and other stakeholders, and to support decision-making processes. In this complex network of different actors, architects and *bygningskonstruktører* define their roles. Their roles and collaborations largely depend on office's management strategy, and yet their responsibilities are defined by their knowledge, skills and competences that have been developed during their professional education. And yet, this division of roles might create a gap within the communication network of different actors, contributing possibly to the *loss of accessibility* during the process of building and planning [6]. In the following, the paper describes how differently these two key actors in the field have been taught accessibility on their education; and how those different education frames their professional responsibilities; and further discusses the consequences, both positive and negative, of this division in practicing to produce inclusive built environments. However, before moving onto these discussions, both method and theoretical framework are outlined in following sections.

3 Method

Two sets of investigations were conducted. The first set of study looked into what and how accessibility and inclusion are taught to educate future architects and *bygningskonstruktører* at four institutions in Denmark. The Royal Danish Academy of Fine Arts, School of Architecture, Design and Conservation (KADK) and Aarhus School of Architecture (AAA) offer BA and MA programmes in architecture[1]. Copenhagen school of Design and Technology (KEA) and Via University College (VIA)[2] offer BA programme in Architectural Technology and Construction Management. The author represents KADK. The investigation includes the study of implemented curriculums

[1] There are only two schools of architecture in Denmark.

[2] There are six institutions for *bygningskonstruktører* in Denmark. These are: Københavns Erhvervsakademi (Copenhagen school of Design and Technology) in Copenhagen; Via University College in Horsens and Aarhus; EASJ - Erhvervsakademi Sjælland (Zealand Academy of Technologies and Business) in Næstved; UCN - Professionshøjskolen Nordjylland (University College of Northern Denmark) in Aalborg; UCL Erhvervsakademi og Professionshøjskole in Odense; and Erhvervsakademi Sydvest in Esbjerg.

and email correspondences and/or interviews for clarification with course responsible persons or programme leaders from these leading institutions in Denmark[3].

The second set was nine qualitative interviews conducted with five practicing architects[4] and four *bygningskonstruktører*[5] who have experience of designing for inclusion. These interviews were to gain first-hand knowledge about the design process and their professional roles, as well as the type of knowledge and information they use when designing accessible and inclusive architecture. The respondents are building architects and *bygningskonstruktører* who were educated at one of the above-mentioned institutions. Landscape architects were not included. To ensure that those architects and *bygningskonstruktører* are familiar with and know different aspects of accessibility, practices selected have considerable experiences in designing publicly accessible buildings such as sport and leisure buildings, cultural buildings, offices, educational institutions or health providing facilities where accessibility and inclusion matter. Those practices are AART Architects, Dorte Mandrup Architects, Force4 Architects, Holscher Nordberg Architecture and Planning and Schmidt Hammer Lassen Architects.

The interviews took a form of semi-structured qualitative interviews. Conversations were left as open-ended, however, within the framework of the pre-defined structure. First, the respondents were asked to explain how they carry out their works in practice, focusing on accessibility and inclusion. How different their roles and responsibilities can be? How does your office organise a team; and how do architects and *bygningskonstruktør* collaborate in a project? Then, the respondents were asked to relate their expertise to knowledge, skills and competences developed during their professional education. How differently these two key actors in the field have been taught; and how those different education frame their professional responsibilities? Lastly, they are asked to reflect on the division of their professional roles. What are positive and negative aspects of this division; and how could the situation be improved in order to produce inclusive built environments? The respondents were guided by those questions, and yet, they were encouraged to divert, tell their own stories and describe their experiences. These conversations were recorded and the quotations in this paper were transcribed and shortened by the author and were approved by the interviewees.

4 Knowledge Field in Designing Inclusive Architecture

What kind of knowledge do architects and *bygningskonstruktører* make use of when designing inclusive architecture? Danish scholar Kirkeby in her paper *Accessible Knowledge – Knowledge on Accessibility* emphasises the importance of context-

[3] Brigitte Geert Jensen from AAA, Claudio Spaziani Testa from KEA and Bo Sørensen from VIA.

[4] These architects are Kathrine Hegner Stærmose from AART Architects, Marianne Hansen from Dorte Mandrup Architects, Andreas Lauesen from Force4 Architects, Mikkel Nordberg from Holscher Nordberg Architecture and Planning, and René Nedergaard from Schmidt Hammer Lassen Architect.

[5] Helga Skovsted from AART Architects, Asbjørn Severin Lütken from Dorte Mandrup Architects, Anders Mortensen from Holscher Nordberg Architecture and Planning, and Kirsten Sennicksen from Schmidt Hammer Lassen Architect.

dependent knowledge in opposition to a general scientific ideal of context-independent knowledge [7]. Drawing on Flyvbjerg who refers to the concept of *phronesis* from Aristotle's description of *praxis* [8], Kirkeby articulates the essential roles of *phronesis* as context-dependent knowledge which enables the practitioner to make responsible choices [9]. In this paper, responsible choices are interpreted as choices for realising inclusive architecture and built environment. Kirkeby develops her arguments based on Aristotle's three types of knowledge – *episteme, techne and phronesis*. *Episteme* forms the body of ideas that is intellectually certain at any particular time and space. It is scientific knowledge gained through analytical rationality. *Techne* is knowledge related to the production of art and craft. It is a type of knowledge used in making and doing. It governs the application of technical knowledge and skills based on practical instrumental rationality. P*hronesis* is knowledge oriented toward action. It is pragmatic knowledge that places ethics centrally and helps persons to make reasonable decisions based on practical value-rationality [8–10].

While *episteme* is context-independent knowledge that concerns theoretical *know-why, techne* is context-dependent knowledge denotes technical *know-how. Phronesis* is context-dependent knowledge specific to each particular circumstance that cannot be reduced to knowledge of general truths [7]. When designing, quantifiable data sets and other rule-based knowledge such as building regulations and building standards are used. Those represent rationality and universality; and, therefore, are context-independent. On the contrary, context-dependent knowledge is a type of knowledge that people learn through their experience and used through the act of designing in reflective practice, as Schön (1983) describes – *reflection-in-action* [11]. It should be noted that these different kinds of knowledge are not to be seen as totally separated. Architects and *bygningskonstruktører* should freely mobilise knowledge while coordinating site, architectural, structural, building service, production and cost information in order to produce inclusive architecture. When designing, they must use accumulated knowledge from previous and current tasks, negotiating with a whole range of design parameters. And, user's experience – like every other aspect of design – should carefully be integrated into the design process as one of essential design generating parameters [12].

Another important distinction is made in-between the imagined space of designers and the lived space of users. This paper recognises that the experience of users that architects and *bygningskonstruktører* imagine when designing and the actual experience of users are never the same. The act of designing demands an ability of designers to imagine how others may use and experience designed buildings, products and services. And yet, no matter how actively one participates to make understanding of situations, the design process involves designers' personal interpretation of gained understandings. In addition, there is tendency of designers to conceptualise the human body as a biological object, which often has average dimensions and ideal proportions [13]. This abstraction suppresses the differences of which human bodies possess and simplifies the complexity of bodily interactions with objects, buildings and their surrounding environments. Accessibility requirements and related guidelines are based, to a large extent, on those standardised bodies. They objectify the body as primarily comprising of physical impairments in which a wheelchair becomes central to facilitate access and movement in and around buildings. Therefore, the roles of accessibility requirements and related guidelines in design process can continuously be questioned [14, 15].

The above-mentioned distinctions between context-dependent and context-independent knowledge, as well as the imagined space of designers and the lived space of users form an analytical model presented in Fig. 2. This model adopts Kirkeby's analytical model that describes the understanding of knowledge used in the making of architecture [7, 9, 10], and is further developed in order to describe a systematic understanding of knowledge learned and used in the production of inclusive architecture. The developed new model specially focuses on the aspect of use and users, and helps to understand how gained knowledge of users is translated into useful resources and mobilised in the design process. This focus on users reflects critical discourses on the codification of accessibility, and aligns with recent attentions towards people-centred practices that are integrated into architectural process which encourages members of project teams to adopt a practical approach to involving users throughout the different development phases of a project.

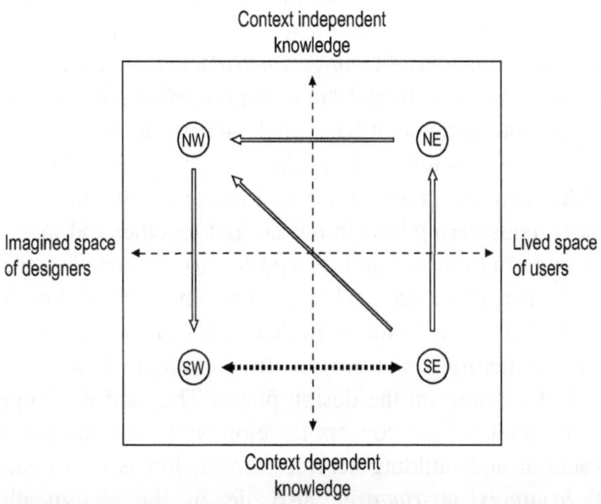

Fig. 2. A plane of knowledge for inclusion that architects and bygningskonstruktører make use of while building works are carried out.

South-east quadrant (SE) in the *plane of knowledge for inclusion* is an area where embodied experiences of individual users are situated. *North-east quadrant (NE)* is an area where users' experiences are translated into resources. This field corresponds with the rationality of science, representing in the type of knowledge such as a data sets that quantify human experience and behaviour. In the design process, designers operate in both fields of *North-west quadrant (NW)* and *south-west quadrant (SW). NW* is an area that available knowledge of users is translated and organised so that architects and designers can directly work with. These can be in the form of rules and guidelines; explicit answers to problem raising in the design process – what is the minimum width of doors for a person in a wheelchair to pass, for example. *SW* is the primal area where design methods and design thinking are employed in their creative process. In this field, architects and *bygningskonstruktører* coordinate gained knowledge but also develop new knowledge in the act of designing. The analysis made below draws on this *plane of knowledge for inclusion* presented in Fig. 2.

5 Findings and Analysis

The interviews with practitioners were carried out by following three main questions of; how differently do architects and *bygningskonstruktører* carry out their work in practice; how does this division of roles and responsibilities relate to their understanding of own profession in relation to professional education received; and what are the consequences, both positive and negative, of this division in practice in order to produce inclusive built environments? Correspondingly, materials from the interviews are organised into following subsections of: *Two Different Professions; Different Types of Knowledge on Accessibility;* and *Mobilising and Sharing Knowledge.*

5.1 Two Different Professions

Amongst practitioners, there is a strong recognition in professional differences between architects and *bygningskonstruktører.* Architects consider themselves – and are considered by *bygningskonstruktører* – as creative ones who look after formal and aesthetical aspects of design; whereas, *bygningskonstruktører* consider themselves – and are considered by architects – as technical ones. *Bygningskonstruktører* are often responsible for preparing and producing construction details and technical specification, which are coordinated with the cost information. As Asbjørn Severin Lütken, *bygningskonstruktør,* stated: "An architect gives a form to a building. Of course, the form changes given circumstances, considering how things are put together and making it possible to build. In the process, I can influence and set a project to a certain direction. But my job is not to create a form, instead, to make what architects have created or envisioned into a something buildable. I stay true to the form that architects pass on to me."

Architects involve themselves in projects from strategic definition phase to design phase; but typically focussing on the design phase. They are in charge of developing overall design of the project from concept development to the final proposal where all architectural, structural and building service information is coordinated. The professional roles of *bygningskonstruktører* also lie in the design phase. However, they typically join a project team in the developed design stage after outline proposals are approved; and take the project up to technical design stage where at the end of this stage, provided materials can form the basis for tendering, contracting and construction. However increasingly, they are involved through the preparation of production information and tender documentation, but also take a part at the concept design stage on consultancy basis. This earlier involvement of *bygningskonstruktører* typically increases as a project gets bigger in size, considering the complexity that the project might bring. *Bygningskonstruktører* handle technical information coordination and sharing, through a process such as BIM (building information modelling); and accessibility is considered as one type of such technical information that needs to be included in the process.

Furthermore, findings articulate that *bygningskonstruktører*'s involvement in a project changes over time. A team structure can change as Kirsten Sennicksen stated: "A project I am on now, which is in a technical design stage, we have ten persons in our team, and five of us are *bygningskonstruktører.* A year ago, in the developed design stage, we were two *bygningskonstruktører*". Responsibilities within a team might also shift. Marianne Hansen: "When *bygningskonstruktører* join a project in developed design stage, they would likely take responsibility of 20%; and that will grow up to 60 to 70%". In some cases, *bygningskonstruktører* act as a project manager and take responsibility on coordinating with other consultants dealing with structural, service system and cost information. However, experienced architects are who typically manage projects. This is due to fact that the practices involved in this study prefer someone – ideally whole team members – to follow a project from the beginning to the end, in order to assure a seamless transition from a vision to a building. *Bygningskonstruktører* would often not be on early stages of the project. Hence, architects become project managers. This is underlined by their mutual understanding of: "aesthetic and form are expertise of architects, and the highest priority in the office", as stated Asbjørn Severin Lütken.

5.2 Different Types of Knowledge on Accessibility

Architects and *bygningskonstruktører* respect each other's skills and competences, and enjoy the collaborative aspects of their works, although they have very different approaches to accessibility. Interviewed architects consider that overall responsibility is on themselves, since accessibility is very much part of form making. However, this comment is always followed by a line such as "but later, *bygningskonstruktører* come into a project to make sure that a building lives up to minimum requirements specified on legal documents". Correspondingly, *bygningskonstruktører* believe that they are the one who takes care of accessibility – accessibility represented in the format of building regulations. This contrasts with how interviewed architects consider accessibility – accessibility as something opaque, relating to key words such as wheelchair, ramp, lift and circulation. However importantly, these different interpretations of accessibility create a hierarchy between two professions. As stated by one of architects, "*Bygningskonstruktører* know how to put details together by codes, but it may not be the way how we would like them to be. So, it might take a long time, doing *back and forth,* to get to the details that we want".

Both KEA and VIA offer education for *bygningskonstruktør as* BA programme in Architectural Technology and Construction Management, which has a duration of three and a half years and consists of seven semesters of a total of 210 ECTS. Their courses do not provide specific lectures or seminars on accessibility. And yet, accessibility is addressed within students' semester projects. Accessibility takes a part in the subject of design and building law, focusing on the Danish building regulations. Accessibility is understood and taught as one type of legal requirement which they as *bygningskonstruktør*er must comply with when designing. KADK and AAA offer education for architects at both levels of BA and MA programmes, which has a duration of five years in total and consists of ten semesters of 300 ECTS. KADK increasingly provides lectures and workshops on inclusion in some occasions. However, it is fair to state that there are not any coherent courses on accessibility and related legal matters across the school at this moment. This also applies to AAA. In both schools, different design teaching studios might address the theme voluntarily. However currently, there is no obligation nor control by the institutions on how and to which extent the issues of inclusion are addressed. Responding to this uneasy circumstance, KADK started to make positive changes in 2016, and is now in the process of integrating teaching on inclusion into its new overall structure.

These changes reflect not only on the lack of accessibility education, but also the critical view on codified rules that simplify the complexity of bodily interactions and represent lived life of persons in static and quantifiable ways [12, 13, 15]. However, KADK acknowledges the importance – and the limitation – of teaching rules and guidelines on accessibility and inclusion, because they provide essential points of reference for technical and functional solutions. Accordingly, KADK aims to balance between: (1) knowledge in *NW* of Fig. 2 – established design tools and information represented in building regulations; and (2) knowledge in *SE* – gained through people-centred practice that encourage designers to work directly with individuals. In *SW*, knowledge on *NW* and *SE* should be integrated. Design must progress through a reflective conversation with the situation where accumulated knowledge of designers from previous and current tasks as well as spatial experience of users, both systematised (*NW*) and embodied (*SE*), are integrated. In addition, KADK emphasises the importance of *reflection-in-action*, and promotes *leaning-while-designing*. This situatedness in contexts during knowledge production [8] and learning process [16, 17] gives an opportunity for architects to avoid objectifying the people, and therefore, to creating architecture that is more responsive to the manifold complexities of peoples' experience in situations.

5.3 Mobilizing and Sharing Knowledge

With regard to accessibility, *bygningskonstruktører* essentially use knowledge in *NW*. They consult the building regulations and other design guidelines, such as *Danish Standard DS3028*. This means that user's embodied experiences in *SE* come into *SW* via *NW*. Architects also use knowledge situated in *NW*; however predominantly through making use of *bygningskonstruktører's* knowledge on building regulations. *Bygningskonstruktører* become in-house consultants for architects who lead a team as masters of aesthetic and form-making. *Bygningskonstruktører* hold key knowledge on accessibility and are "the link between architects and rådgivende ingeniører (consulting engineers) and/or contractors", stated Asbjørn Severin Lütken. However, it must be reiterated that their knowledge on accessibility is technical information based on standardised body dimensions and proportions, of which exclusive use might lead design towards generic and *add-on* solutions [7, 13, 15].

Over the past 20 years, there has been an increase of involving users and other experts in the design process in order to gain valuable insights from them. The process can be formalised, especially when projects consider older and/or disabled persons as specific target groups. Such projects aim to involve these users and experts in periodic, so they participate in the project throughout. As an example, 1:1 mock-ups of different toilets were even made. A procedure was, and would often be, that users give feedback; then experts, clients and architects reflect these users' opinions and make decisions. In the process, architects take a leading role, coordinating with users and other expert groups. In this process, architects transfer knowledge of disabled users from *SE* into *SW*. It should be underlined that such user involvements provide deeper understanding of users' experience, as opposed to broad knowledge gained from statistics or other quantifiable data. Knowledge obtained in this way brings architects valuable insights; but less of comprehensive understanding of every aspects of a person's life. It produces understandings that are subject-dependent, and therefore can be deeply embedded into architects' creatively in practice: but not *bygningskonstruktørerer's*.

There is a strong hierarchy in two professions. Architects set a vision and make a form. This form-making is the highest importance of the practices studied in this paper. In this process, when possible or requested, architects engage with users directly. This movement of knowledge between *SE* and *SW* adds another layer of questions, which then inform the advancement of design. While architects develop the design, knowledge situated in *NW* is sought and integrated into solutions through a dialogue with *bygningskonstruktører.* This externalised process of consulting *bygningskonstruktører* provides an opportunity for both architects and *bygningskonstruktører* to critically assess and qualify information in *NW* while imagining how design could be. Kathrine Hegner Stærmose: "It is a collaboration. It goes back and forth all the time". This *back and forth* – a dialogue between architects who connect users, experts and a client, and *bygningskonstruktører* who communicate with other consulting engineers and contractors – is an essential key for producing accessible and inclusive architecture. And therefore, this division in their roles and the strong dependency to each other could create a gap of knowledge, when there is a lack of shared understandings or communication.

This challenges architects and *bygningskonstruktører* in two perspectives. The first, the common ground for knowledge on inclusion need to be established. *Bygningskonstruktører* oscillate between *SW* and *NW.* Architects firmly dwell in *SW*; and yet move between *SW* and *SE.* In *SW*, architects and *bygningskonstruktører* put together carried knowledge from their previous tasks and accumulated knowledge from their current tasks. In order to enhance collaboration between those with different skills and competences, they should share the foundation of knowledge. They both need to know the rationale behind technical information based on the systematised experiences of users; and be informed by the embodied experience of users. This challenge should primarily be addressed in educational institutions. However, learning process of inclusion must continue while practicing. They gain knowledge while developing the project; and they develop design while developing new knowledge. This aspect of continuous learning leads to the second perspective. The team structure and communication strategy need to be critically reviewed and frequently appropriated in order to optimise knowledge allocation and dissemination within a teams and office. Kirsten Sennicksen: "This *ping-pong* of knowledge allows us to come up with solutions that regulations might not be able to solve. Our office has gained experience in last 19 years. We become better at handling issues, and understand the importance of having this *ping-pong* to make architecture more inclusive." Yet, can the experience of architects and *bygningskonstruktører* be the only solution for better knowledge allocation and sharing?

6 Conclusions

The paper has presented how architects and *bygningskonstruktører* collaborate in practice, using different types of knowledge on inclusion. They receive different education on accessibility which define their respective roles and responsibilities. Based on the *plane of knowledge for inclusion* developed from the distinctions between context-dependent and context-independent knowledge, as well as the imagined space of designers and the lived space of users, the paper systematically analysed how knowledge on inclusion is gained, mobilised and shared during the actual design process. The first conclusion to be drawn from this study is that there is a need for developing the common ground for knowledge on inclusion between architects and *bygningskonstruktører*: legislative framework, statutory requirements and the rationale behind specific requirements, specific and complex needs of users. The second conclusion to be drawn from this study is that the team structure and communication strategy need to be critically appropriated in order to optimise knowledge sharing within a team and practice at large. Rules and standards, as well as people-centred techniques are useful tools; but are not decisive. Hence, the achievement of inclusive environments depends largely on the capability of architects and *bygningskonstruktører* who gain and share knowledge on inclusion effectively. Situate themselves in the primal field of creation of *SW*, they must reach out – as individuals but also as a team – to *NW* and *SE*, as well as *NE* if necessarily: moving far and wide trough the *plane of knowledge for inclusion*.

References

1. European Commission.: European Disability Strategy 2010–2020: A Renewed Commitment to a Barrier-Free Europe (2010)
2. Danish Ministry of Children, Gender Equality, Integration and Social Affairs: Disability Action Plan 2013: One Society for All (2013)
3. United Nations, General Assembly: Convention on the Rights of Persons with Disabilities and Optional Protocol (2007)
4. Danish Association of Consulting Engineers and Danish Association of Architectural Firms: Description of Services: Building and Planning 2018 (2018)
5. RIBA Plan of Works (2013). https://www.ribaplanofwork.com
6. Frandsen, A.K., Kirkeby, I.M., Ryhl, C., Perersen, L.S.: The accessibility requirements in the danish building regulations and the building process: a mapping of compliance and barriers, SBi 2012(16). Danish Building Research Institute, Hørsholm (2012)
7. Kirkeby, I.M.: Accessible knowledge – knowledge on accessibility. J. Civil Eng. Archit. **9**, 534–546 (2015)
8. Flyvberg, B.: Phronetic planning research: theoretical and methodological reflections. Planning Theor. Pract. **5**(3), 283–306 (2004)
9. Kirkeby, I.M.: Transferable knowledge: an interview with Bent Flyvbjerg. Archit. Res. Q. **15** (1), 9–14 (2011)
10. Kirkeby, I.M.: Knowledge in the making. Archit. Res. Q. **13**(3/4), 307–313 (2009)
11. Schön, D.: The Reflective Practitioner: How Professionals Think in Action. Basic Books, New York (1983)
12. Boys, J.: Doing Disability Differently: An Alternative Handbook on Architecture, Dis/ability, and Designing for Everyday Life. Routledge, London (2014)
13. Imrie, R.: Architect's conception of the human body. Environ. Planning D: Soc. Space **21**(1), 47–65 (2003)
14. Imrie, R., Street, E.: Architectural Design and Regulation. Wiley-Blackwell, Chichester (2007)
15. Kajita, M.: Re-addressing the social dimension of accessibility: towards an inclusive society. In: Lots, K., Simpson, D., Raahauge, K.M., Vindum, K., Jensen, M.J., Bendsen, J.R (eds.) Form til Velfærd/ Forming Welfare, pp. 280–291. The Danish Architectural Press, Copenhagen (2017)
16. Lawson, B.R.: How Designers Know. Architectural Press, Oxford (2004)
17. Cross, N.: Designerly Ways of Knowing. Springer, London (2006)

Green for All. A Didactic Experience on Design for All Applied to Gardening Earthenware Objects

Giuseppe Di Bucchianico(✉)

Università di Chieti-Pescara, Viale Pindaro, 42, 65127 Pescara, Italy
pepetto@unich.it

Abstract. The Design for All approach, enhancing human diversity, promotes the spread of a social conscience oriented towards inclusion. This seems particularly useful for the training of young designers, because it obliges them to consider the complex system of constraints and opportunities arising from the extreme variability of the design targets, that more and more contemporary society requires to design activity. Here are described the development and results of a teaching experience, entitled "Green for all", developed at the University of Chieti-Pescara, in which students designed a product for gardening, referring to specific and original use contexts, through the DfA approach. Starting from the dimensional and productive restriction of earthenware, the projects had the purpose to favor the cultivation of ornamental plants or domestic spices even in contexts of original use (in the kitchen, in the bathroom, in the office, but also on the desk, in gym or on the train).

Keywords: Design for All · Gardening · Earthenware products

1 Introduction

In parallel and in response to the latest global socio-demographic phenomena, they have seen a rapid development and diffusion also different design approaches related to the more general definition of "Design for Inclusion", also applicable to productive sectors and new user contexts, such as gardening, especially when it is related to new healthy therapies for the psyche and the human body.

1.1 The New Socio-demographic Horizons

The main demographic challenge of contemporary society, especially for industrialized economies, is represented by the aging of the population. The global population of 9 billion expected for 2050[1] will be set up for about 2 billion from "over 65"[2].

[1] UNITED NATIONS, DEPARTMENT OF ECONOMIC AND SOCIAL AFFAIRS, POPULATION DIVISION, World Population Prospects. The 2008 Revision, New York, 2009, 9.

[2] R. LEE, The Demographic Transition: Three Centuries of Fundamental Change, in Journal of Economic Perspectives, 2003, vol. 17, n. 4, 3.

G. Di Bucchianico (Ed.): AHFE 2019, AISC 954, pp. 15–21, 2020.
https://doi.org/10.1007/978-3-030-20444-0_2

In particular, the same source estimates that in Europe and in China in 2050 the "over 65" will reach 30% of the population, while in Japan the "over 80" will represent even 40% of the population. This phenomenon is the result of two factors: on the one hand, the increase in life expectancy, which will rise from the current 65 years, as a world average, to 74 years in 2050 (in Italy, the first country in Europe, is already 82 years); on the other, the parallel decrease in the average world fertility rate, which will drop from the current 2.7% to 2% in 2050. The first consequence will be a drastic reduction of people of working age and therefore the need to revise the traditional welfare systems to support a growing "old" society. The second consequence will be the increase in pathologies, even partially invalidating, linked precisely to the aging of the population (chronic-degenerative diseases and mental illnesses, reduction of sensory capacities, reduced mobility, etc.). In fact, most observers expect a society marked by disability for the next decades: a great increase in non-self-sufficiency, with all the economic, social security, welfare and anthropological consequences that this entails.

Another social phenomenon that, for different reasons, instead invests across contemporary society in all its age groups is the lack of social relations, which, for the weaker sections of the population, is transformed into social isolation, which at in turn it can lead to serious consequences also on the cognitive level and physical health. To give an example, a study that examined 479 thousand individuals between 40 and 69 years, shows how healthy but lonely people, without a family, friendship or social network, seem to have statistically a greater risk of cardiac events such as heart myocardium attack and ictus [1].

On the other hand, policies of social inclusion start from the assumption that the condition of social hardship does not only concern the forms of social deprivation and suffering, such as economic or material poverty, but also the shortcomings with respect to family and social ties, housing systems, networks of support services and social integration, as well as training and forms of work or cultural marginalization[3].

Aging, disability, isolation and social suffering: these are therefore some of the most significant phenomena that characterize contemporary society and which will presumably increase in intensity in the coming decades.

1.2 The Horticultural Therapy

Horticultural therapy is one of the healthiest therapies for the psyche and the body.

It is a set of floriculture and horticulture activities, which aim to improve the physical and psychological conditions both of individuals and of entire communities.

Taking care of land, gardening, cultivating plants and vegetables is a great stimulus for treatment, socialization and self-esteem. Attending a tactile and olfactively different environment from the everyday is a way to accelerate healing, or simply allows you to increase your perception of yourself, enhance your skills, shift the focus from your person to an object in need of care, with significant emotional, social and physical benefits. The plants therefore can help in many cases, from stress to old age, from addictions to disability, from depression to psychopathology.

[3] United Nations Development Programme, 2011.

The earliest references to the use of plant cultivation as therapy for people with disabilities were found in an Irish monastery and date back to 1300, while in the late 1700s, Benjamin Rush, father of American psychiatry, wrote that working the land had a beneficial effect on mental health [2]. It is necessary to underline the differences that exist between the gardening activities and the Horticultural therapy that, even if are referred only to a question of points of view, seems to be important. In fact, while gardening can improve the well-being of the community or groups of people living in the same environment and share different interests, Horticultural therapy deals with interactions between human beings and plants in a much more intimate way: its primary objective is to promote the individual health of patients while the plants become, in this way, only a "by-product" of the rehabilitation process. Its peculiarity is that it can be applied in a wide range of contexts: at home, in public or private gardens, in green spaces or in plants for the cultivation of plants connected to hospitals, rehabilitation clinics and hospices.

1.3 Design for All for an Inclusive Gardening

Starting from these assumptions, we understand how there can be a close relationship between the therapeutic approach to gardening and the issues of social inclusion.

In order to promote social inclusion, the idea that all people, regardless of age, sex, culture, physical and psycho-sensorial capacities, must have the same opportunities to develop a dignified life and to decide and satisfy themselves desires and aspirations: it is a challenge that, without a new way of thinking and making a project, it is not possible to face and solve.

In addition, gardening, and more generally, taking care of a plant, as well as requiring specific skills, sometimes also requires the need to perform tasks and gestures that are sometimes extremely technical and attentive.

One of the design approaches developed to respond creatively to this dual challenge is Design for All (DfA), which is the "design of human diversity, social inclusion and equality" (EIDD Stockholm Declaration, 2004). It proposes a holistic and innovative approach to the project, which considers the differences between individuals as an important resource for economic and social development, rather than as a "problem" or a "limitation" for the project. In particular, it is opportune to dwell on some key concepts through which the design approach of the DfA can help to generate truly inclusive solutions.

The first concept concerns the theme of the diversity of people that is seen as a resource rather than a problem. In fact, the assumption that the whole population is made up of deeply and radically different individuals - from the biological, cultural, social, demographic, ethnic, pathological, experiential aspects, etc. - is considered a positive value of society and a resource for improving the quality of life. Through appropriate design methodologies the DfA tries to prevent these differences from turning into disabilities, designing products and services that are able to respond as much as possible to the diversity of human needs. Human diversity, if properly managed, respected and valued, can lead to a greater wealth of the whole of what makes some *ad hoc* production coupled with a pervasive, inattentive and generic production [3].

The second concept concerns the theme of the participation of people and their involvement in the design process. The DfA approach tries to involve, both directly and simply through the observation of behaviors and attitudes, the greatest number of people involved in the design process.

The inclusive society, therefore, is also the place where to see the diversity of people in a new way, that is as a resource and not a problem. And it is also the place where all people can actively take part in the processes of transformation and construction of reality in different aspects: they can not only be considered as people who must be satisfied in their needs (target design), but also as people who can contribute in an active and unique way, through participation, collaboration and co-creation (design tool).

In this perspective, applying some principles of Design for All to the project of equipment and tools for gardening, if on the one hand it promotes the spread of practices of Horticultural Therapy, on the other hand it allows to extend the relationship of individuals with plants also to new and unexplored contexts, through the development of new products useful for the cultivation of plants.

2 Objectives

The principle of conscious participation and awareness on the issues of inclusion, on which the Design for All approach is based, is the basis of the "Green for All" didactic experience presented in these pages, developed in the Department of Architecture at the University of Chieti-Pescara.

The initiative started from the idea that green care can be an effective tool to raise awareness of environmental issues, as well as, in some cases, a real instrument of therapy and rehabilitation through the care and management of the green, in the cultivation of flowers, vegetables and other plants.

By extending these benefits to the categories of users not strictly related to disability and starting from the assumption that everyone can benefit from an intimate and direct relationship with the plants, the design laboratory has solicited the students to deal with the theme of design for all, through the project of an artifact for gardening.

Starting from the dimensional and productive bond of the terracotta, the projects have had the purpose to favor the cultivation of ornamental plants or spices of domestic use even in unpublished contexts (in the kitchen, in the bathroom, in the office, but also on the desk, in the gym or in train), and also with original and unconventional use modalities. A selection of these projects forms the basis for the realization of a collection of prototypes, useful for hypothesizing a collection for a traveling exhibition.

3 Method

The "Green for All" design laboratory, through the direct involvement of a complex system of actors, has been a pretext for raising awareness on the issues of diversity among individuals, equal opportunities to enjoy everyday experiences in an autonomous and comfortable way, favoring in this way social and cultural inclusion, according to the Design for All approach.

Therefore, from didactic experience, through different phases the initiative has all the characteristics to be transformed into a multidimensional promotion event, with moments of reflection and discussion within the working group alternating with moments of public exposure of the partial results obtained.

3.1 Critical Reflection and the Definition of the Briefs

In the Department of Architecture at the University of Chieti-Pescara some teachings related to Industrial Design have been dealing with the issues associated with design for inclusion for several years. In particular, the teaching of Product Design 2 of the Bachelor in "Design", starting from the basics of ergonomics applied to the industrial products, focuses mainly on the innovative approach of Design for All, both with lectures and collective reflections and debates, both through the identification of a specific design issue.

Recently, the field of application of design experimentation has been repeatedly identified in the types of products related to the domestic gardening sector, with the aim of developing inclusive product ideas for the care and cultivation of ornamental plants, spices and vegetables, whose use autonomous, comfortable and pleasant, it is extended to the widest variety of individuals with diversified needs and abilities. The only constraint was that the products should refer to the technology of terracotta, and therefore that the projects made specific reference to the formal and productive constraints of the pressing and turning of the clay, with a precise dimensional constraint for each element that does not exceed 30 cm of maximum size.

The brief was developed in groups by the students, starting from the identification of specific and unusual contexts of use and from the description of the respective limit user targets with respect to the chosen activities: this in order to inspire also original strategies and methods of use of the new products.

Some specialist contributions on the themes of home gardening and of the production process of terracotta, together with specific contributions on Design for All issues, have encouraged the students' collective critical reflection. Subsequently they were engaged in groups in the definition of specific design briefs. This starting from the direct observation of users grappling with the use of products for gardening and of users acting in contexts unrelated to the usual ones of home gardening.

3.2 Development of the Concepts Towards the Construction of Prototypes

On the basis of the briefs defined, each student subsequently independently developed a concept proposal, which had to refer strictly to the formal and productive constraints of terracotta.

The concepts defined in this way immediately highlighted the potential of the design process related to inclusion: the students verified how innovative product concepts can be developed to meet specific needs, determined by the needs expressed by multi-user, especially in relation to new contexts and use.

Thus product concepts have been developed, which are also interesting in their simple denominations: pocket flowerpots, kitchen wall-hung pot, mortar for cultivation and processing of spices, umbrellas-cushion vase, height adjustable pot, "a garden on the legs" for users in wheelchairs, spinning top pot (Figs. 1 and 2), etc.

Fig. 1. The "Spinning top pot" concept

Fig. 2. The "Garden for bedside table" concept

A group of private subjects has recently shown interest in the educational initiative. The hope is that it happens as in a previous similar initiative, entitled "Everyone at the table!" [4], in which some private individuals and public institutions have made available facilities, professionalism and equipment for the concrete realization of prototypes. On the other hand, the extension of the project towards the realization of physical prototypes, allows the diffusion of the "Design for All" approach to a wide and diversified public, not limited therefore only to "insiders" able to interpret the project technical drawings. This also through the organization of public events which can foster a broad reflection and widespread awareness on the issues of social inclusion and human diversity.

4 Conclusions

The orchard or the garden are effective and extraordinary setting for testing the principles of inclusive design. This is through the possibility of autonomously using places, systems and products for the cultivation and care of plants, and the comparison and enhancement of the differences between the different crops (and their cultures).

In this the Design for All, with its attention to the "design process", appears to be the most promising and effective approach to experiment and promote social inclusion and equality between individuals, peoples and different tastes, also through the daily activities related to the care of domestic plants.

References

1. Hakulinen, C., Pulkki-Råback, L., Virtanen, M., Jokela, M., Kivimäki, M., Elovainio, M.: Social isolation and loneliness as risk factors for myocardial infarction, stroke and mortality: UK Biobank cohort study of 479 054 men and women. Heart **104**, 1536–1542 (2018)
2. Tereshkovic, G.: Horticultural therapy: a review. HortScience **8**(6), 460–461 (1973)
3. Accolla, A.: Design for All. Franco Angeli, Milano (2009)
4. Di Bucchianico, G.: Food for All. Sala editore, Pescara (2018)

Designing of Inclusive Learning Experiences: Preliminary Outcomes of a Pilot Project Tailored to SLD

Alessia Brischetto(✉) and Alessandra Rinaldi

Laboratory of Ergonomics and Design, Department of Architecture, University of Florence, Via Sandro Pertini 93, 50041 Calenzano, Florence, Italy
{alessia.brischetto,alessandra.rinaldi}@unifi.it

Abstract. Thanks to the knowledge related to Ergonomics in Design and Inclusive Design, the present work aims at finding and providing an opened and integrated approach to the disciplines of cognitive psychology, pedagogy and human-computer interaction by presenting field experiments on learning and case studies. The latter show novel co-design practices through the use of Human-Centered Design and Inclusive Design methods, which were pointed out by means of experiments conducted in collaboration with professionals, end users and people with Specific Learning Disabilities. Through the presentation of a pilot project dedicated to the SLDs, the final output of the work was to illustrate a developed inclusive learning platform and its operating rules. Moreover, the research highlighted the need for systems based on social networking, supporting not only accessibility but also autonomy, socialization and the ability to collectively build new meanings.

Keywords: Inclusive Design · Human Centered design · ICT · Social inclusion · Web accessibility · Learning

1 Introduction

Learning in the digital age means having user-friendly and interactive systems capable of creating a series of new relationships between man and technology. In this regard, the Design is called to face, more and more strongly, with different disciplines, from the engineering-informatics to the humanistic-social ones. Therefore, it is necessary to adopt a bottom-up multidisciplinary approach, within which the design field interfaces with increasingly complex design dimensions. In recent years, computer accessibility and aspects regarding the use, distribution and visualization of contents, not only informative but also educational, became issues of great importance within the technology market, particularly within the Educational Technologies sector.

The survey of the Italian National Institute of Statistics (ISTAT), dating back to 2004–2005, reported a percentage of people with disabilities for an absolute value of about 2,600,000 individuals. The latest unofficial estimates have raised this figure to three million. In 2009–10 there were 15,884 students with disabilities enrolled in the University, a figure that has now tripled compared to ten years before. With regard to

G. Di Bucchianico (Ed.): AHFE 2019, AISC 954, pp. 22–34, 2020.
https://doi.org/10.1007/978-3-030-20444-0_3

Specific Learning Disabilities (SLDs), the statistical service of the Ministry of Education, University and Research (MIUR) provides relevant data for the 2014/2015 and 2015/2016 academic years. These data are organized by Region and school system according to the various levels, defined as pre-primary (3 to 5/6 years), primary (usually from 6 years to 11), middle school (age 11–14) and high school (from 14 to 19 years of age). In Italy, the statistical data for the academic year 2014/2015 state an average percentage of students with SLD of 2.1%. Excluding from the calculation the SLD diagnoses attributed to pre-primary school and its population, the percentage is 2.5%. The highest SLD values seem to be recorded in the middle school years. Comparing these results with those of the academic year 2015/2016, which has a predominant value of SLD certifications of 3.0%, an absolute increase of 0.5% emerges compared to the value recorded the previous year [1]. These alarming data are growing both in Italy and internationally. To face with this issue, within the framework of Educational Technologies, assistive and adaptive systems, including a broad range of technological devices (hardware, furniture, computer screens, software and so on), are usually employed to enable anyone to interact with more easily and effectively, regardless of any impairment or disabilities. Similarly, Learning Management System (blended Learning, E-learning, ecc.), born as electronic educational technologies, are largely used nowadays as networking environments to promote the integration and guide learning through flexible architectures based on current standards of web accessibility, platforms and content format [2]. However, it is noteworthy that the main limitation of these systems is ascribed to accessibility. This causes social exclusion, as it makes more difficult the access to information of people with different types of impairments. Despite the significant efforts, the accessibility of e-learning and web platforms and the use of assistive technologies are issues not entirely addressed yet. With this in mind, the present work studies how Design and related operational methodologies may support the process of analysis and development of complex digital products, which should be as much as possible inclusive. In particular, the preliminary outcomes of a pilot project tailored to SLDs, carried out in collaboration with experts and the C.R.E.D. center are presented. The main objective of the research was to design and develop a platform for inclusive learning dedicated to students over 11 years old. Summarizing, the work is therefore finalized to:

- assess the actual inclusive potential of technologies, in particular by analyzing systems designed for SLDs;
- technical aspects related to accessibility and use of contents;
- interaction styles (variability, transcoding, convergence, multimedia, hypertext);
- teaching and pedagogical aspects (organization, creation and sharing of teaching materials);
- didactic and pedagogical aspects for disables (universal access to information, organization of the study, didactic strategies, motivation);
- how to motivate and foster socialization (i.e. User Experience);
- design alternative scenarios.

2 Methodology

The design workflow was focused on the Human-Centered perspective, according to indications reported on the ISO standard (ISO 9241-210: 2010) [3]. The latter defines criteria for user-oriented design in systems based on human-machine interaction. The main criteria concern the clear knowledge of needs, tasks and contexts of use; the involvement of users in the design and development phases; the multidisciplinarity of skills and perspectives; the centrality of the evaluation phase; the iterative design process that provides for continuous steps and revisions ranging from consultation with users, to the analysis of needs, to the (re)definition of system requirements. In order to make available to users (People) who will use the system in specific contexts (Contexts) to perform certain activities (Activities) through appropriate technologies (Technologies), reference has been made to the PACT model (People, Activities, Contexts, Technologies) proposed by Benyon (2010) [4]. To explore all four categories of PACT, the research was structured in three macro-phases:

- Phase 1: Preliminary analysis and Understanding
- Phase 2: Evaluation phase
- Phase 3: Design and Envisoment

The first phase was dedicated to critically reviewing the reference literature and technical standards in order to get preliminary information about the educational sector, learning models and related technologies. It was also taken into account how and when Assistive Technologies (AT) should be used. Aspects related to the effectiveness and inclusive potential of learning technologies were subsequently investigated for evaluating the contribution of a combined approach between Inclusive Design (ID) and Universal Design for Learning (UDL) [5]. At the same time, the existing web accessibility standards and guideline (WCAG) [6, 7] and standards ISO[1] were analyzed and possible integrations with UDL approach evaluated. During the Understanding phase, qualitative procedures and tools were adopted to collect data on users' needs, but also to build a framework of habits and perceptions in terms of technologies. Usually the work of phase 1 precedes the experimental section (phase 2), since in the guidelines and good practices illustrated we find the necessary requirements to design a platform for inclusive learning. The methods applied were User Observation, Interviews and Focus Groups [8, 9]. The object of the evaluation was a hybrid platform developed by the research team through an Adobe XD interface simulator. The decision to create a platform for the simulation of activity was aimed at recreating all the best features identified in phase 1 and to verify the actual possibility, from the design point of view, to apply the technical standards of reference. Phase 2 was planned on the basis of the following objectives: (1) define the reference framework of the most widely used compensatory and dispensatory tools, (2) evaluate the instructional support that these

[1] It addresses software accessibility and usability (see also ISO 9241-110, ISO 9241-11 to ISO 9241-17, ISO 14915 and ISO 13407). Regarding AT: ISO/IEC TR 13066-3:2012 Information technology - Interoperability with assistive technology (AT) - Part 3: accessibility application programming interface (API).

tools are able to offer and their effective degree of inclusion, (3) measure the user experience and investigate organizational and motivational aspects of users (teacher-student-family), (4) define possible design scenarios.

The work activity of the last phase is an ongoing step. Preliminary functionalities for the development of a web platform will be discussed on the basis of the data collected in phase 2.

3 Results and Discussion

3.1 Preliminary Analysis

Currently on the market there are different types of products for SLD, some of these products are dedicated to specific categories of users, denoting a lack of tools effectively transversal. Looking at all that concerns the technological supports intended for subjects with SLD, it is possible to distinguish three categories of systems: compensatory, dispensatory and those used for the evaluation of the SLD degree. In our analysis we have taken more account of the first two. Compensatory systems aim to compensate for deficits due to SLD. In this category we find applications for both computers and mobile devices. Many of them consist of screen readers, software that, mainly in the field of dyslexia, through speech synthesis or with expedients to properly highlight words, assist the reading and suggest the correct decoding of the text. For example, in Italy there is "EdiTouch", a tablet that is proposed as a collection of the main compensatory tools for SDLs, such as an e-book reader, a calculator and a software for creating mental maps. Another Italian product is represented by the software created by Anastasis, which are computer programs mainly dedicated to the creation of mind maps, speech synthesis and support for study. The dispensing systems involve tools for dispensing some performances (reading aloud, taking notes), the possibility of organizing the study (personalized time of realization of the activities) and evaluation (the form is not evaluated but only the content).

These tools are not always available, some of them need to be paid and/or addressed to operating systems such as Windows. In a for all perspective, the possibility of the web to exploit the basic browsers is the most usable dimension and would also offer the possibility to monitor and implement the system over time. The same concept would also serve to improve the adaptability of AT (always and in any case necessary). Another feature found in the web is the ability to connect multimedia (such as YouTube platforms, and/or EVO, Fliker, SlideShare, Scratch by MIT, etc.) to the platform through the import or sharing of content. It was also found the ability of the web to promote socialization as well as to develop cooperative teaching materials. In terms of accessibility, the web also offers many indications at a regulatory level than the software market, which at the moment, especially in Italy, is poorly based on the study of accessibility in the broad sense.

In particular, ICT and network services have long been considered as an opportunity for growth and methodological-disciplinary reflection within educational and training environments. The substratum of ICT allows to significantly transform the articulation of productivity in a way that is no longer solely individual but cooperative and collaborative. To define pros and cons related to the adoption of ICT, it was made reference to the survey carried out by INVALSI [10]. Below are some aspects considered relevant:

- possibility to create customized teaching materials also for pupils with severe disabilities;
- the increase in self-esteem;
- respond to the need for flexible and shared tools;
- customization of contents to facilitate learning;
- motivation to improve the speed of writing autonomy;
- the work shared between teachers and experts from the school and from the university.

The use of ICT for disability carriers in educational environments simultaneously affects educational-methodological and technical issues regarding the selection of devices to be used. ICT, for its characteristics, would seem to have strength for at least three aspects including "motivation, strictness, adaptability." The use of computer, thanks to its flexibility, allows the customization of specific training processes, playing a key role on learning styles and abilities of each individual [11].

As far as the project dimension is concerned, we have witnessed the development of new disciplinary areas and project approaches with respect to the themes of inclusive learning over the years. These mainly refer to the field of Instructional Design and Universal Design for Learning (CAST). In particular, they refer to sectors such as the User Interface (UI), interaction design and User Experience, which are contributing to the growth of this disciplinary sector in qualitative terms [12]. While usability assessment methods allow us to measure the actual level of accessibility, HCD and UX methods can provide us with qualitative parameters on emotions and pleasantness (strategic aspect with regard to the educational dimension). Inclusive design and, above all, in the field of information technology and education, Universal Design, are today treated as disciplines of Instructional Design. Within the field of educational technologies, the UD (Universal Web Design Principles and guideline) approach and the educational studies of the Universal Design for Learning (UDL) emerge more clearly [5]. The UDL approach, combining the principles of UD and neuroscience-related research on different ways of learning, formulates the following principles (CAST, 2011):

- support models of recognition learning and at the same time provide multiple and flexible methods of content presentation;
- supporting strategic learning through flexible methods of expression and training;
- support affective learning and provide multiple and flexible options to stimulate engagement.

The UDL provides an interesting perspective and a solid foundation allowing to state that it is possible to pass from the concept of "special adaptation" focused on disability to universal design (for all), by valorizing individual differences and taking advantage of the potential of inclusive technologies [13, 14].

Identified in the UDL a viable base of operations to define strategies for planning and structuring learning in accessible way, it is noteworthy that WCAG, Universal Web

Design Principles and guide line and UDL have in common the attention to customize how to display the information, the availability of alternatives to the audio and visual content through the use of different media (even in support of understanding), the readability and comprehensibility of texts and compatibility with assistive technologies.

These aspects were verified and considered in the development phase of the platform for the simulation of activities and by the functions of phase 2.

3.2 Evaluation Phase

In the evaluation phase, web platforms for supporting learning at school and at home were analyzed. The sample of users considered foresaw the involvement of users aged 11 to 13 years and teachers involved in formal and informal training. As this is an experimental research, the aim of this work has been to define an intervention strategy implemented over time, which is expected to be extended to schools and universities. The sample analyzed concerned 5 teachers and 25 students (70% of these SLD).

Step 1: Preliminary Accessibility Review (Expert Evaluation): Checking conformance to guideline and Standards (UD, UDL and WCAG, ISO). In this phase, direct observation sessions were carried out with the involvement of a small sample of users and with the support of experts. In this way it was possible to make a preliminary analysis of the most popular platforms for reading and understanding the text. In particular, through the Task Analysis (TA) [15], the interactions on the teacher's side and then on the student's side were modeled. Once the architecture and functionalities of the platforms had been defined, Nielsen's Usability heuristics, Sheinderman's 8 golden principles of goods interface design were applied [5, 9–16]. The objective of this phase was to validate the correspondence and potentiality of the UDL guidelines and the technical and operational standards identified in phase 1 and to define the objectives of the following phases.

Step 2: User-Based Evaluation: Methods applied were Interviews and Focus Groups. Object of the evaluation at this stage was a hybrid platform developed through a simulator of Adobe XD interfaces. The choice of recreating a simulator was made to outline all the best features identified in phase 1 and the actual possibility from the design point to apply the technical standards of reference.

The aspects observed with respect to the process of organizing the teaching and monitoring of the path by the teachers were: (a) organization of the contents and planning of the training course, (b) creation and sharing of the teaching material, (c) monitoring and personalized support.

The aspects that were observed with respect to the students' learning process were as follows: (d) Access to the information to be studied, (e) understanding of the text, (f) re-elaboration of the information, (g) storage, (h) retrieval of the contents.

To evaluate these aspects a platform that allowed to simulate the functionalities present in Udl Editors, UDIO of the CAST was designed. It also includes open source compensatory tools with functions for speech synthesis Clip Claxon, LeggiXme and Balabolka, for text decoding (Memory for images and words by C.R.E.D and Super-Quaderno di Anastasis, LeggiXme_Jr_SP etc.), mental and conceptual map processing tools (Super Maps, Cooperativa Anastasis, Super Maps EVO, C-Map Tools) and for lesson creation (Facile Facile web platform by C.R.E.D). Figure 1 illustrates some of the functions of the platform created.

The hybrid platform has been tested through direct observation sessions, and each session included the support of experts. The test sessions were structured following the same flow of activities.

Subsequently, on the basis of the data emerging from the evaluation sessions, 2 Focus Groups were organized, one aimed at the teaching dimension and the other at the students. Semi-structured interviews were carried out within the Focus Groups. In particular, this article contains the considerations that emerged and the design requirements identified and discussed with the teachers and experts involved in the survey.

(d) Access to the Information to be Studied: 40% of the observed users read the text several times highlighting a difficulty in the legibility of the font. The presence of textual indexes highlighted and subdivided in a coherent and repeated way in the screens (the title, the images, the subtitles, the words highlighted) have emerged as fundamental. The ability to select the level of difficulty was seen as negative before the test of almost all users observed. Although at the end of the session they reported that they had taken advantage of the possibility of being able to select difficulty levels; 25% said that this feature stimulated them to improve. Voice synthesis is 70% a strategic element and a support to understanding, the same for the dictionary. The dictionary should be revised because the size of the text and external links (Wiki, Translator, etc.) have not been perceived as exhaustive.

Design Ideas: Introduce topic without reading the entire text, possibility to customize the font (OpenDyslexic, Dyslexie, Read Me!, Bianconero, EasyReading) and the line spacing. Introduce the possibility to select text and activate speech synthesis in a more dynamic way through an enhanced tool. Possibility to integrate the notes taken in class to the text being studied. Insertion of verification tools in the form of gaming (scaffolding and tutoring system). This would allow the exchange and sharing of the study experience with others or with a tutor. Create the possibility of integrating content with extra material created by the student.

(e) Text Comprehension: For an LSD student, understanding a text read on its own is often very complex. This can be explained by the fact that, considering the problems in the reading process, all his attention is channelled into reading correctly, leaving out the meaning of the text. The observations made by the teachers and experts were as follows:

- the possibility of dividing the text into smaller parts, giving the possibility of analyzing one part at a time of the text;
- presence of the priming, i.e. providing questions to be read before proceeding to the reading of the text, in order to focus their attention on the salient points;
- possibility to insert and/or identify the key words first in the question and then in the text;
- possibility to highlight the keywords in the text during the reading.

Fig. 1. Screen shot of the hybrid platform - student version 2.0: (a) Selection of the levels of support to the reading (b) Multimedia resources (c) Link window to select support levels and text line spacing (d) Text HELP Toolbar (e) Table of Contents.

Remarks: This phase has been linked to the evaluation phase involving teachers (a). The teacher expressed the need (as on Facile Facile) to create the contents himself and not find all the topics already packaged. This presupposes that the platform is conceived on the one hand for the teacher and on the other for the student. The teacher side, in addition to the usual registration phases, should allow the teacher to insert all the chosen elements necessary for the construction of the teaching units. This is especially strategic for the pedagogical aspect and to follow the student throughout his learning path. Consequently, a significant aspect is the possibility to include learning verification functions.

Design Ideas: Create a structured teaching area that allows you to organize the teaching units by topics and objectives. The architecture section of the functions must be connected to the student section, the teacher follows a format and obligatory operational phases. Consider the possibility of creating a register of the activities carried out by the student and the possibility of activating customized tutoring tools. Stimulate collaboration and sharing of material developed by individual teachers through a shared space.

(f) Re-elaboration of Information: Content reworking is fundamental for learning, and completes understanding by preparing for memorization. The deeper the processing of the information received, the easier it is to store and retrieve it. There are many ways to elaborate what you study, that is to reorganize and manipulate the contents of a text, and what certainly gives more results for a DSA is the construction of a conceptual map, that is a scheme, that visually organizes all that is to be studied and memorized. The dimension of the graphic representation of the contents has been tested, in which the key words previously highlighted are inserted, linked together by arrows and enriched by images to facilitate visual memory. The maps were made by the student (this aspect is fundamental) and not already provided by the teacher or the system, precisely because it is the graphic representation that allows the cognitive stimulation of the student. From the focus group activities it emerged that each user has developed his own personal version in the organization of the contents. The users highlighted the need to develop the map on paper before making it on the platform. This aspect opens a reflection on the aspects related to the manipulation and tangibility of learning that should be considered during the development of the system.

Design Ideas: Possibility to create maps simultaneously and provide the possibility to import paper maps convertible into interactive digital format. Today, there are products that in this direction could increase the link between technology and disadvantage (see: Osmo™, Sifteo Cube). If you access the platform via tablet, you can use the drawing tool through digital pensil.

(f) Re-elaboration of Information: Content reworking is fundamental for learning, and completes understanding by preparing for memorization. The deeper the processing of the information received, the easier it is to store and retrieve it. Here are many ways to process, reorganize and manipulate the contents of a text, and what certainly gives the most results for a SLD is the construction of a conceptual map, that is, a scheme, that visually organizes everything that is to be studied and stored. The dimension of the graphic representation of the contents has been tested. The key words previously highlighted are inserted, linked together by arrows and enriched by images to facilitate visual memory. The maps were made by the student (this aspect is fundamental) and not already provided by the teacher or the system. It is the graphic representation that allows the cognitive stimulation of the student. From the focus group activities emerged that each user has developed his own personal version in the organization of the contents. The users highlighted the need to develop the map on paper before making it on the platform. This opens a reflection on the aspects related to the manipulation and tangibility of learning that should be considered during the development of the system.

Design Ideas: Possibility to create maps simultaneously and provide the possibility to import paper maps convertible into interactive digital format. Today, there are products that in this direction could increase the link between technology and disadvantage (see: Osmo™, Sifteo Cube). If you access the platform via tablet, you can use the drawing tool through digital pencil.

(g) Storage: The analyzed users requested to view the information acquired in the previous sections within all the tabs. Users also had difficulty remembering certain words or topics. The navigability for horizontal tabs was more functional than the vertical scrolling of traditional web pages as well as the ability to deepen the content through the viewing or listening to multimedia material.

Design Ideas: To offer a simple reiteration of the information sought during navigation. Support the association between the information to be remembered and familiar elements through the activation of symbols and/or photographic material and/or the possibility of highlighting portions of text or image in an alternative way, compared to the functions hypothesized in point (e). Possibility to organize the information on the basis of those previously stored. To make possible the transformation of a complex information into a simpler one through a keyword in order to generate a process of meta-cognition.

(h) Recovery of Contents: Once users have created their own concept map and therefore started the memorization process, it has not been possible to verify in an exhaustive way the learning, especially if the user does not have the possibility to expose it to others or recover it from the memory. Some students said that they are used to repeating out loud the contents of their map or the text they have read. Repeating the subject they learned to someone else was also useful for getting feedback about their ability to expose themselves.

Design Ideas: Evaluate the possibility of recording your own voice and hearing it again to encourage self-correction within the platform with the possibility of being able to listen to it again, or to make video calls with your friend or school friend.

Note: If the user has difficulty in repeating what he has represented in the map, this means that the map is not adequate to support him in learning and then in exposure; consequently, it will be necessary to make some changes that the student himself will have to make to understand which concepts are missing, which keywords can help him and which links can facilitate the memorization of the speech.

In addition to the above, other considerations emerged, which were fundamental for the development of the pilot platform. These included the need to include tools and functionalities for the organization and management of time, which was found to be critical for SLD students. Moreover, some hardware tools such as the PC keyboard were more difficult to use than touch devices. This opened up an interesting question and the need to investigate more on the interaction modes. Multimedia material and hypertext revealed to be strategic for SLDs, although a current and ordered layout is necessary to prevent cognitive crowding. This is a determining aspect in the phase of content creation by the teacher. Compensatory tools have not been effective if not used consciously. Automation practices must therefore be preceded, the tool is effective only if it takes the abstract form of a communicative bridge, built in a conscious way.

The systems of self-correction and verification of learning favor meta-cognition, at the same time it is necessary to pay attention to how this agents are administered (the user does not like to be corrected by an adult or by a teacher). The subject understands why he is wrong, and through it the sense of what he is doing is amplified (significant learning). Therefore, it is in this phase that the conscious and non-automatic use of the instrument allows to generate strategies in the name of autonomy. The subject thus has control of what he does and consequently experiences modes of learning until the identification of the preferred practice.

The use of collaborative platforms allows to operate in an inclusive perspective. Any student, regardless of their condition, can learn in a dynamic and collaborative way thanks to others.

In conclusion, the aspects related to the design dimension of the teacher-side content have been the subject of the preliminary design phase, while those of the student-side learning support will be explained in the next paragraph.

3.3 Phase 3: Design and Envisoment

On the basis of the data collected in phase 2, a web platform is currently being developed with a horizontal layout for supporting and sharing learning. The objective of the preliminary project was to develop a conceptual version that allows to evaluate from a technical and graphic point of view the project ideas that emerged in phase 2. The platform will aim to support the following functionalities:

- Organization by degrees of learning support (maximum, moderate, minimal) selectable and modifiable at any time by the user
- Presence of pedagogical agents who support readers in understanding the text through contextualized explanations, operational suggestions and guided questions. These are supposed to be customized by the student through the development of features: Create your own Avatar.
- Presence of text-to-speech technology. Through this tool it will be possible to activate on request the translation of foreign language words. This tool will be provided through audio and text formats to facilitate correct pronunciation allowing for in-depth analysis of words and phrases with direct searches on the Google engine;
- Presence of reading supports such as synthesis and images
- Presence of indications on the correct ways to achieve a careful and effective reading at the level of understanding and appropriation. The strategies proposed consist in predicting, asking questions, visualizing and summarizing the contents of a narrative;
- Tools for creating interactive maps. It will be tested the possibility of importing maps made through digital media that allow you to store the gestures of the drawing - tablet/scanner
- Presence of feedback on several levels to promote self-correction and to encourage problem solving
- Insertion of content manipulation tools. Predict, ask questions, view and summarize content.
- Presence of external links for in-depth analysis (teacher side and student side)
- Integrated comprehension support tools, including a multi-media glossary that can be linked directly from the text for more difficult terms
- Collaborative section activated by a specific icon and activated by choice connected to a section that provides for the exchange and collection of processed materials, rather than dialogue through the chat
- teaching area that allows you to organize the teaching units by topics and objectives. The function architecture section must be connected to the student section, the teacher follows a similar format and operational steps for all. This is to avoid the creation of uneven teaching material.

The platform is currently under development and, as mentioned in paragraph 2, the beta platform will be re-tested and implemented on the basis of the criticalities that will emerge during the evaluation phase.

4 Conclusion

The aim of this experimentation was to define, through the development of a hybrid pilot platform, the functionalities that favor the autonomy process and the problem solving strategies in learning. The experimentation has shown that doing "problem solving" means developing a critical sense (also through others). Active interaction with teaching material within a web container favors these practices and is a positive aspect from the pedagogical point of view. In this sense, the most significant tools are those for the development of concept maps that allow to rework concepts by strengthening key aspects, voice reproducers, tools for the organization and structuring of content and programs of temporal organization. Within this framework, the experimentation allowed to identify strategic aspects in order to support SLDs in the learning process, with a broader approach that addresses all regardless of their capabilities. The future objective of this work will be to extend the observations to the dimension of visual, auditory and cognitive disabilities. In perspective for all, the success of this experimentation will lie in its ability to be compatible with assistive technologies as well as supporting all teachers and students, regardless of their condition.

Acknowledgments. The interviews were conducted as part of collateral activities of the LED laboratory of the University of Florence, in collaboration with experts of the C.R.E.D center (Centro Risorse Educative Didattiche). For the interviews and observation Dr. Gianna Teri, Dr. Claudia Durso – expert pedagogue DSA, Dr. Enrico Rialti - Psycholinguist and Learning Tutor, Gianni Fadda - web designer CRED and Alessandro Guarducci (social cooperative Macramè) are kindly acknowledged.

References

1. MIUR-ISTAT. http://www.istat.it/it/archivio/194622
2. Seale, J., Cooper, M.: E-learning and accessibility: an exploration of the potential role of generic pedagogical tools. Comput. Educ. **54**, 1107–1116 (2010)
3. Benyon, D.: Designing Interactive Systems (2: a uppl.). Pearson Education Limited, Harlow (2010)
4. BS EN ISO 9241-210:2010: Ergonomics of human-system interaction, part 210: human-centred design for interactive systems (2010)
5. CAST- UDL Guidelines. http://www.udlcenter.org/aboutudl/udlguidelines
6. Web Accessibility Initiative. http://www.w3.org/WAI
7. W3C provides 'Web Content Accessibility Guidelines 2.0'. http://www.w3.org/TR/WCAG
8. Cooper, L., Baber, C.: Focus Groups in Handbook of Human Factors and Ergonomics Methods. CRC Press, Boca Raton (2005)
9. Hirai, Y., Morita, Y., Elokia, N.: Evaluation of usability methodologies in the universal design process. In: Proceedings International Conference on Engineering Design 'Design for Society: Knowledge, Innovation and Sustainability' (CD ICED), vol. 7, p. 91, August 2007
10. INVALSI: MonVal – Monitoraggio e Valutazione Progetto Nuove Tecnologie e Disabilità - Azione 6, p. 12 (2010). https://www.invalsi.it/invalsi/rn/doc_monval/6.Reportazione6.pdf
11. Brischetto, A., Tosi, F.: Improving learning technologies and social inclusion through human centred design and universal design approaches: novel designing scenarios. In: Advances in Design for Inclusion, pp. 39–50. Springer, Cham (2016)

12. Petrie, H., Bevan, N.: The evaluation of accessibility, usability, and user experience. Univ. Access Handb. **1**, 1–16 (2009)
13. Boucher, A. R., Evans, M., & Graham, S.: Udio: Rich and authentic literacy experiences for struggling middle school readers. In S. A. Crossley & D. S. McNamara (Eds.), Adaptive educational technologies for literacy instruction. New York, NY: Taylor & Francis, Routledge, (2015)
14. Coyne, M., Evans, M., Karger, J.: Use of a UDL literacy environment by middle school students with intellectual and developmental disabilities. Intell. Dev. Disabil. **55**(1), 4–14 (2015)
15. Hollnagel, E.: Task analysis, why, what and how. In: Handbook of Human Factors and Ergonomics, vol. 4, pp. 385–396 (2012)
16. Steen-Hansen, L., Lundh, M.V.G., Chen, W.: How universal design principles can enhance the interface of 3D printing programs. In: DS 78: Proceedings of the 16th International conference on Engineering and Product Design Education (E&PDE14), Design Education and Human Technology Relations, University of Twente, The Netherlands, 04–05 Sept 2014, pp. 123–128 (2014)

Universal Signaling Based on the Articulation Between the Ergonomic Practices and the Perception of the Visually Impaired

Luís Rocha[1,2(✉)], Thaíla Santos Lima[1], and Lays Campos Brito[1]

[1] Instituto Federal do Maranhão - IFMA,
Avenue Colares Moreira, number 477, Renascença, São Luís, Maranhão, Brazil
luis.rocha@ifma.edu.br, thailaslima7@gmail.com,
layxcampox@gmail.com

[2] Cesar School, Cais do Apolo, 77, Bairro do Recife, Recife, Pernambuco, Brazil

Abstract. Every signaling system involves a range of issues of communication, visual perception, semiotics, pictograms, signage, writing, among others, in order to guide and inform, offering the necessary accessibility conditions for anyone, regardless of their condition. The present work proposes to continue to rethink the construction of inclusive educational environments from the integrated reflection of research related to ergonomics, participatory practices in design and sensory experiences that allow the development of alternatives for a universal signaling system centered on the user, including the visually impaired user, of the São Luís - Monte Castelo Campus of the Federal Institute of Maranhão.

Priority to understand the experience lived by the visually impaired in a theoretically open environment for all was the starting point to see the difficulties faced by this public. However, the user's participation in the process consolidates the opening of new paths that achieve the most significant results. Thus, considering the participation of the visually impaired user with the intention of proposing improvements in existing methodologies or in the process of creating the solution gains more and more strength as a strategy to promote intelligent and accessible environments.

Keywords: Accessibility · Inclusion · Ergonomics · Signage · Universal design

1 Introduction

Throughout the world there has been an increase in the number of people who need the help of some resource to develop their daily activities due to some difficulty or sensory incapacity. The deficiencies are affecting individuals, while, unequally, research and possible solutions and technologies are also advancing in order to balance this gap.

Brazil, with almost 519 years, grew under strong social, cultural and economic influence of the European countries, its formation and construction were conceived and constructed from the adaptation of the material and socioeconomic conditions, disregarding the necessities of the deficient minority of the time.

G. Di Bucchianico (Ed.): AHFE 2019, AISC 954, pp. 35–43, 2020.
https://doi.org/10.1007/978-3-030-20444-0_4

Time has passed and the world has opened up to think and design things that can be used interchangeably by society, regardless of their condition. It took many years for the country to open its eyes to see social needs. When this happened, laws and standards were created to guarantee the rights of those with disabilities. Today, ensuring that these guarantees are met is a goal to be achieved.

Gradually the disabled, have been seeking to exercise their rights of care, accessibility and autonomy. For this to happen, there needs to be a change in behavior and especially in the construction and adaptation of projects, with a more human thinking in conceptions.

In this way, design has been gaining strength because it is a tool capable of repairing failures and injustices by balancing solutions that can involve people's needs to the peculiar aspects that involve social and environmental issues. Both participatory design and universal design work so that your products and services meet as many people as possible, that is, with the participation and use of all. For the visually impaired, the design aims to contribute to the elimination of barriers that impede the access of the handicapped, thus allowing a significant improvement in the well-being and the quality of lives of all.

2 Understanding Disability

Deficiency is part of nature, like other animals, the human being is also prone to have problems that can cause impediments in their interaction with the environment.

Thus, disability is any partial or permanent handicap or incapacity, which limits the individual to develop his activities in its fullness.

This definition of disability has always been based on the medical and individual aspects of the people, which are related to the medical model. Since the 1970s, the way in which disabled people are seen has changed, mainly due to the growing number of people who have become part of the disabled group. In order to have more arguments about human conditions and to propose solutions that reduce the segregation of these people, the United Nations began to investigate these aspects and began to adopt another model that is based on the condition of the individual, from a structural and social perspective, "In which people are seen as disabled by society and not due to their bodies" [11], thus beginning the transition to a new model, the social, adopting the CIF, International Classification of Functionality, Disability and Health.

ICF, adopted as the conceptual model for this global report on disability, comprises functionality and disability as a dynamic interaction between health problems and contextual factors, both personal and environmental. Promoted as a "biological-psychic-social model," the report represents a viable compromise between medical and social models. Disability is a comprehensive term for impairments, limitations to perform, and restrictions to participate in certain activities, which encompasses the negative aspects of the interaction between an individual (with a health problem) and the contextual factors of that individual (environmental and personal factors) [11].

This model now treats the diversity of disability as complex, so that people are no longer part of the traditional generalized group of sensory problems and disabilities, which includes hearing, visual, mental and motor deficiencies, being part of a group whose disability is defined by the relation of their problems involving environmental and personal factors, the latter involving "differences in terms of gender, age, socio-economic status, sexuality, ethnicity, or cultural heritage" [11].

The Brazilian State, through Decree Number 3.298 of 1999 [2], considers the disabled and incapable, all those with loss or abnormality of physiological, anatomical or psychological functions, and those that due to some event have not been able to restore the autonomy in the performance of their activities and because of this they need the help of adaptations, equipment or special resources. The Brazilian Federal Constitution of 1988 recognizes the condition and the rights of the disabled, and advances in the creation of other laws that since then have been extending these fundamental guarantees.

Law Number 13.146 of 2015 [3] establishes the inclusion of the disabled person of any nature and has full participation in society, in order to allow equal and unconditional access to spaces and buildings, equipment, furniture, information systems and technologies, as well as services and facilities of a collective nature, whether public or private.

3 Visual Impairment

According to data from the OMS [11], about 40 million people with vision problems worldwide, in Brazil, according to data from the World Health Organization, are deficient visually impaired. last Census [9], the number of people with vision problems reaches approximately 36 million, representing 3.5% of the population (Fig. 1).

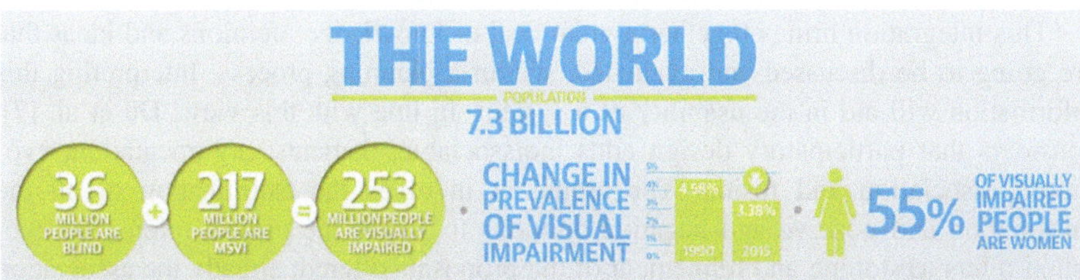

Fig. 1. Data from the international agency for the prevention of blindness.

Visual impairment consists in reducing the level of visual acuity, that is, the visual system, which reacts through stimuli, has the function of converting light energy into neural activity with meanings recognizable by the human being, cannot work correctly, soon, the formation of the images are totally or partially compromised.

According to the International Agency for the Prevention of Blindness [8], problems with the visual system can occur congenital or acquired. The most common causes of visual impairment are due to problems related to cataract, glaucoma, age-related macular degeneration, retinopathies, optic neuropathies and amblyopia, and according to Carletto [6], are presented in several types:

Tubular vision - the individual sees only what is directly in front of him and nothing that is next to him, while others see objects as a puzzle missing one or two pieces; Low vision - the person sees very little, but still they are able to use the vision for the planning and execution of a task; Blindness - when there is no perception of light causing the individual to see absolutely nothing.

Knowing the levels of visual capacity is essential, above all, because the visually impaired individual tends to discover new means and use the other senses to fill the absence of vision, so that they can relate to the environment and carry out their activities.

4 Participative Design

In the last decades, design has undergone several transformations regarding its performance, when it is no longer focused on the project activities with a high degree of complexity, applied to manufacturing and consumption on a large scale, to become a constantly evolving activity, from of the improvement of their practices, which have expanded to the most diverse areas, yet the needs of the human being, have become their center of interest in all these practices.

One of these evolutions was social insertion in its practices, as a way of seeking collaboration for the understanding and resolution of the most varied and complex problems.

Collaboration has a fundamental role in enabling the relationship of different groups with a common interest, having positive effects on the performance and communication of those involved, as well as contributing to the success of the project.

In this sense the participatory design in its planning, it relates to several integrated areas, such as interaction design, interface and product design, architecture and information technology, all of which are user centered.

This integration brings together a great deal of knowledge, opinions and ideas that are going to be discussed and evaluated, within a planning process. Interpreting this information will aid in the usability of a system. In line with this view, Du et al. [7], considers that participatory design adds indissociable elements: argumentation, evolution, association and fusion. The argument involves the examination of all the information and positive and negative opinions to which the project refers; the evolution refers to stoning and refinement of the proposals offered; already the association is when an idea or proposition encourages the other idea, thus conceiving a new one; and finally the merger is the possibility of unification of proposals that are similar.

According to Camargo and Fazani [4] "participatory design assumes that people are always participating, even if unofficially, because the adaptations they make to use their tools are also creative actions, that is, the use is not considered mere consumption". This engagement, which must be democratic in nature, fosters the construction and alignment of perspectives that lead to the construction of solutions.

5 Universal Design

Since its inception, design has always had a clear purpose, "to put order in the mess of the industrial world" [5]. However, the deep transformations that occurred in all social, political, economic and cultural scenarios have meant that within the process of design evolution, it also added to its organism the need to offer more human values, as well as to make its solutions, products for everyone.

Recently, at the center of the discussions, design begins to assume this role of designing things, but things that meet not only needs, but also that can be used by any individual, regardless of their conditions, skills, culture and whatever be their differential for the others. This is how universal design gains notoriety by adding to its solutions elements that extend the possibility of use to most people and can still simplify the life of all of them.

In this way the universal design, it aims at designing objects, equipment and structures of the physical environment intended for use by the majority of people, without recourse to adapted or specialized projects, and its objective is to simplify the life of all, whatever the age, stature or capacity, making the products, structures, communication/information and the built environment usable by as many people as possible, at low cost or without extra costs, so that all people and not only those with special needs, even temporary ones, can fully integrate into an inclusive society [10].

Contrary to popular belief, universal design does not converge with unnecessary actions that are intended to serve specific audiences, such as solutions that target only people with disabilities or an age group. That's why they are basic principles of universal design:

Equality in the possibility of use: comparable use, for people with different abilities;
Flexible: adaptable to a wide range of preferences and abilities;
Simple and intuitive: easy to understand for any audience;
Effective information gathering: communicates effectively the necessary information;
Safe: error tolerant and minimizes the risk of unwanted actions;
Minimized physical effort: requires little physical effort;
Comprehensive: Provides appropriate size and space for access and use.

Thinking about a universal project often does not necessarily involve major interventions, first of all there must be empathy, sensitivity and small attitudes that aim to improve things for everyone. The universal design exists so that things and places are prepared for any individual, be it architectural, functional, or informational questions.

6 The Development of the Project

For the beginning of this project, we sought theoretical basis in articles, laws, legislation and other materials that could feed our repertoire of information, with the purpose of enabling the analysis of a given environment, with respect to an inclusive environment.

Based on the idea that a public teaching environment should be as accessible as possible and always ready to receive any individual, information surveys were carried out at the Federal Institute of Maranhão - Campus São Luís Monte Castelo, seeking to observe the main aspects related to the existence or not of a system of visual communication and accessibility conditions offered, with the aim of offering contributions that can minimize the daily barriers, be they architectural or communication. Assuring thus, the autonomy of the people, and especially those with disabilities, giving them the opportunity to commune with an egalitarian environment. In this way, we focused on experiencing the challenges and difficulties experienced by people with visual impairment.

To do this, a mapping of three main routes used by a blind student was carried out, starting from his arrival on the campus, until his move towards the classroom, library and cafeteria.

For those who use public transportation, whatever the bus line, it is possible to disembark a few meters from the main entrance of the Campus São Luís Monte Castelo, and in this first contact, it is possible to perceive the lack of tactile signaling that directs to its dependencies. Even for those who see it is possible to find some sign, sign or any type of inscription in the surroundings and the facade that identifies the place. However, there is a wide entrance structure, which, although not fully adapted to the ergonomic accessibility standards, allows the entry of handicapped people to move quietly. This initial structure has a ramp with handrail and body guard, which due to its steep slope, should have at least one segment or rest area, following the orientation of the Brazilian Association of Technical Standards – ABNT (Fig. 2).

Fig. 2. View of campus areas. Source: Authors. January 2019.

This standard, NBR 9050 [1], says that only ramps can be considered if the tread surface has a slope of 5% or more. And according to the same standard, for a ramp to be accessible, maximum slope limits, slopes that need to be overcome and the maximum number of segments are set. The value determined for the slope represents the relation between its height and length, represented as a percentage, as it is possible to observe the values arranged in the following Table 1:

Table 1. Calculation of incline ramps. [1].

Maximum slopes of each ramp segment h (meters)	Permissible slope on each ramp segment *i* (%)	Maximum number of ramp segments
1,50	5,00 (1:20)	Without limit
1,00	5,00 (1:20) < $i \leq$ 6,25 (1:16)	Without limit
0,80	6,25 (1:16) < $i \leq$ 8,33 (1:12)	15

In practice a ramp would be as in the image below (Fig. 3):

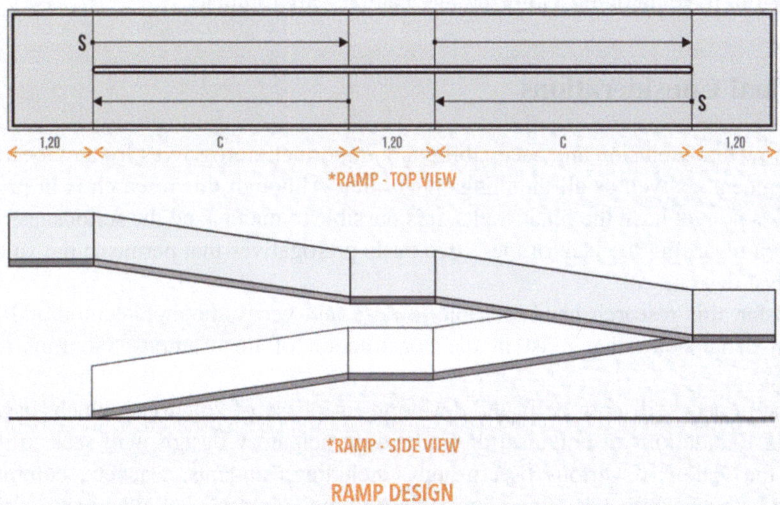

Fig. 3. Ramp design. [12].

Following the course towards the classroom, we come across long corridors, without any type of signage, are ways with insufficient widths for side-by-side passage of a person and a wheelchair, besides possessing several obstacles, buckets and garbage baskets, drinkers, traps for rodents, among others, that can cause some accident for people with disabilities. For arrivals in some rooms, the accesses through the corridors lead to staircases that further hinder the path of the visually impaired student.

In the reverse direction, going to the library and cafeteria, it is not possible to carry out a shortcut on the way, as do the sighted students, since the ramps of access to another level are punctual, and take an even more complicated path.

In many of the rooms observed in the trajectories covered, there were some signs, some paper, others adhesive and some even using more rigid materials, however, all of them are out of the accessibility standards, do not have any of the prerequisites of a signaling, who will say an inclusive communication, that has sensorial resources.

There were also some observations related to other environments, such as the campus restaurant, which even has plates, a table with information on prices and products offered, an ordering desk and shopping receipts, but none of this information is presented in Braille or in the standard communicational communication. The Theater also has only a small plaque, without Braille. Locomotion conditions for disabled people inside the theater are not favorable and seating arrangement makes it even more difficult. In relation to the use of the bathrooms there is also no information and the environment is not able to meet the needs of a student with reduced mobility or wheelchair, and as for access to the upper floor of the theater, it is also difficult to access for those who have visual difficulties and others deficient because they have only stairs, with uneven stairs and different heights, which makes the route dangerous.

As it is a project that is in progress is still being carried out the diagnosis of other environments and in parallel the initial study of a signaling proposal that enables a disabled to have autonomy to reach any campus environment.

7 Final Considerations

We know that inclusion and accessibility are important prerogatives for any location or environment, as well as all the things they offer. Although this research is in progress and does not yet have the final results, it is possible to understand the seriousness of the situation regarding the lack of these two basic prerogatives that permeate the values of universal design.

So far, this research has been able to raise and verify the architectural and information deficiencies that exist in the construction of the Campus São Luís Monte Castelo.

The project will now be in the development phase of solutions, which, based on guiding foundations in universal design and participatory design, will seek to understand the vision of various user groups, including: students, teachers, community, people with and without with a view to developing solutions that effectively fulfill the task of communicating and helping people.

References

1. ABNT: Brazilian Association of Technical Standards. Accessibility to buildings, furniture, spaces and urban equipment (2015). http://www.abnt.org.br. Accessed 01 Jan 2019
2. BRAZIL: Decree Nº 3.298, of December 20, 1999. It establishes the National Policy for the Integration of Persons with Disabilities, consolidates the norms of protection, and gives other measures. Official Gazette of the Federative Republic of Brazil, Brasília, DF, 21 December 1999. http://www.planalto.gov.br/ccivil_03/decreto/d3298.htm. Accessed 01 Jan 2019
3. BRAZIL: Law Nº 13.146, of July 6, 2015. Institutes the Brazilian Law on the Inclusion of Persons with Disabilities (Statute of Persons with Disabilities). http://www.planalto.gov.br/ccivil_03/_Ato2015-2018/2015/Lei/L13146.htm. Accessed 10 Jan 2019
4. Camargo, L., Fazani, A.: Exploring participatory design as an information systems development practice. CID: J. Inf. Sci. Documentation **5**(1), 138–150 (2014). http://www.revistas.usp.br/incid/article/view/64103. Accessed 13 Feb 2019
5. Cardoso, R.: Design for a Complex World. Cosac Naify, São Paulo (2013)
6. Carletto, A.C.: Handbook on Coexistence - People with Disabilities and Reduced Mobility, 2nd edn., enlarged and revised. Realization Institute Mara Gabrilli (2012)
7. Du, J., Jing, S., Liu, J.: Creating shared design thinking process for collaborative design. J. Netw. Comput. Appl. 35, 111–120 (2012). https://www.sciencedirect.com/science/article/pii/S108480451100052X#s0025. Accessed 13 Feb 2019
8. IAPB: The International Agency for the Prevention of Blindness. Blindness and Visual Impairment: Global Facts (2017). https://www.iapb.org/vision-2020/who-facts/. Accessed 13 Feb 2019
9. IBGE: Demographic Census (2010). www.ibge.gov.br. Accessed 02 Jan 2019

10. INR - National Institute of Rehabilitation: Universal design (2014). http://www.inr.pt/content/1/5/desenho-universal. Accessed 13 Feb (2019)
11. OMS - World Report on Disability (2011). http://www.who.int/disabilities/world_report/2011/report.pdf. Accessed 01 Feb 2019
12. Souza, E.: Designing ramps accessible according to NBR 9050 (2018). https://www.archdaily.com.br/br/891636/projetando-rampas-acessiveis-segundo-a-nbr-9050. Accessed 14 Feb (2019)

Industrial Design for Inclusion

Accessibility Evaluation of Automated Vending Machines

Nicholas Caporusso(✉), Kingsley Udenze, Asibi Imaji, Yangyang Cui, Yanjun Li, and Spencer Romeiser

Fort Hays State University, 600 Park Street, Hays 67601, USA
n_caporusso@fhsu.edu, {kzudenze,aoimaji,y_cuill_sia, y_li277,s_romeiser}@mail.fhsu.edu

Abstract. Electronic vending machines are a wide-spread system for delivering and accessing goods: the possibility of operating in a completely automated fashion and independently from business hours makes them extremely convenient for serving public and private locations. However, they fail their purpose in serving individuals with special needs: their lack of accessibility prevents users who are blind or have severe visual impairments from autonomously purchasing products. In this paper, we focus on the ergonomics of electronic vending machines and their automatic payment systems: we discuss their human factors and, specifically, usability to individuals who have severe low vision. Furthermore, we present a user experience analysis of their components and we detail the results of a study that evaluated the performance of different common types of commercially-available vending machines in the three steps of a purchase, that is, product selection, payment, and collection. Our results demonstrate that poor design, lack of tactile information, and absence of non-visual feedback have a significant impact on accessibility and prevent most customers who are blind from being able to independently select, pay, and collect a product.

Keywords: Accessibility · Blindness · eVending · Automated payment systems · Americans with Disability Act

1 Introduction

Electronic Vending machines (eVending) are a convenient method for delivering and accessing different types of goods in public locations as well as in private facilities that cannot be served otherwise, and outside office hours. Modern eVending features a variety of categories of products, such as, tickets for public transportation, electronic gadgets, books, and food and beverage. With specific regard to the latter, other than comforting customers, vending machines provide them with additional choices that include more affordable options. The success of the eVending industry is demonstrated by recent market reports, which estimate that currently there are 4.6 million automated devices in operation, in the United States alone, with an annual spending of 7 billion US dollars in food and beverage, alone [1]. Also, industry reports show a growing trend in all the diverse sectors. Moreover, the increasing adoption of eVending benefits from

G. Di Bucchianico (Ed.): AHFE 2019, AISC 954, pp. 47–56, 2020.
https://doi.org/10.1007/978-3-030-20444-0_5

and facilitates the introduction of cashless technology: according to recent statistics, vending machines that offer an electronic payment system represent 30% of the existing market [1].

Primarily, studies about eVending devices focused on the type and quality of food offered, on the habits of consumers, and on how to improve interaction by connecting vending machines to Internet-based services. However, human factors received less attention, and there are very few studies that investigate the usability of eVending systems. Specifically, there is a lack of understanding of the accessibility principles that enable people with disabilities to independently buy food and drinks from vending machines. Despite the numerous challenges in supporting interaction for people who are blind [2, 3], addressing and supporting accessibility is an often-overlooked aspect. The Americans with Disabilities Act (ADA) established in 2012, outlines design guidelines and details federal regulations for rendering eVending devices usable and accessible, as well as defining penalties for violations [4]. Nevertheless, ADA rules primarily address physical accessibility for people with limited mobility (e.g., wheelchair users), only, and do not consider sensory conditions, such as, blindness or severe low vision. Unfortunately, a recent article [5] reports that the Supreme Court and the U. S. Department of Justice declared that vending machines are not public accommodations under the Americans with Disabilities Act.

In this paper, we introduce a study that investigated accessibility in the most common types of automated vending machines, with specific regard to people who are blind or have significant visual impairments. We discuss the results of an accessibility assessment of the individual components of eVending devices. Moreover, we detail an experiment in which we studied the usability performance of two different types of automated vending machines in serving a group of people who experienced them as if they were blind.

2 Related Work

In the last decade, the topic of accessibility, with specific regard to individuals who are blind, received increasing attention and has been introduced as a major theme in the design of public and private spaces [6], products and appliances [7], and software [8]. Recently, the lack of accessibility in kiosks [9], eVending, and other types of self-service devices, has been considered by the research community, though there is an ongoing legal debate on whether accessibility of automated vending machines should be regulated [10]. The authors of [11] question Title III of the Americans with Disabilities Act and describes the legal pitfalls that can lead to non-compliance in the design of websites as well as self-service vending machines.

Previous studies discussed methods for supporting individuals who are blind or visually-impaired in several tasks [12], using wearable devices [13], touch- or sign-based communication [14], or implementing Braille [15]. Nevertheless, most of research resulted in new technology. Conversely, others focused on analyzing and improving accessibility of existing devices: the work of [16] investigated the use of touchscreens and other types of input devices that utilize capacitive sensors, demonstrating the impact of systems that do not provide any tactile feedback on the usability of self-service kiosks. Indeed, lack of accessibility in devices for enabling equitable use of public transportation has the largest impact on the blind community [17–19]. Therefore, many are addressing support to mobility from different perspectives, such as, leveraging smartphones or increasing the presence of auditory and tactile feedback. Nevertheless, their applicability still requires further investigation. In this paper, we present the result of an experimental study that aimed at increasing awareness on the impact of poor accessibility in eVending. Moreover, we demonstrate the problem and describe simple and resource-effective fixes that, despite negligible investments, might lead to significant social return.

3 Experimental Study

We designed a preliminary experimental study to evaluate the accessibility of vending machines to people who are blind. The objective was to discover if currently available systems can be effectively used by blind people in each of the steps of a purchase, that is, product selection, payment, and product collection. Additionally, by doing this, we also aimed at evaluating the accessibility of automatic payment systems to individuals who are blind. The experimental study was realized in the library of a state university. We chose a public space because our goal was to evaluate the use of eVending in a real-life scenario that an individual with visual impairments might encounter.

3.1 Protocol

The experimental task consisted in completing the purchase of a product. The task was repeated twice with each of the vending machines: first, they realized it while being blindfolded to simulate the circumstances of being blind; then, they were asked to repeat it without any impairment, to simulate acquisition from a control group. The order of eVending was selected at random. The experimental task was divided into three sub-tasks, that is, product selection, payment, and collection. They were recorded, timed, and analyzed separately. Subjects had a 60-s time limit to complete each sub-task, which was considered as unsuccessful after 1 min.

Sub-task 1: Selection. Subjects were verbally instructed to select a specific product: for machine A, they were asked to purchase the drink at the very bottom left; for device B, they had to select product having code 609. In addition to standardizing the protocol and removing confounding variables, this was to simulate the situation in which they already used the eVending and knew which product to select.

Sub-task 2: Payment. Participants were verbally asked to complete a payment for the selected item. Regardless of the type of vending machine and product selection, product price was set at 1.75 US dollars to standardize the protocol and avoid confounding variables. This sub-task was further divided in two pieces: payment with credit card and with cash. The objective of this task was to realize which automated payment method and system might prove to be easier for a visually impaired person to operate. Before the beginning of each activity in this sub-task, participants were provided with a credit card or with 2 dollars in notes and 1 dollar in coins.

Sub-task 3: Collection. In this activity, participants were verbally asked to collect the product from the dispenser. A sample item was placed in the collection container before the beginning of the sub-task. This was to provide subjects with the same type of auditory feedback they would receive if the product was released after they actually purchased it.

Before the experiment, participants were provided with informed consent form and instructions. Subjects were not allowed to touch the vending machine until the timer started. Experimental data were recorded using a camera and a paper-based experiment worksheet. After completing all the experimental tasks, participants were asked to fill out a questionnaire that collected their accessibility evaluation of each of the vending machines.

3.2 Equipment

For this study, we utilized two types of eVending: one serving drinks (A) and one snacks (B). They are shown in Fig. 1. They were chosen because their components had very different design, with regard to product selection and payment. Specifically, machine A, which offered 12 products, featured a large, embossed press-button for each drink. On the contrary, machine B utilized a flat, touch-sensitive keypad to enable users to access a greater choice of products (approximately 50) by entering their code.

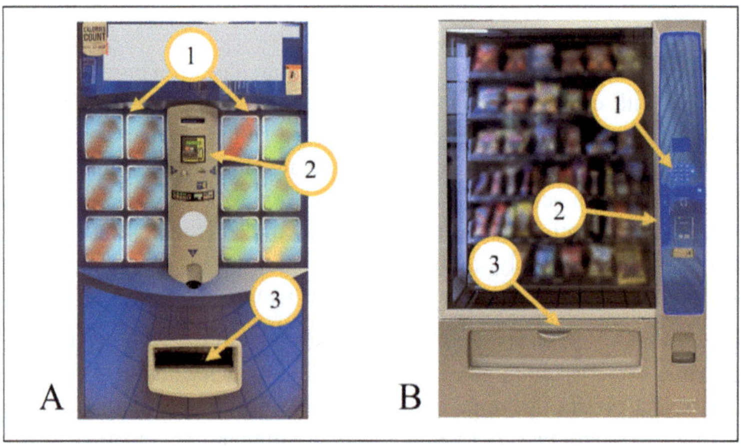

Fig. 1. The eVending utilized for the experimental study. We utilized two standard vending machines for beverages (A) and snacks (B) offering payment with both cash and credit card. Also, machine B implemented contactless payment. The numbers illustrate the different configuration of the devices in regard to: (1) product selection, (2) payment, and (3) product collection.

Furthermore, the payment system of eVending A consisted of an external unit that incorporated a vertical slot for swiping credit cards and a horizontal opening for cash; the opening for coins was located separately, on the bottom of the payment unit.

Conversely, machine B featured 3 adjacent openings: a large vertical inset for credit cards, a horizontal slot for cash and a small vertical opening for coins. Both machines implemented released the product in a container located at the bottom: machine A had a simple opening whereas the compartment of machine B was protected by a sliding door. They are shown in Fig. 2.

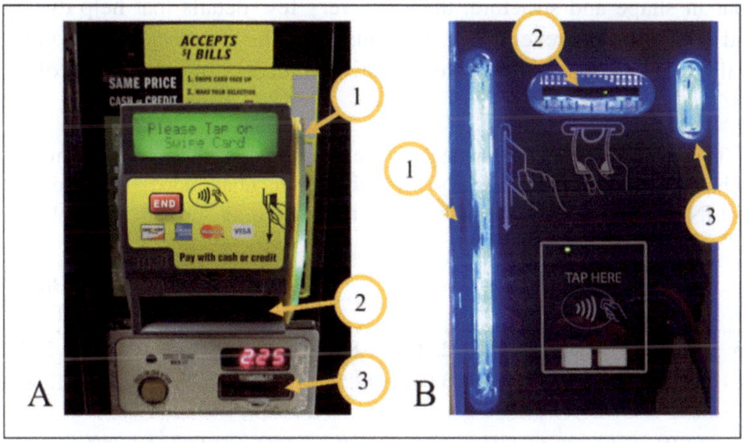

Fig. 2. The automated payment systems for eVending A (left) and B (right). The devices had different configurations. The numbers indicate the different design of the components for (1) swiping credit cards and inserting notes (2) and coins (3).

3.3 Participants

A total of 15 participants were recruited in this experimental study (12 females and 3 males); 12 (80%) aged 18–24, 2 (13.33%) were in the 25–34 bracket, and 1 (6.66%) aged 35–44. As this was a preliminary study, we did not specifically involve individuals with blindness: subjects did not have any visual impairment and their vision was generally good, though it was not assessed with any optometry test.

Participants' familiarity with vending machines was evaluated before the experiment using a self-assessment tool in which respondents described the frequency of use of eVending and their preferred method of payment.

4 Results and Discussion

Our results show that very few subjects were able to complete the experimental tasks when they were blindfolded, despite the presence of the experimental staff, who suggested which product to buy, told participants the price of the item, and prevented actions that could impact the task, such as, inserting the credit card in the slot for notes. Furthermore, subjects would fail every sub-task without this type of guidance. Regardless of the simplified and supervised setting, participants were not successful in finalizing product selection using the touch-sensitive keypad incorporated in machine

B, due to accidental input, lack of auditory or tactile feedback, and to the impossibility of recognizing individual buttons. As a result, in a real-life scenario, they would not be able to move the purchase forward. Table 1 summarizes the experimental result, with specific regard to the success rate (SR) in each task: very few subjects succeeded in paying with credit card, leading to an aggregate success ratio of 17%, on average. In this regard, the design of the payment device is key: the slots for credit card and cash are similar in shape and structure, there are very few details that help distinguishing them, and no measures prevent users from inserting the payment method in the wrong slot. On the contrary, participants were relatively successful in product collection.

Table 1. Timing (T) and success rate (SR) in the experimental sub-tasks using machine A and B. The table shows results for subjects when they were blindfolded (BF) and without any impairment (NBF). Success rate is not shown for the latter as all subjects were able to complete the task.

Sub-task	A/BF T	A/BF SR	A/NBF T	B/BF T	B/BF SR	B/NBF T
Selection	30.32 ± 17.57	92.86%	2.36 ± 1.88	60 ± 0	0%	5.95 ± 3.7
Payment - card	59.19 ± 3.03	7.14%	13.04 ± 12.58	51.87 ± 16.7	28.57%	12.31 ± 9.07
Payment - cash	51.01 ± 13.22	33.33%	11.85 ± 4.86	44.97 ± 13.97	64.28%	17.27 ± 12.8
Collection	16.5 ± 16.07	92.86%	2.91 ± 0.66	22.68 ± 17.94	85.71%	6.79 ± 8.5

The total time for accomplishing the experimental task is described in Fig. 3. On average, subjects experienced a 4-fold increase in purchase time when they were blindfolded. However, as none of the experimental subjects could complete product selection with machine B, the time associated with this sub-task was 60 s.

Nevertheless, in a real-life scenario, individuals who are blind would not be able to select a product and, consequently, move the purchase process forward. Also, the lack of any type of non-visual feedback prevented participants from understanding if the machine was working, if they selected the right product, the cost of the selected item, if they inserted the cash or card correctly, and if there was any issue with the purchase.

In addition to quantitative data acquired during the experiment, we asked participants to complete a survey that featured items on a 5-point Likert scale. Figure 4 shows the overall usability of machine A and B, indicated as perceived frustration and accessibility. Several subjects ranked eVending as accessible, despite their frustration and even if they were not able to complete some of the experimental tasks. This might be due to a lack of familiarity with the concept of accessibility: our future work will address this aspect.

Moreover, users were asked to report the accessibility of machine A and B in each sub-task, considering the experience they had while being blindfolded (see Fig. 5). Overall the accessibility of both devices was considered less than acceptable. Despite product collection had a good outcome, respondents identified selection and payment

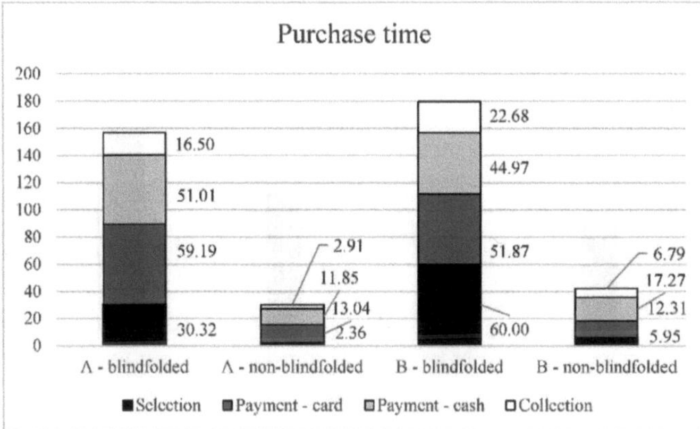

Fig. 3. Time (in seconds) for completing the different steps of a purchase, that is, selection, payment with cash or credit card, and collection. Although customers use only payment method, data are shown in an aggregated form as they describe how design affects each phase.

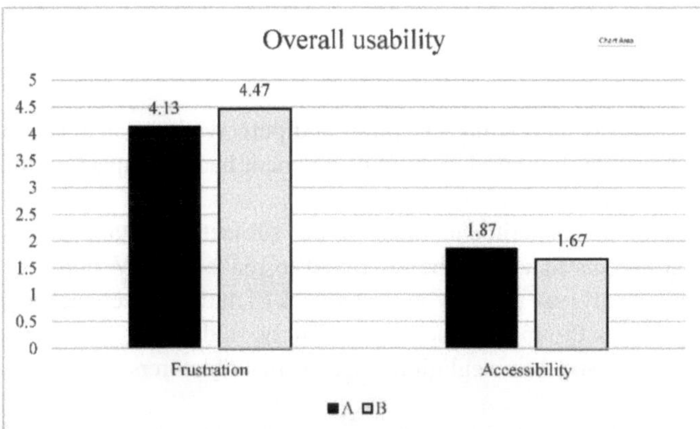

Fig. 4. Overall usability experienced by experimental subjects and displayed as frustration and perceived accessibility. Most of the users felt very frustrated with both machines, and they scored the devices as not being accessible. Unsuccessful product selection was the main reason why eVending B ranked the worst, though both devices scored very low in usability.

as inaccessible. This is supported by other experimental data, which showed high failure rate in these tasks.

Cash was considered a more accessible payment system than credit card. This is mainly due to the lack of tactile indications that help locate and distinguish where to and in which side and direction to insert the card. Also, the design of the slots for inserting cash and credit cards is ambiguous, and in several instances the experiment required the intervention of the assistants, who prevented subjects from inserting the payment system in the wrong slot. We found no significant correlation between

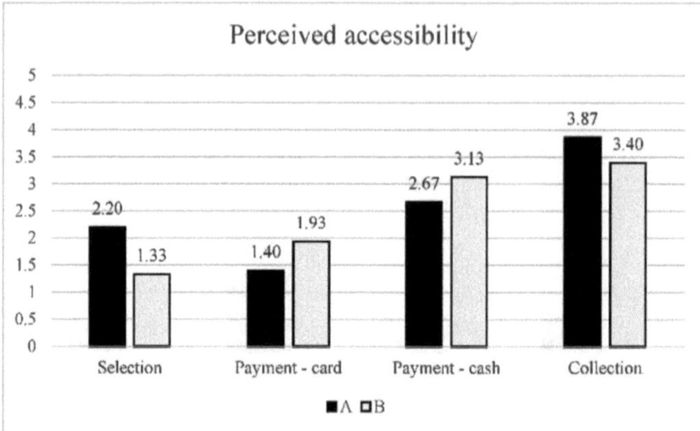

Fig. 5. Detail of perceived accessibility in each of the experimental tasks. Product selection with machine B was the lowest ranked because it was impossible for blindfolded subjects to operate the touch-sensitive keypad due to lack of tactile features and spurious input. Payment with credit card was perceived as the hardest, whereas subjects were able to use cash and coins to complete the purchase. Nevertheless, using either method was indicated as hard. Conversely, product collection was perceived as the most accessible sub-task, though the presence of the sliding door on machine B impacted the user experience.

previous experience with vending machines and perceived or calculated accessibility; moreover, participants' preferred method of payment had no impact on their ability to complete a purchase using cash or credit card.

At the end of the experiment, most of the subjects spontaneously shared their feedback about aspects that could be improved to render the eVending systems more accessible. Specifically, the majority indicated lack of auditory feedback and absence of tactile information as the main issues. Interestingly, addressing the items reported by participants would result in a negligible cost for manufacturers.

5 Conclusion

Automated vending machines are an established technology that experienced a significant growth in the last decades and its diffusion is capillary in public and private spaces. Nevertheless, their accessibility has been addressed only partially: although the sighted use this technology without major limitations and ADA defined implementation criteria for enabling people with limited mobility (i.e., using a wheelchair) to operate them, they are still not accessible to individuals who are blind; also, as regulations do not enforce any accessibility guidelines in regard to blindness, their needs are not taken into consideration. Moreover, there are very few studies on the accessibility of vending machines, kiosks, automated dispensing systems, and devices that serve more fundamental needs (i.e., mobility and personal care).

In this paper, we focused on the accessibility of eVending. Specifically, we were interested in evaluating if people who are blind can independently operate them and successfully complete a purchase. To this end, we realized a study that involved two of the most common types of eVending offering snacks and beverage. Specifically, we realized a quantitative analysis of each task involved in the use of eVending systems, that is, selection, payment with cash- and card-based systems, and collection. Our preliminary study involved 15 participants only. Despite the small sample might limit the applicability of our results, this paper is one of the first attempts to investigate accessibility of vending machines for people who are blind. From our findings, we can conclude that the majority of eVending systems lack features that are crucial for enabling individuals who are blind to independently select, pay for, and collect a product. Current automated vending machines do not support people who are blind at all: although our data show only a 4-fold increase in the time required to blindfolded subjects to complete a purchase, this result was obtained in a setting where participants were guided in every step of the process, that is, which product to select, and the amount to be paid. However, subjects would not be able to even select a product without indications from an assistant. Moreover, our study demonstrates that the adoption of novel technology, such as, touchscreens, which might improve the look and feel of a device, involves the overlooked risk of rendering them less accessible.

Although this study focused on eVending that dispense food and beverage the results are applicable to almost all other types of vending machines as well as to kiosks providing information. In our future work, we will expand our sample and we will involve individuals who are blind, so that we can achieve more accurate results. Furthermore, based on the outcome of our study and on feedback from participants, we will work with appropriate populations on developing design guidelines that allow vendors to incorporate inclusion in their design strategy, or at least to improve the accessibility of their current devices by introducing minimal, low-cost refactoring.

References

1. Anything Research: 2019 U.S. Industry Statistics & Market Research - Vending Machine Operators (2019)
2. Caporusso, N., Trizio, M., Perrone, G.: Pervasive assistive technology for the deaf-blind need, emergency and assistance through the sense of touch. In: Pervasive Health, pp. 289–316. Springer, London (2014). https://doi.org/10.1007/978-1-4471-6413-5_12
3. Caporusso, N.: Issues, challenges and practices in advancing pervasive human-computer interaction for people with combined hearing and vision impairments (2012). http://e-theses.imtlucca.it/87/
4. ADA: Revised Regulations Implementing Title II and Title III (2013). https://www.ada.gov/regs2010/ADAregs2010.htm
5. Launey, K.M., Vu, M.N.: Justice Department Says Vending Machines Are Not Places of Public Accommodation—And So Much More. ADA Title III News & Insights (2017). https://www.adatitleiii.com/2017/07/justice-department-says-vending-machines-are-not-places-of-public-accommodation-and-so-much-more/
6. Tutuncu, O., Lieberman, L.: Accessibility of hotels for people with visual impairments: from research to practice. J. Vis. Impairment Blindness **110**(3), 163–175 (2016)

7. Fusco, G., Tekin, E., Giudice, N.A., Coughlan, J.M.: Appliance displays: accessibility challenges and proposed solutions. In: Proceedings of the 17th International ACM SIGACCESS Conference on Computers & Accessibility, pp. 405–406. ACM, October 2015
8. Caporusso, N., Biasi, L., Cinquepalmi, G., Trotta, G.F., Brunetti, A., Bevilacqua, V.: Enabling touch-based communication in wearable devices for people with sensory and multisensory impairments. In: International Conference on Applied Human Factors and Ergonomics, pp. 149–159. Springer, Cham, July 2017. https://doi.org/10.1007/978-3-319-60639-2_15
9. Veijalainen, A.: Breaking barriers: accessible self-service kiosks for everyone (2017)
10. Darzentas, J., Petrie, H.: Standards, guidelines and legislation related to self-service technologies: developments since 2013. In: International Conference on Computers Helping People with Special Needs, pp. 295–302. Springer, Cham, July 2018
11. Vu, M.N., Sarnoff, J.N.: Websites, kiosks, and other self-service equipment in franchising: legal pitfalls posed by title III of the Americans with Disabilities Act. Franchise Law J. **36**(3), 443 (2017)
12. Caporusso, N., Mkrtchyan, L., Badia, L.: A multimodal interface device for online board games designed for sight-impaired people. IEEE Trans. Inf. Technol. Biomed. **14**(2), 248–254 (2010). https://doi.org/10.1109/TITB.2009.2034739
13. Caporusso, N., Biasi, L., Cinquepalmi, G., Trotta, G.F., Brunetti, A., Bevilacqua, V.: A wearable device supporting multiple touch-and gesture-based languages for the deaf-blind. In: International Conference on Applied Human Factors and Ergonomics, pp. 32–41. Springer, Cham, July 2017. https://doi.org/10.1007/978-3-319-60639-2_4
14. Bevilacqua, V., Biasi, L., Pepe, A., Mastronardi, G., Caporusso, N.: A computer vision method for the italian finger spelling recognition. In: International Conference on Intelligent Computing, pp. 264–274. Springer, Cham, August 2015. https://doi.org/10.1007/978-3-319-22053-6_28
15. Loconsole, C., Leonardis, D., Brunetti, A., Trotta, G., Caporusso, N., Bevilacqua, V.: Braille lab: a new design approach for social entrepreneurship and innovation in assistive tools for the visually impaired. World Acad. Sci. Eng. Technol. Int. Sci. Index Industr. Manuf. Eng. **11**(5), 1252 (2017)
16. Sandnes, F.E.: Improving the robustness to input errors on touch-based self-service kiosks and transportation apps. In: International Conference on Computers Helping People with Special Needs, pp. 311–319. Springer, Cham, July 2018
17. Transit Cooperative Research Program: Accessibility of Automated Fare Vending and Collection for Customers with Disabilities. Transportation Research Board (1998)
18. Paul, A.M., Flynn, C.G.L., Panneton, J., Komissar, J.I.: Improving Accessibility of Public Transport Information for the Blind (2018)
19. Schreder, G., Siebenhandl, K., Mayr, E., Smuc, M.: The ticket machine challenge: social inclusion by barrier-free ticket vending machines. In: Generational Use of New Media, pp. 129–148 (2016)

Human-Centered Design and Quality Function Deployment: Understanding Needs on a Multidisciplinary Automotive Research

Gian Andrea Giacobone(✉) and Giuseppe Mincolelli

Department of Architecture, University of Ferrara, Via Quartieri 8, 44121 Ferrara, Italy
{gianandrea.giacobone,giuseppe.mincolelli}@unife.it

Abstract. The paper reports the activities of a specific design research integrated into multidisciplinary automotive project funded by Emilia Romagna region, Italy. The main goal of the overall project is to design the concept of a high-performance vehicle. In particular, this contribution presents a qualitative user analysis that has been carried out specifically to design the vehicle interface. This work lays on a methodology that points out the importance of inclusive and human-centered design approaches that are used throughout the automotive product development. The Quality Function Deployment (QFD), namely an inclusive decision-making tool, has been utilized into the product planning process in order to obtain measurable design parameters for improving the quality of the expected project through a correlation with the user needs. The outputs derived by QFD tool are used as a strategy to develop a possible design solution for the vehicle interface.

Keywords: Human-center design · Quality Function Deployment · User needs analysis · Automotive product development · Interaction design

1 Introduction

The category of high-performance cars is a particular segment, in which engineering and technical aspects of the vehicle are maximized to provide extreme performances. Given that the sport cars are primary designed either for track or race use, often the product development process prefers to minimize or partially exclude from the project, characteristics of comfort or usability, in favor of technical and mechanical improvements. However, on a different point of view, those parameters are fundamental parts of the total experience that a product provides to its users during its use [1]. In that particular context of automotive, the cited elements are very important to improve, at the same time, human abilities and physical-cognitive performance of all the people who constantly interact with the vehicle, for driving, but also for setting and maintaining its technical parameters in optimal conditions.

Designing an automotive product is very complicated undertaking because it requires many resources and coordinated engineering processes. Due to the difficulty of managing the development of automotive products, there is a possibility to create also a

G. Di Bucchianico (Ed.): AHFE 2019, AISC 954, pp. 57–68, 2020.
https://doi.org/10.1007/978-3-030-20444-0_6

real gap between the level of satisfaction desired by the users and the management utilized by the engineering processes to achieve the requested quality [2]. In this case, for realizing a successful product, the vehicle definition has the necessity to be based on the needs of its final users, in order to improve not only its mechanical properties, but also every physical and psychological aspect elicited by the interaction with the vehicle, which determines the qualitative level of the total product experiences [3].

But often the main problems to produce wrong results come directly from the failure to understand user needs, but also to translate them into technical specifications for designing the vehicle and to validate them, if the final product coincides to the customer expectations [2]. To avoid negative outcomes, it is important to identify user prospective and its main requirements through a research analysis primary focused on users. Every need is immediately recognized, understood and subsequently processed into the automotive development process at an early stage. In this way, clearly defined data from this practice become useful parameters to define the acceptability of the final product and important inputs to the expected concept vehicle.

2 Objectives

Following the consideration touched in the introduction, the paper reports a design research process that is carried out in the field of automotive, in particular focusing on sport car segment[1].

This contribution regards the activities of a specific design research integrated into multidisciplinary project called Automotive Academy: a learning by doing project for innovation engineering automotive, which is financed with public funds by Emilia-Romagna Region, Italy. The peculiarity of this project lays on a collaborative network that involves four universities – University of Modena and Reggio Emilia (which is the coordinator), University of Ferrara, University of Parma and University of Bologna – and many research centers located on the territory.

The main goal of the entire project is to develop several engineering and design solutions in different areas of study that can be applied for developing the future concept of a sustainable sport car. The final vehicle expects to combine, at the same time, engineering solutions that provide high performance with the technology of the hybrid drivetrain. Working on many aspects of the automobile (as weight, aerodynamic shape, propulsion and consumption), the scope of the project is to increase the sustainability of the sport car improving its energy efficiency, in order to reduce its level of carbon emission without compromise its performance. In addition, the research puts the

[1] In automotive industry, every country or region assumes a particular regulatory framework that is used to classify multiple categories of vehicle. In this specific project, the main framework that is taken into consideration is provided by European Commission. The utilization of car segmentation is important to define in which market area a particular vehicle want to be inserted, because it is subjected to technical and administrative regulations that authorize the type-approval and placing on the market. In this case, according to both European segmentation and the specific vehicle attributes – that characterise the entire automobile in terms of driving experience and design forms – the concept vehicle of this research is inserted is S-segment, namely the sport coupés segment.

attention on people needs and for this reason it is aimed to define also functional and ergonomic aspects of both driver package and vehicle interface, in order to provide to the final users, the expected User Experience (in terms of safety, comfort usability and sensation).

Each partner contributes with its specific research in five different areas related to: design methodologies; hybrid powertrain; vehicle body; smart materials; drive-by-wire instruments. This specific research is inserted on the first area regarding design methodologies and it is focused on inclusive and human-centered design approaches. Its aim is to delineate the design guidelines of the vehicle that will be used specifically to define many exterior and interior elements of the final concept. In particular, this contribution presents a research process that will be utilized specifically to design the vehicle interface.

3 Methodology

For this research it has been decided to configure the entire process through the application of a design methodology whit an inclusive character. It is Human-Centered Design (HCD), an approach that puts human needs, capabilities, and behavior first, then designs expected solutions to accommodate those needs, capabilities, and ways of behaving [1].

For a respectable success of the research, this method starts with a good understanding – through observation and direct interviews – of the main users and their needs that the design is intended to meet. Then it processes gathered data to find new possible solutions that can satisfy many unaware problems encountered by people during the interaction with the product. To translate raw information into design outputs and technical specifications for designing the concept vehicle, the overall research uses also an inclusive tool called Quality Function Deployment (QFD). It is a powerful decision-making tool for creating a system of cooperation among different expert divisions (such as decision makers, designers, engineers, etc.) within design process [4].

QFD is based on a bi-dimensional mathematical matrix that represents graphically the correlation between the most important user needs and the product characteristics that are more significant to develop suitable solutions for solving the main problems of people [5]. QFD provides a qualitative product planning process that is used to obtain measurable design parameters, in order to improve the quality of the expected final result. The outputs derived by this tool are used as a strategy to develop a possible design solution for the vehicle interface.

4 User Analysis

According to the goals of the project, the research has identified the main target to which the final concept vehicle – in particular the vehicle interface – is addressed to. In this case, for a good result of the analysis, it is important to recognize different typologies of user who have either direct or indirect relations with the concept vehicle throughout its entire life cycle [6]. The main categories of user identified during the activities of this research analysis are essentially four. Each category has another sub-categories and some of them may be overlapped between them:

- D - Driver: It is the principal typology of user that is directly related to the concept vehicle throughout the usage phase of the product. This category is composed by people who drive the car and interact whit its functions in a period of time much longer that the other classes. Each person can have a different level of driving experience. The relation with the vehicle occurs within the interior package during driving, especially with its vehicle interface. The sub-categories identified for this category are: Owner; Amateur driver; Professional driver; Race driving instructor.
- T - Technician: The members of this category are mainly expert people with a great engineering know-how and many technical and practical skills that are used to set up the vehicle and its performance, in base of the necessities requested by the first category. They accompany and assist drivers in the pit lane during race events. The relation whit the vehicle mainly happens with the engineering elements and technical parameters of the vehicle. The sub-categories identified for this category are: Team manager/director; Race engineer; Mechanic; Tire specialist.
- O - Operator: This category is composed by people who interact with the vehicle in some phases of its life cycle such as production, assembly, distribution. They are mostly employments of the company that realize the vehicle. As the second category, the member of this class are related to the vehicle through a technical point of view, but unlikely the technicians, they are not specialized and their work is essentially linked to manual labor. The relation with the vehicle occurs with its mechanical components during production processes. The sub-categories identified for this category are: Manufacturer; Installer; Truck driver; Repairer.
- S - Decision Maker (Stakeholder): The members of this category are directly and indirectly linked to the vehicle during the early phases of the life cycle of the project. This class corresponds to the decision makers of the company, which manage the entire vehicle development program under its all technical, economic and organizational aspects. Some of the needs derive from marketing analyses, in order to increase the quality of the product compared to its competitors. The relation with the vehicle is connected to the design phase. The sub-categories identified for this category are: Program manager; Commercial director; Technical director; Chief aerodynamicist; Chief designer; R&D officer.

5 Analysis of Needs

As mentioned in the chapter of methodology, the analysis process has been carried out through many direct interviews with the various typologies of user and many observations in the context where people are doing activities. One of the most important activities of needs analysis – reported in this paper – has been carried out at a motorsport event, which it has been taken in exams to acquire as much information as possible about concrete necessities and problems of users (Fig. 1).

Fig. 1. Technician at work while he is analyzing data log of the KTM X-Bow. The vehicle is one of the case studies selected for this research. The image comes from the analysis of needs related to the vehicle interface, held from 27th to 29th of September 2018 at MWC - Misano World Circuit, Misano (Province of Rimini), Italy.

The activity of data gathering was performed within the event organized by KTM company called X-Bow-Battle, during the Italian stage. The event lasted three days from the 27th to 29th of September 2018 and it was held at MWC (Misano World Circuit), Misano (Province of Rimini), Italy. KTM's event has been chosen as a significant step to acquire information because the motor company (in particular its car model) is one of the case studies that have been taken into account to design the expected vehicle and their users are the same of this research.

Given the event was very thigh to the vehicle target, the research activities have allowed to acquire the most important needs, which have been significant to continue the following design activities of this research. Raw data derived from the analysis activities have been evaluated and translated into short sentences that express the needs of users. Every expression of need has been undergone to an abstraction process, namely every phrase has been represented with few simple words that can express only moods, whishes and feelings of users, eliminating any indication about technical characteristics of the product [7]. Indeed, in order to make the phrases as much useful as possible, it is always important to express each problem or expectation of people without anticipate possible design solutions, as they could compromise or affect decision-making choices that will be taken for designing the final project [8].

At the end of data processing, the investigation has gathered more of 200 needs, distributed on each category of user. According to the final goal of this research presented in this paper, the analysis has specifically selected and extrapolated from database, only the main needs related to the vehicle interface. They are:

- D. 054 - I would want to keep me eyes on the road
- D. 069 - I would not want to obstruct the visibility of information
- D. 077 - I would want to see only the essential vehicle information
- D. 079 - I would want to have great readability in all environmental conditions
- D. 068 - I would want to know when the race finish
- D. 074 - I would always want to know the state of vehicle system
- D. 086 - I would want to know if I'm losing time laps
- D. 126 - I would want to shift gear at the right moment
- D, T. 134 - I would want to check the vehicle parameters before a driving section
- D, T. 136 - I would want to track the vehicle on circuit
- T. 157 - I would want to monitor vehicle parameters remotely
- T. 159 - I would want to analyze driving data easily
- D. 007 - I would not want to remove my hands on steering wheel
- D. 060 - I would want to change functions in base of my preferences
- D,T. 135 - I would want to speak with my team on-board
- D. 056 - I would not want to decrease my driving attention
- D. 064 - I would not want to change driving functions accidentally
- D. 081 - I would not want to be stressed by information workload
- D. 082 - I would want to be warned in time in case of malfunctions
- D. 058 - I would want to adjust parameters during driving
- D. 061 - I would want to have a specific driving style based on context
- D. 067 - I would want to be entertained during my trips
- T. 158 - I would want to set up the vehicle tools for race
- D. 083 - I would want to learn from my driving mistakes
- D. 084 - I would want to know every characteristic of the circuit

The list of phrases has been used to create the QFD matrix about the design strategy of the vehicle interface. As noted in the list, the main needs related to the goal of this activity come from driver and technician categories, as they are the two categories that interact with the interface more frequently that the others.

6 Correlation Matrix

This chapter presents the QFD matrix of the vehicle interface. It combines user needs with technical parameters of the product, which can be managed for increasing the overall quality of the expected project. The results of this index are very important for the design phase, because it gives many directions on how the expected project can be designed, by assigning priority to the characteristics with higher score.

6.1 Evaluation System

Each need has received a value in numbers that corresponds to its level of importance. The values are: 5 = indispensable; 4 = very important; 3 = important; 2 = preferable; 1 = negligible. Since the selected sentences may be all important for the final result of the research, during weight assignation, it is indispensable compare that needs to each other in such a way that the rating can be as much distributed as possible. This consideration helps the research to identify concretely what needs are more important that others, without assigning the same weight to all. Instead, each technical attribute has another scale of rating that corresponds to the level of correspondence and relationship with a specific need. The values for this second rating are: 9 = strong relation; 3 = medium relation; 1 = weak relation; 0 = no relation.

The multiplication between the two values (weight of need and relationship between need and technical requirement) shows the effective relevance of a specific technical attribute with a certain need. Then, to understand the total weight of a specific parameter within the entire QFD matrix, it is possible to sum all the values allocated in its own column together. Below each column there is the final level of relevance of each product requirement. During design phase, the values with higher score will be more considered as essential elements to develop the final project. All values are reported both in absolute weight and in relative weight (percentage). The use of color saturation helps designers and engineers to better recognize the most important relations between needs and product specifications.

6.2 Product Parameters

Technical attributes that are utilized for realizing the QFD matrix has been chosen to increase the quality expected by users about the vehicle interface. Driving is a complex, multitask activity [9], consisting of interactions between the driver, the vehicle and the environment [10] and requiring the successful integration and coordination of the driver's cognitive, physical, sensory, and psychomotor skills [11]. Indeed, the aim of these parameters is to provide the adequate driving experience – conforming to the necessities that can be encountered in race contexts – in order to maximize physical-cognitive resources of driver and consequently guarantee to it an acceptable situation awareness [12]. The product elements of the QFD matrix has been identified to improve vehicle interactions taking into account some of usability's general criteria, which are associated to the User Experience. They are: error prevention (low error rate); efficiency (high level of productivity); satisfaction (pleasant to use); learnability (easy to learn rapidly); memorability (easy to remember); flexibility (adaptable to different users) [13, 14]. Moreover, some attributes have been selected in relation with the Hutchins, Hollan e Norman's model of direct manipulation [15], in order to recognize the elements that can facilitate understanding and steps execution of a specific task in Human-Computer Interaction [1].

Technical attributes that concern *viewing distance*, *dashboard form factor*, *dashboard area*, *screen resolution*, *brightness and contrast*, are related mainly to the relation between the driver's field of view and the physical area of the vehicle dashboard. They are necessary to guarantee the execution of different task while continuously monitoring the driving scene [16] (Figs. 2 and 3).

Product characteristics regarding *menu depth*, *font/icon size*, *space of intersection*, *number of information*, *number of options*, *colors*, *border radius ratio* and *personalization* are addressed to the layout of the interface and they are considered because some aspects may affect many visual principles – for example *gestalt* rules, layout grid, basic visual hierarchy, Hick's law or Fitts's law[2] [17, 18] – which are used to increase the entire comprehension of the vehicle interface.

The elements of *frequency of use*, *operation distance*, *operation time*, *operation steps* and *error rate* are connected to some parameters that compose specific tasks, which are fundamental to reach a predetermined goal in a correct way [19].

Technical parameters concerning *input area*, *contour depth difference*, *roughness*, *fingertip force*, *feedback response* and *feedback intensity* are related to both the input elements of the vehicle dashboard and the feedback quality that the interface must guarantee to the driver without producing any type of error.

At the end, the product attributes corresponding *memory*, *number of input* and *number of output* are related to electronic and productive characteristics that are used to define the technical sheet of the final project. In this phase, it is important to report that the QFD is focused on the technical parameters addressed to increasing the quality of User Experience and not to the contents of vehicle interface. The values that must be displayed on dashboard will be selected in an other design research.

7 Results and Discussion

QFD tool has allowed to gather qualitative and quantitative information about the real needs of users. The matrix offers a synoptic view – in a representative form – the most important product characteristics that must be taken in consideration for developing the vehicle interface in next design development.

The close reading of QFD matrix shows that the data for improving the vehicle interface are slightly more concentrated on the quality of operations and their task flow – both provided by the system or performed by the driver – than the others aspects related to readability of information. Indeed, more relevant values (5%) are posed on *frequency of use*, *operation time* and *operation steps*, which are necessary to do a specific task on the system. It means that the project must avoid intrusive and complicated functions that can require a long task flow or a high cognitive workload from driver. On the contrary, it is important to design immediate and easy operations that can be rapidly activated only when it is necessary. In this case, synthetizing operations allows the system to reduce errors during the activities of a particular task.

The QFD matrix shows also that other important values are addressed to *number of information*, namely the parameter that regulates both the correct distribution and the

[2] Hick's Law, says that the time it takes for users to make decisions is determined by the number of possible choices they have. Fitts's Law simply states that the time it takes to move from a starting position to a final target is determined by two things: the distance to the target and the size of the target.

	Weight of needs	% Weight of needs	Viewing distance	Dashboard form factor	Dashboard area	Screen resolution	Brightness	Contrast	Menu depth	Font/Icon size	Space of Intersection	Number of information	Number of options	Error rate	Colours	Border radius
I would want to keep me eyes on the road	5	6%	3	9	9	9	3	9	9	3	3	9	3	3	3	3
I would not want to obstruct the visibility of information	3	3%	9	3	3	9	3	3	1	1	9	1	1	3	3	1
I would want to see only the essential vehicle information	4	5%	3	3	9	9	3	3	9	3	9	9	9	9	9	3
I would want to have great readability in all environmental conditions	3	3%	3	3	9	9	9	9	3	9	9	3	3	3	3	3
I would want to know when the race finish	2	2%	3	1	3	3	3	3	9	9	3	9	1	3	3	1
I would always want to know the state of vehicle system	5	6%	3	1	3	9	3	3	3	3	3	3	3	9	3	3
I would want to know if I'm losing time laps	4	5%	3	1	3	3	3	3	1	9	3	3	3	3	3	3
I would want to shift gear at the right moment	4	5%	3	1	3	3	3	9	1	9	9	9	3	9	3	1
I would want to check the vehicle parameters before a driving section	3	3%	1	1	3	3	3	3	9	3	3	9	9	3	3	3
I would want to track the vehicle on circuit	3	3%	1	1	1	3	1	1	3	3	3	3	3	1	1	1
I would want to monitor vehicle parameters remotely	2	2%	0	0	0	0	0	0	3	0	0	0	0	0	0	0
I would want to analyse driving data easily	3	3%	1	1	1	3	3	3	9	3	9	9	3	1	1	1
I would not want to remove my hands on steering wheel	5	6%	3	9	3	3	3	9	3	3	3	3	9	3	3	9
I would want to change functions in base of my preferences	5	6%	3	1	9	9	1	3	9	3	3	9	9	3	9	9
I would want to speak with my team on-board	2	2%	1	1	1	1	1	3	3	1	1	3	3	3	1	1
I would not want to decrease my driving attention	3	3%	1	3	3	9	3	9	1	3	3	3	3	3	3	1
I would not want to change driving functions accidentally	3	3%	1	9	3	3	3	9	3	3	3	9	9	9	9	9
I would not want to be stressed by information workload	4	5%	1	3	9	3	1	9	3	3	3	9	9	9	9	3
I would want to be warned in time in case of malfunctions	5	6%	3	3	3	9	3	9	9	9	9	3	3	9	9	3
I would want to adjust parameters during driving	3	3%	3	9	3	3	3	3	3	3	3	9	9	3	3	3
I would want to have a specific driving style based on context	3	3%	1	1	1	1	1	1	9	3	3	3	3	3	3	3
I would want to be entertained during my trips	2	2%	1	3	9	3	3	3	9	3	3	3	3	9	1	1
I would want to set up the vehicle tools for race	4	5%	1	3	9	3	3	3	9	3	3	9	9	3	3	3
I would want to learn from my driving mistakes	4	5%	1	1	1	1	1	1	3	3	3	3	1	3	1	1
I would want to know every characteristic of the circuit	4	5%	1	1	1	1	1	1	3	3	3	3	3	3	1	1
TOTAL	91	100%	200	270	382	430	226	422	452	356	386	492	426	408	348	276
Importance of every technical parameter of the final product		➡	2%	3%	4%	4%	2%	4%	4%	4%	4%	5%	4%	4%	3%	3%

Fig. 2. QFD matrix for developing the vehicle interface. First part.

	Weight of needs	% Weight of needs	Frequency of use	Operation distance	Operation time	Operation steps	Input area	Contour depth	Roughness	Fingertip force	Feedback responce	Feedback intensity	Personali-zation	Memory	Number of input	Number of output
I would want to keep me eyes on the road	5	6%	9	3	9	3	9	3	3	9	1	9	1	0	1	1
I would not want to obstruct the visibility of information	3	3%	3	3	3	3	9	0	0	1	1	1	1	0	0	0
I would want to see only the essential vehicle information	4	5%	9	3	9	3	3	3	3	3	3	3	9	0	1	1
I would want to have great readability in all environmental conditions	3	3%	1	1	1	1	9	3	3	3	3	9	3	0	1	1
I would want to know when the race finish	2	2%	9	1	3	3	1	1	1	1	3	3	9	1	3	3
I would always want to know the state of vehicle system	5	6%	9	3	3	3	3	1	1	3	9	9	3	3	3	3
I would want to know if I'm losing time laps	4	5%	3	3	3	3	3	1	1	1	3	3	9	3	3	3
I would want to shift gear at the right moment	4	5%	9	9	9	9	9	9	9	9	9	9	3	0	0	0
I would want to check the vehicle parameters before a driving section	3	3%	3	3	3	9	3	3	1	3	3	3	9	3	3	3
I would want to track the vehicle on circuit	3	3%	9	3	3	3	3	3	3	3	3	3	9	3	9	3
I would want to monitor vehicle parameters remotely	2	2%	3	9	3	3	1	0	0	0	1	1	9	3	3	3
I would want to analyse driving data easily	3	3%	3	3	9	9	1	1	1	3	1	1	3	9	3	9
I would not want to remove my hands on steering wheel	5	6%	9	9	9	9	9	9	3	9	3	3	9	0	0	3
I would want to change functions in base of my preferences	5	6%	3	3	3	9	9	3	1	3	3	3	9	1	3	9
I would want to speak with my team on-board	2	2%	9	3	3	3	3	3	3	3	3	3	9	0	9	9
I would not want to decrease my driving attention	3	3%	1	3	9	9	3	3	1	3	9	9	1	0	1	1
I would not want to change driving functions accidentally	3	3%	3	3	3	9	9	9	9	9	3	3	1	0	3	1
I would not want to be stressed by information workload	4	5%	9	1	9	9	3	3	3	3	3	9	3	0	1	1
I would want to be warned in time in case of malfunctions	5	6%	3	9	3	3	3	1	1	1	9	9	1	0	3	9
I would want to adjust parameters during driving	3	3%	9	9	9	9	9	9	9	9	9	9	3	1	1	3
I would want to have a specific driving style based on context	3	3%	9	3	3	3	3	3	3	3	3	3	9	1	1	1
I would want to be entertained during my trips	2	2%	3	1	9	9	3	3	3	3	3	3	9	9	1	1
I would want to set up the vehicle tools for race	4	5%	3	3	3	9	3	3	3	3	3	3	9	1	9	3
I would want to learn from my driving mistakes	4	5%	3	1	3	3	0	0	0	0	3	3	3	9	3	3
I would want to know every characteristic of the circuit	4	5%	3	1	3	3	0	0	0	0	1	1	9	9	9	9
TOTAL	91	100%	492	340	456	492	412	277	225	326	350	432	484	185	252	303
Importance of every technical parameter of the final product		➡	5%	3%	5%	5%	4%	3%	2%	3%	3%	4%	5%	2%	2%	3%

Fig. 3. QFD matrix for developing the vehicle interface. Second part.

clear visualization of data on the screen. Since the particular race context requires specific configurations to operate on the system rapidly, this specific requirement needs to organize the vehicle interface in an intelligible manner, in such a way that the driver is able to detect and understand information with a few glances. To improve the efficacy of the system, it is also possible to work on the information architecture, as it can give the correct hierarchy to the entire system and consequently display only values or commands that are considered more utilizable for the driver. In fact, the importance of this element is very connected to the first technical parameters, because a good setup of visual data – combined with an immediate task flow of operations – can provide less steps or less time to perform an action. The relevance of information architecture is also highlighted and reinforced by the quite high values (4%) of the attribute *menu depth*.

An other important parameter on which the research must focus the attention is on *personalization*, because the driver needs flexibility on driving functions, in order to use only the information or commands that are considered more significant for comfortable and usable driving experience. Again, the project must provide a functional configuration to personalize the contents of the vehicle interface through an adaptable layout and flexible information architecture. A good setup of this parameter can positively influence the other important characteristics, in order to provide efficient commands to users, based on their needs.

In conclusion, the combination of Human-Centered Design approach with the QFD tool has proven to be effective tools to increase the quality of a design research at the early stages, combining the analysis of a creative process together with the technical aspects of the automotive development process. The results coming from QFD matrix provide a detailed and measurable assessment about user needs, which often report subjective points of view. In addition, the assessments will be utilized as a strategic tool to define concretely the vehicle interface, working on the technical characteristics of the product, in order to satisfy user expectations. Therefore, the most important elements will have a greater influence on the decision-making process of design phase.

References

1. Norman, A.D.: The Design of Everyday Things. Basic Books, New York (2013)
2. Bhise, V.D.: Automotive Product Development: A System Engineering Implementation. CRC Press, Florida (2017)
3. Schifferstein, H.J.N., Hekkert, P.: Product Experience. Elsevier, New York (2008)
4. Akao, Y.: Quality Function Deployment. Productivity Press, Cambridge (1986)
5. Franceschini, F.: Quality Function Deployment. Il Sole 24 Ore, Milan (2003)
6. Mincolelli, G.: Design Accessibile. Esperienze Progettuali e Didattiche sul Tema del Design for All. Maggioli Editore, Rimini (2008)
7. Maritan, D.: Practical Manual of Quality Function Deployment. Springer, Cham (2015)
8. Mincolelli, G.: Customer/User Centered Design. Maggioli Editore, Rimini (2008)
9. Regan, M.A., Lee, J.D., Young, K.L.: Introduction. In: Regan, M.A., Lee, J.D., Young, K.L. (eds.) Driver Distraction: Theory, Effects and Mitigation, pp. 3–7. CRC Press. Florida (2009)
10. Rakotonirainy, A., Tay, R.: In-vehicle ambient intelligent transport systems (I-VAITS): towards an integrated research. In: 7th International IEEE Conference on Intelligent Transportation Systems. IEEE Press, Washington (2004)

11. Young, K.L., Regan, M.A., Hammer, M.: Driver Distraction: A review of the Literature. Monash University Accident Research Centre, Victoria (2003)
12. Camilli, M.: Situation awareness. In: Di Nocera, F. (ed.) Ergonomia Cognitiva. Carrocci Editore, Roma (2011)
13. Nielsen, J.: Usability Engineering. Morgan Kaufmann, San Francisco (1993)
14. Harvey, C., Stanton, N.A.: Usability Evaluation for In-Vehicle Systems. CRC Press, Florida (2013)
15. Hutchins, E.L., Holland, J.D., Norman, D.A.: Direct manipulation interfaces. In: Human-Computer Interaction, vol. 1, pp. 311–338. Taylor & Francis, London (1985)
16. Wierwille, W.W.: Visual and manual demands of in-car controls and displays. In: Peacock, B., Karwowski, B. (eds.) Automotive Ergonomics, pp. 299–313. Taylor & Francis, London (1993)
17. Lupton, E., Phillips, J.C.: Graphic Design: The New Basic. Princeton Architecture Press, New York (2015)
18. Saffer, D.: Designing for Interaction: Creating Innovative Applications and Devices, 2nd edn. New Riders, Berkeley (2009)
19. Diarper, D.: Understanding task analysis for human-computer interaction. In: Diarper, D., Stanton, N. (eds.) The Handbook of Task Analysis for Human-Computer Interaction, pp. 5–47. Lawrence Erlbaum Associates, Mahwah (2004)

A New Approach for an Inclusive Yacht Design

Paolo Ferrari(✉)

Department of Engineering and Architecture, University of Trieste,
via Valerio 6/3, 34127 Trieste, Italy
pferrari@units.it

Abstract. The paper focuses on the definition of a new approach for the design of accessible sailing yachts that takes into account not only the physical problems of users affected by permanent and temporary motor disabilities, but also the presence of architectural barriers on most boats built to date. Through an analysis of the current state of the art of modern sailing yachts is possible to determine a series of cases in which the user, depending on his physical limits, may encounter problems while boarding or using external and internal spaces. The research aims to define the most suitable solutions, in order to create a reference standard in the field of nautical design for accessibility. The research leads to the drafting of a technical instrument useful to promote accessibility and inclusion on board.

Keywords: Design · Yachts · Disability · Accessibility · Inclusion · Ergonomics

1 Introduction

Recent studies, conducted by the OECD [1], show that the world population will be getting older and therefore characterized by different types of disability. For this reason, it is essential that designers focus their attention on projects that meet the needs of all, developing specific design approaches, especially in the context of inclusive disciplines. Inclusive design today is addressed to every human being and it is therefore useful to quote a famous statement by Paul Hogan, EIDD-DfA Europe President Emeritus: "Good design enables, bad design disables [2] ", to underline how much, even objects that arise from a purely aesthetic process, must deal with particular attention to accessibility. Despite its recent introduction, the concepts of Design for All have been widely introduced in most fields of application. Currently, a large public practices recreational activities, which were once considered the prerogative of most prosperous classes, and even individuals with different disabilities are more and more active users. In particular, sailing is one of those activities. In the last few years, some pleasure boats have been modified to favor the inclusion, giving the possibility to form mixed crews with men and women, young and old, able and disabled together. This type of experimentation, even if not widespread, gives an idea of the potential of this new design approach, which would allow to produce boats able to satisfy different types of users simultaneously, creating inclusion, integration and opening new commercial scenarios. This document focuses on the definition of a new approach that takes

G. Di Bucchianico (Ed.): AHFE 2019, AISC 954, pp. 69–78, 2020.
https://doi.org/10.1007/978-3-030-20444-0_7

into consideration the architectural barriers present on every pleasure boat actually produced and tries to find possible solutions suitable for any type of user.

2 Problems and Possible Solutions to Improve the Accessibility of Sailing Yachts

The design of pleasure boats from 33 to 78 feet has, to date, limited to the development of sailing yachts that are almost identical to each other and practically inaccessible both above and below deck (Fig. 1).

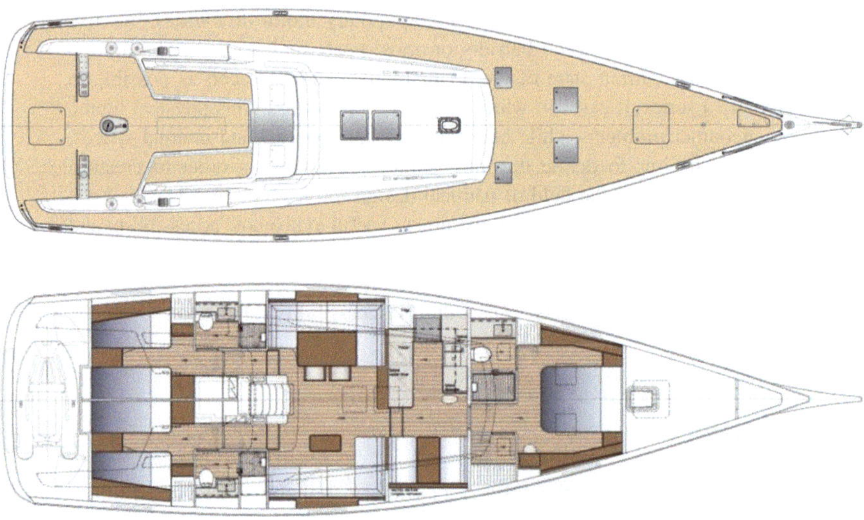

Fig. 1. A modern 65 feet sailing yacht. Both open and lower deck are inaccessible. It is important to note that, even in large boats, accessibility have never been taken into consideration.

Designers must pay close attention to the needs of the person, with or without disability, and try as much as possible to facilitate total accessibility to exterior and interior spaces. This means rethinking any accessory aid, environment or even just part of it, in order to satisfy ergonomic and spatial, safety and autonomy requirements of all those who are not able to overcome certain physical, and sometimes psychological, obstacles. Starting from the modality of accessing on board, a helpful solution consists in a simple electrically controlled scaffolding, permanently fixed to concrete piers, which facilitate access to any type of boat equipped with an open transom. A tilting platform would also ensure the safety of the individual and save the boat from the risk of damage, due to the wave motion. It would be useful for each marina to offer a small percentage of moorings for boats with disabled passengers (Fig. 2).

Analyzing the boat layout from the stern, it is incomprehensible how the vast majority of gangways, produced up to now, presents minimal dimensions. This often causes discomfort even in experienced people, due to the possibility of falling into the

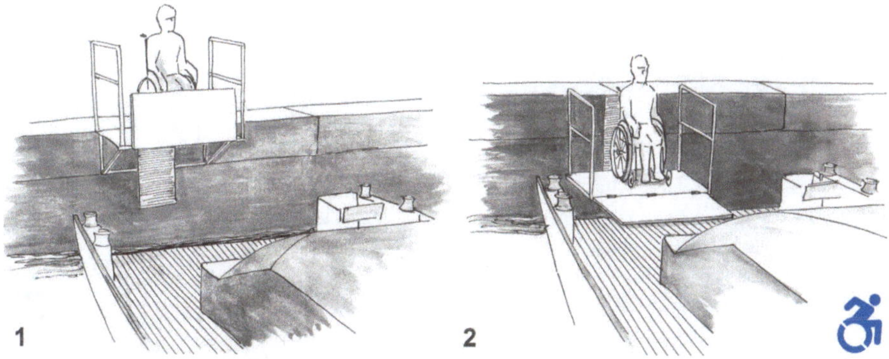

Fig. 2. An innovative system of permanent embarkation for accessible berths, suitable for both harbors and marinas.

water, going over it. It is necessary to create a dimensional standard in the design of gangways that allows all people to be able to embark, without any risk. This solution is also perfectly suitable for all kind of users (Fig. 3).

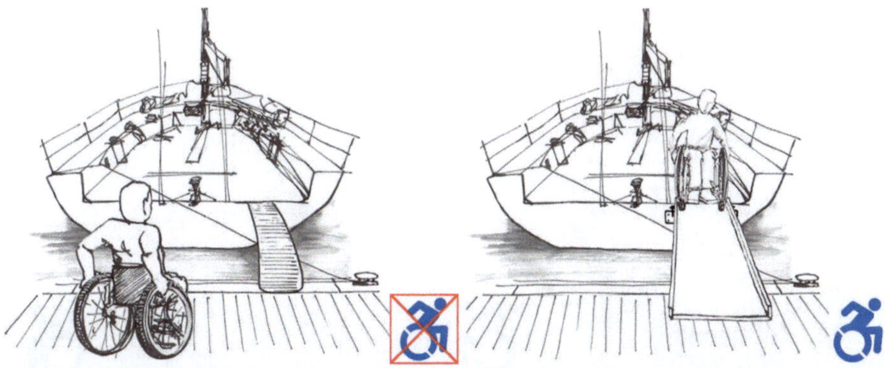

Fig. 3. An inclusive design must be able to facilitate the boarding and disembarkation of the person in a wheelchair, in total safety and, as much as possible, independently.

In order for a boat to be accessible, it is advisable to provide a fully open transom. The design of the boats, thanks to the modern geometry of hull lines, has already led to this result, but it is of fundamental importance that it is not hampered by any kind of obstacle to facilitate the passage for the person in wheelchair (Fig. 4).

Among the major problems of a sailing boat, the most evident are the size and the shape of the helm, normally designed without any consideration about the ergonomics issues for a disabled user. Moreover, it is easily conceivable that, in navigation, the heeling of the hull prevents in any way a person on wheelchair to be able to sit in safety. An innovative system consists in a passing wheel, in which the spokes and the

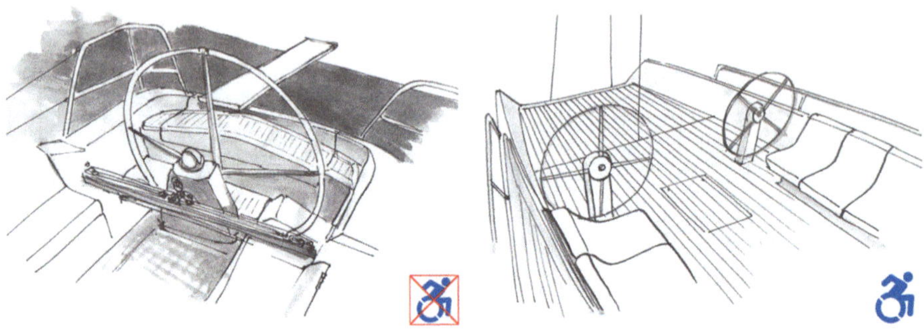

Fig. 4. The transom, completely opened, and the insertion of a helm composed of two wheels, sufficiently distant from each other, are the ideal solution for an easy boarding.

supports are placed below the deck, in order to allow the wheelchair to easily pass through the helm to access the cockpit area (Fig. 5).

Fig. 5. The innovative concept of helm made of a central passing wheel, designed to solve ergonomic issues of people on wheelchair.

Moreover, to ensure the wheelchair in the event of heeling, a tilting platform placed close to the helm is an interesting aid: thanks to actual technologies, it is possible to modify the position of the platform at any time, through an electronic control system based on the principle of the gyroscope. Mechanical systems that fix the wheelchair to the deck and safety belts to secure the person to the wheelchair have to be previously taken into consideration (Fig. 6).

Continuing with the analysis of the open deck problems, the passages that lead to the bow from the cockpit are normally too narrow in most boats. An attentive design must therefore reconsider the entire layout and the spatial relationship between the deck and the interior, in order to be able to obtain larger spaces along the sides, to allow the

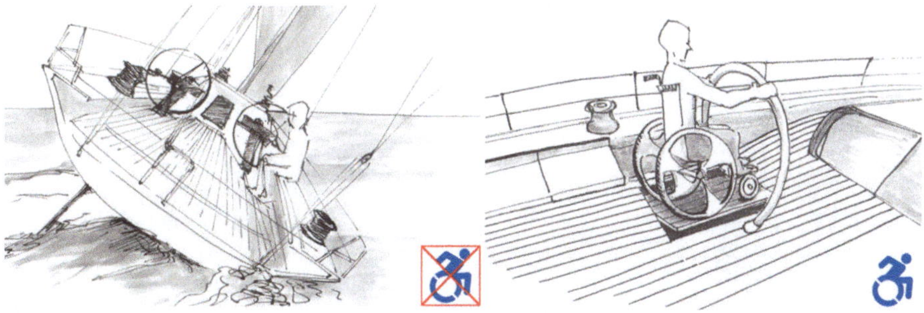

Fig. 6. In order to avoid the overturning of the wheelchair, it is necessary to place it on a tilting platform, electronically controlled.

wheelchair transit. Furthermore, it is necessary that the difference in height between the cockpit and side passages is resolved with moderate slopes, accessible to the disabled, in total autonomy (Fig. 7).

Fig. 7. The step between the cockpit and the lateral passage and the limited size of this one do not allow the wheelchair to access the bow of the boat. A flat deck, with limited slopes on its sides is the ideal solution.

A fully accessible hatch is the first element of the boat on which to set up a design that is attentive to the needs of disabled people. It must have a net light of at least 80 cm, have a one-piece hatch, which can slide sideways, and must not show coamings or obstacles higher than 25 mm. It is also advisable to insert an internal threshold, with a "wheel lock" system, for greater personal safety (Fig. 8).

To access the lower deck and to overcome any difference in height it is suggested the adoption of an electric lifting platform that gives the possibility to the disabled

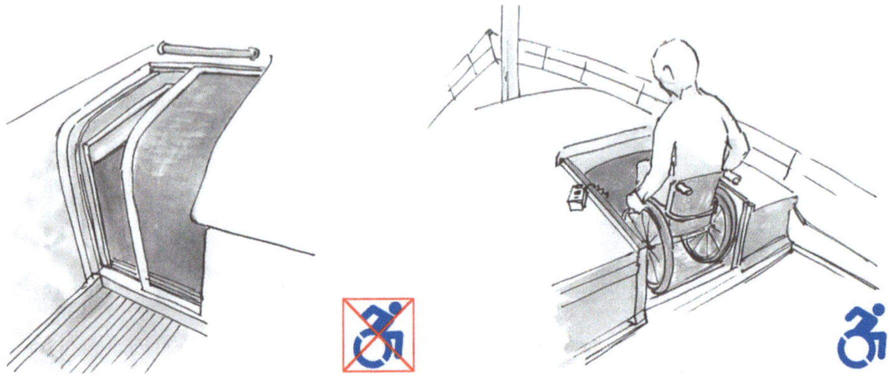

Fig. 8. It is necessary to change the current design approach to allow the best access to the inner spaces of a boat.

person to move between the various levels of the boat in total autonomy. On boats larger than 15 m, it is conceivable to use controlled slope ramps, to replace the classic access ladder (Fig. 9).

Fig. 9. Only by reconsidering the project in its entirety, according to the principles of Design for Inclusion, it is possible to create innovation with solutions accessible to everyone.

The table placed in the central dinette must not have fixed seats along the corridor, either to avoid tightening the passage. Furthermore, it is advisable to plan the interior spaces to give the wheelchair the possibility to make a 360° rotation, without having to force the disabled person to carry out complex maneuvers (Fig. 10).

As often happens in home environments, the main problems in the use of a traditional kitchen are the following: impossibility to use cooktop and sink frontally, impossibility to access the upper units due to their excessive height, difficulties to rotate the wheelchair without colliding with the lower cabinets. An accessible kitchen must therefore offer space for the legs and the wheelchair below the hob and sink, lower cabinets that allow accessing the objects placed on the bottom, large surrounding space for an easy movement of the chair and oven and microwave placed where disabled people can arrive without excessive effort (Fig. 11).

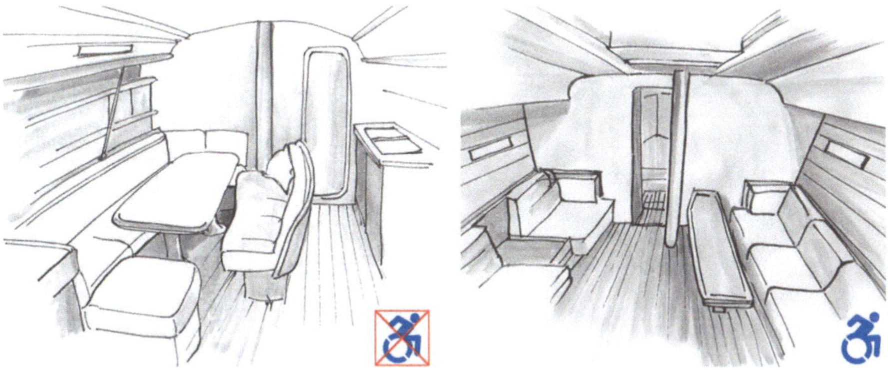

Fig. 10. Removing steps and fixed seats, placed towards the center of the boat, the lower deck space is accessible from stern to bow to people on wheelchair.

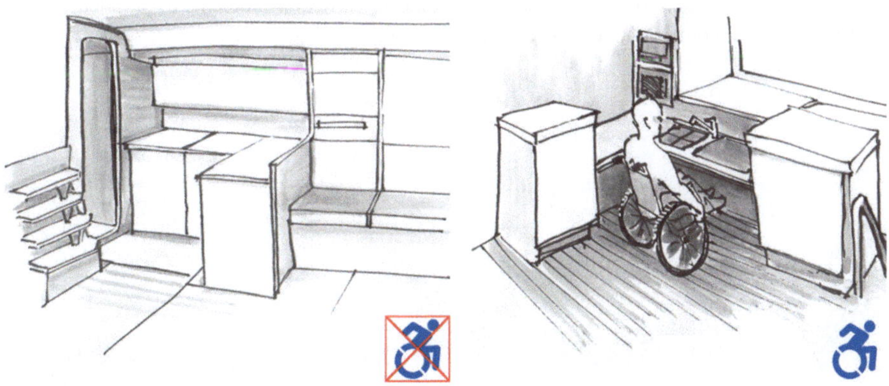

Fig. 11. A design for an extended use does not impact the usability of the kitchen by everyone: it is very important to respect the ergonomic needs that a disabled person needs.

Doors of sailing yacht are usually no wider than 2 feet and often they have a threshold to jump: all these issues make it impossible to access any indoor environment. It is therefore necessary to design wide doors, possibly sliding, and no obstacle to be overcome on the floor. In case of steps between two different spaces, a light slope is the easiest and right solution (Fig. 12).

The height of the beds should never be higher than the height of the wheelchair seat. Therefore, an excessive eight can be a problem for the disabled person when moving from wheelchair to bed. The use of excessively high beds is frequent in the forward cabins of sailing yachts, due to obtain a larger mattress surface. It is therefore clear that a cabin, to be accessible, must be equipped with an easy and sufficiently wide door passage, generous foot traffic to allow the wheelchair to rotate 360° and a bed at the same height of the wheelchair seat. In the case of cabins with single beds, it is advisable to keep a distance between the beds above 3 feet, to allow the disabled to move more simply (Fig. 13).

Fig. 12. The size of the doors must allow an easy passage to the wheelchair: the sliding door is preferable to the swinging one, although does not create particular problems to disabled users; a small ramp to overcome light steps is always advisable.

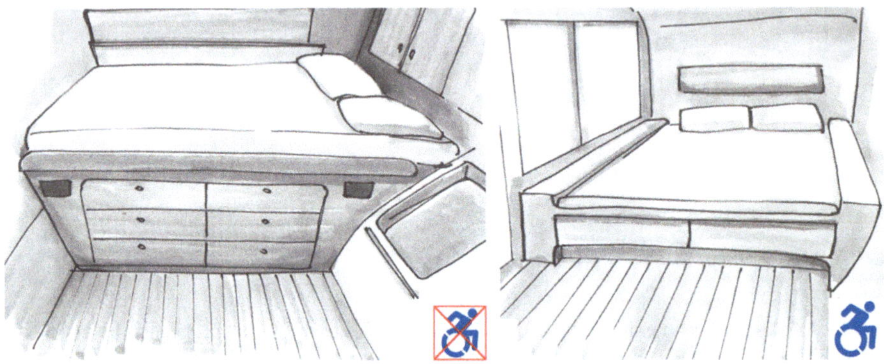

Fig. 13. The bed of a sailing boat must have a height similar to the wheelchair seat and the outer edge free from obstacles, in order to allow the disabled person to move directly on the mattress, without risking to hurt himself.

The bathroom, as the kitchen, follows the rules of ergonomics and common sense, so that disabled people can enjoy the inner space without any discomfort: a good design needs accessible doors, enough space to maneuver the wheelchair, folding seat in the shower, mirrors and fixtures at the correct height. In the case of shower box, it is not necessary to provide the wheelchair rotation, but it is necessary that the disabled person can access it from the front to be able to move on the retractable seat, so as not to wet his wheelchair (Fig. 14).

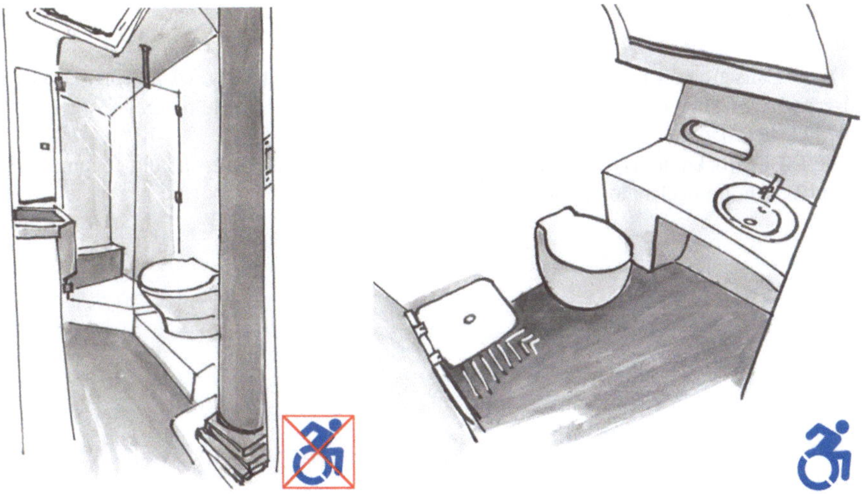

Fig. 14. A well-designed bathroom can meet the basic needs of the disabled individual also being quite small.

3 Conclusions

The research explores the offer of pleasure sailing boats usable by people with disabilities and proposing design solutions inspired by the principles of DfA, a young discipline (about 20 years old) but which starts to have concrete implications, in particular in civil architecture and in the field of product design, where the benefits of its application are immediately tangible with a positive return also on the corporate image. It is a question of translating these principles into the nautical sector: in fact, this analysis revealed that the nautical market is lacking in an offer of accessible boats. The work aims to provide a series of guidelines for the design of inclusive sailing boats, as well as to give solutions also applicable to other kind of boats, as motor yachts. It offers alternative solutions to the traditional design methodology of a boat: innovative proposals studied to improve the accessibility of a boat. The paper also tries to make up for the almost total absence of specific publications on the topic addressed, creating a precedent within the scientific literature of the sector. By adopt-ing integrally or even partially the solutions proposed, it is possible to create a new type of sailing boat and allow everyone not only to navigate autonomously and safely, but also to create new business opportunities such as charter for ex-tended users, inclusive sailing schools. The development of further solutions will be useful for making boats truly "for everyone".

References

1. The Organisation for Economic Co-operation and Development (OECD): Ageing: Debate the Issues, OECD Insights. OECD Publishing, Paris (2015). http://dx.doi.org/10.1787/9789264242654-en
2. The EIDD Stockholm Declaration (2004). http://dfaeurope.eu/what-is-dfa/dfa-documents/the-eidd-stockholm-declaration-2004/

The Effect of Sliding Door Hardware Design on Opening Operation

Satoshi Kose[1(✉)], Yoshiaki Goto[2], Ken Nunota[3], and Shinji Tanaka[4]

[1] Via Parco, 2-9-4, Higashi-Shinbashi, Minato-ku, Tokyo 105-0021, Japan
skose@gakushikai.jp
[2] Okayama University of Science, 1-1, Ridaicho, Kita-ku, Okayama 700-0005, Japan
goto@archi.ous.ac.jp
[3] National Institute for Land and Infrastructure Management, Tachihara 1, Tsukuba, Ibaraki 305-0802, Japan
nunota-k92ta@mlit.go.jp
[4] Comprehensive Housing R&D Institute, Sekisui House, Ltd., 6-6-4, Kabutodai, Kizugawa, Kyoto 619-0224, Japan
stanaka@gz.sekisuihouse.co.jp

Abstract. Usability of sliding door was tested with different door hardware designs, i.e., a grab bar, a vertical door pull, and a recessed door pull. Three levels of force necessary for operation (10 N, 20 N, and 30 N) were compared. Sixteen subjects of different ages participated in the experiments. They reported the ease of use, and their behavior was also observed. Results suggest that no more than 20 N force is desirable for easy operation, and depth of 15 mm would be necessary for a door pull to apply the force effectively.

Keywords: Operating force · Usability · Grab bar · Door pull · Recessed door pull

1 Introduction

Doors that were traditionally smaller, narrower, and lighter are getting larger and heavier in response to recent move toward increased functional requirements such as ease of passage or better thermal insulation. In the case of a sliding door, it often leads to the change of door hardware from a traditional recessed (grooved) pull to a vertical pull or grab bar. However, usability of such new hardware is rarely quantitatively tested. Present paper reports an experimental study examining the usability of the hardware and the effect of design on necessary operating force.

2 Experimental Setup

Three kinds of hardware design were chosen: a vertical grab bar, a vertical pull, and a recessed pull. Overall experimental setup is given in Fig. 1. The recessed pull is just like traditional Japanese hardware with vertical size of 100 mm, horizontal 30 mm and

G. Di Bucchianico (Ed.): AHFE 2019, AISC 954, pp. 79–88, 2020.
https://doi.org/10.1007/978-3-030-20444-0_8

15 mm depth, while the other two types extend 500 mm vertical. The center height was set to be 900 mm, which is customary to traditional Japanese recessed pull design, but vertical extension of grab bar and vertical pull allows different height for force application. The detail of each hardware design is given in Fig. 2. They were attached at the end of a sliding door, and the necessary force to open the door was set with adjusted weights to be 10 N, 20 N and 30 N, respectively.

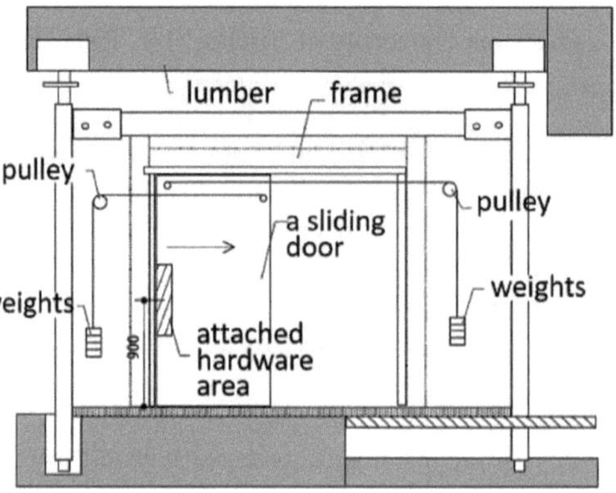

Fig. 1. Setup of simulated sliding door for experiments.

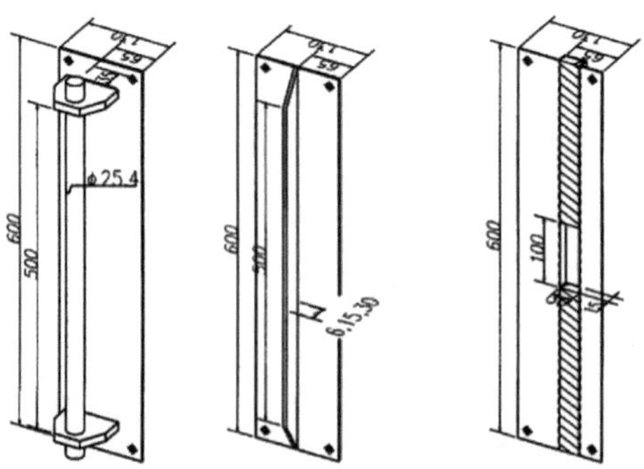

Fig. 2. Experimented door hardware. Grab bar, door pull, and recessed door pull, respectively.

Subjects participated in the experiments were ten seniors and six younger persons. Their basic characteristics are given in Table 1. They were asked to slide open the door and pass to the other side, and five-point sensory evaluation (1: very bad, to 5: very good) was conducted in addition to video recording of the movement. The subjects opened the door as shown in Fig. 3, and they answered their evaluation both ease of opening and ease of closing. No specific instructions were given other than to open the door, pass through, and close, but it turned out that most of the subjects closed the door as they turned to face the door after passing.

Table 1. Characteristics of experimental subjects. Average and range value.

		Number of persons	Age (year)	Stature (mm)	Weight (kg)
			above: average below: range		
Seniors	M	5	75	1579	59.8
			72–78	1485–1667	55.0–72.0
	F	5	80	1420	51.2
			78–83	1340–1516	48.0–54.0
Youngsters	M	2	22	1701	81.5
			22	1680–1721	76.0–87.0
	F	4	27.3	1586	49.1
			21–35	1538–1617	45.0–56.0

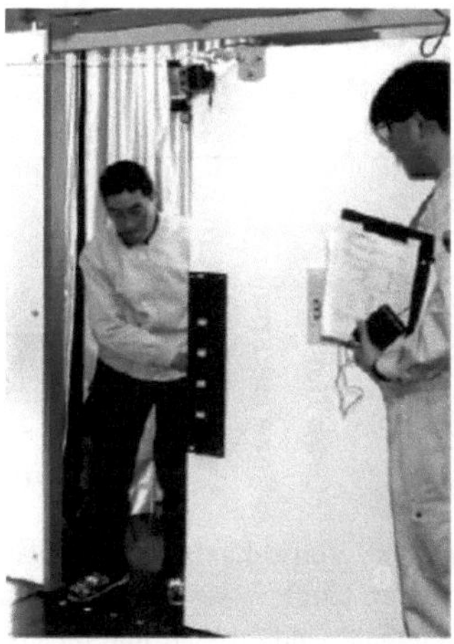

Fig. 3. An example of subjects trying to pass through the door after opening it.

After the experiment, subjects were shown all five kinds of door hardware, and they were asked to give rank order comparison for ease of use (point 5 the best, and 1 the worst). The result is given in Table 2.

3 Results of Force Application Experiment

The averaged points of sensory evaluation for seniors and youngsters, males and females respectively, are given in Fig. 4. As opening force increases, the evaluation gets worse. Regarding the type of hardware, vertical grab bar is evaluated best, and 6 mm vertical pull is the worst. Evaluation of other three types varied depending on the opening force. With opening force 10 N, the evaluation points are around 4 to 5, while with opening force 30 N, the points are around 2. With opening force 20 N, the points were more scattered between 2 and 4 among the subjects and conditions, which suggests usability is easier to differentiate.

To find out the effects of subject characteristics, hardware and opening force, a quantification analysis was conducted. The result is shown in Table 3.

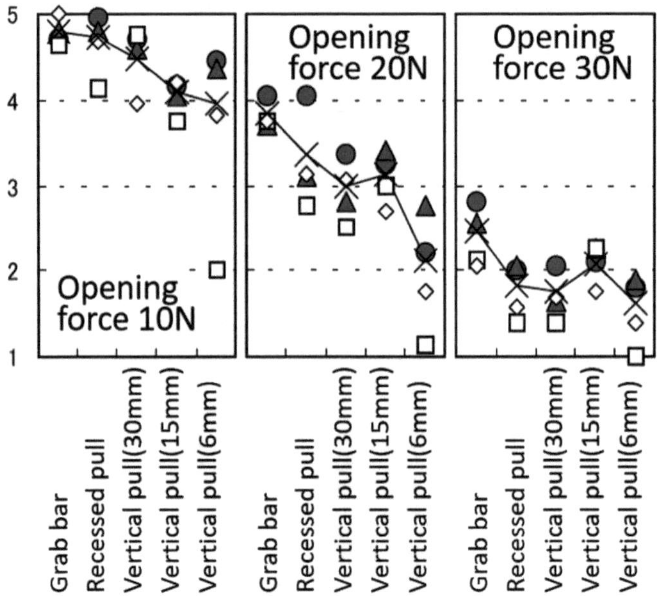

Fig. 4. Results of sensory evaluation. Vertical axis denotes evaluation value, meaning 1 = very bad, 3 = neutral, and 5 = very good. Circle = senior male; triangle = senior female; square = young male; diamond = young female; and crossed line means all averaged.

Table 2. Results of rank order evaluation of various door hardware designs.

Total rank		Pts.	Senior male		Senior female		Youngster male		Youngster female	
			Rank	Pts.	Rank	Pts.	Rank	Pts.	Rank	Pts.
1	Grab bar	71	1	22	2	20	1	10	1	19
2	Recessed pull	63	2	20	1	21	2	7	2	15
3	Vertical pull (30 mm)	56	3	18	3	17	2	7	3	14
4	Vertical pull (15 mm)	33	4	10	4	11	4	4	4	8
5	Vertical pull (6 mm)	17	5	5	5	6	5	2	5	4

Table 3. Quantification analysis of parameters affecting sensory evaluation.

	Category	-2 -1 0 1 2	Category score	Partial correlation
Subject	Senior male		0.3	0.58
	Senior female		0.18	
	Youngster male		-0.37	
	Youngster female		-0.11	
hardware	Grab bar		0.59	0.748
	Recessed pull		0.4	
	Vertical pull (30mm)		-0.03	
	Vertical pull (15mm)		0	
	Vertical pull (6mm)		-0.7	
Opening force	10N		1.25	0.939
	20N		-0.06	
	30N		-1.19	
Coefficient of determination		0.902	Constant term	3.07

4 Discussion on Ease of Use

Opening force was most influential, followed by the type of hardware. Among the hardware tested, vertical pull with the projection of 6 mm was the worst. There was not much difference between 15 mm depth recessed pull, vertical pull of 30 mm and 15 mm projection. Previous studies by Yokobayashi et al. [5], and Tanaka et al. [3] suggested that vertical pull with 15 mm is enough to apply necessary force to operate the door. Overall rank order evaluation (Table 2) also suggested that grab bar was the best, followed by recessed pull, vertical pull of 30 mm, vertical pull of 15 mm and the worst was vertical pull of 6 mm.

Expectation was that the limitation of the recessed pull on its width and placement height from the floor will lead to lower evaluation, but the sensory evaluation during use and comparative rank evaluation after the experiment both revealed relatively high acceptance level. Some seniors replied that they were accustomed to using recessed pull type at home.

5 Analysis of Door Operation Behavior with Video-Recording

Movements of experimental subjects were video-recorded from the side and from above as shown in Fig. 5. Whole activities as they open, pass through and close the door, were recorded. Both slide open the door to the left and to the right were examined. Height of hand application as they open the door was measured from the recording. If they applied both hands to open the door, averaged height was calculated. The opening width as they pass through was also measured.

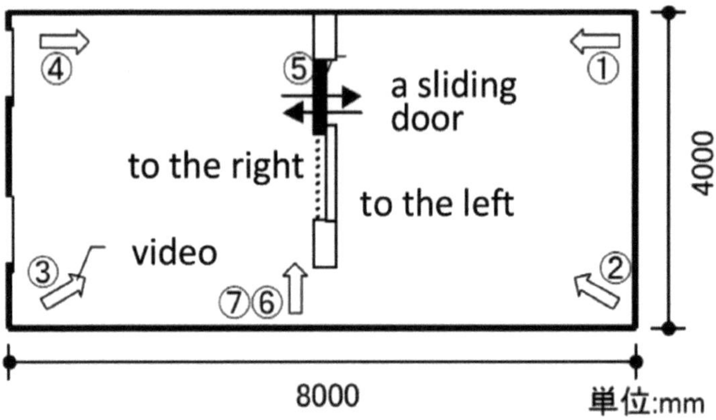

Fig. 5. Settings for video-recording. Operating height was measured from the recordings of cameras 1, 2, and 6. Opening width was measured with camera 5 from above.

6 Results of Behavior Observation

Average height of force application is given in Fig. 6 (recessed pull is not covered since its height is limited by its nature). As a group, seniors and youngsters chose different heights. There was not much correlation between the subject's stature and height of force application, just as reported by Ochiai et al. in 1988 [2]. When the subject applied the force facing backward the door just as shown in Fig. 8, the height was much lower than the case of facing forward the door. A previous study by Endo [1] reported that the force application was best at the height of just below one's elbow. However, there was no tendency among the subjects to change the height depending on the necessary force, and most applied their force somewhat higher than the elbow. No systematic difference was observed among the different hardware.

Average door opening width is given in Fig. 7. The subjects were instructed to open as they might feel wide enough, which basically resulted in an opening of 600 mm or narrower. Their behavior varied among the subjects, and young females showed narrowest opening, twisting their body as they passed. The opening tended to be narrower as necessary operating force increased, while vertical pull of 6 mm resulted in somewhat wider opening.

As to which hand they used, most used their dominant hand: only one subject used both hands according to the situation. Nine subjects sometimes used both hands, but their choice was not related with the hardware type or opening force.

Figure 9 shows the way a subject uses his wrist. It seems to depend on the hardware and opening force (pictures shown here are taken at opening width of 400 mm). With opening force of 30 N, more efforts were observed to ensure force application.

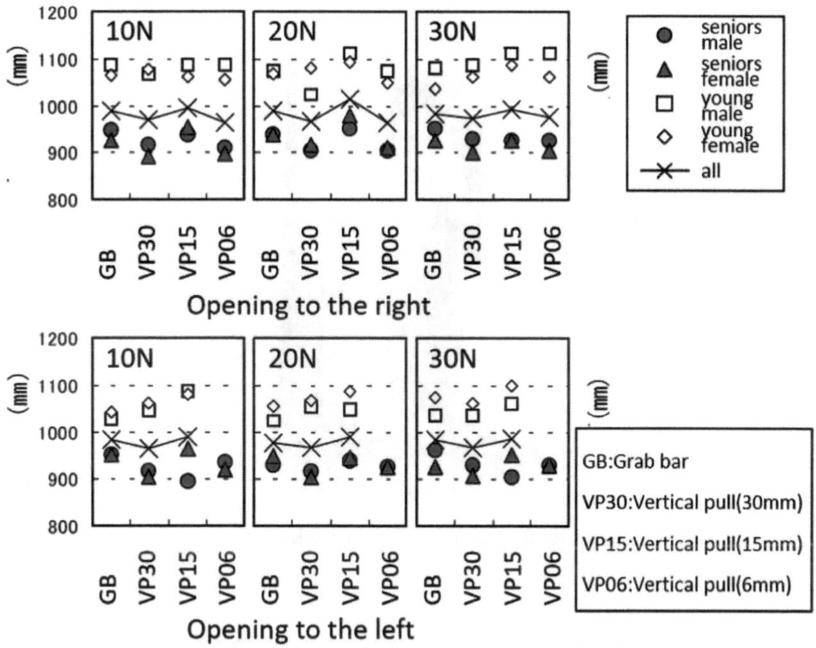

Fig. 6. Height of force application under different conditions.

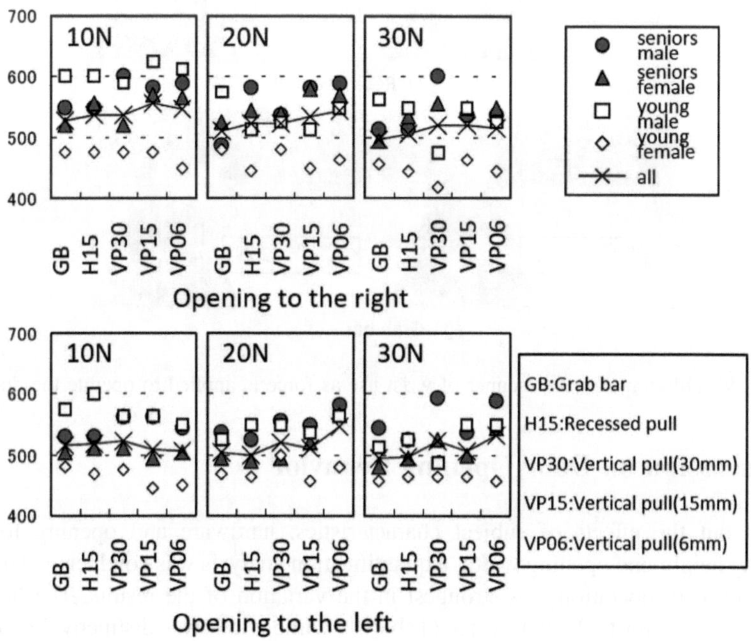

Fig. 7. Width of door opening under different conditions.

Fig. 8. Closing the door backward.

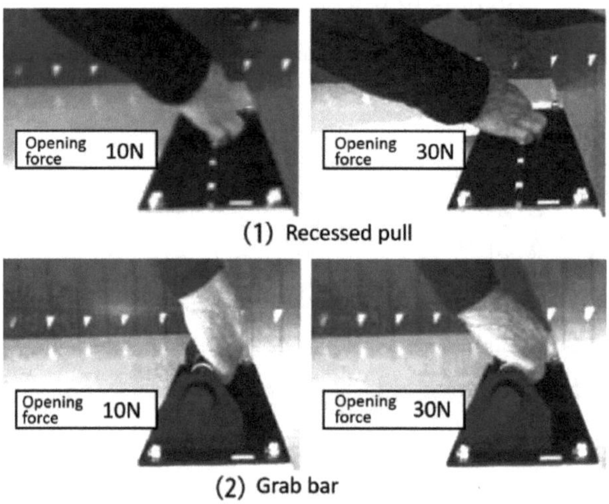

Fig. 9. Difference in the manner of wrist use as force is applied to operate the door.

7 Discussion on Door Opening Behavior

To find out the effects of subject characteristics, hardware and opening force on operating height and opening width, a quantification analysis was conducted. The effect of difference in operation was strongest in the variation of the results. The height of application was most affected by age (Table 4), while width was distinctively narrower for young females (Table 5).

Table 4. Quantification analysis of parameters affecting height of force application.

	Category	Category score	Partial correlation
Subject	Senior male	-61.86	0.986
	Senior female	-62.4	
	Youngster male	69.32	
	Youngster female	54.94	
hardware	Grab bar	2.72	0.617
	Vertical pull (30mm)	-10.17	
	Vertical pull (15mm)	12.02	
	Vertical pull (6mm)	-4.57	
Opening force	10N	-1.72	0.163
	20N	-0.68	
	30N	2.4	
Coefficient of determination	0.972	Constant term	997.91

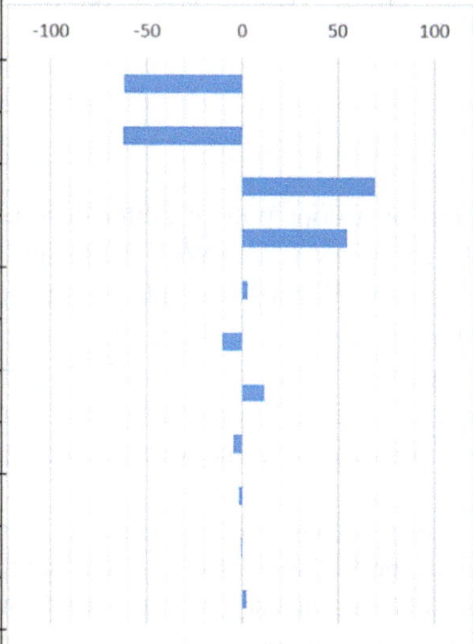

Table 5. Quantification analysis of parameters affecting width of door opening.

	Category	Category score	Partial correlation
Subject	Senior male	29.95	0.884
	Senior female	3.22	
	Youngster male	28.15	
	Youngster female	-61.32	
hardware	Grab bar	-9.22	0.328
	Recessed pull	-5.22	
	Vertical pull (30mm)	0.87	
	Vertical pull (15mm)	3.37	
	Vertical pull (6mm)	10.2	
Opening force	10N	11.05	0.413
	20N	-0.45	
	30N	-10.6	
Coefficient of determination	0.797	Constant term	524.05

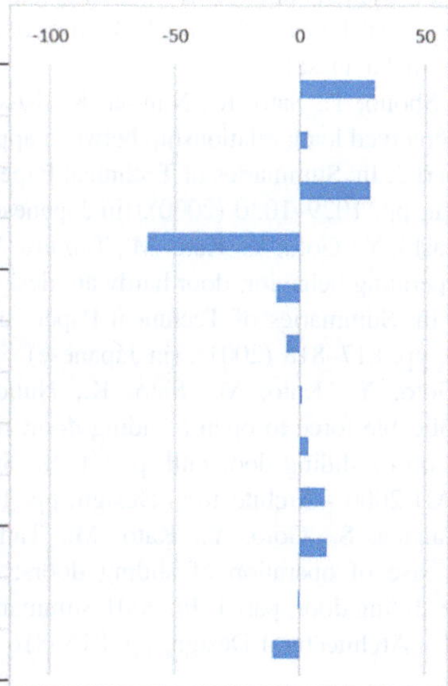

Observation of behavior revealed controlled operation such as trying to prevent the door overshoot under smaller operating force. Quantification results show that hardware with lower sensory evaluation tended to result in wider opening, suggesting the difficulty to control the door movement.

8 Conclusion

With a vertical grab bar or a door pull of better design, usability of the door can be improved, particularly for 20 N force. With 30 N and beyond, however, it will be difficult to reduce the burden. With 10 N condition, the door can be easily operated, but 6 mm projection vertical door pull is still uncomfortable to use. Both operating height and opening width depended more on individuals rather than hardware design or opening force. However, opening force clearly affects the movement of the wrist or elbow angle, which can be used as a measure of door usability.

Acknowledgments. The experiments were conducted as part of a joint research between the Building Research Institute, the Ministry of Construction and the Comprehensive Housing R&D Institute, Sekisui House, Ltd. A summary of this research was reported at the Architectural Institute of Japan Annual meeting in 2001 [4, 6].

References

1. Endo, Y.: A study of force applied to handles by man. In: Summaries of Technical Papers of Annual Meeting of AIJ 1993 - Architectural Design, pp. 935–936 (1993). (in Japanese)
2. Ochiai, O., Tenjin, Y., Naoi, H.: An experimental study on preferrable height of door handles. In: Summaries of Technical Papers of Annual Meeting of AIJ 1988 - Architectural Design, pp. 527–528 (1988). (in Japanese)
3. Tanaka, S., Ogi, H., Shono, T., Sato, K., Nunota, K., Kose, S.: Applicable force to open a sliding door and its perceived load: relationship between applied force and shape & dimension of sliding door pull part 2. In: Summaries of Technical Papers of Annual Meeting of AIJ 2000 - Architectural Design, pp. 1029–1030 (2000). (in Japanese)
4. Tanaka, S., Yokobayashi, Y., Goto, Y., Kato, M., Tamura, Y., Nunota, K., Kose, S.: Effect of hardware design to operating behavior: door hardware design and force necessary to open the sliding door, part 2. In: Summaries of Technical Papers of Annual Meeting of AIJ 2001 - Architectural Design, pp. 817–818 (2001). (in Japanese)
5. Yokobayashi, Y., Goto, Y., Kato, M., Sato, K., Nunota, K., Kose, S.: Experimental measurements of applicable force to open a sliding door: relationship between applied force and shape & dimension of sliding door pull, part 1. In: Summaries of Technical Papers of Annual Meeting of AIJ 2000 - Architectural Design, pp. 1027–1028 (2000). (in Japanese)
6. Yokobayashi, Y., Tanaka, S., Goto, Y., Kato, M., Tamura, Y., Nunota, K., Kose, S.: Experiments on the ease of operation of sliding doors: door hardware design and force necessary to open the sliding door, part 1. In: 2001 Summaries of Technical Papers of Annual Meeting of AIJ 2001 - Architectural Design, pp. 815–816 (2001). (in Japanese)

Designing for Inclusion in the Information Society

Collaborative Quality Function Deployment. A Methodology for Enabling Co-design Research Practice

Giuseppe Mincolelli[1(✉)], Silvia Imbesi[1], and Matteo Zallio[2]

[1] Dipartimento di Architettura, Università degli Studi di Ferrara, Via della Ghiara 36, 44121 Ferrara, Italy
{mncgpp,mbsslv}@unife.it

[2] Center for Design Research, Mechanical Engineering, Stanford University, 424 Panama Mall, Stanford, CA 94305, USA
Matteo.zallio@stanford.edu

Abstract. The Quality Function Deployment, was born in Japan around 1965 as a method to deploy functions forming quality and measurable parameters concerning products, service or processes based on user needs. QFD is normally used in business contexts by multidisciplinary teams that, after significant training, collaborate to compile and to interpret user requirements embedded in a matrix with characteristics and attributes. However, the design discipline is changing, along with methodologies, approaches, multidisciplinary teams and working practice. The rise of recent co-designed methodologies, flexible and holistic approaches in to design research lead to a decreased use of QFD among new designer generations. It has been noticed that when people are not adequately trained, and there is no opportunity to have face to face meetings, some methodological issues can arise while using QFD. The aim of this explorative work is to investigate the key factors that limit the use of QFD in the current multidisciplinary design research practice. A methodological literature review along with holistic experience fostered the development of a manifesto for a collaborative QFD methodology that stimulates collaborative multidisciplinary design research. This research impacts on developing a stimulating approach in to design research and practice, which shall be collaborative, inclusive, flexible, adaptable and open source, following the recent paradigms in cross-disciplines research practice.

Keywords: Human factors · Inclusive design · Quality Function Deployment · Design research

1 Scientific Background

The cross contamination among applied methodologies in the ideation and development phases of industrial answers to human needs and requirements is fundamental in the design process.

Those under the domain of the theoretical approach of scientific research has given birth, during the last decades, to the practice of design research aiming to improve the life quality of people.

G. Di Bucchianico (Ed.): AHFE 2019, AISC 954, pp. 91–99, 2020.
https://doi.org/10.1007/978-3-030-20444-0_9

According to Archer, "Design research is systematic enquiry whose goal is knowledge of, or in, the embodiment of configuration, composition, structure, purpose, value, and meaning in man-made things and systems (…)" [1].

The concept of Design research affects the convergence of two different approaches and attitudes regarding the analysis of the actual reality and the experimentation and validation of theories with different levels of complexity. Different typologies of Design research have been developed and even more are still being proposed, within the dualism between science and industry, quality and quantity, theoretic models and prototypes.

Different research processes offer a variety of outcomes: methodologies aiming to develop shareable solution to human problems provide, generally, results like configurations, processes, and designs. When the aim of the research is to understand the causes and the contexts in which those problems are generated, Design research can provide closer results to the ones related to the scientific approach, such as theories, publications, experiments [2].

It is commonly recognized that the development of innovation in scenarios such as the Internet of Things, autonomous vehicles, assistive environments, eHealth, due to the complexity and variety of the competences required, necessarily implicates a multidisciplinary approach [3].

This exploratory work, very much focused on the Human Centered Design research and very close to industrial products and processes, can provide support for the creation of collaborative methodologies, services and other approaches that can offer new human needs-driven solutions. The term "new" recalls the concept of "innovation" referring to the major drivers in the development of contemporary industrial design research.

The design research team in these cases is composed of experts coming from Industrial Design and Engineering, Social Sciences, Ergonomics and Human Factors, Law and Policy, Economics, Computer Science, Communication and Robotics. The work of such research teams requires instruments and methods to enable them to organize and harmonize quantitative and qualitative approaches.

The analysis phase, mutual understanding, the language used to communicate ideas among different specializations have to be driven towards the direction of shared objectives derived from the real human experience.

The Quality Function Deployment (QFD) is a methodology that aims at fostering the ideation of innovative solutions based on a qualitative human-centered analysis through quantitative instruments of evaluation, like matrices and algorithms and the use of graphic [4].

It is interesting to see how the complexity and accuracy of this methodology in addressing human needs has been used for many years, since 1965.

However, the increased use of collaborative platforms along with different design approaches is nowadays overwhelming the structured, and not up-to-date QFD methodology. This research work, based on observation of design research practice in different teams, is positioning a hypothesis of applying the concept of collaborative platforms to further develop a collaborative QFD methodology able to cope to the new demand of flexible, ubiquitous design approaches for the design community.

2 The Quality Function Deployment Methodology

The Quality Function Deployment, also known as QFD, was born in Japan in 1965 as a method to deploy functions forming quality and measurable parameters concerning products, service or processes, based on user needs [4].

It was developed primarily by Yoji Akao, that described the tool as a "method to transform qualitative user demands into quantitative parameters, to deploy the functions forming quality, and to deploy methods for achieving the design quality into subsystems and component parts, and ultimately to specific elements of the manufacturing process" [5].

The Quality Function Deployment was further defined and rationalized by Yoji Akao, Shigeru Mizuno and Yasushi Furukawa, members of the JUSE (Union of Japanese Scientists and Engineers), in the early '70s after a applying the QFD for designing an oil tanker by Mitsubishi Heavy Industries [6].

QFD has largely been applied over half a century as a tool to literally capture the "voice of the customer". The voice of the customer is collected and analyzed in a variety of different ways: direct discussion or interviews, surveys, focus groups, customer specifications, observation, field reports. QFD is normally used in business contexts by multidisciplinary teams that, after significant training, collaborate to compile and to interpret user requirements embedded in a matrix with characteristics and attributes [7] (Fig. 1).

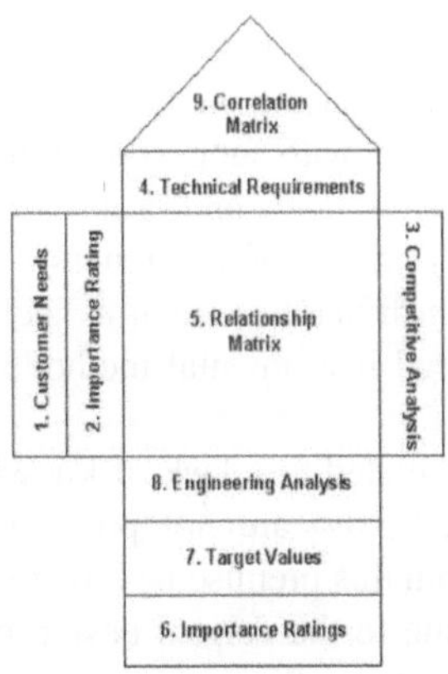

Fig. 1. Graphical representation of the traditional matrix of the Quality Function Deployment.

These parameters are embedded in a structure called the "House of Quality" that allows teams to envision solutions as an output to measure the relevance of the features for the satisfactions of the identified needs by using a graphical approach and predefined algorithms. Matrices are used as a communication tool at each step of the procedure, and the added value is within the process of communication and decision-making [8].

At the end of the 90s, when QFD has been largely diffused across the world, Professor Yoshizawa during the sixth symposium on QFD pointed out two fundamental observations on the significance of QFD in industry [9].

"QFD has changed what we have known as quality control in manufacturing processes, and established quality control for development and design. In other words, QFD has established quality management in product development and design. QFD has played a significant role when the focus of TQC shifted from process-oriented QA (Quality Assurance) to design-oriented QA and creation of a new product development system".

"QFD has provided a communication tool to designers. Engineers, positioned midway between the market and production, need to lead new product development. QFD renders a powerful arm to engineers as they build a system for product development".

After Akao, many other researchers contributed to the evolution of the QFD methodology that is currently in use. Different practitioners across the world took inspiration from different approaches such as: Conjoint Analysis, Theory of the Resolution of Invention-related Tasks (TRIZ), Conflict Management, Taguchi methods and different others [10]. A relevant fact is also shown in the application of QFD to software design other than product design [11]. Considering the perception of the design discipline among practitioners and researchers, it is easily recognizable a paradigm change that rapidly brought innovation across the years.

Methodologies, approaches and working practice are getting contaminated by cross discipline collaboration of teams. The rise of recent co-designed methodologies, flexible and holistic approaches in to design research lead to a decreased use of the originally deployed QFD methodology among new generations of designers. During the past years the authors have been using the QFD in several design research projects, with the aim of bringing innovation with support of Human Centered Design approach, including researchers from different fields and professionals from related markets [12].

It has been noticed that scarcity of awareness for designers and researchers, associated with communication difficulties, such as lack of face-to-face meetings and information exchange, developed instrumental methodological issues while using the QFD methodology.

One of the major issues is related to a lack of knowledge in interpreting matrices, not easily readable by subjects who are not proficient with QFD. An interesting research question that arises from this premise is: how to improve and update the use of the QFD methodology according to the current design practice and envision a further sustainable development for the future?

How collaborative models, platforms and technologies can foster the use and revitalize the application of the QFD tool across disciplines, countries and practitioners?

3 The Rise of Collaborative Platforms

Collaborative design has become a predominant approach in different research and working fields [13].

Collaborative design is an approach that allows a group of people to design a solution together. It helps teams to build a shared understanding of both design problem and solution [14]. It provides the means for people to work together to decide which functionality and interface elements best implement the feature they want to create [14]. Collaborative design is also described by Wang et al., as "a design process where a product is designed through the collective and joint efforts of many designers" [15].

It helps to think out of the box, brainstorms ideas with multidisciplinary teams and engages various stakeholders with different perspectives in the creative process. One of the most direct examples of how collaborative design is becoming popular among different fields is in the use of collaborative software, groupware or online platforms.

A collaborative groupware is an application software designed to support team members involved in pursuing a common task that enables them to achieve a particular goal [16]. During the process, team members use different tools and software to easier the decision making, management and creative thinking procedure.

Collaborative platforms gained popularity around 1990 when the US Government began to use them for different purposes [17]. One of the first robust applications was the Navy's Common Operational Modeling, Planning and Simulation Strategy (COMPASS), allowing up to six users to create connections with each other [18].

As the potential of Internet connection grew up, the number of users increased, the cost diminished and the continuous demand improved. Nowadays we are pervaded by an increased number of software, offering online collaborative services across the world. A short list of examples includes: electronic calendars/time management software, project management (CRM) solutions, workflow systems, knowledge management systems, enterprise bookmarking, online spreadsheets and collaborative documents, client portals, only to mention some of them.

Collaborative platforms, have been successful and used across different fields such as: Urban Planning (facilitated workshops with various communities), Community Building (such as the Detroit Collaborative Design Center (DCDC) [19]), Architecture (with different participatory methods called EDP "Equal Design Partners" and IDP "Integrated Project Delivery").

The success of these platforms is consistently described by Shea et al. in three main tasks: task interdependence (how closely group members work together), outcome interdependence (whether, and how, group performance is rewarded), and potency (members' belief that the group can be effective) [20].

To be successful, a collaborative project must establish a definition of the team, identify their outcomes, ensure there is a purpose of the collaboration and clarify the interdependencies of the members [21]. It is possible to interpret an overlap of the topics already part of the QFD methodology.

On the other hand, there are different design and implementation issues. Workers that collaborate through such groupware do not only have to address technical issues, but also carefully consider organizational aspects [22] and team processes that should be supported with the software application.

Some of the identified issues are: persistence is needed in different cases; online voice and video meetings disappears after the session, leaving behind a lack of information that cannot be stored, if not by accurately recording the sessions. Multiple input and output streams can improve concurrency issues into the shared software applications. Motivational issues are important, especially in settings where no predefined group process was in place. Design patterns in the methodological process can drive design issues and choices in a way that all team members can participate in the development process.

While the QFD has been largely used over decades, the lack of a fresh, innovative and collaborative information sharing among multidisciplinary teams working on solution development represents one consistent issue in the current spreading of QFD to a wider audience.

4 The Case Study: A Methodology for Enabling Co-design Research Practice

The research team during various years of design and research practice, highlighted the majority of constraints while using the QFD methodology, particularly when the communication among teams was weak and inconsistent.

In details, during the research project "Habitat", aiming to design inclusive IoT-based solutions for older people, it has been noticed how the use of QFD methodology was not performing well across different multidisciplinary teams working from remote.

One of the major aims of the preliminary work of analysis of user needs and definition of design requirements, was to combine available technologies on the market with the design requirements expressed by the users by using the QFD methodology [23].

The Habitat research was developed around a User Centred Design approach [24] to design smart objects and systems that control them.

The applied methodology was aimed at optimizing the combination between constraints deriving from the technological choices and requirements derived from the analysis of the needs carried out with users.

During this research the inputs are the "Needs," and are defined to reflect users' necessities and the "Features", as measurable performances of the object to be designed. Both needs and features have been indexed in correlation through a matrix chart, which supported designers to evaluate the degree of relationship.

Starting from hierarchy resulting from expressed evaluations, designers developed through brainstorming the feasible solutions for answering user needs. One of the major issues encountered during this ideation process was related to the lack of communication when combining topics related to quality (emotions and necessities of selected categories of users), with others related to quantity (measurable features of the smart objects to be designed regarding both design and technologies).

As a result, the iteration among teams of designers, engineers and caregivers was not optimized, despite the rigid, functional structure of the QFD. In practice the methodology was performing well, but the connection among teams was not strong enough. This means unclear communication, time not optimized and the most important aspect, inefficiency during the design and optimization stages.

This helped to change the strategy behind the QFD, by organizing different meetings among designers, engineers and caregivers, by using online platforms as well as in person communication strategy.

In a further stage, when the prototypes were defined, a test phase was performed in order to collect feedback for mainly evaluating the usability.

At the end of the test phases, information gathered through observation, questionnaires and focus groups were further elaborated through collaborative platforms and online database.

As a result, since the process of design, implementation, prototyping and test was implemented with collaborative platforms and online database, the team performance as well as the efficacy of the design was perceivable among the team members.

The implementation of the QFD methodology with collaborative platforms available online and data sharing options fostered the inner collaboration between designers of different disciplines by improving their sense of responsibility as well as the group challenge in performing better.

5 Envisioning the Future of Collaborative QFD Platforms

The design process is evolving with a pace which is proportional to the technology development. The rise of recent co-designed methodologies, flexible and holistic approaches in design research may lead to a further decreased use of QFD methodology. This would translate in a big lost in terms of scientifically proven design methodology used in the research and development scenario.

Knowledge and training are also important in order to spread out the concept of QFD and the case study of the Habitat project proved how those topics are extremely relevant in current design research.

As a preliminary result, it is possible to state that the benefit of a ollaborative QFD platform are boosting the design outcomes.

The collaborative QFD platform could support practitioners but also researchers in interpreting matrices, analyzing workflows and foster the mutual understanding process among multidisciplinary teams, who are not physically connected.

As a preliminary statement of the Collaborative QFD platform manifesto, some of the inputs might be taken in to account:

- Allow multidisciplinary teams to foster understating of user needs;
- Enable team players to make meaningful brainstorming;
- Unobtrusively improve the quality of working practice;
- Foster a long-lasting Human Centered Design approach during different stages of the project;
- Connect mindsets, strategies and approaches with as less as possible constraints;
- Allow data sharing, information sharing, idea sharing in an accessible way;
- Increase awareness by allowing practitioners to adjust the system according to their needs.

A future impact of this work will lead towards the creation of a collaborative web-based platform that uses the QFD methodology to enable new generations of designers and engineers to intuitively understand customer needs, translate in to meaningful requirements and foster innovative designs.

This research will impact on developing a stimulating collaborative approach in to design research and practice, which shall be collaborative, inclusive, flexible, adaptable and open source, following the recent paradigms in cross-disciplines research practice.

References

1. Archer, L.B.: A view of the nature of the design research. In: Jacques, R., Powell, J.A. (eds.) Design: Science: Method, Guilford, Surrey. IPC Business Press Ltd., pp. 30–47 (1981)
2. Mincolelli, G.: Financed design research made by Universities: some considerations about the protection of results. In: How to Face the Scientific Communication Today. International Challenge and Digital Technology Impact on Research Outputs Dissemination. Firenze University Press, Firenze (2017)
3. Zallio, M., Berry, D.: Design and planned obsolescence a review of theories and approaches impacting on enabling technologies. Des. J. Int. J. Aspects Des. 20 (2017). In: Proceedings of the 12th European Academy of Design Conference, Design for Next, Sapienza University of Rome, 12–14 April 2017, pp. S3749–S3761 (2017)
4. Akao, Y.: Development history of quality function deployment. In: The Customer Driven Approach to Quality Planning and Deployment. Asian Productivity Organization, Minato, Tokyo (1994)
5. Akao, Y.: QFD: past, present, and future. In: Proceedings International Symposium on QFD 1997 (1997)
6. Hauser, J.R., Clausing, D.: The House of Quality. Harvard Business Review, May 1988 (2016)
7. Akao, Y.: QFD (Quality Function Deployment) - Integrating Customer Requirements into Product Design. Productivity Press, Portland (1990)
8. Akao, Y., Mizuno, S.: Quality Function Deployment, an approach to total quality control (1978)
9. Yoshizawa, T.: Origins and development of internationalization of QFD. In: Proceedings of the Sixth Symposium on QFD, Tokyo (1997)
10. Tadashi, O., Cristiano, J.J., White, C.C.: Comparison of QFD status in Japan and the U.S. In: Proceedings of the JSQC 25th Anniversary 52nd Research Presentations, 1 June (1996)
11. Maritan, D.: Practical Manual of Quality Function Deployment. Springer, Switzerland (2015)
12. Mincolelli, G., Marchi, M., Imbesi, S.: Inclusive design for ageing people and the Internet of Things: understanding needs, in advances in design for inclusion. In: Proceedings of the AHFE 2017 International Conference on Design for Inclusion, pp 98–108 Springer (2017)
13. Dètienne, F.: Collaborative design: managing task interdependencies and multiple perspectives. Interact. Comput. **18**(1), 1–20 (2006)
14. Seiden, J., Gothelf, J.: Lean UX. O'Reilly Media, Inc. (2016)
15. Wang, L.H., Shen, W.M., Xie, H., Neelamkavil, J., Pardasani, A.: Collaborative conceptual design - state of the art and future trends. Comput. Aided Des. **34**(13), 981–996 (2002)
16. Johnson-Lenz, P., Johnson-Lenz, T.: Post-mechanistic groupware primitives rhythms, boundaries, and containers. Int. J. Man Mach. Stud. **34**, 395–417 (1991)
17. Bullen, C.V., Bennett, J.L.: Learning from user experience with groupware. In: Proceedings of the 1990 ACM Conference on Computer-Supported Cooperative Work (CSCW 1990). ACM, New York, NY, USA, pp. 291–302 (1990)
18. U.S. Department of Defense: The Dictionary of Military Terms. Skyhorse Publishing, NY (2009)

19. Detroit Collaborative Design Center (DCDC). http://www.dcdc-udm.org/about/design/
20. Shea, G.P., Guzzo, R.A.: Group effectiveness: what really matters. Sloan Manag. Rev. **28**(3), 25–31 (1981)
21. Kvan, T.: Collaborative design: what is it? Autom. Constr. **9**(4), 409–415 (2000)
22. D'Atri, A., De Marco, M., Casalino, N.: Interdisciplinary Aspects of Information Systems Studies, pp. 1–416. Springer, Germany (2000)
23. Mincolelli, G., Marchi, M., Chiari, L., Costanzo, A., Borelli, E., Mellone, S., Masotti, D., Paolini, G., Imbesi, S.: Inclusive design of wearable smart objects for older users: design principles for combining technical constraints and human factors. In: Advances in Design for Inclusion Proceedings of the AHFE 2018 International Conference on Design for Inclusion, 21–25 July 2018, Loews Sapphire Falls Resort at Universal Studios, Orlando, Florida, USA (2018)
24. ISO 13407: Human-centred design processes for interactive systems, ISO (1999)

Inclusive Design Methodology in Practice: Turning a Prison into an Inclusive Civic University

Jasmien Herssens(✉)

Hasselt University, Agoralaan Gebouw E, 3590 Diepenbeek, Belgium
Jasmien.Herssens@UHasselt.be

Abstract. This paper outlines the importance of contextual theoretical, empirical research and research by design in order to create inclusive concepts at the start of a design process. 26 master students were invited to reflect upon an old prison site in order to create a new entrance for Hasselt University representing an open university for all. Whereas many student excercises in Design for All (DfA) often focus on the needs and disabilities of people, this excercise focused on the abilities of people on site and on the opportunities of the architectural site itself. By means of linking both results, the excercise demonstrates to be very inspirational to both students and staff. The paper concludes with a categorisation in contextual analysis from an experiential as well as an architectural point of view.

Keywords: Architecture · Inclusion · Design methodology · Education · Context · Concept

1 Introduction

How can an architect work towards inclusive solutions in an extreme context of adaptive reuse in which the original typology aims at the principles of exclusion and experiences of isolation and stigma are the main goal: a prison? An old building, originally designed for prisoners and guardians, was the challenge for architects when Hasselt University asked them to reflect upon a design proposal by means of turning the old prison of the city of Hasselt into a civic university. More precisely, architects were invited to create a design proposal for the old prison in order to host the Faculty of Law. This paradox was the main goal for NoA-architects in Belgium when they were invited to renovate the old prison in Hasselt into the University's Faculty of Law. For a long time, an architectural typology was historically defined by a formal defined language and its architectural elements. This approach originates upon the common belief that the architectural language is dominant, and that people should always adapt to these by law and religion determined creations.

Consequently, a formal language outlined the main foundations for public buildings and urban creations. Moreover, the historical and cultural values formed the original

G. Di Bucchianico (Ed.): AHFE 2019, AISC 954, pp. 100–109, 2020.
https://doi.org/10.1007/978-3-030-20444-0_10

differences in between the formal languages of Architectural Schools, like Ecole des Beaux Arts and Ecole Polytechnique. Currently, however, a democratic and inclusive approach in design has positively evolved the architectural language by means of including experiences rather than forms. These experiences are defined by economic, political, cultural and social evolutions within a historical defined society. The architectural language has been challenged because there is no limitation in the needs of people as it is a relative approach based on evolution [1]. Thus, the current challenge is to use the designated formal language of this prison in order to meet the aimed experiences and needs. Designing a building as diverse as possible, makes both the architectural language as well as its outcome more democratic in expression. This revolutionary shift can lead to a diversity of forms and style in architectural language these days. Moreover, it may indicate the level of inclusion of the buildings based on typology. This challenge in historical evolution offers new opportunities to architects and urban planners both for new and old buildings.

This paper outlines insights in design methodology based on an architectural master project for students in interior and architectural design who were challenged to reflect upon the paradoxical question to reverse the site of the old prison into the main entrance of the university. Making an inclusive entrance in an old prison's typology requires a design approach and awareness that examines the right design questions and decisions. By means of generating the different design steps, this paper evaluates and informs upon a workable educational approach by means of using the right methods and awareness in order to transfer insights to design methods.

2 Historical and Architectural Context

In the city center of Hasselt, next to the ring road -designed as a Green Boulevard- the old prison dated 1859 is now one of the main buildings of Hasselt University. Until 2005 this building hosted prisoners and guardians who counted their days in small cells connected by lateral corridors leading towards a center that reminds us of a panopticon. Using a panopticon as a prison was one of the main architectural concepts developed by Jeremy Bentham [2]. Bentham defined the panopticon as a circular place that embraced all the cells with a view. In this circular place was a circular room in which the guardians could watch all the prisoners, whereas they didn't know if they were watching or not. It was a by Foucault (1975) [3] invoked metaphor for a modern "disciplinary society" on "seeing and not to be seen" and it supported the guardians in their efforts of control. This way this structural pattern could act as a powerful object in the environment. Whereas in other prisons, guardians still had to walk all the corridors to check upon their hosts, this typology reduced the number of walking lines and made it possible to check from one control tower. Moreover, the five arms that linked the central body of the building and acted as corridors were placed in a pentagon (Fig. 1).

Fig. 1. Aerial view upon the old prison, the Faculty of Law at Hasselt University

By means of using this form, people who walked the corridors were confused and distracted and consequently by this formal language, orientation was undermined. This design aimed at creating disorienting architecture in order to distract its main users – prisoners- and avoid them to escape. At Hasselt's old prison, part of this concept was built but the cells were still part of five main wings and the wings themselves were connected by the center that wasn't used for a control post but instead offered place for a pulpit. The pulpit was used by a priest and guardians could listen to them. It was a star form developed by Edouard Ducpétiaux, a Belgian sociologist and reformer of the Belgian prison system [4].

In 2005, the prison moved out and the building was bought by Hasselt University. The rectorate invited NoA architects to redefine the site into the Faculty of Law and the main building for the rectorate and part of the University's administration. Moreover, the principles of a civic university, a human centred approach in terms of university vision, are crucial for the rectorate and consequently the representation of these principles into the architectural language requires an inclusive approach since the idea of creating distraction in this protected monument would be unsuccessful in order to create an accessible university.

The prison is a partly protected building by Monuments and Services and parallel to the paradoxical question to make a prison inclusive, the challenge to turn this architecture into an inclusive academic environment was challenging too.

Currently, the university exponentially grows, and this requires new buildings. Some have been built close to this site and others will soon be renovated or erected in the city center as the university has bought old buildings in order to establish its city campus. This evolution requires a vision at urban level as well. These challenges for a future university campus formed the basis for the research and design exercise in the research seminar "Designing for More" at the Faculty of Architecture and Arts at Hasselt University.

Fig. 2. Picture of the old prison, the Faculty of Law at Hasselt University, (Picture: Liesbeth Driessen)

3 Designing for More: Educational Design Research in Order to Unravel Opportunities for Inclusion

During the research seminar "Designing for More", invited lectures as well as workshops and traditional lecturing took place. Guest speakers with different expertise and from different areas and disciplines were invited (CIFAL, staff at Hasselt University, Inter, Doof Vlaanderen, Blindenzorg Licht en Liefde, …). This way students were involved with the challenges of social sustainability defined by different people, approaches and perspectives. A speaker of the United Nations came to raise awareness for the Sustainability goals, an expert in echolocation gave a workshop to train students in listening and hearing differences in architecture based on sound experiences, an architect in a wheelchair explained her experiences and approaches in work, an administrator from Inter, the Flemish expertise center for Accessibility and Universal Design, gave an extended presentation and experts in the experiences of deafness and blindness joined the students during their design research workshops in practice. For 14 weeks, weekly one day courses were built up with lecturing theory and challenging research exercises by 3 professors.

26 students from the master's in interior architecture and the master's in architecture subscribed for this seminar. The group was invited to organize themselves as an architecture office in which several groups of people were responsible for several topics and two student coordinators oversaw communication and delegation of tasks. For the research part we asked half of the group to do research on the experiences of users of the building (interviews, observations, performances); whereas others could focus on the architectural language of the context and analyzed the buildings' and site's context at different levels (history, form studies, statistics of program use of the buildings, traditional modelmaking, stringmodels, typology, architectural plans, architectural offices). At the start students didn't understand this division and they literally asked

why some of them didn't have to do research, considering the architectural model making and analysis not as research. By means of questioning them we created an awareness for the importance of analyzing both the human experiences as well as the architectural language evoking these experiences. Students were invited to question and reflect upon their actual actions and what they were doing. For example, the group that made a traditional model of the context was challenged to remove certain blocks and by means of visualizing and making the physical context by hand, they revealed certain aspects that contributed to the handicaps of the environment. Indeed, we consider the buildings and site to be impaired when people don't experience it as originally planned or intended by the designers. This is the core focus of "Design for All" or "Inclusive Design", more precisely that it is the outcome of a design strategy resulting in an environment, product or service in which users should not adapt but are supported in their actions and experiences in a positive and elegant way [5]. In other words: "Good Design enables, bad design disables" [6].

If we want a building to be as inclusive as possible, we need to know the "translation" and "representation" of different architectural languages and their impact on experiences. This personal impact in terms of spatial experience can be physical as well as social, cultural and cognitive. The physical impact is often more explicitly recognizable for outsiders as it is visually comparable with personal actions. Indeed, every person can be invited to empathize with the experience imaging him- or herself in that situation by means of thinking about the exclusion of functions or abilities linked to their physical bodies. The experiences are often valued based on human standards. The more visible and readable the differences in performance are to others, the easier it is to predict experiences. Because you cannot hide the lack of certain abilities, it is perhaps therefore the most stigmatizing experience as it is a literal and personal confrontation between the physical body of the user and the physical bodily environment. For example, if you only have one arm you will perform differently when having diner or lunch. Results can be easily made by observation and often solutions are made based on principles of accessibility and standardization. The more invisible the experiential impact is to others, the more difficult it becomes to empathize as a designer. Cognitive functions support our daily experiences but are invisible to others. Similarly, cultural and social preferences are more intangible and based on personal preferences and knowledge. Whereas in the domain of accessibility these differences are considered as "impairments" or "handicaps", in Design for All this difference is an expertise because every human is considered as a different and unique being in the world. Focus is not put on the adaptation of people or users but on how design can support the diversity of users. Therefore, it is necessary to inform designers in the best way and inform them about diversity in experiences created by design. Currently, we notice in the design world a lack of insights in this translation. In order to bridge this gap between experiences and design outcome. It takes awareness for diversity into account when designing. During the courses and lectures students were informed and invited to read related scientific articles. By means of linking theoretical knowledge with their empirical research, they were asked to reflect upon the context of the old prison, and they were invited to design concepts for future improvement.

4 Outcome Results

More precisely, students were asked to link their results to the Universal Design principles [7] and goals [8] and motivate their concepts based on the link between actual and aimed experiences and current architectural language and the supportive language. Architectural design supports and shapes interaction with the environment and architecture shows us how we use our body and senses. However, for long architecture has been dominated by a visual culture that particularly has focused on visual languages [9, 10]. In this seminar we stimulate students to design architecture that supports all senses and can contribute to human culture in all means. A panopticon or a derivate, however, was originally planned and designed to support and stimulate vision. It was designed for prisoners to be seen and for guardians to stay invisible. As part of the concept, other senses were given less attention to during the design process. Partly because it was not prioritized in terms of program. For example, the main goal was to watch the prisoners and guard them in a visual way. Secondly, in terms of programming and use, the prison was not designed to be a comfortable place because in the Victorian time it was no longer designed as a temporary transit but as a place meant to feel like a punishment and so as unpleasant as possible [11]. Finally, this architectural language has been inspired on a 19th century architectural language defined by the military design style of the Ecole Polytechnique where focus was put on the constructive and defensive translation of the building and its aesthetic visual perception. The use of the building was designed based on the prior use of the concept: visual guardiancy.

Since the current program requires just the opposite performance of the original program, there is a challenge in renovating this "handicapped" building language into a civic university. Hasselt University aims at being closely linked with society, culture and its users in general. The rector generates a common vision for the identity of Hasselt University as "an open house of knowledge in action" so the opposite of a 19th century's prison concept. Consequently, in order to be inclusive for its users an approach and façade can represent a building's identity as part of a welcoming function and cognitively accessible entrance. In this exercise we invited students to analyze the context of the users as well as the actual building. For this exercise we started from a positive approach and opportunities instead of focusing on impairments for specific groups or disabilities. The abilities and experiences of users have been analyzed as well as the existing language of the architecture on site. The research results of the analysis of both approaches are explained in the next subsections.

4.1 Architectural Language and Context of the Site

Seven years after opening, the building renovated by NoA underlines the principles of an open civic university for all, a university that links with the city and city life. NoA's concept was based on the metaphor of a bird cage where outsiders can follow and connect to what happens insides and students can stay in a protective biotope. Therefore, gateways are carefully chosen, and the powerful typology of the building is transformed into a cognitive representation of safety for students and knowledge creators. The second conceptual layer was a "city in the city" approach where the university acts as a city and connects with its paths and squares to the actual city and is easily accessible. However, due to choices during the design process – for example to protect the outside façade- the entrance of the Faculty of Law is still unclear to unfamiliar visitors or users (Fig. 2). The rectorate was designed as a completely new green building that was stickered with the UHasselt-logo and represented the beating nerve system of the university. By connecting two new buildings to the old prison and creating a colorful green building as a landmark, people often enter this building to search for the reception desk or information. The original mason made prison-façade was originally not designed to support and welcome visitors, because it once was to keep everyone out and in. Now, the site has five possible entrances, and this is confusing as well. Based on the analysis with the expert in echolocation the main gates are the most "visible" and considered as entrances.

At urban level the ringway literally creates a boundary between the city and the university's city. The model and the photo-ethnographic studies demonstrate that the approach of the university is unclear to unfamiliar visitors. The landscape architecture with trees and rocks leads people to the main open gates in the wall and consequently visitors follow roads that don't immediately lead to the main entrance. The idea of the green building as a landmark is even stronger underlined in the concept of the architects as a contrast between the red brick building. Green stands for transparency and openness, the vision of the rector and consequently his office place.

Based on maps and time registration, the university's site is close to the railway station and close to several parking lots and a bus station. A string diagram shows that the walking lines, driveways, and crossing points should need a revision in terms of the university's approach. The program of the functions of the buildings is unclear to many people as well as the signage of the rooms and buildings.

4.2 Context of People's Abilities and Experiences

Moreover, the idea to respect the pathways in the old prison as a labyrinth was theoretically aimed for by the architects but in practice interviews with administration officers, students and staff prove that this undermines the usability and experience of the place. Hence, the flow of users does not align the objectives of the daily practice because, despite the prison holds the most public functions with the reception desk and auditoria, the green rectorate building has been defined by a welcoming façade, underlined by its architectural language. Since there is no reception desk in this building, it may lead to confusion and a less inclusive welcome which doesn't align with the civic university vision of the university. Observations demonstrate that many people use the old prison and the Faculty of Law as a walkthrough from a parking lot and other campus sites towards the city and vice versa. This demonstrates the idea of a city in the city, however, once inside the building people are lost. The green building is also for visually oriented people like people who are deaf the most obvious entrance. They suggest using small landmarks in and around the site to guide people through. Experts in visual impairments confirm the need of a good design of the university's approach and entrance. The stones in front of the building are now placed in a rhythm but can be placed in order to lead the visitors towards the main entrance.

The experience of the buildings in terms of elegance and aesthetics are very positive. The reception managers only ask for a more visible location and a more usable and practical desk. Moreover, they feel that it would decrease their workload in offering guidance, information and support to users if the building was more readable.

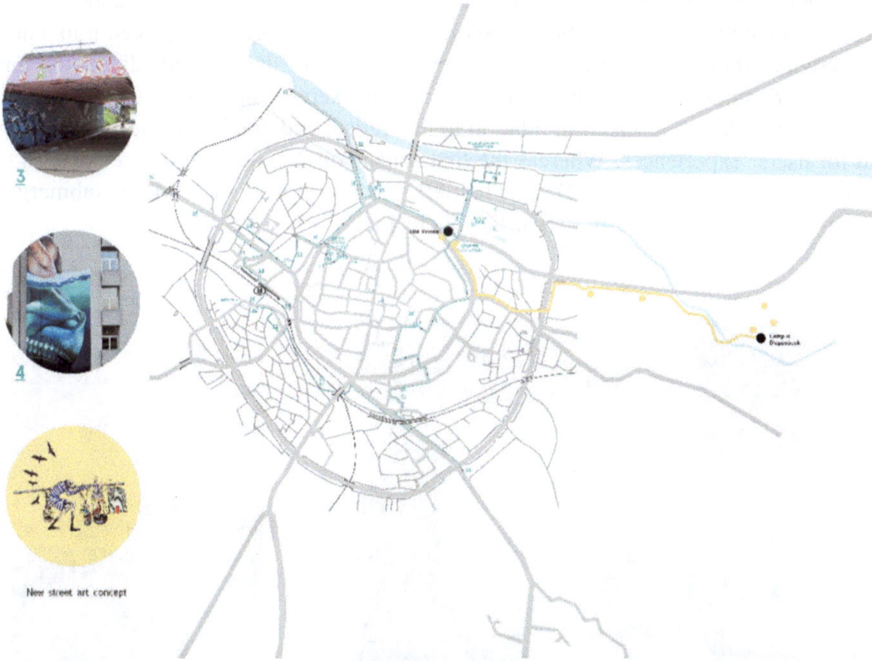

Fig. 3. Design map of contextual analysis at urban level, "New street Art concept", (design by Charlotte Neyens, Maud Orudowski, Femke Steensels, Emma Vanderwegen)

4.3 Conceptual Design Proposals

Based on all the findings in their research and design analysis, students were invited to create a conceptual design that would meet future requirements and expectations of the university. They were not obliged to design but they had to reflect about how they would link the UD-principles with UD-goals by means of their concept. Again, the students worked together in seven different design teams and their results were based on the contextual layers of inclusion of both experiences and architecture. Depending on their preferences students started with the experiences as part of their concept or the architectural characteristics but they all related to the concept of inclusion. For example, one group started from the result that the program of the green building did not align with its representation and instead of designing a new building they reprogrammed the green building with more public functions and making its ground floor serving as an entrance for the whole site. Another group also started from an architectural point of view and used the idea of the "city in a city" to design at urban levels several landmarks that support the orientation towards the university. Similar another group designed artworks or used buildings as art in order to mark the university campus in the city (Fig. 3). Other groups started from the cognitive experience of the site and wanted to extend the signage by means of architecture. One team literally pulled the "panopticon" out of the old building by means of creating a huge surface in front of the building that also acted as a ramp. The wooden entrance door was replaced by glass that is detectable by means of echolocation. Several groups also gave attention to the location of the reception desk. Some took it out of the building and replaced it in a new or renovated building whereas others created columns that served all senses and physical orientation. It was only after the design exercise that students realized the importance of design research in relation to theoretical and empirical research with, by and for users' experiences. Whereas the first seven courses, focused on the creation of awareness for the concept of Design for All, the last part of the courses submerged them into the power of design regarding inclusion.

Fig. 4. Left: picture of the jury presentation of the results, right: picture of one of the design models (pictures by Yasmine Detavernier)

5 Discussion

This master project demonstrates the importance of a decent contextual analysis by means of research by design as well as theoretical and empirical research on experiences. These different layers of contextual information help designers to create a concept that puts priorities ahead and offers insights on inclusive experiences. We define contextual layers of inclusion both in terms of experience as in terms of architectural language. More precisely, an experience can be physically, socially, cognitively or culturally defined and from an architectural point of view a context can be historically, geographically, culturally and socially layered. Future research on how experiential as well as architectural layers can be linked is necessary.

Within the context of this project new insights arose on possible improvements for the university campus. For example, the architectural concept to define contrast on site between transparency and safety undermines the actual practical use in terms of accessibility. Consequently, the general vision of a civic university has not been completely achieved. It would be good to consider this concept not only at an architectural but also at an urban level in which the "transparency" of the university's site is placed into the red brick city of Hasselt. The idea of the labyrinth can be kept but needs extra support in order to work as it is meant to be. Labyrinths in the Renaissance were part of the landscape architecture and parks. Considering the University's new city campus as a Renaissance park could put strength to the original concept of NoA architects. Similarly, new ideas and concepts can underline the university's vision and representation. Currently the staff administration investigates possible implementation of some of the suggestions. They were invited as jury members and the fact that this the exercise was a concrete and real time question contributed to the attitude and awareness of students and can be considered as a positive pedagogical approach (Fig. 4). Furthermore, this paper demonstrates that designers can learn from extreme conditions both in terms of experiences as in terms of architecture. A building typology that represents power and undermines comfort is an extreme situation in order to turn it towards an inclusive building for all and it shows that focusing on experiences instead of forms contributes to inclusive representations in buildings [10]. Similarly, inviting people with an extreme expertise in perception like people who are blind or deaf increases the insights in experiences for all. This is the common vision of Design for All and underlines the positive approach in order to respect people's abilities and diversity.

Acknowledgments. The author would like to thank all the "experiencers" on site, the experts, guest lecturers, students and teachers involved in the course seminar Designing for More as well as the technical team and staff of Hasselt University to give us the required support in order to make this exercise possible and successful.

References

1. Aymonino, C.: El significado de las Ciudades. Blume, Madrid (1981)
2. Johnston, N.: Evolving function: early use of imprisonment as punishment. Prison J. **89**(1), 10S–34S (2009)
3. Foucault, M.: Surveiller et punir: Naissance de la prison, Gallimard, Paris (1975)
4. Aubert, R., Ducpétiaux, E.: Biographie Nationale, **32**(suppl.) 154–176, Deel IV (1964)
5. Herssens, J. Design for health is designing with care, Lecture Interregiomeeting Hasselt, 4 December 2014
6. Hogan, P.: EIDD Design for All Europe (1994). www.dfaeurope.eu
7. Story, M.F.: Maximizing usability: the principles of universal design. Assistive Technol. **10** (1), 4–12 (1998)
8. Steinfeld, E., Maisel, J.: Universal Design: Creating Inclusive Environments. Wiley, New Jersey (2012)
9. Pallasmaa, J.: The Eyes of the Skin. Wiley-Academy, Chichester (1996)
10. Herssens, J.: Designing Architecture for More: A framework of Haptic Design Parameters with the Experience of People Born Blind. Lulu Publishers, Hasselt-Leuven (2011)
11. Willis, J.J.: Punishment and the cultural limits to state power in late 18th century Britain. Punishment Soc. **10**(4), 401–428 (2008)

Defining a Shared Platform in China to Propose a Social, Cultural and Technological Bridge

Avril Accolla(✉)

Design and Innovation, Tongji University, Shanghai, People's Republic of China
avril@avrildesign.com

Abstract. China's continuous and swift transformations pose relevant challenges to a sustainable development in all three areas: social, ecological and economical. Several national policies are implemented to guide these transitions within supported, non-disruptive solutions, seeking to keep an overall balance in the very dynamic socio-economic framework. These policies are generally implemented separately. This paper presents a possible proposal to link an economical policy to a social one through a design driven approach for social innovation: how to create a potentially beneficial link between the major social challenges of the elderly and the family in rural areas and the "Mass Innovation Entrepreneurship", one of the policies to support the transition from "factory of the world" to Made in China and Design in China. This could impact millions of families and offer a new social implementation for a tech driven phenomena.

Keywords: Left behind children · Elderly · Design driven social innovation · Human diversity · Sustainable business · Makers · Generations' bridge · Cultural bridge

1 Introduction

With an increase of production costs and a decrease of productivity, China's role as the "cheap factory of the world" is decreasing and the Country is implementing the transition to the high tech production model of the Made in China 2025 initiative and ultimately, to Designed in China.

One of the solutions China is pursuing is the development of a "Mass Innovation Entrepreneurship" (a policy purposely created to support the phenomenon), where innovation means mostly technology. The maker movement is seen as an important source of technological innovation and a trigger or catalyzer for capillary micro innovation's behaviors, especially for what concerns electronics, mechatronics, robotics and artificial intelligence.

Through a development which growth is not linear nor constant, the social impact of the Chinese makers is still to be demonstrated, yet it holds quite a relevant potentiality both with the big corporations and the small business.

The management of elderly is a growing challenge on a national level. The rural areas are mostly populated by elderlies and their grandchildren, left behind by their

G. Di Bucchianico (Ed.): AHFE 2019, AISC 954, pp. 110–116, 2020.
https://doi.org/10.1007/978-3-030-20444-0_11

parents, who moved to the cities, looking for better life conditions. The technological knowledge absorbed by the middle generation does not contribute to increase the quality of life of the rural areas where the elderly and children live.

The questions are: who is going to take care of the elderly? And who is going to bring forward their traditional knowledge as such and through innovation?

Empowering the elderly in their natural social role and bringing forward their traditional knowledge while at the same time finding a possible answer to the inverse generational gap created in the children growing with their grandparents, with the aim of bringing back the middle generation to the country side and re-tightening the link to their parents and children (as for the traditional "three generations under a roof") requires a design driven approach.

This research aims to evaluate the possibility of deploying a social design driven approach to find a unique combined solution to these two apparently far issues, with the creation of a pilot sustainable business, managed also with the elderly, which can integrate tradition in craft and production of food with technological innovation provided by the makers, serving several clusters of small towns.

2 Elderlies, Families, Policies, Projects

In China, more than 230 million people are over 60 years old. It is the only country in the world with an elderly population of over 200 million. More than 90% of the elderly are taken care of at home, following the traditional "filial pity" model.

However, the one child policy created the inverted pyramid of the Chinese population (4 grand-parents, 2 parents, one child) which is seriously challenging the traditional model. The China Family Development Report, since the 1980s, shows the average size of households has been progressively reducing. In 1990, it was reduced to 3.96 person, and in 2010 it was reduced to 3.10 person. According to the National Bureau of Statistics, the average size of households in 2012 was 3.02 person. China is already a country with a small average family size.

The 2015 Report on the family development in China, published by the National Health and Family Planning commission highlights as the proportion of socialized aged care services received by the elderly is very low, being mainly for physical health checkups and consultations.

On top of that, millions are elderly have been left behind when over 270 million people left their rural villages to move to the multimillion people cities.

Many of them left their children with the other parent, grandparents, relatives or other adults. By 2010 the number of "left-behind children" overcame 60 million The government is fully aware of this issue and in February 2016 released a directive recognizing the very negative impact of the distance had on millions of families [2] (Fig. 1).

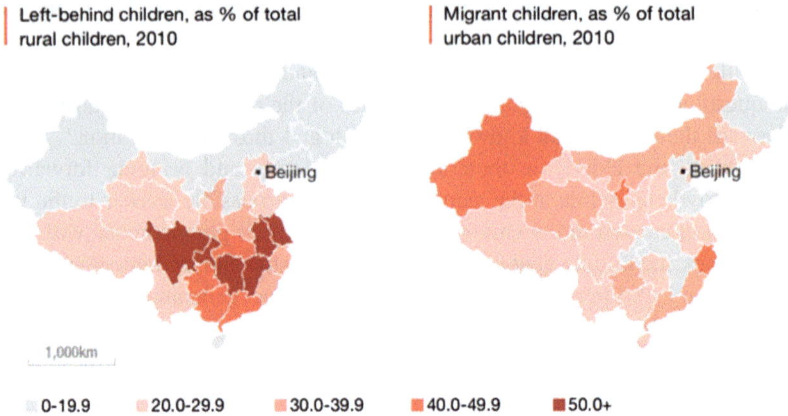

Fig. 1. Left behind and migrant children in China [2]

While China was able to dramatically reduce the number of people living in poverty conditions, many inequalities are still present. The so-called poverty traps is still present, and persistence and immobility are the dominant characteristics in the income distribution dynamics [4].

Brining urbanization to the extreme is not the solution "China cannot turn all the countryside into cities and it must pursue a new path to sustain rural development economically and environmentally," said Du Zhixiong of the Chinese Academy of Social Sciences [5].

The part of population mostly affected are the elderly and the children. A very high percentage of them belong to the above mentioned "left behind".

As an extreme consequence of aging and urbanisation, the pattern of suicide in China changed, with a much higher number of suicides among elderly and specifically those living in rural areas [6].

As already mentioned, the Chinese Government recognized the need for policies aimed to improve the economic situation of the rural regions. This theme was consistently included in the last 15 years No. 1 documents[1]. The establishment of local cooperatives was one of the initiatives supported by several policies and funding. In few years, more than 600.000 farmers' cooperatives were created, covering in theory more than 18% of total rural population [7].

Many projects have been supported with funding dedicated by the Central Government and deployed at local level.

However, many of them failed to have an important impact on the living conditions of the rural residents.

The lack of a real contextualization of any project in the local environment if often a failure factor. In fact, the approach to contextualization is most of the time translating in "treating the social and ecological environment as exogenous forces to which

[1] The first national document the Chinese government issues every year, outlining the top priorities of the nation for the coming year.

organizations must respond to while pursuing profits, competitive advantage and other more central objectives" [8].

The case of YuHu, described by Yang and Hung [9] shows how, when the competition started to seriously challenge the prosperous tourism cooperative, the whole venture was taken over by a private entrepreneur which was much less interest to the involvement of the local residents and merely aimed to increase its income.

One more reason for the failure of these local initiatives is, as underlined by the UNESCAP report [10], "the lack of infrastructures. and an underdeveloped environment of intermediaries and services providers" which "threatens the evolution of the impact investment sector and, ultimately, its ability to realize its potential to achieve social and environmental impact".

Again in 2017, United Nations Under-Secretary-General and Executive Secretary of ESCAP, Dr. Shamshad Akhtar, stated "As infrastructure provides wide economic, social and environmental benefits, it constitutes an essential part of the 2030 Agenda for Sustainable Development".

However, infrastructures are not only streets, sewage or railways. ICT plays a very important role as well.

3 China's Makers and Mass Entrepreneurship Innovation

With the first high-tech business incubator born in 1987[2], the first makerspace in 2010[3] and the second one in 2011[4], China seems to have experienced, from 2015 to now, an incredible growth, reaching the biggest number of incubators and makerspaces in the world and planning to grow further. According to data from the Torch High Technology Industry Development Center of the Ministry of Science and Technology, at the end of 2016 China owned 3.255 incubators and 4.298 makerspaces triggering the creation of 223.000 SMEs. By 2020 the plan is to reach 10.000 between incubators and makerspaces, plus 100 overseas.

As stated in its 13th Five-Year Plan (2016–2020), China aims to become an "innovation nation" by 2020, an international leader in innovation by 2030, and a world powerhouse in scientific and technological innovation by 2050.

The Chinese maker movement started typically with people, makers and hobbyists, experimenting, and creating from scratch mainly tech prototypes in shared workshops which offered shared components, machines, processes and training for research and invention. The makerspaces were funded by rent and membership fees, also with sponsorships from schools and companies.

In 2015 the policy Made in China 2025 is published and indicates a push for innovation led economic growth, changing China's focus from low-cost and Shānzhài[5] manufacturing to innovation, creativity, and design.

[2] in Wuhan, Hubei province.

[3] Xinchejian in Shanghai, founded by David Li, a Taiwanese startup entrepreneur.

[4] Chai Huo in Shenzhen.

[5] Shānzhài, 山寨, refers to counterfeit consumer goods, including imitation and trademark infringing brands and/or particularly electronics, in China.

In the same year the Government policy "Mass Entrepreneurship Innovation" pushes the maker movement to another level and a different significance: makers became synonymous of entrepreneurs and makerspaces of startups. The maker movement is subsidised to boost china's capillary capability of innovation and entrepreneurship, to generate wealth and workplaces.

Today there are 39 registered fablab[6], approximately 100 makerspaces nearer to the original spirit[7] and, according to the data from the Ministry of Science and Technology, more than 5.500 more than 4,000 incubators for technology companies more than 500,000 incubators and enterprises.

Within a specific "Chinese Character" the technological innovation potentiality is being considered and exploited both from the business and the tech educational point of view. Considering the social challenges mentioned earlier and the social potentiality of the maker approach, it is worthwhile to explore a different area of implementation and application to both experiment new opportunities and give a new thrust to the movement. A makerspace with a social innovation focus in rural areas could (i) collect the elderly A.N.A.s, input, practice and transfer their traditional knowledge; (ii) offer to younger adults to use their acquired tech knowledge locally; (iii) facilitate the coming back of young parents, thus the reunification of rural families (three generation under the roof); (iv) enhancing the quality and efficiency of small farms, thus enhancing the living standards of the farmers and their income.

4 The Project: A Mobile Shared Platform

The project foresees the creation of a mobile Fab Lab, having the same capabilities of a fix Fab Lab but providing the advantage of being able to reach different locations. While the choice of the location is a key success factor when planning to install a traditional Fab Lab, as there may be no local need for the Fab Lab neither local human resources able to create it and manage it, the mobile Fab Lab can be moved among different locations, offering several advantages:

1. Make possible assess the appropriateness of a location in order to evaluate to the possibility of eventually installing a stationary Fab lab.
2. Reach a much larger number of users.

The first mobile Lab was created by a group of MIT students in 2007, to offer the very successful making course to a larger base of students, starting from the a high school in Cleveland.

Mobile Fab Labs are presently a growing reality with several pilot experiences deployed in the United States Europe, Turkey, Saudi Arabia, and Brazil. This approach find its best application as the first step to bring digital fabrication technologies to the rural context and clustered locations.

[6] https://www.fablabs.io/labs?country=cn.

[7] https://media.nesta.org.uk/documents/made_in_china-_makerspaces_report.pdf.

In the agriculture sector, there has been some interesting initiatives. Most of them were very specifically dedicated to issues related to precision agriculture (Smolka 2016).

In early cases, such as the of smart irrigation of peanuts fields in Alabama described by Hoills [16], the involvement of the University was a key component to success although the lab was still centralized in the university.

However, our project is the first in China and it is peculiarly aimed to involve the elderly in the use of the technologies developed by the makers of the mobile fab lab.

In this way we aim to create a cultural and inter-generational bridge.

5 Conclusions

A design driven approach to social innovation brought to a possible way to combine a tech-driven government policy, derived from an economic and production challenge, to different socially focused government policies addressing a pervasive and multi-layered social challenge, with economical and health large implications, due to the huge demographics.

*D*esign is a social act, it can define and help to achieve a social-economical goal that could otherwise be difficult to achieve for a solely tech driven phenomena which tend to assume that technological micro innovations will bring China from a production-led to a design led economy.

"Designer's ability to envision and give form on material and immaterial products that can address human problems on broad scale and contribute to social well-being" [17].

"Design has always played a role in the process of production, transformations in society and the economy, shifts in technology and impacts on the environment […]. The post-industrial era is creating complex projects for technology, service, systems, strategy and products. Clients are even becoming undefined stakeholders, and this can be extended to the entire community and the environment. The rise of digital technology and the knowledge society are introducing a new culture, which can be open, participatory, shared and collaborative. Here the designer is acting as a researcher, always questioning the character of the project, its outcomes and process. Open access, co-design, crowdfunding, digital manufacturing, open-source, DIY, enabling systems and networking can be included in the toolbox of the designer and can create opportunities to drive the change towards sustainability, equity and democracy. Social innovation is leveraging forms of collaboration and co-production in which designers, innovators, users and communities co-create knowledge and solutions for a wide range of social needs, exploiting the networking technologies" [18].

The proposal of this research is to explore the feasibility of a design driven approach to a social challenge not with the consolidated assistive approach but with a participative policy, proactive, creative, holistic system integrated locally.

References

1. National Bureau of Statistics. http://www.gov.cn/xinwen/2014-05/14/content_2679681.htm. Accessed 11 Dec 2018
2. A slow awakening; left-behind children. Econ. **418**(8977), 37 (2016)
3. National Health and Family Planning Commission. http://www.cssn.cn/dybg/gqdy_ttxw/201505/t20150513_1793503.shtml. Accessed 12 Oct 2018
4. Wu, J.X., He, J.Y.: Urban-rural gap and poverty traps in China: a prefecture level analysis. Appl. Econ. **50**(30), 3300–3314 (2018)
5. China seeks better rural, urban integration – Xinhua, 3 February 2015. https://search-proquest-com.hult.idm.oclc.org/docview/1650268231. Accessed 11 Nov 2018
6. Li, M., Katikireddi, S.V.: Urban-rural inequalities in suicide among elderly people in China: a systematic review and meta-analysis. Int. J. Equity Health **18**(1), 2 (2019)
7. Li, Y.X.: 600,000 farmer co-operatives with 46,000,000 farmers on agriculture markets. Farmers' Daily, 25 September 2012 (2012)
8. Di Gregorio, D.: Place-based business models for resilient local economies. J. Enterprising Communities People Places Global Econ. **11**(1), 113–128 (2017)
9. Yang, X., Hung, K.: Poverty alleviation via tourism cooperatives in China: the story of Yuhu. Int. J. Contemp. Hospitality Manag. **26**(6), 879–906 (2014)
10. Ratnayake, R.: From Corporate Social Responsibility to Corporate Sustainability: Moving the Agenda Forward in Asia and the Pacific. United Nations Publication Copyright © United Nations (2013)
11. Asia-Pacific Countries with Special Needs Development Report 2017: Investing in Infrastructure for an Inclusive and Sustainable Future. United Nations Publication Copyright © United Nations (2017)
12. National Health and Family Planning Commission: Report on the Family Development in China. China Population Publishing House, Beijin (2015)
13. May, P., Ehrlich, H.C., Steinke, T.: ZIB structure prediction pipeline: composing a complex biologic
14. Foster, I., Kesselman, C.: The Grid: Blueprint for a New Computing Infrastructure. Morgan Kaufmann, San Francisco (1999)
15. Smolka, M., et al.: A mobile lab-on-a-chip device for on-site soil nutrient analysis. Precis. Agric **18**, 152–168 (2017)
16. Hoills, P.: Timeliness, details trademarks of Southeast peanut champions. Southwest Farm Press **32**(15), 12–16, 3 (2005)
17. Margolin, V.: The Politics of the Artificial. University of Chicago Press, Chicago (2002)
18. Imbesi, L.: Design for Social Innovation in Canada. Common Ground Pub, Arlington (2016)

Methods and Techniques Used in the Evaluation of Interfaces in Brazilian Studies Focusing on Visually Impaired People

Dominique Leite Adam(✉), Maria Lúcia Leite Ribeiro Okimoto, Kelli C. A. S. Smythe, and Caelen Teger da Silva

Federal University of Paraná, Curitiba, Brazil
domiadam@gmail.com, kellicas@gmail.com,
caelenteger@gmail.com,
lucia.demec@ufpr.br

Abstract. The aim of this paper is to map the state of art of Brazilian researches focused on the accessibility of interfaces, identifying the methods and techniques used on the evaluation of digital interfaces based on the approach of the visually impaired user. An exploratory search was made, filtering studies between 2013 and 2018, at the Catalog of Theses and Dissertations of CAPES and at the Digital Library of Theses and Dissertations. The found documents were filtered and analyzed, aiming to list the procedures used on the evaluation process. Subsequently, the results were synthesized and compared with the literature. The study is justified by the lack of evaluative research with an inclusive approach. As a result, this paper indicates that usability tests, task analysis, verbal protocols and satisfaction questionnaires are relevant procedures for the interface evaluation from a user's centered approach.

Keywords: Methods and techniques for evaluation · Interfaces · Visual impairment

1 Introduction

Newspapers, books and magazines can be easily accessed with a few clicks. Digital inclusion in parallel with technology advance enables people to participate actively in society, performing professional, personal and entertainment tasks independently.

In contrast, due to a large amount of digital content currently produced, there is a loss of information quality and, consequently, a lack of accessibility. For digital inclusion is necessary that system developers consider the user's needs and limitations to choose the best decisions for interaction [1]. Accordingly, they will be able to reach a larger number of people and come close to a universal interface.

Visually impaired people, for example, use computers, smartphones as an important Assistive Technology (AT) for their communication. Screen readers, interface customization, enlarged and high contrast fonts are adaptive features that intermediate the Human-Computer Interaction (HCI). However, these features are not enough to provide accessibility to the user. Interfaces must comply with the Web Content

G. Di Bucchianico (Ed.): AHFE 2019, AISC 954, pp. 117–128, 2020.
https://doi.org/10.1007/978-3-030-20444-0_12

Accessibility Guidelines – WCAG 2.0 and the standards provided by World Wide Web Consortium – W3C [2]. Although, comply with the established standards do not guarantee the accessibility [3]. In order to enable a better interaction, is necessary to develop a simple and intuitive interface, considering coding, graphical and textual information balancing usability, accessibility and user-experience [1, 4].

From this context, this paper aims to map the state of the art of Brazilian research focused on the accessibility of interfaces, detecting methods and techniques used on interfaces evaluation with a user-centered design approach. For this purpose, an exploratory search of studies was performed at the Catalog of Theses and Dissertations of CAPES and at the Digital Library of Theses and Dissertations, from January 1st, 2013 until December 1st, 2018. The found documents were filtered and analyzed, aiming to identify the procedures used for the evaluation of interfaces. Subsequently, they were compared with the literature in order to recognize the methods and techniques of evaluation. This study is justified by the lack of research in the area and the need to map methods and techniques with a user-centered approach that can be used on the interface's evaluation, based on the visually impaired user's opinion. Consequently, this paper intends to help researchers and interface developers to select evaluative procedures to include the end user on the development process.

2 Accessibility, Usability and User Experience (UX)

According to the Web Accessibility Initiative – WAI, everyone can contribute to web interaction. Therefore, making an interactive system accessible is more than enhance access to content. This promotes inclusion, offering to the population the same quality of information independent of age, limitations and disabilities [2].

Attached to accessibility is the concept of usability which is defined as [the extent to which a system, product or service can be used by specified users to achieve specified goals with effectiveness, efficiency and satisfaction in a specified context of use] [5]. [6] establishes that usability is related to de level of easiness in which the user can perceive the interface, the time spent for this perception, and linking with five components showed on Table 1.

Table 1. Usability components.

Components	Definition
Learnability	Indicates how easy is to perform a task on the system
Efficiency	Measures the time it takes to perform a task after learning it
Memorability	Indicates the ability to remember how to perform a task on the system
Errors	Indicates the number of errors, how severe they are and how the user recovers from them
Satisfaction	Indicates how pleasant is to use the system

Nowadays, the information society is prioritizing the use of digital resources instead of analogical practices. Consequently, is important to take care of principles to facilitate and integrate the HCI, providing effectiveness, efficiency and satisfaction.

Balancing fundamentals of usability and accessibility into the design process tends to promote the universal access of digital information [4, 7, 8]. If developers do not consider these principles, the final product tends to be incomplete and not inclusive for people with disabilities [9]. These authors suggest that accessible and inclusive interfaces and systems have a considerable impact not only on the HCI, but also on the socioeconomic development of the population.

User Experience (UX) is another important practice that influences HCI [10]. The term encompasses all aspects of how end user perceives and makes decisions to interact with systems, interfaces, services and products. These decisions are influenced by *superficial elements* - how it looks, *structural elements* - how it behaves, and *functional elements* - what it allows to do. In this way, it is necessary to consider the user's perception in the design process (*strategy*), the functional attributes and presented content (*scope*), the information architecture and possible interactions (*structure*), elements that facilitates the interaction (*skeleton*), and finally, but not less important, the graphic and visual appearance (*surface*) [10].

Considering these needs, is indispensable to use User-Centered Design (UCD) approach to understand the user's behavior in front of the interface and propose inclusive solutions. UCD can be defined as [*a design approach based on user engagement in an active, participatory or non-participatory way, using iterative processes, methods, techniques and procedures to use the user experience in expanding design requirements to meet the needs of these users, the tasks and context of use*] [11].

Several studies [1, 4, 12, 13] point out that the use of user-centered design methods presents more assertive, accessible and inclusive solutions, mainly in qualitative approaches. Therefore, when developing structurally accessible interfaces, it is fundamental to consider the user's opinion, and follow the accessibility standards for the web content [2].

2.1 Methods and Techniques for Evaluating Interfaces

In scientific research, there are several methods and techniques that can be combined to structure the design process and guide specialists on the research, development and evaluate stages of the process, whether in a user-centered or non-centered design approach [13]. It is important to point out that methods can be understood in terms of their *research approach* (deductive, inductive, hypothetical-deductive, dialectical), or *procedure* (concrete steps of the research) [14]. From this definition, this paper will discuss method as a procedure, a set of systematic actions that uses mechanisms (techniques) that help the researcher construct the path to achieve the result and hence materialize the research.

When considering the user-centered approach (empirical method), it is possible to propose meaningful and useful solutions through observation of people's actions and their experience [15]. [16] state that there are several procedures that can be used to develop, update, redesign and evaluate an interface: from understanding the concepts of usability, information architecture, to prototyping and testing with users. The authors mention that, on iterative and interactive process, two kinds of evaluation can occur: *formative evaluation* (iterative or developmental), when is conducted before the system's implementation, to identify characteristics of development; and *summative evaluation*, when is carried out after the system's implementation, to propose improvements. Significant data come from the environment in which phenomena occur and can be collected through field or laboratory research [17]. In terms of interfaces, the evaluation of this step requires user-system interaction to provide requirements for refinement and development. The evaluation process [16] is related to the ergonomic aspects and the level of usability of the interface, in other words, from the ability to adapt the interface to the user's profile and the interface's potential based on levels of effectiveness, efficiency and satisfaction.

From the methodological approach of the study, qualitative or quantitative, it is possible to select the most appropriate instruments for the user-centered evaluation. It is relevant to mention that there is no consensus on literature on the procedure's classification on methods or techniques, nor of its indication to a specific target audience. In this paper, the evaluation procedures will be classified as shown in [14, 17], in which evaluation can be organized in systematic processes (methods) using mechanisms (techniques) that help in these processes.

Based on methods, ergonomics and HCI's experts [15–20] Tables 2 and 3 present the methods and techniques for evaluating interfaces, its objectives in the computational context, the data obtained, as well as other procedures that may be associated. According to [21, 22] it is important to merge data collection techniques in order to evaluate the usability and accessibility focusing on the visually impaired, with an user-centered approach.

Table 2. Methods and techniques used on interface evaluation [15–20].

Methods	Objective	Data	Associations
Heuristic evaluation [17, 19, 20]	*Usability inspection with experts to identify interface, interaction and structure problems before the end user's evaluation (end user may have expert knowledge)*	Qualitative, quantitative	Interviews, questionnaires
Experiment [15, 20]	*Measure the effect that an action causes in an interaction with the manipulation of controlled variables*	Quantitative	Task analysis, questionnaires
Focus group [15–17, 20]	*Identify user's opinions, feelings and attitudes about the interface and interaction*	Qualitative	Observation, interview, questionnaire
Usability test [16, 17, 19, 20]	*Observe the quality of the user-system interaction experience (effectiveness, efficiency and satisfaction)*	Qualitative, quantitative	Observation, task analysis, verbal protocol, interview, questionnaire
Task analysis [18, 20]	*Analyze the HCI based on the pathway performed, decision-making and system's feedback*	Quantitative	Observation, questionnaire

Table 3. Techniques used on interface evaluation [15–20].

Techniques	Objective	Data	Associations
Interview [15, 17, 20]	*Collect user's opinions, perceptions and experience of interaction, face to face*	Qualitative	Observation, questionnaire
Observation (participatory or non-participatory) [15, 17, 20]	*Observe passively or actively (in a structured or non-structured way) in order to identify user behavior during interaction*	Qualitative	Interview
Questionnaire [15–17, 20]	*Collect user's opinion, perceptions and experience of interaction, without researcher's interference*	Qualitative, quantitative	Observation, usability test
Think aloud (verbal protocol, verbalization) [15, 20]	*Collect information from the user's verbalization of feeling seized and actions performed during the interaction*	Qualitative, quantitative	Observation, questionnaires

3 Method

With a theoretical-conceptual and exploratory approach, this paper aims to map the state of the art of master's and doctorate's Brazilian researches that utilize methods and techniques to evaluate the accessibility of interfaces under a visually impaired user's perspective. As a result, this study intends to list procedures to assist researchers and developers of interfaces and systems on the evaluation process with a user-center approach.

The search strategy that guides this study is derived from the literature review proposed by [23]. According to the authors, from a research question is possible to establish the key words that represents the scope, as well as relevant databases to conduct the research. In parallel, is necessary to establish the inclusion and exclusion criteria, in order to limit the results based on the scope of the research. Figure 1 shows the strategy of this study.

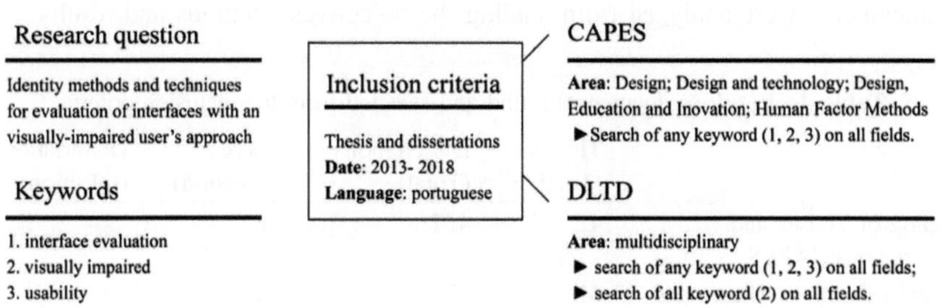

Fig. 1. Search strategy: research question, keywords, database, inclusion criteria.

Based on the research question, representative keywords (interface evaluation, visual impairment and usability) were defined to map the documents. In parallel, two databases: the Catalog of Theses and Dissertations of CAPES, because it is a search system that disseminates theses and dissertations produced by postgraduate programs in Brazil, recognized by the National Council of Education/Brazilian Federal Government [24], and the Digital Library of Theses and Dissertations), also an Federal Government agency which aims to provide visibility to national scientific production in partnership with institutions of education and research [25].

The inclusion criteria to select the studies, were defined from the search strategies of each database. The CAPES Thesis and Dissertation Catalog allows a comprehensive search of indexed documents. In this Catalog it is not possible to define the location of the keyword (e.g. title or abstract), however the website presents filters by areas of knowledge (Applied Social Sciences), date, among others. In this way, the search in the Design area was delimited, since it is an area related to HCI. The Digital Library of Theses and Dissertations allows an advanced search, in which it is possible to select whether all, any or none of the keywords should be tracked, including the search fields in which they can be inserted (e.g. abstract, title, all fields). In this library it is also possible to search by date, however, it is not allowed to select the documents by area of knowledge, this is automatically defined by the system from the results obtained. Consequently, the search was carried out in multidisciplinary areas. Due to the fact that the scope of this study is related to technology, and consequently, in constant evolution, we considered the studies published in Portuguese in the last 6 years, from January 1st, 2013 to December 1st, 2018.

4 Results

According to the presented search strategy, a total of 592 documents were found, among them 93 theses and 499 dissertations. An initial screening was carried out from reading the titles, abstracts and if they were related to the scope of this study were selected for a detailed analysis. Based on this screening, only one dissertation was repeated in both databases, resulting on 3 theses and 11 dissertations (Table 4). This result demonstrates the lack of studies with an evaluative approach from the perspective of the visually impaired, justifying the present study. Based on this selection, the documents were analyzed from reading the objectives, methods and results.

Table 4. Result of documents found and selected from the inclusion criteria.

Database	Theses (Total)	Dissertations (Total)	Theses (selection)	Dissertations (selection)
Catalog of Theses and Dissertations|CAPES	42	427	2	2
Digital Library of Theses and Dissertations|DLTD	51	72	2	9
TOTAL	93	499	**3***	**11**

**As mentioned, one document was duplicated.*

With a multidisciplinary approach, the selected researches correspond to human, social, applied and exact sciences. The studies present a qualitative approach and insert the user into several stages of the design process. It is reiterated that, as seen in the literature review, there is no consensus on the classification of procedures into methods or techniques. In general, searches address terms as synonyms. For the purposes of this article, Tables 5 and 6 organize the procedures (methods and techniques) based on [15–20].

Table 5. Synthesis of *Theses* and corresponding methods and techniques found.

Area/Researcher	Evaluation goal	Evaluation methods	Evaluation techniques
Education [26]	*Evaluate WebSonora application under aspects of ergonomics and usability with the visually impaired*	Usability test, heuristic evaluation, focus group, course analysis	Questionnaire, verbalization
Engineering and knowledge management [27]	*Propose a template for a Thematic Social Network, accessible for disabled people*	Usability test	System Usability Scale (SUS) questionnaire
Science education and mathematics [28]	*Evaluate the usability of a video game about geometry with visually impaired students*	Usability test	Questionnaire, participatory observation, interview
Design [29]	*Usability evaluation of the Brazilian Health Portal, from the perspective of blind people, through WCAG-EM 1.0.*	Usability test, task analysis	Questionnaire, observation

Table 6. Synthesis of *Dissertations* and corresponding methods and techniques found.

Area/Researcher	Evaluation goal	Evaluation methods	Evaluation techniques
Design [30]	*Measure the blind user's level of satisfaction when using an haptic equipment*	Experiment, usability test	Systematic observation, interview
Media and technology [31]	*Develop an evaluation process of screen readers focusing on the visually impaired*	Usability test	Questionnaire
Nursing [32]	*Develop and evaluate an instructional health online course for blind people*	Task analysis	Experience report

(*continued*)

Table 6. (*continued*)

Area/Researcher	Evaluation goal	Evaluation methods	Evaluation techniques
Computer science and mathematics [33]	*Investigate if gestural interactions improve accessibility of videos during the visually impaired user's interaction*	Task analysis	Questionnaire, interview
Social communication [34]	*Evaluate the accessibility of applications from the perspective of the visually impaired*	Usability test, task analysis	Task report
Public policy and management of graduation [35]	*Evaluate the interaction of visually impaired servers with a Federal Academic system*	Activity analysis	Questionnaire, interview, *in loco* observation
Design [36]	*Evaluate ergonomic solutions for designing inclusive interfaces for people with visual and/or auditory impairment*	Experiment, usability test	Questionnaire
Information science [37]	*Identify the behavior of blind people in interactions with interfaces through the screen reader*	Interaction test, task analysis	Experience report
Computer science [38]	*Evaluate the interaction of visually impaired people in a mobile application*	Usability test, heuristic evaluation, task analysis	Direct observation
Information systems [39]	*Evaluate the interaction of visually impaired people in a mobile application*	Experiment, functionality test	Questionnaire

5 Discussion

Considering the literature review and the analysis of the documents, it was possible to observe initially that there is a lack of studies focused on interface evaluation from the perspective of the visually impaired user. The literature defends the relevance of the subject and the results of this search integrate multidisciplinary areas and confirm its importance, even though is still scarce. Another aspect identified corresponds to the nomenclatures of the procedures. In the literature there is no consensus about its classification in methods or techniques. The theoretical approaches and methodological basis differ according to the areas of knowledge, and this can influence the understanding of methods and techniques, as well as their distinction.

Researches that focus on interfaces evaluation from the user's point of view can use several methods, techniques and their associations [13, 16]. There is no specific method or technique for interface evaluation directed to a profile of end user, for example, visually impaired. However, it was possible to observe that the procedures pointed out in the literature are adjusted according to the methodological approach, context, objectives and target audience of the study, being thus dependent on these factors.

In general, the analyzed studies used the usability test, also known as interaction test, as an interface evaluation method. As mentioned [16], the potential of using an interface is an essential factor to make it efficient and provide a satisfactory interaction. Regardless of whether the research demands the analysis of qualitative or quantitative data, the method helps the identification of the quality of HCI [19].

As shown in literature [21, 22] depending on the approach, the association of methods and/or techniques is indicated to guarantee a better control of the data evaluation, as well as to reach a greater level of specificity. Among the studies analyzed, it was possible to identify the recurrence of the use of more than one evaluation procedure. In addition to the usability or interaction test, the task analysis [26, 28, 30, 34, 37, 38] was indicated as a complementary procedure that aims to verify the user's decision making, as well as the difficulties encountered in the interaction [18, 20].

[30, 36, 39] include both usability test and task analysis as complementary methods to experiments. It is worth emphasizing that this method has a quantitative approach, since it aims to control the manipulation of the variables and to allow the replicability of the research. However, the analyzed studies present a qualitative approach, which limits the use of the method. Therefore, adaptations of the procedure can be used, for example: field experiment, quasi-experiment, among others [15, 20] and the association with other methods.

[26] associates four methods for evaluating an application. The author calls one of the methods as a course analysis, which can be understood as task analysis, since there is a to-do list to be carried out and, in the sequence, the user narrates his actions. [35] indicates the same procedure as activity analysis and associates it with interview, questionnaire and *in loco* observation. The mentioned observation technique cites the intervention of the researcher, when necessary, at the moment task accomplishment. This procedure is like participant observation, according to [16, 17, 20].

Narrative evaluation techniques (e.g. experience report, task report, verbal or oral report) are cited [26, 28, 32, 34, 37] as a complement to the usability test and task analysis. These narrative techniques equate to verbal protocol, think aloud [15, 20] or verbalization [19]. These procedures aim to identify the sensations seized during or after the HCI.

Interviews, observation and questionnaire were also cited on the studies. [28, 33, 35] use interview as a complementary evaluation technique to task analysis or activity analysis. Observation was cited by [35, 38] and aims to identify user behavior during interaction [15, 17, 20]. [35] mentions *in loco* observation, or a real environmental observation [17] that aims to evaluate the task with the minimum of external interference. [38] points out direct observation, which seeks to understand the facts and their interpretation, by the users, in a participatory or not participatory way [17].

Questionnaires were used more frequently, as satisfaction questionnaires, which aim to measure and understand the user's reactions about the interface. It was often associated with usability test [26–29, 31, 36] and task analysis [26, 28, 29, 33, 35, 36].

[36] uses a Questionnaire for User Interaction Satisfaction model (QUIS) [40] associated with a questionnaire based on the predictive model GOMS - Operators,

Methods and Selection rules. The last indicates the time of the actions performed to complete a requested task [19]. [27] complements the usability test with a questionnaire based on the System Usability Scale (SUS), composed of 10 questions with five response options (Likert scale), developed in 1986 by John Brooke [41].

Finally, [39] presents the term functionality test referring to the evaluation of the functional requirements of its product. According to [10] functional specifications are initial abstract concepts related to users' needs, which are analyzed separately from the interaction, interface and visual aspects. In this way, it is possible to infer that the functionality test is an evaluation method of the scope of the interface, either in relation to the desired functions or its contents.

6 Considerations

There are many researches in the area of interface evaluation, however a lack of studies that measure its accessibility from the visually impaired user. It is possible to infer, from the literature review, that there is no specific method or technique directed at this target audience. From the exploratory search of this paper, it was possible to identify that the Brazilian researches present a qualitative approach, able to evaluate the interfaces through the association of procedures, regardless of their classification in methods or techniques.

Among the studies found, user-centered evaluation is a requirement for the validation of a digital interface. Regardless of the area in which the study is allocated, the researchers consider the end-user's opinion to guide usability and accessibility improvements for interaction. It was possible to observe that usability tests and task analysis are recurrent methods used on interfaces evaluation by visually impaired. Associated with these, verbalization techniques and satisfaction questionnaires are frequently used. However, it is important to verify the possibilities of using methods and techniques and their associations according to the scope of the study, available resources and research expectations [19]. From this inspection, it was possible to map the procedures that best meet the assessment objectives, providing relevant data to adapt, restructure or redesign an interface with a user-centered approach.

Acknowledgments. Special acknowledgments to UFPR, PPGDesign and CAPES.

References

1. de Carvalho, J.O.F.: O papel da interação humano-computador na inclusão digital (2003)
2. W3C: Web Content Accessibility Guidelines (WCAG) 2.0. https://www.w3.org/TR/WCAG20/
3. Bataliotti, S.E.: Da acessibilidade à autonomia do usuário com deficiência visual em ambientes virtuais de aprendizagem (2017)
4. Petrie, H., Bevan, N.: The evaluation of accessibility usability and user experience. In: Stephanidis, C. (ed.) The Universal Access Handbook, pp. 1–30. CRC Press, New York (2009)

5. ISO - The International Organization for Standardization: ISO 9241-11:2018(en), Ergonomics of human-system interaction—Part 11: Usability: Definitions and concepts. https://www.iso.org/obp/ui/#iso:std:iso:9241:-11:ed-2:v1:en
6. Nielsen, J.: Introduction to Usability. http://www.nngroup.com/articles/usability-101-introduction-to-usability/
7. Lazar, J., Allen, A., Kleinman, J., Malarkey, C.: IJHCI_blind_user_frustration. Int. J. Hum. Comput. Interact. **22**, 247–269 (2007)
8. Power, C., Freire, A., Petrie, H., Swallow, D.: Guidelines are only half of the story. In: Proceedings of the 2012 ACM Annual Conference on Human Factors in Computing Systems, CHI 2012, p. 433 (2012)
9. Brophy, P., Craven, J.: Web Accessibility. Springer, London (2008)
10. Garret, J.J.: The Elements of User Experience: User-Centered Design for the Web and Beyond, 2nd edn. Pearson Education, London (2003)
11. Smythe, K.C.A.S.: Proposta de método de obtenção de dados sobre comportamento informacional dos usuários no processo de wayfinding em ambientes hospitalares (2018)
12. do Carmo Nogueira, T.: Estudo comparativo da experiência de usuários cegos e videntes no design web responsivo e não responsivo (2016)
13. Rogers, Y., Sharp, H., Preece, J.: Design de interação: além da interação humano-computador. Bookman, Porto Alegre (2013)
14. de Andrade Marconi, M., Lakatos, E.M.: Fundamentos de metodologia científica. Editora Atlas, São Paulo (2003)
15. Jordan, P.W.: An introduction to usability. Taylor & Francis, London (1998)
16. Santa Rosa, J.G., de Moraes, A.: Avaliação e projeto no design de interfaces. 2AB, Teresópolis (2008)
17. de Marconi, M.A., Lakatos, E.M.: Técnicas de Pesquisa. Atlas, São Paulo (1986)
18. Crystal, A., Ellington, B.: Task analysis and human-computer interaction: approaches, techniques, and levels of analysis. In: AMCIS 2004 Proceedings, New York (2004)
19. de Abreu Cybis, W., Betiol, A.H., Faust, R.: Ergonomia e usabilidade : conhecimentos, métodos e aplicações. Novatec (2010)
20. Martin, B., Hanington, B.M.: Universal Methods of Design: 100 Ways to Research Complex Problems, Develop Innovative Ideas, and Design Effective Solutions. Rockport Publishers, Beverly (2012)
21. Gonçalves, R., Rocha, T., Martins, J., Branco, F., Au-Yong-Oliveira, M.: Evaluation of e-commerce websites accessibility and usability: an e-commerce platform analysis with the inclusion of blind users. Univers. Access Inf. Soc. **17**, 567–583 (2018). https://doi.org/10.1007/s10209-017-0557-5
22. do Carmo Nogueira, T., Ferreira, D.J., de Carvalho, S.T., de Oliveira Berretta, L., Guntijo, M.R.: Comparing sighted and blind users task performance in responsive and non-responsive web design. Knowl. Inf. Syst. 1–21 (2018). https://doi.org/10.1007/s10115-018-1188-8
23. Dresh, A., Lacerda, D.P., Júnior, J.A.V.A.: Design Science Research: Método de Pesquisa para Avanço da Ciência e Tecnologia. Bookman, São Paulo (2015)
24. CAPES: Catálogo de Teses e Dissertações. https://catalogodeteses.capes.gov.br/catalogo-teses/#!/
25. IBICT, I.B. de I. em C. e T.: Biblioteca Digital Brasileira de Teses e Dissertações. http://bdtd.ibict.br/vufind/
26. Castadelli, G.A.: Estudo da Usabilidade de Software Telemático em Dispositivos Móveis com Interface Háptica e Acústica para Deficientes Visuais (2017)
27. Kaminski, D.: Redes Sociais Temáticas Inclusivas (2014)

28. Menezes, A.L.D.S.: Investigação do Desenvolvimento do Pensamento Geométrico por meio do uso de um Videojogo por Estudantes Cegos (2017)
29. De Souza, E.R.: Avaliação da metodologia de conformidade com o Web Content Accessibility Guidelines (WCAG-EM 1.0) (2015)
30. Honorato, S.: Percepção de Imagens Através de Frequências Vibratórias Captadas pelas Mãos de Pessoas Cegas (2013)
31. Campana, A.R.: Análise da Qualidade e Usabilidade dos Softwares Leitores de Tela Visando a Acessibilidade Tecnológica às Pessoas com Deficiência Visual (2017)
32. Vieira de Carvalho, L.: Construção e Avaliaçao de Curso Online para Pessoas Cegas Sobre a Prevenção da Hipertensão Arterial (2015)
33. Funes, M.M.: Interação gestual para acessibilidade de vídeos na Web por pessoas com deficiência visual (2018)
34. Goss, B.M.: Informação Móvel para Todos: Acessibilidade em Aplicativos Jornalísticos para Dispositivos Móveis (2015)
35. Juvencio, V.L.P.: Contribuição das tecnologias da Informação e Comunicação (TICS) para a Acessibilidade de Pessoas com Deficiência Visual: O caso da Universidade Federal do Ceará (2013)
36. Marcos, J.R.: Usabilidade, acessibilidade e desenho universal para aprendizagem: a experiência de usuários na educação a distância (2013). http://tede.udesc.br/tede/tede/1941
37. Rocha, J.A.P.: (In) Acessibilidade na web para pessoas com deficiência visual: um estudo de usuários à luz da cognição situada (2013)
38. da Silva, J.C.: Recomendações de Projeto de Interface de um Aplicativo Móvel de Registro de Aula para Pessoas com Deficiência Visual (2016)
39. Veriscimo, E.D.E.S.: Interação 3D com retorno de áudio para a identificação de objetos virtuais acessível a usuários com deficiência visual (2016)
40. Chin, J.P., Diehl, V.A., Norman, L.K.: Development of an instrument measuring user satisfaction of the human-computer interface. In: Proceedings of the SIGCHI Conference on Human Factors in Computing Systems, CHI 1988, pp. 213–218. ACM Press, New York (1988)
41. Sauro, J.: MeasuringU: measuring usability with the system usability scale (SUS). https://measuringu.com/sus/

Usability Analysis of the Payment Method with an Application of Digital Parking Meters with Younger and Older Users

Paula González-Torres[1(✉)], Ileana Chávez-Sánchez[1], Andrea Tejada-Gutiérrez[1], Claudia M. Fernández-Rivera[1], and Carlos Aceves-González[2]

[1] Master in Ergonomics, Universidad de Guadalajara, Guadalajara, Mexico
pau_reichel92@hotmail.com,
ilchasa@gmail.com, alejandra.teg@gmail.com,
claudia.fernandezri@gmail.com

[2] Ergonomics Research Center, Universidad de Guadalajara, Guadalajara, Mexico
c.aceves@academicos.udg.mx

Abstract. This research aimed to identify usability issues in the interaction with a payment application for a digital parking meter system located in Guadalajara, Mexico. A usability test was carried out by twenty participants divided into two groups—ten younger and ten older adults—. The main task evaluated was making a payment using the application. Participants were asked to do the task on two consecutive occasions. Once the interaction with the application finished the usability questionnaires A and B were administered. The results show that younger and older participants found problems and made mistakes when using the system. Younger and older participants suggested the need for modifications to the application. The paper discusses how understanding the experience and needs of both groups of users can help to develop a more accessible and usable app.

Keywords: Usability · Parking meters · Smartphone · Inclusive design · Ageing population

1 Introduction

The continuous technological development, mainly known as Information and Communication Technologies (ICT), has stimulated the emergence of a new economy with an impact on the way people use public services, manage businesses and communicate every day. Smartphones play an essential role among the general population because it is one of the first accesses to the ICT, which has been developed mainly through the use of applications [1] that aims to facilitate the execution of the daily activities.

The National Institute of Statistics and Geography (INEGI, for its designation in Spanish) reports that 63.9% of Mexicans aged 6 or over are Internet users. The same report mentions that of people over 55 years old are Internet users, 36% are women and 31.1% are men. Although these represent a smaller group of users compared to younger

G. Di Bucchianico (Ed.): AHFE 2019, AISC 954, pp. 129–139, 2020.
https://doi.org/10.1007/978-3-030-20444-0_13

ones, this number is continuously increasing [2]. It also matches with data provided by the National Statistics Institute (INE) in Spain, which states that the use of technologies by older adults has increased by 85% [1]. As well, it is necessary to consider that the older segment of the population is continually increasing. It is expected that by 2050 this group will represent 22.3% of the total world population; while in Mexico it is calculated to be 21.5% [3].

Similarly, INEGI reported an increase in the number of users of smartphones in Mexico, from 60.6 million people to 64.7 million from 2016 to 2017. Particularly, during the same dates, the number of users with access to the Internet through a smartphone increased from 89% to 92%. A higher percentage (69.7%) of these people downloaded some application [2].

However, although the applications are aimed at improving everyday activities, they have not yet managed to capture all sectors of the population, as is the case for older adults. This group of people often experiences some disadvantage about concerning the use of electronic devices because their capabilities are often reduced (i.e. reduced visual and auditory acuity, motor difficulties and cognitive deficits), factors that are not taken into account during the design of the interface of some applications [4–6]. Also, it is necessary to keep in mind that the pace of learning, the adaptability, and consumer habits constitute the main barriers to the older population regarding using the ICT [1].

Another critical aspect to consider about the adoption of new technologies by older adults is the lack of familiarization, considering that they have not been raised with these technologies around and thus they might ignore how these systems operate [1]. This situation could promote the presence of anxiety and the refusal to use these digital platforms [4, 7]. However, a study conducted by González-Oñate and Fanjul-Peyró in 2018 shows that 48% of its 256 participants over 65 years stated they would use ICTs if someone taught them compared to 15% who were not interested [1].

Therefore, it is necessary to develop applications that respond to the requirements and needs of older adults. Specific parameters must be considered in the design process of smartphone applications, i.e. bigger typographies, the elimination of double-clicks to show more functions, easy access to the menus and the reduction working memory loads by limiting the number of options available to users [5].

1.1 Parking Meters in Guadalajara

In 2017, a new virtual parking meter program, called “Parkimovil”, was implemented in the Metropolitan Zone of Guadalajara (ZMG) Mexico. The intention was to replace the old parking meters that operated with coins with a self-service payment system in which users can opt for three different options: (1) paying with cash in at designated store; (2) sending a text message; and, (3) paying through an application [8]. According to the local government, the use of the app is considered the most important feature in the modernization of the service.

Previous analysis of the digital parking meter system detected that only two out of nine older adults decided to interact with the App of this parking meter system and only one of them was able to complete the payment successfully [9]. It is also important to emphasize that nine of ten younger participants in that study qualified their use of the service as a miserable experience, which could suggest specific difficulty on the interaction with the app.

2 Method

The participants in this research were twenty people divided into two groups; group A included ten older people aged 60–72.9 with an average of 67.3 years (SD $\pm$ 5.8319); group B consisted of ten younger people between the ages of 20 and 29 with an average of 22.9 years (SD $\pm$ 3.3482). Three aspects were taken into account in the inclusion criteria: the participants had to have their smartphone, they had to have previously interacted with an application before, and they should not use the Parkimovil application yet.

The method used in this research was a usability test of the virtual parking meter, in which four tools were applied. (1) A questionnaire to collect sociodemographic data in which were included questions about the type of cellphone and if it had credit at that time, as well as a Facebook account or e-mail because it is necessary information to carry out the registration in the application. (2) A usability questionnaire "A" where the participants were asked about previous experience with similar applications, preference concerning the use of previous parking meters and whether they wished to use the app again. (3) Usability questionnaire "B" which included open-ended questions about its ease of use in comparison with the first attempt. They were also asked about modifications they considered convenient for the interface as well as a Likert scale to rate the application in general, how the information is distributed, ease of use for data entry and ease of use for payment. The scale consisted of five points which were rated from 1 = very good to 5 = very bad. (4) The After Scenario Questionnaire "ASQ" [10] to know the satisfaction of the participants when interacting with the application in which it was considered a seven-point scale where 1 = strongly agree and 7 = strongly disagree.

2.1 Parking Meters in Guadalajara

The following instruments and materials were used to undertake the usability analysis:

- A Reflex camera, to record on video the interaction of users with the app.
- A tripod.
- A smartphone with a stopwatch.
- The ASQ Questionnaire.
- Usability questionnaires A and B of the digital parking meter application.
- An exact reproduction of the Poster whit the same information and measurements as the posters placed on the streets, printed on a rigid material, shown in Fig. 1.

Fig. 1. Picture of the instructions for the digital parking.

2.2 Procedure

The payment process was divided into three main tasks: (1) Downloading the application to the smartphone following the payment instructions. (2) Introducing the data required by the application and finally (3) Making the payment for 30 min of parking.

Each main task was divided into sub-tasks that included the steps required by the system in order to complete the parking meter payment.

The participants were visited at home for the test where it was verified that they had an Internet connection. The activity they were asked to do was to pay through the application 30 min of parking only with the help of the printed information poster that was presented to them; however, the researchers provided help in case the participant needed it. The general procedure followed is shown in Fig. 2.

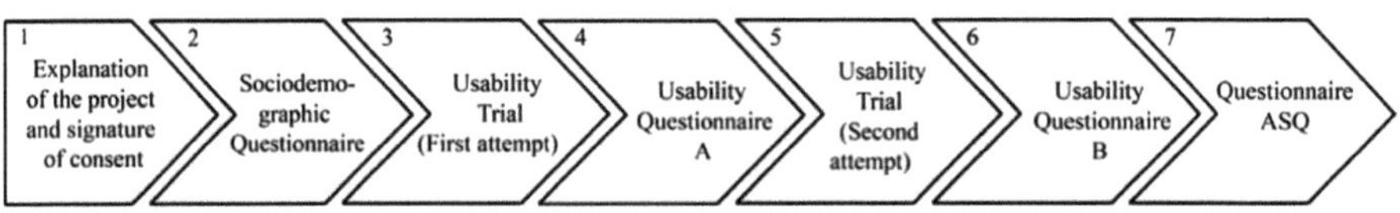

Fig. 2. Usability test procedure

The payment process during the first attempt was divided into three main tasks: (1) Download the application to the smartphone following the instructions on the poster. (2) Enter the required data for the account, vehicle, and credit card registration. (3) Make the payment of the 30 min of parking (Selection of the city, area, vehicle, time and mode of payment). It is important to mention that previously a Hierarchical Analysis of tasks (HTA) was carried out to be able to divide the three steps into sub-tasks and later to be able to make a table of errors in order to have a better follow up when registering them.

On the other hand, the payment process for the second attempt had only one main task; to make the payment for the 30 min, since the previous steps that were made during the first attempt are saved in the cell phone of the users. In both attempts, the time was taken, and the number of errors was recorded.

3 Results

Concerning the characteristics of the smartphones, seven of the younger participants and seven of the older adults confirmed that they have credit on their cell phones, which is necessary to be able to use the application on the street. Also Besides, the operating system used by the majority of the participants was Android, while only two participants from each group had a phone with an iOS operating system.

Regarding the mistakes during the first interaction with the "Parkimovil" application, the most frequent mistake in both groups corresponded to the filling of data of the vehicle where the older adults obtained 13 errors and the younger participants nine.

In addition to this in payment confirmation sub-task, the older adults had made ten mistakes and the younger participants made nine mistakes. In Table 1 the mistakes made during the interaction can be seen with more precision.

Table 1. Mistakes made by the participants during the interaction with the parkimovil app

Task	E* Older group		E* Younger group		Error description
	1A	2A	1A	2A	
1. *Downloading the application to the smartphone*					
Search app in the virtual store	5	0	0	0	Search for the application in a different site than the virtual store
• Select the "Parkimovil" app icon	0	0	0	0	Select a different icon to the one of the Parkimovil app
• Download the app on the cell phone	4	0	0	0	Download another application than Parkimovil
• Open the App	1	0	0	0	Do not find the icon in the cell phone menu
• Register via Facebook	0	0	1	0	Do not allow the access of the app to their Facebook account
2. *Introducing the data required by the application*					
• Select the option to create an account	1	0	0	0	Select login instead create an account
• Register via e-mail	6	0	2	1	Do not register the data required correctly

(*continued*)

Table 1. (*continued*)

Task	E* Older group		E* Younger group		Error description
	1A	2A	1A	2A	
• Link account with the cell phone number	5	0	0	0	Do not enter the cell phone number correctly
• Request the access code to the app via:	0	0	0	0	Do not request a code
– Whatsapp	5	0	0	0	Do not register correctly the code sent via Whatsapp
– SMS	1	0	0	0	Do not register correctly the code sent via SMS
• Scan QR code	2	2	0	0	Do not scan QR
• Enter the option of vehicle registration	13	0	9	0	Do not enter the vehicle data menu.
• Save the data of the registered vehicle	1	0	0	0	Do not select the option to save vehicle registration
• Log in to the account (this step is the start of the second attempt)	0	0	0	0	Register again in the app
– Enter email address	0	1	0	1	Do not register the email correctly
– Enter the password	0	0	0	0	Do not register the password correctly
3. *Making the payment for 30 min of parking*					
• Select the city	4	1	0	0	Do not select the city/Select another city
• Select zone	2	0	4	1	Do not select zone
• Select vehicle	0	0	1	1	Do not select a saved vehicle
• Select parking meter time to pay (30 min)	2	0	1	0	Do not select parking meter time to pay
• Select the payment method					
• "Parkimovil" credit	0	0	0	0	Do not select the option of "Parkimovil" credit
• Select the payment icon	10	1	9	0	Select to pay without entering all required data
• Wait for payment confirmation	1	0	0	0	Quit the app without confirmation of payment
Total number of mistakes	**63**	**5**	**29**	**4**	

Note: *E number de mistakes made by the participants *1A = fist attempt, 2A = second attempt

A comparison was made between the results obtained among the younger people and the older adults regarding concerning to the number of errors presented during the interaction with the application and the total time recorded in the two attempts that were made; this results can be observed in Figs. 3 and 4.

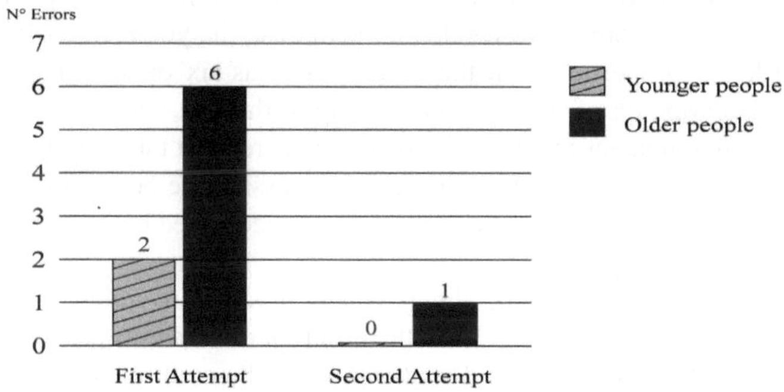

Fig. 3. Comparison of the number of errors recorded in both groups

In Fig. 4 it is shown that the average number of mistakes made by the group of older adults was higher than the mistakes made by the group of younger people in the first attempt, whereas in the second attempt the differences in mistakes made between the two groups decreased.

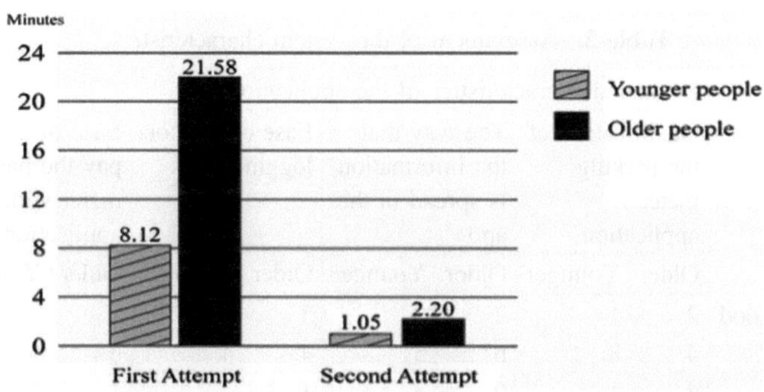

Fig. 4. Comparison of the time required for payment of the parking meter recorded in both groups.

In the same way, in Fig. 4 a similar behavior can be observed to the results obtained in Fig. 3 since the registered difference of total times between both groups decreases in the second attempt, 19.38 min in the group of older adults and 1.44 min in the younger group. It is important to consider that this decrease in time can be explained by the fact that in the development of the second interaction with the application the participants do not have to download the app and register, because these data were already registered.

It should be noted that the lower time recorded by an older adult (19 s) was lower than any other time registered in, the younger group, in which the minimum time was 27 s.

During the interaction with the application, the participants could ask for help from the researchers. Considering this, it was observed that nine of the ten older adults requested assistance in the first payment attempt, while in the second attempt just five of them asked for help from the researchers. In contrast, the younger participants asked for help at a lower proportion than the older adults, as six of the ten younger users requested the intervention of the researchers only at the first attempt.

Referring to the moment in which the participants request the help, it was observed all the users of the group of older adults asked for assistance before they had made a mistake; on the other hand, four of the younger users requested help once they failed in the interaction with the app.

In relation with the section of the payment process in which the participants requested assistance, eight members from the group of older adults asked for help from the researchers in the sub-task related to the entry of the confirmation code. Also, six older adults asked for assistance in scanning the QR code and searching for the application in the virtual store.

In comparison, the younger participants asked for help in fewer amounts than the older adults; this group of users requested assistance mainly in the sub-task related to the QR code scanning with two of the ten participants.

Regarding the overall assessment of the parking meter payment application, it can be found that the majority of the participants rated the application positively. The results are shown in Table 2.

Table 2. Assessment of the system characteristics.

Scale	Evaluated characteristics of the application							
	Qualification of the parking meter application		The way that the information is spread in the app.		Ease of use for logging data		Ease of use to pay the parking meter with the application	
	Older	Younger	Older	Younger	Older	Younger	Older	Younger
Very good	2	1	2	2	3	3	2	2
Good	4	8	6	5	4	4	6	6
Average	4	1	0	3	1	2	1	2
Bad	0	0	2	0	2	1	1	0
Very bad	0	0	0	0	0	0	0	0

Note: older group n = 10, younger group n = 10

Concerning the previous experience with the use of other payment applications, none of the older adults expressed that they had used this kind of application before, they only mentioned having interacted with social networks. Meanwhile, five participants from the younger group indicated that they had used similar applications, mostly for bank operations.

Regarding the results obtained in the usability questionnaire, in the question on whether the new payment method is better than traditional parking meters, half of the participants in the group of older adults considered the application to be more effective, as did seven of the younger ones. Also besides, in both the first and second attempts, younger and older adults expressed a desire to use the parking meter payment application again. Finally, the main suggestion for modification concerned for to menus to facilitate data entry.

In Table 3 shows the results obtained from the ASQ questionnaire by each user based on the average of the three questions that constitute the questionnaire, in which the group of older adults obtained a total average of 3.1, while the group of younger people had an average of 1.9. Thus it can be established that the perception of user satisfaction is high.

Table 3. Results from both groups of the ASQ questionnaire.

Statement	Average	
	Younger group	Older group
1.Overall, I am satisfied with the ease of completing this task	1,9	3,2
2.Overall, I am satisfied with the amount of time it took to complete this task	1,9	3,2
3.Overall, I am satisfied with the support information (on-line, messages, documentation) when completing this task	1,9	3
Total average	**1,9**	**3,1**

Note: older group n = 10, younger group n = 10

4 Discussion

At the end of the interaction process, 60% of the participants from both groups considered that using the app to complete the parking meter payment was proper. This may be related to the perception of ease of use mentioned in The Senior Technology Acceptance and Adoption Model (STAM), which attributes this to the level of effort required to interact with these technologies [11].

In contrast to the results, difficulties were observed among the participants according to the sequence of actions necessary to complete the task correctly. Mainly referring to the vehicle data entry; which highlights the need to design applications with more straightforward menus, in order to reduce the difficulty of their use and avoid the rejection of these applications by older adults [4].

At the same time, we can identify that both, in the number of errors and in the time in which to complete the first task, the group of older adults considerably improved their performance during the second attempt. This finding seems to reduce the gap with respect to the results of the group of younger people; a factor that agrees with González-Oñate and Fanjul-Peyró who affirm that currently there is no element that limits the use of technologies according to age, but instead that it is necessary to adapt the teaching of these technologies [1].

Therefore, ease of learning should be considered as a factor that influences the acceptance and acquisition of smartphone devices because since older adults have greater difficulty in terms of processing and selecting information, which has an impact on the learning process of new technologies [4].

It is important to highlight that the majority of older adults expressed their discomfort when they did not understand what they should have to do next within the application and they justified themselves by affirming that they did not have enough familiarity with this type of technology; which is consistent with the arguments presented by González-Oñate and Fanjul-Peyró, they mentioned that an important factor for which these new technologies have presented difficulties to be accepted in the elderly population is the lack of familiarity, since they have not grown with them and are unaware of their functioning [1].

Additionally, some many participants expressed their dissatisfaction with the information contained in the poster because when they attempted to follow the payment instructions, they pointed out that it did not explain the whole process.

A further factor to highlight is that none of the participants registered their credit card in order to make the payment, although the app asked for it as a mandatory registration since most people expressed distrust of apps based on previous negative experiences with other similar applications.

During the interaction process with the app, there was an inconsistency observed in the design layout of the two different operating systems that were used. Such differences were detected in the data input task alike as time selection, time zone, state, city and in the task concerning the payment finalization; in the android systems all the options previously selected by the participants were deselected when redirected to a new menu (i.e. vehicle data input), while in the iOS systems these options remained saved.

A second problem presented by the application was that some participants did not receive the message with the activation code necessary to complete the registration, which caused a delay in the process. Also in the cell phone of one of the participants, the application presented an error, so it had to be reinstalled.

A limitation of this research was that the usability test was carried out in a controlled environment in order to guarantee internet access to the participants. Therefore an interaction of use could not be observed in the place where the parking meters are located; added to this is the poster used, where the area to which it belongs is not appreciated and the signpost with the information that accompanies this poster in the public street was not included.

5 Conclusion

In contrast to the good ratings obtained by both groups, it was evident that the times obtained especially with older adults, who had reached 33 min, were quite elevated especially if we consider that they were continually guided in order to carry out the activity; This indicates the need for a redesign of the menu interface of the application in order to facilitate and simplify the registration process, a step in which a higher number of errors occurred.

However, it is important to consider the process of familiarization of the users with new technologies; as the participants experienced the payment process with this new application. In the second interaction a better execution of the task was registered, with a the reduction of errors, a diminution of the request for help and in the decrease of the time invested in making the payment; This is consistent with that proposed by Fanjul-Peyró [1] in which it is established that older adults will be receptive to these new technologies as long as they would be instructed in the use of them.

Finally, it is important to emphasize that users find in the application many advantages and conveniences once it is installed in their smartphones. However, there are still many improvements that will help new users to register more efficiently and less confusingly so that they can understand how the application works more efficiently.

Acknowledgments. We want to thank our research participants for their contributions to this research. We would also like to thank the staff of the Ergonomics Research Centre and the Master of Ergonomics for their support and advice during this project.

References

1. González-Oñate, A., Fanjul-Peyró, A.: Aplicaciones móviles para personas mayores: un estudio sobre su estrategia actual. Aula Abierta **47**, 107–112 (2018). https://doi.org/10.17811/rifie.47.1.2018.107-112
2. Instituto Nacional de Estadistica y Geografia. Comunicado de Prensa Núm. 105/18 20 de Febrero de 2018 (2018)
3. González, K.D.: Envejecimiento demográfico en México: análisis comparativo entre las entidades federativas. Conapo 129 (2015). http://www.conapo.gob.mx/es/CONAPO/Envejecimiento_demografico_en_Mexico
4. Petrovcic, A., Rogelj, A., Dolnicar, V.: Computers in human behavior smart but not adapted enough: heuristic evaluation of smartphone launchers with an adapted interface and assistive technologies for older adults. Comput. Hum. Behav. **79**, 123–136 (2018). https://doi.org/10.1016/j.chb.2017.10.021
5. Díaz-Bossini, J., Moreno, L.: Accessibility to mobile interfaces for older people. Procedia Comput. Sci. **27**, 57–66 (2014). https://doi.org/10.1016/j.procs.2014.02.008
6. Zhou, J., Rau, P.P., Salvendy, G.: Older adults' use of smart phones: an investigation of the factors influencing the acceptance of new functions. Behav. Inf. Technol. **33**(6), 552–560 (2014). https://doi.org/10.1080/0144929X.2013.780637
7. Meuter, M.L., Ostrom, A.L., Bitner, M.J., Roundtree, R.: The influence of technology anxiety on consumer use and experiences with self-service technologies. J. Bus. Res. **56**(11), 899–906 (2003)
8. Gobierno de Guadalajara: Aquí hay lugar, parquimetros virtuales, 20 Octubre 2017, de Gobierno de Guadalajara (2017). https://guadalajara.gob.mx/bienestacionado/
9. Tejada-Gutiérrez, A., González-Torres, P., Chávez-Sánchez, I., Aceves-González, C.: Inclusive smart parking: usability analysis of digital parking meter for younger and older users. In: Di Bucchianico, G. (ed.) Advances in Design for Inclusion. AHFE 2018. Advances in Intelligent Systems and Computing, vol. 776. Springer, Cham (2019)
10. Lewis, J.: IBM computer usability satisfaction questionnaires: psychometric evaluation and instructions for use. Int. J. Hum.-Comput. Interact. **7**(1), 57–78 (1995)
11. Renaud, K., Van Biljon, J.: Predicting technology acceptance and adoption by the elderly: a qualitative study. In: Annual Research Conference of the South African Institute of Computer Scientists and Information Technologists, Wilderness, South Africa, 6–8 October 2008, pp. 210–219. https://doi.org/10.1145/1456659.1456684

The Influence of Screen Brightness and Moving Speed of Visual Icons on Visual Acuity

Linghua Ran[1(✉)], Xin Wu[2], Hong Luo[1], Chaoyi Zhao[1], and Xin Zhang[1]

[1] China National Institute of Standardization, Beijing, China
ranlh@cnis.gov.cn
[2] Beijing Huiyunyiting Technical Company, Beijing, China
89427978@qq.com

Abstract. In this experiment, static visual acuity and dynamic visual acuity were tested with standard visual acuity charts and vision testing software for three groups of people in different age groups under different lighting conditions. The results showed that the dynamic visual acuity and static visual acuity had a certain correlation with each other and the dynamic vision was always lower than static vision. Static vision and dynamic vision were gradually decreased with increasing of the age. As the brightness of the screen increased, the individual's vision acuity increased. The moving speed of visual icons had great effect on the visual acuity.

Keywords: Screen brightness · Visual icons · Visual acuity

1 Introduction

Vision is the most important sensory channel for human beings, and it is also the most important means for human beings to obtain external information. Vision plays an important role in distinguishing physical properties such as light and shade, color and shape of external objects, as well as distinguishing spatial attributes such as the size and distance of objects. In the event of visual movement in the outside world, for example, the vehicle runs at a certain speed on the road, and the driver is also in constant motion relative to the external environment. At this point, the dynamic visual characteristics of the personnel are very different from those at rest. Dynamic vision refers to the visual acuity measured when a person and a fixation object are in relative motion.

In daily life, the movement of the human head is very frequent and complex, which will cause the continuous movement of the image on the retina. Studies have shown that even when standing still, the main frequency of the head movement can even reach 5 Hz, the maximum frequency can reach 10 Hz; the maximum speed of the head during running can reach 90°/s, and the main frequency in the horizontal direction is about 2.7 Hz. The vertical direction is 8.2 Hz. However, if the image motion on the retina exceeds 2°–3°/s, it will cause blurred vision. The fixation of the target by the human eyes in different body movements involves complex physiological reflexes, including visual tracking mechanisms, vestibular eye reflexes, smooth tracking, saccades, and neck-eye reflexes [1]. Therefore, dynamic vision is a comprehensive

G. Di Bucchianico (Ed.): AHFE 2019, AISC 954, pp. 140–148, 2020.
https://doi.org/10.1007/978-3-030-20444-0_14

manifestation of static vision and physiological reflection of the human body in different motion states.

Under normal circumstances, dynamic vision will be lower than static vision by 10%–20%, and by 30%–40% in some special cases. This is because when the human eye is in motion, the fixation point may move from side to side, the imaging of the target on the retina is also unstable, and the human eye has very short observation time on the object. Sometimes it is too late to discern it and the object leaves the field of vision. Therefore, people's visual acuity is always lower than static vision. Dynamic vision decreases rapidly as the speed increases. When the speed increases, with the reduction of dynamic vision, it will affect people's perception and observation of the external environment.

The purpose of this experiment is to investigate the visual threshold in different visual velocities for different age groups under different light conditions. The test results are expected to provide an important reference for industrial design.

2 Experimental Variable

2.1 Test Subjects Selection

In this experiment, participants with a total of 191 participants aged from 18 to 75 years old were divided into three age groups Table 1 shows the basic information of participants in each age group. The research goal of this experiment is to provide basic data to support the industrial design which is different from the visual inspection in medicine to evaluate the health of human eyes. Therefore, this experiment is to test the subject's corrected vision.

Table 1. Three age groups

Age group	Number	Age range	Average age	Standard deviation
Youth group 18–35	67	19–35	25.5	5.3
Middle-aged group 45–55	65	45–55	49.4	4.8
Elderly group 60–75	59	58–72	63.2	3.8
Total	191	–	–	–

2.2 Test Screen Brightness

Screen Brightness are divided into 3 levels, which are respectively 10 cd/m^2, 100 cd/m^2 and 300 cd/m^2, representing low brightness, medium brightness and high brightness.

2.3 Visual Target Speed

Visual tests were performed on the visual target under static, low-speed, medium-speed, and high-speed motion conditions respectively. Due to the different line-of-sight used in related research, the linear velocity corresponding to angular velocity also varies. In dynamic visual acuity tests, angular velocity is generally used to represent the unit of velocity. In many studies [2, 3] dynamic visual acuity checks generally have a range of speeds of 0°–150°/s, basically covering the range of motion speed of common objects in daily life. The maximum inspection speed can be determined according to the test conditions, such as checking the viewing distance, and the maximum time that the visual icons can display at the maximum speed on the display screen is not less than 100 ms.

3 Experimental Equipment and Environment

The standard logarithmic vision chart, vision testing software, 50 inch of TCL displayer with a resolution of 1920 * 1080, Z-10 smart illuminometer were used in this experiment. Vision testing software is now available on TCL 50-inch monitors.

The experiment was conducted in a 3 m * 5 m visual ergonomic laboratory with adjustable lightings. According to the national standard, "common indoor and outdoor illumination standards" in the provisions of 100–200 lx between the common office environment light level, the test environment lighting is set to 200 lx, and the test lighting environment was adjusted using a brightness adjustment software system to achieve the required illumination level.

4 Experiment Process

4.1 Preparation

In the preparatory step, the subject first introduced the subject's purpose, the experimental process and related requirements to the subject, and demonstrated the use of the test instrument in the experiment. And the subjects performed training operations and solved the problems existing in the operation of the subjects. After the subjects were completely familiar with the whole experiment, formal testing began.

4.2 Test

(1) First, to use a standard vision chart to test the visual acuity under 300 lx light conditions, and record the left and right eye vision, binocular vision. The test method refers to the standard eye chart usage specification.
(2) Visual acuity testing software was used to test the static visual acuity of the subjects under 3 different screen brightness (10 cd/m^2, 100 cd/m^2, 300 cd/m^2).
(3) Visual test software was used to test the dynamic vision of the subjects under three different screen brightness (10 cd/m^2, 100 cd/m^2, 300 cd/m^2). Dynamic eye movement speeds were: low speed (10°/s), medium speed (50°/s), and high speed (100°/s), for a total of 9 tests equaled by 3 multiplying 3.
(4) Vision testing software testing process interface, as shown in Fig. 1. First, the subjects stared at the red gaze point in the center of the screen to ensure that the initial eye gaze position was uniform, and clicked on the gaze point screen to

randomly present the visual target. In the static vision test, after the subject has finished identifying the visual target (generally not more than 2 s), press the "pace bar", the screen presents the visual target opening direction selection interface, the system records the test results and test results are correct; In the dynamic vision test, when the subject is waiting for the target to be displayed on the screen, the target opening direction selection interface is automatically presented, and the system records whether the test result and the test result are correct.

(a) Staring at the interface (b) Mission interface (c) Judgment interface

Fig. 1. Visual test software vision test process

4.3 Record of Data Results

The visual acuity of the subjects under different screen brightness under different speeds was recorded.

5 Statistical Analysis of Data

5.1 Compare and Analyze Whether There Are Differences Between the Test Results of the Standard Eye Chart and the Vision Test Software

The visual acuity results of the visual acuity of the different age groups by the standard visual acuity chart and monitor are shown in Table 2.

Table 2. Test results of corrected visual acuity under two methods in different age groups

Age group	Visual chart (300 lx)		Vision test software (300 lx)	
	Mean value	Median	Mean value	Median
Youth	1.2	1.2	1.3	1.2
Middle-aged	1.3	1.2	1.3	1.2
Elderly	1.0	1.0	1.0	1.2

The corrected visual acuity and illumination of the subjects were visual acuity values of 300 lx. There was no significant difference in visual acuity between the two genders and three age groups ($p < 0.05$), indicating that the visual acuity of the monitor was consistent with that of the standard visual acuity chart.

5.2 Correlation Analysis of Static Vision and Dynamic Vision

The visual acuity at 0°/s, 300 cd/m^2 as static visual acuity was compared with dynamic visual acuity at three different speeds: low speed (10°/s), medium speed (50°/s) and high speed (100°/s) (Table 3) .

Table 3. Correlation coefficients of static vision and dynamic vision

Age group	10°/s, 300 cd/m^2	50°/s, 300 cd/m^2	100°/s, 300 cd/m^2
Youth	.689**	.777**	.676**
Middle-aged	.658**	.714**	.607**
Elderly	.535**	.600**	.560**

5.3 Analysis of Dynamic Vision Influencing Factors

(1) Visual acuity values of different age groups, different screen brightness and different moving speeds
Figures 2, 3 and 4 show the visual acuity values of different age groups of young people, middle-aged and elderly people under different screen brightness and different moving speeds.

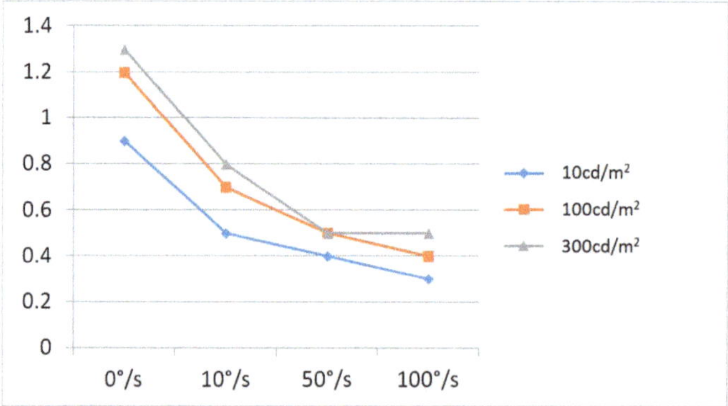

Fig. 2. Visual acuity of young people under different screen brightness and visual speed

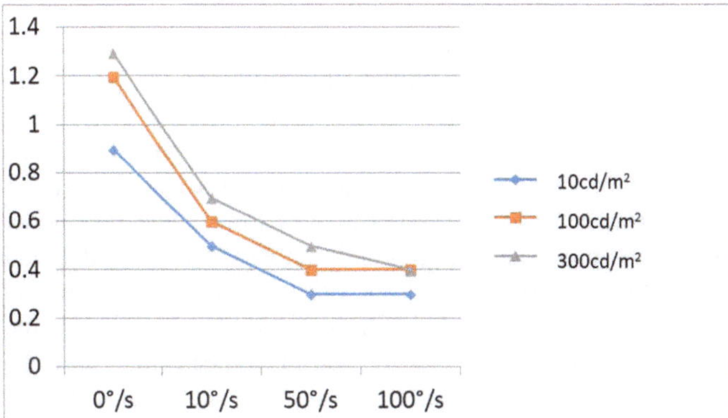

Fig. 3. The visual acuity of middle-aged people under different screen brightness and different moving speed

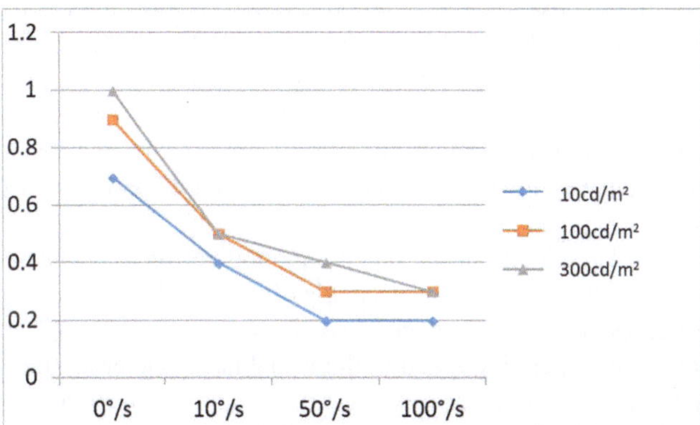

Fig. 4. The visual acuity of the elderly under different screen brightness and different moving speed

(2) Effect of visual speed on dynamic vision
From Figs. 2, 3 and 4, it can be seen that the young people's vision gradually weakens with the increase of the speed of the target. Visual acuity decreased by about 44% when the moving speed of the visual target was 10°/s, and visual acuity decreased by approximately 56% when the moving speed of the visual target was 50°/s. When the moving speed of the visual target was 100°/s, the visual acuity decreased by about 67%. The sight of middle-aged people gradually weakens with the increase in the speed of the visual target. Visual acuity decreased by about 44% when the moving speed of the target was 10°/s, and decreased by about 67% when the moving speed of the target was 50°/s and 100°/s. The visual acuity of the elderly gradually weakens with the increase in the speed of the visual target. Visual acuity decreased by about 43% at a visual movement speed of 10°/s, and decreased by about 71% at a visual movement speed of 50°/s and 100°/s.

From the statistical difference data, with the increase of the speed of the visual target, the individual's visual acuity showed a downward trend. For youth, the vision of visual speed under conditions of low speed (M = 0.67, t(66) = 15.96, $p < 0.0001$), medium speed (M = 0.46, t(66) = 21.36, $p < 0.0001$) and high speed (M = 0.41, t(66) = 15.96, $p < 0.0001$) was lower than that under static status (M = 1.12); vision under medium-speed (t(66) = 12.50, $p < 0.0001$) and high speed (t(66) = 15.68, $p < 0.0001$) was lower than that under low-speed status; and vision under high-speed condition was lower than that under medium-speed status (t(66) = 4.08, $p < 0.0001$. For the middle-aged, the vision of visual speed under low-speed (M = 0.59, t(64) = 14.99, $p < 0.0001$), medium-speed (M = 0.41, t(64) = 19.44, $p < 0.0001$) and high speed (M = 0.37, t(64) = 18.86, $p < 0.0001$) conditions was lower than that under static status (M = 1.06); vision under medium-speed (t(64) = 10.39, $p < 0.0001$) and high speed (t(64) = 10.85, $p < 0.0001$) was lower than that under low-speed status; and vision under high-speed condition was lower than that under medium-speed status (t(64) = 3.92, $p < 0.0001$). For the elderly, it went the same. The vision of visual speed under low-speed (M = 0.46, t(58) = 15.00, $p < 0.0001$), medium-speed (M = 0.31, t (58) = 20.21, $p < 0.0001$) and high speed (M = 0.28, t(58) = 21.00, $p < 0.0001$) conditions was lower than that under static status (M = 0.89; vision under medium-speed (t(58) = 10.43, $p < 0.0001$) and high speed (t(58) = 10.84, $p < 0.0001$) was lower than that under low-speed status; and vision under high-speed condition was lower than that under medium-speed status (t(58) = 3.93, $p < 0.0001$).

When the angular velocity of the relative motion between people and the fixation object increases, the minimum visual angle of the eye also increases, and the visual acuity decreases, resulting in a decrease in the perceptive ability of dynamic vision. In addition, as the operating speed of the visual target increases, the visual field of persons gradually becomes narrower, and the fixation point moves forward, and the ability to distinguish objects near and on both sides is greatly reduced, resulting in a reduction of human recognizable space. In other words, as the speed increases, the human visual acuity is reduced, the spatial recognition range is reduced, and the dynamic visual characteristics become worse. Therefore, during the process of people's progress, it is necessary to avoid the accidents caused by the difficulty of recognizing the obstacles in front and the lack of reaction time due to the drop of the dynamic vision function during high-speed movement.

(3) The influence of age on dynamic vision

Dynamic vision has a lot to do with people's age. The study found that the older the age, the worse the visual acuity. The analysis of variance with age group (variation between groups), visual speed (intragroup variable) and screen brightness (intragroup variable) as independent variables and visual acuity as dependent variable showed that visual acuity of individuals of different groups was significantly different. It showed better visual acuity in youth group (M = 0.67, t(58) = 4.94, $p < 0.001$) and middle-aged group (M = 0.63, t(58) = 3.21, $p < 0.01$) than in elderly group (M = 0.49), and youth and middle-aged group vision did not differ significantly (t(64) = 1.02, $p = 0.31$).

Table 4. Variance analysis results for variables of age group, visual target speed, and screen brightness

Effect item	F Value	df1	df2	P Value
Age group	13.39	2	188	<0.0001
Visual target speed	919.23	3	564	<0.0001
Screen brightness	391.46	2	376	<0.0001
Age group × Visual target speed	3.89	5	564	<0.0001
Age group × Screen brightness	2.24	4	376	0.06
Visual target speed × Screen brightness	45.25	6	1128	<0.0001
Age group × Visual target speed × Screen brightness	0.94	12	1128	0.50

Human dynamic vision is performed by the eyes and the entire visual system, and the influence of age on the visual system is very obvious. As people grow older, their visual function gradually deteriorates, and their dynamic visual characteristics decline rapidly. Moreover, as the age goes up, the chance of suffering from eye diseases will also increase. If you suffer from eye diseases, you may experience amblyopic or loss of vision, affecting the effective movement of the eyes and causing a decrease in visual function (Table 4).

(4) Effect of background target brightness on dynamic vision
From Fig. 5, we can see that as the brightness of the screen increases, the individual's vision increases: For youth group, the visual acuity under medium brightness ($M = 0.70$, $t(66) = 12.49$, $p < 0.0001$) and high brightness ($M = 0.77$, $t(66) = 16.07$, $p < 0.0001$) is higher than low brightness ($M = 0.55$), and visual acuity under high brightness is higher than medium brightness ($t(66) = 5.22$, $p < 0.0001$); For middle-aged group, the visual acuity under medium brightness ($M = 0.66$, $t(64) = 11.44$, $p < 0.0001$) and high Brightness ($M = 0.72$, $t(64) = 13.25$, $p < 0.0001$) is higher than low brightness ($M = 0.49$), and the visual acuity under high brightness is higher than visual acuity under medium brightness ($t(64) = 4.30$, $p < 0.0001$); the elderly group has the same result.

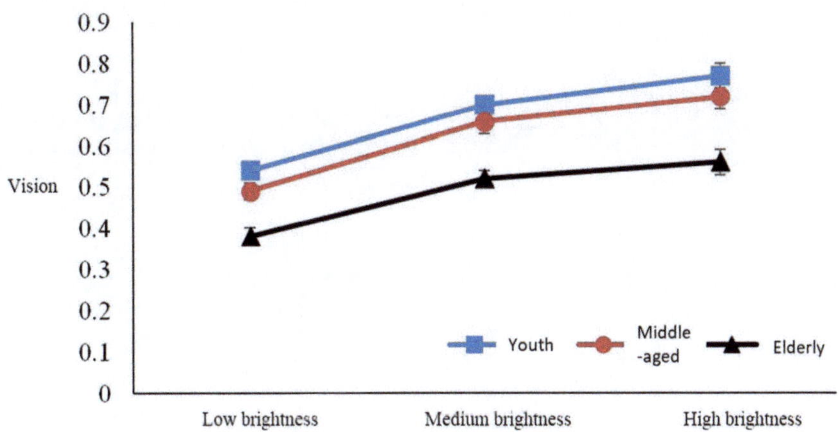

Fig. 5. Effect of background target brightness on dynamic vision

6 Conclusion

In this experiment, static visual acuity and dynamic visual acuity were tested with standard visual acuity charts and vision testing software for three groups of people of different ages under different lighting conditions. In the experiment, it is found that dynamic visual acuity and static visual acuity have a certain correlation with each other. The level of dynamic vision is based on static vision, that is, dynamic vision is always lower than static vision; with increasing of age, and static vision and dynamic vision are gradually decreased.

The quality of dynamic vision is closely related to whether the task can be completed under exercise or not. Therefore, dynamic vision detection plays an important role in checking various types of drivers and improving the training effect of drivers. At the same time, dynamic vision testing is an important part of selecting various types of special workers, which plays a significant role in improving work efficiency.

Acknowledgments. This research is supported by "Presidents Fund Project" (522017Y-5277), National Key R&D Program of China (2016YFF0201701), National Science and Technology Basic Research (2013FY110200).

References

1. An overview of dynamic vision detection methods for pilots enterprise technology development **31**(11) (2012)
2. A new device for dynamic visual acuity inspection. Chin. J. Nautical Med. **5**(3), 185–186 (1998)

Participatory Design with Older Adults: Exploring the Latent Needs of Young-Old and Middle-Old in Daily Living Using a Universal Design Approach

Alex Pui-yuk King(✉)

Faculty of Design and Environment,
Technological and Higher Education Institute of Hong Kong,
Kolwoon Tong, Hong Kong
aking@vtc.edu.hk

Abstract. In 2017, global population aged 60 years or over reached nearly 963 million, becoming twice the figure recorded in 1980. Not surprisingly ageing population will continue to accelerate due to continuing decline in fertility and improvement in survival in major diseases. When people who are suffered from cognitive or physical impairment, they often feel alone and experience different degrees of social loneliness. This paper discusses co-design experiences with various stakeholders to explore latent needs of older persons in their daily living using a universal design approach. Through iterative use of creative methods, freehand sketching and physical models, older adults can express their needs in a more accurate, tactile format. Findings reveal that commonality of interest among older persons are important in building rapport among other participants. It also helps designers develop assistive design related to health care, mobility, dining and leisure activities involving older persons, benefiting society as a whole.

Keywords: Participatory design · Universal design · Elderly · Activities of daily living · Rapid prototyping

1 Introduction

A recent report from the United Nations (UN) indicated that by 2030, the number of older people around the world will reach 1.4 billion. This is expected to double by 2100, putting pressure on current social protections, health care systems, housing and intergenerational solidarity [1]. The rise in the number of older persons is also driving the cost of health care services all over the world, suggesting a need to address social exclusion in both physical and mental health. As the global population aged 60 and over has been growing, population ageing in Hong Kong has become more pronounced than in any other major Asian city. By 2024 Hong Kong will be expected to have 397,000 people over the age of 80, constitutes a 24.8% increase from 2014 [2]. Among those aged 65 and above 12.7% will live alone, constituting a 9.5% increase from 2006 [3]. People who are cognitively impaired are particularly vulnerable because they often feel alone and are socially isolated. They may be unable to communicate with friends,

G. Di Bucchianico (Ed.): AHFE 2019, AISC 954, pp. 149–160, 2020.
https://doi.org/10.1007/978-3-030-20444-0_15

or lose them entirely [4]. People can experience loneliness in various ways, and may even be lonely when they have regular contact with family members. Having said that, research has shown that 17% of older people have contact with family and their neighbours less than once a week, and 11% have contact less than once a month [5]. When people lack social networks, they experience different levels of social loneliness.

2 Participatory Design

Action research has played a crucial role in Scandinavian Participatory Design projects since the 1970s, leading social researchers to follow its guiding principles [6]. Local accountability has been crucial to these studies, which have found that to produce reflective 'designerly' design research interventions, social and design researchers must be accordant with local needs, contribute to the knowledge pool and serve as a stimulus for later research. The purpose of this type of research is to collect, organise and improve on existing design and its operation, and to accumulate knowledge in the design participatory process. 'Design intelligence' offers a new and distinctive view from the scholarly and scientific approach, contributing to a greater understanding of design research [7]. Participation design is driven by social interactions because users and designers work together to create and develop their ideas [8]. The evaluation of possible ideas and insights also plays a critical role in this process and has a strong influence on how design uncovers unforeseen opportunities and desirable outcomes. Participatory design involves all stakeholders in the collective action of creativity, understanding and mutual learning opportunities, offering constructive solutions in terms of products, services and systems. When the idea of participatory design began in the early 1970s, it involved a process of understanding between users and designers representing the two principal participating roles. This practical research method has allowed participants to realise their genuine needs through collective thinking and production. The co-produced outputs have value because they have contributed to the professionalism of the workforce in addition to meeting its wider needs. Researchers have begun to realise that user experience is becoming more important, and should be embedded into the core of the design process, unlike conventional system-centred design [9]. The participant design approach considers the 'user as partner', rather than the 'user as subject' [10]. It involves several key attributes that differ from other typical design methods, producing tangible design ideas through processes in which design insights are gained, and a collaborative design experience is achieved by addressing unmet and unrecognised needs [11].

A recent review of the literature has shown that participatory design practice and theory are becoming more vigorously applied to health care [12]. The shared decision making of participatory design allows common goals to be attained, benefiting every participant when they work together to make improvements. Inviting every user into such a collective process requires a trusting relationship between each of them. This creative method relies on brainstorming, freehand sketching, quickly generating models to facilitate participation and promoting the idea of design as a process. It has been suggested that the elderly could become engaged in product and system testing in addition to being actively involved in the iterative application design process to improve functionality [13]. The idea would be to equalise power between vulnerable and abled persons, allowing them to share their thoughts. The platform would facilitate a mutual learning experience in which designers and older persons work together to explore the problems the latter face in their day-to-day activities. A range of techniques, including paper-based dirty models, print-outs and foam models would be used to identify the form factors, physical sizes and proportions, individual preferences, product details and the use of products. The use of technology by older persons in both their workplace and home would become an expression of equality, because such technologies improve the quality of their lives and standard of living.

3 Universal Design

Universal design has already become a design philosophy in a global design context, influencing our day-to-day activities and the quality of life for people of all ages and abilities. The term 'Universal' refers to a wide range of human diversity where the application of design should be based on user-centered approach, while 'Design' is the creative solution underpinned by the empirical investigation [14]. Universal designing requires designers a deeper consideration and understanding of the extensive range of human abilities throughout a person's life during the development process [15]. Adopting the principles of universal design could make the design of environmental elements accessible to their greatest extent as it addresses all range of functionality. It serves as a fundamental condition and requirement of good design, offering an accessible and convenient way of using things. It plays a vital role in enhancing the potential for better and the significant part of living for every individuals [16]. Universal design promotes inclusiveness in design, to shape products, places and systems by reducing the need for expensive assistive devices and special accommodations – a method of designing that focuses on all ages, sizes and dimensions and thus lowering the economic burden that products and services need to be specially designed to cater people in need. The definition of universal design as a process enables and empowers a wide range of population by enhancing human performance, wellness and social participation [17]. By fully utilizing the principles of universal design, it provides an opportunity for better social support so that people in need become more self-reliant and socially engaged [18]. As the self-image of people becomes more important, the social status of people who are suffering from functional limitations will be largely determined by assistive resources, clothing and living environment. When the assistive design becomes universal elegance, it will have a higher degree to remove stigma and discrimination. By applying the principles of universal design, product design can reduce social stigma and promote social engagement by putting people with functional disabilities on a level playing field with fully functioning people [19]. In short, it is an integrated concept that emphasises accessibility, adaptability, aesthetics and affordability [20]. As the health care cost including long-term caring is rising, in search of universal design solutions to health related issues is becoming both a social responsibility and a financial necessity. The seven principles of universal design adopted to serve as a guideline during the design process in the later stages of research include (1) equitable use; (2) flexibility in use; (3) simple and intuitive use; (4) perceptible information; (5) tolerance for error; (6) low physical effort; (7) size and space for approach and use [21].

4 Aims and Objectives

This study is funded by the Social Welfare Department, Hong Kong Special Administrative Region, to promote social inclusion and active ageing. Its specific objectives are (1) to establish and reinforce mutual support network between older neighbours, (2) to develop community-based education activities that promote social inclusion, (3) to facilitate active ageing through a series of workshops aimed at eliminating social isolation, (4) to offer a collective learning experience with older persons using information technology, (5) to promote social inclusion through design participation with tertiary students and people with disabilities, (6) to preserve local and traditional culture and indigenous intelligence through an iterative production process, (7) to offer tangible product design solutions that improve the quality of living of older persons.

5 The First Day

Forty-two Hong Kong residents (3 males, 39 females) who were 70 years old and above, were invited to join this study. These individuals were placed into seven groups to participate in design workshops with two sessions for each group. Each group included one graduating product design student (designer), one social worker and six elderly participants, all of whom were facilitated by the principal investigator (PI). The duration of the brainstorming session in the first session was limited to 20 min. All of the participants, including the older persons, were required to jot down keywords related to (1) mobility, (2) personal hygiene, (3) dining experience, (4) health care, and (5) leisure activity. They were assisted by student helpers whenever necessary. A sorting process took place after each session, delivering up to 60 fresh ideas or keywords according to specific topics. An unstructured interview was conducted by the PI with leading questions corresponding to different research areas. All of the key concepts, dialogues and answers to the questions had been written down on a long roll of paper covering four units of long benches. This 'knowledge pool' in paper format was accumulative, (i.e., words in context were added according to each unstructured interview). Interview sessions lasted one hour, followed by the first attempt at the production process in the participative design workshop, with full support from Food Grace Centre and Action Care Centre. Both are well established non-profit organisations that support the development of active ageing and community education activities in Hong Kong. All sessions were conducted in a natural setting, a mini hall at the Food Grace Centre and a communal activity room at the Action Care Centre. All of the participants were familiar with the actual location because they visited their corresponding centres to collect free vegetables and lunch boxes twice a week. The first production process lasted 60 min. The designer and the other participants were seated three in a row on both sides of a table. The designer and the social worker sat diagonally from each other, allowing the senior participants to actively participate. Among the 42 older people who joined the workshop, very few failed to show up because their assigned schedules clashed with their regular medical check-ups. The pace of the workshops was controlled by the PI to make sure the process remained on track and the procedures were fully understood by all of the participants. Marker pens, cardboard, clay, modelling foam, felt sheets and thick paper were given to the participants to build a sketch model for testing and evaluation following an iterative approach. No detailed models were expected because the data collection method was new to the older persons. In addition, many of them did not have a fundamental educational background. Therefore, if the participants could express their thoughts freely it was deemed relevant to the study. In this study, the PI decided to adopt unstructured interviews with five pre-set groups of questions relating to the key themes: (1) mobility, (2) personal hygiene, (3) dining experience, (4) health care and (5) leisure activity. The main objective was to explore the potential problems older persons might have in their day-to-day activities.

6 The Next Day

The PI and the designer scrutinized the data collected during the brainstorming sessions, unstructured interviews and the first attempt to produce several testing mock-ups. Thereafter, they prepared a digital model to deliver a true-scaled three-dimensional output for an advanced physical trial with older persons and stakeholders. The tangible format allowed the participants to comment according to what they saw and felt, so that the optimum size and proportion could be determined through collective contribution. Unsurprisingly, the printed objects offered a new approach for users to test, not only in terms of cosmetic and ergonomic considerations, but also in selecting the actual appearance of the materials, colour variants and the simple mechanism to be applied to the prototypes. The first set of physical models was printed on the second day of the participatory design workshop, after the digital models had been prepared by the designer after the first day's participation. Three-dimensional (3-D) files were built using a freeform modeller in true scale, visualised in a photo-realistic renderer to ensure all surfaces were in excellent condition. The models were then converted into another machine language that could be recognised by the 3-D printer. The model was built up layer by layer using a fuse deposition modelling technique with preferred plastic-like substance and biodegradable Polylactide (PLA). This process was executed at least two times during this session to ensure the printed mock-ups fulfilled the fundamental requirements of universal design. Detailed feedback was recorded and the physical models were immediately modified with the aid of light clay if the gripping size and overall proportion did not match most of the models.

Alternation occurs only when the first attempt does not satisfy the user needs proposed in the process of data collection. The production process was new to the older persons even though they were assisted by the designers and research assistants. It required knowledge to synthesise their creative ideas and put them into practice through the interactive process, cross-fertilising ideas and applying technologies to resolve problems with day-to-day activities. In the first attempt, the participants used corrugated boards and a handheld wire cutting machine to model the preferred shapes and modify the existing design. They were asked to form groups of two to make dirty models according to their thoughts and beliefs. Both paper and foam models could effectively express the genuine needs of the elderly, however, there was a limitation in that some tiny details could not truly be reflected in the first attempt using these basic materials. In the second attempt, the designer prepared the digital model and delivered a few possible variations in each category according to the evaluation of the first attempt. The digital model took time to be printed through fuse deposition, therefore, prior preparation was crucial to maintaining the participative workshops within the two-hour schedule. The iterative development helped the PI collect tangible data in a valid and reliable way. The designer circulated the mixture of the print models and foam components to collect immediate feedback from the participants. Modifications were iteratively made using computer simulation if the testing samples were far from the needs of the elderly people. In the end, several testing models were produced for further evaluation in terms of their aesthetics, ergonomics, ease of use and design innovation, with the goal of promoting healthy attitudes, active ageing and an inclusive society.

7 Discussion

7.1 Mobility

This study revealed that over 20% of the participating elderly used walking canes to assist them most of the time. WL, a 74-year-old woman, said she had asked the owner of a local fruit shop to drill a hole in her cane to help her put a typical cotton cord through it because she refused to spend extra money to do it. WL preferred a holding cord because she could manage her cane with her left hand and choose her fruit without putting aside her belongings. This cane was a valuable asset to her and she had been using it for years. She lost a similar cane when she could not hold it with her hands and left it at the fruit market. WL remarked that the overall structure of the cane should be sturdy and without a foldable function. Her motto was 'safety first'. Another priority was that the cane should be light-weight and easy to carry. It was also easy to observe the personalised name plate attached to the upper part of the aluminum cane to make it easy to see, and to make sure everyone collected their own cane, because the typical design of canes was similar. SF, a 65-year-old female participant, mentioned that the hand grip was important to elderly people, and that it should fit for both directions. Although she agreed that most users would locate the cane outwardly, she deliberately kept it closer to her body so that the centre of gravity was closer to her. In this case, the product should have a certain degree of 'tolerance for error', to reduce accidents caused by mistakes or misuse due to different interpretations. The replaceable cane foot also played an important role in this participative design development. Ruby, a social worker from Food Grace, commented that many of the older participants did not want to pay extra to purchase a replacement even though they realised the rubber foot had already lost much of its physical function. Having an optimum level of material softness could facilitate human movement and the user experience with such a product design. KW, a 72-year-old female participant said the developers and designers should pay extra attention to the position of the walking cane, which should not be at a ninety-degree angle to the floor, but in a tilted position. The development of the walking aid was given the highest priority because it affected the social activities of the older persons. They indicated it could be a hurdle to attending social gatherings if it had an unfavourable design, and they also wanted to hide its social stigma. Not surprisingly, nearly half of the participants said they preferred an umbrella cane.

7.2 Personal Hygiene

The dialogue began with female participant, SM, aged 68, who had an unpleasant experience trimming her fungal toenails. She related that the appearance of infected toenails was different from normal nails. It required additional force to cut the toenail, and both the trimming size and actual shape should be considered to allow for exploration. A well-designed trimmer should only require 'minimal physical effort' so that older users benefit without needing a tool specifically designed for them. A sample with an oval magnifier attached to it was not viewed favourably by most of the participants because the pre-defined focus did not match each individual user. At the beginning of the study, the research team believed that the bigger the trimmer, the easier the trimming process would be. However, it was observed that the size of the trimmer should be optimal for older persons to master without having to guess the area to be trimmed. In addition to the trimmer, two shower heads were circulated on the floor to gain insight on the existing product design. SJ, age 66, said she preferred a smaller shower head because she believed it would reduce her water consumption when she showered a few times a week. Larger shower heads with adjustable water outlets were believed to consume more water, which was wasted. Contrary to SJ, CL, a 68-year-old retired female pointed out that large shower heads were useful because they had more space to hold them in a firm position. The turntable on the shower head also allowed users to adjust the volume of water. Thus, water control become easier. Clear instructions should be shown in a graphic presentation so that they can fulfil the basic requirement of product semantics and be easier to understand. They should be *'simple and intuitive to use'* to enable older persons to operate them on their own.

7.3 Dining Experience

FF was a 65-year-old disabled female who had limited ability when she tried to pick up a piece of a cherry tomato with assistive chopsticks. She could master the assistive chopsticks on her own, however, her holding direction was completely upside down, which was unexpected. Even if there had been an in-depth study of the ergonomic considerations of this tool, she could not have felt the curves and textures with her fingers due to a series of minor strokes. The failure to properly use the assistive tool even made the picking-up process improper and impolite. A few of the older participants refused to use it even when they reported it as being useful and reliable. They did not want the social stigma in this regard. If the older persons could pick up the chopsticks during their meal time, they would not opt for assistive tools even if such tools could make it easier for them. More importantly, the typical tool could only help its user pick up chopped food piece by piece. It could not facilitate rice eating due to the wide gap between each chopstick. This study also revealed that some of the Hong

Kong elderly still preferred their knives to be polished before carving their meat, but realised their cutting tools were not in good condition. CL reported that when she polished her knife with a sharpening stone, she put the stone on top of a wet towel. She added that she had stored the stone under the sink for years because of its size and weight. Even though she realised that there was an easier way to do the sharpening, she preferred the traditional method, which gave her the best result and made her cooking experience easier. The overall food preparation experience should blend the chopping process with sharpening, forming an effective tool to fulfil the actual requirements of the users. This study shows that many existing kitchen utensils have failed to satisfy the basic needs of the elderly in Asian cities, without any deep consideration being given to the entire cooking experience most face in their day-to-day activities.

7.4 Health Care

CL was concerned with the pill case and said it could be more user-friendly if the depth of each compartment was carefully designed. According to her, it should allow human fingers to grasp the small pills or turn it upside down to take out the exact amount. The PI passed around a few samples to the participants to collect feedback based on their past experiences and selection criteria. Ruby commented that the existing pill case, designed as a souvenir by the Social Welfare Department, was a good solution except for its oversized and elongated shape, which was not good for carrying pills. The pill case with four compartments (for four distributions in a typical day) seemed to be the best fit for the majority, so they could either take the pills by themselves or with assistance from family members. In this case, the design should follow the 'equitable use' principle of universal design so that older persons suffering from stroke and disabilities, or related diseases, could still operate it on their own. In addition, designers should pay extra attention to 'perceptible information' so that the information appearing on the pill case is easily read, regardless of the sensory abilities of different users. A participant sitting next to CL was unable to operate the sample because the design was totally different from hers. False pressing could be needed to go through one cycle and return to the correct direction. After the workshop, two older participants said they did not know where they could buy the traditional pill cases that are crucial to their daily living. This study revealed that a few of the older participants were living alone and lacked social activities, which might have presented barriers to them keeping up with current information and the news around them.

7.5 Leisure Activity

This area explored the social and leisure activities that older persons do on a regular basis. SF remarked that over the past few years she had visited a local community centre twice a week, where she took training courses, received health checks and networked with people of a similar age. Every Friday afternoon she collected assorted vegetables for free, after she enjoyed a two-hour social gathering. SF shared her views with CL on occasionally attending Chinese opera. The classic opera house was located far away from her residence, therefore, SF needed to rely on a modified walking cane. This study showed that some older persons did not want to carry a walking cane.

Instead, they preferred to use an umbrella cane or even a shopping cart to serve the same purpose. WL preferred to play card games with companions instead of going out in the summer, due to unfavourable weather. The games, she said, made her feel like she belonged to a group and sometimes she could even meet new friends.

8 Recommendations

8.1 Deeper Encouragement

Deeper encouragement to facilitate overall discussion and participation in the participative workshops was needed. Given that all of the senior participants invited to the two-day consecutive workshops did not have similar experience, they might not have been familiar with the workshop setting even though it was organised in a familiar venue. The data collection process seemed to be new to most of the participants. In general, when informing the study group, only one to two senior participants could actively offer personal advice and were eager to share their experiences with the research team. A few needed the PI to open the dialogue on their day-to-day living to obtain research insight for a typical day. Because most of the participating elderly did not have a strong educational background or even a foundation, they needed further explanations in terms of the production process, including the objectives of the study. Concepts such as 'digital printing', 'ergonomic consideration' and 'physical model' were new to these senior participants. For those who did not speak Cantonese, but only spoke their indigenous language, they were either assisted by the translations of teammates or through body language and written words. When the older persons saw the tangible product samples being demonstrated by the research team to facilitate discussion, they became more actively immersed in the data collection process.

8.2 Clear Instructions

This study involved the use of both low-cost materials and digital printing to support two sessions of idea exploration and the development of a product design. The goal was to obtain a genuine understanding of users' needs and how to enhance the quality of their lives. However, even though the physical modelling skillset was not mandatory in this research, the adaptability of the senior participants did not meet expectations. Demonstration was necessary to help them understand what we were doing, what we were going to do and what we expected. Clear guidance was the key to ensuring that every participant, including the social worker and design entrepreneur, fully participated and understood the meaning of the activities in the participative workshop. In this case, a seminar on the effectiveness of participative workshops associated with older participants and stakeholders, followed by a series of videos with working examples among Taiwanese and Chinese would be beneficial. This could provide a holistic picture for the senior participants to help them become familiar with the actual practice. Researchers should make clear instructions, break down the procedures and explain each step one at a time.

8.3 Timing and Time Management

The research schedule was deliberately designed for workshops to be held on Fridays, when Food Grace Centre and Action Care Centre regularly distributed free assorted vegetables and lunch boxes to needy older people. The participation rate was expected to be higher than 92% in this study. Very few participants could not attend the workshops because the schedule clashed with their medical appointments or they decided to attend other free community activities. A public seminar on the participatory workshop intended to generate design experience with older persons familiar with information technology had been organised a week before the first workshop. Both the first and the second sessions consisted of two hours each, which might have exceeded the limits of the majority. It is suggested that the duration of each session could be shortened to one and a half hours to coincide with the typical activities in day care centres. If vulnerable participants are familiar with the regular time schedules within their community centres, it enhances the efficiency of data collection. The longer the time spent in the participative workshops, the worse the attention span. The ideal number of participants was between six and no more than eight in each session.

8.4 Common Interests

The commonality of interest among the older persons played a vital role in building rapport among the other participants, such as the facilitator, social worker and design entrepreneur. Because the project aimed to explore the genuine needs of the elderly in their daily activities through brainstorming, interviews and design participation, the areas that created common interest between the older participants helped to facilitate the on-going research process. When the participants were asked to discuss their dining experience in the past, for example, they might share their experience with the research team and other participants they otherwise seldom paid attention to, due to different generations and living habits.

8.5 Social Benefits

This study found that a few older participants were not satisfied with their existing living environment. Most of them preferred a safer cooking environment when they used a gas stove. The location of the pipe that transfers the gas should have a sturdier design and be in a more reasonable location to make sure it will not easily be damaged by fire or extremely hot temperatures. SF explained that she paid extra attention when she fried vegetables in oil every day. Her interior environment was the standardised typical design of a public estate where the existing interiors still have room for improvement. Although SF had been aware of the unsafe condition for years, she did not know how to improve it, and even she had realised there was a gas leak, she would have to cook in the same way as she had from the first day she moved in. SF was living alone, and her near relatives were living in northern China. She spent thirty minutes with the PI discussing the current transportation system in her district. After suffering from a stroke and cancer she had been required to walk with a cane to travel between the community centre and her residence. By sharing their stories with the other participants, the elderly living alone could meet new friends and interact with neighbours living near the community centre. The participative workshops offered opportunities for them to establish deeper relationships with other older participants and centre staff.

9 Conclusion

Failed product design affects people's initiative to deal with it and removes the incentive to give it a try. In many cases, misuse of the cutlery and walking aid could induce serious injuries that the care givers and even the end users had not thought about. There is a rising trend in social isolation among the elderly. Following universal design principles in assisted living and health care design makes a significant contribution to product development by reducing the social stigma and empowering vulnerable people to become more capable of living independently. Loneliness and social isolation can be mitigated by using the proper assistive tools and daily products that facilitate social interaction and physical activity. If researchers could put designers and end users together to participate in friendly workshops so that new ideas and interpretations could be achieved, it would be meaningful. The contributions of the other stakeholders, such as the nursing home people, care givers and social workers, helped to encourage the older persons to participate in the group. Representatives from a manufacturing company also actively participated in the group to learn the seniors' demands and improve the usability of their existing products and services. This was the main objective of this study.

Acknowledgments. The author would like to express our gratitude to the reviewers for their valuable comments. We would also like to thank the participants and supporting staff who took part in the main study. This research work was supported by The Hong Kong Polytechnic University and Technological and Higher Education Institute of Hong Kong.

References

1. United Nations: World Population Prospects. United Nations, New York (2015)
2. Census and Statistics Department: Hong Kong Population Projections. Hong Kong Special Administration Region Government Printer, Hong Kong (2016)
3. Census and Statistics Department: Hong Kong Population Projections. Hong Kong Special Administration Region Government Printer, Hong Kong (2011)
4. Alzheimer's Society. Dementia 2012: A National Challenge. Alzheimer's Society. London (2012)
5. Victor, C.: Loneliness, Social Isolation and Living Alone in Later Life. Economic and Social Research Council, London (2003)
6. Suchman, L.: Located accountabilities in technology production. Scand. J. Inf. Syst. **14**(2), 91–105 (2002)
7. Cross, N.: Designerly ways of knowing: design discipline versus design science. Des. Issue **17**(3), 49–55 (2001)

8. Simonsen, J., Robertson, T.: Routledge International Handbook of Participatory Design. Routledge, Taylor & Francis Group, London and New York (2013)
9. Johnson, R.: User-Centred Technology: A Rhetorical Theory for Computers and Other Mundane Artifacts. SUNY Press, Albany (1998)
10. Sanders, E.B.-N., Stappers, P.J.: Co-creation and the new landscapes of design. CoDesign **4**, 5–18 (2008)
11. Hunter, M.: What design is and why it matters. http://www.thecreativeindustries.co.uk/uk-creative-overview/news-and-views/view-what-is-design-and-why-it-matters. Accessed 10 Jan 2019
12. Chamberlain, P., et al.: The state of the art of design in health: an expert-led review of the extent of the art of design theory and practice in health and social care. Sheffield Hallam University, Sheffield, UK (2015). designinhealthnetwork.org
13. Alaoui, M., Lewkowicz, M.: A living lab approach to involve elderly in the design of smart TV applications offering communication services. In: Online Communities and Social Computing, OCSC 2013. LNCS, vol. 8029, pp. 325–334 (2013)
14. Froyen, H.: Universal Design, a Methodological Approach. The Institute for Human Centered Design, Boston (2012)
15. Mace, R.: Definitions: Accessible, Adaptable, and Universal Design. The Center for Universal Design, Fact Sheet# 6, Raleigh, NC (1990)
16. Russell, L.: The Future of the Built Environment. The Millennium Papers. Age Concern England, London (1999)
17. Steinfeld, E., Maisel, J.: Universal Design: Creating Inclusive Environments. Wiley, Hoboken (2012)
18. King, A.P., Siu, K.W.M.: Participant observation in cognitive gameplay as a rehabilitation tool for living alone elderly with dementia in Hong Kong: a pilot study. Des. J. **20**(Supl), S2426–S2438 (2017)
19. Danford, G.S., Maurer, J.: Empirical tests of the claimed benefits of universal design. In: Paper Presented at the Proceedings of the Thirty-sixth Annual International Conference of the Environment Design Research Association, pp. 123–128. Environment Design Research Association, Edmond, OK (2005)
20. Behar, S.: A design solution for "aging in place". The ASID report, pp. 6–9 (1991)
21. Mace, R.: Universal design: barrier free environments for everyone. Designers West **33**, 147–152 (1985)

Public Spaces, Building Environment and Communities

A New Inclusive Housing Prototype

Francesco Spagnoli(✉)

IEL Architects, Rome, Italy
info@ielarchitects.com

Abstract. A new model of sustainable housing has been designed, thanks to a research conducted with different kind of users. A ready-made wood residences prototype is proposed as a project able to guarantee a height home environment quality. The research aims to improve people daily Life related to multiple factors. A study has been conducted on materials quality to guarantee a safe environment and a high air quality and at the same time, also a research on colors and contrasts between the volumes. The structural elements are already built in the factory equipped of flexible interiors furniture (multiple layers): this technology permits an ergonomic use of the tables, the armchairs, the windows, handrails, kitchens and toilets [1]. Plugs and switches have a minimum range to be moved and adapted. The house project results anti-seismic, ergonomic and sustainable and it makes people interact inclusively with the outside landscape thanks to a movable windows system that allows to the different users of the house to change the windows height. The flexible structure is able to contrast the earthquake vibrations because of its flexible structure and at the same time the home environment becomes safe thanks to the integrated furniture. Guides for the wheelchair users are integrated in the ready-made panels and can be customized for the clients [2]. The prototype is also eco-friendly and detains all the requirements to be constructed in NZEB (Nearly Zero Energy Building) [3]. It has to be pointed out that the interior walls panels and furniture are made of recycled materials, mixed with chipboards, to increase the project sustainability and to reduce the expenses. The economic factor has been considered as a fundamental element of the research and the already-made technology permit sustainable costs. Users Expert have been involved in the design process thru workshops and seminars organized to investigate needs and measures of people housing thanks to the collaboration of a consultant that works with public administrations to improve people home environment.

Keywords: Ergonomics · Flexible furniture · NZEB

1 Introduction

A new model of sustainable housing has been designed, thanks to a research conducted with different kinds of users. A ready-made wood residence prototype is proposed as a project that guarantees a large home environment quality. The research aims to improve people's daily life, related to multiple factors. The quality of the domestic environment depends on the space quality, the air conditioning, the light, the attention dedicated to user's dimensions, the ergonomic factors of furniture and details. The design project

G. Di Bucchianico (Ed.): AHFE 2019, AISC 954, pp. 163–175, 2020.
https://doi.org/10.1007/978-3-030-20444-0_16

consists in the experiment of a new connection technology between the structural elements. It allows the installations of structural and final layers ready-made in the factory, with an integrated system of conditioning, home automation, electrical and air systems. A study has been conducted on materials quality to guarantee a safe environment and a high air quality. At the same time, has been conducted a research on the effect of colors and contrasts between the volumes in a domestic context. The structural elements are already built in the factory equipped of flexible interiors furniture (multiple layers). The house project results anti-seismic, ergonomic and sustainable and it makes people interact inclusively with the outside [4]. The flexible structure is able to contrast the earthquake vibrations because of its flexible structure and at the same time the home environment becomes safe thanks to the integrated furniture. Guides for the wheelchair users are integrated in the ready made panels and can be customized for the clients. The prototype is also eco-friendly and detains all the requirements to be constructed in NZEB (Nearly Zero Energy Building) [5]. It has to be pointed out that the interior walls panels and furniture are made of recycled materials, mixed with chipboards, to increase the project sustainability and to reduce the expenses. The economic factor has been considered as a fundamental element of the research and the already-made technology permit sustainable costs. Users Expert have been involved in the design process through workshops and seminars organized to investigate needs and measures of people housing thanks to the collaboration of a consultant that works with public administrations to improve people home environment [6].

2 Paper Preparation

Nowadays the development of timber construction models is based on the use of two main technologies: the X LAM construction systems, and the frame and massive system named *BLOCKBAU*. In particular, the United States market has always used the frame system such as the *balloon-frame* and *platform-frame*; in Europe during these last years the X LAM has become the most popular system used, which provides an excellent compromise between tradition techniques and prefabrication, with a significant time-saving concerning the construction time, but it's still expensive technique. In particular, the application of wood in the construction field is fundamental for two different aspects:

1. Environmental/energy aspects. Thanks to the possibility to build several new green buildings made of eco-compatible nature of materials and of high performance envelopes, it is possible to have a minimal bearing impact on the planet.
2. Structural/anti seismic aspects. The excellent response of wooden buildings to the stresses due to seismic events, allows the use of this technology to all the geographical areas in the world vulnerable to this risk, such countries as Italy, California (USA) Japan and Mexico.

3. Sustainable indoor life. Wood is an organic material, and if it's not treated with harmful glues, its release of substances can be relaxing for the smell, and it can capture the pollution contained in the indoor spaces [7].

In particular, in 2007 the *CNR-IVALSA (Italian Research National Centre for timber application)* and the Province of Trento, in cooperation with the National Institute for Earth Science and Disaster Prevention (NIED), the Building Research Institute (BRI), Shizuoka University and the Centre for Better Living in Japan, have worked on a research program named *SOFIE Project 2007.*

The program focuses on the verification of the seismic resistance of multi-level buildings built with the X-LAM system. The seismic tests were carried out on the largest vibration platform of the world in Kobe, Japan. In particular, the timber building (seven floor height) has passed the test positively, thanks to the specific qualities of the material (lightness and strength) and because of the construction system adopted (ductility and capacity to dissipate energy). After more than ten consecutive earthquake simulations at variable intensity there was no residual displacement, indeed, the building has always returned to its right position, keeping the structure safe from the damages [13].

IEL research project, is in continuity with the proposals of the current worldwide market and with the latest scientific applications used in the field of timber housing. In particular, the project consists in the development of a prototype of a single-family house with adaptable dimensions that provides a *360°* integration of all the technological, social and environmental aspects that characterized the development of each design process.

The main goal is to place on the market a wooden model of ergonomic residential houses that allows to give the right answers to the performing requirements, from the performing point of view to the shell. A maximum flexibility of the interior spaces had been designed according with customer needs. The study of interior spaces is fundamental to create inclusive solutions applied to the furniture, walls and flexible partitions. Any object can be designed with multiple functions. All the spaces have a system of adjustable guides that increase both the inclusion in the domestic environment and the adjustment of the light spots. Long handrails can be integrated in furniture guides or for example towel-holders. The same kitchen surface can be adapted for all family members in different moments.

The use of sophisticated technology combined with the use of innovative assembly systems, can be solutions economically advantageous and competitive on the global market for wooden constructions. The particular external structure of the envelope and the way with the wooden elements have been assembled, make possible the integration of photovoltaic systems on the roof, to realize a building totally self - sufficient from the energy requirements point of view (stand alone system) (Fig. 1) [8].

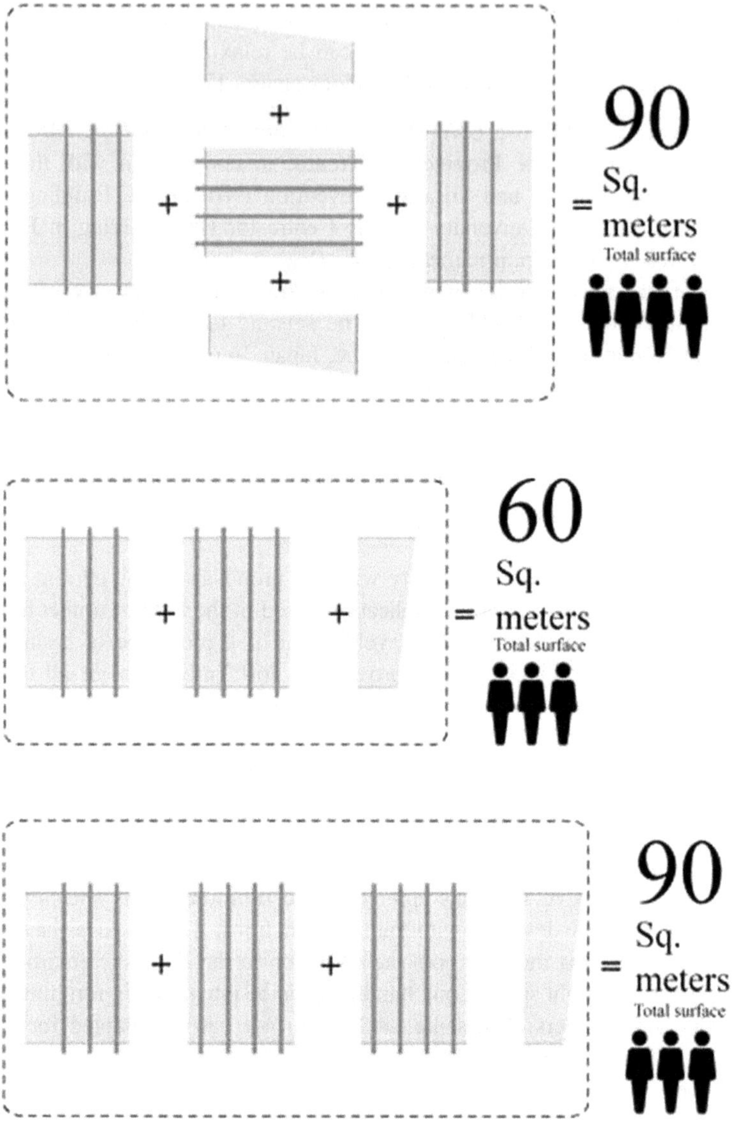

Fig. 1. Different kind of assembly. Those are just a few type of modular possibility of assembled.

Costs reduction is achieved by maximising the production process of prefabricated modules from the production to transportation and assembly on worksite. The wooden modules are designed to be contained and transported thru trailers (according with standard dimensions for conventional trucks), and we've designed these modular elements in easy shape to be easily operated. These types of houses are designed for different kind of users, thanks to the easy way in which they can transform the interior spaces and the possibility of adjusting equipment and furniture. The research has studied a new sliding guides system with rails connections that allow to fix the furniture and all the equipment of different kinds to the walls, to prevent, in case of earthquake any damages caused by the furnishment (Fig. 2).

Fig. 2. External view of prototype's rendering. A multi-storey villa 125 m^2 large with two balconies modules.

Research has investigated also the LCA Life Cycle Assessment of materials Buildings have been designed to have a life cycle of 100 years, wood components and other materials used are almost totally recyclable (90%).

Flexibility, already mentioned in indoor environments, is at the same time a fundamental characteristic of the external design of the prototype; indeed, although the elements are standardised according to the characteristics and its adjustability, the building can be customizable, according to customs requirements.

The project is divided into different areas, each of these stands out to find out new innovative solutions applied to the prototype, such as:

- Structure and Safety of the shell;
- Environmental Sustainability and the use of materials with a natural origin
- Spatial Flexibility and integration of systems,
- Inclusive Design of interiors, according with a large number of potential users (Elders, Families with children, Young couples, Singles Disabled) [9].

2.1 Structure and Safety

The aspect related to the structural safety is one of the main aspects of the design of the prototype. Our engineering team has worked to develop a new type of connection between the different structural elements, something that has not been developed yet in the field of timber constructions. Thanks to the use of these connector systems consisting in a special springs and steel cables, we aim to improve the structural seismic response without significantly affecting production of costs and at the same time improving the prefabrication of modules.

This technology has been developed favouring the possibility of installing the prefabricated modules in any type of parcels and location, and the method of installation and assembly has been designed with the aim to realize a product that can be easily dismantled and reassembled in another different place.

Users well-being is an important characteristic of the project referring to the choice of materials that maintain a good level of *IAQ Indoor Air Quality* standards, this has been possible thanks to the use of wood in particular for the furniture and interior finishes, coming from certified woods according with *FSC© Forest Stewardship Council* and treated without the use of chemicals or dangerous product toxics for the humans health [14].

For us, it is very important to realize liveable residences both from the comfort and the society-economic point of view [10].

2.2 Environmental Sustainability

Another fundamental aspect that this prototype underlines is the model of living the indoors spaces. It's a sustainable model because of the materials but also thanks to the efficient use of energy. The wood is used for the structures, which has itself high environmental standards, cardboard, made of recycled sources, as the filling for the cavities that permit thermal and acoustic insulation in addition to vegetable fibres. This particular combination makes possible the creation of light buildings than the norm, favouring both a better seismic response and a greater ease of manoeuvre during transportation and assembly on site. The model is designed to be easily integrated with roofing systems for the production of energy from renewable sources such as photovoltaic or solar thermal systems. This is because the model is designed to be a *NZEB Nearly Zero Energy Building*, so it will have satisfied the requirements and performance standards according to the *LEED* (for USA and CA system) and *KLIMAHOUSE* (for IT and EU systems) certifications. The possibility of setting up the model with alternative energy systems allows to the houses prototype to be realized in different contexts:

- urban areas (such as towns and villages)
- isolated areas (such as country sides, and all these areas where there is not a direct connection to the infrastructures/networks - electricity, gas, telecommunication, internet and water).

In order to overcome the problem of water supply, a rainwater collection tanks system has been designed, which can be extended if required, also to the treatment of waste water, in the absence of a suitable sewage system (Fig. 3) [11].

Fig. 3. Interior view of living area. Mobil walls flexible system integrated with spot lights.

2.3 Flexibility and Interaction of Spaces

To this prototype, the flexibility of interior spaces is a fundamental element, that permit to extend the housing model to a large market share. The model can actually be adapted for several user needs, considering that in the life of everyone needs can change and our capability can suddenly change, for a temporary timing or permanently. The flexibility has been ideated thanks to the use of special devices and technological equipment that allows the flexibility of the walls and consequently of the interior spaces, especially the regulation of the furniture. This technology permits an ergonomic use of the tables, the armchairs, the windows, handrails, kitchens and toilets. Plugs and switches have a minimum range to be moved and adapted. The house project results anti-seismic, ergonomic and sustainable and it makes people interact inclusively with the outside landscape thanks to a movable windows system that allows to the different users of the house to change the windows height.

Nowadays, indeed, the phenomenon of the high requests of characteristic of sustainability and flexibility is increasing. It's necessary to design ergonomic products that can be commercially attractive for an enlarged target of audience.

Until now, we've mostly considered the inside aspects of the houses. The flexibility of the building can be explored also from the outside of the structure, where the customization of the external skin allows to the module to be placed in different environments and contexts, from the city to the countryside, ensuring a maximum integration. The external shape can be creating depending on the customer needs related to the interior room position. The interior spaces can be composed according to the user's needs, both during the assembly phase that later.

For special categories of users, such people with reduced mobility, a special vertical and horizontal guides allow to change furniture height and are physically integrated with light system and he ceiling.

The lift system is provided on request, and it allows also to persons to go up and down from the ground to the upper floors, in totally autonomy and in few minutes, without the use of the regular elevators or stair lifts [12].

2.4 Production Costs

At present, despite the prefabrication and the short construction time, the use of wood technology is not convenient, because of the high cost of the certified raw material, such as coniferous timber. Currently prices to build prefabricated houses in wood are approximately between 1,200.00 and 1,500.00 €/m^2 (standardized in the EU zone). The challenge of our team consists in the propose of a new prototype for the international market with an estimated cost of 800.00–900.00 €/m^2 (EU currency) or 900.00–1,000.00 \$/m^2 (USA currency).

The achieve of this new prototype is the rationalization of the use of raw material (wood), exclusively for structural partitions and internal finishes such as false ceilings and various fittings. This allows a significant reduction of costs. The masonry package is also filled with wood fibres (less expensive) mixed with organic materials fibres and recycled rigid cardboard. The innovation in the use of this material allows - not only a reduction of costs- but also a significant reduction in the mass of the wall, or rather a reduction in the its weight, maintaining optimal levels of quality and climatic standards.

3 Customers, Users, Developers, Universities - Research Institutes

IEL Architects team is developing the research project thanks to the collaboration of a multidisciplinary staff made of professionals, academics, engineers and trainers. We consider at the same level all people and customers, because we believe that all categories must have the same opportunities to buy a good design level products.

Within the research project the team has started several partnerships with leading companies in the field, based in Italy and abroad. Research institutes and universities are also potential partners during the development and testing phases of structures and materials if they pursuit the same ideals, or they'd like to train students through practical activities. We are constantly looking for new future collaborations with academic-scientific sector, research and development companies, to reach new goals for a better living (Fig. 4) [15].

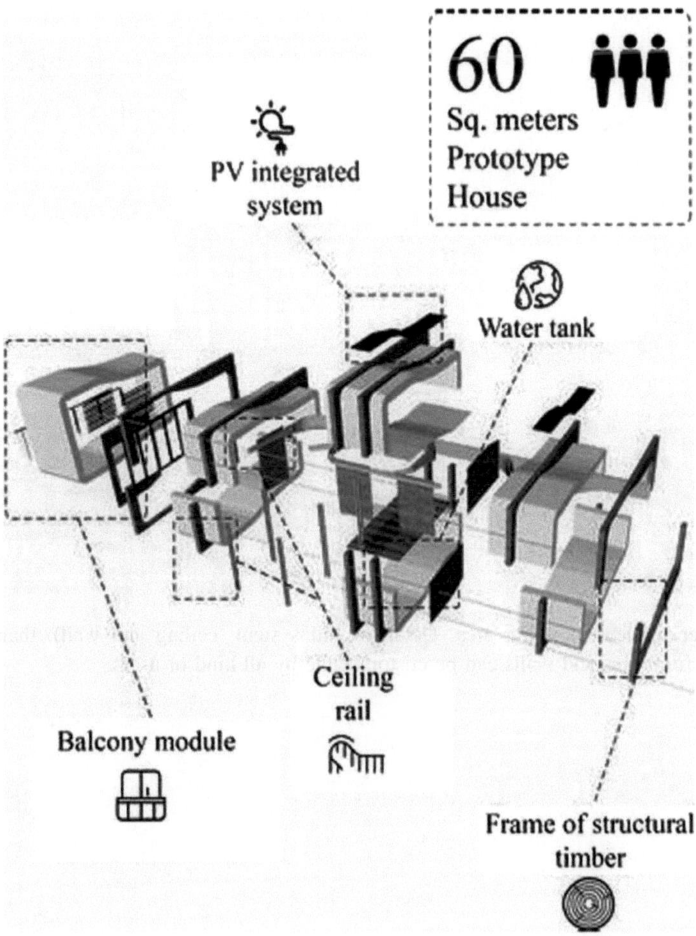

Fig. 4. Exploded axonometry of a housing module, (60 m^2 house).

Since 2018 have been activated new training activity at IEL headquarter in Sabina, a region close to Rome, in Italy, characterised by a significant number of fields where to test structures and where to locate training courses. In particular the training offer is addressed to technicians, workers and companies in specific field related to sustainable principles starting from the ergonomic design, to the use of organic materials or techniques (Adobe), prefabrication of wood and innovative construction solutions, oriented towards the construction of NZEB buildings.

3.1 Customers, Users, Developers, Universities - Research Institutes

See Fig. 5.

Fig. 5. Interior view of living area. Detail of rail system (ceiling and wall), thanks to this technology furnitures and walls can be customizable by all kind of users.

Customers
- Elders (75 +) - Families and young couples - Singles - Disabled people - Younger (18 +)

Developers- Companies
- General constructors - Building construction companies - Timber construction companies - Social housing developers - Public administration and municipalities

Research Institutes & Universities
- Research institutes (tests and labs analysis) - Universities (research and developer) - Trainings classes and activities - Design and research programs

4 Green Housing | for All

The research project has been convinced starting from the principles that sustainability is a necessary condition to build a better world for our future generations, from a social to an environmental point of view. It's not a "slogan", it's a possible objective to perceive if we try extend it to as much people as possible.

The project of ergonomic prefabricated wood housing aims to reach a sustainable design without economic or social barriers. To make it possible, developers, administrations, designers, architects have to be open to critics, suggestions, changes.

So we are (Fig. 6) [16].

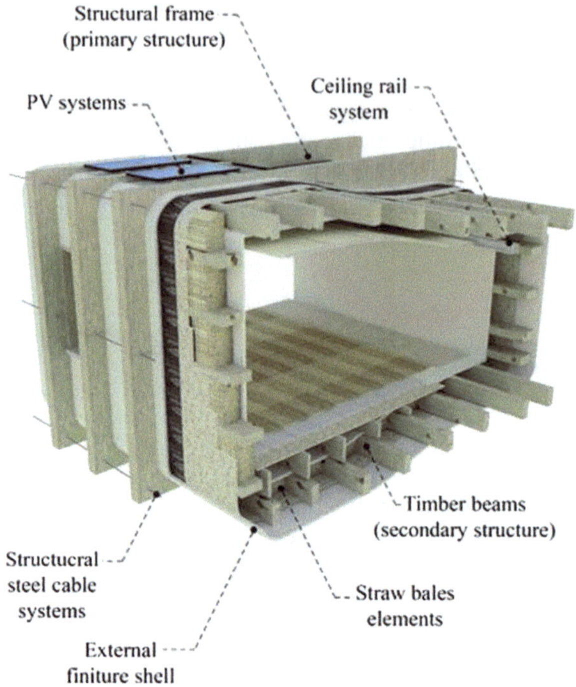

Fig. 6. View of prototype. Stratification of the building package, specification of each different layers.

References

1. Bounin, K., Eckhard, M., Hammerl, D., Krämer, G., Letsch, B., Nutsch, W., Nutsch, T., Schlatter, K., Siebert, S., Willgerodt, F.: Fachkunde Holztechnik. Verlag Europa-Lehrmittel Nourney, Volmer GmbH & Co., Haan Gruiten
2. American Hardwood Export Council: Progettazione strutturale con latifoglie americane. EC3 V 9DS, London (2003)
3. Bendetti, C., Dal Vera, C., Gavioli, M., Gigli, D., Glorioso, M., Pasquale, C., Pezzucchi, S., Terzi, E.: Costruire in legno – edifici a basso consumo energetico. Bozen University Press, Bozen (2009)
4. Minke, G.: Building with Bamboo, Design and Technology of a Sustainable Architecture. Birkhäuser, Basel (2012)
5. Schickhofer, G., Bernasconi, A., Jobstl, R.A., Katzengruber, R.: Pannelli di legno. Promo Legno, Milano (2008)
6. Bernasconi, A., Piazza, M.: Il calcolo delle strutture di legno. Promo Legno, Milano (2010)
7. Guttmann, E., Isopp, A.: Massiv uber Kreuz, Zuschnitt nr. 31.2008- Pro Holz, Wien (2008)
8. Albrecht, B., Bergmeister, G., Bernsconi, A., Binder, G., Carbone, A., Carlet, F., Montuori, L., Ritter, A., Sacchi, L., Scagnol, M., Weger, S., Zweifel, K.: Materia Legno 01. Promo Legno, Milan (2009)
9. Albrecht, B., Bernsconi, A., Carbone, A., Carlet, F., Montuori, L., Ritter, A., Scagnol, M., Weger, S.: Materia Legno 02. Promo Legno, Milan (2009)

10. Albrecht, B., Bernsconi, A., Carbone, A., Carlet, F., Montuori, L., Ritter, A., Scagnol, M., Weger, S.: Materia Legno 03. Promo Legno, Milan (2011)
11. Albrecht, B., Bernsconi, A., Carbone, A., Montuori, L., Ritter, A., Scagnol, M., Weger, S.: Materia Legno 04. Promo Legno, Milan (2012)
12. Bonomo, U.: Felipe Assadi, pp. 62–91. Casabella nr. 892, Milan, Italy, December 2018
13. http://www.ivalsa.cnr.it/foto-e-video/foto/progetto-sofie-test-sismico-2007.html
14. https://it.fsc.org/it-it
15. https://new.usgbc.org/leed
16. http://www.gbcitalia.org/leed

Bioclimatic Design Approach for Low-Income Dwelling at Monte Sinahí, Guayaquil

Boris Forero(✉), Jesús Hechavarría, and Robinson Vega

School of Architecture and Design,
Catholic University of Santiago de Guayaquil, Guayaquil, Ecuador
{boris.forero, jesus.hechavarria,
robinson.vega}@cu.ucsg.edu.ec

Abstract. This document presents the procedures used to carry out architectural design of a housing prototype for low-income people under bioclimatic criteria in Monte Sinahí, at Guayaquil. Bioclimatic design starts with the characterization of the local climate and, based on inclusive design, considers the needs of a family that currently lives in that territory. With climate data obtained from a weather station located on site, region climate is described, which is characterized in two climatic seasons: hot-dry and hot-humid. The analysis of the micro-climatic particularities of the place allowed defining cooling requirements and proposing passive design strategies into architectural project for interior spaces. It also outlines recommendations for dwellings with bioclimatic criteria taking into account architectural morphology, urban form, building elements and solar and wind flow control devices. The use of Ecomaterials as construction elements in the efficient management of local resources is the result of a research program whose objective is to provide solutions in order to improve the constructive quality of low-cost dwellings using bamboo and other natural resources as fibers, leaves and parts of the trunk of plant species. Climatic characterization, added to future dwellers requirements, citizen participation as an example of social inclusion in the decision-making design process and the use of Ecomaterials, will contribute to the appropriation of the prototype by its users to be used as a housing solution that contributes with the care of the environment and the improvement of the quality of life of the inhabitants of Monte Sinahí.

Keywords: Bioclimatic design · Concurrent engineering · Ecomaterials · Participatory design · Low-income housing

1 Introduction

In Ecuador's National Development Plan within its goals to 2021, it is indicated to increase from 53% to 95% the number of families with proper and dignified dwelling units that are in extreme poverty; where access to dignified home and safe and healthy habitats within it is still a fundamental pillar, the *Casa para Todos* Plan, where "the integral guarantee of this right will be achieved through the implementation of coordinated and articulated actions between the Technical Secretariat of *Plan Toda una Vida*, the Ministry of Urban Development and Housing (MIDUVI), the National

G. Di Bucchianico (Ed.): AHFE 2019, AISC 954, pp. 176–185, 2020.
https://doi.org/10.1007/978-3-030-20444-0_17

Secretariat of Planning and Development, the Public Company *Casa Para Todos*, BanEcuador, the Development Bank of Ecuador and the Social Sectoral Council" [1].

The production of social interest housing carried out in Guayaquil since 1942 can be divided into two large groups: formal and informal housing. Within the first one are those of public promotion, which until 2017 registered more than 158 thousand single-family housing solutions [2] being the MIDUVI and the Municipality of Guayaquil the institutions that recorded higher production with 44.7% and 28.8% respectively, followed by the *Junta Nacional de Vivienda* with 22.6%; thus covering 96.1% of formal public housing production. This production is complemented by private production sponsored by the NGO called *Hogar de Cristo* since 1971 [3] with wooden housing types using bamboo and in recent years they have included new construction systems with different materials. These dwellings, which amount to about 185,000 units, were built in areas of low economic resources such as *Isla Trinitaria*, *Flor de Bastión* and *Monte Sinahí* in Guayaquil.

Traditionally these formal and informal Social Interest Housing (VIS) dwellings [4] are not designed to respond coherently to the climatic reality where they are located. Both materials used in its architectural envelope and its construction systems have high rates of thermal transmittance (U-value). Nor are considered aspects such as textures and colors from its thermal performance, as well as the proportions of its interior spaces and designs of windows and doors, configure different scenarios where humidity and temperature present throughout the year, generate absence of any sense of comfort; that is, they have an inadequate thermal performance that affects its dwellers [5, 6].

2 Bioclimatic Design Methodology

The bioclimatic design starts from the collection of information related to the climatic data of the area where the project is located to determine its most relevant characteristics. Within the reality of a specific locality, the construction materials and systems of the building must take into account, because its decisive influence on thermal performance that accompanies the habitability conditions of a dwelling.

Determination of thermal comfort conditions is a fundamental step within the bioclimatic design process, since it establishes the reference of design temperatures inside a dwelling to meet that demand. They are established and visualized in Givoni's psychrometric chart, which also allows us to understand which strategies to develop for the different types of scenarios throughout the year and hours of the day [7].

In this way, main seasons must been described, which gives rise to their diagnosis and characterization to establish cooling requirements and propose passive cooling strategies to condition interior spaces. Bioclimatic recommendations according to urban environment, construction elements and design resources, within which orientations respect to the sun and the prevailing winds, layout, thickness and combination of materials and construction systems, spatial arrangements are also taken into account in relation to the design of the plan, as well as the finishes and textures of the architectural envelope.

Once the architectural project is elaborated integrating the described variables, it is subjected to a thermal validation or simulation with specialized software to be able to evaluate its thermal behavior according to its materials, construction systems, finishes and other essential characteristics of the architectural envelope. Finally, the construction phase allows having an in-situ learning laboratory to analyze and feedback this bioclimatic design process [8] as shown in Fig. 1.

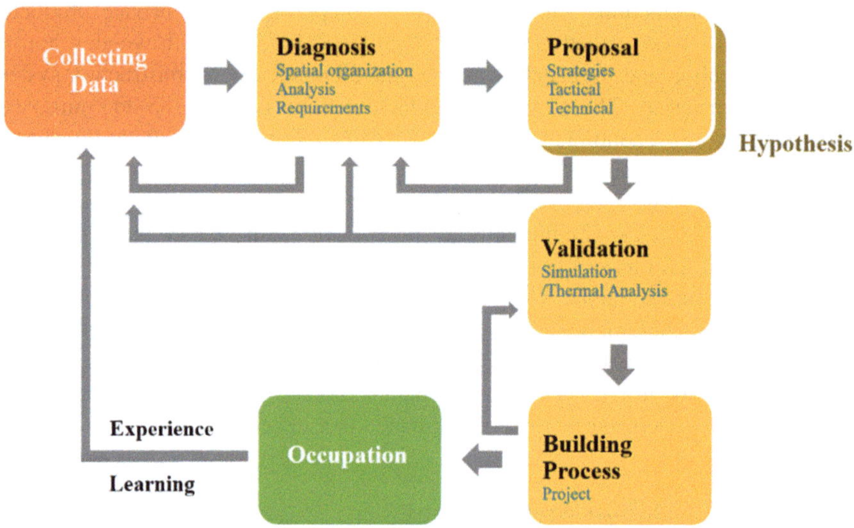

Fig. 1. Bioclimatic design methodology.

2.1 Characterization and Climate Analysis

Guayaquil is located on a plain beside Guayas River, nest to the foothills of the Chongón-Colonche mountain range that is located to the northwest of the city in whose area Mount Sinahí is located. This zone expose to winds of moderate intensity at night, due to the combination of topographic and climatic effects added to the humidity coming from the coast and which converge in this area of the city [5]. Its warm-humid climate, has two clearly differentiated seasons: the dry season, from May to November, where average temperatures range between 23 °C to 25 °C with relative humidity between 67% and 73% to register maximum thermal sensations average of 29 °C and fresh nights with averages of 22 °C. Described scenario is configured due to the presence of the cold Humboldt Current that runs along the Pacific coast of South America from the south. On the other hand the rainy season, generated by the presence of the El Niño ocean current, coming from central Pacific in the opposite direction to Humboldt from December to April. This generates average temperatures between 26 °C and 28 °C with humidities between 68% and 80% with average thermal sensations of 32°C.

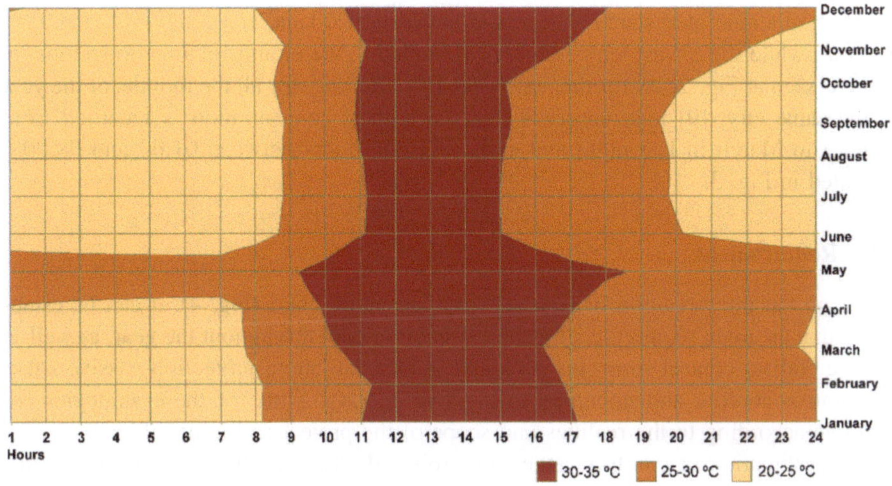

Fig. 2. Isothermal chart for Guayaquil. Hour and monthly temperature graphic behavior displayed indicates different temperature averages. Data has been acquired from an *epw file generated with Meteonorm software [9].

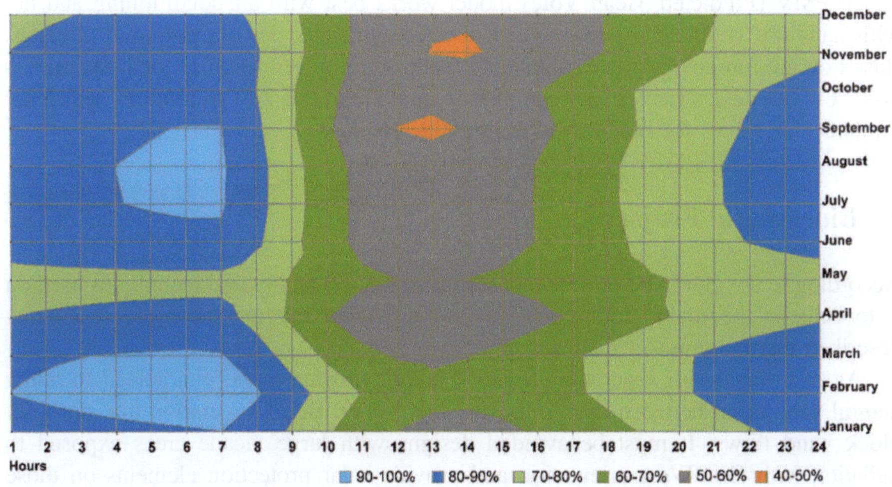

Fig. 3. Isohygral chart for Guayaquil. Hour and monthly relative humidity (RH) and its average percentages over a year [9].

Because it's located close to the equatorial line, solar radiation is fairly strong throughout the year, in addition to its incidence of approximately six months on north and south facades respectively [7, 10]. The hourly thermal behavior obtained according to the method described by Tejada [11] and analyzed through Gómez-Azpeitia's application [9] plots the isothermal chart for each month of the year in Guayaquil.

Highest temperatures are concentrated between 11:00 and 16:00 h for all months of the year, while the lowest temperatures are limited to the months of June to October at night as seen at Fig. 2. As for the hourly relative humidity for all the months of the year of the same city, [9] it is observed that the highest concentration is recorded from January to March in late night and early morning hours between 01:00 and 08:00 h indicated in Fig. 3.

2.2 Requirements

Based on the interpretation of Givoni's psychrometric chart (Fig. 4), it can be established that the main requirement for the project is to cool throughout the year, as well as passive climate control strategies such as natural ventilation, protection passive solar, evaporative cooling and nocturnal convective cooling. Some of these strategies are selective according to the realities and scope of the project and its inhabitants.

The estimation of comfort ranges are projected based on the adaptive model proposed by De Dear [12] with very similar results to Auliciems [13] and compiled in the ASHRAE55 standard [14] which is a dynamic approach related to physical and psychological processes that generate the interaction of the dwellers with their environment in order to achieve comfort.

Other approaches like thermal equilibrium proposed in the seventies by Fanger [15] which PMV (Predicted Mean Vote) model works best with air conditioning and not with naturally ventilated spaces, which are recommended for warm-humid climates. Based on the foregoing, it is established that the comfort temperature for Guayaquil is 26.4 °C, with design ranges between 28.9 °C maximum and 23°C minimum, which are fundamental values for bioclimatic design [13, 16, 17].

3 Bioclimatic Project

According to the cited, the project focuses on strategies; tactics and techniques used on it to achieve the main requirement of cooling down interior spaces to reach that described thermal comfort ranges.

At an urban level, separation between buildings need to be guaranteed to allow natural ventilation between dwellings, and the proper use of vegetation that does not block wind flows. It must be avoided designs with large façade areas exposed to radiation with East-West orientation and provide solar protection elements on those facing North and South where solar incidence will be approximately of six months on each of them.

It is essential to perform solar geometry analysis in order to design windows with good performances according to requirements stated in the previous statements. In addition, spatial arrangement of dwellings must foresee that different uses respond to orientations related to temperatures and direct solar radiation as shown at Fig. 5. Spaces occupied at night, must be located at East side, so indirect heat gains of afternoon hours through walls and roof will not affect those spaces.

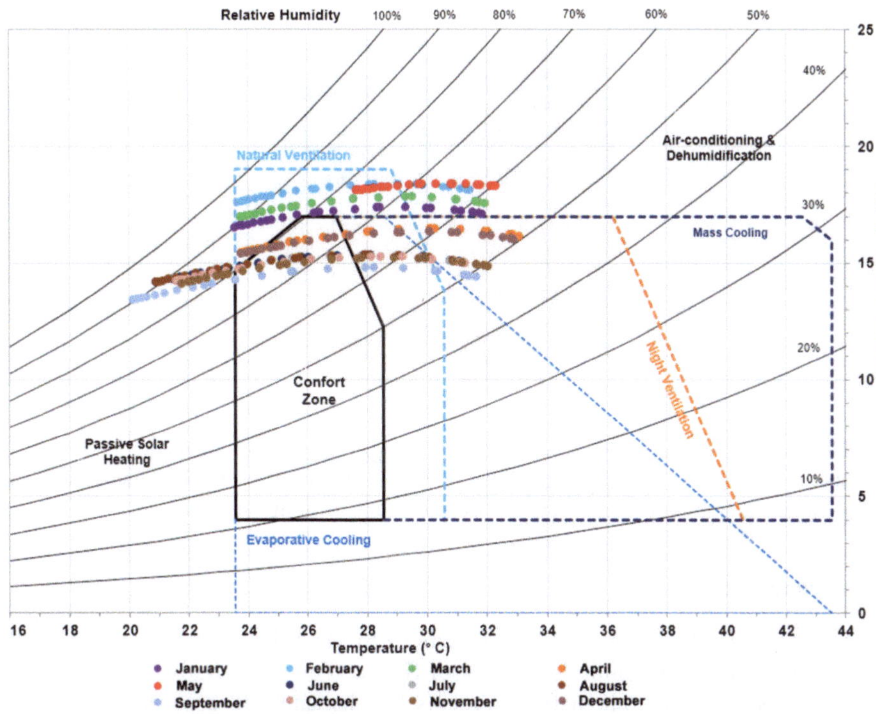

Fig. 4. Psychrometric chart for Guayaquil. Dots indicates hourly mean *temperatures* (°C) for each month of a year. *Natural* and *Night Ventilation* strategies are needed through all the year and it is even needed some heating at night and early morning during the hot-dry season as well as *Confort Zone* temperatures (*center*) at morning and night. During the hot-humid season, some *Air Conditioning and Dehumidification* recommended as well [9].

It is also required that the internal volume is the necessary one that hot air layers are not in contact with the users. This can be achievement with minimum heights of 3 m duly ventilated. The compactness of the home should be in the 1:2 ratio range to ensure adequate ventilation of its interior spaces, although authors like Olgyay recommend 1:1.3 to 1:1.7 with slender volumes [18].

Thermal transmittance of construction materials must be low to avoid as much as possible indirect solar gains within dwelling. In this order color, texture and finishes also play an important role. It is recommended finishes and highly reflective materials (albedo index) as well as low thermal emissivity [19].

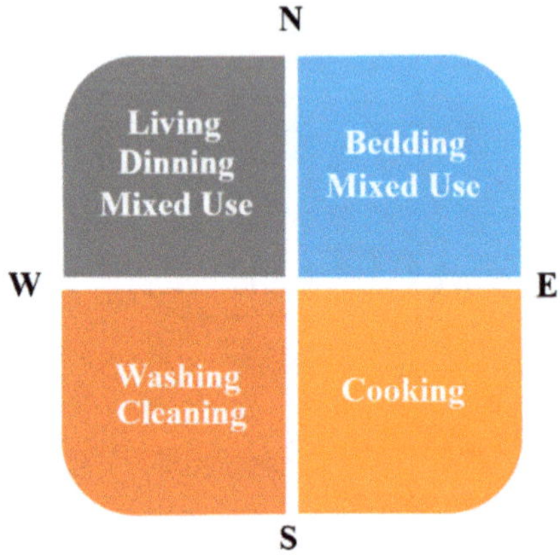

Fig. 5. Spatial arrangement for low-income dwellings. Mixed Use refers to a variety of domestic, economic and productive activities carried out by dwellers. Warm-humid equatorial coastal climate defines this scheme and is part of bioclimatic design resources.

3.1 Ecomaterials

The Ecomaterials are the product of a research agreement between SENACYT (now SENESCYT) and the Catholic University of Santiago de Guayaquil (UCSG), whose objective was to provide alternative solutions for elements and construction components based on the efficient use and management of local resources as seen at Fig. 6.

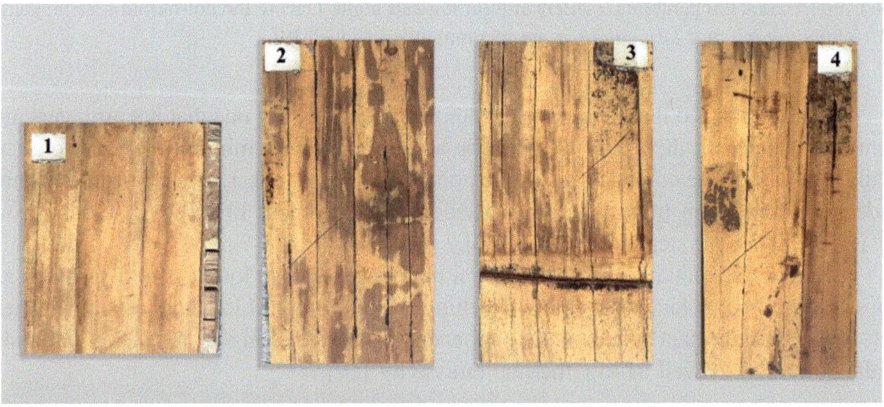

Fig. 6. Different types of ecomaterials made from natural fibers and bamboo (*Guadua Angustifolia Kunth*) as a result of the Project "Pilot plant for research, production and technology transfer in the use of innovative ecomaterials for low-cost housing construction", leaded by School of Architecture and Design (FAD) at Catholic University of Santiago de Guayaquil (UCSG). Triple bamboo horizontal and vertical layered panel (*one*); three types of bamboo sandwich-type panels: with coir fiber (*two*); with internal bamboo thread (*three*); and with internal layer of rice husk (*four*).

With this, we want to improve the constructive quality of low-cost housing, especially using bamboo [20] of the type *Guadua Angustifolia Kunth*, considered one of the three best bamboos among about 1400 existing species worldwide due to its high resistance.

Within the research process, four types of bamboo boards with dimensions of 2.40 × 1.20 m developed, which differ from each other in the manufacturing processes, orientation of the raw material and finishes. Its use is oriented to the construction of houses of social interest (VIS), as floors components, walls, division panels, coatings, ceilings, furniture, among others [20].

Over the years, research has focused on the study of other natural resources such as fibers from agricultural waste from coconut straw, banana and palm leaf fibers. This diversifies the solution options of the Ecomaterials to the different components of the dwellings.

Design for the architectural envelope and its systems of walls and roofs are fundamental in equatorial latitudes, as they are elements that receive a large amount of solar radiation. It is precisely here where the Ecomaterials play a key role to achieve optimal thermal performances in designing dwellings at Monte Sinahí.

4 Conclusions

Design of low-income dwellings must include bioclimatic criteria to respond adequately to a given climate reality. It should be designed with this focus not only the architectural envelope integrating materials and construction systems of good thermal performance, designs of windows and doors, but also spatial arrangements and zoning of interior spaces that respond to different climate scenarios.

Ecomaterials development has a primordial importance at present because they are natural and healthy products, free of chemical compounds and do not harm the environment. Its use in the design of dwelling spaces allows criteria of bioconstruction to be considered with a minimum impact on the ecological footprint, thus preserving the health of living beings and the planet.

The success consists of achieving a creative combination that takes into account technical, social and economic factors, using the existing raw material in the place and combining it with known technologies. By promoting the use of Ecomaterials and the application of bioclimatic design in the construction of social housing, it contributes to the care of the environment and the improvement of the quality of life of the inhabitants of Monte Sinahí, at Guayaquil, Ecuador [20].

Archetypes design of low-cost housing begins with the study of the social, economic and environmental context, thus characterizing the territory where the project will be built. In addition to dwellers, who are protagonists in the decision-making process.

With this knowledge, it is possible to give coherent answers integrating the architectural design, as well as the interior furniture to the social interest houses, taking into account the spatial analysis [2], ergonomic and anthropometric of the community; all as part of a systemic design process [21], where participatory design methodologies in which the inhabitants of the sector or direct beneficiaries are part of the decision making, guarantee the sense of belonging with the developed solutions [4, 16, 22–24].

Acknowledgments. The authors would like to thank to School of Architecture and Design, at Catholic University of Santiago de Guayaquil for the financial support of this research.

References

1. SENPLADES: Plan Nacional de Desarrollo 2017-2021. Gobierno Nacional de la República del Ecuador, Quito (2017)
2. Almeida, S., Anaya, J., Hechavarría, J.: Caracterización dimensional de modelos constructivos de Vivienda de Interés Social construídos en Guayaquil, Ecuador. In: Libro de Memorias Científicas del V Conrgreso Internacional de investigación y actualización en ingenierías, pp. 234–249. CIDE, Guayaquil (2017)
3. De Cristo, H.: En el corazón de la Esperanza. Memoria Institucional, Guayaquil (2012)
4. Forero, B., Hechavarría, J., Alcívar, S., Ricaurte, V.: Systemic approach for inclusive design of low-income dwellings in popular settlements at Guayaquil, Ecuador. In: Ahram, T., Karwowski, W., Taiar, R. (eds.) Human Systems Engineering and Design, IHSED 2018. Advances in Intelligent Systems and Computing, vol. 876. Springer, Cham (2019)
5. Forero, B.: Mejoramiento de las condiciones térmicas de las viviendas del complejo habitacional Socio Vivienda II Etapa I, en la ciudad de Guayaquil, Ecuador. Master's thesis, Universidad de Colima (2015)
6. Dick, S., Hechavarría, J., Forero, B.: Systemic analysis of bioclimatic design of low-income state-led housing program "Socio Vivienda" at Guayaquil, Ecuador. In: Ahram, T., Karwowski, W., Taiar, R. (eds.) Human Systems Engineering and Design, IHSED 2018. Advances in Intelligent Systems and Computing, vol. 876. Springer, Cham (2019)
7. Forero, B., Hechavarría, J., Sandoya, R.: Diseño bioclimático de viviendas de bajo costo. Una propuesta metodológica para estudiantes de la Carrera Arquitectura, Guayaquil. Opuntia Brava, vol. 9 (2017)
8. Gómez-Azpeitia, G.: "Método para el diseño bioclimático". Master's thesis, Universidad de Colima (1990)
9. Gómez-Azpeitia, G.: Caracterización climática. Diagnóstico de Confort, Colima (2016)
10. Johansson, E., Wasim, M., Arroyo, I., Bengs, C.: Outdoor thermal comfort in public space in warm-humid Guayaquil, Ecuador. Int. J. Biometeorol. **62**(3), 387–399 (2018)
11. Tejada, A.: Simulación de la marcha diaria de confort higrotérmico. In: Primer Encuentro Diseño y Medio Ambiente, Colima (1991)
12. De Dear, R., Brager, G.: Developing an adaptive model of thermal comfort and preference. In: ASHRAE, American Society of Heating, Refrigerating and Air-Conditioning Engineering, pp. 145–167 (1998)
13. Auliciems, A., Szokolay, S.: Thermal comfort. Passive and Low Energy Architecture International design Tools and Techniques, vol. 3. University of Queensland (1997)
14. ANSI/ASHRAE: Standard 55. Thermal Environmental Conditions for Human Occupancy. American Society of Heating, Refrigerating and Air-Conditioning Engineering, Atlanta (2010)
15. Fanger, P.: Thermal Comfort; Analysis and Application in Environmental Engineering. Danish Technical Press, Copenhagen (1970)

16. Hechavarría, J., Jaramillo, R.V., Forero, B.: Multi-objective optimization applied to the bioclimatic design of dwellings with ecomaterials. In: Ahram, T., Karwowski, W., Taiar, R. (eds.) Human Systems Engineering and Design, IHSED 2018. Advances in Intelligent Systems and Computing, vol. 876. Springer, Cham (2019)
17. Szokolay, S.: Introduction to Architectural Science. Elsevier, Oxford (2004)
18. Elizondo, M., Gómez, A., Alcántara, A., Esparza, C.: Criterios bioclimáticos de diseño en edificaciones para el turismo en la costa de Jalisco. El caso de El Tuito, municipio de Cabos Corrientes (2018)
19. Forero, B., Hechavarría, J.: Análisis de las condiciones de confort térmico en el interior de las viviendas del complejo habitacional Socio Vivienda 2, Etapa 1, en la ciudad de Guayaquil, Ecuador. In: III Congreso Internacional Tecnología, Universidad y Sociedad (TUS). Universidad de Especialidades Espiritu Santo, Guayaquil (2015)
20. CDB, Centro de Documentación del Bambú (2019). https://goo.gl/xaZbgF
21. Leyva, M., Hechavarría, J., Batista, N., Alarcón, J., Gómez, O.: A framework for PEST analysis based on fuzzy decision maps. Rev. Espacios **39**(13) (2018)
22. Hechavarría, J., Forero, B., Bermeo, P., Portilla, Y.: Social inclusion: a proposal from the University of Guayaquil to design popular housing with citizen participation. In: World Engineering Education Forum, WEEF 2018. Albuquerque, NM (2018)
23. Hechavarría, J., Forero, B.: Aplicación de la metodología de Análisis y Síntesis de Sistemas de Ingeniería en la búsqueda de soluciones a problemas de la sociedad. In: III Congreso Internacional Tecnología, Universidad y Sociedad (TUS). Universidad de Especialidades Espiritu Santo, Guayaquil (2015)
24. Hechavarría, J., Forero, B., Al-Terkawi, J.: Enfoque sistémico como propuesta metodológica para el diseño de Viviendas de Interés Social en estudiantes de arquitectura de la Universidad de Guayaquil. Opuntia Brava, vol. 9 (2017)

Codesign of Public Spaces for Intercultural Communication, Diversity and Inclusion

Alessandra Rinaldi[1(✉)], Leonardo Angelini[2(✉)], Omar Abou Khaled[2(✉)], Elena Mugellini[2(✉)], and Maurizio Caon[3(✉)]

[1] Department of Architecture, University of Florence, Florence, Italy
alessandra.rinaldi@unifi.it
[2] HumanTech Institute, University of Applied Sciences and Arts Western Switzerland (HES-SO), Fribourg, Switzerland
{leonardo.angelini,omar.aboukhaled, elena.mugellini}@hes-so.ch
[3] School of Management, University of Applied Sciences and Arts Western Switzerland (HES-SO), Fribourg, Switzerland
maurizio.caon@hes-so.ch

Abstract. The consistent phenomenon of immigration from countries with different cultures, that characterized the last decade, has changed the use of public spaces in European cities. The need to develop innovative solutions aimed at improving intercultural inclusion and mutual understanding of all citizens emerges, so that urban spaces become inclusive, safe and sustainable. This research project is focused on two main questions: (1) can design and technology enable the creation of a different way of living the public space to catalyze interpersonal interaction and to promote a shared sense of citizenship? (2) is it possible to support intercultural communication and to facilitate social inclusion through the design of innovative and interactive public spaces equipped with smart urban furniture? This paper presents three Design Orienting Scenarios as results of the research conducted by several experts, professionals and people with different cultural backgrounds during a focus groups and two different workshops where they made converge competences from design, business, engineering and social sciences to find innovative solutions for the social inclusion of migrants in urban areas.

Keywords: Social innovation · Design for inclusion · Public spaces · Smart technologies · Interculturalism

1 Introduction and Background

The current society is progressively becoming more and more complex. Indeed, the increasing number of businesses or other organizations developing international influence or starting to operate on an international scale changed how value is created and perceived while strengthening globalized economic relationships. At the same time, the bettering of international diplomatic relations and the raising of new policies for the free movement of people in widening geographical areas, such as the European

G. Di Bucchianico (Ed.): AHFE 2019, AISC 954, pp. 186–195, 2020.
https://doi.org/10.1007/978-3-030-20444-0_18

Union, are changing the shape of our society. These changes of the global socio-economic scenario caused the emergence of the so-called post-multicultural era, which defines a society presenting intertwined effects of different phenomena: transnationalism, superdiversity, cosmopolitanism and solidarity [1]. Transnationalism indicates a social phenomenon due to the increased interconnectivity between people and the fading financial and cultural significance of boundaries among different countries. The term "superdiversity refers to the phenomenon of diversification of diversity, which consists of some current levels of population diversity that are significantly higher than before. Vertovec observed that superdiversity is distinguished by a dynamic interplay of variables among an increased number of new, small and scattered, multiple-origin, transnationally connected, socio-economically differentiated and legally stratified immigrants" [2]. It denotes increased diversity not only between immigrant and ethnic minority groups, but also within them. Cosmopolitanism refers to a shared feeling of belonging to a unique community based on an inclusive morality, a shared economic relationship, or a political structure that encompasses different nations. Solidarity refers to the ties in a society that bind people together as one based on kinship and shared values or other shared interests or goals.

The complex structure of contemporary society requires new policies able to recognize the inevitability of diversity in modern cities while supporting integration and inclusion. Such policies should adhere to the methodological interculturalism, which aims at incorporating diversity and difference into society's public culture rendering it object of affirmation. Methodological interculturalism indicates also the contact between people as the most important and appropriate way to drive integration [1].

Physical spaces can be designed to support such contact and, in the context of cities, public spaces can play a crucial role in catalyzing social encounter, also with strangers. Urban architecture already provided some examples of solutions for social inclusivity, e.g., public art interventions as shown in Sharp et al.'s paper [3]. Unfortunately, to our knowledge, there are not many examples of urban design dedicated to intercultural integration and inclusion. Although some research has already been conducted on the use of technology for the integration of migrants in new countries [4] and some preliminary work explored also the adoption of codesign techniques with immigrant teens [5], there is still an evident lack of work that focuses on intercultural integration in urban areas integrating digital technology.

As stated by De Lange et al., new technologies can enable smart cities to engage citizens and promote cocreation of new solutions [6]. We also believe that smart public spaces, i.e., public spaces augmented with new technologies, can be designed to foster intercultural integration and for this reason we started working directly with migrants to codesign novel concepts of smart spaces to respond to citizens' needs while valorizing diversity [7]. To this purpose, we started a project to explore the possibility to codesign smart public spaces for intercultural integration. The first phase was composed of two steps: a focus group and a "World Café" workshop. During the focus group, researchers and professionals working in different domains, mainly from design, political and social science and computer science, met operators and social workers in inclusion projects and integration of migrants, for analyzing together the current state of the art in urban architecture for multicultural integration, use of technology in public spaces, and design solutions for social inclusion. Thanks to this debate and exchange between the participants, it was possible to generate the research questions to be answered during the following workshop. During this workshop, which adopted the World Café methodology, the participants (including researchers, migrants and cultural mediators) tried to identify the critical points of multicultural integration and inclusion for migrants and relative strategies to cope with these challenges (these results can be found in Rinaldi et al.'s paper [7]).

In this paper, we will present the second workshop, conducted after a few months, where researchers, migrants and cultural mediators tried to co-create different solutions to tackle the previously identified challenges to intercultural inclusion applying the strategies emerged during the previous workshops (Fig. 1). The result consists of three different concepts, depicted in Figs. 2, 3 and 4, which respectively present a nomadic gym, an itinerant escape room and a smart bus shelter. These concepts will be presented in detail in the remainder of this paper.

Fig. 1. One of the tables during the codesign workshop.

2 Methodological Approach

According to the codesign methodology the research questions were identified as a primary element of the research. The main questions were: (i) can design and technology be strategic factors in creating a different use of public spaces and facilitating citizens' interaction and sense of citizenship? (ii) is it possible to support intercultural communication and to facilitate social inclusion through the design of innovative and interactive public spaces by incorporating digital technologies in urban furnishings?

As already mentioned in the previous part, European cities are changing due the immigration of people from different countries bringing a variety of heterogeneous cultural backgrounds. There are many issues related to the integration and dialogue between cultures. Urban design plays a key role in how migrants participate in their host community and it is an important driver for the inclusion process. Promoting in public spaces the interaction of different cultures becomes a crucial element to facilitate social cohesion and living together [8–10].

This codesign workshop aimed at investigating innovative solutions for facilitating migrants' integration and the dialogue between different cultures, through the design of innovative urban furniture and services, integrating smart technologies.

With this objective, researchers shared the design process with the other stakeholders, including users from different cultures, carrying out an activity of ideas' generation and of design concepts' development.

As Rizzo [11] and Sanders [12] wrote, this methodology have to be aimed more at identifying than at solving problems and focuses on the opportunities for innovation, emerging in the phase of ideas' generation, which will be used by designers to define innovative but plausible Design Orienting Scenarios.

The choice of codesign workshop derived from our desire to develop a project activity of a participatory nature together with the users by actively involving them, as this method allowed for designing products and services that are more useful and meaningful for people. For the development of the research, we decided to actively involve some immigrant citizens, coming from five different countries extra-UE, in the codesign workshop. In this way, researchers, expert of design, technology and social issues, were able to identify three different innovative scenarios of digital product-service systems for urban spaces, aimed at encouraging and supporting integration and intercultural dialogue.

Fifteen people participated in the codesign workshop, including designers, engineers, social scientists, professionals and operators working in the social field, educational scientists, immigrants and cultural mediators. During the co-design workshop, for understanding needs and expectations of all the stakeholders, involved and for stimulating the creativity of participants, the researchers exploited their skills and some of the tools normally used in the design discipline, such as design probes, storytelling, design orienting scenarios, prototypes, and design thinking models.

The development of project scenarios took into account the strategies, which were previously identified during the focus group, as fundamental to promote dialogue and intercultural inclusion.

The strategies we referred to are the following: (i) creating a network of interactions to stimulate intercultural exchanges between locals and migrants; (ii) educating for diversity; (iii) tackling of the gender differences in the provision of services or projects *(i.e., migrant women are wary of leaving their children in healthcare structures, and have problems of autonomy)*; (iv) working to maintain identity; (v) bringing of the different cultures closer together *(i.e., through themes such as school, work, food, dance, music and sport)*; (vi) use of catalysts, such as the linguistic tandem that allows two people of different cultures to confront each other and learn not only at a linguistic but also a cultural heritage level; (vii) creating of informal situations, which foster inclusion and direct human relationships [7].

The process adopted has been divided into three main steps: (i) the "*say*" phase, that means understanding needs and expectation of both local and immigrants; identifying the main factors of exclusion and/or inclusion in cultural heritage; exploring how different cultures use public spaces; (ii) the "*do*" phase: moving on to the discussion, collaboration, empathic relationship with the co-designers, for the generation of ideas, concepts and possible solutions creative and summarizing insights; (iii) the *"make"* phase: discussing the insights and identifying a set of design domains for inclusive product-service systems in urban spaces, representing them through innovative and plausible Design Orienting Scenarios.

During the cognitive and exploratory activities, related to the "say" phase, participants told their experiences, documenting them through annotations, images, photographs, using different tools and methods that were made available by designers.

Once the previous phase was analyzed and the co-project brief was well identified, together with the participants, we moved on to the codesign activity: to the "do" phase.

Through discussion, collaboration, and empathic relationship with the co-designers, multiple ideas, concepts and possible creative solutions were generated. Finally, we arrived at the "make" phase, in which concepts of interactive urban place were developed and prototyped, represented by the designers in the form of three main Design Orienting Scenarios, better described in the following section.

3 Results

The workshop, which saw users involved as co-designers in the design process, led to defining and experimental prototyping some innovative, interesting and viable solutions concerning urban spaces and furnishings, in the form of product-service systems, for facilitating social inclusion of immigrants and intercultural communication among citizens.

Migrants have been given the position of 'experts in their own experience', playing an important role in developing knowledge, generating ideas and developing concepts of the products and services that they themselves will use. The ideas have been translated into three main Design Orienting Scenarios, which will play an important role in inspiration of design, as described below.

3.1 Scenario 1: Health and Wellness

The first solution is focused on sports and wellbeing (Fig. 2). Promoting good health and well-being is one of the key research areas promoted by the EU [13] as well as one of the sustainable development goals of the World Health organization [14]. In a superdiverse society, promoting health might imply additional challenges that cannot be solved through traditional interventions. This first scenario tries to address this challenge through the identification and the design of a public space installation for promoting and supporting fitness activities. Such solution could consist in a set of gym equipment to be installed outdoor in parks, squares and neighborhoods or indoor, in schools. As sport in public space could be an opportunity for socialization, this scenario aimed at identifying a common playground for younger people, both millennials and kids ac-companied by their parents. For the first target user group of this wellness installation, technology was identified as an opportunity both for socialization and for increasing motivation in engaging in physical activity, giving an opportunity to the youngest generation to track their physical performance, to compare it with other users and to challenge each other. Dance and music were also seen as an opportunity to foster engagement in physical activity while also valorizing cultural heritage. While the opportunity to have fun and enjoy dance was considered as appealing also for parents with their kids, a possible barrier was individuated for a part of female migrants, whenever their culture would hinder them from enjoying physical activity in fitness outfits. To facilitate their participation to such activities, the identified design solution was a capsule to be installed in the public space, close to the other fitness installations. The proposed capsule would allow female migrants enjoying in fitness activities in a discreet space, with their kids and/or with other female migrants.

A digital platform will augment the capabilities of this nomadic gym providing information, about fitness classes and their schedules, and allowing registration for a selected class. At the same time, this platform can enable facilitated sharing of these events in order to motivate other people in the participants' social network and to catalyze the establishment of new relationships through the social media.

3.2 Scenario 2: Learning and Cultural Heritage

Promoting interculturalism and social inclusion implies a reciprocal understanding of cultural habits and traditions, as well as a good knowledge of local policies and bureaucratic processes of the hosting country (Fig. 3). As both locals and migrants might have difficulties to find (and often little interest in looking for) such information, the challenge of this scenario was to identify a solution to motivate people to get interested in and retain information about different cultures and about local policies. Indeed, policies and bureaucratic processes are often provided by institutions in a format that is complicated and difficult to understand even for locals, while little information is often provided about the culture of the different communities inhabiting an urban space. The solution identified for this scenario is an escape room, a social game that is particularly popular among millennials, but that is also enjoyable by people of all ages. The escape room could be realized in an itinerant format, to be moved from time to time in different public spaces of the city, such as parks, squares, neighborhood, schools and shopping centers. The content of the escape room should be varied and should be targeted at acquiring knowledge about uses and traditions of the different communities living in the surrounding of the public space where the escape room is installed. At the same time, the content should also facilitate learning of the bureaucratic processes that citizens of all communities might encounter during their daily life in the city. As entering the escape room with a multicultural team might facilitate the solution of the riddles required to win the game; socialization between different communities and learning through problem solving will be particularly facilitated. Some challenges have been identified for the sustainability of such solution: rich and varied content is required to raise the engagement of participants and a moderator is required to ensure the best experience during the game. Moreover, such installation should be easy to move to different places of the cities in order to reach the maximum number of citizens.

3.3 Scenario 3: Free Time, Work, Movement

The third scenario was intended to address a more practical need of people with different backgrounds to exchange ideas, services and skills (Fig. 4). In this case, the urban space is transformed in order to allow everyone to easily share information with other citizens, in an accessible and decentralized manner. This sharing info point could be physically materialized into either a multi-ethnic interactive totem, or a table where people can sit with real and virtual guests (for example, with chairs that integrate interactive screens in the backrest), or a smart interactive glass (for example, integrated in a shelter). In all these cases, video-messages can be shared by any citizen looking for particular skills or support or offering particular services or help. Users looking for a job or skill offer could browse the content and listen to messages left by other users, eventually meeting in person with other people sharing ideas at the info point. This solution might foster synergies among citizens and promote the inclusion of people with different needs and backgrounds. Challenges that should be taken into consideration in this scenario are a secured authentication of the users leaving video-messages, as well as a moderation of the contents left by the users. A risk to be avoided is transforming the public space into a ghetto of migrants, where locals would be afraid to gather. To this purpose, such info points should be installed in strategic places where all citizens are already gathering, for example in a park where kids can play together, in a bus stop, in a shopping center or in a school. At the same time, accessible and attractive interactive systems may lower the barriers for citizens to approach the service and to share and access such information.

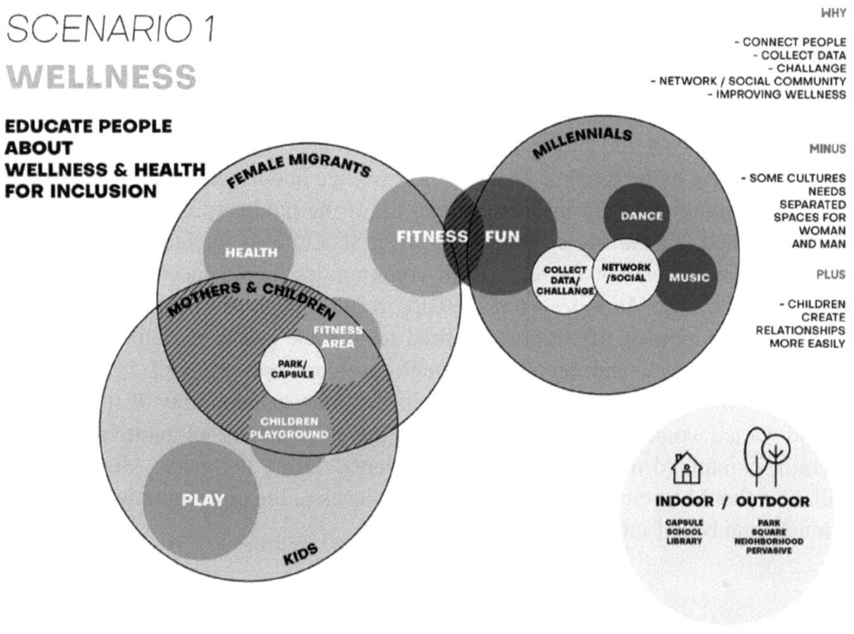

Fig. 2. Design Orienting Scenario to present the concept of nomadic gym

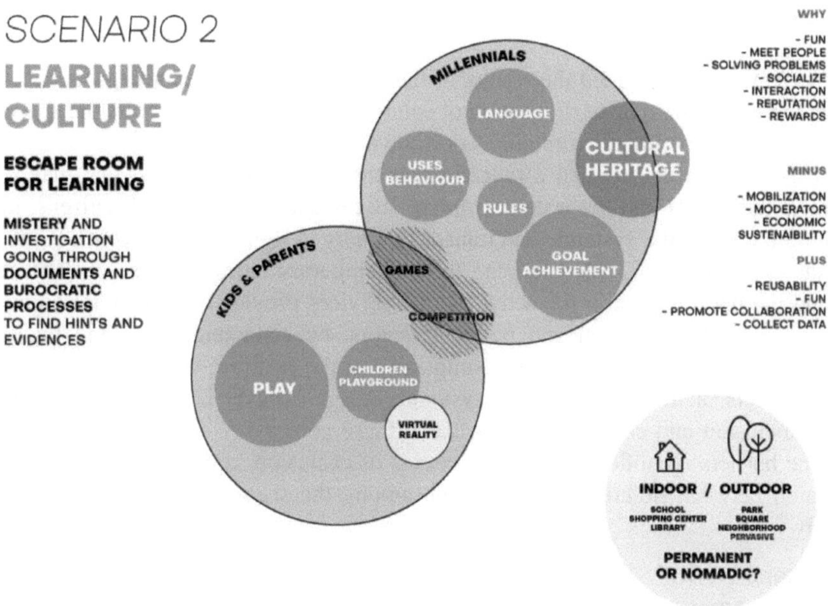

Fig. 3. Design Orienting Scenario to present the concept of itinerant escape room.

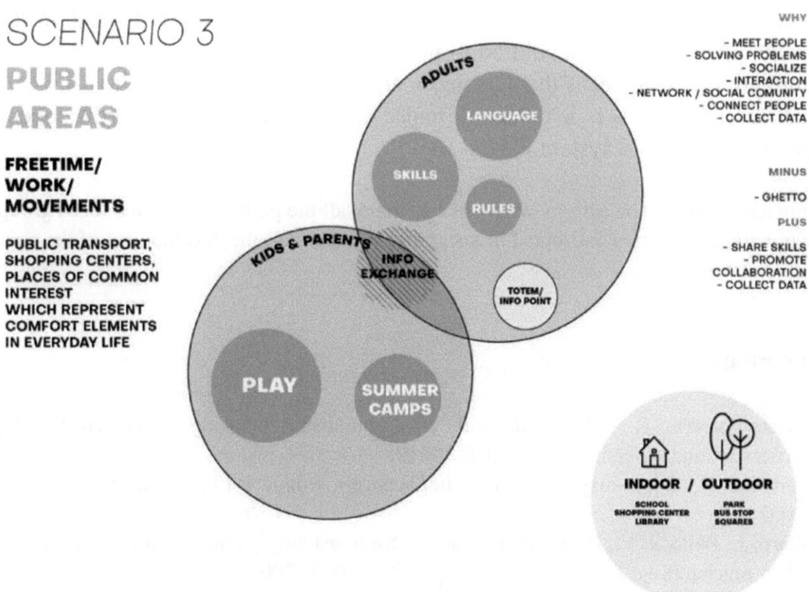

Fig. 4. Design Orienting Scenario to present the concept of smart bus shelter

4 Conclusions

In this paper, we presented the results of the second workshop adopting the codesign approach for the participatory creation of solutions that leverage smart public spaces as a catalyst for intercultural communication, diversity and inclusion.

This research has opened a reflection on the factors of exclusion that may compromise the use of and the access to urban spaces and services by all citizens, highlighting the need of a systemic and transdisciplinary design approach, which considers the intercultural factor, characterizing the contemporary cities. The use of, the access to, and the participation in the activities and services provided in urban areas may be affected by cultural, social and political factors, by economic status and linguistic difficulties of the users. Due to the complexity of the relations and the interactions of these aspects, a multi-domain and systemic design attitude is required in applying social inclusion and cohesion approaches in these contexts.

The barriers identified as the main causes of exclusion, discrimination and lack of cohesion between different cultures, even among the second young generations, are mainly four:

- Cultural barriers
- Social barriers
- Economical barriers
- Legal and political barriers

These barriers must be considered within the usability assessment and the design process of public spaces for intercultural communication, diversity and inclusion, where the main objective should be to prevent any kind of potential exclusion.

Furthermore, this means that it is necessary to adopt a systemic approach where the product-service design is a dynamic, under changeable, human-centered, open and codesigned part of the system [15].

Acknowledgments. The authors would like to thank all the participants of the focus group and the two workshops, who contributed in a significant manner to the development of the concepts presented in this paper.

References

1. Zapata-Barrero, R.: Methodological interculturalism: breaking down epistemological barriers around diversity management (2019)
2. Vertovec, S.: Super-diversity and its implications. Ethnic Racial Stud. **30**(6), 1024–1054 (2007)
3. Sharp, J., Pollock, V., Paddison, R.: Just art for a just city: public art and social inclusion in urban regeneration. Urban Stud. **42**(5–6), 1001–1023 (2005)
4. Coles-Kemp, L., Jensen, R.B., Talhouk, R.: In a new land: mobile phones, amplified pressures and reduced capabilities. In: Proceedings of the 2018 CHI Conference on Human Factors in Computing Systems, p. 584. ACM, April 2018

5. Fisher, K.E., Yefimova, K., Bishop, A.: Adapting design thinking and cultural probes to the experiences of immigrant youth: uncovering the roles of visual media and music in ICT wayfaring. In: Proceedings of the 2016 CHI conference extended abstracts on human factors in computing systems, pp. 859–871. ACM, May 2016
6. De Lange, M., De Waal, M.: Owning the city: new media and citizen engagement in urban design. First Monday **18**(11) (2013). http://firstmonday.org/ojs/index.php/fm/article/view/4954
7. Rinaldi, A., Caon, M., Khaled, O.A., Mugellini, E.: Designing urban smart furniture for facilitating migrants' integration: the co-design workshop as approach for supporting inclusive design. In: Congress of the International Ergonomics Association, pp. 461–470. Springer, Cham, August 2018
8. World Bank: Urban Development (2017). http://www.worldbank.org/en/topic/urbandevelopment/overview. Accessed 15 Mar 2018
9. DESA: Creating an inclusive society: practical strategies to promote social integration (2009)
10. Pinyol, G. (2017). http://instrategies.eu/en. Accessed Apr 2017
11. Rizzo, F.: Strategie di co-design. Teorie, metodi e strumenti per progettare con gliutenti. Franco Angeli, Milano (2009)
12. Sanders, E.B.N.: Generative tool for co-designing. In: Scrivener, S.A.R., et al. (eds.) Collaborative design. Springer, London (2000)
13. European Commission: Health Information on policies, strategy and funding in health research and innovation. https://ec.europa.eu/research/health/. Accessed 5 Feb 2019
14. United Nations: Sustainable Development Goal 3. https://sustainabledevelopment.un.org/sdg3. Accessed 5 Feb 2019
15. Busciantella Ricci, D., Rinaldi, A., Tosi, F.: Supporting inclusive approaches in service design with netnography. In: Advances in Design for Inclusion, Proceedings of AHFE 2018 International Conference on Design for Inclusion, pp. 290–301. Springer, Cham (2018)

Development of an Anthropometric Protocol for Wheelchair Users: Guiding the Decision-Making for Designing Inclusive Spaces

Libertad Rizo-Corona[1(✉)], Adrián Leal-Pérez[1(✉)], John Rey-Galindo[1,2(✉)], Carlos Aceves-González[1,2(✉)], and Elvia González-Muñoz[1,2(✉)]

[1] Universidad de Guadalajara, 44250 Guadalajara, Jalisco, Mexico
libertad.rizo.din@gmail.com, adrian26leal@gmail.com, {john.rey, c.aceves, elvia.gmunoz}@academicos.udg.mx
[2] Centro de Investigaciones en Ergonomía, Universidad de Guadalajara, 44250 Guadalajara, Jalisco, Mexico

Abstract. This paper describes the steps taken toward the development of an anthropometric protocol. The protocol was conceived as a tool for the design of inclusive built environments considering wheelchair user characteristics. The process followed for developing the protocol consisted of two stages. The first one included reviewing the type of dimensions, application context, methods and user profiles within the available literature. The second stage consisted in the corroboration of order and relevance of the measures proposed as well as the measurement instruments and the way of approaching participants. This protocol considers a total of 16 dimensions including height, reach and depth measurements. The paper describes the method, measurement landmarks and the tools used to take the measurement. To guide the application of the protocol a table was developed with recommendations on the use of certain percentiles according to the interaction of the user with the environment.

Keywords: Wheelchair users · Inclusive spaces · Design guidelines · Built environment

1 Introduction

The public built environment should accommodate all users to be a facilitator of daily life activities considering the particular characteristics of all population groups. Providing easy access without any additional adaptation [1]. According to World Health Organization [2] physical barriers are a crucial aspect that hinder service delivery and interferes with the participation of people with disabilities in education, social and productive life causing a disparity when compared to people with no disabilities. Also, the WHO [2] report includes nine recommendations to help in reducing the gap for people with disabilities, among them are: reinforce and support research about disability, improve the data gathering about this condition and allow the access to all conventional services and systems.

In the Mexican national context, some efforts have been made towards creating more inclusive spaces for all groups of the population. Among them are design

G. Di Bucchianico (Ed.): AHFE 2019, AISC 954, pp. 196–205, 2020.
https://doi.org/10.1007/978-3-030-20444-0_19

guidelines for the interior design of public buildings and exterior spaces. Some of the main references are the standards *Manual de normas técnicas de accesibilidad, Gobierno de la Ciudad de México* [3], *Criterios de proyecto de arquitectura para la accesibilidad de las personas con discapacidad* [4], *Normas y especificaciones para estudios, proyectos, construcción e instalaciones* [5], and *Recomendaciones de accesibilidad* [6]. It must be said that the documents mostly fail in providing the origin of the anthropometric data. In some cases data were outdated, or they were from other populations. This situation has been suggested to cause a disparity between the data shown in the handbooks and the current population [7].

Previous protocols reported in the academic literature has considered the characteristic of wheelchair users. Nevertheless, these works were oriented to the design of workspaces [8, 9] or to propose dimensions for other more general applications [10]. Therefore, the interaction with the built environment is not considered with an impact on the way of measuring the user. The interaction of wheelchair users with space is mediated by the wheelchair making it necessary to collect the dimensions of people, but conceiving the user and the wheelchair as a whole.

Although numerous efforts have been made to inform the design of inclusive built environments (IBE), it is considered important to continue in this direction, since there is still a lack of information and standardization for the measurement of wheelchair users for this application purpose.

The objective of this work is 1. to develop an anthropometric protocol that allows standardizing measurement techniques for wheelchair users required for the design of IBE and 2. propose criteria for percentile application in the design of IBE considering wheelchair user. Objectives aligned with [11] recommendations, to reduce the current gap in methodology and application of anthropometry by HF/E professionals and designers to create safe and effective products or environments for a wide range of users, taking steps to ensure well being and human dignity to all users.

2 Methodology for Protocol Development

The procedure was divided into two stages. First, a literature review was made to analyze the current contributions to the subject. Second, a corroboration stage was done. Through the literature review, the landmarks and method references were obtained. In the corroboration stage, the order of the measurement and application of the protocol was tested, as well as the collection of data regarding daily life activities of wheelchair users. This, to incorporate the vision of the active user aligned with the contemporary concept of disability.

2.1 Literature and Documental Review

In the literature review, official Mexican standards were consulted to identify the available information for the design of IBE for wheelchair users. Specific information about the landmarks and the measurement method was missing. Therefore the information obtained from these documents was through the analysis of the graphic schemes regarding wheelchair users in interaction with the built environment. The consulted documents are the following:

- Manual de normas técnicas de accesibilidad (Handbook of technical guidelines for Accessibility; Gobierno de la Ciudad de México, 2016)
- Criterios de proyecto de arquitectura para la accesibilidad de las personas con discapacidad (Architecture project criteria for the accessibility of people with disability; Instituto Mexicano del Seguro Social, 2011)
- Normas y especificaciones para estudios, proyectos, construcción e instalaciones, (Standards and specifications for studies, projects, construction and facilities; Instituto Nacional de la Infraestructura Física Educativa, 2010)
- Recomendaciones de accesibilidad (Accessibility Recommendations; Buenfil-Garza and Gutiérrez-Brezmes, n.d.)

Altogether these documents include exhaustive graphic descriptions of the wheelchair user in interaction with public and private spaces. Even though these are not anthropometric protocols, they allowed identifying some of the user dimensions required for the design of IBE. Which, has been translated to the proposed protocol into the appropriate landmarks and the dimensions (total length of human-wheelchair system and elbow to elbow width, see Table 2), as well as the need for the consideration of the person and the wheelchair as a system. On the other hand, some of the standards did not include the date, reference and population of the data. In other cases, the data was outdated or from other populations. Also, the criteria of data application regarding percentiles were not clearly shown.

With the objective of analysing measurement methods, a search was done in different specialized publications (Applied ergonomics, Ergonomics and Industrial ergonomics). The work related to anthropometric protocols for wheelchair users were considered to analyze the method and the dimensions included. In Table 1 a comparison of the methods, dimensions and consideration of the wheelchair is made.

Table 1. Previous protocols used as a reference for the development of the present protocol.

Author	Measurement method	Dimensions
Nowak [9]	Direct manual measurement on a special chair according to Martin-saller's method cited by Nowak [9] and graphic calculation from data. Subjects were measured without their wheelchair	Stature Eye level Shoulder height Elbow height Knee height Popliteal height Trunk depth Buttock-popliteal length Thigh thickness Shoulder breadth Maximum elbow breadth Elbow breadth Arm overhead reach Arm reach forward Arm reach down Lateral reach Arms span

(*continued*)

Table 1. (*continued*)

Author	Measurement method	Dimensions
Jarosz [8]	Direct manual measurement on a special chair according to Martin-saller's method cited by Nowak [9] and Jarosz [8] graphic calculation and calculation with an average wheelchair. Subjects were measured without their wheelchair, an adjustment was calculated by a mean dimension of the wheelchair, according to Skaradzinska cited by Jarosz [8]	Stature Eye height Shoulder height Elbow height Knee height Popliteal height Trunk depth Popliteal depth Thigh thickness Shoulder breadth Hip breadth max Elbows span Arm overhead reach Arm reach forward And the following calculated measurements Floor to vertex Floor to eye plane Floor to shoulder Floor to elbow Floor to top of thigh Floor to overhead grip reach Floor to upper grip reach Forward reach from front of wheelchair Forward reach from abdomen Lateral reach from side of chair
Das and Kozey [10]	Photogrammetry Subjects were measured on their wheelchairs	Seated stature Eye height Shoulder height Forearm height Knee height Toe height Max reach height Overhead reach Normal reach Radial arm length Maximum reach Trunk depth Bi-deltoid width Acromion width Elbow width Overall length Seat pan height

Other protocols and projects focused on anthropometry were found. Nevertheless, most of them were omitted for this study due to their focus on other populations, such as elderly people [12] or population with cerebral palsy [13], or because their descriptions were focused on data application [14], without specific information about the anthropometric measurement process.

Contributions of the reviewed authors are integrated to the proposed protocol under a systemic approach that considers and measures the user together with its own wheelchair as a human-wheelchair system [10]. Due to its accuracy, the direct manual measurement was selected among the reviewed methods. Furthermore, verification of definitions and landmarks was performed. Among the authors some of the measurements are named differently but measured in the same manner, others are proposed completely different. The work from Jarosz [8] and Nowak [9] focuses mainly on the development of workspaces, and therefore the dimensions are oriented towards reach.

In the reviewed documents (standards and protocols) height dimensions are similar by considering the user and the wheelchair as a unity resulting in some cases in measurements from the ground to the body landmark. For reach dimensions, some landmarks are similar from the graphics shown in the handbooks and the reviewed protocols, although some others and the measurement methods differ. As a final step in the literature review the International Standard ISO 7250-1:2010 [15] was consulted to verify the landmarks and instruments proposed.

2.2 Corroboration

This stage aimed to corroborate the order and pertinence of the measures proposed as well as the measurement instruments and the way of approaching participants. In this stage, a total of 14 wheelchair users participated. Eleven men and three women aged between 19 and 50, most of them had more than fifteen years of experience using a wheelchair. All participants had total or partial dexterity of the upper extremities and were active wheelchair users, having autonomy to perform everyday activities without assistance.

For the anthropometric data collection, the following instruments were used: Martin anthropometer and a Vernier type callipers (700 mm), and a measuring tape. Additionally, a questionnaire was adapted from the Evaluation of Daily Activity Questionnaire [16] including open-ended questions to assess the difficulties of the wheelchair user in everyday life. Specific results from the questionnaire had been published [7].

From this sample was confirmed that this protocol could be used to measure participants in different environments such as work or study place. The dimensions proposed corresponded to some problematic experiences in built environments reported by the participants in the questionnaire, confirming its applicability. Also, the order of the dimensions was adjusted to ease the process. In detail, a change was done in dimension Maximum vertical reach sideways to a wall (see Table 2). Initially, the dimension was extracted from the graphic information in the handbooks, although from the questionnaire to participants it was identified that the real manner for wheelchair users to reach things over the head did not match this description.

3 Protocol Description

This protocol is conceived for the application of its data in the design of IBE. The set of 16 dimensions can be classified into three categories; the dimensions strictly of the person, the dimensions of the human-wheelchair system and the dimensions of the wheelchair alone, even though the latter are not strictly anthropometric dimensions, all are necessary for the design of spaces since the interaction of the person with the space occurs in the wheelchair. The data collection method is direct manual measurement.

After obtaining the consent of the participants, it is essential to consider the following basic guidelines: (1) All measures must be taken in the participant own wheelchair, (2) Participants should wear their normal clothing including shoes, only removing jackets and other bulky clothes, (3) Measurements are taken by a group of two applicants (preferably a man and a woman for the ease of participants). One is in charge of measuring the participant while the other assists in the process and captures the dimensions in a format, (4) For all the dimensions a Martin anthropometer was used, except for 11 and 16, (5) People should not be forced to adopt postures they find uncomfortable or painful, and (6) When possible, all dimensions are measured in the right hemisphere.

Table 2. Proposed dimensions in the protocol

Dimension	References landmarks and measurement postures
1. Total height of human-wheelchair system	Vertical distance from the ground to the highest point of the head (vertex), with the subject's head oriented in the Frankfurt plane. Measuring view: sagittal plane
2. Total length of human-wheelchair system	Length from the rear surface of the human-wheelchair system to the farthest front. Measuring view: transversal plane
3. Eye height	Vertical distance from the ground to the outer corner of the eye (exocantion) with the subject's head oriented in the Frankfurt plane. Measuring view: sagittal plane
4. Shoulder height	Vertical distance from the ground to the upper side of the acromion with hands on thighs. Measuring view: sagittal plane
5. Elbow height	Vertical distance from the ground to the lowest bony point of the elbow (radial humerus joint commissure) with the arm flexed at a 90° angle. Measuring view: sagittal plane
6. Wheelchair height	Vertical distance from the ground to the highest point of the wheelchair. Not including the person. Measuring view: sagittal plane
7. Armrest height	Vertical distance from the ground to wheelchair armrest. Not including the person. Measuring view: sagittal plane
8. Handrim height	Vertical distance from the ground to the top of the handrim. Measuring view: sagittal plane
9. Highest landmark closest to the knee	Vertical distance from the ground to the maximum point of the flexed leg near the knee (upper patellar zone). Measuring view: sagittal plane
10. Toe height	Vertical distance from the ground to the tip of the toe. Measuring view: sagittal plane

(continued)

Table 2. (*continued*)

Dimension	References landmarks and measurement postures
11. Maximum vertical reach sideways to a wall	Maximum distance from the ground to the knuckle of the middle finger with the right arm and body leaning against the wall. A tape measure placed one-meter-high from the ground is used as a reference. The participant can hold a flat object in the palm to ease the measure identification. Measuring view: sagittal plane
12. Elbow to Elbow width	Length from elbow to elbow with hands at the top of the handrims, separating the elbows to the maximum possible point. Measuring view: coronal plane
13. Lateral reach	Length from the middle of the collarbone (suprasternal notch) to the knuckle of the middle finger, with the right arm extended laterally. For better support and accuracy, subject must have the back against the wall. Measuring view: coronal plane
14. Frontal reach	Length from the acromion to the middle finger knuckle, with the right arm extended forward and the fist closed. Measuring view: sagittal plane
15. Thorax-chair depth	Length from the farthest posterior point of the chair to the chest (mesosternal point). The chair must be against a wall and the subject leaning against the backrest. Measuring view: transversal plane
16. Thorax depth	Length from the backrest to the farthest point of the chest (mesosternal point). For this dimension, a Vernier type calliper 700 mm is used. Measuring view: transversal plane

Note: For a graphic description of the dimensions see Fig. 1.

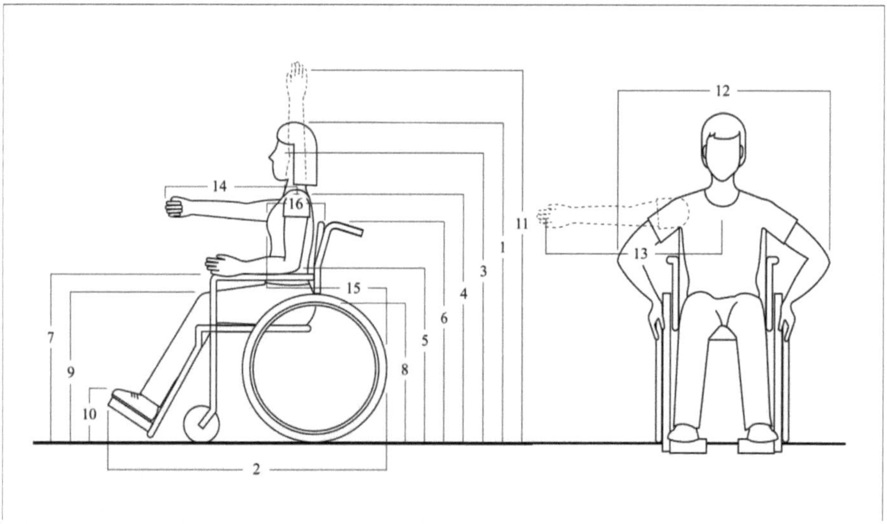

Fig. 1. Anthropometric dimensions proposed in the protocol. The annotation corresponds to the description in Table 2.

From the review of the official documents for the design of inclusive spaces, some gaps were identified. The following criteria were elaborated to respond to these gaps, considering the capacities and limitations of wheelchair users. The development process (Table 3) was nourished with data obtained from the corroboration stage with wheelchair users. The 5th percentile of the female users was considered to define the dimensions regarding reach, surfaces and sight. The 95th percentile of male users for clearances [17].

Table 3. Recommended criteria for selecting the appropriate percentile to apply in the design of public spaces considering wheelchair users.

Application	Criteria
Minimum width of the pedestrian zone	Maximum width of elbow to elbow while grabbing the handrims of the wheelchair, of the 95th percentile, multiplied by two plus clearance
Minimum size for waiting area before pedestrian crossing	Area determined by the maximum width and length of the 95th percentile human-wheelchair system
Minimum corridor width	Length from elbow to elbow while grabbing the wheels of the wheelchair, of the 95th percentile, multiplied by two, plus clearance
Minimum door width	Length from elbow to elbow while grabbing the wheels of the wheelchair, of the 95th percentile, plus a tolerance
Elevator operating controls, contact, window or button	Frontal reach of the 5th percentile, height at eyes. 5th percentile
Information guides for spatial orientation. Except emergency signals.	Eye height of the 5th percentile. (This is a general recommendation and dimensions of the signals must be considered, determining height case by case)
Maximum height for counter surface	Height at elbow in seated position, 5th percentile
Maximum sink height	Height at elbow in seated position, 5th percentile
Sanitary spaces (dimensions of toilet cubicle, dimensions of door, height of handrails)	Area determined by the maximum width and length of the 95th percentile human-wheelchair system, plus the estimation of their companion, plus a tolerance and the path needed for the use of the facilities
Height of handles	Height at shoulder in seated position. 5th percentile
Public emergency phone	Dial pad should placed at height at eyes. 5th percentile
Lifting platform surface	Area determined by the maximum width and length of the 95th percentile human-wheelchair system, plus the estimate of their companion, plus a tolerance

4 Discussion

The proposed protocol is suitable for the purposes of spaces, workstation and furniture design; it is acknowledged some measures may not be useful for other purposes. For example, if dimensions of the user or the wheelchair alone are needed. Although the data collection is easier and can be performed in several environments such as the work or study place of the participants, making data collection easier to obtain and feasible with fewer resources.

Some of the dimensions needed for the design of spaces do not appear in the protocol but, they can be inferred from the rest of the dimensions. For example, to determine the width of a corridor with circulation in both directions, the authors inferred the dimension from the measurements *elbow to elbow width* and *total length of the human-wheelchair system* (see Table 2).

Other authors [8, 9] measure the user separately and then consider the wheelchair dimensions, which results in a data set suitable for multiple purposes, but being less specific regarding the variation of wheelchair types and more complicated to obtain. The particular characteristics of each context will determine the relevance of considering certain dimensions.

As well, other protocols, for more accurate measurements, establish that participants must wear a very light layer of clothing [15], although for the nature of data application, in the design of IBE, clothing is not considering a main factor to interfere with the dimensions.

Due to resource limitations, this research focused only on wheelchair users. However, efforts to include those who use physical aids, visually impaired people and all those whose travel requirements are not considered with the current infrastructure are still missing. It is acknowledged the importance of collecting data from all user groups to inform the development of truly inclusive environments.

5 Conclusion

The development of the present anthropometric measurement protocol is a proposal that integrates the contributions of researchers, design guides, and the view of wheelchair users for the application in the context of IBE.

The insights gained from this study may be of assistance to architects, urban designers, human factors and ergonomics professionals and stakeholders to provide criteria to guide decision-making for the design of IBE.

References

1. British Standards Institution: Design management systems. British Standards Institution (1999)
2. World Health Organization: World Report On Disability. World Health Organization, Malta (2011)

3. Gobierno de la Ciudad de México: Manual de Normas Técnicas de Accesibilidad. México (2016)
4. Instituto Mexicano del Seguro Social: Criterios de Proyectos de Arquitectura para la Accesibilidad de las Personas con Discapacidad. México D.F, México (2011)
5. Instituto Nacional de la Infraestructura Física Educativa: Normas y especificaciones para estudios proyectos construcción e instalaciones (2010)
6. Buenfil-Garza T, Gutiérrez-Brezmes J.L.: Recomendaciones de accesibilidad
7. Leal-Pérez A, Rizo-Corona L, Rey-Galindo J, et al.: Assessment of dimensional needs for designing spaces for wheelchair users (2019)
8. Jarosz, E.: Determination of the workspace of wheelchair users. Int. J. Ind. Ergon. **17**, 123–133 (1996)
9. Nowak, E.: Workspace for disabled people. Ergonomics **32**, 1077–1088 (1989). https://doi.org/10.1080/00140138908966875
10. Das, B., Kozey, J.W.: Structural anthropometric measurements for wheelchair mobile adults. Appl. Ergon. **30**, 385–390 (1999). https://doi.org/10.1016/S0003-6870(99)00010-1
11. Dianat, I., Molenbroek, J., Castellucci, H.I.: A review of the methodology and applications of anthropometry in ergonomics and product design. Ergonomics **61**, 1696–1720 (2018). https://doi.org/10.1080/00140139.2018.1502817
12. Molenbroek, J.F.M.: Anthropometry of elderly people in the Netherlands; research and applications. Appl. Ergon. **18**, 187–199 (1987). https://doi.org/10.1016/0003-6870(87)90003-2
13. Hobson, D.A., Molenbroek, J.F.M.: Anthropometry and design for the disabled: experiences with seating design for the cerebral palsy population. Appl. Ergon. **21**, 43–54 (1990). https://doi.org/10.1016/0003-6870(90)90073-7
14. Reed, M.P., Van Roosmalen, L.: A pilot study of a method for assessing the reach capability of wheelchair users for safety belt design. Appl. Ergon. **36**, 523–528 (2005). https://doi.org/10.1016/j.apergo.2005.04.002
15. British Standards Institution: ISO 7250-1:2010 Basic human body measurements for technological design (2010)
16. Hammond, A., Tennant, A., Tyson, S., Nordenskiold, U.: Evaluation of daily activity questionnaire: v1 (parts 1 and 2), outcome measure. Manchester (2016)
17. Pheasant, S., Haslegrave, C.M.: Bodyspace: Anthropometry, Ergonomics, and the Design of Work, 3rd edn. Taylor & Francis, Boca Raton (2005)

Designing a Pilot System for Sustainable Villages for All

Avril Accolla(✉)

Tongji University Design and Innovation, Shanghai, People's Republic of China
avril@avrildesign.com

Abstract. The rapid change of the Chinese society in the last 20 years, with the biggest urbanization process ever observed, the growing economic and social gap between the remaining rural population and the much more sophisticated resident of the large cities of the coastal area of the East, has been pushing the Chinese government to intervene with several policies aimed to reduce the west-east migration. However, the different policies designed with this aim are not yet reaching the desired impact and their implementations are often not focusing on the need to preserve also the cultural heritage of the Chinese countryside. This paper discusses the applied research on the points mentioned above, focusing on design solutions for a sustainable and innovative pilot system which promotes economic and ecological sustainability, social innovation and re-defines marginality through innovation within tradition.

Keywords: Design driven social innovation · Design for all · Sustainable business · Cultural bridge · Elderly · Rural · Marginality

1 Introduction

Despite the average increase of the income per-capita and the very strong decrease of the number of people living under the poverty threshold, the inequality in China (represented by the GINI index) is still higher than that one of the most developed economies (Fig. 1).

According to the 2016 Statistical Report on Urban and Rural Construction issued by the Ministry of Housing and Urban-Rural Development of the People's Republic of China in 2017, there were approximately 2.617 million natural villages in China in 2016. The natural village (Chinese: 自然村) is a unit of daily life and communication of farmers, often belonging to the same family, but it is not a social management unit. Several natural villages constitute an administrative village (Chinese: 行政村), a rural grassroots management unit established by the state in accordance with the law, with a village committee and a party branch committee. In China there are more than 650.000 Administrative Villages, out of which approximately 5000 are recognized as traditional villages.

Any intervention to improve the quality of life and the economic and social environment in the rural areas will face a very high level of complexity.

G. Di Bucchianico (Ed.): AHFE 2019, AISC 954, pp. 206–214, 2020.
https://doi.org/10.1007/978-3-030-20444-0_20

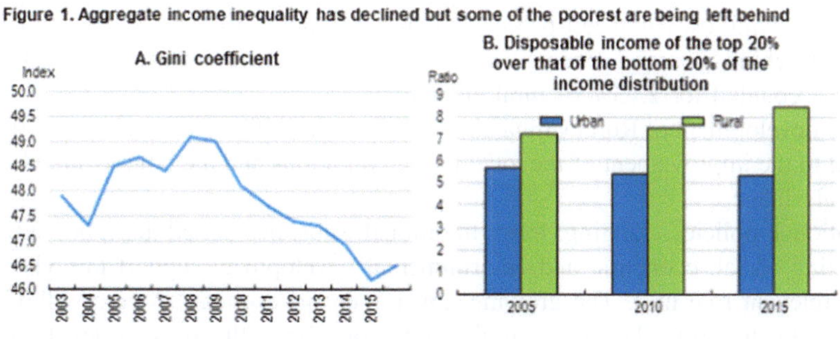

Fig. 1. GINI index trend and allocation of income in China (OECD 2017)

2 The Difficult Life of the Rural Population

The mass migration from the vast rural areas of the west to the more developed areas in the east (Fig. 2), with the resulting progressing reduction of the number of the farmers (and of the food produced by the family farms), although increased the average income of the previous farmers, now transformed for the largest majority in migrant factory workers, created an even bigger divide between the urban population and the countryside.

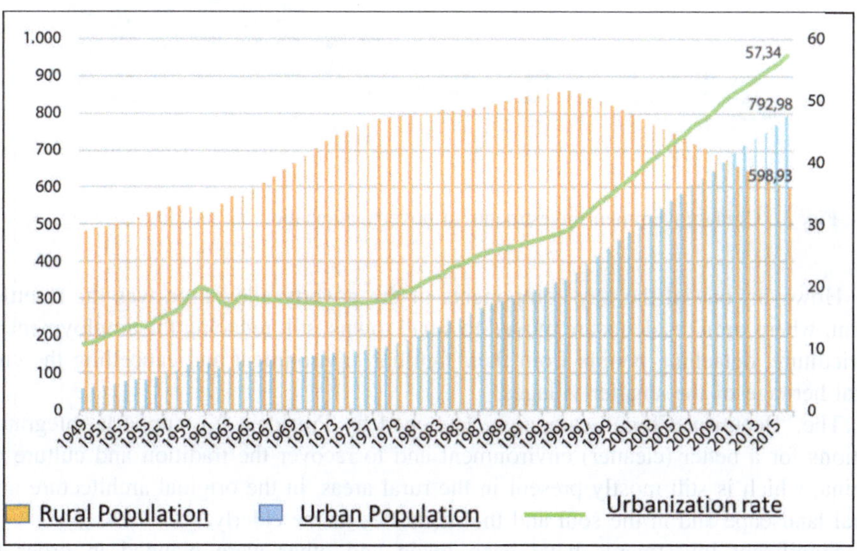

Fig. 2. Evolution of urbanization in China [2]

Furthermore, in the process of urbanization of Chinese society, two new types of families emerged, mobile families and left-behind families. There are 70 million left-behind families in the country, involving 240 million people, of which 77% are rural left-behind families. Rural left-behind children accounted for 35.1% of all rural children, left-behind women accounted for 6.1% of rural married women, and left-behind elderly accounted for 23.3% of rural elderly.

The development of Rural China aims to focus on its diversity and cultural richness parallel to the improvement of the local living conditions and the poverty alleviation in the rural villages.

National policies, ranging from traditional subsidies based to more advanced integrating social, economic and environmental development support this vision and play a relevant role in it. The growing government investment "Poverty Alleviation" policies aim to bring the entire rural population above the poverty level by 2020 (Fig. 3).

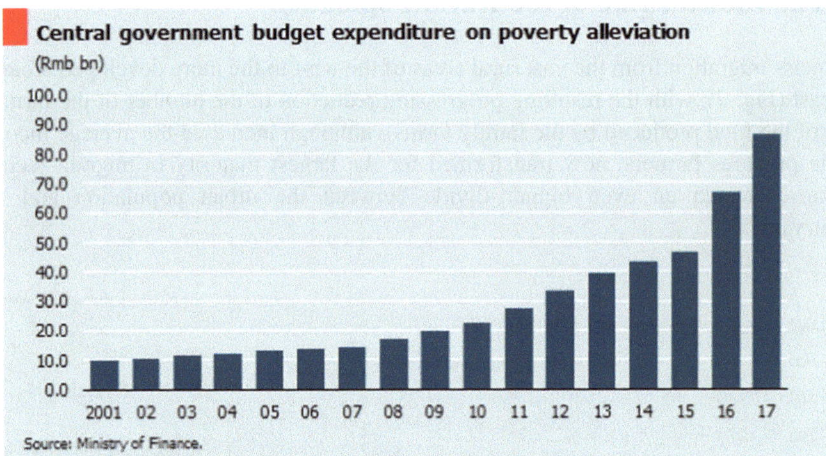

Fig. 3. China government investments in poverty alleviation are steadily growing [3]

However, one of the key components of the poverty alleviation was the resettlement, where rural residents are moved to small towns, still reducing the employment in agriculture, detaching people from their familiar environment and cancelling the cultural heritage of the smaller villages.

The "Beautiful China" campaign, launched by President Xi, supports integrated actions for a better (cleaner) environment and to recover the tradition and culture of China, which is still mostly present in the rural areas, in the original architecture and rural landscape and in the soul and the memories of the elderly.

Significant investments have been made and other were planned in order to improve the quality of living in the rural areas. As jointly stated by General Office of the Communist Party of China (CPC) Central Committee and the General Office of the State Council in February 2018, "although progress has been made, the living environment in rural China is very unbalanced with many areas in the grip of dirt and disorder. It falls far short of people's expectations and marks a key weakness in social and economic development". The three years action plan unveiled in that occasion includes garbage sorting, toilet upgrades, residential sewage treatment and village management [4].

The "Characterizing small town" policy supports systemic interventions where design can impact social, culture, economic and business interactions.

3 Rural Elderly Issue

China's population is becoming more and more aged. In 2017, China's elderly population over 60 years old has exceeded 230 million. It is the only country in the world with an elderly population of over 200 million.

There are more than 88 million households in China with over 65 years of age, accounting for more than 20% of households in the country.

In 2013, the number of disabled elderly people in the country reached 37.5 million. It is estimated that the number of disabled elderly people will exceed 40 million in 2015. In sharp contrast to the rapid growth in the needs of family support, family capacity is weakened due to shrinking family size and time resources.

Among them, the rural left-behind elderly families and the elderly living alone are the most prominent (Fig. 4).

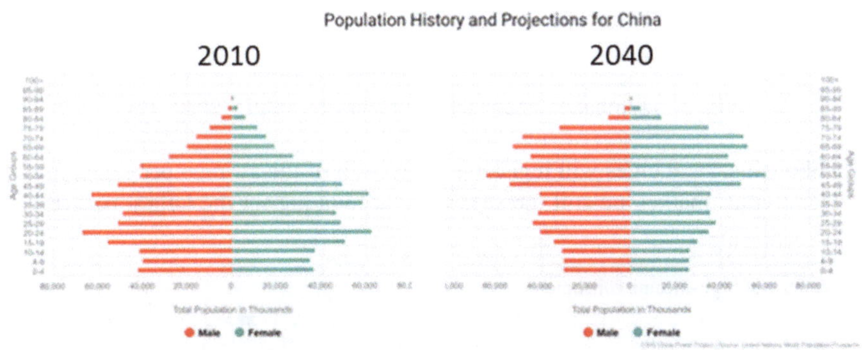

Fig. 4. Demographic composition of China in 2010 and 2040 projection [5].

One of the effects of the solitary life of the rural elderly is the dramatic increase of the rate of suicides, presently one of the highest globally [6]. Policies and specific activities aimed to slow down this process are quickly needed.

4 The Evolving Needs of the Urban Residents

The population of the urban areas is showing a strong increase of attention to health, environment and healthy food, also driven by the many food scandals Chinese customers faced in the last few years. A McKinsey research [7] shows "a decrease in consumption in categories deemed less healthy and a willingness to spend significantly more on health and more environmentally conscious categories…….(omissis) … consumers seemed more concerned with these issues in 2015 than they were before".

This is driven by several forces:

(a) The continuously increasing spending capacity, with the expansion of the upper middle class (Fig. 5)

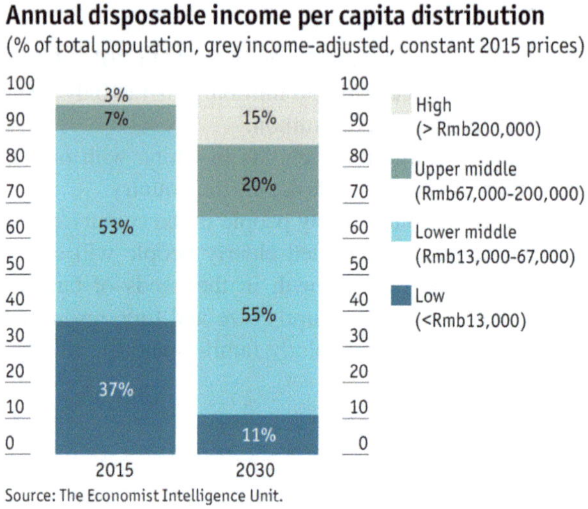

Fig. 5. Distribution of annual income in China 2015–2030 [10]

Although several cities will register a big growth of the average per capita income, the Tier 1 cities market are still foreseen to be accounting for the largest part (Fig. 6).

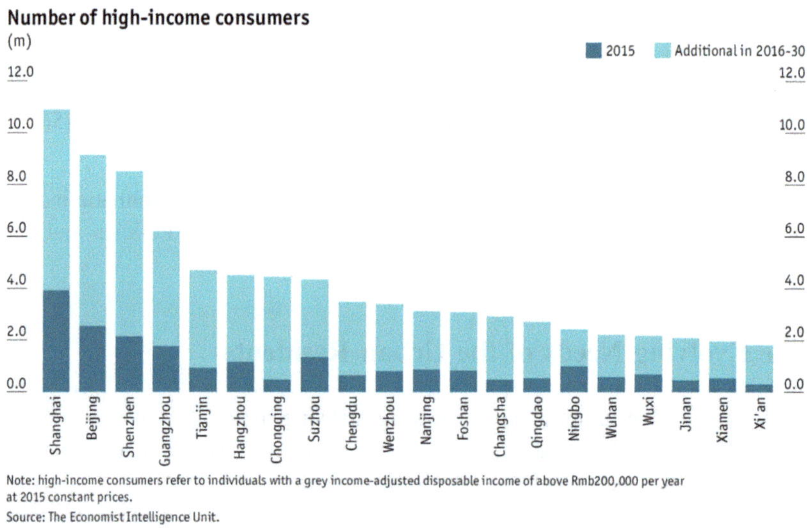

Fig. 6. The 20 most important Chinese cities in terms of income per-capita in 2015 and 2030

(b) The increase attitude to indulge in personal consumption. McKiney (2016) reports an increase of the willingness to spend to get better quality and of the attention to a better quality of life. In 2013, health care and nutrition products jumped to number 2 among 15 product categories in which Chinese consumers plan to spend more, up from number 11 in 2011 (BCG 2014). A survey performed in 20 Chinese cities in 2014 showed that 80% of respondents were not satisfied with food safety and thought the major responsible are food companies and food industry association. A recent research shows that 88% of consumers are willing to spend more in order to get healthy foods, organic and without food additives [9]. The same source reports that

- "Over 30% of consumers in the cities of 1st and 2nd rows often buy organic food. Consumers seek foods with fewer pesticides that can harm their health. Having a child will influence on how to consume the foods, parents are more vigilant about food. Families with high income are willing to pay more for organic food".
- Although the experts' projections [10] show a decline of the % expenditures in food, the healthy food does belong more to the categories of health products and healthy and organic food is often seen as a luxury product.

(c) The strong connected of food with health in the traditional Chinese culture, which expects meals to be nutritionally balanced and to use natural ingredients. From our research, in Shanghai only there is more than 15000 vegetarian belonging to one or the other organization. Wechat (the most common social network in China, with more 800 million of users) hosts more than 20 chats only in the Shanghai area, focused on healthy food/vegan/vegetarian/organic, etc. The Beijing Farmers Market has a network of 30,000 WeChat followers. International market research company Organic Monitor estimated that in 2013 the market for organic foods in China had reached about 2.4US$ billion, compared with less than US$100 million of 10 years before [11]. On top of many companies providing "organic" food sometimes supplied by rural farms, cooperatives of farmers, focusing on fair and organic farming, are starting to emerge and represent an opportunity to establish sustainable business models for small scale. This is made much easier by the e-commerce that is now the standard way of purchasing for hundreds of millions of Chinese, which generally trust more online platform than traditional shops and which is becoming the standard way to purchase food in China [12].

5 Proposed Project

5.1 Rationale

Eco and gastronomic tourism is perceived as a possible effective driving force for the development of the villages that kept traditional characteristics. The Chinese government strongly encouraged the formation of farmers' cooperative to support agriculture and agritourism. There is a growing number of projects promoting the enhancement and development of villages and small towns in China. Among these, the initiatives focusing on the preservation of the ancient charm of villages to support poverty alleviation through tourism are the most renowned and many pilot experiences of rural tourism have been implemented. From an urbanization perspective, to improve the countryside standards of living, some small villages are merging with other villages or towns, creating clusters concentrating scattered realities into higher density residential areas, so facilitating the provision of public services. This process also helps saving agricultural land.

However, rural residents often got a limited benefit from the tourism, as they participated in tourism activities in most cases only because they were recruited as workers because their limited skills and experience prevented them from being involved in the management [13].

Furthermore, as observed by Yang and Hung [14]:

"Sustaining the success of the tourism cooperative remains difficult because of complex factors, such as management ability of leaders, attitude of other interest groups and market competition".

In addition, since it is estimated that nowadays only less than 2% of Chinese villages have maintained their traditional features, the initiatives able to account on a "heritage related tourism" are inevitably focused on a limited number of settlements. Therefore the implementation of innovative development models based on a different pattern of values is likely even more crucial for all those villages that have not preserved their traditional layout and characteristics.

The agricultural sector remains key to provide employment and income opportunities for the rural population. A gradual restructuring for the agricultural sector, from an innovative and environmentally safe perspective, is necessary to allow productivity, efficiency and farm income to increase. This process will have to involve the management of elderly which, as explained before, represent the largest population of the rural villages.

The Project aims to explore innovative ways to improve local quality of life and establish a sustainable maintenance system, creating a favorable living environment that can attract new inhabitants (whether returning natives or newcomers), and visitors.

These objectives would be achieved by increasing local farmers' income and improving local public services, enhancing and promoting the quality of agricultural products, developing innovative and energy-efficient practices in farming and building construction, and by training and capacity building initiatives in agriculture and eco-tourism sectors.

5.2 Project Components

The Project will include three components:

1. Restructuring of agricultural sub-sector, through:
   (a) Technical assistance for restructuring the current agricultural production patterns, introducing new or improved systems of organic production and low emissions, pre-processing/processing technologies for agricultural products, storage and marketing facilities;

2. Strengthening of farm-to-market linkages, including upgrading of quality standards, labelling and certification, branding, product tracing, logistics, and food safety;
3. Support of existing and new farmers' cooperatives and enabling institutions, providing training and capacity building in management and business operation, and start-up investments for new and innovative technologies.

5.3 Facilities Improvement

(a) Technical assistance for residential upgrading, renovation and restoration of traditional buildings, and new buildings construction using traditional materials and modern technology with efficient energy solutions;
(b) Technical assistance for the provision of improved facilities for the agricultural sector, including new facilities for processing, storage and marketing;
(c) Subject to the availability of funds, possible technical assistance for public infrastructure improvements including wastewater treatment and environmental sanitation facilities (solid waste collection and treatment), small-scale agricultural infrastructure (such as production irrigation, drainage, field tracks), security facilities (fire extinguishing equipment).

5.4 Territorial Linkage and Urban-Rural Cooperation

(d) Technical assistance for enhancing small-scale sustainable territorial infrastructure, including restructuring and upgrading existing pathways, and developing new trails connecting settlements;
(e) Technical assistance for cultural and natural resource participatory inventory of the villages in the region;
(f) Technical assistance for drafting protection measures/regulations for landscape and architectonical integrity and consistency;
(g) Technical assistance for drafting urban planning layout for the integration of new buildings, facilities and infrastructure;
(h) Technical assistance for improving eco-tourism services and facilities, and soft mobility offer;
(i) Capacity building for sustainable cultural and eco-tourism development and community self-development (guiding local residents to conduct profitable tourism-related transactions); English-language training; training on service standards and hygiene practices;
(j) Enhancement urban-rural linkages promoting education courses on innovative/organic farming, food-related events and facilities.

The Project will be divided in three phases (1. Research and Evaluation, 2. Design and Research Implementation, 3. Project Implementation)

The project received the interest of a county in the west part of the Hunan Province and we are planning to start the research activities before summer 2019.

References

1. Westmore, B.: China Desk, OECD Economics Department. An immediate Chinese challenge: further addressing vast income inequality. https://oecdecoscope.blog/2017/03/21/an-immediate-chinese-challenge-further-addressing-vast-income-inequality/ Accessed 17 Dec 2018
2. Rossi, A., Fasulo, F.: Cina 2017. Scenari e Prospettive per le imprese. Fondazione Italia-Cina, 12 May 2017
3. The Economist Intelligence Unit. China's plan to eliminate poverty by 2020, 5 January 2018. http://country.eiu.com/article.aspx?articleid=536285437&Country=China&topic=Economy. Accessed 18 Sept 2018
4. Liang, J.: China unveils action plan for improving rural living environment. Xinhua, 06 February 2018
5. China Power Team: Does China have an aging problem? China Power, 15 February 2016. Updated 11 August 2017. https://chinapower.csis.org/aging-problem/. Accessed 1 Feb 2019
6. Vanderklippe, N.: How China's rural elderly are being left behind and taking their lives. The Globe and Mail, 11 March 2016, Updated on 12 November 2017. https://www.theglobeandmail.com/news/world/how-chinas-rural-elderly-are-being-left-behind-and-taking-theirlives/article29179579/. Accessed 17 Nov 2019
7. Atsmon, Y., Magni, M.: Chinese consumers: revisiting our predictions. Mc Kinsey, October 2016. http://www.mckinsey.com/industries/retail/our-insights/chinese-consumers-revisiting-our-predictions. Accessed 12 Nov 2018
8. Boston Consulting Group: From Insight to action. Capturing a share of China consumer health market, February 2014. https://www.bcgperspectives.com/Images/From_Insight_to_Action_Feb_2014_tcm80-155115.pdf Accessed 27 Oct 2017
9. Gentleman Marketing Agency: Focus on new Food habits and consumption in China, 9 June 2016. http://marketingtochina.com/new-food-trends-chinese-consumers/. Accessed 27 Feb 2018
10. The Economist Intelligence Unit Limited: The Chinese Consumer in 2030 (2016). http://www.eiu.com/public/topical_report.aspx?campaignid=Chineseconsumer2030 Accessed 28 Oct 2017
11. Research Institute of Organic Agriculture (FiBL) (2013). http://www.fibl.org/en/media/media-archive/media-release/article/growth-continues-global-organic-market-at-72-billion-us-dollars-with-43-million-hectares-of-organic.html . Retrieved on February 26 2018
12. Zheng, Y., Jin, S., Zhang, J.: The 21st century agribusiness in China: E-commerce, consumer preference, and competition, 14 November 2018. https://doi.org/10.1002/agr.21587. Accessed 25 Jan 2019
13. Scheyvens, R.: Tourism and Poverty. Routledge, New York (2011)
14. Yang, X., Hung, K.: Poverty alleviation via tourism cooperatives in China: the story of Yuhu. Int. J. Contemp. Hospit. Manag. **26**(6), 879–906 (2014)

Assessing Accessibility and Safety Conditions in an Urban Environment: What Do Pedestrians Perceive?

Carlos Aceves-González[1(✉)], Libertad Rizo-Corona[2], Rosa Rosales-Cinco[1], John Rey-Galindo[1], Karthikeyan Ekambaram[3], and Maricela Ramos-Tachiquín[4]

[1] Centro de Investigaciones en Ergonomía, Universidad de Guadalajara, Guadalajara, México
{c.aceves,rosa.rcinco,john.rey}@academicos.udg.mx
[2] Maestría en Ergonomía, Universidad de Guadalajara, Guadalajara, México
libertad.rizo.din@gmail.com
[3] Institute for Future Transport and Cities, Coventry University, Coventry, UK
Karthikeyan.ekambaram@coventry.ac.uk
[4] Departamento de Ciencias de la Salud Poblacional, CUTonalá, Universidad de Guadalajara, Guadalajara, México
maricela.ramos@academicos.udg.mx

Abstract. This study aimed to assess pedestrians' perceptions about accessibility and safety with an aim to identify factors that influence their choice of mobility in an urban environment in the Municipality of Tlaquepaque, Mexico. A survey was conducted of 299 people aged 18 to 72 years old. The study was undertaken in situ while the users were at transit on the place. Most of the participants pointed out the presence of obstacles to free access and over than 60% of the participants considered that the design of the road could imply problems to people with disabilities. In terms of safety, almost 80% of participants consider the lack of respect to pedestrians as the main fact that represents safety risks for them. Overall, the study allows recognising a series of elements of the urban space might be considered to create an age-friendly – safe and accessible - environment.

Keywords: Urban environment · Safety · Accessibility · Ageing · Pedestrians' perception

1 Introduction

Walking is an integral part of every travel chain and is an important component of one's mobility. It is recognised as the cheapest, most natural and sustainable mode of travelling [1]. In addition, walking comprises a series of benefits ranging from physical and psychological to social aspects [2]. Alfonzo [3] proposes a socio-ecological model consisting of a hierarchy of five levels of needs in the decision-making process of walking namely: feasibility, accessibility, safety, comfort, and pleasurability. This study stands on the view that a feasible, accessible and safe journey needs to be

G. Di Bucchianico (Ed.): AHFE 2019, AISC 954, pp. 215–225, 2020.
https://doi.org/10.1007/978-3-030-20444-0_21

guaranteed before pedestrians decide walking. Lack of accessibility and pedestrians' perception of safety risks might affect their decision to walk or even their choice to use some streets [4].

The term accessibility used to refer to the relationship between a person's capabilities and demands generated by the design of the physical environment. At present, it is a term more oriented towards usability, defined as the 'usability of a product, service, environment or facility by people with the widest range of capabilities' [5]. Although accessibility has gained importance and interest to fulfil the needs of both older and disabled people, there are still some gaps to be bridged. The World Health Organization [6] highlights that, in both developed and developing countries, people think that their city was not designed for the elderly. In Mexico, previous studies have drawn attention to the presence of physical and social barriers in urban environments that prevent people from using roads as pedestrians [7, 8].

According to the World Health Organization [9], the number of road traffic deaths on the world's roads is unacceptably high. Accidents involving pedestrians, cyclists and motorcyclists comprise more than half of those traffic deaths. The rate for road fatality varies between countries by income. The risk of road accidents is three times higher in low-income countries than in high-income countries. In 2016, the number of road traffic fatalities in Mexico was estimated to be 16,725, with a rate of 13.1 deaths per 100,000 populations [9]. According to the official statistics, pedestrians account for around 29% in Mexico, but results from a different study suggest that pedestrians represent nearly 48% of the deaths associated with road injuries [10]. Pedestrians are the most vulnerable group of road users and traffic accidents involving injuries to them are a major societal concern.

The local governments in Mexico have made continuous efforts to improve accessibility and safety, especially in urban environments. However, most of those interventions lack participation from road users. Understanding people's needs for walking plays an essential role in designing a better road. Consequently, the objective of this study was to assess pedestrians' perceptions about accessibility and safety in an urban environment with an aim to identify factors that influence their choice of mobility.

2 Methodology

2.1 Site Selection and Participants

The study was conducted in cooperation with the municipality of Tlaquepaque, one of the eight towns within the Metropolitan Area of Guadalajara, Mexico. For this study, five blocks of a road in the headquarters of Tlaquepaque was chosen for evaluation. The specific road is *Juarez* Avenue from *Niños Heroes* Avenue to *Herrera y Cairo* Street comprising of shops, restaurants and tourist attractions (Fig. 1). This specific location was chosen for evaluation because the local government has a plan to improve roads in this area. The Department of Mobility was especially interested in understanding how this built environment could be improved for better pedestrian accessibility.

A total of 299 participants were involved in this study. A random criterion of selection was utilised which consisted of inviting a person every three people passing on the road at the time of the survey. Road users were included after obtaining their acceptance. Additionally, it was required to be aged 15 years old or over with sufficient language and cognitive abilities to allow them to provide informed consent and response to the survey. The age of the participants was registered by ranks between those who were under 18 years of age, from 19 to 29, from 29 to 59 and over 60 years of age. Participants were informed of the study objective, confidentiality of participating, the time needed for the survey, and the right to withdraw at any time if decided. The data were collected in three weeks in July of 2017. The intention was to cover different times of the day. Therefore, collection times were morning and evening hours during weekdays, evening hours on Saturdays and morning hours on Sundays.

Fig. 1. Evaluation site (marked in red)

2.2 Materials

The questionnaire used in this survey was study-specific and was based on previous literature, but some questions were introduced based on the interests of the Department of Mobility. The questionnaire included two sections. The first section was to register the participants' demographics and journey characteristics. The second section aimed to collect participants′ perceptions regarding safety risks and accessibility. This section comprised a series of statements to be answered using tick boxes and two open-ended questions to enter opinions from participants. These questions had the intention to corroborate and complement what participants were responding to the closed-ended questions. A five-point scale with agreement alternatives was used for those statements. These options were: 'strongly disagree', 'disagree', 'neither', 'agree', and 'strongly agree'. The rating scale was from one to five, where one represented the negative values and five the positive ones. For the analysis, ratings of four and five were considered as positive perceptions of safety and accessibility; in contrast of one to three which were considered as the lack of these attributes. The questionnaire is shown in Appendix 1.

Data collection was conducted by three senior researchers and ten junior researchers who had training for two weeks. Two pilot studies were carried out, which served to refine the questions of the survey and to evaluate the rating scale.

2.3 Data Analysis

Quantitative data collected were analysed using, SPSS (v. 21). A process of screening and cleaning was undertaken before analysing the collected data. Some data were missing for each variable mostly due to the participant not responding to a question. Data were analysed descriptively for the preparation of this report. Qualitative data from the open-ended questions were captured and thematically analysed using Nvivo (v11). An inductive approach was used for the analysis, which means the categories identified are strongly linked to the data themselves [11].

3 Results

3.1 Participants and Journey Characteristics

A total of 299 pedestrians were involved in the survey. Majority of the participants were aged between 30 and 59 (41%), male (52%), and Mexican nationality (95%). Eleven per cent of participants reported suffering from chronic illness or disability. Thirty-three per cent of the respondents was in the place for tourism or entertainment. Forty per cent of the participants were either working (27%), or residing (13%) in the survey area, suggesting that they visit the place regularly. Ten per cent of participants visited the area for the first time. Most of the people arrived on foot (40%), by car (32%) or by bus (22%). These results are listed in Table 1.

Table 1. Demographics and journey characteristics of participants.

Participants' characteristics	Frequency (%)	Journeys' characteristics	Frequency (%)
Gender		Frequency of visit	
Female	142(48)	2/3 times per year	38(13)
Male	156(52)	2/3 times per month	51(17)
		2/3 times per week	58(20)
Age group		Daily	121(40)
Less than 18	31(10)	First time	31(10)
From 19 to 29	109(36)		
From 30 to 59	121(41)	Mode of transport to get the place	
60 and over	38(13)	Walking	119(40)
		Bicycle	12(4)
Nationality		Motorbike	4(1)
Mexican	283(95)	Bus	64(22)

(*continued*)

Table 1. (*continued*)

Participants' characteristics	Frequency (%)	Journeys' characteristics	Frequency (%)
Foreigner	16(5)	Car	95(32)
		Other	2(1)
Chronic illness or			
disability		Reason for using the road	
Yes	32(11)	Tourism and fun	97(33)
No	267(89)	Work in the area	80(27)
		Live in the area	40(13)
		Shopping	44(15)
		Other	37(12)

3.2 Pedestrians′ Perception of Safety and Accessibility

Fifty-six per cent of participants in the survey either 'agreed' or 'strongly agreed' that it was safe crossing roads in the area where the survey was undertaken (Table 2). Approximately sixty-two and sixty-seven per cent 'agreed' and 'strongly agreed' that the width of sidewalks was enough and lighting was adequate, respectively. Regarding accessibility, over 57 per cent responded that the design of the roads did not consider the needs of disabled people. Likewise, sixty-three per cent indicated the lack of rest areas in the road.

Table 2. Pedestrians perception regarding safety and accessibility (n = 299)

Statements	Unsafe and not accessible			Safe and accessible	
	1	2	3	4	5
	n (%)	n (%)	n (%)	n (%)	n (%)
Crossing roads in this area is safe	29(10)	43(14)	57(19)	77(26)	88(29)
Sidewalks are wide enough	26(9)	32(11)	53(18)	58(19)	123(41)
Lighting on this road is adequate	17(6)	21(7)	50(17)	82(27)	100(33)
The design of the roads considers people with physical and mental disabilities	65(22)	44(15)	59(20)	75(25)	49(16)
There are enough rest areas on this road (benches, shadows)	63(21)	54(18)	68(23)	67(22)	39(13)
Using this road is comfortable	6(2)	15(5)	45(15)	64(21)	160(54)
There is enough vegetation on this street	27(9)	39(13)	65(22)	73(24)	88(29)
I feel comfortable with the number of people who use this road	10(3)	19(6)	38(13)	80(27)	145(48)
The information provided by signals is adequate	39(13)	26(9)	39(13)	78(26)	108(36)

1 = Strongly disagree; 2 = Disagree; 3 = Neither; 4 = Agree; 5 = Strongly agree
*Percentages may not add due to rounding

3.3 Reasons Behind Safety Perception

Pedestrians were asked to state the main reason based on what they responded to the question regarding the extent of agreement in being safe crossing the roads in the area. Participants expressed a total of 321 reasons, approximately seventy-five per cent (239) were on safety problems. The fact that drivers do not respect pedestrians was the most mentioned negative issue (80 mentions), followed by lack of traffic lights (34 mentions) and the presence of too much traffic (31 mentions). It must be highlighted that negative reasons were raised even by participants that agreed with the statement "crossing the road in this area is safe". Positive comments on safety were led by the fact of drivers respecting pedestrians (39 mentions), followed by cars travelling at a low speed (15 mentions) and the presence of traffic lights (9 mentions) (Table 3).

Table 3. Positive and negative reasons behind of perceived safety

Positive reasons	Number	Negative reasons	Number
Drivers respect pedestrians	39	Drivers do not respect pedestrians	80
Cars at low speed	15	Lack of traffic lights	34
There are signals	9	Too much traffic	31
Presence of police officers	7	A poor culture of road safety	21
Space to cross the roads	4	Cars parked in places that obstruct the visibility and mobility of pedestrians	19
Others	8	Cars in high speed	16
		Lack of signals	15
		Lack of police officers	7
		Lack of bumps	3
		Presence of holes or obstacles on the pedestrian way	3
		Others	10
Total	**82**	**Total**	**239**

3.4 Pedestrians' Suggestions for Improving Safety and Accessibility

At the end of the survey participants were asked for any suggestions to improve this road. More than fifteen categories were raised from the thematic analysis. Table 4 shows the top five recommendations. Pedestrianisation of the road and/or widening the sidewalks was most frequently (48) suggested followed by the proposal to ban parking of cars in the street (41) and removing any obstacles on the walkway and crossing points (31).

Table 4. Pedestrians' top five recommendations for improving safety and accessibility.

Level	Recommendation	# mentions
1	Convert the current road to a pedestrian road and/or widening the sidewalks	48
2	Do not allow cars to park in the road	41
3	Remove obstacles from the walkways and crossing points	31
4	Install pedestrian traffic lights (with audio)	25
5	More signals and information	25

4 Discussion

4.1 Perception of Accessibility Issues

This research has focused on investigating the pedestrians′ perception of accessibility and safety in an urban area in the Metropolitan Area of Guadalajara, Mexico. Results indicate that over half of participants (168) did not agree with the statement of "the design of the roads consider people with physical and mental disabilities", which suggests that the obstacles might prevent or impose difficulties for these group of users. In accordance with the results, a previous study has found accessibility and usability problems for older people when using the roads in Guadalajara [8]. Removal of obstacles from the walkways and crossing points is one of the top five suggestions for improvements, which reveals the relevance of accessibility issues for all pedestrians and not only for older and disabled road users.

In addition, two-thirds of the pedestrians reported that there are not enough rest areas on the road. Spaces and furniture for resting along with enough vegetation have been suggested as an important physical factor promoting walking behaviours [12]. This finding is consistent with that of Alfonzo′s model [3] regarding the level of comfort required for walking. These facts suggest that the current state of the road may prevent some pedestrians avoid using the facility either for the lack of convenience or because some pedestrians - especially most vulnerable users - may have a self-perception of not being able to move through the road.

4.2 Perception of Safety Risks

Results in this study indicate that 57% of pedestrians (129) did not agree with the statement of "crossing the roads in this area is safe". This finding contrast with the percentage of negative reasons explaining the safety risks (75%). The difference may be explained by the fact that some pedestrians who "agree" with this statement also provided negative comments on safety. Only some of those who "strongly agreed" (88) did not provide any negative comments. Beyond these differences, the results are consistent in showing pedestrians concerns on safety. These results may be explained by the fact that the number of road traffic injuries and fatalities is too high in Mexico. It may be due to the reason that the perceived safety is highly associated with the number of road fatalities. Perceived safety concerns have been reported previously in the context of Guadalajara [8].

At least four of the negative reasons for safety concerns are directly related to the interaction between pedestrians and car drivers. In general, the interaction between pedestrian and vehicle drivers are mediated and affected by several factors including social and behavioural, for example, compliance of rules and social norms which have been identified to play the central role compared with formal rules [13, 14]. Having ease of movement without any challenging situation with other motorists found to be an influential factor in the perception of safety. This in line with Šucha [15] who found following behaviours from the car drivers: approaching a crossing at an inappropriate speed, not yielding to pedestrians and not yielding when turning right/left at a junction, among others have a negative influence on pedestrians' perception of safety. In contrast, car drivers perceived mainly pedestrians to have the following unsafe behaviours before an accident: an infraction of some traffic rule, not respecting traffic lights and not crossing in the appropriate zones. At any case, car drivers' respect for pedestrians could be manifested by attending to speed limit, observance of the right of way and use of the designated parking spaces, also, by the consideration to pedestrians with diminished mobility which are more vulnerable while crossing the street.

Another factor described in the literature as part of pedestrian-car drivers interaction and identified by participants among the negative reasons for safety perception is the physical infrastructure, manifested in the presence or absence of signals, traffic lights, road geometry and type of pedestrian crossings [13, 16]. Furthermore, when asked for suggestions for improving safety and accessibility, participants' most recurrent answers were related to having exclusive areas for pedestrians, whether that be pedestrian-only roads or extended sidewalks, as well as forbidding parking of cars on the sidewalks. These results suggest that there is a need to improve interactions between the pedestrians and vehicle in a holistic way, including social aspects of the interaction and improving the infrastructure.

4.3 Potential Interventions

The suggestions from the participants seem to be an excellent starting point for possible countermeasures. The recommendations comprise ideas for both accessibility and safety improvements. Pedestrianisation of the road along with widening the sidewalks was the most voted idea for intervention. This means no access to motorised traffic and the prioritisation of the use of the road only for pedestrians. It should be noted that pedestrianisation has an overall positive environmental effect [4].

Interestingly, the suggestions made by the road users corroborate well with the literature and are within the international guidelines for pedestrian safety [4], which indicates that pedestrians have a good sense in recognising the problems and ability to see the solutions. Furthermore, this study highlights the relevance of including pedestrians' participation for a more human-centred design of our cities.

5 Conclusion

The overall aim of this study was to assess pedestrians' perceptions about accessibility and safety in an urban built environment with a view to identifying factors that will encourage walking as a choice of the commute. The study found that walking needs in terms of accessibility and safety are not fully guaranteed in the evaluated location. A considerable proportion of participants had the perception that crossing roads in that site is not safe, but they also pointed out that the design of the road does not consider the needs of older and disabled people. Car drivers' action towards pedestrians was one of the main reason for the negative perception of road safety. Participants suggested several recommendations that have the potential to impact positively on the decision-making for walking. Finally, it is relevant to state that more research is needed for getting a better understanding of accessibility and safety conditions of urban spaces. However, the collaboration between the road users and local authorities for developing a safe and accessible zone is even more needed for increasing pedestrians walking rates.

Acknowledgements. We want to acknowledge Martín Rubén Corona González, who was the head of the Department of Mobility in Tlaquepaque, for his vision of including an Ergonomic analysis for safety and accessibility in the municipality. We appreciate his input and the facilities provided to undertake the study. We also thank Azucena Hernández, Daniela Hurtado, Alondra Yépez, Andrea Gómez, Yaqueline Rios, Fernando Navarro, Fabiola Zitlapopoca, Cecilia Pérez, Mauricio Alcaraz and Alejandro Ángel. They helped in preparing materials and applying the survey as part of the studentship programme *Verano de Investigación Delfin* 2017.

Appendix 1. Questionnaire

Aspects included on the questionnaire

Gobierno de TLAQUEPAQUE

1. Internal control		
a. Questionnaire number:	b. Time:	c. Date:
d. Place:	e. Name of the researcher:	

2. Demographics and travel characteristics						
a.Gender______________		b. Nationality		1) Mexican	2) Foreigner	
c. Age	1) Less than 18	2) from 19 to 29		3) from 29 to 59		4) 60 and over
d. Any chronic illness or disability?		1) Yes		2) No	Name it ____________	
e. How frequent do you use this road?	1) 2/3 times per year	2) 2/3 times per month		3) 2/3 times per week	4) Everyday	5) First time
f. How did you get to this place?	1) Walking	2) Bicycle	3) Car	4) Bus	5)Motorbike	6) Other

3. User perceptions on safety and accessibility					
a) Please tell us how much you agree or disagree with the following statements:					
Statements	1 Strongly disagree	2 Disagree	3 Neither	4 Agree	5 Strongly agree
Crossing roads in this area is safe.					
Why?					
Sidewalks are wide enough					
Lighting on this road is adequate					
The design of the road considers people with physical and mental disabilities.					
There are enough rest areas on this road (benches, shadows).					
Using this street is comfortable					
There is enough vegetation on this road.					
I feel comfortable with the amount of people who use this road.					
The information provided by signals is adequate.					
Do you have any suggestions to improve this road?					

References

1. Tournier, I., Dommes, A., Cavallo, V.: Review of safety and mobility issues among older pedestrians. Accid. Anal. Prev. **91**, 24–35 (2016)
2. Demerath, L., Levinger, D.: The social qualities of being on foot: a theoretical analysis of pedestrian activity, community, and culture. City Commun. **2**, 217–237 (2003)
3. Alfonzo, M.A.: To walk or not to walk? The hierarchy of walking needs. Environ. Behav. **37**, 808–836 (2005)
4. World Health Organization: Pedestrian safety: a road safety manual for decision-makers and practitioners. WHO (2013)
5. BS 18477: BS 18477:2010. Inclusive service provision – Requirements for identifying and responding to consumer vulnerability (2010)
6. WHO: Global Age-friendly Cities: A Guide. WHO (2007)
7. Aceves-Gonzalez, C.: The application and development of inclusive service design in the context of a bus service (2014). https://dspace.lboro.ac.uk/2134/16265
8. Aceves-González, C., Cook, S., May, A.: Bus use in a developing world city: implications for the health and well-being of older passengers. J. Transp. Heal. **2**, 308–316 (2015)
9. World Health Organization: Global Status Report on Road Safety. World Health Organization, Geneva (2018)
10. Bartels, D., Bhalla, K., Shahraz, S., Abraham, J., Lozano, R., Murray, C.J.L.: Incidence of road injuries in Mexico: country report. Int. J. Inj. Control Saf. Promot. **17**, 169–176 (2010)
11. Braun, V., Clarke, V.: Using thematic analysis in psychology. Qual. Res. Psychol. **3**, 77–101 (2006)
12. Mehta, V.: Walkable streets: pedestrian behavior, perceptions and attitudes. J. Urban. **1**, 217–245 (2008)
13. Rasouli, A., Tsotsos, J.K.: Joint attention in driver-pedestrian interaction: from theory to practice, pp. 1–111 (2018)
14. Wilde, G.J.S.: Social interaction patterns in driver behavior: an introductory review. Hum. Factors J. Hum. Factors Ergon. Soc. **18**, 477–492 (1976)
15. Šucha, M.: Road users' strategies and communication: driver-pedestrian interaction. Transp. Res. Arena (2014)
16. Kaplan, S., Bortei-Doku, S., Prato, C.G.: The relation between the perception of safe traffic and the comprehension of road signs in conditions of ambiguous and redundant information. Transp. Res. Part F Traffic Psychol. Behav. **55**, 415–425 (2018)

Global Perspectives on People-Centered Design and Cultural Heritage

Understanding Visual Information Processing for American vs. Saudi Arabian Users

Yahya Alqahtani(✉), Joyram Chakraborty(✉), Michael McGuire(✉), and Jinjuan Heidi Feng(✉)

Department of Computer and Information Sciences, Towson University, Towson, USA
{yalqahtani,jchakraborty,mmcguire,jfeng}@towson.edu

Abstract. Cultural differences may affect user experiences in many different ways. This may range from having to adapt to different data formats to understanding novel directions of reading and writing. Analyzing data presented in various forms are crucial for both information assimilation and dissemination. Different information presentation methods can substantially affect task performances. To date, limited research has been reported that investigates how users from different cultural backgrounds respond to various information presentation methods while conducting visual information processing tasks. To fill in that gap, we conducted an experiment to better understand the relationship between American and Saudi Arabian users when interacting with different information presentation methods in the context of visual information processing. The specific information presentation methods investigated in the study are textual presentation, graphical presentation, and tabular presentation. The results suggest that information representation has significant impact on task efficiency for both groups of participants. However, information representation only affected the answer quality of the American group.

Keywords: Cross-culture · Visual information processing · Information presentation methods

1 Introduction

The interaction of cultural markers with technology is responsible for the differing perceptions and user experiences [1]. According to Heimgärtner, cross-cultural variables such as globalization and internationalization, can determine the layout of a system's user interface and affect its usability [2]. Systems designed for the global audience are infused with cross-cultural cues in the design process to enhance the target user experience.

User interfaces are increasingly becoming globalized and culturally sensitive at the same time, thereby advancing the development of cross-cultural applications while influencing the design of user interface. Fernandes defined the steps for designing international user interfaces. First step is globalization, which is building basic functions that can be easily modified when used in different culture. Second step is localization, which is adapting the globalized product to preferences of a culture [3].

G. Di Bucchianico (Ed.): AHFE 2019, AISC 954, pp. 229–238, 2020.
https://doi.org/10.1007/978-3-030-20444-0_22

Processing visual information is influenced by the culture of reading and writing. Notable cultural difference that might affect visual information processing exists in various forms such as languages, measurement systems, format of numbers and dates, and the use of colors. The direction of reading and writing is another important factor. For example, Americans have a culture of reading and writing from left-to-right (LTR) while the Arabs read and write from right-to-left (RTL). It is important to understand how the cultural difference affects the way Arabs and Americans process visual information.

In this study, we investigated how users from the United States and Saudi Arabia interact with three information presentation methods while completing visual information processing tasks. The three information presentation methods examined are textual presentation, graphical presentation, and tabular presentation. We also collected eye gaze data in order to examine the visual scanning patterns and strategies. Better understanding of the relationships between cultural difference and visual information processing may help researchers and practitioners design visual contents that better fit the unique needs of people from different cultures.

2 Related Research

Integration of cultural differences into the design of technology systems is essential in users experience. Marcus asserts that a degree of cultural relativism is critical in avoiding technology bias. Most of the United States websites capitalize on visitors and their objective [4]. Mauney et al. conducted a study to define touch-screen gestures' preferences among nine different countries. Three hundred and forty individuals participated to use 28 gesture actions. The actions were recorded on a touchscreen simulator. All participants showed high-level of agreement on all proposed gestures except on symbolic gestures [5].

Researchers also compared different countries and cultures when using web-based tools and websites. In a recent study, Benaida compared Arab universities with US universities using the Hofstede's cultural dimensions. He chose three university websites from each culture. The result shows lower level of individualism within the Arab culture than the American culture. However, Arabic culture is more masculine than the American culture [6]. In another study, a new analysis technique was proposed to maximize the benefit of e-government websites. Aladwani compared British and Kuwaiti citizens' views on e-government interface quality. His technique incorporated content analysis, persuasive quality gap, and user perception. The findings showed a significant difference between the two groups in perceiving the performance of quality features. No significant variation was observed between the two groups in perceiving the importance of quality attributes [7].

In another study, Marcos et al. conducted an eye tracking experiment to determine if there are cultural differences between users from the Middle East and those from Western Europe. The study suggested that Middle Eastern users spent longer time on search engine results pages compared to European users and read more results. The results are found to be consistent with holistic cognition theories, and cultural variations may stem from the fact that the Middle East participants were non-native English speakers [8]. As such they may have had reduced confidence and slower performances in engaging with the pages.

Kralisch et al. conducted a study to identify the role of different cultures on search behavior, specifically information seeking behavior. They recorded the navigation behavior of users from more than 185 countries. People within similar culture have the same search behavior and it might be due to the inheritance of thinking patterns [9].

Al-hamouri et al. reiterated that Arabic reading and writing styles differ considerably from the Western scripts. The authors evaluated the impact of written Arabic on the sequence of language-associated brain activations by scanning a group of Spanish speakers and a group of native Arab speakers. The result showed that the brain processing was mainly left lateralized in both groups; however, the left and right hemisphere of the brain reached similar activation point in Arabs but differently in Spanish participants [10].

Other scholars have given their input on the subject of information processing. According to Helmut, the processing of visual information is different among cultures because of the system of characters. The author specifies that Arabic consume extra horizontal space compared to English due to character variations. Arabic characters are written from right-to-left while the numbers are written from left-to-right; therefore, experts have been careful while translating an Arabic mobile application into English [11]. Better understanding of both American and Arabic visual information processing patterns is important from both the linguistic and the system application perspectives.

According to Chokron et al., writing and reading direction has been demonstrated as a key determinant of visual-spatial performance, such as facial affect perception, line bisection, aesthetic judgment, straight-ahead pointing, and movement perception. The findings show that visual information processing is largely based on the hemispheric interpretation of objects based on directional modes, performance asymmetries, and other factors [12].

3 Method

3.1 Experiment Design

In our study, we adopted a split-plot design that consists of both between-group and within-group variables. The between-group independent variable is the cultural background of the participants: American and Saudi Arabian users. Two within-group independent variables were examined. The first one is the information presentation method with three conditions: textual graphical, and tabular. The second within-group independent variable is the difficulty of the tasks, also with three conditions: easy, medium, and difficult. We developed three sets of webpages that presented the same amount of information in three different methods. Presented on each page was statistics about four health-related conditions (smoking, high blood pressure, cholesterol, and Diabetes) in four imaginary countries.

We collected data regarding the time it took to complete the tasks and the quality rating of the answers to measure task performance. The participants' subjective satisfaction and preference ratings were gathered through a post-test questionnaire. In addition, we used a Tobii X2 60 eye tracker to track the participants' eye movements to help understand the visual scanning paths when interacting with the different information presented methods.

3.2 Participants

Twelve participants from United States (the American group) and twelve participants from Saudi Arabia (The Saudi Arabia group) took part in the study. All participants in the Saudi Arabia group grew up in Saudi Arabia and were pursuing a bachelor or graduate level degree in the United States. All participants were undergraduate or graduate students from different majors. They were recruited through email announcements and flyers. In the American group, seven participants are female and five are male. In the Saudi Arabian group, two are female and ten are male. The age for the American participants was between 20 to 30 years old and for the Saudi participants was between 20 to 37 years old. All participants received a $20 gift card at the end of the study.

3.3 Scenario and Tasks

Throughout the study, each participant observed health-related information presented in different methods and answered questions based on the information presented. We selected health-related information to be our core data because people are constantly looking for such information on the Internet. For instance, the number of visitors of WebMD is estimated to be 75 million each month [13]. We created four imaginary countries with imaginary data including high cholesterol, smoking, high blood pressure, and diabetes and their occurrence rate among four ethnic groups: Caucasian, African American, Hispanic, and Asian. The volume of data presented for each country is the same. For each country, we developed three questions for each participant to answer based on the data presented. The level of difficulty of the questions varies based on the amount of information needed to answer the question. The easy questions required the participant to find the information about a specific ethnic group in order to retrieve the answer. For example, the participant only needed to track the data about the males in the Caucasian group in order to answer the following question:

"Among Caucasian males, which of the 4 health-related conditions has the lowest percentage of occurrence?"

For the medium level questions, the participant had to track the information about a specific gender. For example, the participant needed to track the data about all male ethnic groups in order to answer the following question:

"Among all male groups, which ethnic group and condition has the lowest percentage of occurrence?"

For the difficult questions, the participant needed to investigate all ethnic groups to find the answer. For example, the participant needed to track the data about all ethnic groups, both male and female, in order to answer the following question:

"Among all groups, which ethnic group and condition has the biggest difference between male and female?"

We used numeric ratings to measure the quality of answer with three possible values: 0, 0.5, and 1. The answers to the easy questions only contain one specific health condition, with a correct answer being rated 1 and incorrect answer being rated 0. For the medium and difficult question, the answers consist of two elements: a health condition and an ethnic group. Answers containing both correct elements would receive a rating of 1. Answers containing only one correct element would receive a rating of 0.5. Answers containing no correct elements would receive a rating of 0.

3.4 Information Representation

We designed the pages locally using asp.net and Visual Basic on Visual Studio 2017. Nine webpages were developed for four countries: A, B, C, and D. Countries A, B and C were used for the formal tasks and Country D were used for the training tasks. The pages for all countries consist of the same amount of imaginary data regarding heart disease- related statistics presented in three information presentation methods: textual, graphical, tabular (Figs. 1, 2 and 3). At the top of each page, a question was presented and the participants answered the question based on the data presented on that page.

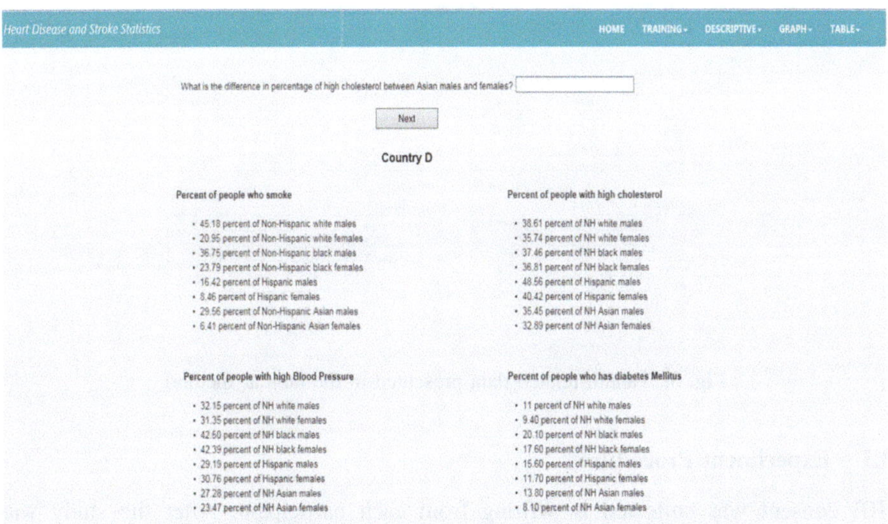

Fig. 1. Health-related data presented in the textual method

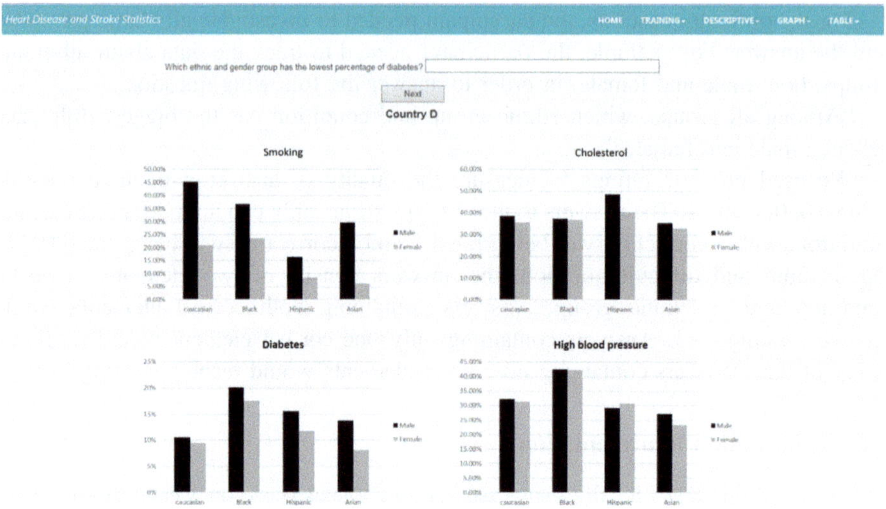

Fig. 2. Health-related data presented in the graphical method

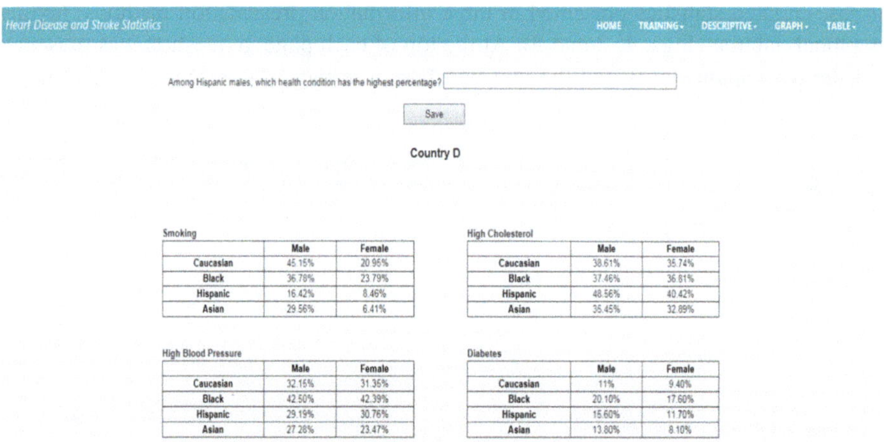

Smoking

	Male	Female
Caucasian	45.15%	20.96%
Black	36.78%	23.79%
Hispanic	16.42%	8.46%
Asian	29.56%	6.41%

High Cholesterol

	Male	Female
Caucasian	38.61%	35.74%
Black	37.46%	36.81%
Hispanic	48.56%	40.42%
Asian	35.45%	32.89%

High Blood Pressure

	Male	Female
Caucasian	32.15%	31.35%
Black	42.50%	42.39%
Hispanic	29.19%	30.76%
Asian	27.28%	23.47%

Diabetes

	Male	Female
Caucasian	11%	9.40%
Black	20.10%	17.60%
Hispanic	15.60%	11.70%
Asian	13.80%	8.10%

Fig. 3. Health-related data presented in the tabular method

3.5 Experiment Procedure

IRB consent was collected in writing from each participant. After the study was explained, participants completed a training session that exposed them to all three information presentation methods. Following the training session, each participant completed information retrieval tasks under all three information presentation conditions: textual, graphical and tabular. Under each condition, they viewed the data of a specific country and answered three questions (one easy question, one medium level difficulty, and one difficult question) about the data. Both the order of the information presentation methods and the country are counterbalanced to control the learning effect. Participants completed a demographic and satisfaction questionnaire at the end of the study. Data was collected from all participants over the course of one semester.

4 Result

The data collected was checked for completeness, cleaned and formatted for analysis using the SPSS software. We completed preliminary analysis of the data using the Repeated Measures Analysis of Variance (ANOVA) test with task completion time as the dependent variable and information presentation method, question type, and participant condition as the independent variables. The results suggest information presentation method and type of questions have significant effect on task completion time ($F(2, 44) = 46.08$, $p < 0.001$; $F(2, 44) = 5.03$, $p < 0.05$). The textual method required significantly longer time to answer the questions than the tabular method and the graphical method for both groups (Fig. 4). There is no significant interaction effect between information presentation method, types of questions and task completion time ($F(4, 88) = 3.90$, n.s.).

There is significant difference between the two groups of participants in task completion time ($F(1, 22) = 11.094$, $p < 0.005$). Participants from Saudi Arabia spent significantly longer time completing the tasks than participants growing up in the United States. As indicated in Fig. 4, participants from Saudi Arabia spent significantly longer time to complete the tasks than the American group when the information was presented in all presentation methods.

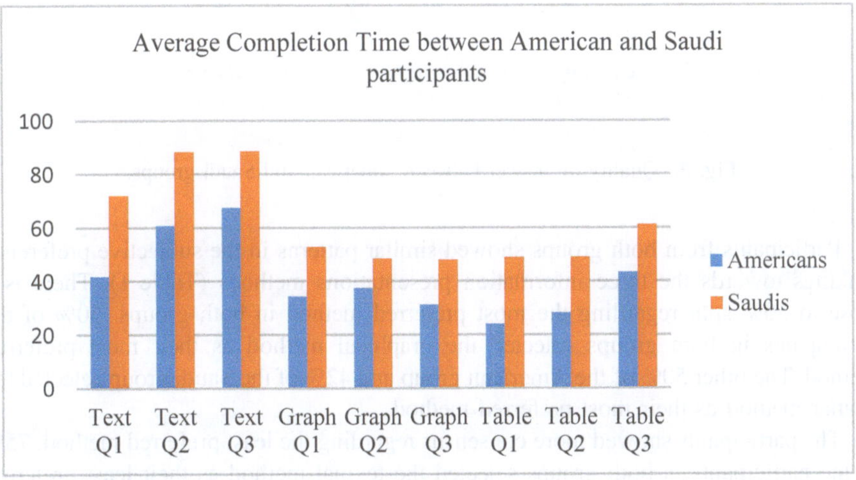

Fig. 4. Task completion time (in seconds) of both participant groups

We conducted a Repeated Measures Analysis of Variance (ANOVA) test with quality rating of the answers as the dependent variable and information presentation method, question type, and participant condition as the independent variables. There is no significant difference in quality of answers between the two groups of participants ($F(1, 22) = 1.733$, n.s.). Information presentation method and type of questions have significant effect on the quality of the answers ($F(2, 44) = 5.04$, $p < 0.05$; $F(2, 44) = 14.82$, $p < 0.001$). There is no significant interaction effect between information presentation method, types of questions and quality of answer ($F(4, 88) = 2.49$, n.s.). Participants from the United States had significantly lower-quality answers when the information was presented in text as compared to graph and table. There is no significant difference in the quality of answer among the three information presentation methods for the participants from Saudi Arabia (Fig. 5).

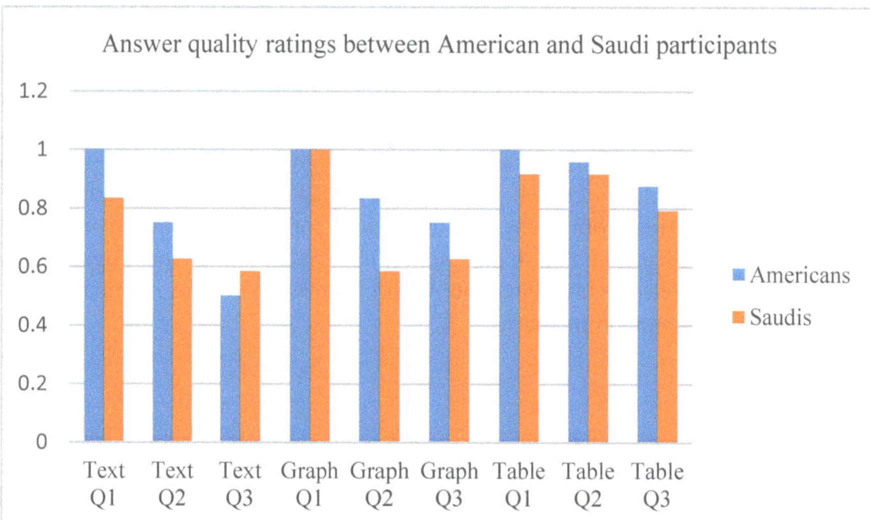

Fig. 5. Quality of answer between American and Saudi groups.

Participants from both groups showed similar patterns in the subjective preference rankings towards the three information presentations methods (Table 1). There is a close to 50% split regarding the most preferred method in both groups. 50% of the participants in both groups selected the graphical method as their most preferred method. The other 50% of the American group and 42% of the Saudi group selected the tabular method as their most preferred method.

The participants showed more consensus regarding the least preferred method, 75% of the participants in both groups selected the textual method as their least preferred method.

Table 1. Number of participants from each group who selected a specific presentation method as their most or least preferred method.

	Group	Textual	Graphical	Tabular
Most preferred	American	1	6	5
	Saudi Arabia	0	6	6
Least preferred	American	9	2	1
	Saudi Arabia	9	3	0

5 Discussion and Conclusion

We studied the influence of three information representations on visual information processing for users from two different cultures through a controlled experiment. While quality of answers between American and Saudi participants was the same, the Saudi participants spent longer time to complete the tasks. The impact of the information presentation methods on efficiency followed the same pattern for the two groups of users. That is, the textual method required significantly longer time to answer the questions than the tabular method and the graphical method for both groups. However, there was a difference in the impact of information representation on the quality of answers between the two groups. The quality ratings of the Saudi participants were similar among the three information representations while the American participants had higher ratings in the graphical and tabular formats than the textual format.

The results indicated that the Saudi participants spent longer time completing the tasks. This could be attributed to many reasons. For instance, Saudi cultures read Arabic text from right-to-left. This may result in a delay in processing the presented information from left to right. Another possible factor contributing to the lower efficiency in the Saudi group is the language barrier. Since the information was presented in English, it might have taken the Saudi participants longer time to read and understand the information. It is interesting that the information presentation methods affected the answer quality of the American participants but did not affect that of the Saudi participants. We are analyzing the eye movement data to study the visual scan patterns of the participants. We hope the visual scan patterns can help explain the differences observed in efficiency and quality of the answers between the two groups.

Our study provided preliminary understanding on how people from different cultural backgrounds interact with different information presentation methods. The results suggest that, for participants in both groups, the textual condition is the least desirable method according to both efficiency and subjective ranking. However, the results should be interpreted with caution due to the small sample size and the specific nature of the participants. All Saudi participants were students studying in a university in the U. S. and therefore, did not represent the general population of the country. We plan to replicate the study with a larger sample size in Saudi Arabia to validate the finding.

References

1. Mushtaha, A., De Troyer, O.: Cross-culture and website design: cultural movements and settled cultural variables. In: International Conference on Internationalization, Design and Global Development, pp. 69–78. Springer, Heidelberg (2009)
2. Heimgärtner, R.: Intercultural user interface design - culture-centered HCI design - cross-cultural user interface design: different terminology or different approaches? In: Lecture Notes in Computer Science (Including Subseries Lecture Notes in Artificial Intelligence and Lecture Notes in Bioinformatics), LNCS (PART 2), vol. 8013, pp. 62–71 (2013). https://doi.org/10.1007/978-3-642-39241-2-8
3. Fernandes, T.: Global interface design: a guide to designing international user interfaces (1995)
4. Marcus, A.: Cultural dimensions and global web user-interface design. Interactions **7**(4), 32–46 (2015)
5. Mauney, D., Howarth, J., Wirtanen, A., Capra, M.: Cultural similarities and differences in user-defined gestures for touchscreen user interfaces. In: CHI 2010 Extended Abstracts on Human Factors in Computing Systems, pp. 4015–4020. ACM (2010)
6. Benaida, M.: Cross-cultural web design and education: a comparison between arab universities and us universities based on hofstede cultural dimensions. IJCSNS **18**(10) (2018)
7. Aladwani, A.M.: A cross-cultural comparison of Kuwaiti and British citizens' views of e-government interface quality. Govern. Inf. Q. **30**(1), 74–86 (2013)
8. Marcos, M.C., García-Gavilanes, R.O., Bataineh, E., Pasarin, L.: Cultural differences on seeking information: an eye tracking study (2013)
9. Kralisch, A., Berendt, B.: Cultural determinants of search behaviour on websites. In: IWIPS, pp. 61–75 (2004)
10. Al-hamouri, F.A., Maestú, F., Del Río, D., Fernández, S., Campo, P., Capilla, A., García, E., González-Marqués, J., Ortiz, T.: Brain dynamics of Arabic reading: a magnetoencephalographic study. NeuroReport **16**(16), 1861–1864 (2005)
11. Helmut: Right-to-left vs left-to-right. LingoHub (2012). https://lingohub.com/blog/2012/07/right-to-left-vs-left-to-right/
12. Chokron, S., Kazandjian, S., Agostini, M.D.: Effects of reading direction on visuospatial organization: a critical review. Quod Erat Demonstrandum: From Herodotus' ethnographic journeys to cross-cultural research, pp. 107–114 (2009)
13. WebMD Facts Sheet (2019). https://www.webmd.com/corporate/press-center-fact-sheet#1. Accessed 10 Jan 2019

An Application for Mobile Systems Developed with the Community Involvement. An Open Source Mobile Application Project of a Perceptive City Map

Francesca Bozza(✉)

DIAP, Dipartimento di Architettura e Progetto, Sapienza Università di Roma, Rome, Italy
francesca.bozza@uniroma1.it

Abstract. The paper describes the hypothesis of an applicative part of the author's research *Design for the sensitive experience* (The research has allowed the author to organize interviews and discussion with different city users and psychologists permitting the definition of parameters and categories described in this paper. Anyway these parameters are questionable and have to be implemented thru the second part of the research, thru the comparison with results taken from investigation that are planned to be made also with technological advices.). It's an experimental project aimed to create the bases to develop an open source mobile application where the *city users* can directly communicate with planners, designers and administration. The 'app' would consist in an interactive city map, a model that can be extended to any cities. In particular, the paper displays the experiment conducted within the PhD research, applied to four itineraries of Rome city. The experiment can be tested on other urban realities and it illustrates the parameters identified to map the city areas and the tools that can be used. The method adopted to monitor the sustainable-mobility, is the development of an application that has to be used especially during the planning phase of the visit, before to start a city tour, not only in the while of a walk.

Keywords: Application · Open sources · Mapping · Design with community · Well-being · Emotional map · Urban contexts · Historical city · Inclusive itineraries · Perceptive · Sensitize the people · Accessibility · Life quality · Inclusive parameters · Data collection methods

1 Introduction

A space could be defined as 'inclusive' if people don't notice the inclusiveness at the first approach. To better understand the concept, people can think about a condition in which they have everything; in this case, they can't perceive the feeling of missing something. This happens with a good design: an inclusive space doesn't make you feel

G. Di Bucchianico (Ed.): AHFE 2019, AISC 954, pp. 239–250, 2020.
https://doi.org/10.1007/978-3-030-20444-0_23

excluded, and you don't reflect on the exclusiveness, because they just feel good. If a place is accessible people don't think that it could be built as inaccessible.

In the meanwhile, we have to consider that the aspiration towards a project "for all" (*Design for All*), is still a distant purpose. In the past it has represented an abstraction as much as today the idea of a design oriented towards a "standard" user, has to be considered outdated. The recipient of the Architectural Project and Design should be, therefore, an extended user, or at least extended to an enlarger number of people.

Nowadays, how can we do not include in our idea of 'community' the so-called "invisible" minorities which are together with children and elderly (which in Italy are no longer a minority of the population) - the physically impaired person with cognitive and sensory problems?

To design welcoming architecture or spaces and urban projects, the specialists have to consider during the design process the different user's needs, and the perceptions that the environment stimulates on the visitors.

To better understand people's needs, should be open new direct links between the community and the planners (productive connections). We have to consider, indeed, that depending on the age, necessities change, and that disability can be temporary or permanent. Some people with less perception of a sense might have developed another capability connected to another particular amplified sense. A collaboration with them can help designers in having different vision of the space and new intuitions.

In 2001, the International Classification of Functioning, Disability and Health developed the definition of 'accessible place' by moving the attention away from personal incapability, to address it rather to the living space: it's the environmental context that plays a fundamental role in the inclusion or exclusion of individuals, as it is characterized by elements inclined in "producing" disability, independently of the physical and perceptive conditions of people.

The ICF transferred the concept of disability from the notion of "*deminutio*" to the definition of "particularity" in the perceptive relationship between person and environment: it has elevated for the first time the concept of equality to a fundamental social value.

2 Design for Sensitive Experience

The research within the author's PhD thesis, in particular, aims to support the principles of Inclusive Design applying them to the stratified urban context, by favoring a line of "participated and shared" choices. The research goal is to point out that in the last decades, architects, planners, developers and designers have lost the real dimension of a design project and the related aspect of "responsibility" own of their job. A public design project can't be self-referential; it has to be designed for people well-being, to emphasize the environment, to be sustainable in all its parts.

An inclusive space is designed with the concept to be lived in by the community and to be capable to offer a comfortable environment for as many people as possible.

The research question is, "Does Inclusive Design haves necessarily to be an indoor space?" Architects and administrations consider it an impossible challenge. Yes, it is, but Designers, planners, architects and engineering started the carrier with the aware to have chosen a challenging job.

Henri Lefebvre in *The Right to the City*[1], described the need to reflect on the essential needs of men, and to the right of people in the city. We should consider the fact that everything passes through the single person, before to arrive to the plurality.

Everything comes from the individual.

The contemporary city is implicitly asking a dialogue with citizens. Because each of us should be able to contribute to the built prevention and care, to work together to take possession of the well-being and of the belonging feeling that once, was living in every citizen.

The Italian architect Giancarlo De Carlo wrote that in the past eras, when there were no magazines and books on architecture, and the architecture was represented only through painting, people were not eliminated from the scene, but they were often represented in the foreground, in the act of use the *architectural event* that was intended for them[2].

The real subject of all created objects and urban landscape is the man, and De Carlo perceived the idea that architecture was not only made up of buildings, but of people and buildings together, related in a relationship of a mutual needed.

The emotional factors arising from a context are determinant in the behaviors, judgments and choices of people. The environment and the architecture influence mind and people mood. They are therefore key elements in the development of cities and in the conservation of our historical heritage.

Citizens with physical difficulties or perceptive sensory diversity are often excluded from city life and from the enjoyment of common heritage; the research proposes to pursue a fundamental and unexceptionable principle of the Universal Declaration of Human Rights of 1948. "*Participation in the cultural life and scientific progress by the community is a right of every individual who must be put in condition to take part freely, to enjoy the arts and to participate in progress and benefit of the heritage properties*"[3].

These are the premises that lead the research. Administrations and designers should have the chance to communicate with the *city users*, citizens and tourist, but not only. The research aims to propose one of the many feasible ideas to start a collaboration helping each other, sharing information, analysis and not only proposals.

---

[1] Henri Lefebvre in *The Right to the City. Philosophy of the City and Planning Ideology.*

[2] Richards J.M., De Carlo G., Blake P., L'architettura degli anni Settanta, Il Saggiatore, Milano 1973, p. 91.

[3] Art. 27 (Universal Declaration of Human Rights 1948).

3 The Feasible Role of a Digital Applications in the Society

A digital application is a tool that can hold different utilities and goals.

In this paper and within the research the final users of the application are supposed to be the citizens, the administrations, the organizations, the tourists and the architects with the planners. The author proposes the development of a concrete tool that is the result of the analysis and studies carried out on the perception of citizens, children, users with sensory, motor or cognitive difficulties in urban areas. The tool is an application for mobile systems that allows the use of an interactive map of some historic center itineraries that pass thru areas of cultural interest.

The tool for mobile systems holds the potential to offer support also to people with disability difficulties in knowing the degree of accessibility and the satisfaction of other users on the perceptual quality and use of a context, before choosing it as a destination of cultural interest. Even tourists, as well as citizens, could enjoy this instrument because often they, if they present sensory or motor difficulties, feel hindered in planning a trip to the Capital.

The research is currently ensuring the development of a plan where the areas of high historical interest and their degree of inclusion àre indicated, to obtain a map of users' perceptions and emotions[4].

The digital application should allow the use of an interactive map of few itineraries which classifies, area by area, the accessible levels as well as perceptions and emotions of people in the interaction between the user and the city.

The idea is to develop a tool that hold the potentiality of an open source system, to implement and to update the information (Cf. C. Ratti, Architettura Open Source). In this way it could be used by the Municipality for the information acquisition to be presented and consulted during the town council, to define priorities, to make decisions related to the street routines maintenance, exceptional maintenances, and renovation of urban furnishing.

The application shouldn't contain only objective data, but also people emotions. People can insert a feedback on a place creating a new point of feedback. The new point of feedback can be related to an emotion.

4 The Hypothesis of an Application as a Partial Result of a PhD Research with the Community Involvement

The application could also be an interesting tool for the partial publication of research results and could represent a first step towards the open source extension of the graphic drawings for the involvement of users in the mapping of the city. The author has received for two years a grants from Sapienza, Università di Roma, (2017–18 and 2018–19)[5], to develop an application for mobile systems to promote a concrete tool that

[4] G. Bruno, Atlas of Emotions, Verso Books, Milano, 2007.

[5] Research Grant del. DR n. 1052/2017 Prot n. 27698 7th April 2017 and n.278/18 del18/09/2018 within the EU Research Horizon 2020.

permit the users to live inclusively the city. The project goal is to open the application to the users' acquirements, so that the accessibility status can be implemented and constantly monitored and updated. The research is based on the information and knowledge acquired by the author thanks to the research work carried out in synergy with the Municipality I ° in the years 2016–2018 in relation to the study of accessibility of the historic center of Rome and the needs of citizens in the field of inclusion.

The tool applied to the mobile systems holds the potentiality to offer a support to all the interested people, tourist, citizens, more than everyone people with perceptive and emotional difficulties because they can be aware of the accessibility level of a path before to arrive in front of the "obstacles".

The 'app' gives the possibility to know in advance which road to take depending on the kind of tour that the person needs to do. The app can assist the user in the walk related to parameters like the pavement, the presence of natures, benches, shadows, cycle lane, urban furnishment and so on.

A practical example: The app support in the tour planning; if a user knows before starting the visit, if to take the right or the left of a sidewalk, he is not obliged anymore to cross the street various times to reach the ramp to go up the sidewalks. Even tourists with special needs, as well as citizens, can enjoy this instrument overcoming the fear of planning a visit to historical centers like Rome. At the same time the application permits to the person to publish comments and information detected, to insert an emotion in the atlas. Four colors help to distinguish the level of satisfaction of the four categories:

Physical factors, sensitive factors, well-being factors and in the end, emotional factors.

4.1 The Application's Users and the Mapping Methodology

The questionnaires surveys filled by some citizens in the frame of the researched have pointed out that they were not informed about some accessibility project that have been occurring for at least three years. The application permits an immediate update by the citizens who can be an active part of the process and not only the final users. They can easily find information about road accessibility status, suggestions about paths and itineraries and also, they can report in real time any urgent discomfort, breakdown, or also, to share nice surprised, why not.

Consulting the digital application, the urban planners can immediately define the SWOT analysis[6], have an overview of weakness and strengthens of the urban area, continually updated. It helps having an overview of the conditions, to be able to detect in reduced time possible interventions. The urban planner will be able to conduct less survey on site, knowing in advance people needs.

The map of the application, defined by different colors that characterized the area, will permit at the first approach to pinpoint the places where to go to have a relax, to avoid if you're with a person with physical impairments, where to stay with children

[6] SWOT is the acronym standing for Strengths, Weaknesses, Opportunities and Threats. It's an analysis method used mostly used in Urban Planning.

where to go if you're by bike. Institutions and Public administrations can easily find out the areas where people don't like to go or where they don't feel in comfort. The places more appreciated can be valorized and it's possible to define new connections that can link the isolated places that have received positive feedbacks.

The main users of the application are:

1. Public Administration
2. Citizen
3. Planners

4.2 How to Map an Urban Context?

This paper aims to share few Mapping advices in the urban context that consider the multisensory experience and people wellbeing the aim of the analysis. The following points are the first results of the research started in 2014[7].

The research investigates only 4 itineraries, and it's an experimentation of a tool that can be applied to a larger scale.

Note: the methodology applied and all the categories and parameters defined by this paper are the result of a study within a thesis research.

The advice is to define itineraries of max. 3 km (Miles) lengths each. The itineraries start from point A and ends to point B.

The research aims to give a contribution to people everyday life, so the starting points have to be characterized by centrality as markets, stations, parks, and public services. The itineraries give the chance to people to pass by the most beautiful and livable parts of Rome city and its archaeological paths and squares.

The first investigation is conducted every 100 m (328 ft), and then the Open Source system can be implemented by user's feedbacks.

To each point the mapping can give one or more categories information (see the Table 1 content).

Table 1. Categories of investigations

Category	Sense/perception
Category 1	Physical description
Category 2	Sensory description
Category 3	Perceptive/Cognitive description
Category 4	Emotional feedback

[7] Between 2014 and 2015 the author has attended the elective course "Inclusive Design" at the KU LEUVEN University in Belgium and she joined the research group Research [x] Design with whom has collaborate for one year (2014–15) at the ASRO Department of Architecture.
(KU LEUVEN Tutor Prof. Ann Heylighen):

– "Rent-a-Spatialist" (ERC Proof-of-Concept)
– "Mobilizing Disability experience to inform design practice" (IOF leverage Project).

Table 2. The relation between the category and the response.

	RESPONSES			
Category 1	Ideal entrance Welcoming	Accessible	Not enough accessible	Not accessible at all
Category 2	Positively Stimulated	Sometimes/or just one or few sense have been considered	Not Considered	Problem/ discomfort felt
Category 3	Understandable e inclusive	Wayfinding Sometimes Present	No Wayfinding	Not understandable
Category 4	Positively surprised	Regular situation	Not Satisfied	Feeling of Discomfort

To each category the digital application will give a written description and a color to have an immediate response on the map.

The interval investigated are ascribed by an opinion codified by a color:

1. Red
2. Orange
3. Yellow
4. Green

To define the colors in the attribution of the places on the map, supports the reading of the plan.

4.3 Mapping

Where to start to investigate the senses? The research has individuated the first part factors to consider during the mapping related to the senses stimulations. The list can be always implemented. The factors are the following, collected for each sense:

Table 3. The five senses stimuli

THE VIEW
Colors
Light/Shadows
Shapes
Degradation
Disorganized/tidy
Clean /Dirty
Brightness
Harmony in the elements juxtaposition/contrasts
Presence of green area/Nature/trees/woods

THE SENSE OF SMELL & Breath
Clean/Dusty
Pollution
Garbage bins in the open air
Factories nearby
Stations nearby
Split conditioning
Rivers, see, marsh nearby
Gastronomic activities
Presence of green area/Nature/trees/woods

THE SENSE OF TOUCH
Stimulation by porosity/shape of materials
Dust or shiny
Presence of Object designed for play/touch/ seat
Presence of shrub evergreen/smelly shrub
Presence of any natural elements to interact with

THE SENSE OF HEARING_SOUNDSCAPES
Noises
Music
Children's laughs
Presence of factories
Presence of stations/trains (elevated railway)
Level of traffic
Restaurant bar with open doors/tables outdoor
Airport nearby
Presence of natural filters (Parks- Tall trees)

4.3.1 Well Being and Emotions, Two Connected Factors

The study conducted thru the analysis of case study, the interviews, and the study of the physical environmental technology disclose the importance of people's the wellbeing associate at the accessibility of the spaces.

The following table shows how the well-being is connected to two factors, physical and psychological (Table 4).

Table 4. Beyond the five senses

Well being	
Phisical factors	Psycholocigal factors
Cold/warm perception	Socially welcoming
Humidity percentage	Presence of fences
Air conditioning flux/directions	Prohibitions
Brightness-luminance phenomena	Wayfinding[a]
Emotional factors - Individual memories - Historical collective memories - Social relations - Relax/Stress - Comfort/discomfort (conf. PANAS Scale parameters)	

[a]K. Lynch, The Image of the City, The Technology Press Harvard University Press, Cambridge, 1960.

4.3.2 Accessibility Physical Factors to Consider During the Mapping Phase

Referring to the four kind of responses (cf. Table 2) it's possible to define a first set of parameters for the classification of the spatial accessibility:

Table 5. First set of Parameters

PHYSICAL BARRIERS_URGENT PRIORITY
No ramp to reach second/first a stretch of the sidewalk
No sidewalk
Cobblestones pavements not flat
Sidewalk less large than 0,90 m (3ft) or without possibility to turn with a wheelchair
Sidewalk large but occupied by restaurant/coffee bar with tables and chairs
Sidewalk large but occupied by market boxes lot
Large roots and big holes on the sidewalk
Pedestrian crossing
Road slope more than 8%
Steps not reported
Barriers on vertical space (branch of the trees…)
No alarm alert in correspondence of sliding doors, traffic lights, obstacles…
No LOGES paths

PHYSICAL BARRIERS_MEDIUM PRIORITY
Sidewalk of a length between. 2 ft-3 ft.
Roots and small holes on the sidewalk
Very few pedestrian crossing
Pavement in gravel/ clay court (in a dry place). Asterisco per quando piove. Se il giorno prima ha piovuto, non andare.
Garbage bin at 0,90 meters height
Road slope not more than 8% (people with wheelchair and not strengthens in the arms have to be pushed by someone)
No tactile maps

PHYSICAL BARRIERS_URGENT PRIORITY
Sidewalk of 1,m
Garbage bin at ground floor level/until 2 ft.
Road slope not more than 5%
Pavement in concrete /asphalt/small crushed gravel better Ramp for each sidewalk stretch

PHYSICAL BARRIERS _MORE THAN SUFFICIENT
Sidewalk < 1,m
Road slope between the flat and the 4%
Presence of Pedestrian crossing each Regular pavements No hills- flat road …

The green defines the status of the art that goes beyond the common idea of "accessibility".

5 Conclusions

As the architect Roberto Secchi observed, at present we underestimate the effects that individualism has produced on the generation of new social conflicts. The loss of feelings of citizenship and sharing of shared values and rules for a free coexistence in the mutual recognition of all rights.

The space represents what has been before, what it is today, and holds the potential of what can become in the future.

Architects and planners must be able to plan with the awareness of the past, but listening to the needs of the inhabitants of the city to work together on the future, involving whom, will really live this future[8] (Cf. Gregotti and Wright 1945).

For this reason, the research aims to communicate to the architects that nowadays more than ever, they hold a social position. They are the bearer of a key role in the link between politics and citizens: a support for the integration and the union among the inhabitants, aid to plan regeneration in different contexts. We have to be aware of the fact that all the people constitute a community, made up of a multitude of different identities, with individual needs and stories, starting from children's involvements.

Social networks are not always the best way to communicate. Wrong messages can pass thru this virtual world, or good messages can also be misunderstood or surrounded by advertising and an infinite multitude of media messages, and become just one of the many. Nowadays we can not do without technologies, aids, media and social networks; what we can do is to use them in the best way to amplify their good effect and to find social roles.

At the same time the research is founding out the difficulties and the contradictions to develop tools able to measure accessibility factors, as well as the sensitive and the emotional ones. At least it's interesting starting from a point and propose valid methods. The paper doesn't pretend to find solutions, but the staring point of a work held in a complex urban context, that once collected to the other contemporary researches, will be able one day to certificate the public spaces and to guarantee an enlarged inclusiveness and an increased community well-being.

References

1. Buti, L.B.: Ergonomia olistica. Il progetto per la variabilità umana. di Serie di architettura FrancoAngeli, vol. 51, Milano (2008)
2. Bruno, G.: Atlas of Emotions. Verso Books, Milano (2007)
3. De Carlo, G.: An Architecture of Participation. Royal Australian Institute of Architects (1972)
4. Gregotti, V.: Come le città italiane potranno cambiare. In: Purini, F., Marzot, N., Petranzan, M., Sacchi, L. (eds.) a cura di la città nuova-y-2026, Compositori, Bologna (2006)
5. Heylighen, A., Nijs, G.: Studying (architecture) in dialogue with disability. reflections on the public role of the University. In: Simons, M., Decuypere, M., Vlieghe, J., Masschelein, J. (eds.) Curating the European University. Leuven University Press, Leuven (2011)
6. Lynch, K.: The Image of the City. The Technology Press Harvard University Press, Cambridge (1960)
7. Romano, A.: Giancarlo De Carlo. Lo spazio, realtà del vivere insieme. Testo e Immagine, Torino (2001)
8. Ratti, C.: Architettura Open Source: Verso una progettazione aperta. Giulio Einaudi Editore, Torino (2014)
9. Richards, J.M., De Carlo, G., Blake, P.: L'architettura degli anni Settanta, Il Saggiatore, Milano (1973)

[8] Cf. Francesco Tonucci, CNR, speech within the Conference Spazi Innovativi Inclusivi Partecipati organized by the Department of Educational Sciences of Università di Bologna, 23th Februarty 2019.

10. Watson, D., Clark, L.A., Tellegen, A.: Development and validation of brief measures of positive and negative affect: the PANAS scales. J. Pers. Soc. Psychol. **54**, 1063 (1988)
11. Wright, F.L.: Architettura e Democrazia, Maggioni, Milano (1945)
12. Wright, F.L.: La città vivente, Einaudi, Torino (1966)
13. Henri Lefebvre in the right to the city. In: Philosophy of the City and Planning Ideology

Inclusive Design Practices for Natural Parks. Products and Services for Experience-Focused Solutions in Places of High Naturalistic Value

Ivo Caruso(✉), Vincenzo Cristallo, and Carlo Martino

Department of Planning, Design and Technology of Architecture (PDTA), Sapienza University of Rome, via Flaminia 72, 00196 Rome, Italy
ivo.caruso@uniroma1.it

Abstract. The research investigates the possibilities deriving from inclusive design methods in contexts of high naturalistic value. Natural parks, reserves and public gardens play a strategic role from an ecological, social and cultural point of view. Therefore it is necessary to place them at the center of a new scientific debate on methods for the study, the development and the governance of these assets. They are particular and multi-functional contexts in which today it is necessary to operate both in terms of environmental and landscape protection (ecodesign), and in relation to the evolved needs of their users (user-centered design). After a deep analysis concerning international case studies, the research proposes a new possible model of inclusiveness for design practices oriented to territories enhancement. A model capable of acting both in the metaproject phases (introducing interdisciplinarity and co-design practices) and in the properly design phases (according to the principles of design-for-all).

Keywords: Design for territories · Natural parks · Inclusive design · Integrated services · Eco-social development

1 Natural Parks as Inclusive Places. Design Models and Methods for Territories Enhancement

Starting from the evidence that the parks are protected areas characterized by high naturalistic and landscape values, the so-called "design for the territory" - in its vocation to promote initiatives for local development and for territorial protection - is the preliminary instrument to operate in these places through actions and methods

The research "Inclusive design practices for natural parks. Product and services for experience-focused solution in places of high naturalistic value" derives from the scientific activities of the II level Specialization Course "Natural capital and protected areas, planning, design and management", promoted by the Department of Planning, Design and Technologies of Architecture (PDTA) – Sapienza University of Rome. The research is coordinated by Professor Carlo Martino and Professor Vincenzo Cristallo – with the scientific support of PhD Ivo Caruso. The general aim of the research is to analyse the system of physical and intangible tools which implement, give value and new functions to naturalistic areas, through specific "integrated services", able to create connections between products and communication design.

G. Di Bucchianico (Ed.): AHFE 2019, AISC 954, pp. 251–262, 2020.
https://doi.org/10.1007/978-3-030-20444-0_24

which promote models of widespread and inclusive planning. These models are targeted to the use and enjoyment of environmental assets by combining the ecological and social priorities with the services provided though products [1]. For this reason, the enhancement of a choral way to project by an extensive system of stakeholders (local administrations and institutions, public or private organizations, companies, researchers and professionals, private citizens) is an indispensable condition for every project with a territorial vocation. It means a project in which the territory is concretely and rhetorically co-authored and in which it is required a network of different skills. Transversal and complex skills which can concretely give value to the specific typicalities of a territory heritage [2]. This is why to design starting from the territorial point of view means recognizing and preserving the identity of a "local culture" that can generate "local products" through its "local resources", of any species. And, "local product" is everything that has a close relationship with the territory and with the community that generated it, even casually. The protection of a territorial vocation therefore becomes the starting point of any planning hypothesis which provides, depending on the case, the relocation, revaluation and reactivation of material and immaterial resources of a given eco-system. However, these are resources (skills and qualities) that must be recognized and shared. Design actions for the territory are aimed to: "read" the territory heritage, interpretate it, visualize it, share meanings and visions, promote forms of participatory planning, transform visions into feasible initiatives, design specific products and the interfaces of the services deriving from them, promote an effective communication of the whole processes. That is to say that design and its system can direct actions towards an environmental sustainability that has a strong relational and social component on a territorial scale. As a consequence, sharing becomes a form of co-definition of the project plan through cooperation actions between proponents and recipients. It is therefore important to recognize that when the goal is to introduce innovative processes on the territory - fostering the connections between places and people, the continuous learning, the enhancement of the material elements, knowledge, culture and local traditions - it is necessary to adopt an approach that supports a community-based form of project; so able to act simultaneously on strategic and services levels. Levels that, in order to generate appropriate and long-lasting actions, must consider the theme of the project of the identity of places as an indispensable value that requires constant monitoring and planning to act in the direction of development and protection of natural heritage. However, the concept of identity is often misunderstood by a pre-constituted representation of its value, considered above all as belonging, defense, opposition. Nonetheless, the objective to be pursued - also understanding the difficulty of many to adhere to a more complex and variable meaning of the term identity - is to ensure that designing identity is equivalent to proposing an updated vision of the word "localism". Consequentially it is fundamental to contemplate territory's multiform nature, its plurality of meanings referable to cultural, normative, economic and political aspects. These aspects need a circular and extended design attitude, focusing on the experiences of the last years that have pushed on themes such as the access to resources, the protection and the use of goods, the environmental enhancement and the sustainable development. Finally, it is necessary to define an idea of a localism that has been updated in the contents by examining the value of typical resources according to a definition ascribable to the sequence of relationship between product/context/identity. Such a linking indicates that what we sometimes identify generically as territorial-environmental resources are assets to which we can recognize such complexity that can fully distinguish the area of origin and its main identifying properties. Moreover, the economic studies referring to the productions that belong to the typical resources consider the territory as a diffused company - or the environment as a systemic enterprise - and recognize in it economic models, often informal, where the originality and the quality of local products, including those of an environmental nature, is given by the particular combination of territorial resources and the typicality of the goods resulting from a specific "Productive Habitat" [3]. This "Habitat", considerable as a place composed of parts in reciprocal stability (balance of relationship and position), results in a peculiar balance between material and immaterial elements. These elements represent the structural, organizational and dimensional typicality of a given context. These arguments are even more evident if the products in this habitat can be considered as identity marks. In this case we can state that the contribution of the design can activate a process of configuration of the product/context/identity relationship acting with bottom-up strategies, with appropriate enhancement actions for the different levels of territorial identity. For this type of interpretative paradigm the design disciplines, starting from the analysis of the existing typicalities, have the task to take advantage from these initiatives of recognition and networking, turning them into competitive factors for a territorial emancipation on economic and social bases [4]. In conclusion, identity, understood as a cultural matrix of places and territories, has a crucial value for local development. It is a resource to be recognized, protected and communicated. By joining the theme of identity and the design action it is possible to increase the relationship between development and enjoyment of environmental assets, and it is possible to enhance the dynamics within scenarios of inclusive and sustainable use of local resources. Design culture, with its aim of manipulating the shape of the cultures and knowledge present in the territory, becomes a real engine of development for goods and services and helps to recognize the leading role that cultural-environmental assets have in the development of local systems. Therefore, developing the territory through the design principles means modifying the reference center of any planning activity from the subjects to the contexts. Through the use of different disciplinary levels and on different fields of action (social, economic, cultural), the action of design on a territorial scale must therefore offer solutions to increase social, economic and technological innovation processes, taking into account the specificity of local resources. This result can be pursued above all by recognizing that the "project for the territory" always plays in a general key a "narrative" role as it strives for a sort of "staging" the territory. Creating a "script" is equivalent to making a subject fit to be represented. It means to be able to portray the complexity and originality of the different components-actors that are present in the scene-territory with different roles.

2 Natural Parks as "Spaces for the Experience". Practices and Artifacts for Inclusive and Sustainable Fruition

In reference to the complex global challenges facing humanity today, nature and man turn out to be fatally interdependent. With regard to this, in the era in which is taking place what Jeremy Rifkin defines "Third Industrial Revolution", we are facing an important change of scientific paradigm, which considers nature no longer as a sum of objects, but as a sum of relations. "Old science is characterized by the values of detachment, expropriation, dissection and reduction; the new one from responsibility, integration and holism. Old science is focused on making productive nature; the new one on making human activities sustainable. Old science seeks power over nature; the new one aim to a partnership with nature. (…) The new science makes us go from a colonialist vision of nature as an enemy to despoil and force to slavery to a perception of nature as a community to feed" [5]. To have the possibility to live in a salubrious place, or to have (at least occasionally) the possibility to stay in direct contact with places characterized by a great naturalistic value, increases the quality of a person's life, and increases the quality of people's health (both from physical and psicological point of view). Moreover it has a positive impact on social skills, local economies and citizens' behavior towards the environment. Making nature simultaneously protected and usable therefore put in connection people and the environment, thus making people more aware and empathetic towards ecological issues. This condition is fundamental for the definition of the territory as a "choral work of art produced through the dialogue between living entities, man himself and nature, in the long time of history. That is, as a work that arises from fertilization by culture" [6]. The challenge (concerning planning, management and cultural aspects) can not therefore be limited to the ergonomic requirements of accessible tourism, but it must aim at creating the real conditions of sustainable coexistence between the emergencies of nature and the needs of the widest range of possible users. In this way it is possible to experience a model of inclusiveness that becomes a fundamental factor for the creation of places that themselves become "ecological tools through socialization". According to John Thackara one of the reasons why the so-called "new economy" is failing is the lack of interest in the contexts. In the book "In the Bubble", Thackara defines in the concept of "territorial patrimony"; a set of natural beauties, cultural heritages, unique specificities of the territory, social cohesion, civic sense, memories [7]. All the actions oriented to "localization" so contribute to the redefinition of social structures and relationships between resources, activities, nature and communities. Natural parks, as we said, are aimed to maintain the environmental balance of a specific place, increasing or protecting its biodiversity, and at the same time providing services (scientific, educational, touristic, recreational) to ordinary or occasional users. "Accessibility is a function of the interaction between the user, with his skills and demands, and the environment, consisting of natural resources and the organization of equipment, personnel and tools aimed at the enjoyment of the natural environment" [8]. The model of the "sanctuary" precluded to anthropogenic contaminations has to be substituted by places where ecology and ecotourism can coexist. These are places able to provide visitors with high-quality experiences. In this sense, the park can be a place in which natural and artificial factors interact; and its fruition is managed by a "complex and articulated system of elements, interconnected or interdependent, which allows the user to tend to the targets of autonomy and self-sufficiency" [9]. The design of these specific products and services is not limited to solve problems of accessibility, interaction and usability, but includes the study of needs, expectations and behaviors of different kind of people. It is important to take into consideration the fact that the theme of diversity has been affected in recent years by a substantial re-reading. "Individuals traditionally considered disabled (of which only a part - less than 50% - are mobility impaired) constitute a significant and growing percentage of the whole population, if we consider, in addition to permanent disabled people, temporary ones, those who are not considered part of the statistics as they are not registered as disabled, those whose handicap is a natural condition (pregnant women, parents with young children, children themselves, elderly people)" [10]. This "integrated landscape" can be composed by: supports for communication and wayfinding (vertical signs, physical information panels, nowadays "augmentable" with virtual contents); furniture and equipments (seats, covers, lighting equipment, delimiting and monitoring systems, instruments for waste management, playgrounds, tools for sports or leisure); equipped areas for rest breaks and reception (so capable of foreseeing and put in connection all the functions necessary for the various activities related to welcome services for visitors, including information, programming, security, lounge or refreshment areas, logistical support); intermodal mobility solutions; permanent or temporary micro-architectures that can be more or less capable of mimesis in the context (observation towers, paths, shelters, exhibition places, research centers or structures hosting initiatives for enhancement of the territory).

Micro-architectures for naturalistic areas are particularly interesting design practices. Although sometimes presenting important sizes, they are conceived in an intermediate conceptual dimension between the sectors of building and set-up design and, often, they experiment, with a particular attention to sustainability and to integration with natural elements, constructive practices based on the standardization of the components, on the use of innovative materials, on prefabrication, on modularity, on lightness, on reversibility, on spatial adaptability. These techniques and instances are usually adopted in the sectors of exhibit and industrial design. Moreover, these constructions are real examples in which a responsible design approach can adopt the guidelines and the rules of inclusiveness as an opportunity to optimize costs (avoiding future adaptations) and to achieve a widespread approval and use deriving from the elimination of obstacles and possible conditions of discomfort, danger and exclusion. Since there are often large areas available, many of these structures are characterized by the use of suspended ramps and walkways that solve the accessibility problems of traditional towers. Long "stripes" (treetop walks) can thus elevate the visitor to allow him to enjoy magnificent bird's eye views that give the sensation of being in direct contact with nature, being able to almost caress the trees, being able to observe the landscape from a privileged and immersive position. One of the first examples of inclusive design for natural parks structures is the Clingmans Dome Observation Tower, designed by Bebb & Olson (Hubert Bebb) in 1959 and built at the highest point of the Great Smoky Mountains National Park, on the border between Tennessee and North Carolina (USA). The designer refused the exclusive use of local materials and the "rustic style" (adhering to a certain traditional imaginary) and outlined a modernist-inspired reinforced concrete walkway that leans on cylindrical columns and leads, above the treetops, to a "circular covered terrace" from which the visitor can enjoy a 360° view. The Shark Valley Tower (1964) in the Everglades National Park, the Look Rock Tower (1965) in the Great Smoky Mountains National Park and, on a smaller scale, the Wooden Birdwatching Observatory designed by the LJB studio in 2017 in Herdla (Norway) have been designed on the same composition scheme (Fig. 1).

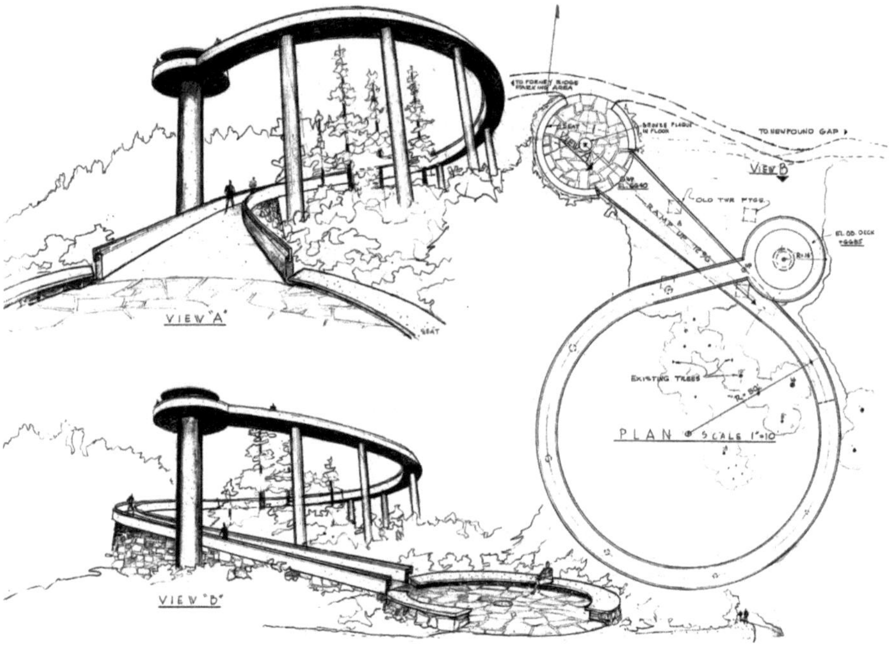

Fig. 1. Clingmans Dome Observation Tower, original drawings by Hubert Bebb, Bebb & Olson, 1959 Great Smoky Mountains National Park (USA). This pioneering project can be considered a fundamental archetype of modern observation tower strongly oriented to accessibility and standardisation of processes.

The project "A Path in the Forest" designed in 2011 by the Japanese studio Tetsuo Kondo Architects and temporarily set up in Kadriorg Park during the LIFT11 (Urban Installations festival in Tallinn - Estonia), pursues the idea of creating the possibility of a new perceptual relationship between people and the secular forest through a suspended path capable to provide alternative points of view. The project consists of a simple structure, 95 m long, and light enough to be able to anchor itself in a symbiotic way to the trees themselves, thus renouncing the need for additional support pillars. In 2012 the Tree Top Walk was inaugurated in the Bavarian Forest National Park, in Neuschonau (Germany). A wooden spiral ramp of 500 m allows visitors (even those with motor disabilities) to reach a height from the ground of 44 m. A similar case study is the Camp Adventure Park - The Treetop Experience, designed in 2017 by EFFEKT - ARUP for the Glisselfeld Kloster Forest, in Haslev (Denmark). An iconic hyperbolic tower made of corten steel can be reached by a system of inclusive paths designed to define diversified functional segments. The user, walking through the foliage of the trees, explores bridges, spirals, aviaries, observation balconies, ramps and common seats. The Wild Walk, designed by Charles P. Reay in 2015 for Adirondack Park (Upstate New York - USA), was also conceived on the model of the "treetop walk". A large system of suspended paths that foresee various activities and create a context of strong visual impact that, at the same time, aspires to provide recreational experiences, but also to encourage people of all ages to explore and understand the surrounding natural environment through exhibitions and scientific events. In 2017 the Latvia Nature Conservation Agency launched an international design competition for an observation tower for the Kemeri National Park (Latvia). All the selected and mentioned projects (the results have been announced in June 2018) propose solutions strongly oriented to inclusivity. This fact confirms that today designers can become the spokespersons of a shared awareness according to which the contemporary natural park can be conceived as a "space for experience"; a place to experiment, to acquire information, to make discoveries, to share activities (and therefore socialize), to rediscover the pleasure of the contact with nature, stimulating virtuous behaviours (Fig. 2).

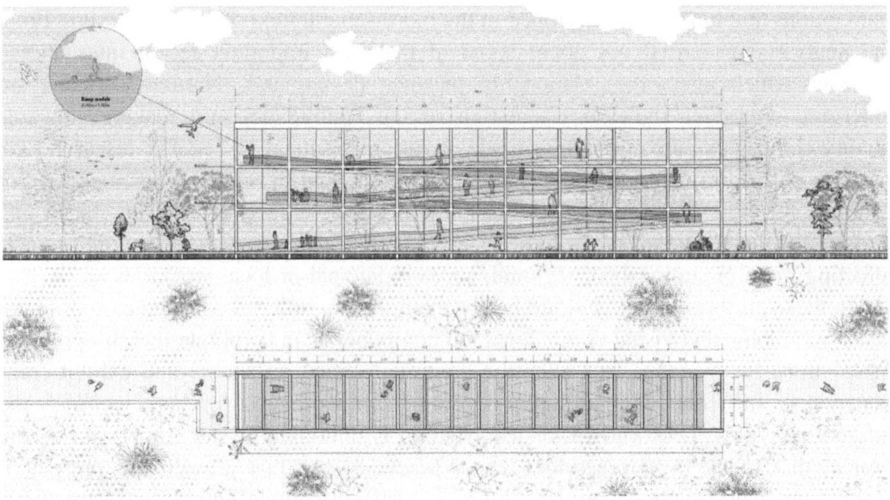

Fig. 2. Funambulist Landscape, Ernesto Urquízar Quesada, María Cervantes Lardón, 1st prize project of the International Architecture Competition Kemeri National Park Observation Tower, 2017. The observatory has been conceived as a sensorial and inclusive promenade, a multiple viewpoint to establish a continuous relationship between nature and visitors. The components of the structure are designed to be easily prefabricated offsite and to be assembled without the use of heavy equipments. It is important to note that all the awarded projects consider as key factors wheelchair accessibility, durability, ease of construction, and budget optimization.

The critical objective consists in conceiving the park no longer as a "sanctuary", a "container of wonders", or even worse as a "themed lunapark" only dedicated to recreational activities. The "park for experience" must instead be a values-amplifier, a determining instrument for the narration of the territory, for the strengthening of the local identity and of the environmental awareness of its users.

3 Natural Parks and Inclusive Services, Design Practices in Communication and Tools for Wayfinding

In the last 10 years, the international system of park authorities and the network of the numerous stakeholders involved in the management of the "Natural Capital"[1] of our planet, has shown to understand the relevance of design in the strategic project of protection and enhancement of the main heritage of humanity, demonstrating, in comparison to the past, a real "paradigm shift".

As the Italian Ministry for the Environment clarifies: "The concept of 'Natural Capital' has been instrumentally borrowed from the economic sector to indicate the value in physical, monetary and welfare terms offered by biodiversity to the human race, also in order to guide the choices of public decision-makers"[2], and it is perhaps precisely this new induced awareness that led this change. As how it happened for urban design practices [11], the "integrated tools" for parks fruition derive from innovations that have been developed in the more general field of outdoor design. From the analysis we carried out about some of the most important recent international projects, which affect the relationship between design and "Natural Capital", are emerging important signs of orientation of the "public decision makers" towards quality choices that are possible to made at various political scales. The research took into consideration good practices affecting a national level, as demonstrated for example by the systemic and coordinated projects developed in Norway, Japan and Croatia for their national parks networks; but also highly innovative specific projects affecting more circumscribed, regional, supra-municipal or local territories.

In these projects, design culture has been applied in order to contribute on different tasks: in graphic and visual communication components, in corporate-identity projects, in solutions for signages and scientific communications, in spatial and exhibit components, in the conceiving different kinds of infrastructures for both external and internal activities. Less known, or less visible, is undoubtedly the component of the project that today invests service design practices, but that are already present in embryonic ways in those projects both of visual identities and of infrastructures, designed in close synergy with the services they provide. The identity projects, for example, are today often integrated with specific digital Apps, so with tools for

[1] The term 'natural capital' was first used in 1973 by E.F. Schumacher in his book Small Is Beautiful and is closely identified with Herman Daly, Robert Costanza, the Biosphere 2 project, and the Natural Capitalism economic model of Paul Hawken, Amory Lovins, and Hunter Lovins. Recently, it begun to be used by politicians, notably Ralph Nader, Paul Martin Jr., and agencies of the UK government, including its Natural Capital Committee and the London Health Observatory.

[2] http://www.minambiente.it/pagina/capitale-naturale-e-servizi-ecosistemici.

exploring the park which provide scientific information and guidance services, as well as the design of infrastructures - towers, birdwatching cabins, etc. - it is developed with full awareness of the "performance" [12] that they can provide to their users, in terms of accessibility and hybridization with digital services.

Since the highest international institutions for environment protection conceive the Natural Capital as the set of "ecosystems which provide essential goods and services" to humanity, and since the same concept of Capital, borrowed from the economic sciences, encourages users and decision-makers to be well-aware of the value and wellbeing that derives from the natural heritage, it results that the design sciences are called today, more than before, to develop increasingly integrated solutions able to connect systems of products and services.

As previously said, this need strongly emerges from the analysis of some of the most successful projects of branding and scientific communication for national parks and protected areas. These case studies have in common that they are fully conceived as systems able to integrate the visual narration with users experiences, starting through the careful study of the peculiarities of each park or system of parks, and admitting the possibility of offering both traditional and innovative services.

3.1 Visual Identities

"Every organization is unique: its identity can only derive from its roots, its personality, its strengths and its weaknesses" [13].

As stated above, if it is possible to conceive of nature as a capital, the theme of identity design for parks must be able to make an association between company branding and place branding, and then imagine the institution of management, protection and enhancement of parks just like an organization that, to express its identity, has to be based on its specific peculiarities. The design of park identity becomes therefore more effective when it succeeds in carrying out this transposition, i.e. translating characteristics and peculiarities of supranational, national and/or local natural heritage into "visual stories" and integrating them into an overall project that offers unparalleled services and experiences.

The identity-experiential storytelling is in fact at the base of the 2015 project of Snøetta Design Studio for the coordinated identity of 44 Norway's National Parks. The logo is characterized by a squared portal, resting on the variable silhouette of a hill. It represents the visual translation of the story of a passing experience: "A portal is an entrance, or a gate, which symbolizes the transition between two dimensions; the traversing between the cultivated and the natural. The new, unifying visual identity opens a gate to these new experiences, it lowers the barriers for visiting, and facilitates the increased knowledge of our precious surroundings"[3].

Japan faced the problem of defining a national identity of the park system in 2017. The project have been developed by Yamaguchi Moeko of the Nippon Design Center. The designer cited the identity theme of the Rising Sun, the secular "logo" well-imprinted in the global collective consciousness, actualizing it through chromatic gradients that move

[3] https://www.domusweb.it/en/news/2015/04/17/norway_s_national_park_s_sn_hetta.html.

away from the full red of the national flag sun. The logo in this case symbolizes the sunrise experience, also sustained by an animation clip, with the shape of the Rising Sun in a typical foggy landscape of Japan: "The logo represents a unique Japanese landscape, the misty moment the sun rises"[4]. These two experiences demonstrate how the identity of a complex system of parks can integrate national themes and symbols with dynamic and experiential elements, able to reach a great emotional impact (Fig. 3).

Fig. 3. Norway's National Park, visual identity, graphic design, signage and wayfinding systems, Snøhetta Design Studio, Miljødirektoratet (The Norwegian Environment Agency), 2015. Photographer Øivind Haug.

3.2 Wayfinding

Wayfinding is an integral part of the communication project. If it is designed according to the logic of consistency and iteration of the identity codes, it plays a fundamental reinforcing role of the identity system and it has a significant functional role in terms of accessibility and security. In fact, the orientation and the "signaling" of predefined paths assumes a central function in a naturalistic context that offers great possibilities for the fruition of the different ecosystems (landscapes, views, etc.) so making users safe from potential risks. In the natural parks context, signaling has more than elsewhere the primary task defined by Arthur and Passini as "Problem Spatial Solving" [14], meaning the ability of the wayfinding to favor the achievement of the goal

4 https://www.ndc.co.jp/en/works/national-park-vi_201711/.

through the resolution of a spatial problem. The most recent wayfinding projects for parks, that we analyzed in the research, have assumed all the functional constraints defined by Massimo Vignelli[5] in his manuals, such as the "Hierarchy", the "Repetition" and the "Sequence" [15], surpassing them also in favor of an extended audience. For example, the recent wayfinding project for the Parc Riu Llobregat by the Spanish studio Claese Bcn elegantly included tactile maps in the whole signage system. Fitzpatrick Woolmer in 2012 designed the signages for the Brecon Beacons National Park, in Wales. In this case the designer used the surfaces of the wooden benches as communication surfaces, so engraving tactile maps on the pieces of furniture that have been installed along the path.

3.3 Interpretive Panels

It is common practice in the functional project of parks and green areas of particular value, to provide some significant points of the paths with graphical panels carrying informations on the ecosystems to be observed, with the emphasis on the faunal and floral species present on the site. The interpretive panels are fully part of the coordinated communication project of the park, often adopting the same codes and identifying characters. These panels are particular examples of infographics. They descent from an analogical world that wanted to bring scientific information, managed with a manual approach and with a simple and informative language, to the public of visitors, often imagined as it was overall composed by children. In some of the most recent communication projects for the parks, the interpretive panels have been conceived as hybrid interfaces - analog/digital - able to increase inclusiveness of communications. So from the simple integration of tags in the graphics of the panels that refer to the online world, we witness the transformation of the same into three-dimensional tactile maps also including short braille texts, so usable by an extended range of users. Important examples are the ones designed by Atlantex Creative Works for the Canadian Fundy National Park, or the ones installed in the botanical garden of Porto Caleri in Italy.

4 Conclusions

This ongoing research demonstrates the need of a paradigm shift in redefining the "eco-social" role of parks that, if redesigned through scientific and interdisciplinary actions, can become places for promotion of health, culture of respect for nature, socialization and valorization of the territories. Natural parks can contribute to the contrast of the fragmentation of the territories and can act as a mitigation instruments between urban, periurban and rural areas. Tools of diffused ecology, economy and education, markers of sustainability and wellbeing. Design sciences play a strategic role in redefining priorities, anticipating transformations, giving identity and new meanings to places, artefacts and services, involving different actors in research-action paths oriented to

[5] Massimo Vignelli, graphic designer, in 1977 he realized, in collaboration with his wife Lella, the Ungrid project; one of the first integrated projects of corporate identity, wayfinding and communication for parks. Client: Natural Parks Service (USA).

sustainability. Design do not only has to propose "constructive practices", but "strategies of resemantization and refunctionalization" to gain true models of inclusiveness and social development. With this in mind, the design of new solutions for parks has the task of updating and redefining the rules of coexistence between nature, people and artefacts and therefore the possible degree of anthropization of nature; and planning new models of use that include new users, new priorities, new sensibilities and new awareness.

References

1. Cristallo, V., De Simone, C., Cappucci, S., Donati, S.: Medonia. Il design per la salvaguardia della Posidonia oceanica. In: Bisson, M. (ed.) MDA 2017. Mediterranean Design Association, II International Conference on Environmental Design. MDA - Mediterranean Design Association, pp. 229–240. De Lettera Publisher, Milano (2017)
2. Parente, M.: Il design per la valorizzazione territoriale. Il caso del Rione Sanità a Napoli. Tafter J. **22**, 1–6 (2010). http://www.tafterjournal.it/2010/04/01/il-design-per-la-valorizzazione-territoriale-il-caso-del-rione-sanita-a-napoli/
3. Cristallo, V.: Habitat produttivo. In: Villari, B; Castelli, A. (eds.) STAR, Sistema Topologico Argomentativo della ricerca, Costruire l'organizzazione della conoscenza, Il caso Me. Design. Edizioni Poli.Design, Milano (2003)
4. Cristallo, V.: Argomenti e opportunità per lo sviluppo locale sostenuto dal design. In: Lotti, G. (ed.) Interdisciplinary design. Progetto e relazioni tra saperi, pp. 244–259. DIDA, Dipartimento di Architettura Università di Firenze, Firenze (2016)
5. Rifkin, J.: The Third Industrial Revolution: How Lateral Power is Transforming Energy, The Economy, and the World, p. 257. Arnoldo Mondadori Editore, Milano (2011)
6. Magnaghi, A.: Il progetto locale: verso la coscienza di luogo, p. 17. Bollati Boringhieri, Torino (2013)
7. Thackara, J.: In the Bubble: Designing in a Complex World. MIT Press, Cambridge (2006)
8. Monzeglio, E.: La natura accessibile. In: Astrusa, F., Rustichelli, R., Zampicini, F. (eds.) Barriere architettoniche. Un progetto per l'uomo, pp. 136–158. BE-MA, Milano (1991)
9. Vescovo, F.: Obiettivo:progettare un ambiente urbano accessibile per un'utenza ampliata. In: Paesaggio urbano n.1. Maggioli Editore. Rimini (2002)
10. Matteucci, E. (ed.): Il verde per tutti. In: Progettare con il verde. Mariella Zoppi & Co., no. 6, p. 13. Alinea Editrice, Firenze (2000)
11. Martino, C.: Il design per la città. Da segno di accentuazione qualificativa a strategia sociale. In: DIID (Disegno Industriale Industrial Design), no. 59, pp. 90–95. RDesignpress, Roma (2015)
12. Pacenti, E.: Design dei servizi. In: Bertola, P., Manzini, E.: Design Multiverso. Appunti di fenomenologia del design. Edizioni Polidesign, Milano (2004)
13. Polano, S., Tassinari, P.: Sussidiario: grafica e caratteri moderni. Mondadori Electa Spa, Milano (2010)
14. Arthur, P., Passini, R.: Wayfinding: People, Signs, and Architecture. McGraw-Hill Publisher, New York (1992)
15. Vignelli, M.: Segnaletica e Contesto. In: XY Dimensioni del Disegno, nos. 35/36/37, January/December. Edizioni Officina, Roma (1999)

User Interaction and Scenario-Based Experience Design for New Media Technology in Museum Spatial Experience

Jian Yu[1(✉)] and Yongbin Wang[2]

[1] Guangzhou Academy of Fine Arts China, Guangzhou, China
619077255@qq.com
[2] Guangdong University of Finance and Economics China, Guangzhou, China

Abstract. This article aims to study the interaction and scenario-based experience design of new media technology in the museum, improve people's sensibility to museum-theme space information, and facilitate propaganda and education. It mainly: (1) summarizes the expression of new media human-computer interaction techniques in the museum; (2) discusses the relationship between museum-them space shaping based on new media technology and humans' visiting experience; (3) summarizes the experience in human and museum space storyline with new media intervention. The article focuses on theoretical analysis and practical significance study from the 3 aspects. Meanwhile, it outlines the influence humans perceive in museum spatial experience.

Keywords: Museum space experience · New media technology · Interactive experience design · Scene experience design

1 Introduction

A museum is an interactive place that takes space as its medium for the purpose of disseminating information. Museums continuously grow as the humans' material and cultural lives become increasingly rich. It satisfies the needs of human information exchange in multiple fields, such as information, knowledge, technology, education, culture, as well as trade and business, while constantly driving forward human development and progress [1, 3]. The development of the exhibition space in museums is dependent on multiple elements, including space architecture, media, and exhibits.

The so-called "new media" refers to a form of technologically advanced medium that emerged relatively late and presents new forms of interactivity and connectivity in information transmission. Born out of the computer and network technologies, the new media breaks the molds of old-fashioned presentation and language of the traditional museum exhibition. Moreover, storytelling techniques in museums are no longer perpetually rigid, and in terms of the spatial expression of a museum, new media act as a kind of emerging medium of narrative expression that creates new spatial plots meaningfully and purposefully, instead of simply using those known or unknown technologies to construct the medium of space, thus influencing the "old" media and people's way of thinking, their behaviors, and their perception of the exhibits in an

G. Di Bucchianico (Ed.): AHFE 2019, AISC 954, pp. 263–270, 2020.
https://doi.org/10.1007/978-3-030-20444-0_25

exhibition space, giving the exhibits a new form of impression expressible in its exhibition space.

The so-called spatial experience is a visual expression-based forming principle of arranging each element of space with media materials as the elements, and the principles of mechanics and modern technology as the support, in accordance with certain aesthetic rules to achieve a beautiful visual experience.

Museum exhibition design is interwoven with space, and museum design can even be understood as an artistic activity of space planning. Then, rational spatial experience and spatial interaction are the keys to determine whether the concept of museum design and the essence of design can be expressed in depth [2, 3]. The spaces of museum exhibition can be categorized into "tangible" and "intangible" spaces by enclosure, as well as physical space and virtual space by form. Hence, the arrangement of space that features the new-media-technology-driven interactive experience among spatial components in museum exhibitions should also be completed in accordance with the form, size, and enclosure of different spaces (see Figs. 1, 2).

Fig. 1. Interactive new media installation of South Korea pavilion at expo 2014 milan, Italy

Fig. 2. Immersive space interactive experience of new media

2 Method

2.1 Forms of Expression of the New Media Human-Computer Interaction Technique in Museums

Space is a "container" of exhibits, and its nature is determined by the theme and characteristics of the exhibits in the museum. Museum space utilizes architectural entity and new media in space to convey information presented according to the exhibition theme, through which the process of human-computer interaction experience is formed.

The use of the new media-based human-computer interaction techniques is an important tool and a necessary medium for information dissemination, and the human-computer interaction involving the new media in museums is a way of using integrated media, such as texts, images, videos, audios, dramas, interactive demonstrations, commentaries, and various physical models [4–6]. Due to the use of new media, exhibited information can exist in a new state and be perceived by the audience in a new form. The new media gives traditional museum spaces the option of digital presentation and incorporates the concept of human-computer interaction, giving rise to unidirectional, bi-directional, and even multi-directional information dissemination (see Fig. 3).

Fig. 3. Interactive multimedia installation at Shanghai museum of glass, China

2.2 New-Media-Technology-Based Exploration of the Relationship Between Museum Space Narration and Visiting Experience

Visiting and experiencing museum spaces in conjunction with the use of new media are activities that are proactive and participative in nature, while museum space is a space of information exchange where the audience input and output information at the same time in the experiential process.

The use of the new media further prompts the audience to integrate intuitional experience in the process of visiting the museum. Information reception in a museum, comes from not only the visual experience, but also from the stimulation of any other sensory terminals throughout the whole body. Only by making the audience feel the information in person can they be more able to perceive the information content, and hence complete the visit and experience process via their senses [7, 8]. At first, vision is the most important channel to collect information from that provides the most intense and clear feeling of the information, and what "seeing is believing" stresses is exactly visualization; second, sound draws the attention of people most easily, and auditory experience often can make space more mysterious; next, tactile sense is the most real, and one feels this world all the time, ranging from air motion and temperature variation to material texture and object shape, so in a manner of speaking it is the first instinct for us to identify surroundings; lastly, normally, olfaction and gustation are rarely used in medium experience and are mainly for the functional needs of daily life, but once used properly, they will greatly arouse the memory nerve of the viewer. When all senses are selectively used in the narration of a museum space, it is highly likely to arouse plot experience featuring life memory and emotional resonance with the people [9, 10]. Therefore, in a narration design of a museum space, the purpose of exhibition design is to utilize new media technology to convey information and receive feedback, and how to utilize new media technology to constitute better content to intervene in the overall visiting route and achieve good visual effects is the focus of the studies conducted by designers (see Figs. 4, 5).

Fig. 4. Human-computer interaction device in Germany pavilion at expo 2014 milan, Italy

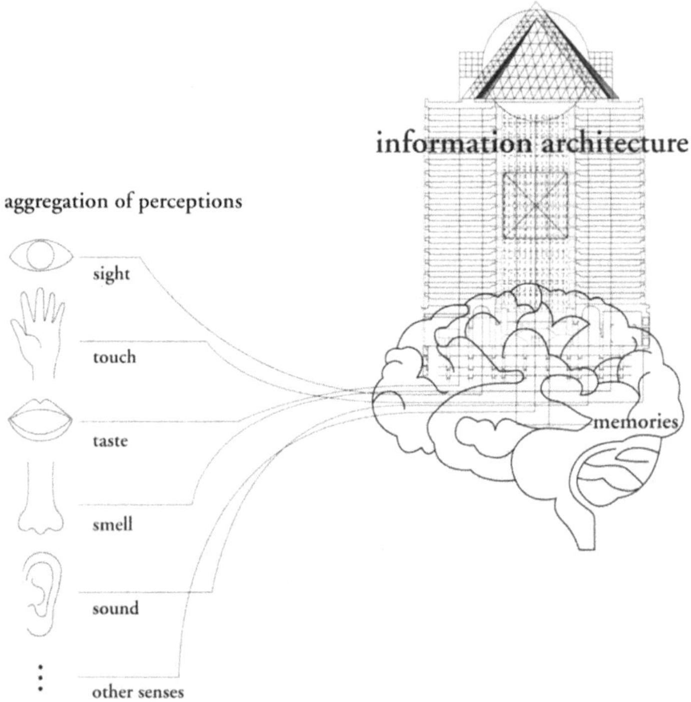

Fig. 5. The five sense organs of the human body, this is from Hara Kenya: Design in design, Shandong people press, 72 (2006)

2.3 People and Museum Space Plot Experience as Intervened by the New Media

The arousing side of plot experience often comes from thematic storytelling, personification of stories, plotting of figures, and orientation of the plot towards everyday life. Thematic storytelling is more likely to generate suspense and moving content, which is why it is best at evoking emotions of the audience. The new media convergence is manifested by the fact that traditional narration with relatively closed and complete narrative text will be gradually replaced by a loose network structure. The latter can develop a plot by expanding the scope of narrative possibility, and offers help for the expression of museum information with people's action or their interaction with others. Personification of stories makes it more vivid, because figures are the foundation to further discuss experience, conflict, and suspense. Correspondingly, plotting of figures can bring people more surprising details [11, 12].

To study the plot experience in a museum under the influence of the new media technology, we need to start from two aspects - humans' physiological dimension and psychological dimension. The experiential transition from physiological to psychological is a progressive change directed by the change in human emotion, as emotional involvement is the result or effect of watching and the result of the dialogue between the audience and the object in the museum space. Humans have a variety of emotions, such as surprise, joy, anger, excitement, irritation, satisfaction, depression, and many other indescribable emotional characteristics [13, 14]. Thus, all kinds of emotions surface in the process of information exchange with the objects in a museum by using auditory sense, visual sense, tactile sense, and even through the gustatory and olfactory organs. Mobilizing these five senses requires the human's active participation, from a supporting role to a leading role, from a spectator to a controller, but this high involvement and sense of control are hard to achieve in story plot experience of a traditional museum (see Fig. 6).

Fig. 6. Multimedia interactive experience device for Germany pavilion at expo milano, Italy

3 Results

In a museum exhibition space, new media-based exhibits and installations have been applied into the thematic expression at a higher and higher proportion, from 15%–30%, to even as high as 80%. Highly-sophisticated digital display methods have completely expanded and overturned the narrative language of traditional exhibition space, new media's display props, and installation arts, such as touch screens, projection systems, interactive systems, holographic images, and 4D cinemas, adding a new language to the exhibition space narration with increasingly diverse design approaches, influencing the speed and breadth of information reception by the audience. New-media-technology-driven information dissemination mode tends to be a way to connect those who share the same opinion and idea [15–17]. In other words, in a sense or to some degree, its forms of expression can transfer some spatial information in someone's idea to the perception of another to make the information interaction possible. In effect, it is just like the rhetorical device called "synaesthesia" in literature, which uses feelings that are originally perceptible only by organ A, to express the feelings perceivable by organ B, for instance, the ability to "see coldness", "smell color", "hear flavor", and the like. Such a figurative language is able to transfer human senses, link up, and interweave different feelings, and hence makes the information in need to be conveyed more vivid, intriguing, and perceptible. This is also an important form of the new media in museum space design.

People change from a spectator to a controller in museum space narration, and this change goes through a long historical period. From technological development, change of concept, the elevated power of speech of the audience, to the power delegation by the exhibition designer, all reflect the importance and subjectivity of the audience, namely the people in space. The audience no longer watches from afar, quietly accept, and passively approve, as they can control the theme of the exhibits or the development of the spatial plot, and choose when to quit and when to change in real time [18].

Museum space experience does not exist alone in a spatial plot. Hence, new media convergence needs to allow the audience to acquire more real feelings through perceptual experience integration and hence immersively and comprehensively feel the content of information. In the process of practicing new media convergence, on the one hand, we can make use of the characteristic of the object to lead people to think in a direction; on the other hand, we do not need to deliberately describe the characteristics of this object, rather, we should tempt the audience into putting together the form of information with abstract experience. Supplement and integration of different perceptual experience allow people to fully dig out the meaning of museum space information, while the omission of some space plot experience can make people throw away some of their long-held concepts and ideas, and re-explore the feeling that this information brings to people (see Fig. 7).

Fig. 7. Multimedia holographic installation at the museum of AC milan, Italy

Fig. 8. Multimedia human-computer interaction device at Japan pavilion at expo milano, Italy

4 Conclusion

The enhancement, improvement, or emergence of every technology endow art with a whole new mode of expression. Therefore, artists and designers are all incessantly seeking and exploring all possible technical media and materials, and trying to get new answers to change and innovate new thinking, new experience, or new artistic possibility. Hence, for museum design, innovation and revolution of science and technology undoubtedly inject critical nutrients in the continuous improvement of museum space design, making it grow and become full-fledged (see Fig. 8).

Acknowledgements. This paper is funded by the youth innovative talents program (humanities and social science) of guangdong provincial department of education. The project title is "theory and practice research on visual design elements". The project number is 2016WQNCX079.

References

1. Smith, T.F., Waterman, M.S.: Identification of common molecular subsequences. J. Mol. Biol. **147**, 195–197 (1981)
2. May, P., Ehrlich, H.C., Steinke, T.: ZIB structure prediction pipeline: composing a complex biological workflow through web services. In: Nagel, W.E., Walter, W.V., Lehner, W. (eds.) Euro-Par 2006. LNCS, vol. 4128, pp. 1148–1158. Springer, Heidelberg (2006)
3. Foster, I., Kesselman, C.: The Grid: Blueprint for a New Computing Infrastructure. Morgan Kaufmann, San Francisco (1999)
4. Czajkowski, K., Fitzgerald, S., Foster, I., Kesselman, C.: Grid information services for distributed resource sharing. In: 10th IEEE International Symposium on High Performance Distributed Computing, pp. 181–184. IEEE Press, New York (2001)
5. Foster, I., Kesselman, C., Nick, J., Tuecke, S.: The physiology of the grid: an open grid services architecture for distributed systems integration. Technical report, Global Grid Forum (2002)
6. National Center for Biotechnology Information. http://www.ncbi.nlm.nih.gov

Inclusive Participation Design Methodologies for Digital Cultural Heritage

Giuseppe Mincolelli(✉) and Michele Marchi

TekneHub – Dipartimento di Architettura, Università degli Studi di Ferrara, via Saragat 13, 44124 Ferrara, Italy
giuseppe.mincolelli@unife.it

Abstract. The European cultural and historical heritage contains a surprising richness and variety that places it, thanks to its multi-diversity and heterogeneity, among the most interesting territories at international level. As a result of the scarcity of accessible technical equipment [1] and lack of time or availability to physically visit each Cultural Heritages, the knowledge of the European cultural heritage is not as widespread as it deserves. For this reason, a highly skilled work team with heterogeneous skills and approaches, has been created. Through the INCEPTION project, this team is developing inclusive methodologies with the aim to increase the dissemination and the knowledge of the historical cultural heritage through a web platform. The paper describes how it has been possible to model and apply some of the principles of the Human Centered Design in an international design research project in order to promote the accessibility of the digitized cultural heritage, through the involvement of institutional subjects. Moreover, the authors illustrate how co-design represents a valid method of work planning and a methodological design tool that allows a productive cooperation between different demanding profiles in the field of Digital Cultural Heritage.

Keywords: Digital Cultural Heritage · Accessibility · Human Centered Design

1 Introduction

The INCEPTION project has been applied under the Work Programme "Europe in a changing world - inclusive, innovative and reflective Societies" (Call - Reflective Societies: Cultural Heritage and European Identities, Reflective-7-2014, Advanced 3D modeling for accessing and understanding European cultural assets).

This research project has received funding from the European Commission's H2020 Framework Program for Research and Innovation under Grant agreement no 665220.

The main purpose of INCEPTION is to make understand and spread the knowledge of the European cultural heritage [2]. Thanks also to the multidisciplinary, heterogeneity and skills of the project partners, it is planned to outline appropriate strategies for the dissemination of information regarding cultural heritage to a wide and diversified audience, also trying to find solutions that are oriented towards an accessible and inclusive knowledge of it. The project therefore aims to define innovative and inclusive

G. Di Bucchianico (Ed.): AHFE 2019, AISC 954, pp. 271–281, 2020.
https://doi.org/10.1007/978-3-030-20444-0_26

methodologies for the development of 3D [3] models of Cultural Heritage, providing information that can increase the perceived quality and interdisciplinary knowledge of cultural identity. The main objectives of the project, which are already partially developed, are related to the deployment of innovative strategies for the acquisition and implementation, through 3D models of buildings and historical contexts, of integrated data as complete as possible on the built cultural heritage, and to the proposal of specific methodologies for the semantic structuring [4] of data for Cultural Heritage, all organized through a standard open web platform for the dissemination and sharing of the harvested information.

The project, thanks also to the geographical heterogeneity of the Partners, aims to apply to nine case studies with different locations and natures, different digital acquisition tools and then, for each case, to make data, 3D models and other informations like pictures, documents, available and easily accessible for the end user.

This paper, edited by the team that is responsible for the design of the modes and tools to access and interact with the platform, focuses on the application of the Human Centered approach to the design of digital instruments and interface, like the web site. The platform is intended to be accessible for a number of user as big as possible, and to make its content available through a collection of spontaneous crowd-sourced contributions, but it also requires the availability of more specific tools, intended for experienced users, for the upload, management, validation and maintenance of the Platform.

The INCEPTION project is under development by a consortium of fourteen partners from ten European countries led by the Department of Architecture of the University of Ferrara. Academic partners of the Consortium, in addition to the Department of Architecture of the University of Ferrara, includes the University of Ljubljana (Slovenia), the National Technical University of Athens (Greece), the Cyprus University of Technology (Cyprus), the University of Zagreb (Croatia), the research centers Consorzio Futuro in Ricerca (Italy) and Cartif (Spain). The clustering of small medium enterprises includes: DEMO Consultants BV (The Netherlands), 3L Architects (Germany), Nemoris (Italy), RDF (Bulgaria), 13BIS Consulting (France), Z + F (Germany), Vision and Business Consultants (Greece).

2 Methodology

The design research shown here starts with the understanding of the needs of the different categories of users and specialized stakeholders involved, to get to the co-design of the interfaces and instruments for the navigation of the platform.

This methodological step was important to try to incorporate all the needs and expectations of the people who will use the final web/mobile platform [5].

To obtain this, two meetings of participatory work were held between the research team, the stakeholders and the users to plan together the best strategies for the conception of the structure, the organization of the contents and the interfaces of the INCEPTION web platform, through some co-design project tools, such as User Experience Map, brainstorming, card sorting and others.

The INCEPTION platform is already structured through a general system architecture. Therefore, a first prototype of the platform has been designed and developed which responds both to the first needs of users and stakeholders (previously involved in participatory days and according to the literature) and of technological and system invariants and constraints determined by software and hardware developers [6].

The first prototype was submitted to a first test through a co-design workshop [7] (held in Ferrara on 11th June 2018). This was the first event in which some strategies were developed together with stakeholders with the aim to improve the accessibility to and of the platform. The goal was not only to receive users' comments and suggestions on the current state of the INCEPTION Platform development, but above all, to have the opportunity and privilege to participate with them in the project of contents and interfaces through experimentation and re-elaboration of the opportunities offered by the platform [2].

The aim of this activity, named co-design [8], was to understand all together – designers, users and stakeholders – if the structure of the first prototype had favored the understanding of the platform and if it had managed to give concrete and tangible solutions to the needs developed during the carried out needs assessment [9]. Therefore, the activity has represented a day of participatory work which has confirmed and improved the work done up to now, integrating it with different or more inclusive investigative tools [10] (Fig. 1).

Fig. 1. View of the first co-design workshop.

The final goal that the team had set to get at the end of the first day of participatory work was not so much to deepen the requests expressed by users and stakeholders, but rather to design with them – involving and intriguing them – in order to plan a new set-up of the platform, in such a way that it also met the expectations that have not been reached yet (Fig. 2).

	Contents	User experience	Interface
Tourist and specialized users	Hyperlinks to external platforms	Storytellers to relate the real site and the virtual model	Accessibility and inclusion even to people with special needs or disabilities
	Update the models with related data and information	Possibility to print virtual models into 3D physical models	Intuitive and hierarchical interface
	Examine the model in different historic periods	Knowledge about building by showing historical events	Information available even on smartphone and by AR
Governments and decision makers	Information about events and exhibitions and info about the physical place	Bridge the gap between real and virtual experiences	Do not waste time to find data during researches
	Download models for related activities	Make users updated on last modifications of their visited models	Provide users the % of browsed elements of a visited model
	Reliable data for scientific researches	Provide augmented reality during a visit	Give users few and clear information
Technicians and professionals	Split the elements of a building in order to visualize different historic phases	Keep in memory information already visited Real-time work sharing	Private area for not accessible models and filter research bar
	Store data of previous interventions and documents available even on mobile devices	Download 3D models with different scales of details	Timeline of interventions and last updating
	Interoperability among several external databases	See missing parts of the buildings	Comprehension of different parts that compose the building

Fig. 2. Summary of the results gathered during the first co-design workshop [7]

Once the first workshop that helped the project team to develop new strategies for the accessibility and inclusiveness of the contents of the web platform was completed, planning a further meeting with the participation of all stakeholders – with different skills and approaches – was necessary [11]. The purpose of the following meeting was to discuss and reflect mainly on the tangible characters (color, font, graphic interface, contents, navigation strategy) of the web platform.

3 Second Co-design Workshop

The second workshop took place on 26 November 2018 at Art'otel in Cologne (Germany).

At the event there were about 15 stakeholders and 5 moderators to better speak to users, for a total of about 20 people involved.

The final goal was to involve all the people in a participatory work day, trying to focus the discussion on what was the most relevant, complete and accessible graphic interface for them. The purpose was not banal and obvious, since the INCEPTION platform so far designed has two main ways to access it: one is to upload (manager, institutions who has in charge the cultural sites, etc.), the other is accessing for visiting places (general users, visitors) and browse available documentation. Institutions can upload 3D models and information to increase knowledge and number of visitors. Access can be done on site or in remote.

Therefore the available material is of multiple nature and interface: navigation of BM models, photos, videos, general information with historical description and location of each Cultural Heritage.

Currently, the platform has 9 case studies:

- National Archive - Museum of Hydra in Greece (inserted in the category Museum)
- Technical-Museum-Nikola-Tesla in Zagreb (inserted in the category Museum)
- The Church of the Assumption of the Blessed Virgin Mary, situated in Mirlovic Zagora (inserted in the category categoria Historical sites)
- Istituto degli Innocenti in Firenze (inserted in the category Historical sites)
- The Saint Nicholas Chapel, also known as the church of Obergum in Groningen province in the north of the Netherlands (inserted in the category Historical sites)
- Villa Klonaridi is located in Patisia District of the Municipality of Athens (inserted in the category Heritage buildings)
- Church-Panayia-Phorviotissa in Cyprus (inserted in the category Heritage buildings)
- The Castillo de Torrelobatón (Torrelobatón Castle) is situated in the province of Valladolid in Castile, Spain (inserted in the category Heritage buildings)
- The Akropolis of Erimokastro, Rhodes, Greece

To date, users are also divided into three categories: tourists and visitors (general information); scientist or scholars (going in depth and more detailed level of information); professionals and technicians (specific login, can upload technical info checking condition assessment etc.) (Fig. 3).

Fig. 3. View of the second co-design workshop.

The working day was structured in two distinct phases. The objective of the first phase was to improve the interface and the accessibility to the platform. In a Human Centered Design approach, the people were involved in a playful and creative activity in order to allow to the different users to customize, as best they consider, the graphical interface of the INCEPTION platform navigation tools. Specifically, the discussion focused on the following "screenshots" of the platform:

- Home page (card 1)
- Geographic search (card 2)
- Most recent Models/Most complete models (card 3)
- Categories (card 4)
- Models (card 5)

In the five cards are presented almost all the tools of the platform. The specific goal of the workshop was not only to understand which color was more suitable for the background or which font was more readable, but to understand which communication strategy made the platform more intuitive and captivating for tourists or visitors, more complete, scientific and reliable for technicians. Within a playful discussion, the aim is to understand what are the latent needs; to be able to recognize what are the needs not yet expressed by the people; to understand the needs and expectations that the people have towards a web platform.

In order to do this, the card sorting [13] method was used. It was chosen because it is particularly suitable to facilitate the processes of prefiguration and shared construction through a visual and tactile experience of possible interaction tools [14]. This tool allows to represent and organize in a coherent way the functional and interactive structure of a complex web platform, focusing it on the needs of the final users.

With the aim to facilitate the communication, participants were divided, according to their knowledge and skills [12], into two different groups. In this phase, stakeholders with diversified skills were appropriately chosen. In this way the range of feedback illustrates a broad target, with different experiences, skills and needs.

Each card has been printed in large size for the two tables in order to allow a correct readability even if not very close to the main table. Each paper card depicts the current state of the platform's graphical interface (with some strengths and weaknesses).

The workshop is useful to highlight the tools and functions that a visitor might need for a first interaction with the INCEPTION platform. To do this, some cards were processed and printed, representing every single operation that the user can perform inside the Platform. On cards, there are alternative set up of different info. Groups choose alternatives, and give suggestions by scratching, rearranging the contents, explaining opinions (with post-it on images or other comments directly on paper).

Furthermore, in this phase the contribution of the moderator was crucial. The moderator, leaving free the discussion, without setting limits, obstacles or empty moments, has addressed it towards sensitive issues for the final design.

This session was very helpful to collect inputs and suggestions on Platform interface, directly testing by final users contents more suitable to their different needs.

Therefore, it was possible to elaborate a database of new and latent needs and strategic choices on how to illustrate or position some contents rather than others in the webpage.

Specifically, for example, we discussed on what was the faster and more effective navigation mode for the choice of buildings. Currently there are three different ways:

- search button
- through the drop-down menu Categories
- Geographic search

Regarding the latter navigation mode, two options (screenshot) of geographic search are presented. Below are some suggestions extrapolated during the workshop, which comment this phase of work.

- Option 1 is desirable. On the left grey column it should be helpful to have a list of different kind of information (e.g. Pop up with historic info; climate);
- In Option 2 the ring in which the picture is included is too strong;
- Dynamic pointer (hand) should address to info on building (Fig. 4).

Fig. 4. In the picture above are shown some of the different cards prepared for the event; specifically, those regarding the geographical search of buildings in the INCEPTION platform.

Once the work divided into two groups was completed, a plenary meeting was organized, where all the stakeholders examined the work carried out on the other table. This last phase of work led to the drafting of a shared final document, in which some strategies or choices were identified in order to improve accessibility to the web platform.

The second part of the workshop wanted to investigate the current and future progress of the INCEPTION platform, with reference to the observations resulted during the participatory events organized in June 2018.

A questionnaire has been developed in order to find out whether the adopted solutions were considered accessible, shared and inclusive for all stakeholders.

This document delivered to each stakeholder, was divided into two parts:

Part 1 - The first part of the questionnaire was focused on the solutions adopted to give answers to the observations which came out during the previous workshops. These observations have involved aspects that will necessarily be shown within the web platform. The document has been graphically divided into two parts. The adopted graphic and textual solution was on the top; the lower part has been left empty with the aim to provide the opportunity to write appropriate advice or suggestions. The graphic part was used by each stakeholder to suggest the navigation mode, however each user had the real opportunity to visit directly the platform through an individual PC; and therefore to observe directly the solution adopted in multimedia format (Fig. 5).

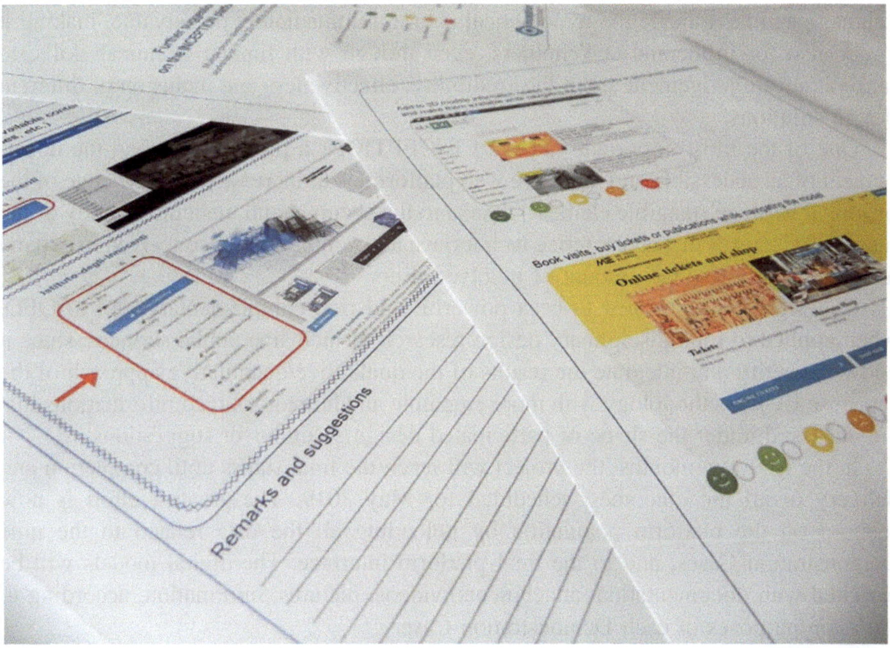

Fig. 5. Picture showing the individual questionnaire delivered to each stakeholder. Both parts are shown: the first (on the left) has been studied to understand the level of pleasantness, inclusiveness and sharing of the solutions adopted; the second (on the right) to investigate which aspects of the future platform are more shared and crucial for the involved stakeholders.

Part 2 - The second part of the questionnaire concerned the possibilities and future suggestions to be applied to the INCEPTION platform. For this reason the document did not illustrate a graphic part of real solutions, instead it contained evocative or suggestive images of how the new future platform could change and be configured.

In this part, Stakeholders are invited to express preferences on a multiple-choice chromatic scale (Green = preferable, yellow = medium; red = bad choice) on several issues emerged during the last meeting in Ferrara, to be implemented.

4 Conclusion

The participatory work days, structured in co-design workshops, have obtained important results on several aspects. Firstly, they allowed the involved stakeholders (with different sectors and specific competencies) to be able to understand other people's technical and emotional needs and therefore improve their awareness about inclusive design configurations. Secondly, they have increased the empathy between the design research and development team, the stakeholders and users of the INCEPTION project web platform. Thirdly, the applied methodologies, such as card sorting, have allowed both a fast and reliable comparison through a shared graphic language and have made the hypothetical solution immediately perceivable, making it easier also for influential stakeholders, even though with limited technical skills, to make an aware judgement about the credibility, effectiveness and inclusion of different design solutions.

One of the most important values of INCEPTION, a project regarding the development of an accessible and inclusive web platform, is to increase the knowledge of the historical heritage (tangible element) thanks to the accurate and meticulous study of the intangible features that characterize its interface and configuration, conceived to expand as much as possible the amount of involved users.

Codesign has confirmed to be a powerful and rich complementary tool to QFD, when applied to multidisciplinary design research of inclusive public services, since it allows to verify and integrate the results of the qualitative/quantitative approach of the Japanese UCD methodology with more empathic and human-sourced informations that are delivered under the shape of participated design solutions or suggestions.

In the next few months, the project will reach the final stages until completion and delivery of all the outcomes, scheduled for May 2019. The greatest effort is now focused on the platform population by uploading all the data related to the nine Demonstration Cases, and to the final platform interface. The digital models will be enriched with documentation, attachments, videos, pictures, information, according to the main purposes of each Demonstration Case.

In addition to the nine pilots under uploading, an INCEPTION Course will be held between April and May 2019 in order to present the platform to Institutions, Public bodies, Research Centres, etc., involved in the Heritage field, as a motivation to propose Heritage digital models to be uploaded on the platform. In this way, digital models currently unknown to the public or unexploited could be shared in an inclusive way.

References

1. Di Giulio, R., et al.: Integrated data capturing requirements for 3D semantic modelling of cultural heritage: the INCEPTION protocol. Int. Arch. Photogr. Remote Sens. Spat. Inf. Sci. **42**, 251 (2017)
2. Di Giulio, R., Maietti, F., Piaia, E.: 3D documentation and semantic aware representation of cultural heritage: the INCEPTION project. In: Proceedings of the 14th Eurographics Workshop on Graphics and Cultural Heritage, pp. 195–198. Eurographics Association (2016)
3. Maietti, F., et al.: Digital memory and integrated data capturing: innovations for an inclusive cultural heritage in Europe through 3D semantic modelling. In: Mixed Reality and Gamification for Cultural Heritage, pp. 225–244. Springer, Cham (2017)
4. Pauwels, P., et al.: Integrating building information modelling and semantic web technologies for the management of built heritage information. In: Digital Heritage International Congress (DigitalHeritage), pp. 481–488. IEEE (2013)
5. Jennings, E., Dobreva, M., Devreni-Koutsouki, A.: Towards user engagement models for citizen science: initiatives in the digital cultural heritage domain. In: Cultural Heritage Communities, pp. 98–115. Routledge (2017)

6. Maietti, F., et al.: Roadmap for IT research on a heritage-BIM interoperable platform within INCEPTION. In: Borg, R.P., Gauci, P., Spiteri Staines, C. (eds.) Proceedings of the International Conference SBE Malta, pp. 283–290. Gutenberg Press, Malta (2016)
7. Maietti, F., Piaia, E., Mincolelli, G., Di Giulio, R., Imbesi, S., Marchi, M., Giacobone, G.A., Brunoro, S.: Accessing and understanding cultural heritage through users experience within the INCEPTION project. In. Ioannides, M., Fink, E., Brumana, R., Patias, P., Doulamis, A., Martins, J., Wallace, M. (eds.) Digital Heritage Progress in Cultural Heritage: Documentation, Preservation, and Protection, EuroMed 2018, pp. 356–365. Springer (2018)
8. Goodman, J., Langdon, P., Clarkson, P.J.: Formats for user data in inclusive design. In: Universal Acess in Human Computer Interaction. Coping with Diversity, Springer, Heidelberg, pp 117–126 (2007)
9. Mincolelli, G.: Customer/user centered design. Analysis of an application case, Maggioli, Rimini (2008)
10. Kankainen, A., Vaajakallio, K., Kantola, V., Mattelmaki, T.: Storytelling group- a co-design method for service design. J. Behav. Inf. Technol. **31**(3), 221–230 (2010). Services and HCI: New Opportunities
11. Pallot, M., Kit, B., Senach, B., Scapin, D.: Living lab research landscape: from the user center to design and user experience towards user cocreation. Hal archives-ouvertes (2011)
12. Davis, F.D.: Perceived usefulness, perceived ease of use, and user acceptance of information technology. MIS Q. **13**(3), 319–339 (1989)
13. Hudson, W.: Playing your cards right: getting the most from card sorting for navigation design. Interact. - HCI High. Educ. **12**(5), 56–58 (2005)
14. Nielsen, J., Sano, D.: SunWeb: user interface design for sun microsystem's internal web. Comput. Netw. ISDN Syst. **28**(1–2), 179–188 (1995)

User Attitudes Towards Wheelchairs: A Mexican Case Study with Special Focus on Vulnerable Wheelchair Users and Wheelchair Assistants

Paulina Manzano-Hernandez[1](✉), Maria Giovanna Trotta[1], Carlos Aceves-Gonzalez[2], Alberto Rossa-Sierra[1], and Fabiola Cortes-Chavez[1]

[1] Facultad de Ingeniería, Universidad Panamericana, Álvaro del Portillo 49, 45010 Zapopan, Jalisco, Mexico
{paulina.manzano, gtrotta, lurosa, fcortes}@up.edu.mx
[2] Centro de Investigaciones en Ergonomía, Universidad de Guadalajara, 44250 Guadalajara, Jalisco, Mexico
c.aceves@academicos.udg.mx

Abstract. In Mexico, wheelchair users face discrimination and attacks on their dignity. To improve this situation, products that consider the needs of the whole population, while highlighting those of vulnerable groups, need to be developed. Nevertheless, little efforts have been made to acknowledge how vulnerable Mexican users and wheelchair assistants perceive this device. Hence, this research aims at gathering a deeper understanding of how the latter users feel towards current wheelchairs and where they perceive areas of opportunity. Semi-structured interviews were made to retrieve information about variables like wheelchair characteristics' prioritization, pain-points and wearing. Some of the most relevant results include the need to implement elements that consider natural body functions like sweating, and the misfit of wheelchair-user dimensioning. With these study's findings and by considering vulnerable groups, we aspire to provide insights for wheelchair design, that if satisfied, could help to create more inclusive devices and thus, more inclusive environments.

Keywords: Inclusive design · Wheelchair users · Wheelchair assistants · Older adults · Children and adolescents · Vulnerability

1 Introduction

Wheelchair users are varied and can be defined as those who already use a wheelchair or those who could be benefited by one due to limited mobility [1]. Regardless of their age, gender, place of birth, lifestyle and overall situation, these users require a dignifying mean of mobility.

In Mexico, vulnerable groups are more prone to suffer attacks that damage their dignity due to factors that as gender discrimination, low levels of schooling and low

G. Di Bucchianico (Ed.): AHFE 2019, AISC 954, pp. 282–291, 2020.
https://doi.org/10.1007/978-3-030-20444-0_27

income. Moreover, if we add that some of them also suffer from health vulnerability and that they are wheelchair users, this situation becomes worse, translating into higher degrees of discrimination and dignity loss [2]. In order to improve and establish effective health and development programs, products that consider the needs of the entire population, while highlighting the needs of vulnerable groups, need to be developed [3, 4].

Considering these vulnerable groups and their needs when designing wheelchairs will result in greater empowerment and more participation in daily activities, this will allow them to have a more independent, inclusive and fulfilling life [1]. Also, having a proper wheelchair/mobility tool has the potential to benefit other sectors, besides healthcare, such as economy and education because of the user-society integration that a wheelchair provides [1, 4].

For this article, we have identified two further vulnerable users when it comes to wheelchair use: older people (55 years or more), and children and adolescents (0 to 17 years), which are sometimes considered in the same group [5–8]. In Mexico between 2014 and 2015, there were about 17.28 million older people and 43.5 million children and adolescents [9, 10]. Out of these, around 3.155 million and 0.307 million respectively had a disability when walking or going up or down the stairs when using their legs [10, 11]. According to [12] around 1% of the total of the population in any given country needs a wheelchair, this would mean a need of 1.19 million wheelchairs in Mexico (considering the latest official data available from 2015) [13]. As reported by [10, 14] wheelchairs were the most used technical aid during 2014 with an estimation of 1.24 million users, which, in theory, would satisfy the national need. Nevertheless several clarifications need to be made: (a) the exact amount of wheelchairs needed is not entirely precise as stated by [12]; (b) the numbers from the Mexican population stated above were from surveys with self-reported answers, therefore people that actually need a wheelchair but do not state it could be higher; (c) there is a lack of more updated data that will not be fulfilled until the next couple of years, and (d) it is not specified how these wheelchairs are distributed amongst the population. Hence some people might be lacking one, including the vulnerable population.

Also, the people helping the user that sits on the wheelchair were contemplated as key users in this study, since they can sometimes be the primary user depending on the level of disability of the seated user.

It has been documented that current aids in Mexico do not adapt to the user as they should, making them feel unsafe and uncomfortable, and this situation could worsen the quality of life of all users (including the most vulnerable ones) [15]. In addition, user experience regarding the manoeuvrability of wheelchairs has been negatively rated on a daily use basis, thus creating an opportunity to improve this device [16].

Considering the importance of designing for vulnerable groups and assistants, and the lack of information about the latter, this research aims at gathering a more in-depth understanding about how wheelchair users and assistants feel towards current wheelchairs and where they perceive areas of opportunity for product improvement. A particular focus was made on the groups mentioned before as further vulnerable to make their needs visible, and hopefully in the future, design better wheelchair proposals.

2 Methods and Materials

2.1 Participants

Four wheelchair users - two children/adolescents and two older adults -, and seven wheelchair assistants were interviewed. From the wheelchair users, three were females aged 16, 58, and 84 and one male aged 16 years. The females were independent full-time users and the male was a part-time user, in both cases, they were assisted from time to time.

Four assistants were females aged 21, 24, 42 and 57; with five, one and a half, two and 20 years of experience respectively. Three assistants were males aged 23, 31 and 74. The youngest and the oldest assistants had two years of experience, whereas the assistant aged 31 expressed his years of experience as "a lifetime". All of the assistants were providing help to wheelchair users at the time of the study.

2.2 Materials

For the evaluation, two interview layouts related to the users' attitudes towards the wheelchair were developed based on questionnaires and suggestions of [16–18]. One of the two layouts (which include open-ended and open questions) was applied to the wheelchair users that sit on the wheelchair and the other one to the wheelchair assistants. The questionnaires had the content shown in Tables 1 and 2 respectively. The slight differences between them were mostly due to the variation of the typical activities that they perform and their user condition (seated vs assistant).

2.3 Procedure

To apply the interviews to the participants, an online request via social media (to gather as many participants as possible), a visit to a public hospital in Guadalajara (Mexico), and a visit to a meeting center for older adults were done. All of the users that sat on the wheelchair agreed to a face-to-face interview at a place of their preference, except for the participants interviewed at the hospital and the meeting center. Out of the seven assistants, five agreed on a face-to-face interview and two answered an online form with the same questions. When the interview was made face-to-face, the questions' answers and any other relevant comment(s) were written by the interviewer. Else, the users were encouraged to write as many experiences as they wished through the online form. The users were not limited on their answers and could provide any extra information that they considered relevant. This freedom was agreed so they could feel in a comfortable environment, and when convenient (and if applicable), the interviewer returned to the previously made questions. When needed, questions were restructured, or sections were asked in a different order to make the interview as natural as possible to the user, mostly to children as suggested by [22].

For the first seven interviews (two with seated users and five with assistants), the questions were answered in their vast majority and the answers provided information about the participants' feelings, preferences and areas of opportunity. Plus, the questions elicited valuable feedback that was not considered on the questionnaires (like challenging scenarios for the users). Nevertheless, users seemed uneasy at the end of

Table 1. Seated user interview's sections and the correspondent data that was retrieved from them.

Section	Data retrieved
0-General data	Age, sex, studies, residence (city/town), time for which the user has used the wheelchair, the reason for using the wheelchair, and type of user*
1-General wheelchair data (as a product)	Type of wheelchair, equipment of the wheelchair, and element-related issues (e.g. seat, wheels, backrest)
2-Wheelchair use	Whether the user was trained to use the wheelchair and how, difficulty of performing daily activities (e.g. eating, showering, sopping), and pain related to body parts when using the wheelchair**
3-Wheelchair perception	Liked and disliked aspects of the wheelchair, whether the wheelchair was adapted to the user's body and if user felt identified with the wheelchair, importance and current users' wheelchair rating of features like comfort, stability, ease of use, safety and price***
4-Suggestions	Any other suggestion that the user could think of to improve the wheelchair

*User types based on the National Accessibility Scheme by [19] and [20]. **Body map based on the Nordic questionnaire's section that evaluates the anatomical areas and that has been used in different populations [21]. A scale that ranged from 1 (little pain) to 5 (intolerable pain), with the option of 0 (as no pain at all) was used. ***A 5-point Likert scale where importance was rated from "Very important" to "Not important", and their wheelchair status was rated from "Very good" to "Very poor" was implemented.

the interview. Furthermore, the term "robustness" and the question "Do you consider that the wheelchair represents you or do you feel identified with it?" needed additional clarifications since they were somewhat confusing. Therefore, to make the interview less confusing and with a faster pace for the other participants, Section 3 (in the seated user interview, see Table 1) and Section 4 (in the assistant interview, see Table 2) were reframed in order to be better understood by a broader population and to shorten the interview. Those sections were specifically chosen since they posed the most extended duration and were the most tiring when the participant was answering. To do so, preference evaluation methods that do not require a significant amount of cognitive effort and that minimize the artificial rise of the evaluation score due to over-enthusiasm were considered so the children and older adults could answer with ease. The format of the *Again-Again* tool that is part of the Fun Toolkit was considered to substitute the 5-point Likert scale. This tool gives a magnitude of preference, shows areas for development, and still provides open choices when compared to other methods like *This or that* [23, 24]. At the same time, the sub-section that reflected the importance of some of the wheelchair's features was left out since it measured the same variables as the sub-section rating the current wheelchair's features and users were tired of answering both. Besides, users tended to get confused and answered the first sub-section as if it were the second.

Table 2. Assistants interview's sections and the correspondent data that was retrieved from them.

Section	Data retrieved
0-General data	Age, sex, studies, residence (city/town)
1-General questions regarding the assistance	The time that the person has been assisting users, assisted people's profile*, the frequency of assistance and location (indoors/outdoors)
2-Wheelchair use	Whether the user was trained to use the wheelchair and how, movements performed when assisting (e.g. lifting the person or allowing them to place their hand on the shoulders), and pain related to body parts when using the wheelchair**
3-User relationship	Daily coexistence issues, user resistance to sit on the wheelchair, and solutions proposed
4-Wheelchair perception	Liked and disliked aspects of the wheelchair, whether the wheelchair was adapted to the user's body and if user felt identified with the wheelchair, importance and current users' wheelchair rating of features like comfort, stability, ease of use, safety and price***
5-Suggestions	Any other suggestion that the user could think of to improve the wheelchair

*Based on whether they were children, young adults, adults or older adults (55+) [5, 8]. **Body map based on the Nordic questionnaire's section that evaluates the anatomical areas and that has been used in different populations [21]. A scale that ranged from 1 (little pain) to 5 (intolerable pain), with the option of 0 (as no pain at all) was used. ***A 5-point Likert scale where importance was rated from "Very important" to "Not important" and their wheelchair status was rated from "Very good" to "Very poor" was implemented.

The interview was also shortened by removing questions and options that were found to be repetitive, or that could be answered with another question. For instance, the word "robustness" was related to stability. Thus it was removed, and "stable", which was less complex, was kept. Terms like "aesthetic appearance" were replaced to more comprehensive ones like "beautiful". Also, the question "Do you consider that the wheelchair represents you or do you feel identified with it?" was restructured as "Do you think that the wheelchair reflects your personality?". When possible, this sub-sections: (a) causes of using the wheelchair, (b) equipment of the wheelchair, and (c) part-related issues, were left to be answered by the interviewer (when it was apparent or visible) or a carer/supervisor that could answer to avoid the user's fatigue.

2.4 Data Analysis

The retrieved answers were analyzed and grouped in terms of similarity and repetition. After that, general variables that described the answers were established. Answers that involved numeric scales were averaged and prioritized. The questions that remained the same over the whole study were grouped with no distinction. When needed, clarifications were made about whether a question or section belonged to a specific questionnaire version. The early version of the survey, which corresponds to the first seven interviews, was referred to as "EV". The latest version of the survey, the refined version after the first seven interviews, was referred to as "LV".

3 Results

The results were divided into two parts, one for the seated users and the other one for assistant users due to each interview's structure. Tables 3 and 4 show the results for the wheelchair users and the assistant users' interviews respectively.

Table 3. Results from the wheelchair users' interview.

Variable	Results
Training	None of these users was trained on how to use the wheelchair. However, one user said that she used videos as a way to learn how to manoeuvre the wheelchair
Pain zone	The lower back was considered as the place where they felt the most pain with a mean score of 4.50 (SD = 1.00), followed by the upper back and shoulders with mean scores of 3.75 (SD = 2.50) and 3.5 (SD = 2.38) respectively
Difficult activities	Showering was considered the most challenging activity to perform. Moreover, although it was not previously considered, getting in and out of the wheelchair was also mentioned as a challenging task
Element- related issues	The lack of a break that assures a complete wheel stop and forgetting to place the break were considered as risks that could lead to accidents. Three out of four participants had issues with the break during the last year, making it one of the most problematic elements alongside seat and backrest wearing and the opening of the fabric's weave
Equipment and features	The four wheelchairs could fold. Three wheelchairs had a cushioned seat (in one case it was added after a while), of those, one had an adjustable footrest, and another one included a cushioned backrest. Footrests were one of the elements that in occasions resulted so inefficient that the users opted not to use them or even removed them since their feet kept on sliding
Adaptation	Opinions were divided on whether the chair was adequately adapted to their body with positive (n = 2), moderately positive (n = 1), and negative (n = 1) ratings
Experience	When asked about their experience while using the wheelchair that they were currently using, answers tended to go to extremes depending on their experience and expectations of the device. This was accentuated with the answers from the "EV" questionnaire. For instance, a user stated that her experience started negatively in a public hospital where wheelchairs were not even available and when she finally got her wheelchair (which was borrowed and previously used) the experience was further degraded. She even stated that it made her look older On the other hand, there was a case where the objective was to use it as little as possible since the user only saw it as a tool. Hence the relationship wheelchair-user was not as dependent and the facilitation that it provided were above any negative consideration. Still, comfort, cheap to repair, portability and easy to maintain and transport in public transportation were features that were poorly rated. It was also commented that a fixed seated position was tiring, and for one older adult's case, it even resulted in blood circulation issues
Shared priorities	Participants highly agreed on the need to improve the seat, since it was not comfortable and it became too hot during summer

Table 4. Results from the assistants' interview.

Variable	Results
Training	Four assistants were trained on how to use the wheelchair and worked as full-time assistants, taking care of various user profiles including adults (n = 3) older adults (n = 2) and children (n = 1). Three of them were trained with demonstrations, oral explanations and written instructions. The other one was trained with demonstrations and oral instructions only. Three assistants acknowledged the demonstrations as the best training method, and one claimed the theory and practice combined as the best
Pain zone	The lower back was considered the place where they feel the most pain with an average score of 3.07 (SD = 1.30) followed by the neck and knees with an average score of 2.14 (SD = 2.12) and 1.86 (SD = 1.77) respectively
Adaptation	Three assistants agreed that the wheelchair was not adapted to their bodies, with remarks as it being designed for the person that sits on it, but not for the person that assists. Two stated that the wheelchair was somewhat adapted to their bodies, and two said they were adapted correctly (one of the wheelchairs was customized to the assistant). The lack of proper adaptation compromises the manoeuvrability negatively
Experience	Heavy wheelchairs (e.g. steel vs aluminium) were harder to manoeuvre. The fact that the wheelchair allowed them to move the seated users (the wheelchair accomplishing its most basic function) was valued. Still, the wheelchair's features barely accomplished an overall rating of "average", with a weighted mean score of 3 out of 5 in the "EV" and 2.25 out of 3 in the "LV"
Shared priorities	Greater adaptability was highlighted, both for them and for the people that sit. It is interesting to note that although aesthetic appearance was given little importance, it was rated between poor and average in both the "EV" and the "LV". One assistant expressed that, although this is not important for them, it is for the users. The latter comment matches the seated users' opinions which expressed comments referencing the aesthetic appearance as important but rated their wheelchair appearance poorly

4 Discussion and Conclusion

Wheelchair users and assistants find value in the services that the wheelchair provides, mostly in the one concerning the mobility of users. As [25] have suggested in product usage, the facilitation of the wheelchair's goal (getting a user from point A to B) elicited positive emotions. This happened even when the wheelchair had many opportunity areas that were clearly stated by the users. The idea of "having this wheelchair is better than having none" was very latent. Still, this device should be improved. The need for a better seat and backrest design that considers natural body functions like sweating or even urinary incontinence must be complied. Similar to [16], discomfort in the seats were reported by the seated users. Failure to tackle this problem could lead to severe problems, including sores in the users' bodies and even more discomfort, thus hindering the possibilities of a better realization of daily routines [1]. The wheelchair's breaks and footrest are also elements that need to be redesigned. The breaks tend to fail, and when the user is on the move, they most likely either make a full stop that is too abrupt or they do not work at all. For that reason, the implementation of progressive breaks that can be used to lower the speed and that can make a full stop when needed (e.g. when the user moves from the wheelchair to the bed) is suggested. For the footrests' case, structures that can be put away, but that do not wiggle while moving the wheelchair should be made. Overall, the wheelchair should be less static and more dynamic when it comes to part adaptation. Even if an element were to be designed considering anthropometric features, if it is fixed, it will most likely cause discomfort to the user. For instance, a low backrest would be good for active wheelchair users so they can manoeuvre the wheelchair with ease. However, even so, there are moments where their need is to be static, in this case, a high backrest would work better. The consideration of materials that diminish discomfort due to weather conditions (like heat or rain) is essential since the beginning of the designs. It is relevant to note that these weather conditions vary from place to place and from time to time. Thus material and design proposals that do not represent a significant cost for the user would be ideal.

Unfortunately, seated users did not receive any training, although it has been reported that these users would appreciate it and could potentialize their abilities and lower the possibilities of injury [1, 16]. In agreement with [1], training programs should be more widespread so the users can have a more independent and safe life.

5 Future Work

Although the selected population aimed to retrieve meaningful information through qualitative data, to have a representative sample of the Mexican population using a wheelchair further work is required. This paper represents the initial step of more extended research. Despite certain uncertainty related to the adequate representation of patterns, which is regularly recognized for population samples, the recommendation for future research is to select a larger adequate sample of the population to reduce the possible errors and generate more accurate knowledge. On this side, the authors suggest, as convenient for future research, to consider further vulnerable conditions that affect seated wheelchair users. This, based on the suggestions reported by pregnant women, which advice that the wheelchair does not project their ability to have a suitable pregnancy, and later on to take care of the child. This is a group that has been highly stigmatized and even labelled as "irresponsible" when they are not [26]. For that reason, their needs, attitudes and opportunity areas to improve the wheelchair should also be taken into account. Mostly in places like Mexico where literature about them is scarce.

References

1. World Health Organization: Guidelines on the provision of Manual Wheelchairs on less resourced settings. https://www.who.int/disabilities/publications/technology/wheelchairguidelines/en/
2. Juárez-Ramírez, C., Márquez-Serrano, M., Salgado de Snyder, N., Pelcastre-Villafuerte, B. E., Reyes-Morales, H.: La desigualdad en salud de grupos vulnerables de México: adultos mayores, indígenas y migrantes. Rev. Panam. Salud Pública. **35**, 284–290 (2014)
3. World Health Organization: Closing the health equity gap: policy options and opportunities for action. http://www.who.int/iris/handle/10665/78335
4. World Health Organization: Health in the post-2015 development agenda: need for a social determinants of health approach (2015)
5. Curvers, N., Pavlova, M., Hajema, K., Groot, W., Angeli, F.: Social participation among older adults (55+): results of a survey in the region of South Limburg in the Netherlands. Health Soc. Care Community **26**, e85–e93 (2018)
6. Waisel, D.B.: Vulnerable populations in healthcare. Curr. Opin. Anaesthesiol. **26**, 186–192 (2013)
7. World Health Organization: Definition of key terms. https://www.who.int/hiv/pub/guidelines/arv2013/intro/keyterms/en/
8. Avila-Chaurand, R., Prado-Leon, L.R., Gonzalez-Muñoz, E.L.: Dimensiones antropométricas de la población latinoamericana, México, Cuba, Colombia, Chile. Universidad de Guadalajara, Centro Universitario de Arte, Arquitectura y Diseño, División de Tecnología y Procesos, Departamento de Producción y Desarrollo, Centro de Investigaciones en Ergonomía, Guadalajara (2007)
9. Instituto Nacional de Estadística y Geografía: Encuesta Intercensal 2015, Tabulado de Población (Data file) (2015). http://www.beta.inegi.org.mx/proyectos/enchogares/especiales/intercensal/
10. Instituto Nacional de Estadística y Geografía: Encuesta Nacional de los Hogares (ENH) 2014, Tabulado de Población con discapacidad o limitación (Data file) (2014)
11. Instituto Nacional de Estadística y Geografía: Encuesta Nacional de la Dinámica Demográfica 2014, Tabulado de Tabulados Básicos (Data file) (2014)
12. World Health Organization: Report of a Consensus Conference on Wheelchairs For Developing Countries. http://www.who.int/disabilities/technology/WheelchairConsensusConferenceReport_Jan08.pdf
13. Instituto Nacional de Estadística y Geografía: Principales resultados: Encuesta intercensal (2015). http://www.beta.inegi.org.mx/contenidos/programas/intercensal/2015/doc/eic_2015_presentacion.pdf
14. Instituto Nacional de Estadística y Geografía: La discapacidad en México, datos al (2014). http://internet.contenidos.inegi.org.mx/contenidos/productos/prod_serv/contenidos/espanol/bvinegi/productos/nueva_estruc/702825090203.pdf
15. Herrera-Saray, P., Peláez-Ballestas, I., Ramos-Lira, L., Sánchez-Monroy, D., Burgos-Vargas, R.: Problemas con el uso de sillas de ruedas y otras ayudas técnicas y barreras sociales a las que se enfrentan las personas que las utilizan. Estudio cualitativo desde la perspectiva de la ergonomía en personas discapacitadas por enfermedades reumáticas y ot. Reumatol. Clínica, vol. 9, pp. 24–30 (2013)
16. Carneiro, L., Rebelo, F., Filgueiras, E., Noriega, P.: Usability and user experience of technical aids for people with disabilities? A preliminary study with a wheelchair. Procedia Manuf. **3**, 6068–6074 (2015)

17. Soares, M.M.: Translating user needs into product design for the disabled: an ergonomic approach. Theor. Issues Ergon. Sci. **13**, 92–120 (2012)
18. Soares, M.M.: Translating User Needs into Product Design for Disabled People: a Study of Wheelchairs (1998). https://dspace.lboro.ac.uk/2134/7050
19. Visit England: National Accessible Scheme: For Serviced, Self-Catering, Hostel and Campus Accommodation. https://www.visitbritain.org/sites/default/files/vb-corporate/visitengland_national_accessible_scheme_serviced_standards.pdf
20. World Health Organization: International Classification of Impairments, Disabilities, and Handicaps. World Health Organization, Geneva (1993)
21. Mendonça, C.R., Noll, M., Silveira, E.A.: Adaptation and validation of body maps for musculoskeletal pain location in patients with severe obesity. Korean J. Pain. **31**, 268 (2018)
22. Irwin, L.G., Johnson, J.: Interviewing young children: explicating our practices and dilemmas. Qual. Health Res. **15**, 821–831 (2005)
23. Sim, G., Horton, M.: Investigating children's opinions of games. In: Proceedings of 11th International Conference on Interaction Design and Children – IDC 2012, p. 70 (2012)
24. Read, J.C., MacFarlane, S.: Using the fun toolkit and other survey methods to gather opinions in child computer interaction. In: Proceeding 2006 Interaction Design and Children – IDC 2006, p. 81 (2006)
25. Desmet, P., Hekkert, P.: Framework of product experience. Int. J. Des. **1**, 57–66 (2007)
26. Iezzoni, L.I., Wint, A.J., Smeltzer, S.C., Ecker, J.L.: "how did that happen?" Public responses to women with mobility disability during pregnancy. Disabil. Health J. **8**, 380–387 (2015)

Inclusion of Children with Down Syndrome Through the Creation and Use of a "Learning Object"

Rita Assoreira Almendra(✉) and Mariana Elvas

Faculdade de Arquitetura, CIAUD, REDES, Universidade de Lisboa,
Rua Sá Nogueira, Polo Universitário da Ajuda, 1349-063 Lisbon, Portugal
almendra@fa.ulisboa.pt, marianaelvas1995@gmail.com

Abstract. This paper addresses the inclusion of children with Down Syndrome in school environments. It is based on the work of one of the authors developed as a Master dissertation in the area of Communication Design. The research integrates the study of the problem in context, with the participation of practitioners involved in the education process of these children. Framing the context implied studying not only this specific pathology but also the used learning-teaching methodologies, communication and interaction design principles and practices among other issues. Moreover, it was made a comparative critical analysis of three case studies of learning objects/systems to acquire knowledge and guidelines to support the creation of a learning object that would favor these children (aged between 5 and 7 years old) inclusion in regular schools. The result was an interactive printed book to mainly support reading and speaking activities in a playfull and shareable way. This book was evaluated by experts and was after improved through a design iterative process.

Keywords: Inclusive design · Down syndrome · Children's teaching-learning process · Interaction design

1 Introduction

Trisomy 21 (T21) or Down Syndrome is a chronical condition that imposes several challenges not only to children that have it but also to their families. It is then necessary to invest in order that these children can mature their neurological functions, practice daily routines and consequently learn and develop themselves.

By having a proactive behaviour towards their learning processes one can contribute to a fairer world, a better and more equal society, free of prejudice and one that constantly pursues information and knowledge to fight all types of exclusion situations.

In this particular area of education, we have chosen the age period between 5 and 7 years old since this is the moment children initiate their learning process as well as they reinforce their social interaction and engagement.

To avoid the difficulties children with Down Syndrome have in their scholar environment it is mandatory to submit them to continuous stimulation and to make them feel equal and included.

G. Di Bucchianico (Ed.): AHFE 2019, AISC 954, pp. 292–300, 2020.
https://doi.org/10.1007/978-3-030-20444-0_28

The creation of materials that will help all children including these ones to acquire the competencies to read and to better speak, to execute daily tasks among other things in a playful and enjoyable way was the challenge we committed ourselves with having had the help of educators and after a field study that included the observation of two schools environments and the study of existing materials supporting these children development.

2 Scholar Integration vs. Scholar Inclusion of Trisomy 21 Children

Inclusion and integration are two different things. The first concept has a broader definition and signifies a total and unconditional insertion of the person. On the other hand, integration implies a partial and limited insertion of a person. In this particular case it means that the difficulties are on the part of children with "impairment" that can be integrated in the normal education system, but only when their characteristics allows it [1].

So, the option is to include, meaning to embrace, understand and reach and that is what we propose to these children: to proportionate them the same experiences that a child with a typical development has, "to accept the different and learn with him". To include is "to abandon stereotypes" [2] and to be happy to do it.

Learning is a process that requires interaction of the individual with the world; being so it is determinant as Bassani proposes [3] to assume T21 children school's inclusion as something natural and desirable since this way one can contribute to the improvement of their self-esteem and support their social development.

Although T21 children can experience some frailties and limitation while learning, the pedagogical work to develop with them must respect the rhythm of each child and to guarantee an adequate stimulation of his/her abilities since it is insufficient to just place the child in a suitable environment [4].

Thus, inclusion means a directed work based on the potentialities the child has instead of his/her weaknesses.

As Castro [5] acknowledges through several inclusion practices of these children in school environments it has been demonstrated that it is possible to stimulate in meaningful ways their abilities having a natural coexistence with the differences.

The inclusion gives a T21 child the possibility to interact with "normal" children through a collaborative apprenticeship process that allows a better quality of general development as well as of language ability.

It is also important to attain that the development of T21 children is better succeeded in ludic environments since it is easier for them to take advantage of a pleasant experience while learning important contents.

3 Role of Game/Ludic in the Teaching-Learning Process of Children with T21

Children are born with the potential to learn but learning occurs as a function of all the experiences children have during their development so, to have a stimulating childhood with appropriate play linked with each of the development phases will contribute to a healthier, integrate and complete personality's shape [6].

Since birth children experience an instructed world that starts with gestures, looks, words, drawings until writing is approached and reached. As Silva [7] states, to live this process through ludic activities proves to be the better option for T21 children's learning and development.

Consequently, the stimulation of T21 children through playful activities and gaming permits these children to access the necessary development of concepts such as identity, autonomy and in general terms allows them to discover the world surrounding them [8].

Friedrich Froebel (1782–1852) was the first educator to value toys and games as important elements in teaching-learning processes assuming that ludic activities allows the children to build up their imaginary world being the toy the vehicle of their self-expression [7].

Since then the ludic has been used to the development and education of children with special needs since "in classroom it is a facilitator of practical knowledge in group or individual activities" [9]. Despite that fact it is crucial to understand that educating by the use of ludic activities it is not "use packaged lessons so the student can consume it in a passive way" instead it is a conscious and planned act to seduce children to engage with the pleasure of knowing, of wanting to learn in a truthful assumption of the real meaning of "school" as a "place of joy, intellectual pleasure, satisfaction and development" [10] (p. 8).

4 Designing an Inclusive Teaching-Learning Object

4.1 Designing Interaction

A designer has an important role and an ethical responsibility towards people since through his/her actions and projects he/she determines their exclusion or inclusion in terms of their daily activities [11].

Being so, to create solutions to diverse problems is to design a complete and enduring answer that magnifies users' abilities and not their limitations, facilitating the acquisition of knew knowledge and helping the development of psychological functions [11, 12].

One of the ways of facilitating development is Interaction that serves the purpose of connecting and promoting development. It has to do with interactivity that gives respect to the result of interaction itself plus the result of the action of one of the parts involved in some event that presupposes a reaction.

As proposed by Casaca [13] the concept of interactivity must be used to improve the relationship a user has with an object, so it can create new opportunities of use and interaction what is reflected on the quality of the object that can become more attractive and useful to the user.

Design process is supported by experience and constant iteration among the various process phases, so the designer can create the better answer to a given problem. It is then a flexible process, adaptative and open to transformation along its development [13].

In the case of a printed object one must create a narrative in which time and space are manipulated so the information is not entirely uncovered giving space to user's motivation to access the whole of it. In terms of printed objects, the disclosure can be guaranteed by the construction of folded elements, overlapped parts, that appeal to the curiosity of the reader and pulls him/her to act with it [14]. When an object presents interaction elements it excels other objects that only have plane pages since an interactive printed object gives "a closer and customizable dimension to its user having physical, tactile and manipulability characteristics that react to user's participation" [14].

4.2 Analysing Case-Studies

To support the design of an interactive printed object to support the teaching learning process to all children aged between 5 and 7 years old it was necessary to analyse existing solutions in the market. Three case studies where analysed and their strengths and weaknesses where acknowledged. Criteria used to select the case studies where the focus on the teaching-learning process of children with any kind of cognitive impairment.

The first case-study was a computer game entitled "Games of Mimocas" and it is targeted to children with mental deficiencies. It is an educational software to be used in a precocious intervention, so these children can have a better inclusion in school. It is a programme that allies education and entertainment increasing children's motivation towards learning and it is adequate to group activities.

The second case study was an activities' book addressed to parents of T21 children to help them along the development of their kids.

Finally, the third case-study was a method "The 28 words method" that departures from the general to the particular and that has good results in T21 children's apprenticeship although it is not targeted to them.

Regardless age target and the themes being approached these three case-studies allowed us to perceive positive aspects of educating tools and materials as well as aspects to be improved and enhanced. They served as inspirational material to the interactive object created (book + cards) that combines characteristics of them adding new and improved ones.

4.3 The Teaching Learning Object

As proposed by Hansen [15] inclusive design is the creation of products and environments used by all without the need of adaptation or specialized design.

A product should be designed to function in different contexts but more than that to make them part of the inclusive system it integrates. That brings us to Hansen [16] definition of inclusive environment, one in which each user, independently of his/her abilities can live everyday life comfortably, with efficiency and security, without being limited by a bad design. Nevertheless, it is proved that in many cases it is impossible to offer a "one size fits all" solution, so, some individuals will always be excluded. Bearing that in mind it should always be part of a good design thinking process the goal of creating a product and/or service that fits the needs of the large majority [16]. An inclusive design approach that understands a product or system as part of a wider system of products and services comprehends the mental switch in design thinking that Atkin [17] synthesizes in a methodology that provides guidance to designers while developing their inclusive design solutions. In the case of an inclusive design approach one can either have an user-centred or a participatory methodology since both focus in users in the structuring of the design/research process. The process we focus on this paper is a user-centred one.

The main goal of the designed interactive kit (book + cards) was to contribute to a better performance of all children (including T21 ones) in terms of Reading and writing. The option was to do it based on activities that would captivate and deliver to children conquest and joy sensations along their learning process. The interactive book can be used autonomously but it is recommended its use together with cards (being more challenging for that reason) and it as a strong presence of images and daily words.

The focus of the developed solution was never the condition of specific children but instead the fact that all the children can learn and overcome difficulties achieving good results.

As previously said the designed kit (book + cards) is based on some elements of the case-studies mainly on the "28 words method" – in which we have 28 words whose syllables allow children to write other words – but also on the "Global method" – a more personal method using words that are more common on children's day-by-day life.

One of the premises of the project was that each word should relate itself with the proposed activity, so children would understand it easily and would have a joyful activity to perform.

The solution was an interactive kit with a main object – a book – that is used with 5 sets of cards – lowercase words; uppercase words, images, syllables, letters – that will allow not only to learn how to read and write but also some daily activities like make shoelaces (see Fig. 1).

Being an interactive kit children have to cut, draw, glue, associate, feel, build, to pair, all activities that are available in the kit never limiting the way of using it and above all supporting the attempt-error approach in each of the tasks.

In what concerns language this one was adapted to the target ages (5 to 7) and complemented with the use of colours, patterns and drawings to facilitate the concept and ideas dissemination.

As previously said besides Reading and writing the kit enables children to work fine motor through the use of textures, fittings and insets etcetera (see Figs. 1, 2, 3 and 4).

One of the main advantages and goals of this project is the fact that children can repeat the activities as many times as they want until they have a 100% assimilation of what they are learning and experiencing. Also important is the fact that each activity can have multiple answers.

Fig. 1. Images of the interactive Kit (book + cards) with some of the activities as well as of the cards that are to be used complementarily to the book (author, 2018)

4.4 Evaluating the Interactive Kit

The interactive Kit was evaluated by a group of experts – related with education, T21 pathology, therapists – so we could have feedback regarding its usefulness and impact on children's teaching learning process. This evaluation is essential since "the identification of flaws and problems is of extreme importance because it represents the first step towards the optimization of the object" [13].

Specialists had the opportunity to test the kit, to compare it with existing solutions, to propose improvements and adjustments and an iteration on the design process was done. The suggestions were implemented, and a new prototype was created. The final Kit was received by the expert panel with enthusiasm being described as an excellent tool to support the teaching-learning process since it has highly positive levels of usability.

Fig. 2. Images of the interactive book – using band-aids (author, 2018)

Fig. 3. Images of the interactive book – playing the rooster game (author, 2018)

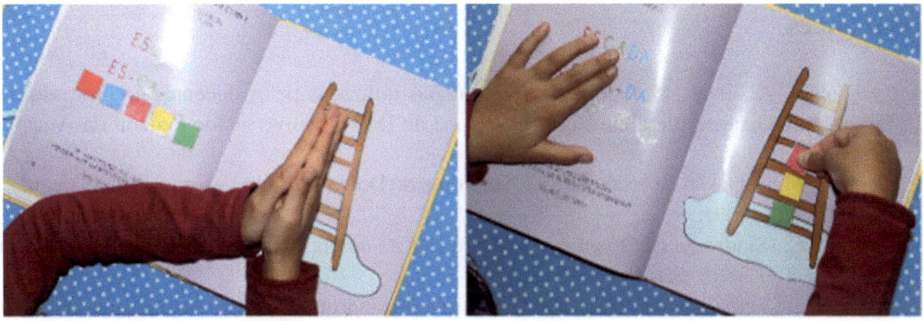

Fig. 4. Images of the interactive book – climbing the stairs (author, 2018)

5 Conclusions

An overview of the themes and audiences focused in designing in an inclusive approach show us that the more common ones are related with elderly and disabled people, and that the contexts of the problems are usually related with the needs of mobility, security, personal basic needs (eat, drink, sleep, dress) health and education.

However, the children group of users is an interesting and important one to be tackled "since they are the future of the nations and working with them in an inclusive way one hopes to create a consistent intervention in societies developing awareness and an effective commitment of young generations with this approach" [18].

Inclusion in the education context is a challenging task especially when we are working with young children because the ethical implications are higher as well as the responsibilities. In the current hyper-complex world, it becomes extremely critical to guide children's development aiming that they will become sensitive, empathic and knowledgeable human beings.

Inclusion practices do facilitate that goal being accelerators of collaborative behaviors, citizenship awareness, and empathic predisposition. It also allows to experience the differences as something positive and rich.

The creation of this interactive kit made clear the need of more tools and equipment supporting inclusion. It also demonstrated that interactivity helps the teaching-learning process of children for the reason that it raises children's attention, enthusiasm and commitment to the process.

Acknowledgments. We want to acknowledge the Research Center in Architecture, Urbanism and Design (CIAUD) and the Foundation of Sicence and Technology (FCT) for funding our research project as well as REDES – a research group in Design Education and Research and mostly all the specialists and parents that had the generosity to work with us in this special project.

References

1. Guimarães, A., et al.: Metodologias e estratégias utilizadas pelos docentes no processo de ensino aprendizagem de alunos com Síndrome de Down. Universidade Federal do Amapá, Brasil (2010)
2. Ferreira, M.: Inclusão de crianças com Síndrome de Down no ciclo I do ensino fundamental. Lins, Brasil (2009)
3. Bassani, C.: Síndrome de Down e as dificuldades na aprendizagem. Faculdade Anhanguera de Taboão da Serra, São Paulo (2012)
4. Silva, M.: A aprendizagem escrita pela criança com síndrome de down inserida em uma sala de aula regular. UFRN, Brasil (s.d.)
5. Castro, A., Pimentel, S.: Síndrome de Down: Desafios e perspetivas na inclusão escolar. EDUFBA, Brasil (2009)
6. Sacchetto, K., et al.: O ambiente lúdico como fator motivacional na aprendizagem escolar. Universidade Presbiteriana Mackenzie, Brasil (2011)
7. Silva, A.: A atividade Lúdica - contribuições para educação das crianças com síndrome de down. Universidade Federal de Paraíba, Brasil (2016)
8. Cintra, R., et al.: As contribuições do lúdico no processo de desenvolvimento das crianças com síndrome de down na educação infantil. Horizontes **33**(2), 159–166 (2015)
9. Nascimento, T., Barreto, M.: A importância de atividades lúdicas na aprendizagem de crianças com necessidades especiais. Faculdade de Edu-cação, Ciências e Letras do Sertão Central, Brasil (s.d.)
10. Dallabona, S.: O Lúdico na Educação infantil: Jogar, brincar, uma forma de educar. Instituto Catarinense de Pós-Graduação, Brasil (s.d.)
11. Maxwell. Design Inclusivo. https://www.maxwell.vrac.puc-rio.br/30055/30055_3.PDF
12. Heidrich, R., et al.: Design Centrado no Usuário para Obje-tos de Aprendizagem Inclusivos. Universidade Feevale, Brasil (s.d.)
13. Casaca, J.: O design centrado no utilizador aplicado ao design gráfico: Interação e participação na comunicação do Património Cultural de Lisboa. Master dissertation, Faculdade de Arquitetura da Universidade Técni-ca de Lisboa, Lisboa (2014)
14. Faísca, M.: O design e a patologia oncológica. Master dissertation, Faculdade de Arquitetura da Universidade Técnica de Lisboa, Lisboa (2015)
15. Hansen, J.: The Inclusive City: delivering a more accessible urban environment through inclusive design. University College, London (2006)
16. Almendra, R.A.: Designing the difference in an inclusive way. In: Proceedings of the 15th International Conference on Engineering and Product Design Education: Design Education - Growing Our Future, EPDE 2013 (2013)
17. Atkin, R.: Sight Line. Designing Better Streets for People with Low Vision. Jeremy Myerson, London (2010)
18. Almendra, R.A., Falcão, G.: Inclusive design for children at the master education. In: Advances in Intelligent Systems and Computing (2016). https://doi.org/10.1007/978-3-319-41962-6_29

Design of a Wheelchair for Low-Income Countries, the Second Stage of a Project

Alberto Rossa-Sierra(✉), Fabiola Cortes-Chavez, David Vidana-Zavala, and Maria Giovanna Trotta Munno

Facultad de Ingeniería, Universidad Panamericana, Álvaro del Portillo 49, 45010 Zapopan, Jalisco, Mexico
lurosa@up.edu.mx

Abstract. In the first phase of the project, it was reported in a previous conference communication, the manufacture of two low-cost wheelchairs developed under this platform, both proposals designed in European countries (Spain and Italy). Following the constructive drawings and instructions available on the open-source platforms, both chairs were manufactured as close as possible to reported instructions. In their manufacture process, both wheelchairs presented some problems, in the construction and during the user testing phase, we report these kinds of problems in the first stage of the project. In the present stage, we start from models manufactured and tested with users, with the aim of improving the product design, since, one of the characteristics of the products designed under the open-source scheme is its ability to be modified and adapted to local needs, materials and production processes.

Keywords: Wheelchair · Product design · Open-source platform

1 Introduction

People with disabilities make up one of the most marginalized groups in the world. These people have worse health outcomes, obtain lower academic results, participate less in the economy and have higher poverty rates than people without disabilities. The figures released by the WHO reflect that 70 million people in the world need a wheelchair, but only 5–15% of them have one. One of the principal causes of this is the high cost that for low-income countries implies the acquisition of a wheelchair. With this information, the low-cost wheelchair design project was launched in 2017 under the open-source platform at the Universidad Panamericana at Guadalajara, Mexico.

The objective of the project is the design of a low-cost wheelchair under the criterion of the open source platform, which will allow its manufacture by any interested party with a minimum of technical knowledge, materials, and tools.

For a product to be considered open source, needs to meet the following parameters:

(a) Free redistribution: The product design must be freely given or sold.
(b) Construction plans: They must be free and easy to understand.
(c) Derived works: The redistribution of modifications must be allowed.

G. Di Bucchianico (Ed.): AHFE 2019, AISC 954, pp. 301–309, 2020.
https://doi.org/10.1007/978-3-030-20444-0_29

(d) No discrimination against persons or groups: No one can be left out.
(e) No discrimination of areas of the initiative: Do not exclude commercial users.
(f) Design distribution: The same rights should apply to everyone who receives the design of the product.
(g) The license must be technologically neutral: Do not require acceptance of the license through access by mouse click or other specific means of support of the product design.

2 Project Development

In the first phase of the project, it was reported in a previous conference communication, the manufacture of two low-cost wheelchairs developed under this platform, both designed in European countries (Spain and Italy). Following the constructive drawings and instructions available on the open source platforms, both chairs were manufactured as close as possible to said instructions. In their manufacture, both chairs presented various problems, both in their construction and during the testing phase, we report this kind of problems in the first stage of the project.

In this second stage, we start from the models manufactured and tested in the field, using as a base the model of the chair built in wood, because the latter proved from its construction to be more robust than the Spanish proposal made from PVC pipes (Figs. 1 and 2)

Fig. 1. User testing the wheelchair proposal made with PVC pipes, this model is economical and complies with being very easy to manufacture, however, it is very unstable and fragile, as well as it is complicated to maneuver by the user.

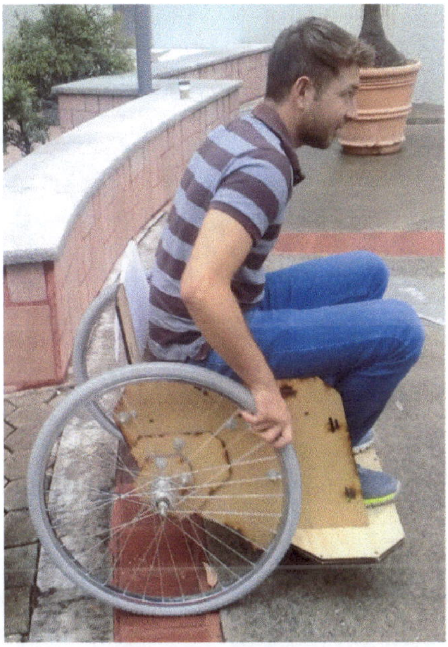

Fig. 2. A user is performing the test on the wooden chair; it seems that the feet are very flexed, as well as the lack of rim to push the wheel among other details.

The manufacturing phase of the chairs was carried out with the aim of improving the design of the same since one of the characteristics of the products designed under the open-source scheme is their ability to be modified and adapted to the needs, materials and local production means. We focus particular emphasis on improving the structural strength of the chair, as well as the use of materials, in order to optimize costs as much as possible, as well as to improve some ergonomic aspects that during the test with users, they showed as areas of opportunity. The overall objective of the project continues to be that the cost of the chair does not exceed USD 80 per unit and that it is feasible to be produced with simple tools and local materials anywhere in the world.

On the other hand, work continued on the search for more proposals to make comparisons, finding several proposals at the global level, some of which are well documented [1, 2], such as the model called "Free wheelchair mission, Gen 1" [3] which reports a cost of only $ 59.20 USD per unit including shipping to anywhere in the world (Fig. 3).

A consideration of this model is that part of the use of a commercial plastic chair reinforced with a metal structure. The proposal resembles the so-called "Kigali chair project", developed by Clara Romani in 2014 [4], where not reports the cost of the wheelchairs (Fig. 4).

A problem with this type of proposals is that they provoke reactions of displeasure in some people considering them to be of low quality in the short life cycle. In Mexico in 2012, the Government of the State of Sonora donated this type of chairs and received sharp criticism, both of the receivers of the chairs and the general public [5, 6]. The

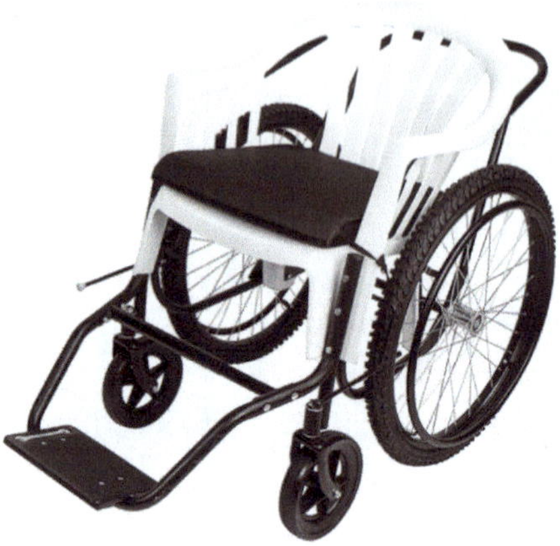

Fig. 3. "Gen_1" is a multipurpose wheelchair design for use in rugged terrain. Free Wheelchair Mission Project (2001).

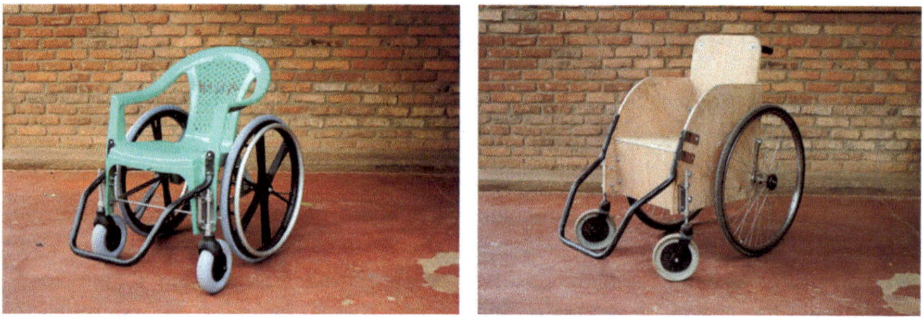

Fig. 4. Two wheelchairs designs from "Kigali chair project" developed by Clara Romani, this initiative has implemented in Rwanda (2014).

same situation of rejection of wheelchairs was experienced in the Argentine Republic in 2014 [7]. This type of reaction to using elements that the public considered "Economic" and adapted as is the case of the plastic chair, although the resistance of the chair that serves as the basis for the product is proven, it is natural, so we must be cautious in the development of our proposals and take into account the quality perceived by users and the general public, understanding by perceived quality that the user believes that the product has (Fig. 5).

This perceived quality can be equivalent or not to the real quality of the product because it is a subjective perception of the client. Also, although it may be influenced by "physical" aspects of the product, it is also affected by intangibles [8], hence the inclusion of this type of chair makes the product perceived as of low quality, which together with "the exclusion and ignorance have been fed back into a vicious circle: the tendency to link people with disabilities with negative stereotypes reverts to the reinforcement of their exclusion". [9], This means that strengthening the differences

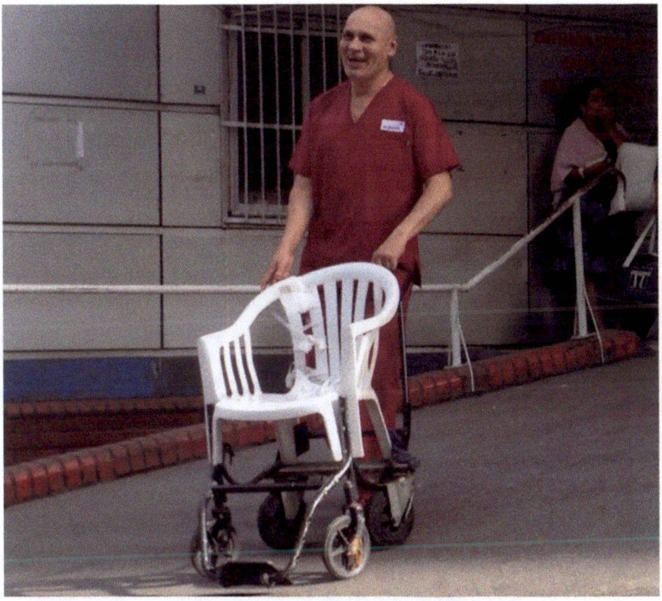

Fig. 5. Adverse reactions have occurred in the past on Latin American countries where used wheelchairs based on plastic chairs, considering as offensive to the dignity of the disabled.

between individuals has resulted in scattered, fragmented and decentralized societies, where the confinement to the isolation of persons with disabilities suggests not only a lack of social cohesion but a possible tendency towards the (growing) suffering of the individual with a disability.

On the other hand there is also a design proposal that does not incorporate prefabricated elements, but instead uses the new 3D printing technologies for the manufacture of parts of the chair and that gives a better appearance and functionality to them. An example of this type of proposal is the HU-GO project, developed by the architect Hugo Riveros [10], and where different components manufactured in a 3D plastic printer are used.

A problem with this type of solution is that this type of reproduction technology, although it has spread rapidly throughout the world, is still inaccessible to the majority of the population, which limits this type of solution (Fig. 6).

Considering the previous issues, we worked on the design of the chair "Too wheels", both in its first version as the version of sports chair badminton that have recently developed (2017) and which we have also made a model for testing [11] (Fig. 7).

The first aspect that we worked in the chair was the stability, since, the original chair is connected to the lateral panels by screws covered by tubular aluminum, which has two consequences. First place, the notably increase the weight of the chair, since, We consider in excess the number of screws and second, it raises the cost of it since the aluminum tubular is expensive and only fulfills a decorative function.

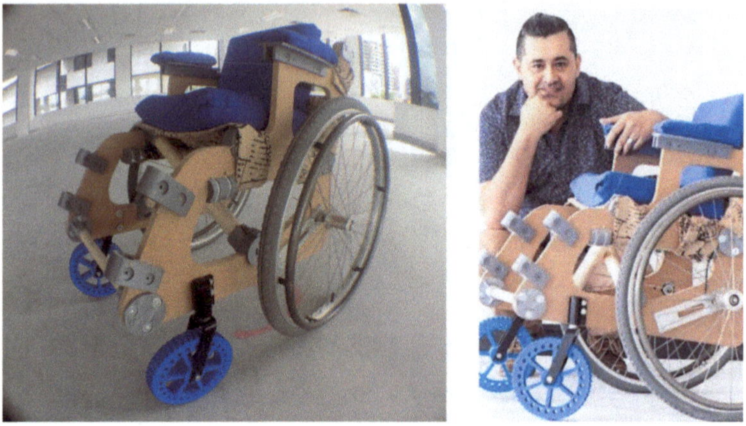

Fig. 6. "Hu-Go wheelchair", a proposal developed by the architect Hugo Riveros made with different components manufactured in a 3D plastic printer.

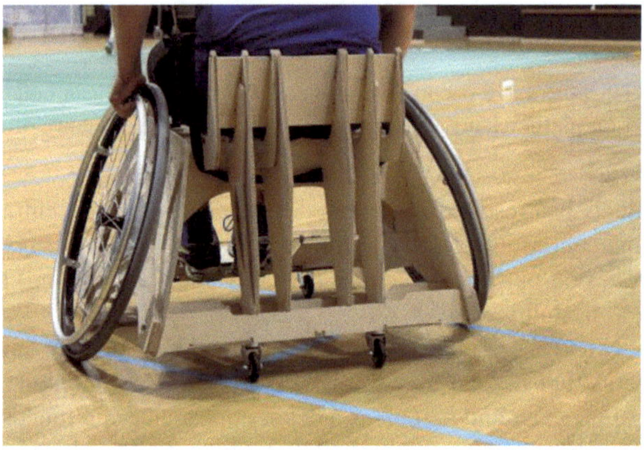

Fig. 7. Badminton wheelchair recent developed by "Too wheels project".

For this, we chose to make a 1/2″ welded steel structure that is occult under the plywood cover. This structure in addition to providing greater rigidity to the chair as a whole makes it lighter than the original (Approximated 11.5 kgs. of our proposal against 18 kgs. of the original project) (Figs. 8 and 9). The distance of the front wheels was also modified to improve stability. Regarding the badminton wheelchair, the chair made following the instructions of the designer, obtaining the product shown below (Fig. 10).

It should be a remark that this chair considerably increases its weight (26 kgs) and uses too much material (4 sheets of MDF wood), so we consider that a re-design of it is easily possible, if compromising its stability or integrity.

Fig. 8. Comparison between the wheelchairs, showing (left side) the number of aluminum tubes and on the other hand (right side) the simplified steel structure.

Fig. 9. Detail of the structure of the redesigned wheelchair.

In all cases, users have tried the chairs and have reported a series of areas of opportunity that we are about to implement in the design of our chair, these observations are:

Fig. 10. Reproduction of badminton wheelchair for evaluation purposes.

(a) The axle of the wheels must be placed a little further back to allow better movement of the user's arms.
(b) The front wheels must be ahead of the current position, as well as have a smaller distance concerning the center.
(c) The backrest of the seat requires more depth for greater comfort and the possibility of accommodating cushioning.
(d) Include at all times the outer ring for a push, the proposals are not to the liking of users without this piece, because, the wheel collects dirt and water that end up in the hands of the disabled.
(e) It should include a brake system for the wheels that block the movement since the user to move from the chair to a fixed surface (toilet bowl, stationary chair, etc.), must be able to push himself from the wheelchair without is displaced by accident.
(f) As much as possible, try to tilt the wheels at a negative camber angle to improve the stability of the chair.

3 Conclusions

We have developed a proposal for low-cost wheelchairs, based on the previous design of the "Too wheels" proposal, considering that, is an open-source product, the model can be re-designed as long as they are substantial improvements and always giving credit to the original author. Likewise, we have saved material in its manufacture, which makes the chair more economical, it has increased the stability and rigidity while we have managed to lower the weight of the said chair. The users report that the use of the new proposal gives them confidence and they report a series of possible improvements that we have listed previously and which we are working to present the final result in the next International Conference to receive criticism and comments from the public assistant.

References

1. Shore, S.L.: Use of an economical wheelchair in India and Peru: Impact on health and function. Med. Sci. Monit. **14**(12), PH71–PH79 (2008)
2. Shore, S., Juillerat, S.: The impact of a low cost wheelchair on the quality of life of the disabled in the developing world. Med. Sci. Monit.: Int. Med. J. Exp. Clin. Res. **18**(9), CR533 (2012)
3. Free wheelchair mission. https://www.freewheelchairmission.org/our-wheelchairs/
4. Kigali chair project. https://cargocollective.com/clararomani/Kigali-Chair-Project-1
5. Redacción: Entregan en Sonora sillas de ruedas de plástico, Aristegui Noticias Sitio web (2012). https://aristeguinoticias.com/2808/mexico/gobierno-de-sonora-entrega-sillas-de-ruedas-de-plastico/
6. El escándalo por las 'sillas de ruedas de plástico'. Noticieros Televisa. http://noticierostelevisa.esmas.com/especiales/493083/el-escandalo-sillas-ruedas-plastico/
7. Polémica por la silla de plástico sobre ruedas que usan para trasladar enfermos. https://www.clarin.com/ciudades/polemica-hospitales-portenos-sillas-ruedas_0_HkiMfucw7l.html
8. Indesit Company. Materials driven design. Dodici Edizioni, Milano (2012)
9. Courtis, C.: Discapacidad e inclusión social. Escuela Nacional de Ciencias en la Tierra. Revista Nexos, México (2015)
10. HU-GO, the Open Source DIY Low-Cost Wheelchair with 3D Printed Parts. https://3dprint.com/106254/hu-go-open-source-wheelchair/
11. TooWheels - the opensource wheelchair. https://hackaday.io/project/25757-toowheels-the-opensource-wheelchair

Media Convergence in Information Transmission in Museum Space

Yongbin Wang[1] and Jian Yu[2(✉)]

[1] Guangdong University of Finance and Economics, Guangzhou, China
[2] Guangzhou Academy of Fine Arts, Guangzhou, China
2366436445@qq.com

Abstract. Museums are important places for people's access to historical and cultural information. Therefore, people's experience and perception of information content shall be the main orientation of museum design. By using multiple media to build the exhibition space and interface, designers aim to form varied spatial effects and thereby combine educational functions with entertaining activities. The research content is divided into three parts: 1. An overview of the collisions between human and spatial scenarios in museums; 2. Observation of the expression of human-computer interaction technology in museum information transmission; 3. A brief description of the relation between media convergence and the spatial visual effects experienced by people in museums. In this paper, based on the aforesaid three aspects, the author unfolded a comprehensive discussion and explored the application of these theories in the design of information transmission in the narrative space of museums.

Keywords: Museum spatial plot · Interactive technology and information dissemination · Media convergence and spatial vision

1 Introduction

The multi-dimensional space of museums constitutes the flow of the discourse in the entire spatial plot. A plot discourse refers to not only the comprehensive experience of the audience, including their understanding and the recalling of their personal memory, while "reading" the information flow presented in the museum, but also the space that the audience see and feel when they are in the virtual world composed of the spatial plot under the theme of the museum. Field of vision is not always an integrated and closed system; while in reading, the shift of readers' field of vision is a natural and flowing process, which is not subjected to a paragraph or a chapter. Therefore, the reappearance of a space is not a composition of numerous unitary scenes, but a mixed and refined spatial complex consisted of a series of flowing scenes which cover multiple factors of the discourse, including those of geographical and spatial-temporal natures. The composition of spatial plot is reflected in the intertwined quotations [1]. The connotation of spatial plot is derived from the origins of the thousands of cultural activities under the theme of the museum. Therefore, the spatial plot is transformed

G. Di Bucchianico (Ed.): AHFE 2019, AISC 954, pp. 310–316, 2020.
https://doi.org/10.1007/978-3-030-20444-0_30

through the media into visible and appreciable information, which is finally passed on to the audience. The thematic discourse of a museum is realized by the presence of media, and the convergence of these media causes the discourse composed of the spatial plots to be infinitely open and even infinitely contradictory. This not only results in the generation of multiple interpretations of the spatial plot, but also the changes in the narrative style.

Media convergence bestows the space an extended impression and produces a multidimensional effect. Such convergence makes people feel that the space is reachable and touchable, despite the lack of any boundary. The feeling of extension and multidimensionality in the space are also embodied in the numerous fragmented and scattered scenes. For example, Archive Dreaming, an immersive art installation created by Refik Anadol, facilitates not only the search and classification of nearly two million documents in SALT Galata art museum, but also the real-time rendering of the corresponding images (see Fig. 1). With Archive Dreaming, space is seemingly composed of hundreds of thousands of visual data. When the data is broken into numerous individual spaces, the concept of boundary will cease to exist. Such shocking immersive space enables people to feel that they are encircled by books as the space becomes multidimensional and relatively virtual. The real-time interaction between people and visual data elements seems to closely resemble a conversation between people and space.

2 Method

2.1 Spatial Plot Experience in Museums

A plot design should attract people's attention; otherwise, it may be difficult to spread. Social development has increased the ease at which people's attention is being diverted from one thing to another. When we walk on the road, advertisements and posters along the road are probably just the complementary cloaks for a spatial scene, whose content cannot be distinguished one by one. We always do not realize that certain hotels or restaurants are located on the road we pass through every day until they are spoken of by others. When we play a video, we often close its window after ten seconds when we don't find it interesting. The audience's attention always shifts this quickly. The first impression or the first feeling always determines whether the entire communication process is smooth or not. While when we successfully draw people's attention, the attraction of the plot is the factor that helps us to focus their "concentration" on our subject for a longer period of time [2].

Therefore, to think about the museum plot experience from the perspective of space enables the information of space construction to be no longer single and separated but coherent and logical. Spatial plot, which plays an important role in museum information dissemination, integrates and transforms mass media of communication. James Carrey also expresses similar opinions in his lecture on the transmission view and ritual view of communication "Media ritual is a kind of activity held based on the themes of major media and relevant fields, which makes 'media as the access into the core of

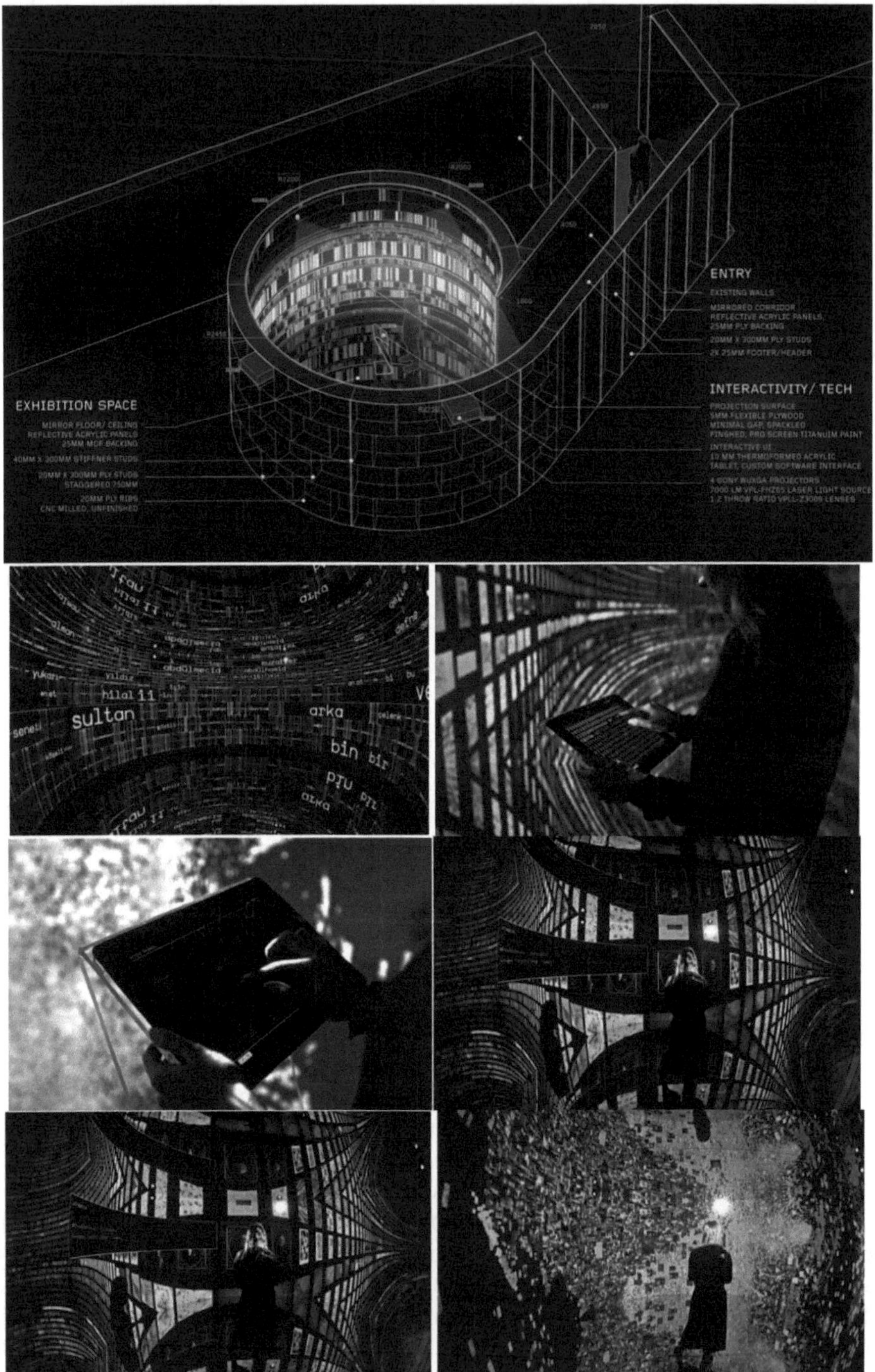

Fig. 1. Archive dreaming – building relations and drawing alt-history with machine learning

society' This values, which is hidden in the communication process, is intensified and rationalized [3]." therefore, the spatial plot experience of a museum becomes particularly important in the interactive experience at the museum.

2.2 The Mode of Conversation Between People and Multimedia Spatial Experience in Museums

In a museum, media convergence can improve the sense of reality and the sense of immersion of spatial experience, which enables the museum to leave a deep visual impression on its visitors. With the upgrading of media technology, new patterns are created in the museum experience design [4]. "Media convergence" becomes a positive developing trend of museum spatial experience. In general, media convergence promotes technology revolution, enriches the perceptual experience, and realizes the free choice of space and time, as well as the exchange between the subjects and objects of experience. However, "media convergence" is neither a simple superposition of media forms nor the development of technology, but the result of multiple attempts and practices conducted by the museums led by new media technologies for the purpose of maximizing the effect of experience by virtue of the advantages of media [5, 6].

2.3 Interactive Technology and Information Dissemination Modes in Museums

The role that the audience plays in museum information dissemination always depends on the following elements: whether the content draws attention; whether the information continues to arouse interests; whether the information is useful; and whether the audience takes action in response to such information. If the information does not draw attention, information dissemination will be impossible; if the information is not in accord with the audience's knowledge and believes, the transfer cannot proceed; if the audience receives information, but do not take action, the transfer will not achieve the expected results [7, 8]. Therefore, a successful museum information dissemination may be swift or circuitous and complex. In the research on the effect of communication, Aristotle focused on the communicators and explored how they persuaded others; while Cartwright researched the information receivers and had deep interests in the change of receivers during communication. The author thinks that it will be easier to draw a conclusion if the audience is taken as the main subject of research.

2.4 The Influence of Human-Computer Interaction Technology on Museum Information Dissemination Mode

Human-computer interaction technology under the background of media convergence is a channel and approach of museum information dissemination. Even with the continuous information flow in a museum, the audience can participate actively, while also perceiving and experiencing the information they are interested in by playing fun games or interactive games, which transforms every individual audience from being a passive receiver to an active participator. Given the support of the development of new media interactive technology, information dissemination can be realized in various aspects, including spatial arrangement, plot design, symbol guidance design, and participation promotion. Human-computer interaction technology should be applied to museum planning, instead of just high-tech production field. After the mutual infiltration and integration of the elements in various aspects like culture, history, and exhibits, the mainstream transfer mode, which adapts to the development of contemporary museums, can be finally created [8].

Museum spatial discourse information dissemination mode should be open. Every media plays its own role in spreading information. Other than spreading the content, the media itself also disseminates information. *Thousands of Miles of Mountains and Rivers* is a silk-based watercolor created by Wang Ximeng, a painter in the Northern Song Dynasty, which is also his only creation. Before being transformed into a human-computer interaction immersive art exhibition, it lies in the Palace Museum quietly. There are always numerous visitors waiting in line for having a look at this scroll painting with a thousand-year history. Nowadays, people "revive" this great creation in another way by integrating art with new media. *Thousands of Miles of Mountains and Rivers* is an about 12-meters-long scroll. In the picture, lies elegant terraces, pavilions, and spiraling smoke, which harmoniously combine motion and stillness. Every stroke of the green hills and clear water depicts a spatial image that, in Guo Xi's words, is "stoppable", "viewable", "traversable", and "livable". Only after appreciating the details inch-by-inch, can the audience explore the underlying secrets. Such "instantaneous change of scenes during motion" is tactfully represented again using the traditional brush and ink technique in conjunction with the new elemental integration of points, lines, and planes via the new media. As the audience walks along, the scroll painting unfolds gradually. When the audience places themselves in the painting, they are not only starting a dialogue with science and technology, but also communicating with the traditional art. The interaction between light-and-shadow and digital technology enables the traditional works to step out of the conventional two-dimensional plane, but also making itself compatible in three-dimensional spaces with an upgraded and more perfect light-and-shadow experience. The multidimensional world of sense makes *Thousands of Miles of Mountains and Rivers* into an audio-visual festivity as an aerial photo, building a connection between the audience and the exhibition through the former's senses. The presentation of the special techniques inherited by the cultural relics and the spiritual connotation of Chinese culture via modern science and technology media convergence just embodies the new narrative mode of spatial discourse. *Qingming shanghe tu* (*Ascending the River during the Qingming Festival*) that was painted by Zhang Zeduan and exhibited in the China Pavilion during the Shanghai World Expo, is another artwork "revived" by similar means (see Fig. 2).

2.5 Results

Spatial scene and spatial exhibit are displayed in front of the audience via media. In the traditional museum exhibition design activity, old media is just regarded as a necessary tool for displaying the appearance of exhibits, whose independence is not explored at all. In the exhibition spaces of contemporary museums, new technologies provide us with great possibilities of spatial expression. Media convergence, as a technique of exhibiting the major visual elements in space and displaying thematic features, has become increasingly common nowadays [9]. The series of visual and symbolic image

Fig. 2. A multimedia display of the riverside scene during the qingming festival at the China pavilion at the Shanghai world expo.

elements extracted from the theme of a museum exhibition is displayed through certain spatial arrangement methods via media convergence, where the change of light, the dynamic images, and hybrid-dimension spatial effect achieve an incomparable visual effect that other spatial elements cannot.

Audience's attention is the emphasis of museum design. Media convergence can expand space, or even make the space multidimensional through optical illusion, so as to "revive" the objects in the museum space and exhibit them by new means of expression. Moreover, when taking the cross-media narrative as an approach, it is necessary to take the audience's need for spatial experience into full consideration, exploit the special disadvantages of each media so as to make contributions to the narrative of the main story of the museum and enrich the information to be disseminated. In a museum information dissemination, promoting the audience's participation by symbols can finally develop the mainstream visual forms of space that can well adapt to the development of contemporary museums.

3 Conclusion

Media convergence sparks a great revolution in museum information dissemination modes and spatial plot expression by interactive technology, network technology, and communication technology in the new era. It is not a simple technology integration; it is a diversified and three-dimensional mode of transmission on the premise of the development of new media technology that is developed under the guidance of the audience's needs and the museum's thematic information. It also has to be considered based on the actual temporal and spatial criteria, after which the execution is to be implemented by the guidance of symbols and through information dissemination and interaction, so that the information offered in the museum can ultimately be spread in conjunction with the experience of entertainment.

Acknowledgments. This paper is sponsored by the characteristic specialty construction project of colleges and universities in guangdong province. Funded by arts program of humanities and social science research and planning fund of the ministry of education, "restoration and construction of lingnan city culture: research on environmental art design method based on geographic spatial information system (GIS)", no.: 15YJA760037; Guangdong province philosophy and social science "12th five-year plan" 2013 art project grant, "sustainable development research on environmental design of the pearl river delta garden city cluster", no.: gd13cys03.

References

1. Raskin, J.: The Humane Interface, pp. 1–8. Trans. by Y. Shi. China Machine Press (2003)
2. Zettl, H.: Image, Sound, Motion, pp. 17–30, 207–216. Trans. by M. Zhao. Beijing Broadcasting Institute Press (2003)
3. Carey, J.W.: As the Transmission of Culture. Trans. by W. Ding. Huaxia Publishing House (2005)
4. Dernie, D.: Advanced Course of British Exhibition Design, pp. 18–40. Trans. by W. Han. Shanghai People's Fine Arts Publishing House (2007)
5. Coper, A.: The Way of Interaction Design, pp. 169–188. Trans. by Q. Ding and F. Jiang. Electronic Industry Press (2006)
6. Gombrich, E.H.: Art and Illusion, pp. 154–217. Trans. by C. Yang, B. Li and J. Fan. Guangxi Fine Arts Publishing House (2012)
7. Martin, D., Jacobus, L.A.: An Introduction to the Art, pp. 395–432. Trans. by H. Bao and S. Huang. Shanghai Social Science Press (2011)
8. Lipson, H., Kurman, M.: 3D Printing – From Imagination to Reality, pp. 268–275. Trans. by the Expert Group of Sadie Research Institute. Citic Press (2013)
9. Hu, X.: Research on the Exhibition Concept of Modern Science and Technology Museum and the Development of New Exhibition Technology, pp. 127–188. Shanghai Science and Technology Academic Press (2006)

Design for Inclusion for Apparel and Textile Design

Formation of the 3D Virtual Models for Clothing Fit Assessment Applied in Human Factors

Hyunjung Lee(✉) and Su-Jeong Hwang Shin

Department of Design, Texas Tech University, Lubbock, USA
prohjlee@gmail.com, su.hwang@ttu.edu

Abstract. The aim of this study is to reveal the inconsistency of virtual model formations manually created by individual technical designers and to identify effective ways of referencing human body dimensions to develop a prototype virtual model in the industry. In this experiment, the technical designers were asked to create two virtual models based on two types of size charts. The results of the experiment show that the problems of inconsistency were associated with the different perceptions by each technical designer on body dimensions and body proportions based on given charts. In addition, the problems were caused by automatic adjustments in the 3D virtual fitting software program. The limitation of adjustable size scales existed in pre-programed built-in features in the program. The result of this study will provide fundamental and referential materials to enhance and supplement virtual modeling technology.

Keywords: 3D · Virtual model · Clothing · Fit · Size · Technical design

1 Introduction

The advancements of 3D technology have changed the way products are produced in various industry sectors. In the apparel industry, the 3D virtual fitting technology is emerging as a new paradigm of changing apparel manufacturing process [1]. In addition, in global outsourcing, the expectations of 3D virtual fitting technology is gradually rising as the new solution of quality control issues. The garment fit issue is a significant challenge in the stage of apparel manufacturing in collaborating with manufacturers that are located overseas [2]. For clothing fit assessment, sending photos of sample garments by email is a typical method. Although some companies communicate with manufacturers through web-based meetings, it is still difficult to handle the fit issues. These communication methods cause the increase of production lead time. As a potential technical solution, a virtual clothing fit assessment using a 3D virtual fitting software program has been introduced. A virtual clothing fit assessment allows manufacturers to save time and cost on producing multiple sample garments by replacing the actual production. Accordingly, expectation of the apparel industry for a virtual clothing fit assessment has led to demand for developing a 3D virtual model similar to human body type and dimension for accurate fit assessment. In previous

G. Di Bucchianico (Ed.): AHFE 2019, AISC 954, pp. 319–328, 2020.
https://doi.org/10.1007/978-3-030-20444-0_31

studies [3–9], creating a virtual model for virtual garment simulation has been explored using sizes of dress forms or human bodies scanned with a 3D body scanner. This way of creating a virtual model might not be appropriate for apparel companies due to the high cost of the equipment. For the 3D virtual fitting process, formation of a 3D virtual model is an important first step in that manufacture's use of the model for clothing fit assessment. In the industry, technical designers usually create a virtual model by using built-in features in the 3D virtual clothing fitting software program. The built-in features allow technical designers to manually create virtual models. However, consistency of the virtual model formation created manually by individual technical designers has been questionable [10, 11].

Thus, the aim of this study was to investigate two scenarios in creating a 3D virtual model by individual technical designers. One is about size consensus of each technical designer's virtual model formation with same body measurements. The other is how consistent the individual technical designers' works are when given two different types of size chart. The investigation will be followed by the suggestion of the effective ways of referencing human body dimensions for technical designers to develop a prototype virtual model in the industry.

2 Methodology

Technical designers were asked to create two virtual models for the one size, which is size 8, by using 3D virtual fitting software program to find the inconsistency of virtual model formations manually created.

2.1 Participants

A focus group for this experiment consisted of twenty-five Korean technical designers who have experiences with using 2D/3D CAD systems in the apparel industry; Korean apparel industry has many apparel companies that cooperate with brands which are located in the U.S. and Europe.

2.2 3D Virtual Fitting Software Program

For 3D virtual model formation, CLO software program was used, which is currently used widely in a garment development and fit assessment process by technical designers in the Korean apparel industry (Fig. 1).

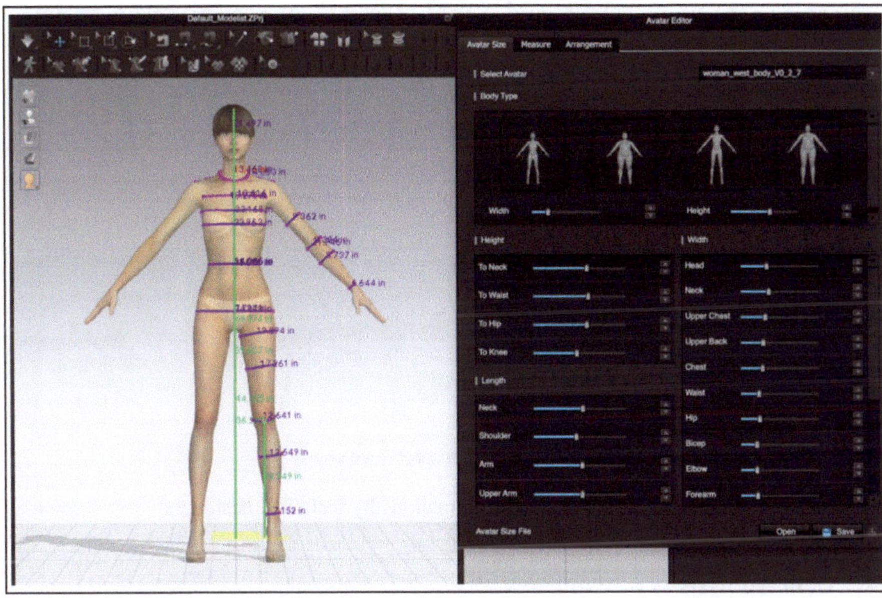

Fig. 1. Manual virtual model

2.3 Sample Size

For references of human body dimensions, we provided (1) a long size chart with 38 measurements from ASTM D 5585 standard (height, weight, bust, waist, high hip, hips, mid-neck, neck base, armscye, upper arm, elbow, wrist, thigh max, thigh mid, knee, calf, ankle, center trunk, crotch length total, cervical height, waist height, high hip height, hips height, crotch height, knee height, ankle height, waist length front, waist length back, across shoulder, cross back width, cross chest width, shoulder length, shoulder slope degree, arm length (shoulder point to wrist point), arm length (center back neck to wrist point), bust point to bust point, neck to bust point, scye depth) and (2) a short size chart with 11 measurements (bust, bust point, under bust, waist, high hip, low hip, waist to floor, back neck to waist, front neck to waist, arm, above bust) from a technical size chart. Accordingly, the technical designers were asked to create two virtual models: Model A-based on the long size chart and Model B-based on the short size chart (Fig. 2).

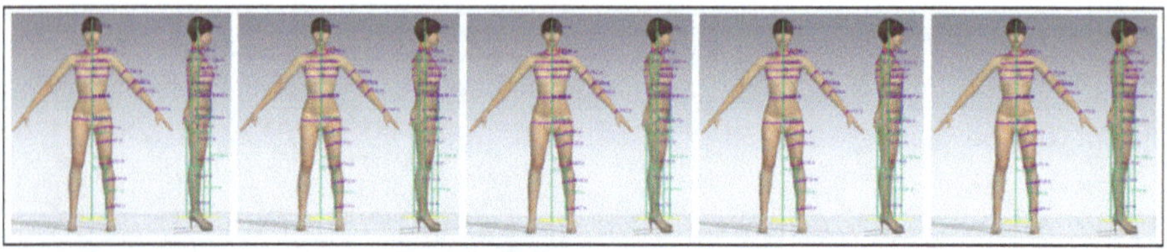

Model A (based on long size chart)

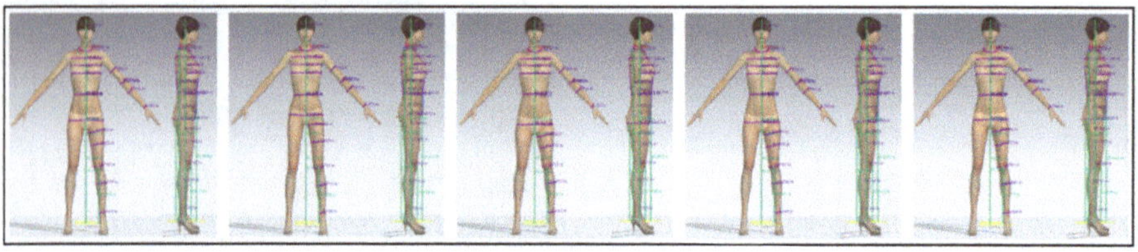

Model B (based on short size chart)

Fig. 2. 3D virtual model formations by technical designers

2.4 Data Analysis

The two virtual models were analyzed with descriptive statistics and t-test.

Descriptive statistics was used to examine size consistency of virtual models created by two types of size chart. T-test was performed to identify effective size chart on creating virtual model formations, comparing size differences between each size chart and virtual model formations.

3 Result

3.1 The Size Comparison Between Virtual Model Formations by Two Types of Chart

Consistency in Size. When technical designers created virtual model, Model A based on the ASTM standard (long size chart) shows that average standard deviation 1.4 and maximum SD 4.22. The average of dimensional error difference from ASTM size chart is −0.34 in. and the dimensional error range is between −3.92 and 0.71 in. When technical designers created the other virtual model, Model B based on the technical size chart (short size chart), results show that average standard deviation 1.98 and maximum SD 8.19. The average of dimensional error difference from the technical size chart is −0.16 in. and the dimensional error range is between −2.54 and 1.02 in. (Table 1). The findings show that body dimensions of virtual models are created differently by different users. When comparing the standard deviation of Model A with Model B, the standard deviation of

Model A is lower than the one of Model B (Fig. 3). In comparing the dimensional errors between Model A with Model B, while the average and range of dimensional errors are larger in Model A than Model B, the dimensional errors in Model B shows more inconsistency (Fig. 4). Therefore, the results show that virtual model formations from long size chart are more consistent than those from short size chart.

Table 1. The standard deviation and the dimensional error range

	Average SD	Max SD	The dimensional error range (inch)
Model A (long size chart)	1.4	4.22 (elbow width)	0.711 (high hip width) ~ −3.922 (cross back)
Model B (short size chart)	1.98	8.19 (hip height)	1.019 (thigh max width) ~ −2.544 (hip height)

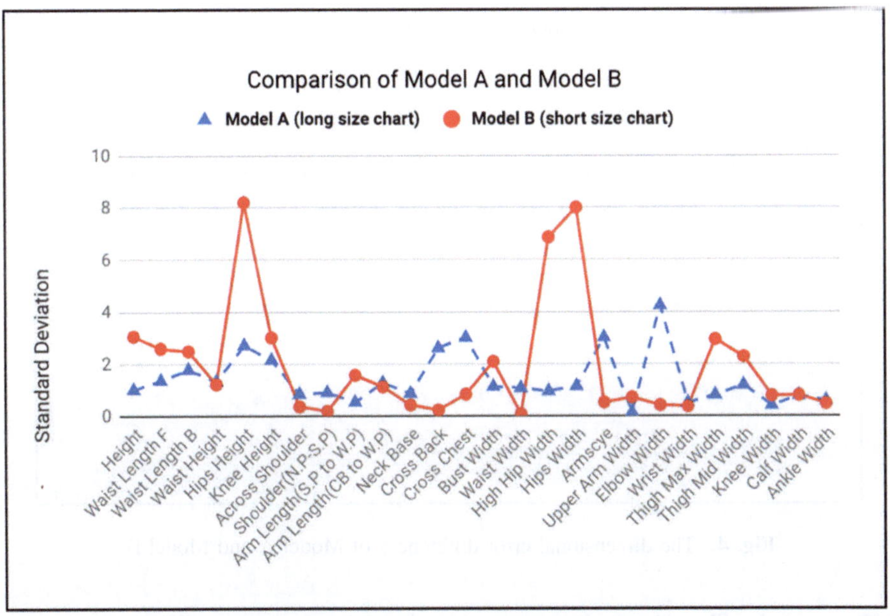

Fig. 3. The size consistency of Model A and Model B

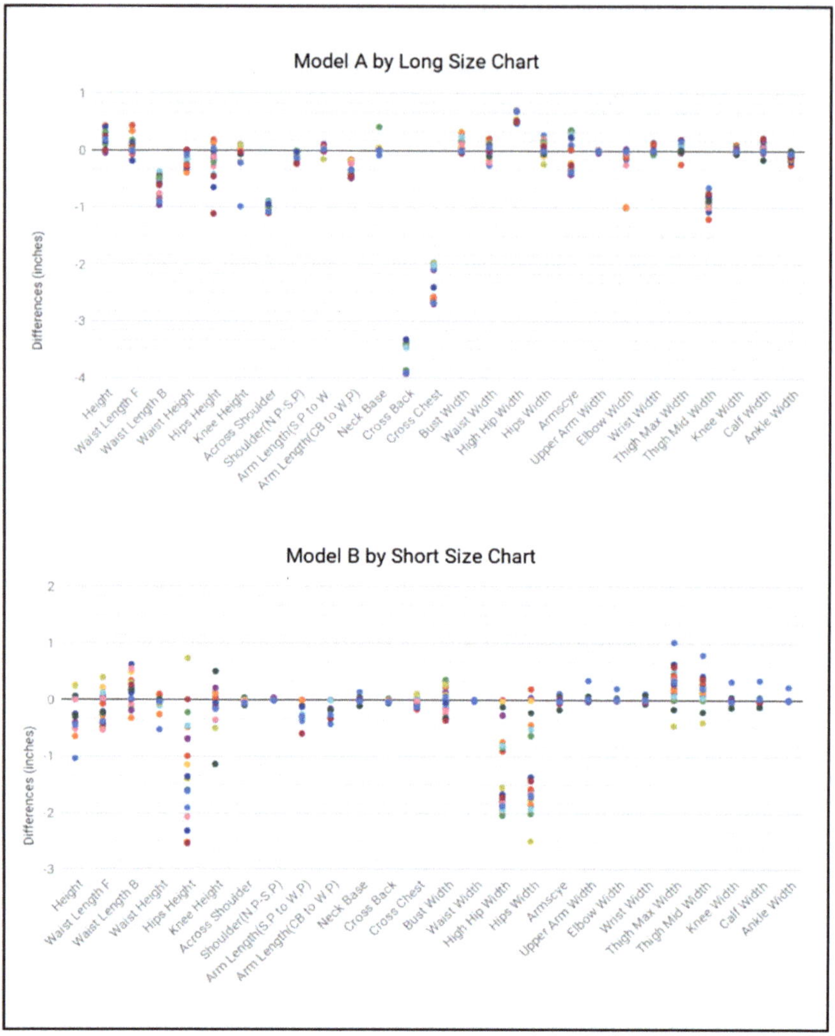

Fig. 4. The dimensional error differences of Model A and Model B

Differences in Size. Apaired t-test was performed to compare differences in size between Model A and Model B. When the same size (size 8) was given, results show significant differences in all body dimensions except for cross chest and armscye ($p < 0.001$) (Tables 2 and 3). This means that the amount of the information given on body dimensions to the technical designers has an effect on a virtual model formation.

Table 2. The size differences (height and length) between virtual model formations by two types of size chart

Height/Length	Mean(SD)		t-Value
	Model A (long size chart)	Model B (short size chart)	
Height	65.20(1.01)	72.70(3.05)	−112.751***
Waist length front	14.29(1.36)	15.88(2.59)	−24.633***
Waist length back	15.59(1.77)	17.47(2.48)	−28.072***
Waist height	39.86(1.32)	45.50(1.22)	−162.765***
Hips height	31.83(2.70)	36.04(8.19)	−25.201***
Knee height	17.93(2.14)	20.00(3.00)	−31.873***
Across shoulder	14.13(0.83)	13.82(0.36)	17.403***
Shoulder (neck point to shoulder)	5.00(0.90)	4.64(0.18)	20.036***
Arm length	23.52(0.53)	24.66(1.56)	−32.03***
Arm length (center back to wrist)	30.72(1.27)	31.50(1.12)	−21.347***
Cross back	10.96(2.60)	10.52(0.22)	8.659***
Cross chest	11.20(3.00)	11.31(0.83)	8.659***

***p < 0.001, **p < 0.01, *p < 0.05

Table 3. The size differences (width) between virtual model formations by two types of size chart

Width	Mean(SD)		t-Value
	Model A	Model B	
Neck base	14.27(0.86)	15.05(0.42)	−38.157***
Bust	35.09(1.13)	35.56(2.08)	8.659***
Waist	27.00(1.07)	26.50(0.07)	24.272***
High Hip	35.11(0.95)	33.86(6.85)	8.307***
Hips	37.55(1.14)	36.46(7.99)	6.498***
Armscye	15.31(3.00)	15.31(0.52)	0.061
Upper arm	10.75(0.13)	9.31(0.71)	97.944***
Elbow	9.48(4.22)	8.28(0.41)	14.207***
Wrist	6.01(0.50)	5.75(0.36)	26.233***
Thigh max	21.77(0.80)	20.69(2.93)	17.678***
Thigh mid	18.86(1.18)	17.88(2.27)	17.989***
Knee	14.13(0.40)	12.76(0.77)	78.104***
Calf	13.64(0.77)	12.76(0.79)	43.116***
Ankle	9.05(0.58)	7.17(0.47)	143.079***

***p < 0.001, **p < 0.01, *p < 0.05

3.2 The Difference of Body Dimensions Between a Size Chart and a Virtual Model Formation

According to the t-test, while the sizes of 50% of virtual model formations are different from the ones of the technical size chart, 57.7% of the virtual model formations are significantly different from the ones of the ASTM body standard ($p < 0.05$) (Tables 4 and 5). This results show that more differences from the body dimensions with the long size chart are made than the short size chart while the consistency between virtual model formations is greater than the short size chart. In the Model A, body dimensions of the virtual model formation display low standard deviation ($SD < 1.00$) with a significance level ($p < 0.001$) of t-Value in the following dimensions: across shoulder, shoulder (N.P-S.P), high hip width, and ankle width. This can be interpreted that there is a limitation of adjustable size scales in inputting dimensions on the body parts and automatic adjustment function causes dimensional errors while creating virtual models.

Table 4. The dimensional difference (height and length) between size chart and a virtual model formation

	Model A			Model B		
	Long size chart (inch)	Mean(SD)	t-Value	Short size chart (inch)	Mean(SD)	t-Value
Height	65	65.20(1.01)	10.075***	72.95	72.70(3.05)	−4.014**
Waist length front	14.25	14.29(1.36)	1.403	16.079	15.88(2.59)	−3.745**
Waist length back	16.24	15.59(1.77)	−18.365***	17.361	17.47(2.48)	2.198*
Waist height	40	39.86(1.32)	−5.492***	45.544	45.50(1.22)	−1.901
Hips height	32	31.83(2.70)	−3.063**	37.248	36.04(8.19)	−7.219***
Knee height	18	17.93(2.14)	−1.628	20.11	20.00(3.00)	−1.796
Across shoulder	15.13	14.13(0.83)	−60.159***	13.85	13.82(0.36)	−4.581***
Shoulder (neck point to shoulder)	5.13	5.00(0.90)	−7.253***	4.638	4.64(0.18)	0.526
Arm length	23.5	23.52(0.53)	1.55	24.881	24.66(1.56)	−7.069***
Arm length (center back to wrist)	31.06	30.72(1.27)	13.483***	31.651	31.50(1.12)	−6.606***
Cross back	14.63	10.96(2.60)	−70.456***	10.535	10.52(0.22)	−4.230***
Cross chest	13.63	11.20(3.00)	−40.690***	11.38	11.31(0.83)	−3.940**

***$p < 0.001$, **$p < 0.01$, *$p < 0.05$

Table 5. The dimensional difference (width) between size chart and a virtual model formation

Width	Model A			Model B		
	Long size chart (inch)	Mean(SD)	t-Value	Short size chart (inch)	Mean(SD)	t-Value
Neck Base	14.25	14.27(0.86)	1.078	15.054	15.05(0.42)	−1.032
Bust	35	35.09(1.13)	3.984**	35.57	35.56(2.08)	−0.261
Waist	27	27.00(1.07)	0.17	26.5	26.50(0.07)	−0.885
High hip	34.5	35.11(0.95)	32.215***	35.213	33.86(6.85)	−9.675***
Hips	37.5	37.55(1.14)	2.218*	37.647	36.46(7.99)	−7.306***
Armscye	15.38	15.31(3.00)	−1.089	15.324	15.31(0.52)	−1.242
Upper arm	10.75	10.75(0.13)	−1.81	9.288	9.31(0.71)	1.511
Elbow	9.75	9.48(4.22)	−3.212**	8.27	8.28(0.41)	1.584
Wrist	6	6.01(0.50)	0.684	5.746	5.75(0.36)	0.32
Thigh max	21.75	21.77(0.80)	1.023	20.406	20.69(2.93)	4.766***
Thigh mid	19.75	18.86(1.18)	−37.592***	17.695	17.88(2.27)	3.996**
Knee	14.13	14.13(0.40)	−0.261	12.763	12.76(0.77)	−0.141
Calf	13.63	13.64(0.77)	2.092	12.763	12.76(0.79)	0.113
Ankle	9.13	9.05(0.58)	−7.265***	7.164	7.17(0.47)	0.935

***$p < 0.001$, **$p < 0.01$, *$p < 0.05$

4 Conclusion

This study provides evidence that the sizes of manual 3D formation are inconsistent. When individual technical designers made virtual sample size model with manual 3D formation, the body dimensions of each virtual model created by the individuals are not consistent. Also, the models are inconsistently created regardless of having long or short size chart of the body dimensions. The inconsistency is partially caused by the different perception by each technical designer on the body dimensions and body proportions based on the given charts. When provided a short size chart to build virtual models, the inconsistency of the sizes is more significant than a long size chart.

On the other hand, when provided a long size chart, the problems are associated with functions for 3D virtual modeling in virtual fitting software program. Automatic scale adjustment and limitation of adjustable size scales cause more difficulties while adjusting a built-in feature pre-programed.

In conclusion, manual 3D formation would create an inaccurate virtual sample size model. This indicates that with the manual formation, there is a high chance that garments with imprecise size and fit will be produced by unreliable fit assessment.

Currently, in the apparel industry, accurate fit and size are essential and should be measured consistently which, in turn, leads to satisfied retailers and customers. Thus, further research should be performed for providing anthropometric references and guidelines for technical designers to have consensus of the sample size model. Further, the developers of 3D virtual technology should improve the limitations of functions for 3D virtual modeling.

References

1. Durupynar, F., Gudukbay, U.: A virtual garment design and simulation system. In: 11th IEEE International Conference Information Visualization, pp. 862–870. IEEE Press, New York (2007)
2. Lee, E., Choi, H., Do, W.: A study on present state of technical designers' work responsibilities and training conditions. J. Korean Soc. Cloth. Text. **37**(3), 292–305 (2013)
3. Allen, B., Curless, B., Popovic, Z.: The space of human body shapes: reconstruction and parameterization from range scans. ACM Trans. Graph. **22**(3), 27–31 (2003)
4. Baek, S.Y., Lee, K.W.: Parametric human body shape modeling framework for human-centered product design. Comput. Aided Des. **44**(1), 56–67 (2012)
5. Charlie, C.L.W.: Parameterization and parametric design of mannequins. Comput. Aided Des. **37**(1), 83–98 (2005)
6. Kim, S.M., Park, C.K.: Parametric body model generation for garment drape simulation. Fibers Polym. **5**(1), 12–18 (2004)
7. Lee, M., Shon, H.: A comparative analysis of the different between CLO 3D avatar sizing and actual body measurement shapes. J. Fash. Bus. **16**(4), 137–151 (2012)
8. Li, J., Chen, J.: A mannequin modeling method based on section templates and silhouette control. Int. J. Cloth. Sci. Technol. **21**(5), 300–310 (2009)
9. Seo, H.W., Magnenat-Thalmann, N.: An example-based approach to human body manipulation. Graph. Models **66**, 1–23 (2004)
10. Shin, S., Robinson, J.R.: An analysis of the 3D sample size model built by individual users. In: International Textile and Apparel Association 71st Annual Conference. ITAA Press, Tennessee (2014)
11. Shin, S.: A practicability of utilizing 3D systems for virtual fitting process in the apparel industry. In: International Textile and Apparel Association 72nd Annual Conference Proceedings. ITAA Press, Tennessee (2015)

Apparels for Comfort. Knitting for People

Gianni Montagna(✉), Cristina Carvalho, and Carla Morais

CIAUD, Lisbon School of Architecture, Universidade de Lisboa,
1349-063 Lisbon, Portugal
g.montagna@gmail.com, cristifig@gmail.com,
carlota.morais@gmail.com

Abstract. In garment and fashion design human and technical factors to take into account are multiple, and are reflected in numerous aspects of the product, process and relationship. Comfort of the objects to be projected and produced are an essential element that in many cases is left aside, favoring the aesthetic aspect and social representation. The knitted textile material is a peculiar element which, because of its characteristics, has reserved a prominent place in the field of fashion design and is characterized by the possibility of creating its own textile surface. Knit material, like any other textile surface, includes and promotes a material and immaterial culture representative of the social and communal world. This paper intends to argue and approximate the different characteristics provided by the knitted textile material to the needs of the users and in particular the possibility of co-creating clothing adapted to the needs of the users.

Keywords: Apparel design · Co-design · Textile design · Knit design · Comfort

1 Introduction

The relationship of users with their own clothing is profound. Textile materials have a history of thousands of years and are part of different levels of our experience and way of living. The comfort of clothing, together with the aesthetic issue, is one of the main elements taken into account when choosing a piece that is increasingly representative of our way to be and well-being. Given that, comfort itself should be understood as a comprehensive issue that manifests itself at the material and immaterial level, its approach must be careful, broad and take into account the objective and subjective aspects of the user and the subject of study itself.

In recent years material developments have evolved rapidly and accompanied the new technologies, allowing users to meet the needs of the users in search of materials that better adapt to their daily needs, more agitated and with higher standards in aesthetic and functional terms [1]. The new possibilities and challenges offered by the new technologies, such as the knit material associated with new products and new structures capable of ensuring greater user comfort and high functional performance, are placed in the textile area. The knit material, due to its intrinsic characteristics, presents high levels of elasticity and consequent adaptation to the body, allowing a high increase in thermal comfort, taking into account its better capacity to allow an

G. Di Bucchianico (Ed.): AHFE 2019, AISC 954, pp. 329–336, 2020.
https://doi.org/10.1007/978-3-030-20444-0_32

easier dissipation of body heat. Since ancient times the knitted textile material has been used to make garments that need to adapt their shape to the body.

To talk about knit is to touch different concepts and variables that, working together and mixing constantly, offer multiple products and structures that are constantly being reinvented [2].

2 The Environment of Knit Material

As an ancient craft, present since 1000BC, knitting was originated from hand knotting or twisting of yarns using hand fingers. The word ‘knit’ originates from the old English term ‘cnyttan’, and from German ‘knütten’, developed to modern term knot. Hand knitting rapidly advanced from fingers, to the use of hand tools – known as pins –, into a skilled and complex craft where, by the fifth century AD, knitted objects such as socks are recorded, combining fashioning, seaming, and circular knitting together with patterning techniques simultaneously [3]. Cross loop knitting (or cross-knit looping) – a single-needle technique – dates back to the 2nd century AD, being found in objects made by civilizations like the egyptians, romans and Scandinavians. Pattern knitting, known in Arabia and India from the 9th century, is introduced in the 13th century in Iberia and southern Italy [4]. As said by Schoeser [4] and Steele [5], the development of knitting has been responsible for the most striking and fundamental changes in textiles and clothing since 1600.

This deep connection with the consumer is one of the strong abilities of the knitted textile over the last decades, and the use of knitted textiles has been increased year by year since the seventies, due to the spreading of the practice of sport and the introduction of specific materials for sports activities. According to Sinclair [6], knitwear will have a massive growth in the fashion industry, related to its intrinsic relationship with casual dress, active wear and sportswear, crescent trends in our society. Recent statistics of the European Apparel and Textile Confederation (Euratex) show that in 2015 more than 1/3 of the fashion market was made of knit garments.

Knitting and knitting can probably be considered as one of the most creative options within the entire fashion and apparel production system given the possibility of offering multiple choices and variables placed in the hands of designers. Due to its constructive characteristics, the designer is allowed to create the textile substrate where it can intervene directly and in a much simpler and direct way in the production of the knit itself, without having to intervene vertically in the textile line, as with the production of orthogonal materials. The construction of knitted textile materials exists essentially in two ways: as a warp knit or as a weft knit. If the warp knit is essentially used for the production of more technical and industrial materials, the weft fabrics are the ones that have more application in apparel and fashion, being able to present with different variables within two primordial types of construction and that depend on different Technologies: Circular knit jerseys more thinner and used for interiors and “V” bed knits mainly for external use, being thicker and heavier (Fig. 1).

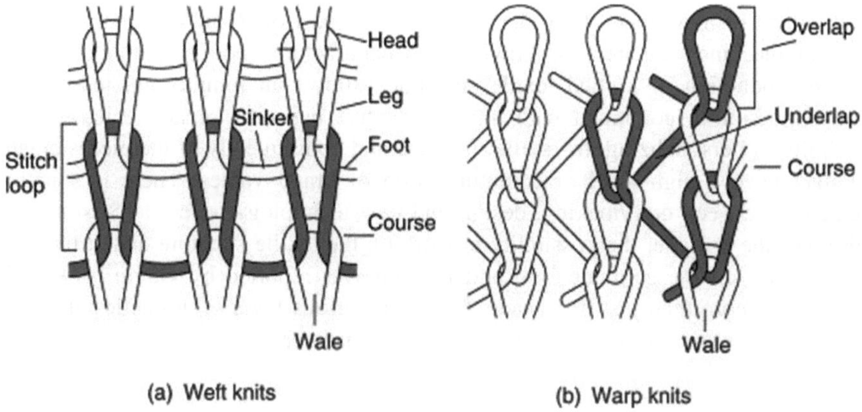

Fig. 1. Weft and Warp knitting structure. https://www.sciencedirect.com/topics/materials-science/knit-fabrics [retrived 15/02/2019]

The production of knit substrate for garments technically obeys the rules for the construction of the weft knit which allows the production of the substrate by interpolation of a new loop of knit from the ring formed in the previous row and that is constructed with the use of a single thread. This textile construction which in general terms is considered simple allows a direct intervention of the designer in its production and the choice of the type of yarn and its finish, color, diameter, yarn effect, composition, etc., is a much simpler choice of rather than intervening in an orthogonal textile row (web and weft) where the processes are much more extensive and complex and where the production and availability times of the materials in question are much longer.

As pointed out by Lindsay [7] referring to her Master Thesis, Jack Lenor Larsen in 1961 described machine knitting as the next industrial revolution, with the possibility of overlapping orthogonal textiles (wowen textiles) in terms of production speed and economic results.

Still far from this reality, Larsen was able to perceive the great potentiality of the knits that had their greater paradigmatic development with seamless technology. Spencer [8] refers that seamless technology knits on spiral, passing from one needles bed to the other working sequentially selvage needles and thus creating a tube. The process of creating garments for the upper part of the human body, refers Black [9], is to knit one tube for the body and two for sleeves, and join it together starting from under the sleeve. The result of the upper body knit garment will be also un unique tube, using the similar garment conception used by centuries for the Norwegian traditional knitting sweaters, with a traditional circular yoke.

In this sense, taking into account that in the production of this type of material the designer can have total possibility of creation and intervention in the production of its characteristics, the parallelism with *haute couture*, where in many cases the textile material used is a poetic construction of the seasonal inspiration is inevitable. In knitwear, the design becomes a basic step in a long industrial chain that starts from the development of new fibers, from the recover of the past ones and comes to the innovation on machines that are now allowing new manufactures more and more often [10].

The production of knitted textile substrates allows an infinite and independent creative possibility combined with high productive capacity, greatly enhancing the possibility of personalizing the textile substrate and consequently of the pieces created that may have a high level of communication of brand values. There is a natural interaction between craft/making, design and new technology, refers to Sissons [11]. Knit gives the designer the possibility to work on the textile structure and at the same time to work on the garment form and function. As reported by Quinn [12] textile surfaces are more than just the façade, but are a curious layer of aesthetics and identities, and a contentious site of exploration and resistance.

3 User's Factors

The human factors involved in the design and production of fashion and apparel are multiple and of different origins. The user/object relationship is a relationship that is based on different stages and involves multiple levels of perception and involvement. The relationship of users with the object of clothing is made not only by the contact with the physical object but especially with the aesthetic and social representation that this object represents and to which it may be associated and consequently represent the user himself.

Despite the great importance and involvement of the aesthetic and social mechanisms of clothing, which gratify and participate in the construction of the psychological well-being of the user, the issue of physical well-being becomes more and more important and contributes to the psychological well-being and the increase of self-esteem, and the garment can be defined as a multisensory interface that communicates constantly with the outside world.

The comfort of the clothing we use on different occasions is increasingly taken into consideration and is one of the determining factors for the end user. In reality comfort is usually measured by its opposite, that is by the level of discomfort that any object or situation may cause the individual. In this sense, Chappells and Shove (2004) [13], says that comfort can be defined as "a feeling of contentment, a sense of cosiness, or a state of physical and mental well-being." In this sense, Sontag [14], Oğlakçıoğlu and Marmaralı [1] and Kamalha [15] add that the main areas of comfort can be summarized as follows: psychological comfort, sensorial comfort and thermo-physical comfort. Psychological comfort can influence the user in their relationship with others and in the way their interpersonal social network can perceive the values that the user wants to communicate. In this sense, it is a matter that, linked to the object of study, is projected to the outside and will help in the construction of greater security of the individual with its social structure. An example of this can be considered the typical motifs of the "fair isle" knits that soon identifies a group of individuals with the small Scottish island. There are small geometric motifs that are repeated throughout the surface of the textile and are often worked in stripes of different and contrasting colors. As stated by Steed [3] knitting has always been the method for expressing oral history, facilitating community engagement and expressing deep personal attachments. In a society where the individual seeks to be heard, knitting emerges as a powerful social and political tool.

With regard to sensory and thermophysical comfort, these two areas of comfort are related more to the individual and less to his social network. The sensorial and thermophysical comfort will eventually be more connected to the user's senses, being more related to the textile material and the garment. The possibility of wearing a comfortable piece, which does not create great thermal oscillations in the entrance and exit of specific environments and that are comfortable in the movement of the user in its interior may be priorities for this area of comfort, being, more and more, today are a necessity to be taken into account by the multifaceted uses of the garments we find on the market. Sportswear, which had a great development of materials from the 70's of last century, with its more comfortable garments, is influencing more and more our everyday clothing, which we require to be practical, due to our overactive life. Knitting, because of its adaptive and comfortable capabilities, has always been used in active wear and sportswear. In the future, refers to Sinclair [6] we will see a growing migration of high-performance knit textile surfaces - used in both of these areas - to everyday life.

4 Design for People

The adaptation of the product of fashion and clothing to the users is often an empirical task that, through advances and retreats, adapts to the market and the needs of the possible customers who, season after season, wait for continuous innovations and proposals of renewed communication codes with its ecosystem. The elements that form part of this equation are multiple and of different levels, ranging from styles and materials, colors and structures, details and volumes, references and significances, hierarchies and recognitions, honors and approaches, among others. The continuous mixing, remixing, alteration, adaptation, forecasting, etc., of these and other slopes and variables are the center of the fashion product that has the capacity to reinvent itself constantly, even if it is found in terrains that are not stable and constantly in balance between the objective and subjective, between material and immaterial, between concrete and abstract, between fashion and tradition, between the individual and society.

Despite all the possibilities presented by the direct actors in the process, we must take into account indirect and subsidiary actors who carry out the activities necessary for the implementation of the product on the market, have specific and specific needs that must be taken into account. It is a multifactorial, interdisciplinary and complex process where the objective needs of the different actors are associated with the subjective ones, in a continuous change of positions and commitments with the aim of presenting a product that satisfies a high number of consumer needs to whom the product or service is directed.

No doubt that design is the discipline that more naturally predisposes as an intermediary of these issues, presenting methods that can help reduce distances between different positions and especially between user and fashion product/textile. Design, subject to different areas of knowledge, combines quantitative and qualitative work methods, subjective and objective, active and observational, etc., which allow the exploration and application of methods and methodologies that provide collaborative work among the different areas disciplinary measures that contribute to the production of the textile object.

Co-design and customization are two typical examples of end-user collaboration in the production processes of the textile products themselves, which for some time have been equated by design teams and companies in order to reduce the distance between those who use and those who make product to its source of revenue.

Although there are different definitions of these concepts, we can broadly agree with those presented here. The co-design process can be defined as a process where user needs and preferences are the basis for the development of the entire design process, improvements and innovation [16, 17]. To Zamenopoulos and Alexiou [18], co-design means that people come together to conceptually develop and create things that respond to certain matters of concern and create a (better) future reality.

This need, already identified by Toffler in his 1980 book The Third Wave when a consumer participates in co-design and at the same time becomes a co-producer of the product he consumes, can be defined as "prosumer" [19, 20]. By customization we can understand the adaptation of a product made available to consumers for a possible adaptation in the technical design phase and the physical production of the service article [21]. Different definitions can be adopted for the development of actions and methods that in general terms differ little from each other, as in the case of tailored product, consumer driven design, made to measure, made to consumption, among many others.

The knit material is a textile structure that has specific characteristics due to its majority of the technology used in its production. While it is true that for the production of any material the technology used in its production has a major influence, for the knitting sector this symbiosis has always been essential and in many cases the new products implemented in the market as fashion products were the result of great advances technological developments. We can state here many, but taking into account the knit/comfort aspect that is the focus of this article, the possibility of seamless knit production has allowed the implementation of a new paradigm for the sector that is still far from exhausting its possibilities. As referred by Spencer 2001 apud Peterson (2016) [22] complete garment technology, also known as the seamless garment technology, was introduced on "V-bed" flat knitting machines in 1995, having evolved form developments in the 1980s. V-bed machines have two needles beds, on a position of an inverted V and equipped with needles. This methodology, initially based on existing technology, allowed the development of garments with different shapes but without seams, which in some cases may represent a problem because of the friction and thickness that may increase the levels of discomfort and in severe could provoke scares on the user's skin surface. The whole garment is made directly in the flat knitting machine without post cutting and sewing processes (Legner 2003 apud Peterson 2016) [22]. At the beginning of the 21st century, another innovative technology for knitting became a reality and a productive one. "Seamless" or "no seam garments" knitting technology allows the production of knit garments that fit the body and use a limited number of seams, usually reserved for hem finishes or hemming applications. This technology was developed in the 80's of the XX century, when a lot of leather shoes were used directly in contact with the skin of the lower limbs, minimizing the use of socks. For this reason, the Seamless technology has been adapted to clothing so as to minimize the ever-increasing decline in stocking markets and to apply the techniques and logics of tubular and three-dimensional fabrics, adapting itself to the production of interior and exterior knit which could be used as any other type of knitted garment, with remarkable increasing levels of comfort for the end user.

The technological possibilities make this type of clothing an example of customization and adaptation to the different needs of use, since it presents infinite possibilities of technical development of the product used different textile structures in the same piece, being able to adopt different levels of compression by zones, application of different intensities of the substrate also by zones/areas at the surface level and in particular, being able to produce finishes that very little interfere in the interaction with the user. This technology is prepared to be able to produce individual pieces adapted to the users producing knitted garments in a mass customization optics. The disadvantage of the high prices of these equipment continues to be a factor that still prevents the large diffusion of these technologies for small shops and small businesses where, despite everything, the competition is also still in the price competition base.

5 Conclusions

If the fashion and clothing product remains one of the central elements of the individual's relationship with the social and professional environment, design processes need better vertical and horizontal integration to help design products to be more representative of consumer needs and its media. The possibility of user participation in the design process can be done in different ways, all of which allow a better efficiency in the interpretation of user needs, increasing the market penetration capacity of brands and designers, decrease in production of garments that, season after season, fill stores of not sold or rejected clothing by consumers for not responding to their needs. The poor sustainability of the garment industry needs to apply co-design strategies in order to produce better and more sustainably, sparing resources that are still wasted today because they do not have a close relationship with their final users.

Participation in the production of the knit garments by the authors themselves may represent the most complete and real scenario of co-design, and may include and extend to the adjacent methodologies of "design by user", "inclusive design", among others, knitted fabric provides this kind of joint production, adapted, customized while providing high levels of comfort and adaptability of the material produced.

With the integration of technological systems, processes, methods, etc., the act of creating products is no longer to be fragmented to become integrated and thought as a joint problem, promoting the interconnection between producer and user and meeting the real material and immaterial needs that make the fashion product a true product of communication and individual expression.

Acknowledgment. The authors of this paper wish to thanks the Centre for Research in Architecture, Urbanism and Design (CIAUD) of the Lisbon School of Architecture of the University of Lisbon and FCT for founding this project.

References

1. Oglakcioglu, N., Marmaralı, A.: Thermal comfort properties of some knitted structures. Fibres Text. Eastern Europe **15**, 94–96 (2007)
2. Montagna, G., Santos, L.: Knit and technology: a long lasting friendship: progress(es) – theories and practices. In: Kong, M., Monteiro, M. (eds.) Proceedings of the 3rd International Multidisciplinary Congress on Proportion Harmonies Identities (PHI 2017), Bari, Italy, 4–7 October 2017, pp. 219–224. CRC Press, Bari (2017). Print ISBN 1351242679
3. Steed, J.: Hand knitting in a digital era. In: Nimkulrat, N., Kane, F., Walton, K. (eds.) Crafting Textiles in the Digital Age (2016)
4. Schoeser, M.: World Textiles: A Concise History. Thames & Hudson, London (2003)
5. Steele, V.: Encyclopedia of Clothing and Fashion. Thomson Gale, Detroit; London (2005)
6. Sinclair, R.: Textiles and Fashion: Materials, Design and Technology (2015)
7. Lindsay, J.L.: Mary walker phillips: the art of knitting. J. Modern Craft **8**, 125–139 (2015)
8. Spencer, D.J.: Knitting Technology: A Comprehensive Handbook and Practical Guide. Woodhead (2001)
9. Black, S.: Knitwear in Fashion. Thames & Hudson (2002)
10. Affinito, L., Maria Conti, G., Motta, M.: New vision on Knitwear design. How the traditional craft methodologies are evolving into fashion international scenarios. Des. J. **20**, S2760–S2770 (2017)
11. Sissons, J.: Knitwear. AVA Academia; Distributed in the USA & Canada by Ingram Publisher Services, Lausanne; La Vergne (2010)
12. Quinn, B.: Textile Futures: Fashion, Design and Technology. Berg, Oxford (2010)
13. Chappells, H., Shove, E.: Comfort: a review of philosophies and paradigms. Unpublished Essay (2004). http://www.lancs.ac.uk/fass/projects/futcom/fc_litfinal1.pdf
14. Sontag, M.S.: Comfort dimensions of actual and ideal insulative clothing for older women. Cloth. Text. Res. J. **4**, 9–17 (1985)
15. Kamalha, E., Zeng, Y., Mwasiagi, J.I., Kyatuheire, S.: The comfort dimension; a review of perception in clothing. J. Sens. Stud. **28**, 423–444 (2013)
16. Sanders, E.B.-N., Stappers, P.J.: Co-creation and the new landscapes of design. Co-design **4**, 5–18 (2008)
17. Burkett, I.: An introduction to co-design. Saatavissa (2012). http://design4socialinnovation.com.au/wp-content/uploads/2014/09/An-Introduction-to-Co-Design-by-Ingrid-Burkett.pdf. Accessed 21 Oct 2015
18. Zamenopoulos, T., Alexiou, K.: Co-design As Collaborative Research. Bristol University/AHRC Connected Communities Programme, Bristol (2018)
19. Kotler, P.: The prosumer movement. In: Blättel-Mink, B., Hellmann, K.-U. (eds.) Prosumer Revisited: Zur Aktualität einer Debatte, pp. 51–60. VS Verlag für Sozialwissenschaften, Wiesbaden (2010)
20. Ritzer, G., Dean, P., Jurgenson, N.: The coming of age of the prosumer. Am. Behav. Sci. **56**, 379–398 (2012)
21. Kaplan, A.M., Haenlein, M.: Toward a parsimonious definition of traditional and electronic mass customization. J. Prod. Innov. Manag. **23**, 168–182 (2006)
22. Peterson, J.: The co-design process in mass customization of complete garment knitted fashion products. J. Text. Sci. Eng. **6**, 1–8 (2016)

Performative Approaches in Designing Costumes: Ergonomics in Immersion and Storytelling

Alexandra Cabral(✉) and Carlos Manuel Figueiredo(✉)

CIAUD – Lisbon School of Architecture, Universidade de Lisboa, Lisbon, Portugal
cabral.fashion@yahoo.com, cfigpt@gmail.com

Abstract. The case studies explore experimental practices and approaches to design based on costumes that lead to the characters' construction and of the plot itself and thus contribute to the immersion of performers and spectators into corresponding fictions. Semiotics and somatics gather in a visual vocabulary based on textiles, colors and garment shapes that tell stories, taking spectators to plausible universes or situations contextualized scenically. Participatory design is considered, with costumes as working tools for performers and directors. User-centered design processes that include the moving body as the mill of a costume's definition promote a staging under construction that turns the plot into a metaphor of the costumes' facets. This brings new symbolic and theoretical frameworks for both performing and designing. Thus, ergonomics is central in the creative process, implying original designs for these staging practices with no preliminary grounds and that rely on costumes as triggers for acting.

Keywords: Performance · Costume design · Ergonomics · Immersion · Storytelling

1 Theoretical Framework

This study aims to understand the interactions between the human element and other elements in a performative context, under an ergonomic framework related to how the body and body expression can be extended by that exchange. It is an approach based on practice that is, nevertheless, "theory-focused, on concepts which imply experimental approaches to performance art that explore body perception through costumes' plastic potentialities and even its constraints" [1].

Costumes are enablers in performance, due to their somatic relationship with the usage of the body as a means of expression, therefore "our quest is to understand the actor on stage and the identities he enacts in costume" [2]. But embodiments in characters imply also the immersion in narratives they belong to, leading us to consider "performances in witch costumes are made to convey specific stories to an audience while moving the spectators emotionally and transform the actors psychologically" [2]. Here, the role of the director gains importance. The choices of a director on the conceptual elements of the narrative complement the visual language telling and the

G. Di Bucchianico (Ed.): AHFE 2019, AISC 954, pp. 337–349, 2020.
https://doi.org/10.1007/978-3-030-20444-0_33

interaction he aims. The design process needs to follow that up, because costumes express an emotional interaction between the character/actor's performance and the storytelling: "Design outcomes had to support in some way the understanding of audience engagement with (…) [the] performance and, if possible, increase that engagement" [3].

Clothing functionality and cognitive readings by the performer are linked to his self-expression and self-image, and also, "this self-image and self-expression tend to be read by spectators not only as an outer display, but also as an inner identity and emotion (…) they blend together with the performer's actions, (…) connecting the artistic expression with the clear notion of [his] selfness. In this process, communication is obviously involved" [1]. But costumes can play a narrative role both in compliance with the characters and other elements of the storytelling and set design, depending on how they are contextualized and emphasized: "The set is the frame and the context for the costume designer's work. (…) Without light the world that the designer makes for the performers and the audience is lifeless" [4]. Thus "A good costume 'tells' the audience about the status, relative wealth, age, profession or trade, attractiveness, temper and underlying psychic state of each character" [5], but they can tell much more about the story if used and framed appropriately, when "the visual effects help place the work in a context, give the characters depth and rigor, and underline and support the emotional content of the work. The nature of devised work lets a designer work with the director and performer to build the characters and the place of the story" [4].

In a designer's perspective, several materials and their combination can highlight or blend the costume and character in the scenic surrounding space and light: artifices like the thorough choice of transparencies, reflection, oversize volumes, neutral colors, can all allow mimetic relationships. Besides those, the perceived quality of fabrics, such as softness, coldness or warmness, can convey sensations to the spectator that he relates to the character's personality. Throughout the play, a costume can then perform a narrative or interactive role, depending on its fixed or dynamic features. Its possibilities of change and interaction can be achieved both with or without the use of smart and technological artifacts, although some art directors support that the use of smart approach is unnecessary: "we believed that fundamental questions about interaction and engagement could still be asked without recourse to 'smart' systems and materials" [3].

Smart textiles or smart design can include microcapsules, LEDs, optical wire, light sensitive prints or sensors in the costumes, textiles or props. These can add smell, light, shapes and color, which can change during the play and be interactively triggered. These changes can be partial or transform the costumes completely. They can be triggered by other characters/actors, by spectators, or by inner factors like scene lighting or sound. A costume change will have narrative intentions, assuming part of the storytelling. A costume can metamorphose during the play, into two or three different characters. These characters can be present simultaneously or be emphasized one by one. In any case, there is a unique set of visual vocabulary in the stage and light design that aims to reinforce the characters' personalities, actions and their fates in the plot. Emphasizing main moments of the scenes contribute to the spectator's immersion: "Spectators may experience interactions as secretive, expressive, suspenseful or magical, depending on the visibility of interface manipulations and their effects, which each may be hidden, partially or fully revealed, and amplified" [6].

Nonetheless, the actor's body, its expression and posture, together with the costume, plays a main role in the character's construction, namely in its personality. The starting point is the body the actor aims to embody: "Performance practice has also been questioning the usage and reading of the body, (…) the imagined body as the starting point of the performer's acting (…)[and] changes in body can as well be done by fashion, being fashion also deeply related to the notion of the self" [1]. Costumes can be involved in the performance creative process and that is even more valid in smart costume design, due to the evinced somatic characteristics and new relationships with the body. Seemingly, "actors should have sufficient time to train with the interactive features and rehearse while wearing the costumes and possibly should also be included in the creation process" [6]. The user-centered methodology used in the costume design process relies on the actors' interference with costumes even from the very beginning, since their inputs are crucial to get to a point where actor, costume and character become one and only.

In live performance, an interactive costume can expand the expressiveness of the body: "A wearable art itself can be an expressive media platform and when it is worn by a performer, it elevates expression of body language" [7]. Art directors fiercely defend or deny the use of smart approaches in performances and play narratives, but both agree that "performing art can be an ideal platform for the wearable media as it provides creative expression for both wearable artist and performers" [7]. Some say that its use impoverishes the role of actors and directors in the making of the storytelling, turning it less authentic for relying more on external processes, rather than on the actor´s body expression, posture and projection. What is certainly true is that directing with «smart» or «non-smart» artifacts differs a lot. Connotations can be given to formal elements of the costumes and setting and the open nature of costumes, often present in simple forms, seems to suggest new possibilities of interpretation, rather than impose boundaries in their meaning. By this way, they become important storytelling elements for creating this close relationship with the plot, and do not necessarily need to be «smart»: "Instead of 'smart' textiles and computational wearables engaging audiences through technology, it seemed that the open nature, or thingliness, of the costumes had a powerful and complex effect on the relationship between performer and audience" [3]. We therefore consider that besides the conceptual approaches, there is a strategic balance between pros and cons in the core of the production related to human, technical and financial resources.

Costumes can also be given attributes linked to the scenic environment they inhabit. They can symbolize a prop or an object and they can stand for themselves alone in terms of meaning: they might become something else but the character, in the fictional space of the play. Their shapes, colors, textures and volumes convey metaphors that can trigger a particular narrative role in the spectator's scenic imagery. The same stands for separating parts of the costume, making them autonomous props in the scene. The performance becomes a dialogue between the physical elements in space and the new possibilities of acting with them, which in the end needs to make sense in the narrative. A mentally and physically trained actor can foresee it, but it takes a lot of rehearsing for testing their true potential, since "an improvisation (…) needs precise rehearsal to happen, and no one may know until strength and skill have developed further, whether it is actually possible. If it is, perhaps it will be incorporated into the narrative" [4]. Furthermore, interactive or shared costumes that involve the acting and that promote the interaction between actors and spectators imply a performance practice allied to a staging process under construction, where those relationships are pre-established in a certain framework. It is, nonetheless, a performance practice led by costumes but also actuated and defined by the chance implied in the duration of the interaction, promoting the creation of new meanings, because "the work revolves around the notion that each body and each body memory, gesture, deliberate and multiply [the] framed staging of the self in performance [and thus] leads to another layering of communication as bodily inscription" [8].

The ergonomics of a costume can then be more functional when linked to gestures or actions; more cognitive when promoting emotions and immersion in the realm of a certain production; more narrative when performing different roles or gaining attributes with new features and contextualization in tune with the storytelling. Semiotics can then play an important role in carrying the spectator through symbolic metaphoric spaces (including post-human) that costumes may suggest, proposing secondary levels of «meaning» telling, because "towards a reading of inhabiting space by wearing costumes, we need to consider the act of wearing in the impact on others, as a decoded language. Hence, a performer (and a costume designer), by deliberately choosing a signifying object for his performance is alluding to a certain social interpretative dimension" [1].

2 Case Studies

The case studies described below were held in 2014, during the exploratory design phase of the PhD research project title "Design of Dynamic and Interactive Costumes for Performance Art", based at CIAUD Research Centre, developed by Alexandra Cabral (the first author) and supervised by Carlos Manuel Figueredo (the second author), together with Cristina Carvalho. The costume design is by the first author.

2.1 Costume Design for *Little Thumb* Theatre Play

The first case under analysis portrays the costume design process for a theatre play by Seara – Companhia de Teatro de Beja (theatre company in Beja, Portugal) directed by David Silva, based on Charles Perrault's *Little Thumb* tale, which started from a financial and human resources restricted set. With three actors for nine characters (Little Thumb, Lumberjack, Lumberjacks' Wife, Lumberjack's Sons, Giant, Giant's Wife, Giant's Daughters, and another two Little Thumbs), the actors enrolled a complex performative context where they assumed three different personalities each. They could not leave the stage to change clothes neither undress them, but they had to transform their costumes and assume different personalities in every garment change, in front of the audience – involving it, rather that distracting it from the plot.

The stage design was limited to a table, a black background and light projections, meaning that the costumes had a narrative role not only in compliance with the characters they represented but also with other elements of the story, as we will see. Before the limited space and the imposed nearness between actors, they faced movement constrains, being the scenic space defined both by their location on set and their confined action fields.

Connotations were given to formal aspects of costumes through acting, which was essential to define the characters' psychology and physiognomy. Early prototypes were requested for improvisation purposes (Fig. 1, top left) although initial instructions had been given by the actors and the director, related to gestures, manners or stature, as well to technical features, such as pockets for carrying props. There was therefore a balance between a pre-established idea of how characters were meant to be and what costumes could allow in terms of presence and body movement.

Fig. 1. Rehearsing with prototypes (top left); using a costume to represent a light in the dark (right); using boots as the reins of a chariot (bottom left). *Little Thumb* theatre play by Seara – Companhia de Teatro de Beja, directed by David Silva. Costume design by Alexandra Cabral.

Conceptual elements in the narrative were strongly linked to a complementary visual language, according to the director. The *Rubik's Cube* imagery had to be present in the color blocking related to the character's changes. There were other references like the layers in a samurai's costume, Oskar Schlemmer's *Triadic Ballet* (1927) geometry and the simplicity in Malevich paintings. The design proposal was therefore based on monochromatic costumes with some hints of color (from the other characters' attire, which could not be hidden) and textures associated to narrative aspects, such as the pleating in the Giant's Daughters' dress, symbolizing them dying horizontally (lying on their own beds), or the vertical pleating in the Lumberjack Sons' trousers, representing the trees in the forest they lived in.

The characters had a color correspondence, according to their griefs and sorrows, destinies or personality traces. The Lumberjack's Sons were dressed in green, due to the forest trees; the Giant's Daughters were dressed in red for dying with their throats cut; orange represented the Lumberjack being antsy; purple was related to the Lumberjack's wife sorrow or idea of death associated with the loss of her kids; blue was as cold as the Giant's nature; white, considered a pure color, belonged to Little Thumb's costume.

There was also a color code given to some aspects of the narrative, so that the costumes could also represent something else apart from the characters. Yellow represented the light in the dark, coming from the Giant's house, thus the Giant's Wife had a yellow dress (Fig. 1, right); before the absence of a physical door on set, her panel skirt could also make a door's turn. The Giant's boots were used to resemble the reins of a chariot (Fig. 1, bottom left). Therefore, shapes, colors and metaphors were hand-in-hand and anticipated the spectator's imagery on the narrative.

The décor had the contribution of the always present costumes that filled the spaces, either on the top of the table or bellow it and sometimes on both. The action in the woods was carried out underneath; the action inside the houses was done on the table top. There were also light projections in the shapes of hats and crowns, indicating the passage from boys to girls (the Lumberjack's Sons vs the Giant's Daughters), guiding the spectator in the quick changes between them and anticipating the tragic moment of one of them dying.

The user-centered methodology used in the costume design process was based on letting the actors interfere in it by wearing prototypes, manipulating them and giving feedback on their usability during the rehearsals – but observing directly how they moved was important to correct mistakes missed before in the costumes' characteristics. Nevertheless, the initial possibilities of the costumes were already tested in small scale (1:25) prototypes that predicted what movements/passages from character to character were possible. Even so, it was important to check if the attired could provide a smooth convincible passage between characters to the audience. And since fastenings were also a little difficult to accomplish by the actors, extra rehearsals had to be done for the actors to practice over and over, until they would perform the passages without stumbling – and those could be then perceived «magically». We can see the transformation from Little Thumb into Lumberjack on stage in Fig. 2.

Fig. 2. Transformation from Little Thumb, in white, into Lumberjack, in orange, on stage (rehearsal). *Little Thumb* theatre play by Seara – Companhia de Teatro de Beja, directed by David Silva. Costume design by Alexandra Cabral.

Each actor had to perform two different characters apart from Little Thumb. There was, nonetheless, one «main» Little Thumb character, therefore two of them could be semi-present, that is, from the hip up. A few steps were needed to achieve the transformations, such: as moving panels sideways (transforming trousers into a skirt – Fig. 3, 1.1–1.2); dressing some parts upwards (turning a hemline into a turtle neck, blouse in Fig. 3, 1.1–1.2); unzipping and pulling some pieces down, revealing white pieces underneath (Fig. 3, 1.3 and 2.3); throwing them to the back and over the head, turning a skirt into a top (Fig. 3, 2.1–2.2) or changing from one vest to another (from the orange one to the red underneath, Fig. 3, 3.1–3.2).

Fig. 3. Costume's transitions: Lumberjack's Sons/green (1.1); Giant's Wife/yellow (1.2); Little Thumb/white (1.3); Lumberjack's Wife/purple (2.1); Giant/blue (2.2); Little Thumb/white (2.3); Lumberjack/orange (3.1); Giant's Daughters/red (3.2); Main Little Thumb/white (3.3).

The actors needed for this play had to be highly skilled and focused, since they could have to move and speak independently during the transformations between characters. The costumes' color blocking minimized the chances of them getting mistaken, being each color related to a certain character. The shapes and textures of each costume were different, also helping the actors to embody them. Besides, the pace factor related to each transformation differed. All these small details mattered in memorizing the steps needed for the costumes' manipulation.

The light design of the play emphasized the central moments of the scenes, meaning that the transformations were not always highlighted, therefore the actors had to memorize how to perform their gestures with almost no light. But the focus on the actor under a low-key lighting helped engage the audience, immersed in the performer's ability to embody constant change into different characters. The costumes encompassed a somatic effect: after a while, the spectators could predict which characters would appear on set, although they could not guess properly how the transformations were done, as they were carried out rather quickly.

Considering the importance of these costumes in the staging process, it can be said that there was co-creation in the staging of the play itself between the costume designer and the director – costumes somehow simultaneously constricted and emphasized the acting process, thus nourishing the director's vision. There was no end for the play and the costumes contributed with the solution: the show was about misery and it was convenient to have three Little Thumbs bowing and begging to the audience. Besides, and as mentioned above, costumes also represented physical elements of the scenic environment (such as houses, trees or lights). In sum, costumes established a broader relationship with the plot, contributing positively for revising the way the narrative was constructed, and of course, the way the story could be told.

2.2 Costume Design for *Epidemia* Theatre Play

Epidemia was a theatre play by David Silva, performed again by Seara, about the new digital technologies and their «epidemic» usage, being the characters an iphone and an ipad. The plot highlighted the dependence people have on this kind of gadgets, making them lose touch with reality. The director had particular requests regarding the direct relationship between lighting and costumes, making them resemble the electronic devices. They had also to be easy to dress and undress on stage.

Transparent and shiny fabrics were used to make the costumes resemble screen surfaces. To emphasize the technological sensation, they had some volume and distance from the body, allowing light projections to fill into the «gaps», adding the impression that costumes were embedded with their own light. Altering between opaque and transparent parts also added the impression of depth in that distance. Moreover, their neutral colors allowed the stage lights to be incorporated, creating a consistency in the décor. Everything was flooded with light of the same color, since costumes could «change» color along with the scenery (Fig. 4).

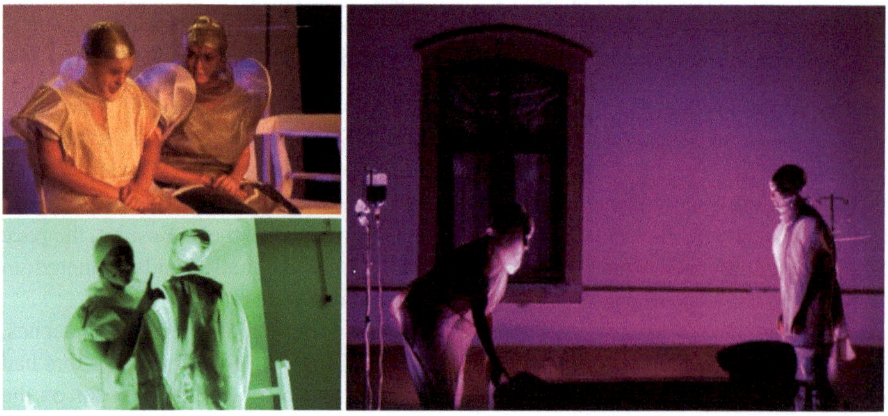

Fig. 4. Costumes transformed by the light design. *Epidemia* by David Silva, performed by Seara – Companhia de Teatro de Beja. Light design by Paulo Santos. Costume design by Alexandra Cabral.

The technological resemblance in the costumes made the audience and the light designer wonder on the presence of an actual electronic technology in them. This led us question how we can make-believe in performance practices through materiality. Performance practices comprehend both strategies, technological and non-technological, being the choice itself a purely strategic decision that considers conceptual and logistic dynamics, according to the global design of the theatre play. The use of technology, electronic in this case, would represent more investment, both financially and collaboratively. As the staging occurred outside at night, the need for a technologically advanced costume was not relevant, because the impact of the lights on the costumes were enough to create a good visual impact. Immersion in the narrative would then be assured. This analysis emphasized the spectators' point of view on the visual grammar and narrative coherence of the play and into what extent the costume design would follow their expectations.

There is a semiotic facet in this theatre play that carries the spectator along the symbolic metaphoric spaces the characters live in. There is some human reminiscence in the costumes, the characters still had arms, legs, heads. They were humans turning into gadgets, or gadgets becoming human-like entities. This dichotomy suggested that post-human dimensions were being brought up. By resembling LED screens and accepting the stage lights with brightness, it was as if the characters were in a mimetic relationship with the background, belonging to a virtual space. Therefore, the light design was leading the acting, with the help of costumes, which were placing the actors «somewhere else». This was also possible to achieve due to the setting: it was a façade from a building, which gave enough room to the action space and established a scale relationship with the small characters. Plus, it was the façade of a church, adding some celestial connotations to the virtual existence. There were only a few props on stage that linked the characters to the real-life setting; a bench was connecting the characters to their mundane existence.

There was no co-creation or collaboration between the actors and the costume designer in this production, but costumes were used in advance to test the possibilities of the light design. All conceptual facets were discussed with the director alone and the experience was rather fruitful because the play exposed how much meaning can be added to a performance, when the visual solutions found for it meet the spirit of the time, being this a play based on very recent technological developments. The costumes had a significant narrative role in the realm of the production.

2.3 Costume Design for *Miss Fruits Go to School* Educational Project

Donas Frutas Vão à Escola (Miss Fruits Go to School) was a draft-project presented to Buzico! Produções Artísticas e Agenciamento Lda. (Lisbon, Portugal) destined to be implemented in primary schools to teach kids to eat more fruit. The director's demands, in this case, were the fruits being peeled, separated and cut, depending on the variety and, most important, with pieces that could be carried around by the child spectators. The banana had to be peeled, the grapes needed their seeds to be shown and the apple should have a slice cut out (Fig. 5).

The project itself had an apparent simplicity, but it was grounded on a few ergonomic topics to be sorted out regarding the users and also the spectators, who could take part in the narrative as collaborative actors – not to mention that the organic matter in the fruits could be correlated to the human body. Apart from peeling a fruit being also undressing in a certain extent, fruit and body alter with time, adding some elements to a possible improvisation in acting and possibilities in design. For example, the fruits being prone to change could lead to an apple oxidizing, making use of LED circuits linked to microprocessors and sensors to create that illusion of a slice being separated from its mother piece.

A few stories could be told, like a monkey eating banana or a caterpillar eating the apple. But what about the banana smell? Technical textiles could be used, adding smell via microcapsules of scent. Fabrics' textures could also add softness and cold or warm sensations to the surface of the fruits, such as velvet for a banana, canvas for an apple and organza for a grape. In other words, other sensorial elements would sum up to the visual elements, like smell and touch.

Fig. 5. *Miss Fruits Go to School* pre-project, presented to Buzico! Produções Artísticas e Agenciamento Lda. Costume design by Alexandra Cabral.

Mechanic features could be considered, such as magnets as seeds in a grape, to help close both halves when cut in two or to get the grapes connected back to the stem. Separating parts of the costume, peeling and unpeeling the fruits by means of velcro or zippers, moving around gracefully in costumes that could change the gravity center of the performer's body, all those facets could represent challenges in the performers' acting. On the other hand, giving the children's the chance to hold separate pieces from a costume would place them as performers telling a story backwards, while looking around for the original pieces of fruit.

Sharing a costume with the spectator implies reviewing the staging process because it becomes a staging under construction. In this case, this would be particularly possible to be put into practice because the storytelling fruits summed up symbolic characteristics to their explanatory features, bordering somehow the possibilities of the narrative to a few predictable topics, allowing the actors to expect, in advance, some reactions from the audience and be ready to deal with them on set.

2.4 Glove Functional Prototype, E-Textiles Art Workshop at Contextile 2014

A simple functional prototype was developed during the E-Textiles Art Workshop held at the contemporary textile art biennale Contextile 2014 (Guimarães, Portugal), based on a few electronic tools: conductive thread, Lilypad Arduino, 5 LEDs, coin cell battery holder and lithium battery. The goal was to accomplish something functional in a very limited time, having to be inevitably small scaled. The challenge represented an introduction to designing with smart materials, hence it sounded appropriate to make an interface someone else could activate, generating an active behavior between sender and receiver. The solution should embody the emotion in a handshake, relating it to lighting up the LEDs in a glove. The limited resources led to a rational choice of where to place them: LEDs along the thumb; switch near it, where the other person's thumb touches; remaining electronic components hidden in the palm (Fig. 6).

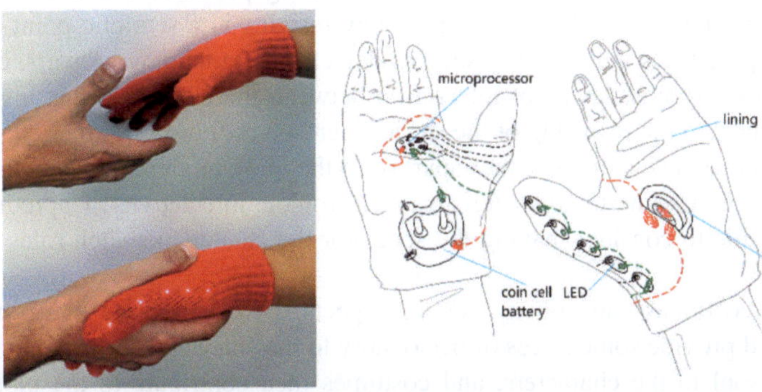

Fig. 6. Glove functional prototype: switching the LEDs on with a handshake. Developed during the E-Textiles Art workshop at Contextile 2014 by Alexandra Cabral.

The prototype surprisingly provoked the urge in the opposite person to keep repeating the gesture of lighting up the glove. Into what extent lighting up the glove meant an immersive feeling into a handshake? The limited functions of the LED circuit proposed other possibilities, such as how to create more correspondence with the intensity of the gesture (harder, with more lights; softer with less), relating the prototype's expressiveness to human emotions in greetings. Colored LEDs could be added, and the electronic circuit could be spread to other parts of the body, accordingly to the desired symbolism.

The ergonomics in this prototype became both functional and cognitive, for being allied to a communicative gesture. What if both users had a glove with lights? What narratives could one compose with such wearables? Into what extend would acting and improvising contribute to open up new possibilities? Chance and intentionality would bring new narratives to a handshake discourse, showing that technological wearables can drive us to explore storytelling.

3 Conclusion

A few performative contexts within the costume design process were analyzed, from including the actors' collaborations from the beginning to counting on the director's inputs only. Anyhow, costumes were used as tools of the staging creative process and proved to perform a significant role in conveying meanings to the audience through their direct relationship with the plot. Accordingly, they helped actors develop both a performative narrative and the characters themselves.

Consequently, questions were posed regarding the acting process about how costumes, by their nature, could either become the core of a narrative or bring possibilities for other stories to be told. The creative process was based on the potentials of the costumes' materiality and their ergonomic relationships with the body itself, considering also the new dimensions added by technological components. Those promoted a transformation of the dynamics of interaction among performers and between performers and spectators, capable of bringing new insights in a semiotic point of view.

Dynamic properties in costumes, such as detachable parts, moving parts, changing colors or lights, promote levels of interaction between the characters that inspire storytelling and lead to the rewriting of the stories themselves. Plus, costumes can become props perceived as new entities, either apart from the body or gaining meaning from it. They can have new «smart» characteristics and promote interactivity. Unpredictability is then a variable to consider, having the spectator the chance to switch roles with the performer.

Visually, costumes can convey certain aspects of the narrative by their plastic expression and provide some traces of personality to the characters they intend to dress. Acting adds soul to the characters, and costumes then contribute to the plot thread, guiding the spectator. Additionally, costumes also influence coordination and interaction among actors and establish their degree of embodiment into the scenic space. Textile materials are relevant for compelling, by their structure, certain body movements, formerly promoted by the costumes' shapes and textures; those features, together with their colors, transparencies or opacities, are prone to either allow them contrast or blend into the scenery, thus creating different meanings.

Designing costumes meets then the staging process. They merge in the intricate ergonomic task, both physical and cognitive, linked to drafting a narrative, narrowing down the results, providing the audience with a convincible recreated fictional world. It's a process obviously framed by an appropriate stage and light design that emphasizes storytelling, promoting the immersion of the spectators in the plot. All cases considered, a user-centered methodology becomes particularly important, not only in the costume design process but also in the process of assuring which aspects should be present and which should be disposable in the construction of a certain narrative, so that the story can be conveyed to the audience in a specific conceptual framework.

Acknowledgments. A special thanks to Maks Tatiana, who manufactured the costumes and to the actors involved in the participatory design in Little Thumb, namely Estêvão Antunes, Maria Lalande and Sandra Maya.

References

1. Cabral, A., Figueiredo, C.: Costume design: ergonomics in performance art. In: Advances in Ergonomics in Design, Usability & Special Populations: Part 1, 5th International Conference on Applied Human Factors and Ergonomics, pp. 511–520. AHFE, Krakow (2014)
2. Shukla, P., Glassie, H.: Costume: Performing Identities Through Dress. Indiana University Press, Bloomington (2015)
3. Glazzard, M., Kettley, S., et al.: Experiential collaborations from garment to costume: play and the thing as design outcome. Craft + Des. Enq. **6**, 137–157 (2014)
4. Bicat, T.: Costume and Design for Devised and Physical Theatre. The Crowood Press Ltd., Marlborough (2012)
5. Clancy, D.: Designing Costume for Stage and Screen. Batsford Ltd., London (2001)
6. Honauer, M., Hornecker, E.: Challenges for creating and staging interactive costumes for the theatre stage. In: C&C 2015 Proceedings of the 2015 ACM SIGCHI Conference on Creativity and Cognition, pp. 13–22. ACM, New York (2015)
7. Kim, Y.: Body graffiti: expressive wearable art through bodily performance. In: The 17th International Symposium on Electronic Art, ISEA, Istambul (2011)
8. Goodman, L.: Introduction, part 1: performance futures: bodies in movement, viewed through multiple screens. Int. J. Perform. Arts Digit. Media **3**(2&3), 101–102 (2009)

Reliability of Anthropometric Reference Data for Children's Product Design

Su-Jeong Hwang Shin(✉) and Mona Maher

Department of Design, Texas Tech University, Lubbock, TX, USA
su.hwang@ttu.edu

Abstract. In the United States, children's body dimensions are significantly different from what they have been in the past. Dissatisfaction with current children's clothing sizes and mismatched furniture are often reported in the industry. Children often show improper sitting posture in class due to mismatched sets of desks and chairs. A reliable reference for children's anthropometric data is needed to improve children's products' safety, comfort, and fit in the environment. The aim of this study was to provide the most up-to-date children's anthropometric data for 6 to 11 year-olds in the United States and examine the reliability of anthropometric reference data for children's products. This study was conducted at Texas Tech University, one of the scan sites for a national sizing survey, Size NorthAmerica (Size NA led by Human Solutions Inc.). A total of 21 girls between the ages of 6 and 11 participated in the study with IRB approval. A set of 52 body dimensions in each child was scanned with a laser-based 3D body scanner. The measuring procedure was compatible with ASTM standard measurement definitions in accordance with ISO 8559 (garment construction and anthropometric survey) and ISO 7250 (basic human body measurements for technological design). Current children's body dimensions were significantly different from the anthropometric reference size charts for children's product design. Results in this study would provide insight into the need for developing reliable anthropometric reference data for children's product design.

Keywords: Children's clothing · Furniture design · Anthropometric · Size · 3D

1 Introduction

The International Ergonomics Association (IEA) defines human factors (ergonomics) as the science which focuses on the interactions among humans and other elements of a system. It also describes anthropometry as a branch of ergonomics that encompasses measuring various dimensions of the human body in relation to products in order to optimize human and system performance. In the United States, apparel and furniture companies utilize anthropometric reference charts to develop their merchandise. In the apparel industry, three major anthropometric references are available, provided by the American Society for Testing and Materials (ASTM), the International Center for

G. Di Bucchianico (Ed.): AHFE 2019, AISC 954, pp. 350–357, 2020.
https://doi.org/10.1007/978-3-030-20444-0_34

Health Statistics (ISO), and the National Center for Health Statistics (NCHS) [1]. However, some of this reference data are outdated for product development [2].

Several national sizing studies have been conducted to update anthropometric references. For example, national sizing studies, CAESAR, SizeUK, and SizeUSA provide accurate anthropometrics using 3D body scanners and reported a significant change in human body dimensions over past decades [3]. However, these surveys excluded children due to the difficult sampling procedures for this age group [4]. Dissatisfactions with the children's furniture sizes have been reported frequently [5–13]. Hwang and Kim [14], for instance, found children's body dimensions are significantly different now than from when anthropometric data reference size charts were collected. Also, based on the National Health and Nutrition Examination Surveys, children are heavier at present. Further, children often show improper sitting posture in classes due to mismatched desks and chairs. Van Wely [15] identified strong relationships between various musculoskeletal disorders and exposure to highly repetitive awkward postures, postures that hardly improve in adulthood [16]. There is a critical need to study children's anthropometrics and to update the references that can help with improving children's products' safety, comfort, and fit. Thus, the objective of this study was to provide the most updated anthropometric data for children aged 6 to 11 in the United States. For children's product development, it also aimed to investigate the reliability of anthropometric reference size charts, including the ASTM D6192-11 (Standard Tables of Body Measurements for Girls, Sizes 2 to 20) [17], and the standard size chart for children's furniture products provided by Panero and Zelnik [18].

2 Methods

In this study, 21 female children aged 6 to 11 were scanned from August to October 2018 using a laser-based 3D body scanner to collect updated data on children's anthropometrics. The measurement was conducted at Texas Tech University, one of the scan sites for a national sizing survey, Size NorthAmerica (Size NA led by Human Solutions Inc.). A set of 52 body dimensions in each child was scanned in four postures: three in standing positions and one while seated. The scanner provided fast and accurate measurements through an optical triangulation process using eight sensors. The measuring procedure was compatible with ASTM standard measurement definitions in accordance with ISO 8559 (garment construction and anthropometric survey) and ISO 7250 (basic human body measurements for technological design [19]. The Institutional Review Board (IRB) at Texas Tech University approved the data collection.

Next, two reference size charts, ASTM D6192-11 for apparel products (Standard Tables of Body Measurements for Girls, Sizes 2 to 20) [17], and the standard size chart provided by Panero and Zelnik [18] for children's furniture were compared. These standards are summarized in Tables 1 and 2. Accordingly, the sets of survey data were compared with the references to investigate their potential differences. The analysis was implemented using SPSS version 25.

Table 1. Summary of ASTM D6119-11 clothing standard size chart [17]

Age	6	7	8	9	10	11
Body measurements (in)						
Body height	46.50	48.50	50.50	52.50	55.00	58.00
Waist girth	23.00	23.50	23.50	24.25	25.00	26.00
Hip/seat height	22.75	24.38	25.63	27.00	28.63	30.13
Waist height	28.00	29.63	31.13	32.88	34.75	36.88
Hip/seat girth	25.00	25.75	27.50	28.50	30.00	32.00
Crotch height	20.25	21.88	23.13	24.38	26.00	27.63
Crotch length	19.13	19.25	20.00	20.88	21.88	23.00
Knee height	12.00	12.50	13.38	14.00	15.13	16.38

Table 2. Summary of Panero and Zelnik furniture standard size chart [18]

Age	6	7	8	9	10	11
Body measurements (in)						
Body height	46.33	48.57	51.00	53.43	55.55	58.22
Knee height	14.14	15.00	15.80	16.88	17.47	18.34
Popliteal length	11.40	12.00	12.75	13.50	14.00	13.00

3 Results

3.1 Current Children's Body Dimensions

Table 3 shows the statistics of the participants' dimensions for body height, waist girth, hip height, waist height, hip/seat girth, hip, crotch height, crotch length, and knee height. These data were subsequently employed for comparison of current anthropometrics with the two reference size charts. Accordingly, Table 4 presents a summary of the participants' characteristics in terms of age, race, and Body Mass Index (BMI). The participants' BMIs were calculated using the BMI percentile calculator for children and teens provided by the Centers for Disease Control and Prevention at the U.S. Department of Health and Human Services [20]. Overall, the sample's average BMI was 19, placing it at the 71st percentile of BMI for girls in the healthy weight category [20].

Table 3. Descriptive statistics of current children's anthropometrics (N = 21, Unit Inch)

Age		6	7	8	9	10	11
Number of participants		5	2	2	1	6	5
Body height	Average	43.75	49.53	51.09	57.35	56.38	58.06
	SD	1.69	0.81	1.81	-	2.98	3.21
Waist girth	Average	23.69	21.58	25.73	39.51	30.31	26.28
	SD	3.24	1.68	-	-	4.35	2.01
Hip height	Average	17.91	21.13	22.54	27.55	26.68	27.42
	SD	1.92	1.62	0.18	-	2.18	1.70
Waist height	Average	25.75	29.44	31.27	34.37	34.53	36.44
	SD	1.36	0.28	0.19	-	1.96	2.60
Hip/seat girth	Average	27.21	27.18	28.93	40.21	35.64	32.81
	SD	1.61	1.29	0.33	-	5.50	2.33
Hip/hip girth	Average	24.68	22.99	26.16	41.33	32.585	28.676
	SD	3.27	1.64	0.89	-	5.16	2.04
Crotch height	Average	18.01	21.24	22.52	23.02	24.70	26.12
	SD	0.79	1.10	0.10	-	1.24	1.86
Crotch length	Average	20.37	22.21	22.51	31.23	26.33	26.23
	SD	2.18	2.23	0.48	-	3.33	1.51
Knee height	Average	11.09	12.63	13.59	14.79	15.15	15.77
	SD	0.56	0.42	0.06	-	0.89	1.07

Table 4. Participants' characteristics

	Count	Average BMI	Percentile BMI	Weight status
Race				
White	10%	18	60	Healthy
Hispanic	71%	20	76	Overweight
African American	5%	32	99	Obese
Asian	14%	16	45	Healthy
Age				
6	23.8%	16	58	Healthy
7	9.5%	15	42	Healthy
8	9.5%	17	71	Healthy
9	4.8%	30	99	Obese
10	28.6%	23	90	Overweight
11	23.8%	19	68	Healthy
Total	**100%**	**19**	**71**	**Healthy**

3.2 Inconsistent Anthropometric References for the Children's Size Charts

One sample t-test was applied to investigate the differences between ASTM D6119-11 clothing standard size chart for apparel [17] and the Panero and Zelnik standard size chart for furniture [18]. Table 5 summarizes the results for body height, popliteal length, seat height, and knee height of children aged 6–11. The results showed no significant difference in body height between the two reference charts. However, popliteal length among all ages (6–11) was found to be significantly shorter in the apparel reference when compared with the furniture reference (MD = −2.28 in, −2.00 in, −2.43 in, −1.50 in, −2.83 in, and −3.48 in, respectively). Similarly, seat height at age 6–9 was significantly shorter in the apparel reference than the furniture reference (MD = −2.40 in, −1.71 in, −1.32 in, −0.94 in, respectively). In addition, knee height was found to be significantly greater in the apparel reference than the furniture reference at ages 6 and 8 (MD = 0.60 in, 0.62 in, respectively) and significantly shorter for ages 10 and 11 (MD = −.57 and −1.02 in, respectively). These differences imply that the body dimensions recommended by the reference size charts for apparel and furniture do not conform in most parameters.

Table 5. Comparison of ASTM D6192 [17] and Panero and Zelnik furniture standard size chart [18]

Age	6		7		8		9		10		11	
Sig./mean difference	Sig.	MD	Sig.	Sig.	Sig.	MD	Sig.	MD	Sig.	MD	Sig.	MD
Body height	0.92	0.16	0.63	−0.11	0.47	−0.41	0.32	−0.93	0.66	−0.56	0.82	−0.23
Popliteal length	0.00*	−2.28	0.00*	−2.00	0.00*	−2.43	0.00*	−1.50	0.00*	−2.83	0.00*	−3.48
Seat height	0.00*	−2.40	0.00*	−1.71	0.00*	−1.32	0.00*	−0.94	0.72	−0.16	0.61	0.05
Knee height	0.04*	0.60	0.23	0.43	0.00*	0.62	0.12	−0.50	0.00*	−0.57	0.00*	−1.02

*Level of significance: $p < 0.05$

3.3 Anthropometric Reference Data for the Children's Clothing Size Chart

Table 6 shows the differences between ASTM D6192-11 [17] and the children's body dimensions measured through the data collection process. Since there was only one participant of age 9, this age category was excluded from the analysis. Comparison of body height showed that the sample at age 6 was significantly shorter than the reference chart (MD = −2.75 in). Also, the sample crotch heights at ages 6 and 10 were significantly shorter than the reference (MD = −2.24 in, and −1.30 in, respectively). In contrast, the sample crotch lengths at ages 10 and 11 were significantly greater than the reference (MD = 4.46 in, and 3.23 in, respectively). However, the sample hip height at age 6 was significantly shorter than the reference chart (MD = −4.84 in). Similarly, the sample knee height at age 6 was significantly shorter than the reference (MD = −0.91 in). Conversely, the sample waist girth at age 10 was significantly greater than the reference (MD = 5.31 in). Finally, the sample waist height at age 6 was significantly shorter than the reference (MD = −2.24 in). The differences of the other values were insignificant.

In summary, children at age 6 showed more significant differences from ASTM D6192. This age group had a significantly shorter body height, crotch height, hip height, knee height, and waist height in comparison with the reference chart ASTM D6192 [17]. In addition, these participants had greater crotch length, hip/seat girth, and waist girth than the reference; however, these differences were not significant.

Table 6. Comparison of participants' anthropometrics with ASTM D6192 [17]

Age	6		7		8		10		11	
Sig./mean difference	Sig.	MD	Sig.	MD	Sig.	MD	Sig.	MD	Sig.	MD
Body height	0.02*	−2.75	0.32	1.03	0.35	0.59	0.31	1.37	0.97	0.06
Crotch height	0.00*	−2.24	0.57	−0.62	0.13	−0.43	0.04*	−1.30	0.15	−1.5
Crotch length	0.26	1.25	0.31	2.95	0.20	2.51	0.02*	4.46	0.00*	3.23
Hip height	0.01*	−4.84	0.22	−3.24	0.49	−1.41	0.81	−1.95	0.24	−2.71
Hip/seat girth	0.07	2.21	0.36	1.42	0.20	1.43	0.54	5.63	0.42	0.80
Knee height	0.02*	−0.91	0.74	0.12	0.26	0.21	0.95	0.02	0.28	−0.65
Waist girth	0.66	0.69	0.35	−1.92	0.41	2.23	0.03*	5.31	0.77	0.28
Waist height	0.02*	−2.24	0.52	−0.18	0.37	0.27	0.80	−0.21	0.73	−0.45

*Level of significance: $p < 0.05$

3.4 Anthropometric Reference Data for the Children's Furniture

The children's body height and knee height were compared to the Panero and Zelnik furniture standard size chart [18]. Table 7 shows the comparison results that the sample at age 6 was significantly shorter than the reference size chart (MD = −2.58 in). This is in accordance with the previous finding where the sample at age 6 was significantly shorter than the apparel reference chart. Further, it was found that the participants at ages 8, 10, and 11 had a significantly longer knee height than did the furniture reference chart (MD = 0.83 in, 1.07 in, and 1.00 in, respectively). The other values did not have significant differences.

Table 7. Comparison of participants' anthropometrics with Panero and Zelnik furniture standard size chart [18]

Age	6		7		8		10		11	
Sig./mean difference	Sig.	MD	Sig.	MD	Sig.	MD	Sig.	MD	Sig.	MD
Body height	0.02*	−2.58	0.44	0.92	0.98	0.18	0.45	0.81	0.86	−0.17
Knee height	0.27	−0.32	0.31	0.55	0.03*	0.83	0.03*	1.07	0.04*	1.00

*Level of significance: $p < 0.05$

4 Conclusions

Children's products help improve body development if designed appropriately. Clothing and furniture are the most commonly-used products and have a great influence on children's bodies by providing comfort and body protection. Manufacturers utilize anthropometric reference data for size charts to develop the children's products. It is necessary for the industry to reconsider the anthropometric references in accordance to changes in children's body dimensions, considering fit and size issues.

This study has revealed a potential problem with the anthropometric reference that would not be applicable for developing children's products. Children's current body dimensions were significantly different from the anthropometric reference size charts. The reference size charts, ASTM D6192-11 for apparel product [17] and the standard size chart by Panero and Zelnik [18] for children's furniture had a significant difference between the children's body dimensions at age 6–11 years and the references with respect to body height, crotch height, crotch length, hip height, knee height, waist girth, and waist height. These finding are in accordance with reports on dissatisfaction with apparel fit and musculoskeletal symptoms due to furniture mismatch.

Importantly, the problem is not only related to children's body changes. The main concern is the use of outdated sizing systems in the industry. Although some manufacturers have attempted to improve their products by considering additional sizes for apparel or producing adjustable furniture, more effort is necessary to better understand children's body proportions and shapes within different ages and stages of development. Also, this study provided evidence for differences among current children's body dimensions both from the past and from reference size charts. However, it had limitations associated with the collected data. For example, most participants (71%) were Hispanic, while this share in the United States' population is about 18% [21]. Due to the sample characteristics, further research is necessary to generalize the findings to the population of children in the United States. A larger sample population may provide a better representation of current children's anthropometric data. The anthropometric data collected on the participants' body heights and weights can be applied to clinical and health studies.

Acknowledgement. This study is associated with the project "*Size NorthAmerica at Texas Tech*", which was a part of the national sizing survey *"Size North America"* led by the Human Solutions Inc.

References

1. Shin, S.-J.H., Istook, C.L.: Sizing systems for children's wear in the United States. In: Proceeding of International Textile & Apparel Association (2008)
2. Lee, J.-H., Hwang, S., Istook, C.L.: Analysis of human head shapes in the United States. Int. J. Hum. Ecol. **7**(1), 77–83 (2006)
3. Apeagyei, P.R.: Application of 3D body scanning technology to human measurement for clothing Fit. Change **4**(7), 58–68 (2010)

4. Shin, S.-J.H., Fowler, D., Lee, J.: Teen size study reveals indefinite figures and fit problems in junior category. In: International Textile and Apparel Association (2012)
5. Parcells, C., Stommel, M., Hubbard, R.P.: Mismatch of classroom furniture and student body dimensions: empirical findings and health implications. J. Adolescent Health **24**(4), 265–273 (1999)
6. Cotton, L.M., et al.: Mismatch of school desks and chairs by ethnicity and grade level in middle school. Work **18**(3), 269–280 (2002)
7. Panagiotopoulou, G., et al.: Classroom furniture dimensions and anthropometric measures in primary school. Appl. Ergon. **35**(2), 121–128 (2004)
8. Gouvali, M., Boudolos, K.: Match between school furniture dimensions and children's anthropometry. Appl. Ergon. **37**(6), 765–773 (2006)
9. Chung, J.W., Wong, T.K.: Anthropometric evaluation for primary school furniture design. Ergonomics **50**(3), 323–334 (2007)
10. Jayaratne, I., Fernando, D.: Ergonomics related to seating arrangements in the classroom: worst in South East Asia? The situation in Sri Lankan school children. Work **34**(4), 409–420 (2009)
11. Van Niekerk, S.-M., et al.: The anthropometric match between high school learners of the Cape Metropole area, Western Cape, South Africa and their computer workstation at school. Appl. Ergon. **44**(3), 366–371 (2013)
12. Castellucci, H., Arezes, P., Molenbroek, J.: Applying different equations to evaluate the level of mismatch between students and school furniture. Appl. Ergon. **45**(4), 1123–1132 (2014)
13. Altaboli, A., Elfituri, O., Alturkey, A., Dogman, M., Almagrhi, M.: Anthropometric evaluation of the classroom desk for middle schools. In: Ahram, T., Karwowski, W., Taiar, R. (eds.) IHSED 2018. AISC, vol. 876, pp. 927–932. Springer, Cham (2019). https://doi.org/10.1007/978-3-030-02053-8_141
14. Hwang, S.-J., Kim, H.-J.: A study on fitness standard sizing systems for children's wear in the United States and South Korea. 패션과 니트 **11**(3), 21–32 (2013)
15. Van Wely, P.: Design and disease. Appl. Ergon. **1**(5), 262–269 (1970)
16. Yeats, B.: Factors that may influence the postural health of schoolchildren (K-12). Work **9** (1), 45–55 (1997)
17. ASTM, ASTM D6192-11, Standard Tables of Body Measurements for Girls, Sizes 2 to 20 (Reg & Slim) and Girls Plus. ASTM International, West Conshohocken, PA (2007)
18. Panero, J., Zelnik, M.: Human Dimension & Interior Space: A Source Book of Design Reference Standards. Watson-Guptill, New York (1979)
19. Human Solutions of North America. http://www.sizenorthamerica.com/. Accessed 11 Jan 2018
20. CDC. BMI Percentile Calculator for Child and Teen: Results. 2018 October 22, 218 December 10 2018. https://www.cdc.gov/healthyweight/bmi/calculator.html
21. USCB, DP05 - ACS demographic and housing estimates, in 2013−2017 American Community Survey 5-Year Estimates, United States Census Bureau (USCB), Editor (2017). https://factfinder.census.gov/faces/tableservices/jsf/pages/productview.xhtml?pid=ACS_17_5YR_DP05&prodType=table

Dynamic Wardrobe from a User's Perspective – A Previous Survey Study for Female Students

Carla Morais(✉), Cristina Carvalho, and Gianni Montagna

Ciaud, Faculdade de Arquitectura, Polo Universitário, Rua Sá Nogueira, Alto da Ajuda, 1349-055 Lisbon, Portugal
carlota.morais@gmail.com, cristifig@gmail.com, g.montagna@gmail.com

Abstract. Evaluating all phases of Fashion Products, it is realized that the linear make-disposal model of Fashion has no longer future. This wasteful and polluting system has been replaced more and more by strategies based on a circular economy which aims to keep the highest clothing value, textiles and fibers during the use and re-enter the economy afterwards, never ending up as waste. The sustainable concepts of the new textile economy can provide new products and new services models that would enhance the benefit experienced by customers. With the aim to know the opportunity of people having a dynamic wardrobe (based on second-hand clothes, rental or personalized clothes) the present study shows a survey results on academic female population in order to investigate which items they most like and would never leave their closets. We realized that in 25 female students there is an interest for the same types of clothes and colors, even finding that the way the Fashion System is currently based is completely out of step with consumer behavior.

Keywords: Female wardrobe · Fashion waste · Wasteful consumption · Sustainable

1 Introduction

The reasons why it makes sense to know what goes on within female wardrobe are important to the future of fashion practice that we hope will become increasingly sustainable. Knowing people and understanding their clothing behavior provides a better development in garments matter [1]. So that is important to ask and evaluate the items they use because the wardrobe is a collection of items accumulated from previous occasions. It can have different orders which may be based upon social roles (work-wear), functionality (activewear, swimwear, leisurewear), type of clothing (such as coats, trousers or shirts), colours, textures or even other type of individual categories such as fun clothing. Taking into account the multiplicity of actions that can occur daily, these orders may help women to organize their outfits and make the act of choosing what to wear easier [2].

Many pieces of garment can be dressed on different occasions, changing the purpose for which they were originally purchased. Woman manage the items and recombined them to construct their appearance through the consideration of social and

G. Di Bucchianico (Ed.): AHFE 2019, AISC 954, pp. 358–364, 2020.
https://doi.org/10.1007/978-3-030-20444-0_35

cultural contexts [3]: a formal black trousers jumper can appear funky when worn with sporty sneakers, for example. And woman can buy a new top to go with something from her wardrobe and yet arrives at the frequent conclusion that she has nothing to wear, a fact which may subsequently encourage her to buy clothing all the time. This attitude is ingrained and linked to all aspects of living in a modern society but the more it is done the more damage goes to the environment. Fashion waste does not directly go to landfill, rather it stays and keep inactive and unused inside people's wardrobes. A garment usually becomes waste when it is superseded by a superior version in style, cut, fit or textile, without describing the high levels of pollution and chemical that all the upstream supply chain has. People do not have the perception of that and who have it probably would express reluctance to sacrifice the habits with which have grown up.

In order to effectively reduce wasteful consumption new strategies, need to be planned so that sustainable development would not be based on the depravation of what people currently value but derived from what is easily accept. Not pushing people that are personally responsible by their consumption pattern, it is needed to create options for consumer having more sustainable opportunities [4].

This research aims to understand what type of clothes people wear more and to know, by user's perspective, what is wasteful consumption or not.

2 Female Wardrobe

2.1 Fashion Consumption

The purchase of clothing has witnessed a historical evolution, since it was made by need until it become a fast and quick impulse. Cultural anthropologists help us to understand its process, describing that people cover or decorate their bodies due to a variety of reasons, such as protection, modesty, and the desire to be sexually attractive [5]. Before and during Middle Ages, clothing characterization, in addition to be cultural and having religious connotations, was also representative showing the power of social classes through conspicuous consumption [6]. When the sumptuary laws were abolished, social discrimination declined, and the use of clothing starts getting broader.

Nowadays, Fashion consumption is a practice of purchasing connected to everyday life in a modern society and a medium to anybody adorn oneself. There are those who described it as an attempt to construct a wearer visual identity to be interpreted by the public [7]. Clothes we use can express our identity telling others something about our gender, class, status and so on, but once they not speak directly they can therefore be open to misinterpretations [8].

Veblen's theory is no longer able to explain the behavior of consumers in a modern and dynamic capitalist society, where the middle class became majority and tastes spread out [9]. The fashion system is no longer characterized by homogenous styles such as those that were represented up to the 50's but rather by fragmentation and uncertainty. At any one time there are multiple definitions appearing no rules, only choices [10]. Images of the fashionable ideal are so universal that women step outside the domain of fashion to go inside in television shows, films, etc. They have balanced so many different concerns, ranging from personal preferences and body shape that a quest for fashionable wear is primarily motivated by their social desires. As identities are socially acquired in relate to socially constructed positions within a society [11] it is much safer aesthetically and less expensive financially to buy social and formal clothing solutions.

A study made by Hamilton [12] revealed that are youngest people (18–35 years) who are affluent or the highest waste-generating consumers. They can buy a lot of clothes and never use them, hence, making many purchasing errors and increasing waste consumption. They are more actively consuming and experimenting more to settle for their identity, once they are going through a transition into adulthood.

Therefore, fashion consumption display characteristics of users' personality, their taste and also their lifestyle but products materiality does not always respect their intentions because others may saw aspects they would have never wanted revealed at that time [2]. Only strong people and self-confident can have more successful impact changes upon others and reduce their environmental footprint because this time we see fashion as offering cheaper and endless possibilities for individual expression.

To examine female fashion consumers buying criteria a study made in 2011 for an active working people showed that they had the habit of buying at least 1 or 2 pieces of clothing from the latest trend. Their most important requirements to buy clothing were "personal taste" and "necessity", and most of them had concerns over reducing garments they do not need effectively. These results fit to the target segmentation theory with personality variables and lifestyles preferences, instead of demographics values [13]. Another issue is that people wear what they like but preferring comfort, an opposite tendency to the Tigert & Ring theory that have characterized the individuals with greater involvement with Fashion who had less relation to the "comfort" [14].

2.2 Wardrobe

A wardrobe can be organized by functionality, by colors, occasions, seasons, etc. There are many collective options to put clothing inside, as long as women know how to negotiate their sense of self, their individuality and their autonomy when choosing what to wear.

Women faced with many possibilities, since being chic or funky until to please family members and loved ones, or even with their position in life, such as being middle-aged, grandmother and not having the body she used to have [2]. These thoughts and concerns are many times restraints that they have to manage on a daily basis. The wide context is so big that sometimes the wardrobe fills up with new items to go with older ones. In turn these items remain inactive, keeping itself as memories from the past, waiting to be dressed on again and acquire a new identity.

Wardrobes can really be clothing collectors and chronological biographies of each owner and as people get older the wardrobe will accumulate clothes.

Analyzing all pieces of garment, and considering that Fashion is a system of signs [15], with greater significance for some people than for others, it has been observed over the years that there are numerous items with different typologies. A study carried out in 2015 in order to find out that basic wardrobe reveals about 20 types of pieces, including black dress, the polo shirt, the t-shirt, the twin-set, the "blazer", the "jeans", the "trench-coat", "black pants", "white shirt", the "skirt", the "day dress", the "training suit", and others like a "sweater", "top", the "cardigan", the "jacket", "shorts", "leggings", "cape", and a "pullover" [16].

Nevertheless, analyzing the typology of each one, since for example in the jeans can exist several models, like skinny, straight leg, jeggings, boot cut, flared, Low rise, boyfriend and *cigarrete*, it begins the need to refine which more common among different users' type.

2.3 Fashion Waste

Garment can become fashion waste when the owner changes in shape it does not fits anymore. Fashion clothes are made in standard sizes and not all facilitate size variation. The users may keep their garment and thinking that will wear it again when they return to their original size, but it is likely that it remains in the wardrobe forever.

The same happens when clothes stops being attractive in Design and another superior version takes its place. The wearer may be attached to its clothes because they were used in very special occasions but as soon as they stop fitting, they start getting inactive and become fashion waste. The wearer may not be conscious of that because their focus is on the new pieces [4]. There are also who store such items of clothing as a significant extension of who they were used to be because they were worn habitually during entire working life [2].

However, saving items that are not used is not as much problematic as those which are bought and never used, such as young generations have done.

This research aims to know if female Portuguese students do the same as Australian young consumers that spend money on clothing they do not wear at all [12] and what happens to the clothes after they through them away.

The fact of having clothes at low prices with good design means a great increase in purchases, getting customers retain a large quantity of clothes in their homes, some stored in wardrobes, drawers, garages, etc. When this storage is exhausted the consumers have the options of selling to second-hand stores. give them as donations to charities or send to clothing banques [17]. In Portugal there are who put next to the waste conventional bin for someone can catch them up.

3 Sustainable Alternatives

3.1 Sustainable Consumption

Environmental concerns around fashion products have provided studies, incentives and partnerships between companies in order to make improvements in processes of production, services and means of communication to change the perception of clothing as a disposable product and reduce environmental impacts during its maintenance and after the first cycle of use [18].

Line of thought, sustainable consumption is an issue to be addressed in terms of resource conservation, waste reduction and reduction of toxic pollutants.

While people are focusing on wealth-creation and consuming commodities satisfaction there are also emerging social groups adopting alternative lifestyles and choosing non-materials behaviors even in the field of fashion. Many of them buy as few clothes as possible and prefers second-hand clothing consumption. There are who called them the minimalist shopper [19], down-shifter 12 or even neo-consumers (Teutsch 2004 in Pears [4]) but all of them have already made a decision to reduce their work, incomes and consumption, having more leisure time for their kids and living in balance with their own values.

3.2 Sustainable Wardrobe

As soon as environment has become a differentiation source and a concern requirement for designers and green companies that they have recognized and promote new tackles for their consumers, adopting an holistic approach «from cradle to cradle» [20].

Design for sustainability focus on the systems that must be worked on several lines of though simultaneously and making a product lifecycle approach, ranging from design to end-of-life stage. There are several methodologies at all stages of the clothing product to reduce the environmental impact. The use stage is the one to which «wardrobe» is connected, thus it is firstly important understand the dynamics of that, which depends on user to user, to verify the best practices of use, care or even reuse. The idea of renting garment instead of buying seems unattractive for many consumers, however, for certain clothing styles (such as social clothing, maternity clothes, school uniforms, etc.) is a fairly common practice [21].

Nowadays our wardrobe can have places of dynamic equilibrium, if there are reworked clothes, shared and reused, without constantly requiring a flow of new goods and resources [22]. The clothing exchange may be a practice as a response to the problem of fashion waste by seeking to reactivate it [23]. However, the potential for exchange, as fashion waste activation, only may be realized if that become a regular event and generate an enthusiastic audience for people who consider it as a reliable method of fashion consumption and a genuine alternative to consumption of new fashion garments [4].

The aim of sharing goods is to increase the intensity of their use while reducing the material intensity and the need to buy new clothing.

3.3 Dinamic Wardrobe

A dynamic wardrobe can be like a house with good maintenance and care that changes when are out of time. It can have rental, shared and bought products, but all are actively used. To know the opportunity of people having one and with the purpose of verifying which pieces of clothing that nobody discards and that all use them, interviews were conducted to women students of higher education.

3.4 Survey Methodology

The interviews were given in the form of an open questionnaire and the students had two months to answer the questions and to illustratively assemble their wardrobe pieces.

The issues were made according to characteristics that have not been studied in the literary review such as:

1. What clothes are on your current wardrobe and what you really wear (coats, skirts, trousers, shirts…)?
2. What style of clothing do you wear more often (style, colors, textures, inspiration)?
3. What kind of store do you usually buy clothing (street store, shopping store, online, etc.)?
4. How many garments do you buy per season?
5. What reasons make you to buy garments (necessity, emotional feeling, to be attractive, etc.)?
6. How many garments do you put away per season and why?
7. Among all existing garments in your wardrobe which do you like most and why?

3.5 Survey Results

Among 25 inquiries, we realize that when we ask people about the wardrobe only a very small minority (8%) included accessories and anybody included bathing suits, gym suits and intimate attire.

Just under half (40%) of people have between 50 and 100 pieces of clothing into their wardrobe while only 8% have less than 30 because they are not living in their country.

64% of people buy their clothes in shopping centers or street shops and only a small minority buy online from foreign brands. About half the people that discard clothes keep them at home and other half give them to friends or send to charity.

Everyone says buying items by necessity, fashion and comfort, although a very small fraction confirms buying pieces of clothes to match what they already have.

In general, there are typologies of pieces and colors the vast majority preferred most, the coats and oversized pieces; the coats for the possibility of composing a personal look and the oversized pieces (especially wool sweaters, large and loose dresses or jumpers) for the ease of handling and comfort, all of them usually in dry, neutral or black colours.

4 Conclusion

All women have a clear sense of which items can be match simultaneously and which clothes go together, involving not only questions of colours and style, but also whether the garment fits on their own body. They know clothes are a medium they use to express their selves and despite the preference for neutral colours and basic styles on complete outfits, the patterns, accessories and specific items (more geometric for some and more floral for others) are aspects of their self-identity that may be externalized in the wider social and cultural context. Women can manage and construct their appearances based on social networks, lifestyle and personal taste as well often going in the opposite direction to what happens in fashion trends. These issues lead us to think that there is great amount of fashion products, more than people need. We have found that although most students have more than 50 pieces of garment, some enjoy wearing items purchased in unconventional stores or inherited by family members. However, this concern is based on differentiation of consumers who are dissatisfied with the same figures of new fashion for mass market rather than environmental matter.

We think that in a more sustainable future the fashion design process will certainly be less massive and more personalized or rotating, where the wardrobes can be dynamic, with exchanges and rental clothes.

References

1. Fletcher, K., Klepp, I.: Opening up the wardrobe (2017)
2. Woodward, S., Miller, D.: Why Women Wear What They Wear. Berg (2007)
3. Kaiser, S.: The Social Psychology of Clothing. Fairchild Books, London (2003)
4. Pears, K.: Fashion re-consumption; developing a sustainable fashion consumption practice influenced by sustainability and consumption theory. School of Architecture and Design, RMIT University (2006)
5. Kawamura, Y.: Fashion-ology - An Introduction to Fashion Studies. Berg (2004)
6. Veblen, T.: The Theory of the Leisure Class: An Economic Study in the Evolution of Institutions. The Macmillan Company, London (1899)
7. Hogg, V.: An Introduction to Social Psychology. Custom Book (1983)
8. Entwistle, J.: The Fashioned Body - Fashion, Dress and Modern social Theory. Polity Press, Boston (2015)
9. Baldini, M.: L'Invenzione della Moda. Le teorie, gli stilisti, la storia. Armando Armando (2005)
10. Muggleton, D.: Inside Sub-culture: The Post-modern Meaning of Style. Berg (2000)
11. Eicher, J.: The cultural significance of dress and textiles. Rev. Anthropol. **30**, 309–323 (2001)
12. Hamilton, C., Denniss, R.: Affluenza: When Too Much is Never Enough. Allen & Unwin, Crows Nest (2005)
13. Solomon, M.: Truth About What Customers Want. Pearson Education Inc, New York (2009)
14. Tigert, D., Ring, L.: Fashion involvement and buying behavior: a methodological study. Adv. Consum. Res. **3**, 46–52 (1976)
15. Barthes, R.: Sistema della Moda (1970)
16. Morais, C., Montagna, G.: Costumized Wardrobe: Clothing According to User. Elsevier B. V, Amsterdam (2015)
17. Gwilt, A., Rissanen, T.: Shapping Sustainable Fashion-Changing the Way we Make and Use Clothes. Earthscan, Abingdon (2011)
18. Ellen Macarthur Foundation: A New Textiles Economy: Redesigning Fashion's Future. Ellen Macarthur Foundation, Cowes (2007)
19. Schor, J.: The new politics of consumption. Boston Review, Boston, USA (2005)
20. Braungart, M., MacDonough, W.: Cradle to Cradle: Remaking the Way We Make Things. North Point Press, New York (2002)
21. Vezzoli, C., Manzini, E.: Design for Environmental Sustainability. Springer, Heidelberg (2008)
22. Fletcher, K., Grose, L.: Fashion and Sustainability - Design for Change. Laurence King Publishing Ltd., London (2012)
23. Mont, O.: Institutionalization of sustainable consumption patterns based on shared use. Ecol. Econ. **50**, 135–153 (2004)

Parametric Design for the Construction of a Corset Surface Based on Historical Female Bodies

Felipe Zapata-Roldan[1,2](✉) and Blanca Echavarria-Bustamante[2]

[1] Department of Technology Management, School of Engineering, University of Bridgeport, 126 Park Avenue, Bridgeport, CT 06604, USA
fzapatar@my.bridgeport.edu

[2] Escuela de Arquitectura y Diseño, Universidad Pontificia Bolivariana, Circular 1 No. 70 01, Medellín, Colombia
{felipe.zapata, blanca.echavarria}@upb.edu.co

Abstract. This work shows the results of a research project which goal was to make a geo-historical study on clothes used by four civilizations: Rome, Greece, Egypt and India. A bibliography search was done, online museum image databases and other sources were accessed in order to have a photographic record of statues and female tridimensional figures corresponding to the selected historical periods and parts of the world. The method included a partial reconstruction of the body, referencing the sculptural representation of female beauty in those ancient civilizations. After the reconstruction an indirect extraction of anthropometric measurements was made over 2D images, and those data were fed into a generative design software to model a tridimensional torso. The outcome is a 'historical-customizable' 3D model that can be applied to the production of clothing and historical costumes.

Keywords: Parametric design · CAD · History research · Image analysis

1 Introduction

Based on the results of the first stage of a project named 'Geo-historical reading of garment' done with a team of undergraduate students from the School of Architecture and Design of a Colombian higher education institution, a set of morphological characteristics of women from four ancient civilizations were categorized. Based on those key morphological characteristics, a parametrization was possible on a second stage of the project.

To obtain both characteristics and parameters, a novel method was applied to collect data from secondary sources, in this case photographs taken by different individuals in different parts of the world. The interpretation of the anthropometrical measurements is key to include the historical force of the concept of beauty, divinity and perfection on each civilization, and their semiotic reading on the selected images that could include reproductions or photographs of sculptures and drawings. Semiotic readings in this context often include cultural, religious and social elements permeating

G. Di Bucchianico (Ed.): AHFE 2019, AISC 954, pp. 365–371, 2020.
https://doi.org/10.1007/978-3-030-20444-0_36

the artistic representation of beauty, and such elements could insert distortions in such representation.

Historical clothes reproduction demands inference tools and indirect extraction of structured body measurements based on aesthetic canons linked to beauty and divinity, because it is not possible to measure subjects. All artistic or crafted representations are the best source possible and that reflects on both dressed and nude bodies what surrounded the maker, with a certain degree of objectivity defined by the cultural agreements on what was correct and divine.

For any of the four civilizations selected, the most long-lasting object to record the shape of bodies, excluding those of deities and mythological figures, are statues and figurines, regardless the scale. These objects can be found in both public and private collections around the world, and there are photographic archives of a large number of these tridimensional objects. An ideal scenario supported by new technologies like CAD and CAM applied to clothing design, could be having access to the physical object and make the measurements directly on it, however, that compromises the integrity of each piece and requires access to many facilities around the world, which may be restrictive for a research project. On a more practical perspective, overcoming these research challenges might help to have accurate historic anthropometrical models to be used in the construction of clothes, covering in a more precise manner the base body cylinder and the layout for cutting patterns and pieces to be sewn.

The main contribution of this work is the creation of algorithms fed with historical data extracted from bi-dimensional images, that can be used in the tridimensional modeling of a body-fitting surface. Although precise, most technical solutions to apply anthropometrical measurements in technical solutions in the clothing design process are rigid, and don't include ethnic nor body morphology factors.

2 State of the Art

Different instruments have been developed in the last two decades to make anthropometrical measurements to diverse groups, regardless they are segmented by region, ethnicity, gender and other characteristics, that help to draw conclusions about their body typology to be included in the identification process of sizing models used in the technical development of clothing. Body scanning technology complements all traditional instruments like rulers, tapes and calipers. Measurements obtained by these methods are called 'naked measurements' in which the subject uses underwear or a tightly-fitted bodysuit to avoid any error in data collection.

Nonetheless, the extraction of anthropometrical data of already deceased individuals is challenging. Some extrapolation can be made from forensic anthropology methods using bones, genotypes and comparative anatomy, however is not an exact measurement. Also, these methods lack the cultural and historical factor analysis that may explain the way clothes were used or why a certain body feature was admired. However some attempts to link body changes and economic cycles have been made [1].

Anthropometrics in clothing design have a direct influence in how the clothes are created and manufactured and any technology that supports data extraction and application can streamline this process. New technological opportunities and other industrial applications have opened new fields of research related to the parametrization of the body, whether is for analyzing body shape variability for product design [2], or creating avatars with accurate body proportions [3], designing products [4, 5], or benchmarking measurement methods [6].

3 Methods

This research project is based on an inductive empirical method, in which early stage results of the project were used to refine the software solution on a later stage. First stage of the project, the geo-historical reading of clothing, included identification of dressing patterns for Rome, Greece, India, Egypt, Sumer, Persia, China and Japan. This was key to identify the body segments that were considered components of 'beautiful' proportions, and this concept is related to three concepts in clothing elements: access, cover and support. Body typology was a selection criterion to refine the search for digital images from which measurements could be made, as the interaction of clothing is relevant to the perception of beauty according to the time and place. The original list of eight civilizations was narrowed down to for, to continue with data extraction, as they had a higher number of tridimensional representations of the body (statues and figurines) compared to the number of paintings and drawings. For each one of the four civilizations eight online-available photographs of statues or figurines were selected, prioritizing those in which the posture was natural, legs were extended, and it was a frontal view. Over those images, a set of main axes were marked digitally as seen on Fig. 1.

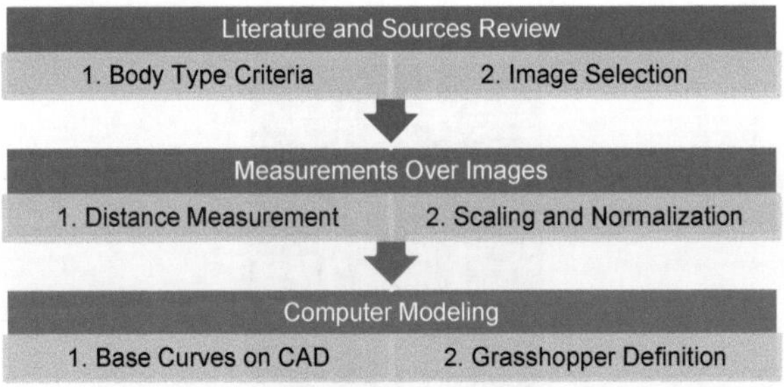

Fig. 1. Methodological process of the project

All images were placed horizontally in a row, and horizontal guidelines were added marking the top of the head and the end of the spine, using the distance between them to normalize and make the images comparable. A vertical guideline was added coaxial with the line that passes through where the spine is estimated to be, when possible and according to the posture to define the rest of dimensions. Finally, the rest of the key vertical levels were marked: chin, waist, hips and heels. These additional markings allowed a comparison between proportions of the beauty canons for each civilization as seen on Fig. 2.

Fig. 2. Statues from five civilizations: Greece (Aphrodite, Getty Museum, USA), Rome (Aphrodite -Venus- of Knidos, National Roman Museum, Italy), Egypt (Cleopatra, Hermitage Museum - Saint Petersburg, Russia) and India (Didarganj Yakshi, Indian Government). Guidelines were added to mark heights of interest: top of the head, chin, waist, hips and Heels.

Next step after marking relevant height levels, was measuring distances defining body proportions. Bust level was included, and for the purpose of this study any spherical lens distortion was considered minimal, also, all distances considered linear. After measuring all distances of interest over letter-size printed versions of the photographs, values were recorded in millimeters for further processing. One of the limitations of the method is to estimate or extrapolate measurements that cannot be made directly due to the existence of 'obstacles' like the garment of the statue representation or a body part hiding key points of reference as seen on Fig. 3.

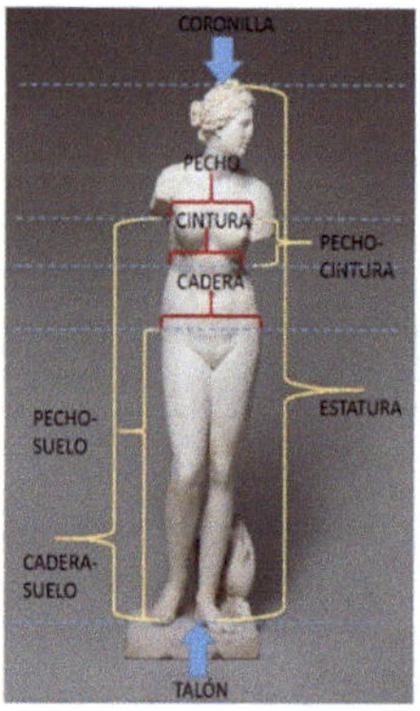

Fig. 3. Original training image handed to students to collect dimensional data. The five points of reference are marked with horizontal dashed lines (top of the head, bust, waist, hips and heels).

A second limitation for the method was handling images from diverse sources, with different printed resolutions and different qualities. To work with the best available printed image, each photograph was printed on a single sheet of paper and measurements done over it, therefore a scale standard was to be implemented. Used on anthropometry, the height (or vertical dimension) of the figure from top of the head to the heels was selected as the dimensional standard.

4 Results

Even though there were limitations, the method allowed to obtain proportion parameters based on body type, and indirect anthropometry of body representations, from different eras and parts of the world. Those historical-based parameters were applied to a customizable algorithm and modeling result generated a large number of versions of the surface modeled. To support the parametric modeling process, measurements on paper must be translated into the unit system used by any digital modeling software, however, those parameters must be useful to the design process.

Measurement data from the five body points of reference (top of the head, bust, waist, hips and heels) of the bodies was processed and derived dimensions were calculated: total body height, distance between bust and waist, bust point height referenced to the heels, and hips height referenced to heels. Normalization and dimension-to-height ratios main advantage is that such data is independent of the measuring method, whether is millimeters on a piece of paper obtained with a ruler, or pixels on a digital image obtained with an image editing software.

Same normalization method was used to establish the ratios for standardized widths or horizontal extension measurements for bust, waist and hips. A sample of data based on the average of the eight sets of measurements for each civilization is shown in Table 1 below.

Table 1. Average height for each civilization and dimension-to-height ratios

	Width ratios				Height ratios			
	Height	Bust	Waist	Hips	Bust	Bust-Waist	Waist-Hips	Hips Height
Rome	168.1	0.189	0.173	0.223	0.754	0.132	0.250	0.508
Greece	172.4	0.169	0.174	0.221	0.750	0.128	0.250	0.497
India	169.0	0.175	0.132	0.233	0.743	0.129	0.257	0.487

After calculating averages for all four civilizations, dimension-to-height ratios were calculated, and this data were used in Rhinoceros 3D (by McNeel and Associates) and its generative design plugin Grasshopper (by McNeel and Associates). As a proof of concept, only height offsetting was modeled, although, contour scaling follows the same logic although that would require an expansion to the Grasshopper definition. To simplify the model a free-form curve resembling the contour of the waist was scaled and used for the bust and the hips contours as the main goal is to obtain a surface that might be used as the base of a corset cutting pattern.

This visual algorithmic definition has four parameters that can be changed: Height on a floating-point slider and three ratios that can be selected from a dropdown list. Not only this allows to change in real time the proportions of the surface, but also, select in out of the different permutations a mix of the proportions from different civilizations, expanding the possibilities for design and clothing adaptation to historical shapes.

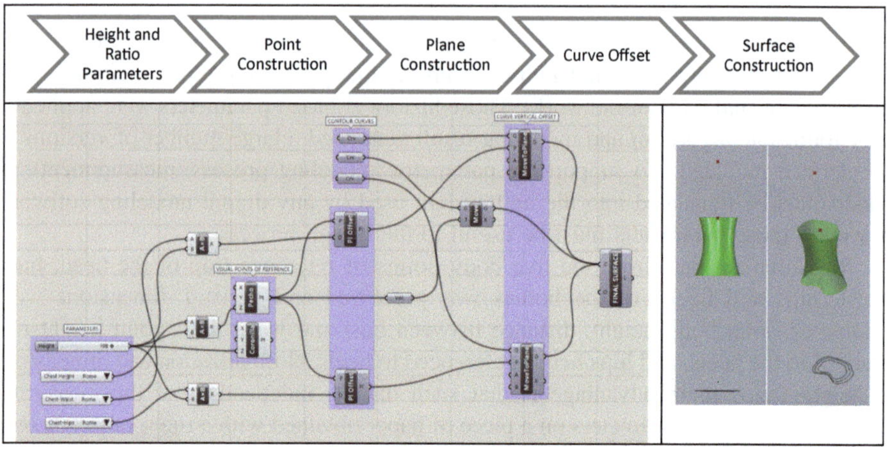

Fig. 4. Algorithmic process and Grasshopper definition

An interesting finding can be seen on Fig. 4 for a coaxial comparison of the three models. There are small variances between the Roman and the Greek beauty canons for the same height, although results suggest exact same waistline with slightly longer legs and upper back. Indian beauty canon for the same height suggests shorter legs, lower waistline and lower shoulders. Geo-historical interpretations of such difference may include religious reasons or deity shape emulation for India, to bone structure for Greece and Rome as seen on Fig. 5.

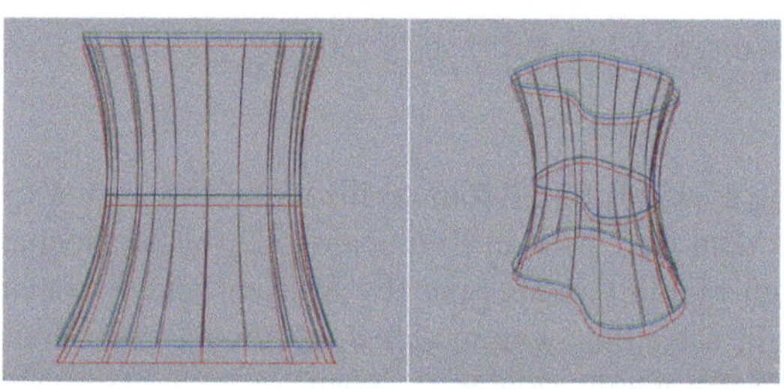

Fig. 5. Comparison of the surface obtained for three civilizations: India (red), Greece (blue) and Rome (green).

5 Conclusions

The use of parametric modeling brings technological opportunities for clothing and costume designers. With a subset of ancient civilizations and a small sample of images of body representations, cultural, genetic and religious aspects of sacred or divine proportions were included in parametric CAD models as a novel method. The results of this work can be expanded integrating other measurements of the body that can be normalized with a dimension-to-height ratio.

Acknowledgements. The authors want to acknowledge the work of the group of undergraduate students who worked on the project, and were part of the garment design research group between 2015 and 2016.

References

1. Cuff, T.: Historical anthropometrics (2004). https://eh.net/encyclopedia/historical-anthropometrics/. Accessed 01 Mar 2019
2. Baek, S.-Y., Lee, K.: Parametric human body shape modeling framework for human-centered product design. Comput. Des. **44**(1), 56–67 (2012)
3. Reed, M.P., Park, B.-K., Kim, K.H., Raschke, U.: Creating custom human avatars for ergonomic analysis using depth cameras. Proc. Hum. Factors Ergon. Soc. Annu. Meet. **58**(1), 1590–1594 (2014)
4. Ellena, T., Skals, S., Subic, A., Mustafa, H., Pang, T.Y.: 3D digital headform models of Australian cyclists. Appl. Ergon. **59**, 11–18 (2017)
5. Lacko, D., et al.: Ergonomic design of an EEG headset using 3D anthropometry. Appl. Ergon. **58**, 128–136 (2017)
6. Simmons, K.P., Istook, C.L.: Body measurement techniques. J. Fash. Mark. Manag. Int. J. **7**(3), 306–332 (2003)

Exploring Fashion Design Methods: Understanding Human Factors in New Society Contexts

Luís Ricardo Santos(✉), Gianni Montagna, and Maria João Pereira Neto

CIAUD - Lisbon School of Architecture of the University of Lisbon, Rua Sá Nogueira, Alto da Ajuda, 1349-055 Lisbon, Portugal
luisricardosantos@gmail.com

Abstract. In the last decades the progressive technological advances have been changing dramatically the way we experience life, shaping the relationships that we establish with what surrounds us in new unprecedented ways. By changing who we are and what we perceive to be, this phenomenon gradually created an individual that is self-aware and self-sustainable.

Technological advances change communication forever and the individual sees its existence split between two different worlds: the actual and the virtual. In this new era of existence new necessities arise from a body that finds new ways of experiencing its limitations (and enhancements).

In a time of growing individuality and with much more technological power at finger disposal, all users and especially new designers and design students, are allowed to play the "individuality game" even if dominated by rules of big corporations and globalized interests. It's a quest for self-identity and affirmation that we seek on the things we do, the things we buy and the people we connect, in order to make a statement on our person and nature.

Fashion design as a privileged means of communication is a great way for creativity and self-expression of the individual, playing and acknowledging his individual role.

This paper is part of a PhD research on the influence of the digital tools and digital environments in human creativity and concept development, in the specific case of fashion design products. With this paper we intend to comprehend how human beings have evolved culturally and socially so we can identify the most reliable paths to an efficient generative experience, regarding human factors and the users necessities and expectations. It's an attempt to map the actual human positioning on how we connect with each other and with the environment.

The digital environments induce a drastic change in the way research, concept development and project communication are made, changing forever the practice of design. The full understanding of the creative potential of digital languages is a real necessity to those who want to be prepared to the uncertain future of fashion design. Regarding this, the results of this study will help to understand how the fashion designer of the future will need to be, how he will act and how he will think.

Keywords: Fashion · Contemporaneity · Human factors · Society · Culture

G. Di Bucchianico (Ed.): AHFE 2019, AISC 954, pp. 372–381, 2020.
https://doi.org/10.1007/978-3-030-20444-0_37

1 Introduction

Design is an ever-changing concept that seeks to respond and adapt to the new social and technical circumstances dictated by the contemporary world. It is a form of culture, of expression and can never be dissociated from the social context. As such, in an increasingly digital world, it will be normal to see design requiring different media and processes [1].

McLuhan and Fiore [2] in their book "The Medium is the Massage" speak about the deep sociocultural changes that contemporary society is suffering, reflecting on the way everything around us is reconfiguring, and how, the human being adaptively evolves. According to McLuhan [3] we are currently in a state of dormancy - which he calls the *Narcissus Narcosis* syndrome - that makes us to be unaware of the social and psychological effects that new technologies have on us. The human sensory balance is altered and the individual changes gradually without realizing it. Because of this unconscious evolution, the need to identify and reflect on the changes infused in the individual becomes urgent.

2 Modernity

The modernity, also called heavy modernity or early modernity, is a social phenomenon that starts in Europe from the XVII century, that changes the local social-cultural panorama and gradually the global landscape [4]. It's a phenomenon that changes radically the way the individual connects with the world.

This period, which Bauman calls "heavy modernity", was characterized with a great solidity in the strategies of growth carried out by countries and enterprises, that privileged a physical and territorial approach. Material assets were directly connected with the growth of financial capital so, the maintenance of work relationships emerges as an economical strategy [5].

In the Twentieth century, modernity becomes a global phenomenon characterized by the following dimensions and consequences:

2.1 Technological Advances

There is a generalized use of machines in the production processes and in war. This industrialization process leads to the emergence of production processes increasingly cheaper and more efficient. As a result, the human impact expands and most of the landscapes (urban or not) become controlled by man [5].

The industrialization and consequential technological advances of the world leads to an exponential acceleration and dynamism of society, resulting in:

The Split Between Time and Space
Human action ceases to be limited by space and information is able to travel freely and increasingly fast, between distant parts of the globe. The local and global realities overlap and the concept of "*globalization*" emerges. Social activity can now be established at a global scale, through big time-space distances [4, 5].

The Reflexive Appropriation of Knowledge
Traditional practices are gradually replaced by more effective ones, as the outcome of a review process whose focus is the knowledge that is constantly produced. This reflexive process of constant enhancement is also appropriated by individuals, generating a state of mental uncertainty that tends to lead to the construction of a trust system as a way to deal with instability [6, 7].

The Growth of Communication Technologies
First the printed text and then the electronic signal, lead society to an exponential growth in terms of the information available. This avalanche of information from all over the globe is gradually incorporated in the everyday consciousness, leading to the juxtaposition of stories and items that have nothing in common except being opportune. Narratives disappear and single elements are separated from their initial referents, turning into sterile signs. These signs go beyond the transmission of information, forming their own reality, disconnected from the actual world [4, 6].

2.2 Market

Capital and property is privatized. The commodification of the work force allows the individual to accumulate capital, boosting the competitiveness of the market. This phenomenon marks the emergence of capitalism as the new class system regulated by individual possessions.

The possession of things as a way of class distinctions is rooted in the exclusiveness of the objects. This exclusiveness is no longer strictly connected to materiality but also to the authenticity that comes from the gesture of who made it. Artistry becomes marketable [8].

2.3 Society

Administrative power is centralized in *nation-states*, in order to become more efficient, being able to mobilize larger amounts of economic and social resources than pre-modern institutions. The access to population data allows these nation-states to control and supervise population, directly or indirectly. Territorial and military power, allow them to control the multi-national companies that expand in economic and social power during this era [4].

2.4 The Individual

The division of work in the modern societies results in the emergence of people as individual entities, fully recognized and valued.

With the ability to make his own choices, the individual sees the future as full of possibilities and no indication on which path to take. This uncertainty exposes them to high levels of psychological insecurity. The individual is judged socially by his choices, and appearance becomes the central element of the construction of the identity of the self. Emergent questions of good taste label the produced goods and their users, positioning them accordingly to the capitalist class system.

Apart from individuality, another phenomenon that emerges in modernity is the emerge of the "lifestyle". More than the objects possessed, it is the practices that an individual embraces that give shape to a particular narrative of self-identity.

Lifestyle ceases to be adopted as a product of tradition or habit and starts to be embraced as a life's project where style is assumed to be synonymous of individuality. The adoption of a lifestyle is influenced by group pressures, by socioeconomic models and by circumstances. It is the lifestyle and not the physical items that are going to establish the class structure and the mechanisms of differentiation in society [4–7, 9].

3 After the Modernism

After the 1950's, the numerous crisis and wars give rise to a growing discredit in the universal and timeless principles of modernism. Particular visions gained a new importance and social regimes become anti-totalitarian. Democracy loses its idealistic grandiosity and accepts heterogeneous, marginal and quotidian discourses.

This period acquires several nomenclatures, depending on the specific focus of different authors. Jean-François Lyotard speaks of this era as a "post-modernity", as a way to mark the change from an industrial to a post-industrialization era. He acknowledges a change in society that goes beyond the values and practices of modernism. To Ulrich Beck, this period arises as a "second modernity" or "reflexive modernity", highlighting the way as society becomes aware of the processes carried out during the early modernity and start to think its conduct in a reflexive, rational way. The anti-totalitarian aspect of modernity leads to the concept of "alter-modernity" proposed by Nicolas Bourriaud. This view states that the same modern phenomena happens but, we have new, alternative views over them, deeply rooted in multiculturalism. To Zygmunt Bauman the uncertainty and instability of the end of the twentieth century gives place to the "liquid modernity" or "light modernity" where solidity of views and relationships become more fluid. Anthony Giddens takes a different stance, calling this era of "high modernity", "radicalized modernity" or "late modernity", since, according to the author, there was no break towards the paradigms of modernity but, instead, we see its consequences being greatly radicalized and globalized. Gilles Lipovetsky, takes a similar approach (with emphasis in the digitalization of society), stating that we now live in an highly radicalized and accelerated modernity called "hyper-modernity" [4, 5, 7–10].

None of these visions are antagonistic and, together, they are able to characterize this period of social history where modernist dimensions still apply, but with different, more amplified consequences:

3.1 Technological Advances

Modernism based its growing production of objects and signs in the thrust of infinite energy and in the projection of present into the future. But with the energy crisis that starts to be felt by the late twentieth century, a discredit in progress arises and ecological concerns emerge. The posture towards technology becomes reflexive, and humanity starts to comprehend the consequences of the rapid technicalization of society.

Digitalization spreads through all the sectors of society, promoting deep socio-cultural changes that raise questions about man's dominion over technology. This highly digitalization of society causes a nostalgic search for physicality (real values and real people) [1, 4, 6, 8, 10].

The Transformation of Reality in Images and the Acceleration of Life
The gradual digitalization of the world since the 1980's leads to an exponential generation and transmission of images. This phenomena stimulate mechanisms of obsolescence where images are constantly consumed and replaced.

The concept of "*tabula rasa*" becomes impossible to conceive because every personal concept or creation is sustained by this chaos of products, media and comments experienced by the individual in his everyday life.

For Zygmunt Bauman, it's this acceleration of the daily life that differentiates "liquid modernity" from the early modernity of the twentieth century. Acceleration achieves its natural limit giving place to the instantaneity (of time and space) characteristic of our era. The surrounding ambient is constantly updated and reformulated, and what was "lasting" in our lives becomes "temporary". The concept of "possession" loses meaning because of the volatility of things and is substituted by the concept of the "access" to the experience.

Life becomes a never-ending race in which the individual tries to stay by all means. The precariousness that comes from this condition leads to everything being perceived as ephemeral and disposable after fulfilling its purpose. This principle encompasses not only objects but also other human beings that also become replaceable.

Progress ceases to be a collective goal and is now centered in the individual that is now responsible to his own upgrowth in life. The excessive amount of choices available generates anxiety in the individual who constantly questions themselves whether he has made the right choices. Therefore, he chooses to live of the immediate pleasure of the present because future is uncertain, and setting long-term goals seems a bad strategy. And since today's commitments can interfere with tomorrow's opportunities, the tendency is to keep commitments open, light and superficial.

If on the one hand this accountability leads the individual to isolate themselves and focus on his path (taking an exasperated stance towards everything that might obstruct his goals), on the other hand it leads to an unquestioned submission to social norms and institutions, in search for mental stability [4–6, 8, 11].

3.2 Market

At the end of the twentieth century, economy evolves towards a "light capitalism", characterized by the growth of technological innovation and by the disconnection from the physicality of the geographical place through a digitalization of the economy.

The economic system based on the manufacture of material goods shifts its focus to the commercialization of information and ideas.

The market is increasingly globalized and is ruled by the values of competition that create relentless policies of capitalist profitability. The automation of the functions of regulation and reproduction, which are no longer made by human administrators, accelerates the production process and increases the accessibility of products [5, 8, 9].

According to Lipovetsky and Serroy,

> "Social relegation of large sections of the population grows, wages are degraded, old labour solidarities disappear and, trapped in the jungle of total capitalism, the individual is left to the uncertainty of the future, to a process of atomization that leaves they alone before himself and towards an unprotected existence, without class culture, without a collective framework, without a political project of world transformation." [10]

The Expansion of Production Capacity

The constant renewal of goods gives the consumer the illusion of complete substitutability of goods and unrestricted access to them. This illusion generates enormous pressures of obsolescence where the consumer gradually detaches himself emotionally from what he possesses, focusing on the immediate and superficial pleasure of consumption, rooted in novelty and not in affectivity.

Science becomes a tool of capitalism, making room for improvements in the performance of production and its products. Knowledge becomes commodified as a way of competitiveness in the world's capitalist scenery [7, 9].

Sign-Goods

The perception of the status or hierarchical position of the individual becomes a complex problem because of the existence of the constant renewal and increasing accessibility of products. Trends are brief, transitory and coexist simultaneously making the distinction process even more difficult to perform.

With this accelerated process of renovation, goods lose their real use value. Through advertising, mundane consumer goods turn into images (of romance, exoticism, achievement, etc.) becoming "sign-goods". The acquisition of things becomes primarily the consumption of signs and not only the purchase of material utilities.

Product rarity and uniqueness resurfaces as an effective way of breaking down the barrier of standardization and to transmit an aura of (marketable) individuality [7, 8].

The Emotional Pleasures of Consumption

Capitalism stimulates the celebration of individual dreams and desires in specific places of consumption. These constructed scenarios are designed to produce various types of physical excitement in the shape of an "aesthetic hallucination of the real" in which image and reality merge. Everything becomes performative, emotional and imaginary in order to constantly reactivate the desires of the consumer.

With the spread of these practices, everything gains a superficiality and instantaneity and the pleasure that comes from its consumption has to be immediate and intense. The "pure" taste of high-aesthetic, related to a cognitive and controlled appreciation is replaced by a "fast-aesthetic" controlled by immediate, direct, emotional and physical pleasure, widespread in and by the masses. Pleasure becomes the purpose of all human action itself [7].

3.3 Society

There is a constant affirmation of singularity, in a world where the right to be different has become a fundamental value. This concept overcomes pre-existing models and stereotypes, giving space to a multiplicity of new lifestyles.

Society moves beyond the concept of the universal (characteristic of the modernism) to embrace the idea of a fluid plurality built by systems (cultural groups, subgroups or single individuals) with relationships between them [9, 10].

The Plural Globalization

In contemporary society, cultural diversity is identified as a patrimony to be preserved, in response to the standardization of imaginations felt during modernism. The process of globalization becomes plural and starts to support several images of what the world is (or should be).

Global culture tends to follow in the opposite direction of homogeneity and of a common culture. In this plural cultural scenery, no discourse assumes itself as "\the true" or "the dominant one" and conflicting trends coexist in harmony.

Globalizing pressures strengthens social feelings of regional cultural identity and local empowerment. This fact results in the recognition of local, regional and subcultural differences, by the *nation-states,* through the process of decentralization.

Society begins to see narrative knowledge (originating from popular cultures) in a more dignified way, accepting it as a method of legitimation in the form of social consensus [4, 7–9].

The Abolition of the Border Between Art and Everyday Life

Art ceases to be seen as a superior form of experience, derived from the creative genius or special qualities of the artist. Artistic and intellectual goods become capitalizable, becoming popular and accessible to larger audiences. Art migrates to industrial design, advertising and the industries associated with the production of symbols and images. This migration makes the separation between "art" and "non-art" doubtful, so that any object of everyday use can be aesthetized.

The mass-produced culture that results from this democratization of art ends up being reduced to the lowest common denominator so it is accessible to a greater public, in the form of extremely limited discourses. Eclecticism and the mixing of codes in the form of parody, pastiche, irony, amusement and the celebration of the "lack of depth" are favored. The volatility of public memory leads to the emergence of "new" information and "new" elements that in fact are recycled or plagiarized from pre-existing ones [7, 12].

3.4 The Individual

A focus on the individual arises after modernism. According to Lipovetsky and Serroy, this new individualism, also known as the "second individualist revolution," is characterized by some fundamental traits: "body worship, psychological and relational worship, consumerist-hedonistic worship and autonomy worship (as a consequence of the disappearance of faith in the great historicist ideologies: Nation, Revolution, Progress)" [10]. Constant concerns with self-development develops in the individual a growing narcissistic, hedonistic and egocentric posture. Compared to early modernity individualism, the contemporary approach is less frivolous, less playful and more reflexive.

The rediscovery of physicality emerges as a counterbalance to a dematerialized and disembodied culture. Contemporary life is divided between these two realities, equally important to the individual. New opportunities for social relations arise through new contexts (such as the internet), which prevents the modern world to become highly impersonal. We do not live in a "world of strangers," we have just changed the ways we relate [4].

Aestheticization of Everyday Life

We tend to see other people's lives as a work of art and thus aestheticize our own lives. To the practice of aestheticization Bauman calls "identity" [5]. We tend to have the illusion of confidence and solidity of these "identities," believing that they are constant in time, covered by the cloak of authenticity. Only with the use of fantasy mechanisms (dreamlike strategies that consumer brands have learned to capitalize) can these "identities" be cohesive. These "identities" are now purchased on commercial surfaces in the form of mass objects that point the way to a costumed self. This is a "false self" that becomes an object manipulated by the "true self". This dissociation generates an anxiety that leads the individual to question his true authenticity and existence, and to increasingly confuse the two types of "self".

Another form of fantasy arises through the proliferation of the image, with special focus on the screens that have spread throughout society. The fantasy presented through images sets standards for the reality world that is now expected to be more intense and pleasurable. The image becomes the only possible reality and our bodies accompany this mentally constructed ideal. But this surrender to image supremacy is not without its consequences. After experiencing the overwhelming intensity of images, the bigger is the sense of poverty felt towards reality and, the more we feel disappointed with our daily lives.

According to Bourriaud, the contemporary individual becomes a "radicant," an entity that only through constant movement manages to form an identity [8]. These "radicants" refuse to settle in a certain space, time or aesthetics and borrow elements, quotes and proximities to construct their identity. If during modernism the individual is transformed into an industrial machine, contemporaneity aims to equate the individual with the dominant technology: the computer and its reticulated networks. The individual becomes a network, establishing several bidirectional connections (influencing and being influenced), both in the personal and professional spheres [5, 6, 8].

4 Conclusions

The contemporary individual carries within themselves the result of more than a century of intense technological advances and deep social-cultural changes. The concepts introduced by modernism evolved over the past decades in an exponential way, resulting in an amplified version of its social impact, as identified by the multiple authors (previously identified in this paper) through their particular perspectives.

At the center of these social changes, like McLuhan [3] notes, is the individual that reflexively evolves and adapts to this new environment.

The levels of stress, chaos and uncertainty of the social environment are absorbed by human beings that incorporate them and generate internal mechanisms of defence. As a result, identities are no longer stable and the individual defines himself through a constant movement that can only capture loose fragments of reality. It's with these fragments of hollow cultural signs that they construct their fragile identities.

The obsolescence introduced by capitalism becomes the normality and the concept of ephemerality is accepted in life. Nothing is stable or lasting so options are kept open and commitments are avoided. Everything is perceived as disposable so the individual lacks connectivity with things in their life.

Even identity is an open concept and individuals change them at ease, in order to reflect their present mood and context. The boundaries between the "outer identities" and the inner selves are blurred and authenticity gradually fades away. Individuals become the images of a self-created world. These images are signs with no reference, just like all the images proliferating in the world. These hollow images create alternative, pleasurable realities, easier to control, and where the individual can escape from the cruelty of actual reality.

In an unstable, unreliable world, the only concern of the individual is with themselves because everything else can fail or collapse. Everything else becomes irrelevant, even their action as citizens. This egocentric approach is intentional, reflexive and acts as a protection towards the adversities of facing the world at their own risk.

This focus on the self leads the individual to pursue immediate pleasure in the form of easy consumable objects, relationships or situations. As a result, everything in life becomes measured by the pleasure it provides. Novelty is the drive of this pleasure, which, in the contemporary social world, can only be obtained by a constant movement and self-reinvention. Like Lewis Carrol said:

> "it takes all the running you can do, to keep in the same place. If you want to get somewhere else, you must run at least twice as fast as that!"

Acknowledgment. The authors wish to thank Research Centre for Architecture, Urbanism and Design (CIAUD) of ULisboa.

References

1. Newson, A., Suggett, E., Sudjic, D.: Designer, Maker, User. Phaidon Press Limited, London (2016)
2. McLuhan, M., Fiore, Q.: The Medium is the Massage: An Inventory of Effects. Gingko Press, Corte Madera ([1967] 2001)
3. McLuhan, M.: The playboy interview: Marshall McLuhan. Playboy Mag. **158**, 53–74 (1969)
4. Giddens, A.: As Consequências da Modernidade. Editora UNESP. São Paulo ([1990] 1991)
5. Bauman, Z.: Liquid Modernity. Polity Press, Cambridge ([2000] 2006)
6. Giddens, A.: Modernidade e Identidade. Jorge Zahar Editor, Rio de Janeiro ([1991] 2002)
7. Featherstone, M.: Cultura de Consumo e Pós-Modernismo. Studio Nobel, São Paulo ([1990] 1995)

8. Bourriaud, N.: The Radicant. Sternberg Press, New York (2009)
9. Lyotard, J.F.: A Condição Pós-Moderna. José Olympio Editora, Rio de Janeiro ([1979] 2009)
10. Lipovetsky, G., Serroy, J.: O Ecrã Global. Edições 70, Lisboa ([2007] 2010)
11. Lévy, P.: Cibercultura. Instituto Piaget, Lisboa ([1997] 2000)
12. Bauman, Z.: Liquid Modern Challenges to education. Padova University Press, Padova (2011)

Impact of a New Designed Automation Process in Textile Natural Fibers and Yarns Dying

Alexandra Ene(✉), Carmen Mihai, Emilia Visileanu, and Cristian Jipa

The National Research and Development Institute for Textiles and Leather, 16, Lucrețiu Patrașcanu str., 030508 Bucharest, Romania
alexandra.ene@certex.ro

Abstract. The fierce competition in the textile sector, especially due to the producers in Turkey, China, India, etc., implies the development of new competitiveness strategies based on cost reduction in the dying process at the same time as ensuring the quality of the products to the required standards.

The solution developed by a team of researchers within the National Research and Development Institute for Textiles and Leather and successfully implemented in the economic environment consists in the design and realization of an electronic and automation system that ensures the process of textile fiber dyeing under conditions efficiency and increased productivity that responds to the technical and quality performance requirements imposed on end-user products and by consumers.

The proposed technological solution is based principally on generating the turbulence stream of the solution in the dyeing bath by driving the main pump motor and maintaining the flow rate in the dye vessel constantly by actuating the float pump motor.

Using the theories of fluid mechanics, optimization of the recirculation pump speed regime in under pressure textile fiber dyeing processes has demonstrated that the implementation of the technological solution leads to a turbulent regime that ensures the uniformity of the solution in the dying bath, while reducing energy consumption and energy costs, increased productivity through better process control, increased equipment lifespan, and decreased maintenance time.

Keywords: On-line management · Natural fibers · Fluid mechanics

1 Introduction

In the context of continued concern for maintaining high quality standards, attention is paid to the textile fiber and yarns dyeing process, which is one of the most important phases of the technological process and which, unfortunately, is currently being achieved with high consumption of materials and utilities.

The technical and economic efficiency of the textile yarn dyeing process is conditioned by ensuring the proper homogeneity of the dying bath (by generating turbulent motion) while maintaining a constant flow rate in the dying vessel [1].

The improvement of the natural fiber and yarns dyeing parameters is based on the intrinsic connection between the process controller, speed variator and pump-adding system.

G. Di Bucchianico (Ed.): AHFE 2019, AISC 954, pp. 382–386, 2020.
https://doi.org/10.1007/978-3-030-20444-0_38

2 Materials and Methods

The novelty in this dying process is to maintain the bath flow at constant pressure and variable temperature depending on the dying diagram.

Based on the information received from the temperature and pressure sensors, the process controller controls the pumping system by means of the speed variators, so that the dying diagram to ensure a constant pressure throughout the technological process.

The electrical signals from the sensors that monitored the functional blocks are transmitted to an automation module that signals the occurring events [2].

At constant pressure in the dyeing vessel, it is ensured a uniform color absorption in the bath, having as result a uniform dyeing process of the material (Fig. 1).

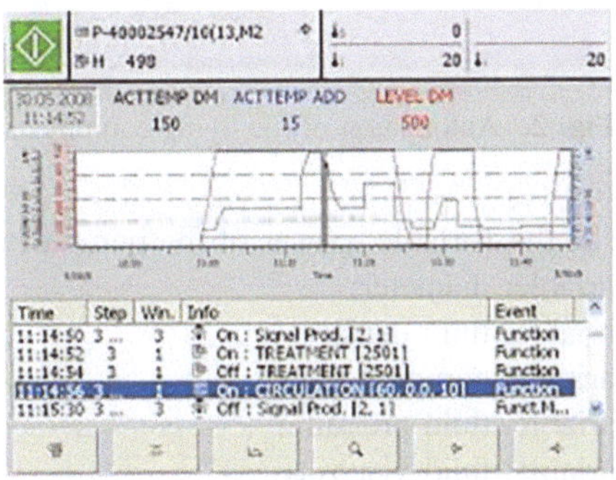

Fig. 1. Management of the dying diagrame through the process controller

The control of functional parameters is based on a sensor system that monitors temperature, pressure, flow as well as the proper functioning of the other functional modules.

Information from these sensors is transmitted to the process controller.

The flow and quantity of the bath are monitored via an analog output flow sensor so that the ratio between the amount of dye and afferent bath being according to the recipe (dying diagram).

Using the theories of fluid mechanics, for the optimization of the recirculation pump speed regime in textile fiber pressure dyeing processes, it has been demonstrated that the implementation of the technological solution leads to a turbulent regime that ensures the uniformity of the solution in the dying bath, while reducing energy consumption and energy costs, increasing the productivity through better process control, increasing the equipment runtime and decreasing the maintenance duration (Fig. 2).

The phenomenon analyzed using the hydrodynamics equations revealed the tendency of the response of the random variables to the turbulent movement (mean and turbulent fluctuation) [3].

To predict the phenomenon, a specialized software was used, starting from the following hypotheses:

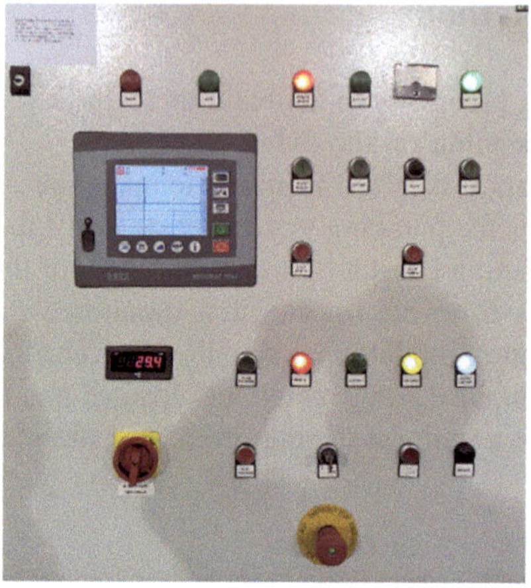

Fig. 2. Automation of the dying equipment

$\ddot{Y}$ – predicted value of the variable: power of the pump of addition;
X_1 – independent variable: diameter;
X_2 – independent variable: fluid density;
X_3 – independent variable: pumping height;
X_4 – independent variable: fluid pressure;
X_5 – independent variable: fluid viscosity.

Using a dedicated software, the coefficient values of the regression equation were determined as follows:

$$\ddot{Y} = A_0 + A_1X_1 + A_2X_2 + A_3X_3 + A_4X_4 + A_5X_5 \quad (1)$$

The distribution form was verified through non-parametric Kolmogorov-Smirnov test (initialized sequential commands).

The values of the Kolmogorov - Smirnov test demonstrate that the data complies with the normality criterion for the population of which they belong [4].

From the analysis of the values (Table 1) it follows:

(i) for a significance threshold p = 0.05 - the values obtained show that there is a relationship between the studied variables, namely "fluid pressure" and "diameter of the feeding device" - Pearson Correlation: −0.762; "Pumping height" and "fluid density" - Pearson Correlation: −0.684; "Pearson Correlation: 0.712"; "Pump power" and "fluid flow" - Pearson Correlation: −0.635;
(ii) for a significance threshold p = 0.01 - there is a relationship between the studied variables, namely: "feed device diameter" and "fluid density" - Pearson Correlation: −0.837; "Fluid density" and "fluid pressure" - Pearson Correlation: 0.848; "Pump power" and "fluid density" - Pearson Correlation: 0.834

Table 1. Experimental values

Nr. crt.	d feed device	h pumping height	P fluid pressure	ρ fluid density	P power of the add. pump
1	60,00	4,70	,140	,1000	1,8542
2	55,00	4,20	,150	,0800	1,8697
3	55,00	4,58	,158	,0790	1,8622
4	50,00	4,56	,159	,0780	1,8797
5	50,00	4,55	,150	,0790	1,8799
6	60,00	4,68	,140	,1000	1,8509
7	45,00	4,33	,160	,0700	1,8834
8	45,00	4,38	,161	,0710	1,8833
9	50,00	4,55	,148	,0782	1,8388
10	60,00	4,66	,150	,0900	1,8552
Total	10	10	10	10	10

(iii) for the “fluid pressure” and “feed device diameter”, “pumping height” and “fluid density” variables, “additional pomp power” and “fluid flow”, the correlation sign is negative, 95% can be said that the link between variables is inversely proportional

(iv) for the “feed device diameter” and “fluid density” variables the correlation sign is negative, so with 99% probability it can be stated that the link between the variables is inversely proportional

(v) for the variables “power pump additive power” and “fluid pressure” the positive sign for correlation was obtained, so with a probability of 95% it can be stated that the link between the variables is directly proportional.

(vi) For the “fluid density” and “fluid pressure” variables, respectively “additional pump power” and “fluid density”, the positive sign for correlation was obtained, thus, with 99% probability, is directly proportional.

(vii) The absolute magnitude of the coefficient (which describes the strength of the link between the variables) shows that there is a strong link between all the variables subjected to the correlation analysis ($r > 0.50$).

3 Results and Discussion

The regression equation is: Power of the addition pump = 1.86 − .002 * Diameter + .001 * Pumping height –247 * Pressure − .049 * Flow rate

This highlights the followings:

- The value of the multi-patch correlation coefficient R^2 reveals that 97.0% of the variance of the dependent (predicted) variable power additive pump is explained by the model, the prediction power being 93.1%;
- The standard error of the predicted dependent variable (addition pump power) demonstrates that the standard power output deviation is 0.041 kW, if the value of the independent variables in the model is known: the feed device diameter, the pumping height, the fluid pressure and its flow.

- The value of 0.970 demonstrates that R^2 does not change when variables are added to the model.
- The variant analysis highlights, with a probability of error of 5%, that the model significantly explains the variance of the dependent variable. The values of the F mark demonstrate that the variance explained by the model is significantly higher than the residual, so the created model is effective in prediction.

The differences between the predicted value of the model and those determined experimentally are: for experiment I: with .002 kW higher, for experiment II: with .002 kW lower and for experiment III: with .003 kW higher but all values are falls within the standard deviation of the .0413 kW dependent variable.

4 Conclusions

The proposed technological solution is based principally on the generation of unsteady vortices (vortex stretching) of the bath into the dye vessel by driving the main pump motor and maintaining the constant value of the pressure in the dyeing bath by driving the addition pump motor that assure the circulation of the bath.

Using the theories of fluid mechanics, for the optimization of the recirculation pump speed regime in textile fiber pressure dyeing processes, it has been demonstrated that the implementation of the technological solution leads to a turbulent regime that ensures the uniformity of the solution in the dying bath, while reducing energy consumption and energy costs, increasing the productivity through better process control, increasing the equipment runtime and decreasing the maintenance duration.

Acknowledgment. This scientific paper is funded by the Ministry of Research and Innovation within Program 1 - Development of the national RD system, Subprogram 1.2 - Institutional Performance - RDI excellence funding projects, Contract no. 6PFE/2018.

References

1. Johnels, A., Crawshaw, G.H.: Measurement of pressure and flow during dyeing of worsted yarn packages. J. Soc. Dyers Colour. **85**(3), 101–107 (2008)
2. Iannetti, A., Stickland, T., Dempster, M.: A CFD and experimental study on cavitation in positive displacement pumps: benefits and drawbacks of the 'full' cavitation model. Eng. Appl. Comput. Fluid Mech. **10**(1), 57–71 (2016)
3. Shankhadeep, D., Bai, H., Wu, C., Barney, B., Kidd, M., Kuettel, M.: Improving the performance of industrial clarifiers using three-dimensional computational fluid dynamics. Eng. Appl. Comput. Fluid Mech. **10**(1), 130–144 (2016)
4. Voineagu, V., Titan, E., Ghita, S.: Statistica. Baze Teoretice si Aplicatii, Editura Economica, Bucuresti, pp. 43–446 (2007)

Lean Manufacturing Model for the Reduction of Production Times and Reduction of the Returns of Defective Items in Textile Industry

Yanira Andrade[1(✉)], Leslie Cardenas[1(✉)], Gino Viacava[1(✉)], Carlos Raymundo[2(✉)], and Francisco Dominguez[3(✉)]

[1] Escuela de Ingeniería Industrial, Universidad Peruana de Ciencias Aplicadas (UPC), Lima, Perú
{u201319340,u201316812,gino.viacava}@upc.edu.pe
[2] Dirección de Investigaciones, Universidad Peruana de Ciencias Aplicadas (UPC), Lima, Perú
carlos.raymundo@upc.edu.pe
[3] Escuela Superior de Ingeniería Informática, Universidad Rey Juan Carlos, Mostoles, Madrid, Spain
francisco.dominguez@urjc.es

Abstract. Textile garment industries must be readily aware of the system's current situation, respond quickly to changes in style to process orders on time, and manage an appropriate method of work to avoid production delays. In addition, the textile sector plays an important role because it presents a significant weight in the world economy. Likewise, this sector is considered influential when defining an international business agreement. Therefore, adequate management is required in the sector. Otherwise, it will generate a competitive disadvantage because the Peruvian industry is approaching an export recession. This paper proposes using the lean manufacturing methodology to reduce production times. The project is validated through a simulation of the current and proposed processes with the application of lean thinking tools to reduce waste (manufacturing process times).

Keywords: Lean manufacturing · Textile · Production time · Kanban · 5S

1 Introduction

Currently, the textile sector reports production delays due to inefficient and lengthy manufacturing processes. According to Grimaldo, Silva, and Molina [1], in a textile company, manufacturing process times delay production since different parts of the product are assembled in this activity.

According to Güner and Ünal [2], "Garment industries, especially those which rely on small batches and orders, must respond quickly to changes in style. It is essential to be aware of the current situation of a system to process orders on time." Therefore, the critical issue is the high times in the manufacturing area, as they evidence a slow and inefficient delivery service. Likewise, exports have decreased by 16.5%, from 2015 to

G. Di Bucchianico (Ed.): AHFE 2019, AISC 954, pp. 387–398, 2020.
https://doi.org/10.1007/978-3-030-20444-0_39

2016, for the Peruvian textile market. This decrease is due to the preference of foreign buyers for Central American products.

The environment of the research project is manufacturing and its sector is textiles. According to the National Peruvian Institute of Statistics, the manufacturing environment represents 13.30% of all economic activities, and in recent years, its gross domestic product has been decreasing, since there is a percentage variation of −0.6%, for 2015 and 2016. Then, in light of these problems affecting textile companies, new techniques that may help reduce high manufacturing times are required. For this reason, the diagnostics VSM tool was used to identify waste in each production process. Likewise, the Kanban and 5S techniques were used, obtaining a favorable impact on the returns of defective products and penalties for late deliveries, generating higher revenue and increased sales for companies. In addition, the Deming Cycle will be applied to secure continuous implementation and future improvements.

2 State of the Art

2.1 Lean Thinking

According to Kovács [3], the contribution of lean thinking is that it helps improve productivity and quality to achieve long-term success to maximize the competitive advantage in the organization. The following results were obtained in different case studies:

For offset printing, cycle times were improved, obtaining 65% of added value. This translates in less inventory and waste [3].

In a chemotherapy health process, average patient waiting times were reduced by 42% [4].

In a manufacturing company, the application of the methodology helped to reduce polishing process times to 4.15 s. Likewise, the activities without added value and cost of overtime hours were reduced by 66.53% and USD $1,764, respectively [5].

The problem of the manufacturing company is that operators take a long time to find information. The main results were the reduction of lead time by 78%, the cycle time by 3.3%, and in-process inventories by 77% [6].

In a health company, the lean manufacturing methodology was used to increase admissions and consultations by 33% and 23%, respectively [7].

In conclusion, according to the aforementioned studies, the implementation of lean thinking is significant for all types of companies, since it focuses on reducing or eliminating factors that do not add value to processes; the latter is executed with the ultimate objective of achieving an efficient productive chain.

2.2 Lean Manufacturing Philosophies

According to Sundar and Kumar [8], using lean manufacturing philosophies may contribute in the improvement of several types of companies. To illustrate this point further, the results obtained in different case studies are mentioned below.

For example, in one case, lean manufacturing was used to reduce workspace by 22.3% and average times by 61.5%. Likewise, in another case, workspace was reduced by 53% and profits of USD $21,350 were obtained [9, 10].

A manufacturing company managed to reduce in-process inventory levels by 71.43% through the Kanban tool. In addition, this company increased its daily production to 105 products [11].

In a textile company, through the application of total productive maintenance, machine downtime was reduced by 25%, increasing machine performance (OEE) by 15% [12].

In a company, by means of the Standard Work tool, cycle times and configuration times were reduced to 350 s and 1,500 s, respectively [13].

In a production process, through the SMED tool, all activities that do not add value were eliminated, reducing average changeover times by 61.29% [14]. In a metallurgical company, the Kaizen technique reduced in-process inventories by 77% and generated savings of USD $37,600 [15].

In conclusion, according to the aforementioned studies, the implementation of lean manufacturing philosophies helps reduce unwanted activities in the production process. In addition, they seek to reduce internal activities within the company.

2.3 Time Studies

According to Jaramillo [16], time studies are an important technique for measuring work. The results obtained in the different case studies are described below.

In a flat weaving plant in Mexico, manufacturing times were reduced by 5%, with an average time of 233.50 min [1].

In a manufacturing company, idle times in the preparation process were reduced by 85%, with an average time of 3981.50 min. In this way, the company was able to save $1,983,682.84 per month [17].

In another example, in a company that produces printed shirts, production bottlenecks were reduced to 21.29 min. In addition, the standard shirt production was reduced to 74.68 min, which equals 1.24 h per unit produced [18].

In a sewing production line, it was observed that productivity increases significantly if order preparation time is reduced. Here cycle times and lead times were reduced by 3.3% and 78%, respectively [19].

In another company, an average reduction of 136 s in production times was reported. In addition, at the end of the day, the company was able to produce 455 more sweatshirts [20].

In conclusion, from the set of time study articles reviewed, reducing production times is a critical factor for the garment assembly (manufacturing) process, since it allows companies to correct errors and assess the cost of bottlenecks.

2.4 Simulation

According to Segovia and Pozo [21], simulations help visualize the possible behavior of the solution to be implemented. Below are the results obtained in different case studies.

In a service company, a Bizagi simulation was used to reflect the average production times, obtaining 30 days for the current scenario, against a proposed ideal time of 12 days [2].

In a manufacturing company, its current diagnosis with respect to production is 432 units per week and the proposed solution scenario, using the Arena simulation tool, is 480 units per week, thus improving productivity by 11.1% [22].

In conclusion, from the set of simulation-related articles reviewed, it is clear that Bizagi is oriented to business processes and Arena to manufacturing processes. In addition, Arena seems to generate results that are more accurate.

3 Collaborations

3.1 Rationale

Once the information and the selected articles were reviewed, five models were selected. Each set had different dimensions, with each one directly contributing to the objectives set forth (Table 1).

Table 1. Comparison matrix

Models	Dimensions				
	Labor	Machine	Method	Communication & planning	Environment
Paper 1: Zakarial and Nik Mohamed et al. [6]		x	x		
Paper 2: Cheng and Cheng [19]			x		x
Paper 3: Randhawa and Ahuja [9]	x	x			
Paper 4: Millstein and Martinich [11]	x	x			
Paper 5: Nallusamy and Saravanan [13]					
Proposal	**x**	**x**	**x**	**x**	**x**

Through the comparative analysis, it is evident that each selected model shows a different approach, depending on the type and sector of the organization under study. This evidences how valuable our proposal is, since the background information greatly contributes to validate our model. This model consists of integrating the Kanban and 5S techniques to find an adequate level of work for operators and maintain efficiencies when performing tasks in the organization. Likewise, this model is applicable to SMEs, and at the same time, it is attractive for future research with broader analysis due to its simplicity and flexibility for the simulation of management models.

3.2 General Model

Based on the research study, opportunities for improvement have been identified. The general production model (Fig. 1) seeks to integrate the dimensions from the comparison matrix. Furthermore, the Deming Cycle will be used to integrate dimensions, thus guaranteeing constant improvement after applying the different lean manufacturing methodology tools.

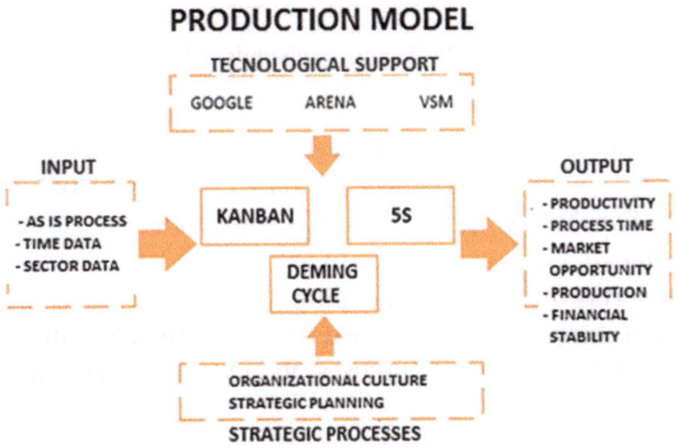

Fig. 1. General proposal model

This proposal then integrates lean manufacturing techniques (i.e., 5S, Kanban) to provide an adequate flow of information, planning, work method, and reduce production times. In addition, this model involves the Deming Cycle to secure proper production model operations through the continuous improvement of its processes. Therefore, the contribution of the Deming Cycle to the proposal is unique and will have a great impact for the project.

3.3 5S Technique

For the 5S technique, several steps were performed. Below, a detailed description of each step has been provided (Fig. 2).

- Step 1: A training schedule for the 5S technique was developed, which focuses on educating the operators on the benefits of applying the tool and providing success stories that support the information. Next, the 5S training schedule is presented.

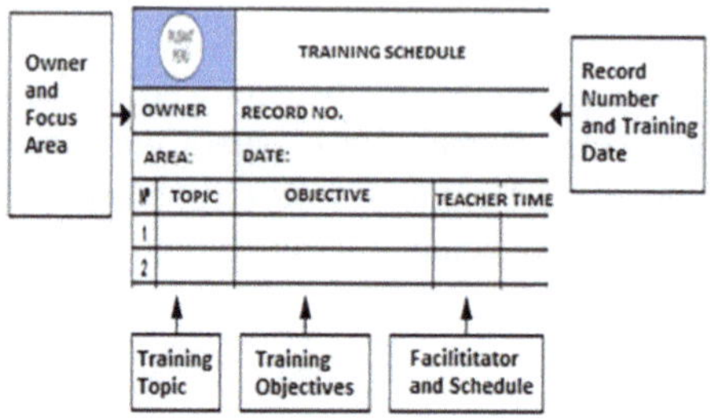

Fig. 2. Training schedule - 5S

- Step 2: This step represents the first stage—classification. In this stage, anything that is not needed should be discarded, and only the required tools should remain in the work area. For these purposes, the following guidelines are followed:
  - *Recognize the area:* Its objective is to identify the area and the number of tools required in terms of their corresponding usefulness.
  - *Identify unnecessary objects:* It lists all process elements to identify any unnecessary items. For these purposes, the red card shown in Fig. 3 is used.

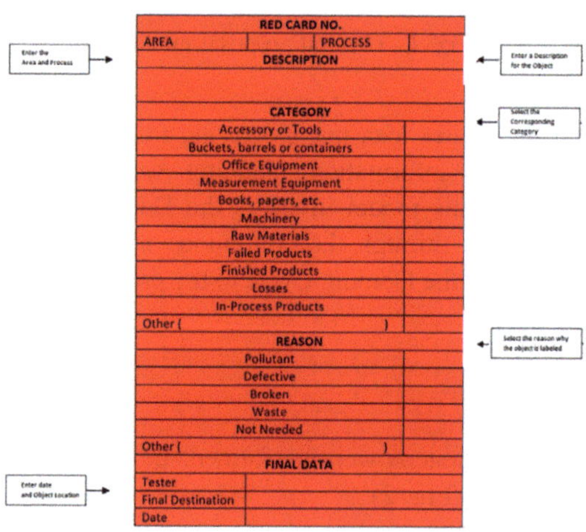

Fig. 3. Red card

  - *List of selected objects:* Once the red cards are made, the list of objects deemed unnecessary for the production process is prepared.
  - *Evaluate the selected objects:* Finally, a diagram is developed for the selected objects, where objects that are still useful for production are separated and sorted for future use. Unnecessary objects are also separated for evaluation.
- Step 3: This step represents the second stage—order. Here, the purpose is to facilitate work for the operators. In this stage, all objects and tools must be in order and in the appropriate position. For this, the following steps must be followed:

- *Tools and quantity to be used:* If production tools are located in strategic places, they can be found and returned quickly. For this reason, keeping a daily tool usage control record is recommended, using tool inventory forms.
- *Specify work area*: It seeks to identify the most important areas within the process. For these purposes, work area forms will be used to describe area specifications and tools required.

– Step 4: Represents the third stage—cleaning—that strives to keep equipment, tools, and work areas clean and tidy. For this, the following steps must be followed:
  - *Determine cleaning responsibilities:* The company under study uses a cleaning service; however, it does not have a specific daily cleaning schedule. Therefore, a cleaning schedule is proposed. See Fig. 4.

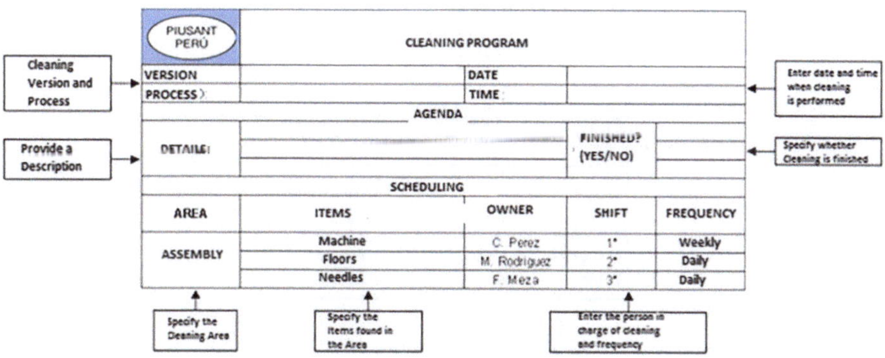

Fig. 4. Cleaning schedule

  - *Record schedule compliance:* A form is developed to guarantee compliance with the proposed cleaning schedule. This form is similar to the previous form, with the addition of one box, where the status on which the area is found before cleaning is recorded.
  - *Evaluate work area:* Finally, work areas were evaluated to identify whether materials may be donated, sold, or transferred. If they can still be used, they are returned to the corresponding area.

– Step 5: The work area is standardized. This process aims at normalizing and maintaining order and cleanliness in work areas through visual controls. In addition, it guarantees compliance with the first 3S. For this reason, this step is also assigned to the same person in charge of enforcing compliance with cleaning schedules and other instructions.

– Step 6: The last step of 5S implementation is discipline. This stage comprises motivating operators. For this reason, management should implement incentives within the organization.

3.4 Kanban

Once the implementation of the 5S technique has started, the Kanban technique will also be implemented. The implementation steps for this technique are:

- *Operator training:* First, training must be provided to operators to disseminate the changes that will be implemented in their work methodologies.
  According to Leonard and Sereno [23], this training should focus on the importance of the Kanban technique for operators and thus foster their commitment of improving the company. Next, Fig. 5 shows the training schedule for the Kanban technique.

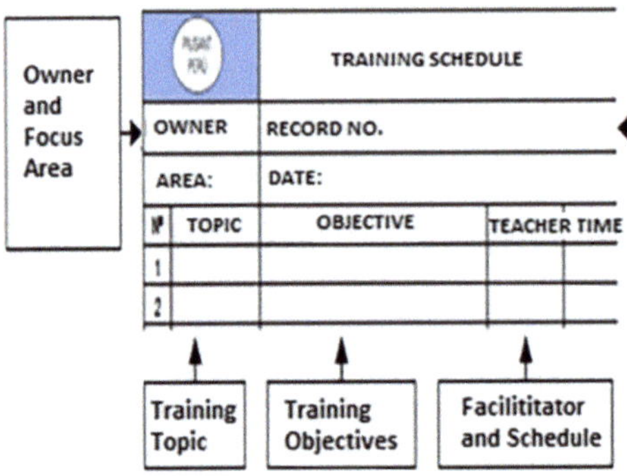

Fig. 5. Kanban training schedule

- *Develop the card:* The second Kanban implementation step is to develop the card, which will prevent workers from making production mistakes. This card promotes better understanding and greater organization, and for this reason, it includes a section for instructions from customers. Next, the Kanban cards are presented (Fig. 6).

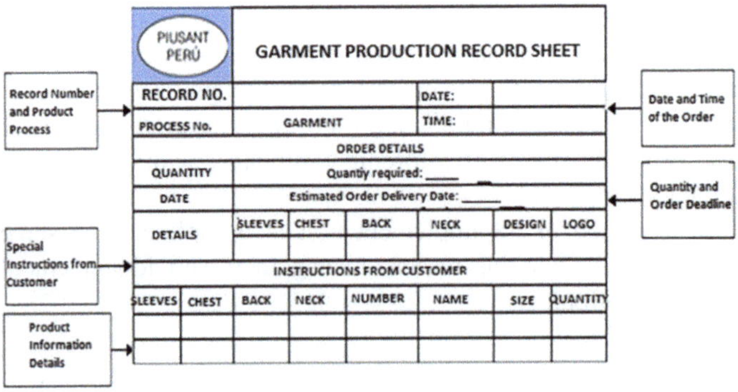

Fig. 6. Improved record sheet

- *Kanban system review:* The last implementation step is review. This stage consists of monitoring the implementation of the technique to see what can be improved and to adopt new prevention measures.

3.5 Deming Cycle

The contribution of this project will be the interaction of lean manufacturing techniques with the application of the Deming Cycle. This Cycle comprises four stages.

- *PLAN:* Determines the objectives and processes necessary to reduce production times. Here the development of Kanban and 5S techniques is described in detail.
- *DO:* Implements the processes proposed in the previous item. Each step and the proposed flows are applied for an adequate management of the techniques.
- *CHECK:* Reviews processes, policies, or objectives, provides information on the results achieved, provides focus, and identifies possible improvements to the process.
- *ACT:* Uses methods to improve the processes conducted, for example, the continuous application of the Kanban and 5S techniques in the company's processes.

The proposal includes linking the techniques to a complete Deming Cycle. This way, implementation may be continuous and may be repeatedly improved (Fig. 7).

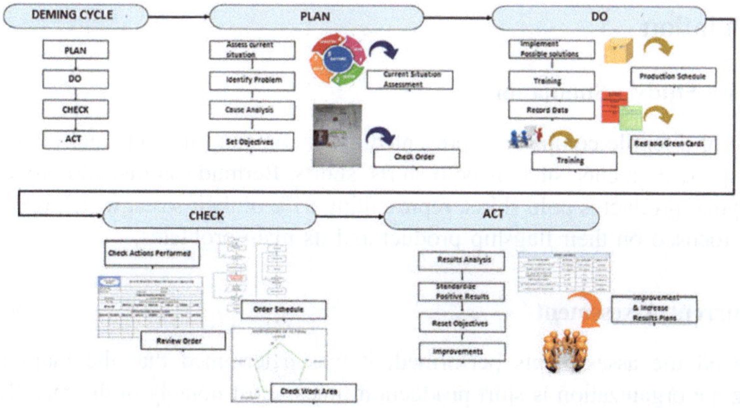

Fig. 7. Proposed model details

3.6 Implementation Process

One of the problems that generate the greatest economic impact for companies in the textile industry is returns due to product defects. The causes for this problem are different for each company, since each company uses a different management style and different processes. For this reason, the following flow is proposed. This flow allows for a complete analysis of the company, from the collection of information to solution simulation. Likewise, follow up and control will be performed with the help of the Deming Cycle (Fig. 8).

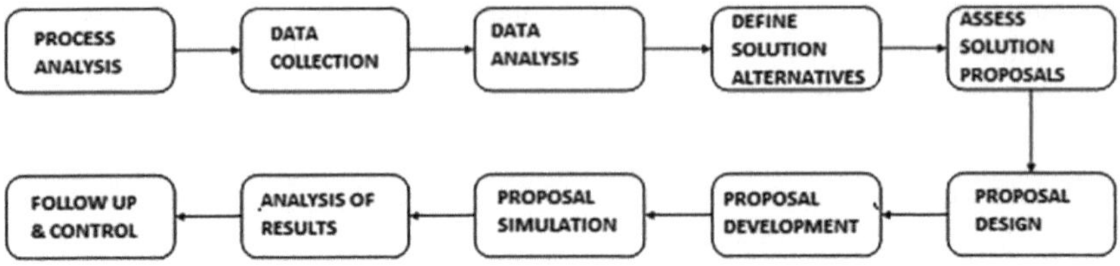

Fig. 8. Model implementation phases

3.7 General Indicators

Primarily, organizations need to improve three indicators: return of defective products, high manufacturing times, and penalties. These indicators are characterized as they hinder companies from meeting their organizational objectives.

- Demand: It has to be fulfilled by the established deadline. Hours are used to complete tasks that were not performed in the established period.
- Percentage of defective and late products: Percentage of defective products and products delivered late due to process inefficiencies per month.
- Percentage of income losses: Reduce income losses per month.

4 Validation

4.1 Case Study—Simulation

The case of a textile company is presented. This company markets and manufactures various sports garments, such as polo shirts, shorts, Bermuda shorts, and gloves. Their most popular product is polo shirts, representing 91% of their sales, which is the reason why we focused on their flagship product and its main problems.

4.2 Current Assessment

Through all the assessments performed, it was determined that the main problem afflicting the organization is shirt production times, most notably in the manufacturing process, since manufacturing times exceed the Takt Time.

MANUFACTURING = 487 sec. > TAKT TIME = 320 sec.

In the current scenario, 58 polo shirts are produced per day.

4.3 Proposed Model Application

After the proposed scenario was implemented, daily production increased to 81 polo shirts, that is, polo shirt production increased by 40% (Fig. 9).

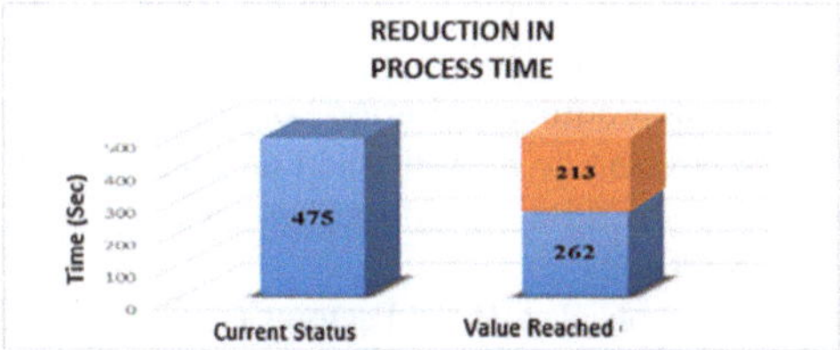

Fig. 9. Decrease of manufacturing time

4.4 Results

The results from the proposed model are as follows (Table 2):

Table 2. Current and achieved indicators

Indicators			
Factor/time	Actual	Reached	Reached
Production per month	1392 units/mo	1968 units/mo	Increased
Production time	671 sec/unit	380 sec/unit	Decreased
Defective/late products	8.14%	2.75%	Decreased
Income losses	6.32%	1.96%	Decreased
Non-fulfilled orders	35.5%	10%	Decreased
Percentage of penalties	9%	2%	Decreased

5 Conclusions

- In the case study, the integration of the Deming Cycle, 5S, and Kanban techniques increased the company's income by 84%.
- By integrating the Deming Cycle, 5S, and Kanban techniques, the proposed model may be adapted for any other company area or even another company.
- Polo shirt production was increased by 41%.
- The number of defective products and non-fulfilled orders was reduced by 25% and 26%, respectively.
- In addition, polo shirt production times and penalties were reduced by 77% and 35%, respectively.

References

1. Grimaldo, G., Silva, J., Fonseca, D., Molina, J.: Análisis de métodos y tiempos: empresa textil Stand Deportivo [Analysis of methods and times: textile company Stand Deportivo], Colombia. Magazine I3+ **3**, 120–139 (2014)
2. Güner, M.G., Ünal, C.: Line balancing in the apparel industry using simulation techniques. Fibres Text. East. Eur. **16**, 75–78 (2014)

3. Kovács, G.: Application of lean methods for improvement of manufacturing processes. Acad. J. Manuf. Eng. **15**(2), 31–36 (2017)
4. Tapia Coronado, J., Escobedo Portillo, T., Barrón López, E., Martínez Moreno, G., Estebané Ortega, V.: Marco de Referencia de la Aplicación de Manufactura Esbelta en la Industria [Reference framework for the lean manufacturing application in the industry]. Cienc. Trabajo **19**(60), 171–178 (2017)
5. León, G.E., Marulanda, N., González, H.H.: Factores Claves de éxito en la implementación de Lean Manufacturing en algunas empresas con sede en Colombia [Success key factors on lean manufacturing implementation, at some companies based in Colombia]. Tendencias **18**, 85–100 (2017)
6. Zakarial, N., Nik Mohamed, N., Ab Rahid, M., Mohd Rose, A.: Success key factors on lean manufacturing implementation, at some companies based in Colombia. Sci. Work **19**(60), 171–178 (2017)
7. Panwar, A., Jain, R., Rathore, A.P.S.: Lean implementation in Indian process industries–some empirical evidence. J. Manuf. Technol. Manag. **26**, 131–160 (2015)
8. Sundar and Kumar
9. Randhawa, J.S., Ahuja, I.S.: "Evaluating impact of 5S implementation on business performance. Int. J. Prod. Perform. Manag. **66**(7), 948–978 (2017)
10. Davis, S.M.: Office efficiency: managing information flow in a virtual workspace with 5S. Qual. Prog. **32**, 154–173 (2016)
11. Millstein, M.A., Martinicha, J.S.: Takt time grouping: implementing kanban-flow manufacturing in an unbalanced, high variation cycle-time process with moving constraints. Int. J. Prod. Res. **52**, 6863–6877 (2014)
12. Junior, P., Alves, M.J., Mendes, J.V.: Operational practices of lean manufacturing: potentiating environmental improvements. JIEM **3**, 134–156 (2016)
13. Nallusamy, S., Saravanan, V.: Lean application in a medium company with standardized work technique. Rev. Electron. Ind. Distrib. Ind. **42**, 145–162 (2016). INEI
14. Manfredsson, P.: Textile management enabled by lean thinking: a case study of textile SMEs. Prod. Plan. Control **27**, 132–148 (2016)
15. Santosa, W.A., Sugarindra, M.: Implementation of lean manufacturing to reduce waste in production line with value stream mapping approach and Kaizen in division standing upright piano. In: MATEC Web of Conferences, vol. 67, pp. 189–198. EDP Sciences (2018)
16. Jaramillo
17. Cui, H., Yan, Y.: The optimization design of uniform's hanging production line. Int. J. Cloth. Sci. Tech. **27**(3), 370–389 (2015)
18. Wickramasinghe, G.L.D., Wickramasinghe, V.: Implementation of lean production practices and manufacturing performance: the role of lean duration. J. Manuf. Technol. Manag. **28**, 531–550 (2016)
19. Cheng, C.Y., Chang, P.Y.: Implementation of the lean six sigma framework in non-profit organisations: a case study. Total Qual. Manag. Bus. **23**, 431–447 (2014)
20. Bevilacqua, M., Ciarapica, F.E., De Sanctis, I., Mazzuto, G., Paciarotti, C.: A changeover time reduction through an integration of lean practices: a case study from the pharmaceutical sector. Assembly Auto. **35**, 22–34 (2015)
21. Segovia and Pozo
22. Kursun, S., Kalaoglu, F.: Simulation of production line balancing in apparel manufacturing. Fibres Text. East. Eur. **17**, 68–71 (2017)
23. Leonard and Sereno

Ancient Handcraft Improved by New Material

Cristina Carvalho(✉), Isabel Bieger, and Gianni Montagna

CIAUD, Faculty of Architecture, University of Lisbon, Lisbon, Portugal
cristifig@gmail.com, isabelbieger@hotmail.com,
g.montagna@gmail.com

Abstract. The bobbin lace is considered an ancient handcraft. It already exists for more than 500 years as we can see it at the city of Peniche in the country of Portugal. Of course, it hasn't a linear story to write about the bobbin lace. It was weave as the same manner, using the same technique over the years, and the same material was been used, the cotton wire.

Following the new tendency of the world and using the skill that the technology brings to us, we are improving this ancient technique. With an interventionist methodology we introduce a new material. Replacing the cotton thread, we proposed to weave with the ceramic wire. Maintaining the same technique of weaving we got different results what bring to us the conception of new other objects.

Keywords: Design · Bobbin lace · New materials · Innovation

1 Introduction

In this paper we focus on an important handcraft, the bobbin lace, that we can see normally in the seaside. In Portugal we have an important focus localized in the little peninsula named Peniche, located in the Midwest of the country. The bobbin lace is considered a part of culture and identity for the people who live there. They refer to it like an ex-libris from the city.

Even tough, we found it in a difficult situation. Less people searching it, lack of demand is nowadays the status of this handcraft. The same way of weaving it and using the same materials that they used a hundred years ago can bring this result.

Worried about it we think to propose change the materials that the lace makers use to employ. That can open new possibilities of products.

2 The Bobbin Laces

The technique of bobbin lace holds up long time. It's necessary having specifically materials for practice it. Because of this it's so called. For to do this are been used both hands that moving at the same time for twisting the wire [1]. It's indispensable use a cylindrical cushion that stay over a stool, we need also the pike, the thread, the pins and the bobbin that is the instrument that gives name to the technique. The numbers of bobbins depend of the type of the lace that we want to do. The bobbins support the

G. Di Bucchianico (Ed.): AHFE 2019, AISC 954, pp. 399–405, 2020.
https://doi.org/10.1007/978-3-030-20444-0_40

thread and always work in pairs [2]. The thread used it's normally one hundred cotton in the white color [3]. "Bobbins are made of bone, wood or plastic; some are highly decorated, others are plain and inexpensive. It is essential that the slim North Bucks bobbins are weighted with rings of beads" [4] (Fig. 1).

Fig. 1. Producing the traditional bobbin lace. Photograph taken on 13 April 2017. Source: Isabel Bieger

"A piece of lace is an artistic composition expressed in twisted thread, just as a piece of wood-carving is the expression of the artist's idea in chiseled wood. Lace is not, like embroidery, an ornamental fabric; it is itself ornament. It is not the application of art to a craft; the whole pattern is the fabric, and the fabric is the pattern. It is this peculiarity that distinguishes lace from needlework and woven work" [5].

The bobbin lace survived through many years, we believe it's because her exceptional characteristics. Many women do it because believe in a tradition handcraft like a true cultural heritage, for increase the family yield or do it like a hobby. Women who produced bobbin lace, the lace makers, were usually the wives of fishermen. To be a lace maker reveals itself in the way of life of the women of the fishermen. These women developed the activity very early. Some before they went to traditional school already knew how to produce the bobbin lace. Activity that demarcates the natural life flow of many Peniche's women [2].

The lace makers of Peniche, mostly had the first contacts with the bobbin lace in their childhood, normally through family. Many of them learned from their mother, from someone they were producing, or from relatives. Others have had the opportunity to learn at schools, such as the Industrial school for example, and learn the bobbin lace in a professional way. In Peniche the bobbin lace, in the early 1900s, was considered the second largest industry in the city [6].

The traditional yarns used in the confection of bobbin lace are cotton, linen and silk. The most important feature is that the yarns used have low elasticity. Preferably thin thickness and in colors of white or beige. For to achieve the mastery of the technique they suggest using the cotton yarn.

The entire fabric process of bobbin lace is made manually "… a craft object must be made substantially by hand, utilizing the hand itself, hand tools, and to some degree, power tools" [7]. The lace makers and the learnings are taking care for maintaining the bobbin lace as a real handcraft. And so, they are producing it with the same ancestral technique.

3 Changing Stereotips Through Design

The handcraft must follow the trends of the world and for not to be forgotten. Especially such an ancient craft. The author Ethel Leon [8] warns of the neglect of artisanal culture. Craftsmanship is not just an archaic knowledge "devoid of great utilities, reserved for the market of futilities or tourism". Craft and manual making encompasses most of the creation of a product. The author also refers to the manufacture of automated objects where the stages of creation are produced with artisanal procedures. Moreover, the author conveys the notion that artisanal culture is an integral part of the industrial culture, validated by the Bauhaus.

For the bobbin lace not be only a reminder of previous generations, strategies are necessary, in the sense of their use, in another profile and at the same time a re-signification of their use.

"This is not to say craft is somehow equivalent to materiality in itself (that is, the extension of matter in space). It would be better to say that craft always entails an encounter with the properties of a specific material (…) craft involves direct engagement with specific material properties. The normative idea of modern art, by contrast, involves the transcendence (which, as in the case of autonomy, is also a repression) of just this encounter" [9].

For changing the way and be able to show other side, we convoke a designer. A designer brings a different vision of the craftsman, has technical knowledge about materials and productive techniques, serves as a facilitator between the product and the maker, building tangible and improving local identity products. According to Lima, the relationship and the work carried out jointly between the designer and the designer are important because "the artisan must be seen as a protagonist and not an auxiliary, that is, labor without creative autonomy, which would lead to the proletarianization of the craftsman" [10]. In this relationship with the bobbin lace of Peniche and the introduction of new materials, it is necessary to consult a specialist in the production of bobbin lace. However, the material to be chosen for its execution needs to be sorted by the lace makers. Considering that not all materials are possible to be produced, and others, we try not to change the identification of the bobbin lace of Peniche.

Changing the material that they usually use for producing the bobbin lace, may be a possibilities for to give another perspective and creating new products. And so, expecting that the practive of doing bobbin lace don't fall into oblivion, like are happening nowadays with this ancestral activity. Handcraft and design have a dichotomy of values as innovation and handmade. Though design the things can be reinvented, specially using new technology to put them to use the language of post-modernism [11].

Innovation requires constant work in search of solutions, of research, of search. The products cease to be innovative very quickly, this is due to the constant competition of the market, where everyone struggles to take their space and captivate customers. To innovate, one projects the thought for the future, for a change of life or for a better quality of life. Continuous innovation is a way of rationalizing creativity in the projection of design into the future of the new. Carrying creativity into material object is to launch the scientific platform, in the strict and precise demand of the new for the future project [12].

"Innovation is about knowledge – creating new possibilities through combining different knowledge sets. These can be in the form of knowledge about what is technically possible or what configuration of this would meet an articulated or latent need. Such knowledge may already exist in our experience, based on something we have seen or done before. Or it could result from a process of search – research into technologies, markets, competitor actions, etc." [13].

For the author Lalkaka [14] the change can cause a great rupture or, to increase, modernize, boost or refine a certain product. Innovation is the dynamic and integrated chain of activities, from the identification and creation of a new concept to meet a market opportunity, followed by its search for commercial achievement. The challenge is to identify a market opportunity followed by the development and implementation of a new concept.

According to Papaneck, the main function of the designer is to solve problems. However, the lack of demand of the bobbin lace makes the designer more sensitive to understand what problems exist. Easily a designer will discover the existence of a problem that no one suspected before. It will define it as a problem and try to solve it. This can be read as a definition of the creative process [15].

4 Introducing New Materials

A product should arouse emotional and deep interest in the user/consumer. Regarding the bobbin lace and Peniche is not difficult to think about. Bobbin lace is imbedded in the culture of the local population. To extend this interest about bobbin lace is necessary that some emotional meaning could be developed for the populations that do not know it. "The engineered product can sustain in the external world only when the meaning and measure of the utility of this product is derived from those contexts that belong to the subjective world of the customer" [16].

Giving a new context to certain materials and objects are marks of this time. Innovation ambitions the search for new forms for products and at the same time gets new solutions to existing problems. The author Blaine Brownell points out that no traditional product is saved from this system, so the materials are in permanent study to improve their function or replacement of existing ones. In this way, the reuse of materials and the new way of using them, accompany the evolution of other technologies, developing a positive attitude towards this development [17].

The path to the choice of new materials and their specification is rarely linear. "As always, it is devious, assembling bits of information by one method, other bits by others, and combining them into a solution that - although based on existing information - can be entirely new" [18]. It is in research and construction, between advances and retreats that innovation arises.

"In this context, the "new materials" are not only the new polymers, new alloys, new ceramics, or new advanced composites, that is, the cutting-edge materials that are emerging from the most advanced research centers. We may define as "new" even materials that are the result of creative combination of familiar materials. This combination, thanks to more sophisticated management of matter and energy, leads to greater information content and higher density of performances" [19] (Figs. 2 and 3).

Fig. 2. Ceramic wire. Photograph taken on 28 March 2017. Source: Isabel Bieger

A yarn to produce bobbin lace need to be flexible enough for to do the knots. So, we thought at the ceramic fiber. The ceramic fiber is frequently employed in the textile industry mainly in the weaving. "These yarns have opened up new possibilities for lacemakers, taking the craft into their rightful place in the twenty-first century" [20]. The ceramic fiber yarn, with a cross section around 2 mm, has like benefits:

- Good mechanical resistance
- Extremely lightweight

Fig. 3. Ceramic wire test. Test performed with ceramic thread. Photograph taken on 28 March 2017. Source: Isabel Bieger

- Very flexible and moldable
- Supports up to 1.260 °C (around 2.300 Fahrenheit)
- Composed of silica and calcium

5 Conclusion

Research and applications of materials in other possibilities compared to those usually applied have led to the abandonment and transposition of barriers between material, technique and specialties. More and more the division between the professional, the material and the product is becoming more tenuous.

As a result, new products as well as new and potential markets could be identified. Therefore, highlights the need for a series of tests that will give increase to the experimental results, being these able to alter the processes of use of materials and consequently change of the variables of the systems.

Trough the test we can confirm that is possible change material for weaving the bobbin lace and so producing new objects. The new objects can be made with pieces of bobbin lace inlaid. The material of outside can be plastic/polymer or even glass. We thought on them because it's materials that support high temperatures like the ceramic fiber. Joining these different materials, we can create decoration pieces with innovation.

We can observe and withdraw this conclusion according to the delicacy of the result of the bobbin income obtained. The lace is quite flexible which can acquire the shape of a mold for example. The joining of the parts was practically perfect, resulting in an integrated and unique piece.

References

1. Mincoff, E., Marriage, M.S.: Pillow Lace: A Practical Hand-Book. E. P. Dutton and Company, New York (1907)
2. Calado, M.: História da Renda de Bilros de Peniche, Autor edn. Gráfica Torriana, Peniche (2003)
3. Vargas, A.: As Guardiãs da Renda. Instituto Estadual do Patrimônio Cultural, Rio de Janeiro (2004)
4. Nottingham, P.: The Technique of Bobbin Lace. Dover Publications, London (1995)
5. Wright, T.: The Romance of the Lace Pillow. H. H. Armstrong, London (1919)
6. Calado, M., Peniche na História e na Lenda. Gráfica Silvas, S.L. (1991)
7. Dormer, P.: The Culture of Craft - Status and Future. Manchester University Press, Manchester (1997)
8. Leon, E.: Design e artesanato: relações delicadas. Revista D'ARTE **12**, 63–67 (2005)
9. Adamson, G.: Thinking Through Craft. Bloomsbury, London (2007)
10. Lima, M.F.: Design e Artesanato: relações de poder. 5º Simpósio de Design Sustentável, Rio de Janeiro, pp. 11–20 (2015)
11. Rees, H.: Patterns of making: thinking and making in industrial design. In: Dormer, P. (ed.) The Culture of Craft, pp. 116–136. Manchester University Press, Manchester (1997)
12. Hitcher, W.: The Innovation Paradigm. McGraw-Hill, New York (2005)
13. Tidd, J., Bessant, J., Pavitt, K.: Managing Innovation - Integrating Technological, Market and Organizational Change, 3rd edn. Wiley, Chichester (2005)
14. Lalkaka, R.: Fostering Technological Entrepreneurship and Innovation. Millennium Book Chapter (2001)
15. Papanek, V.J.: Design for the Real World. Thames & Hudson, Califórnia (1982)
16. Karamchedu, R.: It's Not About the Technology - Developing the Craft of Thinking for a High Technology Corporation. Springer, New York (2005)
17. Brownell, B.E.: Transmaterial. Princeton Architectural Press, New York (2006)
18. Ashby, M., Johnson, K.: Materials and Design - The Art and Science of Material Selection in Product Design, 2nd edn. Elsevier, Oxford (2010)
19. Manzini, E.: The Material Invention - Materials and Design. The MIT Press, Milão (1989)
20. Dye, G., Thunder, A.: Beginner's Guide to Bobbin Lace. Search Press, Tunbridge Wells (2007)

Solutions in Designing of the Composite Textile Structure Destined to Block, Intake and Storage of the Petroleum Residues Due to Natural Disasters

Carmen Mihai(✉), Alexandra Ene, Emilia Visileanu, and Cristian Jipa

The National Research and Development Institute for Textiles and Leather, 16, Lucreţiu Patraşcanu Str., 030508 Bucharest, Romania
carmen.mihai@certex.ro

Abstract. To floating dams for limiting the expansion of the hydrocarbon film to the sea surface, different physicochemical processes are used which consist in the gelation of the marginal portion of the oil film by polymerization or by the dispersal of various products that prevent the expansion of the oil pan by altering superficial tension contact oil.

For the design of the composite structure destined to the blocking, capture and storage of the oil fractions, several structural analyses has been performed by considering the initial conditions differentiated as a function of wind force measured on Beaufort Scale.

In this respect, a specialized software that allows the setting of calculation parameters, performing actual calculations, processing, visualization and export of numerical data was used.

The structural analysis was based on the theories of continuous medium mechanics.

The values resulting from the structural analysis for: structural deformation, Von Mises stress values, displacement vectors, as well as those resulting from the use of membrane theory, allowed the determination of the structural parameters necessary for the design of the textile structure.

Keywords: Disasters · CAD/CAM · FEM

1 Introduction

Water pollution with petroleum residues is an extremely important and difficult problem to be prevented and fix it that affects both surface and underground waters.

The pollution of the marine environment with petroleum residues is a worrying phenomenon that has taken an unprecedented scale since the 1960s. Sources and causes of pollution have multiplied year by year proportionally with the occurrence and proliferation of some risk factors.

Incidents in drilling, extraction, transport, transfer operations, loading - unloading, refining, storage, etc. have generated imminent risks due to the hazardous properties of oil and petroleum products.

G. Di Bucchianico (Ed.): AHFE 2019, AISC 954, pp. 406–410, 2020.
https://doi.org/10.1007/978-3-030-20444-0_41

Marine pollution with petroleum residues may be the cause, also, because of the acts of war targeted at shore-side oil installations [1].

The limitation of the oil fractions pellicle extension can be achieved by the use of dams. The most used dams are pneumatic and rigid ones.

Special firefighting, filtering or recovery dams are rarely used due to their high cost.

In addition, the special dams involve difficult exploitation technologies.

Supplementary, to floating dams for limiting the expansion of the hydrocarbon film to the sea surface, different physicochemical processes are used which consist in the gelation of the marginal portion of the oil film by polymerization or by the dispersal of various products that prevent the expansion of the oil pan by altering superficial tension contact oil.

Antifire dams are intended to protect the pellicle of petroleum fractions that have been incinerated under controlled conditions and in well-defined situations.

Filtering or absorbent dams are those dams that have a semipermeable structure that retains the hydrocarbon film. These dams are used in cases of low-carbon hydrocarbon pollution, situations where classic dams and high-capacity recovery systems would be uneven.

Recovery dams combine the functionality of classic floating dams with those of film-collecting devices.

The design and implementation of petroleum residues blocking, intake and storage systems must take into account several factors, such as: the type and amount of pollutant recovered, the flow rate of the recovery units, the area of use, the hydro-meteorological conditions in the field, transport, way of positioning.

2 Materials and Methods

In order to describe the phenomena that arise when the fluid is delimited by a floating structure that blocks the expansion of a mixture of water and petroleum residues in open space, and in order to clearly predict the efficiency of the blocking, capture, storage of oil fractions, it is necessary, from a mathematical point of view, to properly consider the theories of the type of the wave, the shape of the sea bottom and the dimensions of the floating superstructure.

For the design of the composite structure destined to the blocking, capture and storage of the oil fractions, several structural analyses has been performed by considering the initial conditions differentiated as a function of wind force measured on Beaufort Scale.

In this respect, a specialized software that allows the setting of calculation parameters, performing actual calculations, processing, visualization and export of numerical data was used.

In the preprocessing stage the following structural parameters were defined: the unit system; the reference system; geometry of the structure; the material from which the structure is to be made; element type for mesh structure; the type of analysis to be performed; contour conditions.

Through the included solver, during the post processing process the phenomena that occur on the textile structure, which allowed the determination of the variation intervals of the structural parameters, were evidenced [2].

The structural analysis was based on the theories of continuous medium mechanics. It was considered that the textile composite structure is a continuous, waterproof filling environment that fills a certain area of space, so that at every geometric point there is an environmental material point.

In these conditions, the geometry of the structure was discretized in a series of finite elements, with the elemental mass Δm, in which the continuity property imposed the existence of mass density.

It has to be emphasized from the beginning that the actual exploitation conditions have been taken into account. Thus, the elasticity of the structure was considered and included in the analyses.

The calculation and simulation were performed using FEM, and by setting the values of the constituent elements of the mesh network, namely:

- for discretisation (mesh):
  finite element dimension: 20 mm;
  maximum tolerance between the real model and the discretised model: 9 mm;
  minimum size of the finite element: 0.1 mm;
  tolerance: 0.01 mm;
  element type: linear (4 nodes, 1 Gauss point, 3 degrees of freedom).
- constraints: - for each face of the geometry (Fig. 1). For the horizontal face was used the surface constraint.

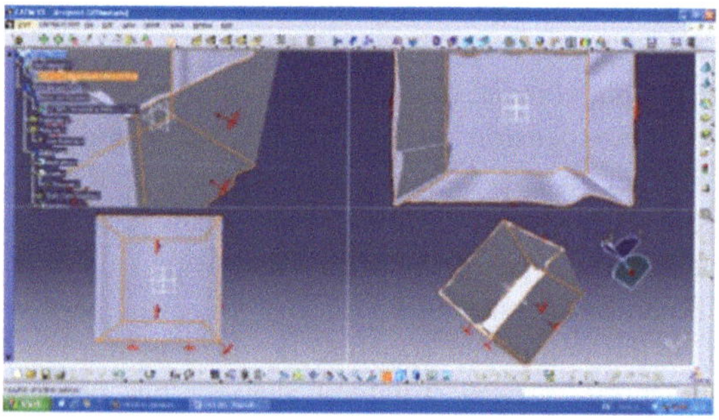

Fig. 1. Constrains of the composite system

Since the textile structure is under the action of an arbitrary medium of mechanical loading, the physical bonds between the material particles are subjected to mechanical stresses that depend on the type and level of the existing demands in the environment in which this system is defined, but also on the position and the direction of the point where the analysis is to be performed.

Thus, the phenomena must be interpreted by its components (normal stress and tangential stress), bearing in mind that the stress values will be different when they are calculated on different directions (the state of the stresses - Fluid Mechanics) [3].

Under these conditions, the description of the composite structure during the movement is accomplished through tensors (mathematical entities).

The software used allowed the visualization of the vector distributions displacement (Fig. 2) as well as of the deformation of the structures as a result of the demands from the real environment of use.

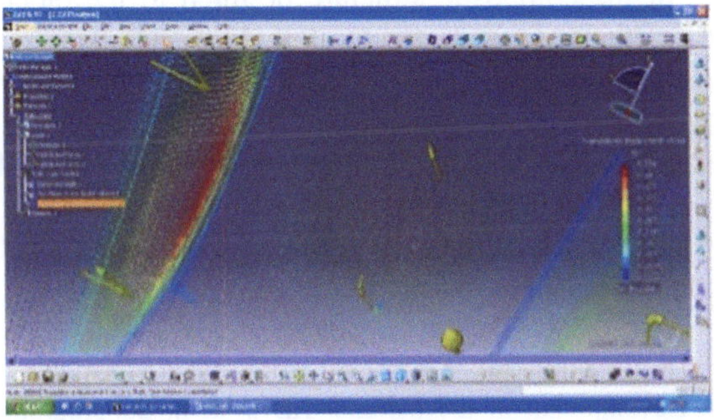

Fig. 2. Vector distributions displacement: wave direction: to the seashore, wind direction: South East

3 Results and Discussion

In order to determine the values of the design parameters of the textile structure, the following specific conditions were considered:

- linearisation of the solution for low amplitude waves (B = 4) which implicitly neglects the slope of the free surface;
- hydrodynamic equilibrium equations (for trochoidal waves);
- free surface degeneration (sinusoidal wave Fourier series); constant temperature and intensity of solar radiation;
- system homogeneity at the trophic level (for closed systems);
- membrane vibration (boundary conditions): the thickness of the membrane that can be neglected compared to the other two dimensions of the fluid interaction surface - the structure (length and width).

In addition, was considered the following conditions:

- the impermeability of the wall to the external surface interface; the equilibrium of the membrane (wall);
- the internal elastic forces determined by the fluid produce an equal mechanical effect of contrary sign with the one determined by the external forces produced by the wall;
- Work of an external force for a membrane sector appearance is proportional to the deformation;
- the proportionality coefficient T is positive and independent of the shape and position of the sector.

4 Conclusions

The theoretical fundamentation of the design of the studied systems was based on the theories of Fluid Mechanics, which allowed the determination of the equations of potential for small amplitude waves, as well as of the equilibrium equations for the case of the trochoidal waves (with solutions for the Stokes waves).

For the design of the composite structure destined to the blocking, capture and storage of the oil fractions, several structural analyses has been performed by considering the initial conditions differentiated as a function of wind force measured on Beaufort Scale.

The values resulting from the structural analysis for: structural deformation, Von Mises stress values, displacement vectors, as well as those resulting from the use of membrane theory, allowed the determination of the structural parameters necessary for the design of the textile structure.

Acknowledgment. This scientific paper is funded by the Ministry of Research and Innovation within Program 1 - Development of the national RD system, Subprogram 1.2 - Institutional Performance - RDI excellence funding projects, Contract no. 6PFE/2018.

References

1. Brown, L.: Probleme globale ale omenirii. Starea lumii, pp. 27–68. Editura Tehnica, Bucureşti (2001)
2. Ghionea, I.C.: Proiectare asistata in CATIA V5. Elemente teoretice si aplicatii. Editura Bren, Bucureşti (2015)
3. Das, S., Bai, H., Wu, C., Barney, B., Kidd, M., Kuettel, M.: Improving the performance of industrial clarifiers using three-dimensional computational fluid dynamics. Eng. Appl. Comput. Fluid Mech. **10**(1), 130–144 (2016)

Changes in Physical and Technical Characteristics of Raw Materials in the Process of Making a Bra

Ana Filipe(✉), Gianni Montagna, and Cristina Carvalho

CIAUD, Lisbon School of Architecture, University of Lisbon, Rua Sá Nogueira, Pólo Universitário, Alto da Ajuda, 1349-055 Lisbon, Portugal
anabrigida84@gmail.com

Abstract. The amount of raw materials possible to be found in the construction of a bra is huge. Such diversity in this single product makes its execution very complex. This study focuses on the relationship that exists between the processes that the materials experience and their result in the fitting moment. When the bra samples are developed in the development office, the tolerances allowed and the repression that the materials experience are different from the results obtained when the same product is run in a mass production. It is important then to understand the discrepancies that exist during both processes and how they can be taken care of. For that, it is necessary to identify and handle those several factors to minimize the suffering of raw materials.

Keywords: Bra design · Raw materials · Bra process · Importance of fitting

1 Introduction

In the construction of a bra, how the elements are taken care of is paramount for the success of its execution and for it's great purpose, the support of the woman's breast.

When it comes to building a supportive brassiere, the materials still acquire a superior exponent of importance so that more naturally their function is achieved. Controlling rigorously such diverse materials is a work that begins from the development of the design of the model. However, it is through the pattern maker that all the perspectives are fine-tuned so that the production of the bra reaches the determined requirements.

In the diversity of materials we can find metallic wire, ribbons, elastic of various sizes, microfibers, powernet, silicone tapes, foam, laces, embroideries, powernet knits and some fabrics, elastic bands, elastic straps, hook and eye, lining that can be of cotton, charmeuse, marquisate or meshes, etc.

This results in a rather complex equation in the sew execution, both in the control of all the machines involved, and in the stability of the materials throughout the process, while safeguarding their characteristics.

The process of building a brassiere will always have to start point with the design office, which defines the development according to the needs of the target. In the case of this type of product, the fitting in the process is of extreme importance, since it will

G. Di Bucchianico (Ed.): AHFE 2019, AISC 954, pp. 411–421, 2020.
https://doi.org/10.1007/978-3-030-20444-0_42

allow faster detection of error in time before the next steps are taken. It is also a product that lives a lot of the subjectivity of women who do the fitting of the bra. Some of the factors are due to the direct relation of the level of comfort with the changes in the patterns; to the type of breast of the woman, since several breast sizes can wear the same bra size.

Consequently, the control of the bras fitting goes well beyond dressing just to make sure everything fits and sits in place. It is through dressing and using the bra that it will prove that has achieved the function for which it was developed. However, the fitting bra may have differences in the process of before and after washing, as well as the user's perception of the level of comfort, and the communication of this information to the designer and/or pattern maker, and finally in the sewing process of that bra.

Without declared center stage, the materials end up by playing much of the problem vs. solution to the understanding and success of the fitting of a bra. If the samples taken during the brassiere study are not with the materials of mass production or if they do not guarantee the same results, there will certainly be significant discrepancies between the sample of the design office and the mass production of the model in question.

Each brand is governed by measures that it considers to be the most appropriate to its target, for that purpose, when subcontracting to the sewing process, they deliver a whole set of rules and parameters that they consider essential to guarantee the viability of the production. The sample made by the design office will have to correspond to all the items in this brand order book, but in addition it must predict the scenarios in which in series production it is more complex to meet all the requirements determined by the brand. Is it correct, then, to affirm that follow-up of the design office during series production allows full control of success?

The tech pack is very unique and personalized according to the characteristics implemented by each brand, however authors such as Jonsson [1], transcend this secrecy and demonstrate the requirements, specifications and standards that can be more common on some brands to standardize the universe of this kind of product.

Understanding how to ensure maximum reliability according to the approved sample is an arduous task, which is often forced to go back to previous steps so that the next steps are not compromised. Testing with real examples of samples (from the office sample and from the production area) and confronting with the tolerance charts (ordered by clients), of the same bra, will reveal the consistency achieved throughout the process.

The patterns of the bra are the starting point when each of the materials takes shape and coordinate with each other; the cutting period is a time of use of several techniques that depend very much on the type of material that is being used. The precision achieved is of extreme importance, whether in manual or automatic cutting. The confection is a fundamental part for the pattern maker to succeed and when it is verified if the cut of the diverse materials was well executed; the allowed tolerances behave as controllers of the whole process. In quality control, the measures are all reviewed but it is undoubtedly in the fitting moment that everything is proven. With the fitting it can be understood whether the tolerances are (not) excessively high or not, and if the different materials have also reacted within the tolerances provided.

The way that all the processes worked out at each stage, through which the bra goes, will provide success according to the tech pack.

All the process reveals that many materials are more vulnerable of being changed in their physical and technical characteristics as result of the effort that they are submitted throughout the process. Such may culminate in the substitution of materials, when the style bra goes into production, to minimize the problems that have occurred.

2 Review of the Bra Process

2.1 Main Discrepancies Between the Creation of a Final Sample by the Design Office and the Mass Production

The Sewing. One of the first differences is the team. Who produces the samples from the design office, is not who will make the mass production. Hence the importance of the specifications that will define all sewing parameters, from stitches per centimeter, types of seams, colors of coating lines, outer lines, tolerances in measurements, changes of materials by grading, etc. However, we will always have to contemplate that even though they are always the same person performing these samples, they will have derivations that may be associated with concentration, stress, speeding of the worker, among other factors. The machines where the samples are made may have oscillations. Would simplify the process if the machines in which the samples are sew, were exclusively for making the samples of the design office. Ideally, don't change the working parameters of the machines until the sample is completely tuned and revised to go into production. As the sew of new samples and new developments are constant, the adjustments to making a style are changed to make another and that can make some information of the entire sewing process losted.

In mass production, there are various ways of organizing such as in-line or by cells, but in any one of them, different workers will do the style. This factor is very important for the chart of tolerances, in order to provide a greater precision and working parameter. So that at each sewing operation it is confirmed that it is within the tolerance gap. To achieve those goal it is necessary that the sewing machines are tuned in perfect harmony with those used for the sew of the samples of the design office.

The Materials. Other of the discrepancies that can affect more deeply what was done as final sample, and what will be the production of this style, are the materials used.

In the development stage of a new bra sample, it's made with the materials chosen from suppliers that come in small quantity, as they will serve only to sew some style, in the design office. In this stage, work is done with materials that do not have industrial finishing. This affects many characteristics such as elasticity, hardness, touch, color, opacity, which may be of extreme relevance for the whole process of creating a new style and it's mass production. Without the guarantee of the materials having an industrial finish, the final sample of the design office will not be correct nor valid to serve as a production sample. Even if it was approved in terms of fitting, the sample should be repeated with the arrival of the purchased industrial quantities of all materials. This will allow adjustments in design, patterns, cutting, molding, sewing and fitting, on the style made, with the correct materials that will enter on the mass production. However, this procedure is not always possible, those with tight deadlines and with high fines if the orders do not leave in the correct time. They risk producing that style by looking at a sample design office made from materials without industrial finishing.

For better understand the problem the research team brought some examples of their experience in the industrial activity in this sector.

If a color requested from suppliers through a *Pantone Textile*, this color will have to be worked in the laboratory and the design office will have to receive a several color shade options ordered on the material in which it will be produced. Designated by *lap dips*, they are small samples (about 20 cm if it is an elastic, about 10 × 10 cm if it is a mesh) representative of what will be the color in the production. The design team will select one of them. It will be the swatch color for to control the bulk (industrial sample representative of the industrial item). After the *bulk* validation allows the suppliers can send the industrial quantity purchased for the beginning of the mass production of the style bra. In the middle while, it is necessary to make the final sample of design office, make the graduation of the style and approve the size range according to the intended fitting. It turns out that, if in the design office, the materials available to sew the final sample and the graduation of the style, are the standard colors available from the suppliers.

Dyeing is probably the method most used by suppliers, since having the yarn in the exact Pantone color required is not economically regular. The industrial quantity purchased of an elastic is always with their technical sheet, which includes tolerances of elasticity percent. Because of the dyeing process, the material can expose a lower elasticity. It will be sewn with others like a mesh or lace or even to several different materials together, that can result in a more rigid behavior, different from what would have been expected and different from the sample of the. Therefore, it is extremely important to make a fully made sample of industrial production materials.

The Patterns. Is a walking that can last for a long time but that allows the viability and the correct fitting to create a successful product. The importance of a discipline such as Patterns is present throughout this creative and productive process. According to the National Geographic documentary - Secret history of the bra, there are six months separating the initial creation until its conclusion and entry into mass production. According to the company Helix - Indústria de Confecções SA, this time can be reduced thanks to the know-how that the team has, since it does not always need to go to the first steps and the others styles already created and validated can be adapted according to the materials and fitting. Nothing can ever enter into production that is not dressed, commented, corrected, and dressed again until the bras conforms to the intended fitting [2].

Just with the accuracy of patterns the bras can be successful. The more we specify the functions we want in a bra, the more important is the patterns of them. When we make the patterns of a D+ cup, the demands of its construction are greater. It is often the job of the patterns maker to change the visual aspect of the bras in favor of a better fitting. This may even result in changes in the visual appearance of the brassiere. This study is done in detail, from size to size, from cup to cup, to understand where it will be necessary to add a lining, to remove the cutlets cup, to reinforce the wings, to increase the strap elastics or even their quality, etc. The limitations found in some materials may require major changes in the patterns, such as the width of a lace. This material may even work in terms of cutting up to a certain size, but when developing graduation to larger sizes, certainly it will need extra cuts that are only made because of the lace width. After approval and validation of the graduation style, the specificities of the style are gathered in its technical sheet. In this way, any pattern maker can pick up and work according to the specifications that the brand requires for their style bras.

From the very beginning, the creation of a style bra just achieves success with the presence of the patterns process. It is the stage to materialize an idea, the design and the function through the patterns that interpret what the woman's breast needs in order to have the support and adequate level of comfort. Often, the patterns are identify as the most responsible in the entire process of creating and developing a new style.

The Cutting. The cut consists of several specialties. Cutting elastic is not the same as cutting lace, therefore it's necessary need to understand the characteristics of different materials to produce the bra. There are machines for the automatic cutting of elastics, which facilitates the quickening of the process, the same goes for the elastics of the loops, in which also the necessary components (ring and slide) are placed for the adjustment of the loops when sewn in the bra. As a rule, all materials are layered and cut into the automatic cutting table, with either saw or laser. The lace and embroidery cannot be extended to be cut automatically, without the presence and support of a worker. They require much more attention to position the molds in the right way so that the seams are correctly sewn and to be positioned in a way that saves as much material as possible, from size to size. The study, the machines and the solutions for cutting lace and embroidery are very particular. It can always be done manually; precision is also possible by cutting machines. However, they require a time-consuming and meticulous process of schematizing all patterns, drawing cuttings to be placed on a tray. It's necessary the constant presence of a worker, to place the lace or embroidery, on the reference points for the board and its cutters, to execute the cutting of the pattern. The great advantage of this method is the precision achieved (which can be better than achieved when cutting manually) and the speed in the cutting stage, which recoup for the entire preparation stage.

When cutting samples for the design office, due to the needs in terms of time and the low amount of raw material available at this stage, it is certain that manual cutting becomes the most appropriate. The extreme precision of cut has to be achieved as early as this first stage because all detour can dictate the measurements out of tolerance.

It is sometimes necessary to unseam to understand whether the cutting of the material it's correct to the patter draw.

Another great focus is the study of the selvage fabric for cutting in the correct position. Something that must already be provided by the pattern maker, but it's always has to be confirmed by the cutting worker, by the position in which the material is intended to have elasticity.

The Molding. The complexity of molding is extreme. The research's team has already saw the difficulty of many molding machine's workers in understanding what to do without having the parameters of the molding machines completely defined by the technical team. The biggest problem of molding is always work with the transformation of the raw material. Something that had 2D characteristics and that by the intervention of the machines originate patterns in 3D. With the creation of volumes, it will allow to increase the versatility of using the same material in different ways in the same in the bra. In addition to the great advantage of removing any seam in the cup which is makes the bras more invisible over the outer garment. It is also the bra molded it's the preferred and bestselling style worldwide [3].

In the opinion of the researcher's team, when it is necessary to mold a new material, a new fabric, it will have to take two steps. One it's to confirm by the technical sheet of this material that it can be molded; other it's make tests not only for the size in which the final sample is made, but also in all the size range that is intended to produce.

The researcher's team perceive that different batches or colors of the same material may have different behaviors in the molding. Which forces adjustments. Whenever there is a change of batch of material, it is necessary to test if the molding parameters need some adjustments to according to the range defined by the tolerance. To confirm it should be necessary to line the canopy with the molded fabric and sew it in the size range.

The Fitting. "Just because a prototype looks good in the bust form, don't mean it can stand up to the demands of a real woman." [1].

Dressing a rigid mannequin will be only a guide to identify easily some structural problems, and serves as a support to attest that the visual image is being achieved. However, it is necessary to understand if the fitting is correct with the requirements and the goals defined by the design office.

The standard industrial size is the 34B. A woman that wear it must to control all the time if her body size change it, like weight, menstrual cycle or others, that can influence the dressing correctly.

In the fitting session with the first prototype there are many conclusions that can be drawn. One of the first is whether the materials, that were used, are most suitable for the intended visual image. Like the lining and the position of the selvage, in which way the fabric has been cut. Have to be reviewed to improve a better fitting. Checked the first measurement charts, for example the distances between the straps, both front and back, distance between cups, etc. At this time the user will indicate how she feels wearing the bra style. The user must have some knowledge about this kind of product. Normally are employees of the company or models that have some training. Some of the parameters that the user will have to evaluate is the cradle and the cup adjustment, in the breast volume, the tightening and comfort level of the bust contour, the stability of the straps, etc. Often neglected, the basic movements and their relationship to the stability of the bra, makes it possible to understand changes in the patterns that will have to be done.

It is very common to repeat two, three time the sewing process of a bra style, to guarantee all the desired settings. During this process its necessary that more than one user woman wear it and check it (in the same size), to refute or add the necessary changes. According to the *Maidform* brand, the prototype goes through five real models to be validated. This practice is very personal, according to the organization of each production company and each brand. With this method, the design and technical office are only validating a single size and it being necessary to understand the graduation that the bra will suffer and its reaction in the fitting test. For this reason, many brands produce all sizes that will move on the mass production and, in the same way, will have users to wear and comment the fitting. Through all the tests in different users and in different sizes of the size range, it will be possible to find most of the problems and resolve them after the mass production.

However, the fitting process does not end in the prototyping stage. In a pre-production stage it is advisable to confirm that everything is like expected. Repeating the process and understanding if there are significant changes to what was defined in the prototype stage. Being considered one of the parameters of the Control and Quality of the product.

2.2 The Importance of a Technical Pack for Raw Materials Control Throughout the Sample Bra Process

In a production working, all the procedures, rules and specifications are according with the Design and Technical Office. This information must be transmitted in a very clearly way to mass production.

A technical or specifications pack is not only for the production stage, but also for the prototyping stage, to follow the rules so that always styles are regulate by the same parameters defined by the brand. Will provide a noise-free information for the creation stage of a new style and in the patterns stage. The guarantee that a technical pack offers will allow new styles to be developed or produced anywhere in the world, having a universal reading, so the results in different places should be the same. Almost like a cookbook, it is a DNA structure in the brand, which is shared only with those who have work partnerships, the technical pack are rich in documentation intended to achieve the intended target in an almost customized way.

It should indicate the wires used their graduation, thickness and dimensions. In what type of models and styles should have lateral bones and in what materials, the ring and slide dimension; their equivalences from cup-to-cup sizes and their singularities. It should indicate in all sizes of cups and waistbands/wings where you will take lining and how they will be reinforced, their formats, the type of material; these lining can be found for example in waistbands/wings, center gore, in the cup, etc. It should indicate when it's necessary to change the elastic widths on straps, waistbands/wings elastics, armhole, neckline… It should show the image about were to open the cup to place the cutlets, show the types of preformed cups, foam thicknesses for cups and cradles. It should show the image of different widths for hook & eye tape according with the size range; stitches for different width of elastics according with the size range; position of each elastic tape.

A more complete technical pack also controls the type of graduation of each pattern and the growing table for every patterns of the style. It is common to present changes of patterns because of the graduation. Things like transform a regular wing to a U-backwards, for example. Jonsson [4] considers that different companies can provide different measurements with which they want their models to be graduated.

Each brand has a technical pack structured according to the characteristics it considers indicated to reach and satisfy its target. Therefore, its uniqueness makes these pack unique and confidential information.

"The bigger the company, the more detailed a tech pack usually is, for the simple reason that it'll pass through many hands before it reaches production".

The production pack is different from a technical pack. The production pack is an archive and knowledge base of the decisions made for the design technical team of that unique style bra. It will have information about fabric tests, color approval, washing tests of the style made, fitting comments, etc. It is a pack that will become more complete as the style gets approved to go into production. Jonsson [4] organizes a list of the documents that should be part of the production pack:

1. Production Order sheet
2. Comment sheet
3. PP Meeting sheet
4. Approved Fabric Swatch sheet
5. Approved Strike Off sheet
6. Trim Card sheet
7. Label Approval sheet
8. FPT/GPT Report sheet
9. PP Size Set Sample sheet
10. Shipment Sample sheet.

The Technical Sheets for Bras. The technical sheets establish a relationship between the design team, the technical team and the production team. It may seem too pretension, says but we consider that there is a higher specificity in the technical sheets of lingerie.

Firstly, it is necessary to make a distinction. While the technical pack is the procedures, rules, tables, agreements to use during the creation of the styles. Serving as support for who will produce the style bra, knowing exactly all the following rules of the brand in a technical analysis. The technical sheets is relative to the details of the style according to the intended concept. It's about all the answer for the aesthetic points, all the specificities details for the style in that particular collection, especially to understanding the graduation. In the technical sheets it is necessary that there is information of size of the style. Yu [5] considers that there are two ways of elaborating the technical sheets:

- The Designer creates the sheet with the specifications of measurements and the pattern maker follows the directions to get to the design wanted.
- The Designer creates only a technical draw for the first sample and only after it first prototype sewed and measured, define the directions and measurements.

In both situations will result in the PP sample (pre-production sample) or sealer. It means that the measures may change from the initial directions due to the adjustments and changes necessary, to agree for the result that the designer wanted to achieve. After this development approved, proceed to the seal sample, which represents what will be the final sample. It is very common to be a label with an inviolable seal, usually a golden tag Jonsson [6].

The technical sheet is changed with the exact measurements according to the seal sample and sent with this physical sample to the factory where it will be produced. It should also be followed by the measures intended for the size range.

The walking of a technical sheet is the same as the sample. Since the creation stage, the technical draw or a simple sketch, with or without measurements, it is forwarded to the pattern maker. After the preparation of the final sample, it is necessary for the purchasing team to understand with the Designer team, if the prices are within the intended or it is necessary to rethink in some decisions taken.

The production team based on the technical sheet, to parameterize the measures for the sewing workers of the style bra to guarantee that the measurements will be inside of tolerances. Finally, the quality control also needs the measurement sheets to ensure that all the parts are in tolerance, for their success in the place of sale.

3 Conclusion and Discussion

Looking at the structure of a bras it's possible to understand the various processes, machines, materials, that are present for its execution. In a bras creation the raw materials are determinants for the success of that style.

According to some authors such as Van Jonsson [1] the bra is have a particularized structure that needs to be dissected so that each element reveals its individual importance and its collective function.

There must give an intelligent support, which certainly has to go through choosing more properly the materials to be used for the intended purpose of the bras.

The anatomy of the bra or the parts of the bra can vary, however some basic parts in the structure are constant. The structures vary depending on the support that intended for a breast that oscillate in size. Understanding the anatomy of the bras can help in their creation, to make it perfect in fitting and in design.

The materials used for the construction of a bra result in its perfect adaptation to the woman's body. The opening of the wires, the typologies and thicknesses of the cups are the fundamental elements, which used according to the needs of the breast that intended to support. With that, it will be possible to provide a very high level of comfort and a greater degree of satisfaction of the woman who uses it.

Often, patters identified as the most responsible in the entire process of creating and developing a new style. It's present in all the process to guarantee the design and function of the model, thanks to the satisfaction in the use, which should be a prime objective to achieve.

The raw materials that most relate to lingerie increasingly used in outerwear. Many of these materials were until recently used exclusively for the clothing used for underwear. However, fashion design began to become more and more interested in their delicacy, specificity, and their hidden side. We consider that the relevance since study is in the diffusion of the type of work that carried out for the construction of the bras, its procedures and specificities. It is a closed sector, where each brand works differently and for the target that considers most appropriate for its brand. However, it is possible to find a uniform and universal structure in the creation of the bra. We believe, therefore, that it is important to insert more knowledge within the professional training and even the academic universe. All of these factors: sewing, sample materials and production materials, fitting, molding, molding, cutting influence on the behavior of the raw materials, up to the visual result and your bras life (Fig. 1).

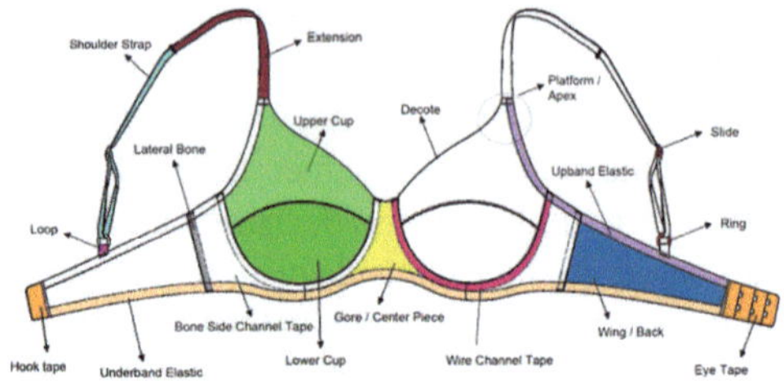

Fig. 1. The bra anatomy, Filipe (2014)

The loyalty of the user to a brand that really enjoy the style bra, want it in all future collections. That is also one of the consequences of a successful process.

Allow to the production pack and the technical pack, information on the entire process, materials, construction, labeling, packaging, can be easily controlled in an exemplary way, unambiguous information for the manufacturer, whether in Portugal, Turkey or other place at the world.

Through the classes Fashion Technology, include in the Master of Fashion Design, there was opportunity to input some issues related to lingerie and more specifically about the execution of a bra. Since it has been discovered what knowledge masters students have. From the fragmentation of the bra to understand its structure, to interaction of the most diverse materials present in this structure. Elastic behavior, molding, the type of cutting and how to do it in different materials, the students did not have the knowledge bases to enter the market and work and to design a style bra as they would easily do for a dress or a sweater. It is understandable that they do not have this knowledge. It is a niche market in production in Portugal, but it already represents a product that manages a world industry equivalent to that of outerwear, and thought to have a production that exceeds 16 billion dollars per year [7]. We can verify that one of the great positive aspects of having introduced some knowledge about the materials, the processes and the product, is the huge surprise and interest that the subject created. The interest was really growing and we had to witness the bras production in loco. Following the progress in each of the different stages, interacting to understand the methods in detail. Increasingly interested in the process of creating a bra, will allow the construction of products with more quality, that are suitable to the user and taking advantage of the maximum of the capabilities of all the raw materials that are bound to this product.

References

1. Jonsson, L.V: How to Become a Lingerie Designer (2012)
2. Filipe, A.B.: Lingerie e Imagem Visual no Mercado Português. Faculdade de Arquitectura Lisboa, Universidade Técnica de Lisboa (2010)
3. Bruce, M., Daly, L.: Buyer behaviour for fast fashion. J. Fashion Mark. Manag. **10**(3), 329–344 (2006)
4. Jonsson, L.V: How to write a bra and a brief Tech Pack (2015)

5. Yu, W., Ng, S.-P.: Innovation and Techonology of Women's Intimate Apparel. Woodhead Publishing Limited, Cambridge (2006). T. T. Institute
6. Jonsson, L.V.: The Anatomy of the Bra (2013)
7. Luciani, J.: The Bra book: The Fashion Formula to Finding the Perfect Bra. Benbella Books Inc, Dallas (2009)

Use of 3D Printing and Nano Materials in Fashion: From Revolution to Evolution

Javed Anjum Sheikh, Muhammad Faisal Waheed(✉), Ahmad Mukhtar Khalid, and Ijaz A. Qureshi

University of Sialkot, 1 Km, Main Daska Road, Sialkot, Pakistan
{Javed.sheikh, faisal.waheed, am.khalid, vc}@uskt.edu.pk

Abstract. The purpose of 3D technology is to transform the most challenging concepts into reality. In Fashion, 3D printing is continually evolving with the designers' interest and professionals in technology. It opens new avenues of fascinating and endless possibilities for fashion designers. In this process the object is created layer by layer without any mould, based on its design which was developed through Computer Aided Design (CAD) and use powder particles materials, like plastics, metals, and resins etc. 3D printing offers different perspectives in almost every field of life. This innovation has impacting the new industrial revolution and creating vibes, in majorly all the industries. Including high fashion apparels and could change the future of both Haute Couture and Pret. Regarding three dimensional printing there are various ways and technologies have been developed by the professionals, but this outrageous technology is having an obstacle in fashion apparels, like materials, which are mainly synthetics and lacking the natural properties to produce desired silhouettes.

Keywords: 3D printing · Computer Aided Design · Sustainable · Nano materials · Environment-friendly · Fashion · Design

1 Introduction

3D is a familiar word. However, 3D movies are more familiar and easily understandable for the layman. However, 3D printing or Additive Manufacturing (AM) is entirely a unique term for most of the people who are not related to this domain. 3D printing refers to a computer-controlled fabrication process, which enables the creation of a three-dimensional object from the 3D model, designed using computer-aided-design software [1]. The ultimate success of the 3D printing process is making three-dimensional objects with extreme intricate, geometric and structured details. Currently, this is a unique technology related to producing anything. In other words, you name it, or you can have it, with the help of 3D printing or AM. 3D printing got the attention of designers because of its sustainability in the form of wastage and environmentally friendly, as no waste of water or any harm to the environment takes place. Technologies have already changed our routine life, like Artificial Intelligence, Robots, and World Wide Web etc. As per data by a specialist of 3D printing market research, it has

G. Di Bucchianico (Ed.): AHFE 2019, AISC 954, pp. 422–429, 2020.
https://doi.org/10.1007/978-3-030-20444-0_43

recorded that financially it has grown amazingly from 3.07 billion dollars in 2014 to 21 billion dollars by 2020. It is approximately 32% growth annually [2].

In the beginning, designers used rigid and synthetic surfaces to create apparel collections. However, 3D printing was having challenges like stiffness, rigidness and non-flexibility. Practicality, comfort and feel of materials were also missing. Designers and professionals related to 3D printing overcome rigidness and stiffness issues by producing materials, which are flexible. They have successfully produced the body-hugging outfits, which move as the body move and helps to perform the routine task without any issue. Nevertheless, the use of NANO materials improves the properties making it durable, sustainable and with no colour constraints. Nanotechnology is itself is the industrial revolution. In which the properties of any material can be changed making it more towards performance-based in all aspects. By advancements of Nanomaterials, possibilities in terms of design and practicality are endless [3].

Using NANO materials through 3D printing technology in Fashion for mass production is the concept of discussed and explored in this paper. If we look into the history of 3D technology which is not very old, and it seems that this technology has already set the strong footsteps in the area of design and manufacturing not only in Fashion but almost all the aspects of life, rather Fashion comes quite after some time. Few of the fields where 3D technology has made its debut; are Medical and dentistry, Aerospace, Luxury Automotive, Jewelry and accessories, Art & Design, Architecture, Food etc. 3D Printing or AM technology has not only changed the apparel design and methodologies of manufacturing dramatically but has moved it forward to limitless boundaries. This unique way of manufacturing has given the new paradigm to conceptualize and materialize the creative design process in the Fashion, textiles and accessories, which ultimately lead to resolve many problems related to textile and fashion sector. It is a well-known fact that the textile industry is one of the most polluting industries in present times [3].

2 History of 3D Printing

The history of 3D Printing technology indicates that it first came in the picture in the 1980s. When the first patent application for rapid prototyping was filed in the 1980s, albeit the first patent was issued and registered in 1986 for "Stereo-Lithography Apparatus" (SLA) [1]. In 1987 SLA-1, appeared, it was the first commercial 3D printer produced and after strong critique on its operations and possibilities the first unit was sold in 1988. [4, 5]. In 1989, the Selective Laser Sintering (SLS) came into existence and was subsequently patented [4, 5]. In 1992 the patent was issued for another technology within 3D printing, it was Fused Deposition Modeling (FDM) [4, 5]. These were the initial developments and technologies of this unique mode of printing, and after mid-nineties, the race is still going on about to reduce the price of 3D printing apparatus and bring it down to be within the approach of individuals. The advancement of this technology has reached a certain level with the passage of time, as its main purpose is to develop this technology was to produce prototyping in almost all the aspects and areas of life. The objective of 3D printing is to create objects which can be conceptualized but difficult or next to impossible with the bare hands. The prototyping or artworks which have been developed with the help of 3D printing technology through its design software's are very intricate and detailed [2, 4, 5].

Using a computer-aided design program (CAD), designers create and modulate a 3D model to generate by computer [6]. From this model, it is necessary to create a "build path" which specifies each movement to be made by the printer, when to start printing out the specific material and at what rate. To do so the designer must make use of a separate program. Through the later, the 3D model is transformed into a series of 2D slices thus informing the printer where to solidify the starting material for each slice [6]. Every layer printed out represents one of the digital slices generated through CAD. Making use of a binder, a laser or an electron beam, the 3D printer will solidify the material along the programmed build path [7]. Today, several different printing processes as described earlier exist such as stereolithography (SL), Polyjet, fused deposition modelling, laminated object manufacturing (LOM), Prometal or selective laser sintering (SLS) [8].

3 Literature Review

At the outset, it is important to lay the framework for the proposed study through a detailed review of the existing literature to identify key theories and concepts deemed relevant to the research topic.

Above mentioned materials, from liquid resins to Powder formulation in the shape of plastic, aluminium, titanium to PLA, ABS, TPU type of materials are being used in the process of prototyping or commercial purposes [1, 9, 10]. If we go in details all these materials are based on synthetic structures, which are somehow, rigid, rubber type flexible materials, and metal like strong. wIf we talk about textiles and in a more specific way wearable textiles, it requires comfort and practicality which are somehow lacking in the currently developed clothing using 3D Printing or AM. As per saying by famous Dutch designer Iris Van Harpen, that now you can wash and Iron the 3D printed garments [12, 13]. Her collections related to 3D printing are the next level of development in the domain of apparel design. She has merged the Design and technology in such a way where every creation is art itself. In other words, 3D printing is a correct and ultimate tool as the expression of designers.

Another company "Continuum" has developed Bikini Dress named N12, Nylon 12, which is entirely developed by the 3D Printing using SLS technology [14]. This outfit is purely for the swimming purpose and its uniqueness is that while getting wet, it is amazingly comfortable and waterproof because of the properties of Nylon. An interesting fact is that it is solid, flexible and strong as it is only 07 mm thick and not gets de-shaped. No hand or machine sewing is involved in its construction, as it is developed entirely with the process of 3D printing including closures and hooks [14].

In current development a graduate student, Lingxiao Luo, who is currently in process of her Masters in Women wear Design from the Royal College of Art(RCA), she has used this unique 3D printing technology and merged with her knitwear project called it AddiToy collection. She has mainly used knitwear fabric and applied 3D printing mesh patterns directly onto knitted fabrics in order to attach another knitted yarn piece. She also used the felting method to use 3D printed patterns directly onto knitted fabric, and then wet and shrinks the knitted fabric to produce three-dimensional structures [15].

Fashion Designer Frances Murphy also studying in Royal College of Arts (RCA) is experiencing the Bioplastic and cellulose-based materials to examine the possibilities to use them for 3D printing. Her idea is to use traditional structures of textile, and experiments on materials like silk, which is cellulose, based [16].

4 Materials and Methods

4.1 Liquid Resins

Stereo-lithography (SLA) technology uses liquid resins, which converts the liquid Resin into solid materials using high strength of ultraviolet UV light. The process of high intensity of light is to harden the liquid layer drawn by the printer as per the design layer by layer [1, 9].

4.2 Powder Based Materials

Powder-based materials used in the Selective Laser Sintering (SLS) technology, which is a rapid prototyping technique. It is more like (SLA) technique but without the ultraviolet light. Powder materials like, Glass, Aluminum, Titanium, Nylon, and Polystyrenes replaced with Resins and spread on the base of machine evenly and the prototype is developed with the help of a laser, which fuses the plastic materials into layers. This process explains the objects developed with this technology, the process is mostly in the rigid forms, and they are fragile. If we use these materials in the form of apparel, yes it would be an enormous and creative achievement, but fragile. It shall be a perfect piece of art but in rigid form. Same materials also use in Selective Laser Melting (SLM) technology. In this process rather fusing the powder, it is melted with the high temperature [1, 6, 9].

4.3 Paper Materials

By using paper material and applying adhesive coating on at lower side of the object as layer-to-layer. The Laminated Object Manufacturing (LOM) is another type of additive manufacturing technique. In this process, the heated stainless-steel roller applies the pressure to place another layer on top of the already laid down layer. The excess material, which covers at the surrounding of the object, removes manually at the end of the process [1, 9].

4.4 Thermoplastic Materials

Thermoplastic materials like ABS, PLA, and HDPE, are types of materials, which are being used in Fused Deposition Modeling (FDM), which is another additive manufacturing technology. A temperature-controlled nozzle develops the object layer by layer. In this process, the material becomes solid form, semi-liquid form during the extrusion process to form an object. As these materials sound, they require high temperature by melting them in the extruder just before printing. In one experiment in which they have designed the 3 different types of plain weave structures [1, 7, 9, 10]. To avoid any complexity basic geometrical shapes like straight line and circles uses for the fabric design. The printing material used in this project was ABS (Acrylonitrile Butadiene Styrene) which is a lightweight filament. It is high impact resistant and high creep resistant polymer.

The result of these samples is quite promising as this shows about the possibilities to develop the weave like structure, which is also wearable and functional through 3d printing or AM [1, 9, 10]. The same FDM technology also used by the Houston Based designer "Maria Alejandra Mora-Sanchez" who has developed the "LOOM" dress by using Push Plastic Flexible TPU material. She was inspired by the "Wayuu" tribe patterns and developed the flexible, stretchable material, which basically quite stretchable and moves along with the body [11].

4.5 Nano Materials

In parallel with 3D printing, Nanomaterials is another key area, which has revolutionized the Textile and Fashion sector. The concept of this unique technology is to treat materials related to textiles like cotton, silk, and other organic based with the coating of Nanomaterials, this procedure improves the properties and making them sustainable, durable, and enhances their performance and they become biodegradable. Nanotechnology has changed the entire perspectives in almost every field of life including, Agriculture, Medicine, Automotive Industry, Fashion and Textile. To examine the Nanoparticles with the naked eye is not possible, as one has to use the proper equipment to see them at the nanoscale level. At this stage, particles can even change their colour, which is great news for the fashion and textile designers as they can create wonders by using such properties, which are helpful for them to create their products [3, 17]. By using Nanomaterials can create wonders in the process of 3D printing technology. As described earlier, Nanomaterials coating apply to any type of material in particular organic and synthetic can change their properties. In addition, when we say organic, we need to look into Biotechnology, which is based on DNA technology heading enzymes synthesis to save natural resources, most importantly water [3, 18].

History of Biotechnology is based on centuries ago when people were not even aware of this word, but it actually came into existence in 1953 when American Biologist James Watson and English Physicist Frances Crick discovered the structure of DNA. Biotechnology is the performance-based technology, which has actually directed the manufacturing industry into a new paradigm. Through, Biotechnology a new range of hi-tech and eco-friendly materials have revolutionized the Textile and Fashion sector and ushering them into the future. These materials will lead the fashion into more sustainable way [19–21]. Few biodynamic materials like, 'Zelfo', which is cellulose, based organic material, like, Hemp, Flax and Paper. This substance uses renewable sources for production without using any inorganic form. The interesting thing is there is no wastage of material in the process of making it, as waste is reusable too [20–23]. Another substance is 'Sorona' polymers, which drive from plants. It is a high-performance fibre made it through plant-based ingredients. An interesting fact is that this fibre is stretchable without any use of stretchable yarn or substance, and its outcome is marvellous after using and washing it numerous times [20, 24]. These materials are the combination of agriculture and biotech science capabilities to give bio-based solutions to the needs of humankind while protecting the environment. Another interesting fibre is 'Cocona', which is another performance-based fibre harvested from nature. These kinds of fibres can be used for woven and knitted fabrics, which promote evaporation of moisture, and protection from ultraviolet rays, giving comfort to the wearer in hot climate regions. An excellent fibre is also derived from coconut husk, which is waste from the water filtration industry. These sort of unusual materials, which are results of Biobased technology, are the great source of inspiration to the industry, which is at the verge of disaster for nature [25].

5 Limitations

Despite there being great, hope about the future of 3D printing, the existing body of literature agrees that 3D printing still faces many challenges. The marginal costs of production continue to remain higher with 3D printing than with conventional technology. This is linked to the cost of material used and the consumption of energy by the machines [26]. The lack of material also raises questions in terms of quality and can discourage designers and firms to opt for 3D technologies. However, researchers advance that once more suppliers enter the market for 3D printing such costs will begin to decrease.

In addition, 3D printing poses problems in terms of designers' abilities and capabilities [27]. If on one hand, the ability to design viable products and bringing them to manufacture is enhancing the demand for qualified designers, faster and simpler prototyping as well as easy to use 3D software's are opening up doors for automation and non-professionals [27]. Another issue linked to 3D printing technology relates to intellectual property.

Today, many blueprints are distributed freely online and can be replicated and printed out by all those having access to the file [28]. So, just as streaming enabled cheap counterfeiting for the music, publishing and movie industries, 3D printing enables decentralized piracy, meaning that consumers can obtain counterfeits cheaply without having to pass through commercial counterfeits [28]. Today, 3D printing is not submitted to IP laws. In the current digital age, copyright laws are under revision, with policymakers under pressure to reevaluate the role of copyright protection [29]. With innovations being rapid and unpredictable, it takes legal systems time to adapt to newly introduced technologies and the consequences they entail in diverse industries. What is more, because of the aforementioned unpredictability of new technologies, it is difficult for legislators to predict or anticipate the effects of given technologies [28].

6 Conclusion

Things are moving quickly. 3D printing will ultimately become supportable. Its beauty is its limitation is dependent upon materials. It is not so difficult to understand that the evolution of the human brain, who invented technology for the betterment of the human kind but ultimately disastrous for the natural world in which they lived. Now the advancement of technology has forced the human kind to develop the solutions to prevent further disaster and apply the reverse engineering method to resolve such problems. Through technologies like Biotechnology and Nano Technology, organic and performance-based materials can transform the 3D technology. Materials can be easily transformed into powder and liquid form to be used in the additive manufacturing capacity. A blend of different fibres like manmade and organic can further enhance the process of 3D printing. Using this unique technology, the mass production of apparel is not very far from existence. Rapid prototyping like its name can set the new horizons for the mass production of clothing by using bio, technological based hi-tech, materials for the sustainability of this advanced unique domain.

Acknowledgements. We are thankful to USKT's Board of Governors, Mr Faisal Manzoor, the Chairman its Members, and Mr Muhammad Rehan Younas, the Executive Director for their keen interest and encouragement during this work.

References

1. i.materialise, 3D Printing Materials. http://i.materialise.com/en
2. Kwon, Y.M., Lee, Y.A., Kim, S.J.: A case study on 3D printing education in fashion design coursework. Fashion and Textiles **4**, 26 (2017). https://doi.org/10.1186/s40691-017-0111-3
3. Waheed, M.F., Khalid, A.M.: Impact of emerging technologies for sustainable fashion, textile and design. In: Karwowski, W., Ahram, T. (eds.) IHSI 2019. AISC, vol. 903, pp. 684–689. Springer, Cham (2019). https://doi.org/10.1007/978-3-030-11051-2_104
4. Richardot, A.: Everything you need to know about Additive Manufacturing (2018). https://futur404.com/3d-printing/
5. Gürcüm, B.H., Börklü, H.R., Sezer, K., Eren, O.: Implemented 3D printed structures as the Newest Textile Form (2018)
6. Valtas, A., Sun, D.: 3D printing for garments production: an exploratory study. J. Fashion Technol. Textile Eng. **4**(3) (2016). https://doi.org/10.4172/2329-9568.1000139
7. Petrick, I.J., Simpson, T.W.: 3D printing disrupts manufacturing: how economies of one create new rules of competition. Res.-Technol. Manag. **56**(6), 12–16 (2013)
8. Wong, K.V., Hernandez, A.: A Review of Additive Manufacturing, International Scholarly Research Network, pp. 1–10 (2012)
9. 3D Printing Materials. https://www.sculpteo.com/en/materials/
10. Partsch, L., Vassiliadis, S., Papageorgas, P.: 3D printed textile fabric structures. In: 5th International Istanbul Textile Congress, Istanbul, Turkey (2015)
11. https://3dprintingindustry.com/news/reshape18-3d-printing-sensing-materialities-calls-for-final-submissions-138841/?utm_source=Twitter
12. Holgate, M.: Meet Iris van Herpen, the Dutch designer boldly going into the future (2016). https://www.vogue.com/article/iris-van-herpen-dutch-designer-interview-3d-printing

13. Iris Van Herpen debuts wearable 3D printed pieces at Paris Fashion Week (2013). https://www.materialise.com/en/cases/iris-van-herpen-debuts-wearable-3d-printed-pieces-at-paris-fashion-week
14. Etherington, R.: N12 3D-printed bikini by Continuum Fashion and Shapeways (2011). http://continuumfashion.com/N12.php
15. Vialva, T.: Interview: RCA graduate Lingxiao Luo creates 3D printed knitwear collection (2018). https://3dprintingindustry.com/news/interview-rca-graduate-lingxiao-luo-creates-3d-printed-knitwear-collection-135434/
16. Innovation in textiles (2018). http://www.innovationtextiles.com/rethinking-textiles-with-3d-printing/
17. Application of Nanotechnology in the Textile Industry. http://textilelearner.blogspot.com/2012/12/application-of-nanotechnology-in.html
18. Application of Biotechnology in Textiles | Importance of Biotechnology in Textile Processing. http://textilelearner.blogspot.com/2012/10/bio-textile-application-of.html
19. Thomas, L.: Nanotechnology and Biotechnology-Similarities and Differences (2018). https://www.azonano.com/article.aspx?ArticleID=4862
20. Fernandez, C.R.: How Biotech is changing the way we make clothes. https://labiotech.eu/features/biofabrication-fashion-industry/
21. Colwell, B.: Biotechnology timeline: Humans have manipulated genes since the 'dawn of civilization' (2017). https://geneticliteracyproject.org/2017/07/18/biotechnology-timeline-humans-manipulating-genes-since-dawn-civilization/
22. Quinn, B.: Fashion Future. Merrell Publishers Limited, London (2012)
23. Material district. https://materialdistrict.com/material/zelfo-2/
24. Sorona, Polymer by DuPont. https://www.apparelsearch.com/sorona.htm
25. Hemmings, J. https://www.jessicahemmings.com/cocona-natural-technology/ (2009)
26. Weller, C., Kleer, R., Piller, F.T.: Economic implications of 3D printing: market structure models in light of additive manufacturing revisited. Int. J. Prod. Econ. **164**, 43–56 (2015)
27. Michalik, J., et al.: 3D opportunity for product design: Additive manufacturing and the early stage, Deloitte University Press, July 2015. http://dupress.com/articles/3d-printing-product-design-and-development/#sup-3
28. Depoorter, B.: Intellectual property infringements & 3D printing: decentralized piracy. Hastings LJ **65**, 1483 (2013)
29. Towse, R.: Creativity, copyright and the creative industries paradigm. KYKLOS **63**(3), 461–478 (2010)

Cognitive Behavior Difference Based on Sensory Analysis in Tactile Evaluation of Fabrics

Jun Zhang[1,2], Hui Tao[1,2(✉)], and Xuewei Jiang[1,2]

[1] School of Fashion, Wuhan Textile University, No. 1, Fangzhi road, Hongshan District, Wuhan 430073, Hubei, China
{zhangjun, maggietao, xwjiang}@wtu.edu.cn
[2] Wuhan Textile and Apparel Digital Engineering Technology Research Center, No. 1, Fangzhi road, Hongshan District, Wuhan 430073, Hubei, China

Abstract. The hand evaluation of fabric is the psychological cognition and experience when people interact with fabrics and clothing. In this study, the subjective evaluation method was used to evaluate the tactile properties of 11 fabrics to find out the cognitive behavior difference in tactile evaluation. 30 subjects were selected for tactile sensory evaluation. The results were collected by a 5-level scale sheet, and were analyzed by Excel and SPSS. Based on the evaluation results, in sensory evaluation, there was a short-term memory effect, which led to a difference in subsequent illusory memory (DIM), it affected the results of sensory evaluation. There was a cognitive behavior difference in tactile evaluation of fabrics. In a sensory evaluation sequence for thickness test, the evaluation value was influenced by the previous sample. Cognitive evaluation of thickness of fabric was also affected by its surface property.

Keywords: Fabric hand · Cognitive behavior · Sensory analysis · Tactile evaluation

1 Introduction

Skin of body is the soft outer tissue covering of human with three main functions of protection, regulation, and sensation. It is the largest organ of human body with a total area of 1.5–2 m^2 and a thickness of 0.5–4 mm [1]. The skin interfaces with the environment and is the first line of defense from external factors. Hands and fingers are very important organs of human beings, sensation is one of the significant functions of hand for temperature, pressure, roughness, hardness, friction and so on, because of the dense distribution of the sensory nerves at the hands, fingertips in particular.

Fabric hand or fabric handle is the "feel" of the fabric, the way the fabric feels when it is touched. The fabric hand is everything you can tell by touching a fabric with your hands: texture, drapability, stretch, wrinkle resistance, etc. it is the sum total of the sensations expressed when a textile fabric is handled by touching, flexing of the fingers and so on [2]. It is the comprehensive evaluation of the mechanical properties of fabrics. The hand evaluation of fabric is the psychological cognition and experience

G. Di Bucchianico (Ed.): AHFE 2019, AISC 954, pp. 430–437, 2020.
https://doi.org/10.1007/978-3-030-20444-0_44

when people interact with fabrics and clothing. The tactile evaluation of fabrics is an indispensable part of fabric hand evaluation. There are two methods used for fabric hand evaluation: objective evaluation and subjective evaluation. The objective evaluation refers to a mechanical equipment measurement, professional instruments are applied to measure the physical properties of fabrics directly. The comprehensive hand value of fabrics is obtained from a data analysis. The measured value is objective and accurate relatively. However, the implementation is complicated, and requires special equipment and professional technicians, in specific occasions with great limitations. Sensory analysis experiment is used as a subjective evaluation method. Subjective cognitive behavior is evaluated in the interaction of subjects and experimental target. Psychological scale is used to reflect the subject's cognition of the experimental target. Subjective analysis is simple and feasible, it is also susceptible to external conditions, leading to inaccurate experimental results. Both methods have its advantages and disadvantages in different application.

Objective evaluation is usually conducted using a special equipment with high accuracy of results. Kawabata Evaluation System (KES) and Fabric Assurance by Simple Testing (FAST) are the widely used devices for objective evaluation. Tokmak investigated the relationship between KES and FAST [3]. Some researchers developed some new method for the objective evaluation of textile hand, Strazdienė and Gutauskas proposed a new method for tensile property testing by pulling a disc-shaped specimen [4]. An artificial finger was developed to evaluate the tactile perception by Chen et al. for tactile evaluation [5]. Some researchers using the two methods for fabric hand evaluation, such as tactile properties of wool [6], wetness sensation [7], and thermal sensation of wet fabrics [8], to obtain the accuracy and reliability of fabric's hand subjective evaluation [9, 10].

New intelligent technologies were developed to predict fabric hand value based on objective or subjective evaluation. An intelligent method was proposed for fabric hand evaluation and prediction of fabric formability for men's suits [11], and a fuzzy neural network (FNN) prediction system is explored with a prediction accuracy over 80% in the past report [12]. Subjective evaluation method is easy for implementation without some special equipment, which is widely used for various evaluation. Kandzhikova developed a methodology to determine the weight coefficient of the factors involved in the evaluation for terry fabrics handle [13], subjective evaluation also can be used to assesses the comfort of functional fabrics for smart clothing [14] and fabrics for protective clothing [15].

Tactile sensation is a common method to evaluate textile products, such as stiffness, roughness, thickness and other properties of fabric. Friction of fingertip skin was measured to study the tactile perception [1] in previous report. Tactile sensation also was used to evaluate quality and aesthetic of woo fabrics [16]. Subjective cognitive behavior is evaluated in the interaction of subjects and experimental target, it is influenced by some test condition, subject situation and the experimental target. In the paper, we investigated the cognitive behavior difference in tactile evaluation of fabrics based on sensory analysis. It was found the surface property of fabric sample and different evaluation sequence would influence the sensory evaluation. And the Subsequent illusory memory (DIM) would cause a cognitive behavior difference.

Table 1. Characteristics of samples

(a) KES-FB3 Compression properties

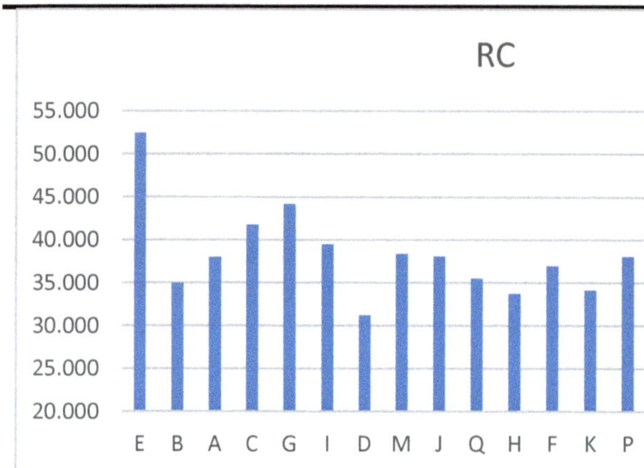

Sample	LC	WC	RC	Tm	To
	—	(gf/cm2)	%	mm	mm
A	0.	0.16	34.	0.	0.
B	0.	0.18	52.	0.	0.
C	0.	0.18	36.	0.	0.
D	0.	0.16	44.	0.	0.
E	0.	0.18	39.	0.	0.
F	0.	0.22	38.	0.	0.
G	0.	0.30	34.	0.	0.
H	0.	0.24	32.	0.	0.
I	0.	0.34	31.	0.	0.
J	0.	0.27	38.	0.	0.
K	0.	0.33	33.	0.	1.

(b) KES-FB4 Surface properties

Sample	Surface				Back			
	Warp		Weft		Warp		Weft	
	MIU	SMD	MIU	SMD	MIU	SMD	MIU	SMD
	-	(μm)	-	(μm)	-	(μm)	-	(μm)
A	0.12	4.26	0.12	2.82	0.13	4.44	0.13	2.70
B	0.18	4.56	0.18	4.20	0.18	4.13	0.18	4.32
C	0.14	8.28	0.15	2.24	0.14	2.37	0.15	2.52
D	0.15	3.32	0.14	2.23	0.14	2.22	0.15	2.47
E	0.15	2.78	0.16	3.15	0.16	2.32	0.18	4.26
F	0.13	13.40	0.14	15.53	0.14	15.70	0.14	15.98
G	0.15	7.95	0.16	8.82	0.16	6.59	0.16	8.23
H	0.17	5.88	0.16	3.45	0.16	3.61	0.16	7.42
I	0.15	1.53	0.16	3.32	0.16	4.31	0.16	2.01
J	0.15	12.54	0.14	10.76	0.14	13.09	0.15	11.53
K	0.16	1.95	0.16	4.07	0.15	5.55	0.14	3.00

2 Experiments

To investigate the cognitive behavior difference in tactile evaluation of fabrics. In our experiment, it was focused on the relationship and interaction between the surface characteristics (roughness and friction) and the compressive properties (perceive thickness) of the fabrics, to find out the correlation and interaction between the two aspects.

In this study, the subjective evaluation method was used to evaluate the tactile properties of 11 different fabrics named A, B, …, J, K, including plane, twill and satin weaves. Samples were cut into a square with sides that were 20 cm in length. The basic properties of the samples were tested using Kawabata Evaluation System for Fabrics (KES-F) after placing into a thermostatic chamber for 24 h under condition of 21±1 °C, RH65±5%. Each sample was tested 3 times to get its average parameter, and the characteristics of the samples are shown in Table 1 (a) and (b). 30 subjects were selected for tactile sensory evaluation with the average age of 21.2 including 20 female and 10 male university students. Tactile sensory evaluation experiments were conducted in a room with controllable temperature and humidity set as 21±1 °C, RH65±5%.

The critical issue discussed in this paper is the perceive thickness of the same fabric sample was evaluated in different evaluation sequences to confirm whether there was a cognitive difference in different evaluation sequence. To investigate this issue, we randomly arranged the sensory evaluation sequence, and randomly repeated some fabric samples in the test sequence to verify the hypothesis of cognitive difference in tactile evaluation of fabrics. In this experiment, sample A, C, F, G, H, I, J, K were repletely evaluated 2 to 4 times randomly. The evaluation sequence is shown in Table 2. For example, the 2nd sample ${}^{K}I$ in the sequence means the I sample is tested after K sample.

The subjects are asked to use fingertips of one hand to pick up the sample and sliding with thumb and other fingers as shown in Fig. 1. Subjects "feel" the 11 fabric samples to understand the coverage of 5-level scale of the sample thickness with a patch over his eye, and then have a three-minute break. After that, they give an evaluation value of thickness in the assessment table whit a ten-second time interval between two samples.

The results were collected by a 5-level scale sheet, and were analyzed by Excel and SPSS.

Table 2. Tactile evaluation sequence

Sequence	1	2	3	4	5	6	7	8	9
Sample	K	${}^{K}I$	${}^{I}G$	B	${}^{B}A$	${}^{A}H$	D	${}^{D}C$	${}^{C}K$
Sequence	10	11	12	13	14	15	16	17	18
Sample	${}^{K}F$	E	${}^{E}G$	${}^{G}A$	${}^{A1}C$	${}^{C}I$	${}^{I}F$	${}^{F}J$	${}^{J}K$
Sequence	19	20	21	22	23	24	25		
Sample	${}^{K}H$	${}^{H}G$	${}^{G}J$	${}^{J}I$	${}^{I}K$	${}^{K}A$	${}^{A2}C$		

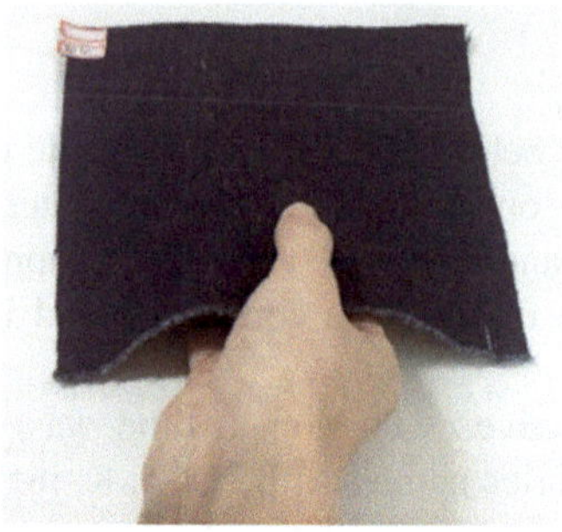

Fig. 1. Tactile evaluation of fabrics

3 Results and Discussion

We measured the surface property and compression property of the samples. Standardizing the collection of data helps to analyze and compare the date differences. Standardized data is shown in Table 3 and Fig. 2. The surface roughness of warp direction is marked as *SMD-swa*, measured thickness is marked as *To*, and evaluated thickness is marked as *Toe*, as shown in Table 4, the average value of thickness collected from 30 subjects.

Table 3. Standardized characteristics data of samples

Sample	Surface		Back		Thickness
	Warp	Weft	Warp	Weft	
	SMD-swa	SMD-swe	SMD-bwa	SMD-bwe	To
A	1.92	1.18	1.67	1.20	1.00
B	2.02	1.59	1.57	1.66	1.57
C	3.27	1.00	1.05	1.15	1.94
D	1.60	1.00	1.00	1.13	2.03
E	1.42	1.28	1.03	1.64	2.38
F	5.00	5.00	5.00	5.00	2.95
G	3.16	2.98	2.31	2.78	3.18
H	2.47	1.37	1.42	2.55	3.40
I	1.00	1.33	1.63	1.00	3.63
J	4.71	3.56	4.27	3.73	3.78
K	1.14	1.55	2.00	1.28	5.00

Table 5 shows the significant difference between *Toe* and *To*, except sample H and J, H and J repeated 2 times in the tactile evaluation, other samples repeated 3 or more times show significant difference at 0.05 level ($P < 0.05$). It indicates that the results of subjective sensation show deference to the objective measurements, and the same sample shows different value if it has a different position in the tactile evaluation sequence.

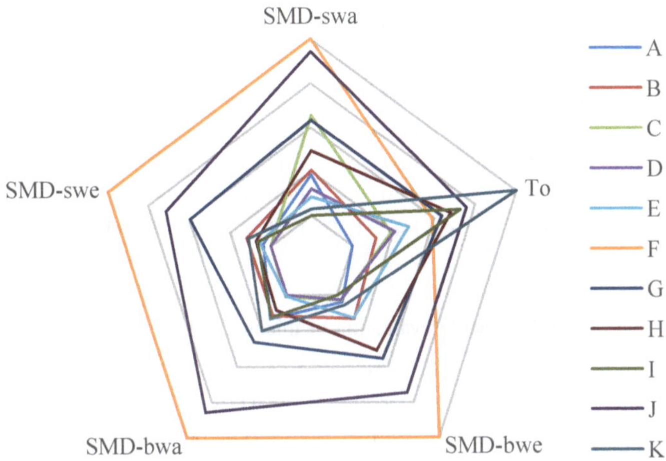

Fig. 2. Standardized roughness and thickness of samples

Table 4. Tactile evaluation of thickness of samples

Sequence	1	2	3	4	5	6	7	8	9
Sample	K	^{K}I	^{I}G	B	^{B}A	^{A}H	D	^{D}C	^{C}K
Toe	3.47	3.16	3.47	2.00	1.68	3.74	1.84	2.74	3.42
Sequence	10	11	12	13	14	15	16	17	18
Sample	^{K}F	E	^{E}G	^{G}A	^{A1}C	^{C}I	^{I}F	^{F}J	^{J}K
Toe	4.68	2.26	3.63	1.42	2.84	2.89	4.74	4.21	3.16
Sequence	19	20	21	22	23	24	25		
Sample	^{K}H	^{H}G	^{G}J	^{J}I	^{I}K	^{K}A	^{A2}C		
Toe	3.42	3.68	3.68	2.84	2.58	1.32	2.79		

Table 5. Significant difference between *Toe* and *To*

Sample	A	C	F	G	H	I	J	K
Sig.	.000	.050	.011	.023	.463	0.22	.645	.003

In a tactile evaluation sequence, the evaluation value is influenced by the previous sample. If the previous sample has a greater thickness, the evaluation value of the current sample will get a stronger influence and shows lover value, according to Tables 6 and 7. The surface property of fabric sample also affects the cognitive evaluation of fabric thickness. Sample has coarser surface, the evaluation value of thickness tends to be smaller than reality and vice versa, according to Tables 3 and 4.

Table 6. Tactile evaluation of thickness of samples A

Sample	^{B}A	^{G}A	^{K}A	A	B	G	K
SMD-swa		1.92		1.92	2.02	3.16	1.14
SMD-swe		1.18		1.18	1.59	2.98	1.55
SMD-bwa		1.67		1.67	1.57	2.31	2.00
SMD-bwe		1.20		1.20	1.66	2.78	1.28
To, Toe	1.68	1.42	1.32	1.00	1.57	3.18	5.00

Table 7. Tactile evaluation of thickness of samples C

Sample	^{D}C	^{A1}C	^{A2}C	A	C	D
SMD-swa		3.27		1.92	3.27	1.60
SMD-swe		1.00		1.18	1.00	1.00
SMD-bwa		1.05		1.67	1.05	1.00
SMD-bwe		1.15		1.20	1.15	1.13
To, Toe	2.74	2.84	2.79	1.00	1.94	2.03

From the perspective of cognitive psychology and medical neuroscience, in sensory evaluation, there was a short-term memory effect, which leads to a difference in subsequent illusory memory (DIM), it affects the results of sensory evaluation. This effect is more significant when the time interval of continuous experiment is shorter. The cognitive behavior difference of tactile evaluation caused by a DIM. The time interval is a key factor which is cannot be ignored in sensory evaluation.

4 Conclusions

In this study, we conducted a sensory evaluation of fabrics, 30 subjects were selected to evaluate the thickness of fabrics. To investigate the cognitive behavior difference in sensory evaluation, we setup a distinctive evaluation sequence with some fabric sample repeated to confirm the cognitive difference.

Based on the experiment and the results, it was found that the results of subjective sensation showed deference to the objective measurements. In a sensory evaluation sequence for thickness test, the evaluation value was influenced by the previous sample. Current evaluated sample showed lower value if the previous sample had a greater value. Cognitive evaluation of thickness of fabric was also affected by its surface property, the evaluation value trends to smaller if the sample had a coarser surface, and vice versa.

Subsequent illusory memory (DIM) caused cognitive behavior difference of sensory evaluation. In sensory evaluation, the time interval between two sample is an important factor affecting evaluation result. Appropriate time interval is essential to offset the short-term memory effect in sensory evaluation.

Acknowledgments. This work is supported by Wuhan Textile University Fund, and the National Natural Science Foundation of China (Grant No. 61802285)in part. The authors wish to acknowledge the Wuhan Textile and Apparel Digital Engineering Technology Research Center.

References

1. Chen, S., Ge, S.: Experimental research on the tactile perception from fingertip skin friction. Wear **376**, 305–314 (2016)
2. Bishop, D.P.: Fabrics: sensory and mechanical properties. Text Prog. **26**(3), 1–62 (1996)
3. Tokmak, O., Berkalp, O., Gersak, J.: Investigation of the mechanics and performance of woven fabrics using objective evaluation techniques. Part I: the relationship between FAST, KES-F and Cusick's drape-meter parameters. Fibres Text. East Eur. **18**(2), 55–59 (2010)
4. Strazdienė, E., Gutauskas, M.: New method for the objective evaluation of textile hand. Fibres Text. East Eur. **13**(2), 35–38 (2005)
5. Chen, S., Ge, S., Tang, W., Zhang, J., Chen, N.: Tactile perception of fabrics with an artificial finger compared to human sensing. Text. Res. J. **85**(20), 2177–2187 (2015)
6. Bacci, L., Camilli, F., Drago, S., Magli, M., Vagnoni, E., Mauro, A., Predieri, S.: Sensory evaluation and instrumental measurements to determine tactile properties of wool fabrics. Text. Res. J. **82**(14), 1430–1441 (2012)
7. Tang, K., Kan, C., Fan, J.: Assessing and predicting the subjective wetness sensation of textiles: subjective and objective evaluation. Text. Res. J. **85**(8), 838–849 (2015)
8. Niedermann, R., Rossi, R.: Objective and subjective evaluation of the human thermal sensation of wet fabrics. Text. Res. J. **82**(4), 374–384 (2012)
9. Grinevičiūtė, D., Daukantienė, V., Gutauskas, M.: Textile hand: comparison of two evaluation methods. Mater. Sci. (Medžiagotyra) **11**(1), 57–63 (2005)
10. Valatkienė, L., Strazdienė, E.: Accuracy and reliability of fabric's hand subjective evaluation. Mater. Sci. (Medžiagotyra) **12**(3), 25–257 (2006)
11. Xue, Z., Zeng, X., Koehl, L.: An intelligent method for the evaluation and prediction of fabric formability for men's suits. Text. Res. J. **88**(4), 43–452 (2018)
12. Yu, Y., Hui, C., Choi, T., Au, R.: Intelligent fabric hand prediction system with fuzzy neural network. IEEE Trans. Syst. Man Cybern. Part C (Appl. Rev.) **40**(6), 619–629 (2010)
13. Kandzhikova, G., Germanova-Krasteva, D.: Subjective evaluation of terry fabrics handle. J. Text. Inst. **107**(3), 355–363 (2016)
14. Tadesse, M., Harpa, R., Chen, Y., Wang, L., Nierstrasz, V., Loghin, C.: Assessing the comfort of functional fabrics for smart clothing using subjective evaluation. J. Ind. Text. **107** (3), 355–363 (2016)
15. Vasile, S., Malengier, B., De Raeve, A., Louwagie, J., Vanderhoeven, M., Van Langenhove, L.: Assessment of sensorial comfort of fabrics for protective clothing. In: 7th European Conference on Protective Clothing, pp. 111–113. Meta Basim Press, İzmir (2016)
16. Kitakuchi, S., Kumazawa, M., Morita, H., Endo, M., Sato, T., Sukigara, S.: Quality and aesthetic of wool fabrics through sensory evaluation. J. Text. Eng. **63**(2), 33–41 (2017)

Human Factor in Apparel and Fashion Exhibition Design

Catarina Carreto[1(✉)] and Rui Carreto[2,3]

[1] Faculdade de Arquitetura, Centro de Investigação em Arquitetura, Urbanismo e Design, Universidade de Lisboa, Rua Sá Nogueira, Polo Universitário, Alto da Ajuda, 1349-055 Lisboa, Portugal
catarinaferreira.design@gmail.com

[2] Escola Superior de Tecnologia e Gestão de Viseu, Instituto Politécnico de Viseu, Campus Politécnico, 3504-510 Viseu, Portugal
ruifccarreto@gmail.com

[3] Escola Superior de Tecnologia e Gestão, Instituto Politécnico da Guarda, Avenida Dr. Francisco Sá Carneiro, n.º 50, 6300-559 Guarda, Portugal

Abstract. The need to preserve past cultures, textile and clothing creations, identities of a society in a certain time period, made the development of exhibition design area relevant. Exhibition spaces must give innovative answers to the aspirations of the society, which are the visitors, reason of being of its existence. In this framework, human factor is essential in the exhibition design field.

The present article intends to address essential points regarding human factor in its relation with the design of exhibitions of fashion and clothing products. Thus, we intend to equate and emphasize the role of the exhibition designer in solving issues related to human factors, both with regard to exhibition design projects and also with regard to the visitor/user experience of the exhibitions. In this sense, through the literature review and the presentation of examples it is our intention to clarify concepts and modes of action.

Keywords: Human factors · Exhibition design · Apparel and fashion design · Design · User/visitor

1 Introduction

The relationship between the disciplinary areas of design and museology is increasingly important, particularly in the museography area and in the problematic of exhibition and fashion activities and projects. The movement of the new museology opened dimensions and fields of work with multiple applications and senses. With the new concepts and purposes, new museographic practices are now necessary as a response to a society of cultural consumption. The need to preserve past cultures, in particular textile, fashion and clothing creations, the identity of a society of a certain period, made the development of the exhibition design area relevant.

G. Di Bucchianico (Ed.): AHFE 2019, AISC 954, pp. 438–448, 2020.
https://doi.org/10.1007/978-3-030-20444-0_45

The museum ceased to be a hermetic and delimited space, became an open process in permanent interaction with the community and the territory of belonging. It is a condition of the museum to preserve and communicate. The communication takes place through various means, among which the exhibition gain special relevance. Everything can be musealized in terms of social, cultural, scientific and educational objectives. The museum is a space of knowledge and enjoyment, a place of citizenship, where design can intervene with its thinking, methodologies and instruments. Thus, the museums field find in design a fundamental support to make visible and understandable textile objects.

As a multidisciplinary area, design relates to the most varied disciplines, acting as a foundation and complement. The design of exhibitions emerges as a field of action, specific and directed to the display area. Thus, the exhibition spaces must give innovative answers to the society aspirations, which are the visitors, reason of being of its existence. In this field, human factors, in its relation with space and objects, are essential because without visitors the exhibition objects are lost. In order to work on these assumptions, by designing creative, coherent, functional, attractive spaces that maximize the visitor's experience, in view of the exhibition of fashion and clothing products, we highlight the discipline of exhibition design, which acts as an intermediary between the exhibition and the visitor. The importance of exhibition design as a means of communication between the object/art and the observer/visitor has come to play a fundamental role over the years.

2 Exhibition Design and Human Factor

Exhibition design is an activity that designates and plans in order to produce or install a communicative expository method through exhibits that use the fundamentals of interior design, equipment design and graphic design, as Fernández and Fernández [5] says.

The realization of exhibition spaces is increasingly recognized as an area of creative expression that uses a process of interaction between the most varied disciplines, such as architecture, design, museography and museology, stage design, lighting, audiovisual and multimedia. According to Dernie [4], in a global way, the term exhibition means *"(...) proposing, offering, displaying the result of the work"* (p.6). Belcher [1] says that an exhibition is like a show. In this sense, *"(...) everything in the world is an exhibition of one kind or another"* (p. 51). They can be commercial or cultural, from exhibition fairs, brand exhibitions, launch events or consumer pavilions, to museums and galleries of all kinds or artistic installations. What they have in common is the fact that they have at their base *"The idea of explaining a story in a three-dimensional space"* [7] (p. 10).

Analyzing the expositions according to their typology or exhibition mode, Fernández and Fernández [5] consider four types of functions: symbolic, commercial, documentary and aesthetic. The first is aimed at political or religious elevation or just the show of objects. The commercials are oriented towards the value of the product to be exposed. The documentaries are directly linked to informative and scientific value of the exhibition products, in order to disseminate knowledge. Finally, the aesthetics are sedimented in the artistic value of the works. Regardless of type, the design of exhibitions is done with specific objectives in the sense of social and cultural inclusion. Design competencies in this area must be bi-dimensional and three-dimensional. For, what is exposed must be explained and well exposed, requiring good communication to be well seen and well interpreted by visitors. The exhibition projects are extremely important for promoting, showing, teaching and disseminating past cultures and knowledge of the present [3].

In the past, the most important was the object and its information. Nowadays, the visitor is attracted more and more by expositions whose expositive method belongs almost exclusively to the design world. The light, the color, the audiovisuals, the touch and the relation of the senses make the visitor himself part of the exhibition, never losing sight of the educational and cognitive. For better consolidation, exhibitions tend to be interactive and didactic. In this field, there is a strong intervention of the visitor with the exhibition contents and with the space [2].

The exhibition space configures in practice all the elements, including the objects. It is a determining factor of the exhibition visitor's experience, a determination that has had different interpretations over time, accompanied by the evolution of exhibition methods. Since the twentieth century, the public gained a place as a protagonist, as a catalyst for cultural consumption in museums. The growing number of audiences in museums led to a series of circulation studies by the architects of the modern movement. Le Corbusier, Mies van der Rohe and Frank Loyd Wright were pioneers, creating flexible, modular and extensible exhibition solutions in their architectural projects. There are several studies on the organization of space, graphically expressing different expository paths through diagrams, but the human dimension should be central to defining these aspects. Therefore, in an exhibition of fashionable objects it is necessary to take into account the minimum measures of the body width, front and side, as well as its height, this with respect to the static body. Then it is also necessary to consider the rotation angles of the head, trunk and eyes.

Inclusivity in terms of exhibition spaces is also one of the essential premises of design and museology. Thus, the exhibition contents must be placed within the limits of the visitors visual fields with special mobility needs, as well as the visitors that accompany them. The exhibition organization should allow the visualization of the displayed objects by the citizen with reduced mobility in two different ways: front and side. In order to be able to make a 90° rotation, the circulation space must have the necessary measures to allow this movement (Figs. 1, 2, 3 and 4).

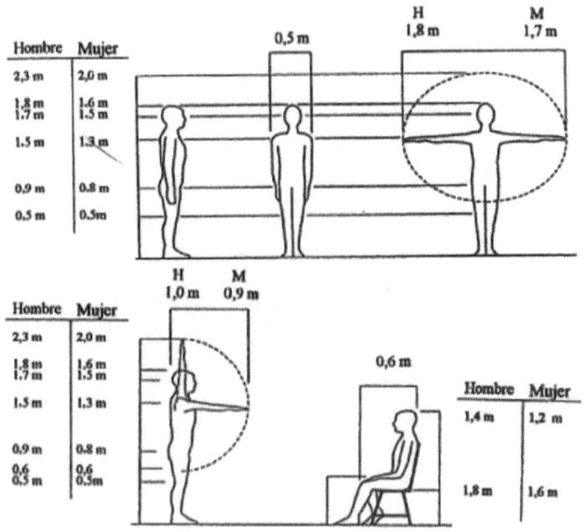

Fig. 1. Human dimensions [Dean, 1994 apud Fernández & Fernández, 2007:52].

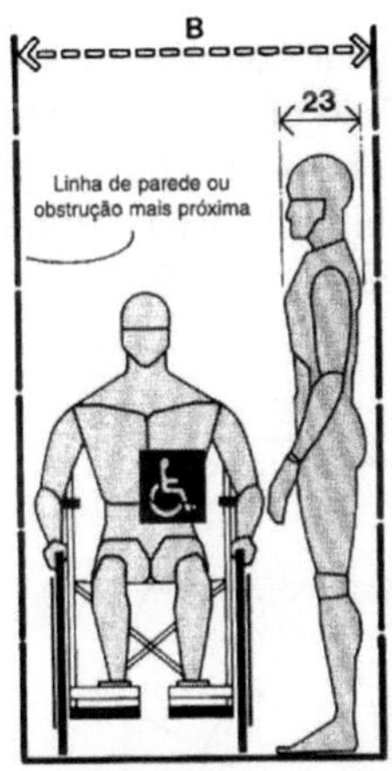

	cm
A	152,4
B	106,7
C	30,5 mín.
D	91,4
E	142,2 mín.
F	63,5
G	213,4
H	91,4 mín.

Fig. 2. Partial circulation for two people [Panero & Zelnik, 2002:270].

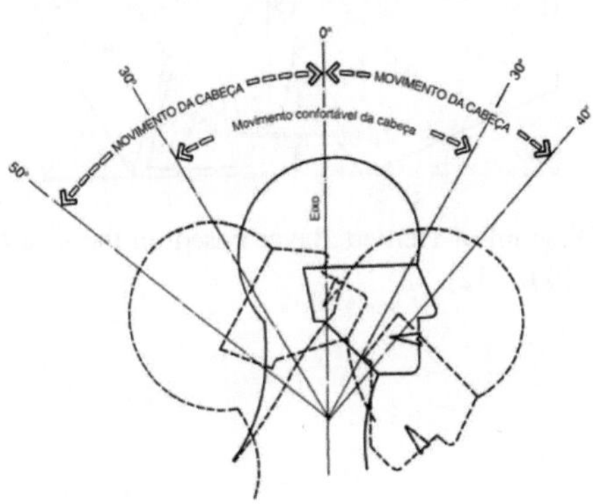

Fig. 3. Movement of the head in the vertical plane [Panero & Zelnik, 2002:286].

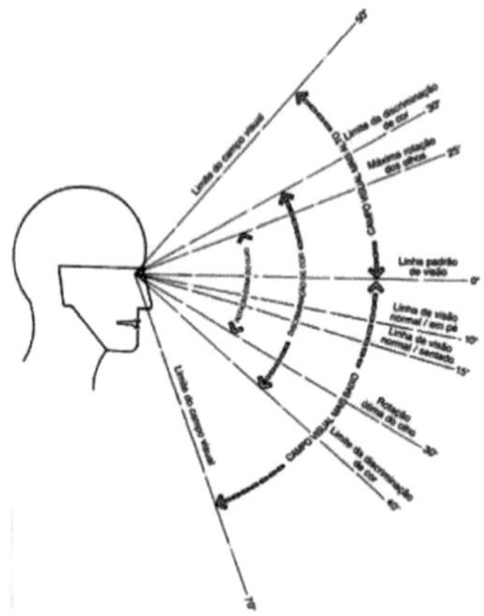

Fig. 4. Field of view in the vertical plane [Panero & Zelnik, 2002:287].

The studies of Herbert Bayer make a great contribution to the conceptual and technical foundation of the designers. The "Field of Vision" diagram demonstrates the principle that graphic support surfaces should be oriented towards the viewer, in view of an optimized effect [3]. The available exhibition space is delimited according to the angle obtained by the eyes, head and neck (Figs. 5 and 6).

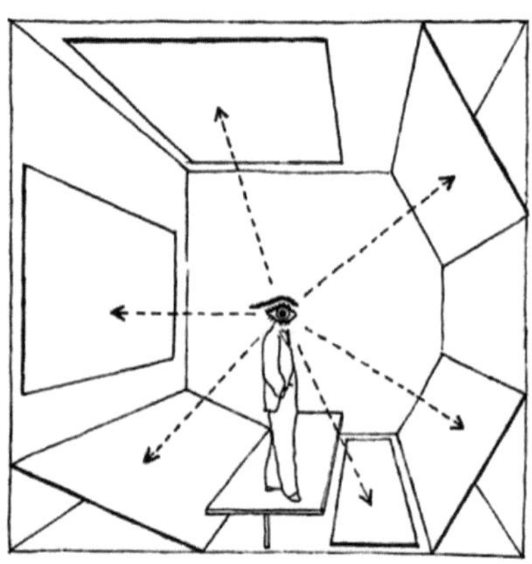

Fig. 5. Field of vision. Diagram of Herbert Bayer based on the idea of the human visual field [Fernández & Fernández, 2007:112].

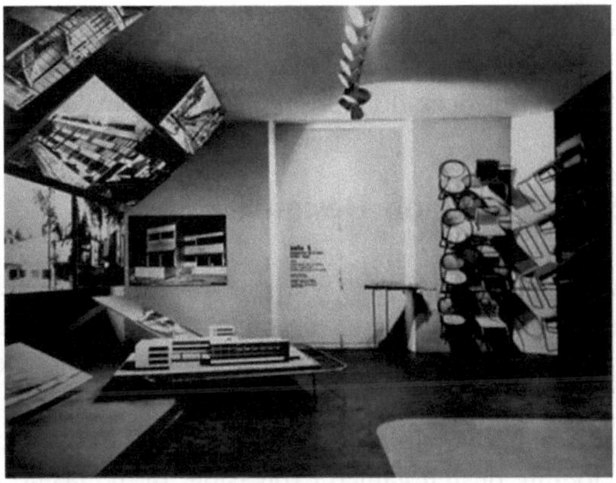

Fig. 6. Deutscher Werkbund Exhibion, Paris, 1930 [Fernández & Fernández, 2007:113].

Not only space, but also the type of objects will condition human behaviour. The specifications when two-dimensional objects are exposed differ from three-dimensional objects, since the height component (Z in the cartesian coordinate system) will assign depth and volumetry to the pieces. We can consider painting, photography, textiles, drawings, posters as two-dimensional objects, as well as three-dimensional objects, such as sculpture, utilitarian objects, furniture, fashion and clothing, among others. Regardless of type, objects should be placed within the human visual field, where the average height of the eyes of a child, an adult and a person in a wheelchair should be considered. In the case of hand-held small objects, the exhibition equipment should allow the visitor's approach for better observation. The distance from the eyes to the object will directly interfere with your enjoyment, as well as with your perception.

In all the exhibitions there is a story to tell or a message to be transmitted that is not limited only to the exhibition of the objects. It is also necessary a whole process of creating texts, based on the interpretation of the object exposed. Hughes [6] stresses the importance of letter size and text, which may be perceptible at a distance and not at all. In many cases, designers develop a hierarchy of signals at different scales, according to the informative and communicational nature: large size for signage, medium size for thematic identification of zones and small size for text body and for object legends. The author also highlights the fact that the text is created based on the behaviour of the visitor and their perception of space.

The quality and type of lighting will also influence the way we perceive the exhibition space and objects. The way the space is illuminated can even indicate the exhibition route to follow by the visitors. The light is directly connected with the vision, so the light intensity should be sufficient to ensure good visibility. In exhibitions, "*The approach to lighting is very similar to that of theater, but there is a significant constraint: lighting equipment must be hidden so as not to destroy the illusion of actual experience"* [8] (p. 161). However, the lighting imposes some care of the parts conservation. Some (organic) textile materials are damaged by exposure to light and, so that they do not degrade more rapidly, their exposure should be limited.

3 Human Factor in Fashion Design Exhibition

Fashion and textile design is designed with the user in mind. Like the design of exhibitions, fashion, clothing and textile products are designed based on anthropometric and ergonomic measures. It is increasingly permeable to the world around it, accompanying the evolution and needs of man and the context in which he was inserted. The technical reproducibility opened up the cultural and creative industries, where fashion and clothing design gained expression. In the fashion world, where the inspirations are diverse and the limits of creativity are infinite, textile productions have gained special relevance for museums, bringing to light exhibition, conservation and restoration needs. Several fashion, apparel, costume and design museums (with fashion and clothing collections) were created in an attempt to preserve the past, educate and stimulate, presenting the new. According to Valencia and Andújar [9], *"Today's society demands to seek new models of exposure, demand, in part, provoked by a new culture or new models of understanding culture"* (p. 21). The incessant search for differentiation sometimes leads to the search for concepts or recurrent practices in parallel creative areas.

The iconic temporary exhibit "China: Through the Looking Glass" by the Metropolitan Museum of Art (MET) in New York, in 2015, clearly portrays different levels of human factor intervention in exhibition design. Having been one of the MET's most visited exhibitions, the large number of visitors created some imponderables in the circulation, as well as in the minimum dimensions necessary for movement in the space, taking into account the high number of people. This aspect interfered both in the visual relation with the displayed object and the set, as well as in the questions of the proxemia. Of the distances defined by the anthropologist Edward T. Hall (intimate, personal, social and public), the one that prevailed more was the intimate distance, causing some discomfort between the visitors/users during the visit. Although the space is not small, the large number of people, in the same space, has made it limited. Considering that the intimate distance presupposes an approximation of up to 40 cm, which, therefore, easily allows the physical contact.

Fashion and clothing allow the use of exhibition techniques applied to two-dimensionality and three-dimensionality. However, the trend is to expose the most important pieces in three-dimensional mode since it allows their visualization to fullness. Mirrors can also be used to allow 360° viewing (Fig. 7).

Fig. 7. "China: Through the Looking Glass" Exhibion, MET, 2015 [Foto by Carreto & Carreto, 2015].

Concerning the rotation of the eyes, head and trunk, the exhibition *"China: Through the Looking Glass"* has several levels of observation. Smaller and more minutia objects were placed at eye level, while larger objects were placed at ground level, simulating, as in the figure below, the object in an original context. Video projections were also used as an expository complement. The projection occurs at a higher level, above the height of the visitor, in order to allow the image not to be obstructed and that all visitors can see. The size of the gallery allows the detachment required for an ergonomically comfortable viewing, not putting the neck into tension (Figs. 8 and 9).

Fig. 8. "China: Through the Looking Glass" Exhibion, MET, 2015 [Foto by Carreto & Carreto, 2015].

Fig. 9. "China: Through the Looking Glass" Exhibion, MET, 2015 [Foto by Carreto & Carreto, 2015].

Display cabinets control the environment of the piece, in terms of conservation and safety, and also allow the installation of lighting equipment. In this type of exhibitors it is necessary to take into account the glare reflected by the exterior lamps, whose incidence angles are not well calculated will interfere with human visual perception [8] (Fig. 10).

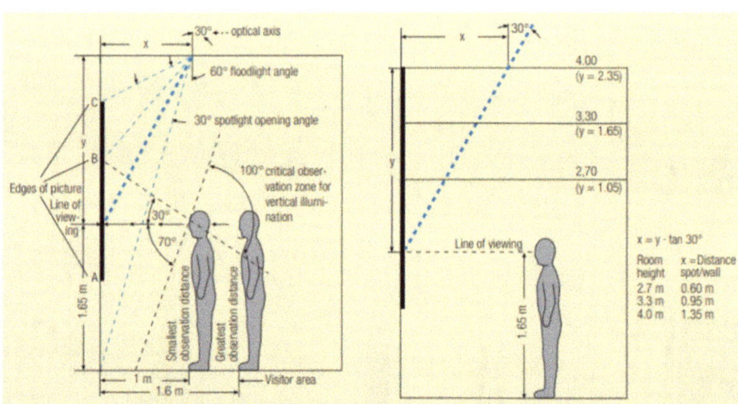

Fig. 10. Calculation of the optimal positioning of a luminaire for pictures on a wall [Retrived from: http://en.licht.de/fileadmin/Publications/licht-wissen/0703_lw18_E_light_museums_galleries_web.pdf, access date 2nd June, 2018].

In the figure below, the lighting intervention transports the visitor along the pieces. Here, in addition to lighting, sounds were used that contextualize the original environment of the piece. In this example, the Guo Pei dress is placed in the center of a base that not only lifts the piece, allowing for its full enjoyment, but also delimits access to visitors (Fig. 11).

Fig. 11. "China: Through the Looking Glass" Exhibion, MET, 2015 [Foto by Carreto & Carreto, 2015].

4 Conclusions

Exhibition design, such as fashion design and apparel, is a constantly updated creative area. The need for the preservation and dissemination of past cultures, in particular textile, fashion and clothing creations, the identity of a society of a certain time period, makes relevant the development of the area of exhibition design. The fashion, although it may be considered artistic creation, is to be worn/worn by people, where the various human factors must be guarded, in terms of dimensions, textures, colors and materials.

When designing an exhibition of fashion and clothing products, the human factor in its relationship with space and objects is essential, since visitors are the main elements and without them, the objective of the exhibitions is lost. In this area, it is essential to take into account the type of lighting and circulation in space, the colors used, the height and distance to which the exhibition object is placed, the minimum dimensions of the human body and the rotational movements of the trunk, head and eyes, as well as the size and legibility of the letters and texts that support the exhibition. With the change of the exhibition paradigm, visitors are increasingly looking for didactic and interactive exhibitions. Today they visit exhibitions not only for their fruitive character, but mainly for the educational and cultural.

References

1. Belcher, M.: Organización y diseño de exposiciones: Su relación con el museo. Ediciones Trea, Gijón (1997)
2. Carreto, C.: Design para a democratização do consumo cultural: A loja do museu. Tese de doutoramento, Faculdade de Arquitectura da Universidade de Lisboa, Lisboa (2018)
3. Carreto, R.: Design para a Mobilidade Expositiva: O Museu Móvel como Museu Reutilizável. Tese de doutoramento, Faculdade de Arquitectura da Universidade de Lisboa, Lisboa (2013)
4. Dernie, D.: Espacios de Exposición. Blume, Barcelona (2006)
5. Fernández, L.A., Fernández, I.G.: Diseño de Exposiciones: concepto, instalación y montaje. Alianza Editorial, Madrid (2007)
6. Hughes, P.: Exhibition Design. Laurence King, London (2010)
7. Locker, Pam: Diseño de exposiciones. Editorial Gustavo Gili, Barcelona (2011)
8. Treguenza, P., Loe, D.: Projecto de Iluminação. Bookman, Porto Alegre (2015)
9. Valencia, P.P., Andújar, A.B.: Joven museografía: La exposición autoportante. Ediciones Trea, Gijón (2011)

Distribution of Clothing Pressure in Support Knee-High Socks and Its Evaluation

Tamaki Mitsuno[1(✉)], Aya Yamagishi[2], and Mayuko Takenaka[2]

[1] Faculty of Education, Shinshu University, 6-ro, Nishinagano, Nagano, Japan
mitsuno@shinshu-u.ac.jp
[2] Former Faculty of Education, Shinshu University, 6-ro, Nishinagano, Nagano, Japan
{6e1632k, 8e1618a}@shinshu-u.ac.jp

Abstract. The relationship between clothing pressure and the evaluation of tightness of the lower leg and foot was examined. Participants were nine young Japanese women in their twenties. Clothing pressures were measured adopting a hydrostatic-pressure balanced method as participants wore three SKSs while standing. The pressure feeling was evaluated employing a ratio scale. The order of the average clothing pressure was sample A > sample B > sample C. However, the feeling of pressure was not in the same order (B > A > C). Our results show that the feeling of tightness depends on the position on the body, and the distribution of clothing pressure is thus important in the development of SKSs. A preferred SKS provides higher pressure for the foot than for the calf/ankle.

Keywords: Clothing pressure · Support knee-high socks · Pressure feeling

1 Introduction

Compression wear for the lower half of the body is expected to reduce swelling and improve the leg shape but its effects on human health and stress remain controversial. Tamura [1] reported that the level of clothing pressure affected not only wearing comfort but also health. The relationship between swelling of the lower half of the body and the current purchasing trends of support knee-high socks (SKSs) was examined in an online questionnaire of 1425 female Japanese consumers aged 10 to 60 years. The questionnaire revealed an awareness that swelling is closely related to the total time spent walking and sitting, and respondents were particularly aware of swelling at evening and night. Among participants who were aware of swelling of their bodies, 79.3% reported swelling in the lower halves of their bodies, such as the calf, ankle, and instep [2]. We therefore expect that knee-high socks are a desirable product to develop. The situation of swelling of the lower leg and foot has been examined for people not wearing SKSs. Relative volumes have been measured for healthy young females sitting for 8 h. The volumes of the lower legs and feet increased slowly from morning to evening whereas the volume of the ankles rapidly increased around noon [3, 4]. The relative volumes of the lower leg and foot, the clothing pressure of SKSs, and the feeling of the tightness of SKSs have thus been measured for three conditions: condition A in which loosely fitting SKSs are worn, condition B in which tightly fitting SKSs are worn, and condition C in which loosely

G. Di Bucchianico (Ed.): AHFE 2019, AISC 954, pp. 449–456, 2020.
https://doi.org/10.1007/978-3-030-20444-0_46

fitting SKSs (condition A) are worn with added support of the calf. The clothing pressures of the calf and foot under condition A and the clothing pressure of the foot under condition C were lower than 5 hPa, and a loose feeling was felt. Meanwhile, the clothing pressures of the calf and foot under condition B and the calf under condition C exceeded 20 hPa, and a tight feeling was felt. The calf relative volume increased 4% under condition A but did not increase for condition B or C. The foot volumes did not differ significantly among conditions A, B, and C. The relative volume of the tightened part of the calf significantly decreased for both partial and total tightening [5]. When the ankle is constricted and the clothing pressure increases, the volume of the ankle decreases whereas the volume of the foot increases. However, the foot volume does not increase when the ankle is constricted, resulting not in a tight-fit feeling but in a perfect-fit feeling. Our final goal is to make comfortable SKSs that prevent swelling. Using three types of SKS, comprising two commercial SKSs and SKSs that we made, the present paper investigates the relation between comfortable clothing pressure and the feeling of pressure.

2 Experiment

2.1 Participants and Samples

An experiment was conducted for nine Japanese female participants in their twenties, whose physical dimensions are given in Table 1. Participants were nonsmokers, had no history of systemic disease, and were not engaged in physical training or dietary programs. Their figures ranged from slender to slightly overweight. Clothing pressures on the abdomen depend on the menstrual cycle of the participant, with the pressures in the luteal phase being higher than those in the follicular phase [6]; the former phase was thus chosen in the present study. Experiments were carried out over four days from the luteal phase latter half to the menstrual phase former half. Clothing pressures were measured for the participants wearing three types of SKS, and the tightness feeling was evaluated as follows (See Sect. 2.4).

Table 1. Physical constitutions of participants

Age (Year)	Hight (cm)	Weight (kg)	Under knee girth (cm)	Calf girth (cm)	Ankle girth (cm)	Arch girth (cm)	BMI (kg/m^2)
22.1 ± 0.4	157.1 ± 5.8	53.5 ± 5.7	21.7 ± 2.0	35.0 ± 1.7	20.9 ± 1.6	22.3 ± 1.3	21.7 ± 2.0

The three types of SKS used in the present study are shown in Fig. 1. Samples A and B were commercial SKSs while sample C was our own product. Each sample came in two sizes; i.e., foot lengths of 23 and 25 cm (corresponding to sizes medium and large). All samples were made from nylon, cotton, and other materials. Each sample was stretched using an AUTOGRAFH (AGS-H, Shimadzu, Japan) at a pulling speed of 300 mm/min, maximum load of 3.5 kgf, holder width of 10 mm, and exposed core of 10 mm in a climate-controlled room (environmental temperature, 20 °C; relative humidity, 65%), and the relative stretching rate was calculated in six directions as shown in Fig. 2.

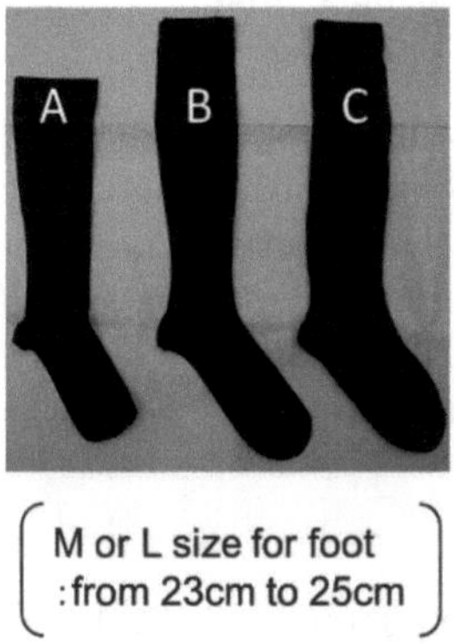

Fig. 1. SKS samples

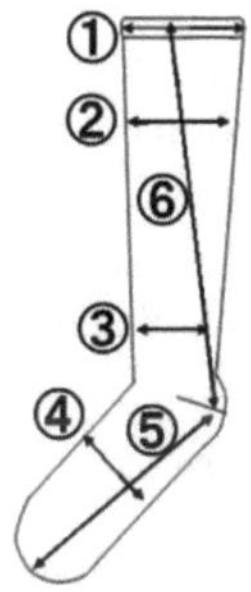

Fig. 2. Measuring positions for stretching

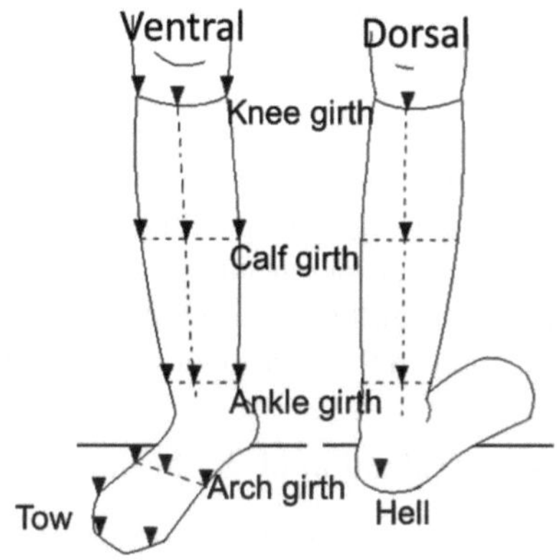

Fig. 3. Measuring points for clothing pressure

2.2 Experimental Schedule

Two experiments, Experiment I (control) and Experiment II (in which participants wore samples) were carried out following the schedule shown in Fig. 4. Participants awoke at 05:30 after 7 h of sleep and ate breakfast until 06:00. They then entered a climate-controlled room (environmental temperature, 24.5 ± 0.3 °C; relative humidity, 50.3 ± 2.5%; lumination, 827 ± 27 lx; air current, 8.0 ± 0.1 cm/s) at 07:00. Participants remained in a sitting posture from 8:00 to 12:00, and the skin temperature of their foot was measured using a thermotracer (TH-3102MRS, NEC, JPN) before and after the 4 h of sitting in Experiments I and II. In experiment II, the clothing pressure and its feeling of tightness were measured after participants wore SKSs.

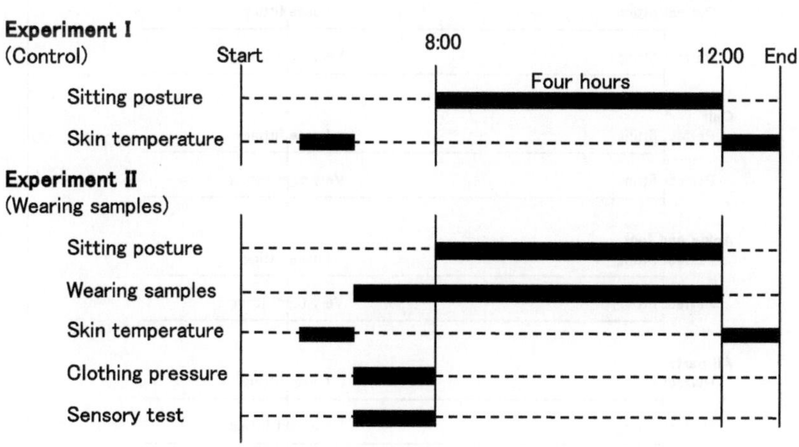

Fig. 4. Experimental schedule

2.3 Measuring System for Clothing Pressure and Measuring Points of the SKS

Clothing pressures were measured adopting a hydrostatic-pressure balanced method [7] for participants wearing SKSs while standing. There were 19 measuring points on the lower right leg corresponding to the intersections of four vertical lines (anterior/posterior midlines, inside/outside lines) and horizontal lines (under-knee girth, maximum/minimum girths, arch/tow girths), and heel as shown in Fig. 3. The average clothing pressures in the horizontal direction were calculated for all subjects.

2.4 Evaluation of the Feeling of Pressure

The pressure feeling was evaluated employing a ratio scale [8]. Participants evaluated the pressure they felt when wearing SKSs by completing a questionnaire (Fig. 5).

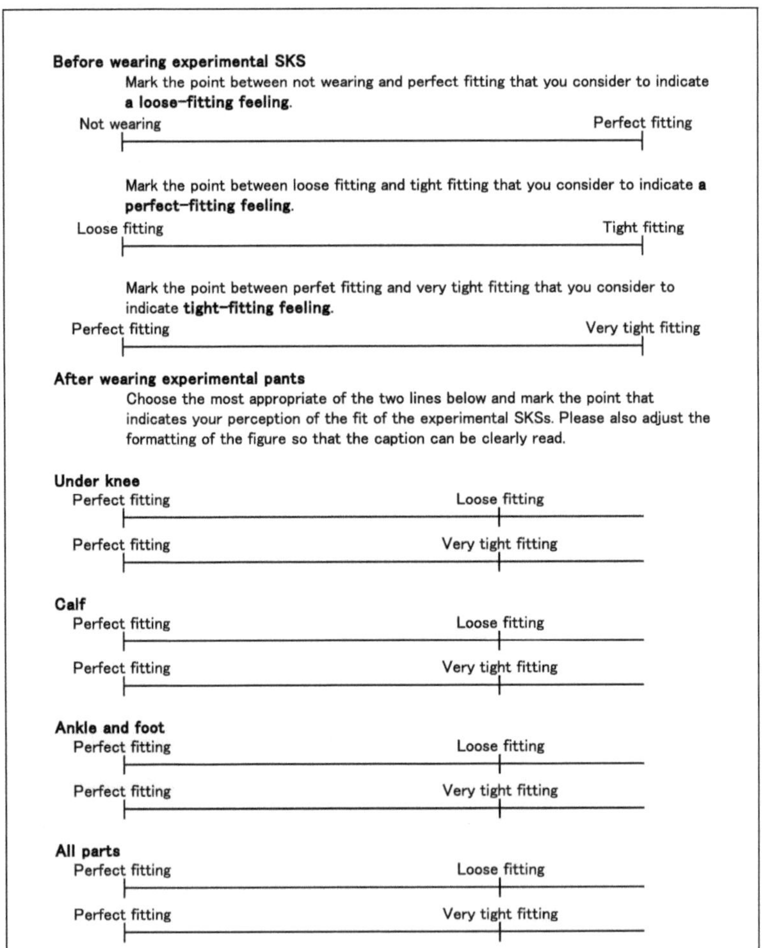
Before wearing experimental SKS

Mark the point between not wearing and perfect fitting that you consider to indicate **a loose-fitting feeling**.

Not wearing — Perfect fitting

Mark the point between loose fitting and tight fitting that you consider to indicate **a perfect-fitting feeling**.

Loose fitting — Tight fitting

Mark the point between perfet fitting and very tight fitting that you consider to indicate **tight-fitting feeling**.

Perfect fitting — Very tight fitting

After wearing experimental pants

Choose the most appropriate of the two lines below and mark the point that indicates your perception of the fit of the experimental SKSs. Please also adjust the formatting of the figure so that the caption can be clearly read.

Under knee

Perfect fitting — Loose fitting

Perfect fitting — Very tight fitting

Calf

Perfect fitting — Loose fitting

Perfect fitting — Very tight fitting

Ankle and foot

Perfect fitting — Loose fitting

Perfect fitting — Very tight fitting

All parts

Perfect fitting — Loose fitting

Perfect fitting — Very tight fitting

Fig. 5. Questionnaire of the pressure feeling when wearing SKSs

Before putting on SKSs, participants estimated what they considered a loose-fitting feeling, perfect-fitting feeling, and tight-fitting feeling. The participants wearing each SKS were asked to indicate individual pressure feelings to compare with the base feelings, and to mark points on a line in the questionnaire. The length of the line connecting the points marked by the participants to indicate the loose-fitting feeling, perfect-fitting feeling, or tight-fitting feeling was measured. Numerical values for the pressure feeling were obtained by supposing that a feeling of no tightening had a score of zero and a perfect-fitting feeling had a score of 1. In other words, the numerical value exceeded 1 if participants felt the SKS was tight while it was less than 1 if they felt the SKS was loose. The data for each body part (i.e., the lower leg, ankle, foot, and overall) were analyzed by one-factor repeated-measures analysis of variance and, once significance was confirmed, analyzed through multiple comparison employing the Tukey method.

3 Results

3.1 Development of Clothing Pressure of the SKS on a Leg Model

Figure 6 presents clothing pressures of SKSs recorded for the leg model of a Japanese female having an age of 20–30 years (shown on the left). The average clothing pressures along horizontal lines are shown for each SKS. The clothing pressures of sample A were significantly higher than the clothing pressures of samples B and C, while the clothing pressures of B were significantly higher than those of sample C (i.e., A > B > C).

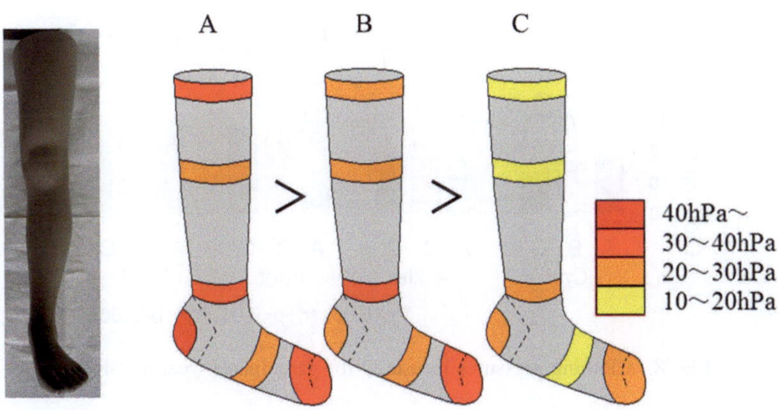

Fig. 6. Clothing pressures of SKSs on a leg model

3.2 SKS Stretching Rate

The stretching rates of SKSs are shown in Fig. 7. The SKSs stretched easily in directions 1–4 and 6 but not in direction 5. All SKSs were produced with greater horizontal stretch than vertical stretch. Sample C had the greatest stretch under the knee

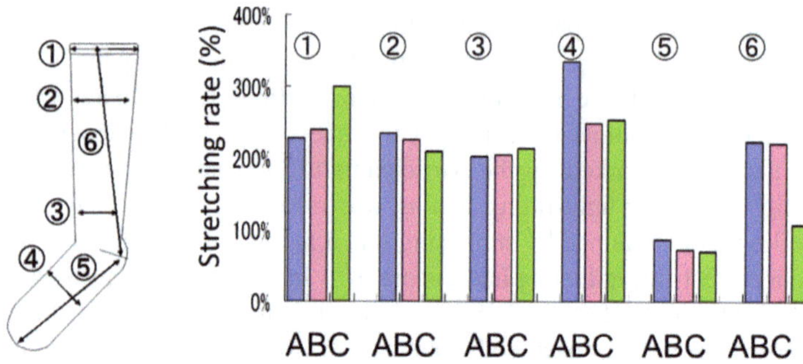

Fig. 7. Stretching rates of SKSs

(i.e., direction 1) and little vertical stretch (i.e., directions 5 and 6). In contrast, sample A/B had grater stretch in the six directions than in sample C.

3.3 Measured Clothing Pressures of SKSs Worn by Participants

Clothing pressures measured for participants wearing the samples are shown in Fig. 8. The overall clothing pressure decreased significantly in the order of samples A, B, and C. The pressures for samples A and C decreased in the order of the foot, ankle, and calf while, for sample B, the clothing pressures of the ankle and foot were almost the same and higher than the clothing pressure of the calf.

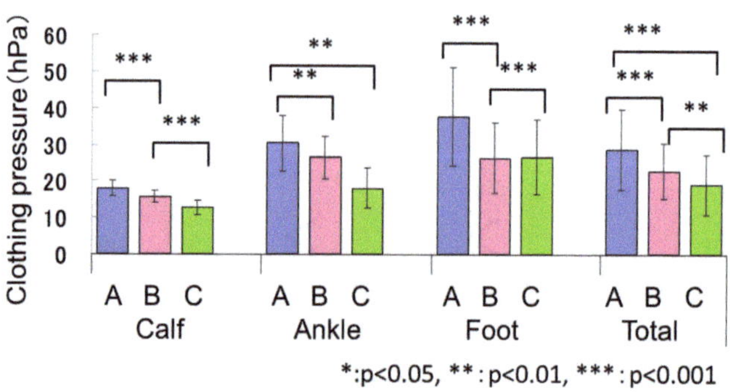

Fig. 8. Clothing pressures measured for participants wearing SKSs

3.4 Evaluation of the Pressure Feeling

The reported feelings of pressure when participants wore SKSs are shown in Fig. 9. The feelings of pressure did not have the same order as the measurements of pressure. There was no significant difference in the feeling of pressure between samples A and B (sample B was slightly tighter than sample A) while sample B felt significantly tighter than sample C.

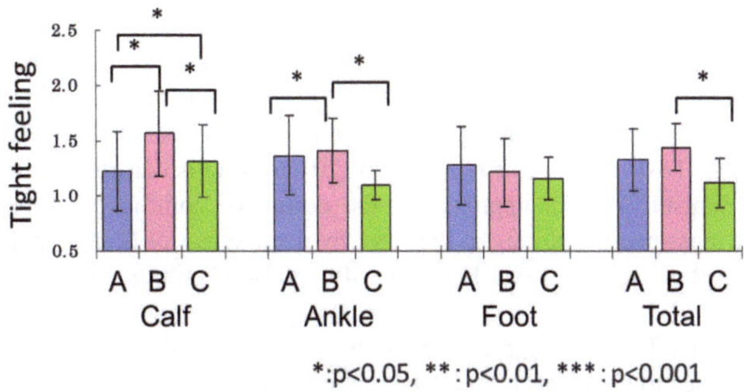

Fig. 9. Pressure felt while wearing SKSs

4 Discussion

The feeling of tightness depends on the position on the body. The distribution of clothing pressure is thus important in the development of SKSs. The skin is not equally sensitive over the body because mechanoreceptors are not equally distributed over the body. Preferred clothing pressure thus depends on the body part, with high pressure being most preferred on the lower leg [7]. The present study found that the preferred pressure order on the lower leg is calf < ankle < foot. That is to say, the preferred clothing pressure becomes higher centrifugally. The order of the overall clothing pressure in the present study was sample A > sample B > sample C; however, the order of pressure feeling was not the same (A/B > C) because the clothing pressures on the ankle and foot were almost the same for sample B. The sequence of pressure that should become higher centrifugally was disturbed. An overall evaluation can thus be established anew by evaluation of the tightness of the ankle. The body surface from the knee to the ankle will have more mechanoreceptors than the foot and the preferred SKS thus provides higher pressure for the foot than for the calf/ankle.

Acknowledgments. Part of this work was supported by the Japanese Society for the Promotion of Science KAKENHI Grant Number 17H01954. We thank Glenn Pennycook, MSc, from Edanz Group (www.edanzediting.com/ac) for editing a draft of this manuscript.

Trial Registration. This study was carried out after the review by the Ethical Review Board for Humans of Shinshu University (approval number 173).

References

1. Tamura, T.: Ifukuatsu no Kouzai (Clothing pressure for merit and demerit). J. Home Econ. Jpn. **51**, 1089–1092 (2000). (in Japanese)
2. Mitsuno, T., Watatani, C., Shinga, I.: Consumer consciousness for swelling in the lower half of the body. Jpn. Res. Assoc. Text. End-Uses **53**, 934–941 (2012). (English abstract with Japanese text)

3. Mitsuno, T., Matsui, M.: An approach to the design of comfortable knee-high socks to support lower legs (part 1): in special reference to the volume changes in lower legs and feet. Jpn. Res. Assoc. Text. End-Uses **49**, 197–206 (2008). (English abstract with Japanese text)
4. Mitsuno, T., Kamijo, M.: Time changes of the volume of the feet in young Japanese women. Jpn. Res. Assoc. Text. End-Uses **51**, 321–326 (2010). (English abstract with Japanese text)
5. Mitsuno, T., Matsui, M.: An approach to the design of comfortable knee-high socks to support lower legs (part 2): aiming at the most hygienic effect of beauty shape in lower legs and feet. Jpn. Res. Assoc. Text. End-Uses **49**, 207–216 (2008). (English abstract with Japanese text)
6. Niwa, H., Mitsuno, T.: Clothing pressure and pressure feeling developed by wearing yukata-in special reference to the phase of seasonal and menstrual cycle in comparison with wearing waistband. Jpn. Res. Assoc. Text. End-Uses **47**, 731–739 (2006). (English abstract with Japanese text)
7. Mitsuno, T., Kai, A.: Distribution of the preferred clothing pressure over the whole body. TRJ, **89**, 2187–2198 (2019). https://doi.org/10.1177/0040517518786272journals. Sagepub. com/home/trj
8. Mitsuno, T., Hibino, A.: Which support pants are the most effective for burning body fat? J. Fiber Bioeng. Inform. **9**, 87–199 (2016)

Woven Fabrics Specially Designed for Thermal Protection and System Recovery of Air Vehicles

Carmen Mihai(✉) and Alexandra Ene

The National Research and Development Institute for Textiles and Leather, 16, Lucreţiu Patraşcanu str., 030508 Bucharest, Romania
carmen.mihai@certex.ro

Abstract. The textile industry is an innovative field, a generator of competitive products, absolutely indispensable to many other fields (aeronautics, medicine, shipping, agriculture, environmental protection etc.)

In this philosophy, important researches have been carried out in order to develop new generations of textile products for strategic applications.

The paper presents important issues regarding: the role of helicopter thermal protection system as well as parachute recovery in the main system (supersonic aircraft); design of thermal protection woven structure for thermal protection of the helicopters as well as the woven structure for main parachute of recovery systems for landing of the supersonic aircraft, based on aerodynamic characteristics and areas of use; the main fabric structure characteristics used in the two systems; description of the phenomena occurring on the main parachute canopy at landing of the supersonic aircraft; structural analysis of the main parachute canopy of the aircraft deceleration system; the interdependence between the aerodynamic, geometric and mass characteristics of the considered systems and the characteristics of the fabric. The design and realization of these product types are based on the specific requirements imposed by its fields of application, which governs the system and the selection of components. Typical design and delivery criteria have been considered: safety in operation (the criterion of the greatest importance); mass and volume in order to efficiently use the loading value; stability; reproducibility of performance, etc.

Keywords: Helicopters · Aircraft landing · Supersonic aircraft

1 Introduction

The researches dealt with the identification and solving of some specific objectives substantiated on the exigencies represented by aeronautics industry performances. Base on relevant accomplishments in the aeronautic and textiles fields, a new generation of woven structure with special characteristics for the main parachute surface of the aerodynamic decelerator of supersonic aircraft and thermal protection of the helicopter installation for heating and ventilation was realized, experimented and implemented at the beneficiaries.

G. Di Bucchianico (Ed.): AHFE 2019, AISC 954, pp. 457–461, 2020.
https://doi.org/10.1007/978-3-030-20444-0_47

To explain the phenomena occurring on the main parachute canopy, for supersonic aircrafts deceleration at landing, both the theories of the Fluid Mechanics were used, referring to: - the rapidly variable transient movement, - the continuity hypothesis, - the physical properties of the fluids, - the operations the motion of a continuous medium, the limiting layer, the turbulent shuffling, etc., and those specific to the aeronautics field related to the flow of fluid around a sphere [1].

2 Material and Methods

By the scientific design of an original fabric, there was identified and solved, the technological problems represented by deceleration at landing of the supersonic aircrafts and by the thermal protection of the helicopter installation for heating and ventilation.

For establishing the main directions for mechanical processing in order to obtain the textile structures for the two types of products, the constituent filaments of a para-aramid filaments were analyzed (Fig. 1).

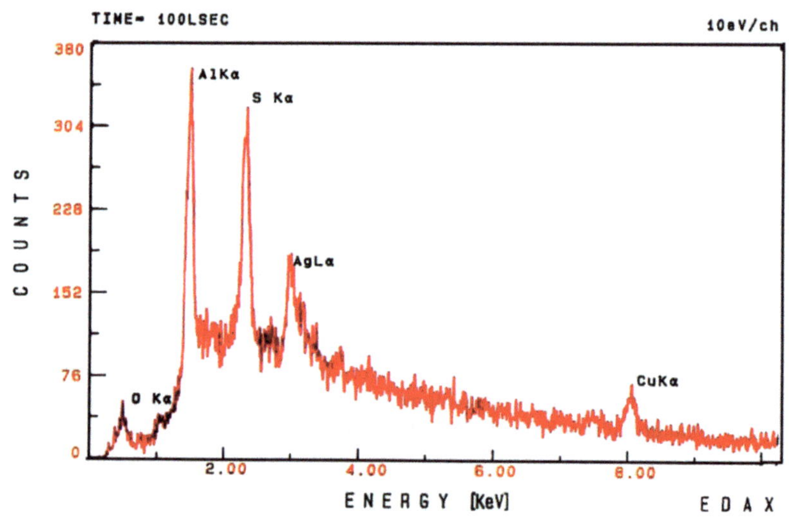

Fig. 1. X-ray spectrum for para-aramidic filament

Considering that this yarn will be used for the thermal protection of the helicopters and the supersonic aircraft deceleration parachute, it has been found that the surface treatment of the filaments revealed by the X-ray spectrum confer to the fabric the necessary characteristics represented by: high thermal conductivity for rapid heat transfer, flame resistance for a long exposure time, good resistance to friction wear, very good resistance to microorganisms during storage.

Designing and accomplishment of the woven structured for the corrugate tube of the thermal protection of the helicopters' installation was based on the specific requirement represented by weight: max.100 g/m^2; tensile strength warp/weft: min.300/320 daN/5 cm; elongation: max.7,5%; tearing resistance warp/weft: min.14/16 daN; permeability: max.100 l/m^2/s [2].

The final product was tested in real conditions and demonstrate its functionality after at least 10.000 flight hours.

The theories of the fluid mechanics enabled to explain the complex phenomena occurring during the supersonic aircrafts landing that enabled us to design and accomplish the textile structure of the parachute according to the requirements imposed by the application domain.

Thus, in order to discuss the motion equation at the contact with the moving canopy surface was considered that its surface is impermeable and on both sides of the fabric there is a pressure difference comparable to the dynamic current pressure.

Upon landing the aircraft, the fluid is initially in stagnation, then, after a very short time, when the braking parachute enters into operation, the fluid in which this phenomenon occurs is in motion.

At one point (t = 0), the braking parachute enters fully in deceleration mode. In this situation, the fluid particles at boundaries are suddenly stopped, causing the kinetic energy to be transformed into pressure energy and thus a Δp increase in the pressure near it and implicitly an increase in the density of the fluid, causing swelling of the canopy.

Thus, a turbulent motion occurs and the air velocity in front of the canopy is decelerated to zero at the stagnation point [3]. Behind this point there is a turbulent flow, which causes a very high static pressure inside the parachute compared to the static pressure in the undisturbed fluid around the canopy.

The positive pressure inside the canopy and the negative pressure outside it form a high-pressure gradient allows the inflation of the canopy.

Designing of the geometry of the recovery system was carried out by using a dedicated software that allowed to preview the specific phenomena-taking place during the landing.

Thus, the unitary effort on the parachute surface was determined, and the obtained value was considered in counting the density of the fabric used for the decelerating parachute surface.

The usage of this program permitted to the highlight of the fact that, for the landing deceleration of the supersonic aircraft, a fabric made of yarns of para-aramid type should be used in order to correspond to the requests imposed by the final application.

The membrane theory was used in order to determine the specific parameters for woven fabric design (i.e. weft and warp density, breaking force and elongation in static and shock conditions etc.).

3 Results and Discussions

The necessary air permeability of the woven structure of the canopy from the deceleration system was determined by performing a complex programme of experiments in subsonic aerodynamic tunnel (Fig. 2).

Fig. 2. Experimental model of brake parachute during the experimentation in aerodynamic tunnel

Testing programme demonstrated that the optimum interval for the air permeability of the woven fabric is [400,700] $l/m^2/s$ that will fulfil the following requirements:

- low oscillation angles of $\pm3°$, recommended for cross deceleration parachute, and therefore good stability of the recovery system at landing;
- low opening shock coefficient of 1.1–1.2.

Ground testing of the recovery system was conducted under the temperature between 4 and 8 °C; atmospheric pressure: 750 mm Hg and wind speed between 4 and 6 m/s.

The obtain results demonstrated that the accomplished system is functional, ensuring the safety opening of the brake parachute at the supersonic aircraft.

This conclusion was validated by the results obtained during the testing programme performed at landing conditions. Thus, after the trials and verification of flights and landing in wet and dry conditions, the deceleration parachute demonstrated to be operational (Fig. 3).

Fig. 3. Deceleration at supersonic aircraft landing with the main parachute

The estimation of the running distance of the open parachute landing was made by the pilots, ranging from 680 to 800 m (depending on opening speed, track condition, wind direction and intensity).

The aerodynamic braking was very good, reaching a normal speed for actuating the mechanical brakes.

4 Conclusions

A new generation of woven structure with special characteristics for the main parachute surface of the aerodynamic decelerator of supersonic aircraft and thermal protection of the helicopter installation for heating and ventilation was realized, experimented and implemented at the beneficiaries.

During the researches was identified and solved, the technological problems represented by deceleration at landing of the supersonic aircrafts and by the thermal protection of the helicopter installation for heating and ventilation.

The complex phenomena occurring during the supersonic aircrafts landing was explained using the theories of the fluid mechanics and aeronautics.

Designing of the geometry of the recovery system was carried out by using a dedicated software that allowed to preview the specific phenomena-taking place during the landing.

Typical design and delivery criteria have been considered: safety in operation (the criterion of the greatest importance); mass and volume in order to efficiently use the loading value; stability; reproducibility of performance, etc.

The specific parameters for woven fabric design (i.e. weft and warp density, breaking force and elongation in static and shock conditions etc.) was determined based on membrane theory.

A complex programme of experiments in real condition validated the functionality of the accomplished systems.

Acknowledgment. This scientific paper is funded by the Ministry of Research and Innovation within Program 1 - Development of the national RD system, Subprogram 1.2 - Institutional Performance - RDI excellence funding projects, Contract no. 6PFE/2018.

References

1. Lien, F.S., Kalitzin, G.: Computation of transonic flow with the v2-f turbulence model. Int. J. Heat Fluid Flow **22**, 53–61 (2001)
2. Bogdanov, V.V., Zalogin, G.N., Zemlyanskii, B.A., et al.: Experimental study of the efficiency of flexible thermal protection of inflatable braking systems. Kosmonavt. Raketostr. **22**, 22–32 (2001)
3. Pilyugin, N.N., Khlebnikov, V.S.: Development of a parachute system for deceleration of flying vehicles in supersonic regimes. J. Appl. Mech. Tech. Phys. **51**(5), 623–632 (2010)

A Pilot Study of a Hygienic Mask Sheet Pattern to Prevent Air Flow Through Gaps

Mika Morishima[1(✉)] and Tamaki Mitsuno[2]

[1] Institute of Human and Social Sciences, Kanazawa University, Kakuma-Machi, Kanazawa, Ishikawa 920-1192, Japan
morishimam@staff.kanazawa-u.ac.jp
[2] School of Humanities and Social Sciences, Shinshu University, Nishinagano, Nagano 380-8544, Japan

Abstract. Hygienic masks are highly useful in healthcare. However, when a commercially available mask is worn, the air gap at the upper edge of the sheet can allow contaminants to reach the wearer's nose and mouth. Here, we aim to identify a novel pattern for a hygienic mask sheet that does not form an air gap. Three-dimensional (3D) scans of the facial shapes of a female mannequin and female human subjects were obtained using a 3D analyzer. A two-dimensional sheet pattern was derived based on the facial coordinate data and a prototype sheet was fabricated using the optimized pattern for the mannequin. In a fitting test, the sheet was well fitted to the face, especially on the upper edge. Moreover, the facial coordinates relative to the nose tip in the subjects were similar to those in the mannequin. Thus, the proposed approach is expected to be applicable to other females.

Keywords: Mask · Fit · Pattern

1 Introduction

A hygienic mask is worn over a wearer's nose and mouth to prevent exposure to external particles such as pollen [1], dust, volcanic ash [2], and atmospheric particulate matter (PM) 2.5 [3]. The mask is one of the most useful textile personal protective measures in healthcare. However, an excellent fit is required to effectively protect the wearer's nose and mouth from harmful substances in the outside environment.

There are three types of facemasks: industrial, surgical, and hygienic. The industrial type of respiratory equipment includes particulate respirators (JIS T 8151:2018) and gas respirator (JIS T 8152:2012). For such respirators, adhesion between the mask and the wearer's face must be measured qualitatively or quantitatively using a fit check or fit test according to national or international standards. For surgical type masks, the fit performance is tested according to international standards. Contrastingly, there are no standards or specific test methods for hygienic masks.

The effectiveness of various commercial hygienic masks has been verified in previous studies [4, 5]. However, in actual use, there is often an air gap between the hygienic mask sheet and the wearer's face. Further, it has been reported that the air gap

G. Di Bucchianico (Ed.): AHFE 2019, AISC 954, pp. 462–468, 2020.
https://doi.org/10.1007/978-3-030-20444-0_48

between the face and the upper edge of the sheet of a commercially available molded mask allows contaminants to reach the wearer's nose and mouth [6].

To overcome this issue, the objective of this study was to develop a novel pattern for a hygienic mask sheet that does not exhibit such an air gap, thus, effectively prevents contaminants from reaching the wearer's nose and mouth. Herein, a three-dimensional (3D) analyzer is used to map the facial shapes of a female mannequin head and several participants. To prevent air flow beneath the upper edge of the sheet to the wearer's nose wing, points along the contour from the nose to the cheek are included in a set of points. Based on the measured coordinates, two-dimensional (2D) patterns are calculated using a triangular combination with the tip of the nose as the origin [7]. A prototype mask sheet based on the analyzed pattern for the mannequin was created and evaluated. Moreover, the mannequin pattern was compared with those of the subjects to verify the adequacy of the proposed approach for human subjects.

2 Experimental Methods

Here, facial shapes of a female mannequin with average head characteristics and young female participants were measured using a 3D analyzer. Then, 2D patterns were generated based on the acquired 3D data. Subsequently, a prototype sheet was fabricated based on the identified pattern. Next, the fit of the mask on the mannequin face was visually evaluated. Finally, patterns for the mannequin and those for the human participants were compared to verify the applicability of the approach to other females.

2.1 Mannequin and Participant

In this pilot test, a mannequin head and five human participants were evaluated. The mannequin was built using a 3D printer (FORTUS360mc; Stratasys) from data about the average head shape of an adult Japanese female published by the National Institute of Advanced Industrial Science and Technology (Anthropometric Database of Japanese Head 2001, H16PRO-212 as collected by the National Institute of Advanced Industrial Science and Technology).

Five young female subjects (Mean age: 22.6 years; Standard deviation (SD): 2.5) were recruited for the study. The head circumference of each subject was measured using a measuring tape and the head length, head breadth, and bigonial breadth of each subject were measured using a spreading caliper; on average, these measures were 557.6 mm (SD 20.4), 181.8 mm (SD 8.1), 150.0 mm (SD 3.7), and 103.6 mm (3.2), respectively. These measures were in the mean ±standard variation ranges for Japanese head characteristics. The study protocol was approved by the ethics committee at the Institute of Human and Social Sciences Ethics Committee of Kanazawa University (IRB permit no. 2017–33).

2.2 Measurement of Facial Shape

The facial shapes of the female mannequin and participants were measured using a 3D analyzer. At each of the landmark points, 3 mm-diameter hemispherical reflection markers were fixed with double-sided medical tape. Figure 1 shows the landmark points. The first ring of points comprised 18 landmark points on the nose and mouth and the second, third, and fourth rings comprised 24, 26, and 26 points, respectively, in concentric circles spaced at 10 mm intervals. To ensure proper fit of the upper edge of the sheet to the wearer's nose wing, the points along the contour from the nose to the cheek were included in the set of points. The facial shape was evaluated based on the positions of the markers in the closed-mouth position, as observed by three infrared cameras. The 3D coordinates (x, y, z) of the markers were computed using the 3D analyzer system.

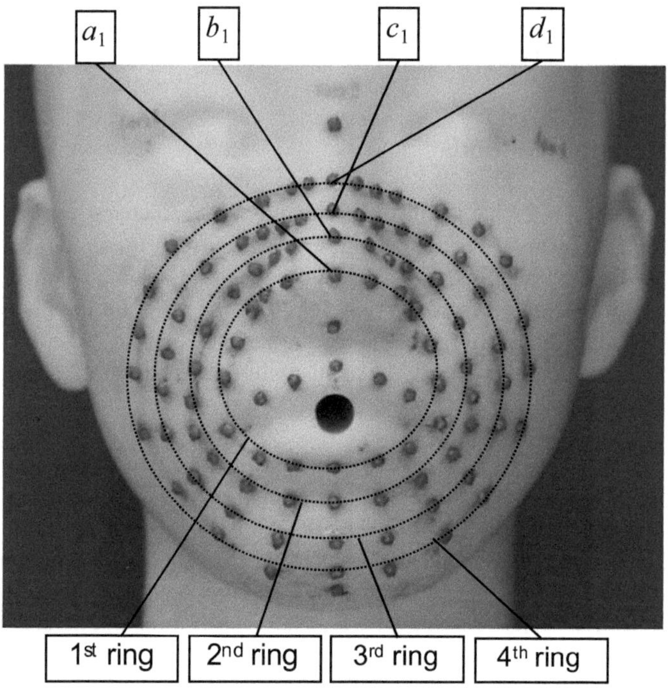

Fig. 1. Measurement point.

2.3 Determination of Mask Sheet Patterns

Using the computed 3D coordinates, 2D patterns were derived by the combinations of triangles method. Briefly, the 2D coordinates of the triangle between the origin point (the nose tip, a_1 (x_0, y_0, z_0), as shown in Fig. 1), each point (x_n, y_n, z_n), and the next point (x_{n+1}, y_{n+1}, z_{n+1}) on each ring were computed based on the distance calculated according to Eq. (1) and the interior angles derived based on the cosine formula.

$$d = \sqrt{(x_n - x_0)^2 + (y_n - y_0)^2 + (z_n - z_0)^2} \tag{1}$$

Then, half of the mask sheet (i.e., the right or left side) was formed by an affine transformation using the integrated inner angle between the nose tip and center of lower lip as an axis for points on the first ring or the angle between the nose tip and the center of the jaw as the axis for points on the other rings. The points in the right side were rotated clockwise based on the integrated inner angles and those on the left side were rotated counterclockwise based on the corresponding integrated inner angles.

3 Result and Discussion

3.1 Analyzed Pattern for the Mannequin

Using the acquired 3D coordinates for the mannequin, 2D mask sheet patterns of four different sizes corresponding to the four rings of points were derived. Figure 2 shows the identified pattern for the mannequin (broken lines). In the pattern reaching the first ring, the ideal pattern required to cover the participant's nose and mouth was found to be nearly elliptical. In the larger patterns reaching the second, third, and fourth rings, the dorsum points of the nose (b_1 in the second ring, c_1 in the third ring, and d_1 in the fourth ring as shown in Fig. 1) were located in the first and second quadrants. When the two points in each quadrant were combined, the 2D pattern evolved into a 3D elliptical cone. The shapes of the analyzed patterns were not awkward.

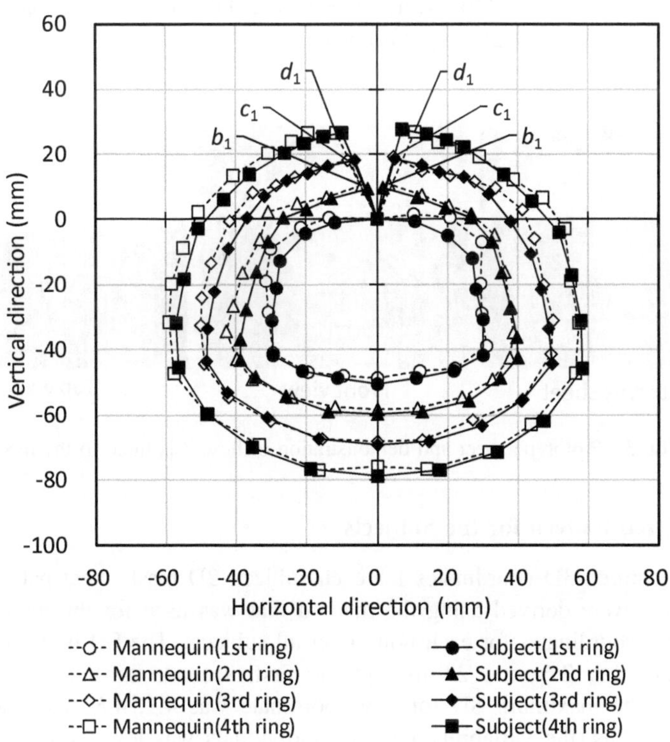

Fig. 2. Analyzed patterns for subjects and the mannequin.

Furthermore, the patterns for the mannequin showed left–right symmetry. The distances between the nose tip, a_1, and each point in the pattern were calculated for both sides. Mean and SD of absolute values of the difference between them were 1.0 ± 0.9 mm in the pattern corresponding to the first ring, 1.6 ± 1.2 mm that corresponding to the second ring, 1.6 ± 1.4 mm in that corresponding to the third ring, and 1.9 ± 1.0 mm in that corresponding to the fourth ring.

In a previous study, fewer measurement points were used in the second, third, and fourth rings compared to the present study. However, the differences between the shapes were smaller in each ring compared to those in the results of the previous study. Further, the outer perimeters of the rings exhibited small differences compared to those in the previous study: −2.6, 0.4, and −0.3 mm in the second, third, and fourth rings, respectively.

3.2 Fit Performance of the Prototype Sheet on the Mannequin

Based on the derived pattern, a prototype sheet extending to the fourth ring was fabricated from non-woven rayon fabric (Torakurosu 30301; Taketora) and put on the mannequin as shown in Fig. 3. It was visually observed that the sheet followed the contour of the mannequin's face very well. The top view shows that the sheet was along the boundary between the nose and cheek. From these results, it was confirmed that the analytical method resulted in a mask sheet pattern that provides a close fit on the upper edge of the sheet. Moreover, it was revealed that the selected markers yielded an effective pattern with high fit performance.

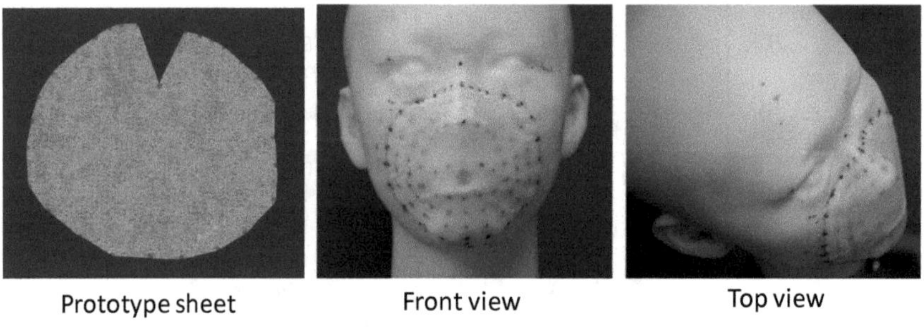

Fig. 3. Prototype sheet and demonstration of how it is fitted to the face.

3.3 Analyzed Pattern for the Subjects

Using the acquired 3D coordinates for each subject, 2D mask sheet patterns in four different sizes were derived using the method that was used for the mannequin and averaging the coordinates for each point over all subjects. The SD in each coordinate averaged 2.0, 2.3, 2.6, and 2.9 mm in the first, second, third, and fourth rings, respectively; the maximum SD for any coordinate was 4.1, 4.5, 5.1, and 6.0 mm, respectively. Thus, the variability between individuals was low so average patterns could be evaluated.

Figure 2 shows the average patterns (solid lines). The shape for each ring was nearly the same as the corresponding shape for the mannequin. Moreover, the analyzed patterns for the subjects exhibited left–right symmetry. The distances between the nose tip, a_1, and each point in the pattern were calculated for both sides and the averages were 0.3 mm in the first ring, 0.6 mm in the second ring, 0.8 mm in the third ring, and 0.6 mm in the fourth ring.

3.4 Comparison of the Analyzed Patterns for the Mannequin and Subject

To evaluate the applicability to a variety of wearers, the analyzed pattern for the mannequin was compared to the pattern for the five subjects. The differences (±SD) were 0.4 mm (±1.2 mm) in the first ring, −0.6 mm (±1.7 mm) in the second ring, −1.1 mm (±1.9 mm) in the third ring, and −1.2 mm (±2.4) in the fourth ring. These small differences indicate that the proposed analytical method using a female mannequin is applicable to the design of masks for human subjects. In future studies, we expect the adjustment of shape and the length of the ear strings will reduce the difference between the mannequin and human subject patterns and the differences between patterns for different individuals.

4 Conclusion

In this study, a 2D pattern for a mask sheet was derived that does not exhibit an air gap between the upper edge and the face. The pattern was derived based on the 3D coordinates of a female head mannequin with average features and the faces of female participants as a pilot test as measured using a 3D analyzer. 94 landmark points on the face were radially set with 10 mm intervals covering the nose and mouth. Patterns of four different sizes for the mannequin and the subjects were compared and the differences were found to be small. Moreover, a prototype of the sheet that was designed for the mannequin was fabricated and shown to fit the female mannequin well, especially, around the nose. Therefore, it was concluded that this approach is appropriate to develop a novel mask sheet and can be applied to human subjects. In future studies, a larger number of participants will be evaluated and the fit will be investigated further.

References

1. Gotoh, M., Okubo, K., Okuda, M.: Inhibitory effects of facemasks and eyeglasses on invasion of pollen particles in the nose and eye: a clinical study. Rhinology **43**, 266–270 (2005)
2. Steinle, S., Sleeuwenhoek, A., Mueller, W., Horwell, C.J., Apsley, A., Davis, A., Cherrie, J. W., Galea, K.S.: The effectiveness of respiratory protection worn by communities to protect from volcanic ash inhalation. Part II: total inward leakage tests. Int. J. Hyg. Environ. Health **221**, 977–984 (2018)
3. Onishi, K.: Health impact assessment of asian dust/cross-border air pollutant and necessary preventive measure. Nippon Eiseigaku Zasshi (Jpn. J. Hyg.) **72**, 43–48 (2017). (in Japanese)

4. Mueller, W., Horwell, C.J., Apsley, A., Steinle, S., McPherson, S., Cherrie, J.W., Galea, K.S.: The effectiveness of respiratory protection worn by communities to protect from volcanic ash inhalation. Part I: filtration efficiency tests. Int. J. Hyg. Environ. Health **221**, 967–976 (2018)
5. Bin-Reza, F., Lopez Chavarrias, V., Nicoll, A., Chamberland, M.E.: The use of masks and respirators to prevent transmission of influenza: a systematic review of the scientific evidence. Influenza Other Respir. Viruses **6**, 257–267 (2012)
6. Morishima, M., Mitsuno, T.: Adhesion distribution between a hygiene mask and a female dummy face via multiple approaches. In: Duffy, V., Lightner, N. (eds.) Advances in Human Factors and Ergonomics in Healthcare and Medical Devices: Proceedings of the AHFE 2017 International Conferences on Human Factors and Ergonomics in Healthcare and Medical Devices, 17–21 July 2017, The Westin Bonaventure Hotel, Los Angeles, California, USA, pp. 324–330. Springer, Cham (2018)
7. Morishima, M., Shimizu, Y., Fujiyama, T., Mitsuno, T.: Analysis of the fitting of patterns of hygiene masks for the closed mouth. Text. Res. J. **85**, 828–837 (2015)

Research on the Characteristics of Hand Shape in Different Countries

Jing Zhao, Fan Zhang, Gang Wu, Chao Zhao, Haitao Wang, and Xinyu Cao(✉)

China National Institute of Standardization, Beijing, China
caoxy@cnis.gov.cn

Abstract. For a long time, science has been dedicated to revealing various indicators of the human body, including height and weight. The same purpose is also for this paper. Based on data on the hand of people from ten different countries, this paper do some research with the help of multi-variate analysis tools. According to the different characteristics of the data, this paper first carried out the analysis of variance on hand length. Then, according to the different dimension of hand shape, including the vertical length of the palm, the metacarpal and the length of the index finger, factor analysis was carried out, and two influence factors were extracted. Then, multiple regressions were performed on these variables, at the same time, stepwise regression method was adopted, and the main independent variables were chose from the constructed model. Finally, gender is set to a dummy variable to study the impact of gender differences on the hand length.

Keywords: Hand · Analysis of variance · Factor analysis · Multiple regression · Dummy variable

1 Introduction

Most scientific research is about people's intuitive physical data, indicating height and weight. However, research on details, such as the study of the hands and feet, is more conducive in other disciplines. For example, reference [1] mentions the improvement of suggestions and opinions on the production of Chinese shoes by analyzing the data of the collected feet. The measurement of the hand has different measurement methods. Reference [2] gives two methods in detail. Overall, there are little literatures discussing the hand characteristics. [2] also gave some comparison of hand anthropometry of females in three ethnic groups: Western Europeans, Indian and West Indian. Additionally, appreciation of the gender differences in hand shape is essential to the proper design of both men's and women's hand tools, gloves. Traditionally, women's gloves are often made using a small version of a men's glove with all dimensions proportionally scaled according to hand length. However, if women's hands differ in shape from men's hands, this is an inappropriate model and could lead to improper gloves fit in women. Courtney [3] carried out an anthropometric study of hand dimensions of Hong Kong Chinese female workers compared to other ethnic groups. Twenty-three hand dimensions were measured and compared with data from the United Kingdom,

G. Di Bucchianico (Ed.): AHFE 2019, AISC 954, pp. 469–479, 2020.
https://doi.org/10.1007/978-3-030-20444-0_49

Japan and the United States of America. Other research referred to hand anthropometric study, please see literatures [4–6].

The data information we analyzed is from ISO/TR 7250-2:2010 "Basic human body measurements for technological design—Part 2: Statistical summaries of body measurements from national populations". This technical report is intended to serve as a continually updated repository of the most current national anthropometric data. It also provides statistical summaries of body measurements together with database background information for working age people in the national population.

2 Method

At first, we conduct the direct comparison among these countries based on analysis of variance (ANOVA). It is because we want to know whether hand length are significantly different among these countries as well as between male and female in these countries. We take hand length as response variable, gender and country as two factors. Analysis of variance is used to analyze the effects of factors, which may have impact on the result of an experiment.

Secondly, we performed factor analysis on the five variables that affect the hand length to extract more intuitive data features. The five variables are the vertical length of the palm, the metacarpal, the length of the index finger, the width of the proximal end of the index finger, and the width of the distal end. We want to know if there are any internal dependencies between these five variables.

Thirdly, we construct a multiple regression analysis of the above five variable opponents, and study how these five variables affect the human hand length, because the stepwise regression method is adopted, resulting the fifth independent variable removed from the model.

At last, we also want to know how much influence the gender has, so we set gender as a dummy variable. The study found that men's hands are about 2% taller than female hands.

3 Results

3.1 Results of ANOVA for Hand Length

The general linear model procedure is used to do analysis with the help of SPSS software. The general linear model procedure has the following advantages: (1) it could be used in unbalanced designs; (2) it can evaluate differences between individual level means; (3) it allows covariates, random factors and nested factors.

As we know, ANOVA based on three assumptions: (1) the dependent variables of each population are normally distributed, (2) the variance of each population is equal, and (3) the samples of each population are randomly sampled independently. Then, first we perform homogeneity tests through Levene's method. The results are in Table 1. From the p-value we can conclude that the data has equal variance.

Table 1. Levene's test of equality of error variances

Dependent variable: hand length			
F	DF1	DF2	P-value
.073	19	40	1.000
Test the full hypothesis that the error variance of the dependent variable is equal across groups			

In Table 2, DF stands for degree of freedom. Without loss of generality we take nominal significance level as 0.05. That means if P-value of one factor is less than 0.05, we think it is significant; otherwise it is not significant.

Table 2. Tests of between-subjects effects

Dependent variable: hand length					
Source	Sum of square	DF	Mean square	F	Sig.
Corrected model	5463.400[a]	19	287.547	1.221	.290
Intercept	2040939.267	1	2040939.267	8663.343	.000
Country	2289.067	9	254.341	1.080	.399
Gender	3024.600	1	3024.600	12.839	.001
Country * gender	149.733	9	16.637	.071	1.000
Error	9423.333	40	235.583		
Total	2055826.000	60			
Corrected total	14886.733	59			

[a]R squared = .367 (Adjusted R squared = .066)

Table 2 provides the results of the analysis of variance for the model. The first line (Corrected Model) is the overall test of all the effects of the model. The significance of two factors, country and gender, is 0.399 and 0.001, respectively. Gender is significant, but country is not significant. However, the test result of the interaction effect between the two factors (country * gender) is extremely insignificant. The table also provides that determination coefficient (R Squared) of the model is equal to 0.367, indicating that the main effect and interaction effect of the two factors in the model together account for 36.7% of the total variation of the dependent variable. However, although the sample analysis shows that the effects of the model have some explanatory ability for the dependent variable, the statistical test of the effects shows that only one effect is beyond the random fluctuation range but that cannot be inferred to the whole.

Figure 1 is the interaction plot for hand length. From Fig. 1, we can obtain the following information: (1) Overall, the hand length of men is larger than hand length of women in all ten countries. (2) The ten countries fall into about three groups: The Netherlands and United States had significantly larger hand length than the other eight countries; The India had the smallest hand length. (3) On average, the hand length of male in the Netherlands and the United States are largest and the hand length of female in United States and Kenya are the largest. The hand length of people in Germany and Italy is similar.

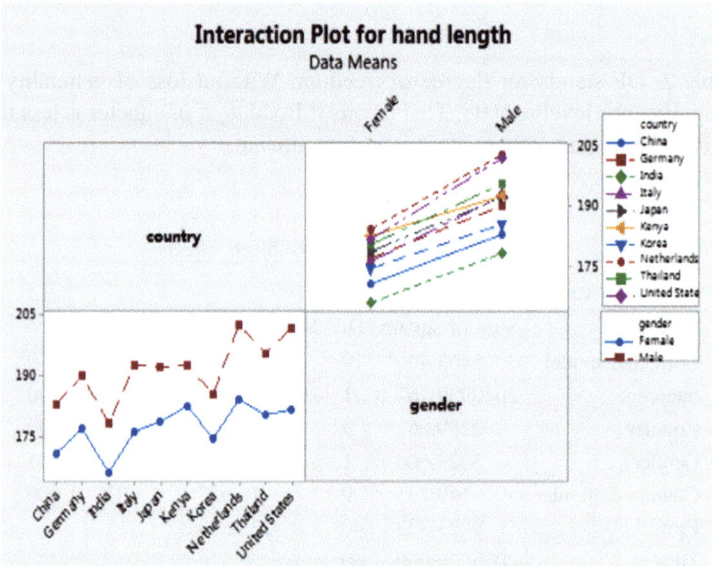

Fig. 1. Interaction plot for hand length

3.2 Results of Factor Analysis

As the original data has some missing values, especially the United States, we do some basic data pretreatment under the help of R software. At first, in order to make sure the accuracy of the data, we delete the information about USA's hand data, then, we make use of R software to multiple interpolate missing values. Furthermore, we try to find some common factors which effect the length of hand in nine countries (except the United States) no matter women or men. Aiming at above purposes, we take the method of Factor Analysis followed.

The purpose of factor analysis is to simplify the data or find out the basic data structure. Therefore, the premise of factor analysis is that there should be a strong correlation between observed variables. If the correlation between variables is small, they cannot share a common factor. Therefore, we produced the correlation coefficient matrix of the five variables in the table and conducted Bartlett's test of sphericity and KMO measures.

Table 3. Correlation matrix

		x1	x2	x3	x4	x5
Correlation	x1	1.000	.843	.838	.754	.750
	x2	.843	1.000	.867	.785	.816
	x3	.838	.867	1.000	.710	.957
	x4	.754	.785	.710	1.000	.650
	x5	.750	.816	.957	.650	1.000

x1 indicates Palm length perpendicular
x2 indicates metacarpals
x3 indicates Index finger length
x4 indicates Index finger breadth, proximal
x5 indicates Index finger breadth, distal

Table 4. KMO and bartlett's test

KMO measure of sampling adequacy		.823
Bartlett's test of sphericity	Approx. chi-square	325.618
	df	10
	Sig.	.000

The correlation analysis of five variables (Table 3) shows that there is a strong correlation between Index finger length and Index finger breadth, distal, which reaches about 0.957. The other three variables have a mutually strong correlation as well. All of these correlation values reach above 0.65. Looking at Table 4, the KMO measured value reaches above 0.8, and the significance level of Bartlett spherical test is extremely high, which indicates that factor analysis can be conducted on this data.

According to the eigenvalues criterion, only one factor can be selected for the hand data including nine countries (except USA). In order to illustrate explicitly, we have selected two factors. From Table 5, it shows the concrete information. The first column "Component" indicates the different kinds of factors, the following three columns are the initial eigenvalues, "Total" column refers to the eigenvalues, "% of Variance" is the percentage of Variance, and "Cumulative %" is the cumulative percentage of variance. For instance, the first two factors explain the data's variance 83.902% and 8.425% respectively. While these two factors explain the vast majority of the total data variance accumulatively, reaching more than 90%. The middle three columns show the two factors selected, as well as their explanatory variance ratio and cumulative ratio. The last three columns are the variance explanations of each selected factor after factor rotation.

Table 5. Total variance explain

Component	Initial eigenvalues			Extraction sum of squared loadings			Rotation sums of squared loadings		
	Total	% of variance	Cumulative %	Total	% of variance	Cumulative %	Total	% of variance	Cumulative %
1	4.195	83.902	83.902	4.195	83.902	83.902	2.546	50.914	50.914
2	.421	8.425	92.327	.421	8.425	92.327	2.071	41.413	92.327
3	.215	4.307	96.634						
4	.138	2.764	99.398						
5	.030	.602	100.000						

Extraction Method: Principal Component Analysis

Communalities is also one of the important judgment principles for determining the number of factors. Table 6 provides information in this respect. In the table, "Initial" is the Initial value, while "Extraction" is the communalities of each variable. The results show that the communalities of the five variables is very high, and the information of each variable contain more than 84%.

Table 6. Communalities

	Initial	Extraction
x1	1.000	.849
x2	1.000	.891
x3	1.000	.976
x4	1.000	.940
x5	1.000	.960

According to the order in which factors are extracted, we portray the screen plot about the change of factor eigenvalues (vertical axis) with the number of factors (horizontal axis). Actually, the number of factors is also judged according to the shape of the figure. It can be seen from the scree plot: from the first factor, the curve goes down rapidly, then becomes gentle, and finally becomes an approximately straight line. The factor load after rotation with the "varimax" method shows in Fig. 2 as well. We can conclude that "Index finger breadth, proximal" (x4) is close to the second factor, and "Index finger length" (x3) and "Index finger breadth, distal" (x5) are closer to the first factor.

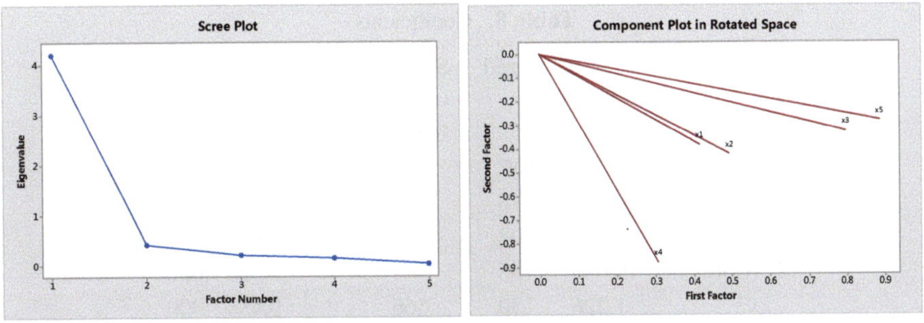

Fig. 2. Screen plot and component plot

3.3 Results of Multiple Regressiom

In order to study how the five variables in the data affect the hand length including nine countries, we build a linear regression model with the stepwise method. Since more than one variable is included in the independent variable, we take multiple linear regression model. The model structure is as follows:

$$y = b_0 + b_1x_1 + b_2x_2 + b_3x_3 + b_4x_4 + b_5x_5 + \varepsilon \tag{1}$$

Among them, y indicates hand length, x_1 = Palm length perpendicular, x_2 = metacarpals, x_3 = Index finger length, x_4 = Index finger breadth, proximal, x_5 = Index finger breadth, distal. $b_1, b_2.\ldots.b_5$ are model parameters, b_0 is intercept value.

Table 7. Model summary

Model	R	R square	Adjusted R square	Std. error of the estimate	Durbin watson
1	.941[a]	.886	.884	5.304	
2	.972[b]	.946	.944	3.694	
3	.979[c]	.958	.956	3.271	
4	.981[d]	.962	.959	3.152	1.192

[a]Predictors: (Constant), x3
[b]Predictors: (Constant), x3, x1
[c]Predictor: (Constant), x3, x1, x2
[d]Predictor: (Constant), x3, x1, x2, x4
[e]Dependent Variable: y

Table 7 shows the interpreting proportion of each model by means of stepwise regression. The order of taking independent variables into the model is Index finger length (x3), Palm length perpendicular (x1), metacarpals (x2), Index finger breadth, proximal (x4). The last model has accounted for hand length more than 95%, but the Durbin-Watson test value of the model is 1.192, which cannot support the premise of error independence, so the regression model is only for reference.

Table 8. Coefficients

Model		Unstandardized coefficients		Standardized coefficients	t	Sig.
		Beta	Std. error	Beta		
1	(Constant)	40.186	7.178		5.598	.000
	x3	2.039	.102	.941	20.083	.000
2	(Constant)	19.086	5.738		3.326	.002
	x3	1.226	.130	.566	9.458	.000
	x1	.741	.099	.448	7.494	.000
3	(Constant)	23.016	5.180		4.443	.000
	x3	.936	.137	.432	6.834	.000
	x1	.580	.097	.351	5.983	.000
	x2	.413	.106	.249	3.882	.000
4	(Constant)	21.093	5.068		4.162	.000
	x3	.931	.132	.430	7.055	.000
	x1	.635	.097	.384	6.568	.000
	x2	.503	.110	.303	4.557	.000
	x4	−.604	.275	−.103	−2.200	.033

Dependent Variable: y

Table 8 shows the coefficients of each model after stepwise regression. In the first step, there is only an independent variable, and the regression coefficient is a large positive value (2.039), reflecting that the larger "Index finger length" (x_3) is, the larger hand length is. The second step regression model increase the independent variable "Palm length perpendicular" (x_1), whose standardized coefficient is 0.448, that is similar to "Index finger length" (x_3). The third step takes "metacarpals" (x_2) into regression model, the fourth step increases the "Index finger breadth, proximal" (x_4).

$$y = 0.384x_1 + 0.303x_2 + 0.43x_3 - 0.103x_4 \tag{2}$$

The above formula is the standardized performance of stepwise regression in the last model. From the standardized coefficients respectively, "Index finger length" (x_3) has the greatest influence on the hand length, each unit it increases, the hand length increases by 0.43 unit. In addition, both "Palm length perpendicular" (x_1) and "metacarpals" (x_2) have positive effects on hand length. From their standardized coefficients (0.384 and 0.303, respectively), the effects of the two variables are similar. Finally, "Index finger breadth, proximal" (x_4) has a negative effect on hand length, whose coefficient is -0.103.

Then, we draw four-in-one residual plot (Fig. 3), we explain the figures in the following:

Normal Probability Plot – Because the points on the normal probability plot roughly follow a straight line, we can assume that the residuals do not deviate substantially from a normal distribution.

Histogram – Use the normal probability plot to make decisions about the normality of the residuals. With a reasonably large sample size, the histogram displays compatible information.

Versus Fits – The constant variance assumption does not appear to be violated, because the residuals are randomly scattered about zero.

Versus Order – The plot of the residuals versus order does not show any pattern. Therefore, there is no time dependence in the residuals.

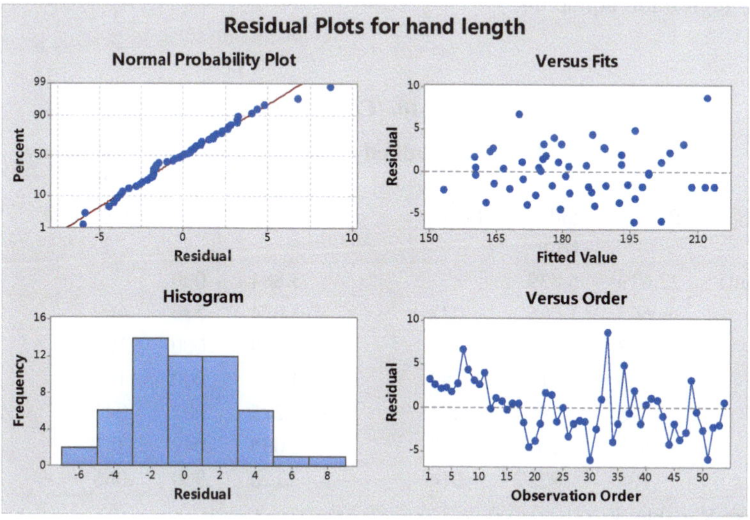

Fig. 3. Residual plots for hand length

3.4 Results of Adding Dummy Variable

Next, considering the impact of gender on hand length in the data of these nine countries, we take gender as a dummy variable in the regression equation, and the other independent variables do not change. Building a multivariate regression equation with dummy variables, Table 9 provides a global test of the regression equation, showing that the F value of the regression equation is 199.989, which is statistically significant at the 0.000 level, indicating that there is at least one independent variable which owns a very significant linear relationship with dependent variable in the equation.

Table 9. ANOVA

Model		Sum of squares	df	Mean square	F	Sig.
1	Regression	12324.111	6	2054.019	199.989	.000
	Residual	482.722	47	10.271		
	Total	12806.833	53			

Dependent Variable: y
Predictors: (Constant), x5, gender, x4, x1, x2, x3

Table 10 provides the regression coefficients for the respective variables after adding dummy variables, including unstandardized coefficients and standardized coefficients. Among them, the last two columns of the table also provide collinearity test items for multiple regression. When the independent variables are completely linearly related, the estimation of the regression model no longer has a unique solution, so the least squares method is no longer applicable, neither. It can be seen from the test that the tolerance of x_3 and x_5 is less than 0.1, which means their variance inflation factors are all greater than 10 but less than 25, indicating that there is a certain degree of correlation between independent variables, and it is necessary to carefully apply multiple regression equations.

Table 10. Coefficients

Model	Unstandardized coefficients		Standardized coefficients	t	Sig.	Collinearity statistics	
	B	Std. error	Beta			Tolerance	VIF
(Constant)	22.679	5.839		3.884	.000		
Gender	−.665	1.066	−.022	−.624	.536	.669	1.495
x1	.623	.106	.377	5.888	.000	.196	5.111
x2	.480	.118	.289	4.061	.000	.158	6.325
x3	.978	.299	.451	3.275	.002	.042	23.695
x4	−.588	.281	−.100	−2.095	.042	.351	2.853
x5	−.104	.825	−.014	−.126	.900	.066	15.053

Dependent Variable: y

We give the following regression equation for the dummy variable according to the column labeled the standardized coefficients in Table 9:

$$y = -0.022gender + 0.377x_1 + 0.289x_2 + 0.451x_3 - 0.1x_4 - 0.014x_5 \quad (3)$$

From the degree of significance of the coefficient, the t-statistic value of gender is -0.624, whose p-value is equal to 0.536 greater than 0.05, rejecting the null hypothesis. The index finger breadth, distal (x_5) is also not significant, rejecting the null hypothesis at the level of 0.9, so we should be careful when interpreting variables with the multiple regression model results. When constructing dummy variables, we use 0 for men and 1 for women. Therefore, the regression coefficient of gender shows that on average, men's hand length are 0.022 units higher than women, which is also a common sense that men's hands are larger than women's.

4 Conclusion

This paper mainly discussed the characteristics of body hand in ten countries based on the statistical data in ISO/TR 7250-2:2010 "Basic human body measurements for technological design—Part 2: Statistical summaries of body measurements from national populations". Based on the average data and percentile values of hand length we can conclude that the ten countries fall into about three groups: The Netherlands and United States had significantly larger hand length than the other eight countries, the India had the smallest hand length, other countries are in a same condition. We do a factor analysis for five variables, which indicates that we can extract two main factors. "Index finger breadth, proximal" is close to the second factor, "Index finger length" and "Index finger breadth, distal" are closer to the first factor. Finally, we do a multiple regression to find out what extended gender and other variables' impact on hand length. The results are "Index finger length" has a great significance and on average, men's hands are about 2% taller than female hands. The obtained results could be a reference for designing the export gloves to these countries. What's more, the obtained regression equations have a great reference value for forensic science and anthropology.

Acknowledgments. This research was supported by National Key Technology R&D Program (2017YFF0206503, 2017YFF0209004, 2016YFF0204205) and China National Institute of Standardization through the "special funds for the basic R&D undertakings by welfare research institutions" (522018Y-5941, 522018Y-5948).

References

1. Bandini: Statistical analysis of foot anthropometric data of young chinese male population. Tsinghua University (2012)
2. Davies, B.T., Abada, A., Benson, K., Courtney, A., Minto, I.: A comparison of hand anthropometry of females in three ethnic groups. Ergonomics **23**, 2 (1980)
3. Courtney, A.J.: Hand anthropometry of Hong Kong Chinese females compared to other ethnic groups. Ergonomics **27**(11), 1169–1180 (1984)
4. Imrhan, S.N., Sarder, M.D., Mandahawi, N.: Hand anthropometry in Bangladeshis living in America and comparisons with other populations. Ergonomics **52**(8), 987–998 (2009)
5. Imrhan, S.N., Nguyen, M.T., Nguyen, N.N.: Hand anthropometry of Americans of Vietnamese origin. Int. J. Ind. Ergon. **12**(4), 281–287 (1993)
6. Wunderlich, R.E., Cavanagh, P.R.: Gender differences in adult foot shape implications for shoe design. Med. Sci. Sports Exerc. **33**, 605–611 (2001)

Research on Human Factors Engineering for the Innovative Design of Work Clothes for Couriers

Longlin Luo(✉) and XiaoPing Hu

School of Design, South China University of Technology, Guangzhou, China
linlongluo@163.com

Abstract. This research focuses on the investigation based on human factor engineering, and explores the relationship between human body and clothing from different aspects. It discusses that clothing materials, sizes, styles, and colors can better meet the physiological and psychological needs of couriers when they work, takes into account the relationship among courier, clothing, and working environment, points out the limitations of courier's work clothes at present, and proposes new design methods. Finally, according to the principle of design optimization, a more comfortable, multi-functional, healthy, safe, and fashionable couriers' work clothes is pointed out, which provides guidance for the improvement of the couriers' work clothes with multi-functional and innovative design.

Keywords: Couriers · Work clothes · Innovative design · Human factor engineering

1 Introduction

With the overall progress of human society and the continuous improvement of people's living standard, comfortable and suitable work clothes play an extremely important role in the designated work type and specific work environment, especially in the express transportation industry that frequently requires long hours of working.

At present, most express companies do not design or customize a standardized uniform for couriers, and currently the work clothes for couriers is also lack of innovation to a certain extent. In addition, in view of the urgent need to reduce costs, the overall form of courier's clothing gradually reveals its shortcomings, mainly due to the lack of consideration of the principle of human factors engineering. The work content of the courier is mainly to place and transport the parcels and other objects, which consumes a lot of energy from the human body. One is long-term work, and the other is because the distance they move about for carrying articles is relatively long. Therefore, considering that a complete set of work clothes can be designed on the basis of meeting the requirements of various indicators, the innovative design of work clothes for couriers has entered a new stage. And for how to effectively obtain the size information of the Courier and combine the requirements of its work content, the key is to apply the new concept into the design.

G. Di Bucchianico (Ed.): AHFE 2019, AISC 954, pp. 480–487, 2020.
https://doi.org/10.1007/978-3-030-20444-0_50

The main research methods adopted in this paper are the observation method and the investigation method. Taking human and clothing as the research objects, this paper focuses on the study of the behavior changes of couriers in the process of investigation. For example, the frequency of repetitive movements, the range of movement changes and the time of continuous operation of a certain movement are frequently carried out in particular time by operators in different positions. And in combination with the inquiry method, the content obtained when talking with the courier is recorded as a data so as to study the new suggestions of the young male couriers on the color, style, size and materials of the work clothes, and take this as the entry point for the design so that the clothing can fully satisfy the principle that the human body can feel comfortable in a variety of motion. This paper mainly analyzes it from the following aspects. The first is the relationship between the physique of a courier and his clothes. And the second one is the analysis of clothing size and courier's body shape. The third point is the interaction between body movements, postures and work clothes. The fourth aspect is the analysis of the structure design and material composition of clothing. On this basis, this paper summarizes the data and extracts the important influencing factors for targeted analysis.

2 Investigation of the Couriers' Work Content and Specific Environment

2.1 Characteristic of Body Shape and Structure of the Couriers

The survey was conducted mainly from the aspects of gender, shape, and structure, taking several districts in Guangzhou as an example, combining with network data information. Among the types of personnel engaged in express delivery, 80% are male and only 20% are female. Usually, the work content of female may be related to customer service jobs such as information entry and consultation, but in the survey, female couriers are also responsible for classifying packages. According to the survey results, most of the young male couriers are between 18 and 30 years old, and their physical characteristics are mostly muscular and strong. The work clothes should form a unified design style, with male body characteristics as the main size for standardized design, which includes the proportional distribution of bright and dark chroma, and warm and cold chroma. The clothes should be inclined to simple and neat style, so as to convey the unique personality culture of the courier profession. In order to make the courier comfortable when wearing work clothes, it is necessary to study the characteristics of human body structure and shape in the process of design, and fully consider the size information of various parts of the human body, especially the key parts. It should focus on observing and measuring the following parts vertically: neck, shoulder, upper arm, forearm, chest, spine, buttock, thigh length, and shin length; and focus on observing and measuring the following part horizontally: neck, shoulder width, axilla, chest, bust, waist, abdomen, buttock, hip line, and other parts [1].

2.2 The Relationship Between the Angle of Movement Posture Change of Courier and Garment Size Space

In the process of packing, sorting, moving, and transporting parcels, the body posture of the couriers will change in many ways. Because most of the couriers are in the state of movement while working, besides standing, sitting, and squatting posture, there are body twisting postures such as bending, rollover, valgus, etc. The range of body changes is relatively large. The body has to bear the load for a certain period of time. Different natures of working posture determine that the design of work clothes should follow a variety of principles to meet the needs of changes in body posture and movement [2].

Although couriers do not have a very fixed commuting time, but their daily work duration is longer than other occupations, usually 10–12 h a day normally. According to different positions, the working time is also different. Figure 1 shows the frequently-used posture changes of the couriers in a high frequency in a fixed-point express store in a day. They repeat nearly similar movements every day. This paper analyzes the composition and structure characteristics of each part of the body of the courier and the relationship among them. When the body is in the active state, the upper part of the body extends more, but the range of changes around the neck, wrist, waist, and so on is smaller. The lower thighs and footsteps are more extensible horizontally. Shoulder is the flattest part. When doing chest enlargement movement, it is the most variable part. Axilla is the most changeable part. The axilla changes correspondingly according to the movement and angle. When the chest moves or the arms are lifted, the muscle of the chest also extends or contracts (Fig. 1).

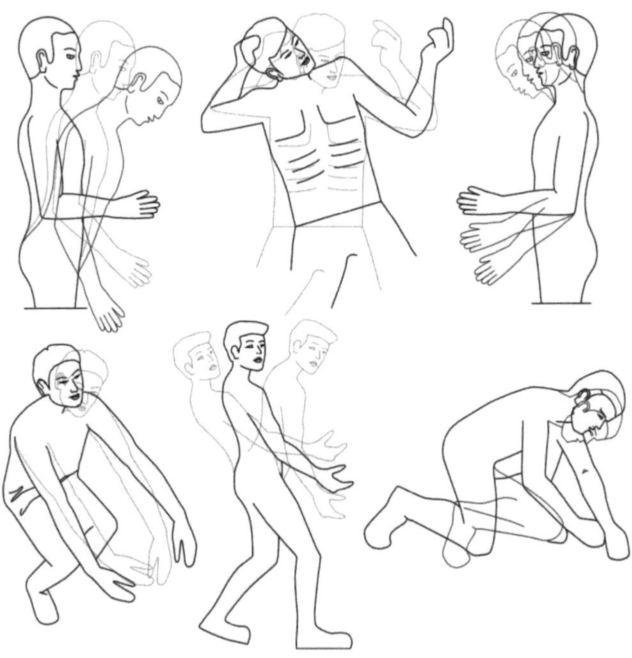

Fig. 1. Frequently-used posture changes of express workers.

According to comprehensive analysis, the couriers have a wide range of physical activities and varied postures, so the design of work clothes should focus on local design. The load caused by suits is often due to the inadequate design of local details of existing work clothes, such as the movement space under the armpit, the difference between shoulders, the width of cuffs, the type of collar, the location of pockets, the type of zipper, design of the pants' elastic band, design of the length of trousers gear, flexible size design of the trousers feet, etc. Local defects of the work clothes can produce obstructive force, which affects the efficiency of the couriers and brings them a sense of fatigue.

2.3 Couriers' Opinion and Expectation Index of Work Suits

This part applies method of inquiry and interview, conducts field visits and small-scale interviews, and extracts part of the personnel engaged in express delivery work in different positions for communication. In the interview, most of the couriers said that a comfortable functional work clothes is naturally the best. Generally, according to the company's requirements, some are free, some are paid, and some are additionally purchased due to the needs of their own work. Generally large express logistics companies will provide unified suits for employees.

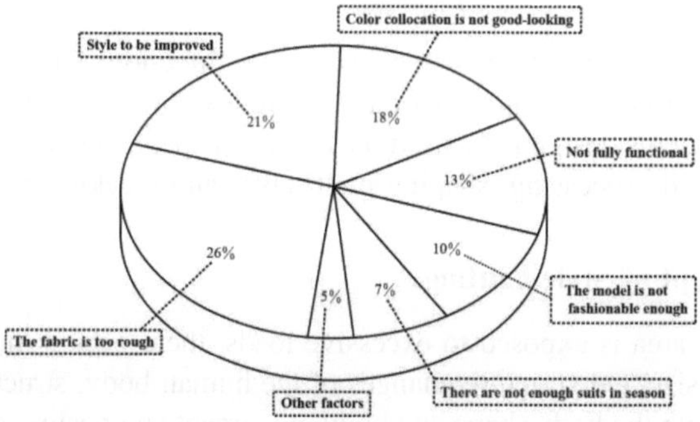

Fig. 2. Proportion of factors influencing the degree of dissatisfaction of couriers to the existing work clothes

The proportion of influencing factors pointed out by couriers to the existing work clothes is as followed. First, the material is too rough, and the fabric is impermeable. Secondly, the style and size are not in conformity, the number of pieces is incomplete, and the type is not fashionable enough. Comprehensive analysis of the impact index results show that: material, style, size, and color are the biggest reasons, and the main characteristics of the couriers' work clothes are comfortable and fit, followed by health, safety, and it is the best if upgraded to a high fashion taste.

3 Research and Analysis on the Orientation of Innovative Clothing Design

Throughout the whole process, the emphasis is laid on the relationship between people and clothing. The purpose of design needs to take the needs of users as the primary starting point. Since the functional and technical aspects required by reality are the main factors indispensable, but more importantly, it needs to be considered comprehensively from multiple aspects [3]. Based on the survey data, detailed rules for innovative design of couriers' work clothes are put forward.

3.1 Function and Decorative Design

For self-detachable hats, the relationship between the body and the hat is connected by zippers. When handling materials, driving, or doing other collective work in the warehouse, they can be dissembled whenever necessary. In front and back of upper and under suit, the pockets are designed with decoration and functionality, at the position where pens can be conveniently placed. In addition, the size of pockets, the types of shapes, and the direction of opening, are designed to meet the needs of the actual size of human hands, so as to make it more convenient and faster. For detachable glove design, the cuffs are connected by hidden buttons. In the specific parts of the cuff or trousers foot, flexible and adjustable size is designed, using anthropometric data to improve the design.

While designing the position of the logo, size, direction, location, identifiability factors such as the decoration and function of the reflex bar, the location of the logo are considered. Key details design is focused on, and local spatial details processing should focus on design and processing, shaping multi-layer anti-cracking process.

3.2 New Concept of Style Cutting

If a specific body area is exposed to excessive loads, there occurs muscle fatigue [4]. According to the size and structure changes of the human body, structural lines which are more in line with the body shape and body movement and posture change are found in the design and tailoring of the suits, and the position and size data are adjusted so as to form a reasonable spatial form with the human body while wearing the suits, reduce the load that hinders the work of the human body, and thus reduce more fatigue. There are many forms of garment line segmentation, such as line segmentation, curve segmentation, free segmentation, variability segmentation, and so on. The relationship between curve and line is fully used to achieve comfort and decorative visual effect. The courier has a large range of activities and consumes more energy. Therefore, the design of the suit style is loose but comfortable.

3.3 Principle of Color

Summer is suitable for light color. From the perspective of sunscreen, the darker the color, the more light and heat are absorbed, the hotter we feel. On the contrary, winter needs to consider protection from wind and cold. In addition to the standard requirements of specific express companies, the choice of color should also meet people's daily visual aesthetic needs to a certain extent. The physiological role of color is mainly reflected in the impact on visual work ability and visual fatigue, and the eyes have different sensitivity to different colors of light [5]. Color also plays a psychological role in influencing people, that is, color will produce a series of psychological effects of color perception through visual stimulation. This effect varies with time, place, and condition. People may have different temperature sensations when seeing different colors, that is, interrelated but antagonistic. The distance sensation has the greatest influence on brightness and hue, and the higher the lightness, the stronger the protruding sensation. The weight sensation is concentrated on lightness. Brightness makes people feel lighter, while darkness makes people feel heavier. For the fatigue sensation, the more dazzling the color is, the greater the stimulation to people, and it is easy to make people feel fatigue. Generally, warm and hot make people feel more fatigued than cold does. The sense of attention, that is, attractive color can be clearly recognized in the unconscious state. The sense of personality has the emotional effect of exciting or calming people. We should play the psychological and physiological role of color in reducing people's fatigue and make people happy.

Work clothes follow "Three-color Principle" in a certain range. Its specific concept is that the work clothes on the body, including jackets, pants, shoes and hats, etc., generally speaking, should not exceed three colors. Such color matching can make the couriers' work clothes present a concise and generous overall style. Black, blue, brown, and grey give people a calm, professional, and formal feeling. On the contrary, red, orange, yellow, and white convey people a warm, bright, brave, and active atmosphere. It is suggested that two or three different types of colors be combined.

3.4 Material Types

The fabrics of working clothes must be constantly innovated, explored, and made. Nowadays, there are many kinds of fabrics on the market, and many types of fabrics are also in line with people's selection requirements. If we add creative design to work clothes fabrics on this basis, and design new different forms of fabrics, we may achieve better results. The key is to use them properly and not appear fancy.

Fabric moisture absorption, air permeability, heat preservation, tactility is the most important four factors to determine the comfort [6]. The material used in the design of work clothes should be sweat-absorption, fast-drying, and air-permeability. In summer, most of the materials used for couriers' work clothes are natural cotton, linen, and polyester fibers, which are cool and breathable. In addition, several kinds of fast-drying fabrics commonly used in the market are ACTIVENT (windproof and water-absorption), MONY-DRY (volatility and water-absorption), COOLMAX, (moisture-absorption and sweat-absorption).

In winter, couriers spend a lot of time working outdoors. The design of work clothes focuses on the attributes of windproof, cold proof and warmth keeping. In the winter, the couriers' work clothes are made of synthetic fibers, such as full-process polyester-cotton blended fabrics, which have double layers of wind-proof and warmth keeping, and are antistatic, thermal, dry, wrinkle-resistant and lightweight.

It is suggested that new textile and garment materials should be introduced to strengthen the skin safety. The selected materials should have good elasticity, wear resistance, dirt resistance, stain resistance, and moisture-absorption. One of the most important factors that influence the perception of comfort is the ability of the clothing to handle thermal [7]. For example, the outdoor jackets worn outside in winter have a significant impact on the heat exchange process. They not only should keep people warm, but also should have good moisture absorption, to maintain the balance of human heat. The factors of thermal environment influence and compensate with each other. The isolation effect of clothes reduces the heat loss of human body to environment [3]. In winter, the temperature difference between indoor and outdoor is large. The outdoor jackets can condense the heat generated by the body on the surface of the jackets through the principle of moisture absorption, releasing some heat, which can convert the heat generated by the human body into energy with the function of warmth keeping.

3.5 Complete Parts Number System

It can be worn single or combined. For example, cloth gloves, duck tongue caps, shoes and other accessories provide couriers with a complete set of work suit, which will standardize their professional areas, clear their job responsibilities, so that they feel more convenient when working outdoors in the city, and this will improve consumers' credibility of couriers' dedication.

4 Reflection and Evaluation

4.1 If we want to add more interesting and fashionable elements, we should make the color of work clothes brighter and brighter, the style more and more leisure, novel and diverse, and take active part in exploring and trying innovative work clothes types. In addition, the harmony between people, clothing and environment system constitutes a begin ecology of clothing behavior, in which environment is the condition, clothing is the means, and people are the focus [8].

4.2 For innovative design direction, first of all, the choice of style and type. Modeling sign can distinguish the different characteristics and applicability of work clothes, and highlight their social symbols and identity symbols. The second is the choice of fabrics. Couriers often shuttle around in the outdoors, and they often encounter windy and rainy days, so we should choose comfortable, skin-friendly, moisture-absorbing and breathable fabrics. The most commonly chosen is pure cotton work clothes, as pure cotton fabrics have good characteristics, making us comfortable, cool and breathable while wearing it. On this basis, we also need to collage reflective strips. After all, couriers have to run for a long time every day in the outdoor streets or alleys. There is a certain risk factor. Setting reflective strips can increase the safety factor of couriers. In thermal environment, perspiration is very important to maintain the balance of heat. Impermeable materials will lead to the generation of heat pressure. The permeability of clothing to moisture is an important factor affecting heat exchange. Then is the choice of color. According to each express company's own personalized customization, brighter color can alleviate workers' fatigue. Finally, we need to design our own personalized logo. The corporate culture in the image of clothing is presented through printing or embroidery.

5 Summary

From the perspective of human factor engineering, we know that the starting point of the design of the couriers' work clothes is to meet the following requirements. First is to wear comfortable suits which are easy for couriers to move. The comfort of work clothes is put in the first place. The second is to have certain functions to make couriers more convenient at work. The third is safety, health, and environmental protection. The logo design of the work clothes should be proper, and the fabrics used should be healthy and environmental protection, so as to reduce irritant harm to the skin. In the research of human factor engineering, human is the most basic factor. Humans' physiological and psychological characteristics and ability characteristics are the basis of the whole system optimization. Its goal is to pay attention not only to human health, safety, and efficiency, but also to human value, comfort, satisfaction, and achievement. More attention should be paid to the research of the couriers, so that the design can truly serve consumers, so as to achieve the highest integration efficiency of the system.

References

1. Cao, X.: Ergonomics. Tsinghua University Press, Beijing (2018)
2. Gupta, D.: Design and engineering of functional clothing. Indian J. Fibre Text. Res. **36**, 327–335 (2011)
3. Kamalha, E., Zeng, Y., Mwasiagi, J.I., Kyatuheire, S.: The comfort dimension; a review of perception in clothing. J. Sens. Stud. **28**, 423–444 (2013)
4. Suh, M., Carroll, K., Cassill, N.: Critical review on smart clothing product development. J. Text. Apparel Technol. Manag. **6**(4), 199–212 (2010)
5. Rao, P.: Human factors engineering. Renmin University of China Press, Beijing (2013)
6. Huang, D.: Application of human engineering in fashion design. J. Wool Text. J. **3**, 71–74 (2018)
7. Bragançaa, S., Fontes, L., Arezesa, P., Edelmanc, E.R., Carvalho, M.: The impact of work clothing design on workers' comfort. Procedia Manuf. **3**, 5889–5896 (2015). 6th International Conference on Applied Human Factors and Ergonomics (AHFE 2015) and the Affiliated Conferences
8. Li, Y., Yang, X.: The application of clothing ergonomics in clothing design. J. Henan Sci. Technol. **11**, 32–33 (2010)

Biodigital Design and Functional Visualization of Multi-Class Personalized Compression Textiles for Ergonomic Fit

Rong Liu[1(✉)], Bo Xu[1,2], and Chongyang Ye[1]

[1] Institute of Textiles and Clothing, The Hong Kong Polytechnic University, Hung Hom, Hong Kong
rong.liu@polyu.edu.hk
[2] School of Energy and Power Engineering, Huazhong University of Science and Technology, Wuhan, China

Abstract. Functional compression textiles have been widely applied in fields of medical, healthcare, rehabilitation, sports and personal protection. The global compression wear market is expected to grow at a CAGR (Compound Average Growth Rate) of 5.1% during 2017–2022. The rising of chronic venous disorders (CVD) (e.g. varicose veins, deep vein thrombosis, lymphedema) and the steady rise of fitness and sports industry are the major drivers to push up the potentiality of compression textiles globally. However, user-oriented ergonomic fit and personalized pressure designs for precise individualized treatment have not matured. The aim of this study is to establish a new compression textiles design-development-assessment system including modular structured sections of 3D body scanning, 3D digital seamless knitting and 3D biomechanical visualization, based on medical magnetic resonance scanning images, leg anthropometry and finite element model, to enhance design efficiency, treatment precision, user fit and compliance of compression textiles in practice. The visualized pressure profiles along lower limbs lengthwise and crosswise as well as internal stress and tissue deformations can be quantitatively detected via new system to guide technical design and functional assessment of multi-class compression textiles. The pressure dosages calculated by FE model presented a favorable agreement with the experimental data, demonstrating its practicability. This new biodigital-based compression design approach raised functional performance of ergonomic design of compression textiles and enhanced our understanding on effects of varying textile design parameters on pressure dosages delivery for CVD treatment.

Keywords: Biodigital design · Biomechanical visualization · Personalized compression textiles · Ergonomic fit · Multi-pressure classes

1 Introduction

Textile-based compression modalities (e.g., compression stockings, leggings and bandages), as a significant prophylaxis and treatment approach, have become the mainstay [1] for management of leg heavy, tiredness, discomfort and chronic venous

G. Di Bucchianico (Ed.): AHFE 2019, AISC 954, pp. 488–499, 2020.
https://doi.org/10.1007/978-3-030-20444-0_51

disorders (CVD) [2–4]. The force-elongation behavior of textiles shell and geometric morphologies and tissue stiffness of lower body determine the contact conditions, thus varying resulting pressure profiles and treatment efficacy. Gradient compression stocking (GCSs) has become one of most widely applied approaches for treatment of CVD through exerting controlled pressure from distal to proximal lower limb to counteract the gravity force and to oppose the increased venous hypertension [5]. However, high non-compliance affected their efficiency and use frequency.

A large-scale invegistration on 3144 CVD patients reported that 63% of them indicated inapplicability and noncompliance [6]. Nonuse of GCSs on a regular basis has led to persistence or recurrence of symptoms [7, 8]. In recent surveys, ill-fitting GCSs is highlighted, which is considered to be related to the skills, training, and competence of clinicians carrying out the assessment [9, 10]. Initiating a hosiery fitting service was recommended by 77% of respondents for effectively improving user concordance [11].

To date personalized medicine are highly demanded for realizing customized compression therapy. Most of the commercial GCSs follow standard size charts by manufacturers or brand companies, which commonly ignore the *inter-individual* variability and discrepancies between consumers in different geographic locations. For example, heights of male in UK are generally greater than that in China by approximate 10 cm, and their body weight is greater by 31.7% in kilograms on average. Such differences in body mass and dimensions would impact accurate delivery of pressure in practical wear. Liu *et al.* conducted a digital body scanning on over 200 subjects in Hong Kong and has established a stratified body shape-driven sizing system [S^4] including classified “balanced”, “diamond”, and “inverted trapezoid” leg shapes [12], as to improve fitting of lower-body compression garment. However, to date a holistic compression design and engineering system for ergonomic fit of specific users is still wanting.

Three-dimensional (3D) digital body scanning and advanced digital knitting technologies provide a highly efficient and flexible fabrication approach to deliver tailored pressure for specific users. However, testing on wooden leg model or human body in lab is still hard to accurately assess overall pressure profile via a few pressure transducers. In our previous works, interactions between finite element (FE) lower limb and stocking has been simulated [13, 14]. In recent studies, effects of GCSs on flow and wall shear stress of deep venous system [15] and more sophisticated FE leg model have been developed [16, 17]. However, till now no study illustrates that how textile shells with different mechanical properties influence pressure function, and digital-based personalized design-knitting-assessment system has not been developed.

Therefore, in this study, a new compression textiles design-development-assessment system has been constructed and validated, including modular structured sections of 3D body scanning, digital seamless knitting fabrication and 3D digital simulation based on medical magnetic resonance scanning images, leg anthropometry and FE model. This digitized system not only contributes to improving personalized pressure design and function for raising ergonomic fit and user compliance, but also provide an operable reference for fitting diverse individual consumers’ demands in compression treatment using compression textiles for healthcare, rehabilitation, sports protection and personal care, etc.

2 Methods

2.1 3D Digital Body Scanning for Morphologic Design of GCSs

To design ergonomic-fit GCS for specific subject, precise anthropometric measurement is the first step. A female subject aged 35 years old (height: 158.5 cm; body weight: 48.3 kg; body mass index: 19.2 kg/m^2) with no visible or palpable signs of CVD was recruited. 3D digital body scanning was conducted via a VITUS^{Bodyscan} scanner (Human Solutions of North America, Cary, NC) when the subject was standing still with her feet placed at two standardized footprints on the platform in a scanning booth. A true-to-scale 3D leg model with precise and reproducible capture (point density: 200 points/cm^3, scan time: 6–10 s) was obtained, which quantitatively determined the dimensions of the key anatomic positions along leg as shown in Fig. 1.

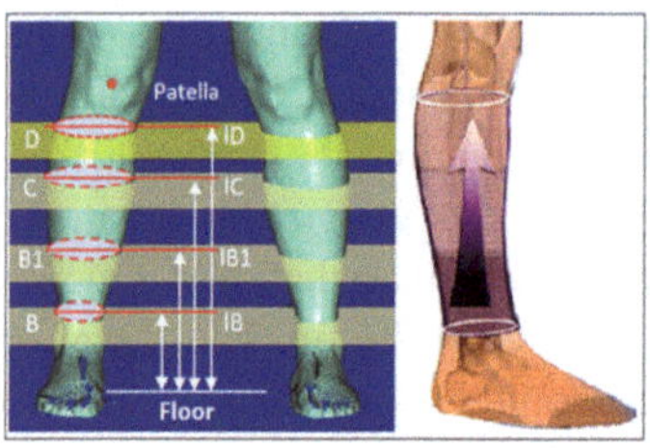

Positions	Height (cm)	Girths (cm)	Indication
Below knee	ID: 35.2	D: 33.9	Below patella center by 5 cm
Calf	IC: 28.0	C: 35.6	Calf with max. girth
Brachial	IB1: 19.1	B1: 28.3	Level at which the Achilles tendon changes into the calf muscles
Ankle	IB:10.8	B: 27.7	Ankle with min. girth

Fig. 1. 3D digital lower limb scanning of specific lower limb of the user

2.2 3D Digital Knitting Design and Fabrication of GCSs with Multi-Class Pressures

Custom-fit knee-high GCSs with multi-pressure levels (including Class I: 18–21 mmHg, Class II: 23–32 mmHg, and Class III: 34–40 mmHg) were designed, fabricated and evaluated. To achieve the proposed gradient pressure from ankle to calf, three graduation zones were designed along the hose, i.e., Zon-1, Zon-2 and Zon-3 (Fig. 2-a) using graphic system of Lonati knitting unit (45ME, Lonati S.P.A., Italy). GCS hoses were knitted by 1X1 laid-in structures via small-circular cylinder with a diameter of $4^{1/2}$ inches and set with a total of 352 needles. The polyamide covered elastomers as ground threads interlooped with polyamide covered inlay threads as illustrated in Fig. 2. GCSs in length of 260 mm were designed to achieve a minimum of 30% stretch in wale direction when being mounted on the subject's lower limb based on standard RAL 387/1.

Classes	Linear density of ground elastomers (set at all 4 feeders)		Linear density of inlay elastomers (set at the 1st and 3rd feeders)	
	Core thread	Covered threads*	Core thread	Covered threads*
I	44 dtex	44 dtex/44 dtex/40F	233 dtex	44 dtex/ 44 dtex /40F
II	44 dtex	44 dtex/44 dtex/40F	294 dtex	44 dtex /44 dtex /40F
III	178 dtex	44 dtex/44 dtex/40F	405 dtex	22 dtex/22 dtex /20F

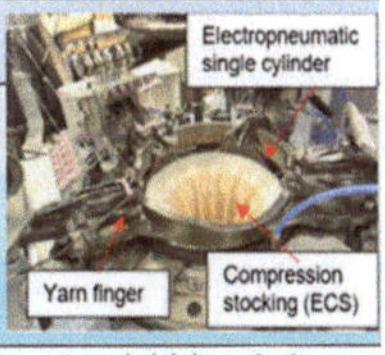

Gradient zones	Pressure residual ratio (%) **	Knitting height of zone (mm)	Mean knitting densities (stitches/inch²)		
			Class I	Class II	Class III
Zon-1	100	86	3858	3561	3032
Zon-2	100-70	84	3116	2612	2574
Zon-3	50-80	90	2632	2522	2152

1x1 laid-in stitches

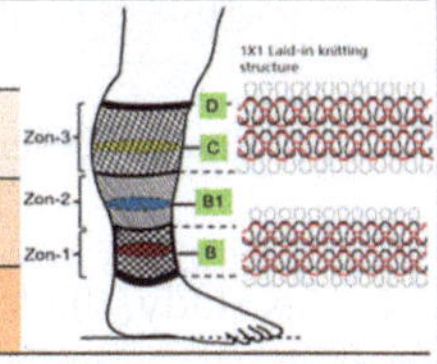

*Double covering: wrapping two spirals of non-elastic covered polyamide elastic threads in opposite directions around the elastic Lycra core. "dtex" indicates "decitex"; F indicates "filament"; **Standard GAL 387/1.

Fig. 2. 3D digital seamless knitting of GCSs with multi-pressure classes

Through adjusting the positive yarn feeding and Yoyo tension controller, the amount of elastomers can be controllably knitted to form GCS tubular with gradient pressure effects. The less the yarn fed, the smaller the loop sizes, thus the shorter the stretch, and higher the compression will be delivered.

2.3 3D Biomechanical Visualization of GCSs with Multi-Pressure Classes

Pressure performances were traditionally assessed via individual pressure transducers on a wooden leg model, which lacked capabilities to holistically reflect surface pressure and internal stress of compression exerted on user's lower limb. In this study, a subject-specific FE model has been developed to digitally visualize and intuitively evaluate GCSs with multiple pressure classes.

(i) Analysis of biomechanical interaction of GCS-lower limb. The interactions between lower limb and compression shell can be differentiated into three fundamental components, i.e., (a) GCS shell is stretched when being mounted to lower limb, exerting pressure on limb surface; (b) pressure compresses soft tissue and induces tissues deformation and internal stress redistribution, and (c) inner stress and skin pressures alter as GCS shell slide upward along lower limb. The contact mechanism between the two bodies (GCS and lower limb) can be depicted in Fig. 3.

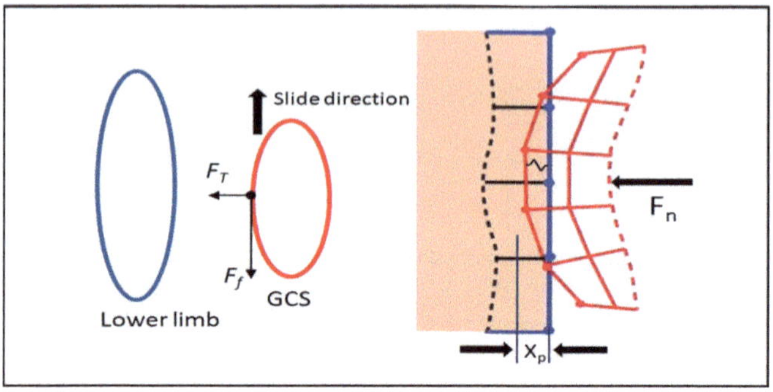

Fig. 3. Description of contact mechanism between GCS and lower limb

In this study, the lower limb and the GCS are assumed as hyperelastic models. The two interactive bodies follow a frictional non-linear connection, and the sliding resistance is proportional to the frictional coefficient. The contact pressure can be computed as

$$u(x,y) = \frac{1-\mu^2}{\pi E}\iint_{\Omega} \frac{F_T(x',y')dx'dy'}{\sqrt{(x-x')^2+(y-y')}} \tag{1}$$

where F_T is contact pressure, u is frictional coefficient, E is Young's modulus, μ is Poisson's Ratio, and u(x, y) is the surface displacement [18]. It is assumed that the interactive contacts of the two bodies do not penetrate each other, then penalty method was used to incorporate the contact condition into the finite element discretization to prevent penetration. The penalty method equation can be expressed as

$$F_T = K_n x_p \tag{2}$$

where k_n is contact stiffness, and x_p is penetration depth [19].

(ii) Reconstruction of geometries of subject-specific lower limb and GCS. The geometry of the subject's right lower limb was reconstructed from coronal images via Magnetic Resonance Imaging (MRI) scanning. The MR image slices in thickness of 2.0 mm and an in-plane voxel size of 0.98 mm × 0.98 mm were taken when the subject was in supine position. The MR images were processed via MIMICS software (v17.0, Materialise, Leuven, Belgium). Soft tissue and rigid bones are the two main components involved in FE model. The boundary surfaces of the two components, as solid sub-models, were processed via SolidWorks (v2013, Massachusetts, US). The geometric shape and gradient zones of GCS were reconstructed using FE Package ANSYS Workbench (v15.0, ANSYS, Inc.) (Figure 4) based on leg morphologies determined in section (i).

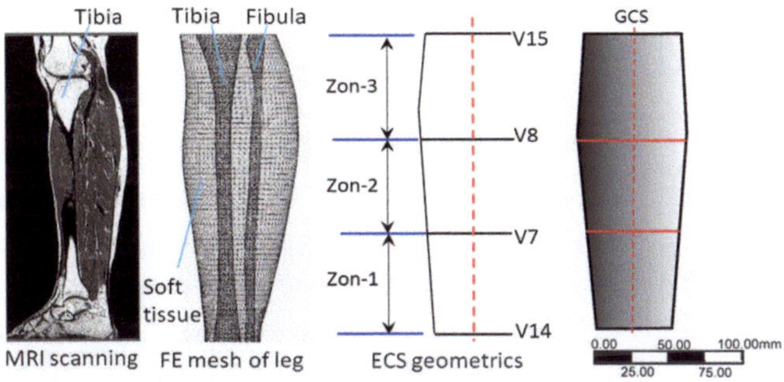

Fig. 4. FE leg mesh and gradient compression stocking (GCS) geometry buildup

(iii) Determination of material properties of FE sub-leg and sub-GCS models. Mechanical properties of soft tissues was modeled via Neo-Hookean hyperelastic model, in which $1/D_1$ and C_{10} were 7.14 MPa and 0.005 MPa, respectively [16], and the bone is designed as homogeneous, isotropic and linear elastic materials. The bone's young's modulus and Poisson's ratio were set to 14.0 GPa and 0.3 [20, 21], respectively. The GCS textile shell was assumed as homogeneous, orthotropic and linear elastic materials. The mechanical properties of GCSs (Young's modulus, shear moduli, mass density, Poisson's ratio and thickness) were obtained through instrumental tests under a standard condition (ISO13934-1:1999). Biaxial tensile tester (KES-G2) and shear tester (KES-FB1) (Kato-Tec Co., Ltd., Japan) were used to determine tensile and shear moduli of GCS fabrics. Table 1 illustrates the mechanical properties of the studied GCSs with multiple pressure classes, which were assigned to the FE model to visualize biomechanical function of different GCSs.

(iv) Finite element mesh and boundary condition. Digital biomechanical assessment of GCS was conducted using ANSYS Workbench modeling system (v15.0, ANSYS, Inc.). The GCS shell was meshed using 4-node quadrilateral shell elements, and the lower limb was meshed using 8-node hexahedron elements. The displacements of bony parts were zero in all degrees of freedom at x, y, z directions. The GCS shell slided onto the leg with displacement of 100 mm from the ankle to top edges of leg in pulling. When GCS reached stable status, the contact between leg and GCS was activated. There was no displacement boundary on the GCS at the end of the simulation.

(v) Validation on the developed subject-specific FE model. Interfacial pressures between GCS and the subject's lower limb were measured *in vivo* to validate the developed subject-specific FE model. PicoPress® pressure sensor was used to test skin pressures at twelve typical locations over three key height levels (ankle, brachial and calf) and around four key directions (anterior, posterior, medial and lateral) underneath GCSs with multi-pressure classes along leg surface in standing position. The average values of the three times of pressure tests were employed for model validation.

Table 1. Mechanical properties of GCS materials with multi-pressure classes

Pressure Classes	Gradient levels	W ($\times 10^{-10}$ tonne/mm^3)	E_1 (MPa)	E_2 (MPa)	G_{12} (MPa)	V	T (mm)
(I) Mild	Ankle-B	3.25 ± 0.12†	0.28 ± 0.02	0.19 ± 0.02	0.11 ± 0.01	0.27 ± 0.02	1.0 ± 0.1
	Brachial-B1	2.75 ± 0.11	0.21 ± 0.02	0.15 ± 0.01	0.08 ± 0.01	0.33 ± 0.03	
	Calf-C	2.30 ± 0.10	0.15 ± 0.01	0.10 ± 0.01	0.05 ± 0.01	0.37 ± 0.03	
(II) Moderate	Ankle-B	3.57 ± 0.13	0.35 ± 0.02	0.21 ± 0.02	0.14 ± 0.01	0.25 ± 0.02	1.1 ± 0.2
	Brachial-B1	3.02 ± 0.13	0.27 ± 0.03	0.15 ± 0.01	0.11 ± 0.01	0.27 ± 0.02	
	Calf-C	2.53 ± 0.11	0.19 ± 0.0	0.09 ± 0.01	0.07 ± 0.01	0.29 ± 0.01	
(III) Strong	Ankle-B	3.74 ± 0.14	0.41 ± 0.03	0.17 ± 0.02	0.17 ± 0.01	0.23 ± 0.01	1.2 ± 0.2
	Brachial-B1	3.16 ± 0.13	0.32 ± 0.03	0.13 ± 0.01	0.13 ± 0.01	0.25 ± 0.02	
	Calf-C	2.65 ± 0.11	0.22 ± 0.02	0.08 ± 0.01	0.08 ± 0.01	0.27 ± 0.01	

Note: W: mass density; E1 and E2: Young's modulus in wale and course directions of fabric; G_{12}: shear modulus; V: Poisson's ratio; T: mean thickness of fabric specimen; †Mean ± standard deviation of the six measured values.

3 Results and Discussion

Figure 5 illustrated the comparison results between the simulated and the measured interfacial pressures at the twelve positions along subject's lower limb beneath GCS with three pressure classes (I, II and III). Most of the testing points presented consistent trends, especially for Class I, which demonstrated the reliability of the developed FE model in biomechanical assessment of GCS. The larger deviations at B1 and C beneath GCS with classes II and III were potentially due to bony anatomic structures at anterior regions of the lower limb.

Figure 6(a–d) quantitatively displayed skin pressures along anterior and posterior sides illustrating more clearly gradient distributions and magnitudes under exertion by GCSs. Skin pressures near anterior tibia crest and posterior Achilles tendon were remarkably higher than those at brachial and calf levels along the same direction lengthwise. Reversed gradient pressures occurred at medial and lateral sides along the lower limb (Fig. 5b–d), where calf pressures were generally higher than those at ankle, largely due to the relative flatter anatomic surface above bilateral tarsus with ellipse cross-section.

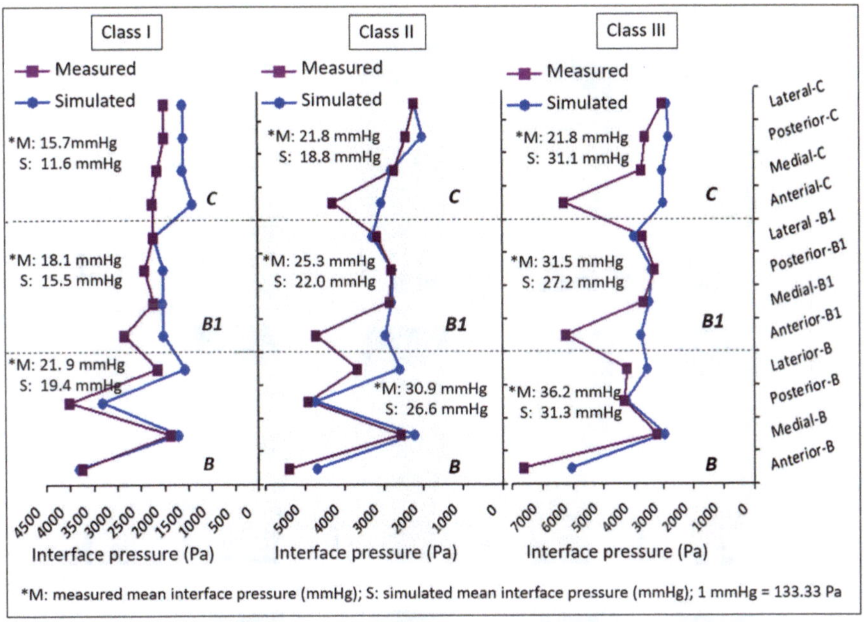

Fig. 5. Comparison on the simulated and measured skin pressures by GCSs with multi pressure classes

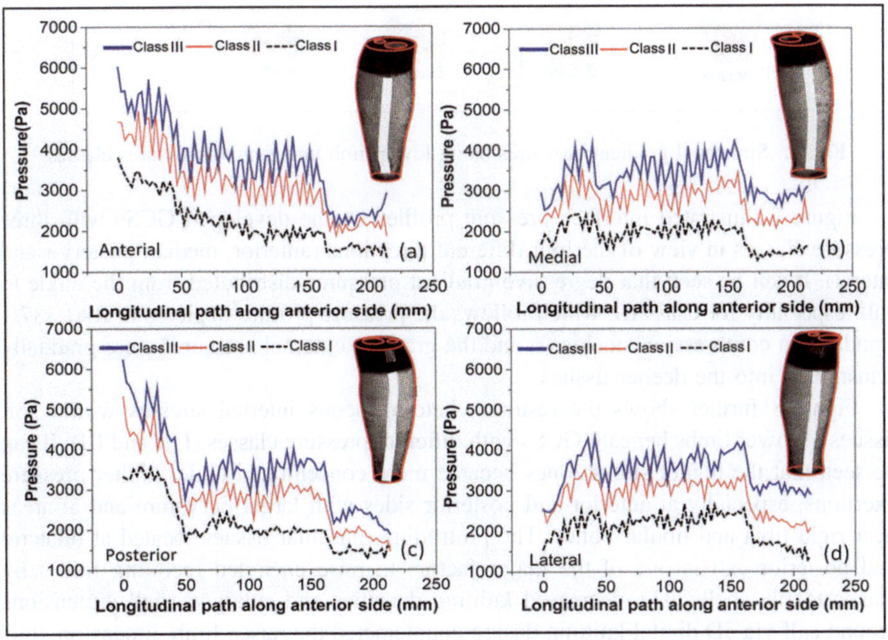

Fig. 6. Quantitative skin pressure gradients along four longitudinal paths of lower limb

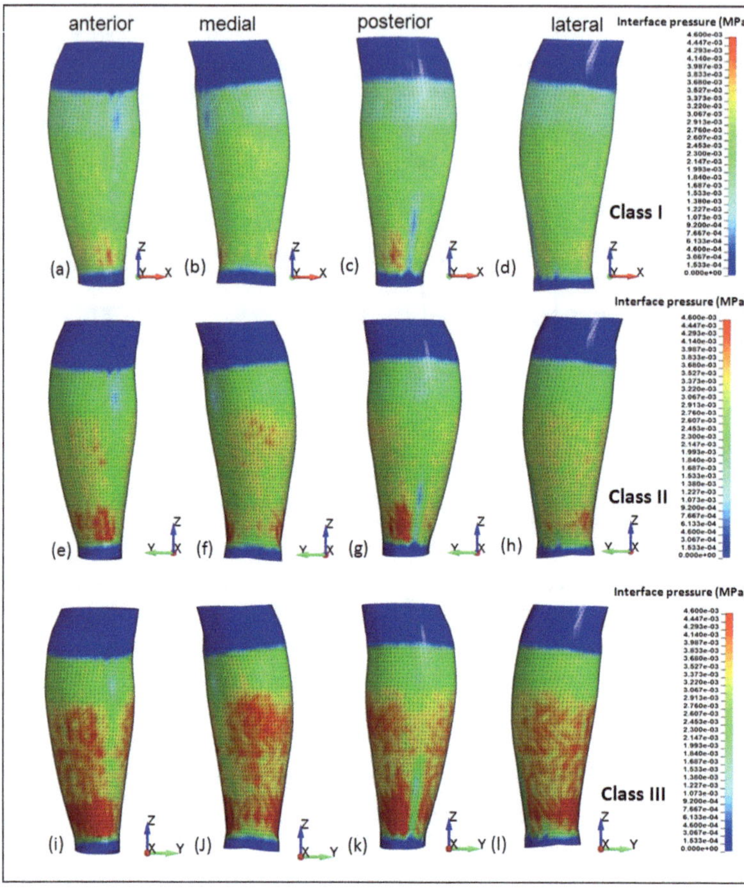

Fig. 7. Simulated gradient pressures along lower limb under multi-pressures classes

Figure 7 illustrated interface pressure profiles by the developed GCSs with three pressure classes in view of the four different directions (anterior, medial, posterior and lateral). It can be seen that degressive gradient pressures distributed from the ankle to calf, especially for Class III, which follows the pressure profiles depicted in RAL387/1 standard on compression stockings, and the gradient external pressures were gradually transmitted into the deeper tissues.

Figure 8 further shows the resultant heterogeneous internal stresses within soft tissues of lower limbs beneath GCSs with different pressure classes (I, II and III). It can be seen that the higher stress zones became more concentrated under higher pressure exertions, especially at anterior and posterior sides with larger curvature and at areas near rigid tibia and fibular bones. The protruding muscular tissues located at bilateral and posterior calf is one of the major factors to raise encircled recoiling forces by elastic textile shells. The decreased knitting densities and enlarged shell dimensions around calf via 3D digital knitting design counteracted the raised limb dimension, thus generating degressive pressure profiles from ankle to calf to prevent tourniquet effect which could impede venous return.

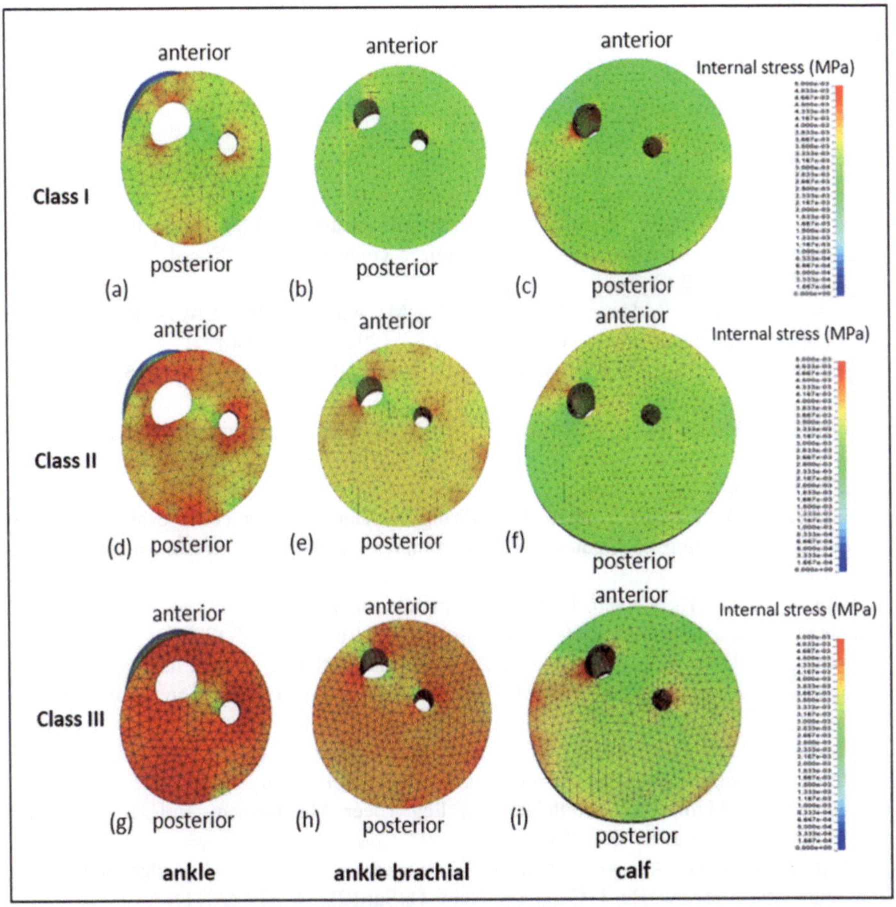

Fig. 8. Internal tissue stress distributions by GCSs with multi-pressure classes

4 Conclusion

Design and development of functional compression textiles is a complex process which requires interdisciplinary cooperation among textile engineering, ergonomic design and biomedical knowledge to cater for personalized compression therapeutic requirements. The present study demonstrated a new approach for designing, fabricating and assessing performances of compression textiles to explore more possibilities of computational-based design and development in modern customized healthcare and medical industries. Applying 3D digital body scanning, hybrid elastic materials, seamless knitting fabrication and finite element modeling, personalized compression textiles (e.g., GCSs) can be developed to realize the expected pressure dosage delivery and ergonomic fit. The digital model visualized pressure and stress profiles with varying textile parametric design, which would largely facilitate pressure design efficiency, and enhance our understanding on effects of material properties on pressure dosage delivery for treatment of chronic venous disorders in practice.

Acknowledgment. This work was supported by General Research Grant (RGC) PolyU 252153/18E and Central Research Grants of the Hong Kong Polytechnic University through Research Projects PolyU 252064/17E and 1-ZVLQ.

References

1. Amsler, F., Blättler, W.: Compression therapy for occupational leg symptoms and chronic venous disorders-a meta-analysis of randomized controlled trials. Eur. J. Vasc. Endovasc. Surg. **35**, 336–372 (2008)
2. Caggiati, A., Rosi, C., Heyn, R., Franceschini, M., Acconcia, M.C.: Age-related variations of varicose veins anatomy. J. Vasc. Surg. **44**(6), 1291–1295 (2006)
3. Perlowski, A.A., Jaff, M.R.: Vascular disorders in athletes. Vasc. Med. **15**(6), 469–479 (2010)
4. Smyth, R., Aflaifel, N., Bamigboye, A.: Interventions for varicose veins and leg oedema in pregnancy. Cochrane Database Syst. Rev. (2015). https://doi.org/10.1002/14651858.CD001066.pub3
5. Liu, R., Kwok, Y.L., Li, Y., Lao, T.T., Zhang, X., Dai, X.Q.: Objective evaluation of skin pressure distribution of graduated elastic compression stockings. Dermatol. Surg. **31**(6), 615–624 (2005)
6. Raju, S., Hollis, K., Neglen, P.: Use of compression stockings in chronic venous disease: patient compliance and efficacy. Ann. Vasc. Surg. **21**(6), 790–795 (2007)
7. Franks, P.J., Oldroyd, M.I., Dickson, D., Sharp, E.J., Moffatt, C.J.: Risk factors for leg ulcer recurrence: a randomized trial of two types of compression stocking. Age Ageing **24**, 490–494 (1995)
8. Mayberry, J.C., Moneta, G.L., Taylor Jr., L.M., Porter, J.M.: Fifteen-year results of ambulatory compression therapy for chronic venous ulcers. Surgery **109**, 575–588 (1991)
9. Moffatt, C., Martin, R., Smithdale, R.: Leg Ulcer Management, vol. 10. Blackwell Publishing Ltd, Oxford, UK (2007)
10. Martin, F., Duffy, A.: Assessment and management of venous leg ulceration in the community: a review. Br. J. Commun. Nurs. **16**(Sup12), S6–S14 (2011)
11. Tandler, S.F.: Challenges faced by healthcare professionals in the provision of compression hosiery to enhance compliance in the prevention of venous leg ulceration. EWMA J. **16**(1), 29–33 (2016)
12. Liu, R., Guo, X., Peng, Q.J., Zhang, L., Lao, T.T., Little, T., Liu, J.D., Chan, E.: Stratified body shape-driven sizing system via three-dimensional digital anthropometry for compression textiles of lower extremities. Text. Res. J. **88**(18), 2055–2075 (2018). https://doi.org/10.1177/0040517517715094
13. Liu, R., Kwok, Y.L., Li, Y., Lao, T.T., Dai, X.Q., Zhang, X.: Numerical simulation of internal stress profiles and three-dimensional deformations of lower extremity beneath medical graduated compression stocking (GCS). Fiber Polym. **8**(3), 302–308 (2007)
14. Liu, R., Kwok, Y.L., Li, Y., Lao, T.T., Zhang, X., Dai, X.Q.: A three-dimensional biomechanical model for numerical simulation of dynamic pressure functional performances of graduated compression stocking (GCS). Fiber Polym. **7**(4), 389–397 (2006)
15. Wang, Y., Downie, S., Wood, N., Firmin, D., Xu, X.Y.: Finite element analysis of the deformation of deep veins in the lower limb under external compression. Med. Eng. Phys. **35**(4), 515–523 (2013)

16. Dubuis, L., Avril, S., Debayle, J., Badel, P.: Identification of the material parameters of soft tissues in the compressed leg. Comput. Methods Biomech. Biomed. Eng. **15**(1), 3–11 (2012). 1025–5842
17. Rohan, P.Y., Badel, P., Lun, B., Rastel, D., Avril, S.: Prediction of the biomechanical effects of compression therapy on deep veins using finite element modelling. Ann. Biomed. Eng. **43** (2), 314–324 (2015)
18. Wang, F., Keer, L.M.: Numerical simulation for three dimensional elastic-plastic contact with hardening behavior. J Tribol-T ASME **127**(3), 494–502 (2015)
19. Carpenter, N.J., Taylor, R.L., Katona, M.G.: Lagrange constraints for transient finite-element surface-contact. Int. J. Numer. Methods Eng. **32**(1), 103–128 (1991)
20. Cheung, J.T., Zhang, M., Leung, A.K., Fan, Y.B.: Three dimensional finite element analysis of the foot during standing – a material sensitivity study. J. Biomech. **38**(5), 1045–1054 (2005)
21. Reilly, D.T., Burstein, A.H.: The mechanical properties of cortical bone. J. Bone Joint Surg. Am. **56**(5), 1001–1022 (1974)

The Application and Development of Smart Clothing

Jia Lyu[1,2], Yue Sui[3(✉)], and Dongsheng Chen[1]

[1] College of Clothing and Artistic Engineering, Minjiang University, Fuzhou, China
[2] Collaborative Innovation Center of Modern Clothing Technology, Minjiang University, Fuzhou 350121, China
[3] Department of Physics and Electronic Information Engineering, Minjiang University, Fuzhou 350121, China
arthur_sui@163.com

Abstract. Smart clothing is a hot research topic in recent years, it is also the application and exploration of intelligent manufacturing in the textile and clothing field. This paper analyzed the design focus and the development trend of the smart clothing through expounding the application, the characteristics, and the domestic and international research progress of smart clothing.

Keywords: Smart clothing · Design focus · Application · Future development

1 Introduction

A prominent feature of the Made in China 2025 strategic mission is the emphasis on upgrading and transforming traditional industries through intelligent manufacturing. As an indispensable part of people's lives, the functional attributes of clothing are expanding constantly with the increasing competition in the clothing industry and the development of modern research technologies. Based on the traditional clothing industry, smart clothing is formed and developed by integrating the concepts and technologies of material science, computer engineering, and fashion design. Its main task is to use new technology to change and improve the traditional functions of clothing.

2 The Basic Concept of Smart Clothing

As an emerging interdisciplinary field, wearable technology mainly explores and creates science and technology that can be worn directly on the human body or integrated into the user's clothing or accessories [1]. Clothing is an essential part of people's daily life. The idea of smart clothing integrates wearable technology into traditional clothing to provide clothing with dual functions of perception and feedback [2], that can not only sense the external environment or internal changes but also react to those changes in real time through a feedback mechanism.

G. Di Bucchianico (Ed.): AHFE 2019, AISC 954, pp. 500–504, 2020.
https://doi.org/10.1007/978-3-030-20444-0_52

Smart clothing is mainly developed in two ways: one is to use intelligent clothing materials, such as shape memory materials, phase change materials, color-changing materials, etc. in clothing design and production. The other is to introduce sensing technology, microelectronic technology and information technology, such as flexible sensors, low-power chip, and low-power wireless communication devices, into everyday wear to perform certain medical testing, track user's vital signs (i.e. heart rate and other vital signs), or to provide fitness guidance (such as calorie consumption, stress level, heart rate).

3 Application of Smart Clothing

Smart clothing was originally created for special occasions, such as military applications to improve combat defense capabilities. With the development of science and technology and the increasing market demand, smart clothing not only plays an essential role in the military field, but also is used in medical field, high-performance sportswear, and high value-added casual wear.

In the past few years, communication and sensors technology have been integrated into clothing and accessories design such as shoes, shirts, belts, and even jewelry or watches to meet the different needs of people or occasions [3–5]. Overall, the application of smart clothing is mainly in two areas: medical monitoring and sports and fitness.

3.1 Medical Monitoring

Smart clothing represents an advance health care system, which attaches mobile devices to everyday wear to provide uninterrupted information transmission between patients and remote telemedical providers through information and communication technologies. Smart clothing can help provide medical support, collect clinical data, and monitor important life characteristics of the wearer. It can be used for collecting important health parameters for the elderly and chronically ill patients and transmitting the information to the smartphone or computer of the wearer or his or her family member(s) via LAN. Therefore, based on the information, remote telemedical provider can decide whether it is necessary to carry out emergency treatment and/or issue a medical alert in time. The home-based e-health care can help the elderly or chronically ill patients to obtain real-time medical monitoring at home, and the diagnosis and treatment can be performed between the patient and the doctor anytime and anywhere. The United States launched the marketing of smart shirts in 1999 [6]. The shirt was equipped with sensors to measure heart rate and lung activities. In addition, it also had an option to transfer the medical data to the control room via the Internet and provide data analysis.

3.2 Sports and Fitness

Utilization of smart clothing in the field of sportswear is mainly embodied in new fiber materials (such as antibacterial fiber, phase change fiber, etc.) and smart devices (such as heart monitoring sports bras). The Athos smart sportswear and shorts developed by the University of Waterloo in Canada [7] incorporated the myoelectric motion sensor into the fabric to analyze the physiological data, such as the movement progression and muscle fatigues of the body part, by measuring the fluctuation range of muscle fiber contraction. The information is then transmitted to the software on the user's smartphone to provide an evaluation of movement progression and monitor the accumulation of stress on muscles throughout training in the form of various charts.

4 Design Focus and Development Trend of Smart Clothing

The design of smart clothing is different from the design of ordinary clothing. In addition to aesthetic performance, factors such as wearability, usability, reliability and durability also need to be considered. The design focus of smart clothing is summarized in the following aspects.

4.1 People-Oriented

In the process of transforming the seller's market into the buyer's market, the organic combination of the various elements of the clothing must satisfy the subjective needs of people's psychological and physiological needs on the premise of satisfying their objective needs. People's acceptance of new technologies is influenced by their sociality, values, gender, and educational experience. As the material carrier of wearable devices, smart clothing is designed to meet the diverse needs of users. Therefore, the people-oriented ergonomic design conducted by observing users' acceptance of smart devices and parameters from their cognition and emotions assessment becomes a problem that needs to be focused on and resolved [8, 9].

4.2 Wearability

In order to broaden the utilization of smart clothing, the overall design shall meet the basic need of comfortability as while as satisfying the functionality of smart clothing. Foreign researchers pay special attention to the comfortability in the design of smart clothing, which include the thermal and humidity comfort of the clothing, the sensory comfort of the contact of sensor and human skin, and the safety of use. Using the flexible sensing technology, the sensor and the apparel fabric are integrated into one, which becomes the basic fabric of smart clothing. It can sensitively understand and react to the changes of various physiological parameters and environmental parameters of human body.

4.3 Enhance Human Body Function

The emergence of smart clothing has also played an important role in improving human body functions to a large extent. A Japanese technology company has recently developed a set of exoskeleton, which provides external strength to the human body when lifting heavy objects through a hydraulic pump device. It is mainly used to help the elderly with limited mobility or partial self-care ability, or people doing longtime physical labor work. The user's physical body function has been improved significantly.

4.4 Accessibility

The accessibility of smart clothing means that the product provided to the end user must be easy to operate and convenient to use. Especially for the elderly and the chronically ill population, the simple and convenient experience is especially important.

4.5 Ergonomics

Smart clothing has many forms to serve different purposes and requirements, but the core factor is the human body. Human has four core elements: shape, movement, physiology, and psychology. Therefore, ergonomic requirements should be considered when designing smart clothing, so that the human body can maintain the most natural movements while wearing smart clothing.

4.6 Provision of Software and Services

Affected by the Internet and Industry 4.0, the development of smart clothing is based on software systems and Internet platforms and is combined with the Internet of Things, big data, and cloud computing to achieve intelligence. The development of smart clothing has also integrated with mobile computing, application software and social media to achieve on-demand customization, such as based on the information transmitted through the LAN to the patient's smartphone and computer, the doctor can decide whether emergency treatment is needed according to the situation.

5 Conclusion

Smart clothing is not only one of the implementation strategies of Made in China 2025, but also one of the means to upgrade and transform the clothing industry. Therefore, an era of rapid development of smart clothing is coming. As one of the manifestations of wearable devices, smart clothing focuses on "wearability". At present, smart clothing has various forms, but the general problem is that the wearability is not high. Therefore, how to improve the wearability and aesthetics of smart clothing as well as how to broaden the applications of smart clothing become very important in the design and the development of smart clothing in the future.

Acknowledgement. This study has been supported by the Fuzhou Science and Technology Project (No. 2018-G-99), the Collaborative Innovation Center of Modern Clothing Technology, Minjiang University (No. MJKFFZ201704) and the Outstanding Young Scientific Research Talents of Universities in Fujian Province (2017).

References

1. Smith, D.: Smart clothes and wearable technology. AI & Soc. **22**, 1–3 (2007)
2. Kiekens, P., Langenhove, L.V., Hertleer, C.: Smart clothing: a new life. Int. J. Cloth. Sci. Technol. **16**(1/2), 63–72 (2004)
3. Linti, C.: Sensory baby vest for the monitoring of infants. In: Horter, H., Österreicher, P., et al. (eds.) International Workshop on Wearable and Implantable Body Sensor Networks, pp. 135–137. IEEE Computer Society Press, Cambridge (2006)
4. Scheermesser, M.: User acceptance of pervasive computing in healthcare: main findings of two case studies. In: Kosow, H., Rashid, A., et al. (eds.) Proceeding of the Second International Conference on Pervasive Computing Technologies for Healthcare, Tampere, pp. 205–213 (2008)
5. Asada, H.H., Shaltis, P., Reisner, A., et al.: Mobile monitoring with wearable photoplethysmographic sensors. IEEE Eng. Med. Biol. Mag. **22**(3), 28–40 (2003)
6. Online Life-Shirt System [EB/OL]. http://www.vivometric.com/
7. Online ATHOS [EB/OL]. https://www.liveathos.com/
8. Ziefle, M., Röcker, C.: Human-Centered Design of E-Health Technologies: Concepts, Methods and Applications. IGI Global, Hershey (2010)
9. Ziefle, M.: Acceptance of pervasive healthcare systems: a comparison of different implementation concepts. In: Röcker, C. (ed.) 4th ICST Conference on Pervasive Computing Technologies for Healthcare, Munchen, pp. 1–6 (2010)

A Classification of Care Apparel Characteristics for Disabled User Requirement

Qilong Feng(✉)

Institute of Textiles and Clothing, The Hong Kong Polytechnic University, Hung Hom, Hong Kong
zoey_feng@163.com

Abstract. Care apparel market for disabled customers in Hong Kong is substantially grown up over last decade. Disabled users could hardly find their required apparel products. It was found that the clothing needs of disabled users tend to have the functional and aesthetic elements of apparel product, which are key factors to determine the feasibility of production method. The study aims at to classify the apparel characteristic based on disability level of users, in order to assist care apparel development in production and market. Observation and interviews were made with the disabled customers. The result revealed that the user expectations for apparel product were classified into three groups based on characteristics of current products.

Keywords: Care apparel · Disabled user · Apparel market · Apparel design

1 Introduction

Nowadays, the apparel industry is converting from manufacture-oriented to customer-oriented, in which consumer requirements has determined the production process. The demand of products from the disabled people is increasing, but the supply of satisfactory specially design apparel is still not efficient, and the situation is even worse for the disabled apparel market in Hong Kong. From the experience in the Care Apparel Center in Hong Kong, which offers special design with both medical requirements and appearance satisfaction to maintain the well-being of disabled users, it is extremely hard for the disabled customers to find proper care apparel products in the market. Questions has been raised about why the accesses to care apparel could not satisfy the demand and requirements of the disabled users. Referring to the previous production of the care apparel center workshop, the lead-time could be as long as several months, depending on the complexity of the product design. Product design and characteristic is the key factor to choose proper manufacture method. Therefore, a study has been designed to identify disabled user requirement, and further investigate the care apparel characteristics, in order to prepare a foundation of proper production selection.

1.1 Care Apparel Definition and Features

The awareness of apparel for the disabled people has been proposed since the 1980s, during which the topic of 'what to wear' was discussed about the apparel design

G. Di Bucchianico (Ed.): AHFE 2019, AISC 954, pp. 505–514, 2020.
https://doi.org/10.1007/978-3-030-20444-0_53

adjustment for disabled elders [1], and it led to the discussion of clothing design. Adaptive clothing, as defined by the 'Disabled world', are designed for people who have difficulty in dressing themselves, including the ones with disabilities, elderly, and patients [2]. 'Inclusive design and product' is introduced in the year 2002 by Keates and Clarkson [3], successfully focuses on the needs, wants, and aspirations of other users (aging population, marginalized minorities, and people with disabilities), and arises conscientious attention to both mainstream market for the abled users, and all other consumers, including the disabled, children, and the older segment [4].

Care apparel, as a part of the concept, has been further identified in Care Apparel Center. It has been defined for users with difficulty in dressing or other special clothing needs [5]. The products of care apparel have three features in common:

(1) The clothing designs permit more independence and enabled the wearers to take care of their personal needs as much as possible.
(2) The clothing can meet functional and appearance requirements, thereby contributing to the social adjustment of the wearers.
(3) Psychological and physical comfort are taken into consideration in order to enhance a sense of well-being for the wearers.

In the review of Ng et al. [6], medical garment (previous name of care apparel) are mainly for three purposes: (1) for protective function purpose, (2) for treatment function purpose, and (3) for care function purpose. Caring function design in the apparel has been the key consideration in care apparel design to maintain the well-being of disabled users.

1.2 Disabled User Classification

The disabled people suffer from different challenge. The evaluation of disability level is used as a measurement in some of studies regarding product development for the disabled users. In general, two approaches were discussed and refined for the measurement. One is to identify the disability level directly into the seriousness of body functions, such as Washington Group general measurement on disability [7]; and the other is to assess the adaptive capability of the disable person into real life activities, as described in Activities of Daily Living (ADLs) [8].

The earliest appeared definition regarding disabled patients' recovery and functional ability is Barthel Index (BI) [9] and modified Barthel Index (MBI) [10, 11]. It has been considered as the golden standard to assess one's performance in self-care tasks, and in the criteria, there would be three to four scales for assessment. Similar index was also developed as Katz's ADLs index [12], in which six elements were examined and scored in difficulty scale and assistance scale separately. The scope covers core activities including bathing, dressing, transferring, using the toilet, continence, and eating [13], and later was extended with more activities such as shopping, cooking, managing money, mobility [14], with more social activities and participation of disabled persons.

Based on the foundation of BI model, Wang et al. [15] have proposed an evaluation system of functional clothing for wheelchair users, referring to a five-scale frame of activity of daily living (ADL) in order to measure the personal capability to handle apparel dressing on and off of the users, which was used in the investigation of trousers trials. The system has divided the personal capability into the levels as: (1) totally dependent, (2) very dependent, (3) partly dependent, (4) slightly dependent, and (5) independent.

1.3 Apparel Characteristics and Production

Assessing the care apparel product should consider both production feasibility and customer expectations. The criteria of apparel design for a variety consumer needs include three elements: functional, expressive, and aesthetic, as the framework combined with the culture background of the target consumer [16]. Especially for the physically disabled user, the functional element and expressive elements determines their reaction in choosing [17]. Function and aesthetics are also identified to be connected in the design of garments for disabilities [18]. Yu and Rahman [19] has proposed the attributes of inclusive apparel for consumer shopping experience, including tangible attributes and intangible attribute. Tangible product attributes include design/styling, comfortability/fit, and material, and intangible attributes include price and product assortment. Product development oriented factors such as comfort, fit, style, value/low cost, and function are most desired attributes to the consumers when accepting the garment [20, 21].

Regarding the various consumers in mainland China and Hong Kong, Yu and Rahman [19] has generated two types of customers based on psychological appeals: (1) practical and convenience-seeking consumers, and (2) fashionable and hedonic consumers, that different groups have various criteria regarding care apparel purchase. Eleven items include utility, comfort, price, appearance, easy to wear, non-outstanding, fitting, easy care, material, durability, and delivery have been identified to be related with disabled user purchase preference [22].

The selection of production strategy is based on the feature of the apparel product. Currently mainstream method in apparel productions include mass production, customization, and mass-customization, which was discussed in many studies. Mass production is low-cost manufacture determined by high demand capability, and also the long runs of identical product based on market orientation and specialization of labor [23]. As the opposite, customization 'uses a flexible production system to deliver a product to our that matches the needs of an individual customer or user', in order to maximize user satisfaction through product design [24]. Mass customization is the strategy combining mass production and customization in order to provide tailored products and service to fulfill the individual consumers based on specific needs and wants [25]. Techniques and design will vary between the production application, which is to be considered with apparel characteristics.

2 Methodology

In this study, a thorough evaluation over care apparels in the Care Apparel Center which is located in the Hong Kong Polytechnic University has been conductions. In order to clearly understand the purpose of choosing care apparel for disability users, their capability in performing daily activities are also considered. First, disability level of different users was defined based on the previous studies, dividing the capabilities into three different categories: low disability level, moderate disability level, and high disability level. Observations about the disability level for each person, individual purchasing preference, and ordered care apparel types and characteristics were recorded for the further classification of care apparel products.

The care apparel characteristics regarding functional design and aesthetic design were evaluated. Details in the apparel design including cutting, structure, opening, closure, accessories were also extracted. Similarities in the deign to fulfill several purposes have been identified, and a classification of the care apparel characteristics from the perspective of production feasibility has been generated. The information from the investigation was summarized to generate an overall classification of care apparel features.

2.1 Disabled User Observation

Disabled clients to the Care Apparel Center have been observed with their preference, personal requirements from care apparel products. The disability level model from Modified Barthel Index (MBI) [10, 11, 15] have been modified in this study, setting the criteria regarding dressing capability of care apparel users in to three levels: (1) low disability level, (2) moderate disability level, and (3) high disability level.

Interviews were conducted after each client having explained about their personal description on disability challenges and product expectations. Open-ended questions were prepared for each client, and the disability level was assessed after they have purchased the order for specific care apparel. The procedure was processed in an experience-base direction, because the clients and their required care apparels were randomly collected from the regular activity in the care apparel center, which was difficulty to control with the amount and detailed requirements.

2.2 Care Apparel Characteristic Classification

Care apparel characteristics were investigated in two directions. Firstly, the description of disabled clients was recorded for analysis. The interviews about each personal requirements and expectations regarding individual conditions have been reviewed in the form of text, and key words have been extracted to identify the determinants of care apparel characteristics for disabled user consideration. Understanding of disabled client types were referred from study of Yu and Rahman [19], and characteristic extraction has integrated the FEA model of Lamb and Kallal [16] and Xie et al. [22].

Secondly, historical orders have been investigated to identify the most frequently ordered products with specific detail modifications. Product type and characteristics in design and application were identified into a general list. Similar types and functions regarding characteristics were classified using the list of all care apparel information. The classification was further applied with the disability level settings of the care apparel users. Finally, a summary of care apparel characteristic regarding different disability levels and feasibility production method was proposed.

3 Result

3.1 Care Apparel Overview

The feature of the garment provided for disabled users are different in design, pattern, material. Application of easy-handling accessories for disabled people are also various. The investigation in the Care Apparel Center in Hong Kong has collected data with totally fifty-four types of products to identify characteristics. Then they were discussed with its functions and production procedures.

In the design of care apparel, structures are adjusted to adapt with the special body movement restrictions. Openings are changed to the side seams in order to add up the easiness in dressing through arms and legs. Accessories such as hooks or buttons are also substituted by sticking fabrics with its soft surface friendly to the sensitive skins of the users, or changed into snap-fastener that could quickly assist the nursing service of the care givers. As a summary, from the perspective of design and material, the care apparel has the following features:

- *Use zippers and VELCRO® fastener as main closure design*
- *Side-open and back-open for easy do-on and do-off*
- *Elastic bands to substitute fixed length waist*
- *Magnetic buttons and zippers to keep the closure oriented*
- *Less structural cutting*
- *Adjusted measurements in tightness or allowance*
- *Knitted fabric to keep warmth and softness*

The similarity in the designs has showed the possibility of production optimization. Even though the structures of the care apparel are different from the regular fashion styles, the large demand from potential consumers still form a foundation of mass production into the market.

3.2 Interview About Care Apparel User Preference and Expectation

An interview with randomly selected disabled people and care givers was conducted to identify the preference and expectations about care apparels. In this study, there were 20 disabled respondents, and 20 care givers. The detailed information of respondents is listed in Table 1.

From the interview, it could be identified that functional-oriented designs are the focus of disabled user expectations. In the responses from the disabled persons, stylish choices are also the main topics about apparel selection. Price and staff engagement in the apparel purchase are also highly discussed during the interview, indicating the relationship between care apparel production cost and related service in the market.

The most participants claimed their requirement in the design of easy do-on and do-off. As in some specific occasions such as toilet or medical examination, disabled users have to manage with self-dressing because their care givers are not allowed to accompany. One of the interviewee as care giver for her mother has mentioned:

Table 1. The statistics of disabled elderly purchase orientation

Disabled users			Care givers		
Items	N	%	Items	N	%
Easy Do-on & Do-off	7	35%	Self-dressing in toilet	11	55%
Easy button up	5	25%	Easy button up	8	40%
Adjusted opening	3	15%	Adjusted opening	2	10%
Fastening	6	30%	Fastening	4	20%
Suitable size & fit	2	10%	Comfort & Function	2	10%
Stylish choice	9	45%	Stylish choice	4	20%
Low price	4	20%	Affordable price	1	5%
Ready-to-wear garment	3	15%	Time-saving	5	25%
Functional design	3	15%	Functional design	3	15%
Physical store	10	50%	Physical store	5	25%
Online shopping	1	5%	Online shopping	9	45%
Require qualified staff (know about special needs)	7	35%	Require qualified staff (know about special needs)	7	35%
Require regular sale	2	10%	Require regular sale	2	10%
Have other to buy clothes	3	15%	Other considerations	5	25%
Average age	74.6 (66 ~ 88)		Average age	39.3 (21 ~ 58)	

Since my mother always need to go hospital for follow-up medical checking. When she carries out X-ray checkups, she needs to change to specialized clothing. It is good that having an easy managed fastening then she can dress and undress by herself, because I cannot to go inside the X-ray clothing chasing room.

From the perspective of psychological impact, the allowance for the disabled to dress themselves could further enhance their confidence, which is essential for them to keep a healthy relationship with families. Some has mentioned his delightfulness of adopting mass customized products for disabled relatives:

They will feel that they are useless when I help them dressing or undressing. Many times, he foes to the toilet by himself. The service you proposed can help them a lot in their daily live. I think it can help them become more confidence.

More disabled interviewees claimed about the unsatisfactory appearance of existing garment that provided from elderly home and rehabilitation centers. Such complaints have expressed the appeal from some of the disabled users who have mild difficulties in daily life:

I do not want to wear the clothing that elderly center promoted, I think those are like the hospital clothing. I don't want wear this just because of missing three toes.

The clothing provided from the elderly center without is aesthetic element. I prefer to buy clothing from the market, but fastening is the main challenge for me.

Psychology for such group of people still weigh much, for the fact that they have enough time for social activities but less choices as entertainments. Whereas the apparel design is limited in the market, resulted with a lack of expectation of them for social participation.

3.3 Care Apparel Characteristics and Classification

Based on the observation of disabled client orders in the care apparel center, and the experience of care apparel procedure, a new classification regarding the marketization of care apparel production has been generated. It has been found out that the users in high disability level tend to pay attention to the functional design of the care apparel. Contrarily, users in low disability level tend to focus on the aesthetic design of the care apparel products. The moderate disability level users have a mixture consideration on both functional performance and aesthetic expression of the care apparel product.

From the perspective of production feasibility, the care apparel can be classified into three categories: a. Highly individualized products, b. Low individualized products, c. Universal products (Fig. 1). The new classification can be refereed as a pre-selection of proper production method before the product is put into the market.

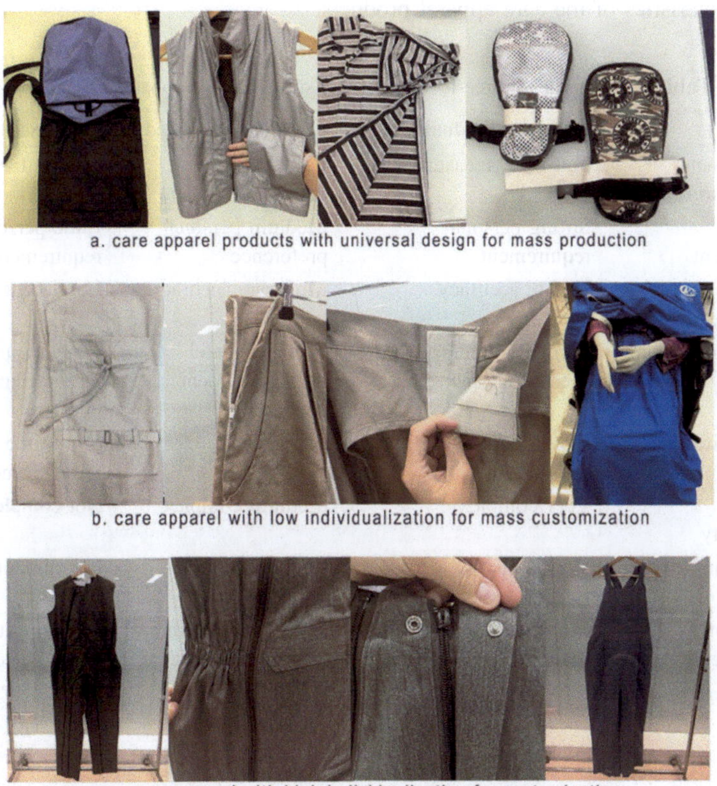

a. care apparel products with universal design for mass production

b. care apparel with low individualization for mass customization

c. care apparel with high individualization for customization

Fig. 1. Three-level classification of care apparel characteristics

Firstly, highly individualized apparel and products require full body measurement, with strong personal preferences, or styles of formal occasions. The highly individualized property of care apparel requires a dedicated manufacture on the product, and different measurements causes the main challenge in production. Therefore, a proper production method could be customization, which will allow for modification on single item based on the body contour or functional innovation specifically. Secondly, low individualized products require partial body measurement, or contain replaceable assembly parts in the design. From the investigation of current care apparel products, key functions remaining at the opening design, closure solutions for self-care, and effective assessors to maintain the easy-wear purpose of the apparel product. The main part of the product which could hold the silhouette could be remained in the mass-customization, with only small part of the apparel requires customization by the users. The universal products require the minimum fitting accuracy in measurement, with a standardized sizing system (or free size) to meet with low personal preference. Such type of care apparel is often in unified size with simple design, in order to maintain the one-for-all purpose for users. The simplified design could be applied into mass production, which could reduce the cost in manufacture and provide the product at a short lead-time to the potential disabled users. Table 2 presents the detailed classification of the characteristics of the care apparel products.

Table 2. Criteria of three-level characteristics for care apparel production

Key items	Highly individualized	Low individualized	Universal design
Measurement modification	Full-body measurement	Partial-body measurement	Standardized measurement
Personal requirement	Strong personal requirement	Medium personal preference	Little personal requirement
Fitting requirement	Highly accuracy	Moderate accuracy	Low accuracy
Wearing occasion	Formal garment, Medical garment	Casual apparel Some treatment purpose	Casual apparel Few medical purpose
Product lifespan	Long-time	Partial duration	Short-time Disposable
Design complexity	Very complex	Medium complex	Not complex
Production difficulty	Very difficult	Medium difficult	Not difficult
Production requirement	Customization Exquisite design	Mass-customization Information system	Mass production High-efficient network

4 Conclusion and Future Study

In this study, the investigation regarding care apparel in Hong Kong has been conducted. The products in Care Apparel Center, which is located in The Hong Kong Polytechnic University has been reviewed and analyzed with its characteristics based on requirements of different user disability levels. Firstly, from observation it has been found that the different disabled users tend to have various clothing requirements regarding care apparel purchase. Functional-oriented garment characteristics were highly related with a high disability level of the users, while aesthetic-oriented garment characteristics were highly related with low disability level of the users. To further identify the relationship and classification of the care apparel features, a thorough investigation has been conducted using the products in the Care Apparel Center, and a three-level classification regarding the feasibility of care apparel production based on its characteristic property could be obtained.

The direction of further study focuses on the investigation of perceived barriers in care apparel application for disabled persons in Hong Kong. Topics regarding care apparel design optimization and specific market business model is also to be proposed and evaluated. Apparel market in Hong Kong has been an international scale platform with high influence to the world industry, however, the development of care apparel in local market still lacks further strength. Hopefully the clothing requirement of disabled customers in Hong Kong could be identified to instruct the market development of care apparel.

Acknowledgements. We sincerely thank all the staff members in the Care Apparel Center from the Hong Kong Polytechnic University, which could provide full support to my research. Their substantial services and extensive experience to the disabled and the elderly have been the foundation of this whole study.

References

1. Reich, N., Otten, P.: What to wear: a challenge for disabled elders. Am. J. Nurs. **87**, 207A (1987)
2. Disabled World.com. News resource website. https://www.disabled-world.com
3. Keates, S., Clarkson, P.: Countering design exclusion through inclusive design. In: ACM SIGCAPH Computers and the Physically Handicapped, no. 73–74, p. 69 (2002)
4. Yang, X., Liu, Y.: Universal Design and its application in housing design. J. Arch. **10**, 7–9 (2005)
5. Troels, H.: Povlsen Care Apparel Center: What is care apparel? https://www.polyu.edu.hk/itc/cac/3_ca.html
6. Ng, S.F., Hui, C.L., Wong, L.F.: Development of medical garments and apparel for the elderly and the disabled. Text. Prog. **43**(4), 235–285 (2011). https://doi.org/10.1080/00405167.2011.573240
7. Washington City Group Statement of rationale for the Washington Group general measure on disability, National Center for Health Statistics, Centers for Disease Control and Prevention (2009). http://www.cdc.gov/nchs/citygroup/rationale.htm. Accessed 15 Dec 2009

8. Palmer, M., Harley, D.: Models and measurement in disability: an international review. Health Policy Plan. **27**(5), 357–364 (2012)
9. Mahoney, F.: Functional evaluation: the Barthel index. Md. State Med. J. **14**(14), 56–61 (1965)
10. Shah, S., Vanclay, F., Cooper, B.: Improving the sensitivity of the Barthel index for stroke rehabilitation. J. Clin. Epidemiol. **42**, 703–709 (1989)
11. Bennett, M., Ryall, N.: Using the modified Barthel index to estimate survival in cancer patients in hospice: observational study. Br. Med. J. **321**(7273), 1381–1382 (2000)
12. Katz, S., Akpom, C.A.: Index of ADL. Med. Care **XIV**, 116–118 (1976)
13. McDowell, I., Newell, C.: Measuring Health: A Guide to Rating Scales and Questionnaires. Oxford University Press, New York (1987)
14. Rodgers, W., Miller, B.: A comparative analysis of ADL questions in surveys of older people. J. Gerontol. **52**, 21–36 (1997)
15. Wang, Y., Wu, D., Zhao, M., Li, J.: Evaluation on an ergonomic design of functional clothing for wheelchair users. Appl. Ergon. **45**(3), 550–555 (2014)
16. Lamb, J., Kallal, M.: A conceptual framework for apparel design. Cloth. Text. Res. J. **10**(2), 42–47 (1992)
17. Freeman, C.M., Kaiser, S.B., Wingate, S.B.: Perceptions of functional clothing by persons with physical disabilities: a social-cognitive framework. Cloth. Text. Res. J. **41**, 46–52 (1985)
18. Quinn, D., Chase, R.W.: Simplicity's Design Without Limits. Drexel Design Press, Philadelphia (1990)
19. Yu, H., Rahman, O.: Inclusive apparel shopping experiences for older consumers in China: product attributes and retail environment. Int. Rev. Retail. Distrib. Consum. Res. **28**(5), 531–553 (2018)
20. Ahmad, R.: The older or ageing consumers in the UK: are they really that different? Int. J. Mark. Res. **44**(3), 337–360 (2002). https://doi.org/10.1177/147078530204400305
21. Nam, J., Hamlin, R., Gam, H.J., Kang, J.H., Kim, J., Kumphai, P., Richards, L.: The fashion-conscious behaviours of mature female consumers. Int. J. Consum. Stud. **31**(1), 102–108 (2007). https://doi.org/10.1111/j.1470-6431.2006.00497.x
22. Xie, Y., Hui, C.L., Ng, S.F.: The evaluation of quality attributes of NPO products: a case in medical garments. Qual. Control Appl. Stat. **56**(4), 393–396 (2011)
23. Senanayake, M., Little, T., King, R.: Mixed Mass Production and Mass Customization Best Practices for Apparel. North Carolina State University, Raleigh (2004)
24. Randall, T., Terwiesch, C., Ulrich, K.T.: Principles for user design of customized products. Calif. Manag. Rev. **47**, 68–85 (2005). http://doi.org/bhfj
25. Fiore, A.M., Lee, S.-E., Kunz, G.: Individual differences, motivations, and willingness to use a mass customization option for fashion products. Eur. J. Mark. **38**, 835–849 (2004). http://doi.org/bgqsfx

Author Index

G. Di Bucchianico (Ed.): AHFE 2019, AISC 954, pp. 515–516, 2020.
https://doi.org/10.1007/978-3-030-20444-0

MIX
Papier aus verantwortungsvollen Quellen
Paper from responsible sources
FSC® C105338
FSC
www.fsc.org

If you have any concerns about our products,
you can contact us on
ProductSafety@springernature.com

In case Publisher is established outside the EU,
the EU authorized representative is:
Springer Nature Customer Service Center GmbH
Europaplatz 3, 69115 Heidelberg, Germany

Printed by Libri Plureos GmbH
in Hamburg, Germany

Torts on Three Continents

Professor Jane Stapleton by Michael Gaskell, oil on board, 2021; By kind permission of the artist and the Master and Fellows of Christ's College, Cambridge

Torts on Three Continents

Honouring Jane Stapleton

Edited by

KYLIE BURNS
JODI GARDNER
JONATHAN MORGAN
SANDY STEEL

OXFORD
UNIVERSITY PRESS

OXFORD
UNIVERSITY PRESS

Great Clarendon Street, Oxford, OX2 6DP,
United Kingdom

Oxford University Press is a department of the University of Oxford.
It furthers the University's objective of excellence in research, scholarship,
and education by publishing worldwide. Oxford is a registered trade mark of
Oxford University Press in the UK and in certain other countries

Published in the United States of America by Oxford University Press
198 Madison Avenue, New York, NY 10016, United States of America

British Library Cataloguing in Publication Data

Data available

Library of Congress Control Number: 2024940350

ISBN 978–0–19–888974–8

DOI: 10.1093/oso/9780198889748.001.0001

Printed and bound by
CPI Group (UK) Ltd, Croydon, CR0 4YY

MIX
Paper | Supporting
responsible forestry
FSC® C013604
FSC
www.fsc.org

Foreword

The Oxford Law Faculty website includes a collection of photographs relating to the legal philosopher Joseph Raz. Among them are photographs taken at summer parties of the student law society at Balliol College, where Joseph was a Fellow. In the photograph for 1984 there appears a young woman, smiling rather shyly at the camera, and wearing a cardigan to keep warm on a sunny summer's day in England. Her name is printed below the photograph: J Stapleton.

Jane Stapleton arrived at Balliol in 1981 to undertake a doctorate on compensation for non-traumatic injuries under the supervision of Patrick Atiyah. If her later work can be seen as having its origins in this research, one can also see in it the influence of the scientific training which she received prior to studying law. The DPhil thesis resulted in her first book, *Disease and the Compensation Debate*.[1]

After graduating from Oxford, Professor Stapleton lectured at the University of Sydney Law School, then returned to Balliol in 1987 as Fellow and Tutor in Law. In 1997 she left Oxford to take up a Research Chair in Law at the Australian National University. She held that chair until 2016, combining it with a chair at the University of Texas at Austin and Visiting Professorships at Oxford, Harvard, Columbia, the European University Institute, and Cambridge. Here, she and her husband, Peter Cane, were the first joint holders of the Arthur Goodhart chair, and she was the first woman to be elected an Honorary Fellow of St John's College. In 2016 she broke through another glass ceiling when she was appointed Master of Christ's College, Cambridge, the first woman to hold that office since the college's foundation in 1505.

Professor Stapleton's books and articles have made an impact on the development of the common law across three continents. How has she come to have such an influence? In the first of her *Three Essays on Torts*,[2] she champions what she calls 'reflexive tort scholarship', that is, a type of scholarship which 'places judges at centre stage and seeks a constructive dialogue with them'.[3] She sees this as a collaborative exercise, in which academics start with the law as the judges have held it to be and then seek to identify connections and patterns, as well as weaknesses and tensions in the reasoning on which it is based.[4]

She has done that in this jurisdiction in relation to a range of issues in the law of tort. In relation to the rule in *Fairchild v Glenhaven Funeral Services Ltd*,[5] in

1 Oxford, 1986.
2 Oxford, 2021.
3 J. Stapleton, *Three Essays on Torts* (OUP 2021) 2.
4 Ibid. 18.
5 [2002] UKHL 22; [2003] 1 AC 32.

particular, the dialogue between successive decisions of our highest court[6] and her commentary in response[7] exemplifies the back-and-forth nature of reflexive tort scholarship. The judiciary acknowledge in their judgments the insights derived from Professor Stapleton's past work. She then comes back to us again to suggest a better account of the law as she finds it, for the judges to consider if an appropriate moment arises. I look forward to seeing how the dialogue develops.

The judiciary and the common law owe Professor Stapleton a considerable debt. Her kind of scholarship is greatly valued by the judiciary and, as she has put it, can offer 'scholars the thrilling prospect of collaborating on the real-world project of identifying and articulating the living common law'.[8] I am very grateful that Professor Stapleton has dedicated so much of her own career to this purpose, to such effect.

Now that she retired as Master of Christ's and returned to Australia, I can only say how much we miss her, and how much we look forward to seeing her back in the United Kingdom when she visits—not only because of her professional achievements but also because of her warmth and generosity as a person.

As far as I am aware, there is not yet a branch of the Ivy in Australia: at least, not the same restaurant as the one Jane likes to go to, in London's theatreland. So I hope that she keeps coming back for her favourite iced berries with hot white chocolate sauce.

Robert Reed
Supreme Court of the United Kingdom

[6] eg *Barker v Corus UK Ltd* [2006] UKHL 20, [2006] 2 AC 572 and *Zurich Insurance Plc UK Branch v International Energy Group Ltd* [2015] UKSC 33; [2016] AC 509.
[7] eg, *Three Essays in Tort* (n 3) 88.
[8] Ibid. 34.

Contents

III. CAUSATION

IV. OTHER TORTS AND LIABILITIES

Abbreviations

AEVA	Automated and Electric Vehicles Act 2018
AI	artificial intelligence
ALI	American Law Institute
APB	Aborigines Protection Board
CJEU	Court of Justice of the European Union
CPA	Consumer Protection Act 1987
DALYs	disability-adjusted life years
GBD	Global Burden of Diseases
GLO	group litigation order
GPSR	General Product Safety Regulation
MIB	Motor Insurers' Bureau
NDIS	National Disability Insurance Scheme
NIIS	National Injury Insurance Scheme
SGA	Sales of Goods Act 1979
VAR	video assistant referee

Contributors

TT Arvind, Professor of Law, University of York

Roderick Bagshaw, Professor of Law, University of Oxford, Fellow in Law, Magdalen College, Oxford

Anita Bernstein, Anita and Stuart Subotnick Professor of Law, Brooklyn Law School

Ellen M Bublick, Dan B. Dobbs Professor of Law, University of Arizona James E. Rogers College of Law

Kylie Burns, Professor of Law, Griffith Law School

Jodi Gardner, Brian Coote Chair in Private Law, University of Auckland

Mark A Geistfeld, Sheila Lubetsky Birnbaum Professor of Civil Litigation, New York University School of Law

Imogen Goold, Professor of Medical Law, University of Oxford, Fellow in Law St Anne's College, Oxford

Genevieve Grant, Associate Professor of Law, Monash University

Sarah Green, Professor of Law, University of Bristol, Commissioner for Commercial and Common Law at the Law Commission for England and Wales

Catherine Kelly, Professor of Law and History, University of Bristol

The Hon Susan Kiefel AC, formerly Chief Justice of Australia

Harold Luntz, Emeritus Professor of Law, University of Melbourne

Nicholas J McBride, Fellow in Law, Pembroke College, Cambridge

Jonathan Morgan, Professor of English Law, University of Cambridge, Fellow in Law, Corpus Christi College, Cambridge

John Murphy, Professor of Law, University of Hong Kong

Philip Sales, Justice of the United Kingdom Supreme Court

Sandy Steel, Professor of Law and Philosophy of Law, University of Oxford, Lee Shau Kee's Sir Man Kam Lo Fellow in Law, Wadham College, Oxford

Jenny Steele, Professor of Law, University of York

Gemma Turton, Associate Professor of Law, University of Sydney

Richard W Wright, University Distinguished Professor Emeritus, Illinois Institute of Technology, Chicago-Kent College of Law

Introduction: Jane Stapleton's Tort Scholarship on Three Continents

Kylie Burns, Jodi Gardner, Jonathan Morgan and Sandy Steel

Jane Stapleton is extraordinary. She is one of the world's leading legal academics and known for her insightful, detailed and fascinating writings. For each of this volume's editors, and many other authors who have contributed to this book, she is also a much loved and valued colleague, friend and mentor. We admire and thank her for her enduring kindness. Her career has been marked by fostering and supporting other private law scholars. Her work has inspired and challenged us. She has been unafraid to engage with other tort scholars (including authors in this book) with curiosity, sometimes critically, but always collegially. Tort law scholarship as a whole is better because of this engagement.

Jane has been a leading contributor in a field often dominated by male voices. She has been a champion of women in private law, particularly in tort. We are pleased that this book includes two female co-editors (including one of her many PhD students) and that half the chapter authors are women from all different stages of their careers. Many of us are the beneficiaries of her teaching, mentorship and support.

Jane's academic career and scholarly contributions have spanned her home country Australia, the United Kingdom, the United States and beyond. This, in itself, is extraordinary. The book includes chapters from authors in each of these jurisdictions. Her eminence in academia in all three is manifest. Jane was, until recently, the Master of Christ's College, Cambridge—the college's first female head since its foundation in 1505. She is an Honorary Fellow of St John's College, University of Cambridge; Emeritus Fellow of Balliol College, University of Oxford; and has been Statutory Visiting Professor of Law, University of Oxford and Arthur Goodhart Visiting Professor of Legal Science, University of Cambridge. She is a Corresponding Fellow of the British Academy, Honorary Bencher of Gray's Inn and was appointed Honorary Queen's Counsel in 2020. She delivered the Clarendon Lectures in Law at Oxford University in 2018.[1] She is also an Emeritus Distinguished Professor of Law at the Australian National University and returned to her country of birth in 2023 as Honorary Professor at the Sydney Law School.

[1] Published as J Stapleton, *Three Essays on Tort* (OUP 2021).

Kylie Burns, Jodi Gardner, Jonathan Morgan and Sandy Steel, *Introduction: Jane Stapleton's Tort Scholarship on Three Continents* In *Torts on Three Continents*. Edited by: Kylie Burns, Jodi Gardner, Jonathan Morgan and Sandy Steel, Oxford University Press.
DOI: 10.1093/oso/9780198889748.001.0001

She is a Fellow of both the Australian Academy of Law and the Academy of Social Sciences.

Jane was formerly the Ernest E. Smith Professor of Law, University of Texas at Austin School of Law, and a visiting Professor at Harvard Law School, Columbia Law School and Washburn Law School. Her immense contribution to tort law scholarship in the United States and more broadly was recognised by her appointment to the Council of the American Law Institute and her award of the American Association of Law Schools' Prosser Award; the American Bar Association's Robert B. McKay Law Professor Award; and the John Fleming Prize in Tort Law. These are extraordinary recognitions, particularly given she is a non-US scholar.

Jane's career has been marked by her close attention and commitment to the need for academic work to be relevant and responsive to the needs of the judiciary and the legal profession. Her commitment to this long pre-dated contemporary university and government agendas which require academic work to demonstrate broader 'impact'. As contributions to this book by the former Chief Justice of Australia, the current President of the United Kingdom Supreme Court and another member of that court, Lord Sales, all demonstrate, Jane's work has been highly influential on the judiciary and the development of law across the common law world. In a time when the legal academy is increasingly seen as distant and disconnected from the concerns of the legal profession, this is extraordinary. Jane's work has always taken judging and judges seriously and, as she describes it, her work is truly 'reflexive' scholarship.[2]

This book reflects on the exceptional contributions Jane has made to tort scholarship, which encompass many of the contemporary and traditional debates in tort law and tort theory. We thank all the authors who have contributed in celebration of Jane and her work.[3] The book is divided into four parts. The first part focuses on general themes that arise in Jane's work, including her academic influence on judges, the role of grand theory in tort law, the role of insurance in the private law and the impact of vulnerability on tort law and liability of public authorities. The second part of the book includes chapters reflecting on Jane's scholarship concerning negligence, including duty of care, economic loss and psychiatric injury. Jane is one of, if not the, world's leading expert on causation. The third part of the book is therefore dedicated completely to causation, with three chapters from authors in three different countries reflecting on the impact of Jane's work in this area. The final part of the book covers a variety of different aspects of tort law and compensation systems including harms committed in the public interest, damage

[2] Ibid 17–20.

[3] Drafts of each chapter were originally presented at a workshop 'Reflecting on Tort Law: Essays in honour of Jane Stapleton' held at Wadham College, Oxford in July 2022. We thank Wadham College and Cambridge Private Law Centre for their support of this project.

in economic torts, statutory product liability reforms and alternative compensation scheme design.

Part I: General Themes of Tort Law

In the first substantive chapter, the Hon Susan Kiefel AC, Chief Justice of the High Court of Australia from January 2017 until November 2023,[4] reflects on the valuable contribution that Jane's tort scholarship has made to courts and the development of tort law. She argues that Jane's 'tort law scholarship has enabled the creation and maintenance of a dialogue with the courts about the law as it is and the direction that it might take'.[5] The chapter discusses the nature of Jane's 'reflexive tort scholarship'. Such scholarship rejects the claims of 'grand theories' such as economic theory and corrective justice theory that they are descriptive of 'what tort law actually is, or what judges do, or say they are doing, when deciding cases'. The chapter presents an overview of Jane's influence on courts, noting the significant citation of her work both internationally and in Australia. The chapter particularly traces the strong influence that Jane's conceptual analysis of causation has had in the development of the law of causation in Australia by courts and by the legislature. Lastly, the chapter discusses Jane's scholarship on economic loss as an illustrative example of her reflexive tort scholarship which has been influential on the High Court of Australia.

In the following chapter, Mark Geistfeld engages critically with Jane's advocacy of 'reflexive' torts scholarship and her criticism of 'grand theory'.[6] Very broadly, Geistfeld seems sympathetic to this project. He too accepts that characteristically, tort judges use 'analogical reasoning based on mid-level unifying normative principles for categorizing cases'.[7] But he points out some significant problems with calls for unqualified 'pluralism' in tort. He also suggests that there is more room for 'monist' theories than Jane allows.

Geistfeld argues that some significant areas of tort law (eg product liability) can be justified both by instrumental (eg economic) theories *and* by non-instrumental (eg deontological) theories. He explains why this is so: convergence in product liability results from the economic linkage between consumer-plaintiffs and manufacturer-defendants. Geistfeld argues that this demonstrates the potential worth of 'monist' theories, including economic analysis of law. The latter, he suggests, provides a better explanation of the non-waivability of product liability

[4] Chief Justice Kiefel was the first woman to serve as Chief Justice of the High Court of Australia

[5] Kiefel, Chapter 1, this volume, p 3.

[6] J Stapleton, *Three Essays on Tort* (OUP 2021) ch 1. See also, eg J Stapleton, 'Comparative Economic Loss: Lessons from Case-Law-Focused Middle Theory' (2002) 50 UCLA Law Review 531.

[7] Geistfeld, Chapter 2, this volume, p 38. cf C Sunstein, *Legal Reasoning and Political Conflict* (2nd edn, OUP 2018).

vis-à-vis consumers than Stapleton's account,[8] whereby tort gives effect to parties' voluntary co-operative arrangements (an avowedly 'incompletely theorized' or mid-level principle).

However, Geistfeld accepts that in many other fields of tort such an 'overlapping consensus' is not available.[9] The law then faces a choice between different, incompatible, 'monist' theories. He questions whether insistence on 'pluralism' can be coherent, or accurately describe the law. If courts are required to commit to inconsistent ideas simultaneously, how can coherent tort law result? If courts are required to defer to democratically elected legislatures to choose between inconsistent principles, on the basis that it is not for judges to decide between values which divide society, why is it that courts routinely decide hard cases in tort (rather than abdicate decisions to legislatures)? As noted, Geistfeld commends the gradual, analogical development of tort law using principles at a lower level of abstraction. This ensures just and consistent treatment of tort plaintiffs and tort defendants. Here too, an overlapping consensus may eventually emerge. But in its absence, Geistfeld seems to accept what he describes as Stapleton's abjuration of 'dependence on foundationalist principles' in favour of 'incompletely theorized principle[s] of general application'.[10]

Anita Bernstein's chapter, 'The Torts Scholar as Disaggregator', also identifies general features of Jane Stapleton's scholarly method. It considers Stapleton's reflexive approach to be part and parcel of being a 'disaggregator'. The chapter begins by explaining the three central characteristics of the disaggregator's method: firstly, a temperamental one, that the disaggregator is 'inclined to focus on divergences more than unities'; secondly, a distinctive emphasis on the primary sources; and thirdly, combining the other two, a characteristic emphasis on the conceptual and normative diversity, and conflict, found in those primary sources.

In the central part of the chapter, Bernstein evidences disaggregationist patterns across the body of Jane's work. A major one concerns Jane's rejection of 'false unities', represented in part by her rejection of 'grand theories' of tort law, such as Law and Economics, Corrective Justice, or Civil Recourse. The fundamental objection is one of 'descriptive inaccuracy'. However, as Bernstein explains, the objection is pursued in various ways, for instance by drawing attention to heterogeneity in tort law's own conceptualisation of its normative justifications and its variance across jurisdictions and time. Another concerns Jane's critique of opaque invocations of 'common sense' in the context of decisions about causation; the disaggregationist seeks to unearth the various normative considerations that lie behind common sense, and to expose them to transparent scrutiny.

[8] J Stapleton, *Three Essays on Tort* (OUP 2021) ch 2.
[9] J Rawls, 'The Idea of an Overlapping Consensus' (1987) 7 Oxford Journal of Legal Studies 1.
[10] Geistfeld, Chapter 2, this volume, p 38.

No collection on Jane's work would be complete without a discussion of insurance. TT Arvind and Jenny Steele engage with two of her significant contributions. First, her 'classic' article of 1995 on 'Tort, Insurance and Ideology'.[11] Secondly, Jane's more recent argument that tort law should, and does, respect 'voluntary arrangements' in tailoring the scope of tort liability.[12] Arvind and Steele argue that there is tension between these two landmark pieces of work.

Stapleton was famously dismissive of viewing tort 'as' insurance in her 1995 article, arguing against the normative relevance, for tort law, of parties being insured. Arvind and Steele note that this was written in response to certain critical approaches to tort law. If tort law is conceived as a form of insurance against certain kinds of accidental injury, that is, as 'insurance' legally imposed in the form of liability in tort, it is in those terms bound to fail. Compared to actual insurance (eg against personal accident), tort's coverage is necessarily incomplete. Tort liability depends on concepts such as fault and responsibility that are irrelevant from the perspective of compensation and loss-spreading. Similarly, as Jane noted, tort's conduct-guiding effect, deterring culpable behaviour, is in tension with any wide loss-spreading goal.

Arvind and Steele accept the force of Jane's 1995 article as a response to such a critique of its social function, and do not dispute her core point that whether society should replace tort with a wider social-insurance scheme is an ideological choice. However, following John Gardner they note that *all* judicial decisions about tort involve distributive choices about how to 'allocate' corrective justice. They question whether in light of this it is helpful to deem the whole of tort law to involve 'political' decisions.

More significantly still, Arvind and Steele contend that Jane's hostility to insurance concerns in tort has been over-generalised. Her outright dismissal of the relevance of insurance is too sweeping. Building on Steele's earlier work (with Rob Merkin), the authors argue that insurance is an intrinsic part of private law.[13] It is woven into the grain of tort. It should not be dismissed as a mere 'makeweight' factor in decisions about tort liability.[14] In particular, as Jane argues in her recent *Three Essays on Tort*, tort law correctly respects co-operative arrangements about risk-allocation—a concern that goes beyond upholding binding contracts between parties in direct privity. It is therefore (in Arvind and Steele's view) misplaced to exclude insurance from this account, since insurability plays a central part in decisions about how to allocate risks of loss—that is, arrangements that re-distribute the incidence of liability that tort law would otherwise, by default, produce. The authors welcome the renewed emphasis on private law's role in facilitating private

[11] J Stapleton (1995) 58 Modern Law Review 820.
[12] J Stapleton, *Three Essays on Tort* (OUP 2021) ch 2.
[13] R Merkin and J Steele, *Insurance and the Law of Obligations* (OUP 2013).
[14] cf J Stapleton, 'Duty of Care Factors: a Selection from the Judicial Menus' in P Cane and J Stapleton, *The Law of Obligations: Essays in Celebration of John Fleming* (OUP 1998).

ordering. The interaction between tort and insurance is an important aspect of this, as Jane's own recent work suggests, contrary to certain aspects of her 1995 critique of insurance 'ideology'.

Part I's last chapter was written by Kylie Burns, one of Jane's PhD students and now a Professor of Tort Law herself. In this chapter Burns considers Jane's important contribution at the Centenary Conference of the High Court of Australia, concerning the role of vulnerability in the development of Australian tort law.[15] Jane argued that the High Court of Australia had identified the 'protection of the vulnerable as a core moral concern of tort law: its golden thread'. Burns considers how 'vulnerability' has been used by Australian courts in developing negligence law since Jane's essay of 2003. It has not proved to be a significant determinative factor in expansion of liability. Given Jane's original argument that 'vulnerability' could form the basis of torts' response to the injustices suffered by the First Peoples of Australia, Burns then focuses on 'Stolen Generation' negligence claims by First Peoples. Finally, she considers public interest litigation on behalf of 'vulnerable' plaintiffs against public defendants, including claims by injured immigration detainees, and climate change litigation. The chapter concludes that protection of the vulnerable is indeed a concern that has the potential to influence the development of Australian negligence law: however, the promise of the 'golden thread' has not (yet) been fulfilled.

Part II: Negligence

In this part, five chapters consider aspects of liability in the tort of negligence. Some contributors engage directly with well-known contributions by Jane: her exploration of the 'policy' factors underlying decisions about duty of care; her scepticism about 'assumption of responsibility' as a criterion for liability; and her blistering criticism of the arbitrary rules limiting liability for psychiatric injury in English law. There are wider reflections on the structure of the tort. All five chapters engage with wider themes of the Stapleton oeuvre: close but critical engagement with judicial reasoning, intolerant of conclusory labels and opaque reasoning—but at heart pluralist rather than mono-theoretical.

In 'The Real Gist of Negligence', Nicholas J McBride rejects the understanding of negligence adopted by many tort theorists which conceives breach of duty as 'carelessly causing harm'. McBride argues that the wrong in negligence law is simply 'carelessly breaching duty' (with causation of harm a separate requirement).

[15] J Stapleton, 'The Golden Thread at the Heart of Tort Law: Protection of the Vulnerable' (2003) 24 Australian Bar Review 135; and J Stapleton, 'The Golden Thread at the Heart of Tort Law: Protection of the Vulnerable' in P Cane (ed), *Centenary Essays for the High Court of Australia* (LexisNexis Butterworths 2004).

McBride suggests that the label for the position he criticises derives from Jane's famous two-part article, 'The Gist of Negligence'.[16] As he accepts though, the substance of the criticised position was not one Jane advocated. Moreover, he argues that an appropriately Stapletonian 'reflexive' approach, sensitive to the understanding and actual practice of the courts, supports his view.

McBride insists that courts do not conflate 'breach of duty' and 'causation of loss' in the way claimed by supporters of 'damage as the gist of negligence'. Telling examples come from difficult English cases on causation. The reasoning in *Fairchild v Glenhaven Funeral Services* proceeds from the *wrongfulness* of the defendant employers carelessly exposing their employees to asbestos.[17] The judges manifestly held this view, which is inconsistent with the theory that their 'wrong' was carelessly *causing* mesothelioma—for that was the very thing which, given the limitations of scientific knowledge, could not be proven. McBride naturally accepts that *Fairchild* and (another key example) *Chester v Afshar* are controversial decisions.[18] But his argument does not assume or seek to demonstrate their correctness. Rather, that without the notion that breach of duty is itself wrongful, arguments for modifying rules on causation become *incomprehensible.*

McBride does not advocate 'pure risk claims'. While there would (on his argument) be a breach of duty if someone drove carelessly too fast near the plaintiff but didn't actually collide with them, McBride accepts that tort law is correct not to recognise an action for ('nominal') damages for a risk of harm that we know never materialised. McBride commends the good pragmatic sense, in most cases, of the legal requirement that causation of 'gist damage' must also be proved for an action in negligence. Sometimes, however, an exception is justified. For example, to justify the award of 'preventive damages' when the plaintiff has successfully taken action to avert a risk created by the defendant from ripening into harm. So long as we discard the conceptual assumptions of the 'loss-causing breach' theory, tort law can coherently award compensation in the absence of any actual harm ever resulting, in such cases. Or (for example) tort could compel a reluctant physician to disclose remote risks of harm, although non-disclosure of those risks would not make injury to the patient very likely (as required by the standard test for a *quia timet* injunction). In short, apart from the particular thesis that McBride advances, the chapter is an example of how 'reflexive' tort scholarship, rooted in doctrinal law, may illuminate debates that otherwise appear purely conceptual and abstract.

Ellen Bublick's chapter 'Duty of Care Factors' focuses less on the conceptual nature of the 'duty' in duty of care, and more on the normative considerations that influence courts' recognition, or refusal to recognise one. Bublick's contribution engages with one of Jane's well-known papers on the diversity of considerations

16 (1988) 104 Law Quarterly Review 213 and (1988) 104 Law Quarterly Review 389.
17 [2002] UKHL 22, [2003] 1 AC 32.
18 [2004] UKHL 41, [2005] 1 AC 134.

which courts have invoked in making duty decisions.[19] Bublick considers state Supreme Court decisions across the United States over a two-and-a-half-year period, aiming to identify the salient considerations that have driven duty of care decisions.

Bublick considers both general 'tests' for duties of care and particular factors. As to the former, the chapter identifies a significant diversity of approaches amongst US state courts, which draw upon the general rules in the *Restatements of Torts*, but with considerable variation. Some approaches, making duty determinations dependent upon liability being consistent with 'public policy', bear obvious similarities to the now discredited *Caparo* test for duties of care in English law.[20] Despite the local variations from state to state, Bublick argues that a 'deep structure' holds across courts. It includes a universal requirement of reasonable foreseeability of harm, a distinction between misfeasance and nonfeasance, some recourse to public policy, a tendency to expand the scope of duties in the face of particularly blameworthy conduct and a tendency to condition novel forms of duty upon especially restrictive criteria.

In relation to particular factors for and against the recognition of a duty, Bublick finds a substantial overlap between those identified in Jane's article and those recognised by US courts. However, the chapter emphasises a more restrictive approach to pure economic loss cases than Jane identified, and finds particular weight being given to a defendant's heightened ability to prevent physical harm. In addition, Bublick identifies a distinctive expansive trend motivated by considerations of equality, and situations in which the parties stand in what the chapter refers to as a 'web of relationships'.

Roderick Bagshaw's chapter asks: 'What is reasonable foreseeability?' Bagshaw argues that lawyers' tendency to treat foreseeability as merely a question of 'fact' is mistaken. The question is not what was in fact running through anyone's mind (eg the defendant's), but a *hypothetical* inquiry (a notional reasonable person's answer had they asked themselves, 'what might happen here?'). While evidence that a given harmful outcome had in fact been foreseen, or had not been foreseen, is relevant, normative judgements are superimposed on the factual evidence. But these judgements take little account (tellingly so, Bagshaw suggests) of defendants' varied cognitive capacities and real-world cognitive limitations.

Bagshaw considers a number of specific controversies in the English case law. Is a defendant excused only if they'd have thought the harm *impossible*—or is it sufficiently exculpatory that it would have seemed 'fantastic or far-fetched'? And how does this relate to the cogency of the defendant's reasons for dismissing any such risk? Bagshaw notes that on this, and other foreseeability issues, courts

[19] J Stapleton, 'Duty of Care Factors: A Selection from the Judicial Menus' in P Cane and J Stapleton (eds), *The Law of Obligations: Chapters in Celebration of John Fleming* (OUP 1998).

[20] cf *Robinson v Chief Constable of West Yorkshire Police* [2018] UKSC 4.

routinely elide distinct approaches, their reasoning being more 'intuitive' than 'transparent'. There are many examples, also, of divergent or inconsistent judicial reasoning: where a given causal system is known to be unpredictable (eg how humans might decide to behave), courts sometimes hold that defendants should be taken to foresee a wide range of 'known unknowns', but on other occasions insist that the particular behaviour must have been especially likely before liability arises.

Referring to Jane's discussion of the 'development risks' defence in product liability, Bagshaw accepts that these labels used may be used to disguise ideological choices, and thereby 'shield inconsistencies from detection'. As Bagshaw notes, an attack on conclusory and evasive legal formulae has been a consistently important theme throughout Jane's tort scholarship. However, he questions how realistic it is to expect legal rules that attain 'moral perfection' by being fully coherent with a particular set of goals. (Thus hinting at the wider debate about judicial reasoning and theories of tort law, discussed by Jane in the first of her *Three Essays on Tort* and more fully here in Geistfeld's chapter on 'pluralist tort adjudication').

Imogen Goold and Catherine Kelly's chapter is entitled 'Time to Cut Ties: Reforming the Secondary Victim "Control Mechanisms" in Pure Psychiatric Injury'. The authors explore Jane's well-known views on psychiatric injury, expressed within a wider critique of 'incremental' development of tort liability. As they observe:[21]

> nearly twenty years ago, Jane Stapleton cut right to the heart of what is wrong with the law's approach to claims for compensation by people who witness events so shocking that they suffer a psychiatric illness as a result, when she said:
>
> 'the area of nervous shock is ... the area where the silliest rules now exist and where criticism is almost universal'.

Goold and Kelly trace the development of tort claims by 'secondary victims' of psychiatric injury. Having outlined the law's historical origins, the authors critically evaluate the problematic control mechanisms developed in recent times by the English (and Scottish and Irish) courts. The chapter argues that the law should take a new direction through 'rejection of the rigid control mechanisms and return to the fundamental principles of negligence'. This would, they conclude, allow the courts to respond on a principled basis with compassion and rationality, avoiding unjust and 'perverse' results.

In 'An "Opaque Conclusionary Label"? Assumption of Responsibility and Pure Economic Loss' Philip Sales highlights Jane's impact on judicial thinking and engagement, as well as the breadth of her scholarship. Lord Sales reflects on the scope of liability for pure economic loss, with an emphasis on Jane's concern about the

[21] See Goold and Kelly, Chapter 9, this volume, p 169.

'opaque' approach often taken by the courts. In particular, she has been particularly critical of the 'assumption of responsibility' test. Lord Sales recognises these concerns, but defends assumption of responsibility as a valuable concept, especially deployed in cases falling between the standard categories of tort and contract. Yet he highlights the practical and normative importance of Jane's criticisms. Her work has stimulated more precise reflection on liability for pure economic loss and particularly on the role of assumption of responsibility; a perfect example of reflexive legal scholarship.

Part III: Causation

Causation in the law has been at the centre of Jane's scholarship, the subject of more than twenty of her publications.[22] This exceptionally influential body of work has contributed to law reform in both the United States and Australia, and frequently cited by appellate courts across the common law world.[23]

Jane's approach to causation in the law can broadly be characterised as involving three main theses. The first we can call, following Hart and Honoré, *causal minimalism*.[24] This is the thesis that the only legal inquiries that are properly understood to be 'causal' in character are those that involve what is conventionally described as 'factual causation'. Doctrines such *novus actus interveniens*, 'proximate cause' and

[22] J Stapleton, 'Law, Causation and Common Sense' (1986) 6 Oxford Journal of Legal Studies 392; J Stapleton, 'The Gist of Negligence, Part 2: The Relationship between "Damage" and Causation' (1988) 104 Law Quarterly Reports 389; J Stapleton, 'Perspectives on Causation' in J Horder (ed), *Oxford Essays in Jurisprudence* (OUP 2000) 61; J Stapleton, 'Legal Cause: Cause-in-Fact and the Scope of Liability for Consequences' (2001) 54 Vanderbilt Law Review 941; J Stapleton, 'Unpacking causation' in P Cane and J Gardner (eds), *Relating to Responsibility: Essays for Tony Honore on his Eightieth Birthday* (Hart Publishing 2001) 145; J Stapleton, 'Scientific and Legal Approaches to Causation' in I Freckelton and D Mendelson (ed), *Causation in Law and Medicine* (Ashgate 2002) 14; J Stapleton, 'Lords A'Leaping Evidentiary Gaps' (2002) 10 Torts Law Journal 276; J Stapleton, 'Book Review: Tort Liability Under Uncertainty by Porat and Stein' (2003) 66 Modern Law Review 308; J Stapleton, 'Loss of the Chance of Cure from Cancer' (2005) 68 Modern Law Review 997; J Stapleton, 'Occam's Razor Reveals an Orthodox Basis for *Chester v Afshar*' (2006) 122 Law Quarterly Review 426; J Stapleton, 'Two Causal Fictions at the Heart of US Asbestos Doctrine' (2006) 122 Law Quarterly Review 189; J Stapleton, 'Choosing What We Mean By 'Causation' in the Law' (2008) 73 Missouri Law Review 433; J Stapleton 'Causation' in P Cane and J Conaghan (eds), *The New Oxford Companion to Law* (OUP 2008); J Stapleton, 'Causation in the Law' in H Beebee, P Menzies and C Hitchcock (eds), *Oxford Handbook of Causation* (OUP 2008); J Stapleton, 'The Two Explosive Proof-of-Causation Doctrines Central to Asbestos Claims' (2009) 74 Brooklyn Law Review 1011; J Stapleton, 'Factual Causation' (2010) 38 Federal Law Review 467;J Stapleton, 'Factual Causation and Asbestos Cancers' (2010) 126 Law Quarterly Review 351; J Stapleton, 'Unnecessary Causes' (2013) 129 Law Quarterly Review 39; J Stapleton, An 'Extended But-For' Test for the Causal Relation in the Law of Obligations (2015) Oxford Journal of Legal Studies 15; J Stapleton and S Steel, 'Causes and Contributions' (2016) 132 LQR 363; J Stapleton, *Three Essays on Torts* (OUP, 2022), Essay 3; J Stapleton, 'Unnecessary and Insufficient Factual Causes' (2023) Journal of Tort Law (forthcoming).

[23] See Kiefel, Chapter 1, this volume, and recently, eg, *Financial Conduct Authority v Arch Insurance (UK) Ltd* [2021] UKSC 1, [183]–[185].

[24] See HLA Hart and T Honore, *Causation in the Law* (2nd edn, OUP 1985) preface.

'remoteness' are properly understood as *non-causal* doctrines that require a normative explanation.[25]

The second thesis—*causation beyond necessity*—is that causal relations include, but are broader than, relations of necessity. Even if the but-for test is *proven* not to be satisfied, this need not exclude the existence of a causal relation. In the well-worn hypothetical, if two bullets strike a victim simultaneously, neither may be necessary for the victim's death, but each may nonetheless be causal. Less obviously, this is also true, Jane claims, for contributions to an outcome that are neither necessary nor sufficient.[26] Across a number of articles, her work has developed heuristics for identifying this broader relation, most recently in her paper on the 'extended' but-for test.[27]

The third thesis is harder to characterise snappily. We might call it the *normative choice* thesis.[28] The basic idea here is that the appropriate causal concept(s) for the law cannot be read directly off the metaphysics of causation. The law faces a choice as to which relation(s) it counts as causal—for instance, relations of necessity, sufficiency, explanation, physical energy transfers—and this choice should be informed by the law's normative aims. This is not the thesis that, in determining whether *x* was a cause of *y*, the law does or should appeal to normative considerations. It is that, *in the design* of the legal rules of causation, a normative choice must be made. This is consistent with *the application* of those rules resting entirely on judgements about non-normative facts.

In 'Dialogues on Causation with Stapleton', Richard Wright, who has also influentially argued for *causal minimalism*, and *causation beyond necessity*, in particular through his development of the NESS analysis of causation, critically probes the points of commonality and disagreement between his and Jane's positions.[29] As Wright explains, there is a broad agreement in relation to *causal minimalism* and *causation beyond necessity*, but still important points of distinction. One example is that Wright's NESS analysis produces different causal conclusions in certain cases involving multiple negative events. For instance, suppose that D1, a mechanic, carelessly fails to repair D2's brakes, and D2 carelessly fails to depress the brakes upon approach to a pedestrian crossing, injuring C in a collision. This is a case in which the but-for test produces the conclusion that neither breach is a cause of C's injury: but for D1's breach, the injury would still have occurred (because of D2's breach), and *mutatis mutandis* for D2's breach. Stapleton's extended

[25] See, most recently, J Stapleton, *Three Essays on Torts* (OUP 2022) Essay 3.

[26] See J Stapleton, 'Unnecessary and Insufficient Factual Causes' (2023) Journal of Tort Law (forthcoming).

[27] See J Stapleton, 'An "Extended But-For" Test for the Causal Relation in the Law of Obligations' (2015) Oxford Journal of Legal Studies 15.

[28] See, especially, J Stapleton, 'Choosing What We Mean By "Causation" in the Law' (2008) 73 Missouri Law Review 433.

[29] See R Wright, 'Causation in Tort Law' (1985) 73 California Law Review 1735, and Chapter 11, this volume, for further references.

but-for test agrees with this result, but reaches a different conclusion in the two bullets hypothetical. In Stapleton's view the latter case is different because each breach adds a *positive* element to the process by which the injury was produced.[30] By contrast, Wright's view, in cases involving multiple failures of preventive causal process, that the cause is properly identified as 'the temporally first lack of instantiation of a required condition in the preventive causal process that irretrievably causes that process to fail', and thus that D1's breach pre-empts D2's breach in the brakes case.[31]

Gemma Turton's contribution to the volume, 'Causation at a Tipping Point: Material Contribution and the "Worse Off" Question' is concerned with the analysis of situations in the tort of negligence in which a breach 'materially contributes' to damage, but is not necessary for its occurrence. Turton agrees with Stapleton's view that a breach may causally contribute to damage despite the absence of necessity, but argues that this is not the appropriate conclusion in cases in which 'a causal process involves an accumulation of a causal agent over time, such that once a tipping point or threshold of exposure has passed [and] the damage becomes inevitable, or in a medical context "untreatable"'. In cases with this structure, Turton contends, further additions of the causal agent are incapable of amounting to causes of the damage. This is because, in line with the NESS analysis, the subsequent breach will not be a 'necessary' element of the sufficiency of a set if the 'tipping point' has already been reached. Turton's chapter also considers the appropriate doctrinal analysis of the lack of necessity when a breach has causally contributed to an outcome, arguing that this issue should be considered to belong to the legal causation inquiry, and not, as Stapleton holds, the definition of 'damage'.

Finally, Sandy Steel's chapter, 'Making a Difference: Liability and Necessity' is focused on understanding when and why the law does and should make liability sensitive to whether a person's conduct was *necessary* for an outcome. It agrees with Stapleton that the role of necessity cannot be understood wholly in causal terms. Sometimes causes are unnecessary for their outcomes. Instead, the chapter proposes that necessity is relevant to liability for various reasons—for instance, as a defeasibly sufficient condition of causation, as determining the existence of loss and as an actionability rule. Conversely, it argues that liability in the absence of necessity is justifiable for various reasons, particularly in relation to remedies that do not hinge upon the existence of loss.

[30] See Stapleton (n 28) above. Stapleton suggests that liability should obtain in such cases, but on a non-causal basis.

[31] See Wright, Chapter 11, this volume, p 226.

Part IV: Other Torts and Liabilities

The final part of this collection is a nod to the impressive and enviable breadth of Jane's scholarship. These chapters show that, rather than just focusing on a single area of law, her research has had far-reaching implications on many areas of tort. This final part covers a variety of different aspects of tort law and compensation systems including harms committed in the public interest, damage in economic torts, statutory product liability reforms and alternative compensation scheme design.

The first chapter in this part is by Jonathan Morgan. He recognises that tort liability is generally conceived as a sanction for wrongful conduct. Morgan however suggests that in some, arguably exceptional, circumstances tort instead *permits* or licenses harmful yet beneficial activities on condition that the actor *pays for all harm* caused, irrespective of fault. This argument is advanced by the analysis of four 'intriguing' cases—*Binsaris v Northern Territory*, *Rylands v Fletcher*, *Vincent v Lake Erie* and *Burmah Oil v Lord Advocate*. Despite the significant factual and jurisdictional differences, Morgan argues that these cases could be brought under a common principle; the 'wrong' in play is not the defendant's original conduct, but that it resulted in *uncompensated* harm. Wrongfulness consists in *not paying* for harm that has been inflicted by a reasonable (indeed, often positively desirable) activity. It is therefore arguable that the 'wrong' is not the actions of the party, but the harm that has arisen. As a result, Morgan suggests that such liability is responding to the harm suffered and not the actions committed.

The 'conditional licence' theory entails a *strict* standard of liability and a *monetary* remedy. Morgan recognises that significant questions are raised by this, grouping them as: (1) legal theory (the connexion between regulatory law and private law; the relationship of rights and remedies); (2) political theory (the compatibility of strict liability with the Rule of Law); and (3) constitutional propriety (the appropriateness of courts rather than legislatures imposing such liability). Despite these challenges, Morgan argues that there are benefits in the approach, particularly in delineating the defence of necessity, and distinguishing between justifications and excuses. He therefore concludes by noting that this area is another confirmation of Jane's reflexive scholarship and that tort law is not a 'seamless web' in that it cannot be explained by any single theory. The conditional licence is another distinct thread within tort's rich fabric.

The next chapter in this part is by John Murphy, who develops his existing work on economic torts and applies it to Jane's seminal article, 'The Gist of Negligence Part 1: Minimal Actionable Damage'.[32] In this article she highlighted that the notion of 'damage' is commonly used but rarely analysed properly, meaning that it

[32] (1988) 104 LQR 213.

tends to lack adequate definition. Whilst there has since been some discussion on the nature of 'damage' in the tort of negligence, Murphy highlights that there is little if any analysis in the realm of intentional economic torts.

Instead of providing a 'grand theory', Murphy emphasises the need to pay 'homage' to Jane's work and focus on what the senior courts are actually doing in decisions on economic torts. He therefore examines the many different types of harm that have been held (or at least suggested) to constitute actionable damage in economic torts cases. This analysis allows a consideration of whether damage constitutes an element of economic torts or instead is better characterised as a condition of actionability.[33] Murphy comes to the conclusion that there are only three clearly rights-based economic torts: passing off, deceit and two-party intimidation. Most, if not all, of the others appear to be loss-based actions. Despite their conventional label, 'economic' torts actually protect a wide range of interests: including damages for personal injury, mental distress, humiliation, property damage and even, perhaps, disappointment. Murphy therefore emphasises what Jane has so eloquently highlighted during her career; that to understand tort law, we must listen to the judges.

Product liability was one of the first areas where Jane made an international contribution. Jodi Gardner and Sarah Green consider the legacy of Jane's 'groundbreaking' work on product liability reform in the United Kingdom. They look at Jane's seminal article analysing the 1985 EEC Directive on Liability for Defective Products, and her 1994 monograph, *Product Liability*.[34] Jane argued that in the United States, where similar reforms had been adopted, ultimately the scope of product liability was not significantly greater than negligence, predicting that similar outcomes could be expected in the United Kingdom under the new statutory regime. Gardner and Green note that Jane ultimately concluded that 'strict liability was to be preferred on a number of levels, including that it is simply a more easily satisfied test for liability'.[35] They discuss recent developments in UK product liability, which they argue strongly support Jane's views. Finally, Gardner and Green discuss the potential future application of product liability law suggesting the need for a review of product regimes to make them fit for purpose for emerging technologies including AI and software-based products.

The final chapter of the collection also focuses on one of Jane's initial research areas, namely injury compensation schemes. Genevieve Grant and Harold Luntz focus on Jane's early work,[36] stemming from her legal doctoral thesis,[37] on tort

[33] Compare also McBride, Chapter 6, this volume.

[34] J Stapleton, 'Products Liability Reform—Real or Illusory' (1986) 6 Oxford Journal of Legal Studies 392; J Stapleton, *Product Liability* (Butterworths 1994).

[35] Gardner and Green, Chapter 16, this volume, p 342 citing J Stapleton, *Product Liability* (Butterworths 1994) 236.

[36] eg J Stapleton, 'Compensating Victims of Diseases' (1985) 5 Oxford Journal of Legal Studies 248; J Stapleton, *Disease and the Compensation Debate* (Clarendon Press 1986).

[37] This was her second doctorate, her first doctorate being in chemistry.

law's failure to compensate people with non-traumatic injury, including through disease. (Professor Luntz was an examiner of Jane's doctorate.) Jane argued that

> the barriers survivors of non-traumatic conditions experience in recovering tort damages 'are not satisfactorily resolved' by the introduction of stricter liability or cause-based, non-tort compensation schemes. A proper and more sustainable solution to the identified limitations of tort's operation requires a needs-based scheme focused on disability (consequences) rather than cause.[38]

The chapter traces the original context for Jane's work and explores its continued relevance, particularly for contemporary Australian compensation schemes. Grant and Luntz examine and evaluate their effectiveness through the lens of available data: including global burden of injury and disease data, public health research on the 'socially determined nature of health' and research relating to propensity to claim. The chapter reviews the expanding landscape of compensation schemes in Australia, and suggests that these schemes replicate the problems Jane identified. Finally, the chapter makes a case for compensation schemes to receive more scholarly attention in the legal academy.

Final Reflection

This collection includes chapters on a wide range of topics from academics across multiple jurisdictions. This reflects the breadth of Jane's impact on academia, the legal profession and, most importantly, the people lucky enough to have known and learned from her, right across the common law world. The editors have been honoured to lead a project celebrating the work of such an intellectual giant. We trust that the people who know Jane Stapleton will read the chapters and think it is a fitting tribute to a truly exceptional person. We also hope they will provide future students, scholars and practitioners of tort law, who have not had the honour of meeting Jane, with a sense of the magnitude of her contribution.

[38] Grant and Luntz, Chapter 17, this volume, p 358.

PART I
GENERAL THEMES OF TORT LAW

1

Professor Jane Stapleton and her Dialogue with the Courts

The Hon Susan Kiefel AC

1. Introduction

Professor Jane Stapleton has had a varied and distinguished academic career: varied because her initial background was in science before she studied law; distinguished because of the wide recognition and the honours which have been bestowed upon her for her contribution to tort law scholarship in Australia, the United Kingdom and the United States of America. Professor Stapleton's scholarship has been especially valuable to the courts in the development of tort law. Her tort law scholarship has enabled the creation and maintenance of a dialogue with the courts about the law as it is and the direction that it might take.

2. Reflexive Tort Scholarship

Professor Stapleton is an advocate of what she calls 'reflexive tort scholarship'. In the first of the Clarendon Law Lectures, which the Professor delivered in 2018, and which were later published,[1] she describes this scholarship in these terms:

> A core feature of this type of scholarship is that it takes the judicial role very seriously. It places at the centre stage what judges do, how they understand their role, the reasons they give in justification of their decisions, and the vital constitutional responsibility they bear to identify and articulate developments in the common law. Such scholarship is capable of both smoothly absorbing legal change signalled by courts and influencing courts with its insights.[2]

This form of scholarship may be contrasted with what Professor Stapleton calls the 'grand theory' approach to torts, which tends to characterise tort law as being 'all about one thing' or to argue for a single coherent justification for tort law and

[1] Jane Stapleton, *Three Essays on Torts* (OUP 2021).
[2] ibid xvii.

The Hon Susan Kiefel AC, *Professor Jane Stapleton and her Dialogue with the Courts* In: *Torts on Three Continents*. Edited by: Kylie Burns, Jodi Gardner, Jonathan Morgan and Sandy Steel, Oxford University Press. © The Hon Susan Kiefel AC 2024. DOI: 10.1093/oso/9780198889748.003.0001

is primarily directed at other academic lawyers.[3] Examples of grand theories are Richard Posner's theory that tort law is concerned with the efficient economic allocation of loss and burdens in society and Robert Stevens' claim that tort law is about the vindication of individual primary rights.[4] Professor Stapleton does not dismiss such theories and considers that they may be useful in other contexts. It is the reality of the law with which she is concerned. She rejects the descriptive claims of grand theories because they do not reflect what tort law actually is, or what judges do, or say they are doing, when deciding cases.

Professor Stapleton's scholarship is directed to judges and seeks a 'constructive dialogue' or a 'creative interactive conversation' with them.[5] She explains that reflexive tort scholarship is 'capable of ... influencing courts to confront tensions in judicial reasoning and doctrinal outcomes, to re-structure precedents and reassess terminology'.[6] She describes the approach as 'reflexive' because it signals a 'two-way conversation between legal academics and the Bench'.[7] Professor Stapleton says that reflexive tort scholarship 'addresses the law as it is'[8] and involves participation in the incremental development of the law by grounding the scholarship in existing principles and precedents. This type of scholarship requires an understanding of the context in which judges operate and the constraints that they are under. One major constraint is the pressure of time. Judges are conscious of timely justice, that justice delayed is justice denied. Time constraints may mean that they are not able to research topics in depth.[9] Judges also determine cases across a broad spectrum of the law, unlike academics who are able to specialise in particular areas. While academics can explore topics from various angles and test different theories, judges must come to a considered conclusion based on the facts of the case before them. They must rely on the parties' legal teams to raise arguments.

Another major constraint for judges is precedent.[10] Judges are not free to change the law arbitrarily but must 'proceed cautiously ... with the rich resource of principle anchored in precedent'.[11] The common law is developed incrementally. Reflexive tort scholarship therefore 'respects the constitutional responsibility on judges to identify and articulate developments in the living common law by evaluating heterogenous social values, expectations and the potential wider social consequences of these articulations'.[12] Reflexive scholarship focuses on judgments. It involves careful analysis and critical evaluation of judgments. In Professor

[3] ibid 2.
[4] ibid 21.
[5] ibid 2.
[6] ibid 2.
[7] ibid 2.
[8] ibid 17.
[9] ibid 3–4.
[10] ibid 4.
[11] ibid 13.
[12] ibid 33.

Stapleton's words, 'reflexive scholars identify connections and patterns, but most importantly weaknesses, tensions and anomalies in judicial reasoning, terminology and doctrinal outcomes'.[13] This enables judges to better incorporate the broader legal context in their decision-making and to build on or confront the connections or tensions identified in the academic work.

This does not mean that reflexive scholarship should avoid normative arguments. Professor Stapleton considers that a reflexive scholar may, for example, assist in identifying 'relevant social facts, values and expectations' in the course of offering policy reasons for adopting or rejecting a particular development.[14] However, she argues that the normative character of such arguments must be acknowledged and be separate from arguments made from doctrinal sources.[15]

3. An Overview of Professor Stapleton's Work

The assistance which can be obtained by judges from academic opinion cannot be underestimated. Judges have long benefitted from the critical analysis and thought of others on a range of topics. In 1955, Sir Owen Dixon, in his paper 'Concerning Judicial Method' said that textbooks and periodicals were often used by judges of the High Court.[16] In the United Kingdom, Lord Dyson has spoken of the 'symbiotic co-existence' of judges and scholars.[17]

Professor Stapleton's work has been cited by courts in Australia and the United Kingdom over a long period of time. Her work appears to have been first cited by the High Court of Australia in 1991, in *March v Stramare (E & MH) Pty Ltd*.[18] The timing is notable, given that she only graduated in law in 1981 and was awarded a D Phil in 1984. The impact of the Professor's work can, to an extent, be gleaned from citation statistics, although it is not possible to definitively count the exact number of times her work has been cited. The Law Cite index records that Professor Stapleton's work has been cited at least 170 times by courts. A search on Westlaw for citations in Australian appellate courts yields 60 cases in which her work has been cited. A search on Lexis Advance for citations in Australian courts yields 137 cases and in United Kingdom courts 71 cases. A search on AustLii of decisions of the High Court of Australia yields 18 cases.[19]

[13] ibid 18.
[14] ibid 20.
[15] ibid 20, 29.
[16] Sir Owen Dixon, 'Concerning Judicial Method' (1956) 29 *Australian Law Journal* 468, 470.
[17] Lord Dyson, 'Academics and Judges' in Lord Dyson (ed), *Justice: Continuity and Change* (Hart Publishing 2018) 35.
[18] *March v Stramare (E & MH) Pty Ltd* (1991) 171 CLR 506.
[19] ibid; *Hill v Van Erp* (1997) 188 CLR 159; *Perre v Apand Pty Ltd* (1999) 198 CLR 180; *Crimmins v Stevedoring Industry Finance Committee* (1999) 200 CLR 1; *Modbury Triangle Shopping Centre Pty Ltd v Anzil* (2000) 205 CLR 254; *Brodie v Singleton Shire Council* (2001) 206 CLR 512; *Henville v Walker* (2001) 206 CLR 459; *Tame v New South Wales* (2002) 211 CLR 317; *Woolcock Street Investments Pty*

Two themes of Professor Stapleton's work are useful to provide examples of her reflexive tort scholarship. They are: first, factual causation and the scope of liability; and secondly, pure economic loss.

A. Factual causation and the scope of liability

One of the earliest and influential contributions made by Professor Stapleton was her argument against what is called 'common sense causation'. In a study published in 1959, HLA Hart and Tony Honoré asserted that 'it is the plain man's notions of causation (and not the philosopher's or scientist's) with which the law is concerned'.[20] This view was adopted in the United Kingdom and in Australia. By way of example, in *Stapley v Gypsum Mines Ltd*,[21] Lord Reid said that the question of what caused an accident 'must be determined by applying common sense to the facts of each particular case'; and in *Naxakis v Western General Hospital*,[22] Gaudron J held that '[f]or the purpose of assigning legal responsibility, philosophical and scientific notions are put aside ... in favour of a common sense approach which allows that "breach of duty coupled with an [event] of the kind that might thereby be caused is enough to justify an inference, in the absence of any sufficient reason to the contrary, that in fact the [event] did occur owing to the act or omission amounting to the breach"'.

Professor Stapleton would have none of this. In a series of articles, she offered two criticisms of the common sense approach. In the first place she argued that the test was unable to serve as a useful precedent and labelled it 'an empty slogan', 'so indeterminate that it is effectively worthless as an analytical guide'.[23] She said:

> It brings the law into disrepute if, when confronted with a hotly disputed complex dispute about the appropriate point at which legal liability should be truncated, a court accepts the 'glib submission' that its resolution rests on nothing more than 'common sense'.'[24]

Ltd v CDG Pty Ltd (2004) 216 CLR 515; *Waller v James* (2006) 226 CLR 136; *Roads & Traffic Authority v Royal* (2008) 82 ALJR 870; *Imbree v McNeilly* (2008) 236 CLR 510; *Tabet v Gett* (2010) 240 CLR 537; *Amaca Pty Ltd v Booth* (2011) 246 CLR 36; *Strong v Woolworths Ltd* (2012) 246 CLR 182; *Alcan Gove Pty Ltd v Zabic* (2015) 257 CLR 1; *Lewis v Australian Capital Territory* (2020) 271 CLR 192; *Tapp v Australian Bushmen's Campdraft & Rodeo Association Limited* (2022) 273 CLR 454.

[20] HLA Hart and Tony Honoré, *Causation in the Law* (Clarendon Press 1959) 1.

[21] *Stapley v Gypsum Mines Ltd* [1953] AC 663, 781.

[22] *Naxakis v Western General Hospital* (1999) 197 CLR 269, 281 [36], citing *Betts v Whittingslowe* (1945) 71 CLR 637, 649.

[23] Jane Stapleton, 'Reflections on Common Sense Causation in Australia' in Simone Degeling, James Edelman and James Goudkamp (eds), *Torts in Commercial Law* (Thomson Reuters 2011) 331, 353.

[24] ibid 334.

Secondly, Professor Stapleton argued that the legal test for causation actually comprised two steps and that conflating the two into one common sense test obfuscates the real issue. The two steps may be referred to as 'factual causation' and 'scope of liability'. In her words:

> I have argued that courts should reject the packaging of disputes in vague causal terminology, it fails to distinguish between the two substantive underlying arguments relevant to the dispute about responsibility: the question of fact about whether the tortious conduct of the defendant, D, was historically involved with the claimant, C, suffering actionable damage; and the normative inquiry into whether the consequence of the tort that C is complaining about should be judged to be within the appropriate scope of liability for consequences of the tortious conduct.[25]

So understood, only the first test involves questions of causation as might commonly be understood by the ordinary person. The second test involves normative, policy and value judgements as to whether the law will hold the defendant liable for the claimant's damage by reference to concepts such as remoteness of damage.

In *March v Stramare*, Professor Stapleton's views were largely adopted by McHugh J.[26] His Honour was in the minority, however, and the majority continued to favour the common sense approach. It would not be until the enactment of legislation, following a review and report in 2002 by Justice David Ipp on the law of negligence,[27] that the law would reflect the views that Professor Stapleton had sought to convey to the courts. The background to the review was complaints from the insurance industry that the common law had developed in a way which rendered persons or bodies liable for negligence on a very broad and uncontrolled basis. Insurance premiums had soared, and many smaller bodies were unable to afford insurance. In the Ipp Report's section on causation, the Negligence Review Panel explicitly recognised the influence of Professor Stapleton. It stated:

> The Panel's consideration of and recommendations about causation have been greatly assisted by the work of Jane Stapleton, especially 'Causation and the Scope of Liability for Consequences'.[28]

The Ipp Report endorsed aspects of Professor Stapleton's criticism of the common sense approach and recommended that a two-pronged test of causation be adopted,

[25] Jane Stapleton, 'Cause-in-Fact and the Scope of Liability for Consequences' (2003) 119 *Law Quarterly Review* 388, 388.

[26] *March v Stramare (E & MH) Pty Ltd* (1991) 171 CLR 506, 534–6.

[27] *Review of the Law of Negligence: Final Report* (2002).

[28] ibid 109 fn 6, citing Stapleton, 'Cause In Fact and the Scope of Liability for Consequences' (n 25), forthcoming as it then was.

consisting of the factual causation and scope of liability steps.[29] It described 'factual causation' as concerning the factual issue of whether the negligence played a part in bringing about the harm and the 'scope of liability' as concerning the issue of the appropriate scope of the negligent person's liability for the harm, once it has been established that the negligence was a factual cause of the harm.[30]

This recommendation was adopted in the uniform civil liability legislation that resulted from the Ipp Report.[31] Subsequently, the High Court recognised that the legislation had departed from the common sense approach to treat factual causation and scope of liability as separate and distinct issues.[32] In *Strong v Woolworths Ltd*,[33] the Court endorsed Professor Stapleton's view that '[i]n particular cases, the requirement to address scope of liability as a separate element may be thought to promote clearer articulation of the policy considerations that bear on the determination'.

Intermediate appellate courts have used Professor Stapleton's work to assist in understanding the recommendations of the Ipp Report and the legislative provisions enacted in response. By way of example, in *Paul v Cooke*[34] it was explained that 'scope of liability' in s 5D(1)(b) of the *Civil Liability Act 2002* (NSW) is 'the means by which what Professor Stapleton described as the "voraciousness of negligence may be controlled by a court"'. On that basis it was held that the negligence of a radiologist who failed to diagnose an aneurysm in 2003 was a factual cause of the appellant's stroke suffered in 2006 during surgery in treatment of the aneurysm, however the misdiagnosis was not within the scope of the liability. This is because had the aneurysm been correctly diagnosed in 2003, the appellant would still have had to have faced the inherent risk of a stroke caused by the surgical procedure to remove the aneurysm.[35]

There is need to add a postscript to the Ipp Report which reflects a concern of Professor Stapleton's about the importance of precision of language in this area of the law. Although the Ipp Report endorsed the two-pronged approach, it continued to describe both steps under the umbrella term of 'causation'. Professor Stapleton was critical of this terminology, regarding it as a barrier to clarity of legal analysis.[36] She encouraged the courts to 'quietly ignore it'.[37]

[29] *Review of the Law of Negligence: Final Report* (2002) 108 [7.25] ff.
[30] ibid 11.
[31] See, eg, *Civil Liability Act 2002* (NSW), s 5D(1).
[32] *Adeels Palace Pty Ltd v Moubarak* (2009) 239 CLR 420, 440 [43].
[33] *Strong v Woolworths Ltd* (2012) 246 CLR 182, 191 [19].
[34] *Paul v Cooke* (2013) 85 NSWLR 167, 187 [86] (Leeming JA, Basten and Ward JJA concurring), citing Jane Stapleton, 'Legal Cause: Cause-in-Fact and the Scope of Liability for Consequences' (2001) 54 *Vanderbilt Law Review* 941, 956.
[35] *Paul v Cooke* (2013) 85 NSWLR 167, 194 [116].
[36] Stapleton, 'Reflections on Common Sense Causation in Australia' (n 23) 336.
[37] ibid 336, 364.

B. Pure economic loss

Turning then to the second theme of Professor Stapleton's work which is illustrative of her reflexive tort scholarship: pure economic loss. Professor Stapleton has written extensively on whether a claimant may recover pure economic loss by way of damage in an action for negligence. Historically the position was that a person did not owe a duty of care in respect of pure economic loss. This changed, to an extent, after the decision of the House of Lords in 1964 in *Hedley Byrne & Co Ltd v Heller and Partners Ltd*,[38] in which it was held that a defendant may be liable for pure economic loss caused by negligent misstatements, where there had been an assumption of responsibility by the defendant and there had been reliance on those statements by the plaintiff. The law developed after that case to allow recovery if a case fell within one of the few recognised categories of cases which were exceptions to the general rule. Professor Stapleton called this approach the 'pockets of case law' approach.[39]

In an article published in 1991 titled 'Duty of Care and Economic Loss: A Wider Agenda',[40] Professor Stapleton analysed decisions of the House of Lords concerning economic loss in order to determine what makes economic loss different from physical loss from the perspective of that court. She found that the principal substantive argument against recognition of pure economic loss as damage is the 'floodgates' argument—where the concern is that defendants might be held liable to an indeterminate number of claimants or for claims of an indeterminate size.[41] The argument is reflected in a well-known statement by Cardozo CJ, that the law should not expose defendants to uncertainty caused by 'liability in an indeterminate amount for an indeterminate time to an indeterminate class'.[42]

Clearly the 'pockets of case law' or categories approach was developed in order to limit the indeterminacy of economic loss claims by restricting liability to only discrete categories of cases. In *Sutherland Shire Council v Heyman*,[43] Brennan J explained it in this way:

> It is preferable, in my view, that the law should develop novel categories of negligence incrementally and by analogy with established categories, rather than by a massive extension of a prima facie duty of care restrained only by indefinable 'considerations which ought to negative, or to reduce or limit the scope of the duty or the class of person to whom it is owed'.

[38] *Hedley Byrne & Co Ltd v Heller and Partners Ltd* [1964] AC 465.
[39] Jane Stapleton, 'Duty of Care and Economic Loss: A Wider Agenda' (1991) 107 *Law Quarterly Review* 249, 249.
[40] ibid 249.
[41] ibid 254.
[42] *Ultramares Corp v Touche* (1931) 174 NE 441, 444.
[43] *Sutherland Shire Council v Heyman* (1985) 157 CLR 424, 481.

This statement was approved by the House of Lords in *Caparo Industries plc v Dickman*,[44] which is still the leading authority in the United Kingdom in this area of the law.

Professor Stapleton criticised the 'pockets of case law approach'. She argued that a process which is based on a particular selection of case law and which imposes or denies liability by analogy with them may sometimes be akin to 'the tail wagging the dog'.[45] That is because the selection of the 'relevant' pocket of cases precludes consideration of factors or policies which the courts have decided should govern the recognition of a duty of care. Consideration of these factors or policies, which she termed an 'agenda of policy concerns', would provide a more coherent approach.[46] The cases which did fall within the policies which the courts decided should govern the recognition of a duty of care would be the natural end product of analysis.[47]

According to Professor Stapleton, examples of policy considerations which might be taken into account include: indeterminacy concerns; whether the plaintiff was able to protect themselves from the risk of loss, which is to say the vulnerability of the plaintiff; whether statute or other areas of law impose conflicting or overlapping duties; and whether imposing a duty would allow the plaintiff to circumvent a prior contractual bargain as to the allocation of risk. From Professor Stapleton's perspective, this approach has the advantage that it better demonstrates what judges in fact do when they reason in this area of the law. Professor Stapleton's approach has not been expressly adopted in the United Kingdom and *Caparo* remains the authority. However, in 2007, in *Customs Excise Commissioner v Barclays Bank plc*,[48] Lord Walker quoted the Professor's criticisms with approval and observed that there had been some progress towards adopting her views.

In Australia, the concern about indeterminacy was historically governed by the test of 'proximity', which, however, was eventually disapproved in the leading case on pure economic loss, *Perre v Apand Pty Ltd*.[49] The defendant in that case negligently introduced a potato disease by causing the supply of infected seed to potato growers in South Australia. The plaintiffs were affected economically because their principal market was in South Australia where regulations prohibited the importation of potatoes which were grown within a certain distance of a known outbreak of the disease. The plaintiffs' farms were within that area. The question was whether the defendant owed the plaintiff a duty of care. The High Court held that it did.

[44] *Caparo Industries plc v Dickman* [1990] 2 AC 605.
[45] Stapleton, 'Duty of Care and Economic Loss: A Wider Agenda' (n 39) 284.
[46] ibid 284–5.
[47] ibid 285.
[48] *Customs Excise Commissioner v Barclays Bank plc* [2007] 1 AC 181, 209 [69]–[71].
[49] *Perre v Apand Pty Ltd* (1999) 198 CLR 180. See also *Sullivan v Moody* (2001) 207 CLR 562.

Three of the justices quoted extensively from Professor Stapleton's works, engaging with her criticism of the 'pockets of case law' approach. Those justices examined her view that there is no one test for a duty of care;[50] accepted her explanation of how indeterminacy may come about in pure economic loss cases;[51] endorsed her view that the emergence of a coherent body of precedent will be impeded by the imposition of fixed categories of cases where recovery will be allowed;[52] and discussed her view that no one clear principle can guide novel cases and that, even if there is some consensus achieved regarding the relevant factors to be weighed, different judges may, quite reasonably, give different weights to competing factors.[53] The features identified as relevant to a duty of care in *Perre* included geographical proximity, the defendant's control over the risk, the defendant's awareness of the risk being created, the vulnerability of the plaintiff and the potential indeterminacy of the claim. It will be observed that Professor Stapleton's 'agenda of policy concerns' bears a similarity to the approach taken in *Perre* to what are salient factors. It is also similar because it is an endorsement of evaluative judgment, with no one factor being regarded as determinative of liability, and it rejects the 'pockets of case law' approach. The main difference between the two approaches is that the High Court regards its approach as based on principle rather than policy.[54]

Consistently, and perhaps characteristically, Professor Stapleton has questioned the terminology of 'principle'. She asks, 'Is a concern that liability not be indeterminate a principle or a policy?'.[55] In her view an appeal to 'principle', so-called, 'simply masks the substance of a judge'-s reasoning process'.[56] And so, the dialogue which is reflexive tort scholarship continues.

4. Conclusion

The well-known comparative lawyer, Professor Hein Kötz has said that 'the weight to be given [to] a view expressed by a legal writer depends on the cogency of the

[50] *Perre v Apand Pty Ltd* (1999) 198 CLR 180, 210 [74] (McHugh J), citing Jane Stapleton, 'Duty of Care Factors: A Selection from the Judicial Menus' in Peter Cane and Jane Stapleton (eds) *The Law of Obligations: Essays in Celebration of John Fleming* (Clarendon Press 1998) 60–2; *Perre v Apand Pty Ltd* (1999) 198 CLR 180, 253 [198] (Gummow J), citing Stapleton, 'Duty of Care Factors: A Selection from the Judicial Menus' (n 50) 60–3.

[51] *Perre v Apand Pty Ltd* (1999) 198 CLR 180, 221 [106]–[107] (McHugh J), citing Stapleton, 'Duty of Care and Economic Loss: A Wider Agenda' (n 39) 255.

[52] *Perre v Apand Pty Ltd* (1999) 198 CLR 180, 254 [200] (Gummow J).

[53] ibid, 325 [404] (Callinan J), citing Stapleton, 'Duty of Care Factors: A Selection from the Judicial Menus' (n 50) 88.

[54] See *Sullivan v Moody* (2001) 207 CLR 562, 579 (Gleeson CJ, Gaudron, McHugh, Hayne and Callinan JJ).

[55] Jane Stapleton, 'The Golden Thread at the Heart of Tort Law: Protection of the Vulnerable' (2003) 24 *Australian Bar Review* 135, 135.

[56] ibid 136.

argument, the reputation of the author and the honesty of [the] scholarship'.[57] Professor Jane Stapleton's work qualifies on all three counts. Her scholarship has guided judges and other lawmakers to critically analyse aspects of tort law. It has encouraged, if not demanded, intellectual honesty and rigour and clarity of expression.

[57] Kötz, 'Scholarship and the Courts: A Comparative Survey' in David S Clark (ed), *Comparative and Private International Law: Essays in Honor of John Henry Merryman on His Seventieth Birthday* (Duncker & Humblot 1990) 189.

2

Unifying Principles within Pluralist Tort Adjudication

Mark A. Geistfeld[*]

1. Introduction

The common law has never had an express rationale, and so scholars have long tried to identify its immanent point or purpose. Tort law provides particularly fertile ground for these efforts, sprouting a number of 'grand theories' as Jane Stapleton puts it, which she defines as 'arguments that the law should ... be conceived as all being about one thing ...; or that it is only normatively coherent if springing from a single integrated justification'.[1]

The connection between normative coherency and a single integrated justification for the law was famously developed by Ronald Dworkin, who defended the conception of law as integrity. 'Law as integrity supposes that people are entitled to a coherent and principled extension of past political decisions even when judges profoundly disagree about what this means.'[2] To develop and illustrate the concept, Dworkin used the construct of Hercules, 'an imaginary judge of superhuman intellectual power and patience who accepts law as integrity'.[3] Rather than imposing his own grand theory from above, Hercules accounts for tort law as it exists and tries to interpret it in a coherent manner based on two dimensions: the extent to which an interpretation fits existing legal materials, and the strength of the justification for interpreting tort law in this particular way.[4]

Employing this interpretive methodology, scholars have debated whether tort liability is best understood as a form of welfarism that formulates tort rules to efficiently minimize the social cost of accidents, or whether tort liability instead instantiates a principle of either corrective or distributive justice grounded in

 I'm grateful for the many helpful comments I received from Jonathan Morgan and Sophia Moreau, from participants in the Oxford symposium honouring Jane Stapleton, and from participants in the law faculty workshop at NYU. Financial support was provided by the Filomen D'Agostino and Max E. Greenberg Research Fund of the New York University School of Law.

[1] Jane Stapleton, *Three Essays on Torts* (OUP 2021) 1–2.

[2] Ronald Dworkin, *Law's Empire* (Belknap Press 1986) 134.

[3] Ibid 239.

[4] Ibid 255–56.

Mark A. Geistfeld, *Unifying Principles within Pluralist Tort Adjudication* In: *Torts on Three Continents.* Edited by: Kylie Burns, Jodi Gardner, Jonathan Morgan and Sandy Steel, Oxford University Press. © Mark A. Geistfeld 2024.
DOI: 10.1093/oso/9780198889748.003.0002

Kantian right or other forms of non-welfarist morality. By deeming these foundational principles to be the products of 'grand theories', Stapleton is flagging a fundamental disconnect between their conceptual purity and empirical reality:[5]

> Judges do not describe the law of torts as being all about the central thing that any of these theories asserts it is about.... On the contrary, it is patent from judicial reasoning and extra-judicial statements that, in identifying developments in the common law of torts, appellate judges take into account a rich, heterogeneous array of social facts, values, concerns, and expectations.[6]

Consistently with Stapleton's observation, eminent judges have complained that the legal academy's theoretical focus on foundational principles is useless for both the bench and bar.[7] Relatedly, two Reporters for the *Restatement (Third) of Torts* have observed that in their efforts to restate tort doctrines and their associated judicial rationales, 'tort theory plays virtually no role'.[8] Insofar as the single integrated rationales for tort law are practically irrelevant, they merit the pejorative label 'grand theories'.

This line of criticism has been further developed by Stephen Smith, who also begins with the observation that judges do not defend rulings with monistic rationales such as allocative efficiency or Kantian right. Judges instead apply 'settled rules' to the case at hand or otherwise 'develop new rules by extending established rules', neither of which typically involves or requires 'a justification for the rules that these practices apply or develop'.[9] Consequently, '[t]he distinguishing feature of judicial justifications for legal rules, at least so far as their form is concerned, is that they are almost always expressed in terms of intermediate moral principles'.[10]

These intermediate principles, such as 'fairness, dignity, honesty, reasonableness, autonomy' and the like, are 'more general than the particular rules that they are employed to justify', while also being 'narrower than the foundational

[5] The term 'grand theory' refers to 'Any theorizing involving a general explanation of society or human experience rather than the study of particular societies or experiences. Often used pejoratively within empiricist perspectives.' Oxford Reference, <https://www.oxfordreference.com/view/10.1093/oi/authority.20110803095903555> accessed 20 February 2024.

[6] Stapleton (n 1) 25–26.

[7] See ibid 34 (quoting Chief Justice John Roberts of the US Supreme Court: 'What the academy is doing, as far as I can tell, is largely of no use or interest to people who actually practice law. Whether it's analytic, whether it's at whatever level they're operating, it doesn't help the practitioners or help the judges.'); see also, for example, Harry T. Edwards, 'The Growing Disjunction Between Legal Education and the Legal Profession' (1992) 91 Michigan Law Review 34, 35 ('[I]t is my impression that judges, administrators, legislators, and practitioners have little use for much of the scholarship that is now produced by members of the academy.').

[8] Nora Freeman Engstrom and Michael D. Green, 'Tort Theory and Restatements: Of Immanence and Lizard Lips' (2021) 14 Journal of Tort Law 333, 371.

[9] Stephen A. Smith, 'Intermediate and Comprehensive Justifications for Legal Rules' in Simone Degling (ed), *Justifying Private Rights* (Bloomsbury Publishing 2021) 63, 66 (sentence structure omitted).

[10] Ibid 69.

principle or principles that define comprehensive moral theories' such as welfarism or Kantian right.[11] Intermediate moral principles are the only ones that expressly factor into judicial decision-making, and so tort 'theories that invoke comprehensive justifications run the risk of transforming the law into something unrecognisable'.[12]

Judges resolve tort disputes with intermediate moral principles for reasons most fully developed by Cass Sunstein, who has long maintained that 'well-functioning legal systems produce stability and agreement in the midst of social disagreement and pluralism' by relying on incompletely theorized agreements that enable judges 'to be clear on the result' for the case at hand 'without agreeing on the most general theory that accounts for it'.[13] An incompletely theorized agreement is based on intermediate moral principles such as fairness or reasonableness that different judges can justify in different ways. The resultant judicial decision does not fully specify or completely theorize the justification for any of the intermediate principles in question, allowing different judges to agree on the case outcome even though they disagree about the single best justification for the decision. According to the prevailing account of pluralism, this type of judicial decision-making is necessary to legitimize tort law in a society riven by a plurality of competing conceptions about what is good or right.[14]

So, too, the ideal judge is no longer exemplified by Hercules, who instead now represents the type of judicial approach that ought to be avoided. By attempting to justify all forms of tort liability with a single integrated justification, a judge who tries to act like Hercules 'would prevent people who disagree on large principles from reaching consensus on particular outcomes. In this way, Hercules could not really participate in ordinary judicial deliberations; he would be seen as a usurper, even an oddball.'[15]

For largely these same reasons, pluralist tort adjudication would seem to entail that tort scholarship should also be pluralist. 'If judicial justifications for legal rules should be acceptable to citizens with different comprehensive views, then academic justifications for legal rules should be similarly acceptable. A justification succeeds or fails regardless of who offers it.'[16]

To be clear, the mid-level principles that largely shape pluralist adjudication are theoretical—they abstract from narrowly specified rules and the particular facts of concrete cases. The principles such as fairness or reasonableness are just not

[11] Ibid (sentence structure omitted).
[12] Ibid 64.
[13] Cass Sunstein, *Legal Reasoning and Political Conflict* (2nd edn, OUP 2018) 4–5 (sentence structure omitted).
[14] Steve Hedley, 'Private Law Theory: The State of the Art' (8 September 2021) 8–9 <https://papers.ssrn.com/sol3/papers.cfm?abstract_id=3917777> accessed 14 October 2023 (summarizing views of different scholars taking this position).
[15] Sunstein (n 13) 49.
[16] Smith (n 9) 81.

completely theorized in the sense that judicial opinions do not justify them with a single integrated rationale for tort law. Hence the practice of pluralist tort adjudication conducted in terms of intermediate principles would seem to imply that tort scholars should engage in mid-level theorizing that straddles between unduly narrow doctrinal analysis and overly general fully theorized comprehensive theories. The scholarly ideal, on this account, is a theoretical mid-point that identifies intermediate moral principles capable of informing and thereby improving pluralist judicial decision-making in individual cases.[17]

For example, Stapleton forcefully defends 'reflexive torts scholarship' that 'seeks a creative interactive conversation with judges'.[18] A reflexive tort scholar responds to the difficulty courts face in 'identifying and articulating the living common law given the complex heterogeneous social values and expectations from which it springs'.[19] Instead of shoe-horning this complexity into a grand theory,

> [t]his sort of scholarship is fully accountable to the cases. By 'accountable' I mean the technique interrogates the facts, the result, and, in particular, the reasons judges give to justify that result. By critically evaluating the reasons judges actually give for their determination, reflexive scholars identify connections and patterns, but most importantly weaknesses, tensions, and anomalies in judicial reasoning, terminology, and doctrinal outcomes. These insights are invaluable to Bench and Bar, whose members rarely have the time to construct such broad analytical perspectives of the evolving common law landscape.[20]

Though appealing when framed in this manner, mid-level theorizing begs important questions; it is, ironically, undertheorized. Most importantly, the claim that judicial decisions must always find justification in a plurality of competing values presents an obvious problem: How do courts proceed when confronted by claims requiring judges to choose among conflicting principles? Although pluralism clearly poses this question, '[m]any of the pluralist tort theorists accept the mixture of objectives without inquiry into the problem of internal consistency'.[21] The problem, moreover, would seem to be intractable. Harmony requires a unifying principle that pluralism apparently rejects, which would seem to imply that the destiny of a pluralist tort system resides in unprincipled resolutions of

[17] Compare Engstrom and Green (n 8) 334 ('The doctrinal level at which Restatements operate and the case law that fuels the production of Restatements—ground level law—is a disjunction from theory, which operates at 30,000 feet.').

[18] Stapleton (n 1) 2.

[19] Ibid 22.

[20] Ibid 18.

[21] Englard, Izhak, 'The Idea of Complementarity as a Philosophical Basis for Pluralism' in David G. Owen (ed), *Philosophical Foundations of Tort Law* (OUP 1995) 183, 184.

conflicting pluralist values—a characterization many leading tort scholars have effectively adopted.[22]

Whether tort law is riddled with these problems depends on how one conceptualizes the practice of pluralist tort adjudication and its relation to tort theory. As I will try to demonstrate, pluralist tort adjudication that proceeds on the basis of incompletely theorized agreements embodies unifying principles that can satisfy the demands of law as integrity.

Most obviously, pluralist tort adjudication is normatively coherent when conducted in the context of an overlapping consensus of the competing pluralist values. To identify an overlapping consensus, one must apply the different foundational moral principles that plausibly describe tort law in order to determine whether their demands coincide with the class of cases under consideration. Doing so is not merely a pedantic exercise. When a case is situated within an overlapping consensus, doctrinal analysis is considerably sharpened by eschewing mid-level theorizing in favour of the more fully theorized foundational moral principle best fitted for addressing the issue at hand.

Wide swaths of tort law find justification within an overlapping consensus and are normatively coherent in that domain, an important point tort scholars have largely overlooked. A focus on this attribute of tort law also reveals important classes of cases in which pluralist values conflict.

Within this space of hard cases, there is an overlapping consensus concerning the reasonable structure of pluralist adjudication. Litigants have an institutional right to be treated equally under the law, which precludes judges from resolving hard cases by simply invoking their preferred value systems when doing so would conflict with other pluralist values. The common law implements this requirement of equal treatment with its characteristic mode of judicial decision-making based on analogical reasoning—treating like cases alike.

Analogical reasoning depends on unifying principles for categorizing cases in order to determine whether one case is sufficiently 'like' another to merit equal treatment. These unifying categorical principles make pluralist tort adjudication a dynamic system for implementing law as integrity. Throughout this entire process, monistic interpretive theories are integral for developing the concepts and

[22] See, for example, Kenneth S. Abraham and Edward G. White, 'Conceptualizing Tort Law: The Continuous and Continuing Struggle' (2021) 81 Maryland Law Review 293, 343 ('[T]ort law is a series of causes of action—the classic set of fragmented "civil wrongs not arising from contract" that it has always been. Any effort to make it more than that, except at the most general level, is bound to fail.'); Charles O. Gregory, 'Trespass to Negligence to Absolute Liability' (1951) 37 Virginia Law Review 359, 359 (observing that 'many' teachers who 'present the law of torts as they find it' 'conclude[] that there is no such thing as the "law of torts" but only a mass of contradictions not teachable as a body of knowledge'); Emily Sherwin, 'Interpreting Tort Law' (2011) 39 Florida State University Law Review 227, 241 (concluding that 'there is no reason to think that tort law is a composite whole rather than a rationally inexplicable heap'); A.W.B. Simpson, 'The Common Law and Legal Theory' in A.W.B. Simpson (ed), *Oxford Essays in Jurisprudence, Second Series* (Clarendon Press 1973) 97, 99 (characterizing the common law as 'degenerate', 'more like a muddle than a system').

doctrines of tort law in a normatively coherent manner that enforces the institutional right of litigants to be treated equally under the law.

2. Mid-Level Theory Illustrated: Pure Economic Loss in Contractual Settings

In both the UK and the US, courts have had difficulty determining how tort law applies to pure economic losses that arise in contractual settings. The disarray in the case law presents exactly the type of problem that mid-level theorists are supposed to address, and Stapleton without doubt is one of the soundest practitioners.

Insightfully moving beyond the categorical confines of black-letter tort doctrine, Stapleton identifies an important analogy: like contract law, tort law embodies 'a parallel principle of support for cooperative human arrangements'.[23] Stapleton illustrates the tort principle of cooperation with cases involving one party (such as a landowner) who gives reasonable notice that she disclaims liability to others, and another party (an entrant on the land) who then chooses to face the risk in light of that notice. The parties do not have a contractual relationship, and yet tort law relies on the disclaimer to bar recovery under the assumption of risk doctrine. Stapleton concludes that the reason for this limitation of liability is based on the principle of cooperation. 'By giving legal effect to the landowner's notice, tort law vindicates this particular cooperative arrangement and gives support to similar future cooperative arrangements, such as future landowners allowing others to take a short cut across their land.'[24]

Based on a nuanced and extended discussion of cases in which commercial parties are situated within a web of contracts, Stapleton concludes that the principle of cooperation adequately explains why courts rely on disclaimers to bar recovery. For example, the principle explains why courts bar recovery for pure economic losses caused by defective products, even when there is no contractual privity between the manufacturer and the injured commercial consumer: 'courts are vindicating a market expectation of economic self-reliance by commercial parties, which the parties to the arrangement are taken to have accepted when they came together to cooperate rather than walk away (and with which expectation the recognition of a duty would be inconsistent).'[25]

Acknowledging that courts are much more willing to recognize the tort duty in cases involving physical harm, Stapleton distinguishes them based on the nature of the loss. 'There is a clear public interest in actors taking care not to injure others physically, so it is understandable that the law tends to judge it as reasonable for us

[23] Stapleton (n 1) 36.
[24] Ibid 37.
[25] Ibid 47.

to rely on others being careful of our physical interests.' By contrast, pure economic losses are not proximately caused by a predicate physical harm, and so tort law's 'modest response reflects a view that, in our society, it is not generally assumed that we are our brother's or sister's economic keeper, at least in terms of avoiding economic loss to them by our careless conduct'.[26]

Having identified the principle of cooperation with robust application in cases of pure economic loss, Stapleton then asks how it applies to 'investment decision' cases that occur within a web of contracts among commercial parties. For reasons her prior analysis elucidates, courts 'emphasise the expectation of economic self-reliance by commercial parties' in denying recovery for plaintiffs who suffer bad investment outcomes the defendant's negligence allegedly caused.[27] However, in another line of cases courts permit tort recovery based on the plaintiff investor's reliance on a defendant who had 'assumed responsibility' for supplying information and was careless in doing so. According to Stapleton, courts distinguish the two lines of cases only with 'opaque notion[s]' that do not rest on 'normatively salient factual differences'.[28] Absent such normative distinction, the two lines of cases cannot be squared with the principle of cooperation, making one set inconsistent with the other—a 'doctrinal discontinuity' unmasked by mid-level theorizing.

Stapleton's analysis illustrates a particular type of pluralist theorizing. Her approach abjures any express dependence on foundational principles and instead relies on an incompletely theorized principle of general application (the tort principle of mutual cooperation) that unifies otherwise seemingly disparate cases (those involving landowners and entrants, defective products, and investment decisions), which in turn enables the analyst to identify inconsistencies that ought to be eliminated. The generalizing mid-level principle of mutual cooperation can easily be justified by a plurality of values, and judges could engage in this type of analysis without ever invoking any type of grand theory and stirring up the associated controversy.

3. Pluralism and Overlapping Consensus in Contractual Settings

The type of theorizing Stapleton employs is not the only possible way to analyse the tort problem of pure economic loss in a pluralist manner. In the case of an overlapping consensus, a category of cases has attributes that satisfy the normative concerns of *all* the plausible comprehensive moral theories for interpreting

[26] Ibid 46.
[27] Ibid 47.
[28] Ibid 64.

tort law.[29] Doctrinal analysis that relies on only one of these foundational principles is still pluralistic in this context. It ensures that the rule conforms to this particular principle, and, by extension, to all other principles within the overlapping consensus.

Although Stapleton's analysis of pure economic loss does not expressly make this move, one could readily justify the underlying principle of cooperation with an overlapping consensus of allocative efficiency (mutually beneficial voluntary exchange) and rights-based concerns of personal autonomy, equal freedom, or liberty (to choose whatever risks one wants to face and thereby incur responsibility for that choice). The open question is whether any of these foundational principles can meaningfully guide the judicial inquiry, or whether the best mode of judicial analysis instead depends on intermediate principles of the type Stapleton invokes, in which case resort to foundational principles would be a form of 'grand theorizing' as she claims.

As Stapleton's analysis illustrates, pluralists who unnecessarily ignore foundational moral principles run a considerable risk of neglecting normatively relevant issues. Comprehensive theories are formulated to canvass the full range of relevant concerns. Different theories depend on different foundational moral principles, but when these comprehensive theories overlap within the relevant category of cases, reliance on their foundational principles provides greater assurance that the pluralist analysis will not miss anything of normative importance.

Consider the mid-level pluralist principle of cooperation within Stapleton's analysis. In her view this principle has less robust application in cases of physical harm, though she recognizes that in some cases disclaimers can bar recovery for these injuries.[30] Hence her formulation of the principle potentially applies to a standard case of strict products liability involving the manufacturer of a defective product that physically harms a commercial consumer. These parties have no direct contractual relationship but are instead situated within a web of contracts running through the commercial chain of product distribution. According to Stapleton's analysis, a manufacturer should be able to disclaim tort liability by giving the commercial consumer reasonable notice that it is doing so.

However, courts deem these disclaimers to be violative of public policy.[31] How, then, can the recognition of such a tort duty be squared with Stapleton's

[29] See John Rawls, 'The Idea of An Overlapping Consensus' (1987) 7 Oxford Journal of Legal Studies 1, 9 ('Yet despite the fact that there are opposing comprehensive conceptions affirmed in society, there is no difficulty as to how an overlapping consensus may exist. Since different premises may lead to the same conclusions, we simply suppose that the essential elements of the political conception, its principles, standards and ideals, are theorems, as it were, at which the comprehensive doctrines in the consensus intersect or converge.').

[30] Stapleton (n 1) 40–44 (applying the principle of cooperation to cases of this type).

[31] American Law Institute, *Restatement (Third) of Torts: Products Liability* (American Law Institute Publishers 1998) §18. As in this instance, the following discussion is based on US law but relies on reasoning that presumably generalizes in the relevant substantive respects to the UK.

formulation of the cooperation principle that relies on disclaimers to bar tort recovery on the ground that 'the parties to the arrangement are taken to have accepted when they came together to cooperate rather than walk away (and with which expectation the recognition of a duty would be inconsistent)'? Is the reason simply that the injury involves physical harm rather than economic loss? If so, why does this distinction matter only in some cases and not others? As these questions indicate, the principle of cooperation needs to be more fully specified.

A pluralist analysis that instead relies on foundational moral principles supplies the requisite normative content. In product cases pitting consumers (and not bystanders) against manufacturers, any fairness or rights-based argument centres on the need to protect the interests of consumer right-holders without unfairly or unjustly burdening manufacturers as duty-bearers. A welfarist argument, by contrast, exclusively focuses on efficiently minimizing the social cost of product accidents. For this class of cases, the concerns for fairness or justice entirely overlap with the minimization of accident costs, making it possible to justify and more fully specify the principle of cooperation with one of these foundational principles.

Because product cases are situated within a web of contracts, consumer right-holders internalize both the costs and the benefits of the manufacturer's tort obligations. For the equilibrium in a market governed by tort rights and their correlative obligations, the manufacturer is entitled 'to generate a sustained profit, above the cost of doing business'.[32] The manufacturer's cost of doing business includes those costs which are imposed as a matter of legal obligation. The ordinary consumer accordingly pays (via higher product prices or decreased product functionality) for both the safety investments and the injury compensation that tort law requires of the manufacturer.[33] The ordinary consumer also incurs the costs of any uncompensated injuries. A liability rule that minimizes the sum of these costs also maximally protects consumer interests. Hence liability rules that minimize consumer costs—the socially efficient outcome (in the absence of bystander harms)—also satisfy any rights-based concern for maximally protecting consumer interests without unfairly disadvantaging manufacturer duty-bearers.[34]

By relying on the overlap among the plausible foundational principles of tort law, this type of pluralist analysis rectifies an important distributive mistake that

[32] *Tincher v Omega Flex, Inc.*, 104 A 3d 328, 385 (Pa 2014).

[33] Costs and benefits cannot be determined until the associated legal entitlements have first been specified, and so that baseline determines how a tort rule distributes costs and benefits across duty-bearers and right-holders. See Mark A. Geistfeld, 'The Tort Entitlement to Physical Security as the Distributive Basis for Environmental, Health, and Safety Regulations' (2014) 15 Theoretical Inquiries in Law 387, 392–98. For the market equilibrium at that baseline, the manufacturer must cover all of its costs, including liability costs. Hence the price that consumers pay for the product covers the full costs of manufacturer tort liability.

[34] Compare Dworkin (n 2) 390 ('[T]he point of constitutional adjudication is not to merely name rights but to secure them, and to do so in the interests of those whose rights they are.'); *id.* at 392 ('Hercules' thesis is at least plausible, that judges have a duty to enforce constitutional rights up to the point at which enforcement ceases to be in the interests of those the rights are supposed to protect....').

underlies Stapleton's analysis. In her view, 'whenever tort recognises a claimant has an entitlement against a defendant, the claimant could be seen as getting a "free ride" because the claimant has not paid for the entitlement. For example, the principle in *Donoghue v. Stevenson* gives consumers physically injured by carelessly made products a free ride.'[35] Insofar as it would be inequitable for tort law to give one interacting party a free ride at the expense of the other, Stapleton provides a further reason why tort law bars recovery under the principle of cooperation. However, consumers do not get a 'free ride' for reasons made clear by economic analysis, the reliance on which does not drag one into controversial positions about the competing foundational principles of welfarism or rights-based concerns of fairness or justice—they all overlap in this category of product cases.

When more fully developed with economic analysis, the tort principle of co-operation has substantive content that is missing from Stapleton's account. As a matter of basic microeconomics, contracting predictably results in unreasonably dangerous products when consumers are not well informed about product risk.[36] Contract law, however, does not adequately protect the expectation interest of poorly informed consumers who suffer product-caused physical harms, creating an important role for tort law.

Aside from extreme cases of unconscionable terms, contract law does not address the fairness of transactions involving uninformed or unsophisticated buyers.[37] An unconscionable term, moreover, only renders the contract (or that particular term) unenforceable.[38] So, too, an uninformed buyer's unilateral mistake only voids the contract and perhaps generates a claim for restitution.[39] None of these remedies adequately protect the expectation interest of a poorly informed buyer across all cases.[40]

For example, the average buyer cannot determine whether a food product is fit for human consumption. Unable to inspect the product for contaminants like germs, the purchaser ordinarily does not have adequate information about this component of product quality. In most cases, a buyer discovers that the food is

[35] Stapleton (n 1) 51.

[36] See Mark A. Geistfeld, *Principles of Products Liability* (3rd edn, Foundation Press 2020) 45–52.

[37] American Law Institute, *Restatement (Second) of Contracts* (American Law Institute Publishers 1981) § 208 cmt c ('Inadequacy of consideration does not of itself invalidate a bargain, but gross disparity in the values exchanged may be an important factor in a determination that a contract is unconscionable.').

[38] Ibid § 208; American Law Institute and Uniform Law Commission, *Uniform Commercial Code* (American Law Institute Publishers 2003) § 2–302.

[39] E. Allan Farnsworth, *Farnsworth on Contracts* (3rd edn, Little, Brown 2004) 623.

[40] Contract law would protect the expectations of poorly informed buyers if it permitted reformation of the contract to embody the agreement that purchasers would have made if they were adequately informed. However, a 'court will not reform a document to reflect an agreement that the court merely *thinks* the parties *would have decided to make* had they not been mistaken'; reformation instead requires 'clear and convincing evidence' showing why the agreed-upon contract did not contain the language in question. E. Allan Farnsworth, *Alleviating Mistakes: Reversal and Forgiveness for Flawed Perceptions* (OUP 2004) 101.

unwholesome only after consuming it and getting sick. Transactions of this type do not satisfy the baseline assumption of contract law that the contracting parties are adequately informed, which explains why the contractual remedies are not adequate. The seller's return of the purchase price or replacement of the defective product with wholesome food would not compensate the buyer for the sickness the contaminated food caused.

Since medieval times, the common law has addressed this problem by protecting the purchasers of food products with the implied warranty of quality.[41] This warranty 'was in its origin a matter of tort liability' that for centuries has subjected the seller of contaminated food to strict liability.[42] By assuring compensation for injuries caused by unwholesome food, the implied warranty protects consumer expectations that commercially distributed food is fit for human consumption. The buyer did not expect to be made sick by the food, and the guarantee of compensation for that injury remedies the buyer's frustrated expectation.

In this respect, the tort remedy functions no differently from the ordinary contractual remedy 'of compensation based on the injured party's expectation'.[43] But unlike a contractual remedy, the tort remedy is formulated to protect the safety expectations of poorly informed buyers by obligating commercial sellers to provide products that the ordinary consumer would purchase or use if well informed.[44] Hence the tort rule supports mutual cooperation in product markets by protecting consumer expectations of safe product performance and correcting the market failure that would otherwise inhibit these voluntary transactions.

Consumers are not always poorly informed, in which case they need not be protected by the tort duty. Indeed, a tort duty that required outcomes different from the voluntary choices of well-informed consumers would be both inefficient and paternalistic for cases that only implicate consumer interests. Consequently, when the ordinary consumer makes an informed choice (such as purchasing a convertible rather than a sedan), the seller is relieved of any obligation to make that identical choice in the opposite manner (as would occur if the court held that the safer properties of a sedan render the convertible defectively designed).[45] As these cases illustrate, the principle of cooperation can sometimes bar recovery in cases of physical harm, but the reason has nothing to do with the important differences between

[41] *Jacob E. Decker & Sons Inc. v Capps*, 164 SW 2d 828, 830 (Tex 1942) (tracing the rule 'as far back as the year 1266 A.D.').

[42] James Barr Ames, 'The History of Assumpsit' (1888) 2 Harvard Law Review 1, 10 (explaining that the implied warranty of quality 'stood anciently upon the ... footing' of tort law).

[43] Farnsworth, *Farnsworth on Contracts* (n 39) 190.

[44] See Geistfeld, *Principles of Products Liability* (n 36) 10–21, 96–97 (explaining why the tort version of the implied warranty requires sellers to provide products that consumers would purchase if they knew the product's 'true character', a rule courts first recognized in cases of contaminated food that they then extended via analogical reasoning to generate the rule of strict products liability).

[45] Ibid 130–40, 320–21 (explaining why this so-called rule against categorical liability is best defined in terms of informed consumer choice that functions no differently from primary assumption of risk).

physical harms and pure economic losses, or with the presence of a disclaimer—the most relevant factors within Stapleton's formulation.

So, too, consumers ordinarily have enough information to protect their economic interests by contracting, explaining why courts generally bar recovery for pure economic loss (lost profits and so on) while also recognizing important exceptions involving the financial costs of asbestos abatement and medical monitoring (monetary expenses for preventing future physical harm).[46] Once again, the case outcomes depend on the ordinary consumer's information about the risk, not the nature of the loss or any notice supplied by the defendant's disclaimer of liability.

The liability rules support the principle of cooperation by ensuring that informational problems do not impede the voluntary interactions between consumers and manufacturers. Disclaimers do not solve this informational problem, and so courts refuse to enforce a product seller's attempt to avoid tort liability via a disclaimer: 'It is presumed that the ordinary product user or consumer lacks sufficient information and bargaining power to execute a fair contractual limitation of rights to recover.'[47] The unenforceability of these disclaimers is clearly explained by the information-based formulation of the cooperation principle, whereas Stapleton's formulation is more ambiguous, asking what 'the parties to the arrangement are taken to have accepted when they came together to cooperate rather than walk away (and with which expectation the recognition of a duty would be inconsistent)'?

All of these points can be recast in rights-based terms of fairness or justice. After all, consent is normatively meaningful only when it is both voluntary and adequately well informed.[48] Recognizing as much, courts conclude that '[s]trict liability for failure to warn [about product risks] is founded on twin principles of social utility and the right of the individual to determine his own fate.'[49] Product cases are situated within an overlapping consensus which makes it possible to justify these liability rules with any of the underlying foundational principles.

Although various norms can guide the analysis in these cases, for reasons the foregoing discussion illustrates, economic methodology is tailored for analysing contracting problems in product or service markets. Though overtly instrumental and welfarist, economic analysis nevertheless supplies the best form of pluralistic reasoning in this context: it does not conflict with concerns for fairness or justice while being considerably more incisive than mid-level theorizing. Pluralism

[46] Ibid 290–300.

[47] American Law Institute, *Restatement (Third) of Torts: Products Liability* (American Law Institute Publishers 1998) §18 cmt a.

[48] See, for example, John Kleining, 'The Nature of Consent' in Franklin Miller and Allen Werthheimer (eds), *The Ethics of Consent: Theory and Practice* (OUP 2010) 3, 13–18 (explaining why the linkage between consent and responsibility for the ensuing outcomes depends on competent agents making voluntary, informed choices).

[49] *Pavildes v Galveston Yacht Basin, Inc.*, 727 F2d 330, 338 (5th Cir 1984).

presumably should embrace judicial reliance on one or more foundational moral principles in cases of an overlapping consensus.

4. Does Pluralist Adjudication Always Require an Overlapping Consensus?

Welfarist and non-welfarist comprehensive theories of tort law overlap within contractual settings due to the distributive properties of these social relations—the parties voluntarily cooperate and expect to mutually benefit by doing so, with all of the normatively relevant costs and benefits being internalized to the cooperative interaction. Tort law, however, also governs social interactions that do not share these normative attributes, creating ample opportunities for conflicts to arise among the requirements of welfarist and non-welfarist foundational principles.[50]

On one view, these inevitable value conflicts can still be accommodated within a pluralist theory of adjudication. According to Smith, 'It is precisely for this reason that, today, it is typically the legislature, not the judiciary, that develops the law in these areas.'[51] For example, Stapleton argues that in cases of negligently caused stand-alone emotional harms, 'there was not even a broad consensus on how relevant social concerns might be distilled into coherent doctrine', making the issue a 'classic illustration' of the situation described by Lord Reid: 'Parliament is the right place to settle issues which the ordinary man regards as controversial.'[52] On this view, if an overlapping consensus cannot justify a tort rule, then its viability turns on controversial value choices that require legislative resolution.

Though elegant, this formulation of pluralism does not persuasively explain why judges do not routinely invoke foundational moral principles to analyse tort problems. If, for example, a judge concludes that the rule under consideration cannot be squared with the welfarist requirements of allocative efficiency, then the analysis identifies the type of value conflict that must be resolved by the legislature, not the court. Foundational principles either form part of the justificatory pluralist principle within an overlapping consensus or otherwise establish an inevitable conflict in the case at hand, giving each one of them decisive importance for *all* tort rules within this account of pluralism.

This formulation of pluralist tort adjudication fails on its own terms. It apparently licenses judges to invoke monistic moral principles when analysing virtually any tort question, even though it seeks to explain why judicial decision-making

[50] For extended discussion, see Mark A. Geistfeld, 'Fault Lines in the Positive Economic Analysis of Tort Law' in Jennifer Arlen (ed), *Research Handbook on the Economics of Torts* (Edward Elgar Press 2013) 149–68.

[51] Ibid 74.

[52] Stapleton (n 1) 14 (quoting Lord Reid, 'The Judge as Law Maker' [1972] JSPTL 23, n 4).

is typically based on incompletely theorized agreements formulated in terms of intermediate principles such as fairness or reasonableness.[53]

This pluralist account is even more troubling than its proponents have recognized. If judge-made tort law must always be justified by the plurality of values embodied in an overlapping consensus of competing foundational moral principles, then courts would never be able to rationally make trade-offs among these pluralist values. This type of pluralism radically undermines the legitimacy of tort adjudication, something that is likely to be anathema for most judges.

In unpacking what it means for adjudication to be rational, William Lucy identifies three different standards. The first is simply that adjudication depends on reasons for action or decision, which 'would seem to be far too lenient'.[54] A more demanding account of rationality 'insists that the decisions judges reach be based upon genuine reasons and thus requires sorting and evaluation of the reasons that might support the decision'.[55] And yet another, still more demanding form of rationality involves weighing the genuine reasons that have first been given and then sorted from the non-genuine reasons. This form of rationality is most germane to the pluralist account. 'Weighing reasons to X only makes sense when there is a plurality of genuine reasons to X. Such plurality also makes sense of those situations, in adjudication and beyond, in which it seems that a choice either way is plausible, supported by reasons, and thus in no sense obviously wrong'.[56]

Although the judicial weighing of reasons is well-suited for pluralist adjudication, the claim that pluralism forecloses judges from choosing among the competing pluralist values prevents tort adjudication from being rational in this respect.[57] If judges cannot decide cases by weighing, balancing, or otherwise trading off incommensurable pluralist values, then tort adjudication would still be rational—it would involve judicial decisions based on genuine reasons (foundational principles) which are then sorted (by an overlapping consensus of the varied foundational principles) in order to make a decision. But in the absence of an overlapping consensus, these pluralists deny that tort adjudication can be rational in the more robust sense that judges are able to weigh the conflicting competing foundational principles to determine how they apply in a particular case.

The incommensurability of conflicting pluralist values would make it quite 'unlikely that adjudication is any more rational than many non-legal and non-arbitrary means of formulating and resolving disputes'—the claim made by critical legal scholars and other sceptics of adjudication.[58] And once adjudication ceases

[53] See *supra* notes 5–12 and accompanying text.
[54] William Lucy, 'Adjudication' in Jules L. Coleman and Scott Shapiro (eds), *The Oxford Handbook of Jurisprudence and Philosophy of Law* (OUP 2018) 206, 229.
[55] Ibid 230.
[56] Ibid 230–31.
[57] Ibid 240.
[58] Ibid 241–42.

to be any more rational than a host of other non-legal, non-arbitrary alternatives, it becomes apparent why some pluralists maintain that judges should defer to the legislature when pluralist values conflict.

Deference to the legislature, however, does not neutrally solve the problem in the manner that pluralists depict. Judicial inaction in the face of value conflict would inevitably protect welfarist foundational principles at the expense of rights-based principles. In general, an individual right protects an interest from invasions or interferences justified solely by collective concerns such as maximizing social welfare. One has a right to be protected against discrimination, for example, even if most members of the community would otherwise benefit by engaging in this behaviour. But under the pluralist guise of legislative deference, common-law judges could never enforce an individual tort right when it conflicts with majoritarian preferences that largely drive aggregate social welfare—the type of case that creates the need for legally protecting the right in the first instance.

Pluralism raises a number of troubling issues when formulated as an approach requiring judicial reliance on an overlapping consensus of the competing moral principles that plausibly explain tort law. If tort rules must be justified by such an overlapping consensus, judges would regularly employ foundational principles either to sharpen the analysis in cases of overlapping consensus or otherwise to show that a particular issue involves a conflict among pluralist values which requires legislative resolution. However, judges do not ordinarily engage in this type of reasoning. They often weigh competing interests and values rather than refusing to resolve the issue on the ground that the legislature must do so.[59] The practice of pluralist tort adjudication is hard to square with the claim that tort law must always be justified by an overlapping consensus of otherwise inherently conflicting pluralist values.

5. Pluralism as Process

The formulation of pluralism we have been analysing is a comprehensive rationale for tort law. It unifies tort law with a foundational moral principle—an overlapping consensus of pluralist values—that obligates judges to defer to the legislature when such consensus does not exist. Like the comprehensive rationales defined in terms

[59] This type of decision-making is perhaps most apparent in cases requiring courts to determine whether they should limit the duty to exercise reasonable care. See, for example, *Rowland v. Christian*, 443 P2d 561, 564 (Cal 1968) ('A departure from this fundamental principle involves the balancing of a number of considerations; the major ones are the foreseeability of harm to the plaintiff, the degree of certainty that the plaintiff suffered injury, the closeness of the connection between the defendant's conduct and the injury suffered, the moral blame attached to the defendant's conduct, the policy of preventing future harm, the extent of the burden to the defendant and consequences to the community of imposing a duty to exercise care with resulting liability for breach, and the availability, cost, and prevalence of insurance for the risk involved.').

of a unifying monist moral principle such as welfarism or Kantian right, the pluralist foundational principle of an overlapping consensus does not adequately describe the practice of tort law. None of these foundational principles plausibly fit tort law, leading to the question whether there are other unifying principles that could make tort law normatively coherent.

This question cannot be adequately resolved without considering the possibility that pluralism is best interpreted as a process of the common law rather than the necessary rationale for all forms of tort liability. Pluralist tort adjudication could be the process the common law employs to identify unifying principles among a host of viable options—*procedural pluralism*—in which case it would be a mistake to interpret pluralism as embodying the endpoint or ultimate objective of adjudication—*substantive pluralism*, the characterization we have thus far been considering.

Procedural pluralism implicates a distinctive type of right 'that individuals hold against courts rather than (as in the case of ordinary rights) other individuals'.[60] Tort adjudication involving the ordinary or primary tort rights that plaintiffs seek to enforce against defendants takes place within an institutional framework that endows the individual litigants with a new set of rights. 'When claimants argue that they have a right to a particular "remedy" from the court, they are arguing that the law requires that the court do something—namely, make an order against the defendant.'[61]

Such an 'action'[62] or 'institutional'[63] right includes the right of each individual litigant to be treated equally under the law. Framed at this level of generality, this institutional or second-order action right can be justified by an overlapping consensus: courts must conduct the adjudication concerning primary or first-order tort rights in a manner that maximally respects the plurality of values while also protecting the right of each litigant to be treated equally under the law. This 'political conception of justice', as John Rawls explains, 'is different from many familiar moral doctrines [such as welfarism or Kantian right], for these are widely understood as general and comprehensive views' that fully specify the content of primary or first-order tort rights and their correlative duties.[64] In contrast, 'a political conception of justice involves, so far as possible, no prior commitment to any wider doctrine. It looks initially to the basic structure [of tort adjudication] and tries to elaborate a reasonable conception for that structure alone.'[65] Litigants can

[60] Stephen A. Smith, 'The Rights of Private Law' in Andrew Robertson and Tang Hang Wu (eds), *The Goals of Private Law* (Hart Publishing 2009) 113, 119.
[61] Ibid.
[62] Ibid.
[63] Ronald Dworkin, *Taking Rights Seriously* (Harvard University Press 1977) 81.
[64] Rawls (n 29) 3.
[65] Ibid 4.

agree about the reasonable rules of adjudication while fundamentally disagreeing about the best rationale for tort law.

The question then becomes, how does this abstract institutional right of litigants shape the actual practice of pluralist tort adjudication? Does the second-order overlapping consensus concerning the reasonable structure of adjudication plausibly provide another unifying principle for the rules of tort law governing primary or first-order tort rights and their correlative duties? If so, this unifying principle would supplement the overlapping consensus that independently applies to litigants' primary tort rights in many tort cases, like those situated within the contractual settings analysed in section 3 above.

6. Procedural Pluralism and Common-Law Adjudication

The common law has always relied on the process of pluralist adjudication for reasons that trace back to its sources in medieval England. According to the legal historian J.H. Baker, 'the unifying force' in this era was 'not a common law but the general social and moral assumptions of the age, or even the natural instincts of mankind at particular stages of development: broad parallels are often found to transcend national and geographical boundaries'.[66] Based on the historical sources now available to us, the 'general social and moral assumptions' of primitive societies were shaped by the normative requirements of reciprocity or equitable balance: An eye for an eye in cases of aggression, the value of an eye for an eye in cases of accidental injury, and the quid pro quo of voluntary mutual exchange.[67] These reciprocity norms generated customs for regulating social interactions that respectively mapped into areas of the early common law now demarcated as criminal law, tort law, and contract law.[68]

But even though the early common-law courts could readily identify customary norms of behavioural reciprocity, enforcing them through the legal system required justification. An enforceable norm, as William Blackstone put it negatively in his synthesis of the early common law, 'must not be unreasonable'.[69] The legal inquiry into reasonableness turned on questions such as whether 'reasonable, honest and fair-minded men' would adopt the custom, whether it was 'in accordance with fundamental principles of right and wrong', or whether the custom might implicate utilitarian or distributional concerns that rendered it unreasonable for

[66] John H. Baker, *An Introduction to English Legal History* (4th edn, Butterworths 2002) 1–2 (emphasis added).

[67] See Mark A. Geistfeld, 'Hidden in Plain Sight: The Normative Source of Modern Tort Law' (2016) 91 New York University Law Review 1517, 1523–48 (describing how the reciprocity practices of primitive societies form the basis of the self-help model of legal development that legal historians widely employ to explain the emergence of centralized governance in England and elsewhere).

[68] Ibid 1548–50 and 1567–68.

[69] 1 BL Comm 77.

having the effect of being 'to the general prejudice, for the advantage of any particular person', or 'where the effects of a custom fell unevenly on a class of people'.[70] Whether courts would enforce a customary norm of behavioural reciprocity accordingly depended on the plurality of values embodied within the legal concept of reasonableness—the same type of pluralistic, incompletely theorized mid-level moral principle that structures modern tort adjudication.

Pluralistic reasoning is an essential component of the 'classical common law jurisprudence' for reasons Gerald Postema describes:

> [L]aw, common lawyers maintained, is not concerned with the moral vision of any individual, however soundly argued it may seem to be, but rather with the convergence of the views and judgments of the larger community, and forging and maintaining a common sense of reasonableness. Salience, not vision, and pragmatic convergence, not theoretical coherence, were its fundamental aims.[71]

To attain these ends, the common law is designed to protect pluralist values. When a ruling can be justified by a plurality of values, a judicial decision that instead invokes only one of the foundational principles would be mere dicta in this respect—the reliance on the single moral principle would be unnecessary for resolving the case and therefore not properly considered to be part of the ruling. By limiting the legal reach of foundational monistic principles, this common-law practice protects a plurality of values in tort adjudication.

However, the 'fundamental aims' of the common law also include the objective of 'pragmatic convergence' as Postema puts it, so how could its pluralist institutional design plausibly attain that end? Most obviously, the common law could attain pragmatic convergence by requiring that liability rules must always be justified by an overlapping consensus of pluralist values. But as we have found, this form of judicial reasoning does not plausibly describe tort practice and is merely another type of grand theory. The only remaining possibility for common-law courts to attain pragmatic convergence is through the process of pluralist adjudication.

The common law is famously dynamic, not static. Courts have incrementally developed the common law with analogical reasoning, 'arguing from one case to the next on the basis of perceived likenesses and differences and the location of the instant case in the landscape of common experience painted by the judge or lawyer in command of the full resources of the common law'.[72] Analogical reasoning relies on unifying categorical principles that make it possible for a pluralistic body of tort rules to pragmatically converge over time into a normatively coherent body of law.

[70] The quotations come from the case law and commentaries as collected in the survey by David J. Bederman, *Custom as a Source of Law* (Cambridge University Press 2012) 33–34 (footnotes omitted).
[71] Gerald J. Postema, 'Philosophy of the Common Law' in Jules Coleman and Scott Shapiro (eds), *Oxford Handbook of Jurisprudence and Philosophy of the Law* (OUP 2002) 588, 594.
[72] Ibid.

To determine whether one case is sufficiently 'like' another despite their myriad differences, courts must ask whether they are in the same category of cases.[73] The construction of these categories depends on the substantive reasons or principles for grouping individual cases in this way.[74] Analogical reasoning accordingly depends on general theoretical components that identify the unifying principle(s) for a category of individual cases that merit equal treatment under the law.

Like pluralist tort adjudication, analogical reasoning does not depend on fully theorized principles for unifying cases within legal categories. 'To say that one case is like another, we need a reason or principle, but we can often offer a reason or a principle that operates at a low level of ambition.'[75] Analogical reasoning has this property, according to Sunstein, because it proceeds from fixed points or widely shared agreements not subject to serious debate, such as deeply engrained social norms or the legal conclusion that a particular precedent or rule is clearly binding in a certain set of circumstances (so-called 'easy' cases). By relying on these fixed points when engaged in analogical reasoning, courts can develop unifying principles for categorizing cases even when they fundamentally disagree about the underlying moral rationale for those principles.[76]

As courts confront more cases over time, the principles for unifying cases within a single category can become more general. For example, cases involving trespasses on land are in a different legal category from those involving conversion (real property cannot be converted). Nevertheless, these early common-law rules, along with a host of others such as assault, battery, and fraud, now comprise the broader substantive category of the intentional torts. Despite their apparent differences, these torts all address aggressive interactions and accordingly cohere into a normatively distinct category of tort law.[77]

The unifying categorical principles embodied within analogical reasoning can drive pluralist tort adjudication towards pragmatic convergence. In the theoretical limit, the dynamic properties of the system point towards a single substantive principle for unifying the entire category of tort law—the basic reason why different forms of liability applicable to different sets of circumstances are nevertheless substantively consistent with one another, ultimately comprising a normatively coherent body of the common law.

[73] See, for example, ibid 604 (observing that because analogical reasoning 'is reasoning by or from example … from like to like', it 'presupposes some degree of (threshold) relevant similarity').

[74] See, for example, Neal MacCormick, *Legal Reasoning and Legal Theory* (Clarendon Press 1978) 186 ('[N]o clear line of distinction can be drawn between argument from legal principles and argument from analogy. Analogies only make sense if there are reasons of principle underlying them.').

[75] Sunstein (n 13) 61.

[76] Ibid 77.

[77] See generally Mark A. Geistfeld, 'Conceptualizing the Intentional Torts (2017) 10 Journal of Tort Law 159, 177–78 (explaining why the intentional torts are a normatively and substantively distinct form of tort liability that coherently address aggressive interactions in which the defendant intentionally invaded or interfered with a legally protected interest of the plaintiff).

7. Analogical Reasoning and the Rule of Law

Analogical reasoning relies on unifying principles because the 'analogizer is committed to consistency'.[78] This commitment does not fetishize formal logic; it instead 'appreciates that the value served by the demand to treat like cases alike is a substantive value of political morality—a kind or form of fairness' that protects the institutional rights of litigants to be treated equally under the law.[79]

This property is most apparent in hard cases requiring judges to choose among conflicting pluralist values. Hard cases implicate what Dworkin calls the 'institutional' right of litigants to receive from the judge 'a justification for a decision' resolving the dispute.[80] Such an institutional right gives one of the litigants 'a right to win' a hard case, leaving the losing party without any such entitlement.[81] Even though one wins and the other loses, each litigant is treated equally under the law whenever the court adequately justifies its decision in a manner that satisfies the requirement of equal treatment.

In a hard case, different pluralist values justify the construction of different legal categories, each of which generates a different legal outcome (resulting either in liability or in no liability). To decide that the case at hand is sufficiently 'like' others in one of these categories, courts must justify their decisions with substantively sound analogical reasoning that strives to identify the categorical principle for unifying cases in this way. Such a justification for the decision satisfies the requirement of equal treatment because it adequately explains why the result in the case at hand is substantively consistent with other analogous cases receiving the same treatment.

A pluralist judge who denied that such a unifying principle is immanent in the case law could adjudicate hard cases only by arbitrarily invoking one of the conflicting pluralist principles. Does it make sense to say that the winning party was 'entitled' to this outcome, and that the loser should be placated by a judicial explanation to the effect that 'there were a number of good reasons for deciding either way, but I had to choose one of them, and you lose'. Maybe you lose because the judge doesn't like your race, gender, or political leanings. Pluralism can invite bias and discrimination rather than giving voice to a diversity of different values and perspectives.

In contrast, the adjudicative requirement that judges must justify their decisions with substantively sound analogical reasoning prevents judges from exclusively relying on their own moral and political views to resolve hard cases. Hence substantively sound analogical reasoning protects the institutional rights of both

[78] Sunstein (n 13) 76.

[79] Gerald Postema, 'Law's System: The Necessity of System in Common Law' (2014) 2014 New Zealand Law Review 69, 100.

[80] Dworkin, *Taking Rights Seriously* (n 63) 81.

[81] Ibid 93.

litigants to be treated equally under the law. Even though one wins and the other loses, each litigant is treated equally under the law because the court resolved the dispute with a substantive principle that tort law has already applied to similarly situated litigants in analogous cases.

The commitment to substantive consistency instantiates other rule-of-law values for reasons Jeremy Waldron emphasizes:

> there is a felt requirement essential to law that its norms make some sort of sense in relation to one another; even apart from formal consistency, we should interpret them so that the point of one does not defeat the point of another. That is why the use of analogy is appropriate in legal argument; it builds on a sense that individual norms are not self-contained and that the point of any one of them may have some bearing on how it is appropriate to think about any other. This broader sense of the systematicity of law helps explain why we think of a body of law as consisting of not just legislation and decisions in particular cases, but also principles whose content reflects powerful themes that run implicitly through the whole body of law and that are reflected in various ways in its explicit norms.[82]

By functioning as a system of interrelated rules, tort law 'can present itself to its subjects as a unified enterprise of governance', thereby enabling individuals 'to make sense of the "big picture", with an understanding of how the regulation of one set of activities relates rationally to the regulation of another.'[83] Hence the commitment to substantive consistency and systematicity 'is a necessary condition of law's normative guidance', according to Postema, 'and since law's distinctive modus operandi is normative guidance, de facto systematicity is essential to law'.[84]

The commitment to substantive consistency explains why pluralist tort adjudication can satisfy rule-of-law requirements across the full set of cases, both easy and hard. This dynamic process is driven by substantively sound analogical reasoning, which depends on an identifiable set of unifying principles for categorizing cases in order to determine whether the case at hand is more analogous to the cases in one category rather than another. These principles for unifying cases within legal categories enable a pluralist tort system to function as a system of interrelated rules, thereby furthering the associated rule-of-law values of fair notice and equal treatment.

[82] Jeremy Waldron, 'The Concept and the Rule of Law' (2008) 43 Georgia Law Review 1, 34.
[83] Ibid (sentence structure omitted).
[84] Postema, 'Law's System: The Necessity of System in Common Law' (n 79) 89.

8. Pluralism and Judicial Reasoning

A commitment to substantively sound analogical reasoning does not entail the need to conduct tort litigation on a pluralistic basis, nor does it guarantee that tort adjudication will ultimately identify the single best unifying principle for justifying this body of the common law. 'The fact that people agree that case A is analogous to case B does not mean that Case A *or* case B is rightly decided.'[85] Consequently, as Frederic Schauer correctly points out, '[t]o believe that the common law works itself pure is to believe that subsequent cases correct the errors of earlier ones far more than they add errors to previously sound doctrines, and that new cases present opportunities for refinement rather than occasions for mistake'.[86]

Once we recognize the obvious problem of judicial error, another property of pluralist tort adjudication becomes apparent. Acknowledging that they can be mistaken about their understanding of the single best interpretation of tort law, judges could defensibly rely on the widest range of rationales that would justify their decisions in particular cases. Prudence can persuasively explain why judges base their decisions on a plurality of values.

To be sure, judges might reason in a pluralist manner because doing so is dictated by their underlying moral philosophy. According to Kenneth Simons, 'Most judges (and ordinary lay people) are not absolutists or monists, even with respect to their foundational beliefs and principles. More likely, they are utilitarians who recognize deontological and distributive justice constraints, or deontologists who acknowledge utilitarian overrides once a threshold of social harm has been exceeded.'[87] Perhaps judges engage in pluralistic legal reasoning as a matter of foundational moral principle rather than prudence.

The types of pluralist foundational moral theories Simons identifies are neither utilitarian nor rights-based. They instead are comprised of welfarist and nonwelfarist moral principles which are inherently conflicting in the sense that there is at least one set of circumstances in which the principles require incompatible behaviours. Both types of principles cannot comprehensively govern all tort cases, and simply invoking one to constrain the other is an ad hoc solution that illustrates the problem of normative coherency within pluralist moral theories.[88]

85 Sunstein (n 13) 9.

86 Frederick Schauer, 'The Failure of the Common Law' (2004) 36 Arizona State Law Journal 765, 777.

87 Simons, Kenneth, 'Justifying and Categorizing Tort Doctrines: What is the Optimal Level of Generality?' (2022) 14 Journal of Tort Law 551, 558 (paragraph structure omitted).

88 See, for example, Alexander, Larry, 'Bad Beginnings' (1996) 145 University of Pennsylvania Law Review 57, 85 ('Correct moral principles will never dictate their own abandonment. They might dictate that we follow bright-line rules, even at some moral cost in particular cases, in return for greater moral benefits generally (moral costs and benefits calculated by reference to the correct moral principles). Correct moral principles, however, will reject all other purported justificatory principles as counterfeit.'); Robert Stevens, 'Rights in Conflict' in Andrew Robertson and Tang Hang Wu (eds), *The Goals of Private Law* (Hart Publishing 2009) 140, 140–41 ('By what further criterion is a judge to determine whether considerations of interpersonal justice can be trumped by concerns for the general welfare, or

Judges presumably are not committed to incoherent decision-making strategies, and so a more plausible reason why they or anyone else might adopt a pluralist moral theory that stems from empirical contingencies—the practical constraints of time and effort that prevent judges and most if not all of us from fully working out a coherent theory of morality governing each and every possible situation we might confront. Unsure of whether a particular comprehensive moral theory is the right one, judges would understandably adopt one highly plausible theory tempered by the others, at least until a judge has had the time or need to figure out which one makes the most sense when their inherent incompatibilities become manifest.

Recognizing that they do not have unlimited time and full subject-matter mastery of each and every legal issue, humble judges in a pluralist society should realize that their interpretations might be mistaken. Judges also know that they (often) are unelected officials who in any event are not authorized to unilaterally shape the law to their own views. Rather than impose their potentially mistaken comprehensive theories of the law to decide the case at hand, prudent judges would instead develop their interpretive theories incrementally on a case-by-case basis, constructing analogies for attaining substantive consistency across cases.

This type of judicial decision-making resolves cases on the basis of mid-level moral principles that satisfy the requirements of substantive consistency and equal treatment. Decisions of this type are capable of the widest range of pluralist justification, including whatever comprehensive rationale each judge has tentatively identified. This kind of decision-making leaves open the greatest latitude for future normative development of tort law in a manner that a judge might have mistakenly missed without unfairly treating the parties in the case at hand.

This interpretation of pluralist tort adjudication explains why judicial reliance on a single comprehensive moral principle would be mere dicta when applied to legal rules that can be more widely justified with pluralist rationales. This practice reflects the concern that justifications reaching beyond the confines of the case at hand run an unnecessary risk of error that does not adequately recognize competing pluralist values, each of which is otherwise sufficiently interpretively plausible to justify tort law.

Thus, in an adjudicative system designed to attain normative coherency on a case-by-case basis through the institutional obligation of treating like cases alike, prudence can explain why judges eschew reliance on the single best foundational principle for tort law. Judges who reason in this way recognize that their interpretive conclusions, like those of their predecessors, are revisable and might not stand the test of time. Prudence in the face of legal error explains why time-pressed judges base their decisions on a plurality of values within an adjudicative

vice versa, when these concerns do not exist on the same plane? One cannot be a bit of a utilitarian or a half-hearted Kantian, and it is certainly impossible to try and be a combination of the two.').

framework designed to produce over time and across cases a single integrated rationale for tort law.[89]

9. Pluralist Torts Scholarship

Having identified the role of unified principles within pluralist tort adjudication, we can now see how legal scholars might help judges in a pluralist society develop tort doctrine in a substantively sound manner. Scholarship based on mid-level theorizing of the type Stapleton and others champion is self-consciously formulated along these lines, although it should be clear by now that pluralist legal scholarship can also embrace monist foundational principles such as allocative efficiency or Kantian right.

To engage in pluralist tort adjudication, judges must know the foundational monist principles worthy of recognition; that is, whether a principle derives from a comprehensive moral theory that adequately fits and justifies tort law. In this obvious and yet basic respect, judges would greatly benefit from the efforts of legal scholars who have the time and subject-matter expertise to more fully work out these interpretations.

To identify a plausible unifying principle, scholars in the first instance must largely focus on well-established doctrines and practices in order to show that the proffered interpretation adequately fits the most important features of tort law. Having largely ignored the kinds of contentious liability questions that vex judges, these first-generation comprehensive theories will strike many as being practically irrelevant. But once these comprehensive theories have been adequately developed, scholars can then apply them to identify cases in which there is an overlapping consensus of the pluralist values. This type of second-generation theorizing—illustrated by the doctrinal analysis in section 3 above—belies the claim that comprehensive monist theories are irrelevant for practice.

[89] After discussing the manner in which Hercules, the Dworkinian construct of a superhuman judge, approaches the resolution of hard cases, federal judge Gerard Lynch addressed 'the humbler question of how a working judge struggles with the many questions that confront him or her in deciding actual cases'.Gerard E. Lynch, 'Complexity, Judgment, and Restraint' (2020) 95 New York University Law Review 621, 623. In doing so, Lynch describes a mode of judicial reasoning in a manner that is largely consistent with the account provided in this section without expressly discussing the roles of analogical reasoning, incompletely theorized agreements, and other related matters. Compare Dworkin (n 2) 257–58 ('No mortal judge can or should try to articulate his instinctive working theory so far, or make that theory so concrete and detailed, that no further thought will be necessary case by case. He must treat any general principles or rules of thumb he has followed in the past as provisional and stand ready to abandon these in favor of more sophisticated and searching analysis when the occasion demands.'). By contrast, Dworkin's depiction of Hercules is only supposed to flesh out 'theoretical disagreement in law', and so his discussions about how Hercules resolves legal questions do not address 'the reasons judges have for tempering their statements' of what they think is the 'most accurate' interpretation of the law: ibid 11–12.

Many, if not most, tort rules are justified by an overlapping consensus about the underlying primary or substantive tort rights. This largely overlooked property of tort law explains why its interpretation is so contestable: each of the plausible foundational moral theories adequately fits and justifies wide swaths of tort law. The resultant overlap of these theories stems from the institutional obligation of judges 'to enforce ... rights up to the point at which enforcement ceases to be in the interests of those the rights are supposed to protect'.[90] Enforcing rights in this manner requires analysis of how different forms of tort liability affect the legally protected interests of right-holders, which in turn shows why an overlapping consensus inheres in so many tort rules.[91]

In the remaining set of hard cases—those in which there is no overlapping consensus of the pluralist values—disagreement is unavoidable. The legal materials, after all, are still open to interpretation. These issues are contentious for reasons comprehensive moral theories make clear. Having identified the underlying policies and principles at stake in these hard cases, tort scholars can try to determine which pluralist value or values would render the case law substantively coherent across the relevant doctrinal domain, thereby helping judges to apply analogical reasoning in a substantively sound manner. Thus, in addition to helping judges improve upon doctrinal analysis in cases of an overlapping consensus, scholars who theorize in this manner can help judges ground their decisions on the proper set of pluralist values in the remaining set of hard cases.[92]

10. Conclusion

The ongoing fact of pluralist tort adjudication calls into serious question whether tort law plausibly coheres into a single integrated moral justification such as welfarism or Kantian right. A healthy political society values diverse viewpoints and their associated moral theories. Recognizing as much, Jane Stapleton and numerous other tort scholars have concluded that tort law is committed to pluralism

[90] Dworkin (n 2) 392 (discussing constitutional rights); see also ibid 390 ('[T]he point of constitutional adjudication is not to merely name rights but to secure them, and to do so in the interests of those whose rights they are.').

[91] The non-welfarist moral theories of tort law limit the ability of the tort system to pursue social welfare in violation of individual tort rights. To identify an overlapping consensus of welfarism and these non-welfaristic moral theories, the maximization of social welfare can be constrained by the requirements of the individual tort rights in question. For cases in which the constraint is not binding, enforcement of the individual right coincides with the maximization of social welfare. This property explains why the cost-minimizing tort rules embraced by welfarism also best protect the relevant interests of right-holders who incur the costs of the correlative duty within either contractual relationships or objectively defined reciprocal risky interactions. For more extensive analysis, see Geistfeld, 'Hidden in Plain Sight: The Normative Source of Modern Tort Law' (n 67) 1564–82. Hard cases are those in which the constraint is binding, and so the pursuit of welfare conflicts with the right in question and requires courts to resolve that conflict.

[92] For my analysis along these lines, see ibid 1582–92.

as a substantive matter. Their reasoning seems incontrovertible: tort adjudication has always been conducted in terms of incompletely theorized mid-level principles such as reasonableness or fairness that can be justified by a plurality of values. Attempts to interpret tort law in terms of a single integrated justification miss the mark on this view, turning them into 'grand theories' that are practically useless for the bench and bar.

Though otherwise valuable, substantive pluralism is problematic in the context of adjudication. If tort law must always be formulated to protect a plurality of competing values, it could not justify judicial decision-making in hard cases, nor could it meaningfully protect individual rights or otherwise provide individuals with requisite guidance on how they should behave when pluralist values conflict. These problems explain why the practice of pluralist tort adjudication is hard to square with the claim that courts must always justify tort liability with a plurality of values.

Rather than entailing a commitment to pluralism as the substantive rationale for tort law, pluralist adjudication is more plausibly characterized as a dynamic process of constructive interpretation. Tort litigation takes place within an institutional structure that gives litigants rights against courts, including the right to be treated equally under the law. To implement the overlapping consensus about this institutional right, tort adjudication is structured so that common-law judges resolve tort claims with analogical reasoning based on mid-level unifying normative principles for categorizing cases. Over time, these lower-level unifying principles have the potential to converge into a single integrated justification for the entire category of tort law.

Although idealized in this respect, the emergence of such a comprehensive unifying principle would not be some sort of instantiation of Platonic justice. It would instead derive from an intersubjective agreement within the political community about the best substantive principle immanent in the extant body of tort law for determining one individual's responsibility for another's injury.

A unifying principle could not survive this process if it embodied an unduly narrow conception about what normatively matters. A comprehensive unified principle recognizes the diversity of social values and weighs each one of them in relation to itself. If the unifying principle were one of self-determination or personal autonomy, for example, then tort law would mediate conflicting interpersonal interests in physical security, liberty, emotional tranquillity, and so on, in terms of their relative importance for the individual exercise of autonomy within the community. This type of unifying principle can still be incompletely theorized in the sense that judges can continue to disagree about the correct understanding of personal autonomy and its legal implications. Nevertheless, a principle of this type enables judges to weigh and balance conflicting interests and values in a manner that is normatively coherent across cases.[93]

[93] To employ the terminology of the US Declaration of Independence, the important tort rules and limitations of liability can be plausibly interpreted as balancing one individual's right to life against

Pluralist tort adjudication can be validly characterized as a dynamic process for attaining such unity even if it will never realistically attain that end. As Oliver Wendell Holmes famously observed, 'By the necessity of its being', the common law 'is always approaching and never reaching consistency. It is forever adopting new principles from life at one end, and it always retains old ones from history at the other end.... It will become entirely consistent only when it ceases to grow.'[94]

another's right to liberty in relation to the importance of these respective rights for the 'pursuit of happiness'. See Mark A. Geistfeld, *Tort Law: The Essentials* (Aspen Publishers 2008).

[94] Oliver Wendell Holmes, 'Common Carriers and the Common Law' (1879) 13 American Law Review 609, 631.

3

The Torts Scholar as Disaggregator

*Anita Bernstein**

1. Introduction

Writing about tort law crafted primarily by judges in published English-language common law decisions, the scholar fêted in this volume chose to build by un-building. Or so I shall contend in this celebration, aware that this claim about a choice to un-build is mine rather than a characterization announced in that work. Jane Stapleton created her corpus by a method just as interesting as the acclaimed substance in it.

The method is disaggregation. As practised by Stapleton, this approach takes aim at diverse targets. Schools of thought, decisional law (mostly from courts in England and Australia), syntheses like Restatements, and predilections of torts scholars have earned scrutiny from our honouree. This chapter reviews half a dozen patterns. Each of them, though separate from the others and emergent over decades in multiple writings, joins a coherent whole.

2. The Disaggregator's Work

Three related but distinct traits characterize disaggregation as a method.

The first is attitudinal, a matter of temperament. A disaggregator is inclined to focus on divergences more than unities. Differences stick out and make a strong claim on attention and analysis.

Second, disaggregation steers the scholar to favour primary sources—statutes, decisional law, constitutions—over secondary materials, which describe or comment on the law rather than constitute it.[1] An implicit hierarchy that places primary above secondary legal sources in a simple ordinal ranking enjoys ample support among readers who are not disaggregators,[2] but the disaggregator brings

* Thanks to Benjamin Smith for an array of inspired contributions.

[1] On the distinction, see Michael Whiteman, 'Appellate Jurisprudence in the Internet Age' (2017) 14 Nw. J. Tech. & Intell. Prop. 255, 271.

[2] See ibid. (observing that American appellate judges cite primary materials much more than the secondary kind); William A. Hilyerd, 'Using the Law Library: A Guide for Educators Part IV: Secondary Sources to the Rescue' (2005) 34 J.L. & Educ. 273, 274 (claiming that 'secondary sources have some serious drawbacks' and function best as instruments that lead readers to primary sources).

Anita Bernstein, *The Torts Scholar as Disaggregator* In: *Torts on Three Continents.* Edited by: Kylie Burns, Jodi Gardner, Jonathan Morgan and Sandy Steel, Oxford University Press.
DOI: 10.1093/oso/9780198889748.003.0003

her own appreciation to the preference. Primary sources are a kind of counterpart to prime numbers. Unlike their secondary counterparts, which necessarily rely on citations to other writing, primary materials resist division.

The third signature-trait in the disaggregator's work combines the first two: disaggregation as temperament intersects with primary materials. Along with other creators of legal scholarship, a central constituent of secondary materials, the disaggregator engages with law as it's found in the 'primary' array. What disaggregation as method brings to theory and analysis is attention to disunity and disharmony present in what statutes have codified and, to quote Stapleton, what 'judges say'.

Disaggregation as method can be applied to secondary sources too, as Stapleton shows most overtly in her *Three Essays on Torts*, the book version of her Clarendon Law Lectures. In its first Lecture, this text speaks against what Stapleton in it calls Grand Theories.[3] Associated with a quarrel that the American sociologist C. Wright Mills had with 'structural functionalism' as propounded by another sociologist,[4] the term 'grand theories' is an outsider's critical-sounding take on what might more neutrally be called aggregation.[5] A grand theory unites; disaggregation questions and resists not just the unity that such a construct proposes but also its claim to importance.

Whereas Stapleton herself coined the phrase 'v. Grand Theories', five other disaggregation stances in her work are labelled for the first time in this chapter. All six resist efforts to unite tort. My second example: Stapleton has said that insurance does not explain what tort sets out to do. Third, recourse to 'common sense', a claim to aggregation that has been applied most often to causation in negligence, meets the same fate. Fourth, restating tort fails. Fifth, comparative study cannot achieve harmonization. Sixth, the boundary that divides tort from non-tort material, which tacitly yet necessarily rests on a unified view of this subject, can shift and disrupt.

3. On False Unities

Disaggregation as expressed by Stapleton in multiple writings on tort offers three broad criticisms of efforts to achieve or assert unity. Such contentions about unity manifest recurring weaknesses, Stapleton has argued. The most fundamental of these weaknesses is descriptive inaccuracy.

[3] Jane Stapleton, *Three Essays on Torts 2* (2021) (hereinafter Stapleton, *Three Essays*).
[4] C. Wright Mills, *The Sociological Imagination* (1959) 35.
[5] Cf. Marcelo Gleiser, 'A "Theory of Everything" Doesn't Make Sense' Big Think (1 December 2021) https://bigthink.com/13-8/theory-of-everything-2/, last accessed 20 February 2024.

The aggregator claims that tort decisions and doctrines hold together in a unitary framework. They don't, says Stapleton. The aggregator can attain unity only by ignoring, denying, or disparaging materials that the theory must by its own commitments reckon with. Contrary doctrines, cases that reach a different result, and disagreements by other theorists all impede the unity pursued. And so when one of these inconvenient items gets in the way, the theorist discards it 'on the basis that it is "wrong" or not really a decision about tort law'.[6]

Efforts to say that tort is about One Thing necessarily crash against the heterogeneity of this field of law, Stapleton contends.[7] Each Grand Theory of tort collides with at least one dissonant condition that it must conceal or minimize. These conditions remind Stapleton of 'the bits of clothes' that one of the Marx Brothers cut with shears when he observed them sticking out of his suitcase.[8] Stapleton uses one of her papers critical of a leading Grand Theory, civil recourse, to identify embarrassments to two other Grand Theories, corrective justice and economic analysis.[9]

Corrective justice, to Stapleton, lacks a 'theory of entitlements' to tell courts whether entering judgment in favour of a plaintiff would actually correct anything.[10] Its focus on wrongdoing by a defendant does not align well with doctrines that tort courts seem to find easy to apply: vicarious liability, occasional variances from the rule about no affirmative duties, and strict liability all provide that defendants can be responsible for injury without having done anything wrong to cause that harm.[11]

Economic analysis, the dominant rival of corrective justice among the Grand Theories of tort, copes better with those doctrines, but is at a loss to explain a bigger topic, the rule that plaintiffs must prove that a defendant's action caused them injury.[12] One might add to the list of smaller topics that challenge economic analysis the negligence doctrine of informed consent, which indulges patient vetoes that

[6] Stapleton, *Three Essays*, above note 3, at 17–18.

[7] Ibid. at 12.

[8] Jane Stapleton, 'Comparative Economic Loss: Lessons from Case-Law-Focused "Middle Theory"' (2002) 50 UCLA L. Rev. 531, 583 (hereinafter Stapleton, 'Middle Theory'). The Marx brother whom Stapleton references must have been Harpo, an actor frequently filmed putting a large pair of scissors to mischief. See 'Harpo Marx, The Silent Comedian, Is Dead at 70' N.Y. Times, Sept. 29, 1964, at 1. Two years ago, having announced the question What is Tort? in an article title, I chose for my answer two traits that appeared to me present in all tort cases and put the pair into one phrase, 'categorical hurt'. Anita Bernstein, 'Q: What is Tort? A: Categorical Hurt' (2021) 14 J. Tort L. 261. (hereinafter Bernstein, 'What is Tort?'). I didn't want to deny, or snip with metaphorical scissors, an aspect of tort that contradicted my answer. Including traits found in some but not all tort cases would have been a source of trouble that Stapleton has helpfully labelled.

[9] Jane Stapleton, 'Evaluating Goldberg and Zipursky's Civil Recourse Theory' (2006) 75 Fordham L. Rev. 1529 (hereinafter Stapleton, 'Evaluating').

[10] Ibid.

[11] Ibid. at 1537–38.

[12] Richard W. Wright, 'Actual Causation vs. Probabilistic Linkage: The Bane of Economic Analysis' (1985) 14 J. Legal Stud. 435; see also Jay Tidmarsh, 'A Process Theory of Torts' 51 Wash. & Lee L. Rev. 1313, 1345–46 (1994) (attributing to the common law 'norms [that] demand a causal connection before reallocation from a defendant is allowed').

could be silly or wasteful and thwarts experts who could otherwise apply what they know about costs and benefits to the delivery of medical care. Additional doctrinal disunity ensues from any Grand Theory that chooses to look forward while trying to explain a retrospective-minded jurisprudence.

Even if decisional law could coherently pursue a forward-looking aggregative goal for tort to be labelled welfare or efficiency or wealth maximation,[13] Stapleton writes, in practice it has not hewed to that destination. Judicial decisions manifest interest in ends that economic analysis cannot readily measure or even understand, 'such as the concern that tort law not positively encourage abortion'.[14] A writer could try to price stakes like these present in tort cases, but the exercise would strain the law-and-economics machine past its capacity.[15] In sum, the methods of economic analysis, like those of corrective justice, can reckon with some of tort law but not all.

Engaging with the other Grand Theory of civil recourse, Stapleton praises its architects, John Goldberg and Benjamin Zipursky,[16] for what she deems a greater degree of descriptive accuracy than what's present in corrective justice and economic analysis.[17] This youngest child of the Grand Theories family improves on its predecessors by going lighter on the embarrassments.[18] Stapleton nonetheless finds doctrinal disunity here as well. For example, Stapleton contends, Goldberg and Zipursky have coined labels—among them 'relational duty', 'relational directives', 'substantive standing'—that in her view misdescribe and obscure how courts rule.[19] Even when new jargon about doctrine is descriptively accurate, until it catches on enough to supersede older terminology it adds more to multiplicity and plurality than to unity.

Another unity problem found in a Grand Theory, according to Stapleton, originates in the diversity always present in judicial decisions. Among fields that case law creates and addresses, tort arguably contains an extra-large share of this diversity because its doctrines originate in pleadings that derive from individual misadventures more than from legislation or regulation. Landing in 'private' rather than

[13] See Anita Bernstein, 'Whatever Happened to Law and Economics?' (2005) 64 Mod. L. Rev. 303, 311 (gathering 'efficiency', 'allocative efficiency', 'wealth maximization', and 'welfare' as approximate synonyms in the literature).

[14] Stapleton, 'Evaluating', above note 9, at 1537.

[15] Ibid.

[16] Though associated with these two scholars writing together, civil recourse theory was first named in an article solo-bylined by Zipursky. Benjamin C. Zipursky, 'Civil Recourse, Not Corrective Justice' (2003) 91 Geo. L.J. 695, 697–98 (announcing an alternative to corrective justice attentive to 'rights, wrongs, and recourse').

[17] Stapleton, 'Evaluating', above note 9, at 1562 (observing that unlike law and economics and corrective justice, civil recourse theory 'confronts the complexity of the contours of the law of torts').

[18] For a more critical take, see Michael L. Rustad, 'Torts as Public Wrongs' (2011) 38 Pepp. L. Rev. 433, 469 (titling part of an article 'Ten Things I Hate About Civil Recourse Theory').

[19] Stapleton, 'Evaluating', above note 9, at 1536–37.

'public' jurisprudence puts tort farther from the public-law quest for unity in the instrumental sense of making like-cases-alike rules of general application possible.

Stapleton observes that Grand Theorists of tort generalize about all the field while writing for readers in multiple countries and working with primary materials produced in several jurisdictions. They don't limit themselves to one intersection of doctrine and geography, such as—let's pick an example from a non-Grand Theory American work of tort aggregation—privity as undone in the highest court of New York State.[20] To tell the truth about tort, a Grand Theory must align with conditions in the varied places it purports to cover. Every Grand Theory also aspires to unite wide stretches of time under its umbrella. What was true about the state of doctrine in year X must also be true about the state of doctrine in year X minus Y, that is, in the past, and useful to guide prediction about the future, year X plus Z.

Unity powerful enough to transcend variations of geography and time could in principle exist,[21] but Stapleton is sceptical. She makes her point about geographic constraint with a couple of rhetorical questions: 'Why would a tort lawyer in Kansas be interested in how tort law operates in England? Of what value might an appreciation of the tort case law of the Tasmanian Supreme Court be to a lawyer in New Zealand?'[22]

On tort disunity of the temporal type, Stapleton notes the truism—in place long before her own theorizing emerged, and entirely independent of her views—that 'the common law is "living", that as society changes and develops, so too does the common law change and evolve in line with this, independently of institutional pronouncements.'[23] The familiar and never controversial notion that 'society changes and develops' in ways that alter the common law necessarily acknowledges impacts from not only the passage of time but differences in settings and locations. 'Society' encompasses a subset of persons rather than the entirety of all humanity, and each of these ecosystems typically exists with reference to a distinct geographic place.[24]

[20] See Edward H. Levi, *An Introduction to Legal Reasoning* (1949) 9-27.

[21] For an example of this unity-by-assertion in the torts literature, see Alex Stein, 'The Domain of Torts' (2017) 117 Colum. L. Rev. 535, 611 (claiming that a 'bottom-up account of tort law' delivers 'a comprehensive descriptive account of how our tort system works'). Aware that Stein wrote *The Domain of Torts* from the post he then held as a law professor in the United States and as a holder of law degrees earned in Israel and England, I read the personal pronoun in 'our tort system' as uniting implicitly some variations of geography and time.

[22] Jane Stapleton, 'Benefits of Comparative Tort Reasoning: Lost in Translation' (2006) 1 J. Tort L. 1, art. 6 (hereinafter Stapleton, 'Lost').

[23] Stapleton, *Three Essays*, above note 3, at 6.

[24] See Anita Bernstein, 'The Communities That Make Standards of Care Possible' (2002) 77 Chi.-Kent L. Rev. 735, 754–55 (identifying 'geography-based community' as central to tort law); Mae Kuykendall, 'Restatement of Place' 79 Brook. L. Rev. 757, 757 (2014) (claiming place is 'a factor that is present in law but often not fully seen').

Stapleton finds a second false unity by digging deeper into description. Unity postures that claim that the object under study, here tort, is 'all about one thing' are assertions about an Is rather than an Ought. Writings that fall under Stapleton's pejorative rubric of Grand Theories claim to describe tort and only incidentally to assert what tort rules *should* provide. The authors' preference for one characterization adjective over another is understandable. Description appears humbler, more credible, and more useful to readers with diverse goals than expositions that want to install a particular Ought. Writers are likely to be sincere, I think, in this description of what they believe about their work. The problem here is not intellectual dishonesty or overt misrepresentation but lack of clarity about 'the claims that scholars actually make'.[25]

Grand Theories are normative, claims Stapleton, more than their creators admit. Where description in this scholarship ends and argument begins is unclear to readers of these works. Not only to readers: theorists themselves are unaware of the wavering lines around what they claim.

It may be helpful to note what about this actually-normative-not-descriptive criticism constitutes disaggregation. My Brooklyn Law colleague Adam Kolber has coined a word that aids in this task. Kolber defines his coinage, 'scholarmush', as 'a tangled combination of claims rooted partly in law and partly in morality that are partly dependent on facts and partly dependent on values'.[26] Even without Kolber's 'tangled combination' definition and his reference to a blend of (descriptive) facts and (normative) values, any word with 'mush' in it adverts to aggregation. Think of how we can see and feel distinct food constituents before we cook them into one of the mushes: porridge, congee, and polenta. Unless both writer and reader stay aware of the contrasts that normative claims add to a descriptive work, undifferentiated inclusions risk softening that work into scholarmush.

The third false unity identified in Stapleton's disaggregation perspective on tort is one of omission. We have noted her reference to the Marx brother who cut away with shears the occasional inconvenient inclusion that trailed from his suitcase. Here the tendency is categorical rather than a one-off: aggregative accounts of tort proceed abstractly without heed to individual choices, predilections, and biographical events. To Stapleton, tort always includes theory but never stops with theory. It exists in actions, which always contain variety, and it cannot exist without these human actors.[27]

Among the many actors that populate tort and make it possible, the cohort of judges repeatedly draws attention from Stapleton. Judges are neglected in tort theory relative to their importance, Stapleton suggests. They write tort law. When tort changes, the record of its change takes form in judicial decisions. Judges as

[25] Adam Kolber, 'How to Fix Legal Scholarmush' (2020) 95 Ind. L.J. 1191, 1212.
[26] Ibid. at 1193.
[27] See Stapleton, *Three Essays*, above note 3, at 2 (referencing 'real-world impact').

human beings with biographies and predilections bring yet another source of disaggregation into tort. The false unity that Stapleton finds here is an implicit belief that although judges create tort decisional law within particular national legal systems, those of them who decide cases in separate English-speaking common law countries and US states operate as part of a transnational whole.

Stapleton not only contends that judges deserve more attention but delivers that attention herself. While other leading torts scholars active today share this interest in the judge as writer-decider,[28] Stapleton talks the talk about judges as indispensable to tort—tort theory too, not just practice or 'the real world' as seen through an academic lens—and walks the walk. I can think of no other torts writer who maintains such steady attention to these people as individuals who have expressed views on doctrine. Attempting a census of the judicial population in this body of scholarship, I count 19 named judges whom Stapleton has written about as writers—and I can't be sure I've found them all. My inexact but long-held impression of the torts literature is that it's rare for any author's judge-tally, even after a long career, to approach double digits.

4. Attention to Insurance Fails

Interpreting tort as an endeavour arrayed around insurance does not qualify as a Grand Theory in the sense that Stapleton has called 'overblown',[29] an adjective that corresponds with the 'grand' in Grand Theories: abstract, high-concept, and confident that one's readers regularly learn from legal scholarship. Though an everyday word, insurance hews to grandeur in that any contention that tort is about insurance aspires to unite the field. Similar to other Grand Theories that have drawn attention from Stapleton, the insurance argument regards tort as having a central goal or descriptor.

Perhaps because she appreciates the un-fanciness of insurance as a putative unifier, Stapleton starts by treating the concept with some kindness. Her principal paper criticizing insurance as a would-be aggregator or unifier of tort, an article in the Modern Law Review,[30] starts with loss or what Stapleton calls 'misfortune'.[31] Misfortune is always present in tort.[32]

[28] See, e.g., Catherine M. Sharkey, Book Review (2021) 134 Harv. L. Rev. 1423, 1423–24 (identifying Brett Kavanaugh of the United States Supreme Court as a judge inclined to impose responsibility on the 'cheapest cost avoider', an academic-analytic category). See also ibid. at 1432–33 and 1436–40 (reviewing scholarly writing on tort by the American judges, Oliver Wendell Holmes, Jr., Richard Posner and Guido Calabresi, coupled with much-studied decisional law by Benjamin Cardozo, Roger Traynor, and Posner and Calabresi writing as judges).

[29] Stapleton, 'Evaluating', above note 9, at 1562.

[30] Jane Stapleton, 'Tort, Insurance and Ideology' (1995) 58 Mod. L. Rev. 820 (hereinafter Stapleton, 'Insurance').

[31] Ibid. at 820.

[32] See Bernstein, 'What is Tort?', above note 8.

Enter insurance. By staying focused on misfortune as loss, Stapleton's analysis puts aside the third-party variety. Having to defend oneself in response to a complaint is burdensome and unpleasant, but not unambiguously 'misfortune' in that tort litigation can be profitable for a defendant in a way that experiencing the impact of tortious conduct cannot.[33] And so for Stapleton insurance is either the 'private' first-party sort, where injured persons can get their bills paid because they paid for this protection in the past, or 'a scheme of social insurance or social welfare'.[34]

Both liability and insurance can be understood as responses to misfortune that set out by design to lessen its impacts. From here, Stapleton can identify another quasi-Grand Theory, a characterization of tort as being about one unifying thing. Misfortune as a descriptor lays a foundation for tort as a kind of peer of two insurance possibilities, the private kind that customers buy before they experience an injury and the public, or government-managed, type of insurance. Liability could be a third type of misfortune insurance if observers are willing to accept 'the conflation of tort with the insurance model of response to misfortune'.[35]

Examining this putative unifier makes it collapse, Stapleton says. She starts by noting the entitlement of bystanders to receive compensation for injury as tort plaintiffs. Bystanders did not 'pay an element in price to cover liability costs',[36] yet collect through court-enforced 'freeloading'. Especially because public insurance is uncontroversial, this result might be 'a tolerable side effect of the pursuit of other goals'—but an observer cannot know what those goals are, given that tort liability withholds money from needy injured persons who cannot fulfil the elements of a claim.[37]

The homespun plainness of insurance in contrast to grander unifiers like corrective justice, law and economics, and civil recourse theory may seem friendly, but for Stapleton it brings danger. Grand Theories fail, Stapleton argues, in that they don't hold up under scrutiny. Consequences of this failure are 'academic' in the sense of idle. Equating tort with insurance has very different effects.

Stapleton writes that once this equating 'is embraced, the normative agenda for the future of tort inexorably points to retrenchment'.[38] By 'retrenchment' Stapleton appears to mean that the reach of tort will diminish. This effect follows because the descriptive assertion that tort is insurance slides easily into a normative

[33] Here I have in mind counterclaims and also court-ordered payments from plaintiffs to defendants, available routinely under 'loser pays' rules followed in much of the common law world and episodically when judges compel plaintiffs to pay monetary sanctions to defendants as a penalty for misconduct. Judicial choices to order (non-tort) plaintiffs to pay money to defendants as sanctions were approved in, *inter alia*, *Roadway Exp. v Piper*, 447 U.S. 752 (1980); *Footman v Cheung*, 341 F. Supp. 2d 1218, 1229 (M.D. Fla. 2004), and *Schutts v Bentley Nevada Corp.*, 966 F. Supp. 1549 (D. Nev. 1997).

[34] Stapleton, 'Insurance', above note 30, at 820.

[35] Ibid.

[36] Jane Stapleton, *Product Liability* (1994) 215.

[37] Ibid.

[38] Stapleton, 'Insurance', above note 30, at 820.

posture: any doctrine inconsistent with this insurance understanding is peripheral to tort, at best, and judges became entitled to repudiate it.

At first blush, Stapleton could be wrong when she says that retrenchment follows the equating of insurance with tort. A contrary result, the enlargement or expansion of tort, could at least in principle ensue when courts choose to give insurance more doctrinal work to do. Take for example *Rowland v Christian*, the 1968 decision where the California Supreme Court widened the duty of care owed to land visitors after reciting 'considerations' for a tort court to balance when it assesses problems of limited duty. The final consideration of this seven-item list is 'the availability, cost, and prevalence of insurance for the risk involved'.[39] Because it ended up changing tort law by replacing a limited-duty rule, *Rowland* implies that adding insurance considerations to the negligence analysis can expand tort. An expansion is the opposite of retrenchment.

It turns out, however, that insurance does not deliver insurance to injured persons. *Rowland*, in ruling for the plaintiff, briefly recited insurance as something for courts to consider, but the decision paid no attention to 'the availability, cost, and prevalence' of any kind of insurance, neither the first-party kind that a land visitor might purchase (or enjoy as a government benefit) nor homeowner's liability insurance paid for to cover this category of claim. As Stapleton observes, citing English cases rather than *Rowland v Christian*, 'on the very few occasions when courts refer to liability insurance they do so as a "makeweight" factor after it has been decided, having balanced other concerns, to impose liability'.[40]

Other plaintiff-favouring decisional law by the California Supreme Court of the mid-century span that yielded *Rowland*, including the slightly older *Greenman v Yuba Power Products*,[41] could be read as equating tort with insurance at a general plane. Roger Traynor, the judge-author of *Greenman*, mentioned insurance overtly in his precursor *Escola v Coca Cola Bottling Co. of Fresno*, a concurrence rather than a court decision but a noteworthy piece of judicial writing that used the word Stapleton chose, 'misfortune'.[42] An injury, said Traynor, 'may be an overwhelming misfortune to the person injured, and a needless one, for the risk of injury can be insured by the manufacturer and distributed among the public as a cost of doing business'.

That expansive-appearing move doesn't look like the retrenchment of tort, yet it is. Tort rules fit poorly at best with the doling-out of money to injured people in response to their needs rather than what a defendant did to them or failed to do. Enlarging products liability along the lines Traynor favoured in *Escola* and

[39] 443 P.2d 561, 564 (Cal. 1968).

[40] Stapleton, 'Insurance', above note 30, at 827. Stapleton identifies Lord (Tom) Denning as prominent among the insurance-minded English judiciary. Ibid. at n.20.

[41] 377 P.2d 897 (Cal. 1963).

[42] *Escola v Coca Cola Bottling Co. of Fresno*, 150 P.2d 436, 441 (Cal. 1944) (Traynor, J., concurring).

Greenman benefits plaintiffs but because it lacks a commitment to tort rules, that expansion comes at a price of lessening tort as doctrine.

Retrenchment marches on. As Stapleton explains, 'tort as insurance' has come to mean not so much that tort judgments furnish injured people with coverage for misfortune but that insurance considerations thwart tort liability as a source of compensation. Insurance as a unifier of torts supports exclusions from what plaintiffs may collect, among them nonpecuniary damages.[43] More fundamental to the retrenchment that Stapleton has found, 'tort as insurance' encourages courts to reject liability altogether on the ground that an alternative path, first-party insurance, does the work of insuring more efficiently.[44]

5. 'Common Sense' Fails

My next item in the catalogue of putative unifiers that try and fail to pull tort together helps make the case for Stapleton as a disaggregator rather than (merely) a foe of scholarly pretension.[45] Stapleton is indeed an unpretentious writer and a reader with little use for vague highfalutin prose about tort, but the trait present in tort writing that she most reliably rejects is the aggregation of disparate materials. She cares less about hauteur.

Economic analysis, corrective justice, civil recourse theory, and even insurance are fancy words or phrases in that they come to the English language from Latin and Greek,[46] and expect readers or listeners to think about tort through an educated lens. Common sense is the opposite of fancy. It sounds like an antonym of grand. Another distinction between insurance and the three Grand Theories that Stapleton condemns is that, whereas insurance and Grand Theories aspire to explain all of tort, 'common sense' as a unifier has a narrower descriptive agenda. The instances of it that Stapleton has found all focus on one corner of tort doctrine, actual or factual cause.[47]

[43] Stapleton, 'Insurance', above note 30, at 833. For support of Stapleton's position from a pro-retrenchment scholar, see George L. Priest, 'The Current Insurance Crisis and Modern Tort Law' (1987) 96 Yale L.J. 1521, 1536 (arguing that 'increased recovery of emotional and other non-economic losses' has harmed the supply of liability insurance).

[44] Stapleton, 'Insurance', above note 30, at 837 (citing the economic analysts William Landes and Richard Posner, scholars who took the view that tort as insurance is extremely costly and justifiable only if it delivers more than just insurance, which to them meant deterrence).

[45] 'Common sense' in this chapter is indebted to analyses ably expressed and defended by Benjamin Smith.

[46] On Latin and Greek as a source of fanciness within tort, see Bernstein, 'What is Tort?', above note 8, at 271.

[47] Disparate terminology for this element of the negligence prima facie case is a problem even within my own country, where courts use 'causation', 'proximate cause', 'legal cause', and other labels inconsistently. Stapleton's unique immersion in the tort law of 'three continents', see Stapleton, 'Lost', above note 22, makes her aware of how this problem gets bigger with the addition of more jurisdictions than the 51 that I live with.

Yet even though 'common sense' looks more modest than the other putative unifiers in that it lacks two of their ostentatious traits, fancy language and an ambition to announce what all of tort is about, Stapleton lines this understanding up against the same wall where she has arrayed grand theories and insurance. It too gets shot down. The trouble with believing that the causation element of tort is about common sense, for Stapleton, is not that this belief is pretentious or aspires to cover too much but that it aggregates that which is diverse and should remain separate.

As she did with insurance, which also sounds plainspoken, Stapleton finds Grand Theory-like trouble with common sense. Observers who claim that common sense can explain legal conclusions about causation no doubt believe that they have rejected lofty abstract thought. Maybe they have: but a problem familiar to Stapleton from Grand Theory, the presence of implicit normative claims in ostensible description, complicates this usage of common sense too. Stapleton quotes with approval the philosopher Judith Jarvis Thomson who, in a festschrift piece of her own, a tribute to H.L.A. Hart, wrote that 'it is very plausible to think that the plain man's thought about causation is drenched in'—wait for another fancy word— 'normativity'.[48]

The festschrift essay by Thomson that Stapleton quotes celebrated a book about causation by the revered English scholars Hart and Tony Honoré. As read by Stapleton, this book, *Causation in the Law*, exemplifies academic error about common sense as a unifier. Yet another festschrift, this one put together when Honoré turned 80 in 2001, gave Stapleton an occasion to praise Hart and Honoré's articulation of how deeply the twentieth century changed understandings about legal cause.[49] Joining acclaim from other experts on causation in tort,[50] Stapleton singled out Hart and Honoré's 'necessary element of a sufficient set' test for particular esteem.[51]

The main trouble with *Causation in the Law*, for Stapleton, is its embrace of 'common-sense principles' as underlying a test to determine liability for tortious conduct.[52] Hart and Honoré argued that the law should not adopt a special meaning for causation as a legal conclusion.[53] For them, the ordinary use of causal language can clarify responsibility: 'it is the plain man's notions of causation', they said, with which the law is and should be concerned.[54]

[48] Jane Stapleton, 'Choosing What We Mean by "Causation" in the Law' (2008) 73 Mod. L. Rev. 433, 480 (quoting Judith Jarvis Thomson, 'Some Reflections on Hart and Honoré, *Causation in the Law*').

[49] Jane Stapleton, 'Unpacking "Causation"', in *Relating to Responsibility: Essays for Tony Honoré on his Eightieth Birthday* (Peter Cane & John Gardner eds., 2001) 145 (hereinafter Stapleton, 'Unpacking').

[50] See Richard C. Ausness, 'Causation and Apportionment Issues in Opioid Litigation' (2021) 49 Cap. U. L. Rev. 535, 547 (2021) (observing that the Third Restatement's test for causation follows this influential development). The leading expositor here is festschrift participant Richard Wright. See Richard W. Wright, 'Causation, Responsibility, Risk, Probability, Naked Statistics, and Proof: Pruning the Bramble Bush by Clarifying the Concepts' (1988) 73 Iowa L. Rev. 1001, 1018–27.

[51] Stapleton, 'Unpacking', above note 49, at 167.

[52] Ibid. at 148 n.16 (citing H.L.A. Hart & Tony Honoré, *Causation in the Law* 1 (1959)).

[53] Ibid. at 145.

[54] Ibid. at 148 nn.13–15.

Disagreeing, Stapleton maintains that the law *does* need a special legal concept of cause.[55] Because a word in 'ordinary' use will mean different things in different contexts,[56] 'common-sense principles' do not achieve the delineating of legal cause that *Causation in the Law* claimed a unitary approach can attain.[57] The topology of causal language that Hart and Honoré supply in their analysis 'neglect[s] the rich diversity of concerns that influence our attributive judgments which we may express in causal language[]',[58] Stapleton continues, and 'ignores the conflicting causal assertions of the parties to the litigation'.[59] More false unity emerges from the failure of these ostensible principles to keep separate 'historical contribution' and 'truncation',[60] or what American readers would call actual cause versus legal or proximate cause. Incoherent and analytically useless with respect to questions of fact,[61] 'common-sense principles' in application offer to Stapleton little guidance for disputes where reasonable minds might differ on moral or policy grounds.[62]

Trouble with common sense emerges in decisional law as well as academic writing like the Hart and Honoré book. Making what Stapleton regards as a prudent choice, courts in the United States and the United Kingdom have avoided an Australian error: the High Court of Australia, Stapleton writes, purports to use a 'common sense test' for causation.[63] Echoing the wisecrack by Voltaire that the Holy Roman Empire that loomed over Europe in the eighteenth century was neither holy nor Roman nor an empire,[64] Stapleton dismisses this Australian judicial favourite as 'an empty slogan, neither a test nor anything to do with common sense'.[65] Like other putative unifiers of tort, Stapleton concludes, common sense fails as both description and normative pursuit.

6. Restatements Do Not Unify

Restating as a failed unifier has long occupied Stapleton's critical reflections on tort. In an essay I published 29 years ago that put together her monograph on products liability with a draft of what would become the American Law Institute's (ALI)

55 Ibid. at 184.
56 Ibid. at 148 n.17.
57 Ibid. at 150.
58 Ibid. at 154.
59 Ibid. at 151.
60 Jane Stapleton, 'Reflections on Common Sense Causation in Australia', in *Torts in Commercial Law* (James Edelman, Simone Degeling, & Tom Goudkamp eds., 2011) 331, 335 (hereinafter Stapleton, 'Reflections').
61 Stapleton, 'Unpacking', above note 49, at 150.
62 Ibid. at 151.
63 Stapleton, 'Reflections', above note 60.
64 François Marie Arouet de Voltaire, *Essai sur les Moeurs et l'Esprit des Nations* (1769) 70.
65 Stapleton, 'Reflections', above note 60, at 331.

first constituent of the third Restatement of Torts,[66] I spoke for the first time in print about this torts scholar as disaggregator, calling her *Product Liability* 'a great work of destruction, an attack not only on the basic dogma that there ought to be a separate law for harms caused by manufactured products but also, more obliquely, on the very idea of restatement'.[67] This theme of disaggregation had also informed Stapleton's pre-*Product Liability* scholarship.[68]

Because 'restatement', especially in torts, can signify an ALI compendium and can also refer to other efforts, a few words on terminology here may be in order. My 1995 working definition, written to describe not just what the American Law Institute produces but also the European product liability directive adopted in 1985 and efforts to enact federal law in the United States, will serve my present purposes: let us say that restatement means 'a codification, produced by a group of persons and subject to a vote, that purports to reconcile and improve the state of legal doctrine in a particular subject, where at least part of this doctrine derives from the common law'.[69] The shift key may be helpful here. I'll say Restatement when I reference an ALI project and restatement to mean the generic kind that other collaborations can produce.

Readers can anticipate what Stapleton will think of any endeavour, whether topped with a capital R or rendered in lower case, that purports to reconcile and improve anything with a unitary new creation. Restating may be extra problematic. Until now, unifiers that Stapleton has criticized left no doubt about the object of their attention. Grand Theories seek to describe all of tort at a general plane. Same for the claim that tort is insurance. The 'common sense test' endorsed by the Australian judiciary addresses causation. Say what she might about the infirmities of these ostensible descriptions, Stapleton broaches no quarrel with their spokesmen about what it is they have described.

Since the launch about 30 years ago of a products liability volume within the aegis of Restatement (Third) of Torts, the object of restaters' attention has splintered. Drafters are now freer to stray from lines that were in place before they started their work. They proceed with confident entitlement to determine not only what the law is now and what it should be, but what counts as pertinent to the document they write. In the new volume they create, they assert unity while also breaking up pre-existing cohesion.

[66] Anita Bernstein, 'Restatement Redux' (1995) 48 Vand. L. Rev. 1663 (reviewing Jane Stapleton, *Product Liability* (1994)) (hereinafter Bernstein, 'Restatement'). Following an American convention, and differing sometimes from Stapleton, I use 'products liability' plural rather than singular unless I am quoting someone else.

[67] Ibid. at 1664.

[68] See Jane Stapleton, 'Product Liability Reform—Real or Illusory?' (1986) 6 Oxford J. Legal Stud. 392 (hereinafter Stapleton, 'Product Liability Reform'). I've noted the influence of this work on two articles of mine: Anita Bernstein, 'A Model of Products Liability Reform' (1993) 27 Valp. U. L. Rev. 639 and Anita Bernstein, 'L'Harmonie Dissonante: Strict Products Liability Attempted in the European Community' (1991) 31 Va. J. Intl. L. 673. See Bernstein, 'Restatement', above note 66, at 1665 n.7.

[69] Bernstein, 'Restatement', above note 66, at 1665.

When James Henderson and Aaron Twerski prepared the blackletter and commentary of *Restatement: Third of Torts: Products Liability* to replace what the Second Restatement had provided on point, for example, they changed the ALI's products liability work in scope and ambition—not just the doctrines that they endorsed, revised, and rejected. Pre-1998 products liability restating by the ALI had occupied a single section in a larger four-volume tome: the entire field reposed in the much-cited § 402A of the Second Restatement under a narrow title, 'Special Liability of Seller of Product for Physical Harm to User or Consumer'. Products liability deserved to become a book-sized topic with rules that extend beyond 'physical harm to [a] user or consumer', as Henderson and Twerski argued expressly in law review articles and implicitly by writing their much fuller Restatement.[70] The ALI chose to replace few words on products liability with many, in short: but terse § 402A still gets cited by American courts,[71] suggesting that the Third Restatement has not generated a consensus to support the expansion that Henderson and Twerski undertook.

A restatement need not replace an old one for disunity to ensue. We see the problem in two sub-volumes of the larger ongoing *Restatement: (Third) of Torts* now in progress, Remedies and Concluding Provisions. Both projects embark on paths not taken in the Second Restatement and neither of them sets out to supersede an ALI predecessor in the products liability mode. Which topics or doctrines belong in these new Third Restatement volumes and which lie outside their scope cannot be known from their titles. These restatements-in-progress will unite only what their author-publishers choose to unite.

Three examples from Stapleton illustrate particulars under the larger banner of disunity in restating. The first two come from law review symposia where Stapleton, invited to comment on draft tort Restatements, found—brace yourself for this revelation—disarray under the banner of unity. The third example I offer moves to a lower-case restatement, the European product liability directive promulgated in 1985.

At a Vanderbilt Law School symposium on Restatement (Third) of Torts: General Principles, which I remember as a lively event,[72] Stapleton used causation doctrine to press her view that understanding tort law must engage with the work of the judge. Recall that Stapleton believes that tort theory writ large has

[70] See, e.g., James A. Henderson, Jr. & Aaron D. Twerski, 'A Proposed Revision of Section 402A of the Restatement (Second) of Torts' (1992) 77 Cornell L. Rev. 1512 (arguing for a new Restatement); 'Symposium, The Products Liability Restatement: Was It a Success?' (2009) 74(3) Brook. L. Rev (reviewing the authors' Restatement 10 years after its publication).

[71] See, e.g., *Bessette v IKO Indus., Inc.*, 30 F.4th 75, 83 (1st Cir. 2022); *Millers Cap. Ins. Co. v Hydrofarm, Inc.*, 340 F.R.D. 198, 224 (D.D.C. 2022).

[72] It gave a home to one of my early efforts to unite tort and feminist jurisprudence. See Anita Bernstein, 'Restatement (Third) of Torts: General Principles and the Prescription of Masculine Order' (2001) 54 Vand. L. Rev. 1367.

the same obligation.[73] The priority is necessarily stronger in a restatement, which purports to examine and influence judicial decisions.

And so Stapleton had occasion to fault Section 105 of the draft General Principles, in which reporter Harvey Perlman excluded foreseeability from duty with the remark that it's juries, not judges, who decide whether an accident is foreseeable.[74] Influenced, perhaps, by her training in a contrary tradition with respect to the jury, Stapleton retorted that 'the notion of reasonable care can only operate against the notion of some perception of foreseeable risk'.[75] Stapleton was saying that the draft General Principles assert unity—that's what 'duty law categorically doesn't care about foreseeability' might mean—and then fail at that goal, generating fragmentation and disarray.

At another symposium, Stapleton found clangour in efforts to restate another fundamental of negligence law, responsibility for risk. She honed in on draft blackletter that went on to be Section 7(a) of Restatement (Third) of Torts: Liability for Physical and Emotional Harm: 'An actor ordinarily has a duty to exercise reasonable care when the actor's conduct creates a risk of physical harm'.[76] After faulting the drafters for not stating what it means for conduct to create this risk and thereby forcing her into 'a close and labor-intensive trawl of the whole text',[77] Stapleton followed this lapse beyond the duty element of the negligence prima facie case into breach,[78] then causation.[79]

Her disapproval of the trawl emphasized the disunity in it. To Stapleton, this volume of the Restatement is vexed not only by 'inconsistent and conflated' diction but also 'an inexplicable variety' of verb forms: Actors' conduct sometimes 'creates a risk', sometimes 'poses a risk', on other occasions 'creates' that same consequence.[80] The disarray extends beyond conduct: 'We are told that there may be a danger 'posed' by a chair or conditions on land',[81] Stapleton observed.

Shortcomings in the Liability for Physical and Emotional Harm segment of the Third Restatement are to Stapleton evidence of a larger problem with ALI restating. The American Law Institute, says Stapleton, ought to emphasize 'securing terminological clarity', a challenge that under present methods 'asks too much of even the most gifted reporters'.[82] The Restatement path to publication regrettably 'freeze[s] provisions chunk by chunk' by relying on a vote at the ALI annual

[73] See above note 28 and accompanying text.

[74] Jane Stapleton, 'Legal Cause: Cause-in-Fact and the Scope of Liability for Consequences' (2001) 54 Vand. L. Rev. 941.

[75] Ibid.

[76] Jane Stapleton, 'The Risk Architecture of the Restatement (Third) of Torts' (2009) 44 Wake Forest L. Rev. 1309, 1310.

[77] Ibid. at 1312.

[78] Ibid. at 1320–22.

[79] Ibid. at 1322–24.

[80] Ibid. at 1330.

[81] Ibid.

[82] Ibid. at 1333.

meeting to approve these provisions 'before sophisticated academic evaluations appear and before later sections are even drafted'.[83]

Lest one think that this disapproval of restating as a source of false unity extends only to the capital-R output of the American Law Institute, an early paper by Stapleton finds the same problem in another effort that fits my working definition of a restatement.[84] Readers can predict Stapleton's answer to the question posed in her title, *Product Liability Reform—Real or Illusory?*, once they note that the species of products liability reform examined in that article, the European instrument formally called the Council Directive on Liability for Defective Products 25.7.85 and more colloquially 'the EEC Directive' or just 'the Directive',[85] purported to unify the law.

The Directive starts by emphasizing its attention to uniformity in tort within a shared economic sphere:

> [A]pproximation of the laws of the Member States concerning the liability of the producer for damage caused by the defectiveness of his products is necessary because the existing divergences may distort competition and affect the movement of goods within the common market and entail a differing degree of protection of the consumer against damage caused by a defective product to his health or property.[86]

As used here, 'approximation' refers to the obligation of member nations to align their national laws with the supranational law of the European Economic Community (as it was known in 1985; it has been the European Union for more than 30 years).[87]

Stapleton identifies a fundamental problem with this ideal, familiar from the ALI Restatement tension between reconciliation and improvement. On one hand, the Directive pursues what Stapleton calls a level playing field for businesses in Europe; on the other hand, as elaborated in her 1994 monograph on products liability, 'we now have a second competing formal objective of the Community', which is 'a social union with explicit emphasis on consumer protection'.[88] For Stapleton, what the Directive called 'existing divergences' in tort do not go away at the command of an instrument that calls for both levelling and progress.

The Directive brings its own divergences while purporting to reconcile. To the extent it takes intelligible postures on products liability controversies, Stapleton

[83] Ibid. at 1332–33.

[84] See above note 68 and accompanying text. Issuance of the Directive came from the Council of Ministers of the European Communities; 'European Union' arrived in 1993.

[85] Stapleton uses 'the EEC Directive'. Stapleton, 'Product Liability Reform', above note 68, at 392 n.2.

[86] Directive, Preamble.

[87] See https://ec.europa.eu/environment/archives/guide/part1.htm, last accessed 20 February 2024.

[88] Stapleton, *Product Liability*, above note 36, at 62.

says, the Directive has favoured negligence-like attention to 'defect' with its implicit commitment to cost-benefit balancing—a move that in her view does not install unity even at the level of asserting it,[89] and makes case outcomes less predictable.

Stapleton avoids expressing an overt preference for the untaken path she calls 'full strict liability',[90] or indeed any new transnational legal regime. Her criticism instead faults the Directive as a lowercase-r products liability restatement. This instrument achieves neither reconciliation (or, in European jargon, approximation) of existing law nor improvement. Oneness fails.

7. Transnational or Comparative Study of Tort Has Other Virtues, But Does Not Achieve Unity

Readers willing to accept my characterization of the European products liability Directive as a kind of restatement can move to a nearby locus where Jane Stapleton has maintained her commitment to disaggregation: transnational or comparative study.

The Directive ordered member states of the European Union, autonomous nations that enact their own laws, to adopt legislation aimed at one products liability reform goal. Each member state had its pre-1985 products liability law to reconcile with what the Directive directs. The implementing legislation that member states adopt do not copypaste the entire Directive without change: they have an opportunity to reject and rewrite.[91]

Faced with instruction from Brussels that calls for judgement as well as deference, then, separate-yet-united participants need to know not only what this document says but how their peers are reading it. National lawmakers complying with the European Union order to adopt the Directive necessarily engage in harmonization or approximation and also a comparative-transnational study of tort law. Stapleton approves of the study that this work entails but not its dream of unity.

Differing with comparativists like Basil Markesinis, Christian von Bar, Gert Bruggemeier, and Helmet Koziol, who have written that European tort law either (a) is 'converging', (b) already holds together, or (c) could *mutatis mutandis* be put into a unitary code,[92] Stapleton finds more divergence than commonality when she examines tort in different nations. Her admittedly 'extremely selective' perspective,[93] oriented as it is towards 'the practical value to practitioners and judges

[89] Stapleton, 'Product Liability Reform', above note 68, at 393.
[90] Ibid. at 403.
[91] See Ernst C. Stiefel, 'Resolution of International Products Liability Disputes: An Emerging Procedural Framework' (1990) 16 Brook. J. Int'l L. 267, 269–72 (reviewing divergencies in implementation).
[92] Stapleton, 'Lost', above note 22, .at 41–43
[93] Ibid., n.p.

in the court of final appeal in an English-speaking jurisdiction of paying attention to how tort issues are analysed in a different jurisdiction',[94] excludes considerable swaths of comparative tort law. Yet Stapleton locates division and disunity on tort even within the European continent.[95]

Moving to Stapleton's 'shuttle between three continents',[96] or what comparing law can achieve by way of unity for a lawyer based in Britain, 'Australasia',[97] or North America, the trouble with trying to impose aggregation continues. 'Tort law cannot simply be "exported"', Stapleton writes, 'because there will always be structural and cultural features which cannot be ignored and which have no equivalent in the potentially 'importing' system'.[98] An example: 'certain deliberate self-interested uncooperative behaviour will not be judged dishonest or unconscionably contradictory in England, whereas it might well be on the Continent'.[99]

This example adverts to Europe. Does its disunity abate for a lawyer based in one of Stapleton's three continents who sticks to common law jurisdictions? Narrower comparative study may be understood as principled rather than (or in addition to) lazily monolingual after January 2020: when Brexit took Britain out of the countries whose laws had to unite under one transnational banner, the European Union's common law outlier no longer needed reconciliation with a civil law tradition,[100] and could join the more compatible kin network of Australia, New Zealand, Scotland, Canada, and the United States. Stapleton acknowledges the presence of torts commonality in Britain-Australasia-North America but also says that in separate nations judges reach divergent results and scholars have different priorities. Furthermore, one of the three continents includes a large country with more than 50 tort jurisdictions each topped by its own little high court. More disaggregation.

The comparative study of tort has virtue for our disaggregator, however. Stapleton chose to title the paper of hers containing the most detailed criticism of this turn as 'Benefits of Comparative Tort Reasoning'—that's praise of the endeavour, not a pejorative. Benefits of comparative tort reasoning vary depending on what the reader-consumer wants from torts material that originated in the

[94] Ibid., n.p.

[95] Ibid. at n.176. So too does Helmut Koziol, one of the comparativists whom Stapleton cites, although Koziol's Principles of European Tort Law project aspires to lessen this disunity. See Michael L. Wells, 'Harmonizing European Tort Law and the Comparative Method: A Review of Basic Questions of Tort Law from a Comparative Perspective (Helmut Koziol ed., Sramek 2015)' (2016) 9 J. Civ. L. Stud. 539, 544.

[96] Stapleton, 'Lost', above note 22, n.p.

[97] Stapleton uses this word when describing tort decisions by Australia and New Zealand courts in Stapleton, 'Middle Theory', above note 8, at 552.

[98] Stapleton, 'Lost', above note 22, n.p.

[99] Jane Stapleton, 'Good Faith in Private Law' (1999) 52 Curr. Leg. Probs. 1 (hereinafter Stapleton, 'Good Faith').

[100] Cf. Wells, above note 95, at 545 (suggesting that trying to unify European tort law may become 'largely moot' once Britain, the source of the most significant tort divergences, leaves the European Union).

jurisdiction of Someplace Else. For example, 'the outcomes of similar fact situations in other jurisdictions' offer value to a reader-consumer who has reasons and opportunities to forum-shop,[101] but presumably will do less for others.

The benefit of comparative tort reasoning that provides the most gain for the most readers, Stapleton says, is support for *argumentation*.[102] Foreign and transnational materials give advocates and judges ideas that can both support and profitably challenge what they want to claim. When writers may cite only sources limited to what has been published in or about their own nation, their opportunities to persuade are unduly narrowed. Problems that tort tries to remedy or lessen are found in multiple jurisdictions. Foreign decisions and secondary writings do not and should not determine how cases come out, but they can often 'enrich the palette of ideas, concerns, perceptions—in short, arguments—that a judge brings to bear on the matters in dispute'.[103]

8. The Boundary around Tort Can Move

My final example of disaggregation shows where Stapleton has examined, and now and then implicitly questioned, the boundary that separates tort from other fields of law. Stapleton appears comfortable with tort as a distinct domain. Disaggregation efforts from her do not set out to break down this border. Instead, her attention to nearby realms of private law—especially contracts, restitution, and the European-style law of obligations—inform her understanding. Responsive to the admonition in Rudyard Kipling's aphorism that one cannot know England when one knows only England,[104] Stapleton's tort scholarship stays attentive to where rules and ideas associated with tort flourish outside its confines. This variant on disaggregation simultaneously accepts the aggregation in tort as a legal category and finds constituents of this category elsewhere.

And so when Stapleton takes on civil recourse theory, she looks not only at its core—to her, civil recourse is in this core a Grand Theory 'overblown in its claims, awkward and inconvenient in application, and internally incoherent in its account of the "guidance" it claims that the law of torts sends out'—but also at a perimeter of neighbouring fields. The claim that tort is 'relational' prompts Stapleton tartly to bring up relationality as more central elsewhere. Restitution, for example, 'is based on reversing certain bilateral transfers of wealth; equity ... is based on

[101] Stapleton, 'Lost', above note 22, n.p.
[102] Ibid (emphasis in original).
[103] Ibid.
[104] See Sanford Levinson, 'Foreword: "I Read the News Today, Oh Boy": The Increasing Centrality of Constitutional Design' (2009) 87 Tex. L. Rev. 1265, 1270 n.18 (noting the use of Kipling's line in slight alteration by Billy Bragg: 'What do they know of England who only England know?').

certain bilateral status relationships attracting norms of altruism; and contract norms ... are based on prior dealings/bargains between the two parties'.[105]

Critical reading of other writers is not the only locus of this type of disaggregation for Stapleton, who has moved the boundary around tort in her own claims as well. Let us apply the disaggregation lens to her article on good faith.[106] Good faith is relatively unimportant to tort, peripheral rather than central. Stapleton explains that as a legal standard, good faith typically demands less than negligence's 'requirement to satisfy a standard of reasonable behaviour',[107] while sharing with reasonableness a degree of indeterminacy and cultural construction.[108] This view of good faith may seem dismissive of its value to tort. Not so fast.

Disaggregation as practised by Stapleton builds rather than destroys, and here moving the boundary of tort makes tort bigger and deeper. Stapleton finds underexamined good faith in strict liability, a core tort concept.[109] We don't regularly apply good faith to defamation, Stapleton says; 'in certain circumstances', though not all, 'defamation liability requires breach of good faith'.[110] Stapleton concludes *Good Faith in Private Law* with three congruences between tort and good faith framed as 'lessons from tort for the good faith debate'.[111] The transmittal of these lessons is bidirectional: Tort travels to inform good faith, and in this movement of a boundary shows what it shares with non-tort doctrines.

A final example of boundary moving as disaggregation by Stapleton is her revisit to privity. Observers learned in the early twentieth century that 'the boundary around tort can move' from two great expanders of duty in the law of negligence, *MacPherson v Buick Motor Co.* and, dearer perhaps to Stapleton's heart, *Donoghue v Stevenson*.[112] In both game-changing decisions, common law courts widened the reach of tort.

One prominent scholar read *Donoghue*, soon after it was decided, as having found 'a citizen's duty ... not to turn dangerous or noxious things loose on the world'.[113] In *Product Liability*, Stapleton offers a criticism of the European directive that disaggregates this 'citizen's duty' ingeniously. She writes her criticism by exploring 'the appropriate relationship of contractual and tortious obligations, their boundary and possible concurrent operation' with a simple-on-the-surface hypothetical.[114] A dog is injured after it eats pet food sold in a can.[115] Using a couple of

105 Stapleton, 'Evaluating', above note 9, at 1560.

106 Stapleton, 'Good Faith' above note 99.

107 Ibid. at 8.

108 Ibid. at 9–13.

109 Ibid. at 14, 33. But see ibid. at 26 (observing that good faith is at odds with strict liability as applied in the tort of conversion).

110 Ibid. at 26.

111 Ibid. at 31.

112 111 N.E. 1050 (N.Y. 1916); [1932] A.C. 562.

113 Frederick Pollock, 'The Snail in the Bottle—and Thereafter' (1933) 49 Law Q. Rev. 22, 26.

114 Stapleton, 'Product Liability', above note 36, at 272.

115 Ibid. at 273.

variations on the story, Stapleton shows that liability *vel non* for that injury cannot rest on whether the can of dog food was a (unitary) 'dangerous or noxious' thing. The answer to that question must be It Depends.

Stapleton has found another instance of unity hidden in plain sight, doing underacknowledged mischief. Especially in the United States, a cliché-metaphor about privity speaks of a 'citadel': the word implies impregnability and centralized defence. The privity barrier undone by *MacPherson* and *Donoghue* once purported to unite product sellers behind the constancy of a wall. Cardozo and Lord Atkin achieved a benevolent kind of disaggregation, taking apart what had stood together unsoundly to thwart good claims. Stapleton finds, and deftly faults, the aggregation in that disaggregation. The concept of a defective product that emerged following *MacPherson* and *Donoghue* has brought to tort another confounding unity. As Stapleton shows with her pet-food example, 'defect' in contrast to conduct—a powerful twentieth-century instance of where the boundary around tort has moved—unites what tort principles suggest ought to remain distinct.[116]

9. Conclusion: The Disaggregator's Discipline

Jane Stapleton has been taking on other putative unifiers, not just the Grand Theories engaged in the first of her *Three Essays on Torts*, for decades. In this chapter I've gathered a few specimens of this approach to tort to illustrate a careful method. Disaggregation done by Stapleton isn't potshots or naysaying. It's creation.

Utter or pure disaggregation cannot exist in scholarship—or indeed anywhere. That the understanding and enhancement of tort law calls for both disaggregation and aggregation is a point nicely made elsewhere in this volume. Offering a nuanced partial defence of what Stapleton has labelled Grand Theories, Mark Geistfeld provisionally accepts the pejorative nature of this characterization. Geistfeld agrees with Stapleton that no Grand Theory can state accurately what tort is. Judges, however, do 'synthesize doctrinal analysis with their best understanding of the range of plausible comprehensive moral theories'.[117]

Such theories need not be called Grand, but they make normative claims,[118] and they unify in the sense of putting together 'seemingly disparate cases'.[119] Evoking the agenda announced by John Dewey of pluralism in service to ideals that sound unitary, Geistfeld finds 'mid-level pluralist theorizing' in Stapleton's interpretation of decisional law on one tort doctrine, pure economic loss.[120]

116 See ibid. at 272.
117 Geistfeld conference draft, at 6.
118 Recall that for Stapleton all Grand Theories are normative. See above notes 25–26, 118 and accompanying text.
119 Geistfeld conference draft, at 8.
120 Geistfeld, Chapter 2, this volume, pp. 18–19.

This characterization of Stapleton's methodology as complementary to, not just critical of, aggregation in tort takes me back to a torts gathering of the last century. There I felt spurred to broach a metaphor about disaggregation pertinent to this praise of an extraordinary scholar.[121] One instrument of disaggregation is a centrifuge, a machine that spins fluids to extract and disperse their constituents. From my view that dispersal and unity coexist in tort, the figurative centrifuge I posited works with and alongside a centripetal counterpart.

As illustrated by the scholarship celebrated in this chapter and this volume, a 'mechanism that breaks formations apart' joins 'a force that pulls toward the center. Concepts spin away but also coalesce.'[122] The honouree of this festschrift has explained, improved, and enlarged torts jurisprudence by deploying disaggregation with unequalled prowess.

[121] Anita Bernstein, 'The New-Tort Centrifuge' (1999) 49 De Paul L. Rev. 413.

[122] Ibid. at 415–16. Using different language, Stapleton has remarked on themes that come close to unity and even Grand Theories present in her work. See Jane Stapleton, Thoughts on Torts, Lecture 2, Clarendon Lecture Series at 2:57 (May 1, 2018) https://ox.cloud.panopto.eu/Panopto/Pages/viewer.aspx?id=db469e97-4d23-43b3-a35a-a8cf0099a5e4, last accessed 20 February 2024.

4

Tort and Insurance as Private Law

Jenny Steele and TT Arvind

1. Introduction

In this chapter, we address Jane Stapleton's complete exclusion of insurance concerns from tort reasoning. Jane's rejection of insurance as a relevant consideration in tort is, in our view, misplaced. While the battles she initially fought over tort reform were real, the total exclusion of insurance does not follow. The over-exclusionary approach is based (we will argue) on a misdescription of the nature of insurance and of the degree to which it is embedded in private law. At the same time, it oversimplifies tort principles. Jane's approach is, in fact, at odds with much of her contribution, which views tort law as properly open to a multitude of considerations.[1] Insurance appears to be singled out for exclusion when many other considerations are allowed in. Jane's openness to multiple 'policy' factors is often found unsettling by tort scholars. We come from a different angle and suggest that her unique hostility to insurance as an aspect of tort reasoning is not justified. Two key aspects to the argument in this chapter can be set out.

First, the resistance in Jane's work to recognising a role for insurance is based on the perception that insurance is an alternative to tort. We argue that they are complementary. The chapter does not argue that tort liability should be reformed, cut back, or replaced by insurance. Instead, it points to the ways in which the law of tort, and the nature of insurance, are aligned. There is no question of tort's logic being replaced by 'risk pooling' supposedly derived from insurance. And indeed, while risk *spreading* is naturally an element of insurance, risk pooling is something of a rare 'ideal' type. Rather, insurance should be analysed as a system of risk allocation, both private and public. We can understand insurance as connected both to party interactions and arrangements, and to demonstrably preferred social and commercial arrangements which are relevant to those interactions. In both senses, insurance is vitally connected to the law of tort and its reasoning.

Second, Jane's recent work, in which she emphasises a principle of respect for parties' 'cooperative arrangements', is devised in such a way that the absence of

[1] See, eg, J Stapleton, 'Duty of Care Factors: a Selection from the Judicial Menus', in P Cane and J Stapleton (eds), *The Law of Obligations: Essays in Celebration of John Fleming* (Oxford, 1998) and more recently, J Stapleton, *Three Essays on Torts* (Oxford, 2021), especially Chapter 1.

Jenny Steele and TT Arvind, *Tort and Insurance as Private Law* In: *Torts on Three Continents*. Edited by: Kylie Burns, Jodi Gardner, Jonathan Morgan and Sandy Steel, Oxford University Press.
DOI: 10.1093/oso/9780198889748.003.0004

insurance concerns seems artificial. These 'cooperative arrangements' often involve arrangement of insurance, or at least reflect understandings of the underlying insurance position. The exclusion of insurance obstructs understanding of how risks are allocated. Notably, this is a discussion about parties' *ex ante* allocations and the law's response to them. It is not a discussion about the *ex post* pursuit of an insurance fund. After arguments for the cutting back or reform of tort, the *ex post* pursuit of 'deep pockets' is the most feared source of potential incoherence, should insurance be considered. Paying attention to party risk allocation provides a welcome focus in the context of tort decisions. Its focus is on the facilitatory role of insurance and tort law, rather than its compensatory role. Such an exercise leads to complexity, but complexity is different from incoherence.

The chapter proceeds as follows. Section 2 introduces Stapleton's argument for the exclusion of insurance, also pointing to its broader legacy in her later work. It then appraises 'reflexive tort scholarship', and points out that this approach actually provides a way of addressing the significance of insurance which Stapleton does not take up, continuing the exclusion of insurance through her silence on the question. Section 3 explains both insurance and tort as connected aspects of private law. Section 4 presents examples of the role played by insurance in respecting party risk allocations or (in Stapleton's terms) 'cooperative human arrangements' beyond contracting. Section 5 concludes. Of course, the chapter cannot take up all elements of the relationship between insurance and the law of tort, and is focused on the presence or absence of insurance concerns in reasoning about tort duties.[2]

2. Stapleton's Denial of the Relevance of Insurance and Later Developments

In her influential article of 1995, 'Tort, Insurance and Ideology', Stapleton argued that tort should not be approached 'as' insurance.[3] The impact of the article has been broader than the range of its argument; and the attack on the significance of insurance within tort reasoning seems to have been continued by Stapleton herself beyond the narrow confines of the argument.[4] In fact, the key argument seems to build upon frustration at 'vague' claims that tort was 'about' insurance—so that the key target could be taken to include many rather different offshoots.[5] The enduring

[2] For much more varied discussion, see R Merkin and J Steele, *Insurance and the Law of Obligations* (Oxford, 2013); and, eg, K Abraham, *The Liability Century: Insurance and Tort Law From the Progressive Era to 9/11* (Cambridge MA, 2008); T Baker, 'Liability Insurance as Tort Regulation: Six Ways that Liability Insurance Shapes Tort Law in Action' (2005) 12 Connecticut Insurance Law Journal 1; R Lewis, 'Insurance and the Tort System'(2005) 25 (1) Legal Studies 85–116.

[3] J Stapleton, 'Tort, Insurance and Ideology' (1995) 58 Modern Law Review 820.

[4] This was described by Tom Baker as a 'jeremiad': T Baker, 'Insurance in Socio-Legal Research', University of Pennsylvania Institute for Law and Economics, Research Paper No 10-07, May 2010.

[5] Stapleton, 'Tort, Insurance and Ideology (n 3) 820.

ambition became to exclude insurance from all tort reasoning, and it is this which is described here as mistaken.[6] The very power of the article—combined perhaps with a message which was very welcome to many private lawyers—discouraged proper attention to insurance in relation to the law of tort.

Stapleton linked insurance closely with political tort reform agendas. Insurance stood outside private law, as one alternative to tort. The argument that it should be compared with tort as a response to misfortune was a political argument. Such a shift would indeed have winners and losers. But as we will argue further, all arguments about the extent of tort liability have distributive effects, and such effects are not outside the range of permissible tort reasoning. There is a contrast with the later argument of John Gardner,[7] that the decision as to *which* tort duties to recognise—or to put it another way, when to give effect to claims of corrective justice—is an unavoidably distributive aspect of tort reasoning itself. The decision whether to impose a tort duty has inherently distributive effects, and this is one reason why it is not reducible to a narrow range of principles or reasons.

Stapleton's original argument was a counter to analyses which addressed tort and insurance—including social insurance[8]—as alternative responses to 'misfortune', on which basis tort will appear less than successful or rational. Indeed, the outcome of such an exercise will inevitably be the production of arguments for radically restructuring tort liability, replacing significant portions of it altogether with the (supposedly) more rational and effective system of insurance.[9] Stapleton's original argument therefore objected to theories which would reduce and replace tort liabilities in order to provide more effective compensation, rather than objecting to insurance arguments in the different form of a search for 'deep pockets', which is inclined to *extend* tort liability. It is worth noting however that some arguments for the restriction of tort have used arguments of responsibility and deterrence, aimed chiefly at deterrence of victims who have transgressed in various ways, or of careless parties who are not primary risk bearers.

The ideology of deterrence is also present in tort reform arguments. In Stapleton's article, insurance and tort encapsulated different and inconsistent 'ideologies', based in different values, a fact which she argued was disguised by apparently empirical arguments. While the adoption of different values was legitimate, this should be done expressly and—contentiously perhaps—the 'personal' nature of ideological preference should be acknowledged.[10] It was through

[6] See also J Morgan, 'Tort, Insurance and Incoherence' (2004) 67 Modern Law Review 384–401. Our analysis here of both insurance and tort aims to show that insurance may lead to complexity but that this may be justified.

[7] J Gardner, 'What is Tort Law For? Part 2. The Place of Distributive Justice', in J Oberdiek (ed), *Philosophical Foundations of the Law of Torts* (Oxford, 2014).

[8] The original plan was to contrast tort with social insurance in a separate article.

[9] For example, the arguments put forward by PS Atiyah, particularly in *The Damages Lottery* (Oxford, 1997).

[10] We say this is contentious because, as argued below, there may be more than one core value at play in any institution or practice, including the law of tort, so that departure from what has been posited as

political argument that tort liability might be required to give way to other responses to loss, as the values could then be fought out. There was no clear identification of a core value for the law of tort, rather, it was necessary to leave space for what Stapleton identified as its two core values of deterrence and corrective justice. Arguably, reducing tort to a question of deterrence—since this is notoriously an unproven effect of its liabilities even if it is one of its values[11]—could leave it unnecessarily vulnerable in any such 'political' discussion. But so too does identifying it with corrective justice, since this primarily aims for an internal coherence that is not easily translated into politically intelligible claims for support or protection of tort. This is one reason why courts must themselves determine the proper extent of corrective justice through the law of tort. Corrective justice cannot exist in an unlimited form, but it is hardly the role of the legislature to determine every question about whether and when a claim based on corrective justice should succeed. Courts must, in this particular sense, make distributive decisions—they must determine when and where corrective justice is available (and, thus, its distribution) unless and until the legislature draws a line. Thus Stapleton has on some important occasions been too quick to describe evaluative elements of the law of tort in terms of 'policy' or 'political' values which stand outside legal principles. Her more recent work on 'cooperative arrangements' (risk allocations) is an important counter-balance to this.[12]

It seems unlikely that the survival of tort, and indeed the support offered to tort principles by consecutive legislatures, is based on an appreciation of its deterrence or corrective justice values alone. Indeed, identification of these values as 'inherent' gives the impression that the operation of tort simply cannot be evaluated other than in 'political' (interpreted as extra-legal) terms. If there is to be critique of corrective justice and deterrence, or a drawing of boundaries for its operation, let it be labelled political and not legal.[13] Within the United Kingdom, legislative interventions have made significant contributions to the law of tort but have not sought to cut it back fundamentally. Even more incremental approaches to scaling back tort liability which preserve its essential structure, such as the Ipp reforms in

'the' core defining value of tort need not be personal, but may be an interpretation of the field. This is compatible with Stapleton's general approach in her most recent work, where she identifies a 'principle' of respect for collaborative human arrangements, which is not expressed as part of core values of deterrence or corrective justice.

[11] There are those who reject deterrence as a value of tort law. In particular, there are corrective justice theorists who regard deterrence as an extrinsic goal. Therefore, Jane's identification of corrective justice and deterrence as core values of tort is itself controversial.

[12] Stapleton, *Three Essays on Torts* (n 1).

[13] To reiterate a point above, the extension of corrective justice to new areas is very much a legal question, even if it is at the same time distributive and political. See also J Gardner 'The Negligence Standard, Political not Metaphysical' (2017) 80 Modern Law Review 1–21

Australia, have had limited appeal in the United Kingdom,[14] where most interventions, on the contrary, have been aimed at the better operation of the law of tort as developed in the courts.[15]

The argument here, which is undoubtedly contentious, is that the continuation of tort in the United Kingdom is most simply explained by its social usefulness; but that this social usefulness is not to be captured in terms of the achievement of any single goal. Perceiving tort law as socially useful does not in itself condemn us to preferring other mechanisms which can be argued to achieve specific goals more effectively. The appropriate distribution of corrective justice can be seen as a socially useful function of the judicial role. For this reason, Parliament acts in a range of ways to facilitate the operation of the law of tort, for example through funding regimes, and sometimes acts to introduce new tort liabilities or refine existing ones. But the essential point here is that internal tort reasoning inescapably involves distributive issues about the availability of its remedies, and these are not extrinsic to law to be discussed on policy, rather than legal grounds.[16] The boundaries of 'legal' questions are being drawn too tightly in Stapleton's original argument, and in some of her later work also. Defining principles narrowly leads to a broader remit for what is called 'policy' or 'ideology', where lawyers have no particular claim to expertise. While Gardner argued that many elements of tort were political—notably, the standard of care in negligence—these were at the same time no less legal.[17] The contrast was not between the political and the legal, but the political and the 'metaphysical'. For Gardner, there were few boundaries in what an appropriate legal consideration can be. Reasons valuable in one context are also valuable in another.

Stapleton argues that the boundaries of tort are to be set through 'ideological' choices. In Gardner's argument, these choices are political but as we have seen no less legal. However, Stapleton's argument is presented in decisive normative terms—for example that non-liable defendants would be getting away 'scot-free' should we tinker with tort principles and reduce the coverage of corrective justice.[18] There is no apolitical extent or coverage for corrective justice, for distributive concerns are always at work in determining whether—as a matter of law—demands of corrective justice are to be met. These are legal questions. In other work, Jane argued that 'no duty' solutions are offensive because they appear to amount to an argument that defendants are 'allowed to' be negligent towards the claimant.[19] In much of her work, we see this sort of line-drawing around 'legal'

[14] For Australia, see, eg, P Underwood, 'Is Ms Donoghue's Snail in Mortal Peril' (2004) 12 Torts Law Journal 139; P Cane, 'Taking Disagreement Seriously: Courts, Legislatures and the Reform of Tort Law' (2005) 25 Oxford Journal of Legal Studies 393.

[15] There is of course much variety behind this statement: cf TT Arvind and J Steele (eds), *Tort Law and the Legislature: Common Law, Statute and the Dynamics of Legal Change* (Oxford, 2012).

[16] See n 7.

[17] See n 13.

[18] Stapleton, 'Tort, Insurance and Ideology' (n 3).

[19] J Stapleton, 'Evaluating Goldberg and Zipursky's Civil Recourse Theory' (2006-7) 75 Fordham Law Review 1529, 1531.

questions, leaving fundamental issues such as the 'extent of liability' to be dealt with separately in openly evaluative terms.[20] We think that her later work on the principle of respect for party cooperation has necessarily moved away from this, but this frame of thinking affected the original exclusion of insurance, with lasting consequences.

Jane Stapleton's 1995 article was a hit. While it was ostensibly aimed at the tendency to see tort and insurance as like-for-like equivalents, to be judged in terms of effectiveness as a response to misfortune, the article was couched so broadly that it seemed to justify the exclusion of insurance from all analyses of tort reasoning. And indeed, this was what was argued by Stapleton in other contexts, without the same level of analysis as the original argument. For example, insurance was condemned, of all considerations, as the one which could only ever be a 'makeweight'.[21] Beyond the tort reform argument, there was a secondary fear that insurance arguments would simply be given a primary role, ignoring the relationship between the parties. In any case, the denial of relevance to insurance was a relief to many, not only because of the threat of tort's abolition, but because it was thought that insurance arguments *within* tort reasoning would lead to an unprincipled pursuit of 'deep pocketed' defendants at the expense of legal principle. But this is not the impact of giving due consideration to insurance in terms of *ex ante* arrangements, as explored below. And ironically, Stapleton's recent work has set out the very kind of analysis which shows—or would show, if insurance had not already been excluded as incompatible with the values of tort—the inherent significance of insurance to private law. Insurance is an essential component of the principle of respect for party risk allocation, or of respect for cooperative human arrangements, as Stapleton puts it. Insurance is in and of party relationships, and it is in and of private law.

'Risk allocation' as a factor in the law of tort does inherently acknowledge that the risk of a breach of duty is something to which parties will put their minds when conducting their relationships, contractual and otherwise. Extension of this to 'cooperative human arrangements'—that is, beyond contracting in the strict sense—begins to show how broad the point is. Liability is in principle consistent with a goal of deterrence, but the law will not seek to disturb efforts to deal with risks *ex ante* in order to impose a new attempt at deterrence. This reflects the fact that no amount of tort liability will remove the possibility of negligence from life. While this is also true of intentional acts, it is fair to say that even honest individuals will fail in negligence duties from time to time. When it comes to organisations, the possibility of both negligence, and intentionally harmful behaviour by those other

[20] J Stapleton, 'Cause-in-Fact and the Scope of Liability for Consequences' (2003) 119 Law Quarterly Review 388; J Stapleton, 'Occam's Razor Reveals an Orthodox Interpretation for *Chester v Afshar*' (2006) 122 Law Quarterly Review 426, 433.

[21] Stapleton, 'A Selection from Judicial Menus' (n 1).

than the assured, are recognised risks which private persons cannot wholly control, whatever incentives or deterrents law may impose. Insurance represents a bottom-up, party-led response to this inevitable and intrinsic dimension of social life. Deterrence is only ever part of the story in thinking about harm, risk, and the response to them.

In earlier work, Stapleton was careful to protect the value of deterrence. She argued that party arrangements could only be taken into account when this was consistent with the value of deterrence. Given how slight the evidence is for deterrence as the defining value of the law of tort—or even a primary effect—it is difficult to justify this hierarchy of concern. However, this further underlines the significance of Stapleton's more recent work on party risk allocations, for the later account makes no mention of deterrence and does not set it as a condition on respect for cooperative arrangements. This is more in tune with reality, for those arrangements accept that duties may sometimes be broken.

That, in fact, is something that is disguised when we do *not* pay attention to insurance. One of the important factors here is the principle of subrogation, which allows insurers to stand in the shoes of their assured for the purposes of an action. Questions about deterrence and fault may be raised by subrogated insurers, seeking to use their position of standing in the insured claimant's shoes to shift liability to other parties at fault. The 'duty enquiry' may therefore take place in a context where an insurer is seeking to shift part of their liability to another party. Sometimes, these may be 'peripheral parties', such as local authorities who are alleged to have failed in duties to make the claimant safe. This was the case in *Stovin v Wise*,[22] and it is clear that the House of Lords was aware of this context. Lord Hoffmann observed that while there was an argument in favour of a deterrent against highway authority inefficiency, 'there must be better ways of doing this than by compensating *insurance companies* out of public funds'.[23] The particular context of *Stovin v Wise* is now governed by the Supreme Court's principle that there is no liability for failures to confer a benefit (save exceptionally: for example, following voluntary assumption of responsibility).[24] This appears to remove situations like *Stovin's* case from multi-factorial consideration, although 'assumption of responsibility' as a criterion is itself not simple and may bring some of the same considerations back into play. Subrogated insurers may appeal to principles which seem derived from arguments of deterrence when seeking to shift liability to a party at fault, rather than a party who has an agreed or implied risk-bearing responsibility. But the possible existence of fault was presupposed by the very risk allocation the parties entered into: as Lord Hoffmann further and correctly observed in *Stovin*, and it is far from clear that premiums would in any way be different if a subrogated insurer

[22] [1996] AC 923.
[23] ibid 955 (emphasis added).
[24] Michael; Robinson xxxx.

could recover compensation from the public body. While an assured cannot contract for cover of its own intentional wrongdoing, it is possible for an organisation to insure against *vicarious* liabilities arising from deliberate acts of its employees and others in the absence of a specific prohibition, so that this is not purely a point about negligence, but about the risk of wrongdoing more broadly.[25]

This section has evaluated the initial reasons for Stapleton's argument for the complete exclusion of insurance from tort reasoning, and the designation of all insurance arguments as being at odds with the values of tort in ideological and distributive terms. We question the idea of distributive concerns being necessarily 'external' to tort, and the apparent reduction of tort to two simple values which in any case may not be compatible. The next section appraises the way in which insurance is presented in Stapleton's 1995 article, and the way in which the understanding of tort is structured so as to exclude insurance considerations, taking issue with both.

3. Tort and Insurance as Private Law

A. Perceptions of tort

Tort and insurance are more closely related than is often appreciated. Not because tort should be approached as an 'alternative to' (or substitute for) insurance, but because both can be understood through a private law lens. It is inappropriate to single insurance out for dismissal as irrelevant or at best a 'makeweight' argument. Insurance considerations are often within the frame of private law decision making, including tort decisions: not as a factor to end all debate but as part of the fabric of party relations. Denying reductionism could lead us also to deny narrow frames of interpretation.

Most importantly perhaps, tort responds to a range of different relationships. Risks like those associated with tort (including the risk of harm done through negligence) are often considered by parties within their relationships. Some of these relationships are not purely bipartite. They may operate within a contractual 'matrix'; or they may involve insurance relationships. Understanding of insurance arrangements is often a natural element in the courts' consideration of those relationships and the framework of rights and duties appropriate to them, including the ambit of tort liability.

In her recent work,[26] Stapleton proposes that contract is centrally concerned to respect party risk allocations. Turning to tort, she argues that a parallel concern

[25] Where intentional wrongdoing is one's own, the language of 'risk' may seem misplaced. But the risk of intentional harm by employees and others is a coherent idea.

[26] Stapleton, *Three Essays on Torts* (n 1) Essay 2, 'Cooperation and Economic Self-Reliance in Commercial Arrangements'.

can be found within the law, which she defines in terms of respect for human co-operation. We agree, and will seek to show in the next section how important this move is. A similar approach is taken by Merkin and Steele, who also begin with an analysis of contract and turn to parallel and related thinking in tort. The key difference between these positions is that Merkin and Steele added the point that insurance is in fact an inherent element of these agreed risk allocations, that is, of cooperative human arrangements (to use Stapleton's subsequent term) beyond contract. Insurance is so endemic to risk arrangements that it would be odd to discuss them without any reference to insurance. The 'cooperative human arrangements' explored by Stapleton may be positively vindicated either by a tort duty, or a denial of a tort duty. But it is notable that the parties' own placing of risk through insurance arrangements is again absent from Stapleton's analysis of the cases here, even though these are among the key features of parties' 'cooperative arrangements'. It is suggested therefore that the continued exclusion of insurance from the picture makes no sense when the focus is (quite rightly) on either a principle of co-operation or a principle of economic self-reliance, since insurance arrangements are inherently connected to these very principles.

The similarities and differences between the approaches of Stapleton's recent account, and the approach taken by Merkin and Steele, can be illustrated by reference to the latter authors' analyses of risk allocation practices by commercial parties in particular. The point is not that courts will try to seek an insurance fund as the fairest way to place liability—an *ex post* exercise—but that courts appreciate that parties typically consider the risks to which they are exposed when they enter into relationships, *ex ante*, including non-bilateral relationships in which the law of contract may well need to be supplemented by tort. All of this is common to the respective accounts. The key difference is that Merkin and Steele point to the significance of insurance to party risk allocations, while Stapleton remains silent on this point. Perhaps that silence is a residue of her earlier argument about insurance. The point to be made is that the objections of the 1995 article, about incompatible responses to 'misfortune', are not at all applicable in the context of tort claims which seek appropriate duties in the context of party risk allocations. We are dealing here with losses that were within the contemplation of the parties involved, and anticipated in their arrangements.

Merkin and Steele's analysis emphasised that parties' risk allocations are typically underpinned by insurance. That is, the parties will need to determine what their exposure to risk is, and to insure accordingly. If they are to be content with the likely liability of a party with whom they are entering a contract for example, they may then insist that the party carry insurance to enable any judgement to be satisfied, or consider paying for insurance themselves. The same goes for potential liabilities to third parties on the part of occupiers of land through the activities of contractors—if the contractors cause harm and are not insured, liabilities to third parties may fall on the occupier. Arrangements of these sorts are underpinned by understandings of insurance practices whether implicit or explicit. These arrangements are not unique to individual relationships. Instead, there are recognised patterns of insurance and risk allocation in different contexts. In the 'contractual matrix' cases represented by cases such as *Norwich City Council v Harvey*,[27] a case of construction contracts, there is also an interest in avoiding over-insurance. The negotiations between the parties in this case had resulted in an allocation of liability to the claimant, in the sense that liability to the claimant on the part of the head contractor was excluded. When a subcontractor was negligent, the court found that although the subcontractor was not directly in a contractual relationship with the claimant, the exclusion of liability in the head contract also benefited the subcontract. The majority of commentators, including Stapleton, have interpreted this as a case about the agreement of the parties operating to exclude a claim by the Council, which had effectively agreed to take the loss.[28] The parties' cooperative arrangements fell to be assessed in the light of this risk allocation.

Merkin and Steele's point in *Insurance and the Law of Obligations* is that the parties' arrangements will have been the basis on which insurance was put in place, and the insurance premium paid (an important element in the pricing of contracts). Norwich City Council had agreed to maintain insurance and the main contract stated that the buildings were to be at their risk. The action will have been brought by a subrogated first party insurer (ie the council's insurers). The defendant subcontractor was not a party to the head contract. However, it will have been aware of the risk structures in place, and without that allocation, it would have insured against its potential liability. To allow the claim would thus have created a double burden of insurance, the price of which will be reflected in the cost of the services. Merkin and Steele's analysis, like Stapleton's, leads to the observation that sometimes tort duties will be denied because of a party risk allocation. However their point is that insurance arrangements comprise a significant element. The protagonists will include insurers as (subrogated) litigants, and the defectiveness of a legal response which undermines the parties' structures is still more stark. Its consequences will include an uninsured loss in a context where the parties had chosen who would secure and pay for insurance, priced accordingly, and the impossibility of placing insuring obligations efficiently because of the potential for insurers' tort claims to unravel the arrangement.

The role of insurance was not clearly articulated in *Norwich*, by contrast with later cases such as *John Hunt Demolition Ltd v ASME Engineering*,[29] which interpreted the parties' overall arrangement in terms of an intention to place risk with insurers rather than with parties to the contract. This is not an instance of 'deep pockets liability', but an understanding that parties come together with the

[27] [1989] 1 WLR 828.
[28] Stapleton, *Three Essays on Torts* (n 1).
[29] [2007] EWHC 1507 (TCC).

intention of allocating risk in such a way that insurance—the cost of which is no doubt an element in the pricing of the contracts—is effective. These cases illustrate that the exclusion of insurance from discussion of risk allocations and cooperative arrangements is artificial, and understates the significance of being attentive to those arrangements when determining tort duties. It can also be noted—as indeed Stapleton does note—that these cases are concerned with duties in respect of physical damage (eg destruction of building by fire in the *Norwich* case). In cases of physical damage, a duty of care in negligence has sometimes been said to be subject to no special control devices (in contrast with liability for pure economic loss). This illustrates that considerations about insurance and risk are not in fact 'control devices', but inherent elements of the consideration of tort duties.

But there is more. If we pay attention to insurance, we will see that the language of tort principles can unfortunately be used in a narrow way in order to circumvent parties' allocations of risk. This kind of appeal relies on a sense that tort is 'about' issues of responsibility, and that considerations of risk allocation are less central, maybe even considerations merely of 'policy'. The parties invoking such principles are very often subrogated insurers. The argument that to pay attention to insurance is to change the subject can be turned on its head if we accept the significance of party arrangements, as Stapleton does. From this perspective, the question becomes a radically different one: if the parties have allocated risk either as between themselves, or by transferring it out of their relationship by paying a third party to bear it, under what circumstances is it appropriate for the courts to disrupt their private arrangements? Insurance, from this perspective, is no different from any other voluntary technique by which a third party is engaged to bear a risk that arises in the course of interaction between two parties. To think about insurance is therefore very much to keep our mind on the subject of the actual nature of the parties' arrangements, and indeed to avoid distraction through more apparently simple (acontextual) appeals to basic principle.

Inclusion of insurance within the analysis can also help us to understand the boundaries of this approach. For example, it is connected to the ways in which personal injuries are often treated differently from other forms of damage. Claimants suffering personal injury typically deserve less scrutiny than is sometimes invited in terms of autonomy, dependency, and good behaviour, for example. Insurers sometimes propose a narrow and acontextual reading of tort principles. However, risk allocations may still be at work. This includes risk allocation through public, political decisions which support some aspects of the functioning of the law of tort. This is an instance where political and legal decision making are mutually reinforcing and difficult to disentangle. Attention to patterns of insurance can however help us to understand when it is appropriate to resist this special treatment and recognise that a denial of duty even in relation to personal injury can protect existing risk allocations. That is one potential interpretation of the decision in *Stovin v Wise*, since here the primary defendant was a driver, and the liability will have been covered by compulsory motor insurance. The claim against a local authority for not alleviating a danger can be perceived as an attempt to shift this insured liability onto a third party. English law now deals with such cases in terms of a much more general principle that there is no duty to benefit outside certain exceptions. The previous form of reasoning is not correctly dismissed as 'policy-based'.[30] It can however be rightly identified as complex. It is arguable that the choice of simplicity in *Poole v GN*[31] and related cases, which is in any case somewhat qualified by the exceptions to its essential principles, comes at the cost of comprehension of reasons why the law has developed as it has.[32]

A very different example is the successful resistance of the Motor Insurers' Bureau (MIB) to indemnification of a passenger seriously injured while being carried in the car of an uninsured driver in *Delaney v Pickett*.[33] The driver and passenger were in possession of cannabis. The MIB failed in its argument that a tort claim should fail for illegality—the possession of cannabis was unrelated to the claim for damages—but it is notable that this argument was deployed. In the event, the MIB successfully argued that the claim was outside the MIB Uninsured Drivers' Agreement 1999, since the claimant knew that the vehicle was being used in the course of a crime. This outcome was at odds with the values of the common law in its approach to illegality, where a strong causal link between illegality and injury is required as illustrated by *Delaney v Pickett* itself. Not everyone will welcome the stark distinction between deserving and undeserving personal injury claimants in a sense unrelated to the injury suffered.[34]

B. Perceptions of insurance

So far, we have mainly discussed the perception of tort and of tort scholarship that are at work in Stapleton's exclusion of insurance from tort decision making. Another part of the picture concerns the perceptions of insurance that were at work. Merkin and Steele have also argued[35] that Stapleton's approach seems to

[30] cf Robinson (n 24).

[31] [2019] UKSC 25; [2020] AC 780.

[32] The Supreme Court in *Poole* (n xxx) recognised the value of incrementalism, and the correctness of the large majority of negligence cases decided under *Caparo*, but rejected the reasons given by the courts in many such cases. If those were valid reasons, then incrementalism will become more difficult to achieve if they are not to be discussed.

[33] *Delaney v Pickett* [2011] EWCA Civ 1532; [2012] 1 WLR 2149. See also the discussion of the MIB and its pursuit of tort damages from the pockets of uninsured drivers, often over many years, in R Merkin and J Steele, 'Policing Tort and Crime with the MIB: Remedies, Penalties and the Duty to Insure', in M Dyson (ed), *Unravelling Tort and Crime* (Cambridge, 2014) 22–57.

[34] Therefore, even those who approve of the sentiment of *Tomlinson v Congleton*, which concerned responsibility for one's own risk-taking, need not approve of the non-compensation of the passenger in *Delaney*.

[35] Merkin and Steele, Insurance and the Law of Obligations (n 2) Chapter 2, 'Characterising Insurance'.

proceed from a strongly 'actuarial' perception of insurance. This is associated with the idea that insurance is quintessentially an instrument of loss-spreading across defined risk pools. It tends to focus on the supposed 'technology' of insurance as a means of calculating and distributing like risks. Insurance is thus perceived to be a mechanism which spans both markets and collective purposes. In a sense it is a hybrid achievement of social purpose.

Merkin and Steele suggested that the notion of homogenous risk pools may not reflect the reality of insurance in many contexts. This is linked—as we will explain—to the existence of a very different perception of insurance, which addresses insurance as a factor of *market* arrangements and therefore quintessentially of private law. This approach emphasises that insurance has its source in and is integrated into market activity. Merkin and Steele call this a 'relational' model of insurance. It is concerned with party relationships, primarily market relationships, though it does not exclude the sense of social and political preference which is present in guiding the operation of markets. Market arrangements are fundamentally about coordination. Their role is not just to allocate scarce resources efficiently, but also to create the conditions for entrepreneurial innovation—that is to say, of devising new ways of doing things that have value to other market participants—by enabling other market participants to discover and, thereby, take advantage of, opportunities for mutually gainful collaboration.

Entrepreneurs provide services to other participants that they either could not do for themselves at all, or that the entrepreneur can do more effectively. The precise way in which entrepreneurs do this in a profitable way is not reducible to any one principle: indeed, it is precisely the creation of new transaction opportunities through innovation that is the heart of entrepreneurship.[36] Insurance, seen as a factor of market arrangements, follows precisely this model. An insurer provides an innovative service to other market participants, namely, the insured. Insurance, in this sense, provides an *alternative* to risk-apportionment between the parties *inter se*. Rather than distributing or allocating the risk between parties to the primary transaction or instance of social interaction, an insurer offers market participants the ability to deal with a risk by engaging a third party (ie the insurer) to bear the financial consequences of the eventuation of that risk. The precise techniques an insurer uses to achieve this are from this perspective an aspect of their entrepreneurial approach and are likely to be the subject of market competition wherein competing insurance providers arrive at different ways to manage the risk of loss. But the various methods cannot be reduced to a single reductionist model such as pricing through risk pools.

Even if we want to resist policy-based reform arguments that compare tort's response with the responses of 'risk-pooling' technology, we need not avoid thinking

[36] The classic exponent of this theory was Kirzner. See esp IM Kirzner, *Discovery, Capitalism, and Distributive Justice* (New York, 1989) 87–91.

about insurance in this particular guise. The 'relational' approach focuses especially on the ways that losses are initially shifted by the insurance relationship, rather than necessarily pooled across like risks as the actuarial image would suggest. It is clear that the nature of insurance is to avoid concentration of losses. But it is fundamentally important that losses are then spread in a variety of ways—partly through reinsurance, and through various market activities including diversification of risks—but this does not closely resemble the model of like risk pools. The model of like risk pools is part of a conception of insurance as technology, which has commended itself to governance theories of social science.[37] The recognition that insurance achieves spreading through a variety of risk transfers, and that these are not susceptible to calculation, brings insurance firmly into the fold of private commercial arrangements, and private law.

These features of insurance, departing from the idea of homogenous risk pools, are both a strength and a weakness, in terms of social purpose. It means that apparently incalculable and extreme risks are nevertheless capable of being shifted and spread.[38] The capacity of insurance to ameliorate risks does not stop at the limits of calculable risk pools. But it also means that there is less control over risk and loss-spreading. There is the potential for shocks and disorder,[39] as with any financial arrangement. Law governs insurance in a range of ways, including contract law and legislation. But what it is governing incorporates a set of market relationships. Insurance is often required to deal with the unexpected and, seen in terms of risk pools, this may even appear unfair. Seen in terms of the broader picture including diversification of risks, reinsurance, and so on, this perception is by no means clearly correct.[40] The heart of insurance lies in the insurance provider's commercial bet that they can deploy a mix of different risk diversification and spreading techniques that will let them make a profit by agreeing to bear the financial consequences of the eventuation of a range of uncertain events. Distributive considerations will sometimes be part of the approach the law takes—and, indeed, as the *Fairchild* 'enclave' and the Compensation Act 2006 instantiate, they may even come to dominate the law's approach.[41] But as the discussion in this section has shown, that is by no means the entirety or even the predominant part of private law's concerns in relation to insurance: as Males LJ put it in *Equitas Insurance v MMI*, when the objective of 'victim protection' underlying the *Fairchild* enclave has been satisfied, tort liability will necessarily revert to orthodox principles.[42]

[37] P O'Malley, *The Currency of Justice: Fines and Damages in Consumer Societies* (Abingdon and New York, 2007), is a valuable contemplation of the significance of money in law's response to tort and crime, which also illustrates the inclination to observe insurance as a means of governance or control.

[38] The existence of catastrophe insurance, often requiring collaboration between state and private parties, is an illustration of this. The risks are too big to be spread across a homogenous risk pool.

[39] J Steele, 'Satisfying Claims? Money, Tort, and "Consumer Society"' (2011) 20(4) Social & Legal Studies 516–527, takes issue with the 'insurance as governance' approach.

[40] R Merkin, 'Insurance and Reinsurance in the *Fairchild* Enclave' (2016) 36 Legal Studies 302–325

[41] For discussion, see Merkin and Steele, *Insurance and the Law of Obligations* (n 2) Chapter 12.

[42] *Equitas Insurance v MMI* [2020] QB 418, [91] (CA).

What is in play is a series of complex transactions and a complex mix of private arrangements (contractual and non-contractual, and between private persons *inter se* as well as between them and an insurer); legal obligation including statutory obligation at different stages of the relationships in question (obligations to insure, statutory definitions of insurer liability, and contribution in relation to mesothelioma claims, and so on); and financial risk arrangements. Risk pools are a minor part of this wider picture, being no more than one of a wide range of risk diversification techniques deployed by insurers, and not part of the constitutive social or market purpose of insurance, nor the overarching concern of tort law in its interaction with insurance.

Implications

Once perceived in this latter way, it can be seen that insurance lies within the domain of private law, for better or for worse. To consider the contribution of insurance to private law—and particularly for our purposes the law of tort—is not to 'change the subject'. It does not need to entail imposing a single goal onto a private law analysis, thus ignoring its values. Rather, it can take on a range of guises that are familiar aspects of private law itself. It may focus on the interpretation of party arrangements and the intention behind them, and the implications of either honouring or unravelling those arrangements. These are factors which are relevant to private law decision making in party disputes. Equally, it may be relevant to the question of how far observable social policies, including the policy that personal injury victims should be supported so far as possible by a source of compensation, can be pursued and to what extent arguments couched in terms of private law doctrine may in fact be used to frustrate this purpose, as well as the risk allocations arrived at by the parties.

One conclusion is that insurance arrangements can be approached as inherently part of the relationships to which tort and contract apply. There should be no arbitrary prohibition about consideration of insurance arrangements, of all factors. In fact, they are intimately connected to the cooperative human arrangements to which Stapleton points. Their consideration does not feed into an infinite regression in the form of a 'deep pockets' analysis. Insurance is far more intelligibly approached as part of wider consideration of the parties' *ex ante* arrangements, in determining where responsibility lies. Nor does it imply that the state may as well distribute *all* risks, which some would argue is the implication of accepting that loss-spreading is a goal of private law. Some analysis of insurance relationships in determining tort questions is natural and often indispensable, not because the 'whole picture' is 'really' about insurance, but because it is about party *relationships*—of which one familiar aspect is insurance arrangements. As with any factor that is of relevance to the analysis of a private law relationship, it will sometimes weigh more heavily than others, and will have more salience in determining the outcome in some cases than others. It is not uniquely powerful. But nor is it uniquely irrelevant.

Of course, these observations do not exhaust the relevance of insurance to the way that we think about tort. And as ever, it is important to think critically about assertions that are made in the name of insurance—as Stapleton would have it, vague ideas that tort is 'about' insurance. For example, it has sometimes been proposed that the objective standard of care in negligence—and perhaps also the full measure of damages payable—can be justified by the fact that few people in fact pay for liabilities, and that this is largely thanks to insurance.[43] This sort of argument is contentious, as there are other arguments in favour of objective standards. However, the example does focus the mind on the fact that what is typically thought of as the 'paradigm' case of tort because it is between strangers who are typically individuals—the road traffic collision—is in fact supported by compulsory liability insurance, and by insurer agreements in accidents where there is no insurance in place. So it is fair to say that the paradigm case of tort is in fact operating on a substratum not just of insurance but of law relating to and regulating insurance, and is operated by insurers on a largely routine basis. So while Stapleton's insistence that arguments should be clear and focused is fairly made, there are grounds for saying that in order to understand the operation of the law of tort, we should understand the influence upon it of insurance. For the way in which tort operates and in which its claims are made and governed to have had no effect on applicable principles would be somewhat unexpected.

At the same time, attention to social policy regarding insurance—and especially liability insurance—adds further weight to the notion that personal injury is different from other forms of harm or damage, not because all personal injuries require to be compensated (this would indeed be circular), but because there may be stronger reasons to recognise duties to avoid personal injury and because the satisfaction of liabilities for personal injury through insurance is itself an observable social policy which has weight against some arguments couched in terms of apparent (contestable) principle. For example, it has weight against arguments of personal responsibility and self-reliance where the use of these arguments by insurers would have the effect of enabling them to avoid the risk they had contractually agreed to bear, and of inserting back into the parties' relationship risks that they had sought to remove from the relationship by transferring them to an insurer. This is not to suggest that arguments of personal responsibility and deterrence are irrelevant, but rather that to give them unconditional prominence will give them a disproportionate significance—an effect which is averted if insurance

[43] Insurance of liabilities is one of the reasons given by Lord Denning in *Nettleship v Weston* [1971] 2 QB 691 why an objective standard of the reasonable driver could be appropriately imposed on a learner driver. Notably, criminal law also imposed an objective standard here, which had nothing to do with insurance.

becomes part of the picture. In this specific respect, insurance arguments may be used against insurers themselves who are wielding private law principles, because awareness of the context shows us something significant about the weight of those arguments.

Stovin v Wise serves as a good illustration: as this chapter has discussed, awareness of the insurance context enabled the House of Lords better to contextualise and evaluate the weight of the arguments advanced around deterrence. But in other contexts, insurance arguments may very well lead to a denial of duty, because awareness of the context may reveal attempts at free riding—an issue which Stapleton considered significant, but which can be understood considerably more effectively by considering insurance.[44] The point in both instances is that private law principles are often used by insurers—typically subrogated insurers—in an attempt to avoid liability and shift losses. To the extent that this frustrates private collaborative intentions or observable social policies, then it is as well to be aware of the insurance context.

4. What are the Obstacles?

Some elements of received tort theory are also obstacles to accepting that tort and insurance are 'on the same page'. But there are variations of tort theory which do not share these elements.

A. Tort duties

A key obstacle to the integration of tort and insurance is the supposition that tort law is essentially concerned with prescribing duties of behaviour.[45] These duties therefore have no inherent or intrinsic 'ambit'. Ambit is, rather, imposed as a result of value-judgement.

This implies that tort law's focus is only on behaviour. Tort is in this view a series of prohibitions, and sanctions (remedies). Getting from here to the idea that tort responds to party risk allocations would require a change of perspective. It

[44] Stapleton argued that insurance was uniquely irrelevant in determining free-riding: the failure to purchase first-party insurance, for example, could never amount to free-riding or be a counterpoint to a deterrence argument, regardless of the ease with which it could have been purchased. J Stapleton, 'Duty of Care: Peripheral Parties and Alternative Opportunities for Deterrence' (1995) 111 Law Quarterly Review 301. The argument advanced in this chapter suggests, in contrast, that support for permitting the parties to use market techniques to manage the risk inherent in a transaction, including by transferring it out of the transaction, is of far greater importance to the law, and considering whether widely available first-party insurance was in fact taken out is of obvious relevance to determining what their risk allocation arrangements in fact were.

[45] Compare McBride, Chapter 6, this volume.

particularly requires that duties are understood to be imposed on the basis of a broad range of considerations and reasons, often related to the roles undertaken by the parties and their relationship with others. Equally, that the location of risk in these relationships is one of the key questions to be addressed. Stapleton would therefore appear to have changed sides on this question. In earlier work, she argued that tort duties were simply duties to take care, and argued for a wider conception of minimum actionable damage in the context of personal injury that gave better effect to this duty by abandoning the requirement of physical change in the claimant's person.[46] Now, she places considerable emphasis on the scope of duties, whilst continuing to argue that duties are owed to all the world in respect of positive acts that cause personal injury. We doubt that this is analytically useful—it appears rather to be the remnant of a broader argument that is no longer pursued. But we certainly agree that the majority of risk allocation questions concern damage that is not in the form of personal injury.[47]

We would also agree with Stapleton that the decision in *Marc Rich v Bishop Rock Marine*[48] clearly illustrates the application of risk allocation questions to a case of physical damage rather than (as the Supreme Court has wrongly asserted[49]) economic loss. That is justified because property has commercial value and can be approached in terms of risk to be allocated; equally because there is recognised potential for first party insurance cover to a tolerable level for most. The uninsured element of personal injury, where aspects not covered either because there is no insurance policy at all or because it does not offer full cover, leaves a more bitter taste—though this distaste has not held back the application of contributory negligence principles. These principles are based on a compromise rather than on a clearly stated moral or legal principle of proportionate liability.[50] Again, it is *insurers* who have tried to generalise—with considerable success in a number of jurisdictions—from this example, in order to limit defendants (in fact, defendants' insurers) to a 'proportionate' share of liability. This illustrates the dangers of seeking to develop 'principles' without context. It also demonstrates the willingness of insurers to attempt exactly this. Academics may champion clear and simple

[46] J Stapleton, 'The Gist of Negligence' (1988) 104 Law Quarterly Review 213 (Part I) and 389 (Part II); Stapleton, 'Evaluating Goldberg (n 19).

[47] Rob Merkin pointed out in 'Tort, Insurance and Ideology: Further Thoughts' (2012) 75 Modern Law Review 301–323 that a focus on personal injury was not the most promising basis for an exploration of the relationship between tort and insurance, especially since first party insurance is more widespread in relation to other harms. This reflects the points made earlier, that the concerns of Stapleton's article were with particular 'tort reform' movements which have tended to focus on personal injury.

[48] [1996] AC 211.

[49] *Poole BC v GN* [2019] UKSC 25; [2020] AC 780. This is an incorrect reading of the damage in *Marc Rich*.

[50] K Barker and J Steele, 'Drifting Towards Proportionate Liability: Ethics and Pragmatics' (2015) 74 Cambridge Law Journal 49–77. The argument of that article is, somewhat like the argument of this chapter, that we should remain alert to the source of arguments no matter how 'principled' they may appear.

principles, but should be wary of thereby assisting acontextual readings of these principles.

Adopting a relational view, as explained above, tort duties do not simply state rules of how to behave—they also allocate risks, and the definition of the risk is significant. The consequence is greater scope to determine the ambit of duties through analysis that is party-focused. This includes the parties' various options for anticipating and avoiding risks. This analysis is not apolitical. But it is connected to party relationships and roles. This means the application of tort principles and process for determining the ambit of duties is complex, not simple. But it also means there is more to work with than purely personal values. For example there may be evidence of social and political agreement in relation to a party's roles or the location of risk; or there may be party arrangements which deal with the issue and which may be disturbed by intervention. What is 'political' is a broad field and may be more or less institutional.

5. Insurance and Tort beyond Party Risk Allocation

Having addressed these road-blocks, the way is clearer to consider some of the other roles played by insurance, for example where there are public judgments about where risks should be allocated, or where the whole structure of litigation rests on insurance.[51] The rhetoric of responsibility and deterrence—the latter of which was strongly associated by Stapleton with tort—can be used to avoid solutions which take into account the possibility of the very risks which are then brought home to a party at fault. This argument does not deny the significance of fault. But it does accept the importance of proportionality in response. The institutional capacity of tort and insurance to distribute risks is extremely important not just to its practical role, but also the justification of its remedies. This includes its philosophical justification: proportionality is a part of the philosophy of responsibility and tort lawyers may need to pay attention to the resources for distribution of liabilities, for example, to understand the proportionality of responsibility in their own field—and to address it. Of course, this is a broader undertaking than can be attempted here. But the fundamental connection between the practical operation of the law of tort, particularly its dependence on insurance, and its moral structure, deserves much more attention. This is not to deny the significance of responsibility in the law of tort, but to suggest that responsibility too is sensitive to the potential impact of liability.

[51] As it does in the United Kingdom. See Merkin and Steele, *Insurance and the Law of Obligations* (n 2), Chapter 13. For the United States, see Baker, 'Liability Insurance as Tort Regulation' (n 2).

6. Conclusions

One of the more interesting trends in recent private law theory has been the renewed focus on the facilitatory, or supportive, role of law. Jane Stapleton's work on cooperative arrangements in tort makes an important contribution to the wider body of work on contract law's role in supporting market arrangements.[52] Yet its wider implications in relation to the role of insurance in structuring relations in tort law has received perhaps less attention. The argument of this chapter has been that insurance should be seen in this light in relation to tort law, both in negligence and beyond. Insurance is an important part of the arrangements parties make *ex ante* to structure their relationship and to deal with the presence of risk within it. Insurance complements the rest of tort law, and should be seen as simply another factor in the wider set of considerations on which tort law bases its evaluation of where liability and responsibility sit within the context of the parties' relationship.

It could be argued that none of this is apolitical, since all of it uses disparate arguments to determine distributive questions. But 'political' arguments can take place in a way that is structured by the particular role that private law is asked to play. Political arguments may be a part of institutional argument. This is not invoking the myth of neutrality that allows one ideology to trump another in the name of a supposed rationalism, but is a part of the evaluative process of managing relationships between parties. Those relationships contain insurance concerns, and the parties include insurers. It matters, in a range of ways, that this is the case, because the arguments engaged in by courts are contextual. We can put forth some relatively simple principles about the law of tort, or some more complex arguments, and it is often useful to do so. But they are not forged from simple abstract reasoning. They are forged from responding to a range of relationships and a range of claims.

Perhaps most importantly, insurance is at best poorly understood if we continue to regard it as a risk pooling mechanism in this context, at least if this is taken to confine its operation to the operation of 'like risk pools'. Insurance is, rather, a risk transfer mechanism taking a contractual form, which has all the advantages and disadvantages of market arrangements. Its special features lie in the creation of market mechanisms to transfer risk out of the context of a particular relationship, and the techniques that it uses involve not only risk pooling, which is heavily qualified, but also transfer and diversification of risks, reinsurance, and so on. It remains an essential mechanism for market protagonists, individuals, and societies. Its ability to deal in larger numbers and to abstract from individual responsibility

[52] Examples include J Morgan, *Contract Law Minimalism* (Cambridge, 2013), TT Arvind and J Steele, 'Remapping Contract Law: Four Perceptions of Markets' in TT Arvind and J Steele (eds), *Contract Law and the Legislature: Autonomy, Expectations, and the Making of Legal Doctrine* (Oxford, 2020), D Campbell, *Contractual Relations* (Oxford, 2022).

or misfortune is of course still essential, and it is therefore important to note that the effort to reintroduce personal responsibility is often aimed at removing a loss from insurers that the insurers had contractually agreed to bear—and, indeed, whose transfer to an insurer was the primary purpose of the market arrangements underlying insurance. It is not always wrong, but whether it is appropriate should depend on an awareness of the context. To equate insurance with risk pooling, sharing, or redistribution, and tort with deterrence, is a false move, designed to obscure the connections between them in an attempt to head off a particular attack on the law of tort. But as this chapter has argued, it is neither necessary nor particularly helpful to rely on these false equations in order to do so. Not only this, but we need to make room for complexity in our understanding of private law and the role of the courts. Simplicity of particular principles is of course useful in its own right. But the existence of a range of different considerations, values, and principles is not a sign of incoherence in its own right. Tort deals with a wide range of relationships in numerous different contexts, and a proper understanding of the role of insurance would start with this, rather than by associating insurance with a broad, external policy context.

5

Fulfilling the Promise of the Golden Thread?: Vulnerability and Australian Negligence Law

Kylie Burns

1. Introduction

In October 2003, at the Centenary Conference of the High Court of Australia, Professor Jane Stapleton argued that the High Court had identified the 'protection of the vulnerable as a core moral concern of tort law: its golden thread'. Her hope was that Australian courts could utilise the concern of vulnerability to develop tort law in a way which could right the 'most grievous wrongs in Australia's history', particularly those wrongs suffered by First Peoples. This chapter will consider whether and how 'vulnerability' has been used by Australian courts in developing negligence law during the last two decades.

Section 2 will consider the themes of Professor Stapleton's original paper. Section 3 will overview how the concept of vulnerability has been utilised by the Australian High Court in negligence cases in the last two decades. Section 4 will focus on 'Stolen Generation' negligence claims by First Peoples wrongfully removed from their families as children and will particularly discuss *State of South Australia v Lampard-Trevorrow ('Trevorrow')*.[1] Section 5 will consider public interest litigation on behalf of 'vulnerable' plaintiffs against public defendants including claims by injured immigration detainees and climate change litigation. The chapter will conclude that despite signs that protection of the vulnerable is a concern that has the potential to influence the development of Australian negligence law, the promise of the golden thread has not (yet) been fully fulfilled.

[1] *State of South Australia v Lampard-Trevorrow* (2010) 106 SASR 331.

Kylie Burns, *Fulfilling the Promise of the Golden Thread?: Vulnerability and Australian Negligence Law* In: *Torts on Three Continents*. Edited by: Kylie Burns, Jodi Gardner, Jonathan Morgan and Sandy Steel, Oxford University Press.
 DOI: 10.1093/oso/9780198889748.003.0005

2. 'The Golden Thread at the Heart of Tort Law: Protection of the Vulnerable'

Professor Stapleton's paper at the Centenary Conference of the High Court of Australia in October 2003 came at a particularly critical moment in the history of Australian negligence law and of the Australian High Court.[2] Changes in the composition of the High Court from the late 1990s were followed by marked retraction in negligence liability favouring defendants from around 1999.[3] In *Sullivan v Moody*, a 2001 case concerning whether a novel duty of care could be owed to two fathers in relation to investigations related to child sexual abuse, the High Court held that the test for duty of care did not encompass a unified concept of proximity and the three-stage test from *Caparo Industries Plc v Dickman*[4] was not the law of Australia.[5] The High Court (notionally) turned away from policy-based reasoning holding 'there are policies at work in the law which can be identified and applied to novel problems, but the law of tort develops by reference to principles, which must be capable of general application, not discretionary decision-making in individual cases'.[6]

In October 2002, some months before being appointed as a Justice of the High Court by the then conservative Federal Government, Justice Dyson Heydon gave a controversial speech at a Quadrant dinner criticising 'judicially activist' judges and advocating a strictly legalistic approach to judging.[7] This sparked a further period of public, political, and academic debate in Australia about the nature and limits of the judicial role particularly on the High Court. Finally, the early 2000s in Australia saw the so-called insurance crisis where liability insurance became unavailable and unaffordable to many potential defendants, including medical practitioners, public authorities, and others. There was much media and political commentary at the time that suggested this insurance crisis occurred as a result of a 'compensation crisis' with too many plaintiffs getting too much compensation from out-of-control

[2] Jane Stapleton, 'The Golden Thread at the Heart of Tort Law: Protection of the Vulnerable' (2003) 24 *Australian Bar Review* 135; and Jane Stapleton, 'The Golden Thread at the Heart of Tort Law: Protection of the Vulnerable' in Peter Cane (ed), *Centenary Essays for the High Court of Australia* (LexisNexis Butterworths 2004). References to the paper in this chapter will refer to the version published in the *Australian Bar Review*.

[3] Harold Luntz, 'Torts Turnaround Downunder' (2001) 1 *Oxford University Commonwealth Law Journal* 95; Harold Luntz, 'Round-up of Cases in the High Court of Australia in 2003' (2004) 12 *Torts Law Journal* 1. Murray Gleeson CJ was appointed as Chief Justice in May 1998. Hayne J was appointed in 1997. Callinan J was appointed in 1998. Heydon J was appointed in February 2003.

[4] [1990] 2 AC 605 (HL).

[5] *Sullivan v Moody* [2001] 207 CLR 562, 578–79 (Gleeson CJ, Gaudron, McHugh, Hayne, and Callinan JJ).

[6] Ibid 579.

[7] Later published as Dyson Heydon, 'Judicial Activism and the Death of the Rule of Law' (2004) 10 *Otago Law Review* 493. Quadrant is a magazine which has a 'founding bias towards cultural freedom, anti-totalitarianism and classical liberalism'. See *Quadrant Online*, https://quadrant.org.au/ (accessed 2 February 2023).

courts in unwarranted negligence actions.[8] Despite this commentary, there was no convincing empirical evidence that there was a compensation crisis or that negligence principles had become too expansionary.[9] The insurance crisis was likely predominantly the result of other factors such as the Global Financial Crisis, the collapse of several large insurers, and lack of regulation in the insurance market.[10] Nevertheless, in a highly unusual cooperative display of federalism, all states and territories and the Commonwealth commissioned an inquiry into the law of negligence, which provided recommendations for law reform.[11] During 2002 and 2003, statutory tort law 'reforms' were passed in all Australian jurisdictions,[12] changing the landscape of Australian negligence law to one dominated by legislation.[13] Against this background of shift and change in the Australian tort landscape (and the political and legal landscape more broadly), a number of key themes emerged in Stapleton's paper at the High Court conference.

A. Courts of ultimate appeal make law

Stapleton commenced by acknowledging that courts of ultimate appeal in a common law system make law.[14] While it might have been thought that by 2003 this was uncontroversial in Australia, as discussed above, debate had been reignited at the time by Heydon J's speech prior to his High Court appointment.[15] Stapleton suggested:

> recent pro-legalism critics of the open lawmaking by the High Court in tort law seem to believe that using the terminology of 'principle' will convey judicial probity, a back-ward looking reliance on settled precedent and a somewhat mechanistic, at most incremental, process of judging, that seeks to keep itself to itself, and well out of the province of the legislature and media controversy.[16]

[8] See discussion in Kylie Burns, 'Distorting the Law: Politics, Media and the Litigation Crisis: An Australian Perspective' (2007) 15 *Torts Law Journal* 195; Harold Luntz et al, *Torts: Cases, Legislation and Commentary* (9th edn, Lexis Nexis, 2021) 21.

[9] Luntz et al (n 8) 6–9.

[10] See discussion of factors which drive insurance crisis' in Peter Cane, 'Reforming Tort Law in Australia: A Personal Perspective' (2003) 27(3) *Melbourne University Law Review* 649.

[11] David Ipp et al, *Review of the Law of Negligence Final Report*, https://treasury.gov.au/review/review-of-the-law-of-negligence (accessed 2 February 2023). For a discussion of the background of the Report, see Cane (n 10). It should be noted that states and territories introduced many more additional reforms that limited liability to plaintiffs that had not been recommended by the Ipp report.

[12] See Luntz et al (n 8) for a discussion of the legislative reforms.

[13] However, the principles of duty of care were left predominantly to the common law and remained relatively untouched by legislative tort law except in relation to public authorities.

[14] Stapleton, 'The Golden Thread' (n 2) 135 and 137.

[15] Stapleton references Heydon J's speech at ibid 136 (fn 3) citing the version published in *Quadrant*: Dyson Heydon, 'Judicial Activism and the Death of the Rule of Law' *Quadrant* (January–February 2003) 9.

[16] Stapleton, 'The Golden Thread' (n 2) 136.

However, as Stapleton had already pointed out in previous work,[17] appealing to 'so-called principles' could simply be a masking exercise to disguise and legitimise the substance of 'a judge's reasoning process'.[18] To this day, one of the best examples of this 'masking' which Stapleton referred to in her 2003 paper,[19] was the very first judgment Heydon J gave in the High Court in the wrongful birth case *Cattanach v Melchior (Cattanach).*[20] Despite the pro-legalism approach he had only recently defended in his Quadrant speech, in *Cattanach* Heydon J 'based his reasoning in a large part on a consideration of matters that would seem better described in moral, social or scientific terms'.[21] Many of these matters were contentious, had no empirical evidence base, or were based on an outdated evidence base.[22] For example, Heydon J referred repeatedly to the so-called 'emotional bastard' argument that wrongful birth litigation would lead to children being emotionally damaged by their parents based on the fact they were unplanned and the subject of litigation.[23] Stapleton particularly critiqued Heydon J's attack on judicial activism and his equation of 'legalism with incrementalism' commenting that she was 'not sure this matches what we see in High Court jurisprudence or even the common law more broadly'.[24] She argued that there was a real danger in using 'a veneer of legalism' to cloak 'violent new advances' with 'undisclosed and unidentified reasons'[25] with some of the 'most admired tort judgments' 'characterised by realism and creativity'.[26] While acknowledging that 'top courts make tort law', Stapleton did not, however, advocate for unfettered judicial expansion of tort law[27] suggesting change need not only be in the 'direction of expanding tort liability'.[28]

[17] eg Jane Stapleton, 'Duty of Care Factors: A Selection from the Judicial Menus' in Peter Cane and Jane Stapleton (eds), *The Law of Obligations: Essays in Honour of John Fleming* (Clarendon Press, 1998) 5.

[18] Stapleton, 'The Golden Thread' (n 2) 136.

[19] Ibid 137.

[20] (2003) 215 CLR 1, handed down on 16 July 2013.

[21] Stapleton, 'The Golden Thread' (n 2) 137.

[22] Kylie Burns, 'The Way the World is: Social Facts in High Court Negligence Cases' (2004) 12 *Torts Law Journal* 215, 233–34.

[23] Ibid.

[24] Stapleton, 'The Golden Thread' (n 2) 138. Stapleton noted (at 140) that whatever the 'motivation for the reform of tort liability by the 2002 Ipp Report' it could not have been 'an acute dissatisfaction' with the Mason and Brennan courts, as the Ipp Report did not suggest 'reversal of any tort decision' of those courts.

[25] Ibid 138.

[26] Ibid. For recent exploration of the creative and living judicial decision-making approach of tort law judges, see Jane Stapleton, *Three Essays on Tort* (OUP, 2021) Chapter 1, 4–10) (which is the published version of the Clarendon Law Lectures delivered by Stapleton in 2018)

[27] Stapleton had identified in previous work areas where liability ought to be ousted or tempered. See, eg, Jane Stapleton, 'In Restraint of Tort' in Peter Birks (ed), *The Frontiers of Liability*, vol 2 (OUP, 1994) 83 and Jane Stapleton, 'Duty of Care: Peripheral Parties and Alternative Opportunities for Deterrence' (1995) 111 *Law Quarterly Review* 301.

[28] Stapleton, 'The Golden Thread' (n 2) 139. She discusses, eg, cases where the High Court had abolished incoherent tort doctrine such as *Burnie Port Authority v General Jones Pty Ltd* (1994) 179 CLR 520.

B. Is there a meaningful difference between principle and policy?

Stapleton also revisited a critique she had made previously, and has developed further in her work since—that is, that there is not a 'compelling' or intelligible difference between principle and policy in negligence law.[29] Many concerns which tort judges take into account are not comfortably described as principle or policy, for example the concept of indeterminacy.[30] Stapleton argued it would be better to 'ditch' the policy/principle terminology and describe the elements judges utilise neutrally as 'legal concerns' while openly acknowledging that the law will take on new values as society changes.[31] In my own work (including empirical content analysis), I have argued that an analysis of Australian High Court cases shows that judges refer to a whole range of concerns and social facts in negligence cases in ways that are not neatly categorised as consequentialist policy nor principle.[32] Despite the Australian High Court in *Sullivan v Moody* determining that only legal principle and 'legal policy' was relevant to determining duty of care, Australian courts have nevertheless continued to refer, from time to time, to consequentialist policy arguments in negligence cases.[33] The salient features or multifactorial test for duty of care that has developed in Australia in the last two decades[34] is an excellent example of what Stapleton has referred to in the Clarendon lectures as tort law 'in all its seeming messiness'[35] in which 'individual judges may place different weight on the variety of social and legal concerns from which that law emerges'.[36]

C. Conceptual advances in tort law: Protection of the vulnerable

The major focus of Stapleton's paper was the identification of three major 'concerns' or 'conceptual advances' which the Australian High Court had utilised 'as part of its tort law-making' which she believed had improved the 'clarity of tort law reasoning generally'.[37] The first two concerns are those that the High Court had used to restrain liability—indeterminacy and coherence. In relation to

[29] Ibid 135–37. See also Stapleton, 'Duty of Care Factors' (n 17); Stapleton, *Three Essays on Torts* (n 26).

[30] Stapleton, 'The Golden Thread' (n 2) 136 and 140.

[31] Ibid 137.

[32] eg Kylie Burns, 'It's Not Just Policy: The Role of Social Facts in Judicial Reasoning in Negligence Cases' (2013) 21(2) *Torts Law Journal* 73.

[33] Ibid.

[34] For a summary of the many 'salient features' the High Court has identified in different categories of case as being potentially relevant in determining duty of care in novel cases, see *Caltex Refineries (QLD) Pty Ltd v Stavar* (2009) 75 NSWLR 649, 675–76 (Allsop P).

[35] Stapleton, *Three Essays on Torts* (n 26) 1.

[36] Ibid 18.

[37] Stapleton, 'The Golden Thread' (n 2) 140.

indeterminacy, Stapleton argued that the High Court, in cases such as economic loss case *Perre v Apand Pty Ltd*[38] in 1999, showed 'innovative appreciation' of the need to identify duty in a way which makes the 'scope of liability determinate'.[39] Second, she identified the then emergent notion of coherence 'as part of a package of restraints on tort liability, especially that of public authorities' being utilised by the High Court.[40] She noted the High Court in a number of cases[41] had begun to stress that duties should not be recognised where they would not be compatible with public duties owed by public bodies more broadly.[42] In the last two decades, the notion of coherence (particularly legislative coherence) has become the predominant factor and fundamental principle in tort law reasoning in novel cases in the High Court including cases which do not concern public authorities.[43] The High Court has repeatedly stressed the importance of coherence of the law of negligence with other areas of the common law, and the necessity of coherence between negligence law and statute.[44] As Fell has suggested, incoherence has generally been regarded as 'a conclusive reason against a proposed ruling or decision'.[45] Despite this increased focus in the High Court since 2003 on coherence, the concept remains under-defined and has not been 'fully articulated' by the High Court.[46] The appropriateness of defining or limiting the ambit of negligence law by reference to other areas of law, including statute, private law, and more generally, has also not been sufficiently critiqued and is under-theorised.

It is the third of the emerging concepts, vulnerability, that was the main focus of Stapleton's paper, and which is the primary focus of this chapter. Stapleton argued that for the two decades prior to 2003, the High Court had identified 'vulnerability' 'as a necessary if not sufficient pre-requisite for protection by the tort of negligence, and as a central organising feature in that tort'.[47] She suggested the Court had 'identified this as a core moral concern of tort law' and 'whatever limits may govern this vulnerability concept it certainly has a powerfully convincing core'.[48] Stapleton's concept of 'vulnerability' was not a broad and unconfined concept.[49] She referred

38 *Perre v Apand Pty Ltd* (1999) 198 CLR 180.

39 Stapleton, 'The Golden Thread' (n 2) 141.

40 Ibid.

41 Ibid 141 (fn 42). Cases cited by Stapleton included *Sullivan* (n 5); *Cattanach* (n 20); *Modbury Triangle Shopping Centre Pty Ltd v Anzil* (2000) 205 CLR 254; *Hill v Van Erp* (1997) 188 CLR 159.

42 Stapleton, 'The Golden Thread' (n 2) 141.

43 See discussion in Andrew Fell, 'The Concept of Coherence in Australian Private Law' (2018) 41 *Melbourne University Law Review* 1161.

44 Ibid 1161.

45 Ibid 1162.

46 Ibid.

47 Stapleton, 'The Golden Thread' (n 2) 142. For earlier discussion by Stapleton of the concept of vulnerability, see Jane Stapleton, 'Comparative Economic Loss: Lessons from Case-Law-Focused Middle Theory' (2002) 50(2) *UCLA Law Review* 531, 554.

48 Stapleton, 'The Golden Thread' (n 2) 142.

49 The 'salient feature' of vulnerability, as referred to in *Caltex Refineries* (n 34) by Allsop P, is 'the degree of vulnerability of the plaintiff to harm from the defendant's conduct, including the capacity and reasonable expectation of a plaintiff to take steps to protect itself' seems broader in scope than the concept favoured by Stapleton.

to the 'most extreme form of vulnerability: namely, where a person is exclusively dependent on another to take care, even if that person is a total stranger'.[50] Her earlier work linked economic loss liability to the inability of the plaintiff to self-protect—in that context, vulnerability was a limiting factor restricting duties owed to commercial plaintiffs.[51] Importantly, she noted the High Court's 'astute use' of the concept of vulnerability to extend tort liability, but in a controlled manner which did not open floodgates to those who could self-protect.[52] Similarly, she suggested that the concept of vulnerability was 'confined' and could be 'outweighed by other concerns' such as incoherence or indeterminacy.[53]

Having identified the emerging 'golden thread' of vulnerability in the tort jurisprudence of High Court, Stapleton considered what future best contribution the High Court might make to social and public challenges. She identified, in particular, the failure of the Australian legal system to properly address the wrongs perpetuated against Indigenous people—the First Peoples.[54] This included issues such as over-incarceration, deaths in custody,[55] and wrongful removal of Indigenous children from their families (the 'Stolen Generation').[56] The Australian Institute of Health and Welfare estimates that in 2018–19 there were 33,600 surviving members of the Stolen Generations in Australia,[57] many of whom were victims of abuse. Survivors suffer a high level of poverty and impaired health.[58] Indigenous children continue to be removed from their families in Australia at disproportionate rates and the rate of Indigenous children in out of home care has increased in recent years.[59] At the time of Stapleton's address in 2003, there had been no private law[60] or public redress in relation to the unlawful and racist removal of Indigenous children from their families on the basis of assimilation, or for the physical, emotional,

[50] Stapleton, 'The Golden Thread' (n 2) 142 which refers (fn 46) to Stapleton, 'Duty of Care Factors' (n 17) 86–87. For recent discussion of vulnerability in negligence law in English Courts, Canada, and Australia, see Carl Stychin, 'The Vulnerable Subject of Negligence Law' (2012) 8(3) *International Journal of Law Context* 337. Stychin discusses (at 337) vulnerability as both a descriptive and broader 'normative device by which to make sense of the law and to guide courts in the future'.

[51] Stapleton 'Comparative Economic Loss' (n 47) 531.

[52] Stapleton, 'The Golden Thread' (n 2) 142.

[53] Ibid.

[54] Ibid 143–44. For further contemporary discussion of the failure of Australian tort law to respond to the wrongs against the First Peoples of Australia and the racism inherent in the tort compensation system, see Luntz et al (n 8) 94–95.

[55] See *Royal Commission into Aboriginal Deaths in Custody*, National Report, vol 1 (AGPS, 1991).

[56] Human Rights and Equal Opportunity Commission, *Bringing Them Home: Report of the National Inquiry into the Separation of Aboriginal and Torres Strait Islander Children from Their Families*, 1997.

[57] https://www.aihw.gov.au/reports/indigenous-australians/indigenous-stolen-generations-50-and-over/contents/summary (accessed 14 June 2024).

[58] Ibid.

[59] Between 2017 and 2021, the rate of Indigenous children in out of home care increased from 51 per 1000 to 58 per 1000: https://www.aihw.gov.au/reports/child-protection/atsicpp-indicators/contents/about (accessed 20 February 2024).

[60] Several cases had been litigated and failed. See *Williams v The Minister, Aboriginal Land Rights Act 1983* (1994) 35 NSWLR 497; *Cubillo v Commonwealth* (No 2) (2000) 103 FCR 1; 174 ALR 97, (2001) 112 FCR 455, 183 ALR 249.

and sexual abuse suffered by many children after their removal.[61] Stapleton identified potential tortious wrongs arising from Stolen Generation cases including false imprisonment, battery, assault, and negligence and argued tort law had a 'substantial potential to provide individual redress'.[62] She suggested that the 'special vulnerability' of children, together with other important factors such as government's ultimate legal control and power to protect and knowledge of individual children's past appeared to constitute a 'strong case' that cases concerning abused Indigenous children could fit within 'criteria set by the High Court for' duty of care.[63]

D. Potential of human rights legislation

Finally, Stapleton canvassed in her paper whether given the failure of tort law to address the grievances of vulnerable parties, domestication of human rights legislation in Australia could provide an adequate remedy. At that time, no Australian state or territory, nor the Commonwealth, had introduced dedicated human rights legislation. Despite the potential that human rights legislation may allow some remedy, Stapleton was 'not yet convinced that a bill of rights would be a good thing. My hope is that, in contrast to our past, the next 100 years will deliver the promise of our existing legal system to be 'a living instrument'.[64] Since 2003, only three Australian jurisdictions have introduced human rights legislation, and the protection in that legislation is limited in scope.[65] None of the relevant Acts allow a direct private action or remedies for damages for breach of a human right.[66] There has been little appetite, until recently, for a Federal constitutional or legislative Bill of Rights.[67]

[61] Stapleton, 'The Golden Thread' (n 2) 144. Stapleton noted that there was evidence that a few plaintiffs had obtained compensation from other sources, for example in statutory criminal compensation.
[62] Stapleton, 'The Golden Thread' (n 2) 144. Stapleton also identified the potential of tort law to respond to deaths in custody and racial harassment.
[63] Ibid 145. She also suggested (at 144) that problems of limitation could be overcome and problems of proving causation could be addressed with creative solutions and (at 146) that in many other areas where Australian law had failed Indigenous peoples, more could be done to 'sensitise' tort law to respond to its 'racial, cultural or religious blindness'.
[64] Ibid 148–49.
[65] *Human Rights Act 2004* (ACT); *Charter of Human Rights and Responsibilities Act 2006* (VIC); *Human Rights Act 2019* (QLD).
[66] See discussion of the approach of the 'dialogue model' of Human Rights legislation in Australia in Bruce Chen, 'Revisiting s 32(1) of the Victorian Charter: Strained Constructions and Legislative Intentions' (2020) 46(1) *Monash University Law Review* 174. This may be compared to the Human Rights Act 1998 (UK) which allows remedies against public authorities, eg see s 8.
[67] See however the recent report of the Parliamentary Joint Committee on Human Rights, Inquiry into Australia's Human Rights Framework, https://www.aph.gov.au/Parliamentary_Business/Committees/Joint/Human_Rights/HumanRightsFramework (accessed 18 June 2024).The recent decision of the Constitutional Court of South Africa in *AK v Minister of Police* [2022] ZACC 1 is an example of a private law action in jurisdictions with extensive constitutional human rights which can integrate those human rights into negligence actions to protect vulnerable plaintiffs. The Court upheld a negligence (delict) action by a victim of sexual assault who was repeatedly raped following the failure of the police to search for her and properly investigate. The Constitutional Court (by majority) held that the actions

E. Responding to Stapleton's critics

Following her address to the High Court Centenary Conference, Stapleton was criticized by some media commentators and members of the academy. Critique included that she had a 'brazen agenda for conscience-driven judicial decision-making';[68] a vision which 'represents the worst kind of rule by personal conscience';[69] called 'at the conference for courts to make more, not less, novel extensions of principle in the area of negligence' which was 'classic ivory-tower stuff, which explains why some academics are best kept off the bench';[70] thought that judging was 'politics' or was advocating a 'openly instrumental style of judging'.[71] Gava went further, in a defence of strict legalism, asking rhetorically after vigorously criticising Stapleton's paper:

> Is it unfair to suggest that Stapleton has let emotion override her intellectual judgment, that she has allowed her commendable concern for Aboriginal disadvantage to temporarily overcome her powers of analysis by advocating a role for law that she has elsewhere demonstrated to be implausible and undesirable?[72]

In response to these critiques, I make several observations about what Stapleton's identification of vulnerability as a core moral concern of tort law in Australia in 2003 was and was not. First, this was not a normative 'Grand Theory' of tort law (a vulnerability tort theory if you will),[73] of the type that she has critiqued in her later work including in the Clarendon lectures.[74] In her address to the High Court Centenary Conference and her earlier work, Stapleton's discussion of vulnerability was in the nature of what she later called 'reflexive tort scholarship'.[75] It emerged from her close analysis of judicial reasons of the High Court in Australian tort cases. It was a descriptive rather than normative account of the role of vulnerability

of the police had to be based on their positive obligations and duties to protect and fulfil the rights of women and to combat and investigate crime pursuant to the Constitution of South Africa. The judgment of Theron J at [137]–[146] distinguishes this approach to the law of police liability in South Africa from other common law jurisdictions such as the United Kingdom.

68 Janet Albrechtsen, 'Leave Policy to the Law-Makers', *The Australian*, Canberra, 15 October 2003, 15.

69 Ibid.

70 Ibid.

71 John Gava, 'Unconvincing and Perplexing: Hutchinson and Stapleton Judging' (2007) 26(1) *University of Queensland Law Journal* 67, 67.

72 Ibid 79.

73 eg, it was not based on the well-known vulnerability theory work of Martha Fineman, https://web.gs.emory.edu/vulnerability/ (accessed 20 February 2024). For a discussion of the theoretical significance of vulnerability in tort law, including feminist theory and the work of Fineman, see Stychin (n 50).

74 See Stapleton, *Three Essays on Torts* (n 26).

75 Ibid 17–20.

in tort judicial reasoning. Second, as discussed above Stapleton did not call for some unrestrained, unconstrained 'activist' use of vulnerability to expand the boundaries of tort liability. This is unsurprising, given that her previous work, prior to 2003, had very carefully examined where boundaries of tort law liability should be drawn and her previous recognition that in economic loss cases, vulnerability (or lack thereof) could be a limiting liability factor. Stapleton very clearly recognised in her paper that even where a plaintiff is vulnerable, and exclusively dependent on the defendant, vulnerability may still properly be outweighed in the determination of liability by countervailing factors such as indeterminacy or incoherence. Vulnerability is a 'delicate' tool.[76]

3. Vulnerability in the Australian High Court in the Last Two Decades

In the last two decades since Stapleton's 2003 paper, it cannot be said that 'vulnerability' has emerged in the Australian High Court as an expansionary or significant force in negligence cases or in tort cases more broadly, or in cases concerning First Peoples. It has not played a dominant role in righting public wrongs. There has been no key tort case before the High Court concerning tortious wrongs suffered by First Peoples. The High Court of the last two decades has remained an essentially conservative tort court. Vulnerability has remained a 'salient feature'[77] considered by the High Court when determining novel duty cases, but rarely (if ever) has it been a positive determinative factor in recognising new categories of liability or expanding liability. It has most often emerged (perhaps more influentially) as a key limiting duty factor, particularly in economic loss cases where commercial parties who had an ability to self-protect (eg through contract) have been found not vulnerable[78] and hence have been owed no duty of care.

Outside of the context of economic loss, plaintiffs who may have been thought to have been especially 'vulnerable' or particularly dependent on the defendant have been unsuccessful. For example, in *Stuart v Kirkland Veenstra*,[79] a case concerning whether police officers owed a duty to prevent suicide, Crennan and Kiefel JJ stated that vulnerability may be one relevant factor in determining the duty of

[76] Stapleton, 'Duty of Care Factors' (n 17) 86–87, 149.

[77] See Stychin (n 50) 345–46.

[78] eg *Brookfield Multiplex Ltd v Owners Corporation Strata Plan 61288* (2014) 254 CLR 185; *Woolcock Street Investments Pty Ltd v CDG Pty Ltd* (2004) 216 CLR 515. See critique of this approach in Alexander Jackman 'Vulnerability, Autonomy and Protection: The Role of Actual and Hypothetical Contracts in the Duty of Care to Protect against Pure Economic Loss' (2020) 49 *Australian Bar Review* 82. Jackman argues (at 89) that in the economic loss cases vulnerability appears to have 'three layers'—ie an evaluation of 'social, political and economic standing'; inability for self-protection; and may encompass consideration of other salient features as indicators such as reliance, knowledge, control, reasonable reliance.

[79] (2009) 285 237 CLR 215.

public authorities, although 'it has not been universally accepted as a useful analytical tool'.[80] The plaintiff's autonomy may be inconsistent with a finding of vulnerability, and ultimately coherence with any relevant legislation (and the way the legislation defined the powers of the public authority) were key factors in defining liability. Liability was refused in cases concerning an array of plaintiffs who may have been thought to be immensely vulnerable, such as: plaintiffs affected by alcohol including where the alcohol was provided by the defendant;[81] employees especially vulnerable to vicarious trauma;[82] pedestrians injured by the actions of council contractors working on a footpath;[83] unborn plaintiffs in wrongful life cases;[84] and a plaintiff killed by a mentally ill friend who was released from hospital.[85] In public authority cases (and more broadly) it has been clear that even where a plaintiff may be considered to be immensely factually 'vulnerable' to injury, liability will be constrained by tightly identifying the confines of relationships of special dependence.[86] 'Coherence' rather than vulnerability has emerged as the 'trump' factor in negligence cases. The High Court will not find a duty of care exists (despite the presence of vulnerability) where such a duty would be inconsistent with a legislative framework.[87] Tort reform civil liability legislation in all Australian states and territories has also restricted the ability of 'vulnerable' plaintiffs to bring actions, particularly against public bodies.[88] In addition, civil liability

[80] Ibid 260–61.

[81] *C.A.L. No 14 Pty Ltd v Motor Accidents Insurance Board* (2009) 239 CLR 390 390, 405– 06 (Gummow, Heydon, and Crennan J) who stressed that the plaintiff had autonomy over his actions and his vulnerability did not arise from the defendant's actions and a duty would not be coherent with other torts.

[82] eg *New South Wales v Fahy* (2007) 232 CLR 486. See Kirby J's dissent at 537–39 where he held that the police service employer should have taken further proactive steps to protect an employee in circumstances where the employer was aware of the 'well known vulnerability and special risk of police work'.

[83] *Leichhardt Municipal Council v Montgomery* [2007] 230 CLR 22. Kirby J held, at 65, that non delegable duties rest on relationships of particular vulnerability and special dependence in closed identifiable categories.

[84] *Harriton v Stephens* (2006) 226 CLR 52; *Waller v James* (2006) 226 CLR 136.

[85] *Hunter and New England Local Health District v McKenna* (2014) 253 CLR 270.

[86] The High Court did have an opportunity to consider the scope of public duty owed to intensely vulnerable plaintiffs who were injured as children when child protection authorities failed to notify police, pursuant to their statutory power, about ongoing sexual abuse by a family member. However, special leave was ultimately withdrawn on the basis that the case was not an appropriate vehicle to consider the scope of duty question; see *New South Wales v DC* [2017] HCA 22. The NSW Court of Appeal had held that the scope of the duty of care extended to notification to police, noting: 'As Sackville AJA observed when this matter was previously before this Court … it is difficult to think of a more vulnerable class of persons than children subjected to sexual abuse by parents or guardians.' See *DC v State of New South Wales* [2016] NSWCA 198 [274]–[276] (Ward JA).

[87] *Hunter and New England Local Health District* (n 85). See, however, the recent case, *Electricity Networks Corporation v Herridge Parties* (2022) 406 ALR 1, where the High Court held that a duty of care was owed by a statutory corporation to take reasonable care in relation to fire risk arising from its operation of its electricity distribution system to people and property in the vicinity. While stressing that the 'starting point' for determining whether a duty of care is owed by a public authority was the statutory framework, the High Court held (at [55]) that in this case the statutory framework was not inconsistent with the relevant duty of care and scope of duty.

[88] Luntz et al (n 8) 463–64.

legislation heavily impacts actions by parties who might otherwise be thought to be vulnerable, such as those who are alcohol or drug impacted or prisoners.[89]

4. Vulnerability and the Stolen Generation

While the Australian High Court has not considered a Stolen Generation torts case in the two decades since Stapleton's paper, there has been a single successful Stolen Generation case in an Australian appellate court in *State of South Australia v Lampard-Trevorrow (Trevorrow)*.[90] Bruce Trevorrow, an Indigenous man, was admitted to hospital on Christmas Day 1957 when he was 13 months old.[91] He had been taken to hospital with stomach issues which were thought to require medical attention.[92] He recovered quickly. Although his condition was not serious and was not indicative of neglect, he was removed from his family by the Aborigines Protection Board ('APB') without notice nor parental consent and placed, on 6 January 1958, with an untrained white foster family who wished to adopt or foster a child.[93] The APB made inadequate inquiries and did not follow their own procedures in relation to removal and fostering.[94] Mr Trevorrow was placed in an inappropriate foster home, was not adequately checked upon, and adequate steps were not taken to return him to his family, or to prepare him for the ultimately unsuccessful return to his family many years later.[95]

For 'a number of years' following his removal, Bruce Trevorrow's parents were not told where he was even though they were in periodic contact with the authorities asking for information.[96] Bruce Trevorrow did not become aware that he was not the biological child of the foster family until 1965.[97] Signs of distress and emotional damage emerged from when he was a young child, including separation anxiety, depression, behavioural deterioration, psychiatric disorders, and physical disorders such as speech defects and hair loss.[98] Mr Trevorrow had a 'troubled' life, spending time in and out of institutions from age 11.[99] He claimed his

[89] Ibid 392–95 for discussion of civil liability legislative provisions limiting liability where the plaintiff is intoxicated.

[90] (2010) 106 SASR 331. Causes of action included misfeasance in public office, false imprisonment, breach of fiduciary obligations, and negligence.

[91] Ibid 336–37.

[92] Ibid.

[93] Ibid 337, 340. The APB was a statutory body constituted by the *Aborigines Act 1934* (SA). See a discussion (at 374–79) of the racist policy intent of this Act.

[94] Ibid 341.

[95] Ibid 342–50. See also discussion (at 345) of a legislative change in 1962 where (if it had indeed been lost) legal guardianship reverted to Bruce Trevorrow's mother but the State took no steps to inform her or return the child to her custody. It was only when she raised it in 1966 'that things began to happen'.

[96] Ibid 342–44.

[97] Ibid 346.

[98] Ibid.

[99] Ibid 337.

removal from his natural family and the manner in which he was ultimately re-united with them, caused psychiatric injury and 'all contributed to a loss of cultural identity, depression, alcoholism, poor health, poor relationships and erratic employment history'.[100] At trial,[101] Mr Trevorrow was found entitled to substantial damages for misfeasance in public office; wrongful imprisonment; breach of a fiduciary duty (although no damages were assessed); and in negligence.[102] Sadly, Bruce Trevorrow died after judgment and before the determination of the appeal by the State of South Australia to the Full Court. The Full Court of the Supreme Court of South Australia upheld the trial judge's judgment in relation to the actions of misfeasance in public office[103] and breach of duty of care in negligence but reversed the trial judge's findings in relation to false imprisonment and breach of fiduciary duty.

The Full Court (Doyle CJ, Duggan and White JJ) commenced its consideration of whether a duty of care was owed by the APB (in relation to the removal; process of determining care after removal; supervision of foster care; and process of return) by considering whether a duty would be consistent with, or excluded by, relevant legislation.[104] This demonstrates the primacy of the factor of coherence or legislative consistency in determination of novel duties of care in Australia. The Court held that if the APB lacked statutory power to act, then a case for imposing a duty of care was strengthened.[105] In addition, under the relevant legislation, the best interests and needs of the child were the paramount considerations, and the aim of the legislation was 'the protection and welfare' of Aboriginal children.[106] The postulated duty was held to be consistent with the statutory requirements and objects, and complementary (rather than opposed) to them.[107] The Court did 'not consider that the imposition of the duty would cut across or undermine the statutory scheme for the protection of Aboriginal children'[108] or 'any other legal obligations to which the APB is subject' and would not 'subvert other principles of law'.[109] The Court also rejected any argument that a duty would result in incoherence of law, for example because of inconsistency with the absence of a duty owed by parents to their children, or because of changed standards over time.[110] The

[100] Ibid.

[101] *Trevorrow v State of South Australia* (No 5) (2007) 98 SASR 136 (Gray J) (*Trevorrow Trial*).

[102] *Trevorrow* (n 90) 337

[103] On the grounds that evidence showed the relevant officers of the Aborigines Protection Board knew that they had no legal grounds under the relevant legislation to assume guardianship and custody of Aboriginal children without the consent of their parents.

[104] *Trevorrow* (n 90) 404–05. The Court noted at 405 [360] that this case was in relation to the exercise of a statutory power as opposed to a failure to exercise a power.

[105] Ibid 405.

[106] Ibid 406.

[107] Ibid 407.

[108] Ibid 409.

[109] Ibid. The Court particularly noted (at [385) that it did not anticipate 'chilling' or 'flow on' effects from the decision to impose a duty and noted that there have been 'shifts in judicial attitudes over time' since previous cases which rejected a duty based on policy reasons such as indeterminate liability.

[110] Ibid 410.

Court accepted the trial judge's findings that expert evidence and contemporaneous research about the state of knowledge about child attachment and the rupture of attachment in the 1950s and 1960s, showed that it should have been reasonably foreseeable to the APB that the removal of Bruce Trevorrow from his mother may cause him psychiatric harm.[111]

In their consideration of the relevant salient features in support of a duty of care, the Full Court particularly stressed the vulnerability of Bruce Trevorrow:

> Bruce Trevorrow was in a vulnerable position. He was an infant. He was quite unable to protect himself until well after the injury had been suffered and the harm done. The APB knew that he was vulnerable. The APB had the power to act, and decided to act. It was, in a sense, in control of the situation, and it made the important decisions. It placed Bruce Trevorrow in a situation of its making, and the APB accepted a general responsibility, having fostered him, to supervise his welfare. It assumed a particular responsibility to him to consider his welfare, having exercised its powers in the manner it did.[112]

The Full Court noted that the trial judge had identified matters that showed Bruce Trevorrow's innate vulnerability and his increased vulnerability over time.[113] The matters identified by the trial judge included that he was placed by the APB in an acutely vulnerable position; he was removed from his natural family; he was 13 months old at removal and a 'defenceless infant in need of support, supervision and care'; during infancy, childhood, and adolescence he 'was a child at risk'; his attachment to his mother was ruptured and he 'was isolated from his natural family'; there was no attempt to maintain a bond with family or Indigenous culture; he showed 'early signs of emotional distress' and depression; he was given antidepressants and tranquilisers at aged 10 showing acute vulnerability; and he 'sustained organic brain damage' and additionally had cerebral palsy and epilepsy.[114]

While the *Trevorrow* case was welcomed as showing the potential of the common law to respond to the historical wrongs suffered by First Peoples, nevertheless there remained constraints to future litigation including litigation based on claims in negligence.[115] This may explain the lack of reported cases or successful Stolen Generation litigation over the last decade. There were distinctive factual and legal features of the *Trevorrow* case that may distinguish it from other

[111] Ibid 412–13.
[112] Ibid 413.
[113] Ibid.
[114] *Trevorrow Trial* (n 101) 359–60.
[115] See discussion in Honni van Rijswijk and Thalia Anthony 'Can the Common Law Adjudicate Historical Suffering?' (2012) 36 *Melbourne University Law Review* 619. For a discussion of the difficulties in Stolen Generation litigation more broadly, see Chris Cunneen and Julia Grix, 'The Limitations of Litigation in Stolen Generations Cases' (Research Discussion Paper No 15, Australian Institute of Aboriginal and Torres Strait Islander Studies, 2004).

Stolen Generation cases.[116] This includes the 'specific legislative framework' which was found in *Trevorrow*, as discussed above, to be consistent with (rather than to contradict) a duty of care.[117] In addition, compared to previous unsuccessful Stolen Generation cases,[118] there was a strong written and oral evidence base in the *Trevorrow* case including medical and departmental records, archival records, expert evidence, and oral evidence from Bruce Trevorrow's siblings.[119] An extension of limitation period was also granted in *Trevorrow* on the basis that injustice to the plaintiff was not outweighed by inconvenience to the defendant of delay, as the delay was contributed to by the defendant and it was in the public interest to have Stolen Generation claims determined by Courts.[120]

As foreshadowed by Stapleton in 2003, the factual vulnerability of the plaintiff in the *Trevorrow* case was a significant factor in both the trial and appellate decisions in support of a duty of care. However, vulnerability was not the predominant factor or even perhaps the 'golden thread'. The issue of legislative consistency and coherency was the threshold issue for determination by the Court. The implication of this is that even where a plaintiff is intrinsically vulnerable and dependent on the defendant, a duty of care will not exist where it would be inconsistent or incoherent with legislation. This places potentially tight boundaries on a duty of care based on vulnerability. While there have been no successful Stolen Generation tort cases since *Trevorrow*,[121] the case formed part of the impetus for the introduction of statutory compensation and reparation schemes across Australia. It has been a slow journey to compensation for survivors, however. The government of South Australia did not introduce a reparation scheme until 2015, seven years after Bruce Trevorrow died, and a number of other jurisdictions have only introduced schemes in the last several years.[122]

[116] Antonio Buti, 'The Stolen Generations and Litigation Revisited' (2008) 32 *Melbourne University Law Review* 382; See also van Rijswijk and Anthony (n 115) 626–27.

[117] *Trevorrow* (n 90).

[118] See n 60.

[119] van Rijswijk and Anthony (n 115) 626. van Rijswijk and Anthony also suggest (at 628–29) that there was a change in *Trevorrow* in the approach to judicial interpretation of the evidence compared to previous unsuccessful cases.

[120] Ibid 636. Future cases may benefit from legislative changes made in Australia in recent years to reduce procedural impediments in civil cases relating to child abuse; see Luntz et al (n 8) 746 [12.4.24].

[121] There are an unknown number of settled cases. For a successful settled claim, see https://www.news.com.au/breaking-news/victorian-stolen-generation-man-neville-austin-gets-compensation-and-apology/news-story/427d152e45b1f7c466f5b8584e201f54 (accessed 20 February 2024). A class action against the Commonwealth of Australia by Northern Territory survivors was settled in 2022 for $50.45 million and was approved by the Supreme Court of NSW in April 2023; see https://www.shine.com.au/service/class-actions/northern-territory-stolen-generations-class-action (accessed 20 February 2024).

[122] The government of South Australia did not introduce a reparation scheme until 2015, seven years after Bruce Trevorrow died: https://www.agd.sa.gov.au/aboriginal-affairs-and-reconciliation/reconciliation/stolen-generations-reparations-scheme (accessed 20 February 2024). The Commonwealth government did not open a scheme to compensate survivors from the ACT and NT until 2022; see https://territoriesredress.gov.au/ (accessed 18 June 2024). QLD and Western Australia have not introduced dedicated Stolen Generation reparation schemes, however members of the Stolen Generation were able to claim through reparation schemes established to compensate children harmed in state care.

5. Vulnerability and Public Wrongs in Australia

Over the last decade, tort law has been used, particularly in the Federal jurisdiction, in a series of cases[123] as the basis for actions to protect vulnerable people including those in off-shore immigration detention[124] pursuant to Australia's punitive offshore processing policy towards 'unauthorised maritime arrivals'[125] and as discussed below in climate change litigation.

A. Vulnerability and immigration detention

In *Plaintiff S99/2016 v Minister for Immigration and Border Protection*,[126] a young woman with severe epilepsy was placed by the Australian government in a regional processing centre offshore on the island Republic of Nauru. She became pregnant after being raped while unconscious after suffering a seizure.[127] She required an abortion, which was not 'straightforward', and complex specialist medical support due to her medical condition.[128] An abortion on Nauru was neither safe nor legal.[129] The Minister for Immigration and Border Protection refused to bring the woman to Australia, due to an Australian government policy that detainees would not be brought to Australia other than in 'exceptional circumstances'. The Minister transferred the woman to Papua New Guinea where abortion was also illegal and evidence suggested it would also not be safe due to limited medical resources.[130]

Similarly, members of the Stolen Generation who experienced institutional child sexual abuse are eligible to claim through the National Redress Scheme established in response to the *Royal Commission into Institutional Responses to Child Sexual Abuse*, https://www.nationalredress.gov.au/ (accessed 20 February 2024).

[123] eg *RX17 as Litigation Representative for FRM17 v Minister for Immigration and Border Protection* [2018] FCA 63; *AYX18 v Minister for Home Affairs* [2018] FCA 283 (6 March 2018); *CQ18 v Minister for Home Affairs* [2018] FCA 918 (Robertson J); *BAF18 v Minister for Home Affairs* [2018] FCA 1060 (Bromberg J); and *DJA18 v Minister for Home Affairs* [2018] FCA 1050 (Murphy J); *FRM17 V Minister for Home Affairs* [2019] FCAFC 148; *AKU21 v Minister for Home Affairs* [2022] FCA 1061.

[124] Since August 2012, the Australian government has sent 4,255 people (including children) to 'offshore processing' in Nauru and Papua New Guinea. It was the policy of the Australian government that people who were transferred offshore could never settle in Australia. As of 31 February 2024, the Refugee Council of Australia reports that there are 54 people left in Nauru (including recent arrivals) and 64 people left in Papua New Guinea, living in the community. The majority of people were returned to their country or have been resettled in third countries including in the United States; see https://www.refugeecouncil.org.au/operation-sovereign-borders-offshore-detention-statistics/2/ (accessed 17 June 2024). For a discussion of the legislative regime, see Anna Talbot, 'The Constitutionality of Australia's Offshore Detention Regime' (2019) 153 *Precedent* 30.

[125] For a discussion of Australia's approach, see Ebony Birchall, Hannah Lewis, and Laura Majewski, 'The Cost of Immigration Detention' (2017) 138 *Precedent* 4; Ebony Birchall, Laura Nigro and William Zerno, 'Child Sexual Abuse in Australia's Immigration Detention (2019) 155 *Precedent* 34.

[126] (2016) 243 FCR 17.

[127] Ibid 30.

[128] Ibid.

[129] Ibid 31.

[130] Ibid.

The applicant argued that the Minister had a duty of care to 'procure a safe and legal abortion', and sought orders that injunctive relief be granted to prevent a breach of that duty.[131] Bromberg J held a duty of care was owed to exercise reasonable care 'to discharge the responsibility he assumed to procure for her a safe and lawful abortion' which could not be discharged in Papua New Guinea, although this did not require she be brought to Australia.[132] An injunction was granted to restrain breach of this duty. Bromberg J extensively considered the principles of duty of care concerning public authorities[133] including the relevant statutory framework[134] and the salient features test.[135] In relation to vulnerability, Bromberg J stated:

> The applicant is obviously in a position of vulnerability should the Minister fail to procure a safe and lawful abortion for her. The applicant has no capacity to protect herself from that potential for harm other than not to have the abortion. But, there is no issue that not having the abortion will also be harmful to the applicant, and the evidence, which I later address, shows the potential for harm to be grave. In those circumstances, no reasonable expectation can be imposed on the applicant to take steps to protect herself against the harm, nor could she. As the Minister admitted, she is entirely reliant upon the Commonwealth to procure the abortion. She is acutely vulnerable to the risk of harm in question and this factor strongly tends in favour of the existence of the putative duty.[136]

Contrast *Ibrahimi v Commonwealth of Australia*,[137] a case arising from a tragic incident in December 2010. A boat (SIEV 221) carrying people bound for Australia from Indonesia ran into a severe storm and was destroyed on the rocks at Christmas Island. Fifty people died and forty-one people were able to be rescued.[138] The NSW Court of Appeal held that there was no duty of care owed by the Commonwealth of Australia to rescue the asylum seekers as the risk of harm was not directly increased by positive acts of border protection; interception operations contained choices about exercise and non-exercise of powers; a duty may promote incoherence with legislation; and there had been no reliance which would give rise to vulnerability.[139]

[131] Ibid.
[132] Ibid 32.
[133] Ibid 79–83.
[134] Ibid 83–85.
[135] Ibid 87.
[136] Ibid. The other 'Stavar' salient features such as reliance, assumption of responsibility, and control were also held to present.
[137] (2018) 366 ALR 321.
[138] Ibid 343.
[139] See ibid 389–90 for a discussion of vulnerability, with the NSWCA finding those on board had not proved that they could not self-protect from the consequences of any want of care by the Commonwealth; they were not reliant on the Commonwealth to protect them from the risk of

The COVID-19 pandemic raised further risks for vulnerable people detained in immigration detention. During the early stages of the pandemic in 2020, there was litigation to prevent breaches of duty exposing vulnerable people in immigration detention in Australia to COVID 19. For example, in *BNL20 v Minister for Home Affairs*,[140] Murphy J made an interlocutory injunction order that the Minister for Home Affairs cease to detain a 68-year-old man with multiple health conditions in an immigration detention centre in Melbourne, on the basis that the Minister was in breach of the duty of care owed to the detainee as there was foreseeable risk 'to his health and safety ... arising from the risk he may contract COVID 19'.[141] At that time, Melbourne was 'within a hot spot for transmission' and expert evidence revealed the applicant was at 'extremely high risk' of contracting COVID 19 due to conditions in the detention centre (including the movement of staff, lack of physical distancing, and lack of masks).[142] The applicant was also a diabetic with a 15 per cent risk of death if COVID 19 entered the immigration detention centre and he contracted the virus.[143]

Murphy J held:

> The respondents accept that the Commonwealth has a duty to exercise such care for the applicant's health and safety as is reasonable in all the circumstances. That duty is well established. By reference to the salient features identified by Allsop P (as his Honour then was) in *Caltex Refineries (Qld) Pty Ltd v Stavar* [2009] NSWCA 258; (2009) 75 NSWLR 649; at [102]-[103], the foreseeability of harm in the present case, the nature of the harm alleged, the degree and nature of control able to be exercised by the respondents to avoid harm, the degree of vulnerability of the applicant to harm resulting from the respondent's conduct, the incapacity of the applicant to take steps to protect himself from the harm, the degree of the applicant's reliance upon the respondents, and the respondents' assumption of responsibility in relation to the risks to detainees posed by SARS-CoV-2, all support the existence of such a duty.[144]

shipwreck; and could have protected themselves from the risk of shipwreck by not embarking on the risky journey.

140 [2020] FCA 1180. See further discussion of immigration detention litigation during COVID 19 in Sara Dehm, Claire Loughnan, and Linda Steele, 'COVID-19 and Sites of Confinement: Public Health, Disposable Lives and Legal Accountability in Immigration Detention and Aged Care' (2021) 44(1) *UNSW Law Journal* 60, 65–79.

141 Ibid [1].

142 Ibid [6].

143 Ibid [6]. The Commonwealth accepted, for the purposes of the interlocutory application, that it owed a duty of care to the applicant but denied there was a breach on the basis the CTH had acted reasonably in the circumstances, see [20]–[24].

144 Ibid [29].

The Department of Home Affairs did not release the applicant to home detention with family in Melbourne, however, but rather complied with the order by transferring him to an immigration detention centre in Western Australia thousands of kilometres away. There was a very low COVID infection rate in Western Australia given that Western Australia had closed borders with other Australian states.[145] A further application to the Federal Court and the Western Australian Court for an injunction to stop this transfer failed, with the Western Australian Court noting that Western Australia was a very safe place at that time for people vulnerable to the threat of COVID 19 and much safer than in the community in Victoria.[146]

B. Vulnerability and climate change litigation

Recent climate change litigation has also sought to proactively utilise negligence actions as part of a strategy to respond to the risk of climate change-caused injury, particularly to vulnerable people. In 2022 the Full Federal Court in *Minister for the Environment v Sharma (Sharma)*[147] considered an appeal against an earlier ground-breaking decision which had recognised the State could owe a duty of care in relation to the risk of injury from catastrophic climate change.[148] A representative group of eight children (on behalf of all Australian children) had sought an injunction, based on breach of a duty of care, against the Commonwealth Minister for the Environment to restrain an approval of the expansion of a coal mine in New South Wales under environmental protection legislation. They argued this would cause them and other children future injury due to climate change. In the original Federal Court decision, an injunction was not granted by Bromberg J on the basis it was undesirable to pre-empt the Minister's decision or to assume any breach of duty was imminent.[149] However, Bromberg J held, referring to the adaptability of tort law to contemporary social conditions and changing values,[150] that the Minister owed the applicants and other Australian children a novel duty of care 'to take reasonable care to avoid causing personal injury to the children when deciding' under the relevant legislation, to approve or not approve the project.[151] On balance, salient features were said to favour a duty.[152] In particular, Bromberg

[145] Dehm, Loughnan, and Steele (n 140).

[146] See *BNL20 v Dawson in capacity as State Emergency Coordinator and Commissioner of Police* [2020] WASC 315, [70]-[75]; Dehm, Loughnan, and Steele (n 140).

[147] (2022) 400 ALR 203 (*Sharma).*

[148] *Sharma v Minister for the Environment* (2021) 391 ALR 1 (*Sharma Trial*). See discussion in Rachel Pepper, 'Environmental Law: Quartet of Recent Landmark Climate Change Cases (2021) 95(11) *Australian Law Journal* 861.

[149] *Sharma Trial* (n 148) 117. Following the initial Federal Court decision, the Minister did subsequently make a decision to approve the coal mine.

[150] Ibid 37.

[151] Ibid 113. See discussion in Wendy Bonython, 'Tort Law and Climate Change' (2021) 40(3) *University of Queensland Law Journal* 421.

[152] *Sharma Trial* (n 148) 113. See discussion of positive and negative salient features at 110.

J made extensive (and moving) factual findings in relation to the vulnerability of children to injury from catastrophic climate change in the future.[153]

The Commonwealth subsequently appealed the decision to the Full Federal Court. The Full Federal Court upheld the appeal and were unanimous in their view that a duty of care was not owed by the Minister to the children, although the three Justices' reasons differed somewhat.[154] Allsop CJ held that the content and scope of the duty would involve evaluation of 'high public policy' unsuited to judicial evaluation;[155] would be inconsistent with the decision-making powers of the Commonwealth under the relevant environmental protection legislation;[156] that there was a lack of control of the risk of harm of climate change (apart from a tiny contribution);[157] and a lack of special vulnerability, indeterminacy of liability, and proportionality of risk to possible damage.[158] Beach J held that there was insufficient 'closeness and directness' between the exercise of the statutory decision-making power and the likely risk of harm to the class of children; and there would be indeterminate liability.[159] Whelan J held that the legislation did not 'erect or facilitate' a relationship between the children and the Minister that would support a duty; there would be incoherence between the proposed duty and the public focus of the discharge of the 'statutory functions'; and personal injury to the children (or class of children) caused by extension of the coal mine was not reasonably foreseeable'.[160]

Allsop CJ rejected the trial judge's findings regarding vulnerability, holding that the children were not 'vulnerable in the relevant sense'.[161] Vulnerability 'is special vulnerability', and 'children are in the same position as everyone in the world who is or will be alive at the future times'.[162] There was no 'relevant control'[163] and no constitutional or legal principle (such as *parens patriae*) or legislative foundation which gave rise to a 'protective relationship' between the Minister and Children.[164] The children had no reliance on the Minister which was different to any other Australian—at best it was 'general political reliance' not 'legal reliance'.[165] Beach J held that vulnerability 'is not any generalised sense of vulnerability to climate change' but rather 'narrower' vulnerability to the specific harm of personal injury

[153] Ibid 72–73.
[154] Note, however, that the Full Court accepted all of the factual findings made by Bromberg J in relation to climate change and the dangers climate change posed to humans.
[155] *Sharma* (n 147) 210, 280–83.
[156] Ibid 283–84.
[157] Ibid 292, 297.
[158] Ibid 297.
[159] Ibid 301, 360–70.
[160] Ibid 371.
[161] Ibid 297.
[162] Ibid.
[163] Ibid.
[164] Ibid.
[165] Ibid.

caused by the defendant's particular breach.[166] In addition, the risk of injury due to climate change was speculative:

> Many members of the claimant class are or will be relevantly vulnerable to personal injury. But in my view one cannot reasonably ascertain their parameters or likely number today. Who knows how or where members of the claimant class will be living in 80 years and the then technological and adaptive ingenuity of humanity to deal with the consequences flowing from any increases in temperature?[167]

There was also no 'special protective relationship' between 'all members of the Executive and Australian children' either due to the legislative framework or any other proximity factor.[168] Despite the loss in the *Sharma* case, a separate climate change class action is currently proceeding based on duty of care. Torres Strait Islanders, who are Native Title holders on Boigu and Saiburu Islands with a deep spiritual connection to land and waters, are alleging the Commonwealth government failed to take pre-emptive action to prevent loss and damage to their property and cultural rights.[169] While there are a number of features that distinguish this case from *Sharma*,[170] it may be difficult to overcome the principles of the *Sharma* case which precluded a duty. Nevertheless, as Beach J suggested in *Sharma*, it is within the power of the High Court to ultimately develop new negligence solutions for the emergent catastrophic challenges of climate change injury:

> ... it is for the High Court not us to engineer new seed varieties for sustainable duties of care, modifying concepts such as 'sufficient closeness and directness' and indeterminacy to address the accelerating complexity, multiple links and cross-links of causal relations. Such concepts in their present form may have reached their shelf life, particularly where one is dealing with acts or omissions that have wide-scale consequences that transcend confined temporal boundaries and geographic ranges, and where more than direct mechanistic causal pathways are involved.[171]

[166] Ibid 355.
[167] Ibid 356.
[168] Ibid 356–57.
[169] https://www.fedcourt.gov.au/services/access-to-files-and-transcripts/online-files/pabai-v-australia (accessed 16 June 2024). The judgment in the case is pending. See also the New Zealand Supreme Court case of *Smith v Fonterra Co-operative Group Limited* (2024) 2 NZSC 5 a further tort case concerning liability for climate change-caused harm, involving a Maori elder arguing a cause of action in tort (including negligence) for contribution by large corporations to climate change, on similar grounds to the *Sharma* case. The Supreme Court of New Zealand refused to strike the action out, and held the action should be allowed to proceed to trial.
[170] eg, a legislative regime which protects the Torres Strait and interests of Torres Strait Islanders, the existence of specific damage and Native Title, and cultural interests of the claimants.
[171] *Sharma* (n 147) 371.

It is true to say that, as originally suggested by Stapleton in 2003, the High Court could ultimately develop the law of negligence in a way that utilised vulnerability to provide a response to harms suffered by the First Peoples of Australia. This will increasingly include harm suffered because of climate change. However, as I have argued in Section 3 of this chapter, this has not been the path taken by the High Court over the last two decades.

6. Conclusion

In 2003, Stapleton argued that the 'protection of the vulnerable' was a core moral concern of tort law, its 'golden thread'. Her hope was that Australian courts, recognising the protection of the vulnerable as a key concern, could develop tort law in a way which might right the 'most grievous wrongs in Australia's history' particularly those wrongs suffered by First Peoples. While protection of the vulnerable has been a factor in reasoning in novel duty cases in the High Court of Australia, other concerns such as coherency and consistency with legislative frameworks have dominated decision making. Vulnerability has not been an expansionary influence on Australian tort law in the High Court or otherwise. The *Trevorrow* Stolen Generations case in South Australia showed the promise of tort law to respond to vulnerable victims of public wrongs like Bruce Trevorrow when other factors (or salient features) such as legislative coherence are also present. However, the *Trevorrow* case has not given rise to further successful Stolen Generation tort cases, and compensation has generally been provided by statutory reparation schemes. Public interest litigation based on breach of duty of care owed by the State to vulnerable people, such as those in immigration detention, has met with some success. However, recent cases in areas such as climate change litigation demonstrate that the legal concept of vulnerability is confined to especially close protective type relationships, and even where vulnerability is present it can be outweighed by other duty factors particularly lack of coherence with legislative framework. Despite some signs that protection of the vulnerable is a concern that has the potential to influence the development of Australian negligence law, the promise of the golden thread has not (yet) been fully fulfilled. However, perhaps recent constitutional cases in the High Court of Australia that recognised the powerful and deep spiritual connections that First Peoples have with country and culture give some hope that this may one day influence the development of Australian tort law in the same way it has influenced other areas of law.[172]

[172] See *Love v Commonwealth of Australia; Thoms v Commonwealth of Australia* (2020) 270 CLR 152 which recognised that Aboriginal and Torres Strait Islander people cannot be 'aliens' (and thus subject to immigration detention or deportation pursuant to constitutional powers) even when they are born outside Australia and are not an Australian citizen. See, however, the subsequent related case of *Thoms v Commonwealth of Australia* (2022) 401 ALR 529 which held that an action of false imprisonment could not be maintained in relation to that detention.

PART II
NEGLIGENCE

6

The Real Gist of Negligence

Nicholas J McBride[*]

1. Introduction

It is given to few academics to alter legal language with their writings. However, Jane Stapleton's two-part article 'The Gist of Negligence'[1] puts her in that select company. By the 'gist of negligence', Jane was—of course—referring to 'damage', and the fact that a claim in negligence cannot be brought by a claimant unless they can allege that they have suffered actionable damage at the defendant's hands.

I think Jane can be fairly said to have coined the phrase that 'damage is the gist of negligence'. The proposition that damage is the gist of *an action in* negligence first appears in a couple of House of Lords cases near the start of World War II,[2] and the same is also said in a number of the cases that Jane was commenting on in her 1988 article.[3] This is almost certainly what put Jane in mind of the phrase 'damage is the gist of negligence', but this contraction—saying not that damage is the gist of *an action* in negligence but that damage is the gist of *negligence*—originates with Jane, and when in *Gregg* v *Scott* Baroness Hale said that 'damage is the gist of negligence'[4] she was quoting Jane. That she might have been quoting Jane unconsciously is all the more tribute to the way this phrase has cemented itself into the English legal imagination since 1988.

It might seem unduly fussy to distinguish between saying that (1) 'damage is the gist of an action in negligence' and (2) 'damage is the gist of negligence', but I think the distinction is important.[5] If we identify negligence with the breach of a duty of care owed by the defendant (D) to the claimant (C), (2) suggests, in a way that (1) does not, that a duty of care owed by D to C *will not be breached* unless C has suffered actionable damage at the hands of D. In other words, a duty of care that D

* My thanks to Jonathan Morgan for comments on an earlier draft of this chapter.

1 J Stapleton, 'The Gist of Negligence: Part 1 Minimal Actionable Damage' (1988) 104 LQR 213; J Stapleton, 'The Gist of Negligence: Part 2 The Relationship between "Damage" and Causation' (1988) 104 LQR 389.

2 *Owners of SS Heranger* v *Owners of SS Diamond* [1939] AC 94 (HL) 104; *Caswell* v *Powell Duffryn* [1940] AC 152 (HL) 164.

3 *Cartledge* v *E Jopling & Sons Ltd* [1963] AC 758 (HL) 783–84; *Sidaway* v *Board of Governors of the Bethlem Royal Hospital* [1985] AC 871 (HL) 882; *Hotson* v *East Berkshire AHA* [1987] AC 750 (HL) 765.

4 *Gregg* v *Scott* [2005] UKHL 2, [2005] 2 AC 176, [217].

5 Jane herself did not make anything of the distinction, routinely and repeatedly referring in the main body of her 'The gist of negligence' articles to 'the gist of a negligence action'.

Nicholas J McBride, *The Real Gist of Negligence* In: *Torts on Three Continents*. Edited by: Kylie Burns, Jodi Gardner, Jonathan Morgan and Sandy Steel, Oxford University Press. © Nicholas J McBride 2024.
DOI: 10.1093/oso/9780198889748.003.0006

owes C always takes the form of a duty not to cause C to suffer damage of some kind by acting in some way. Many tort law theorists believe this to be the case.[6] I do not, and I want to spend the first part of this paper explaining why (2) is untrue. In the second part, I want to turn my guns on (1): the idea that C should be disabled from suing D in negligence if they cannot establish that they have, or (in a *quia timet* case) are likely to suffer, actionable harm as a result of D's actions. While this is usually a sensible limit on allowing C to take up the time and resources of the courts in dealing with a complaint that D has breached, or is in breach, of a duty of care owed by D to C, there are occasions when this limit will prove counter-productive.

2. Damage is Not the Gist of Negligence

The proposition under scrutiny here is the *Popular View* (among tort theorists) that when D owes C a duty of care under the law of negligence, that duty always takes the form of a duty not to cause C to suffer certain harm by acting carelessly. *My View* is that that duty will take the form of (i) a duty to take care not to act in a particular way; or (ii) a duty to take reasonable steps to see that a particular outcome does not happen; or (iii) a duty to perform a particular task with reasonable skill or care.[7] Whatever form the duty takes, it can be breached by D without causing C to suffer any kind of harm. How can we determine which view is correct?

A. Dead ends

Let's begin by dismissing a couple of arguments that do *not* help us determine whether the *Popular View* or *My View* is correct.

First, if we think that duties of care really exist—that the law means what it says when it says D owes C a duty of care and is not saying that if D carelessly causes C harm, D will be held liable for that harm—that does not tell us anything about whether the *Popular View* or *My View* is correct.[8]

One might argue that if duties of care really exist, then they—like other legal duties—are intended to guide people's conduct, and because only duties of care that conform to *My View* are capable of doing that, *My View* is correct. However, proponents of the *Popular View* could argue that duties of care—like other private law duties—are intended to mark out boundaries that cannot be crossed without violating other people's rights, and duties of care that conform to the *Popular View*

[6] For a nice summary (and rebuttal) of some of these other tort law theorists' positions, see J Oberdiek, 'The Wrong in Negligence' (2021) 41 OJLS 1174, 1180-93.

[7] See NJ McBride and R Bagshaw, *Tort Law*, 6th ed (Harlow, Pearson Education, 2018), § 4.1(A).

[8] See Oberdiek (n 6), 1181, n 15.

are perfectly capable of doing that. As a result, there is no reason why someone who accepts that duties of care really exist must accept *My View*.[9]

Second, one of the arguments in favour of thinking that duties of care really exist is the potential availability of injunctions requiring D to comply with a duty of care that D owes C.[10] Such an injunction might require D to provide C with a reasonably safe working environment; or to disclose to C material risks attendant on a medical treatment that D is proposing that C should undergo at D's hands; or to provide C with food and water while C is staying at D's medical establishment.

The fact that such injunctions might be awarded against D does not tell us that *My View* is correct. This is because even if the *Popular View* is correct, C would not need to establish that she *has suffered* harm as a result of D's actions in order to obtain an injunction against him. All C would need to show—to obtain an injunction on a *quia timet* basis—is that D is *likely to cause her* to suffer actionable harm if an injunction is not awarded. After all, the cases acknowledge that damage is (almost always) the gist of claims in private nuisance[11]—and that does not prevent the courts routinely awarding injunctions in such cases.

B. Four arguments

I will now present four arguments in favour of *My View*—at least so far as the English law of negligence is concerned. These arguments fit well with Jane's preference for what she calls 'reflexive tort scholarship': none of these arguments are based on 'grand theoretical claims' about the nature of tort law or the function of legal duties but are instead rooted in taking 'seriously what judges say are the concerns to which they afford weight in their reasoning in tort cases'.[12]

Making causal judgements against wrongdoers

Jane once told me she realised the defendants' case in *Fairchild* v *Glenhaven Funeral Services Ltd*[13] was lost when, in the course of argument, Lord Bingham asked counsel for the defendants: 'Is it your position that if a claimant lives an equal distance from two factories, both of which produce and release huge quantities of asbestos filled-smoke, and the claimant subsequently develops mesothelioma, he should not be able to recover from either factory owner?' Lord Bingham

[9] Cf D Nolan, 'Deconstructing the Duty of Care' (2013) 129 LQR 559, 561–63. Nolan's later piece 'Rights, Damage and Loss' (2017) 37 OJLS 255 seems to endorse the 'boundary' view of legal duties, but he fails to see that doing this opens the way to his endorsing the view that duties of care really exist without accepting *My View*.

[10] NJ McBride, 'Duties of Care—Do They Really Exist?' (2004) 24 OJLS 417, 427–30.

[11] *Network Rail Infrastructure Ltd* v *Williams* [2018] EWCA Civ 1514, [2019] QB 601, [19], [23].

[12] J Stapleton, *Three Essays on Torts* (Oxford, Oxford University Press, 2021), 13.

[13] *Fairchild* v *Glenhaven Funeral Services Ltd* [2002] UKHL 22, [2003] 1 AC 32.

evidently thought the law would be in a sorry state if it were to deny recovery in his hypothetical.

In *Fairchild*, a defendant employer who had negligently exposed a claimant to excessive quantities of asbestos was held liable for the claimant's subsequent mesothelioma even though other, equally negligent, employers had done the same thing in relation to the claimant—thus making it impossible to say whose negligence had resulted in the claimant suffering mesothelioma. One of the reasons for the House of Lords' decision in *Fairchild* was a common feature of the English law of negligence: the courts will not allow a defendant to rely on their own, or another's wrong, in order to escape a finding that the defendant's tort caused the claimant to suffer a particular harm where doing so will result in the claimant's being undercompensated for suffering that harm.[14] We see this feature at work in cases such as *Baker* v *Willoughby*[15] and *Bolitho* v *City and Hackney Health Authority*.[16]

In *Baker* v *Willoughby*, the defendant driver negligently caused severe damage to the claimant's left leg, which same leg had to be amputated three years later as a result of its being shot in a post office robbery. The defendant was held liable for loss of the use of the left leg for the rest of the claimant's life: he was not allowed to rely on the robbers' wrong in shooting the claimant in the left leg to say that he should only be liable for the loss of the use of the leg for the three years before it would have been amputated anyway.[17]

In *Bolitho*, a child died in hospital of a heart attack brought on by respiratory failure. The child had had two previous episodes of breathing difficulties while in the hospital and despite being told of the child's problems, in neither case did the doctor on duty attend to the child. It was negligent of the doctor not to do this, but the doctor gave evidence that even if she had looked at the child herself, she would not have ordered that the child be intubated, which was the only treatment that would have saved the child from the heart attack that eventually killed him. The House of Lords held that if it would have been negligent of the doctor not to intubate, the doctor could not rely on the fact that she would not have intubated the child to establish that her negligent failure to attend personally to the child did not cause the child's death.[18]

The results in these cases are compatible with either the *Popular View* or *My View* being correct. On the *Popular View*, the defendant driver in *Baker* had obviously committed a wrong in damaging the claimant's left leg. The only issue in that

[14] ibid, [9], [33] (per Lord Bingham), [38]–[40] (per Lord Nicholls), [114] (per Lord Hutton), [155], [160], [170] (per Lord Rodger).

[15] *Baker* v *Willoughby* [1970] AC 467 (HL).

[16] *Bolitho* v *City and Hackney Health Authority* [1998] AC 232 (HL).

[17] *Baker* (n 15), 495–96 (per Lord Pearson). See also *Jobling* v *Associated Dairies* [1980] AC 794 (HL) 804 (per Lord Wilberforce), 807 (per Lord Edmund-Davies), 810 (per Lord Russell of Killowen), 813, 815 (per Lord Keith of Kinkel).

[18] *Bolitho* (n 16), 240. See also *Wright* v *Cambridge Medical Group* [2011] EWCA Civ 669, [2013] QB 312, [57].

case was how to quantify the damage caused by that wrong, and in doing that, the defendant was not allowed to rely on an equally obvious wrong (a battery) committed by someone else (the robbers) to reduce their liability. And *Bolitho* can be explained compatibly with the *Popular View* on the basis that the doctor's failure to intubate the child would have amounted to a wrong (assuming the failure to intubate would have been careless) because the failure to intubate would have resulted in the child suffering actionable harm (a heart attack).

But it is impossible to give a similar explanation of the outcome in *Fairchild* if the *Popular View* is correct.[19] If the *Popular View* is correct, there was no basis for finding that anyone in *Fairchild*—either the defendant or the other employers—had done anything wrong to the claimant. This is because on the *Popular View*, the defendant in *Fairchild* could only be said to have done something wrong in relation to the claimant if their actions caused harm to the claimant; and there was, of course, no proof of this. The same was true of the claimant's other employers. So if the *Popular View* is correct, there are two problems with explaining *Fairchild* (or Lord Bingham's hypothetical) in the same way as *Baker* v *Willoughby* and *Bolitho* can be explained: (1) if the defendant did nothing wrong, why is he being held liable? and (2) why is the defendant disabled from relying on what other people did to absolve himself of liability to the claimant if those other people also did nothing wrong?

It goes without saying that the Law Lords who decided *Fairchild* would have emphatically rejected the idea that the defendants in that case—or the factory owners in Lord Bingham's hypothetical—did nothing wrong. Instead, they proceeded on the basis that *My View* is correct. According to *My View*, in both *Fairchild* and Lord Bingham's hypothetical, the claimant is wronged by the defendant when the defendant carelessly exposes the claimant to excessive quantities of asbestos dust,

19 Of course, other explanations of *Fairchild* are possible and such explanations had to be adopted when the House of Lords decided in *Barker* v *Corus UK Ltd* [2006] UKHL 20, [2006] 2 AC 572 to extend *Fairchild* to cases where only one wrongdoer might have caused the claimant's mesothelioma. Such alternative explanations of *Fairchild* are weakened by the subsequent decision of the UK Supreme Court in *Sienkiewicz* v *Greif (UK) Ltd* [2011] UKSC 10, [2011] 2 AC 229 which made it clear that: (i) in non-mesothelioma cases, *Fairchild* would only apply in cases where multiple wrongdoers might have caused the harm suffered by the claimant (at [105] (per Lord Phillips); see now *Heneghan* v *Manchester Dry Docks Ltd* [2016] EWCA Civ 86, [2016] 1 WLR 2036 (applying *Fairchild* in a lung cancer case)); (ii) the extension of *Fairchild* to a single wrongdoer case in *Barker* was regrettable ([189] (per Lord Mance)), and *Barker* was only not overruled in *Sienkiewicz* because of fear that Parliament would react as adversely to *Barker*'s being overruled as it did to the limits placed in *Barker* on how much a defendant to whom *Fairchild* applied would be held liable for ([167] (per Lady Hale)); (iii) the original case applying *Fairchild avant la lettre* to a single wrongdoer case— *McGhee* v *National Coal Board* [1973] 1 WLR 1 (HL)—was probably wrongly decided ([92] (per Lord Phillips), [177] (per Lord Brown)). In any case, even if—all things considered—an alternative explanation of *Fairchild* better fits both it and the subsequent history of its application, it is undeniable as a matter of historical fact that one of the reasons that led the House of Lords to decide *Fairchild* in favour of the claimants was a concern not to allow multiple wrongdoers to rely on each other's wrongs to escape liability (see, eg, Jonathan Morgan's contemporaneous case note 'Lost Causes in the House of Lords: *Fairchild* v *Glenhaven Funeral Services Ltd*' (2003) 66 MLR 277, at 280–81), and that fact can only be explained on *My View*.

and in assessing whether that wrong caused the claimant's mesothelioma for the purpose of determining what damages should be payable to the claimant, the defendant is not allowed to rely on anyone else's equal wrong in carelessly exposing the claimant to excessive amounts of asbestos dust to argue 'You can't show it was my wrong that caused your harm'.

Manipulating causal rules in order to sanction breaches of duty

One aspect of English law which shows that when the courts find that D owes C a duty of care, they mean to bring D (and those like D) under a legally binding obligation for the benefit of C (and those like C) is the courts' willingness to bend the normal rules on causation where applying those rules threatens to allow D to breach the duty of care that he owes C without incurring any kind of sanction for doing so.[20] There are some *dicta* in *Fairchild* that indicate that *Fairchild* might be regarded as one case where the courts bent the rules on causation to ensure that the defendants in that case could not breach the duties of care they owed the claimants without paying a price for doing so.[21] But much clearer examples of this tactic at work are provided by *Reeves* v *Commissioner of Police of the Metropolis*[22] and *Chester* v *Afshar*.[23]

In *Reeves*, the claimant's husband killed himself while in police custody. The police breached the duty of care that they owed the husband to take reasonable steps to see that he did not commit suicide, but the police attempted to argue that they were not liable to the claimant because their breach did not cause the husband's death. He had broken the chain of causation between the police's breach of duty and the husband's death through his deliberate, voluntary, informed and unreasonable decision to commit suicide. The House of Lords gave that argument short shrift. Were it to be accepted, the police's duty of care to prevent prisoners in their custody from killing themselves would become meaningless: every time they breached the duty, and a prisoner of sound mind took advantage of the breach to kill themselves, they could say that they were not liable for the prisoner's death because it amounted to a *novus actus interveniens*.[24]

In *Chester* v *Afshar*, the defendant doctor negligently failed to warn the claimant that an operation he was recommending that she have on her spine carried with it a small (1–2 per cent) and inescapable risk of nerve damage below the sign. The claimant had the operation and the risk of nerve damage materialised. Had she been warned, she would have taken a little while to think about the risk and this would have resulted in her having the operation on a different day when the odds of her not suffering nerve damage would have been overwhelmingly in her favour

[20] McBride (n 10), 430–31.
[21] *Fairchild* (n 13), [33] (per Lord Bingham), [62] (per Lord Hoffmann), [155] (per Lord Rodger).
[22] *Reeves* v *Commissioner of Police of the Metropolis* [2000] 1 AC 360 (HL).
[23] *Chester* v *Afshar* [2004] UKHL 41, [2005] 1 AC 134.
[24] *Reeves* (n 22), 367 (per Lord Hoffmann).

(98–99 per cent). This was enough to establish a 'but for' connection between the doctor's negligence and the claimant's nerve damage—it was overwhelmingly likely that had the doctor warned the claimant of the risk of nerve damage, she would not have suffered that injury—but there was, in fact, no causal connection between the two things. It was just a coincidence that the one patient the doctor failed to warn of the ever-present risk of suffering nerve damage when having her spine operated on was the patient in respect of whom that risk materialised. Despite this, a majority of the House of Lords held that the doctor's negligence *should* be found to have caused the claimant's nerve damage because otherwise a doctor's duty of care to warn their patients of small, but significant (to the patient), risks of harm associated with their treatment would be rendered nugatory: doctors would be free to breach the duty, safe in the knowledge that they could hardly ever[25] be sued should the risk they failed to warn about materialise.[26]

While these cases show that when the courts say that D owes C a duty of care, they mean what they say, they also vindicate *My View*. Cases like *Reeves* and *Chester* v *Afshar* make no sense on the *Popular View*.[27] This is because on the *Popular View* the absence of any causal link between the harm suffered by the claimant (or claimant's husband in *Reeves*) and the defendant's carelessness means that the defendant did not breach their duty of care in those cases; and if the defendant did not breach their duty of care in those cases, it makes no sense to worry that the defendant is being allowed to breach their duty without any kind of sanction. Only *My View* makes sense of that worry. The duty of care the police owed in *Reeves* was a duty to take reasonable steps to stop their prisoner committing suicide. The duty of care the doctor owed in *Chester* v *Afshar* was a duty to warn the patient of a significant risk of harm associated with the operation that it was being proposed the patient undergo. In both cases, those duties of care were breached and operation of the normal rules on causation threatened to allow the defendants breaching those duties to do so without incurring any kind of sanction.

Wrongful death cases

Section 1(1) of the Fatal Accidents Act 1976 provides that:

[25] The only case where they could be is where, had the patient been properly informed of the risks of the operation, they would not have agreed to it. In such a case the patient's injury is more than a merely coincidental result of the doctor's failure to warn: the failure to warn exposed them to the very risk that has now materialised.

[26] *Chester* (n 23), [73], [87] (per Lord Hope).

[27] Of course, it could be argued (and many do argue) that the rules on causation *should not* have been stretched in the way that they were in *Reeves* and *Chester* v *Afshar*—but my point is that the fact that they *were* stretched only makes sense on *My View*. On the *Popular View*, what the House of Lords did in *Reeves* and *Chester* v *Afshar* was incomprehensible. On *My View*, it was perfectly comprehensible; whether it was wise or desirable is an entirely different matter.

> If death is caused by any *wrongful* act, neglect or default which is such as would (if death had not ensued) have entitled the person injured to maintain an action and recover damages in respect thereof, the person who would have been liable if death had not ensued shall be liable to an action for damages, notwithstanding the death of the persons injured.[28]

The requirement that the deceased must have been the victim of a *wrong* before an action for wrongful death can be brought under the 1976 Act is the reason why s 10 of the Animals Act 1971 and s 6(1) of the Consumer Protection Act 1987 set up a fiction that someone who sues under those Acts was the victim of a wrong. Absent these provisions, no claims could be brought under the 1976 Act by dependants of someone killed by a dangerous animal or product. This is because the Animals Act 1971 and Consumer Protection Act 1987 set up *liability rules*, where a defendant is not held liable because they have done anything wrong, but because the law thinks a defendant who is in control of a dangerous animal or produces a dangerously defective product should be held liable for the harm done by that animal/product *regardless* of whether the defendant did anything wrong to bring about that harm.

In light of this, consider the case of *Hicks* v *Chief Constable of South Yorkshire Police*.[29]

This case concerned two sisters—Sarah and Victoria Hicks—who died in the Hillsborough tragedy, at the ages of 19 and 15 respectively. Their estates sued the police in negligence. Their claims were dismissed: it was found that the sisters had suffered no actionable losses before they died because the only physical injuries they suffered occurred at exactly the same time they died. Before that, all they had suffered was distress and anxiety—which losses are not actionable in negligence except in very special circumstances. The House of Lords went on to hold that Sarah and Victoria's parents *could* bring a wrongful death claim against the police under the Fatal Accidents Act 1976, and recover damages for bereavement in respect of Victoria's death, as she was under 18 when she died.[30]

This result is inexplicable under the *Popular View*. While the police at Hillsborough were admittedly careless in creating the conditions where 96 spectators died at Hillsborough, their carelessness had not caused the Hicks sisters to suffer any actionable harm before they died—and so on the *Popular View* the police did no wrong to the sisters in creating the crushing levels of overcrowding that killed them.[31] By contrast, *My View* makes things very simple. The police owed the Hicks

[28] Emphasis added.

[29] *Hicks* v *Chief Constable of South Yorkshire Police* [1992] 2 All ER 65.

[30] ibid, 67. Under s 1A(2)(b) of the 1976 Act, a claim for bereavement damages can only be brought by a parent of a deceased *minor*.

[31] Could it be said that on the *Popular View* the police's wrong to the Hicks' sisters was completed by their death? In other words, by dying did the sisters suffer the sort of harm needed for us to be able to say that the police breached a duty owed to the sisters? But if that were the case, then the estates of the sisters should have been able to sue in respect of the damage suffered by the sisters in dying, when they could

sisters a duty to take care not to usher more spectators into the already full pen of spectators that the Hicks sisters were in—because it was foreseeable that doing so would result in physical injury to the sisters—and they breached that duty of care by opening the gate to that pen. That was the wrong that caused the death of the Hicks sisters, and allowed claims to be made in respect of their deaths under the 1976 Act.

The practice of the courts

The final argument—and one which should provide especial appeal to 'reflexive tort scholars'—is that it is fundamentally alien to the way negligence cases are decided in practice to conflate the issue of whether D breached a duty of care owed to C with the issue of whether D caused any actionable harm to C in the way that the *Popular View* conflates those issues.

No judge would regard any of the cases discussed in the previous three subsections as being 'breach of duty' cases (other than *Bolitho*, where the question of whether the doctor in that case could rely on the fact that she would not have intubated the child to establish there was no causal link between her careless failure to attend to the child personally and the child's death turned on whether it would have been negligent not to intubate).

And the judges hardly ever state what duty of care a defendant owed a claimant in terms that make the question of whether the defendant breached that duty turn on the question of whether the defendant caused the claimant to suffer harm. Instead, they say things like:

> You must take reasonable care to avoid acts or omissions which you can reasonably foresee would be likely to injure your neighbour.[32]

> The duty to take care is the duty to avoid doing or omitting to do anything the doing or omitting to do which may have as its reasonable and probable consequence injury to others.[33]

> doctors [are subject to a duty] to take reasonable care to ensure that a patient is aware of material risks of injury that are inherent in treatment.[34]

And when statute has codified a common law duty of care, the duty is again expressed in terms that conform to *My View* and not the *Popular View*:

not: Law Reform (Miscellaneous Provisions) Act 1934, s 1(2)(c). See, generally, S Hedley, 'Death and Tort' in B Brooks-Gordon (ed), *Death Rites and Rights* (Oxford, Hart Publishing, 2007).

32 *Donoghue* v *Stevenson* [1932] AC 562 (HL) 580 (per Lord Atkin).
33 *Bourhill* v *Young* [1943] AC 92 (HL) 104 (per Lord Macmillan).
34 *Montgomery* v *Lanarkshire Health Board* [2015] UKSC 11, [2015] AC 1430, [82] (per Lords Kerr and Reed).

> the common duty of care [owed by the occupier of premises to a visitor] is a duty to take such care as in all the circumstances of the case is reasonable to see that the visitor will be reasonably safe in using the premises.[35]

The fact that the *Popular View* is so alien to the way English law approaches negligence cases is shown by the fact that when Arthur Ripstein argues that private law can be seen as giving effect to various injunctions that derive from Kantian interpersonal morality, he includes on that list the injunction 'Don't run with scissors'.[36] On *My View*, the law of negligence does indeed tell us not to run with scissors (at least if other people are in the vicinity), but a proponent of the *Popular View*—like Ripstein[37]—would be expected to say that the law tells us 'Don't stab someone with scissors by running with them'. That a thinker as careful and clever as Ripstein cannot avoid slipping into thinking about the law of negligence in a way that makes perfect sense under *My View* but is inconsistent with the *Popular View* that Ripstein endorses, shows that it is not the facts on the ground that are driving the *Popular View* but 'grand theoretical' claims about tort law that demand we accept the *Popular View* regardless of what the facts on the ground tell us.

C. Three Objections to *My View*

Why is damage the gist of actions in negligence?

As we will see, the fact that C cannot sue D in negligence without showing that she has suffered actionable harm as a result of D's breach of the duty of care that he owed C is perfectly compatible with *My View*.[38] There are pragmatic reasons why it is sensible—in most cases—for the law to place such a limit on C's ability to sue D in negligence.

Why can't you sue for damages for being exposed to a risk of harm?

The English courts have set their face against such damages being awarded,[39] and it might be thought that this is inconsistent with *My View*. After all, if D culpably exposes C to a foreseeable risk of being physically injured, D has wronged C on *My*

[35] Occupiers' Liability Act 1957, s 2(2).

[36] A Ripstein, 'Kantian Perspectives on Private Law' in A Gold et al (eds), *The Oxford Handbook of the New Private Law* (Oxford, Oxford University Press, 2020) 71, 73.

[37] Ripstein's views are discussed in Oberdiek (n 6), 1180–86.

[38] See C Webb, 'Duties and Damages' in P Miller and J Oberdiek (eds), *Oxford Studies in Private Law Theory: Volume 1* (Oxford, Oxford University Press, 2020), 5. *Contra*, H Hurd, 'The Deontology of Negligence' (1996) 76 Boston Univ Law Rev 249, 262 (moving from observing that *liability* for negligence is harm-based to a conclusion that negligence *qua wrong* must involve causing another harm).

[39] *Gregg* v *Scott* [2005] UKHL 2, [2005] 2 AC 176; *Durham* v *BAI (Run Off) Ltd* [2012] UKSC 14, [2012] 1 WLR 867 (disapproving the view—potentially encouraged by *Barker* (n 19)—that damages awarded under *Fairchild* are damages for being exposed to the risk of developing mesothelioma).

View—so why should C not recover damages for the fact that D exposed her to the risk of injury?

The reason is that C has not suffered the 'right kind of loss' to maintain an action against D for damages.[40] The duty that D owed C and breached was imposed on D not in order to avoid C being exposed to the risk of injury but in order to avoid C's being injured. As C has not been injured, C has not suffered the kind of harm that the duty breached by D would allow her to sue to be compensated for.

A proponent of the *Popular View* might counter 'If D's duty of care were designed to avoid C's being injured, the duty of care that D owed C would be a duty not to injure C by acting carelessly—which is what the *Popular View* tells us it is.' But this does not follow. There seems nothing wrong with trying to avoid people being stabbed by telling other people 'Don't run with scissors' instead of giving them the more tailored (and more complicated and less intelligible) instruction 'Don't stab someone with scissors by running with them'. In the same way, there seems nothing wrong in the law's being overbroad in the duties it imposes on us and trying to avoid people's being injured by requiring us to take care not to act in ways that are likely to injure others.

Doesn't history establish that damage is the gist of negligence?

The third objection is summed up by Lord Dyson in *Lumba* v *Secretary of State for the Home Department*: 'an action on the *case* (of which a claim in negligence is the paradigm example) regards damage as the essence of the wrong.'[41]

It is doubtful whether this was ever the case: damage may have been essential to *actions* on the case—but that is a very different thing from saying that damage is the essence of the wrong that a claimant complained of when bringing an action on the case. But let's assume for the sake of argument that Lord Dyson is right and that the kind of wrongs that actions on the case covered all involved causing another to suffer some kind of damage. Does that establish that the wrong of negligence as we know it today *necessarily* involves causing the claimant to suffer some kind of damage? Is negligence as a wrong always destined to be tethered to the concept of damage by 'medieval chains'[42] that were forged by decisions hundreds of years ago as to what wrongs would and would not be remedied by the issuing of writs to hear claims for trespass on the case?

The answer is 'Obviously not'. We know the answer is 'Obviously not' because the modern law of contract emerged out of actions in *assumpsit* and actions in *assumpsit* were actions on the case. Despite the origins of claims for breach of contract in actions on the case, no one would dream of saying nowadays that damage

[40] For this limit on C's ability to sue D in tort (not just in negligence), see McBride and Bagshaw (n 7), §§ 4.1(B), 9.3.

[41] *Lumba* v *Secretary of State for the Home Department* [2011] UKSC 12, [2012] 1 AC 245, [64] (emphasis in original).

[42] *United Australia Ltd* v *Barclays Bank Ltd* [1941] AC 1 (HL) 29 (per Lord Atkin).

is the essence of the wrong of breach of contract. If breach of contract *qua* wrong has managed to escape the medieval chains that sought (we are supposing) to tie it to the concept of damage, it is hard to see why negligence *qua* wrong cannot do the same thing.[43] And given the arguments made above, it seems to have already made its escape.

3. Damage as the Gist of Actions in Negligence

Nothing I have said in the preceding section denies that the victim of the breach of a duty of care can only *sue* in negligence if they have suffered (or are likely to suffer) actionable harm as a result of that breach. That is what is meant by saying that 'damage is the gist of actions in negligence'. In this section of the chapter I want to question whether that should always be the case. We begin our exploration of this issue by asking: why is it *ever* the case that a claimant needs to show they have suffered (or are likely to suffer) actionable harm in order to sue in negligence?

A. Why isn't negligence actionable per se?

The question assumes that English private law is religiously wedded to the principle that '*ubi ius, ibi remedium*', with the result that if *My View* is correct and if D does C wrong by running down a corridor carrying some scissors and narrowly misses stabbing C with them, C should be entitled to a remedy. However, this assumption is incorrect.[44] As Robert Stevens—another adherent to the *Popular View*—observes:

> There are torts where the infringement of the right is only actionable upon proof of consequential loss. Four examples are public slander, misfeasance in public office, deceit, and the breach of rights arising from gratuitous undertakings.[45]

So why are these torts only actionable upon proof of consequential loss? Stevens argues that 'In principle, the division between wrongs actionable per se and those

[43] See, to similar effect regarding the tort of misfeasance in public office, Lord Walker's remarks in *Watkins* v *Secretary of State for the Home Department* [2006] UKHL 17, [2006] 2 AC 395, [74]: 'Historically the tort of libel developed, as misfeasance in public office did, from the action on the case. But both have come a long way since then. If … a claimant can recover general damages of over £100,000 for a single libel without proof of any monetary loss whatsoever, ancient legal history cannot be a good reason why misfeasance in public office should not develop in the same direction.'

[44] *Contra*, Y Adar and R Perry, 'Negligence Without Harm' (2022) 111 Geo LJ 187, who see something incoherent about negligence consisting in a breach of a duty owed to another while not being actionable per se.

[45] R Stevens, *Torts and Rights* (Oxford, Oxford University Press, 2007) 88.

only actionable upon proof of consequential loss should reflect a choice between those rights which are, and are not, as a question of social fact sufficiently important to be deserving of protection irrespective of the consequences of violation.'[46] To similar effect, the House of Lords ruled in *Watkins* v *Secretary of State for the Home Department*[47] that the tort of misfeasance in public office should not be actionable per se because 'If public officers behave with outrageous disregard for their legal duties, but without causing material damage, there are other and more appropriate ways of bringing them to book'[48] than allowing them to be sued in tort.

There seems no reason why we cannot give a similar explanation of why negligence is not actionable per se. In the case where D goes running with scissors and narrowly misses stabbing C with them, D has breached a duty of care owed to C (on *My View*) but there seems little point in allowing C to bring a claim against D for nominal damages. The courts have more important things to do with their limited time and resources than to deal with C's claim against D. This is, after all, why damage was originally the gist of actions on the case: expanding the number of claims that the royal courts were able to hear under writs of trespass on the case brought with it a need to place a limit on what cases would be heard under this novel procedure. The limit was the claimant had to allege they had suffered some kind of harm at the defendant's hands.

But just as the requirement of damage was relaxed in certain actions on the case that were socially important to deal with regardless of whether the defendant had caused the claimant harm—such as claims for breach of contract or libel—it follows from the above explanation of why negligence is not actionable per se that the 'actionable harm' limit on a claimant's ability to sue a defendant in negligence should be dispensed with where it is socially important that the claimant should be allowed to sue the defendant regardless of whether they have suffered (or are likely to suffer) actionable harm as a result of the defendant's negligence. I can think of at least two situations where this will be the case.

B. Duties to avert danger

It is well-established that where D has (perhaps completely innocently) put C in danger of suffering some kind of physical harm, D will owe C a duty to take reasonable steps to avert that danger.[49] But what if C spends money averting the danger herself? In such a case, C has a problem with suing D in negligence in respect of the money she has spent. She has not suffered the 'right kind of loss' in order to be able

46 ibid 89.
47 *Watkins* (n 43).
48 ibid [9] (per Lord Bingham).
49 McBride and Bagshaw (n 7), § 6.3.

to sue D. The duty of care D owed C was designed to protect C from being physically injured, not to save C from having to spend money on her own protection.

It seems that all those who have considered this issue agree that C should be entitled to sue D in this kind of case.[50] The case law indicates that the courts are also eager to provide C with a remedy here.[51] The social importance of allowing C to sue D in this situation was identified by LaForest J in *Winnipeg Condominium*: not allowing C to sue 'serves to encourage, rather than discourage, reckless and hazardous behaviour. Maintaining a bar against recoverability for the cost of repair of dangerous defects provides no incentive for plaintiffs to mitigate potential losses and tends to encourage economically inefficient behaviour.'[52] The difficulty lies in finding an appropriate cause of action under which C's claim can be brought. For example, C might try to sue D in restitution, on the basis that she has spent money discharging a legal obligation of D's and therefore D has been enriched at C's expense.[53] However, it takes more than that to bring an action for restitution, and there are difficulties around establishing that D has been *unjustly* enriched, and that D's enrichment is sufficiently 'direct' to count as being at C's *expense*.[54]

The most straightforward approach is to allow C to sue D in negligence—but for that to happen the 'actionable harm' limit on C's ability to sue D in negligence has to be dropped and C must be allowed to sue D even though she has suffered no such harm.[55] C would instead be allowed to sue D in negligence for the cost of curing D's breach in failing to take reasonable steps to save her from the danger of

[50] Stapleton, 'The Gist of Negligence: Part 1' (n 1), 221–24; R Cooke, 'An Impossible Distinction' (1991) 107 LQR 46; J Fleming, 'Preventive Damages' in N Mullany (ed), *Torts in the Nineties* (Sydney, Law Book Co, 1997); M Moran, 'Rethinking *Winnipeg Condominium*: Restitution, Economic loss, and Anticipatory Repairs' (1997) 47 UTLJ 115; J Goldberg and B Zipursky, 'Unrealized Torts' (2002) 88 Va Law Rev 1625, 1701–15; M Geistfeld, 'The Analytics of Duty: Medical Monitoring and Related Forms of Economic Loss' (2002) 88 Va Law Rev 1921; D Nolan, 'Preventive Damages' (2016) 132 LQR 68; J Morgan, *Great Debates in Tort Law* (Oxford, Hart Publishing, 2022) 202–07.

[51] *Murphy* v *Brentwood DC* [1991] 1 AC 398 (HL) 475 (per Lord Bridge) (arguable building owner can claim for cost of repairing dangerous feature of building where it is potential source of danger to others); *The Orjula* [1995] 2 Lloyd's Rep 395 (QB) (arguable can claim for cost of cleaning up acid spill); *Winnipeg Condominium Corp No 36* v *Bird Construction Co* (1995) 121 DLR (4th) 193 (Can SC) (can recover for cost of making dangerous building safe); *Friends for All Children* v *Lockheed Corp*, 716 F 2d 816 (DC Cir 1984) (establishment of fund for medical monitoring of children endangered by plane crash); *Ayers* v *Jackson Township*, 525 A 2d 287 (NJ 1987) (establishment of fund for medical monitoring of children exposed to environmental toxins); *Potter* v *Firestone Tire and Rubber Co*, 863 P 2d 795 (Cal 1993) (ditto).

[52] *Winnipeg Condominium* (n 51) 213.

[53] Moran (n 50); Stevens (n 45) 31–32.

[54] Nolan (n 50) 91–93.

[55] Morgan (n 50) suggests (at 203) that the 'actionable harm' limit could be retained but C could be held to have suffered such actionable harm if she has merely incurred a risk of suffering in the future the kind of harm that D's duty was imposed on D in order to avoid. This would then allow C to make claims in respect of losses flowing from C's being exposed to that risk. However, adopting such an approach would require a host of cases to be overruled or revisited, including *Gregg* v *Scott* (n 39), *Sienkiewicz* (n 19) (insofar as it does not allow *Fairchild* to apply to claims against a single wrongdoer in non-mesothelioma cases), and *Rothwell* v *Chemical & Insulating Co Ltd* [2007] UKHL 39, [2008] 1 AC 281. Paradoxically, simply getting rid of the actionable harm limit in this context is less revolutionary than modifying it.

physical injury to which he exposed her. A claim for cost of cure damages can be routinely made in contract but is alien to claims in negligence because of the idea that damage is the gist of actions in negligence. Drawing an analogy with claims for contractual claims for cost of cure damages makes it easy to see when C should be allowed to sue D for damages representing the cost of averting the danger to which he exposed her: (i) D must be in breach of his duty to take reasonable steps to save C from harm (so D's duty must require him to do more than simply warn C of the danger she is in,[56] and D must have been given a chance to perform that duty himself and failed to do so); and (ii) the amount spent on averting the danger to which D has exposed C must be reasonable (so it must not be more than D could have been expected to spend himself on saving C from danger).

C. Duties to warn

Imagine that C is a candidate for specialist spinal surgery that is routinely performed by D, where D is the only consultant in the country who performs this surgery. C's friend warns her that there is a risk—which he assesses at one in a thousand—of paralysis associated with the surgery that D is proposing to carry out on her. When C asks D about this, D denies that there is any such risk, dismissing what her friend says as 'fake news'. In fact, if he consulted the relevant medical literature, he would discover that the friend is correct, but D does not bother to keep himself up to date with the latest literature in this area. D's response to C is based purely on his personal experiences with carrying out his surgery, where (so far as he knows) none of his patients have ended up paralysed. C's friend is adamant that he is correct and D is misinformed. C—who was unimpressed with D's blithe dismissals of her worries—does not know who to believe. She is clear in her own mind that if her friend is right, she will not have the surgery; and if her friend is wrong, she will.

Could C seek an injunction, requiring D to check the relevant medical literature and in light of that inform her whether or not there is a 0.1 per cent risk of paralysis associated with the surgery he is proposing to perform on C? The idea that damage is the gist of actions in negligence will stand in C's way. This is because C will fall a long way short of the current standard for obtaining a *quia timet* injunction in relation to wrongs that are not actionable per se. Under that standard, it has to be shown that it is *very likely* that C will be injured if she is not properly warned of the risks associated with her operation.[57] If there *is* a 0.1 per cent risk of paralysis, it is overwhelmingly likely that C will—like D's other patients—pass through the surgery unscathed. Moreover, it will not be clear to the court deciding C's case that

[56] This is unlikely to be the case where D was not at fault for creating the danger that C is now in.
[57] For discussion, see J Murphy, 'Rethinking Injunctions in Tort Law' (2007) 27 OJLS 509.

there *is* even a 0.1 per cent risk of paralysis associated with C's operation. Instead, in the eyes of the court, there will simply be a hard to assess risk that there is a 0.1 per cent risk of paralysis—which takes C even further away from the threshold of likelihood of injury that C needs to cross to get a *quia timet* injunction.

The social importance of ensuring that patients like C are properly informed of the risks associated with their treatments suggests that in deciding C's case, the courts should focus not on the probability that C will be *injured* if she is not granted an injunction but on the probability that D will *breach the duty of care* he owes her to warn her of material risks associated with her treatment if she does not obtain an injunction. In other words, in a case like this, the likelihood of a breach of duty and not injury to C should be regarded as the gist of C's claim against D.

D. Other cases?

Generalising from the previous two examples, we might say that we should be more willing to dispense with the 'actionable harm' limit on C's being allowed to sue D in negligence: (i) the more urgent it is that D comply with the duty of care he owes C, or (ii) the more important the duty of care that D owes C is, or (iii) the more likely it is that that D's breach is reflective of a more systemic pattern of breaches of duties of care committed by D or the organisation D works for.

4. The Truth

'Damage is the gist of negligence' is a wonderfully seductive but ultimately misleading phrase. The real gist of negligence is the breach of a duty of care owed to another, and breaching a duty of care owed to another does not involve causing that other damage. While damage is the gist of an action in negligence, there are occasions when we have good reason to think that it should not be—such as in cases where it is socially important that C be allowed to sue D in negligence regardless of whether D's breach of duty has caused (or is likely to cause) C to suffer harm. While my conclusion threatens the contribution Jane has made to the language of the law, I don't think she will care. To my mind, Jane's finest quality as a legal scholar, and the one that I most associate with her scholarship, is her fierce commitment to the truth in talking about the law. As Jane herself has observed, a key influence on her work has been her

> early scientific training. For scientists, reality is the primary material of study to which they refer back as the corrective to any theoretical account... The scientist also cannot be selective about which phenomena his or her theoretical model has to accommodate: in the end, 'best fit' approaches have no value. Reality always

> trumps the model. So, if firm data emerges that does not fit it, the scientist's reaction is to reassess the model, not to condemn that data as somehow 'wrong'. Finally, the scientist seeks to confirm the strength of a model by its capacity to resolve real problems and to predict outcomes.[58]

By these admirable yardsticks, 'damage is the gist of negligence' comes up short and Jane would be the first to embrace the more complex picture of the law that should take its place.

[58] Stapleton, *Three Essays on Torts* (n 12), 1.

7

Duty of Care Factors: Principle and Policy Decisions in the United States

Ellen M Bublick[*]

1. Introduction

All law tells the story of a particular place and time. The Texas Deceptive Practices Act, for example, forbids a laundry list of fraudulent schemes.[1] The Act bars 'taking advantage of a natural disaster, declared by the governor, to demand exorbitant prices for food or fuel'.[2] It prohibits 'using a foreign language translation of the title "attorney" or "immigration law expert", on a business card to advertise someone who is not an attorney'.[3] It outlaws 'resetting the odometer of any motor vehicle so as to reduce the number of miles indicated on the odometer gauge'.[4] By these, and other restrictions, the Act makes visible deceptions to which the state has (and has not) responded over time.

So too does common law tell a story. As honoured distinguished scholar Jane Stapleton observes, the common law is living: '[a]s society changes, so too does the common law change'.[5] Tort law 'moves with changes in social facts, values and expectations'.[6] Stapleton, who held posts at Oxford, Harvard, Texas and the Australian National University, all in a similar time frame, is well situated to understand the extent to which law is influenced by its environment.[7]

Tort law navigates values that not only vary and evolve, but also conflict. In a single jurisdiction, residents have interests in rights, efficiency, deterrence, incentives, and certainty—all at once. Conflicting values are inevitable, and accommodation required. As Judge Posner vividly illustrated with respect to accommodation of the conflicting values of informational privacy and disclosure, 'If Congress required airline passengers to fly nude in order to reduce the risk of a

* Thanks to participants in the Oxford Festschrift for helpful comments on this work.

1 Texas Deceptive Trade Practices Act, Vernon's Texas Bus. & Comm. Code § 17.46 (a statute from Professor Stapleton's one-time US home).

2 Ibid.

3 Ibid.

4 Ibid.

5 Jane Stapleton, *Three Essays on Torts* (Oxford 2021) at xvii

6 Ibid. at 10.

7 Ibid. at 12.

Ellen M Bublick, *Duty of Care Factors: Principle and Policy Decisions in the United States* In: *Torts on Three Continents*. Edited by: Kylie Burns, Jodi Gardner, Jonathan Morgan and Sandy Steel, Oxford University Press.
DOI: 10.1093/oso/9780198889748.003.0007

terrorist incident, one imagines that the law might well be held to infringe a constitutional right to privacy even though there is a substantial social interest in airline safety.'[8]

It is into this complex and conflicting mélange of interests and values that common law courts write. The story of duty of care in tort law is a story of substantive values. In her ground-breaking article, *Duty of Care Factors*, penned 25 years ago in celebration of another tort luminary, John Fleming, Stapleton articulated this uncomfortable truth. She then unmasked a number of 'complex moral and policy concerns underlying [the British and Commonwealth courts'] resolution of novel claims for a duty of care in the tort of negligence'.[9] Stapleton identified a list of factors that influence judicial decisions, and also counselled that the list is 'by no means complete or closed'.[10]

Though courts might wish to avoid an open-ended list of substantive values, Stapleton denied the power of any 'test' to set non-substantive doctrinal rules. At the same time, as reflects her understanding of the practical work that judges must do, Stapleton has also examined how courts might best employ standards to guide their substantive analysis.[11]

My aim in this chapter is simple: to follow Professor Stapleton's lead and review judicial opinions in order to describe important substantive factors that influence duty analysis. I focus on US cases today. This article examines both the substantive factors at work in recent US decisions about the duty of care, and the doctrinal tests employed to frame these choices. To tell this American story, I have reviewed state supreme court decisions on the subject of duty during a two and a half year time span—more than one hundred opinions in total. While there are many more substantive factors in the decisions than this short chapter can begin to name or describe (norms are as pervasive as the air we breathe), in this chapter, I outline a few that are salient in judicial explanations.

The takeaways of the analysis are these: First, state courts draw on common standards and language to decide duty. They have also devised many more tests than US commentators have previously enumerated. These tests include both general standards for determining duty and specific standards for duty in particular contexts. Nevertheless, this chapter identifies a deep structure that pervades.

Second, many of the substantive factors Stapleton notes in her article *Duty of Care Factors* still actively inform duty determinations—in the United States no less than in the Commonwealth countries Stapleton described. And yet the tenor of recent US decisions adds a slightly different cast. In terms of local flavour, US courts

[8] *Wolfe v. Schaefer*, 619 F.3d 782, 786 (7th Cir. 2010).

[9] Jane Stapleton, *Duty of Care Factors: A Selection from Judicial Menus* in *The Law of Obligations: Chapters in Celebration of John Fleming* (Peter Cane and Jane Stapleton, eds, Oxford 1998) 59.

[10] Ibid. at 87.

[11] Ibid. at 87.

are much more concerned with protecting against physical harm than against economic loss. Within the physical injury sphere, harm-prevention rationales are particularly important to US contexts and decisions. A critical area for fostering the physical security of persons arises in the context of people who do not have a previously recognized special relationship—duties to strangers. In this area, state supreme courts are hesitant to embrace a 'duty to the world', but increasingly recognize duty within 'webs of relationship'—in supply chains or shared work environments for example. Often, expanded duties in webs of relation arise in contexts in which one party's actions control structural factors that foreseeably affect the care, risk and harm to which others will be subject. A defendant's special relationships with a harm-causer fit into this group. Another area of increased substantive attention is equality. Duties of police to citizens, and proprietors to sexual assault victims feature in this set of cases. But concerns for inclusion extend beyond groups defined by race and gender. Expanding duties *to* police and to workers suggest a more inclusive focus of individual recovery overall.[12]

But this summary provides just a small glimpse into the larger story of duty in the recent American case law. By way of roadmap, Section I provides background on the importance of duty determinations in the US system, outlines the inadequacy of formulaic tests to decide normative issues and embraces the importance of substantive factors to the project of examining US decisions. Section II reveals some of the commonalities among state supreme court tests for duty, describes many variations, identifies a deep structure of commonality. Section III then describes some oft-addressed substantive factors at play in US opinions. Finally, the chapter concludes with ideas for future study with respect to the normative dimensions of courts' decisions.

2. The Boundary-Line Function of Duty of Care: The Project of Articulating Tests and Factors

Duty of care is the main doctrine through which US courts control the boundaries of negligence liability. When courts resolve whether a duty exists, they mark which harm-producing careless conduct may give rise to tort liability and which may not.[13] Parties who owe a duty of reasonable care may be accountable for failures to live up to that standard. When parties owe no duty, they face no accounting, even when their unreasonable risks cause harm to others.

In the United States, it is judges who decide when a duty, typically a duty of reasonable care, exists. Duty is the single most important question judges decide in

[12] Ellen Bublick, *Tort Common Law Future: Preventing Harm and Providing Redress to the Uncounted Injured* (2021) 14 Journal of Tort Law 279.

[13] Stapleton, *Duty of Care Factors* (n 9) at 59–60.

the negligence case. Indeed, in the United States, duty is the *only* element of the negligence case that is wholly entrusted to judges rather than juries. Because actors cannot be subject to liability in the absence of duty, as determined by a judge, the duty question is a critical gateway through which all negligence suits must pass.

Yet how courts determine whether a duty exists is a mystery—to litigants, scholars, and courts themselves. Even after the shift from formalism to legal realism, judges and scholars long for a self-executing test from which to determine the existence of duty. Professor Stapleton provides wise counsel against the search for such a salve. 'There *is* no test for the duty of care. There *can* be no "duty test" given what it is that judges do under the cloak of this analytical label.'[14]

What judges do under cloak of duty analysis, it turns out, is to evaluate varied substantive factors that militate in favour of, or against, liability in different types of cases. As Dean Prosser famously wrote, 'duty is not sacrosanct in itself, but is only an expression of the sum total of those considerations of policy which lead the law to say that the plaintiff is entitled to protection.'[15] Working under a realist impulse similar to Dean Prosser's, Stapleton thoughtfully articulates many important principle and policy factors underlying duty analysis in the United Kingdom, Australia, Canada and New Zealand.

For example, Stapleton argues that the power of Lord Atkin's famous judgment in *Donoghue v. Stevenson*,[16] lays not in intoning the case's innovative neighbour principle over a factual situation to resolve the duty question.[17] Rather, the neighbour principle merely flags a 'particular gate through which the plaintiff must pass.'[18] Other determinants direct the outcome. Specifically, the substantive import of *Donoghue* is its expansion of liability in the absence of privity—a step beyond the recognized pockets of duty and liability at the time. Also important is the case's recognition of foreseeability as a necessary but not sufficient foundation for liability.

Across many cases, Stapleton pinpoints the substantive concepts doing actual work in courts' duty decisions. She finds that courts invoke the following 'convincing factors *in favour* of recognition of duty': The law is concerned with the physical security of persons and property.[19] The law wants to influence risks that are in the exclusive control of the defendant, on whom plaintiff is thereby dependent.

[14] Ibid. at 60.

[15] W. Page Keeton and Dan B. Dobbs et al., *Prosser and Keeton on Torts* (5th ed., West Publishing Co. 1984) § 54, at 358.

[16] [1932] AC 562 (HL) ('The rule that you are to love your neighbour becomes in law, you must not injure your neighbour; and the lawyer's question, Who is my neighbour? receives a restricted reply. You must take reasonable care to avoid acts or omissions which you can reasonably foresee would be likely to injure your neighbour. Who, then, in law, is my neighbour? The answer seems to be – persons who are so closely and directly affected by my act that I ought reasonably to have them in contemplation as being so affected when I am directing my mind to the acts or omissions which are called in question.')

[17] Stapleton, *Duty of Care Factors* (n 9) at 60–61.

[18] Ibid. at 62.

[19] Ibid. at 93.

The law does not want to discourage rescue or encourage abortion, in part because it wants to foster socially desired conduct and avoid bringing itself into disrepute.[20]

Stapleton also uncovers a long list of 'convincing factors' that courts employ '*countervailing to* the recognition of duty'. This list, six times as long as the list of factors in favour of a duty, includes concerns *not to*: undermine individual liberty by requiring affirmative acts towards a stranger; assign liability to causally peripheral parties; aid a plaintiff who had adequate means of avoiding a risk; produce unattractive socio-economic impacts; create indeterminate liability; disrespect separation of powers; bring the law into disrepute; create systemic evidentiary difficulties; threaten orderly administration of justice or control of public order; encourage exploitation; deter free speech; undermine legislative policy; frustrate the public authority's role; chill valued communications; encourage abortion; adversely affect people in the position of the plaintiff; deter involuntary and undeterrable conduct; embrace liability in an undefinable category; place onerous burdens on disadvantaged groups; substantially depart from traditional categorization; invade fundamental liberties; or discourage socially beneficial hospitality.[21]

Stapleton's factors well explain judicial rationales in many state supreme court opinions in the United States, as well as in opinions from the Commonwealth. But before examining the substantive list of factors favouring and disfavouring duty determinations, an outline of this chapter's project is in order. As US readers will be aware, telling a story about US tort law presents a special challenge. There is no US tort law per se. Instead, US tort law is state common law, and the United States has 50 states. Unless the US Constitution or federal statues are involved, the US Supreme Court does not resolve, or even examine, state tort law.

Accordingly, to examine duty an observer must look at US tort law in its 50-state context. The task is daunting. To capture a snapshot of these many cases but keep the project to a manageable level, this chapter examines decisions from only each state's highest court. It also limits the period under review to the 2-and-a-half years before this symposium—from 1 January 2020, until 1 July 2022. Even so, this small slice of fora and time span yields more than 100 duty cases. If the project were to add back opinions issued by federal courts applying state law when the parties reside in different states, as well as state appellate courts, the data set would grow by a factor of 8 or more. This set of nearly 1,000 opinions would itself provide only a sample of the US case law. Those published opinions are drawn from less than 1 per cent of filed cases that make their way from filing to trial to appeal. All of which is to say that the duty determinations evaluated here are but a small sample of a large, fluid and important doctrinal boundary in US negligence law.

[20] Ibid.
[21] Ibid. at 93–95.

The approach I take in this chapter, examining state supreme court opinions on their own terms, honours Stapleton's guiding work on 'reflexive analysis'.[22] As defined by Stapleton, reflexive analysis 'places judges at center stage and seeks a constructive dialogue with them'.[23] The word reflexive signals a two-way conversation between bench and bar.[24] To engage in this honest conversation, scholars must look at 'law as it is', rather than law as a theorist wishes it to be.[25] The method takes judicial explanations seriously, and examines both judicial outcomes and reasons.[26] As a part of the process, reflexive analysis accepts judicial differences in values and outcomes as valid and inevitable.[27] Yet reflexive analysis does not simply take decisional outcomes and reasoning at its word. The approach notes patterns and tensions among cases, evaluates and criticizes decisions and suggests normative arguments that should be taken into account.[28]

As envisioned, the reflexive scholar's approach has much in common with the American Law Institute's (ALI) stated mode of inquiry for developing suggested common law. The guiding principle of the ALI has been articulated in this way: 'We should feel obliged in our deliberations to give weight to all of the considerations that courts, under a proper view of the judicial function, deem it right to weigh in theirs.'[29]

3. Judicial Tests

When analysing duty, factors and outcomes are evaluated through the lens of doctrinal tests. This section examines those tests. The section first explains how duty fits with the other elements of the US negligence case. It then outlines the most frequently deployed tests of duty in the United States. Next, it describes the much wider variety of tests for duty that courts employ. Finally, the section identifies a deep structure that underlies courts' numerous tests.

A. Duty in context: elements of the US negligence case

For those outside the United States, a bit of context about the role duty plays in the elements of state negligence cases may be of help. The common law negligence causes of action in the United States are generally carved into five elements—duty,

22 Stapleton, *Three Essays on Torts* (n 5) at 1.
23 Ibid. at 2.
24 Ibid.
25 Ibid. at 17–18.
26 Ibid.
27 Ibid. at 18.
28 Ibid. at 19–20.
29 Herbert Wechsler, *Report of the Director*, The American Law Institute: Annual Reports (1967) 5.

breach, actual harm, actual cause and proximate cause. Some state courts enumerate only four elements because they combine actual and proximate cause (a combination which can lead to confusion).[30] A few courts have named just three elements of the negligence case by combining actual harm with proximate (and presumably actual) cause.[31]

A comparison of the US action with the three elements of French negligence law—fault, causation and harm—elucidates the role of the US elements. Duty (obligation), breach (failure to use reasonable care) and proximate cause (harm stemming from the type of risk that made the breach tortious) are but three parts of the question of fault towards the plaintiff.

The UK question of actionable fault seems to be subdivided somewhat further. I leave it to British scholars to sort out how inquiries about actionability, breach, nexus, legal responsibility, proximity and remoteness all fit together.[32] As an outsider to the British system, all of these queries seem to be subparts of issue of actionable fault. I cannot tell whether the fractionation of actionable fault increases the doctrinal hurdles claimants face and obscures courts' focus on substantive factors and reduces the soundness of ultimate outcomes. US negligence law contains subtests that have those effects.[33]

B. Standard tests for duty

The best-known tests for duty in the United States stem from the widely used Restatements of the law, which serve as model common law standards. From the First Restatement onwards, definitions of duty have been a central part of the blackletter law. Indeed, the First Restatement of Torts mentioned the term duty in over 450 of its 951 sections. According to that Restatement, duty denotes that an 'actor is required to conduct himself in a particular manner at the risk that if he does not do so he may become liable to another to whom the duty is owed …'[34] That Restatement structures its duty framework around acts involving risks to others and omissions.[35] This structure aligns with the English common law, from

[30] Dan Dobbs, Paul Hayden, and Ellen Bublick, *The Law of Torts* (2d ed. 2011).

[31] *Keith v. Health-Pro Home Care Svcs., Inc.*, 873 S.E.2d 567, 574 ¶ 24 (N.C. 2022) (elements are duty, breach and 'injury proximately caused by the breach'); *Harwood v. Ardagh Grp.*, 2022 OK 51, ¶ 26, 522 P.3d 473, 480, as corrected (19 September 2022) (elements to a claim based upon negligence include (1) a duty to the plaintiff, (2) a violation of that duty and (3) injury resulting from that violation).

[32] James Goudkamp and Donal Nolan, *Robinson v. Chief Constable of West Yorkshire Police: Taking Duty Back to Basics* (2023) 16(1) Journal of Tort Law 125.

[33] For example, in the doctrine of open and obvious danger, a focus on whether the danger was visible can detract from the principled question—could the defendant reasonably expect that people encountering the danger would be able to perceive and avoid it? Dobbs, Hayden, and Bublick, *The Law of Torts* (n 30) § 276.

[34] Restatement (First) of Torts (1934) § 4.

[35] Ibid § 291 ((1) Where an act is one which a reasonable man would recognize as involving a risk of harm to another, the risk is unreasonable and the act is negligent if the risk is of such magnitude as to

which it came, and the philosophical tenet that action may be wrongful 'because it involves an unreasonable risk of invading one of a particular species of interests, such as an interest of personality'.[36] However, a person generally need not take positive action to protect others.[37]

This dual structure of duty with respect to acts and omissions has remained constant throughout Restatement iterations yet has been characterized by slightly different language. The Second Restatement of Torts distinguished between misfeasance and nonfeasance. According to that distinction, 'anyone who does an affirmative act is under a duty to others to exercise the care of a reasonable man ...'[38] By contrast, 'the duties of one who omits to act are more restricted ...'[39] On the omissions, or nonfeasance, side of the scale, the Restatement Second presents a number of exceptions to the no-duty rule and articulates specific tests for establishing these duties.[40]

The latest Restatement would have courts ask whether the defendant's conduct created a risk of physical harm.[41] Under the Third Restatement standard, two opposite default rules apply based on the answer to that central question. If an actor creates a risk of harm, § 7 applies, and a duty of care is presumed. If an actor *does not* create a risk of harm, § 37 applies, and an actor is presumed *not* to have a duty of care. Both default rules are subject to exceptions based on principle and policy.[42] Indeed, the Third Restatement enumerates many of the traditional exceptions on the nonfeasance side.[43] Duties in nonfeasance can stem from statutes, prior risk-creating conduct, special relationship with the person harmed or the harm causer, undertakings and taking charge of another.[44]

Under this 'risk architecture', a term coined by Professor Stapleton, the whole machinery of tort law—duty, care and potential liability—hinges on a judicial determination of risk creation.[45] But, Stapleton cautioned, the Restatement provides almost no guidance about how courts should determine when an actor has created a risk. 'It is not self-evident what it means to say that conduct "creates a risk' of physical harm. It is not a term of art.'[46] Courts need a substantive definition of

outweigh what the law regards as the utility of the act or of the particular manner in which it is done. (2) The duties to take positive action for the protection of another stated in §§ 316 to 325 are limited to situations in which the risk involved in inaction is of such magnitude as to outweigh what the law regards as the utility of permitting the actor to remain inactive).

36 Ibid. § 281. See also Goudkamp and Nolan, *Robinson v. Chief Constable* (n 32) at 128.
37 Restatement (First) of Torts § 291.
38 Restatement (Second) of Torts (1965) § 302 cmt a.
39 Ibid. § 302.
40 Ibid. § Topic 7 Title A and B.
41 Restatement (Third) of Torts: Physical and Emotional Harm (2010) § 7.
42 Ibid. §§ 7, 37.
43 Ibid §§ 38–44.
44 Ibid.
45 Jane Stapleton, *The Risk Architecture of the Restatement (Third) of Torts* (2009) 44 Wake Forest Law Review 1309.
46 Ibid. at 1310.

'risk creation' to decide which default rule applies. As such, the risk standard, like *Donoghue*'s neighbour principle, can't do much work on its own to resolve duty questions. As predicted, the Restatement's risk architecture is not easy for courts to deploy in the absence of fuller explication of the features of risk creation.

C. Tests employed by state supreme courts

The tests that state supreme courts routinely use to determine duty often employ Restatement of Torts principles and yet are far more varied than those standards. Alongside analysis of their own case law, state supreme courts cite a number of additional national authorities. The secondary authorities that courts rely on most heavily are the Torts Restatements. Although courts need not follow, or refer to, the Restatement of Torts, most of the state supreme court cases from the 2-and-a-half-year sample cite to either the Second Restatement, the Third Restatement or both. Specifically, of the 127 cases in the sample, 82 referenced the Restatement at least once. The Restatement Second was the most cited national authority with mentions in 68 of the state supreme courts' judicial opinions. The Restatement Third was the second most cited source, which courts mention in 27 cases. 23 judicial opinions reference both Restatements. In terms of treatises, 21 courts made note of Dean Prosser's hornbook,[47] 15 noted Professor Dobbs' work (with 13 of the citations to the Law of Torts).[48] Three courts mentioned the views expressed in *Harper, James and Grey*.[49]

In terms of standards themselves, there are many commonalities among state court decisions. 100 cases mention principle, fairness or morality. 61 of these cases use the term in the same sentence as duty. 87 cases mention policy (56 'public policy'). 64 cases mention policy and duty closely together. 92 discuss preventing or deterring harm, injury or loss. 6 cases remind readers of Dean Prosser's advice that duty is 'an expression of the sum total of those considerations of policy which lead the law to say that the particular plaintiff is entitled to protection'.[50] 53 cases refer to factors and duty within the same paragraph. 10 of these cases, almost all in Western states, cite California's specific *Rowland* factors, although not always by name.[51] 79 cases address foreseeability, with 8 advising that 'foreseeability alone' does not determine duty. 54 cases address risk creation, 18 mention 'affirmative

[47] *Prosser and Keeton on Torts* (n 15).
[48] Dobbs, Hayden and Bublick, *The Law of Torts* (n 30).
[49] *Harper, James and Grey on Torts* (3rd ed. 2009).
[50] *Brown v. USA Taekwondo*, 483 P.3d 159, 172 (Cal. 2021) (Cuellar concurrence).
[51] *Weir v. Expert Training, LLC*, 507 P.3d 442, 451 (Wyo. 2022), determining whether such a duty exists involves balancing the factors originally set forth in *Gates v. Richardson*, 719 P.2d 193, 196 (Wyo. 1986).

acts' and 9 misfeasance. 66 cases refer to omissions, 26 talk about 'affirmative duties' and 10 discuss nonfeasance.

From this shared terminology, commonalities in state supreme court language and standards are easy to see. Yet the cases also expose the limits of relying on Restatements as a shorthand for American common law. Courts use a large variety of tests—both general standards and specific standards for particular cases. The diversity of analyses is an important reminder that variation is a feature and not a bug of state common law.

One need look no further than the dizzying array of judicial policy factor tests to see the differences. For example, in Maryland, 'the threshold question of whether the alleged tortfeasor owed a common law duty of care in a particular case depends on whether: (1) a qualifying special relationship or affirmative undertaking existed or occurred under the circumstances at issue; (2) the harm at issue was of a type reasonably foreseeable under the circumstances; and (3) imposition of liability comports with public policy under the circumstances.[52] Connecticut, after an analysis of foreseeability of harm, considers the final step in the duty inquiry, 'to make a determination of the fundamental policy of the law, as to whether the defendant's responsibility should extend to such results'. To make that determination, the state's courts consider four factors: '(1) the normal expectations of the participants in the activity under review; (2) the public policy of encouraging participation in the activity, while weighing the safety of the participants; (3) the avoidance of increased litigation; and (4) the decisions of other jurisdictions.')[53]

Meanwhile, Wyoming adopts California's 8-part list of factors from *Rowland*. In that state, determining whether a duty exists involves balancing the following factors:

(1) the foreseeability of harm to the plaintiff,
(2) the closeness of the connection between the defendant's conduct and the injury suffered,
(3) the degree of certainty that the plaintiff suffered injury,
(4) the moral blame attached to the defendant's conduct,
(5) the policy of preventing future harm,
(6) the extent of the burden upon the defendant,
(7) the consequences to the community and the court system, and
(8) the availability, cost and prevalence of insurance for the risk involved.[54]

[52] *Maryland Cas. Co. v. Asbestos Claims Ct.*, 2020 MT 70, ¶ 28, 399 Mont. 279, 301, 460 P.3d 882, 895.
[53] *Raspberry Junction Holding, LLC v. Se. Connecticut Water Auth.*, 263 A.3d 796, 806 (Conn. 2021).
[54] *Weir v. Expert Training, LLC*, 507 P.3d 442, 451 (Wyo. 2022), determining whether such a duty exists involves balancing the factors originally set forth in *Gates v. Richardson*, 719 P.2d 193, 196 (Wyo. 1986).

D. Underlying structure

While there is no universal or essential test of duty that applies in all state courts, some core similarities in US duty standards stand out.

First, to establish duty, courts almost always require some form of risk creation, affirmative acts working positive harm (misfeasance), or on the nonfeasance side of the equation, an affirmative duty based on statute, special relationship, contract, undertaking or the like.

Second, despite the Restatement Third's advice that courts should avoid discussing foreseeability in the duty analysis, foreseeability remains a mainstay in duty analyses. The locus of the Third Restatement's concern—that judicial forays into foreseeability analysis usurp juries' rightful role in determining breach—is often addressed by a caveat that foreseeability in the duty sphere must be generalized rather than specific. Even so, the level of generality of foreseeability analysis on the misfeasance side is far broader than that on the nonfeasance, or affirmative duty, side.

Third, almost all state courts engage in some examination of public policy, either guided by specific policy-analysis factors, or in the form of a large-scale costs/benefit analysis.

Fourth, although duty, breach and actual cause are all separate elements, most duty analyses look at the moral blame attached to the actor's conduct and the level of certainty that the actor's conduct caused plaintiff's harm. Wrongful conduct that causes harm is a powerful motivator for courts to extend duty—it is the moral heart of negligence liability.[55]

Finally, when liability is seen as problematic—either because it is new or potentially difficult to confine, courts often create additional specific-factor lists, require heightened levels of certainty or increase judicial involvement in the determinations.

4. Duty of Care Factors in the United States

Whether substantive factors are the focus of judicial tests or not, they animate duty decisions. Identifying these factors is an elusive, if critical, project.

Many of the factors Professor Stapleton addressed in her article about the Commonwealth are very much at play in decisions in the United States today. For example, protecting physical security is a central concern of recent decisions in the United States—in fact, *the* central concern. For example, in the case of a snapchat

[55] *Ho-Rath v. Corning Inc.*, 275 A.3d 100, 107 (R.I. 2022) (weighing factors that include '(1) the foreseeability of harm to the plaintiff, (2) the degree of certainty that the plaintiff suffered an injury, [and] (3) the closeness of connection between the defendant's conduct and the injury suffered').

'speed filter' that records the photographer's speed of movement at the moment a photo is taken, the Georgia Supreme Court found that the product designer had a duty of care because it could reasonably foresee that the product's design created a risk of car accidents.[56] Relatedly, risk in 'the exclusive control of the defendant on whom the plaintiff was, therefore, "exclusively dependent"' is another pro-duty factor that resonates strongly with US decisions. In the case of seniors who relied on a home healthcare agency for their caretaker, the North Carolina Supreme Court found the agency had a duty to hire competent actors for these vulnerable plaintiffs, whose safety was dependent on their promised care in hiring.[57] Many duty claims regarding negligent risks towards workers reflect a measure of positional vulnerability of the employees dependent on employers as well.[58]

Factors that Professor Stapleton identified as countervailing to recognition of duty are very much a part of US decisions as well. US courts, like their Commonwealth counterparts, are hesitant to mandate affirmative acts. For example, the Wyoming Supreme Court declined to recognize a duty of care that would have required an actor to maintain or inspect an attic when no contract requires him to do so.[59] In addition, US courts are reluctant to use the tort system to aid a plaintiff who has adequate means to protect herself. The Rhode Island Supreme Court ruled against an adult trespasser at an electric substation for such a reason.[60] Furthermore, US courts share Commonwealth courts' sensitivity for imposing costs on public entities. As the Oklahoma Supreme Court wrote when a cemetery headstone fell and injured a child, 'to impose a continuous duty on Town to maintain every single headstone in the cemetery is a tremendous burden'.[61] Indeed, every one of the duty factors that Professor Stapleton identified from Commonwealth opinions resonate with recent US decisions as well.

Having identified the wisdom in Professor Stapleton's work, one might sensibly end the inquiry there. But the usefulness of Commonwealth factors to US analysis provides reason to believe that US decisions might also be of use abroad. One caveat is in order. Professor Stapleton marked certain duty factors extant in

[56] See, e.g., *Maynard v. Snapchat, Inc.*, 870 S.E.2d 739, 747 (Ga. 2022) (duty reinforced by fact that Snap knew that other drivers were using the Speed Filter while speeding at 100 miles per hour or more as part of 'a game', purposefully designed its products to encourage such behaviour, knew of at least one other instance in which a driver who was using Snapchat while speeding caused a car crash and warned users not to use the product while driving).

[57] See, e.g., *Keith*, 873 S.E.2d at 582 ¶ 69 (in-home health services company had a duty with respect to criminal acts by employee suspected of prior thefts where aide worked unsupervised in the home of elderly and vulnerable adults).

[58] See, e.g., *Boynton v. Kennecott Utah Copper, LLC*, 500 P.3d 847 (Utah 2021) (second-hand asbestos exposure at work); *Harwood v. Ardagh Grp.*, 522 P.3d 473 (Okla. 2022) (employee adequately alleged that employer had duty to provide safer crosswalk for access to parking lot).

[59] See, e.g., *Weir v. Expert Training, LLC*, 507 P.3d 442, 452 (Wyo. 2022) (maintenance staffing company that had no contractual obligation to maintain or inspect a shopping centre attic could not be held to a duty to do so).

[60] *Borgo v. Narragansett Elec. Co.*, 275 A.3d 567, 575 (R.I. 2022).

[61] *Kamphaus v. Town of Granite*, 510 P.3d 181, 186 (Okla. 2022).

Commonwealth law 'unconvincing' there.[62] If jurisdictional variation is to be anticipated and celebrated, it may be no surprise that factors thought unconvincing in one jurisdiction may well carry the day in another.

A. Economic loss

Although Stapleton marks as 'unconvincing' to the rejection of a duty, and the bald assertion that the loss involved is pure economic loss, many American courts find that explanation fairly persuasive. American tort liability is perhaps more bidirectional than its Commonwealth cousins in broadly recognizing a duty of care with respect to negligently caused physical harm and yet eschewing a duty with respect to negligently caused economic loss. In terms of economic loss, court rationales see economic losses as less important than physical injury, easier for plaintiffs to protect against or insure, more likely to undermine contractual agreements, harder to confine in terms of scope and better handled by a legislature. In all but one state supreme court case to address negligently caused stand-alone economic loss during the study period, courts declined to find a duty of care.

Many courts reviewing negligence cases that caused economic rather than physical loss mentioned the lesser importance of a duty of care when physical safety was not at issue. For example, in *Raspberry Junction Holding, LLC v. Southeastern Connecticut Water Auth.*, the Connecticut Supreme Court held that a municipal water authority owed no duty to a hotel operator to use care to avoid economic losses arising from an interruption in hotel water service for several days. This was in part because recognition of a duty 'fails to provide a corresponding increase in safety' given the nature of the service and the potency of existing regulation.[63] Similarly, when a subcontractor placed survey stakes in the wrong place, resulting in a building being built in the wrong place with resulting economic loss, the Arizona Supreme Court wrote, 'this case does not implicate "safety" or "health", and as included alongside those interests, "welfare" most reasonably connotes physical welfare, not economic welfare'. As such no duty pertained.[64] The Wyoming Supreme Court employed similar reasoning in a case in which a bailee-driver entered an off ramp on an interstate highway resulting in a head on collision. The Court held the bailee owed no duty to the bailor-car owner for the increased insurance costs following the bailee's fatal automobile accident. Said the court, since drivers already have a duty of ordinary care while driving, '[i]t is therefore unlikely that imposing an extended duty on those who borrow vehicles would have any measurable effect on preventing negligent driving'.[65] Indeed, as to the 'increased insurance costs, we

[62] Stapleton, *Duty of Care Factors* (n 9) at 92.
[63] 263 A.3d 796, 806 (Conn. 2021).
[64] *Cal-Am Properties Inc. v. Edais Eng'g Inc.*, 509 P.3d 386, 390 (Ariz. 2022).
[65] *Moses Inc. v. Moses*, 509 P.3d 345, 356 ¶ 40 (Wyo. 2022).

believe that [the car owner] is in the best position to prevent the harm since it has control over whether to loan its vehicles and to whom'.[66]

Sometimes the reason for denying a duty with respect to economic loss is a desire not to disrupt private ordering. For instance, the California Supreme Court ruled that a mortgage broker did not have a duty to process, review and respond carefully and completely to a borrower's loan modification application when the contract's terms did not require it.[67] Respecting the parties' bargain and the rights afforded to the lender under the contract made the court unwilling to assign a common law duty. However, where a state had assumed a fiduciary-type relationship, cultivating foreign investment and promising oversight of those investments, a special relationship was established such that a duty would attach.[68] Where parties have no relationship to act on behalf of another, as in the case of hospitals' suit against pharmacies for uncompensated care related to opioid abuse, the Arizona Supreme Court wrote '[c]ourts have not recognized a general duty to exercise reasonable care for the purely economic well-being of others'.[69] This court, and others, have also noted that legislation would be required to address the economic losses.[70]

B. Physical security

In the United States, harm-prevention rationales are particularly important. In contrast to the area of economic loss, in the area of physical injury, many courts are recognizing duty of care to a wider range of persons. In US jurisdictions, the duty of care arises in one of two broad categories—those involving a special relationship, or those involving strangers. An illustration of the first group would be the duty a common carrier owes to its passengers. An example of the second group would be the duty a driver owes to other drivers or to pedestrians.

In recent special relationship cases, results are mixed, but courts seem to be modestly increasing duties. For example, in one Oklahoma case an employer had a nearby parking lot for executives and a parking lot across a busy highway for other workers. The workers were instructed to park at the lot and cross the highway to get to work. The employer said it would make the intersection as safe as possible and put up lights on its lot to illuminate the crossing. When a worker was injured crossing the highway (and workers compensation did not apply because the worker had not yet 'come to work'), the court held that the employer had a duty of care to the employee. The duty was not based on the employer's status as a landowner—the

[66] Ibid. at 356 ¶ 41.
[67] *Sheen v. Wells Fargo Bank, N.A.*, 505 P.3d 625 (Cal. 2022).
[68] *Sutton v. Vermont Reg'l Ctr.*, 238 A.3d 608, 616 (Vt. 2020).
[69] *CVS Pharmacy, Inc. v. Bostwick, in & for County of Pima*, 494 P.3d 572, 580 (Ariz. 2021).
[70] Ibid. and see *Sheen* (n 67).

street was a public way. Instead, the court held that the employer had voluntarily assumed the duty. A vigorous dissent argued that the employers should have no duty to employees off of employer property and outside work hours.[71]

Although an 'unconvincing factor' in the United Kingdom, preventing or deterring physical injury is a strong pro-duty factor in the United States. Often courts look at which party can best protect against injuries. The area in which duties are most increasing is in the area of 'stranger' cases. Often the cases involve third-party physical harm in contexts in which a more remote actor is considered more capable of promoting safety. For example, a social worker may be best able to prevent an attack by a patient with a history of violence who is suffering from hallucinations.[72]

If we think of relationships on a sliding scale—from no duty at all, to duty owed to the world—US courts are not at either extreme. Rather, they are beginning to articulate a middle category, which I will call—recognizing a duty based on defendant's web of relationships. I borrow this term from the economic loss sphere in which courts sometimes say that contractors and subcontractors are in a web of contractual relationships. In the economic loss setting, the interdependence can be used to show that an economic loss plaintiff not in privity could have protected itself from losses through contract.[73]

On the physical injury side, when a defendant is in a web of relationships with the plaintiff, courts may say that the defendant is in a superior position to protect others. Harm prevention is a key rationale here. Broader harm prevention by defendants in webs of relationships may be typified by supply chain liability. But the cases provide dozens of other examples, for which a few will have to suffice here. In the second-hand asbestos exposure context, for example, a number of the recent state cases have recognized a company's duty of care, not only to its employees, but also to their household members sickened by asbestos.[74] The theory in such cases is that a jobsite can better prevent injury from second-hand asbestos exposure to workers' household members. Duty in asbestos products liability cases follows along a similar rationale. The New Jersey Supreme Court determined that defendants had a duty to provide warnings 'given the foreseeability that third parties would be the source of asbestos-containing replacement components'. Its reasoning: 'Warnings on defendants' products would have provided a reliable form of protection for the ultimate user.'[75]

[71] .*Harwood v. Ardagh Grp.,* 2022 OK 51, 522 P.3d 473, as corrected (19 September 2022).

[72] *Coleman v. Martinez*, 254 A.3d 632, 650 (N.J. 2021) (social worker did not treat patient-attacker delusions and inexplicably identified child protection worker as the source of information adverse to patient regaining custody of her children).

[73] *Indianapolis-Marion County Public Library v. Charlier Clark & Linard, P.C.*, 929 N.E.2d 722 (Ind. 2010).

[74] *Boynton v. Kennecott Utah Copper, LLC*, 2021 UT 67, 500 P.3d 847 (premises operator's duty to wife exposed to asbestos through husband's work clothes).

[75] *Whelan v. Armstrong Int'l Inc.*, 242 N.J. 311, 340, 231 A.3d 640, 657 (2020), as revised (4 June 2020).

Situations in which the defendant's relationship is with the person who caused the harm, rather than the plaintiff, can be considered special relationships with the defendant, or these sorts of webs of relation with the plaintiff. For example, the Louisiana Supreme Court held that a company that misidentified the cause of a work system problem which led to an explosion and deaths of workers, owed a duty not only to the employer company, but to its injured employees as well.[76] The Georgia Supreme Court held that snapchat had a duty to use reasonable care to prevent foreseeable risks and harms from a filter that clocked drivers' speed as a photo was taken.

In contexts involving crime prevention, duty with respect to third-party criminal conduct may be based on foreseeable crime, the defendant's control of factors that can reduce the risk of criminal harm and a duty based on relationships with either the plaintiff, the defendant or both. One example here is a home health services company that said it had reviewed employee criminal histories when it had not. The home health aide had a history of drug arrests, threats and dishonesty, but the healthcare services provider never checked to find out about it. When the aide had accomplices rob her elderly charges at gunpoint using information she had gleaned from her work, the court upheld a jury verdict against the home health company for damages stemming from the crimes—saying that a jury could have found them foreseeable.

One situation in which ability to reduce harm to workers has not dictated a duty is in the case of some independent contractors. The California Supreme Court has said that: 'A hirer's failure to correct an unsafe condition is insufficient, by itself, to establish liability for injuries suffered by an independent contractor; to be liable, a hirer must instead exercise its retained control over any part of the contracted-for work.'[77] Specific doctrines about retained control are important to no duty findings in this area.[78] In this case, a factor counter to a finding of duty may also be that the defendant was an individual rather than a corporation.

C. Equality

Another rationale that is gaining increased recognition in US courts is the equality of value of all persons' lives. In the US social context, equality means in part racial equality. Several recent US cases expand tort duties owed to residents by the police. One New York case, for example, held that New York city might have a special

[76] *Malta v. Herbert S. Hiller Corp.*, 2021-00209 (La. 10/10/21), 333 So. 3d 384 (fire inspection company duty to worker).

[77] *Gonzalez v. Mathis*, 12 Cal. 5th 29, 493 P.3d 212 (2021).

[78] *Webb v. Giant of Maryland*, LLC, 477 Md. 121, 266 A.3d 339 (2021) (grocery store did not retain sufficient control over Pepsi employee's work in its store to have a duty to store's customer injured by the employee's negligence).

duty to people in the home in which a no-knock warrant was being planned and executed.[79] In that case, an unarmed person who was sitting on the couch was shot, when the person police sought to locate was not home. However, police had not surveilled the home to be sure the suspect was there, before coming in.

There are also cases animated by issues of gender equity. In a Minnesota case in which the court asked, 'whether the specific danger was objectively reasonable to expect', sexual assault was foreseeable such that a duty was established.[80] Similarly, a programme that worked with aspiring Olympic athletes had a duty of care to female athletes with respect to sexual assaults by coaches.[81] Although some of the state courts still take a narrow view of the foreseeability of sexual assault. For example, a Massachusetts Supreme Court Case denying a university's duty with respect to a student's sexual assault by a classmate said: 'nothing in the record indicates that [the perpetrator] had a history of sexual assaults.'[82] Thus the duty requires foreseeability of injury by a specific assailant—a higher hurdle to meet.

But the equality of persons in recent tort cases is far broader than race and gender. Many states have chipped away at firefighter rules that had denied a duty of care to professional rescuers.[83] Courts have expanded duties to injured police officers, though it remains to be seen if third-party liability can be upheld in the context of protest given constitutional free speech concerns.[84] In one case that will test the question, the Louisiana Supreme Court said that a black lives matter protest organizer who told marchers to block a busy highway outside the police station in a nighttime 'days of rage' protest, could be liable when the marchers threw bottles and rocks, injuring police. Though not the subject of a final decision yet, tort suits have also been filed by injured police against those who directed the insurrection at the US Capitol.[85]

5. Conclusion

In terms of substance, the chapter of the US story most important for addition to factor lists are the importance of deterring physical, rather than economic, injury; growing contemporary norms of harm prevention in webs of relationship; and an expanded sense of equality of persons.

These factors are themselves driven by other legal and cultural factors. For example, most US states have adopted various forms of splitting, rather than

[79] *Ferreira v. City of Binghamton*, 38 N.Y.3d 298, 194 N.E.3d 239 (2022).
[80] *Abel v. Abbott Nw. Hosp.*, 947 N.W.2d 58, 78 (Minn. 2020).
[81] *Brown v. USA Taekwondo*, 483 P.3d 159, 172 (Cal. 2021).
[82] *Helfman v. Ne. Univ.*, 485 Mass. 308, 322, 149 N.E.3d 758, 772 (2020).
[83] *Ipsen v. Diamond Tree Experts, Inc.*, 2020 UT 30.
[84] *Doe v. McKesson*, 2021-00929 (La. 3/25/22), 339 So. 3d 524.
[85] Ellen M. Bublick and Jane R. Bambauer, *Tort Liability for Physical Harm to Police Arising from Protest: Common-Law Principles for a Politicized World* (2024) 73 DePaul Law Review 263.

all-or-nothing liability. Under comparative fault and comparative apportionment systems, parties can be assigned small percentages of responsibility for damages. Though unarticulated in judicial opinions, this factor may well be a more important driver of expanded duty than other visible pro-duty factors.

US judges and scholars also practice in a framework of instrumental thinking, much more so than in other parts of the Commonwealth. As such, rationales about the functions of liability—harm prevention, equality or other substantive goals—may be more expected.

At present, it is difficult to see these unseen factors. However, newer technologies such as artificial intelligence (AI) may yield a deeper analysis, particularly in a country such as the United States, in which decisions from so many states are a part of the analysis. Evaluating judges' reasoning with conventional tools can be illuminating. However, AI may show that patterns of what courts actually do may be more powerful predictors than what they say. And even what courts say may be more intelligible through AI than through old-fashioned human articulation.

Duty is an unruly, changing and contested area of US case law. It may well be that AI is the right actor to tell US duty factors' next story.

8
What is 'Reasonable Foreseeability'?

Roderick Bagshaw

1. Introduction

In October 1947, Cecil Henry Roe, a farm labourer, was left paralysed after an operation on his knee at the Chesterfield and North Derbyshire Royal Hospital. The anaesthetic nupercaine was supplied in glass ampoules; the outside of these ampoules was not sterile, so that year the hospital's anaesthetists had adopted a new method to address the risk of sepsis; they immersed them in a solution of phenol. The trial judge, McNair J, held that Roe's paralysis was caused by phenol contaminating the nupercaine before it was injected into his spine, apparently because it percolated through cracks in the glass ampoule, or molecular flaws, that could not be detected by the visual and tactile examination usually conducted in an operating theatre.[1] Roe's negligence claim failed: the Court of Appeal[2] upheld McNair J's conclusion that neither the Minister of Health, as the successor in title to the hospital's trustees, nor the anaesthetist, was liable: anaesthetists in England were only alerted to the risk of phenol contaminating anaesthetic through invisible cracks in ampoules by a book published in 1951;[3] so, in October 1947, the possibility that an apparently undamaged ampoule could contain enough phenol to paralyse a patient was not reasonably foreseeable.

In her book on product liability,[4] when addressing the thinking behind the inclusion of a 'development risk defence' in the final Product Liability Directive,[5] Professor Stapleton explains why it is not self-evident that a *producer* of a product that has proved to be dangerous should avoid liability for undiscoverable—including unforeseeable—risks: legitimate goals may be served by reallocating the cost of responding to the fruition of such risks, and the consequences and fairness of particular allocations can be debated; on the other hand, the cost of losses associated with undiscoverable risks *can* be left to lie where it falls, and, if so,

[1] *Roe v Minister of Health* [1954] 1 WLR 128.
[2] *Roe v Minister of Health* [1954] 2 QB 66 CA.
[3] Findings of McNair J ([1954] 1 WLR 128, 133), as quoted by the Court of Appeal: [1954] 2 QB 66, 77.
[4] Jane Stapleton, *Product Liability* (Butterworths, 1994).
[5] Council Directive 85/374/EEC of 25 July 1985 on the approximation of the laws, regulations and administrative provisions of the Member States concerning liability for defective products. The defence, in Article 7(e), was transposed into English law as Consumer Protection Act 1987 s 4(1)(e).

Roderick Bagshaw, *What is 'Reasonable Foreseeability'?* In: *Torts on Three Continents*. Edited by: Kylie Burns, Jodi Gardner, Jonathan Morgan and Sandy Steel, Oxford University Press. © Roderick Bagshaw 2024.
DOI: 10.1093/oso/9780198889748.003.0008

producers will not have to *plan* for absorbing this cost, by insurance or otherwise.[6] She concludes:

> How appropriately to make the trade-off between the values represented by efficiency or a moral enterprise liability argument and ... real-world concerns about R[esearch] & D[evelopment] etc, is necessarily an ideological one. It depends on the weight to be given to desiderata such as the protection of choice, the security of supply of certain products, minimum levels of physical security and the long-term protection of a free enterprise culture. These are not susceptible to further legal analysis.[7]

This assessment invites the question whether the decision to leave Roe uncompensated was also 'necessarily an ideological one'.

Perhaps our answer should be 'yes', even if most participants in tort practices would baulk at this description,[8] and even without evidence of courts *consciously* negotiating such a 'trade off' between desiderata; probably, the Court of Appeal's intellectual process in *Roe* involved *unconscious* application of a well-established condition of such 'fault' liability. But such liability ensures that some proportion of the unwanted costs of innovation is allocated to unlucky victims, rather than to those who *chose* to innovate. Moreover, since 1954, the significance of 'reasonable foreseeability' has extended further, without any detailed judicial defence of the underlying ideological project.[9] Thus, currently, outside exceptional situations, a defendant cannot be held liable for having committed the tort of negligence unless a reasonable person in their position would (could?) have foreseen that their act (or omission) was sufficiently likely to cause the relevant type of harm to the claimant, or, at least, to someone in a similar situation to the claimant.[10]

[6] Jane Stapleton, *Product Liability* (Butterworths, 1994) 225–229.

[7] Ibid 228. The ellipsis replaces 'these', which referred back to the previous two paragraphs.

[8] The Protean nature of the adjective 'ideological' may explain their reaction: Michael Freeden observes that 'ideology' is 'polysemic' and: 'Both scholars and non-scholars have invested in it not only purposive, reflective, and critical thought but strong emotions': Michael Freeden, *Ideologies and Political Theory* (OUP, 1996) 13. Linguistically, difficulties are compounded, see *OED* 3rd edition (online, March 2022), by the noun 'ideology' frequently being used neutrally ('A systematic scheme of ideas, usually relating to politics, economics, or society and forming the basis of action or policy') but the adjective being used to impart a judgement ('based on a principle or set of *unshakeable* beliefs' (emphasis added)).

[9] The principal extension has been through an increased role in delimiting the scope of liability for the consequences of torts, aka 'remoteness of damage': see, eg, *Overseas Tankship (UK) Ltd v Morts Dock & Engineering Co Ltd (The Wagon Mound)* [1961] AC 388 PC. A well-known defence of negligence law's ideological project was provided by Andrew Burrows, 'In Defence of Tort', ch 6 in Andrews Burrows, *Understanding the Law of Obligations* (Hart Publishing, 1998) *before* he joined the Supreme Court.

[10] In English law 'reasonable foreseeability' is a pre-condition for establishing a duty of care. As James Plunkett has pointed out, 'Foreseeability of harm to the claimant ... has formed an integral part of every general duty test since *Heaven v. Pender* (1883)': James Plunkett, *The Duty of Care in Negligence* (Hart Publishing, 2018) 2. Thus, the House of Lords ruled in *Bourhill v Young* [1943] AC 92 that the motorcyclist John Young was not liable for the injuries that he caused to Mrs Euphemia Bourhill because—after he had motored several metres past the tram from which she had recently descended—there was

Further, as *Roe* illustrates, whether a duty-owing defendant will be held to have taken sufficient precautions—or to have breached their duty—will depend, in part, on what sources and degrees of risk (to someone such as the claimant?) a reasonable person in the defendant's position ought to have foreseen. And, in negligence, and several other torts, the extent of a defendant's liability for consequential loss is limited in accordance with whether each of the types of damage resulting from the defendant's wrong was a reasonably foreseeable consequence of such a wrong. In summary, in English tort law the concept of reasonable foreseeability plays established roles in seeding duties, setting standards of care, and limiting liability for 'remote' types of damage.

Tort theorists generally associate the 'foreseeability' concept with a *moral* demand:

> ... there is little dispute that the foundational element of negligence – wrongdoing, breach of duty, falling below a community behavioural norm – necessarily rests first upon an actor's ability to contemplate, at least abstractly, the possibility that his or her contemplated behaviour may cause the type of hazard to the type of person that actually does result.[11]

There is more dispute, however, as to why the law asks about *reasonable* foreseeability, rather than about what each defendant actually foresaw, or *could* have foreseen; disputants have identified (at least) four possibilities, which may supplement each other: *objective* standards may be an essential element in ensuring that negligence law promotes equal external freedom;[12] it may be *practically advantageous* to employ an objective standard, because less time will need to be spent investigating any particular defendant's state of mind and characteristics; an objective standard may facilitate the task of *communicating to potential defendants* how they ought to behave (concrete applications of an objective standard are *more* generalizable, since they are expressed as applicable to similarly situated persons); and an objective standard may better reflect claimants' expectations (I am entitled

no reason for him to have had her in contemplation as a person who was sufficiently likely to be injured by his continued speeding.

[11] David Owen, 'Figuring Foreseeability' (2009) 44 *Wake Forest Law Review* 1277, 1291. It appears that the justification for using 'foreseeability' as a condition for liability is also part of the justification for its use as a *limit* on liability: eg *Overseas Tankship (UK) Ltd v Morts Dock & Engineering Co Ltd (The Wagon Mound)* [1961] AC 388, 425–426: 'It is irrelevant to the question whether B is liable for unforeseeable damage that he is liable for foreseeable damage, as irrelevant as would the fact that he had trespassed on Whiteacre be to the question whether he has trespassed on Blackacre.'

[12] Richard Wright, 'The Grounds and Extent of Legal Responsibility' (2003) 40 *San Diego Law Review* 1425, 1433: 'To have sufficient security, each individual's person and property must be secured not merely against harm caused by subjectively blameworthy conduct of others, but also against harm caused by objectively specifiable conduct of others that, if generally allowed to occur, would (contrary to the supreme principle of Right) foreseeably reduce, rather than enlarge, everyone's external freedom.'

to rely on whichever motorist is indicating to turn across my lane meeting the objective standards of driving competence,[13] and entitled to rely on whichever taxi driver or surgeon I am allocated to—as passenger or patient—meeting the same basic standard of competence).[14] But these four possibilities hint at a difference of opinion as to the appropriate *moral* demand: does morality require a subjective approach, with the law adopting an objective approximation for practical reasons, or does *morality itself* require an *objective* approach ('to promote everyone's equal external freedom')? Moreover, any elaboration of 'reasonable foreseeability' *could* represent a trade-off between goals that appeal to rival ideologies—say, a compromise between the allocation of accident costs recommended by 'fault' liability (*perhaps*, subjective) and the allocation supported by pursuit of some *other* goal (*perhaps*, objective to reduce spending on dispute resolution).

In this chapter, I do not purport to provide any superior explanation for *why* English tort law uses an *objective* standard rather than allowing defendants to invoke their idiosyncrasies. Instead, I intend to investigate some of the practical challenges raised by asking whether damage to a claimant—or a particular degree of damage—is 'reasonably foreseeable' when applying the tort of negligence. My expectation is that these challenges will illuminate the nature of the 'reasonable foreseeability' element; in particular, whether 'reasonable foreseeability' might be a *designed* concept, reflecting judicial choices.

Professor Stapleton has methodically interrogated many recurrent elements in tort law reasoning—such as 'proximity', 'floodgates', 'novus actus interveniens'—and exposed them as 'labels' or 'smokescreens' for under-analysed normative judgements. 'Reasonable foreseeability', however, appears to have avoided similar questioning: scholars regularly present it as simply 'factual'.[15] Is that presentation plausible? My contention is that this can be investigated by considering the *sort* of inquiry that courts undertake to assess if Euphemia Bourhill's wrenched back was a 'reasonably foreseeable' consequence of John Young's speeding,[16] or if Bessie Stone's head injury was a 'reasonably foreseeable' consequence of the Cheetham Cricket Club continuing to allow cricket to be played on their ground after the construction of Beckenham Road,[17] or if the burning of the *Corrimal* was a 'reasonably

[13] In English law the proposition that drivers are *entitled to expect* others to drive competently is false: a reasonable driver must take steps to anticipate the negligent mistakes that others will foreseeably make: *Lang v London Transport Executive* [1959] 1 WLR 1168, following dicta in *London Passenger Transport Board v Upson* [1949] AC 155.

[14] No doubt there is scope to debate cause and effect: does the objective standard *establish* claimant expectations or *reflect* them? Further, I can only 'rely' on other drivers *either* meeting the objective standard or becoming liable to compensate me for the injuries caused by not doing so.

[15] Donal Nolan, 'Deconstructing the Duty of Care' (2013) 129 *Law Quarterly Review* 559 consistently refers to the question 'whether damage to someone in the claimant's position was a reasonably foreseeable consequence of the defendant's faulty conduct' as 'the factual duty question' (eg at 560).

[16] *Bourhill v Young* [1943] AC 92.

[17] In *Bolton v Stone* [1951] AC 850, the claimant was injured whilst in Beckenham Road; the defendants were the committee and members of the Cheetham Cricket Club; in 1910 the club had given up

foreseeable' consequence of the *Wagon Mound's* engineers allowing bunkering oil to overflow into Sydney Harbour.[18] What sorts of evidence or submissions did the courts receive from the parties in support of their contention that the wrenched back, head injury, and burnt ship was, or was not, 'reasonably foreseeable'? My thesis is that the methods used cast doubt on the proposition that the foreseeability inquiry is 'factual'.[19] If this is correct, there could be consequences for *procedural* issues, such as whether the question of what was reasonably foreseeable can be dealt with on a strike-out or summary judgement application. But procedural issues are not my focus. If the inquiry is *not* factual, or *not exclusively* factual, then 'reasonable foreseeability' may be exposed as *another* 'smokescreen' for under-analysed normative judgements.

2. Foreseeability Fundamentals

A. A retrospective question about what would have been believed to be possible?

In tort claims, the relevant question for the courts is what was reasonably foreseeable at a past point in time, when the defendant allegedly committed a tort,[20] rather than what is *now*—at trial—reasonably foreseeable. Being a question about the past, it might seem to be about 'fact', since what events have occurred is part of the realm of fact. 'Adam foresaw that Brian might break his wrist' appears to be a proposition about Adam's state of mind at a point in the past with respect to a (then) future event, equivalent to 'Adam *believed* that Brian might break his wrist'.[21] But

some land at one end of its ground to facilitate the building of the road, in exchange for additional land at the other end, but had not moved the wicket (ie where the batsman stood).

[18] *Overseas Tankship (UK) Ltd v Miller Steamship Co Pty (The Wagon Mound (No 2))* [1967] 1 AC 617, PC. (The *Corrimal*, a 1,100-ton collier, had been undergoing a refit at the Shearlegs Wharf, Mort's Bay when the oil caught fire: Malcolm Park, 'Jean Who? (Or Good Lord, Your Place Has Gone up in Flames!)' (1991) 65 *Law Institute Journal* 1026. The ship's rusting hull is now a tourist attraction, 'the floating forest', in Homebush Bay, Sydney: https://www.wrecksite.eu/wreck.aspx?202970.)

[19] My primary goal is to consider whether the question is one of 'fact' in the sense of: about items that should be listed in an inventory of what there is in the world. (The phrase 'inventory of what there is' is from Kevin Mulligan and Fabrice Correia, 'Facts' in Edward N Zalta (ed), *The Stanford Encyclopedia of Philosophy* (Winter 2021 edition).) At this point, English tort lawyers' broader use of 'question of fact' to mean a 'question that will not usually to be assessed *de novo* by an appellate court' can be noted and set aside.

[20] Those who believe that the tort of negligence is not committed until a claimant has suffered actionable damage may prefer, eg, 'when the defendant allegedly engaged in the conduct that formed part of his or her commission of the tort'.

[21] 'Adam *knew* that Brian might break his wrist' might be accepted by many, but: (1) one can only *know* facts that are true; (2) it seems events that do not occur can have been reasonably foresee*able*; (3) the question whether it can be *true* that such events *might have occurred* requires consideration of how far events are pre-determined. To simplify matters, I will assume that 'Brian breaking his wrist' is a sufficient specification of the event that must have been foreseeable; in practice, the law may take an interest in the foreseeability of the *method* by which it was broken.

the proposition 'at time T it *was* reasonably foreseeable that Brian might break his wrist' is distinct from 'it *was* foreseen …' since a court can hold that the former is true even if nobody at time T had formulated any beliefs about the future state of Brian's wrist. For example, it seems reasonable to doubt whether any police officer supervising the crowd at Hillsborough on 15 April 1989 turned their mind to whether negligently allowing overcrowding in one part of the stadium could cause psychiatric injuries to relatives present in other parts of the stadium, or watching on television.[22] Similarly, when courts accept propositions such as it having been reasonably foreseeable that the defendant's negligent driving might cause a physical injury to a rescuer responding to an accident, this does not appear to be on the basis that any proportion of drivers are believed to have *actually contemplated* such scenarios; certainly no defendant in a reported case has thought it worthwhile to survey what proportion of drivers have, when thinking about the risks associated with careless driving, turned their minds to the dangers that might be faced by emergency services employees.[23] The proposition 'it was *reasonably* foreseeable (at time T) that Brian might break his wrist' is about the possibility of someone coming to believe in that possibility *if* they went through a particular mental process (at time T). To conclude: the essential question is not about a simple past fact; it is about what beliefs an artificial, constructed person (the *reasonable* person) would have held, given their understanding of the world, *if* they had asked themselves 'what might happen?'

This question is, however, one that will often turn on findings of fact. Simple facts existing at time T *can* be decisive: for example, proof of the actual beliefs of those who *did* ask themselves 'what might happen if … ?' Suppose Brian's friend asked a trampoline operator: 'Isn't it dangerous to let him on when he is obviously intoxicated?' Logically, proof that a possibility was foreseen seems to demonstrate the *possibility* of foresight.[24] (A rare exception—where this logic might fail—would be where the risk had only been identified through a 'lucky guess', or some method that a reasonable defendant could not have been expected to replicate (eg divine revelation).)

The primary method of negating evidence of *actual* foresight involves seeking to establish that the person who *actually* foresaw the event was someone it would be unfair to equate with 'a reasonable person in the defendant's shoes', perhaps because of their exceptional expertise. But even if evidence of actual identification of the risk by experts (such as regulators) is not decisive, it can *assist* a party

[22] See *Alcock v Chief Constable of South Yorkshire* [1992] 1 AC 310.

[23] It would be interesting to know if physical injuries to emergency personnel after road accidents are more frequent than psychiatric injuries to unrelated bystanders (such as was held unforeseeable in *Bourhill v Young* [1943] AC 92).

[24] In *Haley v London Electricity Board* [1965] AC 778 there was no direct evidence of other contractors identifying the risks to those challenged, but Lord Reid drew attention, at 792, to the Post Office's use of *taller* light fences for excavations, and the Electricity Board's ownership of such fences.

seeking to prove 'reasonable foreseeability': the sub-issue may become whether there is a plausible basis for claiming that such expert beliefs *ought* to have come to the attention of a reasonable defendant inclined to ask, 'what might happen?' Clearly here the issue of what was 'reasonably foreseeable' blends fact (experts *knew*...) with a normative point (so people like the defendant *should have known*). For example, in the mesothelioma case of *Abraham v G Ireson & Son (Properties) Ltd*,[25] Swift J assessed what, in the early 1960s, would have been reasonably foreseeable by defendants exposing a plumber to small amounts of asbestos intermittently, and constructed the answer by referring to the Annual Reports of the Chief Inspector of Factories and general guidance about asbestos published by the Home Office and the Ministry of Labour;[26] no evidence was presented of what proportion of building firms, if any, paid attention to such Reports; instead, the evidence of *expert* foresight was absorbed into a constructed account of what a reasonable employer in the general building trade *ought* to have learned if they had asked appropriate questions.

Proof of actual *non-prediction* may also be relevant. Thus, evidence that those similarly placed to the defendant, or the defendant themselves, sought to assess the risks and did not come to believe that the relevant event was possible, can be presented as tending to disprove reasonable foreseeability: for example, in the first *Wagon Mound* case it was treated as relevant that the claimant's works manager stopped welding at the Sheerlegs Wharf until after he had contacted the manager of the Caltex Oil Company, at whose wharf the *Wagon Mound* was berthed, and had been reassured.[27] Frequently, such non-foresight is inferred from the consistent failure of a class of potential defendants to address a risk: 'recognised and general practice at the material time is of relevance and importance ... [F]oreseeability of injury must not be judged with the benefit of hindsight, rather by reference to the standards of the time.'[28] Clearly, however, there is a gulf between proof that

[25] [2009] EWHC 1958 (QB). See also *Cuthbert v Taylor Woodrow Construction Holdings* [2022] EWHC 3036 (KB) and *White v Secretary of State for Health and Social Care* [2022] EWHC 3082 (KB).

[26] The judge, Swift J, left out of account official letters that had been directed to those in sectors with high asbestos exposure, such as shipbuilding, and points only found in expert journals.

[27] At the initial trial in the *Wagon Mound* litigation, (1958) 61 State Reports (NSW) 688 both parties introduced evidence showing that it was widely believed in 1951 (the time of the spillage) that furnace oil could not be ignited on open water. The claimant was seeking to support its claim that the likely cause of the fire was a contemporaneous spillage of petrol from the same ship, whilst the defendant's expert testified (at 697) that he had believed furnace oil not to be a fire hazard before conducting 300 experiments in preparation for the litigation. (These led to his conclusion that a sufficiently thick layer of furnace oil could be ignited on seawater if there was a 'wick', such as an oily rag.) In *Overseas Tankship (UK) Ltd v Miller Steamship Co Pty (The Wagon Mound (No 2))* [1967] 1 AC 617, PC, 640–641, Lord Reid suggested that the claimant did not seek to prove that it was reasonably foreseeable that such oil would ignite because such a finding might have led to a conclusion that there was contributory negligence (then a complete defence in New South Wales) in re-starting the welding.

[28] *Cuthbert v Taylor Woodrow Construction Holdings* [2022] EWHC 3036 (KB) at [75]. The judge cited *Baker v Quantum Clothing Group Ltd* [2011] UKSC 17, [2011] 1 WLR 1003 as authority for the proposition that foreseeability must not be judged with the benefit of hindsight. An alternative route to proving non-foresight might be to prove that in the immediate aftermath the defendant could not understand how the events had occurred: see, eg, *Doughty v Turner* [1964] 1 QB 518, where evidence

an event was not foreseen and proving that it was not foresee*able*, and hindsight may expose that the lack of recognition of a risk flowed from a failure to identify appropriate questions or to employ suitable methodologies. Here again, however, *normative* judgements are superimposed on the factual matrix.

More must be said about the nature of the relevant normative judgements. Although what is reasonably foreseeable is a question about what beliefs a *reasonable* person would have reached *if* they asked themselves 'that might happen?', and this is sometimes expressed as a question about what a defendant *ought to have foreseen*,[29] it is important to distinguish, and deny, the notion that English law believes that real people in the defendant's position (always) *ought to* have asked themselves 'what might happen?'. At least, it must be doubted that courts' use of this normative language is intended to promulgate that *real defendants* in dynamic situations ought to constantly recalculate who their acts might be capable of affecting so as to *predict* the sorts of accidents they might cause and who could be injured; such a diversion of mental capacity would frequently amplify dangers. For example, encouraging drivers to catalogue mentally all the possible road accident scenarios whilst driving would almost certainly *increase* the number of road casualties.[30] Thus, when Bankes LJ stated, in *Hambrook v Stokes Brothers*,[31] that 'upon the authorities as they stand, the defendant *ought to have anticipated* that if his lorry ran away down this narrow street, it might terrify some woman to such an extent, through fear of some immediate bodily injury to herself, that she would receive such a mental shock as would injure her health', it seems unlikely that he believed that those parking motor lorries on steep streets in Folkestone should have really been *thinking about* the prospect of terrifying pedestrians whilst doing so. Such an understanding of 'ought to have anticipated'—equivalent to: *should have actually had in mind*—would create an extraordinary gulf between the behaviour of the *most* cautious and responsible drivers and the *judicial prototype* of a reasonable driver. Indeed, it is unclear that anyone—judges included—would think it appropriate to *criticize* a lorry driver for a *moral* failing if they admitted that they had never thought about the prospect of terrifying nearby pedestrians when

pointing away from it being reasonably foreseeable that an eruption could be caused by immersion of a sindanyo (asbestos and cement) cover in a cauldron of molten sodium cyanide included the workforce not moving away when the cover was knocked into the cauldron and ICI Ltd having to conduct experiments after the accident to determine the chemical mechanism (at 519–520).

[29] This language appears in Lord Atkin's reasoning in *Donoghue v Stevenson* [1932] AC 562, 580 (emphasis added): 'You must take reasonable care to avoid acts or omissions which you can *reasonably foresee* would be likely to injure your neighbour. Who, then, in law is my neighbour? The answer seems to be — persons who are so closely and directly affected by my act that I *ought reasonably to have them in contemplation* as being so affected when I am directing my mind to the acts or omissions which are called in question.'

[30] The effects on driving performance of concurrent tasks such as selecting music or talking on a telephone are generally known; the relative roles of variables such as limited short-term memory, constrained processing capacity, and attention control are controversial topics in psychological research.

[31] [1925] 1 KB 141, CA, 151 (emphasis added).

parking. Anyone tempted to disagree must consider how a judge ought to respond to a defendant's offer to prove that the realities of human cognitive capacity are such that real human beings *could not* contemplate such risks when deciding how to act (particularly in any context where a decision had to be made swiftly)—because there would be *more* risks than the human brain can simultaneously use in computation, or, perhaps, because risks of a particular type (eg terrorizing pedestrians) would frequently be 'masked' by the prominence of more conspicuous dangers (eg damaging vehicles or buildings), or even because almost all *real* human beings do not undertake exhaustive audits of possibilities but employ heuristics when decision-making. The absence of references in either judgments or practitioners' guides to the usefulness of evidence about the realities of human cognitive capacity and decision-making practices is, I contend, significant.

This said, in less dynamic situations, it may be more plausible to claim that judges believe reasonable defendants ought to have actually identified (some of?) the relevant risks, perhaps through some sort of 'risk assessment' process. However, even in those contexts where risk assessments are mandatory courts have emphasized that more risks can be held to be reasonably foreseeable than those actually identified by a competent risk assessment exercise.[32] For example, even where a competent workplace risk assessment identifies a risk of physical injuries to employees, it is very unlikely also to identify the (supposedly foreseeable) risks to the physical safety of emergency responders and the psychiatric well-being of close relatives who witness the 'immediate aftermath': the judicially constructed, normative account of what is reasonably foreseeable does not simply replicate competent professional risk assessment.

B. The types of knowledge that make a future event foreseeable

The previous section contended that when a court determines what beliefs a reasonable person would have reached if they asked themselves 'what might happen?' (at the relevant time) it is not, usually, making a straightforward finding about a past event (fact). What has actually been foreseen, or actually not foreseen, can be relevant; but often *no one* will have asked themselves the essential question at the relevant time, or, indeed, at any previous point in time. For example, before John Young sped his motorcycle across the junction on 11 October 1938 it seems unlikely that anyone had asked themselves when, if ever, the *noise* of a road traffic accident can cause injuries, let alone whether it could injure a person behind a tram eighteen metres away.[33]

[32] *Uren v Corporate Leisure (UK) Ltd* [2011] EWCA Civ 66, [2011] ICR D11 at [45].
[33] *Bourhill v Young* [1943] AC 92. Lord Thankerton (at 99) also suggested that a reasonable motorcyclist would not have contemplated the motorcycle ricocheting and hitting the claimant, or broken glass reaching her position.

Setting aside evidence of what was actually foreseen, or not foreseen, the most important *facts* when assessing what a reasonable person ought to have foreseen are that person's 'knowledge'.[34] But what types of knowledge might be relevant? Professor Cane proposes:

> Foresight is a function of knowledge, and the relevant questions are: at the time when it is alleged that the defendant ought to have taken precautions against the risk, did anyone have the knowledge required to foresee it? If so, ought the defendant to have had that knowledge? The first question is an empirical one, but the second question can be answered only by making a value-judgement.[35]

This highlights an important division between cases where a defendant seeks to establish that a risk was not reasonably foreseeable because of *general* ignorance of facts at the time of the alleged tortious conduct, and those where, instead, a defendant submits that—regardless of what others knew—there was no reason why they *personally* ought to have known about the necessary facts. My analysis will track this distinction; the second question—the distribution issue—is addressed in a subsequent section;[36] here, I want to sub-divide the types of knowledge that are necessary before an outcome will become foreseeable; to categorize what sorts of things a person must know before it will become *possible* for them to foresee that their proposed conduct could cause damage of a relevant type to a person such as the claimant.

Consider an unfortunate dog-sitter:

> Don agrees to stay in Clem's house and look after her dog in the week before Christmas; Clem instructs him to prevent the dog getting into the dining room. Nonetheless, one night, Don leaves the dining room door open; the dog enters, eats chocolate decorations from a Christmas tree, and when, early the next morning, it has begun to exhibit the symptoms of serious methylxanthine poisoning, Clem returns to collect some forgotten item and witnesses its final hour. Clem subsequently claims against Don for both the value of her dog and a recognised psychiatric injury developed as a result of witnessing its demise.

This hypothetical is designed to illuminate the types of facts Don—or any notional reasonable dog-sitter—would have to know before being able to foresee that such injuries might be caused by his failure to shut the door: specifically, (i) that

[34] See fn 21.

[35] Peter Cane, *The Anatomy of Tort Law* (Hart Publishing, 1997) 39. In the original a footnote is attached to 'anyone': 'But this is a vague term. It probably does not mean "anyone" in the world, but something like "anyone situated similarly to the defendant" (whatever "similarly" is taken to mean).' This imperils the claim that only the second question requires a value judgement.

[36] Section 4, below.

chocolate contains chemicals which can poison dogs, and (ii) that the dining room might contain dog-accessible chocolate, additionally—for the psychiatric injury claim—that (iii) seeing their dog dying from chocolate poisoning can cause a recognised psychiatric injury to a dog owner, and (iv) if he allowed her dog to be poisoned that night, Clem might witness its death. To generalize, the relevant knowledge can be divided into that relating to: (i) causal rules (eg chocolate can poison dogs; seeing a dog die can cause psychiatric injury to its owner) and (ii) the possible existence of conditions for the applicability of those causal rules in the situation concerned (eg Clem's dining room might contain chocolate; Clem might come into a position to perceive the death of her dog if it was poisoned).

Considering causal rules, employing a simplified model of scientific method, there is a difficult question about the point—between the first formulation of the hypothesis that x *might* cause y and confirmation of that hypothesis through reliable experimentation—at which the law should conclude that it became 'reasonably foreseeable' that x might cause y.[37] In less scientific terms: is a person who testifies that 'I vaguely remember having heard something about not giving chocolate to dogs' someone capable of foreseeing that chocolate can kill a dog?

Professor Stapleton has identified the parallel problems that arise when determining whether a 'defect' was 'discoverable' at a particular moment when applying the statutory product liability regime.[38] Importantly: it is frequently impossible to say that the fact that a type of product[39] causes a particular form of harm was *absolutely* undiscoverable at any time *before* this was discovered, because barriers to previous discovery, such as that no one had formulated a hypothesis, or no one had thought it *worthwhile* to collect the necessary data or to investigate a particular hypothesis experimentally, frequently reflect human choices rather than *physical* impossibility;[40] *different* choices will always have been *possible*.[41] Stapleton illustrates these issues by discussing the drug thalidomide: at its launch 'no-one had yet suggested testing pregnancy drugs on pregnant animals from the species which would have manifested birth defects', perhaps because it was widely thought that any drug that was non-toxic for a mother would also be non-toxic for her foetus;[42]

[37] Clearly real cases might involve complex intermediate situations: for example, where experiments have produced inconsistent and inconclusive results, and multiple, incompatible, hypotheses have gained a degree of support.

[38] Jane Stapleton, *Product Liability* (Butterworths, 1994) 236–242.

[39] I have used the phrase 'type of product' to avoid confusion with the *different question* whether it was possible for a producer to know exactly which product (within a batch) had the capacity to cause harm.

[40] Even where a barrier appears to involve physical impossibility, such as no experimental test for 'x' being available at the time, closer investigation usually implicates a human choice (such as, no one believed it to be *worthwhile* to develop such a test). In practice, *past choices* frequently create situations where it would have been impossible to discover a problem with the product *immediately* at the relevant time; but the *temporal issue* is a matter of degree—almost all *discoverable* (but not yet discovered) defects would have taken *some* time to discover.

[41] With apologies to any determinist reader; though if you are right, my mistake was unavoidable.

[42] Jane Stapleton, *Product Liability* (Butterworths, 1994) 241. It is unclear whether thalidomide was tested on *any* pregnant animals before being marketed: compare Jack Botting, 'The History of

but experimentation after concerns were triggered by a sudden increase in prevalence of human birth defects swiftly provided support for the drug's teratogenic character.[43] Standing back, the obvious conclusion is that the point in a scientific process at which causal rules become sufficiently accepted to make a possibility 'reasonably foreseeable' is highly contestable, and—perhaps for that reason—has never been *fixed* by judicial authority in England and Wales.[44]

Turning to beliefs as to the possible existence of the *conditions* for the application of a particular causal rule, here, in the dog-sitting hypothetical, foreseeability of Clem's psychiatric injury depends on whether someone such as Don would believe that Clem's dining room *might* contain dog-accessible chocolate and that Clem *might* return home (unexpectedly) during the time window in which a chocolate-poisoned dog might die. A recurrent difficulty in assessing what a reasonable person could be expected to *foresee* in such cases is that the question can be about the beliefs a person might reach without *any* reliable *direct* evidence. To illustrate: if we asked Don, as a reasonable person, to estimate the likelihood that Clem might not go through with her plan to be away, and instead return home to collect a forgotten item, what evidence would he have? No doubt people—that is, people other than Clem—have sometimes set off on trips then returned home having forgotten an important item, but what basis would *Don* have for assessing the likelihood of such circumstances in *Clem's* case?[45]

Thalidomide' (2002) 15 *Drug News and Perspectives* 604, 605 ('no') and Ray Greek, Niall Shanks, and Mark J Rice, 'The History and Implications of Testing Thalidomide on Animals' (2011) *Journal of Philosophy, Science and Law* 1, 5–6 ('probably, yes'). A critical review concluded that subsequent experiments produced effects similar to those in human beings in rabbits, monkeys, and apes, but not mice, rats, cats, dogs, armadillos, ferrets, or pigs: James L Schardein, Kit A Keller, and Bernard A Schwetz, 'Potential Human Developmental Toxicants and The Role of Animal Testing in Their Identification and Characterization' (1989) 19 *CRC Critical Reviews in Toxicology* 251, 276 (Table 21). In the United States, it was evidence about the incidence of a side effect *for mothers* (peripheral neuritis) that played a decisive role in preventing thalidomide's licensing: Frances O Kelsey, 'Thalidomide Update: Regulatory Aspects' (1988) 38 *Teratology* 221, 221. (Frances Kelsey was the medical officer in the Food and Drug Administration assigned the task of reviewing the application in 1960.)

[43] The question whether the effects of thalidomide were foreseeable remains contentious. For example, Professor Widukind Lenz, who played a key part in raising the alarm in Germany (see Anthony H Lipson (1992) 46 *Teratology* 411, 411: 'Lenz's contribution was monumental, even heroic.'), believed that problems could have been anticipated, because medical literature had identified a shortfall in investigation of the possible teratogenic effects of many widely used chemicals: Widukind Lenz, 'A Short History of Thalidomide Embryopathy' (1988) 38 *Teratology* 203, 203.

[44] Judges have expressed relevant opinions: for example, in *Independent Broadcasting Authority v BICC Construction Ltd* (1980) [1955–1995] PNLR 179, HL, the defendants were held negligent in designing a cylindrical television mast that collapsed under the *combined* force of 'vortex shedding' (a known problem) and asymmetrical icing on its stays (wrongly assumed not to be a problem *because* ice falls off the stays of *lattice* masts below the wind speeds likely to produce problems for *lattice* masts), and Lord Fraser (at 200) explained that this conclusion was justifiable because the defendants' 'error arose not from difficulty of calculation but from the omission of what seems to me to have been a simple piece of reasoning about known facts'. (The necessary 'simple' reasoning: 'vortex shedding' problems for a *cylindrical* mast would arise at a lower wind speed than for a *lattice* mast, and that wind speed might be insufficient to dislodge ice, thus meaning that the stresses could combine.)

[45] Even if Don had evidence of Clem forgetting items in the past, it is not easy to see what basis he would have for elevating this to a constant element in Clem's character.

From a defendant's point of view, their best prospect of establishing that an outcome was *not* foreseeable might involve showing that it was reasonable for them (and other similarly situated people) to believe that it was *impossible* for the particular factual conditions to exist in the situation concerned. An example of a case turning on belief that the conditions for application of a causal rule were impossible might be *Cambridge Water Co v Eastern Counties Leather plc.*[46] So far as this case turned on a belief as to the conditions: one of the reasons why the contamination of the claimants' borehole could be held to be 'not reasonably foreseeable' was that the defendants' spillages of perchloroethene took place in circumstances where the defendants believed that it would evaporate.[47]

Can a defendant establish that an outcome was not reasonably foreseeable because of (their reasonable belief in) the extreme unlikelihood (rather than impossibility) of the necessary conditions existing for the application of a known causal rule? Lord Reid's claim that 'If I buy a ticket in a lottery or enter a football pool it is foreseeable that I may win a very large prize—some competitor must win it'[48] tends to suggest that *only* a belief that it was impossible for the necessary conditions to exist will be sufficient to prevent a defendant who was aware of the applicable causal rules from foreseeing an outcome. Other cases, however, have allowed courts to conclude that outcomes were 'not reasonably foreseeable' even when the defendant did not seek to prove that reasonable people believed the necessary conditions *impossible*; instead their submission was that they would have believed them to be sufficiently unlikely to ignore for practical purposes: thus, in *The Wagon Mound (No 2)*,[49] Lord Reid bracketed situations that a reasonable person would think were impossible with those 'so fantastic or farfetched that no reasonable man would have paid any attention to [the possibility]', and moved on to address the situation of risks so small that a reasonable person could disregard them *if* he or she had a good reason to do so. Importantly, he treated the risk in *Bolton v Stone*[50] as falling into the latter situation, rather than being 'fantastic or farfetched', despite his quantification of the *ex ante* likelihood of a ball hit from the cricket ground striking a person in the street as 'once in so many thousand years'.[51]

46 [1994] 2 AC 264.

47 [1994] 2 AC 264, 292. The contamination was *also* unforeseeable because relevant causal rules were unknown; for example, there were gaps in knowledge with respect to the 'behaviour of groundwater in a fissure-flow, anisotropic (ie where permeability is higher in one direction rather than constant in all directions) chalk aquifer': see Lord Goff at [1994] 2 AC 264, 293.

48 *Dorset Yacht Co Ltd v Home Office* [1970] AC 1004, 1028. Lord Reid's proposition pre-dates the current UK National Lottery 'Lotto' draw, which does not guarantee that someone will win the jackpot each week.

49 *Overseas Tankship (UK) Ltd v Miller Steamship Co Pty (The Wagon Mound (No 2))* [1967] 1 AC 617, PC, 641–642 and 643–644.

50 *Bolton v Stone* [1951] AC 850.

51 *Overseas Tankship (UK) Ltd v Miller Steamship Co Pty (The Wagon Mound (No 2))* [1967] 1 AC 617, PC, 642.

This account presents three situations where a defendant can plausibly contend that the relevant matter was not reasonably foreseeable: where they reasonably believed that the conditions for the outcome were impossible, where they reasonably assessed the prospect of the conditions existing as 'so fantastic or farfetched that no reasonable man would have paid any attention to [the possibility]', and where they reasonably believed the prospect to be sufficiently unlikely for it to be reasonable to ignore it in light of their reason for doing so. Judges, however, rarely distinguish between these situations, despite the second and third requiring normative judgements different from the first. For example, it is unclear how we should classify several cases where defendants have convinced courts that it was not reasonably foreseeable that their driving could injure someone such as the claimant *because of the claimants' unpredictable behaviour*: thus, in *Scott v Gavigan*,[52] a defendant motorcyclist observed a pedestrian walk past a crossing point, and concluded that he did not want to cross, just before he ran into the road and directly towards the motorcycle; the Court of Appeal upheld a trial judge's conclusion that risk of physical injury to the claimant pedestrian was not reasonably foreseeable, so the defendant avoided liability despite having driven faster than he ought to have done in the vicinity of such a crossing. Was this a conclusion that reasonable drivers could regard the pedestrian's behaviour as impossible, farfetched, or merely insufficiently likely to justify any delay?

Three concluding observations: First, the point at which causal rules become sufficiently 'known' ('discovered'?) to make outcomes foreseeable is contestable; but the tort of negligence seems capable of operating in practice without this matter being settled, perhaps because relatively few cases raise matters at the frontiers of scientific discovery. Secondly, the tipping-point at which a defendant can no longer convince a court that their ignorance of the conditions for application of a (known) causal rule made the outcome unforeseeable is beyond what was reasonably believed to be impossible and fantastically far-fetched, and also beyond what was sufficiently unlikely to (reasonably) ignore given the reasons for doing so; in practice, it appears that defendants who are 'repeat players', such as cricket clubs allowing *many* batsmen to hit *many* balls, may find it more difficult to establish that they reasonably believed that conditions would *never* arise, than defendants involved in 'one off' events, such as a motorist dealing with the *extraordinary* behaviour of a *specific* pedestrian. Thirdly, this discussion of relevant types

[52] [2016] EWCA Civ 544 at [12]–[18]. See also *Little v Glen* [2013] CSOH 153, where the first defender's taxi collided with a pedestrian at night, but he contended—successfully—that it was not reasonably foreseeable that anyone would cross in front of him at night on *that road*, since it was unlit, had no pedestrian pavement, and his headlights would have been visible from a significant distance: at [138]. In *Wright v Lodge* [1993] RTR 123, Parker LJ doubted whether a motorist whose vehicle was stationary in the nearside lane on the eastbound side of a dual carriageway would owe a duty to motorists on the westbound side, on the basis that such a vehicle would not foreseeably endanger those on the westbound side (at 126–127), but the Court upheld the decision that she was not liable on the basis of her failure to move her vehicle to the hard shoulder not having caused the claimants' injuries.

of knowledge suggests that the significant boundaries are influenced by normative judgements, sometimes overtly, as when courts evaluate whether a possibility was sufficiently unlikely that it could be ignored *for good reasons*; but there are few signs of transparent ideological debate: for example, the decisions that allocated the accident costs caused by unpredictable pedestrian behaviour to an (uninsured) pedestrian rather than an (insured) motorist proceeded without express discussion of what social goals might be advanced (or impeded) by such allocations. Most probably the power of 'intuitive' impressions about 'fault' has sidelined the factors that might otherwise feature in transparent rationalization of these regular normative judgements; if these 'intuitive' impressions were consistent between cases then concerns about non-transparency might abate, but the analysis necessary to demonstrate this has not been undertaken, perhaps because of the resilient misconception that reasonable foreseeability is a 'question of fact'. The next sub-section provides an example of the difficulties confronting any assertion that 'consistency can be assumed'.

C. Imagination

Many tort claims involve unique combinations of circumstances: and foresight of the possibility—*before* the events occurred—would require *more* than *knowledge* of causal rules and possible conditions: a *reasonable* defendant would also have to be able to identify *which* causal rules, from amongst thousands of candidates, to use in predicting what might happen. To illustrate, in *Donachie v Chief Constable of the Greater Manchester Police*,[53] the allegation was that the defendant had negligently endangered the claimant police officer by supplying vehicle-tagging devices with batteries that frequently failed. The basis for the submission that unreliable batteries would foreseeably endanger police officers was the risk of them being physically attacked whilst *fitting* the devices onto suspects' vehicles and amplification of this risk if re-fittings became *more frequent* to replace defective batteries.[54] The point here is that an after-the-event assessment of the degree of likelihood a person would have attached to a risk before-the-event cannot tell us whether a *real* person would have actually been able to *imagine* the events. Bluntly, there is no reason to believe that a *real* person asked to list every sequence of events that could happen if a police employer supplied vehicle-tagging devices with defective batteries would succeed in imagining every possible narrative short of the 'fantastic or farfetched'.

Abouzaid v Mothercare (UK) Ltd,[55] may provide an example of a case of failure to imagine the scenarios which might arise rather than ignorance of basic causal

[53] [2004] EWCA Civ 405.
[54] This was the basis on which Donachie's appeal succeeded.
[55] [2000] EWCA Civ 348.

rules or belief that the conditions for their operation were impossible or farfetched. The defendants produced a child-sized sleeping bag, which they intended to be attached to a child's pushchair by joining two buckled elastic straps behind the pushchair's seat, and submitted that they could not have reasonably foreseen that a metal buckle on one of the straps might recoil so as to injure the eye of a person trying to join the straps from in front of the pushchair: that submission effectively reports that no one involved in assessing the safety of consumer products had appreciated that the (well-understood) physics of elastic was relevant to the assessment of what dangers *might* face users of a sleeping bag attached with such straps.[56]

So far as a pattern can be detected, it seems that courts are more likely to hold that an outcome was *not* reasonably foreseeable where the accident prompts research that exposes a previously unknown causal rule than where the accident reveals a general failure to draw an 'obvious' conclusion.[57] What is unclear, however, is *why* judges would choose to allocate the cost of accidents associated with the first sort of failure to claimants and the cost of accidents associated with the second, a lack of imagination, to defendants. Perhaps 'there was no data from which a causal hypothesis could have been identified' seems more *rational* as an explanation for non-prediction than 'nobody had put two and two together', but it would be *bizarre* to believe that scientific discovery held up by a shortfall in imagination has usually been delayed by the *culpable lassitude* of scientists.

3. Specific Challenges

So far, I have claimed that the 'reasonable foreseeability' question asks whether a 'reasonable defendant' would have found it *possible to forecast* the relevant outcome at the relevant past point, if he or she had asked the appropriate questions, and that courts' answers take into account knowledge of causal rules and beliefs

[56] In *Abouzaid*, the trial judge imposed liability under the Consumer Protection Act 1987 on the basis that there was 'an obvious danger' (quoted by the Court of Appeal at [15]) but held that the producers had not been negligent. In the Court of Appeal, Wright J (at [52]) insisted that this conclusion on negligence was correct because 'in 1990 the occurrence of such an accident as this would not have been within the reasonable contemplation of the defendant in the light of his then state of knowledge'. Pill LJ, however, *appears* to have upheld the conclusion on negligence on the more convincing basis that a reasonable producer would not have assessed the *magnitude* of the risk as sufficient to require modification of the design (at [33]).

[57] Unique situations where the accident stimulated research include *Doughty v Turner Manufacturing Co Ltd* [1964] 1 QB 518 CA and *Cambridge Water Co v Eastern Counties Leather plc* [1994] 2 AC 264. Situations where defendants failed to 'put two and two together' and were held liable include *Independent Broadcasting Authority v BICC Construction Ltd* (1980) [1955–1995] PNLR 179, HL, and, arguably, *Dorset Yacht Co Ltd v Home Office* [1970] AC 1004 (Lord Reid asserted, at 1026, 'The three officers knew or ought to have known that these trainees would probably try to escape during the night, would take some vessel to make good their escape and would probably cause damage to it or some other vessel', but it seems highly unlikely that they *knew* any such thing: no evidence of previous attempts at nocturnal navigation by trainees.)

about the likelihood of conditions existing, but insist that the imagination needed to put these together is universal, thereby ignoring the cognitive limitations of real human beings. Under this sub-heading I discuss some further challenging scenarios courts can face when attempting to construct what a reasonable defendant could have foreseen.

A. Unpredictable elements

Assessing what a reasonable defendant could have foreseen becomes more difficult where the potentially applicable causal rules include elements that behave unpredictably: a good example is provided by *Jolley v Sutton London Borough Council*,[58] where the question was what injurious consequences a defendant occupier of land could reasonably foresee as possibilities if it chose to leave an abandoned boat on grassed amenity land where children were likely to play: the Court of Appeal thought a reasonable defendant would only anticipate children climbing on the boat and having its rotten planking collapse under them; but the House of Lords upheld the trial judge's opinion that a reasonable defendant ought to have anticipated a far wider range of forms of children's 'meddling with' the boat and endangering themselves. For the trial judge, the breadth of the range of things that *might happen* reflected the range of extraordinary ways in which children might—predictably?—choose to play.

But judicial treatment of causal systems that include unpredictable elements—usually human action or animal behaviour—is not consistent. In contrast to *Jolley*, courts *sometimes* respond to the unpredictability of human conduct by demanding foreseeability of an *increased* level of likelihood before imposing liability. For example:

> where human action forms one of the links between the original wrongdoing of the defendant and the loss suffered by the plaintiff, that action must at least have been something very likely to happen if it is not to be regarded as novus actus interveniens breaking the chain of causation. I do not think that a mere foreseeable possibility is or should be sufficient...[59]

Can we construct a plausible explanation for (sometimes[60]) leaving the cost of *rare* behaviour by unpredictable elements with claimants, whilst demanding that

[58] [2000] 1 WLR 1082, HL(E).

[59] *Dorset Yacht Co Ltd v Home Office* [1970] AC 1004, 1040 *per* Lord Reid. Lord Reid explains his opinion in terms of *when* human behaviour will amount to a new cause: but *how frequently* a form of behaviour follows a particular act is not conclusive evidence of whether it was *caused by* the act.

[60] There are exceptions: eg when such behaviour is the very thing that the defendant was obliged to protect the claimant against.

defendants generally address equally rare risks of their behaviour where no such elements were involved? Perhaps the case for holding a defendant not liable because of the *rarity* of the unpredictable behaviour is stronger when different types of unpredictable behaviour will require *different* types of precautions, as in a driving case like *Scott v Gavigan*,[61] and weakest when a variety of types of unpredictable behaviour can be addressed by the same precaution, as in *Jolley*. It is also arguable, however, that courts are more likely to find *unpredictable behaviour* unforeseeable where the *claimant* had a measure of control over the unpredictable element, suggesting a situation where the judge's assessment was influenced by a view about relative culpability.[62]

B. Cognitive overload

In some circumstances the notion of a reasonable defendant trying to predict the ways in which their behaviour might injure others seems wholly artificial because there are *so many* possibilities that no real person would ever try to think about them. What driver, for example, would ever attempt to catalogue all the ways in which his inattentive driving might injure his passenger, and all the possible injuries his passenger might suffer? To an extent, tort law addresses these problems by 'lumping' together injuries of the same 'type' or 'kind' for the purposes of assessing whether such outcomes were foreseeable.[63] The law appears to take a different approach, however, in dealing with *secondary* effects. As already noted, some dangerous acts (eg careless driving) are capable of causing a wide range of types of accident; and beyond this, each of those types of accident may be capable of causing a range of secondary effects, such as injuries to relatives and rescuers. In the field of secondary psychiatric consequences, the solution to the proliferation of *ex ante* possibilities appears to be the adoption of an exceptional rule, granting a greater role to *hindsight*: the foreseeability of the secondary consequence is

[61] [2016] EWCA Civ 544. Simplifying somewhat, the unforeseeable behaviour of the claimant involved him appearing to deliberately reject an opportunity to use a crossing before running across the road at a point where he should have been able to see the defendant's vehicle. Slowing down to address the prospect of each and every pedestrian behaving in all similarly unexpected ways would require all urban motorists having to reduce speed at many points.

[62] *Scott v Gavigan* [2016] EWCA Civ 544. An inclination to hold *some* claimants wholly responsible for their misfortune has withstood the habituation of apportionment under the Law Reform (Contributory Negligence) Act 1945: James Goudkamp and Donal Nolan, *Contributory Negligence: Principles and Practice* (OUP, 2018), para 3.36, treat *Anderson v Newham College of Further Education* [2002] EWCA Civ 505; [2003] ICR 212 at [19] as confirmation that a 100 per cent reduction in a claimant's damages is impermissible under the statutory regime; this may partly explain the use of 'not foreseeable' (and 'break in chain') doctrines to achieve full allocation of accident costs to *some* culpable claimants.

[63] *Hughes v Lord Advocate* [1963] AC 837 and *Page v Smith* [1996] AC 155 are good examples of 'lumping' decisions; though the principles applied in both are frequently exaggerated: see, eg, the unsuccessful submissions based on *Page v Smith* in *Rothwell v Chemical & Insulating Co Ltd* [2007] UKHL 39, [2008] AC 281.

assessed *in light of the type of accident that actually occurred.*[64] (It is probably appropriate to assume that this *type of* accident must itself have been reasonably foreseeable.[65]) Introduction of this *hindsight* element, however, involves a move to a question different from anything that a defendant (of ordinary cognitive capacity) could have plausibly asked themselves at the moment of *breach* (when a wide variety of types of accident could have occurred).

C. Claimant variability

In the context of the remoteness rule a defendant is expected to 'take their claimant as they find them', meaning that they will be held liable for unexpected forms of injury *whenever* the deviation from the predictable was caused by a physiological, psychological, or social feature of the claimant.[66] Where, however, the claimant is only endangered by the defendant's behaviour because of an idiosyncratic feature the position is different: sometimes a defendant will only be held to owe a duty to the claimant if injury to a reasonably robust claimant would have been foreseeable, or if the defendant had some basis for predicting that the claimant was not reasonably robust.[67] Thus the default position is that a defendant is entitled to assume that those he interacts with have an average degree of resilience against psychiatric injury: 'customary phlegm'.[68] Similarly, a defendant will rarely, if ever, be held to

[64] In *Page v Smith* [1996] AC 155, 197, Lord Lloyd noted: 'In claims by secondary victims, it may be legitimate to use hindsight in order to be able to apply the test of reasonable foreseeability at all.' More controversially: 'Hindsight, however, has no part to play where the plaintiff is the primary victim.' This was criticized by Lord Goff in *Frost v Chief Constable of South Yorkshire Police* [1999] 2 AC 455, 477.

[65] Without such an assumption the hindsight rule might *sometimes* extend the situations where a defendant is held liable (ie where the type of accident was extraordinarily improbable, but the secondary consequences a familiar sequel to it).

[66] The precise dividing line between social features—such as adherence to a religious creed which forbids certain forms of beneficial treatment—and failure to mitigate damage—such as choosing to prioritize a pre-booked family holiday over treatment—is unclear.

[67] The more individualized approach is used in the context of an employer's duty to protect employees from 'workplace stress': *Hatton v Sutherland* [2002] ICR 613, CA at [23] (emphasis in the original) '*the threshold question is whether this kind of harm to this particular employee was reasonably foreseeable*. The question is not whether psychiatric injury is foreseeable in a person of 'ordinary fortitude', and at [29] (emphasis in the original) '*Unless he knows of some particular problem or vulnerability, an employer is usually entitled to assume that his employee is up to the normal pressures of the job*. It is only if there is something specific about the job or the employee or the combination of the two that he has to think harder.' An appeal in a case heard alongside *Hatton* was successful: *Barber v Somerset CC* [2004] 1 WLR 1089, HL(E). In explaining this, Lord Walker, at 1109, described a passage containing the second part of the quoted reasoning as: 'useful practical guidance, but it must be read as that, and not as having anything like statutory force'.

[68] *Bourhill v Young* [1943] AC 92, 117: 'The driver of a car or vehicle, even though careless, is entitled to assume that the ordinary frequenter of the streets has sufficient fortitude to endure such incidents as may from time to time be expected to occur in them, including the noise of a collision and the sight of injury to others, and is not to be considered negligent towards one who does not possess the customary phlegm.' It is less clear whether drivers are entitled to make the same assumption about other motorists: compare *Page v Smith* [1996] 1 AC 155.

have assumed a responsibility to look after an interest of a claimant unless a 'reasonable claimant' would have relied on them:

> If it is not reasonable for a representee to have relied on a representation and for the representor to have foreseen that he would do so, it is difficult to imagine that the latter will have assumed responsibility for it. If it is not reasonable for a representee to have relied on a representation, it may often follow that it is not reasonable for the representor to have foreseen that he would do so.[69]

Standing back, it seems plausible to diagnose that claimant variability is *sometimes* addressed by *constriction* of the default 'reasonable foreseeability' test: accident costs associated with certain types of vulnerability (eg lack of 'customary phlegm') are allocated to claimants, whilst the costs of other vulnerabilities (eg blindness) are distributed using the ordinary 'reasonable foreseeability' approach. Anyone seeking an explanation for this division might be enticed by the suggestion that English law has absorbed outdated, judgmental views about vulnerability to psychiatric harm and acts harshly towards those who rely on advice 'gullibly' or 'opportunistically'. But this explanation may discourage continued use of non-transparent normative judgements.

4. Difficulties with Distribution

A. Inaccessible expertise

I noted above that instead of alleging *general ignorance* of a causal rule a defendant may submit that knowledge of the risk was unevenly distributed, such that people *like him or her* were unaware of what others may have known. The parallel discussion of when a risk is 'discoverable' for the purposes of product liability law has—perhaps unexpectedly—concentrated on such a distribution issue: in particular through discussion of hypothetical cases where a causal rule has been 'discovered' by someone, but the knowledge has not reached the defendant, despite being 'published' in some form—for instance, in Manchuria in Chinese.[70] In this context the pivotal question has become under what conditions such information is 'accessible'.[71]

[69] *NRAM Ltd (formerly NRAM plc) v Steel* [2018] 1 WLR 1190 at [19].

[70] The Opinion of Mr Advocate General Tesauro in Case C-300/95, *Commission of the European Communities v United Kingdom* [1997] ECR I-2649 at [23], attributes the 'Manchuria' example to the Commission. (Electronic cataloguing of scientific papers has overtaken the assumptions behind the example.)

[71] Case C-300/95, *Commission of the European Communities v United Kingdom* [1997] ECR I-2649; *A v National Blood Authority* [2001] 3 All ER 289 at [53] and [64].

In the context of what is 'reasonably foreseeable' for the tort of negligence 'accessibility' has not been openly discussed, probably because of the tort standard's express reference to 'reasonableness'. Where the issue is whether *expert* knowledge ought to have percolated to less-expert (but reasonable) defendants, judges generally go no further than reciting Swanwick J's familiar guidance in an employer's liability case:[72]

> where there is a recognised and general practice which has been followed for a substantial period in similar circumstances without mishap, he is entitled to follow it, unless in the light of common sense or newer knowledge it is clearly bad; but, where there is developing knowledge, he must keep reasonably abreast of it and not be too slow to apply it.

This makes pivotal what would be forecast by a reasonable defendant who had kept 'reasonably abreast' of 'developing knowledge'. But rulings on such points are frequently *normative*: as noted above, in *Abraham v G Ireson & Son (Properties) Ltd*,[73] Swift J assessed what a small firm of general builders ought to have foreseen about asbestos risks by referring to the Annual Reports of the Chief Inspector of Factories and general guidance published by the Home Office and the Ministry of Labour, but disregarded letters written to employers in 'high exposure' sectors, such as shipbuilding, and expert journals: there was no evidence that this reflected the reading practices of building firms in Northamptonshire (or nationally).

B. Misinformed defendants

One special sub-set of cases where specific defendants (rather than a whole class) have failed to foresee a possible consequence is where they have been *misled* as to the validity or applicability of a causal rule. For example, in *Wagon Mound (No 1)*[74] it appears that the claimants anticipated that the oil *might* catch fire, but were misled through being re-assured by the manager of the Caltex wharf (where the *Wagon Mound* had been loading the oil that was spilled) that ignition was not possible once the oil had been spread on water. In circumstances where the supplier of the misinformation was associated with a party, we might think that this party ought to have found it difficult to submit that the risk was reasonably foreseeable by their opponents, and hence they were contributorily negligent.[75] Such reasoning,

[72] *Stokes v Guest, Keen and Nettlefold (Bolts and Nuts) Ltd* [1968] 1 WLR 1776, 1783.
[73] [2009] EWHC 1958 (QB). See also *Cuthbert v Taylor Woodrow Construction Holdings* [2022] EWHC 3036 (KB) and *White v Secretary of State for Health and Social Care* [2022] EWHC 3082 (KB).
[74] *Overseas Tankship (UK) Ltd v Morts Dock & Engineering Co Ltd (The Wagon Mound)* [1961] AC 388, PC.
[75] See fn 27.

however, would incorporate a normative element: a defendant misled by someone *not* associated with the claimants might find it harder to invoke this misinformation to deny foreseeability.

C. General variability in knowledge

What about situations where the fracture is not between experts and non-experts: how can a test of 'reasonable foreseeability' operate where individuals in the same class of potential defendants have significantly *different* levels of knowledge? For example, whilst most defendants who drive, or occupy premises, will have basic awareness of cause-and-effect physics, understandings of how traumatic experiences can cause injuries are unevenly distributed. Similarly, knowledge about types of behaviour that are not part of general experience, such as drug use and some sexual practices, can vary dramatically.[76]

It seems plausible that variability in general knowledge helps to explain why courts (must) sometimes *rule* on what is reasonably foreseeable, so that it becomes—in effect—a question of *law*: for example, in *Alcock v Chief Constable of South Yorkshire Police*[77] Lord Keith began: 'As regards the class of persons to whom a duty may be owed to take reasonable care to avoid inflicting psychiatric illness through nervous shock sustained by reason of physical injury or peril to another, I think it sufficient that reasonable foreseeability should be the guide.' But continued by insisting that where there were 'close ties of love and affection' between a claimant and another then it would be 'reasonably foreseeable that those bound by them may in certain circumstances be at real risk of psychiatric illness if the loved one is injured or put in peril', but that: 'Psychiatric injury to [a bystander unconnected with the victims of an accident] would not ordinarily, in my view, be within the range of reasonable foreseeability.'[78] He left open the possibility of an unconnected bystander falling 'within the range of reasonable foreseeability' 'if the circumstances of a catastrophe occurring very close to him were particularly horrific.'[79] But the Court of Appeal subsequently rejected this: in part, because it would be impossible to offer any *general guidance* on *when* psychiatric injury to a bystander would be reasonably foreseeable:

[76] An extreme example is provided by *MM v RF* (1997) 52 BCLR (3d) 127: the defendant was alleged to have negligently failed to protect her foster daughter against her paedophile adult son, and successfully contended that her reaction to 'warning signs' was influenced by 'her age, her limited education and experience of the world, her strict religious beliefs, her aversion to discussing sexual matters, and her lack of sophistication with respect to them' (at [133]); what a dissenting judge summarized as her 'exceptional naivete or aversion to sexual matters' (at [40]).

[77] [1992] 1 AC 310 HL.

[78] [1992] 1 AC 310, 397.

[79] This possibility was also left open by Lord Ackner and Lord Oliver, [1992] 1 AC 310, 403 and 416.

> Reactions to horrific events are entirely subjective; who is to say that it is more horrific to see a petrol tanker advancing out of control on a school, when perhaps unknown to the plaintiff none of the children are in the building but are somewhere safe, than to see a child or group of children run over on a pedestrian crossing?[80]

In the absence of a *ruling* on what is reasonably foreseeable, like that in *Alcock*, how could courts deal with discrepancies in knowledge? In the context of children as defendants the courts accommodate variations as to what they know, but only between age-based bands.[81] By contrast, with (non-specialist) adults, any attempt to distinguish between, for instance, what should be foreseen by those who know about shock-induced psychiatric injuries and those who do not, would be in tension with several of the general objectives of an *objective* approach: predictability, efficiency, equal freedom, and communicability are all enhanced by clear, consistent rulings. But *rulings* as to what is reasonably foreseeable can predictably be over-extended to advantage classes of expert defendant, such as medical practitioners, who are aware that the *Alcock* rules are out-of-line with whose psychiatric well-being is *really* endangered.

5. Conclusion

'Reasonable foreseeability' is *not* a simple question of fact. Some may be tempted to classify it as a 'question of evaluative judgement', to signal that appeal courts should be hesitant, but some authorities appear to settle the issue as a matter of *precedent*, both where foreseeability is *deemed* (eg endangering of rescuers[82]) and where goals such as predictability require trial judges to be instructed that certain things are *not* reasonably foreseeable (eg psychiatric injury to bystanders). Further, since conclusions about 'reasonable foreseeability' frequently depend on non-transparent normative assessments the concept can be criticized: there could be gains in judges disclosing and defending the reasoning behind their choices. Above, I noted Professor Stapleton's opinion that, in the parallel field of allocation of the costs of unanticipated product defects, the choices were driven by ideology: and a lack of transparency impedes attempts to discover whether similar (undisclosed) drivers explain the inconsistent patterns within negligence's concept of 'reasonable foreseeability'.

[80] *McFarlane v EE Caledonia Ltd* [1993] EWCA Civ 13, [1994] 1 Lloyd's Rep 16.
[81] eg *Mullin v Richards* [1998] 1 WLR 1304, 1308.
[82] *Baker v T E Hopkins & Son* [1959] 1 WLR 966, 975–976: 'There is, happily, in all men of good will an urge to save those who are in peril. Those who put men in peril can hardly be heard to say that they never thought that rescue might be attempted ...'

It may be worth emphasizing, however, that judicial use of concepts that appear to be 'labels' or 'smokescreens' for non-transparent normative judgements may not be pathological. Sometimes the burden of identifying all the reasons that would—ideally—be taken into account, and generating the data necessary to weight them properly, would be inconsistent with the practical realities of judicial decision-making. If so, reliance on non-transparent devices like 'intuitions about what is fair' or 'intuitions as to what would best serve the values of the system (not further specified)' can have the substantial advantage of *enabling* decision-making. A judge who felt unable to decide a case before, for example, identifying at what point between 'first formulation of the hypothesis' and 'confirmation through reliable experimentation' the moral project of securing 'everyone's equal external freedom' would regard a causal rule as sufficiently known to allow a defendant to be held liable, might find they had undertaken a Sisyphean task. Non-transparency impedes understanding, and may shield inconsistencies from detection, but this does not mean we should necessarily insist that human judges strive to achieve coherence with a particular scheme of goals, or moral perfection; perhaps it would be better for us to tolerate them doing not-quite-perfect, not-quite-consistent, justice in more cases more swiftly. These final suggestions, however, may—foreseeably—be contentious.

9

Time to Cut Ties: Reforming the Secondary Victim 'Control Mechanisms' in Pure Psychiatric Injury

Imogen Goold and Catherine Kelly

1. Introduction

Nearly twenty years ago, Jane Stapleton cut right to the heart of what is wrong with the law's approach to claims for compensation by people who witness events so shocking that they suffer a psychiatric illness as a result, when she said:

> the area of nervous shock is ... the area where the silliest rules now exist and where criticism is almost universal.[1]

Two decades later, her words ring just as true, perhaps even more so. For over that period, the 'control mechanisms' on such claims ossified into rigid rules that now draw hard, but incoherent, lines between which claims will succeed and which will fail.

Her observation was part of a wider critique of the impacts of incrementalism on the development of the law. She argued that 'doctrinal developments have now produced areas of intolerable embarrassment with which incrementalism—in particular, incrementalism of the "pre-judged pockets" form—is inadequate to deal.'[2] As she put it, in her inimitable style:

> P.R. Glazebrook has already eloquently attacked the 'affront to public decency' represented by awards of damages for [95] the upkeep of 'unwanted' children. But a class of claims which even more obviously offends this principle is that concerning nervous shock.[3]

[1] Jane Stapleton, 'In Restraint of Tort' in Peter Birks (ed), *The Frontiers of Liability (Volume 2)* (Oxford: Oxford University Press, 1994) 95.
[2] ibid 94.
[3] ibid 94–95.

Imogen Goold and Catherine Kelly, *Time to Cut Ties: Reforming the Secondary Victim 'Control Mechanisms' in Pure Psychiatric Injury* In: *Torts on Three Continents*. Edited by: Kylie Burns, Jodi Gardner, Jonathan Morgan and Sandy Steel, Oxford University Press. © Imogen Goold and Catherine Kelly 2024. DOI: 10.1093/oso/9780198889748.003.0009

Even Lord Oliver, she noted, lamented that the law in this area was not 'logically defensible'.[4] In Stapleton's view, it is 'no small irony' that this area of silly rules is also 'the best historical example of incremental development'.[5] We agree. And we agree that a solution to this might well be, as she has suggested, to wipe out the current rules and start afresh, potentially via a legislative approach.[6] In line with this general position, we explain here why the control mechanisms for secondary victim claims for negligently induced psychiatric injury should be fundamentally reformed, and indeed potentially abolished altogether. Incrementalism, we suggest, has led the law away from the early approach to such claims, wherein the courts drew on the fundamental logic of reasonable foreseeability in determining who might claim for the harm caused by seeing the dreadful impact of another's negligence. We demonstrate how over time, and particularly in response to some highly unusual events, the courts have moved away from the general principles grounding the duty of care inquiry, and replaced them with what is essentially a line-drawing exercise undertaken without reference to defensible principle. Such an approach is problematic not only because it is inherently indefensible, but also because the law in this area has consequently lost its ability to respond to new challenges. This has occurred precisely because the option of drawing on the broader principle of reasonable foreseeability has been effectively closed off by the rigid 'control mechanisms' approach. In short, incrementalism in this area represents a series of retrograde steps away from the broad, open approach of *Hambrook v Stokes* (1925),[7] and towards the blunt inflexibility of *Alcock v Chief Constable of South Yorkshire Police* (1992)[8] and *White v Chief Constable of South Yorkshire* (1999)[9] that now binds the courts so unhelpfully.

We start, then, by returning to the very beginning, outlining the origins of the law on negligently induced psychiatric injuries. We trace this back to *Dulieu v White* (1901)[10] and *Hambrook* and offer a brief overview of the original thinking in the English (Scottish and Irish) courts on secondary victim claims particularly. From here, we explain the current position, and then detail the problematic implications of that position generally, before critically evaluating each control mechanism. As part of this evaluation, we tease out early thinking on each mechanism, and draw on a detailed examination of many unreported decisions from the early period. These offer insights into the general views among the various courts at the time as to how such witnesses to tragedy ought to be treated. We conclude with

[4] *Alcock and Others v Chief Constable of South Yorkshire Police* [1992] 1 AC 310, 418 (Lord Oliver).

[5] Stapleton (n 1) 95.

[6] We have made the argument for starting afresh elsewhere: Imogen Goold and Catherine Kelly, 'Time to Start *De Novo*: The *Paul, Purchase and Polmear* litigation and the temporal gap problem in secondary victim claims for psychiatric injury' (2023) 39(1) *Journal of Professional Negligence* 24.

[7] [1925] 1 KB 141.

[8] [1992] 1 AC 310

[9] [1999] 2 AC 455.

[10] [1901] 2 KB 669.

our suggestions for the direction in which the law should move, namely a rejection of the rigid control mechanisms and a return to the fundamental principles of negligence. Such an approach would remove the overly simplistic distinction that has grown up between physical and psychological harms, and bring the law on psychiatric injuries more in line with the rest of negligence. It would also afford courts greater discretion and flexibility, allowing them to respond to situations of the kind they have faced in the twenty-five years post-*Alcock* with compassion and rationality, rather than binding them to controls that produce perverse results and considerable injustice.

2. Defining Secondary Victims and Controlling Their Claims

Until the early twentieth century, the English courts had refused to permit recovery for nervous shock occasioned without physical impact.[11] Psychological injury was accepted as a real injury, but was seen as in some way connected to a physical injury. Psychological injury arising without any physical impact was recognised in *Wilkinson v Downton* in 1897,[12] but it was not until 1901, following the decision in *Dulieu*, that a claimant who had been shocked by fear for their personal safety due to the negligence of another and where there was no physical injury could make a successful claim for damages. The Irish courts had made this step some decades earlier, in *Byrne v Great Southern and Western Railway Co*[13] and *Bell v Great Northern Railway Co of Ireland*,[14] and these decisions were influential on the English courts.[15]

Nearly a quarter of a century later, the Court of Appeal extended the law in *Hambrook* to allow compensation for nervous shock occasioned by fear for the safety of another if the claimant perceived the shocking event with their own eyes or ears. This position was applied in *Owens v Liverpool Corporation* (1939)[16] and in a number of unreported cases[17] but was then restricted in *Bourhill v Young* (1943) when the House of Lords refused recovery for a bystander who had merely heard the sound of an accident nearby.[18]

[11] *Victorian Railways Commissioners v Coultas* (1888) 13 App Cas 222. For a detailed discussion of the early cases and the complex contemporary understandings of the relationship between physical and psychological injury in the early to mid twentieth century, see Imogen Goold and Catherine Kelly, 'Who's Afraid of Imaginary Claims? Common Misunderstandings of the Origin of the Action for Pure Psychiatric Injury in Negligence 1888–1943' (2022) *Law Quarterly Review* 58.

[12] [1897] 2 QB 57.

[13] Unreported 1884 Court of Appeal (Ireland).

[14] (1890) 26 LR Ir 428.

[15] The Scottish courts had also accepted that a psychological injury could arise without a physical injury: *Pugh v London, Brighton and South Coast Railway Co* [1896] 2 QB 248; *Wallace v Kennedy* (1908) 16 SLT 485.

[16] [1939] 1 KB 394.

[17] *Henderson v Middlesbrough Corporation*, unreported (1928) 72 *Solicitors' Journal* 229.

[18] [1943] AC 92.

Over the course of the next forty years, the courts continued to allow recovery for claimants who suffered injury due to fear for others, and over time they developed positions on how the principles of reasonable foreseeability and proximity would apply. We explore this development in later sections, but for now it suffices to say that by the early 1980s, the more open and flexible approach of the early courts had been replaced by what would come to be known as the 'control mechanisms' for limiting liability to secondary victims. While they had been emerging for some time, they were most clearly crystallised into 'controls' or 'elements' in the first case dealing with the claims of psychiatric injury brought by the families of those injured and killed in the Hillsborough disaster—*Alcock*. The Law Commission reviewed the law on psychiatric injuries in the wake of the decision, and in 1998 described the controls as unnecessarily tight, producing 'arbitrary results'.[19] Despite this, the House of Lords confirmed the 'control mechanism' approach in the second Hillsborough case—*White*—a short time later.

Lord Oliver explained the controls in *Alcock* as follows:

> [F]irst, that in each case there was a *marital or parental relationship between the plaintiff and the primary victim*; secondly, that the injury for which damages were claimed *arose from the sudden and unexpected shock* to the plaintiff's nervous system; thirdly, that the plaintiff in each case was either personally present at the scene of the accident or was in the *more or less immediate vicinity and witnessed the aftermath shortly afterwards*; and, fourthly, that the injury suffered arose from *witnessing the death of, extreme danger to, or injury and discomfort suffered by the primary victim*. Lastly, in each case there was not only an element of physical proximity to the event but *a close temporal connection between the event and the plaintiff's perception* of it combined with a close relationship of affection between the plaintiff and the primary victim. (our emphasis)[20]

It is also a requirement that the secondary victim's injury is a 'recognised psychiatric illness'. These requirements are necessarily interrelated, and perhaps the most helpful way to break them down is the structure proposed by the Law Commission in its 1998 report, *Liability for Psychiatric Illness*.[21] Explaining that they form part of the proximity inquiry, it separated them into three main elements:

1. The claimant's ties of love and affection with the immediate victim;
2. The claimant's closeness in time and space to the incident or its aftermath; and
3. How the claimant perceived the event or its aftermath.

[19] Law Commission, *Liability for Psychiatric Illness* (Law Com No 249, 1998) iii.

[20] *Alcock* (n 4) 411F–H (Lord Oliver), where Lord Oliver described the common features of the reported cases at that time. Sir Geoffrey Vos' emphasis included.

[21] Law Commission (n 19) Chapter 2.

As Stapleton has pointed out, 'a popular but obvious criticism' of these controls is that they do not represent 'sine-qua-non factors to the causation of actionable damage'.[22] People can perfectly foreseeably come to suffer psychiatric illnesses or pathological grief in situations where there are no relational ties, or where they are not physically present at the event, or slowly over time. Indeed, the cases in which the courts have struggled with these control mechanisms offer excellent examples of exactly such factual scenarios.

For example, *W v Essex County Council* (2001) concerned a couple whose children were sexually abused by a boy to whom they had given foster care after the local authority and social worker negligently failed to warn them of his past history of sexually abusive behaviour.[23] The parents felt considerable guilt as a result of the role they inadvertently played in exposing their children to this risk, and as a consequence of this both suffered psychiatric injuries. Their situation fell well outside the bounds laid down by the control mechanisms, as they neither witnessed the event(s), nor their 'immediate' aftermath. But the harm to the parents was certainly foreseeable, particularly as the defendants *knew* that the parents did not want to expose their children to such a risk, and were aware of the boy's history. The House of Lords found for the parents by taking a broad view of the concept of 'aftermath', but it had to do so *in spite of* the recognised stringency of the control mechanisms. In doing so, the court placed emphasis on the foreseeability of the risk of harm.[24]

In other contexts, the courts have had to resort to arguably quite problematic reasoning to avoid the otherwise unjust impact of the control mechanisms. In the case of *Yearworth v North Bristol NHS Trust* the claimants were all men who had stored their semen to preserve their fertility before undergoing treatment for cancer.[25] All suffered various forms of psychiatric harm upon learning that their semen had been negligently destroyed by the Trust, as this left them potentially unable to conceive genetically related children. Despite having clearly suffered a psychiatric injury as a result of the Trust's negligence, the control mechanisms created insurmountable barriers to their claims. This, coupled with the fact that the Court of Appeal refused to regard the damage to their semen as 'personal injury', left them with no clear avenue for redress. Only because the court was prepared to take a broad approach to whether the semen was a form of property (and then, further, a rather open-minded attitude to the principles of recovery for a breach of bailment) were they able to obtain a remedy. Such is the stringency of the control mechanisms, then, that the court had to stretch another area of law to achieve

[22] Stapleton (n 1) 95.
[23] [2001] 2 AC 592.
[24] *W v Essex* (n 23) 598F.
[25] *Jonathan Yearworth and Others v North Bristol NHS Trust* [2009] EWCA Civ 37.

what it considered a just outcome. In other situations, where the court has not been willing to 'stretch' principles, the stringest control mechanisms have denied relief to claimants who have clearly suffered harm as a result of the defendant's negligence. This was the case in *A and others v Leeds Teaching Hospital NHS Trust* (2004), in which a property-based argument in support of the claims of parents distressed at the mishandling of their children's remains failed.[26]

Even in cases that present more straightforward negligence claims, the control mechanisms have acted as an impenetrable barrier. The cases that turned on shock and aftermath, such as *Sion v Hampstead Health Authority* (1994)[27] and *Liverpool Women's Hospital NHS Foundation Trust v Ronayne* (2015),[28] are, in our view, two good examples where the defendant's negligent actions caused entirely foreseeable harm to the claimants but no remedy could be given. In both cases, the claimants witnessed harm to their close relation in hospital resulting from medical negligence, but the courts considered the sight to be not 'a shock' or temporally proximate enough to the negligence as required to meet the criteria. The logic in both seems to be that there is a point shortly after medical negligence occurs where one should almost expect to see horrible things and be sufficiently prepared to do so.

There has been some recognition in the courts that the controls should not be rigidly applied, such as Lord Scarman's assertion in *McLoughlin v O'Brian* that the 'elements' he outlined should be seen only as factors that might point to when an injury was foreseeable,[29] and Lord Slynn's invocation to the same approach in *W v Essex* many years later.[30] But the usual approach of the courts reflected in the cases cited above demonstrates the very reason *why* they felt the need to emphasise this—because the courts do, in fact, apply them rigidly, and they do so because they have been repeatedly framed in a rigid manner in cases that have come before. This makes it exceptionally difficult for any court to now take an open-minded or inclusive approach to cases that do not fit in these controls and certain narrow categories. This is precisely what Lord Slynn was concerned about in *W v Essex* when he emphasised that the courts should not treat those categories as closed,[31] yet his plea has made little difference in the twenty years since, with the law remaining highly constrained.

We are now seeing the impact of those constraints once more, in the recent joint appeals in *Paul v Royal Wolverhampton NHS Trust; Polmear and another v Royal Cornwall Hospitals NHS Trust* and *Purchase v Ahmed* (2022).[32] In each case, the

[26] [2004] EWHC 644 (QB).
[27] [1994] 5 Med LR 170.
[28] [2015] EWCA Civ 588.
[29] [1983] 1 AC 410, 431 (Lord Scarman). This is not, of course, to suggest that Lord Wilberforce considered foreseeability as in and of itself sufficient to ground a duty.
[30] *W v Essex* (n 23) 599–601 (Lord Slynn).
[31] ibid.
[32] [2022] EWCA Civ 12; *Polmear v Royal Cornwall Hospitals NHS Trust* [2021] EWHC 196 (QB); *Purchase v Ahmed* (Birmingham County Court, 6 May 2020).

deleterious impact of the negligence on the victims could certainly have been foreseen, but it was triggered sometime after the initial event of medical negligence. Hence arguably in each case, the injury suffered by the claimants fell outside even an expanded aftermath understanding of the time and space requirement, and so the Court of Appeal denied the claims. The court was bound to do so, but in each case the result, as shall be outlined later in this chapter, is arguably unjust. Each claimant was denied only because their psychiatric injury was triggered by a later event in a chain of events set off by the original failure on the part of the defendant. In each, the claimant was understandably traumatised by seeing a loved one die or discovering them just after their death. These deaths were the result of much earlier clinical negligence. However, because the claimants observed horrifying consequences of that negligence temporally removed from the event of that negligence, their claims had to fail, regardless of how foreseeable their reaction, and how otherwise proximate their outcome to the negligence. These cases are discussed in detail in Section 3.

Andrew Burrows and John Burrows note in a critique of the decision in *Ronayne* that numerous policy arguments for restricting psychiatric injury claims (by primary and secondary) victims have been articulated. They note the concerns about 'opening the floodgates of litigation', and worries about potentially crushing liability that might fall on defendants without some control mechanisms to limit the impact of claims, as well as the burden these would place on courts, insurers, and in medical negligence claims, the NHS.[33] They explain that the control mechanisms for secondary victims are a means to address these concerns, but as they rightly argue '[n]one of the four requirements are easy to defend from a medical perspective'.[34]

Lines must be drawn, that much is true. But they must be drawn in places that can be justified by more than mere expediency, or a desire simply to limit claims. It cannot be, as Stapleton puts it '*simply* a line-drawing exercise' (our emphasis).[35] As members of the Australian High Court asserted in *Tame v New South Wales* (2002),[36] such justification is largely absent in relation to the control mechanisms deployed in English law. In particular, Kirby and Gummow JJ referred to the controls as '[u]nprincipled distinctions and artificial mechanisms' which 'bring the law into disrepute', asserting that 'the "nervous shock" cases predicate elusive distinctions with no root in principle and which are foreign to the merits of the litigation'.[37] The impact of this, they suggest, is that controls have operated in an

[33] Andrew Burrows and John Burrows, 'Commentary: A Shocking Requirement in the Law on Negligence Liability for Psychiatric Illness: *Liverpool Women's Hospital NHS Foundation Trust v Ronayne* [2015] EWCA Civ 588' (2016) 24(2) *Medical Law Review* 278, 280.

[34] ibid 282.

[35] Stapleton (n 1) 95.

[36] *Tame v New South Wales; Annetts v Australian Stations Pty Limited* [2002] HCA 35.

[37] ibid [90].

'arbitrary and capricious manner', as well as creating a barrier to 'the emergence of a coherent body of case law'.[38] We now critically examine each of the controls in turn to demonstrate the lack of justification and the problems to which its application gives rise.

3. The Control Mechanisms—A Critical Analysis

The control mechanisms are variously grouped or categorised by different legal commentators, but most commonly and usefully can be divided into requirements of: a recognised psychiatric injury; a close tie of love and affection; a shock; and finally proximity in both time and space to the event or its immediate aftermath.

The requirement that the injury must have arisen due to shock is longstanding. The term 'nervous shock' itself is the original name in law for what we now refer to as 'psychiatric injury'. The requirement is necessarily entangled with the need for spatial and temporal proximity to the scene or its aftermath, and with the need to witness the injury to another. We attempt to tease out the shock element first, then move to spatial and temporal proximity including the notion that this can be achieved only when the claimant perceives the event with their 'own eyes and ears'.

A. Caused by sudden shock and the need for spatial and temporal proximity to the scene or its aftermath

We lack space to trace the origins of the shock requirement here, but others such as Harvey Teff, Peter Handford and Danuta Mendelson have already done so in valuable detail.[39] The key features of this development are worthy of note, however, as they give some sense of why the law is in its current state. The need for some *immediate* fear that precipitates the mental injury derives from the cases that we would now call 'primary victim' cases—where the claimant feared for themselves. The first of these in English law is *Dulieu*. When the law expanded to include fear for *others* in *Hambrook*, the immediacy requirement was not removed. This may be part of why the 'shock' dimension was retained, and it also is to some extent inherently related to the requirement at this point that the accident to a loved one must be seen or heard (as a limiting factor on claims).[40] This was also congruent

[38] ibid [188]–[190].

[39] Harvey Teff, *Causing Psychiatric and Emotional Harm: Reshaping the Boundaries of Legal Liability* (Oxford: Hart Publishing, 2008); Nicholas J Mullany and Peter R Handford, *Tort Liability for Psychiatric Damage: The Law of 'Nervous Shock'* (London: Sweet & Maxwell, 1993); Danuta Mendelson, *The Interfaces of Medicine and Law: The History of the Liability for Negligently Caused Psychiatric Injury (Nervous Shock)* (Dartmouth: Dartmouth Publishing Co, 1998).

[40] See further Teff (n 39) Chapter 2.

with medical understandings of this type of injury at the time. Whether the notion of something acting quickly and abruptly on the mind was also an aspect of judicial thinking is not entirely clear, and the few judgments from the period do not delve into the complexities of *how* nervous shocks were thought to produce injury. However, even if it were the case that the judges believed a shock to be necessary for injury to arise, developments in psychiatric medicine mean that such a foundation is no longer tenable.

Indeed, this is precisely how the 'shock' requirement is now regarded by mental health professionals. The Law Commission consulted the Royal College of Psychiatrists Mental Health Law Group as part of its inquiry. That group's views were clear:

> For psychiatrists the 'shock-induced' requirement causes serious problems. The term is vague, has no psychiatric meaning and is emotively misleading. The requirement should be abandoned. Psychiatric evidence should require demonstration, or not, of a psychiatric disorder distinct from a normal mental reaction and, if present, its relationship with the index event.[41]

Despite these developments in our understanding of the causes of psychiatric illness, even in the late 1990s and beyond, we can see the judiciary focusing on the need for 'a sudden assault on the nervous system' or 'a sudden and unexpected shock to the plaintiff's nervous system'.[42]

The Law Commission was excoriating in its critique of the shock requirement.[43] It pointed out that it is not considered a causal requirement for psychiatric illness by the medical profession. Nor is it even required in other areas of law, such as stress-induced illness, claims for mental illness consequent on a physical injury, nor even in the more tenuous area of claims for breach of contracts geared towards enjoyment or prevention of distress. Further, even by the time the Law Commission reported, the requirement was clearly leading to what it referred to as 'some harsh and seemingly arbitrary decisions'.[44] The most stark example of such arbitrariness is arguably the gulf between the decisions in *Sion* and *North Glamorgan NHS Trust v Walters* (2002).[45] In *Sion*, the claimant father alleged that his son had been subjected to negligent treatment, including a misdiagnosis of bleeding in his kidney. He claimed for his own psychiatric injuries induced by the negligence to his son. The claim was denied principally due to the father's lack of shock.[46] The father had sat by his son's bedside for more than a week as he slowly succumbed to his injuries

41 Law Commission (n 19) [5.29].
42 *Alcock* (n 4) 398, 411 (Lord Keith and Lord Oliver respectively).
43 Law Commission (n 19) [5.28]–[5.33].
44 ibid [2.63].
45 [2002] EWCA Civ 1792.
46 [1994] 5 Med LR 170.

and eventually died. It was found that because some time had passed between admission to hospital and the son's death, that the death was not itself shocking, but rather expected. The court felt itself bound by the need to find a shock to the father that led to the injury, and without such, considered his claim had to fail. As Staughton LJ explained, the shock requirement was 'authoritatively and conclusively fixed by the House of Lords in *Alcock*'.[47] Peter Gibson LJ agreed that it was a 'sudden awareness' of what had occurred (rather than that slow process in the fact before him) that was 'the crucial ingredient of shock', and that while this was certainly possible in medical negligence cases, it was not the case in this situation.[48]

By stark contrast, in *Walters*, the Court of Appeal was prepared to regard the period from the start of a child's illness (arising due to negligence) to their death in their mother's arms thirty-six hours later as a 'shock'. In that case, the events were understood as 'a seamless tale lasting for a period of 36 hours which for W was undoubtedly one drawn-out experience'.[49] This event was, taken together, productive of a sufficient shock, whereas the longer experience of the father in *Sion* was not. Nor was that of the parents in the sadly similar case of *Taylorson v Shieldness Produce Ltd* (1994).[50]

It might be tempting to think that as *Sion* and *Taylorson* pre-date *Walters* by nearly a decade, the difference can be explained as a liberalisation in the courts' approach to shock. But the decision in *Ronayne* suggests otherwise. In that case, the claimant saw his wife in an unexpectedly very poor state when she had been admitted to the Accident and Emergency department after becoming ill following a hysterectomy. Again, the claimant was denied compensation despite suffering a psychiatric illness that was clearly caused by the defendant's negligence solely because it was not induced by shock. Again, the refusal rested on the fact that the events were not shocking because they were drawn out over time.

This is the point at which the relationship between the shock element and the aftermath doctrine become importantly intertwined. Of all the controls, the need for the claimant to have been 'either personally present at the scene of the accident or … in the more or less immediate vicinity and witnessed the aftermath shortly afterwards' has been the one with which the courts have struggled the most. It has paradoxically produced both more willingness to stretch principles (as in *McLoughlin* and in *W v Essex* discussed above) and in the absence of 'stretch' some of the most searing injustices. The recent appeals in *Paul, Polmear* and *Purchase* are apt examples.[51]

[47] ibid 173.
[48] Ibid 176.
[49] *Walters* (n 45) H3.
[50] [1994] PIQR P329 (CA).
[51] *Paul v Royal Wolverhampton NHS Trust* [2022] EWCA Civ 12, appealing the decisions in *Paul v Royal Wolverhampton NHS Trust* [2020] EWHC 1415 (QB); *Polmear v Royal Cornwall Hospitals NHS Trust* [2021] EWHC 196 (QB); *Purchase v Ahmed* (Birmingham County Court, 6 May 2020). As in *Paul*, the claimants in *Polmear* and *Purchase* witnessed the death of a loved one, or came upon a loved one very shortly after their death. In *Polmear*, the claimant's 7-year-old daughter Esme died in front of

The Law Commission explained that there are a number of understandable, if insufficient, justifications behind the 'sudden shocking event' requirement. One is that it 'maintains a level of immediacy with the primary event, thereby protecting the defendant from the possibility of claims arising years after his or her negligent conduct'.[52] *Paul*, and the earlier case of *Taylor v A Novo (UK) Ltd*,[53] are good examples of where this concern has some traction.

In *Novo*, the primary victim (Cindy Taylor) had been injured when her employer's negligence led to a pile of boards falling on her. However, she did not die until three weeks later, when she suddenly collapsed due to deep vein thrombosis and consequent pulmonary emboli caused by the accident. Her daughter, Crystal, did not see the original accident but she did see her mother's death as a result of which she suffered post-traumatic stress disorder (PTSD).

Similarly, in *Paul*, Mr Paul (the primary victim) had received negligent treatment from the defendants in November 2012. He had presented at the defendant hospital with a number of coronary symptoms, but due to a negligent failure to perform a coronary angiograph, his coronary artery disease was not diagnosed. As a consequence of that failure the disease was not detected, and went untreated. On 26 January 2014, fourteen and a half months later, Mr Paul died of a heart attack in front of his two daughters. Both children suffered PTSD as a result. The key issue was the relevance of the temporal gap between the negligent act, the resulting damage to the primary victim, and the horrifying event causing psychiatric injury to the secondary victim. Unlike situations such as that in *Alcock* or *McLoughlin*, there was a substantial distance between the negligent act and the consequent damage to the primary victim (the sight of which caused the injury), certainly well beyond the period of hours accepted as 'immediate aftermath' in *McLoughlin*. So, while Mr Paul's daughters were certainly shocked, the event that caused the shock was long after the negligence.

This led in *Paul*, as in *Novo* before it, to an inquiry into which event must be the one that shocks. In both *Novo* and *Paul* there was shock, but it was in response to a later event, considered to be removed from the original negligent event. By contrast, in a case like *Ronayne*, there was a lack of shock, but there was sight of the aftermath. However, in *Walters* the court was prepared to see the passage of only thirty-six hours as meaning that the series of events following the negligence were to be considered still part of the original event.

In *Novo*, Lord Dyson considered that the primary issue was what constituted the 'relevant incident' that must be perceived and shock the claimant—the original

her parents due to a negligent failure to diagnose and treat her pulmonary veno-occlusive disease. In *Purchase*, the claimant came upon her daughter only minutes after she had died due to alleged negligence by the defendant, a GP, in failing to diagnose and treat her pneumonia.

[52] Law Commission (n 19) [5.30].
[53] [2013] EWCA Civ 194.

accident, or the death?[54] In his view, to allow Crystal to recover for witnessing only the second 'would be to go too far' for two reasons.[55] One, which with respect is not convincing, was that to allow her to recover but deny someone who came upon the aftermath of an accident sometime later (as *Alcock* would require) 'would strike the ordinary reasonable person as unreasonable and indeed incomprehensible'.[56] Whether that is actually the case is a question that cannot, of course, be answered, nor is it really the appropriate test. His second reason was, in agreement with Lord Steyn and Lord Hoffmann in *White*, that the law should simply not expand beyond the control mechanisms' limits.[57]

Lord Dyson did not, however, engage with the more convincing concern the Law Commission felt might justify the shock requirement—that having to see the accident and be shocked by it maintains immediacy with the primary event. It maintains a link, and hence protects the defendant from the possibility of claims far into the future once the damage precipitated by the negligence manifests. But even as the most convincing reason, it is not at all a good justification. Why should it be the case that the mere fact that a defendant's negligence takes some time to cause its harm, mean that he or she should then entirely escape the consequences? The matter is not the same as one of *indeterminate* liability or of an indeterminate class of claimants, where the defendant cannot know the possible impact he or she might have in terms of its scope. There is merely a time gap between the event and the outcome, but that gap alone should not be enough to absolve them of all responsibility for the consequences.

In the Court of Appeal in *Paul*, Sir Geoffrey Vos stated he was bound by the decision in *Novo*, but that in his view 'there is no logical reason for these rules'.[58] Rather, he felt it was clear that 'secondary victims in these cases ought to be seen to be sufficiently proximate to the defendants to be allowed to recover damages for their psychiatric injury'.[59] Part of the problem was, he understood, that the immediate perception and shock requirements exclude many claimants who experience psychiatric injury as secondary victims in the clinical negligence context. In such situations, a gap between the failure to take care and the outcome is common, and it would (we agree) be problematic to effectively exclude all such situations. He rightly suggested that the real focus should be on the relationship between the defendant and the secondary victim and on the foreseeable impact of the event on *that* victim.[60] As we have argued elsewhere:

[54] ibid [29].
[55] ibid.
[56] ibid [30].
[57] ibid [31].
[58] *Paul* (CA) (n 51) [12].
[59] ibid.
[60] ibid.

> In doing so, he resisted the temptation to apply convoluted approaches derived from the primary victim's injury, and applied a more standard negligence analysis to the situation, in line with the approach taken in *McLoughlin* and earlier authorities.[61]

The main reasons advanced for retaining the shock and immediacy requirements are weak because they can be met in other ways. A proper causal inquiry can address concerns about whether an illness that derives from long, slow exposure or later events was actually *caused* by the initial negligence. The tendency of the courts to stretch the aftermath and shock requirements in some cases but not others demonstrates that the line itself leads to injustices, which they have responded to by merely creating yet more arbitrary lines. The better approach, we argue, is to return to general negligence principles, look for some form of link between the negligence and the complained of outcome, and temper this with foreseeability. Applying current medical knowledge, rather than outdated notions of 'shock' will support this approach.

B. Perceived the event with unaided senses

It was accepted in *McLoughlin* that the claimant must perceive the event or its immediate aftermath through his or her own unaided senses,[62] a requirement that stems from *Hambrook*. As Harvey Teff notes:

> the majority in *Hambrook v Stokes* favoured one limiting device which ruled out many relational claims, namely, that the injury must have resulted from 'sight or sound of the accident', that is, from experiencing it at the time with one's own unaided senses ... A requirement of direct perception helped to allay concerns about opening the floodgates.[63]

Merely having the news of the impact conveyed by some other means would not suffice. The House of Lords confirmed this view in *Alcock*, holding that viewing via television, even via a live broadcast, would be insufficient.[64]

The Law Commission explored the basis of the perception requirement, and found it had little grounding in medical understandings of the causes of psychiatric illness. Citing the analysis of the medical and scientific evidence offered by Teff, it took the view that direct perception of how the harm comes to a loved one is not

[61] Goold and Kelly, 'Time to Start *De Novo* (n 6).
[62] *McLoughlin* (n 29) 422–423 (Lord Wilberforce).
[63] Teff (n 39) 47.
[64] *Alcock* (n 4) 398D–H (Lord Keith). Lord Ackner (405E–F), Lord Oliver (416H–417B, F–418A) and Lord Jauncey (423E–G, 424B–D, D–E) agreed.

particularly crucial in determining whether a psychiatric illness will result.[65] We agree with their view, and will not rehearse it here, but make the point instead that this analysis fits with our general argument—that a strict requirement of perception does not fit the reality of how psychiatric illnesses can be precipitated.

The requirement, like the other controls, prevents a fully holistic evaluation of the situation. As Teff's work demonstrates, where the harm is to a loved one, then merely *hearing* about it may be sufficient to cause an illness, and in fact, potentially *more* likely to do so. This was recognised by Nolan LJ in the Court of Appeal in *Alcock*, with whom Lord Ackner agreed in the House of Lords when he cited Nolan LJ's example of a live broadcast of the hot air balloon accident. As Lord Ackner suggested:

> Many other such situations could be imagined where the impact of the simultaneous television pictures would be as great, if not greater, than the actual sight of the accident.[66]

Like the other controls, the perception requirement adds little that cannot be captured via a considered application of the foreseeability and proximity concepts, serving only to prevent the courts from being able to consider the *actual* impact and effect of the negligence on the secondary victim. Such a control will, no doubt, come under further pressure, as future cases are likely to involve new technologies that offer ways for a person to perceive events which are at least as affecting as traditional sight or hearing. The requirement is not merely redundant—it is potentially overly restrictive and supported by no clear principle or medical evidence. Of all the requirements, it is the one most in need of rejection.

C. Close ties of love and affection

In *Alcock*, Lord Oliver stated the other control mechanisms must be 'combined with a close relationship of affection between the plaintiff and the primary victim'.[67] He asserted that such a requirement was a vital part of a 'combination of circumstances from which the necessary degree of "proximity" between the plaintiff and the defendant can be deduced'.[68] He accepted that the concept of 'proximity' was

[65] Law Commission (n 19) [6.10]–[6.13] citing Harvey Teff, "Liability for Negligently Inflicted Nervous Shock" (1983) 99 *Law Quarterly Review* 100.
[66] *Alcock* (n 4) 405.
[67] ibid 411F–H (Lord Oliver).
[68] ibid 411.

> an artificial one which depends more upon the court's perception of what is the reasonable area for the imposition of liability than upon any logical process of analogical deduction.[69]

What he was seeking was some 'reasonable' boundary for liability, and one element of that, in his view (and that of the rest of the House), was a degree of personal, emotional or relational closeness between the primary victim and the person who witnessed injury occurring to them or its aftermath. Why would this seem to be a reasonable or relevant boundary? The only explanation can be that it is only in cases where there is a close relationship between the victim and the witness that it is reasonable for the witness to then suffer a psychiatric injury as a result of seeing them harmed. That must be the logic at work for it to operate as a legitimate boundary.

Lord Oliver did not, however, state this so baldly. Rather, he asserted in justification that such a relationship was to be found in 'all the reported cases of this type decided in this country prior to the decision of Hidden J. in the instant case and in which the plaintiff succeeded in establishing liability'.[70] Seemingly, he considered this sufficient to support his point.

However, it is not, with respect, entirely true that all preceding successful claims involved a parental or marital relationship. In the well-known case of *Chadwick v British Railways Board* (1967),[71] the secondary victim had no relationship with the primary victims, but it was entirely foreseeable he would suffer psychological trauma from what he witnessed. Lord Oliver was of course aware of the decision in *Chadwick*, but spent little time addressing this point (despite it being raised by counsel). Rather, he merely said that

> the plaintiff recovered damages for the psychiatric illness caused to her deceased husband through the traumatic effects of his gallantry and self-sacrifice in rescuing and comforting victims of the Lewisham railway disaster.[72]

This suggests he considered recovery to arise due to Chadwick being a *rescuer*, but that position is not well-supported (nor does it any longer hold).

A better reading of *Chadwick* comes from setting it against the background of earlier decisions, including some that were not reported from the lower courts. It has not always been the case that a close relational tie was required, nor that that tie need be marital or parental. In *Hambrook*, Lord Atkin was open to the idea that a bystander might be compensated for seeing something terribly traumatic. Such

[69] ibid 411.
[70] ibid 411.
[71] [1967] 1 WLR 912.
[72] *Alcock* (n 4) 408.

was the interest in the question of whether the duty might extend to bystanders that the issue was the subject of the 1927 Middle Temple moot between Middle Temple and Trinity College, Oxford. The subject of the moot was an accident, in which a car narrowly missed a child but hit a man named Herbert.[73] The fictional claimant was a Mrs Herbert, his wife, who suffered nervous shock from witnessing the incident, and one of the issues was whether her shock arose from fear for her husband, or for the child (who was not known to her). In the facts presented, she answered that it was fear for the child's safety that caused her shock. Interestingly, the moot court found in favour of Mrs Herbert.[74] Thus the question for the moot was, in effect, an application of *Hambrook* extended to bystanders in line with Lord Atkin's *obiter* comments in that case, that if psychiatric harm to a bystander is foreseeable in accident cases, it should be compensated.

Post-*Hambrook*, Lord Atkin's view on bystanders was not widely adopted, however cases can be found in which bystanders succeeded. For example, the *Solicitor's Journal* in 1928 reported the decision in *Henderson v Middlesbrough Corporation* in which the plaintiff was walking along the pavement when a bus swerved and injured a man standing near her.[75] She reported that ever since the accident, she had been 'a victim of nerves' and unable to undertake household duties. She received £10 damages for the extra cost of outside help to cover these duties.[76]

Even setting aside the question of mere bystanders, post-*Hambrook* the lower courts did not limit successful claims to those in marital or parental relationships. In *Benton* (1934),[77] an uncle successfully claimed for seeing his nephew killed in a machine, and in *Owens v Liverpool Corporation* (1939) the successful claimants included an uncle, a cousin and the cousin's husband.[78] The point here is not to debate past cases, but rather to point out that it is not accurate (entirely understandably) to say that a parental or marital relationship has always provided a perfect delineation at law between when it will and will not be reasonable for a person to suffer a psychiatric illness when witnessing something awful happen to someone else.

The reason this is so should be self-evident—it is not merely the relationship between a witness and a victim that determines whether it is reasonable that the witness might develop a psychiatric injury. This is precisely why *Chadwick* is an apt example. His illness derived not from the fact that he saw a loved one in distress, but because he saw *people* in distress, truly dreadful distress, and it was clearly the

[73] 'Middle Temple Moot', *The Times*, 13 May 1927, 5.
[74] ibid.
[75] Unreported County Court decision, reported in *The Solicitors Journal*: 'Our County Court Letter: Damage from Shock' *The Solicitors Journal*, 7 April 1928, 234.
[76] ibid.
[77] This is an unreported judgment, and the full case name is not given in the newspaper report of the decision: '£450 for Shock' *Daily Mail*, 16 March 1934, 5. Source: *Daily Mail Historical Archive*, 1896–2004.
[78] *Owens* (n 16).

case that this *in and of itself* was sufficient to lead to the deleterious state of his mental health in the months and years that followed the disaster. One only has to read a little of what he saw that night to appreciate why his reaction was entirely reasonable—and foreseeable.[79]

The close ties requirement finds no support in the medical understanding of the causes of psychiatric illness. It is certainly not the case that PTSD or other psychiatric illnesses arise only from witnessing harm to loved ones. The diagnostic criteria for PTSD in the DSM-V are not limited to events that occur to family members, referring only to 'traumatic events' more generally, but they do list examples such as 'first responders collecting human remains: police officers repeatedly exposed to details of child abuse', neither of which involve family members.[80] In only one context is there a requirement that the events involve family members (or friends): where the condition is triggered by indirect exposure, where the Manual states

> indirect exposure through learning about an event is limited to experiences affecting close relatives or friends and experiences that are violent or accidental (e.g., death due to natural causes does not qualify).[81]

Indeed, the clue lies even in the old name for psychiatric illness—nervous shock, a term that has its origins in the era in which our current understanding of the causes of post-traumatic shock disorder began to emerge, namely the First World War. Soldiers developed shellshock from many aspects of their war experience, from fear for their own lives, from the sound of gunfire and bombing, and from the generally traumatic conditions in the trenches. Some, certainly, developed from witnessing the death and injury of others, some of whom were friends, but many of whom were merely strangers. Yet we would not suggest that they, nor anyone else who develops PTSD has done so *unreasonably* because their reaction was not to an event that befell a close relative. Indeed, it would be offensive to do so.

The logical (and medical) foundation of the close ties requirement is therefore exceptionally weak. But, we acknowledge, lines may need to be drawn, and in *Alcock*, given the numbers of claims, we can appreciate why in that particular context, the House of Lords felt pressure to prevent the opening of the dreaded floodgates. However, we should ask a few questions of this. First, why should it be the

[79] [1967] 2 All ER 945. Chadwick suffered psychiatric injuries after spending many hours attending the Lewisham rail crash (1957). He attended victims, offering comfort and administering care. One witness described the scene as a 'sea of bodies', and the facts reported in the decision note his particular distress at the fate of a young boy. Accounts of the events also tell us that he saw many people in terrible pain, including one who had been impaled but was still conscious. It is small wonder that he developed an anxiety disorder, including personality changes, as a result of his experience.

[80] 'Post Traumatic Stress Disorder' in American Psychiatric Association, *Diagnostic and Statistical Manual of Mental Disorders* (5th edn, Arlington, VA: APA, 2013) 265.

[81] ibid 266.

case that the causing of *great* harm to *many* people should have a protective effect on those defendants? It seems entirely perverse to say that because so many people might experience trauma as a result of seeing the preventable and terrible deaths of nearly 100 people, those who caused it should be shielded from the volume of that culpability. The logical corollary of that must be that had they injured *fewer* people, then a different approach could have been taken and liability found. Morally, that must be wrong.

Second, should not the question be whether they *ought* to be liable, based on the responsibilities they took on and their capacity to prevent harm? That, fundamentally, is the heart of duty of care. Certainly, that must be limited to some extent, and proximity (in tandem with foreseeability) does that work. But it does so by focusing on reasonable limits, and hence the real question should focus on properly engaging with what a reasonable limit would be. Using the clumsy line drawn by a close relationship, a line with no grounding in logic, human experience or medical science, is not the right way to do this. Rather, we must turn to a contextual and holistic approach in evaluating reasonableness in a given situation. Such an approach is also exactly what the courts do more generally, and can do to prevent the feared influx of 'dubious' or even 'imaginary' claims. Hidden J made precisely this point in the High Court in *Alcock*, when he presented an analysis of past English cases and many from other jurisdictions to offer

> a timely reminder of the way in which courts can deal with such problems when they are presented with them, and the way in which, even when the lines as to particular relationships is extended by a court, that court can nevertheless differentiate between genuine and unestablished claims.[82]

His point, and ours, is that the courts need not draw sharp, simplistic lines, but can instead evaluate situations with nuance. That is, in fact, exactly why the government refused to follow the legislative reforms proposed by the Law Commission—because it considered this judicial approach more able to respond to the varied situations in which psychiatric injuries might arise. Decisions such as *W v Essex*[83] and *Yearworth*[84] are good evidence that the courts can do this work, and demonstrate why rigid rules prevents them doing so (or forces them into problematic manoeuvring to achieve the just outcome).

Such evaluation should be directed at what the control mechanism is attempting to bluntly identify—when it is reasonable and reasonably foreseeable for a person to suffer such an injury. The concept of 'ordinary fortitude' coupled with medical evidence can achieve this in a principled manner, given that both are focused on

82 [1991] 1 All ER 353.
83 *W v Essex* (n 23).
84 *Yearworth* (n 25).

that question much more directly than a crude relational category approach. An approach of this kind also avoids exactly the concern Stapleton raised in 1994, that to require claimants to demonstrate a relationship of sufficient closeness necessarily creates the unpleasant situation of requiring claimants to prove their love for a relative. As she put it:

> In future cases will it not be a grotesque sight to see relatives scrabbling to prove their especial love for the deceased in order to win money damages and for the defendant to have to attack that argument? Moreover, most other feasible control devices which might be used instead of 'ties of love' would still turn on the status of the plaintiff and so produce the same unseemly arguments over meaningless boundaries.[85]

We should return, instead, to the earlier approach, before it hardened into the relationship requirement. The earlier cases are better understood not as having drawn a sharp line around the 'right' kind of relationship, but instead as looking to foreseeability and the medical evidence of whether a sight has (as well as could) trigger the complained of illness. That cannot just be a question of the relationship in play, precisely because it is evident that people suffer psychological trauma from witnessing events where there is no relational dimension. Therefore, a relationship cannot be a *necessary* factor for the precipitation of such an injury, so that cannot be a basis on which to require it. There would need to be some other reason to make it a requirement, or it should be reduced to something merely to *consider* as indicative of when an injury might be foreseen.

The reason to consider these cases is that they, unlike the presumption now to be made post-*Alcock* and *White*, saw the courts asking not 'is the relationship one of the right kind?' but rather 'might we foresee psychological harm befalling a person in this situation?' where the nature of their relationship was a factor in reasoning out the answer. That is, the courts once took the approach that both Lord Scarman and Lord Slynn asserted they should, but which the framing of the relational control makes much more difficult. It is made difficult in particular because of the *presumptive* approach, where rather than the relationship feeding in as a factor, certain categories of relationship are presumed to be sufficient (and others explicitly, not).

Courts need to approach these situations by severing ties with the requirement for a close relationship, and instead starting fresh by thinking about how a claimant might be psychologically injured via the combination of what they see, how they see it and to whom it happens. Just as they do in other contexts, a matrix of these factors working together can be evaluated, supported by evidence of how

[85] Stapleton (n 1) 94–95.

the claimant's condition has arisen. Of course, the involvement of a relative may increase susceptibility to injury, but so too might the particular horror of the events (absent any relationship). Removing the close ties requirement affords the court space in which to consider each situation on its own merits appropriately. It should also be recognised that the concept of reasonable foreseeability can do much of the work in distinguishing those instances that are reasonable, as it will almost always be foreseeable that witnessing trauma to a close relative may trigger psychiatric illness, while in the absence of a relationship, asking whether such an illness arising is foreseeable is a logical first step to determining whether a claimant could be expected to react as they did to highly distressing sights.

4. What Should We Do?

More than a quarter of a century ago, the Law Commission set out why three of the four control mechanisms should be abolished. It considered that only the ties of love and affection requirement should be retained.[86] It stated that there should be no shock requirement, and secondary victims

> should be entitled to recover damages from the negligent defendant in respect of [their] illness, regardless of the plaintiff's closeness (in time and space) to the accident or its aftermath or the means by which the plaintiff learns of it.[87]

It recommended legislative reform to achieve this, but its recommendations were eventually rejected by the Department for Constitutional Affairs in 2007 on the grounds that the courts had adopted a flexible approach, and hence reform was best left to the judiciary. The Ministry of Justice concurred with this view in 2009.[88] Clearly, however, the Department's view was misplaced (and given the lack of flexibility displayed by the courts in the years leading up to it, ill-founded). Far from developing and demonstrating flexibility, the law has stood still and ossified. The approach taken to the claimants in *Paul* in the Court of Appeal is a perfect example of the inflexibility of the control mechanisms which now operate as fixed rules from which the courts do not feel they can diverge. This is far from surprising, given that Lord Hoffman in *White* made it abundantly clear that the House of Lords did not intend to take a reformist approach. As he said then:

[86] Law Commission (n 19) 122–124.

[87] ibid Recommendation 11.

[88] Law Commission, *Annual Report* (2009–10), [3.11]. As will be clear from the foregoing, in our view this was incorrect, and the courts were unable to adopt a flexible approach.

> It is too late to go back on the control mechanisms as stated in the *Alcock* case [1992] 1 A.C. 310. Until there is legislative change, the courts must live with them and any judicial developments must take them into account.[89]

So the law has stood still, despite the clear case for reform articulated by the Law Commission and endorsed by many judges and commentators. The courts feel themselves bound, and the government has chosen not to act for nearly a quarter of a century. Given the government's refusal to act, it must be for the Supreme Court to take the much-needed step towards reform itself— and it has this opportunity in the cases of *Paul*, *Polmear* and *Purchase* before the Supreme Court at the time of writing this chapter.

The key change required is to move away from the concept of 'control mechanisms' entirely, and return instead to the fundamental logic underpinning the duty of care inquiry, namely to ask, 'against which risk should someone be expected to guard?' The second is for this question to be approached by reference to the basic principles of reasonable foreseeability and proximity. Doing so would afford the courts the flexibility they need, but also provide an opportunity to develop *principled* approaches to deciding claims, based on foreseeability and proximity as they can be applied in such cases, rather than either applying, or working around, the current strict controls.

5. Postscript

Shortly before this volume went to press, in January 2024, the Supreme Court handed down the decision in *Paul*.[90] We are grateful to the editors and to Oxford University Press for allowing us to include this very brief postscript summarising that decision. In the context of Jane Stapleton's work and the foregoing, we venture that the emphasis placed by both the majority and the dissenting judgment of Lord Burrows on incrementalism and its importance in the development of the law is notable.

The majority judgment, written jointly by Lord Leggatt and Lady Rose,[91] considered the action for pure psychiatric injury in tort, and framed it as a longstanding exception to the widely accepted principle that no action can lie in negligence for death or injury to another person. The majority declined to revisit the established law, and reaffirmed the decisions in *McLoughlin*, *Alcock* and *White* (*Frost*). They stated:

[89] *White* (n 9) 504.
[90] *Paul and Another v Royal Wolverhampton NHS Trust* [2024] UKSC 1.
[91] With whom Lord Briggs, Lord Sales and Lord Richards agreed. Lord Burrows dissenting.

> Historically, the leading cases have involved accidents (mostly road traffic accidents). The question then is whether the rules which have been developed in those cases either apply already or can *by a permissible incremental development of the common law* be extended to apply to claims of the present kind arising in the field of medical negligence. (emphasis added)[92]

In their discussion of the leading decisions, the majority clarified that the action (or exception) only applies to those who witness an 'accident', defined as 'an unexpected and unintended event which caused injury (or a risk of injury) by violent external means to one or more primary victims'.[93] The tragic circumstances presented by each of the claimants could not be considered accidents within this ordinary meaning, nor would the vast majority of clinical negligence claims that might arise.

In reviewing the control mechanisms, the majority did address some of the most widely criticised of them and rejected both the requirement for a claimant to have suffered a 'sudden shock', and that they must have witnessed a 'horrifying event'. Further clarification of the 'present or at the immediate aftermath' control mechanism was also provided, and it is now tolerably clear that a claimant must be either present at an accident or arrive upon the scene very shortly thereafter. The facts of *McLouglin* were said to lie 'on the margin' of what might be considered the scene of an accident or its aftermath.[94] The ambiguity introduced by the reasoning in *Walters* wherein 'a series of events' perceived by the claimant as a single continuum extending the duration of 'immediate aftermath' was rejected.[95]

In his dissenting judgment, Lord Burrows found for the claimants. He stated 'these appeals present a rare opportunity for this court to move the law forward, *by traditional incremental development*, to a more satisfactory position …' (emphasis added).[96] He joined with the majority in rejecting the 'shocking and horrific' requirements, but fundamentally disagreed on the requirement for an 'accident' to ground a claim. Taking up the question of what constitutes a 'relevant event' posed in *Novo*, he determined that the 'relevant event' must be the death of the primary victim and that, flowing from that analysis, foreseeability and the control or proximity factors were satisfied in the claims in question.[97] More broadly, Lord Burrows argued for the incremental extension of the action to claims not fitting the 'accident' rubric, including where the relevant event witnessed by a claimant resulted from clinical negligence.[98]

[92] *Paul* (n 90) [23].
[93] ibid [24].
[94] ibid [108].
[95] ibid [121].
[96] ibid [202].
[97] ibid [254].
[98] ibid [201]–[204].

10

An 'Opaque Conclusionary Label'?: Assumption of Responsibility and Pure Economic Loss

Philip Sales[*]

1. Introduction

Jane Stapleton's contribution to the law has been immense. Her work is closely read by judges. She is the perfect example of an academic who, in Lord Goff's words, has been a 'fellow-pilgrim with the judiciary in a common search for principle'.

In this chapter I try to carry on the task of reflexive doctrinal scholarship championed by Jane. I do this by setting out my response as a judge to an aspect of Jane's work: her cogent critique of the concept of assumption of responsibility as the basis for a duty of care in relation to pure economic loss, set out in *Three Essays on Torts*.[1] There she questioned whether the concept is anything 'more than an opaque conclusionary label'.[2] I put forward a limited defence. Assumption of responsibility is a viable and justified basis for imposition of a duty of care in tort in a central category of case, but outside that category it is much more difficult to sustain as an explanation for liability in tort for pure economic loss. Accordingly, I will argue that Jane's critique is misplaced in relation to the central category, but is justified in relation to attempts to apply the idea outside that category. If the concept is to be applied outside the central category of case, Jane is right to highlight the need for articulation of specific criteria to govern that application and give the concept determinate content in the other contexts where it is to be used.

This chapter is divided into six sections and a conclusion. The first section locates the ancient roots of the concept of assumption of responsibility; the second discusses pure economic loss in the law of negligence; the third addresses the history of the idea of assumption of responsibility; the fourth identifies how the idea came under pressure from the conception of legal science in the nineteenth century; the

* I am grateful to Jake Thorold for his excellent assistance and to Lord Burrows for comments on a previous draft. The usual caveat applies.

[1] J Stapleton, *Three Essays on Torts* (2021)

[2] Ibid, 60. See also K Barker, 'Unreliable Assumptions in the Modern Law of Negligence' (1993) 109 LQR 461.

Philip Sales, *An 'Opaque Conclusionary Label'?: Assumption of Responsibility and Pure Economic Loss* In: *Torts on Three Continents*. Edited by: Kylie Burns, Jodi Gardner, Jonathan Morgan and Sandy Steel, Oxford University Press.
 DOI: 10.1093/oso/9780198889748.003.0010

fifth explains how the conception of legal science itself came under pressure and assumption of responsibility re-emerged as a significant formulation in the law; and the sixth discusses the modern uses of the idea of assumption of responsibility.

2. Assumption of Responsibility: Ancient Roots

Assumption of responsibility was not a novel doctrine cooked up in *Hedley Byrne v Heller*.[3] It has its roots in old common law forms of action available before the systematisation of the law of obligations into the twin pillars of contract and tort from the mid-nineteenth century. *Hedley Byrne* and the identification of assumption of responsibility as a basis for liability can be viewed as a re-emergence of old strands of common law thinking as the bifurcated structure of contract and tort came under growing strain during the twentieth century. That bifurcated structure was a major intellectual achievement, as the law came to be theorised in terms of substantive rights and obligations rather than as a commentary on the historic forms of action. But the claims for scientific rigour associated with that effort sometimes went too far in suppressing previous insights and aspects of the fair mediation of relationships which the old forms of action, as developed over centuries, were capable of capturing. Human life is manifold and variegated and expresses a plurality of values. Sometimes a certain bagginess in common law doctrine can be helpful in accommodating this.

Examination of the historical roots of assumption of responsibility can help to elucidate its normative justification in the present. The history illustrates the *sui generis* place of assumption of responsibility within our law of obligations. Although commonly taught as part of the tort of negligence or cast as a form of 'contract-lite', in reality assumption of responsibility does not fit neatly within either of the orthodox categories of contract or tort. It might perhaps be better conceptualised as a distinct and significant source of rights and obligations with a particular history and purpose. We should be careful not to obscure the basis and nature of this form of obligation by trying to fit it into a Procrustean framework which insists on pigeon-holing it as solely contractual or as based on limited categories of tort.

In this sense, the idea of assumption of responsibility as restated in *Hedley Byrne* can be seen alongside other mid-twentieth century doctrinal developments which challenged the dominance of contract and tort as legal categories. These developments included proprietary estoppel and unjust enrichment. Such developments were based on a sense that a rigid insistence that claims be framed within the strict criteria of contract and tort carried the risk that the law would become divorced from common and fairly basic conceptions of justice. Assumption of responsibility, as a distinct basis of liability, provides a route to avoid claims foundering on

[3] [1964] AC 465.

the requirement of consideration in circumstances where insisting upon it would produce a morally unjust outcome. In all three areas, of course, spelling out the criteria for imposition of liability carries its own risk of introducing uncertainty into the law. That has been a common criticism of each of these areas as doctrine has deviated from the strict paths of contract and tort, to explore other ways of looking at the creation of obligations. The challenge is for doctrine in such areas to provide sufficiently determinate tests for liability and also to provide an adequately coherent account of their interaction with other areas of liability, in particular contract and tort.

3. Pure Economic Loss in Negligence

In *Spartan Steel v Martin & Co*[4] a contractor carrying out work digging up a road negligently damaged an electric cable which they knew was the supply for power to the claimant's factory. The claimant had to dispose of molten metal out of their furnace to prevent damage to the furnace, which meant that the metal lost value and they suffered a loss of profit when they sold it. They also lost profits from further smelting they could have undertaken had the power not been cut. The Court of Appeal held that the defendant was liable for the loss of profit on the ruined metal as foreseeable financial loss immediately consequential on the foreseeable physical damage to the metal, but not for loss of profit from the further smelting. That was pure economic loss which was irrecoverable.

One rationale for limiting the circumstances in which liability will be imposed in tort for causing pure economic loss, as explained in *Spartan Steel*, is a policy one. If the law allowed recovery for all economic loss the extent of liability in many situations would be enormous. For example, the contractor who cut the mainline electricity cable would have been liable to a range of businesses whose trade had been interrupted. This single mistake would have bankrupted the contractor. It would have been forced to carry an excessive part of the risk associated with a socially useful activity which it carried on. As Lord Denning MR said, in this case, 'in such a hazard as this, the risk of economic loss should be suffered by the whole community who suffer the losses—usually many but comparatively small losses—rather than on one pair of shoulders, that is, on the contractor on whom the total of them, all added together, might be very heavy'.[5]

Cardozo CJ famously stated in *Ultramares Corp v Touche* that to permit recovery of all loss suffered by anyone as a result of having relied on a negligent misstatement which came to their attention would be unacceptable, as it would involve imposing 'liability in an indeterminate amount, for an indeterminate time, to an

[4] [1973] QB 27.
[5] [1973] QB 27, 39.

indeterminate class'.[6] This has an underlying policy basis, but it also hints at something else. To impose liability for economic loss suffered in such circumstances would mean that the ambit of liability would pull apart from the forms of relationship which might justify imposing liability in the first place. To balance interests justly, the law has to focus specifically on the relationship between claimant and defendant, to see whether and to what extent the imposition of liability is justified.[7]

The law has sought to protect defendants from being exposed to an unknowable scope of potential liability. A party like the contractor in *Spartan Steel* could not meaningfully incorporate such an indeterminate risk into the price charged for its services nor realistically could it obtain insurance to protect itself. It is unable to bargain around the risk it takes on, so without a rule against the recovery for pure economic loss it would bear all of the risk. If the risk cannot be managed in these ways, there will be a chill effect potentially deterring the activity from being carried on at all. Or the effect may be that the contractor is not incentivised to take proper care, since if the liability to be imposed if an accident occurs is crushingly great, and there is no way to eliminate all risk of an accident, it may calculate that it should simply cut its own costs to maximise its profits.

By contrast, assumption of responsibility provides a sound justification for imposition of liability in a distinct subset of pure economic loss cases in which the defendant *does* have an effective opportunity to bargain over the extent of the obligation to which it is to be subject. Where the relationship between the parties is such that the claimant's potential loss is both limited and foreseeable by both parties, so there is a meaningful opportunity for them to bargain over the liability to be taken on by the defendant, it is possible to spell out determinate criteria to explain when and to what extent liability should be imposed when something goes wrong. In a context where the defendant had a meaningful opportunity to negotiate over taking on the risk or refusing it, the concept of assumption of responsibility is the appropriate one to use.

In *Hedley Byrne*, for example, the defendant bank provided a credit reference to the claimant which was negligently prepared, and the claimant suffered loss as a result of relying on it. The bank had no contractual relationship with the claimant but knew that it would only be the claimant who suffered loss if their advice was negligent. Their risk exposure was therefore limited and identifiable. Although the bank opted to provide the statement for free, it had the possibility of negotiating the extent of risk it would take on and potentially extract a price for doing so. Indeed, it took advantage of that opportunity by expressly excluding any liability based on the statement. Given this opportunity to bargain, it was appropriate to frame the claim in terms of whether the bank had assumed responsibility to the claimant for the action which resulted in loss to the claimant.

[6] *Ultramares Corp v Touche* 174 NE 441 (1932), 444.

[7] See also *Spartan Steel* [1973] QB 27, 37.

In this context, assumption of responsibility has proved useful for delineating a principled basis for when recovery for pure economic loss is justifiable. Following *Hedley Byrne* and its extension to the negligent performance of a service in *Henderson v Merrett Syndicates*,[8] it is now commonplace for the courts to impose a duty of care in respect of pure economic loss on the grounds that the defendant had assumed responsibility to the claimant. In *Lejonvarn v Burgess*,[9] for example, the Court of Appeal held that assumption of responsibility provided a standalone basis for a duty of care for pure economic loss where an architect had provided her expertise free of charge to friends wanting to carry out extensive landscaping to their garden. Explaining the basis for the duty, the Court explained that the architect had 'voluntarily tender[ed] skilled professional services in circumstances where she knew the [claimants] would rely on the proper performance of those services'.

This provides a good encapsulation of the fundamental rationale of assumption of responsibility as a basis for finding a duty of care. An assumption of responsibility occurs when a person has voluntarily elected to take on a task for the claimant in circumstances where the claimant's intended reliance on the performance of that task is apparent along with the consequences should the defendant fail to perform with due skill and care. Unlike in classic examples of negligence, where a duty of care can exist simply by virtue of a person's general position or status, for example as a road user, the duty of care based on assumption of responsibility only arises because of something that a defendant has committed to do for a particular party and—typically—has in fact done. As pointed out in *Lejornvarn*, the architect 'did not have to provide any such services, but to the extent that she did she owed a duty to exercise reasonable skill and care in the provision of those services'.[10]

In contrast to contract, assumption of responsibility normally provides no remedy against a person who has promised to do something but entirely fails to carry out the promised task. It is the commencement of the task which normally triggers liability.[11] The strongest type of case for use of the concept is one where the defendant says to the claimant, in effect, 'Here, let me do that for you', and the claimant accepts and leaves that task to the defendant. The defendant, in its direct interaction with the claimant, has freely chosen to take on the task for their benefit, involving in ordinary circumstances detrimental reliance by the claimant in being put off from taking other steps.

In terms of modern doctrine, the concept has escaped the confines of pure economic loss. Given its historic antecedents, this is not surprising. But questions

[8] [1995] 2 AC 145.

[9] *Lejonvarn v Burgess* [2017] EWCA Civ 254; [2017] BLR 277.

[10] Ibid, [88].

[11] It is possible that the making of the promise to do something might itself be an act which induces detrimental reliance by the claimant, in the sense that the claimant believes that the promisor has indeed made arrangements to do that thing and therefore takes no steps to do it for itself (maybe until it is too late). It may be that a relevant assumption of responsibility could be found in such circumstances, if the promisor appreciated that its promise would be relied on in this manner by the promisee.

arise as to how far any satisfactory unifying principle can be stated to cover all such cases. The range from cases involving damage in the form of pure economic loss to other kinds of harm is very wide and the relevant factors bearing on the imposition of a duty of care are not uniform. Following *Michael v Chief Constable of South Wales*[12] there has been consideration of the circumstances in which the police can be said to have assumed positive obligations towards particular persons.[13] Similarly, the idea of assumption of responsibility has been used to try to identify when a non-delegable duty of care arises.[14] It is questionable how far a unifying theory can be applied across all the contexts in which the idea of assumption of responsibility has been invoked. The pure economic loss cases have a distinctive rationale based on the availability of an opportunity to bargain which justifies the imposition of liability.

4. Assumption of Responsibility: The History of an Idea

Despite broad judicial acceptance of the idea of assumption of responsibility, its historical roots are often overlooked. As Cardozo J observed, 'it is ancient learning that one who assumes to act, even though gratuitously, may thereby become subject to the duty of acting carefully, if he acts at all'.[15]

The ancestors to modern ideas of assumption of responsibility, termed the 'assumpsit duties', arose in the medieval period. For early lawyers these claims were not defined by substantive principles, but had to be brought using one of the various forms of action. These were usually framed by reference to particular established relationships, but included the more general 'action on the case'. This provided a remedy for the invasion of personal or property interests where the other established forms were not suitable. The criteria for making out an action on the case were not precisely defined.[16]

However, as demonstrated by Paul Mitchell, this lack of precision did not prevent categories of claim emerging within the broad parameters of the action on the case which resemble modern assumption of responsibility approaches.[17] As Maitland put it, from the fourteenth century onwards 'various precedents collect

[12] [2015] UKSC 2.

[13] See, eg, *Tindall v Chief Constable of Thames Valley* [2022] EWCA Civ 25; cf *Poole Borough Council v GN* [2019] UKSC 25.

[14] See *Woodland v Essex CC* [2013] UKSC 66; [2014] AC 537, paras 11–12; but more specific criteria were employed to identify when a non-delegable duty had arisen.

[15] *Glanzer v Shepard* 135 NE 275 (1922), at 276.

[16] See P Winfield, 'The History of Negligence in the Law of Torts' (1926) 42 LQR 184, 197 ('the action on the case has been the life-blood of the English law of torts but is calculated to make any professor of jurisprudence desperate').

[17] P Mitchell, 'Hedley Byrne & Co Ltd v Heller & Partners Ltd (1963)' in C Mitchell and P Mitchell (eds), *Landmark Cases in the Law of Tort* (2010).

in which the allegation is made that the defendant had undertaken to do something and then hurt the plaintiff either in his person or in his goods by doing it badly'.[18]

One significant category of action on the case was where a defendant professed a so-called 'common' or 'public' calling. Various occupations were classified in this way, meaning that the services rendered had to be performed with reasonable skill.[19] For example, if an innkeeper hosted a customer, they were deemed to have assumed responsibility for the safe keeping of that customer's possessions.[20] The duty was imposed by law, but only after the innkeeper had agreed to provide a service. Other professions considered 'common' or 'public' included common carriers, surgeons and attorneys. Usually, of course, the services were rendered in return for payment.

Bailment provides perhaps the clearest antecedent for contemporary understandings of assumption of responsibility.[21] In *Coggs v Bernard*, an action on the case against a carrier for carelessly transporting a hogshead of wine succeeded, even though (on the facts assumed by the court) the carrier agreed to carry the wine without payment. In his classic judgment, Holt CJ examined the nature of the duty owed by the carrier and how the extent of it was affected by the presence or absence of an agreement to pay him. The absence of payment did not mean that no duty was owed where someone agreed to carry out a task for another and that other relied on him. Holt CJ said,

> assumpsit does not only signify a future agreement, but in such a case as this, it signifies an actual entry upon the thing, and taking the trust upon himself. And if a man will do that, and miscarries in the performance of his trust, an action will lie against him for that, though nobody could have compelled him to do the thing.[22]

In giving an undertaking, the bailee led the bailor to believe that he possessed the capability to transport the wine carefully; the bailor would not have entrusted the wine to the bailee if he thought otherwise. The loss was viewed as attributable to the undertaking, or the assumed duty, which was not performed with reasonable care.

[18] FW Maitland, *The Forms of Action at Common Law: A Course of Lectures* (1948), 68; R Sutton, *Personal Actions at Common Law* (1929), 63.

[19] See, eg, *Elsee v Gatward* (1793) 5 TR 143.

[20] P Winfield, 'Duty in Tortious Negligence' (1934) 34 Colum L Rev 41, 44, 45: 'by professing a common calling a man in mediaeval times in effect assumed what we should call a *status* from which the duty sprang without further undertaking'.

[21] Cf Y B Hil. 2 Hen VII, f11 pl 9, concerning a sheep drowned in the care of the defendant: 'When the party has taken upon himself to look after the sheep, the action lies because he has not executed his bargain and has undertaken to do this.'

[22] (1703) 2 Ld Raym 909, 919.

Powell J drew a distinction between non-performance and mis-performance, which persists to this day. If a promise was made without consideration, '[a]n action will not lie for *not* doing the thing for want of a sufficient consideration; but if the bailee will take the goods into his custody, he shall be answerable for them; for the taking of the goods into his custody is his own act.'[23] The carrier could have elected not to carry the wine or decided only to do so for something in return. The reasoning here is essentially indistinguishable from that in modern assumption of responsibility cases such as *Lejonvarn v Burgess*. The fact that the assumption of responsibility was gratuitous in both cases did not matter because the defendants in both cases could have negotiated either a price in return or a different allocation of the risk, or simply walked away from the arrangement. It is the existence of a real opportunity to bargain over the task and the extent of the responsibility to be taken on which makes it appropriate to speak of assumption of risk, and to treat that as the source of liability.

Coggs v Barnard was followed in *Elsee v Gatward*.[24] The claimant brought an action upon the case against a carpenter on two grounds: first, for failing to complete works which had been agreed and secondly for having used new materials rather than the old ones he was instructed to use. Relying on *Coggs v Barnard*, the second claim was allowed because 'if a party undertake to perform work, and proceed on the employment, he makes himself liable for any misfeasance in the course of that work'. The first claim, based on non-performance, failed because no consideration had been pleaded. Kenyon CJ explained, 'If this had been an action of assumpsit, it could not have been supported for want of a consideration … if assumpsit will not lie in such a case, there is no technical reasoning that will support such an action as for tort.' This illustrates the gradual division of the action on the case into distinctive categories of contract and tort during the eighteenth century. The meaning of the word 'assumpsit' had changed since *Coggs v Barnard*, where it meant an undertaking and formed a part of the action on the case. In *Elsee v Gatward* the term had come to signify a distinct cause of action concerned with holding people to their agreements and possessed of its own principles, most centrally a requirement for consideration. Yet still the basis for liability combined elements of modern contract and tort. The significance given to the undertaking mirrored contract, while the imposition of liability for damage caused through negligence resembled tort.[25]

The principle in *Coggs v Barnard* also had application in the context of claims for pure economic loss. In *Wilkinson v Coverdale*[26] the defendant gratuitously undertook to get a fire policy renewal for the plaintiff but, in doing so, negligently omitted

[23] 2 Ld Raym 909, 911. See D Ibbetson, 'Coggs v Barnard (1703)' in C Mitchell and P Mitchell (eds), *Landmark Cases in the Law of Contract* (2008).

[24] (1793) 5 TR 143.

[25] P Winfield, 'The Province of the Law of Tort: Tagore Law Lectures Delivered in 1930' (1931), 96.

[26] *Wilkinson v Coverdale* (1793) 1 Esp 75, 170.

certain formalities, thus rendering the policy unenforceable. The plaintiff's premises burned down. Lord Kenyon 'expressed a doubt whether any action could be maintained on such an undertaking' given that the defendant 'had undertaken it gratuitously on the plaintiff's account' without consideration; but he agreed with the plaintiff's submission that

> though there was no consideration for one party's undertaking to procure an insurance for another ... where a party voluntarily undertook to do it, and proceeded to carry his undertaking into effect by getting a policy underwritten, but did it so negligently or unskilfully, that the party could derive no benefit from it, that in that case he should be liable to an action.

The law was in flux. Consideration was becoming a central concept, but it co-existed uneasily alongside other established bases of liability not dependent upon the defendant having received something in exchange for their undertaking.

The old cases allow one to pick out an underlying logic. A defendant is taken to have assumed responsibility to carry out a task with reasonable care where they have dealt with the claimant in circumstances which allowed the defendant a fair opportunity to refuse to do the task or to bargain, if they wished, to limit or define the extent of their responsibility, and knowing of the importance of the task for the claimant, the defendant chose to take it on. In these cases the defendants could have negotiated to allocate differently the risk that they decided to take on, or could have chosen not to carry out the task. It was not for the courts to save them from the consequences of deciding to proceed.

5. Assumption of Responsibility under Pressure from Legal Science

Assumption of responsibility is a concept with a long history, which shows that our concepts of contract and tort were not preordained systems of organisation. As is well documented, they emerged relatively late as dominant categories in the law of obligations. They were laid over the existing patterns of liability according to the forms of action, and came to mould those patterns in new ways according to a new set of ideas about substantive rights and obligations.

During the nineteenth century the traditional forms of action began to lose traction as a framework for analysis. This was partly due to procedural reforms which made them less important and partly to new ways of thinking about law, in terms of a new science of law. The modern law of tort and contract began to be articulated and developed. They became dominant, apparently leaving little room for anything else. As a crude generalisation, assumpsit morphed into contract while various strands of the action on the case coalesced into the tort of negligence.[27]

The story is complex but, to speak broadly, in light of the prestige of the natural sciences and Benthamite philosophy the law increasingly became seen as a science, entailing a view that it should be taken to consist of definite formulae with intellectual, 'scientific' coherence a priority. The old forms of action were not regarded as a satisfactory way of ordering the law, with many gaps and applying in what had come to seem a fairly random way. More intellectual coherence was required to provide predictability. The common law increasingly focused on bright line rules, epitomised by its strict requirements in relation to the formation of contracts and discreet elements required for a claim in negligence.[28]

Despite this, the ambiguous place of traditional assumption of responsibility cases in this schema was apparent. In *Boorman v Brown*[29] the defendant oil broker was retained by the claimants to sell and deliver linseed oil to customers against receipt of cash. The defendant failed to use reasonable skill and care to comply with this instruction and delivered the oil to a customer on credit, who then became bankrupt so that the claimant was unable to recover the price due for its oil. The claim was for pure economic loss, brought as an action on the case, not in assumpsit. Nonetheless, it was held that the claimant could recover. The duty to exercise care resulted from a contract between claimant and defendant, but an action in tort lay for breach of that duty. It was necessary to refer to the terms of the particular contract, since the detail of what the defendant broker was supposed to do could not simply be derived from their status as broker. The argument was framed in terms of whether there was a misfeasance or a non-feasance and ranged across many authorities, including *Coggs v Barnard* and *Elsee v Gatward.*

Tindall CJ observed that there was 'a large class of cases'—for example claims against attorneys, surgeons and other professionals—'in which the foundation of the action springs out of privity of contract between the parties, but in which, nevertheless, the remedy for the breach, or non-performance, is indifferently either assumpsit or case upon tort… for want of competent skill or proper care in the service they undertake to render'. For such claims, the claimant could elect to proceed in tort or contract. The distinction between misfeasance and non-feasance was said to be 'very fine and scarcely perceptible'. He noted that

> the action of case upon tort very frequently occurs where there is a simple non-performance of the contract… as in the case of *Coggs v Bernard*… The principle

[27] See D Ibbetson, *A Historical Introduction to the Law of Obligations* (1999), Part 3.

[28] The first edition of *Chitty on Contracts* was published in 1826, that of *Clerk & Lindsell on Torts* in 1889.

[29] *Boorman v Brown* (1841) 3 QB 511, 525–526; aff'd 11 Cl & F 1; 8 ER 1003 (HL).

in all these cases would seem to be that the contract creates a duty, and the neglect to perform that duty, or the non-feasance, is a ground of action upon a tort.

Over the course of the nineteenth century the previously diffusely applied principle that a person could be liable for inadvertent harm to someone else crystallised into the distinct tort of negligence, dependent on establishing a duty of care.[30] So, compared to assumption of responsibility, negligence in its modern form was a latecomer. As *Boorman v Brown* shows, duty of care analysis and assumption of duty analysis were not wholly distinct, but overlapped.

Ideas of contract had a much longer history, yet again these were also only rationalised into the contemporary model during the nineteenth century. European Will Theory was especially influential. This conceptualised contractual liability as stemming from the mutual assent of the parties to a bargain rather than the breaking of a promise given freely by one party to another. The aim of contract law became to discover what the parties intended to agree and enforce it, reflecting a moral view that where there had been a meeting of minds it was right to give people what they expected to receive as a result.[31]

Historians have argued that the common law concept of consideration sat uncomfortably with the Will Theory, suggesting that if liability depended exclusively on the union of the parties' wills, then there was no reason to impose any additional element of reciprocity.[32] Yet consideration continued to be regarded as an important element in the law of contract and was the principal dividing line which marked it off from the province of tort.

There was perhaps more of a place for consideration to apply as an aspect of the Will Theory than is sometimes allowed. If there is consideration for an agreement, it can serve as a proxy or marker that the party to it had and made use of a genuine opportunity to bargain over the terms on which they were prepared to accept legal responsibility for the task for which the agreement provides. On this view, what they ended up agreeing to can therefore be regarded as a genuine product of the exercise of their will. As foreshadowed above, the modern assumption of responsibility cases in the pure economic loss context can potentially be explained by reference to a similar rationale based on the opportunity to bargain, albeit this depends on an analysis of the particular facts without using the proxy of consideration to reach that conclusion.

Moreover, a form of assumption of responsibility found a firm place in contract doctrine in the rules governing remoteness of loss. The rule in *Hadley v Baxendale*, limiting legal liability for damages arising from a breach of contract to matters of which the party in breach was aware at the time of contracting, and in relation to

[30] *Heaven v Pender* (1883) 11 QBD 503.
[31] See, eg, WD Evans, *A Treatise on the Law of Obligations or Contracts, by M. Pothier* (1806).
[32] See, eg, Ibbotson (n 27), 236.

which it had a fair opportunity to bargain either as to price or limitation of liability, reflects this way of thinking.[33]

However, the supposed rigours of legal 'science' made it more difficult to maintain a picture of disparate sources for legal obligations springing up in a vibrant and variegated garden. The ambiguous position of assumption of responsibility as between contract and tort made it vulnerable. Its focus on a one-sided undertaking, not necessarily tied to consideration, was at odds with Will Theory's preoccupation with mutuality. Moreover, the imposition of liability for inadvertent fault had coalesced into the tort of negligence, which did not depend on any undertaking being willingly given by the defendant. According to negligence analysis, strangers might owe duties of care to each other. Road accidents were a classic example of this. *Donoghue v Stevenson*[34] was the product of this view of the tort.

The history of proprietary estoppel shows that the concept of detrimental reliance as a basis of liability also faltered in the face of Will Theory, as the law centred on enforcing parties' shared expectations rather than compensating for misplaced reliance.[35] The law of proprietary estoppel emerged to reflect the justice of the case in situations where a property owner encouraged another to act to their detriment in the belief that they enjoyed property rights or would receive property rights in the future, but where the legal formalities to convey or contract to convey property were not observed. There was always a question as to how far the doctrine should be allowed to circumvent legal formalities and the usual rules for formation of a binding contract. Orthodox contract theory tended to push it to the margins, but in the twentieth and twenty-first centuries the doctrine gained new vigour. The parallel fates of assumption of responsibility and detrimental reliance in the law of proprietary estoppel are perhaps unsurprising given the scope for them to overlap.

6. Legal Science under Pressure: Re-emergence of assumption of Responsibility in *Hedley Byrne*

This may help to explain why both assumption of responsibility and detrimental reliance in the form of proprietary estoppel re-emerged in the mid-part of the twentieth century. Aspects of the law of obligations came to seem increasingly untethered from common conceptions of justice. The bright line rules favoured by the nineteenth century jurists—especially in contract—came to be seen as too

[33] *Hadley v Baxendale* (1854) 9 Exch 341; *Victoria Laundry (Windsor) Ltd v Newman Industries Ltd* [1949] 2 KB 528.

[34] [1932] AC 562.

[35] P Sales, 'Proprietary Estoppel: Great Expectations and Detrimental Reliance' in N Mrockova, A Nair and L Rostill (eds), *Modern Studies in Property Law, vol 12* (2023); also P Atiyah, *The Rise and Fall of Freedom of Contract* (1979). The balance between expectations and detrimental reliance is at the heart of the division of opinion in *Guest v Guest* [2022] UKSC 27.

rigid, being overly focused on certainty and predictability at the expense of a more basic sense of doing what is right.

In *Hedley Byrne* the Law Lords strained to find a way around the unforgiving requirements of contract doctrine. This is particularly clear in Lord Devlin's speech, most obviously in his incredulity that

> if a doctor negligently advises a patient that he cannot safely pursue his occupation when in fact he can and he loses his livelihood, there is said to be no remedy. Unless, of course, the patient was a private patient and the doctor accepted half a guinea for his trouble: then the patient can recover all. I am bound to say, my Lords, that I think this to be nonsense.[36]

Lord Devlin's response was to invoke the line of authority from *Coggs v Barnard*, drawing from it the conclusion that 'a promise given without consideration to perform a service cannot be enforced as a contract by the promisee; but if the service is in fact performed and done negligently, the promisee can recover in an action in tort'.[37] Lord Devlin recognised, however, that too much had passed to simply revive the old approach wholesale. The regime of tort and contract was too entrenched, and any reassertion of the assumption of responsibility concept needed to work against this backdrop. This led him to limit assumption of responsibility to 'relationships which … are "equivalent to contract", that is, where there is an assumption of responsibility in circumstances in which, but for the absence of consideration, there would be a contract'.[38]

'Contract without consideration' implies a focus on the face-to-face nature of the concept of assumption of responsibility derived from the old authorities. But it perhaps obscures the more radical leap that the House of Lords was making by reviving a doctrine with a distinct grounding and a distinct set of obligations and entitlements. It is also wrong to see assumption of responsibility as a threat to contract. Contract does not exhaust the category of promises or agreements having legal effect.[39] Contract still requires consideration and remains the only means of securing particular entitlements.

7. The Modern Uses of Assumption of Responsibility

I contend that the historical pedigree and theory behind assumption of responsibility as a concept are sound, as applied in a particular class of case. That class

[36] [1964] AC 465, 520.
[37] Ibid, 526.
[38] Ibid, 528–529.
[39] G Treitel, *The Law of Contract* (11th edn, 2003), 73; R Stevens, *Torts and Rights* (2012), 33.

of case is where the assumption of responsibility is grounded in a direct relationship between claimant and defendant which gave the defendant a real opportunity to bargain over what it was prepared to do for the claimant and therefore the extent of the liability which it was prepared to take on. However, there is a danger of dissipating the moral basis it provides for imposing liability and its determinacy as a concept by trying to use it in an over-extensive way. This has been the tendency since *Hedley Byrne*. This is where there is greatest academic disquiet, Jane's included.

There is force in Roderick Bagshaw's observation that critics of the idea have perhaps been overly wedded to classical-form concepts carrying checklist-style criteria, drawing the conclusion that assumption of responsibility must be defective 'if *all* applications of it do not involve the presence of facts satisfying some other concept, such as "promise" or "invited reliance"'.[40] Bagshaw argues that assumption of responsibility is better understood as a 'category including a cluster of examples, that are similar to each other in overlapping and criss-crossing ways'.[41] This captures the reality that assumption of responsibility becomes difficult to pin down if one moves too far beyond the core category of case.[42]

The central unifying feature of the cases in the core category is that the parties had a relationship in which it was open to the defendant to have bargained in respect of the risk involved in taking on a task for the claimant, in a context in which the defendant invited the claimant to rely on the due performance of the task. It is this feature which—in the absence of other controlling mechanisms such as consideration—provides the principled basis for delineating the circumstances in which recovery for pure economic loss will be available and when it will not.

This is clearest in the cases in which the courts have concluded that there *was not* an assumption of responsibility in the relevant sense. *Commissioners of Customs and Excise v Barclays Bank* illustrates this.[43] The bank did not owe a duty of care to the Commissioners when implementing a freezing injunction because this was not a task which it had agreed to undertake, but one which had been imposed on it by a court. Lord Hoffmann noted that in assumption of responsibility cases 'a duty of care is ordinarily generated by something which the defendant has voluntarily decided to *do*', but this could not be said of the bank because the obligation had been imposed on it.[44]

One can argue that a person may assume a responsibility to another by electing to act in a role to which general obligations attach which are intended to safeguard

[40] R Bagshaw, 'Tort Law, Concepts and What Really Matters' in A Robertson and TH Wu (eds), *The Goals of Private Law* (2009), 248.

[41] Ibid, 240.

[42] Cf *Commissioners of Customs & Excise v Barclays Bank* [2007] AC 181, [5] (Lord Bingham).

[43] [2007] AC 181. Also *Playboy Club London v Banca Nazionale del Lavoro* [2018] UKSC 43 and *Royal Bank of Scotland International v JP SPC 4* [2022] UKPC 22.

[44] [2007] AC 181, [37]; also [14] (Lord Bingham) and [74] (Lord Walker).

others from harm, as here the bank chose to participate in banking activity and was therefore vulnerable to being exposed to freezing orders. But that involves slippage from the core case of voluntary assumption of risk in which imposition of a duty of care occurs and has been found to be justified; the question then is more redolent of whether a general duty ought to be imposed on grounds of the status of the defendant rather than by reference to their voluntary assumption of responsibility by dealing with this particular claimant in this particular context with this particular loss in contemplation: see the distinction in *Boorman v Brown* between the specific undertaking given (akin to contract) and obligations derived from general status (ie obligations generated simply by taking on a particular role or function in society, rather than by virtue of giving any particular assurance regarding behaviour). When issues of obligation associated with general status or role arise, which are more remote from obligations voluntarily assumed to the particular claimant in the defendant's face-to-face dealings with them, the moral foundation for imposition of duty based on the free consent of the defendant is weaker and the scope for introduction of wider considerations of policy is wider. Accordingly, the slippage in usage of the concept carries with it a requirement for further justification if a duty of care to hold harmless from pure economic loss is to be imposed.

This is more or less to reassert the rationale given thirty years earlier by Lord Bridge in *Caparo Industries v Dickman* for when imposition of a duty of care for negligent misstatements was appropriate. Having reviewed the authorities including *Hedley Byrne*, Lord Bridge commented that:

> The salient feature of all these cases is that the defendant giving advice or information was fully aware of the nature of the transaction which the plaintiff had in contemplation, knew that the advice or information would be communicated to him directly or indirectly and knew that it was very likely that the plaintiff would rely on that advice or information in deciding whether or not to engage in the transaction in contemplation. In these circumstances the defendant could clearly be expected, subject always to the effect of any disclaimer of responsibility, specifically to anticipate that the plaintiff would rely on the advice or information given by the defendant for the very purpose for which he did in the event rely on it.[45]

Knowledge on the part of the defendant indicates that they had an opportunity to bargain about the extent of the liability they were willing to take on. Where the defendant knows how and by whom their work will be used, they are able to foresee the risks should the work be deficient. They can then bargain around that risk, and agree via price mechanisms and limitations of liability as to how to allocate it should they wish to do so.

[45] *Caparo Industries plc v Dickman* [1990] 2 AC 605, 620–621.

This is also the force of Lord Hoffmann's reasoning in *Barclays Bank*, where he said:

> The purpose of the inquiry is to establish whether there was, in relation to the loss in question, the necessary relationship (or 'proximity') between the parties … the existence of that relationship and the foreseeability of economic loss will make it unnecessary to undertake any further inquiry into whether it would be fair, just and reasonable to impose liability.[46]

The reason for emphasis on the features of proximity between the parties and foreseeability of economic loss is that these show that the parties were in a position where the defendant had a meaningful opportunity to bargain about the task and the extent of legal liability it was willing to take on and the claimant's reliance on the defendant was direct.

By contrast with *Barclays Bank*, in cases in which an assumption of responsibility has been found it was in a context where it was possible for the defendant to bargain in this way.[47] In *Henderson v Merrett Syndicates*,[48] for example, the defendant underwriting agents took on the task of managing the insurance syndicates of which the claimants were members. As found by Lord Goff, the underwriting agents 'well knew' that their expertise would be relied upon in circumstances where the risks were clear if the service was provided negligently. It was open to the defendants to bargain in respect of those risks or alternatively to walk away from the arrangement. Indeed, what makes *Henderson* such an interesting case is that the defendants did in fact bargain in making a contract, albeit not directly with the claimants, to provide their services. Explaining the relationship between tortious and contractual liability, Lord Goff stated:

> Approached as a matter of principle, therefore, it is right to attribute to that assumption of responsibility, together with its concomitant reliance, a tortious liability, and then to inquire whether or not that liability is excluded by the contract because the latter is inconsistent with it.

In other words, an assumption of responsibility will exist unless the parties elected to allocate the risk differently via contract. The presence of the contract in

[46] [2007] 1 AC 181, [35].

[47] This is debatable in the case of *Spring v Guardian Assurance* [1995] 2 AC 296 (request for reference from prospective employer, not from the employee whose chance of employment was ruined by the negligent preparation of the reference) and *Smith v Eric S Bush* [1990] 1 AC 831 (which was not itself decided on the basis of assumption of responsibility, but is sometimes sought to be explained in that way; the surveyor in one of the appeals did not appreciate that his survey would be shown to the claimant purchasers of a property): see D Nolan, 'Assumption of Responsibility: Four Questions' (2019) 72 CLP 123, 132–133.

[48] [1995] 2 AC 145.

Henderson which did not allocate risk away from the defendant was clear evidence that the defendant could (and indeed did) bargain, and chose to accept the allocation of risk inherent in what it undertook to do. The finding of an assumption of responsibility in these circumstances concurrent with the contractual liability is therefore consistent with the rationale I have set out.

In *Three Essays*, Jane asks what the 'coherent normative justification for a tort entitlement' in the *Hedley Byrne* type cases is, and what 'displace[s] the expectation that each participant in the commercial arrangement bears the risks to its own economic interests *inter se* arising out of the arrangement?'[49] My answer is that an assumption of responsibility will have occurred in circumstances where the defendant agreed with the claimant to take on a task, thereby inviting the claimant's reliance on them, in circumstances where it could have bargained to secure an alternative allocation of risk or could simply have declined to do it.

The question remains, why in such cases it is the defendant rather than the claimant who should *prima facie* bear the risk if there is not an express agreement about how that risk is to be distributed. Absent agreement to the contrary, why should the claimant not bear the risk when choosing to rely on the defendant when it was not required to?

The answer to this is to be found in the old cases on the proper performance of tasks which the defendant volunteers to do for the claimant. By taking on the task, the defendant invites the claimant to rely on its due performance of that task. Imposition of liability is justified by the moral claim that if a person voluntarily undertakes to do something in these circumstances it is right that they be required to do it to a proper standard. Even when the undertaking is given gratuitously, the courts have made a judgement that the force in this moral claim overpowers the countervailing claim that ordinarily claimants should bear their economic losses in circumstances where they have not chosen to contract in positive terms. The voluntary assumption of responsibility by the defendant puts off the claimant from seeking to stipulate terms. If the defendant has volunteered to do the task, the claimant already has a moral claim to expect that it is done with proper care and does not have to negotiate to acquire a right to have that very expectation.

There remains good reason, however, for not over-extending the categories of case in which a duty to protect against pure economic loss will be found to arise. Ordinarily it is not important enough to justify the law imposing general duties on people not to cause other people to suffer that kind of loss.[50] Economic loss consequent upon breach of legal obligations arises in many different ways. In general terms, it is legitimate for the law of tort to seek to provide a focus for liability upon those forms of injury that implicate human interests which have choate and obvious forms, particularly bodily integrity and property (a type of economic

[49] Stapleton (n 1), 60.
[50] N McBride, *The Humanity of Private Law* (2019), 131.

interest, but in a crystallised form recognised in a variety of positive ways by the law). As a consequence, it is also legitimate for the law usually to limit liability for pure economic loss (such as loss of earnings or profits) to that particularly associated with damage to those interests. However, this is a general tendency in the law, not a justification for an absolute prohibition on recovery of pure economic loss in other situations. The law of tort applies across a highly varied set of contexts and protects a highly variegated range of interests. So in principle it remains possible to justify the imposition of liability for pure economic loss in cases where it is not associated with damage to some other conventionally protected kind of interest. However, special reasons of the kind discussed above are required to overcome this general attitude.

8. Conclusion

The concept of assumption of responsibility cannot do the whole work to explain when a duty of care arises in tort in relation to economic loss. We should be wary of single concept explanations for duties of care, whether on the basis of the neighbour principle in *Donoghue v Stevenson*, assumption of responsibility or the 'fair, just and reasonable' mantra.[51] But there is a core category of case in which the concept of 'assumption of responsibility' does real substantive work and cannot be dismissed as a conclusionary label. It is an idea with deep roots in the law of obligations. It continues to serve a useful and principled purpose by providing a means for the common law to avoid the unattractiveness of result which would ensue if strict categories of contract and tort were the only available means of redress.

It is a great virtue of Jane's scholarship in the law of tort that her penetrating critique of the vacuity of the idea of assumption of responsibility in many contexts should prompt more precise reflection upon when it may be capable of doing real normative work. Jane's writing is an important step towards the common law 'working itself pure', in Lord Mansfield's phrase, through the refinement of the ideas which underpin it. It seems to me that this is what Jane's conception of reflexive legal scholarship is all about.

[51] *Caparo* (n 45), 617–618.

PART III
CAUSATION

11

Dialogues on Causation with Stapleton

Richard W Wright

1. Theory and Practice

I appreciate being invited to participate in this festschrift in honour of Jane Stapleton, with whom I have been discussing and debating tort liability issues, especially those regarding causation, since our time together at Oxford in 1995, when she was a fellow at Balliol College and I was a visiting fellow at Brasenose College and a visiting lecturer in the Oxford law faculty co-teaching BCL graduate seminars on Causation, Tort Law, and Jurisprudence with Tony Honoré, John Gardner, and John Finnis. Legal academics who have engaged in in-depth analysis of causation, especially in its core sense of 'factual causation' based on the laws of nature, have been few and far between. Those few who have engaged in such analysis have formed mutually supportive friendships, despite what might seem otherwise based on our published papers. This has been especially true for me and Jane (hereafter Stapleton).[1]

In her recent book, *Three Essays on Torts*, Stapleton criticises 'Grand Theories' of tort law for assuming that law is or should be 'all about one thing'. She offers instead a 'reflexive' approach, which supposedly differs from Grand Theories by 'taking judges seriously' and acknowledging diversity in doctrines and results in different legal jurisdictions.[2] I acknowledge being a Grand Theorist, but only in the sense that I, along with Aristotle, Aquinas, Kant, the founders of the United States, and the great majority of judges and ordinary people, believe that the proper purpose of civil society and its laws is justice: the promotion through law of the equal freedom of each member of society.[3] Contrary to her assertion that rights-based theorists focus on the interests of the individual claimant or parties in a lawsuit and ignore the interests of others in the community,[4] I have repeatedly emphasised the need to focus on the common interest understood as the equal freedom of each

[1] See also Richard W Wright, 'Dialogues on Causation with Puppe' (2022) 1(11) Zeitschrift Für Internationale Strafrechtswissenschaft [ZfIStW] 610 <https://www.zfistw.de/dat/artikel/2022_11_1528.pdf> accessed 17 August 2023.

[2] Jane Stapleton, *Three Essays on Torts* (OUP 2021) xvii, 2, 21–22, 33.

[3] eg Richard W Wright, 'Substantive Corrective Justice' (1992) 77 Iowa L Rev 625; Richard W Wright, 'Right, Justice and Tort Law' in David G Owen (ed), *Philosophical Foundations of Tort Law* (OUP 1995) 159; Richard W Wright, 'The Principles of Justice' (2000) 75 Notre Dame L Rev 1859.

[4] Stapleton, *Essays* (n 2) 11–12, 26, 31–33.

Richard W Wright, *Dialogues on Causation with Stapleton* In: *Torts on Three Continents*. Edited by: Kylie Burns, Jodi Gardner, Jonathan Morgan and Sandy Steel, Oxford University Press.
DOI: 10.1093/oso/9780198889748.003.0011

member of society, taking into account distributive justice as well as interactive justice (which is usually misleadingly referred to as 'corrective justice'), rather than aggregate pleasure or wealth.[5] I have argued that multiple areas and aspects of tort liability generally conform with the principles of justice in their results, although doctrines may vary in different jurisdictions due to their history of legal development and different local circumstances, while the competing utilitarian efficiency theories not only are morally bankrupt but also fail to explain tort law or the law in general, especially the causation requirement for legal responsibility and ultimate liability.[6]

Neither I nor, as far as I know, any other Grand Theorist has argued that the law in practice, which is often affected by misunderstandings and corrupted by the self-interested actions of private and governmental entities,[7] is always in conformance with our respective Grand Theories. While Stapleton criticises Grand Theorists for treating departures from their Grand Theories as mistakes,[8] her 'reflexive' approach allows courts' departures from her analyses to be described as 'inaccurate', 'misguided', 'wrong', and 'flawed'.[9] I have always taken judges seriously, while agreeing with Stapleton and the moderate legal realists on the need to pay attention not only to what judges say, but especially to what they do.[10] Yet, as Stapleton has acknowledged, even this is an insufficient basis for a coherent and comprehensive restructuring of the many different issues that courts have addressed using causal language.[11]

So what separates Stapleton's 'reflexive' analyses from the Grand Theorists' analyses, especially my own? As she stated to me at the end of the symposium at which the papers in this festschrift were initially presented, I believe that a coherent, comprehensive, and useful analysis requires consideration of relevant theory as well as actual practice, while she does not. In her 'reflexive' reconstructions of actual

[5] eg papers cited in n 3.

[6] eg Richard W Wright, 'The Efficiency Theory of Causation and Responsibility: Unscientific Formalism and False Semantics' (1987) 63 Chi-Kent L Rev 553; Richard W Wright, 'Allocating Liability Among Multiple Responsible Causes: A Principled Defense of Joint and Several Liability for Actual Harm and Risk Exposure' (1988) 21 UC Davis L Rev 1141; Richard W Wright, 'Principled Adjudication in Tort Law and Beyond' (1999) 7 Canterbury L Rev 265; Richard W Wright, 'The Grounds and Extent of Legal Responsibility' (2003) 40 San Diego L Rev 1425; Richard W Wright, 'Hand, Posner, and the Myth of the "Hand Formula"' (2003) 4 Theoretical Inquiries in Law 145; Richard W Wright, 'The Principles of Product Liability' (2007) 26 Rev Litig 1067; Richard W Wright, 'Private Nuisance Law: A Window on Substantive Justice' in Donal Nolan and Andrew Robertson (eds), *Rights and Private Law* (Hart 2012) 491.

[7] See, eg, Richard W Wright, 'Allocating Liability Among Multiple Responsible Causes; Principles, Rhetoric and Power' in Kit Barker and Ross Grantham (eds), *Apportionment in Private Law* (Hart 2019) 35, 36, 50–62.

[8] Stapleton, *Essays* (n 2) 17–18, 24–25.

[9] eg ibid 19–20; Jane Stapleton, 'Occam's Razor Reveals an Orthodox Basis for *Chester v. Afshar*' (2006) 122 LQR 426, 429.

[10] See Stapleton, *Essays* (n 2) 18, 80–81; Wright, 'Principled Adjudication' (n 6) 269–72.

[11] Stapleton, *Essays* (n 2) 66–67, 78–79; Jane Stapleton, 'Factual Causation' (2010) 38 Fed L Rev 467, 467–71.

practice, especially when discussing causation, she often denies the relevance and even the existence of any possibly applicable theory, instead focusing on judges' weighing or balancing of discrete 'policy' concerns in different types of situations.[12]

This difference in approach, together with Stapleton's desire to communicate with judges using concepts and language with which they are familiar, may explain her attempts, described in Section 2, to elaborate 'more forensically useful' modified-but-for versions of the philosophically based NESS analysis of factual causation that I have continued to develop. Her attempts, as well as her support and criticism of my work in our written and verbal dialogues across continents and in the American Law Institute (ALI), have made me rethink and attempt to improve and clarify my work and, as I briefly describe in Section 3, have been important in helping to convince academics and judges of the need to distinguish the factual causation issue from attribution of legal responsibility and to provide a clear and comprehensive analysis of factual causation. I look forward to continued dialogue and cooperation.

2. Our Dialogues on Causation

Our mutual interest in causal issues in the law, generated perhaps by our respective scientific educations, began with our readings of Herbert Hart and Tony Honoré's landmark treatise, *Causation in the Law*, the first edition of which was published in 1959[13] and the second in 1985.[14] My first articles focusing on causation were published in 1985.[15] During the next three years I published two more relevant articles,[16] and in 1988 Stapleton published her first two relevant articles.[17] In these papers, we each discussed and disagreed with Hart and Honoré's attempt, supposedly based on analysis of ordinary language usage, to treat as supposedly factual 'common sense' principles of causation not only the core issue of scientific causation based on the laws of nature, often referred to as 'cause-in-fact' or 'factual causation', but also the normative issues involved in determinations of legal responsibility and ultimate liability. We instead agreed with the moderate legal realists that, in order to provide much needed clarity to legal analysis, the factual

[12] Stapleton, *Essays* (n 2) 22, 25–34.

[13] HLA Hart and AM Honoré, *Causation in the Law* (OUP 1959).

[14] HLA Hart and Tony Honoré, *Causation in the Law* (2nd edn, OUP 1985)(used for all subsequent citations).

[15] Richard W Wright, 'Causation in Tort Law' (1985) 73 Calif L Rev 1735; Richard W Wright, 'Actual Causation vs. Probabilistic Linkage: The Bane of Economic Analysis' (1985) 14 JLS 435.

[16] Wright, 'Efficiency Theory' (n 6); Richard W Wright, 'Causation, Responsibility. Risk, Probability, Naked Statistics, and Proof: Pruning the Bramble Bush by Clarifying the Concepts' (1988) 73 Iowa L Rev 1001.

[17] Jane Stapleton, 'Law, Causation and Common Sense' (1988) 8 Oxford JLS 111; Jane Stapleton, 'The Gist of Negligence, Part 2: The Relationship between "Damage" and Causation' (1988) 104 LQR 389.

causation issue should be clearly distinguished from the normative legal responsibility issues.[18]

In the most valuable parts of their book,[19] which unfortunately were overshadowed by their primary focus on developing supposedly factual causal principles for asserting and limiting legal responsibility,[20] Hart and Honoré described three different senses of 'necessity',[21] which I and others subsequently labelled 'weak', 'strong', and 'strict'.[22] Building on John Stuart Mill's ground-breaking philosophical analysis of the nature and structure of natural laws,[23] they described a weakly necessary condition as 'one of a set of conditions jointly sufficient for the production of the consequence: it is necessary because it is required to complete this set',[24] but differed from Mill by applying this weak sense not only to the constituent elements of the laws of nature but also as the proper test, more inclusive than the traditionally assumed strong (but-for) sense, for identifying 'causally relevant factors' in specific instances.[25]

In 1985 I created an acronym, NESS (**n**ecessary **e**lement of a **s**ufficient **s**et), for their weak-necessity analysis, which, stated more precisely, requires for a condition to be a cause that it was necessary *for the sufficiency* of a set of existing antecedent conditions that was sufficient for the occurrence of the result,[26] and in 1988 I provided definitions of weak, strong, and strict necessity and sufficiency:[27]

[18] Stapleton, 'Law, Causation and Common Sense' (n 17) 116–17, 123–26, 129–30; Wright, 'Causation in Tort Law' (n 15) 1737–39, 1745–50; Wright, 'Pruning' (n 16) 1008–09, 1014, 1011–18. See also Richard W Wright, 'The Nightmare and the Noble Dream: Causation and Responsibility' in Matthew Kramer, Claire Grant, Ben Colburn, and Antony Hatzistavrou (eds), *The Legacy of H.L.A. Hart: Legal, Political and Moral Philosophy* (OUP 2008) 165, 165–77.

[19] Honoré agreed. At the Hart conference in Cambridge in 2007, he told me that their most significant contribution was making philosophers aware of the problem of overdetermined (duplicative and preemptive) causation.

[20] See Wright, 'Causation in Tort Law' (n 15) 1788–89.

[21] Hart and Honoré, *Causation in the Law* (n 14) 112–13.

[22] See n 28 and accompanying text.

[23] Hart and Honoré, *Causation in the Law* (n 14) 16–22; see Richard W Wright and Ingeborg Puppe, 'Causation: Linguistic, Philosophical, Legal and Economic' (2016) 91 Chicago-Kent L Rev 461, 466–69.

[24] Hart and Honoré, *Causation in the Law* (n 14) 112.

[25] ibid 112–13, 122–29; see also Ingeborg Puppe, 'Der Erfolg und seine kausale Erklärung im Strafrecht' (1980) 92 Zeitschrift für die gesamte Strafrechtswissenschaft [ZStW] 863, 865–68, 875–76 (English translation available at https://ssrn.com/abstract=2743259); see Wright and Puppe, 'Causation: Linguistic, Philosophical, Legal and Economic' (n 23) 468. Puppe was unaware of Hart and Honoré's prior work.

[26] Wright, 'Causation in Tort Law' (n 15) 1788–95; Wright, 'Pruning' (n 16) 1021–22, 1024–25, 1027, 1041. The required sense of sufficiency is the instantiation of all the conditions in the antecedent ('if' part) of the relevant causal law(s), the consequent ('then' part) of which is instantiated by the consequence at issue. Wright, 'Causation in Tort Law' (n 15) 1789, 1795–98, 1803–04, 1808–09, 1823; Wright, 'Pruning' (n 16) 1031, 1033, 1041–42, 1045–46, 1049–53; Richard W Wright, 'Once More Into the Bramble Bush: Duty, Causal Contribution, and the Extent of Legal Responsibility' (2001) 54 Vand L Rev 1071, 1103 and fns 112 and 113, 1111.

[27] Wright, 'Pruning' (n 16) 1019–22.

> In descending order of stringency, a strict-necessity test requires [for *Q* to be a cause of *R*] that *Q* be necessary for the occurrence of *R* whenever *R* occurs; a less stringent, strong-necessity test requires only that *Q* have been necessary for the occurrence of *R* on the particular occasion, considering the circumstances that existed on the particular occasion [the but-for/*sine qua non* test]; and the least stringent, weak-necessity test requires only that *Q* have been a necessary element of some set of actual conditions that was sufficient for the occurrence of *R*—the NESS test. A strict-sufficiency test requires that *Q* be sufficient by itself for the occurrence of *R*; a less stringent, strong-sufficiency test requires only that *Q* be a necessary element of some set of existing conditions that was sufficient for the occurrence of *R*—the NESS test again; and the least stringent, weak-sufficiency test merely 'requires' that *Q* be a part of some set of existing conditions that was sufficient for the occurrence of *R*.[28]

Hart and Honoré's account had some significant deficiencies,[29] including their only discussing conditions that were strongly necessary or independently strongly sufficient—for example, each of two shots, sufficient independent of each other but in conjunction with the other existing conditions to kill the victim, which simultaneously strike the victim.[30] I explained that the NESS weak-necessity analysis can be used to identify as causes conditions that are neither strongly necessary nor independently strongly sufficient—for example, each of three separately supplied drops of poison or, using subsets or 'at least' descriptions of actual conditions, the single drop supplied by one enemy as well as the two drops supplied by another, when only two are necessary to kill someone.[31]

Stapleton's two 1988 articles barely touched on the analysis of factual causation. In each article, there is a brief reference to the 'unorthodox' Hart-and-Honoré 'argument' that each of two independently strongly sufficient fatal shots 'is properly described as the "cause" of the [victim's] death',[32] but she did not discuss their weak-necessity analysis of 'causally relevant factors' and failed to recognise its relevance in two well-known cases that have often been discussed in her work.

In *Bonnington Castings Ltd v Wardlaw*,[33] the plaintiff was simultaneously exposed at work to harmful dust from a tortious source and a non-tortious source,

[28] ibid 1020, citing Hart and Honoré, *Causation in the Law* (n 14) 112–13, JL Mackie, *The Cement of the Universe: A Study of Causation* (OUP 1974, corrected edn 1980) 38–40, 60–66, 126–27, and Michael Scriven, 'The Logic of Cause' (1971) 2 Theory and Decision 49, 49, 52–53.

[29] See Richard W Wright, 'The NESS Account of Natural Causation: A Response to Criticisms' in Richard Goldberg (ed), *Perspectives on Causation* (Hart 2011) 285, 286–88.

[30] Hart and Honoré, *Causation in the Law* (n 14) 123–25, 206–07, 235–39, 245, 249. An independently strongly sufficient condition is a condition that by itself fully instantiates a required element in a fully instantiated causal law. Richard W Wright, 'Acts and Omissions as Positive and Negative Causes' in Jason W Neyers, Erika Chamberlain, and Stephen G Pitel (eds), *Emerging Issues in Tort Law* (Hart 2007) 287, 295.

[31] See Wright, 'Causation in Tort Law' (n 15) 1791–94.

[32] Stapleton, 'Law, Causation and Common Sense' (n 17) 126; Stapleton, 'Gist Part 2' (n 17) 402 fn 1.

[33] [1956] AC 613 (HL).

which caused pneumoconiosis. She argued then and has since continued to argue that the tortious source was a cause of only a separable or marginal part of the victim's supposedly 'divisible' pneumoconiosis.[34] This is incorrect. As the Law Lords (most explicitly Lord Keith of Avenholm) emphasised, the victim's pneumoconiosis was an indivisible harm, to which each source materially contributed, that was caused by all of the harmful dust in the workplace atmosphere, rather than there being distinct separable or marginal portions of the harm caused by each source.[35]

In *Hotson v East Berkshire Area Health Authority*,[36] the defendant's negligent delay in diagnosis and treatment, which allowed bleeding from blood vessels ruptured by the plaintiff's fall from a tree to compress and block the supply of blood from the intact blood vessels, contributed to the loss of blood supply to the epiphysis and thus to the plaintiff's permanent disability, just as stabbing a person who more likely than not already has been stabbed sufficient times to bleed to death contributes to that person's death.[37] However, although some of the judges in the Court of Appeal and the House of Lords mentioned the material contribution analysis employed in *Bonnington* and (less clearly) *McGhee v National Coal Board*,[38] the case was argued and decided in the trial and appellate courts as a 'loss of chance' injury case, which argument was accepted in the trial court and the Court of Appeal but rejected by the House of Lords, using a myopic combination of the but-for test of causation and the highly problematic 50+ per cent probability interpretation of the standard of persuasion.[39] Stapleton agreed (until 2008, when

[34] Stapleton, 'Law, Causation and Common Sense' (n 17) 127–28; Stapleton, 'Gist Part 2' (n 17) 402–06; Jane Stapleton, 'Scientific and Legal Approaches to Causation' in Danuta Mendelson and Ian Freckelton (eds), *Causation in Law and Medicine* (Routledge 2002) 14, 23; Jane Stapleton, 'Cause-in-Fact and the Scope of Liability for Consequences' (2003) 119 LQR 388, 394; Stapleton, 'Factual Causation' (n 11) 477; Jane Stapleton, 'Unnecessary Causes' (2013) 129 LQR 39, 51–53, 56–57; Jane Stapleton and Sandy Steel, 'Case Comment: Causes and Conditions' (2016) 132 LQR 363, 363–65, 368.

[35] *Bonnington* (n 33) 621, 623 (Lord Reid), 623 (Lord Tucker), 626 (Lord Keith of Avenholm); see also Tsachi Keren-Paz and Richard W Wright, 'Liability for Mass Sexual Abuse' (2019) 56 Am Crim L Rev 185, 202–24.

[36] [1988] UKHL 1, [1987] 1 AC 750.

[37] Richard W Wright, 'Liability for Possible Wrongs: Causation, Statistical Probability, and the Burden of Proof' (2008) 41 Loyola LA L Rev 1295, 1321–22.

[38] *Hotson v East Berkshire Area Health Authority* [1987] AC 750 (HL) 758–59 (Donaldson MR, citing *Bonnington* and *McGhee v National Coal Board* [1973] 1 WLR 1), 765–66 and 768–69 (Croom-Johnson LJ, citing *McGhee*), 782 (Lord Bridge), 786 (Lord Mackay).

[39] Wright, 'Liability for Possible Wrongs' (n 37) 1321–23. For discussion of the irresolvable problems and paradoxes created by probabilistic rather than degree of belief interpretations of the standard of persuasion, see Wright, 'Pruning' (n 16) 1042–67; Richard W Wright, 'Proving Causation: Probability versus Belief' in Richard Goldberg (ed), *Perspectives on Causation* (Hart 2011) 195; Richard W Wright, 'Prueba Jurídica: Fundherentismo y Estadística' (Alejo Giles tr) in Diego Dei Vecchi (ed), *Susan Haack Premio Internacional de Cultura Jurídica 2020* (Marcial Pons 2024) 131, 140–45 (original English version available at https://ssrn.com/abstract=3930330). Steel acknowledges the problems and paradoxes but downplays them and ends up supporting and employing the 50+ per cent probability standard of persuasion. Sandy Steel, *Proof of Causation in Tort Law* (Cambridge University Press 2015) 50–51, 61–66, 85–103.

she read and accepted my analysis of the case[40]) that causation was not proven, but argued for proportional liability based on the probability of causation.[41]

I had noted that denial of liability in *Hotson* might be supported by the 'no worse off' limitation on liability, which denies liability if the injury would have occurred anyway as a result of legally non-responsible conditions, but which is very often confused with the but-for test of factual causation.[42] I identified this limitation as a scope of liability or damages issue distinct from the factual causation issue in my initial 1985 article.[43] In subsequent articles[44] and the internal deliberations of the ALI, I described it as the 'no worse off' limitation, extensively elaborated it, and repeatedly defended it, which efforts, with support from Stapleton and (eventually) Michael Green, have resulted in its recent recognition in the *Restatement Third of Torts*.[45] I have argued that, as a matter of justice, the burden of proof on the 'no worse off' limitation should rest on the defendant, who is a proven wrongful cause of the plaintiff's injury.[46] Stapleton has stated that the burden of proof (on what she has recently renamed the 'no better off' principle) should be on the victim and should employ the highly problematic 50+ per cent probability interpretation of the standard of persuasion.[47]

Stapleton first addressed the analysis of factual causation in a series of papers published between 2000 and 2002, in which she argued that, given the use of causal language to encompass legal responsibility as well as factual causation, any terms including the word 'cause' should not be used even with reference to the 'first-level' (factual causation) enquiry. Instead, supposedly more transparent phrases such as 'how things came about',[48] 'historical involvement in an outcome',[49] or 'playing a role in the history of the outcome'[50] should be used. In 2003, she dropped her opposition to the use of 'cause-in-fact' to refer to the 'first-level' enquiry, while still arguing that the law should 'confine the notion of "cause-in-fact" to the special

[40] Jane Stapleton, 'Choosing What We Mean By "Causation" in the Law' (2008) 73 Missouri L Rev 433, 475 fn 148, citing Wright, 'Liability for Possible Wrongs' (n 37) 1322–24; Stapleton, 'Unnecessary Causes' (n 34) 48–49 and fn 40 (same); Stapleton, *Essays* (n 2) 82.

[41] Stapleton 'Gist Part 2' (n 19) 391–400; see also Stapleton, 'Scientific and Legal Approaches to Causation' (n 34) 25–27; Stapleton, 'Cause-in-Fact' (n 34) 396–97, 399.

[42] Wright, 'Liability for Possible Wrongs' (n 37) 1324.

[43] Wright, 'Causation in Tort Law' (n 15) 1798–1801.

[44] Wright, 'Legal Responsibility' (n 6) 1434–67; Richard W Wright, 'Causation (Contribution) and the "No Worse Off" Limitation on Liability' (2022) 49(1) U Western Australia L Rev 1, 31–55 <https://www.able.uwa.edu.au/__data/assets/pdf_file/0010/3671155/2.-Wright.pdf> accessed 17 August 2023.

[45] *Restatement of the Law Third, Torts: Remedies* (Tent Draft No 2) (ALI 2023) s 12.

[46] Wright, 'Legal Responsibility' (n 6) 1434–38; Wright, 'No Worse Off' (n 44) 32–33.

[47] Stapleton, 'Unnecessary Causes' (n 34) 54–58; Stapleton, *Essays* (n 2) 83–84; cf Stapleton, 'Choosing' (n 40) 453–54, 479 and fn 168 (using 'no worse off' to refer to this limitation on 'damages').

[48] Jane Stapleton, 'Perspectives on Causation', in Jeremy Horder (ed), *Oxford Essays in Jurisprudence: Fourth Series* (OUP 2000) 61, 63–64.

[49] Jane Stapleton, 'Unpacking Causation' in Peter Cane and John Gardner (eds), *Relating to Responsibility* (Hart 2001) 145, 145–46, 157, 166–67, 175; Jane Stapleton, 'Legal Cause: Cause-in-Fact and the Scope of Liability for Consequences' (2001) 54 Vand L Rev 941, 958–61.

[50] Stapleton, 'Unpacking Causation' (n 49) 146 and fn 3, 158, 167, 174; Stapleton, 'Legal Cause' (n 49) 945, 958–60; Stapleton, 'Scientific and Legal Approaches to Causation' (n 34) 21.

unambiguous meaning of "historical involvement in [the victim's] suffering actionable damage"'.[51]

In these and subsequent papers Stapleton declared that the 'first-level' (factual causation) enquiry 'is a question of scientific fact to which there is only one correct answer',[52] which is resolved 'using our knowledge of the physical laws of nature, evidence of behaviour and so on' and thus 'carries the potential for the concept of causation to be untainted by normative controversies',[53] which should be distinguished from the 'second-level' enquiry which, in the law, focuses on the determination of legal responsibility and ultimate liability.[54] I fully agree.

Yet Stapleton claimed that 'this first order task neither requires nor is illuminated by "causal theories"'[55] and, furthermore, that 'Causation is not a concept at all'.[56] She described several factual situations, including in her descriptions some of the antecedent conditions and a subsequent harm, with no explicit statement of a causal connection,[57] which connection she apparently believed, despite Hume's demolishment of this idea,[58] we somehow perceive without employing any explicit or implicit causal concept, test, or analysis:

> It is striking that we do not have difficulty understanding these situations in the sense of 'how things came about' because the facts are known. Indeed, if it was always true that by 'causation' we simply meant this first-level understanding of how things came about, then lawyers and philosophers would be wrong to describe the 'problem' in these four cases as a problem of causation.[59]

This statement is especially puzzling given her including as one of the factual situations the desert traveller scenario, in which an enemy poisons the water in the desert traveller's water barrel (his only source of water), a second enemy (not knowing that the water has been poisoned) empties the poisoned water out of

[51] Stapleton, 'Cause-in-Fact' (n 34) 393; see ibid 388–89, 391–93; Stapleton, 'Choosing' (n 40) 433, 435, 473. For criticism of her proposed language, see Wright, 'Once More' (n 26) 1080–81, 1111 fn 132.

[52] Stapleton, 'Legal Cause' (n 49) 958; see also Stapleton, 'Perspectives' (n 48) 61, 63, 66 fn 13; Stapleton, 'Unpacking Causation' (n 49) 149 fn 23; Stapleton, 'Factual Causation' (n 11) 478; Stapleton, *Essays* (n 2) 67.

[53] Stapleton, 'Choosing' (n 40) 474; see ibid 437, 441, 446, 447.

[54] Stapleton, 'Perspectives' (n 48) 62–65, 71, 80, 82; Stapleton, 'Unpacking Causation' (n 49) 167–68.

[55] Stapleton, 'Perspectives' (n 48) 72 fn 24; see Stapleton, 'Choosing' (n 40) 439 ('it is futile for philosophers to search for a coherent freestanding metaphysical account of "causation" unless a choice of underlying interrogation … is specified'), 439 fn 15 (the 'search for "a broad and nondiscriminatory concept of causation"' is 'doomed') ('the central argument of this article is that [such a concept] is indeed a myth').

[56] Stapleton, 'Scientific and Legal Approaches to Causation' (n 34) 14; see Stapleton, 'Choosing' (n 40) 456 ('causal language can denote whatever we choose it to').

[57] Stapleton 'Perspectives' (n 48) 63–64.

[58] See Wright, 'Causation in Tort Law' (n 15) 1789.

[59] Stapleton 'Perspectives' (n 48) 63–64; see also Stapleton, 'Choosing' (n 40) 474 ('where all the facts of the case are known, there would be no room for disagreement on the issue of "causation"'); Stapleton, *Essays* (n 2) 67.

the barrel (before the traveller, also unaware of the poison in the water, attempts to drink it), and the traveller dies in the desert (of thirst).[60] As the NESS analysis easily explains and the *Restatement*, Mackie, and (eventually) Honoré agree, the second enemy's emptying of the barrel preempted the potential causal effect of the first enemy's poisoning of the water and caused the traveller's death, even though it likely prolonged the traveller's life. The set of conditions sufficient for death by thirst all actually occurred, and the emptying of the water barrel was necessary for the sufficiency of this set, while several elements of the set of conditions sufficient for death by poisoning failed to occur, most notably the drinking of the poison.[61]

Stapleton stated that this conclusion is 'surprising' and claimed that it is based on a 'fiction' or 'rationalisation' that ignores the conduct of the first enemy, which resulted in a lack of 'fresh' water and thus, allegedly, the 'dooming' of the traveller to die by thirst if not by poisoning.[62] She discussed different 'perspectives' by which neither, both, or one or the other enemies could be held to be legally responsible and thus a cause,[63] which I criticised in my 2001 Wade Conference article.[64] She stated that to focus on cause-in-fact in such cases 'is dangerously misleading and generates sterile debates'[65] and trumpeted her 'new understanding'[66] 'of the nature of causation and legal disputes about causation [which] turns tradition on its head':[67]

> The traditional lawyer's view is that usually you cannot be responsible for an outcome without being causally connected to it, and that this connection has to be established before you can construct a case for responsibility. In contrast, I argue that in legal disputes where the facts are agreed, it is notions of individual responsibility that determine the perspective a party adopts when looking at those facts to identify what had made a difference in bringing about the relevant outcome. In such cases, notions of individual responsibility can affect whether we call something a 'cause'.[68]

Stapleton has accused me of trying to resolve legal responsibility issues based simply on factual causation analysis, without providing any normative

[60] Stapleton 'Perspectives' (n 48) 63 (Case 2).

[61] Wright, 'Causation in Tort Law' (n 15) 1802; Wright, 'Pruning' (n 16) 1024–25; Tony Honoré, 'Necessary and Sufficient Conditions in Tort Law' in Owen (n 3) 363, 377–79; Mackie, *The Cement of the Universe* (n 28) 44–46; see also *Restatement of the Law Second, Torts* (ALI 1965) s 435 A comment *a* at 455.

[62] Stapleton, 'Perspectives' (n 48) 84 and fns 59–61; Stapleton, 'Unpacking Causation' (n 49) 175, 178–83.

[63] Stapleton, 'Perspectives' (n 48) 82–84; Stapleton, 'Unpacking Causation' (n 49) 178–82.

[64] Wright, 'Once More' (n 26) 1115–20.

[65] Stapleton, 'Legal Cause' (n 49) 967, 966 fn 60.

[66] Stapleton, 'Perspectives' (n 48) 61.

[67] ibid 81.

[68] ibid 81–2; see also ibid 61.

justification.[69] The reverse is true. I have consistently emphasised the need to distinguish factual causation from legal responsibility, explicitly limited my discussions of 'causation' to the factual causation issue, and provided arguments based on the principles of justice for holding defendants liable for unjustly caused injuries.[70] As exemplified by the prior discussion, it is she and David Fischer, upon whose articles she relied,[71] who have failed to distinguish the factual causation and legal responsibility issues in overdetermined causation situations.[72] While Fischer discussed 'corrective justice' and efficiency theories, Stapleton has merely suggested that failing to hold one or both 'wrongdoers' liable would treat the victim of two wrongdoers worse than the victim of one and allow wrongdoers to shelter behind each other's wrongs, thereby bringing the law 'into disrepute', impairing 'the dignity of the law', 'embarrassing' the law, and 'offend[ing] our social values and expectations of the law',[73] without providing or referencing any normative account of wrongdoing or why it should matter.

Despite her statements regarding the lack of need for—and indeed impossibility of—a comprehensive test or theory of factual (natural) causation, Stapleton until recently accepted and supported the NESS analysis as the best available analysis, while also proposing 'more forensically useful' modified-but-for tests allegedly based on the NESS analysis. Her first attempt was the 'targeted but-for' test, which she claimed, 'builds on the "necessary element of a sufficient set" approach of Hart, Honoré, and Wright, and reflects the methodology of experimental science' and should be used in the 'first-level' enquiry into 'the scientific-derived notion of historical involvement':[74]

> [T]ake all factors existing at the time of the actual transition, including the factor that we are investigating ... If there is a notional sequence of removing factors from that set such that

[69] ibid 66, 80, 84; Stapleton, 'Legal Cause' (n 49) 966–67 and fns 60–61; Stapleton, 'Choosing' (n 40) 474, 476–77.

[70] eg Wright, 'Causation in Tort Law' (n 15) 1737–58, 1764 fn 121, 1774–1813; Wright, 'Pruning' (n 16) 1004–42; Wright, 'No Worse Off' (n 44) 6–10; papers cited in nn 6, 7, 23, 35.

[71] Stapleton, 'Legal Cause' (n 49) 957 fn 38, 966 fn 61; see eg David A Fischer, 'Causation in Fact in Omission Cases' (1992) 41 Utah L Rev 1335, 1348 (stating that the substantial-factor 'test' is useful for resolving difficult cause-in-fact problems in multiple-omission cases because it merges the issue of factual causation and legal responsibility).

[72] See Wright, 'Once More' (n 26) 1101 fn 108, 1111, 1119–22, 1121 fn 172; Wright, 'NESS Account' (n 29) 305–06; text at nn 64–68.

[73] Stapleton, 'Perspectives' (n 48) 81, 83 fn 57; Stapleton, 'Legal Cause' (n 49) 967–68, 1007 and fn 175; Stapleton Cause-in-Fact' 413; Jane Stapleton, 'An "Extended-But-For" Test for the Causal Relation in the Law of Obligations' (2016) 36 Oxford JLS 697, 711; Stapleton, *Essays* (n 2) 80; see Wright, 'Once More' (n 26) 1122–23 fn 177. In 'Unpacking Causation' (n 49) at 147 and *Essays* (n 2) at 100, Stapleton references Honoré's 'outcome responsibility' argument, which is overbroad from a moral and especially legal responsibility viewpoint.

[74] Stapleton, 'Legal Cause' (n 49) 959 and fn 42; see also Stapleton, 'Perspectives' (n 48) 61, 66; Stapleton, 'Unpacking Causation' (n 49) 146, 157, 167, 174–75, 179; Stapleton, 'Scientific and Legal Approaches to Causation' (n 34) 19; Stapleton, 'Cause-in-Fact' (n 34) 393–94.

- a stage is reached where, given the remaining factors the actual transition might still have occurred, but that
- the further removal of the targeted factor leaves a set that would not (in the course of things that we now know happened) have produced the transition,
- then the targeted factor played a role in the history of the original transition.[75]

This supposed 'crystallisation' of the NESS analysis[76] is based on Stapleton's misunderstanding of the NESS analysis as using a double-level but-for inquiry into whether the condition at issue would have been a but-for factor 'had another factor/set of factors not been present'.[77] It fails to distinguish duplicative causes from preempted conditions or to recognise causation by conditions which are neither strongly necessary nor independently strongly sufficient when there is a concurrent independently strongly sufficient condition.[78] After I explained this, Stapleton inserted into her published Wade Conference paper and subsequent papers brief statements that 'the plaintiff is not allowed to remove a factor that had actually been present which had completely preempted the relevant effect of the defendant's tortious conduct'.[79] This is circular and question-begging. It assumes that we have already identified preemptive conditions before applying the 'targeted but-for' test, which is supposed to be used to distinguish preemptive from duplicative conditions.[80]

From the beginning, I had recognised that my original elaboration of the NESS analysis, although analytically sound, is too complex for practical use in some situations, especially those situations in which some or all of the relevant contributing conditions were neither strongly necessary nor independently strongly sufficient, in which case the analysis might require use of subsets or 'at least' descriptions of some actual conditions. I soon realised that we need only require (i) that the abstract elements in causal laws be NESS conditions and (ii) that the actual condition at issue be part of the complete instantiation of the relevant causal laws in the particular situation.[81] In my 2001 Wade Conference article, I suggested a practical three-step analysis of 'causal contribution', considering, in order, whether the condition at issue was strongly necessary, independently strongly sufficient, or (subsuming the first two steps) part of the complete instantiation of the relevant causal (natural) laws.[82]

75 Stapleton, 'Legal Cause' (n 49) 959–60.

76 Stapleton, 'Unpacking Causation' (n 49) 146, 167, 175.

77 Stapleton, 'Perspectives' (n 48) 61; Stapleton 'Scientific and Legal Approaches to Causation' (n 34) 19; Stapleton, 'Cause-in-Fact' (n 34) 393–94.

78 Wright, 'Once More' (n 26) 1112–13.

79 Stapleton, 'Legal Cause' (n 49) 960 fn 43; ibid 1005; Stapleton, 'Unpacking Causation' (n 49) 175, 177; Stapleton, 'Cause-in-Fact' (n 34) 393 fn 24.

80 Wright, 'Once More' (n 26) 1114–15 fn 149.

81 Wright, 'Pruning' (n 16) 1037–39; Wright, 'Once More' (n 26) 1107; Wright, 'NESS Account' (n 29) 303–06.

82 Wright, 'Once More' (n 26) 1107–09.

In 2008, recanting 'to the extent necessary' her prior targeted but-for test,[83] Stapleton re-packaged and supplemented it to create a new 'algorithm' for identifying factual causes, focusing on 'involvement', which again is supposedly based on 'the NESS idea: not as a self-evident "meaning of causation", but simply as an extremely effective algorithm for identifying all of the relationships of involvement … with which the Law must deal'.[84] She stated that she employed the term 'involvement' (no longer modified by 'historical') 'to avoid circular causal terminology'.[85] The circularity is transparent. She described three different forms of 'involvement': 'necessity, duplicate necessity and contribution, though contribution subsumes the others'.[86] The structure and content of this three-part analysis is essentially identical, except for its defective employment of double-but-for 'duplicate necessity' to identify independently strongly sufficient conditions,[87] to the three-step analysis of 'causal contribution' that I had previously suggested in my Wade Conference article.[88]

By 'necessity' Stapleton means what is usually referred to as 'strong necessity' (a 'but-for' condition).[89] She attempts to rescue the 'necessity' and 'duplicate necessity' tests (and her subsequent 'extended-but-for' test) from their inability to distinguish duplicative causes from preempted conditions by including in the description of the injury legally irrelevant details, including 'as much specificity in relation to time and place as the available facts allow',[90] and/or (tautologically) the manner of its occurrence—for example, 'death by electrocution at that instant',[91] 'death by explosion at 1 a.m.',[92] 'death by poison at noon on Friday 13 June 2014 under Dan's palm tree',[93] 'death by thirst on Friday at noon'[94]—while claiming that I fall into 'manifest error' by rejecting such widely discredited dodges,[95] which

[83] Stapleton, 'Choosing' (n 40) 441 fn 18.
[84] ibid 474; see also ibid 444, 471–72, 480.
[85] ibid 436.
[86] ibid 433; see also ibid 436, 441–44.
[87] Stapleton, 'Choosing' (n 40) 441–43, 472–74.
[88] See text at n 82.
[89] Stapleton, 'Choosing' (n 40) 436 and fn 7, 438–39. She created (but subsequently failed to employ) the term 'medium necessity' to refer to what is usually called 'strict necessity' ('the relation that in our natural world … if A does not occur, B also does not occur') and defined (but subsequently failed to employ) the term 'strong necessity' to refer to what might better be described as 'super strict necessity' ('necessity in all imaginable worlds'). ibid 434 and fn 3. She did not suggest any term other than 'necessity' to refer to what is usually described as 'strong necessity' (a but-for condition) or 'weak necessity' (a NESS condition).
[90] Stapleton, 'Extended But-For' (n 73) 700; see Stapleton, 'Unnecessary Causes' (n 34) 42. Steel also includes the usually legally irrelevant time of occurrence and erroneously claims that doing so is necessary in the NESS analysis. Steel, *Proof* (n 39) 18 and fn 16; see Wright, 'NESS Account' (n 29) 292, 297–303.
[91] Stapleton, 'Choosing' (n 40) 452.
[92] ibid (italics removed).
[93] Stapleton, 'Extended But-For' (n 73) 700, 704; see also ibid 710 ('death by a poison').
[94] ibid 724.
[95] Stapleton, 'Choosing' (n 40) 452 fn 42; see also Stapleton, 'Extended But-For' (n 73) 700 fn 9, 702 fn 19.

allow any existing condition to be treated as a cause.[96] As Honoré explained, the precise time and place of some injury or event may be helpful in identifying it, but they are rarely legally relevant.[97] If they are included, how can one justify excluding other legally irrelevant conditions? Even if the time and place are included, the injury may well have occurred at the same time and place as a result of a duplicative or preempted condition,[98] as she sometimes admits for duplicative conditions,[99] but tries to avoid admitting when preemption is involved by employing tautological descriptions of the injury[100] or, without admitting doing so, NESS causal sufficiency analysis.[101]

Her third form of 'involvement', 'contribution' (which word also allegedly does not involve any circular use of causal terminology) employs the NESS analysis. As she recognises, the fully elaborated NESS analysis is needed to identify as causes conditions that are neither strongly necessary nor independently strongly sufficient but nevertheless contributed.[102] She nevertheless criticises my supposed failure to explain why my initially applying the NESS analysis to some conditions, including reasons for a decision, by 'disaggregating' other conditions is 'useful for the Law',[103] and she ignores my previously published articles that avoided the disaggregation approach by merely requiring for a condition to be a cause of some result that it be part of the complete instantiation of the antecedent conditions in the relevant causal laws, which, however, must themselves be composed of NESS conditions.[104]

I have repeatedly emphasised the need to provide a means of legal redress for all unjustly caused harm and the need to employ the full NESS analysis in order to encompass all actual causes.[105] Stapleton, on the other hand, has failed to justify the full employment of the NESS analysis or any of her alternative formulations. She

96 See Wright, 'Causation in Tort Law' (n 15) 1777–80; Wright, 'Once More' (n 26) 1113–14; Wright, 'NESS Account' (n 29) 292–93; Wright and Puppe, 'Causation: Linguistic, Philosophical, Legal and Economic' (n 23) 475–77.

97 Honoré, 'Necessary and Sufficient Conditions in Tort Law' (n 61) 377–79. Stapleton seems to acknowledge this distinction in Stapleton, *Essays* (n 2) 79–80, yet inconsistently still seems to individuate by time and place in her causal analysis: 79 fn 34.

98 See papers cited in n 96.

99 Stapleton, 'Choosing' (n 40) 442 fn 19 (multiple advice or reasons when any one would be sufficient for the decision or action); ibid 453 (simultaneously fatal shots).

100 Stapleton, 'Extended But-For' (n 73) 703–04 and fn 30 (Gayle's persuasion of Harry caused 'the destruction of Beth's car which *actually* occurred', even though otherwise Roy would have persuaded Harry).

101 ibid (the 'hypothetical back-up story' did not actually occur); ibid 709 ('the prevention mechanism ... would have been complete'); Stapleton, 'Choosing' (n 40) 452–53 and fn 45 (When Y replaces X's bomb with another bomb, set to go off at the same time and place, '[t]he Law will not ignore the incontestable physical fact that for a bomb to be involved in the actual outcome ... that bomb had to present at the time of the lethal explosion!').

102 Stapleton, 'Choosing' (n 40) 443–44, 450 fn 38, 472, 474–77, 480.

103 ibid 476–77.

104 Wright, 'Pruning' (n 16) 1037–39; Wright, 'Once More' (n 26) 1107; Wright, 'NESS Account' (n 29) 303–06.

105 See papers cited in nn 3 and 6.

merely states that it is 'most convenient' for 'the Law' to do so, in order to identify 'all the relationships of involvement … with which the Law must deal' 'untainted by normative considerations',[106] while inconsistently claiming that there is no core concept of causation and that the 'Law' is free to choose whatever concept of causation it finds most useful for its purposes, without engaging in any discussion of relevant purposes, especially the principles of justice.[107]

In her 2008 article and subsequent articles, Stapleton has repeated criticisms by others of the NESS analysis of causation:

> [C]ritics have claimed that there are residual problems with Wright's otherwise impressive exposition of the NESS device. Some criticisms are relatively minor such as concern with Wright's occasional appeal to intuitions and his curious assertion that, in as much as NESS requires a 'counterfactual' analysis, this is to proceed not by speculating on what might have happened in a hypothetical world where, inter alia, the specified factor is absent, but by considering what actually happened.
>
> But one criticism of Wright, that by Fumerton and Kress, is fundamental and devastating. They demonstrate that, far from delivering on his claim that NESS 'is not just a test for causation, but is itself the meaning of causation', Wright's reliance in NESS on the idea of 'causal sufficiency' involves 'vicious conceptual circularity'. This attack is indeed fatal to Wright's claim concerning the *character* of the NESS test. Nevertheless, it does not, in my view, detract from the potential practical value of NESS to the lawyer.[108]

Although I initially referred to the analysis of necessity in both the but-for and NESS tests as a counterfactual analysis,[109] I have always insisted that, for both tests,[110] the analysis ordinarily is (or should be) a real-world matching of actual conditions against the required elements of the relevant causal laws/generalisations rather than a counterfactual 'possible worlds' exploration of what might have occurred in the absence of the condition at issue, and I have demonstrated the need

[106] Stapleton, 'Choosing' (n 40) 474–76, 479–80; see also ibid 433, 442, 443.

[107] See nn 55 and 56 and accompanying text.

[108] Stapleton, 'Choosing' (n 40) 472–73, citing Richard Fumerton and Ken Kress, 'Causation and the Law: Pre-emption, Lawful Sufficiency, and Causal Sufficiency' (2001) 64 Law and Contemp Prob 83, 84, 101–04.

[109] Wright, 'Causation in Tort Law' (n 15) 1804–05.

[110] The strong necessity criterion, when properly applied through a real world factual analysis, is a corollary of the NESS analysis that is valid only when there is no causal overdetermination and has two steps, the second of which is causally irrelevant but is satisfied when there is no causal overdetermination: (1) was the condition at issue part of the complete instantiation in the specific situation of the antecedent of one or more relevant causal generalisations and their underlying laws that have as their ultimate consequence the effect at issue (the NESS analysis); and (2) were the other existing conditions insufficient without the condition at issue for such complete instantiation (the strong necessity analysis)?

to employ the former rather than the latter analysis to resolve properly both duplicative and preemptive causation cases,[111] as both Honoré (eventually) and Mackie (a major proponent of but-for analysis[112]) admitted when discussing the desert traveller scenario,[113] and as Stapleton has implicitly admitted when discussing several preemption situations.[114]

However, sometimes counterfactual analysis is required, as when projecting what a person would have decided or done if she had been provided some information (rather than whether provided information contributed to an actual decision)[115] or had been, for example, attentive and not drunk (rather than whether drunkenness contributed to an actual action). Sometimes there may be a clear answer, for example, if there is evidence of habitual behaviour rather than mere propensities. But otherwise the best that can be done is to estimate probabilities, which are never enough to prove what actually would have happened, contrary to what is often assumed using the highly problematic 50+ per cent probability interpretation of the standard of persuasion.[116]

As Stapleton has documented, I have often incautiously claimed that the NESS analysis 'captures the essential meaning of the concept of causation' or 'is itself the meaning of causation'.[117] I made those statements considering only causal explanation. After reading Reinherd Merkel's writings and subsequently conversing with him, I agree with his more restricted view of the NESS analysis as being 'the only correct analysis of the logical structure of the overall causative net and as the basis for causal explanations', while still needing an ontological account of the processes that are actually going on (or being blocked) in single instances of causation.[118]

Stapleton's repetition of the 'vicious circularity' criticism of my elaborations of the NESS analysis, which often refer to '*causal* sufficiency' and '*causal* laws', is ironic, since each of her proposed tests has employed viciously circular criteria, for example, 'involvement' and 'contribution', which are made more definite only insofar as she refers to the NESS analysis for their elaboration. A short answer to this criticism is that these terms simply refer to the laws of nature and related generalisations and their complete instantiation in specific instances. For more demanding readers, I have, without employing any circular causal language,

[111] Wright, 'Causation in Tort Law' (n 15) 1789, 1793–96, 1802–03, 1806; Wright, 'Pruning' (n 16) 1022, 1024–25, 1035–37, 1039–42. The incoherence of counterfactual analysis is discussed in Wright and Puppe, 'Causation: Linguistic, Philosophical, Legal and Economic' (n 23) 470–72.

[112] Mackie, *The Cement of the Universe* (n 28) 38–48, 76–77, 126–27, criticised in Wright, 'Pruning' (n 16) 1023–34.

[113] Mackie, *The Cement of the Universe* (n 28) 45 (distinguishing between completed and uncompleted causal stories); Honoré, 'Necessary and Sufficient Conditions in Tort Law' (n 61) 378 (relying on 'which causal process ran its course and which was frustrated').

[114] See n 101.

[115] See Wright, 'Acts and Omissions' (n 30) 301–02.

[116] See n 39.

[117] Stapleton, 'Choosing' (n 40) 472–73 and fn 145.

[118] See Wright, 'Dialogues on Causation with Puppe' (n 1) 617 and fn 89.

provided NESS-based definitions of natural (causal) laws, causal sufficiency, and individual causes (contributing conditions),[119] which, however, she and others have refused to acknowledge while repeating without discussion Fumerton and Kress's original criticism.[120]

Stapleton has also repeated David Fischer's criticisms of my application of the NESS analysis in overdetermined negative causation ('double prevention') cases.[121] In my initial brief discussion of this issue, I tried to include negative conditions as well as positive conditions as part of the complete instantiation of the relevant laws of nature.[122] Fischer correctly criticised my analysis as being easily used to reach opposing conclusions.[123] After further reflection, I realised that, while the analysis of positive causation requires a condition to be part of the complete instantiation of the causal laws underlying the relevant causal process, the analysis of negative causation requires identification of the condition(s) that prevented the complete instantiation of a preventive causal process and thus contributed to the success of the unprevented causal process. I argued that when the failure of a preventive causal process is overdetermined, the lack of instantiation of a 'causally prior' step (eg applying the brakes) in the physical sequence of steps that constitutes the causal process preempts the potential failure that would have occurred at a subsequent un-reached step (eg the failure of the brakes to operate).[124] Fischer responded, again correctly, that this revised analysis also is able to be used to reach the opposite result, depending on how one identifies the steps in the relevant causal process.[125]

After considering Fischer's and Stapleton's criticisms, I ultimately agreed with Ingeborg Puppe that the correct analysis of overdetermined negative causation situations is a combination of our respective prior analyses, which should (i) (as I have argued) focus on how negative conditions contribute to the ultimate result by causing the failure of a causal process that would have prevented that result, but (ii) (as Puppe has argued) identify the temporally first lack of instantiation of a required condition in the preventive causal process that irretrievably causes that process to fail (eg the failure to have working brakes) as a preemptive cause of its failure and thus as a negative cause of the unprevented result. Just as you cannot

[119] Wright, 'NESS Account' (n 29) 289, 291; Ingeborg Puppe and Richard W Wright, 'Causation in the Law: Philosophy, Doctrine and Practice' in Marta Infantino and Lena Zervogianni (eds), *Causation in European Tort Law* (Cambridge University Press 2017) 17, 28.

[120] eg Stapleton, 'Extended But-For' (n 73) 702, 703 fn 26; Steel, *Proof* (n 39) 27 fn 59; David A Lagnado and Tobias Gerstenberg, 'Causation in Moral and Legal Reasoning', in Michael R Waldmann (ed), *The Oxford Handbook of Causal Reasoning* (OUP 2017) 565, 572.

[121] Stapleton, 'Choosing' (n 40) 477–79. Stapleton oddly claims that specific omissions, such as a specific failure to repair brakes and a subsequent specific failure to attempt to use them, 'do not "occur in sequence"': ibid 478.

[122] Wright, 'Causation in Tort Law' (n 15) 1801–02.

[123] Fischer, 'Causation in Fact in Omission Cases' (n 71) 1357–59.

[124] Wright, 'Once More' (n 26) 1125–26, 1128–31; Wright, 'Acts and Omissions' (n 30) 303–05.

[125] David Fischer, 'Insufficient Causes' (2005–2006) 94 U Kentucky L Rev 277, 311.

kill an already dead horse, you cannot cause the failure of an already failed causal process.[126]

In her two most recent articles, published in 2013 and 2015, Stapleton unfortunately has departed from her prior (inconsistently applied) insistence on leaving normative issues out of the factual causation analysis and her prior support of the NESS analysis of factual causation. She instead has engaged in an extensive attack on the NESS analysis, repeating criticisms by herself and others[127] that I have addressed elsewhere in this paper and in prior papers.[128] She has added a new criticism: that, by 'focusing primarily on physical relations' and describing my account 'as one of "natural (scientific …) causation"'[129] (as she had previously described her own analyses[130]), I failed to recognise 'that, to be adequate, any account of the "causal relation" in the law of obligations must clearly encompass … abstract [legal] relations',[131] including 'doctrinal rules [that] apply within a legal jurisdiction, specific terms govern[ing] the legal relations of two individual contracting parties, a certain voting regime [for determining] ballots of a Board of Directors and so on'[132] and even individuals' own decision rules.[133] However, such humanly created rules, even when instantiated in a particular situation, are not causal relations and by themselves have no causal effect. The relevant causal relations instead involve human awareness of and reactions to those rules and their perceived satisfaction or lack of satisfaction, including misinterpreting or ignoring them.[134]

In her 2013 article, Stapleton set forth the following test for 'factual causation' and claimed, once again, that it has 'greater forensic utility' than the allegedly 'algorithmic' NESS analysis:[135]

> [A]ll that is required before the relation between a factor and the existence of a phenomenon will be recognised as 'causal' is that: either, [i] but for the factor, the phenomenon would have been prevented; or [ii] the factor resulted in some positive contribution to the relevant mechanism by which the phenomenon came about.[136]

[126] Puppe and Wright, 'Causation in the Law' (n 119) 53–54.
[127] Stapleton, 'Unnecessary Causes' (n 34) 49 fn 40; Stapleton, 'Extended But-For' (n 73) 697 and fn *, 700 and fn 9, 701–02 and fn 19, 703 and fn 26, 714 and fn 81. As this festschrift was going to press, Stapleton submitted and had immediately published online a new article, which repeats her prior arguments and promotion of the 'extended but-for' test but fails to address any of my criticisms of those arguments and that test. Jane Stapleton, 'Unnecessary and Insufficient Factual Causes' (2023) 16(2) J Tort Law 1, https://doi.org/10.1515/jtl-2023-0030 (published 8 August 2023).
[128] eg Wright, 'NESS Account' (n 29).
[129] Stapleton, 'Extended But-For' (n 73) 700 fn 7, quoting Wright, 'NESS Account' (n 29) 285.
[130] See text at nn 52–54 and 74.
[131] Stapleton, 'Extended But-For' (n 73) 700; see also ibid 702 fn 19.
[132] ibid 703; see also ibid 699–700, 705, 707, 710; Steel *Proof* (n 39) 32–33, 35.
[133] Stapleton, 'Extended But-For' (n 73) 706.
[134] See Wright and Puppe, 'Causation: Linguistic, Philosophical, Legal and Economic' (n 23) 468–69; Wright, 'NESS Account' (n 29) 307–09.
[135] Stapleton, 'Unnecessary Causes' (n 34) 49 fn 40.
[136] ibid 39 (bracketed numerals added to clarify subsequent discussion); see also ibid 65.

In 2015, without any references to 'factual causation'[137] and again recanting her prior analyses,[138] she applied the but-for restriction to both conditions (i) and (ii) and named it the 'extended but-for' test, which she continues to promote:[139]

> A specified factor is a cause of the existence of a particular phenomenon ... only if, but for that factor alone, (i) the phenomenon would not exist or (ii) an actual contribution to an element of the positive requirements for the existence of the phenomenon would not exist.[140]

She claims that these complex but-for tests can be deployed without any reference to 'the notion of sufficient sets'.[141] However, but-for analysis, whether performed (incoherently) counterfactually or (coherently) factually, requires consideration of the instantiation in the particular situation of the relevant natural laws, which consist of minimally sufficient sets of conditions.[142]

Condition (i) in both tests is meant to be the but-for test, although it is inaccurately worded in the 2013 version. In the 2015 version, does 'but for that factor alone' assume the presence or absence of competing factors? Condition (ii) in both versions, which is deficiently based on my simplification of the NESS analysis,[143] is viciously circular, ambiguous, and incomplete. It fails to define or elaborate 'contribution', 'mechanism', 'element', or 'requirement'. The 2013 version assumes we already know 'the relevant mechanism by which the phenomenon came about'. The 2015 version fails to require instantiation of all of the required positive elements. Depending on how one interprets 'but for that factor alone ... an actual contribution to [a required positive] element' would not exist, the factor must be either sufficient by itself for the required contribution (thus excluding conditions which are not independently strongly sufficient), or is not a cause if there is any other actual contribution to the required positive element, or is tautological and question-begging ('but-for this contribution, there would not be this contribution'—which seems to be how Stapleton interprets it[144]). In her discussion of condition (ii), Stapleton returns to the claims in her initial articles that, without applying any causal test, theory, or analysis, it is 'known exactly what happened and by what agency' or that some condition had an 'indisputable physical role'.[145]

[137] But see the subsequent references to 'factual causation' in Stapleton, *Essays* (n 2) 65, 67, 78, 81.
[138] Stapleton, 'Extended But-For' (n 73) 714 fn 81, 716 fn 89, 723 fn 125.
[139] eg Stapleton, *Essays* (n 2) 81 and fn 39.
[140] Stapleton, 'Extended But-For' (n 73) 713.
[141] ibid 697; see also ibid 714.
[142] See Wright, 'Pruning' (n 16) 1023–34; Wright and Puppe, 'Causation: Linguistic, Philosophical, Legal and Economic' (n 23) 470–73; text at nn 110–14.
[143] See text at nn 81–82.
[144] Stapleton, 'Extended But-For' (n 73) 719; see text at n 158.
[145] Stapleton, 'Unnecessary Causes' (n 34) 43–44, 52, 57; see also Stapleton, 'Extended But-For' (n 73) 704; Steel, *Proof* (n 39) 20 fn 22.

Stapleton's formulations generally preclude recognition of an omission or other negative condition as a cause unless it was a but-for cause.[146] I have countered claims that negative conditions, as supposed 'nothings', cannot be causes by explaining that they are real aspects of states of affairs that usually have a causal effect on some result by preventing the initiation or successful completion of a causal process that would have prevented that result, and that, indeed, recognition of negative conditions as causes and effects (eg death, which is a lack of life) is essential theoretically and practically.[147] Inexplicably, Stapleton claims that I fail to provide any such explanation or to engage with the opposing view.[148] Nevertheless, she repeats my explanation of how negative conditions generally operate causally as 'prevention-foiling factors' (a useful phrase). But when there is more than one such prevention-foiling factor, she refuses to treat any of them as a cause,[149] even though the injury at issue would not have occurred if neither negative condition had existed. She states:

> [R]espect for the separateness of a distinct legal entity grounds a fundamental normative principle: ... an individual defendant is only identified with and held responsible for the impact, relative to the hypothetical world in which he conforms to the legal norm, that his *own* violation of the norm had on the world as he found it.[150]

Stapleton fails to identify or explain the source, content, and contours of this normative principle. Is it based on some principle of justice, some version of utilitarianism, or some other moral theory? She implicitly assumes, incorrectly, that only but-for conditions contribute to some state of affairs. It is true that, from both a factual causation and normative justice perspective, we should not employ an aggregate but-for test, which not only fails to focus on individual legal responsibility[151] but also fails to distinguish duplicative causes from preempted conditions and allows irrelevant conditions to be treated as causes.[152] She fails to explain how

[146] See text at nn 136 and 140.

[147] Wright, 'Acts and Omissions' (n 30) 287–92; Wright, 'NESS Account' (n 29) 311–15; Richard W Wright, 'Causation: Metaphysics or Intuition?' in Kimberly Kessler Ferzan and Stephen J Morse (eds), *Legal, Moral, and Metaphysical Truths: The Philosophy of Michael S. Moore* (OUP 2016) 171, 177–82. Moreover, negative conditions can and often do participate in a positive causal process through a sentient being's observation of the negative condition, which positively affects that being's conduct. See Wright, 'Acts and Omissions' (n 30) 291.

[148] Stapleton, 'Extended But-For' (n 73) 701–02.

[149] Stapleton, 'Unnecessary Causes' (n 34) 39–40 and fn 6; Stapleton, 'Extended But-For' (n 73) 708–09, 715–20.

[150] Stapleton, 'Extended But-For' (n 73) 718.

[151] ibid 717–18.

[152] Wright, 'Causation in Tort Law' (n 15) 1780–81. Steel recognises these problems for overdetermined positive causation, but nevertheless uses the aggregate but-for test for overdetermined negative causation and is thus unable to distinguish duplicative from preempted conditions or to exclude totally irrelevant omissions. Steel, *Proof* (n 39) 32, 34, and fn 82.

this principle justifies (contrary to statements in her prior articles[153]) not treating duplicative or preemptive negative conditions as (responsible) causes while treating duplicative and preemptive positive conditions as (responsible) causes. She has admitted that her arguments that 'a defendant is entitled to take the world as he finds it' and that each defendant's conduct considered by itself had 'no relevant [but-for] impact on the world'[154] apply equally to overdetermination (which she now calls 'over-subscription'[155]) by positive as well as negative conditions.[156] In her 2015 article, she tries to avoid having these arguments—which, contrary to 'taking the world as you find it', she applies using hindsight at the time of trial[157]—undermine her treatment of non-but-for positive conditions as causes with the following tautological and question begging statement: 'a contributing [positive] factor … has such an impact because without it there would be one fewer actual contributions to an element of the positive requirements for the existence of the phenomenon as individuated'.[158]

I explained above that preemptive conditions should be distinguished from preempted conditions in overdetermined negative causation situations by identifying which negative condition occurred first and irretrievably prevented the initiation or completion of a preventive causal process, thereby contributing to the occurrence of the unprevented result.[159] This results in a defendant's failure to provide a warning or safeguard being identified as a preemptive cause of the plaintiff's injury even if the warning would not have been heeded or the safeguard would not have been used, contrary to the usual result and reasoning in the cases, as Stapleton and others have pointed out.[160] However, in the great majority of these cases, the plaintiff was the person who failed or would have failed to heed the warning or use the safeguard, and the better explanation of no liability is the 'no worse off' limitation on liability, since the plaintiff would have suffered the same injury anyway due to her own conduct, rather than the but-for test of factual causation with which this limitation has very often been confused.[161] As Stapleton acknowledges, but treats as a 'special rule', courts and scholars are more inclined to recognise (correctly) one or more positive or negative contributing conditions as legally responsible and liable when all of them were tortious and the plaintiff was an injured third party.[162]

[153] Stapleton, 'Perspectives' (n 48) 82–84; Stapleton, 'Unpacking Causation' (n 49) 178–83; Stapleton, 'Choosing' (n 40) 478–79; Stapleton, 'Factual Causation' (n 11) 474–75. These statements are recanted in Stapleton, 'Extended But-For' (n 73) 716 fn 89, 723 fn 125.

[154] Stapleton, 'Unnecessary Causes' (n 34) 40 and fn 6; Stapleton, 'Extended But-For' (n 73) 718–19.

[155] eg Stapleton, 'Unnecessary Causes' (n 34) 41, 48, 54, 58; Stapleton, *Essays* (n 2) xix, 79.

[156] Stapleton, 'Unnecessary Causes' (n 34) 60 fn 95; Stapleton, 'Extended But-For' (n 73) 713 fn 79, 721–22.

[157] Stapleton, 'Extended But-For' (n 73) 718, 719.

[158] ibid 719.

[159] See text at n 126. If they occur simultaneously, they are duplicative causes.

[160] Stapleton, 'Extended But-For' (n 73) 719–20; Steel, *Proof* (n 39) 29–30.

[161] Wright, 'No Worse Off' (n 44) 29–30; see Stapleton, 'Choosing' (n 40) 479 and fn 168.

[162] Stapleton, 'Extended But-For' (n 73) 713 fn 79, 721–22; see Wright, 'No Worse Off' (n 44) 30 and fn 126.

3. Doctrine and Practice

Hart and Honoré's focus on identifying and elaborating supposedly factual 'common sense' principles of responsibility for treating 'causally relevant factors' as 'causes' encouraged continued failure by academics and courts to distinguish factual causation from legal responsibility and overshadowed and distorted their weak-necessity analysis of causally relevant factors, which received very little notice and even less acceptance by legal academics or courts until 1985, when I published my initial article.[163] The rare exception was John Fleming, who incorporated much of their analysis, including its deficient aspects, into the second edition of his treatise.[164] Beginning with his seventh edition in 1987 (he was one of the reviewers of my 1985 article for tenure purposes), he clearly distinguished the scientific factual causation issue from the legal responsibility issues and adopted my NESS analysis of factual causation.[165]

In 2005, the ALI completed work on the basic liability provisions in the *Restatement Third of Torts*,[166] which had begun in 1999 but was not officially published until 2010. Citing publications by Stapleton, me, and others, the ALI departed from the prior *Restatements* by adopting a clear distinction between factual causation and the further issues involved in attributing legal responsibility ('scope of liability').[167] The sections on factual causation were meant to employ the NESS analysis, which is adopted and elaborated in the comments.[168] Unfortunately, however, the blackletter of section 27 employs (without attribution) Stapleton's since-retracted 'targeted but-for' test,[169] which, as discussed in Section 2, was based on a misunderstanding of the NESS analysis as using a double-but-for inquiry,[170] which misunderstanding was also incorporated contemporaneously in the *Principles of European Tort Law*.[171]

[163] See Wright, 'Causation in Tort Law' (n 15) 1788 and fn 227.

[164] John G Fleming, *The Law of Torts* (2nd edn, The Law Book Co 1961) 178–81 and fn 1.

[165] John G Fleming, *The Law of Torts* (7th edn, The Law Book Co 1987) 172–77 and fns 18 and 37.

[166] *Restatement of the Law Third, Torts: Liability for Physical Harm* (Proposed Final Draft No 1, ALI 2005).

[167] *Restatement of the Law Third, Torts: Liability for Physical and Emotional Harm* (ALI 2010) s 26 and comment *a*, s 29 and comment *g*, and related reporters' notes.

[168] ibid s 26 comments *c*, *i*, and *k* and reporters' note to s 26 comment *c*, s 27 comments (all) and reporters' notes to comments *f* and *i*.

[169] ibid s 27:

> Multiple Sufficient Causes. If multiple acts occur, each of which under s 26 alone [requiring but-for causation] would have been a factual cause of the physical harm at the same time in the absence of the other act(s), each act is regarded as a factual cause of the harm.

Stapleton became a member of the Council of the ALI in 2002.

[170] See text at nn 76–80.

[171] European Group on Tort Law, *Principles of European Tort Law: Text and Commentary* (Springer-Verlag 2005) s 3:102:

> Concurrent causes. In case of multiple activities, where each of them alone would have caused [under s 3:101. Conditio sine qua non] the damage at the same time, each activity is regarded as a cause of the victim's damage.

The authors of the Australian 'Ipp Report', completed in 2002, which led to significant statutory changes in civil liability in all but one Australian state, stated that their work on causation was 'greatly assisted' by Stapleton's scholarship.[172] The report nominally distinguished, as urged by Stapleton, 'factual causation' from the normative issues involved in attributing legal responsibility ('scope of liability'),[173] but it included both under the heading of 'causation' and used the same normative consideration of 'whether (and why) responsibility for harm should be imposed on the negligent party' not only for 'scope of liability' but also to allow but not require treatment of non-but-for conditions as 'factual causes' if there is 'proof that the negligence materially contributed to the harm or the risk of the harm'.[174] Stapleton and others have correctly criticised the report for its failure to clearly distinguish scientific factual causation from the normative issues involved in attributions of legal responsibility and, more specifically, for making recognition of causation in overdetermined causation situations discretionary and dependent on normative consideration.[175]

The acceptance and support of the NESS analysis by Fleming, Stapleton (until recently), and the *Restatement Third* have contributed to its general acceptance by legal academics in common law jurisdictions,[176] and Ingeborg Puppe's initial independent elaboration and defence of the NESS analysis and subsequent joint work with me[177] have contributed to its increasing acceptance by academics in civil law jurisdictions.[178] As far as I am aware, very few academics have adopted any of Stapleton's modified-but-for tests or tried themselves to develop modified-but-for or other alternatives to the NESS analysis.[179]

The NESS analysis has usually been accepted by the courts, including the highest courts in Australia, the United Kingdom, and the United States, when made aware of it in appropriate cases. Australia led the way, starting in 1991 with Mason CJ's and McHugh J's opinions in *March v E & MH Stramare Pty Limited*,[180]

[172] Commonwealth of Australia, *Review of the Law of Negligence: Final Report* (2002) 109 fn 6 ['Ipp Report'].

[173] ibid 11 (Recommendation 29(b)).

[174] ibid 11–12 (Recommendations 29(d), (e), (f), and (h)).

[175] eg Stapleton, 'Factual Causation' (n 11) 471, 474–78.

[176] See eg papers cited in Wright, 'NESS Account' (n 29) 285 fn 1; Wright and Puppe, 'Causation: Linguistic, Philosophical, Legal and Economic' (n 23) 488 fns 110 and 111.

[177] See Puppe 'Der Erfolg' (n 25) and Puppe and Wright, 'Causation in the Law' (n 119).

[178] See eg Wright, 'No Worse Off' (n 44) 23 fn 103; text at n 118.

[179] But see Sarah Green, *Causation in Negligence* (Hart 2015) (proposing a 'necessary breach analysis', critically discussed by Wright in Book Review (2016) 132 LQR 524); Steel, *Proof* (n 39) 36–37 ('draw[ing] upon' Stapleton's 'extended but-for' test when the NESS analysis 'would be needlessly complex', despite acknowledging that it is analytically inferior to and reliant on the NESS analysis for plausible elaboration and being aware of the simplified version of the NESS analysis upon which her test is deficiently based (see 27 fn 58, 35–37)); Elise Bant and Jeannie Marie Paterson, 'Statutory Causation in Cases of Misleading Conduct: Lessons from and for the common law' (2017) 24 Torts LJ 1, 15–16 and fn 70 (proposing a question-begging 'played some role (was "a factor") in the historical process' weak-sufficiency test, which allows any condition to be treated as a cause by adding it to an already sufficient set).

[180] [1991] HCA 12, (1991) 171 CLR 506.

who, citing Hart and Honoré, Honoré, and Fleming, noted academic lawyers' modification of Mill's 'scientific and philosophical' definition of a cause as the entire set of minimally sufficient conditions by treating any necessary member of that set as a cause. McHugh argued for a strict separation between factual causation (based on the but-for test with an exception for independently strongly sufficient conditions) and the further issues involved in attributing legal responsibility and ultimate liability; the other Justices disagreed.[181] Subsequent Australian cases have cited Stapleton and Wright as well as Hart and Honoré and Fleming.[182] All of the relevant citations of which I aware have been references to or illustrations of the NESS analysis; none have referenced Stapleton's modified-but-for tests.

The Supreme Court of the United Kingdom, quoting Stapleton and Wright, recently employed the NESS analysis in *The Financial Conduct Authority v Arch Insurance*.[183] Previously, in *Kuwait Airways Corp v Iraqi Airways Co (Nos 4 and 5)*,[184] Lord Nicholls of Birkenhead cited Stapleton on the distinction between factual causation and scope of liability and Fleming on overdetermination and the overly exclusive nature of the but-for test.[185]

Courts in the United States have cited approvingly the comments to *Restatement Third* § 27, rather than its deficient blackletter statement. The Supreme Court, in *Paroline v United States*,[186] relied on the statement of the NESS analysis in § 27 comment *f* and illustration 3 while describing it as a legal fiction.[187] Otherwise, employing deficient linguistic and historical analysis, the Court has insisted on interpreting all causal language in federal statutes as requiring but-for causation, unless there is a clear expression of contrary intent.[188] Other federal and state courts have been more knowledgeable and astute,[189] although a couple, perhaps

[181] 171 CLR at 508 [5] and [19]–[22] (Mason CJ with agreement by Toohey and Gaudron JJ), 525 [12] and [21] (McHugh, J), 519 [6] (Deane J) (citing Wright, 'Efficiency Theory' (n 6)).

[182] eg *Swan v The Queen* [2020] HCA 11 [25]; *Strong v Woolworths Ltd* [2012] HCA 5 [19]–[20]; *Amaca Party Ltd v Booth* [2011] HCA 53 [48] (French CJ); *Minister for the Environment v Sharma* [2022] FCAFC 35 [305] (Allsop CJ), [438]–[439] (Beach J), [879]–[880] (Wheelahan J); *Allianz Australia Ltd v Sim* [2012] NSWCA 68 [136]–[139] (Basten J).

[183] [2021] UKSC 1 [183]–[185], [189]–[191].

[184] [2002] UKHL 19, [2002] 2 AC 883.

[185] ibid [73].

[186] (2014) 572 US 434.

[187] ibid 451–52; see Keren-Paz and Wright, 'Liability for Mass Sexual Abuse' (n 35) 193–201.

[188] eg *Comcast Corp v National Association of African American-Owned Media* 589 US ___, 140 S Ct 1009 (2020).

[189] eg *June v Union Carbide Corp* 577 F3d 1234, 1242–43 (10th Cir 2009); *United States v Kearney* 672 F3d 81, 98–99 (1st Cir 2012) (citing Wright and *Restatement Third* s 27 reporters' notes to comments *g* and *i*); *Kemper v Deutsche Bank AG* 911 F3d 383, 390–91 (7th Cir 2018); *Owens v Republic of Sudan* 412 F Supp 2d 99, 112 (DDC 2006); *Major v RJ Reynolds Tobacco Co* 14 Cal App 5th 1179, 1198–1200 (2017); *Natural Resources Defense Council v Zinke* 347 F Supp 3d 465, 488–91 (ED Calif 2018); cf *Ford Motor Co v. Boomer* 736 SE 2d 724, 732 (Va 2013) (accepting the NESS analysis in *Restatement Third* s 27 comments *a* through *e* for independently strongly sufficient conditions, but declining to employ comment *g* in mesothelioma cases given the current state of medical knowledge).

misled by the double but-for language in the blackletter of § 27, have misinterpreted its 'causal sets' analysis as a but-for analysis, with an exception for independently strongly sufficient conditions.[190]

As Stapleton has said, much work remains to be done, especially with the courts.

[190] *Wilcox v Homestake Mining Co* 619 F 3d 1165, 1167–69 (10th Cir 2010); *Doull v Foster* 163 NE 3d 976, 987 fn 14, 991 (Mass 2021).

12

Causation at a Tipping Point: Material Contribution and the 'Worse Off' Question

*Gemma Turton**

1. Introduction

Much ink has been spilt over the question of what it means to say that a defendant's negligence made a material contribution to harm suffered by the claimant.[1] This continues to trouble English courts, particularly in medical negligence cases,[2] and the law remains in a confused and unsatisfactory state. Outside the negligence context, however, the Supreme Court in *FCA v Arch*[3] recently adopted a clearer approach to factual causation that recognised the conceptual inadequacy of the but-for test. Lord Hamblen and Lord Leggatt, with whom Lord Reed agreed, recognised that 'the "but for" test is inadequate, not only because it is over-inclusive, but also because it excludes some cases where one event could or would be regarded as a cause of another event'.[4] They provided as an example the well-known two hit hunters scenario, where each hunter negligently fires their gun, bullets from both hit a walker who dies, and each bullet alone would have been sufficient (along with all other external conditions) to kill him.[5] The but-for test leads to the conclusion that neither bullet was a cause of the walker's death whereas, Lord Hamblen and Lord Leggatt recognised, both bullets were factual causes of the death. They also cited Stapleton's work to support the view that a factor may be a cause of an outcome where it is neither necessary nor sufficient by itself to bring about the result.[6] They noted that the insurance context differs from the negligence context

* I am grateful to Stephen Bailey, Donal Nolan, and Sandy Steel for their invaluable comments on earlier drafts.

1 The focus is only on the test of material contribution to harm, derived from *Bonnington Castings Ltd v Wardlaw* [1956] AC 613, and not the test of material contribution to risk/material increase in risk established in *Fairchild v Glenhaven Funeral Services (t/a GH Dovener & Son)* [2002] UKHL 22, [2003] 1 AC 32.

2 See, eg, *Bailey v Ministry of Defence* [2008] EWCA Civ 883; *Popple v Birmingham Women's NHS Foundation Trust* [2012] EWCA Civ 1628; *Williams v Bermuda Hospitals Board* [2016] UKPC 4; *John v Central Manchester and Manchester Children's University Hospitals NHS Foundation Trust* [2016] EWHC 407 (QB); *Davies v Frimley Health NHS Foundation Trust* [2021] EWHC 169 (QB).

3 *Financial Conduct Authority v Arch Insurance (UK) Ltd* [2021] UKSC 1, [2021] AC 649.

4 ibid [182].

5 ibid [182] referring to *Cook v Lewis* [1951] SCR 830 (SCC).

6 ibid [183]–[185].

Gemma Turton, *Causation at a Tipping Point: Material Contribution and the 'Worse Off' Question* In: *Torts on Three Continents*. Edited by: Kylie Burns, Jodi Gardner, Jonathan Morgan and Sandy Steel, Oxford University Press.
 DOI: 10.1093/oso/9780198889748.003.0012

because 'judgements of fault or responsibility are not relevant. All that matters is what risks the insurers have agreed to cover',[7] but went on to say that:

> there is nothing in principle or in the concept of causation which precludes an insured peril that in combination with many other similar uninsured events brings about a loss with a sufficient degree of inevitability from being regarded as a cause—indeed as a proximate cause—of the loss, even if the occurrence of the insured peril is neither necessary nor sufficient to bring about the loss by itself.[8]

This shows clear recognition by the Supreme Court that a factor can be a factual cause of an outcome when it is neither necessary nor sufficient on its own to bring about that outcome, and it is to be hoped that as and when the Supreme Court is required to consider the meaning of material contribution to harm in negligence it builds on this foundation of understanding of the concept of causation to reflect this in the notion of 'material contribution to harm'. Section 2 considers the theoretical basis for this approach as well as identifying issues that have clouded the reasoning in recent decisions.

A potential barrier to this development is the caution expressed about the difference between the insurance context and the tort context. Lord Briggs (with whom Lord Hodge agreed) noted that

> the majority ground their treatment of concurrent causation firmly within the process of construction. The question whether particular consequential harm to a policyholder is subject to indemnity is as much a part of the process of interpreting their bargain as is the identification of the insured peril. It is therefore a quite distinct process from, for example, applying the law about causation and remoteness of loss for the purpose of identifying the harm liable to be made good by tortfeasors to their victims.[9]

Despite this, the hypothetical scenarios provided by Lord Hamblen and Lord Leggatt to illustrate the idea that an unnecessary and insufficient factor can be a cause are all based on wrongdoing causing damage.[10] It is notable, however, that Lord Briggs included reference to the evaluative question of remoteness here. It is desirable that the law should reflect one consistent understanding of *factual* causation, addressing normative questions of responsibility through distinct inquiries

[7] ibid [190].
[8] ibid [191].
[9] ibid [320].
[10] See ibid [184]–[185]. They refer to individuals who combine to push a bus over a cliff causing the destruction of the bus where 20 people push but only 13 or 14 would have been needed, company directors who vote to put on the market a dangerous product that causes injuries where only a majority vote was needed, and multiple polluters who discharge hazardous waste into a river.

such as 'remoteness' or legal causation or scope of liability. Section 3 considers the issues relating to responsibility that may arise in respect of legal causation when unnecessary causes are captured at the factual causation stage.

Stapleton and Steel have argued[11] that the claimant must still prove that they are worse off as a result of the damage caused by the defendant,[12] and Wright makes a similar argument but placing the burden on the defendant to prove that their breach did not make the claimant worse off.[13] Unlike factual causation, this is not solely a factual inquiry; what is the benchmark against which the claimant must prove that they are worse off as a result of the defendant's negligence? When the loss would still have followed but for the negligence, the question of whether the defendant ought to be held responsible for that loss may properly vary depending on whether the factors that would still have caused the loss were tortious/otherwise wrongful or innocent.

In practical terms, many, if not most, causation cases will be resolved on the but-for test because the causal process will not be over-determined.[14] In those that are over-determined, the negligence will often combine with non-negligent factors to cause the damage, so the 'worse off' question will also often be resolved through simple application of the but-for test. This is what Stapleton and Steel argue should have occurred in *Bailey*[15] and in *Williams*.[16] The defendant in *Bailey* negligently failed properly to resuscitate the claimant after surgery, meaning she had to undergo further procedures and was left in a weakened state. She was also weakened by a non-negligently occurring illness, pancreatitis. When she later vomited, her weakened state meant that she aspirated the vomit, leading to cardiac arrest and brain damage. The Court of Appeal accepted that the negligent weakness had materially contributed to the harm she suffered so factual causation was established even though the claimant could not satisfy the but-for test and the defendant was liable for the whole of her condition. The claimant in *Williams* attended the defendant hospital suffering from acute appendicitis, where there was a delay of between 2 hours 20 minutes and 4 hours 15 minutes in operating. During surgery it was discovered that the appendix had ruptured and the claimant had developed sepsis which caused complications to his heart and lungs, leaving him seriously unwell for a period of weeks. Although the innocent and negligent exposures to

[11] See, eg, Jane Stapleton and Sandy Steel, 'Causes and Contributions' (2016) 132 LQR 363; Jane Stapleton, 'Unnecessary Causes' (2013) 129 LQR 39.

[12] Along with proof of the other elements of negligence, ie duty of care, breach of duty, legal causation/remoteness. Section 3 of this chapter considers which limb of the negligence inquiry this belongs in.

[13] Richard W Wright, 'Causation (Contribution) and the 'No Worse Off' Limitation on Liability' (2022) 49 UWAL Rev 5.

[14] See Gemma Turton, 'Using NESS to Overcome the Confusion Created by the 'Material Contribution to Harm' Test for Causation in Negligence' (2014) 30 PN 50, 60.

[15] *Bailey v Ministry of Defence* [2008] EWCA Civ 883.

[16] *Williams v Bermuda Hospitals Board* [2016] UKPC 4. See Stapleton and Steel, 'Causes and Contributions' (n 11).

sepsis occurred consecutively rather than concurrently, the Privy Council held that the delay materially contributed to the process and therefore materially contributed to the injury to the heart and lungs. While factual causation was established on the basis of a material contribution to harm, Stapleton and Steel argue that the claimants should not have been able to recover compensatory damages because they had not proved that some tortious conduct was necessary for the loss to occur; they were no worse off than they would have been as a result of non-tortious causes.[17]

Arguably it is only where the negligence made a contribution to the damage without being necessary for that outcome, but combined with other tortious causes so that the loss would not have been suffered but-for *somebody's* wrongdoing, that a defendant might justifiably be held liable despite their negligence not being necessary for the outcome. Hopefully this helps allay fears that treating material contribution to harm as being broader than the but-for test would lead to an explosion in liability. However, given that the 'worse off' question will generally begin with a test of necessity, thereby reintroducing a but-for analysis, we might ask why it is important for negligence law to capture factual causes beyond those that were necessary for the damage. This chapter seeks a fuller answer to this question.

Section 2 of this chapter focuses on factual causation and further develops the argument that a factor may be a factual cause despite being neither necessary nor individually sufficient to bring about a particular outcome. This is an area calling out to be settled by the Supreme Court, which requires a case with appropriate facts, and part of the problem identified in this chapter is that material contribution to harm is often raised in inappropriate cases. Section 3 then turns to the 'worse off' principle and argues that it should form part of the negligence inquiry, but that it is a liability principle relevant to legal causation rather than simply being a matter of quantification of damages. Noting that the 'worse off' principle reintroduces but-for questions alongside normative questions about responsibility, we cannot ignore the potential for it to add complexity to litigation. In response, Section 3 also seeks to articulate more fully why it is important to capture unnecessary causes at the factual causation stage before focusing the inquiry on but-for causes at the legal causation stage.

2. Causation and Contribution in Theory: Broader Than Necessity

As we saw in the introduction, the Supreme Court has recognised that a condition may be a cause of an outcome where it is neither necessary nor sufficient to bring

[17] Stapleton and Steel (n 11) 367.

about that outcome. This section begins by defending the theoretical basis for this claim, and seeks to allay fears of its expansionary potential by identifying when an unnecessary condition, that is, one that does not satisfy the but-for test, is a factual cause. An important consideration identified in this section is to understand the causal process in any case, notably whether in 'cumulative' cause cases involving a build-up of a causal agent, the process involves a 'tipping point' beyond which further exposures cannot have a causal effect.

A. The NESS test

The NESS test was developed by Richard Wright in response to the weaknesses of the but-for test,[18] and built upon ideas initially developed by Hart and Honoré in *Causation in the Law*.[19] In this account of causation something is a cause if it was a Necessary Element of a Sufficient Set: 'a particular condition was a cause of (condition contributing to) a specific consequence if and only if it was a necessary element of a set of antecedent actual conditions that was sufficient for the occurrence of the consequence. (Note that the phrase "a set" permits a plurality of sufficient sets.).'[20] The first part of this is uncontroversial, since causes are 'complex' any outcome is caused by the coming together of a set of conditions which are together sufficient to cause it. This is reflected in the recent decision of the High Court in *John v Manchester Children's Hospitals* which recognised that there need not be a 'single harmful agent' for a court to conclude that the defendant's negligence made a material contribution to harm.[21] The claimant in that case suffered a head injury and was taken to the defendant hospital where there was a negligent delay in performing a CT scan and, consequently, a delay of several hours in operating on him. The court found that he developed raised intra-cranial pressure during this negligent delay. He also suffered a non-negligent post-operative infection which required further surgery. He ultimately developed cognitive and neuropsychological impairment. The head injury and infection alone would have caused some impairment, but the damage seems to have been regarded as indivisible with Picken J finding that apportionment was not merely difficult but impossible[22] because the evidence showed that one cannot attribute particular functions to particular parts

[18] Richard W Wright, 'Causation in Tort Law' (1985) 73 Cal L Rev 1735 (hereafter 'Causation'); Richard W Wright, 'Causation, Responsibility, Risk, Probability, Naked Statistics, and Proof: Pruning the Bramble Bush by Clarifying the Concepts' (1987) 73 Iowa L Rev 1001; Richard W Wright, 'The NESS Account of Natural Causation: A Response to Criticisms' in Richard Goldberg (ed), *Perspectives on Causation* (Hart Publishing 2011).

[19] HLA Hart and T Honoré, *Causation in the Law* (2nd edn, Clarendon Press 1985).

[20] Wright, 'Causation in Tort Law' (n 18) 1790.

[21] *John v Central Manchester and Manchester Children's University Hospitals NHS Foundation Trust* [2016] EWHC 407 (QB).

[22] ibid [98]–[105].

of the brain.[23] All three factors—the head injury, the negligently raised intra-cranial pressure, and the infection—materially contributed to the damage. In NESS terms we would say that each factor was a necessary element of the set of conditions that was jointly sufficient to cause the damage since causes are 'complex'.

The novelty of the NESS test is that it fully incorporates plurality, so while in most instances one particular outcome will only have been preceded by one of these possible sets, the NESS test recognises that multiple sets may actually occur. It is based on the understanding that an event is caused by a set of conditions that are jointly sufficient, and that more than one sufficient set may occur simultaneously. Of course, the negligence inquiry is not concerned with identifying all factors in the set, just with the question of whether the defendant's negligence was a cause of the damage suffered by the claimant. In many cases there will only be one set of conditions that actually occurred, and the question will simply be whether the defendant's negligence was a necessary element of that set so the NESS test effectively collapses into the but-for test, and we see this reflected in the fact that a material contribution will often also be a but-for cause. For example, if I fail to secure my daughter properly in her stroller and she falls out injuring herself, the injury will be caused by a set of conditions such as her wriggling and the bumpy path we are walking along, but we simply want to know whether my failure to properly strap her in was a necessary element of the sufficient set of conditions that occurred, so the NESS inquiry operates in the same way as the but-for test in simple cases.

The classic case where NESS does assist us in identifying causes beyond those that were necessary is the so-called 'double hit hunters' hypothetical outlined above. The but-for test leads to the conclusion that neither bullet was a cause since neither was necessary for the walker's death; absent bullet *A*, the death would still have occurred and vice versa. The NESS test allows for two simultaneously occurring sufficient sets, so is able to identify both as causes. Here, causation is 'duplicated' because each bullet alone can form a sufficient set (with external conditions).

But the NESS test also allows us to identify as causal a duplicated contribution that on its own cannot form a sufficient set, and Stapleton's bridge example illustrates this:

> On a certain train line the trains weigh 10 units each; a bridge carrying the line was built to withstand a weight of 20 units. A train will pass across the bridge at noon. Before noon *X* deposits a weight of 6 units within the bridge structure, then *Y* deposits another 6-unit weight, then *Z* deposits another 6-unit weight. *X*, *Y*, and *Z* act independently and are unaware of the conduct of each other. At noon

[23] Ibid [104].

the train attempts to cross the bridge which collapses, killing a passenger on the train.[24]

No individual weight was a necessary condition of the passenger's death because, for example, but-for *X*'s 6 units of weight, the death would still have occurred, but the same could be said of *Y* and *Z* leading to the false conclusion that none of them were causes of the passenger's death. The NESS test asks whether *X*'s weight was a necessary condition of *an* actually occurring sufficient set, so setting aside *Y* or *Z*'s weight, *X*'s weight was a necessary condition of a set that contained one of the other weights and the weight of the train. That *X*'s weight was placed there innocently or wrongfully makes no difference to the purely factual conclusion that it was a cause of the death; this consideration is relevant to other limbs of the negligence inquiry. Likewise, whether *Y* and *Z*'s weights were innocently or wrongfully present does not alter the factual conclusion that *X*'s weight was a cause of the death, but would be relevant to the loss question where minds may rightfully differ as to whether liability for compensatory damages ought to follow.

The facts of *Bailey* reflect this hypothetical.[25] The claimant was exposed to negligent weakness and pancreatitis weakness, prior to the arrival of an external stress, her vomiting. The difference is that while we know that 20 units of weight is sufficient to collapse the bridge in the hypothetical, we do not know what degree of weakness would have been sufficient for the claimant to aspirate her vomit and what degree of weakness she would have been able to cope with. What the hypothetical shows us is that this should not matter for proof of factual causation; all weakening that occurred prior to the external stressor (and which was still operative at the time of the stressor) was causal whether necessary or not.

Although Stapleton is right to suggest that recognising unnecessary contributions as causal would have 'no impact at all in most cases because in most cases it is known that the relevant step in the mechanism by which the injury came about was an either/or step, and as we have seen, it is this knowledge that justifies the continued application of the but-for test of causal connection in most cases',[26] it is a scenario that seems likely to arise more frequently in a medical negligence context than in other cases. This is because medical negligence will often involve a patient who is already unwell, and a negligent delay in diagnosis and/or treatment that prolongs the effects of that illness as occurred in *Williams*. The two sources of weakness in *Bailey* operated concurrently, but the Privy Council in *Williams* held that negligence may also make a material contribution to harm when the exposures are consecutive.[27] This is true, but the next section suggests that where the causal

[24] Stapleton 'Unnecessary Causes' (n 11) 48–49.
[25] *Bailey* (n 15).
[26] ibid 46.
[27] *Williams* (n 16) [39].

process involves the accumulation of the causal agent to reach a 'tipping point' beyond which the damage becomes unavoidable, as in *Hotson*[28] and *Davies*,[29] the focus should be on when the tipping point is reached rather than when the damage becomes apparent.

B. Causation and the NESS test in cases of consecutive exposures: is there a tipping point?

Where a causal process involves an accumulation of a causal agent over time, such that once a tipping point or threshold of exposure has passed the damage becomes inevitable, or in a medical context 'untreatable', subsequent additions to that causal agent arguably cannot contribute to the causal process. This is because 'being treatable' is one of the necessary elements of the sufficient set of conditions required for 'failure to treat' to also be a necessary element of that set. Some examples help illustrate the issue.

The claimant in *Hotson* fell from a tree injuring his hip. The defendant hospital negligently failed to diagnose the injury to the hip when the claimant first visited, resulting in a delay in diagnosis and treatment. The claimant suffered avascular necrosis which occurs when there is an insufficient supply of blood to the epiphysis. A number of blood vessels were damaged by the fall. During the period of delay in diagnosis these blood vessels continued to bleed into the joint, blocking the blood vessels that had remained intact. Although the claimant advanced a claim for the loss of a chance of avoiding the injury, the House of Lords adhered to traditional principles, retaining the avascular necrosis itself as the gist of the action and applying the but-for test on the balance of probabilities to decide whether the delay in treatment was a cause of this injury. They held that, on the balance of probabilities at the time of the negligence, there were already insufficient blood vessels remaining intact to keep the epiphysis alive. Effectively they considered that the claimant was already doomed to suffer avascular necrosis.

Stapleton criticises this approach in 'Unnecessary Causes'.[30] She explains: 'Both the weakening of the blood supply due to the fall and the weakening resulting from the breach began to starve the hip joint of blood leading later to an avascular necrosis: that is, the necrosis only happened after Hotson's blood supply had been weakened by both the fall and the breach.'[31] However, a key distinction between this case and *Bailey* is that *Bailey* involved duplication because both sources of weakness operated on the claimant simultaneously, whereas in *Hotson* the damage

[28] *Hotson v East Berkshire Health Authority* [1987] AC 750 (HL).
[29] *Davies* (n 2).
[30] Stapleton, 'Unnecessary Causes' (n 11) 48–50.
[31] Ibid 49.

to the blood vessels operated on the claimant consecutively. This means that in *Hotson* it is possible that later bleeding was unable to have any effect on the outcome because the outcome had already become inevitable. This point is made clearer if we revisit the bridge hypothetical. Asking whether *Z* was a cause of the passenger's death, Stapleton observes that the passenger seems to be doomed once the second weight is placed on the bridge but argues that 'this dooming question is irrelevant to the question of whether the third weight physically contributed to the collapse or not ... The law cannot ignore what it knows to be true: that the third weight was exerting its force on the bridge when the train arrived'.[32] She therefore concludes, correctly, that *Z*'s weight was a cause of the death. This is comparable to *Bailey* where 'a threshold point of physical weakness had become oversubscribed before an external stress, in this case the vomit, came on the scene to complete the mechanism'.[33] It is not, however, analogous to the facts in *Hotson*. In the train example *Z*'s weight is a necessary element of an *actually* occurring sufficient set because the set only becomes sufficient once the train's weight is also added. Until that moment *Z* could choose to remove his weight and it would not form part of an actually occurring sufficient set. His weight therefore forms part of an 'indecipherable mass' when the set becomes sufficient. In *Hotson* the negligent damage to blood vessels occurs after the initial damage rather than simultaneously. The negligently damaged blood vessels are not part of an *actually occurring* sufficient set if the hip was already untreatable at that time because 'being treatable' is a necessary part of such a set. This was not a case where an 'external stress' was needed to complete the sufficient set, so once enough blood vessels were damaged there would be a sufficient set. The question was simply at what point enough blood vessels were damaged, and the House of Lords was satisfied on the balance of probabilities that this moment was reached before the negligence occurred.

The scope of this analysis is confined to cases involving a build-up of a causal agent to reach a tipping point where the exposure becomes untreatable. This is distinct from a situation, for example, where the victim is fatally ill from consuming a poison but is subsequently shot and killed before the poison takes effect. There the question is which harmful agent(s) was causal. Here there is a single causal agent and once the tipping point has passed the damage is inevitable and it is simply the passage of time that makes the damage apparent. The passage of time is not, however, an external stressor in the same way as the passing train or the patient vomiting in *Bailey* because it too is inevitable.

Stapleton and Steel object that 'the starving of the hip took time and did not result in the death of the hip joint until considerably after the breach had adversely impacted blood vessels',[34] so the negligence contributed to the causal process and

[32] ibid 49.
[33] Ibid 50.
[34] Stapleton and Steel, 'Causes and Contributions' (n 11) 366.

'to conclude that the starving of the joint by the initial condition was the "sole" cause ... is to ignore the physical reality'.[35] This argument must be taken seriously, indeed this chapter argues that part of the importance of recognising that causation was established in *Bailey* is because this reflects the physical reality.[36] The difference between the cases lies in the fact that all sources of weakness in *Bailey* were operative at the time the external stressor arrived, completing the sufficient set of conditions for the claimant to suffer cardiac arrest, whereas the process in *Hotson* did not involve an external stressor, so once the threshold was passed whereby the damage to the hip was unavoidable because of the degree of damage to the blood vessels, further damage to blood vessels could not form part of an actually occurring set of conditions sufficient to cause avascular necrosis.

Ultimately the difference in result between this approach and Stapleton's arises from the focus on the tipping point rather than the damage becoming apparent. This approach is based on a slight difference in emphasis between the NESS test and Stapleton's extended but-for test. Under the extended but-for test, 'a breach is a cause of an injury only if, but for it alone, (i) the injury would not exist or (ii) an actual contribution to an element of the positive requirements for its occurrence would not exist'.[37] This emphasises the 'positive requirements', that is, the damage to the blood vessels in *Hotson*. The NESS test focuses on the whole of the sufficient set, so more readily includes 'being treatable' as a necessary condition in cases of this type.

This analysis would have simplified the approach to the facts of *Davies v Frimley Health NHS Foundation Trust*.[38] The claimant's wife attended hospital with severe pain in her ear which was in fact bacterial meningitis from which she ultimately died. The judge explained the process:

> the presence of the bacteria triggers a defensive reaction in the host which, through one or more processes, or a combination of them, lead to inflammation and a build-up of fluid in the brain lining ... It also leads to brain swelling. Because the skull is a rigid enclosed chamber, these phenomena lead to increased ICP [intra-cranial pressure] ... The impact on the brain causing swelling is, itself, a further source of increased pressure, so that the process starts to self-reinforce in a cyclical fashion.[39]

It was found that the defendant hospital was negligent in failing to administer antibiotics by 10.40am on the day that she arrived there, and they were in fact

[35] ibid.

[36] See text to n 89.

[37] Jane Stapleton, 'An Extended but-for Test for the Causal Relation in the Law of Obligations' (2015) 35 OJLS 697, 714.

[38] *Davies* (n 2).

[39] ibid [25].

administered at 1.20pm. Later that evening she deteriorated and was ventilated, and two days later she died. Because of the self-reinforcing process, the court found that the causal process involved a 'tipping point'[40] or 'point of no return'.[41] If antibiotics were administered before that point was reached, they would quickly take effect and prevent death. After that time the self-reinforcing process would unfold and antibiotics could not make any difference; the delay in giving antibiotics could not be a necessary element of an actually occurring sufficient set of conditions to cause her death.

A court addressing causation, particularly in a medical negligence case involving delay, should distinguish between those cases that require an external stressor, and those that are the unfolding of a process where there is a tipping point. If an external stressor is involved, such as the vomiting in *Bailey*, then both sources of prior weakness are causal (as long as they are still operative). If the process involves a tipping point, like in *Davies*, then we need to consider whether the exposures were concurrent so both causal, or consecutive, in which case they may or may not be depending on whether they occurred before or after the tipping point had been reached. This NESS-informed analysis should be used by future courts faced with the 'test' of material contribution to harm which the next section shows is currently in a muddled state.

C. Unnecessary causes: a role for the test of 'material contribution to harm'

The test of material contribution to harm is commonly traced back to the decision in *Bonnington Castings v Wardlaw*,[42] although its origins lie in older nuisance cases.[43] Many attempts to understand the meaning of material contribution to harm therefore start with what was decided in *Bonnington*. This is problematic because, as Bailey has explained: 'There appears to be a fundamental ambiguity at the heart of *Bonnington Castings Ltd v Wardlaw*. It is unclear whether the "material contribution" rule as applied to these facts involved an *application* of the but-for principle or an *exception* to that rule.'[44] The House of Lords simply did not explicitly address the question so it remains unclear whether unnecessary causes would be captured by a test of material contribution to harm. Either interpretation is compatible with the judgments in *Bonnington*, and this chapter advances the normative argument that it ought to capture unnecessary causes.

[40] ibid [166].
[41] ibid [132].
[42] *Bonnington* (n 1).
[43] See Sandy Steel and David Ibbetson, 'More Grief on Uncertain Causation in Tort' (2011) 70 Cambridge LJ 451, 453.
[44] SH Bailey, 'Causation in Negligence: What is a Material Contribution?' (2010) 30 LS 167, 174.

The test of material contribution to harm has been applied in numerous cases since *Bonnington* was decided. Two issues seem to have been settled at Privy Council and High Court level: that it applies to consecutive as well as concurrent exposures,[45] and that it does not require a 'single harmful agent'.[46] A number of issues remain unsettled: whether it is an exception to the but-for test, whether it applies to 'indivisible' damage (as opposed to divisible, ie dose-related, damage), and whether it should lead to apportionment of damages. It is therefore calling out to be addressed by the Supreme Court in an appropriate case. A real challenge in this area is that the uncertainty surrounding the meaning of the test has led to it being raised by counsel and addressed by courts in a number of inappropriate cases.

A source of confusion is that where the but-for test can be applied to establish factual causation, the test of material contribution to harm is often still raised by counsel in the alternative (understandably given the lack of clarity surrounding it), yet these cases do not require a material contribution analysis. Courts are therefore invited to apply the test, and to consider whether it is an exception to the but-for test, on facts where the but-for test is satisfied. This exacerbates confusion about the meaning of the test; if the facts are such that the negligence made a necessary contribution to the causal process, a court cannot meaningfully address whether the test of material contribution to harm is an exception to the but-for test. This appears to have been the case in *Popple v Birmingham Women's NHS Foundation Trust* where the process involved various sources of the harmful agent.[47] Ward LJ applied the *Bailey* principle despite his reasoning indicating that the negligence was a necessary condition.[48]

The Court of Appeal in *Popple* simply applied the *Bailey* principle to the facts, but the wider trend is for courts to consider what a 'material contribution' is and whether it is an exception to the but-for test. In *Davies*, HHJ Auerbach felt compelled to address material contribution to harm where the but-for test was satisfied because it was a matter of doctrinal contention.[49] He proceeded to survey the case law, from *Bonnington*, through *Fairchild*,[50] *Gregg v Scott*,[51] *Bailey*,[52] *AB v Ministry of Defence*,[53] *Heneghan v Manchester Dry Docks Ltd*,[54] *Sienkiewicz v Greif*,[55] and *Williams*.[56] Of these cases, only *Bonnington*, *Bailey*, and *Williams* involved application of the test of material contribution to harm. *Fairchild* established the

45 *Williams* (n 16).
46 *John* (n 21).
47 *Popple v Birmingham Women's NHS Foundation Trust* [2012] EWCA Civ 1628.
48 ibid [78].
49 *Davies* (n 2) [168].
50 *Fairchild* (n 1).
51 *Gregg v Scott* [2005] UKHL 2.
52 *Bailey* (n 15).
53 *Ministry of Defence v AB* [2010] EWCA Civ 1317.
54 *Heneghan v Manchester Dry Docks Ltd* [2016] EWCA Civ 86.
55 *Sienkiewicz v Greif (UK) Ltd* [2011] UKSC 10.
56 *Williams* (n 16).

exceptional material contribution to risk of harm approach, which was applied in *Sienkiewicz* and in *Heneghan*, while *Gregg* was a case on loss of a chance. In *AB* the Court of Appeal did hold that the test of material contribution to harm did not apply to indivisible damage, but this was based on the misconception that the relevant damage in *Bailey* was the claimant's weakness rather than her brain damage.[57]

As *Davies* illustrates, the law on material contribution to harm has reached a stage where its meaning is to be gleaned from *obiter dicta* discussing other *obiter dicta*. This is not conducive to clarity. What is clear from this case is that the law in this area is calling out for determination by the Supreme Court. The Supreme Court recognised in *FCA v Arch* that unnecessary and insufficient conditions may be factual causes,[58] and the theoretical analysis in this chapter both supports this view and indicates a need for any elucidation of 'material contribution to harm' to be based on the NESS test which gives courts the concrete guidance that is lacking in the phrase 'material contribution to harm'.

We saw, however, that Lord Briggs in *Arch* expressed caution about whether the same would be said in the negligence context where courts apply both factual causation and principles of remoteness to identify the damage that the defendant is liable for. The chapter now turns to the role that legal causation or remoteness might play in assessing whether the defendant ought to be held liable for the damage in cases involving unnecessary causes.

3. Legal Causation and the 'Worse Off' Question

Only a small number of cases will involve unnecessary causes, and this chapter has further developed the case for recognising them as causal at the factual causation stage of the negligence inquiry. As Stapleton,[59] Steel,[60] and Wright[61] have argued, however, that is not the end of the matter. They argue that a claimant must also show that some wrongdoing was necessary to bring about the damage,[62] and that courts should then engage with normative questions about the appropriate allocation of responsibility. Stapleton calls this the 'no better off' principle of compensatory damages,[63] while Wright calls it the 'no worse off' limitation on liability.[64] In essence it reintroduces a but-for analysis, asking whether the claimant's damage would have been avoided but for some wrongdoing, and then invites courts to air

57 *AB* (n 53) [150].
58 See text to n 8.
59 Stapleton, 'Unnecessary Causes' (n 11).
60 Stapleton and Steel, 'Causes and Contributions' (n 11).
61 Wright, 'No Worse Off' (n 13).
62 Although Wright would place the burden on the defendant to prove that the damage would have occurred in the absence of any wrongdoing (n 13).
63 Stapleton, 'Unnecessary Causes' (n 11) 48.
64 Wright, 'No Worse Off' (n 13).

evaluative questions about responsibility. This section has two concerns. The first is whether this is a principle of damages or of liability. This is addressed in two parts, the first delineating the issue and the second considering the appropriate label for it. The second concern is a practical concern that litigation in causation cases is already complex, with counsel raising both but-for and material contribution to harm at the factual causation stage regardless of the facts, and risks becoming further complicated and costly unless there is clarity as to the appropriate inquiries at the factual causation stage and the 'worse off' stage.

A. The nature and limited scope of this 'worse off' issue

Drawing on the NESS test to address Stapleton's bridge scenario, this chapter has argued that a factor may still be a cause despite being neither necessary nor sufficient for the damage. Varying the scenario helps illustrate the distinction that is being made between causation and 'worse off'. Suppose *X* was performing maintenance and properly placed 6 units on the bridge, *Y* was not a person but a natural event such as rainfall adding 6 units of weight, and *Z* negligently deposited 6 units of weight on the bridge, again all independently and unaware of each other's conduct. The passenger's estate would still be able to establish that *Z*'s negligence was a cause of their death but would not be able to prove that they were worse off as a result of negligence because the bridge would still have collapsed, resulting in their death, due to the non-tortious weights alone. If, however, two or all three of the actors depositing weights on the bridge were acting negligently, then the victim would be worse off as a result of negligence than they would have been in the absence of any negligence, and normative considerations affect whether it is appropriate for liability of any individual defendant to follow. As Stapleton argues, 'tort law has a choice whether the "no better off" principle ... is that compensatory damages should not make a claimant better off relative to: (i) where he would have been but for the individual defendant's tortious contribution; or (ii) where he would have been but for all the tortious contributions'.[65]

We should, however, be cognisant of the distinction that Nolan draws between damage and loss.[66] In his account, loss is the question of whether the claimant is worse off while damage involves an interference with a person's protected interests. The 'worse off' issue at stake in these cases does not arise in such broad terms as 'loss' though. The question is not whether the damage has left the claimant worse off than they would otherwise have been, the question is whether the damage itself, although caused by the defendant's negligence among other things, would have

[65] Stapleton 'Unnecessary Causes' (n 11) 58–59.
[66] Donal Nolan, 'Rights, Damage and Loss' (2017) 37 OJLS 255, 256.

occurred anyway due to non-tortious causes.[67] The distinction between *Jobling* on the one hand and *Bailey* and *Williams* on the other helps illustrate this point.

The claimant in *Jobling* suffered a back injury as a result of the defendant's negligence but three years later, before trial, was diagnosed as suffering an unrelated condition which left him unable to work.[68] The House of Lords held that the defendant was not liable to compensate the claimant for loss of earning capacity from the point when the claimant became unable to work because of the unrelated illness, explaining:

> When the supervening illness or injury which is the independent cause of the loss of earning capacity has manifested itself before trial, the event has demonstrated that, even if the plaintiff had never sustained the tortious injury, his earnings would now be reduced or extinguished. To hold the tortfeasor, in this situation, liable to pay damages for a notional continuing loss of earnings attributable to the tortious injury, is to put the plaintiff in a better position than he would be in if he had never suffered the tortious injury.[69]

Similarly then, Stapleton and Steel argued, while the claimants in *Bailey v MOD*[70] and *Williams v Bermuda Hospitals Board*[71] were able to establish factual causation on the basis of material contribution to harm where the defendant's negligence was (arguably) neither necessary nor sufficient for the harm suffered, they should also have been required to prove that they were worse off as a result of the negligence.

Stapleton and Steel's argument goes further than *Jobling* however. In *Jobling* the defendant's negligence had caused the claimant's back injury and the contested issue was the defendant's liability for a particular head of damages. In relation to *Bailey* and *Williams*, the argument is that although the defendant's negligence was a cause of the physical injury to the claimant, it is arguable that the defendant should not be liable at all because the claimant would have suffered the same physical injury due to non-tortious causes. They suggest that the physical injury did not represent damage at all,[72] while Steel more recently refers to this as a question of 'loss'.[73] Given that this issue would affect whether the defendant is liable at all, it should be accommodated within a liability-determining aspect of negligence

[67] Stapleton, Steel, and Wright each use different labels for this issue, and their views have evolved over time, so caution is required before placing too much weight on the labels they use.

[68] *Jobling v Associated Dairies Ltd* [1982] AC 794.

[69] ibid 820 (Lord Bridge).

[70] *Bailey* (n 15).

[71] *Williams* (n 16).

[72] Of course the gist damage could have been reframed in much more limited terms in each case, so the claimant in *Bailey* could have sought damages for the weakness and its impact on her recovery prior to her aspirating her vomit, and the claimant in *Williams* could have limited his claim to the pain and suffering during the period of delay in diagnosis, but these would have been very low value claims and likely not worthwhile pursuing.

[73] Sandy Steel, 'Damages without Loss' (2023) 139 LQR 219.

law, otherwise it could result in a defendant notionally being liable to the claimant in respect of the damage suffered, but compensatory damages being calculated as zero. While it is not simply a question of loss, nor is it really a question of damage because the claimants in *Bailey* and *Williams* clearly had suffered a setback to their interests in the form of physical injury, so the next section suggests a role for 'legal causation' or 'scope of liability'.

B. A role for legal causation?

Stapleton refers to this 'worse off' question as the question of whether the injury that the defendant is proved to have caused represents 'damage' for the purposes of the 'no better off' principle of compensatory tort damages.[74] She explains:

> It is a defining and well-known principle of the approach the law of torts takes to the assessment of compensatory damages that it should not make a claimant better off than he would have been had he not been the victim of tortious conduct.[75]

In earlier work Stapleton located this issue within 'scope of liability'.[76] Steel now prefers the label 'loss',[77] while Wright calls it a 'limitation on liability'.[78] Building on the argument developed above that this is not simply an issue of damages but an issue of liability, it is argued here that it can most coherently be accommodated within 'scope of liability', and is an example of why the label 'scope of liability' is more appropriate than 'legal causation'.

Stapleton has long advocated for the separation of the factual causation from normative issues that are captured by 'legal causation'.[79] Where the defendant's negligence was a factual cause of the claimant's damage but was not a necessary condition, the question of whether liability ought to follow may justifiably depend on normative considerations arising from whether the other causes were wrongful or innocent, and whether the claimant would have suffered the same damage solely due to innocent causes or whether some wrongdoing was required. Situating the issue outside of factual causation allows those normative considerations to be openly aired. The Supreme Court in *FCA v Arch* hinted at such an approach, noting, as we saw above, that the negligence inquiry involves 'applying the

[74] Stapleton 'Unnecessary Causes' (n 11) 54.
[75] ibid 54–55.
[76] Jane Stapleton 'Cause in Fact and the Scope of Liability for Consequences' (2003) 119 LQR 388, 412.
[77] Steel (n 73).
[78] Wright (n 13) 32.
[79] See, eg, Jane Stapleton 'Unpacking Causation' in Peter Cane and John Gardner (eds), *Relating to responsibility* (Hart Publishing 2001) 158–160; Jane Stapleton, 'Cause in Fact' (n 76).

law about causation *and remoteness of loss* for the purpose of identifying the harm liable to be made good by tortfeasors to their victims' (emphasis added).[80]

However, Stapleton has also advocated for the label 'scope of liability' over the causal language of 'legal causation', and the need to address this 'worse off' issue further advances the case for change.[81] We tend to think of 'legal causation' as the question where to truncate the identification of a person with the consequences of his or her conduct.[82] This suggests a focus on the relationship between the breach of duty and the outcome where, for example, reasonable foreseeability plays a significant role. As Wright has observed, situations involving overdetermination 'are usually unforeseeable'.[83] The issue is not that the damage is unforeseeable, but that the duplicating causes are unforeseeable. This may be relevant to whether liability should follow, but so may the question of whether the duplicating causes are tortious or innocent, and specifically the question of whether any tortious conduct was necessary to cause the damage. The suggestion is not that the defendant should not be identified as responsible in some sense for the damage suffered by the claimant, but that the defendant should not be liable for that damage in negligence because, in the circumstances, the same damage would have been suffered due to non-wrongful causes. It is a question of liability that seems more readily captured by the label 'scope of liability' given its clear appeal to normative questions about the appropriate scope of a defendant's liability.

C. The value of separating causation from loss

This approach would enable a claimant to prove that the defendant's negligence was a cause of the damage they suffered despite the negligence being neither necessary nor sufficient for the damage, as long as it contributed to the outcome. If, however, but for the defendant's negligence, the damage would still have been caused by non-tortious factors, arguably the defendant should not be considered responsible at the legal causation stage. It is only in the handful of cases where *some* tortious conduct was necessary that there then arises a question as to whether compensatory damages should be available. In a case such as *Bailey* or *Williams* where there is only one tortious factor, the but-for question would re-emerge at the loss stage and the claimant would not recover damages unless they could satisfy that test. We might wonder then whether it is efficient to require the claimant to prove both

80 *FCA* (n 3) [320].

81 Stapleton, 'Cause in Fact' (n 76). The 'scope of liability' label has been adopted in legislative reforms in most Australian states and the Australian Capital Territory, eg s 5D *Civil Liability Act* 2002 (NSW).

82 ibid 411 referring to the work of Tony Honoré, 'Responsibility and Luck: The Moral Basis of Strict Liability' (1988) 104 LQR 530.

83 Wright (n 13) 37–38.

causation of damage, on either a but-for or material contribution to harm basis, if they will also be required to prove they are worse off on a but-for basis. There is real force then to the concern motivating Green, that 'Hard cases make for bad law for the very reason that they favour exceptions over rules. The formulation, therefore, of any rule designed expressly for exceptional circumstances is bound not only to prove an extremely difficult exercise, but also to produce a result which is far more complicated than the vast majority of situations will demand.'[84]

Indeed, it seems to be this concern that motivates Green's 'necessary breach analysis'.[85] That approach is based on necessity, but first aggregates all breaches of duty and asks whether '*at least one* breach was necessary to divert the claimant from "the mandated course of events"'.[86] She explains: 'The crucial objective of this stage is both simple and narrow: to establish whether the claimant's injury is the concern of the law, or whether it is the result of misfortune.'[87] This first stage captures duplicative causes, in other words those where causation is over-determined such as the two hit hunters and bridge scenarios outlined above, then a second stage asks whether the defendant's breach was 'operative'.[88] This approach has the attraction of efficiency. Rather than establishing that the defendant's negligence was a factual cause but later concluding that the same loss would have been caused by non-tortious factors in any event so compensatory damages should not be awarded, the necessary breach analysis first filters out cases of this type, then addresses issues of both factual causation and the appropriateness of liability/damages in a second stage. Such efficiency is compelling given the complexity of the causal issues that arise in many cases and the lack of clarity in the case law surrounding proof of factual causation. Despite this, there are a number of reasons for preferring an approach that separates the evaluative considerations into legal rather than factual causation (so long, of course, as the question of legal causation is not overlooked).

The first is a concern for conceptual neatness and the accuracy of the labels the law places on its reasons for imposing or denying liability. In *Bailey* for instance, the claimant was weakened by negligence and weakened by pancreatitis, and as a result of her weakness she aspirated her vomit, resulting in cardiac arrest and brain damage. The necessary breach analysis would conclude at the first stage that the negligence was not a factual cause of the claimant's brain damage because the claimant could not prove that some negligence was necessary for this outcome. But to say to the claimant that the negligently caused weakness was not a cause of

[84] Sarah Green, *Causation in Negligence* (Hart Publishing 2015) 10.

[85] See ibid.

[86] ibid 13 (emphasis in original), citing Jane Stapleton, 'Choosing What We Mean by "Causation" in the Law' (2008) 73 Missouri L Rev 433, 460.

[87] Green (n 84) 13.

[88] The concept of being 'operative' is explained at length in Chapter 3 of Green's book and incorporates both factual and evaluative questions that would currently be addressed under the label of 'remoteness' or 'legal causation' or 'scope of liability'. A summary here would not do justice to Green's work and is not necessary for the argument being made in this chapter.

her brain damage is simply a fiction. As Stapleton has said, 'This result would, I believe, offend our social values and expectations of the law, which … judges seek to reflect when they are identifying and articulating the common law'.[89] However, to say to the claimant that although the negligence was a cause of her brain damage, the evidence shows that she would likely have suffered that damage as a result of the pancreatitis weakness alone so she cannot prove that she is worse off as a result of wrongdoing is accurate. Admittedly it would still be a difficult result for the claimant to accept because of the unknown and unknowable counterfactual position, in comparison with say *Jobling* where the natural cause of the loss of earning capacity was evident, but that is what the balance of probabilities standard of proof responds to.

Addressing 'worse off' separately from causation would also allow for the responsibility-focused questions of 'worse off' to be articulated more clearly and transparently. In advocating for the law to limit the causal inquiry to the factual question of involvement, Stapleton said 'this ensures that normative concerns are located elsewhere in the legal analysis where traditionally they are more likely to be exposed as normative and evaluated accordingly'.[90] Having a clearer understanding of the evaluative nature of the issues at stake would assist in resolving them. Although only a limited number of cases would require considered analysis of the appropriateness of recognising responsibility, because the issue would only arise where no defendant's negligence was either necessary or sufficient for the damage to occur, by locating the analysis under legal rather than factual causation, tort law would acknowledge that factual causation is established and the case would be treated as 'exceptional' in the sense merely of unusual, rather than 'exceptional' in the sense of imposing liability in the absence of causation. Wright has noted that the reporters for the *Restatement Third: Remedies* rejected the 'no worse off' limitation on the basis that

> the reason courts crafted the multiple-sufficient-causes addition to but-for causation in the first place was to make sure that a plaintiff's loss was compensated … it is hard to imagine that those courts recognized a broader class of factual causes than but-for causation for any reason *other* than to make such defendants liable for their tortious conduct towards plaintiffs. Otherwise, the courts would have been altering the rules on factual causation for no reason at all.[91]

As Wright notes, however, 'it does not follow that all factual causes should be held legally responsible and … such clearly is not the case, even if we set aside the

[89] Jane Stapleton, *Three Essays on Torts* (Oxford University Press 2021) 80.

[90] Jane Stapleton, 'Choosing What We Mean by "Causation" in the Law' (2008) 73 Missouri L Rev 433, 446.

[91] ALI, *Restatement of the Law Third, Torts: Remedies* (Council Draft No 1, 31 August 2021) 194.

NWO [no worse off] limitation'.[92] This is correct. The but-for test fails to capture all factual causes, leaving unnecessary and insufficient causes to be treated as somehow exceptional when, in fact, they are causes. But identifying unnecessary and insufficient causes at the factual causation stage then provides the option to impose liability and opens up the space for the relevant normative considerations to be properly articulated.

Finally, this approach would enable tort law to maintain a single concept of factual causation across torts, beyond negligence, including in torts where loss is not required, such as conversion. Stapleton highlights the examples of unlawful physical duress and fraudulent misrepresentation,[93] and Steel's chapter in this volume considers material contribution in a range of torts.

4. Conclusion

The test of material contribution to harm is currently used, without distinction, in a range of senses. In *Popple* it was used where the negligence made a necessary contribution to a causal process to reflect the fact that causation was indeed a process that occurred over a period of time. In *John* it seems to have been used once again where the negligence was necessary for the damage, to reflect the fact that the causal set was obviously complex because there were other factors capable of causing the same type of damage. It is applied where the damage is divisible so the negligence was a necessary condition of a portion of the overall harm suffered by the claimant, and one interpretation of *Bailey* is that it stands for a principle that, while damages should normally be apportioned in such a case, the claimant can recover for the whole of the harm suffered when they cannot prove the portion that is 'damage', that is, is attributable to negligence.[94] It is sometimes applied where the damage is indivisible. Indeed the damage in *Bailey* was indivisible so, properly understood, that case stands for the proposition that an unnecessary contribution can be recognised as a cause.

It is clear that the test is 'ripe for authoritative review, *at least in a case where it may affect the result*'.[95] As this chapter has highlighted, it is important that these issues are addressed in a case where it may affect the result, because part of the confusion stems from the fact that it is repeatedly raised and addressed in cases where it is simply not required by the facts. This has two consequences: since the facts do not involve unnecessary causes, they do not allow a court to meaningfully analyse whether the test of material contribution to harm is, or can be, an exception to the

[92] Wright (n 13) 39.
[93] Stapleton, *Three Essays* (n 89) 85.
[94] See, eg, *Holtby v Brigham & Cowan (Hull) Ltd* [2000] EWCA Civ 111.
[95] *Thorley v Sandwell & West Birmingham Hospitals NHS Trust* [2021] EWHC 2604 (QB), [151] (Soole J) (emphasis added).

but-for test; and the test is analysed in *obiter dicta* discussing *obiter dicta* which does not make for clear resolution of the issues.

It is understandable that courts use the phrase 'material contribution' in the *Popple* and *John* senses, to reflect the complexity of causes. But just because many, if not all, but-for causes can conveniently be described as having made a material contribution to the damage, not all material contributions need be necessary for the damage,[96] and the NESS approach helps us identify when an unnecessary contribution to a duplicated causal process is a cause. What should not be overlooked is that this approach to material contribution is quite different in nature from the proposition that where the negligence contributed in the sense of being a but-for cause of part of the divisible damage suffered by the claimant, but the claimant cannot prove the extent of the damage attributable to the negligence, they should be allowed to recover in full unless the defendant can establish a basis for apportionment. That is a rule of proof rather than concerning the concept of causation, and it is undesirable to use the same label to refer to it.

It might seem strange that this chapter supports and defends the view that courts should address causal issues at two stages of the negligence inquiry, factual causation and the 'worse off' question in scope of liability, given that the law on factual causation is currently in a muddled state and litigation is consequently drawn out by counsel raising both the but-for test and material contribution to harm test as alternative bases for factual causation. Surely, it might be argued, incorporating causal issues into a 'worse off' question at the scope of liability stage just expands the scope for confusion and for increasing the complexity and cost of cases. This practical concern should not be dismissed lightly, and the aim of Section 3 of this chapter was to defend the separation of causation and 'worse off' at a doctrinal level. It should also be clear at this stage that it is only in a small minority of cases where causation was duplicated, and the negligence made an unnecessary contribution but some negligence was necessary to cause the damage, that the 'worse off' question will not simply be resolved by the but-for test and normative considerations will take centre stage. This means that in most cases parties can continue to focus on a but-for analysis, and that if the case does fall into that limited category the distinction between the factual and normative issues allows for focused argument.

[96] See also *Strong v Woolworths* (2012) CLR 182 [24] (High Court of Australia) citing *Zanner v Zanner* [2010] NSWCA 343, [111].

13

Making a Difference: Liability and Necessity

Sandy Steel[*]

1. Introduction

Making a difference matters in law. Sometimes a defendant's argument that 'it would've happened anyway—I made no difference' is a legally good one. Sometimes, however, it is not: liability may be unaffected by whether the defendant's conduct was necessary for the events that ensued. This raises a question: when does and should making a difference make a difference to liability?

Here is an answer that naturally comes to mind:

> *Causation.* 'It would've happened anyway' is legally relevant only in so far it bears on whether a person is a cause of an outcome. If the fact that the outcome would have happened anyway entails no causation, then it is relevant. If it does not, then it is irrelevant.

On this view, the relevance of making a difference is determined exclusively by the law's approach to causation.

Jane Stapleton has argued persuasively against *Causation.*[1] On her account, there are situations in which a person is a cause of an outcome, but is not liable to a legal response in virtue of the fact the outcome would have happened in any event. Consequently, the relevance of making a difference cannot be accounted for entirely in causal terms. Instead, other, non-causal, facts are needed to explain when and why making a difference matters.

* With thanks to Jonathan Morgan for very helpful comments on a draft, and to Cher-Yi Tan for valuable research assistance (errors are mine). It is a real honour to contribute to this volume. Jane Stapleton has been an intellectual inspiration and a source of kind support and encouragement from the beginning of my academic career.

[1] See especially J Stapleton, 'Unnecessary Causes' (2013) 129 LQR 39; J Stapleton, 'An 'Extended But-For' Test for the Causal Relation in the Law of Obligations' (2015) 35 OJLS 697; J Stapleton, *Three Essays on Torts* (OUP, 2020), 82–87.

Sandy Steel, *Making a Difference: Liability and Necessity* In: *Torts on Three Continents.* Edited by: Kylie Burns, Jodi Gardner, Jonathan Morgan and Sandy Steel, Oxford University Press. © Sandy Steel 2024.
DOI: 10.1093/oso/9780198889748.003.0013

Building on Stapleton's analysis, this chapter aims to improve understanding of the role of necessity—of making a difference to the occurrence of an outcome—to legal liability, and the relationship of this idea to causal requirements in law.

It proceeds in three main sections. First, it begins by considering the role of necessity in the current law, broadening attention beyond the traditional focus on compensatory liability, to consider the role of necessity in relation to liability to other legal responses, such as rescission. This analysis shows the importance, in determining the relevance of necessity, of distinguishing between different kinds of liability. Making a difference sometimes affects the availability of a particular form of legal liability, rather than the existence of liability at all.

Second, it considers how, conceptually, liability without necessity should be understood, arguing that it is sometimes causal, and sometimes non-causal.

Third, it examines possible justifications of liability in the absence of difference-making. In light of this discussion, it identifies various respects in which the current law needs modification. Overall, the chapter provides further support for Stapleton's rejection of *Causation*: it will be shown that the relevance of necessity to liability depends on various normative considerations that are external to whether the defendant was a cause of the outcome.

2. The Role of Difference-Making in Legal Liability

Let's say that an agent's conduct 'makes a difference to an outcome' if and only if, had the agent not engaged in that conduct, the outcome would not have occurred.[2] In this sense the agent is 'necessary' for the outcome.[3] This section (1) explains how necessity is relevant at all to liability, and (2) sets out instances in which liability is decoupled from necessity.

A. The relevance of necessity

Causation

The law's starting point for the assessment of whether *x* was a cause of *y* is the 'but-for', counterfactual, test: *x* is a cause of *y* if, had *x* not occurred, *y* would not have occurred. This is a test of necessity. The law's understanding of causation naturally, then, makes necessity relevant wherever causation is relevant.

[2] On the counterfactual questions this raises, see Section 2.C.

[3] As this implies, it only makes sense to inquire about the role of a person's 'difference-making' in relation to legal rules that are concerned with, or make some legal response contingent upon, the relationship between agents' conduct and outcomes. Not all legal rules are like this. A legal rule that defines when one has a capacity to enter into a legal transaction does not specify any connection between an agent's conduct and an outcome.

Matters are more complex, however. It is clear that the law accepts that *x* can be a cause of *y* even if *x* fails the but-for test and is thus not necessary for *y*. We can distinguish three kinds of case:

Overdetermination. X is held to be a cause of Y despite the existence of another *sufficient* cause of Y. For example:

(i) two large fires, independently set by A and B, join to burn down C's house. Each fire causes the destruction of the house, though neither is necessary for this destruction.[4]

(ii) Covid 19 cases within a 25-mile radius of a person's business can be a 'cause' of loss to the business even if cases outside that radius were independently sufficient to trigger restrictive measures that would have produced the same loss.[5]

Pre-emption. X is held to be a cause of Y despite the existence of another set of circumstances that would have caused Y at the same time (or shortly after). For example:

(i) Assassin A kills a target at *t*, but the target would have been killed by Assassin B, working independently of A, at *t* anyway (A's bullet knocks the victim out of the path of B's bullet).

Unnecessary, insufficient, contribution. X is held to be a (contributory) cause of Y, despite the existence of at least another sufficient cause of Y, and X is not necessary for Y. For example:

(i) A adds 1 drop of poison to C's tea, while B adds 2 drops. 2 drops are sufficient for a fatal dose. C drinks the tea, ingesting all 3 drops, and dies.

In each case, the law holds that A is a cause, or at least a causal contribution, despite being unnecessary for the outcome. This is not a peculiarity of the common law.[6]

In short, then, necessity is not necessary for causality (according to the law). However, it is still true that necessity is *defeasibly sufficient* for causality. Absent some unusual circumstance, such as a break in the chain of causation, being necessary is sufficient for a person's conduct to be a cause (in law). Hence we can still explain why necessity is of importance to the law, even given the possibility of

[4] *Anderson v Minneapolis S P & S S M R Co*, 146 Minn 430, 179 NW 45 (1920).

[5] See *Financial Conduct Authority v Arch Insurance (UK) Ltd & Others* [2021] UKSC 1.

[6] For German and French law, see S Steel, *Proof of Causation in Tort Law* (Cambridge University Press, 2017), ch 1.

causation in the absence of necessity: it provides a defeasibly sufficient condition of causation.

Loss

'Loss' is normally understood to involve a person's being caused to be worse off in some respect, by an event, compared to the position they would have been in had the event not occurred.[7] To determine whether a person has suffered a 'loss' from a breach of contract, then, one asks whether that person's current position is worse than the position they would have been had the breach not occurred. The notion of 'loss' thus has a difference-making component baked into it. Whenever liability is contingent upon, or affected by, loss being suffered, the necessity of a person's conduct for another's present condition is thus naturally relevant: it matters to the existence and quantum of loss.

B. Liability without Necessity

This section identifies examples of liability in relation to an outcome, despite a person's conduct not being necessary for the occurrence of the outcome. A person is 'liable in relation to an outcome', in the sense intended here, when they are situated such that a court has the power to issue a (remedial) order against them in relation to the outcome.

It is worth noting immediately that many legal wrongs can be committed independently of difference-making in relation to an outcome: battery, false imprisonment, direct discrimination, murder, trespass to land and goods, negligent damaging of the body, mind or property, and many others. If I trespass on your land, it is irrelevant to the *existence* of liability that the same interference with your exclusive possession would have occurred in any event.[8] Legal wrongdoing and difference-making in relation to an outcome come apart. Consequently, the relevance of necessity in many cases is not to the issue of whether a person committed a legal wrong, but to whether they ought to be liable to a particular kind of legal response, such as an order to pay compensation.

[7] *Morris-Garner v One Step Support Ltd* [2018] UKSC 20, [36] (Lord Reed).

[8] Strictly, one can have legal wrongdoing in the absence of liability at all in cases in which a person is immunised from liability (eg diplomatic immunity). In these cases, difference-making may be irrelevant to the existence of a *wrong*, even when the wrong does not give rise to liability. For an argument that liability should exist in negligence in the absence of *loss* (and so absent difference making in relation to an injury), see McBride's, Chapter 6, this volume and A Beever, *Rediscovering the Law of Negligence* (Hart, 2007), 433ff.

Declarations

To the extent that a declaration could simply have the content 'X wronged Y', and given that wrongdoing can come apart from difference-making, it is fairly straightforward to see how declarations can be available in the absence of necessity. When liability is grounded largely in the wrong itself, this closes off arguments about whether the same thing would have happened in any event.

However, even declaratory relief might be made contingent upon difference-making. Sections 31(2A) and (2B) of the Senior Courts Act 1981 provide that, in an application for relief in judicial review proceedings, a court must refuse to grant relief, including declaratory relief 'if it appears to the court to be highly likely that the outcome for the applicant would not have been substantially different if the conduct complained of had not occurred', unless there is an 'exceptional' public interest in disregarding this requirement.[9] Clearly, this provision does not involve the idea that public law wrongdoing intrinsically involves difference-making; it is a purely remedial restriction.

Punishment

If Assassin 1 intentionally kills Victim without justification, Assassin 1 is clearly guilty of murder, and liable to punishment, even if Assassin 2 would have intentionally killed Victim without justification at exactly the same time in the same way in any event. This is, again, explicable on the basis that the law takes punishment to be justified by the culpable wrong itself, and the wrong of murder is clearly possible (as this example shows) without making a difference to the occurrence of death.

Profits

If a fiduciary makes profit from a transaction which the fiduciary induced their principal to enter by misrepresentation, the fiduciary is liable to account for the profit, even if the profit would have been made had there been full disclosure of the truth. Thus in *Murad v Al-Saraj*, Al-Saraj had proposed to the Murads that they jointly purchase a hotel, with £500,000 of the price being provided in cash by Al-Saraj, £1 million by the Murads, and the remainder by bank loan.[10] The hotel was purchased and sold later for a substantial profit. Al-Saraj had in fact made his contribution to the purchase by way of set-off of debts owed to him by the vendor of the hotel. The Court of Appeal held that Al-Saraj was in a fiduciary relationship with the Murads and was liable to account for the profit made from the sale of the hotel. This was so even though the transaction would have been entered into anyway, had disclosure been made, with the Murads having a greater profit share—and the same result would have followed even if the identical transaction would have been entered into anyway. If the breach of fiduciary duty consists in

[9] See section 31(1)(b) of the Senior Courts Act 1981.
[10] [2005] EWCA Civ 959.

non-disclosure, then this breach made no difference to the existence of the gain (or part of it). I return later to whether the breach is properly characterised as entering-into-a-transaction-without-obtaining-consent.

In so far as profit-based remedies are available for certain proprietary wrongs, it seems unlikely that the wrongdoer can escape liability by arguing that the right-holder would have consented to the interference. If I sell your thing without permission, I am liable to make restitution of the value received.[11] Suppose that I sell your vase without permission, but you would have permitted me to sell it and retain any profits because you believed that the vase was worthless and not worth trying to sell. If the rationale for profit-based liability here is simply that the profit is the fruit of right-holder's thing, because it was acquired in exchange for it, then the fact that consent to the purported alienation would have been given if sought is neither here nor there.[12]

Rescission

The relevance of whether a transaction would have been entered into 'but for' the existence of a vitiating factor is varied. The current position is not entirely clear, but it appears that necessity matters to the existence of right to rescind in relation to misrepresentation and duress, but not to undue influence or breach of fiduciary duty.

Consider, first, undue influence. A victim of undue influence has a right to rescind even if, had they made a free and informed decision, they would have entered into the transaction in any event. In *UCB Corporate Services v Williams*, the Court of Appeal considered the relevance of evidence that a wife would have executed a charge to secure a loan to her husband, even if the husband had not failed to disclose important information about the loan.[13] It was held that this did not affect the existence of a right to rescind. Parker LJ said this:

> Is that fact [that the transaction was procured by undue influence] sufficient to found a right for Mrs Williams, as against Mr Williams, to have the UCB Charge set aside; or is it necessary for her also to establish that, absent the equitable fraud, she would not have executed the UCB Charge?
>
> Undue influence is exerted when improper means of persuasion are used to procure the complainant's consent to participate in a transaction, such that 'the consent thus procured ought not fairly to be treated as the expression of [the complainant's] free will' ... In such a case, equity proceeds on the basis that the complainant did not consent to the transaction. Is that enough to give rise to an

[11] *Oughton v Seppings* (1830) 1 B & Ad 241.

[12] On the relevance of counterfactuals in claims for restitution of an invalid payment, see R Stevens, *The Laws of Restitution* (OUP, 2023), 59–60.

[13] [2002] EWCA Civ 555.

> equity in the complainant to set aside the transaction as against the wrongdoer? In my judgment, it is.[14]

This was held to follow from Lord Browne-Wilkinson's analysis in *CIBC v Pitt*, which was that, in such a case, the husband's undue influence 'deprived the wife of the opportunity to make [an informed] choice, and ... it is that fact which founds the wife's equity ... to set aside the transaction'. In short, it is possible to be deprived of an informed choice, even if one would have made the same decision in any event, and this grounds the right.[15]

Next consider rescission of a transaction entered into by a fiduciary with their principal, or on behalf of their principal, in breach of fiduciary duty. Although necessity is necessary for equitable compensation—at least in non-fraudulent cases—this requirement appears not to apply in relation to rescission. In the context of a claim for compensation, the Privy Council said this in *Brickenden v London Loan & Savings Co*:[16]

> ... Once the Court has determined that the non-disclosed facts were material, speculation as to what course the [company], on disclosure, would have taken is not relevant.

This was rejected by the Court of Appeal in *Swindle v Harrison* as a general proposition in relation to compensation, but the position in relation to rescission was explicitly distinguished.[17]

Consider, now, duress. It is submitted that necessity—but-for—is required as a matter of substantive law for all forms of duress: if it is *known* that the transaction would have been entered into in any event, then there is no entitlement to rescission, unless the transaction would have been entered into because of another vitiating factor.[18] However, if the physical threat or fraudulent misrepresentation is shown to be 'a factor' or influenced the deliberation of the addressee, then a special proof rule applies. The real difficulty is precisely what kind of proof rule. It has been stated that the legal burden of proof is not altered.[19] Rather, an evidential presumption that a but-for connection exists applies, that presumption being

[14] [85]–[86].

[15] *CIBC Mortgages v Pitt* [1994] 1 AC 200, 209. See similarly *Hewett v First Plus Financial Group* [2010] EWCA Civ 312, [34] (Briggs J).

[16] *Brickenden v London Loan & Savings Co* [1934] 3 DLR 465, 469.

[17] *Swindle v Harrison* [1997] PNLR 641, 656 (Evans LJ), 660 (Hobhouse LJ).

[18] See *Versloot Dredging BV v HDI Gerling Industrie Versicherung AG* [2016] UKSC 45, [29] (itself on the 'fraudulent claims rule' in insurance): 'If he would have done the same thing even in the absence of the misrepresentation, a claim based on it will fail. The same applies in a claim to rescind a contract for misrepresentation, fraudulent or otherwise' (Lord Sumption); *BV Nederlandse Industrie van Eiprodukten v Rembrandt Enterprises Inc* [2020] QB 551, [43]–[45] (Longmore LJ).

[19] *BV Nederlandse Industrie van Eiprodukten v Rembrandt Enterprises Inc* [2020] QB 551, [43]–[45] (Longmore LJ).

'very difficult to rebut'. It is suggested that if 'very difficult to rebut' means 'either proof on the balance of probabilities or to a higher standard that there is no but-for connection', then the legal burden of proof has been reversed on this specific issue. This is also supported by the underlying justification given. The justification is not that fraud is somehow always more likely to have been a but-for cause. It is that the legal system is permissibly less concerned about making errors—false positives—in relation to those making culpable threats against people's bodily rights.[20]

In relation to misrepresentation, the rule in cases of non-fraudulent misrepresentation is that the substantive law requires a but-for connection for the existence of a right to rescind, except perhaps when the absence of a but-for connection is explained by the presence of another misrepresentation or vitiating factor.[21] The cloudier matter is fraudulent misrepresentation. In *Barton v Armstrong*, the Privy Council said that the same rule should apply to rescission for duress to the person and rescission for fraudulent misrepresentation:

> the same rule should apply in cases of duress and that if Armstrong's threats were 'a' reason for Barton's executing the deed he is entitled to relief even though he might well have entered into the contract if Armstrong had uttered no threats to induce him to do so.[22]

Subsequent courts have interpreted this as imposing a 'weaker requirement of causation':

> To rescind a contract for fraudulent misrepresentation, the requirement for causation is weaker, it is sufficient to show that the representation was a factor in the representee's decision and that but for it, they might have acted differently.[23]

These are unfortunate formulations. They blend together the issue of the content of the substantive legal rule—what connection must exist between the misrepresentation and the transaction when all facts are known—with the issue of what proof rule ought to apply when the facts are not all known. However, the upshot of

[20] See *Al Nehayan v Kent* [2018] EWHC 33, [233]: 'where physical threats are made, the interest of the law in deterring violence and the grave nature of such wrongdoing justifies setting a low threshold for inferring that a threat of violence has adversely affected the claimant' (Leggatt LJ, applying this reasoning to the causal issue in the tort of intimidation).

[21] See *Raiffeisen Zentralbank Osterreich AG v The Royal Bank of Scotland Plc* [2010] EWHC 1392 (Comm), [198]–[199] (Christopher Clarke J); *Huyton SA v Peter Cremer GmbH* [1999] CLC 230, 250; *SK Shipping Europe plc v Capital VLCC 3 Corp* [2020] EWHC 3448, [117](ii). The qualification after 'perhaps' is not, so far as I am aware, stated in the authorities but, in principle, it is essential.

[22] [1976] AC 104, 119.

[23] *Instagroup Ltd v David Carroll, Christopher Barclay* [2022] EWHC 464, [75] (Cooper QC); *Cassa di Risparmio della Repubblica di San Marino SpA v Barclays Bank Ltd* [2011] EWHC 484 [232]-[233] (Hamblen J, proceeding on the basis that this is the correct understanding); *Raiffeisen* (n 21) [198]–[199].

these statements is that, when the fraudulent misrepresentation is shown to have influenced entry into a transaction, a special proof rule applies according to which a but-for connection will be proven to exist unless it is shown by the representor clearly not to exist.[24]

Compensation

The final, and most difficult, category is compensation. This is the most difficult category because compensation is normally conceptually tied to loss and, as seen, loss is standardly taken to involve a breach of duty making a person *worse off*. Making worse off is, in turn, normally understood to involve a breach being *necessary* for a worsening in the right-holder's position. Nonetheless, it is possible to identify instances of compensatory liability in which the breach is not necessary for the outcome that grounds the wrong.

Multiple actual or hypothetical wrongs

There is general agreement that D1 and D2 would be jointly and severally liable in respect of C's damage if D1 and D2 independently negligently fired a bullet at C's body, each impact being sufficient to produce the same effect on C. D1 and D2 would be liable to compensate C both in respect of the damage that constituted that violation of C's right—the damage to C's body—but also in respect of the earnings that C would have made had D1 and D2 not breached their duties to C.[25] D1 would be similarly liable to C if D1's breach pre-empted the effect of D2's breach, say, if C had fallen to ground after D1's shot, with D2's shot then missing C. In each case, D is liable despite the fact that D's breach of a duty of care is not necessary for the damage which constituted (or made actionable) the tort.[26]

Damage to chattels

If D negligently damaged C's car and the car would have been unexpectedly damaged in the same way by an unforeseeable earthquake moments later, it seems that D is still liable to compensate C for the market value of the car at the time of destruction. The law distinguishes between 'immediate loss' and 'consequential loss'. 'Immediate loss' is said to crystallise immediately upon the damage to the property, with subsequent events not defeating a person's entitlement to substantial compensation in respect of that immediate loss. The principles were summarised by the Court of Appeal in *Coles v Hetherton*:[27]

[24] See *SK Shipping Europe Plc* (n 21) [117]. Compare E Peel, *Treitel on The Law of Contract* (15th edn, Sweet and Maxwell, 2020), at [9-028], rejecting a distinction between rescission and damages.

[25] See *Financial Conduct Authority v Arch Insurance Ltd* [2021] UKSC 1, [182]–[185]; S Steel, *Proof of Causation in Tort Law* (Cambridge University Press, 2015), ch 1.

[26] For the reason for the proviso in parenthesis, see McBride, Chapter 6, this volume.

[27] [2014] Lloyd's Rep IR 367, [2014] 2 All ER (Comm) 223, at [27] (Aikens LJ).

> (1) where a chattel is damaged by the negligence of another that loss (the 'direct' loss) is suffered as soon as the chattel is damaged. (2) The proper measure of that loss is the diminution in value that the chattel has suffered as a result of the negligence of the defendant [...] (3) If the chattel can be economically repaired, the claimant is entitled to have it repaired at the cost of the wrongdoer, although the claimant is not obliged to repair the chattel to recover the direct loss suffered. (4) Events occurring after the infliction of the damage are irrelevant to calculating the diminution in value measure of damages. Thus, subsequent destruction of the chattel, or a decision to delay repairs, or an ability to have the repairs done at less than cost or for nothing will not prevent the claimant from recovering the diminution in value of the chattel that has been caused by the negligence of the tortfeasor.

Choice-based rights

In claims for user and negotiating damages, counterfactuals that bear upon whether the breach was necessary for the interference that constituted the wrong are of limited relevance.

Consider this example:

> *Parking.* While C is on holiday, D parks D's car in C's driveway. D had asked C if C would permit D to do this, but C refused.

C is entitled to substantial damages, measured *prima facie* by reference to the market value of the use of C's land. This has been rationalised on the basis that C loses an opportunity to exercise control over C's right in relation to the land.[28] This example shows, however, that damages should be payable even when C had an opportunity to exercise control—that is, decide whether to permit the interference—but that decision was not respected. If this is correct, it seems that damages are payable simply because C's choice as to how the land ought to be used was not respected. Whether the interference would have occurred in any event is neither here nor there. If, in a variation, C would have permitted the interference but C's permission was not sought, damages should still be payable: a use of the land occurred that was not licenced by C.

Material contribution to damage

Sometimes the law uses the words 'material contribution to damage' to admit the possibility of a liability-generating relation between a person's conduct and an outcome, when the conduct is not proven to have been necessary for the outcome. In *Bailey v Ministry of Defence*, the defendant's breach was not proven, on

[28] *Morris-Garner v One Step (Support) Ltd* [2018] UKSC 20. For a fuller discussion, see S Steel, 'Damages without Loss' (2023) 139 LQR 219.

the balance of probabilities, to have been a but-for cause of the claimant's brain damage, but it was held to have materially contributed to its occurrence. This was because the breach had added to the claimant's weakness and this weakness was considered to have caused the sequence of events leading to the brain injury.[29] The Court of Appeal held that the defendant was liable on the basis that it had materially contributed to the damage and it was scientifically impossible to prove but-for causation. Although *Bailey* has been re-described as a but-for causation case, the re-description is untenable.[30]

Strictly the rule in *Bailey* is different from most of those so far identified. It is a rule which permits the right-holder to succeed in establishing compensatory liability when the breach 'materially contributed' to their damage, *only if* it is scientifically impossible to establish that the breach was necessary on the balance of probabilities.[31] It thereby combines a rule about the substantive connection that must exist between a breach and damage—namely, that a necessary connection is not necessary—and a rule of proof—namely, that the risk of its being impossible to prove the counterfactual of what would have happened but-for the breach is placed upon the party in breach, when science does not provide an answer on the balance of probabilities. So it is not a rule that permits compensatory liability to be established *when it is known on the balance of probabilities* that the breach was unnecessary. In this way, it is similar to the (*de facto*) burden-reversing rules on necessity in relation to rescission for fraudulent misrepresentation.

Fraud and breach of fiduciary duty

There is evidence that fraudulent breaches of duty and breaches of fiduciary duty attract a special proof rule in relation to the existence of *loss* for the purposes of compensation claims.[32] In some jurisdictions, a reverse burden of proof on the necessity issue is explicitly recognised in claims for compensation for breach of fiduciary duty: if the principal shows that the fiduciary entered into a transaction wrongfully, the legal burden of proof on whether the transaction would have occurred in any event shifts to the fiduciary.[33]

[29] [2008] EWCA Civ 883.

[30] Notably by Lord Toulson in *Williams v Bermuda Hospitals Board* [2016] UKPC 4. The argument (at [47]) is that since the *weakness* was a but-for cause of the claimant's damage and the breach contributed to the weakness, *the breach* is therefore a but-for cause. This is fallacious. The fact that the defendant's breach contributed to something that was a but-for cause does not entail that the breach was a but-for cause. My carbon emissions contribute to the process that is a but-for cause of climate change, but I am not a but-for cause of climate change. See further J Stapleton and S Steel, 'Causes and Contributions' (2016) 132 LQR 363; *Holmes v Poeton Holdings Ltd* [2023] EWCA Civ 1377, [67] (Stuart-Smith LJ, rejecting Lord Toulson's analysis).

[31] For an explanation why the rule still exists, despite recent doubts about its existence, see S Steel, 'Material Contribution, Again' (2022) 138 LQR 545.

[32] This was the position of Evans LJ in *Swindle v Harrison* (n 17). It was endorsed in *Nationwide Building Society v Balmer Radmore* [1999] All ER (D) 95. See also *BHP Billiton v Dalmine* [2003] EWCA Civ 170, at [33], [36].

[33] See *Sim Poh Ping v Winsta Holding Pte Ltd* [2020] SGCA 25; the burden-shifting position in Canada, New Zealand, and Hong Kong is explained therein at [146]–[179].

C. An objection

It might be objected that the alleged examples of liability in the absence of necessity are no such thing. Whether one's conduct makes a difference, the objection runs, depends upon how one constructs the relevant counterfactual in which one's conduct does not occur, and it might be objected that I have misstated the relevant counterfactual.

Consider, again, liability to account for profits made in breach of fiduciary duty. If the specific breach consists of 'failing to disclose a conflict of interest', then it is clear that there can be liability when a gain would have been made anyway even with disclosure. But, an objection might run, this is incorrect: the breach is *entering-into-a-transaction-without-obtaining-consent*, and the correct counterfactual is 'not entering into the non-consented-to transaction'. If this counterfactual is employed, then the breach will generally be necessary for the gain made, since, had there been no transaction, the gain would not have been realised.[34] In *Murad*, if Al-Saraj had not bought the hotel, then clearly no profit would have been made.

My main response to this objection is that, even if successful, it is of limited scope: this strategy will not work, for instance, in relation to cases of causal overdetermination or pre-emption. Consider, again, the assassins example, above: it is not clear how one can alter the counterfactual to arrive at the correct result.[35] If one posits that neither assassin wrongfully fired, then one reaches the wrong conclusion that *both* assassins were causes. Or suppose that a fiduciary makes a profit from a business opportunity that was obtained initially through his fiduciary position. Suppose, however, that the business opportunity would have been presented to him in a non-fiduciary capacity at some time anyway had he not obtained it through his fiduciary activity. It still seems clear that the fiduciary will be liable to give up the profit if the opportunity obtained in the fiduciary capacity is exploited. This is so, even though one is supposing a 'no transaction' counterfactual—that is, a counterfactual in which the fiduciary does not proceed with the initial exploitation of the opportunity at all, rather than one in which consent is sought and the transaction proceeds.[36] If this is correct, the better view is that the fiduciary

[34] See M Conaglen, '*Brickenden*' in S Degeling and J Varuhas (eds), *Equitable Compensation and Disgorgement of Profit* (Hart, 2017), ch 6.

[35] See Section 2.A.

[36] If one asked, in a claim for compensation for breach of fiduciary duty, 'what would have happened had the transaction not proceeded, disclosure being lacking', this would indeed not amount to an exception to the relevance of 'necessity', but simply a different way of specifying 'necessity'. However, it would amount to an exception if one ignored, as a matter of law, what the defendant would have done had the transaction not proceeded: one still needs to replace no-transaction with duty-conforrming conduct, and it may be that the defendant would ultimately have disclosed and the transaction have proceeded. For the approach in negligence to specifying the counterfactual, which looks to what would in fact have occurred had there been duty-conforming conduct, see, eg, *Phethean-Hubble v Coles* [2012] EWCA Civ 349; *Robbins v Bexley BC* [2013] EWCA Civ 1233.

liability to account does not strictly depend upon a counterfactual, but only upon the profit being causally attributable to the impugned transaction.[37]

D. Summary

We can see, then, that the role of necessity varies depending on the nature of the liability. First, in relation to liabilities which respond merely to the fact of the legal wrong, such as a declaration, the relevance of necessity largely depends upon the nature of the wrong itself. If the wrong can be committed in the absence of any difference-making, then the liability may exist without difference-making. Second, in relation to liabilities, such as to rescission, which have no necessary connection with loss, the role of necessity is varied; in some cases, it is substantively not required; in other cases, there is a reversal of the burden of proof in relation to necessity. Third, in relation to compensation, the centrality of loss foregrounds necessity, and putative exceptions sometimes (indeed, typically) take the form of modifications of *proof* of loss, rather than dispensing with necessity entirely. In cases of substantial damages in the absence of loss, these are arguably instances in which the law is not aiming to compensate for loss at all, for instance, when it is providing damages to cure the *damage* constitutive of the wrong.

3. Conceptualising Liability without Necessity

How should liability in respect of an outcome in the absence of necessity be characterised? This section considers whether it is *causal* or *non-causal* in character. In the first subsection, I explain the significance of this question, and respond to a recent objection to the view that such liability is causal in character. The second subsection then considers how the examples in the previous section should be characterised.

A. Causation, contribution, and necessity

The conceptual question of whether liability in the absence of necessity can be causal in character has some importance: if legal rules make reference to a requirement that a person has 'caused' an outcome, should this reference be interpreted as normally requiring a relationship of *necessity*?

[37] On why counterfactuals and causation come apart, see again Section 2.A.

The case for a negative answer—at least as a default interpretation—is that ordinary thought recognises causal relations in the absence of necessity. Consider again the pre-emption example from above:

> Assassin A kills a target at *t*, but the target would have been killed by Assassin B, working independently of A, at *t* anyway (A's bullet knocks the victim out of the path of B's bullet).

Suppose that target's estate sues Assassin A and the relevant statute requires that A *caused* target's death. Intuitively, Assassin A did cause the death, even though it would have happened anyway. Killing is causing death *in a particular way*. If the ordinary understanding of 'causation' justifiably bears on the interpretation of statutory rules utilising the concept, then the statutory rule should be held to be satisfied, even in the absence of necessity.

Justice Edelman has argued against this view:

> An event will only cause an outcome if the event is necessary for the outcome. In other words, if the outcome would have occurred even without the event then the event made no difference to the outcome. It did not cause the outcome. This is commonly known as the 'but for' test.[38]

Edelman's argument is that '[t]he most natural, and most common, description of cause is that E is a cause of O if E was necessary for the occurrence of O'. Additionally, a further motivation for this view is the utility of having a separate word, in law, to refer to the relationship between events which is not one of 'necessity'. This encourages transparency of reasoning: 'It means that departure from a rule of necessity must be justified, not concealed.'[39]

Edelman's claim that necessity is 'the most natural, and most common description of cause' should be rejected. It is over and under-inclusive as an account of the ordinary concept of cause. Over-inclusive because it counts 'oxygen' as a cause of every murder. It is under-inclusive because in pre-emption cases, such as the assassins example, it is correct to the first assassin as a 'cause' of C's death.[40] What *does* the assassin do to the target in the pre-emption case if not *cause* C's death at that time? It would be unusual to describe Assassin 1 as merely 'contributing' to C's death, as if there were something else on the scene that had done most of the causal work. We could justifiably revise or refine ordinary usage for theoretical or legal

[38] James Edelman, 'Unnecessary Causation' (2015) 89 ALJ 20, 20–21. See also R Stevens, 'Causation and Contribution' https://www.youtube.com/watch?v=LL6re4bl8N4.
[39] Edelman (n 38) 25.
[40] See Section 2.A.

purposes, but such examples falsify the specific claim that 'the most natural, and most common' description of causation is necessity.

Edelman's claim about what is 'most natural' has more traction in examples like the other given above:

> A adds 1 drop of poison to C's tea, while B adds 2 drops. 2 drops are sufficient for a fatal dose. C drinks the tea, ingesting all 3 drops, and dies.

It doesn't sound right to say that A *caused* B's death. Here is an explanation: the poison example is different to the assassins example because, unlike in the assassins example, the poison is not *sufficient* for the outcome. If an individual event is not sufficient for an outcome, we are less inclined to say it *caused* it.[41]

But it does seem apt to say: 'A was *a* cause of C's death'. This might seem to be logic-bending: if a person is *a* cause of X, then shouldn't it be possible truthfully to say that they caused X? Not necessarily. Compare the concepts of author and co-author. It would be odd to say of a particular co-author that they *wrote* the co-authored book. But it would be natural to describe them as *an author*.[42] Consequently if a legal rule makes reference to a requirement that X must have *caused* Y, it may be ambiguous whether this encompasses being *a* cause of Y, as well as having 'caused' Y.

There are three further problems with describing relationships of necessity as 'causation' and influencing outcomes without necessity as 'contribution'. The first is that 'contribution' also, in ordinary thought, encompasses relations of necessity. Suppose Scotland beat England 3-2 at football, as often happens in a dream I have. Suppose each of the three goals was necessary for the victory.[43] It seems awkward to say that the first goal *caused* the victory. It contributed. It was part of the cause of

[41] The idea that an event can be *sufficient* for another event is likely to prompt the objection that 'nothing is sufficient for anything else'. It depends on what notion of sufficiency is employed. By sufficiency, I mean 'if *x* occurs, and holding fixed the other features of the circumstances, then always *y*'. The objection is a bit like the point being made by A in the following conversation:

A. Can you do 10 press-ups?
B. Depends what you mean by 'can'.
A. I mean 'if you try to, you will'.
B. Yes, I can, then.
A. No you can't.
B. Why?
A. Because there's a possible world in which when you try to do 10 press-ups, there is a quantum tunnelling event which knocks a piano down on you from a tall building before your 10th press-up.
B. I really enjoyed our conversation.

The point is that in assessing whether one 'can' do something, one holds fixed ordinary circumstances. The same arguably goes for 'sufficiency'. See A Kaiserman, 'Necessary Connections in Context' (2017) 82 Erkenntnis 45.

[42] See, making this point, ibid.

[43] This isn't logically guaranteed since, if the first goal had not occurred, it's possible that another goal would have occurred anyway, and victory would still have been secured.

Scotland's victory, but it was necessary. If so, contribution cannot be restricted to lack of necessity. More evidence: consider the defence of *contributory* negligence. It's in the name. Most of the time, the right-holder's failure to take care of their own safety is a necessary condition of their damage. Still, it is typically said only to *contribute* to their damage.

Second, there is a risk in treating causation and contribution as wholly distinct in forgetting that 'contribution' is *causal* contribution in the sense of contribution *to* causing an outcome. Compare the concepts of author and co-author. Contributing stands to causation as co-authoring stands to authoring. Contributing is being part of the cause (and doing part of the causing). This matters since legal rules that refer to being causative of an outcome might properly be interpreted to include 'causally contributing', even if they do not explicitly refer to this.

Third, there is a danger in treating situations in which a person is liable in the absence of necessity as exceptional departures from 'a rule of necessity'. There is no such rule. Necessity matters to certain kinds of question (eg the existence of loss), but not to others (eg the existence of a right to rescind for undue influence). Edelman's objection entrenches the erroneous impression that departures from necessity are intrinsically, *prima facie*, anomalous.

B. Categorising the examples

Once it is accepted that a person can be a cause of—or causally contribute to—an outcome without being necessary for it, the main reason for considering the examples considered above as non-causal in character falls away. For instance, it has been argued that a fiduciary's liability to account for profits is non-causal on this basis, and instead a question of 'attribution'.[44] As we have seen, however, the fact that necessity is not required is inconclusive, given the possibility of being a cause in the absence of necessity. Consider: Mum gives Son £100 as a Christmas present. Son gains £100 from his Mum even if Dad would have given Son £100 for Christmas if Mum had not. If a fiduciary takes a valuable business opportunity from his principal, but the fiduciary would have non-wrongfully exploited an equally valuable business opportunity if he had not done so, the gains from the taken opportunity are still caused by the breach, even if they would have been made in any event.

[44] eg *Murad* [2005] EWCA Civ 959, [72]: 'There needs to be some link or nexus ... but a link or nexus does not need to be of a causal character' (Arden LJ); *Novoship (UK) Ltd v Mikhayluk* [2015] QB 499, [96]. M Conaglen, 'Identifying the Profits for Which a Fiduciary Must Account' [2020] 79 CLJ 38. Describing the rule as an 'attribution' rule doesn't exclude the relevance of causation (as Conaglen accepts). Causation is one basis on which 'x' could be attributed to 'y'. Clearly, it is not the only basis. A rule which said 'actions of Bill are to be considered the actions of Ben' is a non-causal rule of attribution. See R Stevens, 'Causation in Equity', end of Section 4 (forthcoming).

There are two further possible reasons for resisting a causal understanding of some of the examples above. First, it might be argued that the liability does not, properly understood, depend upon an empirical relation between the defendant's act and an outcome. Rather the defendant's act must *constitute*, rather than *cause*, a certain kind of interference. This is plausibly the case in relation to trespass and user damages. The defendant's act must simply *amount to* a non-consensual interference; it is not that the act must *cause* a non-consensual interference. *X* can't be a *cause* of Y if X is partly constitutive of Y. Eating the whole cake doesn't *cause* you to eat a bite of the cake. Having a bite is partly constitutive of eating the whole cake.

Second, Stevens has argued that rescission for undue influence is analogous to a non-causal rule of incapacity or validity.[45] He gives the example of P who agrees to purchase shares from D at fair market value, and P is unbeknownst to D, a 12-year-old child. As he points out, it makes no difference to the enforceability of the contract that P would have entered into such an agreement even if P had been 18 or older. And it is surely correct that the law is uninterested, here, in whether there is a causal connection between P's *age* and the entry into the transaction. P's age is rather a constitutive condition for the validity or enforceability of the agreement. It seems to me, however, that undue influence is distinctive and there is at least sometimes a causal element. Suppose that a husband sends his wife various bullying emails, with the intention of procuring her consent to a transaction, but she never reads them before agreeing to a transaction. Or suppose that a husband fails to disclose his affair prior to seeking his wife's agreement to execute a re-mortgage over their home, but a friend tells the wife of the affair, and she still agrees to the re-mortgage.[46] In these examples, the transactions would not be voidable for undue influence. The best explanation of this is that the controlling party must have a causal impact on the quality of the consent to the transaction. So the position in undue influence appears to be the following: there is no causal connection required between the 'influence' and the transaction, but there is a causal connection required between the influence and the quality of the transacting party's consent.

4. Justifications for Liability in the Absence of Necessity

This section is divided into three subsections. The first considers five possible justifications of liability in the absence of necessity. The second considers possible justifications of departures from ordinary *proof* rules in relation to the necessity. The third considers respects in which the law on necessity needs alteration.

[45] Stevens (n 44) Section 5.

[46] For this case, but without the friend: *Hewett v First Plus Financial Group plc* [2010] EWCA Civ 312.

A. Substantive justifications

Causation and culpability

If it is accepted that (1) causation extends beyond relations of necessity, (2) causation bears on liability, then there is no reason why culpably causing an outcome, despite not being necessary for it, cannot serve as a ground of punitive liability. Highly culpable, serious, wrongdoing makes a person justifiably liable in principle to punishment, and it is possible to commit highly culpable, serious, wrongs in the absence of a relationship of necessity holding between the wrongdoer's breach and any particular outcome.

There are at least two ways of showing this. First, if it is accepted that *attempted* wrongdoing is sometimes sufficient to justify punitive liability, it is difficult to motivate the idea that necessity should be a necessary condition. In cases of attempted crimes, there is *ex hypothesi* no outcome in respect of which the defendant is being held liable and therefore there can be no relationship of necessity between the breach and any outcome. If liability in the absence of outcomes is justifiable, then so too must necessity-less liability be justifiable. Second, the facts which make a person a highly culpable, serious, wrongdoer are frequently unaffected by whether their conduct is necessary for a particular outcome. Discrimination is a good example. If an employer does not hire a person having weighed in their deliberations the fact that he was of a certain race, but would not have hired him anyway, the employer still plausibly morally (and legally) wrongs the candidate.[47]

The value of choice and the quality of consent

A possible justification of the irrelevance of necessity in certain contexts is that A can impair B's consent to a transaction, regardless of whether the transaction would have occurred in any event. Using threats, false statements, or trust reposed, one can impair another's consent to what follows, regardless of whether it makes a difference to this.

A similar idea underpins the fact that *hypothetical consent*—the fact that a person *would have consented* to an interference with a right-protected interest—is generally considered insufficient to justify the interference: actual consent is required. The fact that I would have consented to your using my driveway is insufficient to permit your parking there without my consent.

The law on rescission for undue influence involves an extension of this idea from the absence of consent to impaired consent. Here the person entitled to rescission has made a choice, but the choice was, for example, made in the absence of information of which they were entitled to be made aware; consent was impaired, not absent.

[47] See Stapleton (n 1) 86 n 59.

There are at least two possible reasons for a person having a right to undo a transaction, even if the impairment made no difference to its occurrence. First, the meaning of a decision to enter into a transaction might vary depending on the reasons for which it was entered into. Even if I would have agreed to guarantee my friend's loan, had he told me that it was high-risk, the guarantee would have had a different significance had I known of this fact: it would have shown my willingness to take on a significant burden for a friend. Second, there is a sense in which a person cannot identify the transaction as fully 'theirs' as would have been possible if they had made the choice with the relevant information. The transaction was not an expression of their free agency, even if it would have been entered into anyway, because the agent was not able to weigh up the relevant reasons for and against entry into the transaction.

It is not obvious, however, why the value of unimpaired choice rationale does not extend to misrepresentation and duress. In these cases, too, a person enters a transaction with impaired consent, even if the transaction would have happened in any event. Yet the issue of necessity, subject to the proof rules discussed, *is* relevant to the latter vitiating factors. The distinction cannot be that choice in the context of relationships of trust and confidence is impaired to a greater extent than in relation to duress; threats of violence are surely a counter-example.

The rationale of certain rights

Arguably certain moral and legal rights do not exist primarily to protect a person from counterfactual worsenings in their positions. Consider the right not to be caused reasonably foreseeable physical damage by another's unreasonable risk imposition. Being damaged is not the same as being caused to suffer a counterfactual loss.[48] In the example given above of a valuable chattel being damaged by negligence, before it would otherwise have been pulverised by an earthquake, the tortfeasor has *damaged* the object, but not caused any *loss*, understood as a counterfactual worsening of the right-holder's position. If our secondary rights to compensation aim to reflect the rationale of the primary right violated, and if that primary right is conditioned on *damage*, not *loss*, then the secondary right ought, *prima facie*, to compensate for *damage*.[49]

Similarly, the trespassory torts are arguably justified by a concern to protect choice-related interests: the ability to choose whether to permit an interference, and to have one's choices about one's body, property, or other right-related goods respected. This idea might have some justificatory power in relation to the availability of user and negotiating damages even when the interference would have been permitted in any event. There is a value in the right-holder being the one who makes the choice to allow an interference.

[48] See D Nolan, 'Rights, Damage and Loss' (2017) 37 OJLS 255.

[49] For further argument, see S Steel, 'Damages without Loss' (2023) 139 LQR 219.

Fairness: avoiding wrongs diminishing rights
Consider:

> *Overdetermination.* D1 and D2 independently fire a bullet at C, in breach of duties of care. Each bullet strikes C in the head at the same time. C dies from the head wounds.

> *Overdetermination 2.* D1 fires a bullet at C in breach of a duty of care. D2's bullet was fired when D2 was being attacked by a bear, by the bear's paw, despite the gun's safety catch being securely fastened. Each bullet strikes C in the head at the same time. C dies from the head wounds.

The law is that D1 will be liable to compensate C for the extent to which C is attributably worse off than if D1 and D2 not damaged C in *Overdetermination.*[50] In *Overdetermination 2*, the law appears to be that D1 will not be liable for any earnings (or other consequential detriments) that C would otherwise have made. It may be that D1 is liable for nothing in *Overdetermination 2. Jobling v Associated Dairies* supports this view: in this case D's breach of duty caused damage to C's back, but C developed, independently, after D's tort, a spinal disability which prevented C from working in any event.[51] D's liability was eliminated to the extent that the earnings would not have been made in any event due to spinal disability. Structurally, *Jobling* is the same as *Overdetermination 2.* C's inability to make the earnings that C would otherwise have made has two sufficient causes: D's tort (a wrong) and C's back injury (a non-wrong). This is also true of *Overdetermination 2*: D1's wrong and the non-wrongful fire of the gun are both causes of C's inability or diminished ability to earn.

Arguably the justification of this pattern of liability is that there is an unfairness in a person's right to compensation being diminished by the presence of multiple actual or potential wrongs. Intuitively, it is unjust that the existence of a person's right to compensation is diminished by the presence of other wrongdoers. This occurs in *Overdetermination*, but not *Overdetermination 2.* It might be objected that wrongdoing often justifiability diminishes a person's right to compensation: when the 'chain of causation' is broken by a highly culpable wrong, that wrong relieves the original actor of liability in respect of the loss; hence, wrongdoing does seem justifiably to diminish a person's compensatory rights. The response to the objection is that, in the example just given, the effect of the intervening wrong is merely to generate a right to compensation against another person, that is, the intervening wrongdoer, not to prevent the existence of a compensatory right *at all.* What is so troubling about the idea that there is no liability to compensate in

[50] *Baker v Willoughby* [1970] AC 467.
[51] [1982] AC 794.

Overdetermination is that the multiplication of wrongs eradicates any entitlement to compensation, when that entitlement would otherwise exist.

Impossibility of proof of necessity

Suppose we believed that, ideally, the law ought only to rescind transactions when the absence of the relevant vitiating circumstance would have made a difference to the entry into the transaction or its terms. It could be argued that, in the context of human decision-making, it is simply impossible to say what would have happened if there had been no breach.[52] In other contexts, the law has developed exceptional proof rules that are applicable in cases in which it is *scientifically* impossible to determine the answer to a counterfactual issue.

Even accepting the legitimacy of such doctrines, it simply seems false that the proof of counterfactuals in the undue influence or breach of fiduciary duty contexts is inherently impossible. The findings in *Murad* demonstrate the contrary.[53] It is more plausible that it is often extremely difficult to prove what the terms of an alternative transaction would have been. If this is true, it nonetheless does not provide a reason for dispensing with necessity: if the right-holder is able to prove that the transaction would have been entered into on different terms, this still shows that the vitiating factor was necessary for the transaction which occurred.

B. Justifying alterations in proof rules relating to necessity

Three justificatory ideas emerge from the proof-related examples discussed in Section 2:

Culpability

In relation to fraudulent misrepresentation, duress to the person, deceit, and intimidation, it seems that the wrongdoer's high culpability is held to justify a relaxation of the standard of proof on the necessity issue.[54] This is an evidential counterpart to the substantive law modifications of the law on remoteness that apply in cases of intentional wrongdoing.[55]

[52] E Bant and J Paterson, 'Statutory Causation in Cases of Misleading Conduct: Lessons from and for the Common Law' (2017) 24 TLJ 1, 18–22.

[53] See also N Venkatesan, 'Causation in Misrepresentation: Historical or Counterfactual? And "But For" What?' (2021) 137 LQR 503, also rejecting this view.

[54] See Section 1. In German law, a reversal of the burden of proof on causation applies in relation to gross fault against medical professionals: see S Steel, *Proof of Causation in Tort Law* (Cambridge University Press, 2017), ch 5.

[55] The directness rule of remoteness applies: see, eg, *Smith New Court Securities v Citibank* [1997] AC 254.

Stakes

Relief in judicial review proceedings is only barred when it is *highly likely* that the same decision would have been reached in any event, unless an exceptional public interest demands otherwise. In so far as such a restriction can be justified at all, the higher proof threshold is presumably a reflection of the premium placed on reaching decisions in accordance with the law by public authorities.[56] Even a small probability that a different result would have been reached should be enough to permit quashing of the exercise of a power. More generally, if there is insignificant harm in undoing an exercise of power, and there is a high premium on arriving at the correct answer, then the exercise should typically be undone, despite significant uncertainty surrounding the counterfactual.

Impossibility of proof

In addition to accepting a conceptual enlargement of causal relations beyond necessity, *Bailey* recognises a proof rule that is equivalent to a reversal of the burden of proof on whether the defendant's breach was a cause of *loss*, when there is impenetrable scientific uncertainty surrounding this question. A burden of proof rule is a rule which allocates the risk of proof being impossible on a certain factual issue. *Bailey* allocates this risk to the person who has contributed to damage in breach of a duty of care. In short, causal contribution to damage, plus scientific impossibility of proof of the counterfactual loss question, was thought to merit shifting the burden of proof onto the person in breach.

As a matter of authority, this was based on a reading of *Bonnington Castings Ltd v Wardlaw*, where the defender was held liable in full for the pursuer's pneumoconiosis on the basis that the wrongful exposure to dust materially contributed to its existence, despite the presence of a non-wrongful source of dust.[57] This reliance was misplaced, however. First, *Bonnington* involved two sources of dust—one wrongful and one non-wrongful—each caused by the defender. The argument for placing the risk of proof of the counterfactual upon the defendant is stronger when the defender's own (innocent) conduct has exacerbated the proof difficulty in relation to the counterfactual of what would have happened but-for the wrong. Second, this proof of necessity issue simply was not considered in *Bonnington*, the decision being focussed upon whether liability existed at all.

The approach in *Bailey* is also difficult to reconcile with the approach taken when the defendant's breach itself affects the evidence available to prove causation. Here the law relaxes the standard of proof on causation.[58] By contrast, when the

56 To be clear, the value in lawful decision-making, and thus declaring publicly the fact of unlawful decisions, speaks strongly against a necessity restriction of any kind.

57 [1956] AC 613.

58 See *Keefe v Isle of Man Steam Packet Company* [2010] EWCA 683, and the cases applying it to causation.

source of the evidential uncertainty is the limits of scientific knowledge, the risk of proof being impossible is allocated entirely to the defendant.

Is there a normative case for shifting the burden of proof to the defendant on the existence of counterfactual loss once a causal contribution is shown? The starting point is surely that, since loss is what grounds a claim for compensation in respect of a breach, the burden of proof ought to be firmly on the right-holder in respect of this question (in so far as the right-holder seeks compensatory damages for loss). The burden of proof is on the right-holder in relation to the facts which ground the claim for the remedy sought. Loss is such a fact, when the remedy sought is compensation.

D. The law in light of these justifications

There are two areas in which the law seems to have taken a wrong turn: (1) the role of necessity in rights to rescind, and (2) the proof rules of necessity when a person's breach has materially contributed to damage.

Rescission

In relation to rescission, the central difficulty is one of consistency (Table 13.1).

If the rationale for the position in undue influence is that the influence deprived the right-holder of an opportunity to make a free decision and so a right to rescind will provide a surrogate for this lost opportunity, by allowing the decision to be re-made, this rationale applies to duress and misrepresentation, too. In each case, if that rationale is accepted, the relevance of necessity should be purely negative: it should operate as a reason for denying rescission, if, for instance, the consequences for the other party are severe, and they have behaved non-culpably.

Table 13.1 The role of (proof) of necessity in relation to rights to rescind

Ground of rescission	Necessity relevant as a matter of substantive law to existence of the right?	A special proof rule applicable to necessity?
Undue influence	No	N/A
Breach of fiduciary duty	No	N/A
Duress	Yes	Yes, in cases of duress to the person
Misrepresentation	Yes	Yes, in cases of fraudulent misrepresentation

Proof rules of necessity
The main issue here is that the law on when culpability is relevant to the burden of proof on necessity is unclear, and potentially inconsistent. It has been applied in relation to duress to the person, fraudulent wrongs, and the tort of intimidation. There seems to be no reason why the rule should not apply whenever a person intentionally commits a wrong which has materially increased the probability of the right-holder being worse off than if the wrong had not occurred, and the right-holder *prima facie* seems to be worse off in this sense.

5. A Necessary Ending

The question with which we began was: When does and should making a difference make a difference to liability? This chapter has shown, through analysis of examples from across the law, that the relevance of making a difference is not reducible to its impact upon whether a person was a cause of an outcome. In Stapletonian spirit, it has been argued that there is no single idea which explains when and why necessity matters, and that an emphasis on the particular type of liability in question is important in determining the legal and normative relevance of making a difference.

PART IV
OTHER TORTS AND LIABILITIES

14

Compensation as 'Conditional Licence' for Harmful Activities: An Exploration

Jonathan Morgan

1. Introduction

Dr Jane Stapleton was my tutor at Balliol in the 1990s. We had an undergraduate tutorial every single week of three consecutive Oxford terms, on constitutional law, contract, and then tort. This unforgettable tuition was a formative experience: lively, challenging, and inspiring. Jane's teaching has shaped all my thinking about, writing about, and in turn my own teaching of law. It is a privilege to offer this chapter in fondest tribute. It mostly avoids topics where Jane herself has written so influentially, such as duty, causation, or product liability (lest I produce but a pale imitation). Instead, the chapter concerns an idea immanent within a group of leading cases in the United States, Australia, and England. Nobody can rival Jane's expertise across those three great common law jurisdictions. But she has inspired me at least to try.

The idea, in short, is that some (exceptional?) instances of tort liability cannot be conceived as sanctions for wrongful conduct. Instead, tort sometimes *permits* or licenses harmful yet beneficial activities, on condition that the actor *pays for all harm* caused, irrespective of fault. The existence, and scope, of the principle are controversial. Still more so is its incompatibility with certain theoretical approaches to tort. It has an undeniably regulatory or instrumental flavour. Yet in my view, we are correct to follow Jane's lead and examine the social functions that tort law might fulfil, such as compensation for injuries,[1] or deterrence of risky conduct.[2] Scholars are necessarily theorists. However, as we were admonished in the recent magisterial Clarendon Lectures, by a former research chemist turned global legal authority, the lesson of natural science is to fit our theories to the empirical data of experience (ie the rules of law as we find them).[3] It would be preposterous to bend the facts to comply with a preconceived Grand Theory.[4]

[1] Jane Stapleton, *Disease and the Compensation Debate* (OUP 1986).
[2] Jane Stapleton, *Product Liability* (Butterworths 1994).
[3] Jane Stapleton, *Three Essays on Torts* (OUP 2021) 1.
[4] Nor desire homogenisation, 'like carrot puree': ibid xvii.

Jonathan Morgan, *Compensation as 'Conditional Licence' for Harmful Activities: An Exploration* In: *Torts on Three Continents.* Edited by: Kylie Burns, Jodi Gardner, Jonathan Morgan and Sandy Steel, Oxford University Press.
 DOI: 10.1093/oso/9780198889748.003.0014

This chapter begins with four leading cases. First, *Binsaris v Northern Territory*, where the distinctive approach of one judge in the High Court of Australia is used to introduce the problem and its possible solution. Secondly, *Burmah Oil v Lord Advocate* focuses on government liabilities. We examine the reasoning of the divided House of Lords, and the decision's rapid legislative reversal. The final two cases scarcely need introduction. *Vincent v Lake Erie*, is briefly described before concentrating on scholarly debates about how *Vincent* liability is best justified (or rather, a selection of those debates). Fourth comes *Rylands v Fletcher*. We turn straight to the debates about its nature and extent. The eponymous case could hardly be more famous—yet its legacy remains vigorously contested.

The chapter then outlines proponents of the 'conditional licence', prior to its critical assessment. Can the 'conditional licence' be reconciled with critiques of strict liability?[5] Is it acceptable to offer the weak protection of compensation only, rather than specific remedies? Finally, what is the proper scope of 'conditional licensing' and should its boundaries be set through the common law process or by legislation?

A final introductory caveat. This chapter consciously draws on cases, as well as scholarship, across common law jurisdictions. Hopefully it describes a common underlying theory. Of course however, there are important differences between legal systems. For example, as mentioned below, whereas *Rylands v Fletcher* was enthusiastically received in US jurisdictions, it has only just survived as an anomaly in England, and was abolished as a separate tort in Australia. Divergent development of the common law rules in different societies is perhaps only to be expected. But there are more profound differences too. Some flow from legislation. The wide-ranging restrictions on tort liability enacted in the various Australian states circa 2002 have been criticised as 'a wholesale attack' on common law tort principles that had been 'gradually and carefully worked out by the courts over time'.[6] Perhaps the most striking difference concerns governmental liability, an important strand in this chapter's argument. England adheres closely to its Diceyan tradition whereby public authorities are presumptively liable in the ordinary way for torts of public officials (or indeed in unjust enrichment etc),[7] just as private defendants would be. In the United States, however, special Constitutional Torts and other public law doctrines have long since supplanted ordinary private law, when it comes to government liability.[8] It may be at this point that the chapter's English presuppositions are most evident.

[5] On the wider debate, cf Sandy Steel, 'Culpability and Compensation' in James Goudkamp et al (eds), *Taking Law Seriously: Essays in Honour of Peter Cane* (Hart 2021).

[6] Peter Underwood, 'Is Ms Donoghue's Snail in Mortal Peril?' (2004) 12 Tort LJ 39.

[7] *Woolwich Equitable Building Society v Inland Revenue Commissioners* [1993] AC 70.

[8] Chaim Saiman, 'The Law Wants to be Formal' (2021) 96 Notre Dame LR 1067, 1106 ('American law begins with the opposite presumption [to Diceyan England]: barring an express waiver, sovereign immunity and other statutory schemes shield the state [from liability]').

2. Four Intriguing Cases

A. *Binsaris v Northern Territory* [2020] HCA 22

When force is reasonably used to restore public order, with the unavoidable consequence that bystanders are harmed, does the usual defence of 'lawful force used in self-defence (etc)' bar a tort claim by the blameless bystanders? In *Binsaris v Northern Territory* a prisoner had escaped from his cell in a youth detention centre in Australia. The detention centre's director decided to quell the escapee's violent rampage by deploying a CS gas 'fogger'.[9] However, the CS gas also unavoidably entered the cells of the four plaintiffs, other prisoners who were not relevantly involved in the disturbance. Those bystanders brought battery claims against the state government. The High Court of Australia unanimously upheld their claims. Four of the judges did so on the straightforward grounds that the 'fogger' was a prohibited weapon under the relevant statute, so that its use constituted unlawful force per se. The fifth, Gageler J, took a different, more intriguing approach.[10] We explore his reasoning that necessity rendered deployment of CS gas lawful, but nevertheless did *not* remove the government's obligation to pay compensation to the injured plaintiffs.

Gageler J disagreed with the other judges that the statute, on its proper construction, outlawed the 'fogger'. On ordinary common law principles, the question was therefore whether using CS gas had been reasonably necessary to prevent a breach of the peace. In Gageler J's decision, it had been. The prison director had carefully balanced the plaintiff bystanders' exposure to CS gas (an 'inevitable consequence') against the need to end the disturbance. The prison director *reasonably* decided that using CS gas, so inevitably harming the plaintiffs, was 'the lesser of two evils' (given the risks of serious harm to the un-subdued escapee, or prison staff, if it were not used). Indeed, Gageler J indicated that a responsible director had had no choice but to act as he did. Nevertheless His Honour held that one who makes 'a calculated choice to do an act which it is known will cause harm to a bystander in order to avoid a risk of greater harm' is liable to compensate that bystander.[11] The privilege to harm was not an unqualified one. While there would be a complete defence to a battery claim brought *by the escaped prisoner*, the plaintiff *bystanders'* damages claim succeeded.[12] They could not be used as 'instruments' upon whom collateral damage could be inflicted 'with impunity'.[13]

[9] Dispensing 'tear gas that disables those who breathe it by inducing uncontrollable burning and tearing of the eyes, and intense irritation of the nose and throat, causing profuse coughing and difficulty breathing' [2020] HCA 22 [51].

[10] It sparked the ideas for the present chapter.

[11] [2020] HCA 22 [42].

[12] On 'defensive' liability and innocent bystanders, cf Steel (n 5) 50–53.

[13] [2020] HCA 22 [38].

Gageler J noted a 'surprising dearth of modern authority'. He drew analogies with two of the leading cases discussed below. Gageler J found academic commentary on *Vincent v Lake Erie* illuminating (although suggested the 'incomplete privilege' concept lacked 'direct utility').[14] More pertinent was *Burmah Oil v Lord Advocate.*[15] It showed that government action can require payment of compensation, notwithstanding its lawfulness: distributive justice had 'moderated' the oil company's sacrifice by making it good at public expense. Gageler J summarised his conclusion: 'the burden of the necessitous infliction of harm on an individual ... in the public interest should in fairness be borne by the public.'[16] (Gageler J suggested that earlier decisions, absolving *individual* public-official defendants faced with hard choices in 'conscientious discharge of official duties', had been superseded by the Crown's *corporate/vicarious* liability. Direct governmental liability rendered earlier concerns about the chilling effect of (individual) liability outdated.)[17] As will be seen, this judgment in the High Court of Australia builds on venerable case authorities, and a century of academic commentary.

B. *Burmah Oil Co Ltd v Lord Advocate* [1965] AC 75

Was there a duty to pay compensation when the armed forces of the Crown destroyed private property to stop it falling into the hands of the enemy? (Oil wells belonging to the pursuer company were torched by the British army, retreating before the Japanese forces' 1942 advance into Burma.) The House of Lords held by a 3-2 majority that the Crown indeed had such an obligation. Paying compensation was a condition for the lawful seizure of property under the Royal Prerogative. Lord Pearce stated: 'It is plainly just and equitable that when the state takes or destroys a subject's property for the general good of the state it shall pay him compensation.'[18]

Indeed, all five law lords accepted the attractiveness of that principle, at an abstract level. What divided them was whether it had authority as a *rule of common law.* As an (anonymous) contemporaneous commentator noted, the court's reasoning followed the 'traditional' British pattern (at that time), being 'apparently based entirely on logical analysis of earlier thinking', ignoring entirely the ruling's consequences and effects.[19] Whatever the general merits of focusing exclusively on precedent, it proved deeply problematic in *Burmah Oil.* There had been few cases, of any kind, about the Royal Prerogative between the English Civil War and the First World War (ie for two and a half centuries before 1914). Indeed, there

[14] ibid.
[15] ibid [45].
[16] ibid [48].
[17] ibid [43]–[44].
[18] [1965] AC 75, 149.
[19] Note, 'The Burmah Oil Affair' (1966) 79 Harvard LR 614.

was no authority whatever on the 'pure' war prerogative. The question had never, in practice, arisen. For centuries prior to 1942, Parliament had invariably granted *statutory* powers for requisitioning property in wartime (these statutes invariably including a duty to compensate the property-owners). *Burmah Oil* exceptionally concerned the 'pure' common law prerogative, because the relevant Second World War legislation did not extend to Burma. This was literally unprecedented: 'no war which has put this country in real peril has been waged in modern times without *statutory* powers of an emergency character.'[20]

The majority held there was insufficient legal authority to displace the 'plainly just' principle mentioned above. For Lord Pearce, 'the matters on which the Crown relies are too slender to support [uncompensated] infringement of the subject's rights.'[21] Whereas the two dissenters (as it were) reversed the burden of proof, demanding positive legal authority to *support* the Crown's putative liability. Despite the longstanding and invariable practice of governments paying compensation, either ex gratia or (more usually) pursuant to statutory obligation, neither of those could evidence an underlying *common law obligation*. In Viscount Radcliffe's dissenting opinion, the fact that recompense had long flown through 'other channels' was not 'adequate reason to open the law courts to such claims': on the contrary, the judiciary were not entitled 'to import into the common law a legal right that has not hitherto been recognised to exist.'[22] Lord Radcliffe sharply distinguished between 'what is expected by public sentiment and what is actually obtainable by legal right.'[23]

Burmah Oil is also remembered for its legislative sequel. The War Damage Act 1965 reversed the House of Lords' decision with expressly retrospective effect. It is a notorious example of Parliament refusing to accept a judicial ruling.[24] Interestingly, however, Her Majesty's government (in resisting Burmah Oil's claim) had warned the company that it would react in this way. Early in the litigation the government informed Burmah Oil that if by some remote chance its compensation claim succeeded (the Law Officers of the Crown confidently believed it bad in law—as the Court of Session had held), legislation would be introduced to reverse the court's decision. In the event this retroaction was, naturally, controversial during debates on the War Damage Bill. The House of Lords (*qua* legislative chamber) initially voted to delete the Bill's retrospective commencement, a revolt led by the Lord Chief Justice, Lord Parker (who had not participated in the *Burmah Oil* litigation). When the House of Commons (supporting the government) insisted on retrospectivity, their Lordships made an expedient retreat—accepting

[20] [1965] AC 75, 101 (Lord Reid) (emphasis added).
[21] ibid 155.
[22] ibid 127.
[23] Ibid.
[24] Compare also *Barker v Corus (UK) Plc* [2006] 2 AC 572 with Compensation Act 2006, ss 3 and 16(3) (retrospectively reinstating joint and several liability for '*Fairchild*' mesothelioma claims).

that during a crisis in the public finances, an oil company's claim for £160 million compensation (equivalent to £2.5 billion in 2023), and peers' rights to delay legislation, were unattractive causes for a general election—which was threatened by the Prime Minister, if the Lords did not concede.

Constitutional lawyers have universally condemned the War Damage Act 1965 for violating both the Rule of Law and the Separation of Powers. Yet the contemporary view from Harvard was more positive. Rather, the judicial decision in *Burmah Oil v Lord Advocate* was criticised as 'essentially inadequate' in failing to consider its implications for the public finances. The court seemed 'functionally unable to consider a major element of the case', the majority law lords citing 'feeble and ancient authority supporting compensation' without pausing to consider whether this was 'sufficiently impressive to outweigh the disadvantages to the British Treasury'.[25] In such circumstances and given that the Burmah Oil Co, having been warned in advance, could not have relied on the court's judgment, the Harvard commentator deemed its legislative reversal 'altogether proper'.[26]

Viscount Radcliffe, in dissent, reasoned that 'sound public policy' required the *absence* of a common law compensatory obligation. Because *Parliament*, rather than the courts, should decide how much compensation should be awarded—if any. Fully accepting the moral desirability of state compensation, it was simply not a question for the judiciary: 'it is for those who fill and empty the public purse to decide when, by whom, on what conditions and within what limitations such compensation is to be made available.'[27] In its reaction to the majority's decision, Parliament emphatically vindicated Lord Radcliffe's point.

C. *Vincent v Lake Erie Transportation Co* 124 NW 221 (1910)

This is one of the most discussed US tort cases. Its facts are relatively simple. Justification of the outcome is much less straightforward. Accordingly, we devote more space to discussions about *Vincent* than to the decision itself.

The defendants moored their ship (*The Reynolds*) to the plaintiff's dock during a violent storm. This did significant damage to the dock, for which the plaintiff successfully claimed damages. Yet the defendants had not trespassed. Necessity, in the emergency, gave them a privilege to moor *The Reynolds*: and indeed had the plaintiff cast the ship off (apprehending damage to the dock), he would himself have been liable for trespass.[28] Nor had the defendants been negligent: 'in holding the vessel fast to the dock, those in charge of her exercised good judgment and prudent

[25] (1966) 79 Harvard LR 614, 633.
[26] ibid 634.
[27] [1965] AC 75, 134 (Viscount Radcliffe).
[28] *Ploof v Putnam*, 71 Atl 188 (1908).

seamanship'. But 'No matter how justifiable' their actions, the defendants had to pay for the damage caused (just as if they had requisitioned valuable property for their use). In O'Brien J's exquisitely concise judgment:

> Theologians hold that a starving man may, without moral guilt, take what is necessary to sustain life; but it could hardly be said that the obligation would not be upon such person to pay the value of the property so taken when he became able to do so.

The case has engendered commentary quite disproportionate to its original brevity. Writing about *Vincent* is a rite of passage for private law theorists. What they write—what they *see*—is arguably more revealing about the explainer than the explanandum, a kind of Rorschach Inkblot Test.[29] It is impossible to attempt an exhaustive account of the scholarly debates here. The core problem is that, as the court accepted, the *Vincent* defendants had acted entirely properly. Thus from the tort perspective, this decidedly *strict* liability requires justification—explaining why the defendant 'can *reasonably* use another's property and *still* incur liability for the property damage'.[30] Some commentators reject any such explanation. Sugarman argues that the strict liability in *Vincent* was inappropriate.[31] Parties like the plaintiff dock-owner have a moral duty of 'easy rescue' which not only obliges them to permit use of their property in an emergency, but should also bar any subsequent compensation claim. (Sugarman suggests that assistance is always 'easy' in situations where the shared/sacrificed property is *insured* against loss). In Sugarman's view, the liability imposed in *Vincent* undermines norms of community solidarity—the moral obligation to share resources during emergencies.

Another perspective emphasises the defendant shipowner's *benefit* (cf O'Brien J: defendant 'availed itself of the plaintiffs' property for the purpose of preserving its own more valuable property'). Bohlen, labelling the plaintiff's right to moor to the dock an 'incomplete privilege', argued that while the law wished to encourage salvage of life and property, the cost of saving it 'shall be put upon him ... whose interests are advanced by the act ... [not] one who derives no benefit from the act'.[32] Bohlen also suggested that one criterion for its application was that the interest protected must be 'of a value greater than, or at least equal to, that of the interest invaded'. Gordley too prefers this explanation of *Vincent*.[33] It answers the challenge laid down above (why does liability arise if the defendant acted entirely

[29] As suggested to me by Nicholas J McBride.

[30] Mark A Geistfeld, 'Necessity and the Logic of Strict Liability' (2005) 5 Issues in Legal Scholarship.

[31] Stephen D Sugarman, 'The "Necessity" Defense and the Failure of Tort: The Case Against Strict Liability for Damages Caused While Exercising Self-help in an Emergency' (2005) 5 Issues in Legal Scholarship.

[32] Francis H Bohlen, 'Incomplete Privilege to Inflict Intentional Invasions of Interests of Property and Personality' (1926) 39 Harvard LR 307, 316,

[33] James Gordley, 'Damages Under the Necessity Doctrine' (2005) 5 Issues in Legal Scholarship.

properly?): because liability in *unjust enrichment* is strict. Its principles require restitution even for *entirely innocent* appropriations of property. This strict liability bolsters the exclusive right of property-owners to decide who may use their property and on what terms. The owner's right-of-control is invaded by all *deliberate* 'appropriations of the benefit of the property', although not in 'standard' tort situations involving merely accidental damage.

Gordley's preferred approach is rejected by one pre-eminent scholar of unjust enrichment.[34] Virgo declares: 'nothing can be gained from resorting to a restitutionary analysis.' The claim in *Vincent* was concerned solely with the plaintiff dock-owner's *loss*. The sum awarded was not quantified according to the defendant shipowner's gain. It would be 'highly artificial' to equate the two (albeit that there could be some connexion: as Gordley suggests, a hypothetical *ex-ante*-agreed 'price' for mooring *The Reynolds* would incorporate both the anticipated benefit to the defendant (of saving the ship) *and* the plaintiff's anticipated loss (the damage to the dock)).[35] As Virgo argues however, that hardly justifies 'capping' a supposedly 'restitutionary' claim (imagine that the saved ship carried a fantastically valuable cargo) at the level of damage done to the dock. As well as this remedial mismatch, pointing to the (undoubtedly) strict nature of unjust enrichment is also unconvincing. As Virgo asks, why is the enrichment deemed 'unjust' if the defendant's behaviour was entirely proper and justified?

Virgo calls for fuller articulation of the necessity defence. (One problem is that opportunities for its judicial consideration rarely arise.) Virgo advocates the well-developed criminal law distinction between circumstances that *justify* conduct (so that it is no longer wrongful at all, but positively welcome), and necessity as merely an *excuse*. Although not punished, an 'excused' act remains impermissible—the criminal law tolerates rather than welcomes it, bestows a pardon but not praise. The former 'justification' category includes medical treatment administered on an unconscious patient in their best interests (which the law would surely not wish to discourage). In the 'excuse' category falls the 'duress of circumstances' defence.

Applying this distinction to tort liability, Virgo suggests that necessitous action in the *public* interest should be seen as *justified*. There should be no liability at all, at least not upon *private* individuals inflicting harm on another for the public good.[36] (Public authorities might have a duty to compensate, by analogy with *Burmah Oil*.) While English law appears to suggest that necessitous action in the defendant's own private interests would also (within its highly circumscribed scope)[37] provide an *absolute* defence, Virgo commends *Vincent's* more nuanced

[34] Graham Virgo, 'Justifying Necessity as a Defence in Tort Law' in A Dyson et al (eds), *Defences in Tort Law* (Hart 2015).

[35] For 'negotiating damages' generally, cf *One Step (Support) Ltd v Morris-Garner* [2018] UKSC 20.

[36] *Mouse's Case* (1608) 12 Co Rep 63. See further Kenneth W Simons, 'Self-Defense, Necessity, and the Duty to Compensate, in Law and Morality' (2018) 55 San Diego LR 357.

[37] *Southwark LBC v Williams* [1971] Ch 734 typifies the narrow approach: accepting a private necessity defence in theory, but robustly rejecting it on the facts (trespass to land by homeless persons)

solution. The *Vincent* approach 'straddles the line between justification and excuse, involving conduct which is half-right and half-wrong'. (As seen, the dock-owner could not lawfully cast-off the ship-owner (showing that the latter was *not a wrongdoer*), but nevertheless the ship-owner was liable to pay compensation.) The 'incomplete privilege' concept usefully captures the 'sophisticated' hybrid nature of *Vincent v Lake Erie*.

D. *Rylands v Fletcher* (1866) LR 1 Ex 265

This case, and its facts, really do need no introduction. Again, it has generated discussion disproportionate to its infrequent application.[38] Again, therefore, our focus is on *debates about* the great case. In particular, whether the defendant committed any *wrong*, and if not, how liability can be justified.

In *Rylands* itself in the Court of Exchequer,[39] Martin B gave the leading judgment *dismissing* liability. The defendants had done nothing wrong: 'making a pond for holding water is a nuisance to no one'. He rejected Bramwell B's dissenting judgment ('The nuisance is not in the reservoir, but in the water escaping.'). Why make the defendants liable for an escape that they could neither have foreseen nor prevented? This would, Martin B reasoned, 'make [the defendants] insurers against the consequence of a lawful act upon their own land'. (There was, moreover, no good reason to give stricter protection to property when there was 'no better established rule of law' than the requirement for *negligence* when personal injuries were sustained in highway collisions.)

Could we justify (rather than condemn) *Rylands v Fletcher* as a kind of 'insurance' to protect persons suffering from hazardous but socially beneficial activities, 'lawful' because properly carried-out? (Like reservoirs: indispensable for storing drinking water (or as in *Rylands*, to power a water-mill), but lethally destructive if they overflow.) McBride and Bagshaw accordingly conceive *Rylands v Fletcher* as a 'liability rule'.[40] The defendant does *nothing wrong*. There is no 'tort' in the true sense. There is certainly *liability*, but it is 'a confusion of ideas' to attribute it to a breach of duty.[41] (Incidentally that means that *Rylands v Fletcher* cannot properly be a subset of nuisance, a true wrong-based tort.) *Rylands* liability is based rather on 'expediency'. McBride and Bagshaw discuss it alongside other examples

(eg Edmund Davies LJ at 745–746 'circumspection' required because 'necessity can very easily become simply a mask for anarchy').

[38] *Transco plc v Stockport MBC* [2004] 2 AC 1 [39] (Lord Hoffmann) ('the intellectual effort devoted to the rule by judges and writers over many years has brought forth a mouse').
[39] (1865) 3 Hurlstone & Coltman 774.
[40] Nicholas J McBride and Roderick Bagshaw, *Tort Law* (6th edn, Pearson 2018) ch 14.
[41] ibid 445 quoting CK Allen, *Legal Duties and Other Essays in Jurisprudence* (Oxford 1931) 194.

of liability-without-wrongdoing. Some were recognised at common law (eg liability for escaping fire, the 'scienter' action for harm done by animals, and absolute liability for trespassing cattle). Others have been imposed by statute (eg water undertakers' strict liability for damaging leaks;[42] or the stringent liability for radioactive contamination).[43]

Although mainstream tort lawyers still classify *Rylands* as a 'tort', this may only reflect unthinking usage. More persuasively than the traditional arrangement of textbooks, Goldberg and Zipursky make a reasoned response to McBride and Bagshaw's taxonomic claims.[44] Most torts involve a mixture of strict and fault-based liability.[45] This mandates caution before incorporating fault into the core definition of 'tort'. Goldberg and Zipursky accept that *Rylands v Fletcher* falls within tort law, albeit marginally. It exemplifies a legitimate regulatory strategy of *conditionally* licensing hazardous yet beneficial activities.

> The permissibility of this kind of extraordinary risk imposition is conditioned on the readiness of the risk imposer to take responsibility for the injuries that flow from it, irrespective of the presence or absence of wrongdoing.[46]

Usually, they note, such regulation is done through *statutory* taxes or charges—with *Rylands* as a common law 'parallel'. Goldberg and Zipursky reject the argument, suggested by Keating, that the 'wrong' in *Rylands v Fletcher* was neither constructing the reservoir (entirely properly), nor the escape of the water (unforeseeably and unavoidably), but rather in *failing to compensate the harm inflicted*.[47] In Goldberg and Zipursky's view, non-payment of damages in 'flawless conduct' cases would be mis-described as a 'wrong'. For liability arises as soon as the harm occurs (not at the point when the defendant fails to pay for it). The authors here contrast situations where non-payment is *itself* the wrong (eg an insurer refusing to indemnify loss under a valid policy).

Thus, Goldberg and Zipursky view *Rylands v Fletcher* as exemplifying 'licensing-based strict liability'. *Rylands* liability has continuities with undoubted torts such as negligence or trespass, from which analogies are drawn in defining its limits.[48] It both belongs and does not belong within the law of torts, conceived as a 'gallery of wrongs'. *Rylands v Fletcher* holds defendants accountable for harm done,

[42] Water Industry Act 1991, s 209 (NB preservation (only) of 'sole fault of claimant' and contributory negligence defences: subs (2) and (4)).

[43] Nuclear Installations Act 1965 (NB solitary defence of enemy action: s 13(4)(a)).

[44] John CP Goldberg and Benjamin C Zipursky, 'The Strict Liability in Fault and the Fault in Strict Liability' (2016) 85 Fordham LR 743.

[45] See further on these 'amalgams', James Goudkamp, 'Rethinking Fault Liability and Strict Liability in the Law of Torts' (2023) 139 LQR 269.

[46] Goldberg and Zipursky (n 44) 763.

[47] Gregory C Keating, 'Strict Liability Wrongs' in John Oberdiek (ed), *Philosophical Foundations of the Law of Torts* (OUP 2014).

[48] eg *Cambridge Water Co v Eastern Counties Leather* [1994] 2 AC 264 (remoteness of damage).

but without seeking to condemn their behaviour. It is a viable institutional choice for the law to locate such 'non-pejorative' liabilities within tort. There is sufficient resemblance. Nevertheless, *Rylands v Fletcher* is ultimately anomalous. It plays a minor supporting role in a department of law centrally concerned with laying down rules to govern how people should behave. Tort is 'an important repository of norms of conduct'. Those norms suffer degradation if pure 'liability rules' like *Rylands v Fletcher* are elevated above subordinate 'sidekick' status. For Goldberg and Zipursky, no wider lessons should be inferred from such special liabilities at the very margin of 'tort'.

Others might welcome *Rylands v Fletcher's* 'regulatory' dimension, so grudgingly accepted by Goldberg and Zipursky (and responsible for its expulsion from the pantheon by McBride and Bagshaw). Simpson identifies it as the historical explanation for the courts' ultimate finding of liability.[49] Simpson notes contemporary debates about whether Parliament should generalise the (sporadic) practice of inserting absolute liability clauses into legislation authorising construction of reservoirs. (*Rylands* however involved a small reservoir built on private land—with no need for compulsory purchase, there was (unusually) no governing Private Act, and the disaster raised a purely common law question.) Simpson speculates that judges who favoured strict liability might have been encouraged by that legislative policy, or indeed may have hoped that *Rylands v Fletcher* would give 'encouragement to Parliament to pass general legislation imposing [strict liability] on all water undertakings'.[50]

There are difficulties with Simpson's explanation. First it was, and remains,[51] heretical to generalise particular statutory provisions into a general (common law) principle.[52] Secondly, there is no trace of ideas of loss-internalisation anywhere in the great judgment of Blackburn J in the Exchequer Chamber, contrary to those who hail *Rylands v Fletcher* as the seminal authority for twentieth-century 'enterprise liability'.[53] As Oliphant argues, particularly telling was Bramwell B's silence. Bramwell too argued in favour of liability using formal interpretation of doctrine and authorities. His failure to discuss ideas of cost-internalisation is striking, and puzzling, compared with his sophisticated explanation of the theory in judgments both before and after *Rylands v Fletcher*.[54]

[49] AWB Simpson, 'Legal Liability for Bursting Reservoirs: The Historical Context of *Rylands v. Fletcher*' (1984) 13 JLS 209.

[50] ibid 252 (in fact the government announced in 1867 that it would not pursue such legislation, ibid 230, which eventuated only with the Reservoirs (Safety Provisions) Act 1930, s 7).

[51] PS Atiyah, 'Common law and statute law' (1985) 48 MLR 1.

[52] This is equally problematic for the argument that late nineteenth-century 'workmen's compensation' statutes also gave a 'terrific impetus' to common law strict liability: cf CO Gregory, 'Trespass to Negligence to Absolute Liability' (1951) 37 Virginia LR 359, 388.

[53] Ken Oliphant, '*Rylands v Fletcher* and the emergence of enterprise liability in the common law' in H Koziol and BC Steininger (eds), *European Tort Law 2004* (Springer 2005).

[54] ibid 108–118 (cf *Bamford v Turnley* (1862) 3 B&S 66; *Powell v Fall* (1880) 5 QBD 597, 601 (Bramwell LJ commenting on *Rylands v Fletcher*)).

Citing Simpson's analysis, Lord Hoffmann thought it 'tempting to see, beneath the surface of the rule, a policy of requiring the costs of a commercial enterprise to be internalised'.[55] However, *Rylands v Fletcher* was 'an isolated victory for the internalisers. The following century saw a steady refusal to treat it as laying down any broad principle of liability'.[56] Mid-nineteenth-century judges had contrasting views on the question and decisions reflected the 'shifting ascendancies' between them, not the 'steady development' of cost-internalisation.[57] Simpson, too, accepts that any opportunity to generalise the ideas behind *Rylands* was lost amid 'incoherent' decisions on whether publicly valuable activities were obliged to compensate those unavoidably harmed, following almost immediately after the great case.[58]

Courts today view extensions of strict liability as a matter for Parliament.[59] Lord Hoffmann emphasised that where the regulatory goal required *absolute* liability Parliament can impose it with the necessary stringency, whereas common law doctrines undermine *Rylands*' strictness, for example causation and defences.[60] (He noted that had the defendant in *Transco* been a water undertaker not a gas supplier, its *statutory* liability would have been much stricter than *Rylands v Fletcher*.)[61] Lord Hobhouse observed that Parliament has regulatory options besides imposing liability. Statute can directly regulate, or prohibit outright, 'dangerous or anti-social uses of land'.[62] His Lordship suggested that such statutory regulation grew throughout the twentieth century owing to cultural changes (the decline of 'laissez faire'). This misrepresents nineteenth-century 'political culture'. As Simpson pointed out, 'the mechanism used to regulate enterprises and cope with the problems of rapid industrial and agricultural change was not the common law but the private act of Parliament'—so even in the 1860s the judicial role had been 'peripheral'.[63]

As discussed below, *Rylands v Fletcher* survives, just, as a relic of strict liability in English law (having been deleted from Australian common law). It seems anomalous within tort law conceived as a response to, and condemnation of, *wrongful* conduct. The 'licensing' conception seems more promising. It advances cost-internalisation, attractively summed up as the 'polluter pays' policy. But however attractive, that idea has never been generalised, whether at common law or in legislation.

[55] *Transco* (n 38) [29].
[56] Ibid
[57] *Wildtree Hotels Ltd v Harrow LBC* [2001] 2 AC 1, 8–9 (Lord Hoffmann).
[58] Simpson (n 49) 258 ('the judiciary seemed hopelessly divided ... some leaning in favor of compensation for loss arising through the operation of public works, others not').
[59] *Cambridge Water* (n 49) 305 (Lord Goff).
[60] *Transco* (n 38) [42].
[61] Water Industry Act 1991, s 209.
[62] *Transco* (n 38) [55]–[56].
[63] Simpson (n 49) 264.

3. Rationalising the Precedents: Compensation as the Price of Permission (or 'Conditional Privilege')

It may be possible to bring these fascinating—or notorious—cases together under this common principle. It holds that the 'wrong' in play is not the defendant's conduct, but that it resulted in *uncompensated* harm. Wrongfulness consists in *not paying* for harm that has been inflicted by a reasonable (indeed, often positively desirable) activity.[64]

In each of the four cases, what the defendant had actually been doing was socially valuable and could not have been carried out in a way that avoided the harm. In *Burmah Oil v Lord Advocate* the British army's conduct was *ex hypothesi* necessary to thwart the invading force; *by design destructive* of the property (to prevent it coming under enemy control). In *Vincent v Lake Erie* it was eminently reasonable for the defendant to moor lawfully to the plaintiff's dock during the storm; the damage done was an *unavoidable* consequence. In *Binsaris v Northern Territory* it was at least permissible, arguably necessary, to use CS gas to subdue the escaped prisoner; the plaintiff bystanders' injury was an *inevitable*, albeit undesired, side-effect. In *Rylands v Fletcher*, the defendant was perfectly entitled to use a reservoir on his land, built by apparently competent engineering contractors. In Victorian times (and perhaps again in a carbon-neutral future?), water storage was a vital source of industrial power. The defendant was not personally negligent, having acted as a reasonable landowner would to make the reservoir safe.[65] It was not a nuisance. Yet the defendant was held liable for harm that he could not reasonably have avoided.

The defendants all acted excusably, or even laudably. Hence it follows that their initial conduct should not be condemned or, if it were legally possible to do so—as through a *quia timet* injunction—prohibited. For obvious sensible reasons, the British army was not required to leave the oil wells unharmed (to be appropriated presently by the Japanese); the dock-owner was not permitted to cast off the distressed ship in the storm;[66] the governor could not be prohibited from using the most suitable means for restoring order in the detention centre; and John Rylands could not be enjoined to drain or abandon an apparently properly constructed reservoir. Yet in each case the defendants' conduct posed the risk (or in the first three cases, the inevitability) of harm to others. The principle that seems to underlie the

[64] Keating, 'Strict Liability Wrongs' (n 47).

[65] Not being vicariously liable for the negligent independent contractor (engineer) responsible for construction. On 'non-delegable' duties and 'ultra-hazardous' activities, see *Honeywill v Larkin Bros* [1934] 1 KB 191; *Burnie Port Authority v General Jones Pty Ltd* (1994) 179 CLR 520; cf *Woodland v Essex CC* [2014] 1 AC 537 [6] (Lord Sumption) ('Many of these decisions are founded on arbitrary distinctions between ordinary and extraordinary hazards which may be ripe for re-examination').

[66] *Ploof* (n 28).

cases is that the defendants were permitted to act in the way they did, despite that risk (or certainty) of harm, *on condition that they paid compensation* for that harm.

The principle has been identified and advocated by a number of US legal scholars. (There is, however, no consensus around its name.) Bohlen analysed *Vincent* as an 'incomplete privilege'—the emergency precluded liability for 'mere *harmless invasion* of the [dock-owner's] right' but the defendant was 'not so far privileged as to relieve him from liability to pay for any *material harm*'.[67] Bohlen saw no reason why the privilege conferred had to be absolute; the 'modern' approach in *Vincent* permitted 'far nicer discriminations'. Nor did the reasonableness of the defendant ship-owner's conduct necessarily preclude liability. Keeton generalised a similar idea as 'conditional fault' in 1956.[68] When a defendant inflicts harm through a risky but valuable activity, in otherwise non-blameworthy fashion, they are required to pay compensation. Failure to pay would be blameworthy according to contemporary moral standards ('though the quality of blame is different from that found in most cases of negligent or intentional torts').[69] This included not only situations of using another's property (as in *Vincent v Lake Erie*) but *accidental* harm resulting from carefully conducted yet risky activities (like the reservoir in *Rylands v Fletcher*). If a party engaging in '*Rylands v Fletcher* activity' (the standard US example is blasting with explosives) refused to insure themselves, or otherwise to make financial provision to pay for unavoidable harm, their conduct could be enjoined: 'it is the moral sense of the community that [blasting] should not occur without an advance provision for compensation'.[70]

For Keeton this was an exception to the rule that absent fault, losses lie where they fall (the 'axiom of adjudicatory inertia').[71] More recently, Keating identifies two major strands of strict liability in tort.[72] First, 'boundary crossing' torts (such as battery, or trespass to land) where *all* physical violations are actionable, irrespective of excusable mistakes or reasonable conduct on the defendant's part. Secondly, 'harm-based strict liability [which] imposes responsibility on harm that should not have been avoided *on the ground that it was wrong to harm-without-repairing*'.[73] In contrast to liability based on fault (ie harm which we ought reasonably to avoid):

> Strict liability speaks distinctively to responsibility for harm which should *not* be avoided ... What it does is to place financial responsibility for harms which

[67] Bohlen (n 32) 313 (emphasis added).
[68] Robert E Keeton, 'Conditional Fault in the Law of Torts' (1959) 72 Harvard LR 401.
[69] ibid 434.
[70] ibid 429.
[71] ibid 401.
[72] Gregory C Keating, 'Is Tort Law "Private"?' in Paul B Miller and John Oberdiek (eds), *Civil Wrongs and Justice in Private Law* (OUP 2020).
[73] ibid 366 (emphasis in original).

> should not be avoided on those who inflict those harms. Negligence leaves those losses on the shoulders of their victims.[74]

Such strict liability does not condemn the conduct that causes harm, but the subsequent failure to repair that harm. By requiring the defendant to pay for it, tort law ensures that the activity takes place on terms fair to both actor and victim.[75] The law holds that that 'payment of damages to those injured by the characteristic risks of an activity is a condition for the legitimate conduct of an activity'.[76]

The American Law Institute has recognised the principle too. The *Restatement (Second) of Torts* (1977) contains this comment on § 520 (c) (strict liability for abnormally dangerous activities):

> The utility of [the defendant's] conduct may be such that he is socially justified in proceeding with his activity, but the unavoidable risk of harm that is inherent in it requires that it be carried on at his peril, rather than at the expense of the innocent person who suffers harm as a result of it.

Hence there is widespread acceptance of a principle of making permission to carry out valuable, harmful activities conditional on paying compensation to those who suffer harm.

4. Controversies

Many questions arise about this proposed basis for the intriguing cases. At its core, the 'conditional licence' entails a *strict* standard of liability and a *monetary* remedy. It raises significant issues of: (1) legal theory (the connexion between regulatory law and private law; the relationship of rights and remedies); (2) political theory (the compatibility of strict liability with the Rule of Law); and (3) constitutional propriety (the appropriateness of courts rather than legislatures imposing such liability). In addition, legal-doctrinal controversy surrounds the 'conditional privilege's' extent.

Space precludes a full exploration of—let alone any definitive answer to—these various questions. We can attempt only to show how thinking about this concept can illuminate deeper controversies in tort law and legal theory, before (tentatively) concluding the chapter.

[74] ibid 369 (emphasis added).
[75] Gregory C Keating, 'Rawlsian Fairness and Regime Choice in the Law of Accidents' (2004) 72 Fordham LR 1857.
[76] ibid 1871.

A. Regulation, morality, and private law

The great divide in contemporary tort theory is between instrumental and non-instrumental approaches. Perhaps any given issue can be approached from those different angles. Certainly, that is true of the 'conditional privilege'. It could be defended as a matter of justice (see, eg, Keaton's 'moral sense of the community' cited above). Alternatively, it could be seen as an instrument to address the problem of valuable-yet-harmful activities.

Non-instrumental justifications

In a conscious reaction to the dominant US approach, Keating has championed the *moral* virtues of *Rylands v Fletcher*.[77] Strict liability's merits can, in part, be advanced by attacking the supposedly superior justice of the fault principle. As Coleman strikingly argues,[78] the result of leaving losses lie where they fall (in the absence of a culpable defendant) is strict liability *for victims*. They bear the relevant losses, without any fault on their part. The 'unfairness' of strict liability is equally present in the fault system (but allocated, by fault, to the sufferer rather than the doer): they are 'structural analogues'.[79] Secondly, one can question whether negligent *inadvertence* is, anyway, morally culpable. For those who deny this, the central case of tortious 'fault' attracts little more than 'superficial moralizing'.[80] Keating too deems negligence a 'very mild wrong' compared to the intentional wrongdoing characterising traditional criminal offences; negligence dominates tort law because accidents are so *harmful*, but not particularly *blameworthy*.[81] Waldron questions the morality of negligence law imposing crushing liability on parties whose inadvertence causes 'massive loss' (a momentarily inattentive driver), whereas it ignores those who chance not to cause injury, however culpable their behaviour.[82]

Keating draws a 'provocative' lesson from *Rylands v Fletcher*: 'moral responsibility does not come to an end because harm is unavoidable ... even unavoidable harm can trigger a duty of repair.'[83] Keating questions the venerable anti-*Rylands* argument that immunity from liability advances the general interests of society ('We must have factories, machinery, dams, canals and railroads') and thus immunity benefits the claimant as much as the defendant ('He [the claimant] receives his compensation for such damage by the general good, in which he shares').[84]

[77] Gregory C Keating, 'Recovering *Rylands*: An Essay for Robert Rabin' (2012) 61 DePaul LR 543.
[78] Jules Coleman, 'The Morality of Strict Tort Liability' in *Markets, Morals, and the Law* (OUP 2002).
[79] ibid 182
[80] Heidi M Hurd, 'Finding No Fault with Negligence' in John Oberdiek (ed), *Philosophical Foundations of the Law of Torts* (OUP 2014).
[81] Keating, 'Is Tort Law "Private"?' (n 72) 353–355.
[82] Jeremy Waldron, 'Moments of Carelessness and Massive Loss' in David G Owen (ed), *Philosophical Foundations of Tort Law* (Clarendon 1995).
[83] Keating, 'Recovering *Rylands*' (n 77) 592.
[84] *Losee v Buchanan* 51 NY 476 (1873) (Earl CJ).

Why, Keating asks, should losses be concentrated on luckless plaintiffs, when the benefits are diffused across society at large (or concentrated in the defendants). *Rylands* should be seen as a landmark, 'one of the principal common law fonts of both modern strict liability and enterprise liability'. In Keating's view, its force has been undermined by the modern obsession with economic analysis. 'Strict enterprise liability cannot be fully understood and appraised until the *moral* reasons in its favor are unearthed and assessed.'[85] Albeit as seen, Goldberg and Zipursky argue that non-payment of damages is not itself 'wrongful', fearing in consequence that the 'conditional licensing's' rationale undermines tort's moral project of guiding conduct.

Instrumental justifications

The conditional licence could alternatively be conceived as a regulatory strategy. In outline, there are various ways for the law to approach harmful activities: (1) outright prohibition; (2) total immunity; (3) fault-based liability; or (4) strict liability including 'conditional licensing'.

First, an activity could be *prohibited* altogether. Sometimes prohibition is clearly optimal. The ideal level of murders is zero: criminalised activities usually have no redeeming social utility (or so Parliament has decided).[86] But many other harm-creating activities are 'socially indispensable'—and it would be against society's interests to prevent such harms by suppressing the accompanying valuable activity.[87] Given their social value, none of the inevitably harmful activities in our four leading cases are plausible candidates for outright prohibition.

Secondly and conversely, some activities of clear social benefit are *immunised* altogether from liability for the harm they cause. The concern is that 'over-deterrence' could bring curtailment, or complete cessation, of valuable endeavours. An example is statutory immunity for US blood banks, reflecting concerns that (strict) product liability could disrupt vital blood transfusions.[88] The absence of a duty of care entails de facto 'immunity' in negligence, which is justified by multifarious public policy questions (whether this is openly acknowledged or not).[89] Nolan proposes the duty of care's disaggregation, to isolate 'public policy immunity' as a discrete stage in the negligence inquiry.[90] It is currently difficult to identify clear examples of such immunity.[91] (There have been prominent attempts

[85] Keating, 'Recovering *Rylands*' (n 77) 594 (emphasis added).

[86] cf Hunting Act 2004; *R (Countryside Alliance) v Attorney-General* [2007] UKHL 52 [1] ('For some [foxhunting] is ...a highly cherished, skilful, healthy and useful form of communal outdoor exercise. Others find [foxhunting] abhorrent...') (Lord Bingham).

[87] Alex Stein, 'The Domain of Torts' (2017) 117 Columbia LR 535, 548.

[88] Stapleton, *Product Liability* (n 2) 309–311.

[89] Jane Stapleton, 'Duty of Care Factors: A Selection From the Judicial Menus' in Peter Cane and Jane Stapleton (eds), *The Law of Obligations: Essays in Celebration of John Fleming* (OUP 1998).

[90] Donal Nolan, 'Deconstructing the Duty of Care' (2013) 129 LQR 559.

[91] cf *Mulcahy v Ministry of Defence* [1996] QB 732 (previously, UK military immunity was *statutory*: Crown Proceedings Act 1947, s 10; cf Crown Proceedings (Armed Forces) Act 1987).

to reconceive 'public policy' immunities as based purely on 'principles', such as the absence of liability for omissions.[92]) There is also the well-known, but controversial, thesis that nineteenth-century judges rolled back tort liability as a deliberate subsidy for nascent industrial development.[93] *Rylands v Fletcher* would seem a significant exception to this pro-industrial reconfiguration of tort.[94] In fact, as seen, it would be hard to draw any general lesson about Victorian judges' attitude to compensation for invasion of property rights. Differing judicial attitudes produced utterly incoherent liability. As so often, to view common law development here as a system of 'social engineering' would evoke 'the spirit of Heath Robinson at his most extravagant'.[95]

At any rate when recognised, immunity allows activities to 'externalise' harm onto the uncompensated victims. The result is higher levels of such activity than if it bore its full social cost. For example, if property could be requisitioned 'for the public good' without compensation, governments would surely expropriate land that would not be requisitioned if they had to pay the full market value. Immunity should therefore be used with great caution. Intermediate strategies are (3) to make actors liable for harms avoidable by reasonable care (negligence), or (4) liable for all harms attributable to the activity (strict liability).

The 'conditional licence' for valuable-yet-harmful activities equates to that last option. It could be praised for enlarging the choices available on the regulatory 'menu'. There is a well-known economic literature on the regulatory choice. It involves two main debates. First, whether to regulate *through tort* in the first place, rather than by public enforcement of regulatory sanctions.[96] Secondly, whether better incentives for efficient behaviour are set by fault-based or strict liability (never better explored in the Anglo-Commonwealth literature than in Jane Stapleton's 1994 monograph).[97]

Strict liability has certain advantages. First, it means that all costs are taken into account. For example, strict liability creates incentives for defendants regarding their 'activity levels' whereas negligence addresses only *how* a given activity is carried out (whether a car was driven carefully—but not whether the driver's initial decision to use the car was 'justified'). As Judge Richard Posner put it, tort imposes strict liability for dangerous wild animals because 'we desire not just that the owner take all due care that the tiger not escape, but that he consider seriously the possibility of getting rid of the tiger altogether'.[98] Secondly, strict liability places

[92] Compare *Robinson v Chief Constable* [2018] AC 736 with *Van Colle v Chief Constable* [2009] 1 AC 225.

[93] Morton J Horwitz, *The Transformation of American Law, 1780–1860* (Harvard UP 1977).

[94] cf initial, hostile reaction of US judges: *Brown v Collins*, 53 NH 442 (1873) (*Rylands v Fletcher* the law of 'immature … undeveloped' agricultural society; a 'serious … obstacle in the way of progress … trade, and productive enterprise').

[95] SFC Milsom, *Historical Foundations of the Common Law* (2nd edn, Butterworths 1981) 7.

[96] S Shavell, 'Liability for Harm versus Regulation of Safety' (1984) 13 JLS 357.

[97] Stapleton, *Product Liability* (n 2).

[98] *GJ Leasing Co v Union Elec Co* 54 F 3d 379 (7th Cir 1995) 386 (cf Animals Act 1971, s 2(1)).

all costs on defendants and simply leaves it to them to decide the efficient (ie the cost-justified) level of accident precautions. In negligence by contrast, the court has to define 'efficient' (ie careful) behaviour itself, deciding directly what cost-justified precautions should be taken using a 'Learned Hand' cost-benefit calculus. It is scarcely controversial to point out that such complex trade-offs require more expertise and information than any court is realistically likely to possess.[99] There are considerable informational advantages in remitting choices about precautions to those carrying out the activity.

Suffice to say that the 'conditional licence' possesses these regulatory virtues of strict liability. As for fears that it might financially over-burden socially valuable conduct, Bramwell B pointed out in a nuisance case: 'It is for the public benefit there should be railways, but it would not be unless the gain of having the railway was sufficient to compensate the loss occasioned.'[100] Strict liability is one way to test the argument that a given activity enhances public welfare overall.

B. Rights—and remedies

Controversy arises also about 'conditional privilege's' emphasis upon *compensation*, as opposed to specific remedies such as injunctions. If harm-inflicting conduct is permitted to continue on condition that the actor pays compensation to those harmed, this could seem to downgrade damages into merely 'the cost of doing business'. To paraphrase the leading English nuisance case advocating the priority of injunctions, would not the award of damages involve *compulsory sale of the victim's rights at the option of the defendant*, permitting tortfeasors to purchase permission to persist in the wrong?[101]

This raises two problematic issues about the nature of remedies. First, is it correct to postulate an unbreakable logical connexion between remedy and right, so that a lesser remedy is necessarily weakening the right that it protects?[102] Secondly, when should monetary remedies be understood as punitive or condemnatory sanctions, and when are they indeed just a 'cost'? In the shorthand, is awarding damages 'a fine or a price'?[103]

[99] Similar criticisms could be made about *legislative* safety regulations, although government information and expertise are hopefully superior to that of a generalist court.

[100] *Bamford* (n 54) 84–85.

[101] *Shelfer v City of London Electric Lighting Co* [1895] 1 Ch 287, 315–316 (Lindley LJ).

[102] Reasons to doubt it include: first, the degree of choice courts enjoy in crafting remedies responding to *wrongs* (eg Peter BH Birks, 'The Concept of a Civil Wrong' in David G Owen (ed), *Philosophical Foundations of Tort Law* (OUP 1995)); secondly persisting, historical jurisdictional differences. Damages' primacy at Common Law (cf Equitable specific relief) is largely historically contingent. A parallel oddity is the *absence* of damages for public-law unlawfulness, a 'tenet' which may derive simply from the prerogative writs' historical scope: Peter Cane, 'Damages in Public Law' (1999) 9 Otago LR 489.

[103] Robert Cooter, 'Prices and Sanctions' (1984) 84 Columbia LR 1523.

The latter puzzle was first noted, of course, by Holmes. He contrasted tort damages with (statutory) liability for overflowing mill-ponds, imposed even though 'the law approves and encourages' the activity. How should lawyers distinguish 'between conduct which is prohibited, and that which is merely taxed?'[104] This is by no means straightforward. It helps to disentangle contract and tort.

As for contract law, Holmes notoriously extrapolated the conclusion that there is no duty to perform contracts, but a free option either to perform *or* pay damages (ie conceiving damages as the *price* for exercising an option not to perform).[105] His aim was doubtless provocative, advancing a theory that 'stinks in the nostrils of those who think it advantageous to get as much ethics into the law as they can'.[106] In tort by contrast, Smith argues that in many significant (if marginalised) instances, tort damages are calibrated according to the *wrongfulness* of the conduct rather than the claimant's loss alone.[107] This is entirely inconsistent with the Holmesian 'price' view. It is, further, arguable that even in negligence, courts condemn a defendant's deliberate willingness to incur tort liability as a 'cost of doing business', paying compensation being cheaper than making safety improvements to avert the injuries/liabilities in question.[108] Although cost-saving breach is tolerated in contract—such tolerance being essential to reconcile absolute contractual obligations with real-world changes of circumstances—this does not translate to other branches of the law. We do not tolerate, by analogy, 'efficient theft'.[109] It is, then, less plausible to view tort damages as mere 'prices'.

A different critique from Holmes's contrasts tort and crime. Even within criminal law, whereas some sanctions clearly articulate society's condemnation (especially imprisonment) others seem less morally freighted (eg traffic violation fines).[110] Coffee argues the 'characteristic' distinction between crime and tort is precisely that between prohibition (criminal law) and 'pricing' (tort law).[111] Coffee locates tort's non-condemnatory approach in its ubiquitous 'soft-edged' standards and 'indeterminate language' (eg 'reasonable care'). It would be morally questionable to *punish* those stepping across such unclear lines. Tort is better seen as a warning that behaviour may be costly ('proceed at your own risk'). Compare: 'True

[104] OW Holmes, *The Common Law* (1881) 148

[105] See generally RE Scott and GG Triantis, 'Embedded Options and the Case against Compensation in Contract Law' (2004) 104 Columbia LR 1428.

[106] OW Holmes, 'The Path of the Law' (1897) 10 Harvard LR 457, 462.

[107] Stephen A Smith, 'The Significance of a Civil Wrong' in Paul B Miller and John Oberdiek (eds), *Oxford Studies in Private Law Theory: Volume I* (OUP 2020).

[108] Gary T Schwartz, 'The Myth of the Ford Pinto Case' (1991) 43 Rutgers LR 1013. NB Stein (n 87) 551–559: claimants' interests are vulnerable to cost-benefit analysis only when *public* benefits are in play, not purely private interests of the defendant. cf *Watt v Hertfordshire CC* [1954] 1 WLR 835, 838 (Denning LJ).

[109] Cf Daniel Friedman, 'The Efficient Breach Fallacy' (1989) 18 JLS 1.

[110] Joel Feinberg, 'The Expressive Function of Punishment' (1965) 49 The Monist 397.

[111] John C Coffee Jr, 'Does Unlawful Mean Criminal: Reflections on the Disappearing Tort/ Crime Distinction in American Law' (1991) 71 BULR 193.

commands, such as those stated by the criminal law, [that] require an unequivocal tone and the use of moral absolutes'.[112] Coffee's distinction appears too stark, given how tort remedies sometimes respond to wrongful conduct per se (most obviously, exemplary damages).[113] But clearly tort does not aspire to the same level of 'moral education' as traditional criminal law (crimes '*malum in se*').

Pace Holmes, the 'price' view, or something like it, need not be conceived in ostentatiously amoral terms. The 'conditional licence', seeking to make defendants bear costs but *not* to condemn, deter, or prevent the conduct that causes the harm, could well (as Keating argues) be a morally attractive compromise. This explains current developments in tort law. Notably, senior English judges have argued recently that while the public interest should not be a defence to *liability* in nuisance, property rights can be reconciled with the public interest (and public planning law) at the *remedy* stage by awarding damages in place of an injunction.[114] Lord Leggatt comments that while private property is 'not absolute' and

> may be subordinated to the general good of the community ... it is fundamental to the integrity of any system of property rights that, in any such case, the individuals whose rights are infringed or overridden receive compensation for the violation of their rights.

Namely: 'the public interest may sometimes justify awarding damages rather than granting an injunction but it cannot justify denying the victim any remedy at all'.[115] On the other hand again, parallel developments in US nuisance law occasioned one of the most famous economic analyses.[116] Perhaps in this area at least, the moral and instrumental arguments are mutually supportive, rather than antagonistic.[117] Despite the intellectual difficulty of reconciling such different values, it may well be that like human life itself, 'the legal Cathedral cannot be structured according to one style—it is by its very essence pluralistic, composed of contrasting parts'.[118]

[112] ibid 225.

[113] Smith (n 107).

[114] *Coventry v Lawrence* [2014] AC 822 [155]–[157] (Lord Sumption).

[115] *Fearn v Tate Gallery* [2024] AC 1 [122].

[116] *Boomer v Atlantic Cement Co* 257 NE 2d 870 (NY 1970); Guido Calabresi and A Douglas Melamed, 'Property Rules, Liability Rules, and Inalienability: One View of the Cathedral' (1972) 85 Harvard LR 1089.

[117] That is not to say (obviously) that commentators universally hold 'liability rules' (compensation) superior to 'property rules' (specific remedies). cf eg Yotam Kaplan, 'The other view of *The Cathedral*' (2023) 82 Maryland LR 479 (preference for compensation perpetuates racial inequalities).

[118] Izhak Englard, 'The Costs of Accidents: A Retrospect View from the Cathedral' (2005) 64 Maryland LR 355, 361. See also Liam Murphy, 'Purely Formal Wrongs' in Paul B Miller and John Oberdiek (eds), *Civil Wrongs and Justice in Private Law* (OUP 2020) 34 (harder to disentangle moral/instrumental foundations of *torts* (cf contract and property) 'because of the hodgepodge factor [in tort]').

C. Strict liability and the Rule of Law

Is it consistent with the Rule of Law to impose *absolute duties*, insensitive to the subject's ability to comply? Should not 'ought' imply 'can'? That basic precept of political morality is violated by absolute duties to succeed (rather than merely 'duties to try').[119] Tort duties cannot play their role of guiding the practical reasoning of actors when the law requires things that actors are simply *incapable* of achieving.[120] This violates the core, 'thin', or 'procedural' sense of the Rule of Law as a criterion for successful legal rules. Namely, laws must be clear, precise, published, and stable so that their subjects can know their obligations in order to comply with them. Having *capacity* to comply—negatively, the avoidance of substantive legal duties that are impossible of performance—is a still more basic requirement of what Fuller called 'the inner morality of law'.

One possible response to this critique of strict liability follows from the points raised above, about the nature of remedies. What if paying compensation is better conceived as a tax—the price of legal permission to continue with a risky-yet-valuable activity? This deflects the criticism. As outlined above, 'conditional privilege' does not involve an absolute duty to refrain from causing harm. On the contrary, the 'privileged' activity is permitted to continue notwithstanding its harmfulness, on 'condition' that payment is made for all the harm it causes. There is no Rule of Law objection to levying a financial charge on absolutely unavoidable 'activities'.[121] The Rule of Law is satisfied so long as the event is specified with certainty.[122] Of course in the cases discussed here, the activity involves some degree of free choice. Mr Rylands was not obliged to have a reservoir on his land; the boat-owner in *Vincent* was not obliged to lash it to the dock. The 'conditional licence' could be seen as putting defendants on notice that should they choose to engage in such activities, the law requires them to bear the costs of all harms caused thereby. While strict, the liability can be avoided by choosing not to engage in the activity at all.[123]

On this view, a crucial question is precisely when tort imposes 'conditional privilege' liability. That all legal systems seem to view strict liability as a series of *exceptional* instances helps reconcile it with the Rule of Law. Precisely because it is exceptional,[124] one can engage in fruitful 'self-effacement' in deciding what

[119] John Gardner, 'Obligations and Outcomes in the Law of Torts' in *Torts and Other Wrongs* (OUP 2019) (misconceived to view strict duties as extraordinarily high standards of care—they are simply *indifferent* to the defendant's conduct (compliance might eg be effortless, or fortuitous), examining only its *outcome*).

[120] Adam Slavny, 'Should Tort Law Demand the Impossible?' in Paul B Miller and John Oberdiek (eds), *Civil Wrongs and Justice in Private Law* (OUP 2020).

[121] Paradigmatically 'inheritance tax', an impost on dying.

[122] LL Fuller, *The Morality of Law* (2nd edn, Yale UP 1969) 70–79.

[123] John Gardner, 'Some Rule-of-Law Anxieties about Strict Liability in Private Law' in *Torts and Other Wrongs* (OUP 2019) 191.

[124] Cf David G Owen, 'The Fault Pit' (1992) 26 Georgia LR 703, 710 ('Fault lies at the very heart of tort law and provides it with a meaningful, moral definition').

activities to undertake, and thereby avoid the liability.[125] Whereas if strict liability were to be *generalised*, it could not be avoided (since people can hardly refrain from *all* conduct whatever; nor could they avoid liability by acting with all reasonable care). The present putative doctrine's uncertain, contested, common law basis is admittedly not the best 'signalling' method for those potentially subject to it. Much clearer 'fair warning'[126] about strict liability should be given up front, as in *statutory* licensing and liability regimes, such as the Nuclear Installations Act.[127] Hopefully, this chapter's discussion of 'conditional privilege' will assist wider understanding of the common law's logic, helping to reconcile it with the Rule of Law.

D. Scope of 'conditional privilege': how wide—and who decides?

Is 'conditional privilege' a vain attempt to build a general principle upon a quartet of curious cases? As just noted, strict liability is treated as 'exceptional' in legal systems worldwide. In Germany and Austria, for example, strict liability is seen as a matter for specific imposition by the legislature in particular situations—and very much not a general delictual principle embodied in the jurisdictions' Civil Codes.[128] The resulting patchwork of liability, and deference to *legislative* strict liability through German law's 'enumeration' principle, closely resembles the English courts' reluctance to extend *Rylands v Fletcher*.[129] Debates about its scope, and even its existence, are longstanding. Even in US law there seems little appetite to extend it.[130] In England, *Rylands v Fletcher* has been strictly confined to escapes from land,[131] whereas in Australia the tort has been subsumed within negligence.[132] *Vincent v Lake Erie* cases (ie 'emergency-necessity' claims) are also rare in practice. Our other two cases concern action by public officials. Whether even here there should be a general (constitutional?) principle is controversial—namely whether damage unavoidably inflicted in the public interest must be compensated at public expense.

There are numerous underlying questions. Can the justifications for strict liability (either moral or instrumental) be *generalised*? If not, what are its boundaries? Strict *enterprise* liability and *government* liability for public-interest-harm

[125] Gardner (n 119) 151.
[126] ibid.
[127] n 43 above.
[128] Johannes W Flume, 'Strict Liability in Austrian and German Law: On the concept of strict liability in the age of technological advancement' (2021) 12 JETL 205.
[129] *Cambridge Water* (n 49).
[130] Keating 'Recovering *Rylands*' (n 77) 553.
[131] *Stannard v Gore* [2014] QB 1.
[132] *Burnie Port* (n 65).

are possible, limited, doctrines of strict liability. But each is highly controversial. We take these in turn.

Universal strict liability in tort would be unacceptable and unworkable. First, as seen, is the Rule of Law objection. Strict liability cannot be avoided by acting with reasonable (or all humanly possible) care. But it can be avoided, when it is imposed *only exceptionally*, by refraining from activities to which it attaches. But such avoidance becomes impossible as strict liability is generalised. Second is the 'Coasian' problem of mutual causation.[133] Which of two equally 'causal' parties would be liable for what, if their culpability were irrelevant? The difficulty was recognised a century before Coase by Bramwell B. In *Fletcher v Rylands* (no less) Bramwell said: 'Where two carriages come in collision, if there is no negligence in either it is as much the act of the one driver as of the other that they meet.'[134] For precisely this reason, US courts have declined to extend strict liability to road accidents.[135] As Keating comments, the apparent (and ethically troubling) affinity of strict liability for *property* torts may not stem from a judicial preference for material wealth over bodily protection. Simply that such strict liability is more *feasible*: 'Property norms enable drawing boundaries between activities without recourse to any criterion of fault.'[136]

Henderson, too, argues that feasibility precludes general strict liability, however desirable it might be in theory.[137] 'Discrete [existing] areas of strict liability ... are viable because courts and legislatures draw the boundaries of each area cleanly and crisply'—that is, precisely because they are 'discrete'.[138] But if generalised, strict liability would present intractable, 'unadjudicable' problems. Multiple activities are contributory causes to every given harm. If fault or concepts like product 'defectiveness' were ignored, how could the law decide whether and how much harm to attribute to which activity? There would be 'conceptual chaos'. The courts 'could not possibly reach consistent, rational outcomes' when apportioning relative liability in contribution proceedings between multiple causally-relevant defendants.[139]

In Henderson's view, this equally rules out broad-based strict *enterprise* liability. Making businesses strictly liable for the characteristic harms of their activities has undeniable theoretical appeal. From a regulatory perspective, profit-making organisations should be particularly sensitive to, and able to act upon, the price signal. It is also morally attractive: businesses are permitted to internalise unexpected gains from their activities; equally, they should bear unexpected losses.[140]

[133] Ronald H Coase, 'The Problem of Social Cost' (1960) 3 J L&E 1.

[134] (1865) 3 H & C 774, 790.

[135] *Hammontree v Jenner*, 20 Cal App 3d 528 (1971).

[136] Keating, 'Recovering *Rylands*' (n 77) 587. For criticism of extension of this strict liability to *intellectual* property rights, see O Bracha and P Goold, 'Copyright Accidents' (2016) 96 Boston University LR 1025.

[137] James A Henderson, 'Why Negligence Dominates Tort' (2002) 50 UCLA LR 377.

[138] ibid 397.

[139] ibid 395.

[140] Stapleton, *Product Liability* (n 2) 187.

It is arguably the idea underlying not merely this chapter's four leading cases, but whole swathes of significant tort doctrine including product liability and vicarious liability. But those liabilities, while broad, are not general. Tort law holds a business strictly liable for its employees' wrongs, and for its defective products, but not for *all* the harms it causes. Perhaps it should. Jane Stapleton brilliantly critiques the arbitrary and illogical boundaries of *product-specific* strict liability.[141] But generalisation would be a momentous step. The European Commission's proposal effectively to extend the 1985 Directive from products to professional *services* (as firmly advocated by Jane Stapleton) was quickly abandoned.[142] It would be controversial for the common law to recognise such a broad extension of strict liability in opposition to such legislative reluctance.[143] Keating, although a champion of strict liability, accepts that specifying the boundaries of enterprise liability 'requires the exercise of essentially contestable, normative, political judgment'.[144]

It is equally difficult to imagine the 'discovery', immanent in the common law, of a general principle that *government* must compensate all those who suffer loss as a result of public-interest activities. Even more clearly, this would involve 'contestable ... political judgments'. The principle admittedly has considerable merits and it underlies numerous specific kinds of public authority liability. But to synthesise from those particular examples a *general* strict government liability would exceed the judicial function. We briefly expand on these points.

The idea that persons incidentally harmed by action in the public interest should have their losses compensated at public expense has strong appeal as a matter of justice. As noted, all the judges in *Burmah Oil* accepted this was a compelling moral principle (but disagreed over whether it was a rule of *law*). But as also discussed, the retrospective reversal of the court's decision shows that Parliament disagreed that there was such a principle in play. (This emphatic legislative reversal was overlooked in the recent suggestion that *Burmah Oil* establishes a 'constitutional principle' that 'the executive cannot exercise prerogative powers so as to deprive people of their property without the payment of compensation'.[145]) As de Smith commented, there many examples of 'administrative compensation' being

[141] ibid chapter 12.

[142] See *Commission Proposal for a Council Directive on the liability of suppliers of services* (9 November 1990) (service-providers would have been required to *ensure* 'in normal and reasonably foreseeable conditions ... the safety which may reasonably be expected'). The lobbying power of the services industry exceeds even that of 1980s farmers (Stapleton, *Product Liability* (n 2) 303–305; cf now Directive 1999/34/EC (repealing agricultural exemption in Directive 85/374/EEC)).

[143] cf now European Commission, *Proposed Product Liability Directive* (28 September 2022) (revision of Directive 85/374/EEC extending liability to software and 'clarifying when a related service is to be treated as a component of a product').

[144] Keating, 'Rawlsian Fairness' (n 75) 1862.

[145] *R (Miller) v Prime Minister; Cherry v Advocate General for Scotland* [2020] AC 373 [40]. Gageler J's reliance on *Burmah Oil* in *Binsaris v Northern Territory* [45] equally overlooks its legislative disapproval.

paid, *ex gratia* or *ad hoc*, by the British government; what is absent is any 'coherent body of general principle' to explain why and when this should happen.[146]

French law offers a telling contrast. English administrative law contains no equivalent to its *egalité devant les charges publique* doctrine.[147] England has many particular instances, but no governing theory. Given the financial implications for the government, the recognition of such a general principle would surely exceed the courts' constitutional role. To quote Viscount Radcliffe's dissent again: 'it is for those who fill and empty the public purse to decide'.

One particular advantage of a 'conditional licence' approach to government liability should be noted. It avoids the constitutional sensitivities of adjudicating on a public authority's 'negligence'. For example, concerns that recognising a duty of care to equip and train soldiers injured during active combat entails re-litigation of sensitive military and political decisions (eg 'negligent' battle strategy, resource allocation, or procurement policy).[148] Such questions are at the extreme end of the spectrum of 'policy' as opposed to 'operational' matters. If this 'non-justiciability' objection is sound, what follows? If it results in total immunity, then (as argued above) this could have unfortunate consequences too. If the government bears none of the costs of its actions, it has no *economic* incentive to limit the scale of harm.[149]

Strict liability avoids each of these objections. It does not require courts to assess whether sensitive political decisions were negligent or otherwise culpable. But nor does it permit the government to 'externalise' the costs of harm onto those who suffer it. For precisely these reasons, Rebecca Crootof proposes that belligerent states should be liable to compensate *all* harm that results from military conflict.[150] Such strict liability would provide 'much-needed incentives for states to mitigate or reduce civilian harms' in wartime. Yet (by excluding questions of fault or culpability) it would not alter 'the range of lawful actions in armed conflict'. The very point is compensation for *lawfully inflicted* harms—in this chapter's terms, that the permissibility of military force should be conditional on compensating all the inevitable harm that results. (We have suggested elsewhere that the UK's riot damage compensation regime, making police forces strictly liable for all property damaged by rioting,[151] similarly sharpens police incentives to keep public order without

[146] S de Smith, *Judicial Review of Administrative Action* (3rd edn, Sweet & Maxwell 1973) 21–22, quoted in C Harlow, 'Administrative Compensation: Bypass or Dead End?' in James Goudkamp et al (eds), *Taking Law Seriously: Essays in Honour of Peter Cane* (Hart 2021).

[147] R Errera, 'The Scope and Meaning of No-Fault Liability in French Administrative Law' [1986] CLP 157.

[148] *Smith v Ministry of Defence* [2014] AC 52: see Lord Mance, dissenting, at [125]–[134].

[149] Obviously, there are moral, reputational, political, and military reasons to train and equip forces effectively.

[150] Rebecca Crootof, 'War Torts' (2022) 97 New York University LR 1063.

[151] Riot Compensation Act 2016, s.1.

requiring judges to decide whether a riot could have been more effectively policed, prevented, or contained.[152])

But again, it is questionable whether extensions (let alone generalisations) of this idea are properly for the courts. To take the riot example, why should its principle not extend from rioting to encompass *all* criminal damage to property? There is an obvious answer of brute politics. As the Home Secretary once confessed (explaining why the Criminal Injuries Compensation Scheme compensates only bodily injuries and not property damage), the government simply could not afford it.[153] Compensation schemes come in many forms. Some, notably, reflect (or ought to reflect) tortious ideas of individual justice if they are properly to fulfil cognate (eg vindicatory) purposes.[154] But design choices, including eligibility criteria, cannot be disentangled from sensitive decisions about the appropriate degree of public funding.

It is not for *courts* to distribute public money. Proposals for 'conditional licensing' can often appear morally and instrumentally attractive. The idea underlies numerous specific instances of government liability. But it cannot be generalised as a common law principle. That would exceed the constitutional function of the courts, departing from the 'traditional judicial prudence with public funds'.[155]

5. Conclusion

Perhaps this chapter should not, after all, have begun with the four intriguing cases. Their attractiveness might ultimately distract our attention. Judicial decisions engage the common lawyer as her stock-in-trade, natural home, and chief delight. Yet the seed of the idea found immanent within our four cases is not likely to germinate. A grand new principle cannot grow from them because it would be unstable, too contentious, and too expensive.

The true home of the conditional licence is in legislation. As noted, leading examples of strict (near-absolute) liability for harms done by socially indispensable risky activities are to be found in statute.[156] As also seen, Simpson maintains that even at the high point of nineteenth century common law creativity, when *Rylands v Fletcher* was decided, making risky-yet-beneficial activities absorb their costs was primarily a decision for Parliament.

[152] J Morgan, 'Strict Liability for Police Nonfeasance? The Kinghan Report on the Riot (Damages) Act 1886' (2014) 77 MLR 434.

[153] Quoted A Ashworth, 'Punishment and Compensation: Victims, Offenders and the State' (1986) 6 OJLS 86, 103 n 66 (cost 'absolutely astronomical').

[154] Simone Degeling and Kit Barker, 'Designing Reparation: Lessons from Private Law' in R Levy et al (eds), *New Directions for law in Australia: Essays in Contemporary Law Reform* (ANU Press 2017).

[155] Harlow (n 146) 167 (noting low compensatory awards under the Human Rights Act 1998).

[156] N 42–43.

Rightly so. The common law struggles with creating 'exceptional' rules which are to be given arbitrary limits; applied 'thus far and no further' according to a notorious psychiatric injury decision.[157] How much better, when a closed list is desired, to enumerate it in a statute.[158] A central lesson of English law's special *Fairchild* rule for proof of causation[159] is that drawing 'bright lines' in judge-made law is unhappily impossible. According to Lord Neuberger and Lord Reed the problems with *Fairchild* arise precisely because, unlike legislation:

> the common law cannot confine itself to a particular situation and deal with it in isolation from the remainder of the law; nor can it resolve problems on a purely pragmatic basis. It is a complex and extensive network of interconnected principles applicable to all situations falling within their scope.[160]

Statute can be an institutionally superior means of implementing policies such as the conditional licence. Line-drawing is inherent to legislation. Compromise is the stuff of politics. Unlike the common law, legislatures are not confined to principles that can be extended by analogy, and ultimately universalised. Legislative tools also offer greater precision. Tort lawyers must shake off their inclination to stigmatise legislation. Statute is not per se illegitimate, an unnecessary and 'semi-arbitrary' intrusion on the principles consistently developed and logically refined by judicial decision.[161]

In any event, legislative liability as a 'conditional licensing' strategy runs with the grain of the common law. The four cases show that torts should not be conceived exclusively as sanctions for wrongdoing.[162] Sometimes strict liability is better viewed as the price paid for continuing with valuable, yet harmful, activities. The conditional licence is fruitful also in delineating the defence of necessity, and distinguishing between justifications and excuses. Tort law is not a 'seamless web' and cannot be explained by any single theory. The conditional licence is another distinct thread within tort's rich fabric.

[157] *White v Chief Constable* [1999] 2 AC 455, 500 (Lord Steyn). cf Jane Stapleton, 'In restraint of Tort' in Peter BH Birks (ed), *The Frontiers of Liability, Volume 2* (OUP 1994) 95 ('unseemly arguments over [these] meaningless boundaries' are 'not only artificial but bring the law into disrepute').

[158] *Alcock v Chief Constable* [1992] 1 AC 310, 419 (Lord Oliver) (commending Law Reform (Miscellaneous Provisions) Act 1944 (NSW)).

[159] *Fairchild v Glenhaven Funeral Services* [2002] UKHL 22; [2003] 1 AC 32 (see Jane Stapleton, 'Lords A-Leaping Evidentiary Gaps' (2002) 10 Torts LJ 276).

[160] *Zurich Insurance plc v International Energy Group Ltd* [2015] UKSC 33, [2016] AC 509 [192]

[161] Peter Cane, 'Taking Disagreement Seriously: Courts, Legislatures and the Reform of Tort Law' (2005) 25 OJLS 393 (citing *Cattanach v Melchior* [2003] HCA [137] (Kirby J)).

[162] cf Owen (n 124) 724 (fault principle's 'moral rectitude ... a wellspring of ethical bounty, providing a touchstone of propriety ... to guide all persons').

15

Damage as an Essential Element in the Economic Torts

*John Murphy**

1. Introduction

Jane Stapleton is almost certainly best known for her work on causation. In this connection, she has been a—perhaps *the*—standout voice in the common law world for several decades. But we ought not to overlook the fact that she has written much more widely than this about tort law, and that her scholarship in the field is more or less bookended by her engagement with the damage component in the tort of negligence. During her second year at Balliol, and focusing exclusively on this issue, she published the seminal article, 'The Gist of Negligence Part 1: Minimal Actionable Damage'.[1] Then, in 2021, the damage element in the tort of negligence was a topic to which she returned—albeit as part of a more wide-ranging project—in the last of her Clarendon Law Lectures in an essay entitled, 'Conceptual Interplay between Elements of the Tort of Negligence'.[2]

In the first of these two essays, Jane rightly observed that, the 'modern law of torts is in need of a coherent doctrine about the notion of "damage". The word is bandied about in a number of different contexts, usually without clear definition.'[3] Yet, in spite of a growing recognition among torts scholars that this is indeed an important matter[4]—after all, it is key when seeking to identify the frontiers of tortious liability—there remains much work to be done. For example, although commentators have now begun to engage seriously with one of Jane's initial questions—namely, where and on what basis we may draw the line on what counts as damage for the purposes of the law of negligence[5]—noticeably less has been written about

* I am grateful to the editors of this volume for affording me the opportunity to contribute, to James Goudkamp for written comments and to all those at the symposium who provided oral comments. Last, but by no means least, I am grateful to Jane for all the learning, friendship and advice.

[1] (1988) 104 *Law Quarterly Review* 213.

[2] J Stapleton, *Three Essays on Torts* (Oxford, OUP, 2021) ch 3.

[3] Stapleton (n 1), 213.

[4] See, eg, D Nolan, 'New Forms of Damage in Negligence' (2007) 70 *Modern Law Review* 59; J Morgan, 'The Outer Limits of Personal Injury' [2018] *Cambridge Law Journal* 461; E Voyiakis, 'Is Asymptomatic Sensitisation Actionable Harm?' (2015) 31 *Professional Negligence* 199.

[5] M Fordham, 'The Protection of Personal Interests—Evolving Forms of Damage in Negligence' (2015) 27 *Singapore Academy Law Journal* 643; D Nolan, 'Damage in the English Law of Negligence'

John Murphy, *Damage as an Essential Element in the Economic Torts* In: *Torts on Three Continents*. Edited by: Kylie Burns, Jodi Gardner, Jonathan Morgan and Sandy Steel, Oxford University Press. © John Murphy 2024.
DOI: 10.1093/oso/9780198889748.003.0015

three other, equally important questions: namely, (1) whether damage is best seen an element of that tort, or a mere condition of actionability;[6] (2) how exactly damage relates to the other components in a negligence action;[7] and (3) how damage comes apart from other concepts frequently used as synonyms for it, such as 'loss', 'harm' and 'injury'.[8]

On top of this, there is another striking gap in the academic literature. To date, scholars have overwhelmingly focused their attention on the concept and role of damage *within the tort of negligence*. Yet, despite its being a moot point for years, the nature and role of damage within the economic torts has received far less juristic attention. This chapter seeks to begin the process of redressing the balance. It does so mindful of two things: not just the lacuna highlighted a moment ago, but also—fittingly, I hope—the fact that, in the second of the essays just mentioned, Jane specifically laid down the gauntlet saying that damage, 'is a research topic that is ripe for exploration'.[9]

The chapter unfolds as follows. In Section 2, I explore the question of whether terminology matters, and in particular whether courts should feel able to use the words 'damage', 'loss' and 'harm' more or less interchangeably. In Section 3, as a *homage* to Jane's own exhortation that torts scholars should take seriously what our most senior courts tell us,[10] I examine the various types of harm that have been held (or at least suggested) to constitute actionable damage within the economic torts in the cases as we find them. In Section 4, I explore the related question of whether damage is best seen as an element of the various economic torts, or whether, sometimes, it is more accurately to be regarded as a condition of actionability. Section 5 concludes.

2. The Need for 'Damage' in the Economic Torts?

In an important article published in 2017, Donal Nolan insisted that distinguishing the concepts of 'loss' and 'damage' was essential to the enterprise of evincing a

(2013) 4 *Journal of European Tort Law* 259; J O'Sullivan, 'The Meaning of "Damage" in Pure Financial Loss Cases: Contract and Tort Collide' (2012) 28 *Professional Negligence* 248.

[6] See, eg, D Nolan, 'Deconstructing the Duty of Care' (2013) 129 *Law Quarterly Review* 559, 561–4; cf NJ McBride, 'Duties of Care: Do They Really Exist' (2004) 24 *Oxford Journal of Legal Studies* 417. For detailed discussion of the difference between the two concepts, see J Murphy, 'Unpicking Torts: Elements, Conditions of Actionability and Standing Rules' (2024) 87 *Modern Law Review* (forthcoming).

[7] J Goudkamp and D Nolan, *Winfield and Jolowicz on Tort* (20th edn, London, Sweet & Maxwell, 2019) 150–1.

[8] See D Nolan, 'Rights, Damage and Loss' (2017) 37 *Oxford Journal of Legal Studies* 255 for a convenient collection of judges and jurists who have treated the concept of 'damage' as being interchangeable with each of 'loss', 'harm' and 'injury'.

[9] Stapleton (n 2) 70.

[10] Stapleton (n 2) ch 1.

plausible rights-based understanding of the tort of negligence.[11] In his view, it is important to eschew the idea of 'loss'—which connotes 'being worse off'[12]—on the basis that it is conceptually impossible to possess a right not to be caused loss. The claim rests on two key points. The first—originally highlighted by Peter Birks—is that that a wrong occurs at a moment in time.[13] The second is that it is impossible to say at any given moment in time that someone has been made worse off.[14] It follows from this, Nolan argues, that we cannot say at any particular juncture that D has caused C a loss. And from this, in turn, he insists, we cannot say that C has a right not to be caused a loss. Put more succinctly, if a loss cannot be said to have been inflicted at a moment in time, and a wrong (which involves the violation of a right) necessarily occurs at a moment in time, then it must—as a matter of logic—be true that there can be no right not to be caused a loss.

By contrast, according to Nolan, no such problem attends the concept of 'damage'. This is because damage entails 'a certain kind of interference with a person's protected interests';[15] and any such interference *can* occur at a moment in time. This, therefore, makes it possible to construct an understanding of negligence that has at its core a right that D not carelessly inflict damage on C.

In order to fortify his argument that thinking in terms of damage is not just preferable, but interpretively accurate, Nolan directs his readers to the rules on relational economic loss. For here, as he rightly points out, the ability to sue in negligence rests with the holder of the relevant right rather than with the person who appears to be made worse off. As he puts it, 'a seller of goods can sue in tort for damage done to the goods before the property in them passed to the buyer, even if the risk had passed to the buyer at the relevant time ... [and] the buyer (the one who actually loses out) has no claim'.[16]

Now, the potential importance of Nolan's analysis for present purposes should be obvious from the fact that cases concerning the economic torts are littered with talk of both 'loss' and 'damage'. Indeed, in the wake of *OBG Ltd v Allan*,[17] the now orthodox way of referring to one of the two most prominent economic torts is as the tort of 'causing loss by unlawful means'. If we accept what Nolan says about loss, then some fundamental questions naturally arise in connection with this and the other economic torts. Among them are the following. Have the courts been insufficiently attentive to terminology in this setting? Ought they to revert to a former

[11] Nolan (n 8).
[12] ibid, 2.
[13] P Birks, 'Rights, Wrongs and Remedies' (2000) 20 *Oxford Journal of Legal Studies* 1, 5: a wrong is 'an event which happens in the world'.
[14] Nolan gives the classic example of A being injured in a car crash by B's negligence, such that A cannot then take a plane which crashes on take-off killing all passengers. His point is that, when the accident caused by B occurs, it is impossible to say with certainty that B has been made worse off.
[15] Nolan (n 8)..
[16] ibid, 17.
[17] [2008] 1 AC 1.

label for this tort, such as 'unlawful interference with trade'?[18] Must the terms 'damage' and 'loss' bear the same meanings in the sphere of the economic torts as Nolan ascribes to them in the context of negligence? Do all of the actions that are generally regarded as members of the economic torts 'family' even deserve to be regarded as torts?

If we were to attempt to answer these questions, it would be inappropriate—perhaps even arrogant—simply to dismiss as irrelevant Nolan's article on the basis that 'the focus of the analysis [in that article] is on the law of negligence'.[19] The fact that negligence law was Nolan's principal concern cannot, in and of itself, be taken to mean that his arguments have no purchase elsewhere in the law of torts. After all, it seems perfectly reasonable to suppose that, all other things being equal, a concept used as a touchstone of liability in one tort ought presumptively to bear the same meaning when it is also used as a touchstone of liability in another one. A general concern for the coherence of the law would seem to dictate as much. Accordingly, if we are to put his analysis to one side, we must have good reasons for doing so.

But before considering any such reasons it is only fair to concede the logic of what Nolan says about the inescapable uncertainty of whether a particular event can be said to have caused someone a loss. For the logic is, I think, impeccable *provided* we accept his definition of loss. But herein lies the rub, since it is not at all clear why we must *as matter of law* accept his definition of a loss. And if another understanding of loss is on offer, the possibility of talking interchangeably of 'loss', 'harm' and 'damage' in the economic torts setting need not be ruled out.

Nolan's conception of a loss is only required for the purposes of his specific enterprise, namely, to evince a plausible *rights-based understanding* of the tort of negligence. More particularly, the rights-based nature of his endeavour is doomed to failure if what he attempts to construct is an understanding of negligence that centres on a right not to be caused loss. But as soon as we recognise that some torts—including most of the economic torts—are not rights-based,[20] the significance of the loss/damage distinction is *to this extent* undermined. For even if loss (as Nolan defines it) remains an awkward concept,[21] the complaint that it is logically impossible to possess a right not to be caused loss clearly has no purchase in

[18] See, eg, H Carty 'Unlawful Interference with Trade' [1983] *Legal Studies* 193. See also, for a slightly broader title, J Neyers, 'Rights-Based Justifications for the Tort of Unlawful Interference with Economic Relations' [2008] *Legal Studies* 215. See too, Lord Nicholls' description of it—in *OBG* (n 17), [141]—as 'the tort comprising interference with a trade or business by unlawful means'.

[19] Nolan (n 8), 256.

[20] See, eg, the difficulties of unearthing a rights-based conception of the unlawful means tort in Neyers (n 18); and the difficulties of identifying a rights violation in cases of inducing breach of contract explored in J Murphy, *The Province and Politics of the Economic Torts* (Oxford, Hart Publishing, 2021) ch 3.

[21] Nolan would maintain that loss, strictly speaking, is still an unusable concept, and that the more accurate term would be apparent loss: 'if the legal process reaches the conclusion that a person has suffered loss … [then] "loss" really means "apparent loss"': Nolan (n 8), 256, n 3.

relation to any tort that falls beyond Nolan's conception of a tort as 'the violation of another's legal right'.[22] Most of the economic torts are loss-based rather than rights-based.[23] And lest it be thought that this absence of a rights-basis renders them anomalous, it might also be noted that even the tort of negligence is sometimes only dubiously anchored to a rights infringement.

For example, in cases that turn on D's assumed responsibility towards C, it is highly questionable whether there is a rights infringement at work. True, Robert Stevens has argued that such cases involve rights infringements, with the relevant right being created by D's 'objective manifestation' of willingness to undertake a responsibility.[24] And there is, I concede, ostensible plausibility in his contention that such a right may be deduced by analogy with the way such objective manifestations are treated as generating rights in both the law of bailment and the law governing express trusts.[25] Yet these are far from conclusive analogies; for it need not follow that cases involving assumed responsibilities in the law of negligence must come into line with the law of bailment or the law concerning express trusts. After all, in contract law—with which body of law tort is generally thought to have a much closer affinity—the making of a gratuitous promise is *not* a rights-generating event. Thus, in the absence of any clear reason why tort should follow equity and/or bailment rather than contract, it seems unsafe to presume that assumed responsibility cases can be cashed out in terms of rights violations.

As to my more central claim that for the majority of the economic torts there is no need for C to show that she has suffered a rights infringement, just consider the following. The actions for causing loss by unlawful means, inducing breach of contract, conspiracy in both its guises and injurious falsehood are all widely accepted—even by dyed-in-the-wool rights theorists—as torts in which C's ability to sue D does not hinge upon the infringement of an *ex ante* right held by C against D.[26] Of course, the absence of any need to show a rights infringement might be

[22] ibid, 5.

[23] The economic torts are typically committed via D's conduct towards an intermediary, T. In these torts D may violate a right held by T against D that results in loss to C (as in the tort of causing loss by unlawful means); or D may procure certain lawful conduct on the part of T that results in loss to C (as in the torts of passing off and injurious falsehood); or D may induce T to violate a right held by C against T (as in the tort of inducing breach of contract). But in none of these torts does D personally violate a right held by C against D. See further, J Murphy, 'Rights, Reductionism and Tort Law' (2008) 28 *Oxford Journal of Legal Studies* 393, 403–4; J Murphy, 'The Heterogeneity of Tort Law' (2019) 39 *Oxford Journal of Legal Studies* 455.

[24] R Stevens, *Torts and Rights* (Oxford, OUP, 2007) 10.

[25] ibid.

[26] Numerous authors have attempted to justify the existence of the general economic torts in terms of rights infringements. In relation to inducing breach of contract, such attempts have typically posited the infringement of a (quasi) proprietary right but they all fail for reasons too lengthy to recite here (but explained fully in Murphy (n 20) ch 3). Neyers has proffered a possible rights-based understanding of the unlawful means tort as well as of the two conspiracy torts based on an 'abuse of rights' idea. But his claims are contradicted by House of Lords authorities. He admits as much, but (unlike me) is happy to put these cases to one side: see, J Neyers, 'The Economic Torts as Corrective Justice' (2009) 17 *Torts Law Journal* 1, 21–34. In cases of unlawful means conspiracy where the unlawful means constitute a

taken by some as a basis on which to call into question their pedigree as torts proper. But that entails a separate debate and it need not unduly detain us here.[27] Instead, it suffices to note that a good *prima facie* case for calling these loss-based actions, 'torts', can be anchored to the fact that they are routinely regarded as such by Bench, Bar[28] and—so far as I can tell—most members of the academic tort law community.[29] And that being the case, Nolan's principal reason for attaching significance to the distinction he draws between loss and damage falls away.

There is also a second reason for putting to one side Nolan's insistence that we should talk of damage rather than loss. It inheres in the fact that those practitioners who work with the law on a day-to-day basis—and I have in mind especially the appellate courts—are perfectly happy to use the language of loss. The fact that our most senior judges are untroubled by the concept of loss counts for a great deal *even if*, ultimately, the idea that a particular tort is animated by the causation of loss is logically indefensible. A simple analogy helps to explain why.

Before the video assistant referee (VAR) system was introduced in games of top tier professional football to help check the correctness of referees' on-field decisions, the fact that penalty and offside decisions had sometimes been given in error still counted for something. A penalty awarded in error still counted as a valid penalty from which a goal would usually be scored. It is just the same in law. If, however illogically, the House of Lords or Supreme Court has declared a particular tort to be enlivened by D's causing loss to C, this matters. For one thing, all inferior courts must henceforth proceed in line with what the apex court has said; and for another, the treatise writers, too—even if they exercise their right to criticise the merits of a particular holding—must also at least acknowledge what the appellate court has said. Herbert Hart captured well the point about the inescapable significance of whatever is said by the highest court in the land when he wrote:

> A supreme tribunal has the last word in saying what the law is and … [although] [t]he decision may, of course, be deprived of legal effect by legislation … the very

tort, there may well be a rights infringement; but any such case could be dealt with on the basis of joint tortfeasors committing that tort.

[27] For doubts concerning the idea that there is plausibly any such thing as a pedigree or pure tort, see J Murphy, 'Hybrid Torts and Explanatory Tort Theory' (2018) 64 *McGill Law Journal* 1.

[28] It is true that inducing breach of contract was described by Lord Hoffmann in his leading speech in *OBG Ltd v Allan* [2008] 1 AC 1 (HL) as a form of secondary liability. But it is equally true that he also referred to it as either 'the tort of inducing breach of contract' or 'the *Lumley v Gye* tort' no fewer than eight times: see ibid, [17], [25], [32], [33], [38], [44], [62] and [69]. Furthermore, it has consistently been referred to in both these ways by our most senior courts in the wake of *OBG*: see, eg, *Kawasaki Kisen Kaisha Ltd v James Kemball Ltd* [2021] EWCA Civ 33 (CA), [6] (Popplewell LJ); *Racing Partnership Ltd v Sports Information Services Ltd* [2021] 2 WLR 469 (CA), [142] (Arnold LJ); *Allen v Dodd & Co Ltd* [2020] 2 WLR 1070 (CA), [21] (Lewison LJ); *Eurasia Sports Ltd v Aguad* [2018] 1 WLR 6089 (CA), [23] (Floyd LJ); *Secretary of State for Health v Servier Laboratories Ltd* [2020] Ch 717 (CA), [55], [68] and [84] (Sir Terence Etherton MR, Longmore LJ and McCombe LJ).

[29] For a much fuller defence of their status as torts, see Murphy (n 20) *passim*.

> fact that resort to this [legislation] is necessary demonstrates the empty character, so far as the law is concerned, of the statement that the court's decision was wrong.[30]

For present purposes, then, even if the courts are mistaken in thinking that it is in fact meaningful to talk of C's being able to sue D in tort for having wrongfully caused C some or other loss, the fact that they say this must still be taken seriously *qua* statement of the law. *Declarant ergo est.* The inferior courts cannot simply disregard as misguided or inapt the interchangeable use of the terms 'loss' and 'damage' *if* this is something that our most senior judges have done. And indeed they have so used the terms. In *OBG Ltd v Allan*,[31] for example, Lord Hoffmann flitted freely between talk of loss and damage in elaborating the ingredients of the unlawful means tort. He insisted that, 'one is not liable for loss which is neither a desired end nor a means of attaining it';[32] and then, just a few paragraphs later (when discussing one of the earlier cases), he went on to observe that, '[t]he damage to the abattoir was neither the purpose of the strike nor the means of achieving that purpose'.[33] Equally, even though the tort of injurious falsehood is frequently described as being a tort that makes 'damages recoverable for pecuniary loss',[34] the orthodox nomenclature for this element of the tort is that of 'special damage'.[35]

Because the notion that all (or even most) of the economic torts are rooted in rights infringements is untenable, and because our apex court has treated the terms 'loss' and 'damage' as synonyms in this setting, I would suggest that we need not be unduly troubled by the way that Nolan seeks to draw a distinction between them. Rather, although the difference to which he alludes seems plausible enough (assuming one accepts his definition of loss), I think the fact that the courts (and even on occasion the legislature[36]) are prepared to treat them as synonyms provides us with sufficient licence to do likewise.

But if the foregoing reasons for side-lining Nolan's complaint about 'loss' being an insufficiently determinate concept fail to convince the reader, there is one final basis on which we might feel justified in putting to one side his analysis. It is that, although terminological accuracy in the law is obviously desirable, it is ultimately inessential. To see why, consider a well-known branch of the law of obligations. Many years ago, the courts considered it appropriate to handle a certain class of cases according to the notion of quasi-contract. Over time, the descriptive

30 HLA Hart, *The Concept of Law* (2nd edn, Oxford, OUP, 1994) 141.
31 Above (n 28).
32 ibid, [61].
33 ibid, [64].
34 *Joyce v Sengupta* [1993] 1 WLR 337 (CA), 348 (Sir Donald Nicholls VC).
35 *Ratcliffe v Evans* [1892] 2 QB 524 (CA), 528 (Bowen LJ), 'special damage … is the gist of such an action on the case'.
36 eg, the Limitation Act 1980 speaks both of 'compensation for loss' (Sch 3, para 9) and 'compensation for that damage' (s 10(4)).

inaccuracy of this term gave way to talk of a law of restitution, which itself (because of its own descriptive inaccuracy) gave way, later still, to today's preferred nomenclature: unjust enrichment. It is by no means clear—and no one to my knowledge has ever suggested—that, during the years in which we were still inching towards terminological precision, the relevant cases in this setting were badly decided by virtue of their being decided according to a legal lexicon that lacked descriptive accuracy. My point, then, is that imperfect terminology is neither necessarily nor obviously inimical to, or obstructive of, the doing of justice.

3. The Forms of Recognised Damage in the Economic Torts

None of the economic torts is actionable *per se*. However, while it is clear that C must always show a tangible loss in order to succeed in a claim for damages in this sphere, the scope of what counts as actionable loss in these torts is unclear. Some writers maintain that recovery is only possible in connection with economic setbacks affecting one's trade, business or employment interests. Simon Deakin, for example, insists that the 'purpose of the economic torts is to protect a person in relation to his trade, business or livelihood',[37] and that their 'fundamental rationale … is to maintain the integrity of the competitive process'.[38] In similar vein, Hazel Carty claims that 'the economic torts, as their name suggests, have as their primary function the protection of economic interests'.[39] Yet, as I have shown elsewhere, the idea that these torts comprise a family of actions united by the common purpose of protecting only such economic interests is a relatively recent one that did not emerge—and then only stutteringly—until the second half of the twentieth century.[40]

The aim of this section of the chapter, then, is to expose the wrongheadedness of regarding all these torts as being inextricably tethered to the protection of only economic interests. It does so in two ways: first, by highlighting various authorities that make crystal clear the fact that most of the economic torts can be invoked in connection with damage beyond interests in one's trade, business or livelihood; and, secondly, by showing that two of the remaining three economic torts may also be put to such uses in the light of various obiter and other judicial indications in some of the leading cases without any risk of contradicting what those authorities establish. In fact, as a final prefatory remark, it should be noted that the only economic tort that is strictly confined to the protection of trade and business interests

[37] S Deakin and Z Adams, *Markesinis and Deakin's Tort Law* (8th edn, Oxford, Oxford University Press, 2019) 458.

[38] S Deakin and J Randall, 'Rethinking the Economic Torts' (2009) 72 *Modern Law Review* 519, 520.

[39] H Carty, *An Analysis of the Economic Torts* (2nd edn, Oxford, Oxford University Press, 2010) 1.

[40] Murphy (n 20) ch 1.

is passing off, since in that tort, damage to goodwill—the 'attractive force which brings in custom'[41]—is firmly entrenched as the gist of the tort.[42]

A. Deceit, conspiracy, intimidation and injurious falsehood: non-economic losses formally recognised

In appellate decisions concerning each of the torts of deceit, conspiracy (in both its forms), intimidation and injurious falsehood, it is possible to identify clear judicial support for the claim that C may obtain a remedy in respect of various forms of harm that cannot be cashed out in terms of trade or business interests.

Starting with deceit, it is noteworthy that the claimants in both *Langrdige v Levy*[43] and *Burrows v Rhodes*[44] were able to claim successfully in respect of the personal injuries they suffered.[45] Neither decision can be seen as anomalous. Indeed, they both serve to make tort law more coherent.[46] They do so because they align the tort of deceit with negligence under which branch of the law it is well established that C's reliance on a *careless* misstatement that leads to C's physical injury is actionable.[47] The law in relation to tortious liability for false statements would be odd indeed if it allowed recovery for, say, a broken leg attributable to a negligent misstatement, but ruled out such recovery for the self-same injury caused by an intentionally or recklessly uttered falsehood. One might even argue that the tort of deceit would be indefensibly narrow of scope if it did not protect bodily integrity as well as financial interests given that the former is, all other things being equal, generally considered to be more worthy of protection than the latter.

Physical inconvenience has also been held to be recoverable in deceit. In *Mafo v Adams*,[48] the claimant was duped into quitting his protected tenancy when the defendant's wife tricked him into thinking he had acquired the lease to alternative premises. The claimant and his family packed their belongings into a van, went to the new premises, and waited for several hours in the cold. Eventually, it became clear that the claimant had been deceived. But this was only after suffering the physical inconvenience involved in an attempt to relocate, including a lengthy wait in the cold outside the premises the claimant was hoping to move into. Among

41 *Inland Revenue Commissioners v Muller & Co's Margarine Ltd* [1901] AC 217 (HL), 224.

42 Strictly, goodwill is a proprietary interest. But it is a proprietary interest the infringement of which invariably results in trade and business losses.

43 (1837) 2 M&W 519.

44 [1899] 1 QB 816.

45 In *Langridge*, C suffered a shotgun wound; in *Burrows*, C's injuries were sustained in battle conditions after C had been duped into joining an illegal fighting force in South Africa.

46 Tort law is only, however, broadly coherent. There are significant aspects of the law that may be regarded as outliers from the perspective of coherence: see, J Murphy, 'Contemporary Tort Theory and Tort Law's Evolution' (2019) 32 *Canadian Journal of Law and Jurisprudence* 413.

47 See, eg, *Perrett v Collins* [1998] 2 Lloyd's Rep 255 (CA).

48 [1970] 1 QB 548 (CA).

other heads of loss, the claimant sought compensation for this inconvenience. And finding in his favour Sachs LJ said:

> The plaintiff apparently still has not got a tenancy with an equally sure protection, and for that he is entitled to compensation. In addition, he is entitled upon accepted principles to *compensation for the physical inconvenience* suffered when put in the position to which I have already adverted.[49]

Even recovery for mental distress cannot be ruled out in deceit. Although no such harm was alleged in *Doyle v Olby (Ironmongers) Ltd*, Winn LJ said that, in addition to relatively easily measured pecuniary losses: 'one considers other consequences … [such as] *strain, anxiety and unhappiness* … [since] in some cases such considerations might well be appropriate.'[50] Furthermore, in *Shelley v Paddock*,[51] the Court of Appeal affirmed a first instance decision of Bristow J that included in the damages a sum reflecting the distress the claimant had suffered when she was swindled in connection with the purchase of a house in Spain.[52]

Taken together, these various illustrations of non-economic losses that have been formally recognised by the courts put beyond doubt the fact that there is no good reason to consider deceit an action that is confined solely to the protection of economic interests.

If we shift the spotlight to conspiracy and start, first, with the lawful means version, there is one fairly obvious point that can be made about the forms of recognised damage for the purposes of this tort. This is that damage of a physical kind—whether to person or property—is more or less inconceivable in this setting. People very seldom consent, beyond the realm of contact sports, to others harming them, or damaging their property. Accordingly, it will overwhelmingly be the case that whenever a defendant inflicts such harm or loss upon a claimant, it will have been without the claimant's consent. And with the absence of consent there disappears any prospect of there having been lawful means used. Such unconsented-to contact will typically trigger liability for battery or negligence or (in the case of chattels) a trespass to goods.

But this does not mean that lawful means conspiracy must be seen as limited to the protection of only economic interests. There is, after all, at least one concrete example of the courts allowing this tort to be invoked beyond the sphere of trade or business interests.[53] In *Gulf Oil (GB) Ltd v Page*[54] it was the claimant's reputation

[49] ibid, 554 (emphasis added).
[50] [1969] 2 QB 158 (CA) (emphasis added).
[51] [1980] QB 348 (CA).
[52] In similar vein, see *Burki v Seventy Thirty* [2018] EWHC 2151 and *Kinch v Rosling* [2009] EWHC 286 (QB).
[53] cf Deakin and Randal (n 38), 523.
[54] [1987] Ch 327 (CA).

that was damaged. The defendants had conspired to fly a light aircraft over a racecourse at which the claimant was entertaining a number of customers, and the defendants attached to the aircraft a banner saying: 'Gulf exposed in fundamental breach' (which was perfectly true). Now, because what was said was strictly true, the first instance judge quite correctly ruled out any action based on defamation.[55] But since the conspirators were actuated by malice in their spreading of a damaging truth about the claimant, the latter sought an injunction based on lawful means conspiracy to restrain the defendants from displaying the sign. In the course of his judgment, Parker LJ noted the potential overlap between lawful means conspiracy and libel.[56] He also noted that, in the instant case, the defendants had 'no immediate interest of their own to protect against Gulf and no interest of their own to further as against Gulf' for which reason there was, 'a strong inference that the purpose of the display was simply to inflict upon Gulf the maximum possible damage … [and thus] a strong prima facie case in conspiracy to injure'.[57] Put concisely, he accepted that an injunction for lawful means conspiracy was permissible.

Oddly, only a few years later, a differently constituted Court of Appeal held in *Lonrho Plc v Fayed (No 5)* that *damages* for loss of reputation should never be available under the auspices of lawful means conspiracy.[58] But given the earlier recognition of the potential availability of injunctive relief, the notion that there should be a blanket ban on the recovery of *damages* seems, with respect, rather odd. For one thing, it is ordinarily far easier to get damages than injunctions in tort. Indeed, it is trite to state that injunctions are typically unavailable unless damages would prove an inadequate remedy. Furthermore, as Gary Chan has convincingly argued, 'the ratio in *Lonrho* that non-pecuniary loss of reputation is not recoverable in the tort of conspiracy should be restricted to its particular facts', and '[t]he case should not serve as a precedent to support a general blanket prohibition against loss of reputation claims in the tort of simple conspiracy'.[59] His argument is fortified by reference to the fact that *Lonrho* appears to have been decided *per incuriam*. (There was a gap of only six years between the two cases, yet *Gulf Oil* was not cited once in *Lonrho*.) More importantly, however, the suggestion in *Lonrho* that defamation

[55] There is no report of the first instance decision, but it is clear from Parker LJ's judgment that this is what the first instance judge had held: see ibid, 332.

[56] ibid, 333–334: '[a] plaintiff in an action against the author and publisher of a newspaper article, for example, might well establish a combination … [but given the predominant purpose requirement in the latter] the court would scrutinise with the greatest care any case where a cause of action in conspiracy was joined to a cause of action in defamation'.

[57] ibid, 332–333.

[58] [1994] 1 All ER 188 (CA), 195–196, 202–203 and 210–211 (Dillon, Stuart-Smith and Evans LJJ, respectively).

[59] G Chan, 'Never say "Never" for the Truth Can Hurt: Defamatory but True Statements in the Tort of Simple Conspiracy' (2007) 31 *Melbourne University Law Review* 321, 342.

law would be undermined if damages for loss of reputation were, instead, available under the tort of lawful means conspiracy is unfounded.[60] It is not hard to see why.

The fact that truth would comprise a complete defence in defamation cannot be translated into the proposition that we all possess an absolute freedom to say to someone else things concerning a third party that happen to be true. Modern developments in the law on misuse of private information[61] would clearly have no basis if there were always an immunity granted to those who disclose a confidential fact about someone else. Furthermore, the House of Lords in *Spring v Guardian Assurance plc*[62] made clear that, just because the facts of a case might support a defence under the law of defamation does not preclude the possibility of liability under a different tort. In that case, the claimant successfully sued a former employer in connection with a negligently prepared reference written by the latter about the former. The reference stated falsely that the claimant was dishonest and not to be trusted. According to their Lordships, the fact that a defence of qualified privilege would have been available had the action been framed in defamation made no difference to the question of negligence liability. One would think, then, reasoning analogically, that defamation law is similarly unable to impose limits on the boundaries of lawful means conspiracy liability.

Unlawful means conspiracy overlaps, to a very large extent, with the ground covered by the joint tortfeasance principle. But the former cannot, *pace* Stevens, simply be folded within the latter and declared 'otiose' *qua* separate tort.[63] This is because the kinds of 'unlawful means' recognised in this tort need not—as they must, by definition, in a case of joint tortfeasance—involve a tortious wrong. As was held in *Revenue and Customs Commissioners v Total Network SL*[64] and *JSC BTA Bank v Khrapunov*,[65] respectively, both crimes and acts amounting to a contempt of court will suffice for the purposes of unlawful means conspiracy. This is an important point to note because, if unlawful means conspiracy were indistinguishable from joint tortfeasance, it would be obvious that 'the tort' could be invoked in connection with damage to interests beyond those in trade or business. Unlawful means conspiracy would be no more than shorthand for D1's participation in a tort—of whatever stripe—committed by D2.

But if unlawful means conspiracy enjoys an independent juridical existence, the question concerning the interests it protects does become salient. The enquiry must therefore be into whether there is any authority for the claim that it can be invoked in relation to losses other than those of a trade and business variety. And the

60 This argument appears in the judgment of Dillon LJ in the *Lonrho* case. He said: 'it would, in my judgment, be lamentable if a plaintiff could recover damages against defendants who had combined to tell the truth about the plaintiff and so had destroyed his unwarranted reputation': ibid, 195.

61 See, eg, *Campbell v MGN Ltd* [2004] 2 AC 457 (HL).

62 [1995] 2 AC 296 (HL).

63 Stevens (n 24) 249.

64 Above (n 28).

65 [2020] AC 717 (SC).

answer, in part, can be gleaned from the same two cases just mentioned despite the fact that in both *Total Network* and *Khrapunov* the claimant sought damages in respect of financial losses. In neither case could it plausibly be said that what was suffered was a *trade or business loss*. In the former, as Lord Hope rightly pointed out in connection with the VAT fraud perpetrated in that case, '[t]he [Revenue] commissioners are not authorised by the statute to carry on a business for profit ... [t]hey have no commercial interests that need to be protected by the tort of conspiracy'.[66] Equally, in *Khrapunov*, the defendant had conspired to conceal and dissipate assets allegedly stolen from the claimant bank by his co-conspirator. There was no damage to the claimant's trade interests involved. But the fact that there were no trade losses in these cases should not be thought troubling, for as Lord Walker made plain in his leading speech in *Total Network*. '[t]he gist ... [of the tort] is damage intentionally inflicted by persons who combine for that purpose ... and *the claimant need not be a trader who is injured in his trade*'.[67]

Such an open-ended description of the damage element in a landmark case constitutes, I think, a firm indication at the highest judicial level that the tort is not, as Carty claims, confined to the role of regulating 'excessive rivalrous behaviour ... in the context of trade competition'.[68] The fraudsters in *Total Network* were not in competition with the Inland Revenue; and the son-in-law of the bank's former chairman in *Khrapunov* was not in competition with the claimant bank.

The fact that there is no need to show that C was 'injured in his trade' paves the way for the suggestion, rooted in logic, that unlawful means conspiracy ought to protect at least as many interests as lawful means conspiracy. To put it another way, if conspirators who use *lawful* means to inflict harm of type *X* upon C can be held liable in tort, they must surely also be liable where they use *unlawful* means to do just the same.

If we turn next to intimidation, it is clear that this tort, too, may be invoked in relation to non-economic losses. If we leave to one side the doubts that some commentators have expressed about the continued vitality of this cause of action (which I have rebuffed elsewhere[69]), the key point to note is that there exists Court of Appeal authority for the proposition that the protection of a claimant's health interests falls within the ambit of this tort. In *Godwin v Uzoigwe*,[70] a woman was cruelly used as a domestic drudge. When she sued in tort, highlighting the way that she had been coerced into performing household tasks as though a modern-day slave, her action for intimidation succeeded. And that she should have succeeded is unsurprising once it is realised that the gist of this tort is coercion of the claimant's

66 Above (n 64) [28].

67 ibid, [100] (emphasis added).

68 H Carty, 'The Tort of Conspiracy as a Can of Worms' in SGA Pitel et al (eds), *Tort Law: Challenging Orthodoxy* (Oxford, Hart Publishing, 2013) 391.

69 See J Murphy, 'Understanding Intimidation' (2014) 77 *Modern Law Review* 33.

70 [1993] *Fam Law* 65 (CA).

will. If someone is coerced into acting in a manner that is financially or physically to their detriment, the damages awarded to compensate for such forms of harm are merely reparative of a consequential loss. The critical wrong in a case of this kind consists of an injury to the claimant's decision-making freedom. Although the authorities do not always spell this out in the clearest possible terms, they have nonetheless consistently placed coercion centre-stage. In *Morgan v Fry*, for example, Lord Denning was insistent that only 'the person damnified *by the compliance* [with the threat] can sue for intimidation';[71] while in another case, Stuart-Smith J said this:

> If a threat is little more than idle abuse and is not to be taken seriously, then it would not be sufficient to found an action for intimidation. Indeed, the tort is not complete *unless the person threatened succumbs to the threat.*[72]

What is clearly central to both these judicial pronouncements is the idea that coercion of the threatened person's will is vital to the commission of the tort.[73] Furthermore, several even clearer statements to this effect were made in the leading case of *Rookes v Barnard* where Lord Devlin said: '*[t]he essence of the offence is coercion* … [and] the plaintiff who suffers by the aversion of the act threatened can legally claim that he is illegally injured.'[74] A little later in his speech he added: '[a]ll that matters to the plaintiff is that, metaphorically speaking, a club has been used.'[75] No less explicitly, Lord Evershed also remarked in that case that, 'the threat, properly understood … *had a real and substantial coercive force*' and '[a]s such … [the] threat itself constituted a cause of action'.[76] A slightly more elaborate iteration of such thinking was also supplied by Lord Denning in *JT Stratford & Son v Lindley*. He said: 'it is essential to the cause of action that the person threatened should comply with the demand … *that the threat should be a coercive threat* … It must be intended to coerce a person into doing something that he is unwilling to do.'[77]

Since the cases consistently stress that it is *coercion achieved via an unlawful threat* that animates this tort, it seems clear that the wrong in play is C's being forced into doing something he or she would not otherwise have chosen to do. But for all that it is easy to identify C's decision-making freedom as the gist of the tort, there is a crucial supplementary point that must also be made. This is that the relevant coercion must always be accompanied by—indeed, cause—a loss. And as

[71] [1968] 2 QB 710 (CA) (emphasis added).
[72] *Newsgroup Newspapers Ltd v SOGAT '82*, [1987] ICR 181 (CA), 204.
[73] However, coercion is not the be all and end all. For the coercion to count, it must have been brought about by an unlawful threat.
[74] [1964] AC 1129 (HL), 1207 (emphasis added).
[75] ibid, 1209.
[76] ibid, 1189 (emphasis added).
[77] [1965] AC 269 (HL), 283 (emphasis added).

regards the types of recognised harm, we can fortify what was held to be actionable in *Godwin* by reference to the fact that, in *Rookes v Barnard*, when considering the scope of two-party intimidation, Lord Evershed specifically opined that actionable damage could include injury to health (as well as other losses). He said:

> [The claimant] might well, as a consequence of B's intimidation, suffer in many ways including health, and, if he did, then, as I conceive, his cause of action against B would be founded on the tort of intimidation and would not be confined, by reference to the damage suffered, to interference with his business.[78]

Turning finally in this section of the chapter to injurious falsehood, it is again true to state that the courts have regarded as actionable, forms of damage that go beyond mere trade and business losses. Of course, in practice, it will normally be pecuniary loss that prompts C to bring an action for this tort because in the typical case, D's untruthful remarks about C (or her business) will lead to C's losing custom. In *Ratcliffe v Evans*, however, Bowen LJ said that the requirement of 'special damage' should be understood in terms of any 'particular damage (beyond the general damage), which results from the particular circumstances of the case'; and that, here, particular damage simply 'denotes the actual and temporal loss which has, in fact, occurred'.[79] Equally, in his description of what needs to be shown by way of special damage, Lord Robertson laid down no requirement more specific than that 'the plaintiff has in fact suffered loss which can be and is specified'.[80]

Against such a background, the Court of Appeal in *Joyce v Sengupta* intimated strongly that damages for injured feelings might be possible in this tort, albeit under the banner of aggravated damages. Sir Donald Nicholls V-C asked rhetorically: '[c]an it be right that the law is unable to give him [the claimant] any recompense for this suffering against the person whose malice caused it?'.[81] Then, just a few years later, in *Khodaparast v Shad*,[82] openly taking a lead from what had been said in *Joyce*, the Court of Appeal went one step further and actually awarded aggravated damages for the claimant's injured feelings after she had been falsely portrayed by the defendant as having appeared in a pornographic magazine advertising telephone sex services. More notable still, however, is the fact that there was a ringing endorsement in *Khodaparast* of something that had been floated only speculatively in *Joyce*, namely, that it may be possible to recover damages for injured feelings without labelling such an award 'aggravated damages'. The Vice-Chancellor in *Joyce* expressly countenanced the idea that such injured feelings (or at least the distress that accompanies them) might constitute special damage pure

[78] *Rookes* (n 74) 1194.
[79] Above (n 32) 528.
[80] *Royal Baking Powder Co v Wright, Crossley & Co* (1901) 18 RPC 95 (HL).
[81] [1993] 1 WLR 337 (CA), 348.
[82] [2000] 1 WLR 618 (CA).

and simple. He said: '[i]f distress can inflame the damages recoverable for pecuniary loss, the difference between awarding aggravated damages for that reason and awarding damages for distress *as a separate head is a difference in words only*'.[83] And the evident willingness to permit this in *Khodaparast* was subsequently approved by the Court of Appeal in *George v Cannell* where Saini J (who delivered the only full judgment in the case) said, in relation to the compensability of distress: 'I would endorse what was said at [42] of *Khodaparast* and I would decline to rule out, as a matter of law, an award in a case where the publication caused no actual pecuniary loss in the event'.[84]

In my view, the combined effect of these cases is to provide a very strong indication that the courts would regard the requirement of special damage satisfied, even if C could only show that she suffered severe distress or injury to feelings.[85] The Vice-Chancellor's hypothetical in *Joyce* certainly provides a plausible enough example of how someone might suffer such distress, as do (with a slight tweak) the facts of *Khodaparast*. For it is easy to imagine that, had the claimant in *Khodaparast* not lost her job as a teacher at an Iranian religious school, she would still have suffered very considerable distress at the thought that others in her community believed what had been said of her. And the general claim about recovery for distress can be buttressed further, I think, by two observations. The first is that this tort—central to which is the presence of malice on D's part—is one that (like deceit and intimidation) typically turns of egregious wrongdoing. The second is that such wrongdoing tends to licence awards of aggravated damages which are closely linked to the feelings of distress and humiliation that typically accompany one being treated contemptuously.[86]

Finally, it is noteworthy that, although speaking obiter in *Secretary of State for Health v Servier Laboratories Ltd*, Lord Hamblen also specifically endorsed the idea that injurious falsehood might be invoked by a patient who suffers the amputation of the wrong leg because the defendant has maliciously misinformed the relevant surgeon that it was the patient's left (not right) leg that was scheduled for the operation. And he thought the same would be true of property damage if D were to misinform airport authorities that certain luggage (belonging to C) had been left unattended such that it be taken away and destroyed.[87]

[83] Above (n 34) 348–349 (emphasis added). This particular part of the Vice-Chancellor's dictum was expressly considered in the later case, with not a single word of disagreement being registered: see *Khodaparast* (n 82) [37].

[84] [2022] EWCA Civ 1067 (CA), [78].

[85] For doubts about distress being compensable, see NJ McBride and R Bagshaw, *Tort Law* 6th edn (London, Pearson, 2018) 672. And for a full rebuttal of those doubts, see J Murphy, 'The Vitality of Injurious Falsehood' (2021) 137 *Law Quarterly Review* 658, 672–4.

[86] See, J Murphy, 'The Nature and Domain of Aggravated Damages' [2010] *Cambridge Law Journal* 353.

[87] [2021] 3 WLR 370 (SC), 85. In my view, it is questionable whether these examples could be treated as injurious falsehood cases *because of the absence of the requisite disparaging claim* about the claimant or her business. But Lord Hamblen's words at least indicate a judicial willingness to entertain the idea that injurious falsehood is a tort capable of protecting personal and proprietary interests.

B. Causing loss by unlawful means and inducing breach of contract: are non-economic losses recoverable?

In *OBG*, there was a strong suggestion in Lord Hoffmann's leading speech that the tort of causing loss by unlawful means is available solely in connection with trade and business losses. He said the tort is 'designed *only* to enforce basic standards of civilised behaviour in economic competition between traders or between or employers and labour',[88] adding that there must be 'a wrongful interference with the actions of a third party *in which the claimant has an economic interest*'.[89] His saying this was taken by Swift J, in *Chalfont St Peter Parish Council v Holy Cross Sisters Trustees Inc*,[90] to mean that 'the tort is aimed only at the protection of forms of economic interest'.[91] But, with respect, that is not quite what Lord Hoffmann said; and nor has the need to injure someone in their trade ever been conclusively established as the remedial limit of this tort in any of the landmark cases. True, there was talk, in the oft-cited case of *Keeble v Hickeringill*, of the liability of 'he that hinders another in his trade or livelihood'.[92] But that was a nuisance case, and not one, properly understood,[93] concerning the economic torts. Indeed, it was precisely because it was a nuisance case that it was specifically held in the seminal case of *Allen v Flood* to have 'no bearing on the present case'.[94] Furthermore, Lord Herschell was openly dismissive in *Allen* of the idea that there was anything special about trade or employment interests. Discussing the ancient case of *Garret v Taylor*,[95] he said:

> I am unable to see ... that the case rests on any principle specially relating to trade. If the plaintiff had not been a tradesman, but the owner of a house, and the same menaces had been uttered to those who came from time to time to visit him, I cannot but think that he would equally have had a cause of action.[96]

Lord Davey found it 'strange that if there be any such right of action [purely] *for interference with trade*, there is not to be found some clear authority in the law books in its favour'.[97] And Woolf LJ made essentially the same point in *Lonrho plc*

[88] Above (n 28) [56] (emphasis added).
[89] ibid, [47] (emphasis added).
[90] [2019] EWHC 1128 (QB).
[91] ibid, [99].
[92] (1706) 11 East 574, 575–576 (Holt CJ).
[93] By contrast, *Keeble* was regarded as an economic torts case in *Allen v Flood* by a number of the puisne judges called upon to assist the House of Lords in deciding that case. For example, Cave J considered it 'a very important case', while Wills J regarded it 'an authority of great importance' and Lawrence J thought the principles on which this case [ie, *Allen*] is to be decided are to be found in the dicta of Holt CJ in the case of *Keeble v Hickeringill*: see *Allen v Flood* [1898] AC 1 (HL), 30, 47 and 58 respectively.
[94] ibid, 133 (Lord Herschell).
[95] (1620) Cro Jac 367.
[96] Above (n 94) 136.
[97] ibid, 174.

v Fayed (No 1) when he said, 'the tort relied upon by Lonrho of unlawful interference is still, in my view, of uncertain ambit'.[98] So the key question is whether this changed in *OBG*.

Notably, all that Lord Hoffmann said there was that there must be some kind of economic relation between C and the intermediary. Elsewhere in his speech, despite the part of his dictum cited already, he also specifically suggested the possibility of the tort being used to protect non-economic interests. He declared: '[i]t is sufficient that the intended consequence of the wrongful act is *damage in any form*.'[99] In similar vein, he also opined in *Hunter v Canary Wharf Ltd* that there is 'no reason why a tort of intention should be subject to the rule which excludes compensation for mere distress, inconvenience or discomfort in actions based on negligence'.[100] Thus, it becomes apparent that, although the need for an economic relationship between the third party and the claimant is now a formal requirement of the unlawful means tort,[101] it is not one that necessarily limits the range of compensable losses to those of the trade or business variety.[102] A couple of examples can be invoked to illustrate the point.

In an adaptation of a scenario suggested by Tony Weir,[103] we may imagine a kidney patient, C, who had a contract for the provision of regular dialysis by a private doctor. If C's sworn enemy, D, kidnaps the doctor (and thus prevents C from receiving treatment), C ought in principle to be able to recover damages when alternative treatment cannot be arranged at short notice and her health deteriorates as a result. The remedy in such a case would be for C's personal injury, rather than for a trade or business loss. And other loss, too, such as distress might be recoverable.

Imagine another defendant, the claimant's evil-minded neighbour, who threatens to injure a comedian engaged to provide entertainment at the latter's forthcoming birthday if that comedian should dare to show up on the agreed occasion. In such a case, the fact that the comedian succumbs to the defendant's threat and refuses to do his act would doubtless cause C disappointment which would be compensable if the claimant were to sue the comedian for breach of contract.[104] To my mind, there is no obvious reason why the same loss caused by D's deliberate unlawful interference with the contract should not be recoverable on the basis of the unlawful means tort.

[98] [1992] 1 AC 448 (HL) 493.

[99] *OBG* (n 28) [8].

[100] [1997] AC 655 (HL), 707.

[101] *Secretary of State for Health v Servier Laboratories Ltd* [2021] 3 WLR 370 (SC).

[102] Whether it will succeed in restricting claims to those in respect of trade or business losses is questionable: see J Murphy, 'Floodgates Fears and the Unlawful Means Tort' [2021] *Cambridge Law Journal* 436.

[103] T Weir, 'Chaos or Cosmos: *Rookes*, *Stratford* and the Economic Torts' [1964] *Cambridge Law Journal* 225, 228. (Weir envisaged the kidnap of a prima donna who would thus be prevented from performing for C.)

[104] See, eg, *Jackson v Horizon Holidays* [1975] 1 WLR 1468 (CA).

To these observations we might also add the fact that a current member of the Supreme Court is on record as saying, albeit extra-judicially, that, for the purposes of this tort, 'it is difficult to see why those who suffer loss as a result of a defendant's deliberate wrongdoing aimed at them should not be able to recover damages merely because they suffer no damage to any trade, profession or business'.[105] And although Lord Hoffmann's insistence upon an economic link—the so-called 'dealing requirement'—was recently confirmed by the Supreme Court, the panel in that later case conspicuously failed to rule out the prospect of this tort being used to recover for losses of a non-pecuniary kind.[106]

The second tort in respect of which there has not yet been *express* judicial recognition of the recoverability of damages for types of loss other than setbacks to trade and business interests is the tort of inducing breach of contract. But even here there are two good reasons to think that the tort's remedial capacity is not so confined.

To start with, the decision in *Lumley v Gye*[107] was in large measure anchored to the ancient action *per quod servitium amisit* which featured in the case of *Winsmore v Greenbank*.[108] There, the defendant persuaded the claimant's wife to remain absent from her husband in breach of the marriage contract. In *Lumley*, Wightman J went so far as to describe *Winsmore* as 'an exceedingly strong authority in the plaintiff's favour in the present case'.[109] This is important because, in *Winsmore*, the husband recovered damages for loss of what Wightman J called the 'comfort and society of the wife and her aid in his domestic affairs'.[110] It was, in other words, a case concerning a non-economic loss,[111] and yet it was treated as foundational in the formulation of the tort of inducing breach of contract. Furthermore, Erle J was equally content in *Lumley* to formulate a new tort that rested in significant part on a case in which it had been held that 'procuring a violation of the plaintiff's *right*

[105] P Sales and D Stilitz, 'Intentional Infliction of Harm by Unlawful Means' (1999) 115 *Law Quarterly Review* 411, 431. One caveat here is *Mbasogo v Logo Ltd* [2007] QB 846 (CA) where Sir Anthony Clarke MR held that mental distress short of a recognized psychiatric condition was not recoverable. His reason (ibid, [99]) was that the Protection form Harassment Act 1997 requires acts inflicting such harm on at least two occasions, thus: '[since] Parliament has chosen not to provide for compensation arising out of a single action … it seems to us impossible for the court to recognise such a right by development of the common law'.

[106] With respect to the tort's being confined to the protection of purely economic interests, Lord Hamblen noted that '[h]istorically the tort has been so confined' but then went on to note that, '[i]t may be may be that in an appropriate case a challenge is made to the established position: *Servier* (n 87) [85].

[107] (1853) 2 El & Bl 216.

[108] (1745) Willes 577.

[109] Above (n 107) 238.

[110] ibid.

[111] That *Winsmore* was genuinely a case about a form of non-economic loss rather than, say, the loss of the economically valuable services that would have been provided by wives at the time comes through loud and clear in another case in which Lord Kenyon stated of the seduced wife type of case that, 'this kind of action materially differs from that for harbouring an apprentice, the ground of the action in that case being the loss of the apprentice's service': *Philp v Squire* (1791) Peake 114, 115. See also *Berthon v Cartwright* (1796) 2 Esp 480 in which the very brief report of the case records that the action in that case was purely for 'seducing the plaintiff's wife, detaining her, and thereby depriving him of her society': ibid, 480.

under the marriage contract is … an actionable wrong'.[112] Nothing said by these two judges suggested that the *Lumley* tort was one that applied only in connection with commercial contracts.

And there is nothing else that was said by the panel in *Lumley* that suggests that the tort that was minted in that case is one that protects only trade and business interests. Rather, everything that was said about inducing breach of contract was said about contracts generally. This is important. For, although it is true that many, if not most, contracts have a commercial purpose (such that an induced breach will ordinarily result in economic loss), this is not universally true. Other types of contract—the breach of which will ordinarily result in other types of loss—are perfectly possible, as the court in *Lumley* specifically acknowledged. Thus, if we tweak the facts of my earlier examples such that the private doctor and the comedian were merely suborned into breaking their contracts, it becomes easy to see how inducing breach of contract might protect interests other than those in trade or business. In those examples, permitting recovery in respect of personal injury and disappointment would seem eminently sensible.

We could even add a third hypothetical involving, say, a contract between a home owner and a company that specialises in the removal of fast-growing Japanese knotweed (the roots of which can easily damage houses and garden walls). If any such company that had been contracted to remove the knotweed were induced to break the contract, and no alternative company were available in the short term, it is easy to see how an action might lie in respect of the resulting property damage.

Perhaps even mental distress could result from an induced breach of contract. If we imagine that, at the eleventh hour, D persuaded the bodyguard of a much-maligned politician, C, to break a personal protection contract,[113] it seems perfectly possible that the politician in question might suffer distress if required to fulfil the terms of a particular engagement without such protection. Of course, whether the courts will, in time, come to recognise the availability of this tort in relation to the range of interests just sketched remains to be seen. But as with causing loss by unlawful means, there is nothing in the relevant case law that obstructs its being used in these ways; and I am by no means alone in thinking that it might well be available.[114]

Naturally, it might be said by way of rejoinder that, for all that there is nothing in the relevant case law on inducing breach of contract or the unlawful means tort that specifically rules out recovery for non-economic losses, nor is there anything there that specifically rules them in. But by way of surrejoinder it can at least be said that the arguments I have advanced in favour of their inclusion are principled

[112] ibid, 232 (emphasis added).
[113] I am grateful to Roderick Bagshaw for inspiring some of these examples.
[114] See PS Davies, *Accessory Liability* (Oxford, Hart Publishing, 2015) 143; McBride and Bagshaw (n 85) 669.

and pay close attention to what the courts have actually stated. The same cannot be said of Carty's much narrower claim that the economic torts 'are all weapons in the policy to prevent "excessive" competition or endeavour. They are all about protecting your competitive edge.'[115] The notion that excessive competition was key in *Total Network* and *Khrapunov* is clearly misplaced, as we noted earlier. But so, too, is it misguided to think in terms of injurious falsehood as a tort confined to a form of wrongdoing that is perpetrated only by one business rival against another. The various cases in which the defendant was a newspaper which printed something disparaging about C show as much.[116] Nor can the brow-beaten domestic drudge, or the ailing patient whose private physician is kidnapped, or the disappointed party giver whose comedic entertainer is persuaded not to show up and perform, be seen, respectively, as the business rivals of an oppressive 'employer', a kidnapper or an evil-minded neighbour who simply wants to cause upset.

In short, there are at least some judicial indications in the relevant case law on causing loss by unlawful means and inducing breach of contract to suggest that these torts may legitimately be used in the ways that I have proposed they might be used. At the same time, though, there is a great deal to be gleaned from virtually all corners of the settled law concerning the economic torts that may be marshalled against the narrow role that Carty (and others) would have them perform.

4. Damage's Status: Element or Condition of Actionability?

A claimant must show that she suffered loss when relying on *any* of the economic torts. But it does not follow from this that such loss can safely be regarded an *element* of each of those torts. In most of these torts, as we shall see presently, it does function as such. But in two of them—intimidation and deceit—it appears to function as a mere condition of actionability. The distinction between the two concepts is not merely a semantic one.[117] It has a number of important practical ramifications that are readily exposed if we consider the issue of what counts as unlawful means within the tort of causing loss by unlawful means.

In *OBG*, Lord Hoffmann stated that the term 'unlawful means' should ordinarily be understood in the context of this tort as some or other actionable civil wrong committed against the intermediary, or at least a potentially actionable civil wrong against that intermediary.[118] In relation to those wrongs that are merely potentially actionable, his Lordship said this: '[i]n the case of intimidation ... the threat will

[115] Carty (n 39) 311.

[116] *Ratcliffe v Evans* [1892] 2 QB 525 (CA); *Kaye v Roberston* [1991] FSR 62 (CA).

[117] It could be argued that talk of a condition of actionability is vacuous, that if C has to prove a certain form of damage that makes it an element of the tort. But I have comprehensively rebuffed this contention elsewhere: see Murphy (n 6).

[118] Above (n 28) [49] (Lord Hoffmann).

usually give rise to no cause of action by the third party because he will have suffered no loss.'[119] But he quickly added that any such intimidation 'is nevertheless unlawful means'.[120] This is vital.

A claim in tort will fail if all the elements of that tort are not present. A missing element means that no recognised legal wrong can be said to have occurred. It must follow, then, that for intimidation to count as unlawful means, all the essential elements of the tort must be present *even though* the person at whom the relevant threat is directed suffers no loss. But if all the elements of the tort are present, and loss is merely something additional that must be shown by the claimant in order to secure an award of damages, it makes no sense to describe the need to show loss as an element, too. Rather, the need to show loss must be seen as some other kind of juridical entity. And in my view, the notion of a 'condition of actionability' best captures what this is.

The status of the damage requirement in deceit would appear to be identical. In one case which clearly chimes with the concept of a condition of actionability, Stephenson LJ said that: '[i]n deceit … the false representation … has to cause damage to be actionable, but no damage to the plaintiff is necessary for the tort to be committed.'[121] A very similar remark was made by Lord Toulson in *Zurich Insurance Co plc v Hayward*[122] in a speech with which Lords Reed, Lord Neuberger and Baroness Hale agreed. He said:

> To establish the tort of deceit… [t]he elements essential for liability can be broken down under three headings: (a) the making of a materially false representation (the defendant's conduct element); (b) the defendant's accompanying state of mind (the fault element); and (c) the impact on the representee (the causation element). Where liability is established, it remains for the claimant to establish (d) the amount of any resulting loss.[123]

The fact that damage is a mere condition of actionability in deceit—just as it is in a case of two-party intimidation—means that A's deceiving an intermediary, B, into acting in a way that harms C may perfectly well ground an action by C for causing loss by unlawful means, even though B will have no prospect of obtaining damages. And this is essentially what happened in *National Phonographic Co Ltd v Edison-Bell Consolidated Phonograph Co Ltd*.[124] There, the defendant, who was a retailer on a suspended dealers list, fraudulently induced a third party to sell it phonographs that had been made by the claimants and which had been bought

[119] ibid.
[120] ibid.
[121] *Diamond v Bank of London and Montreal* [1979] QB 333 (CA), 349.
[122] [2017] AC 142 (SC).
[123] ibid, [58].
[124] [1908] 1 Ch 335 (CA).

wholesale by that third party. The wholesale purchase had taken place on the condition that the third party would not later sell those phonographs to companies or persons on the suspended retailers list. Yet by virtue of the fraud, the third party was duped into making such a sale to the defendant. As things turned out, the third party did not itself suffer any loss as a result of this sale. It was therefore unable to bring an action in deceit against the defendant. But, as Lord Hoffmann explained when he discussed the case in *OBG*, this was no basis for precluding an action by the claimants. He said:

> [where] the defendant intentionally caused loss to the plaintiff by *fraudulently inducing* a third party to act to the plaintiff's detriment. The *fraud was unlawful means* because it would have been actionable if the third party had suffered any loss, even though in the event it was the plaintiff who suffered.[125]

In each of the other economic torts, it is submitted that damage constitutes an essential element. Passing off is a clear-cut case. On House of Lords authority, damage to goodwill forms one of the three fundamental elements of the tort, typically referred to as the classical trinity.[126] The status of the damage requirement in the tort of inducing breach of contract is equally clear. Russell LJ stated emphatically in *Sefton v Tophams Ltd* that, '[d]amage is an essential ingredient of the tort'.[127] In *OBG*, Lord Nicholls described harm as being the gist of causing loss by unlawful means;[128] and the very name of the tort (if we put to one side Nolan's objections) also tends to confirm the correctness of his Lordship's ascribing such centrality to the loss component. In injurious falsehood, damage is again central to the commission of the wrong. For, as noted earlier, Bowen LJ said in *Ratcliffe v Evans* that 'special damage ... is the gist of such an action on the case'.[129]

In the conspiracy torts, too, damage ought to be considered an essential element rather than a condition of actionability. In *Kuwait Oil Tanker Co SAK v Al-Bader (No 3)*, Nourse LJ put beyond doubt the centrality of loss in the lawful means variety. He did this (admittedly by implication) when he pointed out that it is 'sometimes described simply as *conspiracy to injure*' and also when he emphasised the tort's principal component, namely, 'the predominant purpose of the defendant ... [being] to injure the claimant'.[130] In similar vein, Lord Diplock was adamant in *Lonrho Ltd v Shell Petroleum Co Ltd (No 2)* that, '[t]he gist of the cause of action is damage to the plaintiff'.[131] By similar reckoning, we ought, I think,

[125] Above (n 28) [49] (emphases added).
[126] *Conzorsio del Prosciutto di Param v Marks & Spencer Plc* [1991] RPC 351 (CA), 358.
[127] [1965] Ch 1140 (CA), 1206. In an earlier case, Slesser LJ expressed the point differently saying that 'damage is of the gist of the action': *British Industrial Plastics v Ferguson* [1938] 4 All ER 504 (CA), 511.
[128] Above (n 28) [141].
[129] [1892] 2 QB 524 (CA).
[130] [2000] 2 All ER (Comm) 271 (CA), [107]-[108].
[131] [1982] AC 173 (HL), 188.

to reach a corresponding conclusion about the fact that, in *Lonrho v Fayed*, Lord Bridge was keen to place the infliction of loss front and centre in unlawful means conspiracy when he relabelled this sister tort, 'the tort of conspiracy *to injure* by unlawful means'.[132]

Although a lengthy defence of the claim is not necessary here, it might usefully be added that, at least in the case of the intermediary-use economic torts, loss is not just an essential ingredient but the very basis for the liability that is imposed. Of course, rights theorists take a different view, and would happily side-line these actions as anomalies.[133] But it is submitted that they are wrong to take this view,[134] and that it is a mistake to believe that all of tort law can and must be cashed out in terms of the infringement of *ex ante* private rights.[135] And there is comfort in this, too. For realising the loss-based nature of the intermediary-use economic torts saves us from striving assiduously to show that, somehow, C held a right that D not commit a tort against the relevant intermediary, T. Trying to evince or conceptualise such rights sails rather too close to the unpalatable notion that C holds some kind of legally recognised interest in, or perhaps right over, T.

5. Conclusion

Within the economic torts, C must always show that she has suffered a tangible setback before she can sue. Sometimes this measurable setback is referred to as 'loss'. On other occasions it is referred to as 'damage', or 'special damage', or 'harm', or even 'injury'. With the possible exception of 'special damage', nothing insuperably troubling turns on which label is used. The need to distinguish loss from damage is plausibly of importance if what one is trying to do is unearth a rights-based understanding of the law of negligence; but it is not necessary to make this distinction in the economic torts setting. Indeed, there are only three economic torts that appear clearly to be rights-based: passing off, deceit and two-party intimidation. Most, if not all, of the others appear to be loss-based actions. (Nolan would doubtless prefer the term 'liability rules'.)

Much more important than any conceptual difference there may be between loss and damage is the breadth of meaning that can be attached to those terms in this sphere. It is simply not the case that, *pace* numerous commentators, all the economic torts are tied inexorably to the protection of only trade and business interests. Various authorities, drawn from across the full range of these torts, reveal that

[132] [1982] AC 448 (HL), 464.
[133] See, eg, Stevens (n 24) ch 8.
[134] See J Murphy, 'Anomalies in Tort Law: A Cause for Concern' (2023) 86 *Modern Law Review* 872.
[135] For a series of articles in which, in different ways, I have attempted to dismantle this idea, see Murphy, 'Reductionsim' (n 23); Murphy, 'Heterogeneity' (n 23); and J Goudkamp and J Murphy, 'Tort Statutes and Tort Theories' (2015) 131 *Law Quarterly Review* 133.

many other interests besides are protected. Indeed, when we look closely to what the courts have actually said in the relevant cases, there seems no good reason at all to think that, between them, these torts are incapable of providing damages for personal injury, mental distress, humiliation, property damage and even, perhaps, disappointment.

Furthermore, the idea that they should protect these various interests seems intuitively right given that, all other things being equal, 'in the hierarchy of protected values wealth, especially corporate wealth is simply not as important [as virtually all of the others]'.[136]

[136] T Weir, *Economic Torts* (Oxford, Clarendon Press, 1997) 9–10. For an expression of this hierarchy in terms of the relative weight of different rights, see R Stevens, 'The Conflict of Rights' in A Robertson and HW Tang, *The Goals of Private Law* (Oxford, Hart Publishing, 2009). Note, however, that intuition only takes us so far; and for a guide to the circumstances where intuition may be put to one side, see J Murphy, 'Tort's Hierarchy of Protected Interests' [2022] *Cambridge Law Journal* 356, 377–382.

16
Continuing the Product Liability Illusion

Jodi Gardner and Sarah Green[*]

1. Reflections

I first heard about Jane Stapleton when I was a 17-year-old undergraduate student, being taught Tort and Accident Compensation by the fabulous Kylie Burns at Griffith University. Kylie had a photo of Jane on her desk and would tell me all about her incredible PhD supervisor who was an international expert on three continents and also an unfailingly supportive and amazing person. When I finally met Jane in real life, it was clear that she lived up to the praise (and more!). What sets Jane aside is not that she is an intellectual powerhouse who can rub shoulders with the best and the brightest, but that she always has time to mentor and support young people in academia, particularly women in private law. Many generations of young scholars owe Jane a debt of gratitude for leading the way, and then extending a hand to guide others along the path—Jodi Gardner

I first met Jane when she interviewed me for an undergraduate place at Balliol a very long time ago. Jane asked me about Causation rather than Product Liability on that occasion, although over the next three years she did of course interrogate me about most aspects of private law—As I remember it, I rarely impressed. After I returned to academic life a few years later, it was always lovely to see Jane in the audience when I gave a paper or talk, albeit that I was always slightly nervous about what question was forthcoming. Much of what Jane taught me as a student has stuck with me over the years, and I am very grateful for all those tutorials during which my essays were torn apart and rebuilt pretty much from scratch. Few benefit from such intensive training—Sarah Green

[*] We would like to thank the participants of the workshop where a draft was presented for their feedback and Professor John Murphy for his helpful comments on an earlier draft. We are greatly appreciative of Professor Kylie Burns for her patience waiting for a long overdue paper, as well as her insightful and helpful review. The usual caveat applies.

Jodi Gardner and Sarah Green, *Continuing the Product Liability Illusion* In: *Torts on Three Continents*. Edited by: Kylie Burns, Jodi Gardner, Jonathan Morgan and Sandy Steel, Oxford University Press.
DOI: 10.1093/oso/9780198889748.003.0016

2. Introduction

Slightly over thirty-five years ago, Jane Stapleton analysed the Council Directive on Liability for Defective Products (Council Directive), arguing that the reform made by these instructions was—for all intents and purposes—largely illusory. This article has been widely cited across multiple jurisdictions in the USA, Canada, Europe, Japan, and Australia.[1] In her discussion, Jane questioned whether there were any real amendments made by the Directive.[2] This chapter will reflect on three and a half decades of product liability cases, starting with *A v National Blood Authority*[3] and finishing with the *DePuy* cases,[4] considering Jane's analysis of the initial Council Directive. It will outline how so much of what Jane predicted has come true, and how the Consumer Protection Act 1987 (CPA) (which came into force after Jane's article was published) provides little additional protection for consumers. This is particularly evident in the light of recently emerging technologies, the implications that these have for interpretations of 'product', and the developing notions of the definition of 'defective'. The chapter will conclude with an analysis of the product liability challenges for new and emerging technologies, and what we can (and should) learn from Jane's analysis when addressing these consumer protection challenges.

3. Part 1: Jane's Argument on Product Liability

In 1993, Jane herself described the CPA as 'the most important new statutory liability in the post-war era in the UK'.[5] Interestingly, the article which is central to the current chapter was published in 1986 and was based on the Council Directive, as opposed to the CPA which came into force the year after the article. In the article's detailed, thorough, and important analysis, Jane contended that the requirement for a 'defect' and inclusion of the 'development risk defence' weakened the inherent value of product liability reform. She argued that these provisions undermined any

[1] See, eg, Geraint G Howells and Mark Mildred, 'Is European Products Liability More Protective Than the Restatement (Third) of Torts: Products Liability' (1998) 65 *Tennessee Law Review* 985; Marshall S Shapo, 'Comparing Products Liability: Concepts in European and American Law' (1993) 26 *Cornell International Law Journal* 279; John Murphy, 'Contemporary Tort Theory and Tort Law's Evolution' (2019) 32(2) *Canadian Journal of Law & Jurisprudence* 413; Geraint G Howells and Mark Mildred 'Infected Blood: Defect and Discoverability a First Exposition of the EC Product Liability Directive' (2002) 65(1) *Modern Law Review* 95; David G Owen, 'Products Liability Principles of Justice' (1991) 20(3) *Anglo-American Law Review* 238.

[2] Jane Stapleton, 'Products Liability Reform—Real or Illusory' (1986) 6(3) *Oxford Journal of Legal Studies* 392.

[3] [2001] EWHC 446 (QB).

[4] *Wilkes v DePuy International Ltd* [2016] EWHC 3096 (QB) and *Gee v DePuy International Ltd* [2018] EWHC 1208 (QB).

[5] Jane Stapleton, *Product Liability* (Butterworths, 1994) Preface.

significant difference between the scope of liability under the subsequent Act compared with the existing common law of negligence.

Jane referred heavily to academic work and case law from the United States in her discussion, something that she highlighted was not at all common in this area of the law.[6] The article showed that in the USA, where a development risk defence was also accepted, the scope of product liability was not significantly greater than that of negligence.[7] Drawing on the experiences from the USA, Jane argued that a similar outcome could be expected in the UK under the regime foreshadowed in the Directive.

A. Scope and purpose of product liability

Jane commented that even after years of interest and debate in product liability reform and an EU Directive, it still could not be said with any certainty what was the scope, effect, or justification of the Directive. As will be shown by this chapter, we are still struggling to find clear answers to these questions—nearly forty years later.

In attempting to provide a coherent rationale for the reform, Jane contended that there are two potential bases for product liability reform:

(1) Enterprise liability theory—which does not depend on whether the risk of injury could have been foreseen, but only on the fact that it was a risk created by the defendant.[8]
(2) Deterrence theory—which is more complicated and focuses on appropriate liability rules assuming priority is given to the economic goal of efficiency (for achieving the optimum level aggregate welfare and improving the allocation of resources).[9]

Despite identifying these different bases, Jane convincingly stated that 'both the loss distribution idea of enterprise liability, that is sheeting home losses to the enterprise which was responsible for them, and both limbs of market deterrence theory can be used to support the imposition of stricter liability for injuries attributable to the use of products of those enterprise'.[10] She went on to state that reform should take the form of full strict liability, covering both undiscoverable and unavoidable risks. The benefits would be undermined by allowing a test that involves a cost-benefit analysis and a range of defences, as this would no longer be true strict liability.

[6] ibid, 5.
[7] ibid, 5.
[8] Stapleton (n 2) 394.
[9] ibid, 394–395.
[10] Ibid, 396.

Jane proposed different models of product liability, each with its own benefits and detriments. The first was Full Strict Liability (Approach I).[11] This predicates liability on injury being sustained as a consequence of the use of the product. There are a number of benefits of this approach: it is simple and treats injuries equally, it does not draw complex distinctions between discoverable, undiscoverable, avoidable, and unavoidable risks, and it is most in line with the accepted justifications for product liability reform.[12] Jane emphasised that to take a truly strict liability approach, there is no role for defences or negligence-based concepts of fault or reasonableness.[13]

The second approach considered by Jane is a two-pronged one that she defined as 'Costs Balanced Against Benefits to the Victim (Approach II) or Public (Approach III)'.[14] Unlike Approach I, these other two approaches are complex and have a number of difficulties with their application. It is however the approach that is predominant in the Directive (and therefore subsequent Consumer Protection Act). When focusing on balancing costs against benefits, it is unclear whether the 'benefits' considered are specifically those to the victim, or to the public more generally.[15] The Directive unhelpfully does not provide much guidance on this point, so it is unclear what is balanced against what, creating uncertainty and confusion. Jane argued that the balanced methodology is also likely to introduce a negligence approach to analysis when determining the liability of parties, including considerations of risk assessment and policy evaluations.[16]

The final approach discussed by Jane, Approach IV, was labelled 'Costs balanced against Benefits except where Development Risk Defence Applies'.[17] She highlighted that Approaches II and III apply independently of whether the defendant knew or ought to have known of the alleged risk or defect.[18] Again drawing on American literature, Jane argued that case law highlights the harm that this type of defence causes to product liability cases. She contended that it results in a movement away from a strict liability approach and instead continues, albeit to a lesser extent than Approaches II and III, the negligence style of analysis.[19]

Jane's article was heavily critical of many aspects of the Directive, but particularly cynical about the impact of the developmental risk defence.[20] This cynicism was shared by a number of members of the EU, and many member states did not

[11] ibid, 400.
[12] ibid, 402.
[13] ibid, 400–403.
[14] ibid, 403–413.
[15] ibid, 404.
[16] ibid, 408.
[17] ibid, 413.
[18] ibid, 413.
[19] ibid, 414.
[20] ibid, 416–419.

implement the defence into their national law or limited its application to certain types of products. For example, a 1999 EU Report on the Directive noted that:

> Some national authorities ... believe that the provision laid down in Article 7(e) of the Directive needs to be reviewed in order to remove this exclusion of liability. In their opinion, removing this defence would contribute to the internal market functioning better.[21]

Other States advocated keeping the defence, stating that it helped promote innovation and the development of new products. The Report noted that there did not appear to be any differences in innovation between countries that implemented the defence and those that did not, questioning the justification of retaining the defence.[22]

It is interesting to note that Jane's article was written based on the Directive, as opposed to the CPA. The wording of the defence in the Directive is:

> The producer shall not be liable as a result of this Directive if he proves ... (e) *that the state of scientific and technical knowledge at the time* when he put the product into circulation was not such as to allow the existence of the defect to be discovered.[23]

The UK implemented different wording for the relevant defence in the CPA, which stated:

> In any civil proceedings ... in respect of a defect in a product it shall be a defence for him to show ... (e) *that the state of scientific and technical knowledge* at the relevant time was not such that a producer of products of the same description *as the products in question might be expected to have discovered the defect* if it had existed in his products while they were under his control.[24]

The latter is arguably a wider definition of 'development risk defence' as the second italicised section allows a reasonableness defence, with the actions of the producer being considered in the light of the actions of other producers of the same description of the products in question.[25] Despite the understandable concerns about

[21] Fourth report on the application of Council Directive 85/374/EEC of 25 July 1985 on the approximation of the laws, regulations and administrative provisions of the Member States concerning liability for defective products amended by Directive 1999/34/EC of the European Parliament and of the Council of 10 May 1999 under the heading 'Development risk defence (Article 7(e))'.

[22] ibid.

[23] Council Directive on Liability for Defective Products, Article 7(e) (emphasis added).

[24] Consumer Protection Act 1987, s 4(1)(e) (emphasis added).

[25] For further analysis, see Christopher Newdick, 'The Development Risk Defence of the Consumer Protection Act 1987' (1988) 47(3) *Cambridge Law Journal* 455–476. This article was written after the

the potential differences in the wording between the Directive and the CPA, the European Court of Justice held that the UK legislation was not incompatible with the Directive in the 1997 case, *Commission v United Kingdom*.[26] Whilst the Court recognised that there could appear to be a compatibility issue, it emphasised that there was no evidence that the UK judges were reading the defence in a narrow or strict manner in such a way that would breach the Directive.

Jane concluded her article with significant criticism of the Directive, declaring that:

> in short, the history and content of products liability reform often supports the view that it is at best simply a cynical exercise to deal with issues which attract controversy and thereby to defuse momentum for more radical reform; and at worst it appears to be a mere exercise in 'window dressing' these few flashpoints with remedies which will prove nugatory in practice.[27]

Jane continued her close analysis of both UK and USA laws in this area, culminating in the publication of her book *Product Liability* in 1994. This further supported the arguments from the 1986 article contending that strict liability was to be preferred on a number of levels, including that it is simply a more easily satisfied test for liability.[28]

4. Part 2: Development of Product Liability in the United Kingdom

As Jane highlighted in 1994, the expected tsunami of defective product cases did not occur after the passing of the CPA. In fact, the 'general claims environment seem[ed] to remain unchanged' and there was no notable increase in insurance premiums.[29] Jane therefore commented that the CPA, and other consumer-focused Directives, 'allow governments to appear to be active in fields such as consumer protection and to appear to establish substantially greater protection than before even when this is in fact not the case'.[30]

enactment of the CPA. Despite the remarkable similarities in analysis, this article does not refer to Jane's seminal earlier piece.

[26] [1997] All ER (EC) 481.

[27] Stapleton (n 2) 422.

[28] Stapleton (n 5) 236. It is recognised that Jane's 1994 book provided significant further grounds for criticisms of the CPA, including providing liability for parties who had minimal roles to play in the defective product, not sufficiently harmonising the law, and adding additional costs to litigation which could be made on other legal grounds,

[29] ibid, 357.

[30] ibid, 355.

A case analysis appears to strongly support Jane's views. When looking for UK cases that have considered the CPA, fewer than 250 have been reported in both Westlaw and LexisNexis—equating to fewer than 7 cases per year.[31] Considering the number of consumer products sold in the country every year, this represents a miniscule number of cases. A review of the cases is even more concerning, with only a handful going beyond the High Court. A significant portion do not consider the CPA in any detail and are instead decided on another legal ground, including common law negligence, contract, or another legislative regime, such as the Employer's Liability (Defective Equipment) Act 1969.

The Court of Appeal took a more robust approach to defective products in *Iman Abouzaid v Mothercare (UK) Ltd ('Abouzaid v Mothercare')*.[32] Despite being a Court of Appeal decision, the judgment was surprisingly short and did not involve a detailed consideration of the CPA and its relationship with common law negligence principles. In this case, the claimant was 12-years-old and assisting his mother with an attachment to his younger brother's pram. When attempting to join the straps, one of the metal buckles hit him in the left eye. This resulted in a severe penetrating injury and no useful central vision as a result.[33]

The trial judge found in favour of the claimant, holding that the product was defective under the CPA, stating:

> the manufacturer ought to have had in mind that older children, such as this claimant, or teenagers, would be using it. They would probably be bending down into the line of the elastic with the risk of this piece of metal coming into contact with an eye. There was therefore a reasonable chance that older children or teenagers would use it at eye height or level. In my judgment, this was an obvious risk or danger, and the manufacturers should have appreciated it.[34]

This decision was appealed to the Court of Appeal. Whilst refusing the appeal, Pill LJ was nonetheless quite critical of the trial judge's decision, stating that it 'has a flavour of common law negligence rather than liability under the statute'.[35] The Court of Appeal held that even though the case was 'close to the borderline', the product was defective under the CPA as the design allowed a risk to arise without providing a warning to consumers. It was also held that the development risk defence did not apply, as there was no 'scientific or technical knowledge' related to the risk at hand[36] Whilst providing an interesting case study and showing a movement

[31] A search was conducted looking for UK cases considering and/or applying the CPA on both databases.
[32] [2000] All ER (D) 246.
[33] ibid, [5].
[34] Cited in ibid, [15].
[35] ibid, [16].
[36] ibid, [28]–[30].

towards stricter liability—Pill LJ's discussion in *Abouzaid v Mothercare* on the application of the CPA (including the defence) was nine paragraphs long and did not refer to any other cases.

The (defective product) world however appeared to change when the decision in *A v National Blood Authority*[37] was handed down a year later, in 2001. This was a group action of 114 claimants for damages arising out of their infection with Hepatitis C following transfusions of blood and blood products from 1 March 1988 (when the CPA came into force) and 1 April 1991 (when testing for Hepatitis C became available). Despite being a High Court decision, Burton J's judgment provided a detailed, comparative, thoughtful, and insightful analysis of strict liability for defective consumer products.[38] It was also the first extensive analysis of the CPA and its impact on product liability in the UK. His Honour confirmed the lack of case law in relation to the CPA, stating that he had only been referred to two UK cases, *Abouzaid v Mothercare*[39] and *Richardson v LRC Products Ltd.*[40]

After reviewing the legal obligations, Burton J held that the hepatitis-infected blood was a defective product under the CPA. He came to this conclusion despite the fact that, as outlined above, there was no effective test available to distinguish between blood that had been infected with Hepatitis C and blood that had not until 1 April 1991. This meant that the only way the defendant could avoid the risk of Hepatitis C infection completely was not to provide any blood transfusions at all. This created a difficulty eloquently described by Eisler as,

> a classic 'hard case' in the Dworkinian sense, because it reflected a conflict between two foundational values and how they could be reconciled as a matter of legal and community norms. Namely, the urgent need to ensure blood was supplied for medical use for the welfare of all patients; and the right of individual patients not to be exposed to blood that was contaminated.[41]

The judgment resolved a number of difficult questions regarding interpretation of the CPA. Burton J stated that the following should *not* be factors in determining whether a 'product is less safe that consumers can legitimately expect' under the CPA: (1) the unavoidability of harmful characteristics; (2) the impracticability of measures that would eliminate the harm; and (3) the benefit of the product to society. This moved the relevant considerations away from standard negligence

[37] [2001] EWHC 446 (QB).

[38] For a consumer protection-focused discussion on this case, see Jacob Eisler, 'A. v. National Blood Authority: An Experiment in Radical Consumer-centric Strict Liability for Products' in Jodi Gardner and Iain Ramsay (eds), *Landmark Cases in Consumer Law* (Hart Publishing 2023) 283.

[39] [2000] All ER (D) 246.

[40] [2000] Lloyds (Med) 280.

[41] Jacob Eisler, 'A. v. National Blood Authority: An Experiment in Radical Consumer-centric Strict Liability for Products' in Jodi Gardner and Iain Ramsay (eds), *Landmark Cases in Consumer Law* (Hart Publishing 2023) 287.

principles. He also departed from the approach in the USA discussions by drawing a distinction between 'standard' products (which operate as expected) and non-standard products (which perform in a manner that consumers would not expect). Burton J defined the difference as:

> [a] standard product is one which is and performs as the producer intends. A non-standard product is one which is different, obviously because it is deficient or inferior in terms of safety, from the standard product: and where it is the harmful characteristic or characteristics present in the non-standard product, but not in the standard product, which has or have caused the material injury or damage.[42]

It was held that Hepatitis C infected blood was non-standard, which meant that issues of avoidability could not be considered. Burton J's analysis and strict liability approach excluded traditional law and economics concepts for allocating risk, such as risk-benefit utility analysis, in favour of robust assurances that consumers could expect their products to not harm them.[43]

Burton J also considered whether the development risks defence applied to the infected blood transfusions. He found that the defence did not apply in this situation. This was because the defendants were aware of the risk of Hepatitis C in the blood, even though there was nothing that they could do about avoiding that risk. Burton J drew a distinction between an *unknown* risk (covered by the defence) and an *unavoidable risk* (not covered by the defence).[44] In doing so, he drew heavily on the discussion in *Commission v United Kingdom*[45] and academic literature, including Jane's work[46] (albeit disagreeing with her analysis). In this case, the defendants admitted that they knew there was a risk of infected blood (and chose not to inform recipients about the risk unless specifically asked) and therefore they were not covered by the defence.[47]

Despite being a victory for consumers, Burton J's decision was widely criticised by academic commentators, including Jane.[48] Whilst she did not agree that the decision was 'necessarily wrong',[49] she commented that it appeared to be 'determined to give the Directive "work to do" in the United Kingdom' by adopting the construction of the statute urged by the claimants.[50] Jane noted in particular that comparing a non-standard product to a standard one seems to have led the judge

42 *A v National Blood Authority* (n 3) [36].
43 See, eg, discussion at ibid, [36].
44 ibid, [50].
45 [1997] All ER (EC) 481.
46 Stapleton (n 5) and Jane Stapleton, 'Restatement (Third) of Torts: Products Liability, an Anglo-Australian Perspective' [2000] 39 *Washburn Law Journal* 363–403.
47 *A v National Blood Authority* (n 3) [55].
48 Jane Stapleton, 'Bugs in Anglo-American Products Liability' (2002) 53 *South Carolina Law Review* 1230.
49 ibid, 1252.
50 ibid, 1249–1259.

inexorably to a finding of a defect, and that the judge ignored public-policy considerations of the importance of blood supply.[51]

Burton J's analysis and decision in *A v National Blood Authority* therefore posed a quandary for Jane's scholarship. Her early work had raised concerns that the CPA would not provide consumers with additional protection as it was drafted in a manner to allow concepts of fault-based negligence to be considered. The first case to undertake a truly detailed analysis however proved Jane's fears had been misplaced, taking a consumer-centric and strict liability approach to defective products. The outcome about which Jane expressed concern did not arise, but she strongly disagreed with the legal analysis that was used to reach the outcome. As outlined in the next section, the consumer victory (and the analysis that went along with it) was short-lived.

5. Part 3: A Movement Away from Strict Liability: Jane's Analysis Realised

Despite the significant publicity about and academic criticisms of *A v National Blood Authority*, the case was not widely cited and later cases considering the CPA either ignored it or did not embrace its fully consumer-focused approach.[52] This area of law therefore remained largely out of the academic limelight until a pair of cases (the '*DePuy* cases'), *Wilkes v DePuy International Ltd*[53] (*Wilkes*) and *Gee v DePuy International Ltd*[54] (*Gee*) challenged the approach in *A v National Blood Authority* and gave renewed salience to Jane's initial fears about the Directive.

Both the *DePuy* cases involved artificial hip implants that resulted in harm and suffering by the recipients, who then sued under the CPA. Owing to the complexity of the product, causation could not accurately be determined, and it was not possible precisely to pinpoint the cause of the fractures. The claimants commenced a class action against DePuy arguing that an artificial hip implant that leads to fractures is defective, despite an inability to determine the exact cause.

The first case, *Wilkes*, was brought by a single claimant whose hip replacement fractured three years after it was implanted. The decision was handed down by Hickinbottom J, who provided the first detailed analysis of the *A v National Blood Authority* by the courts, despite the fact that fifteen years had passed between the two decisions. As outlined above, there had been limited case law considering the

[51] ibid, 1252.

[52] Duncan Fairgrieve and Geraint Howells, 'Rethinking Product Liability: A Missing Element in the European Commission's Third Review of the European Product Liability Directive' (2007) 70 *Modern Law Review* 967. See, eg, *Bogle v McDonald's Restaurants Limited* [2002] EWHC 490 (QB) and *Tesco Stores v Pollard* [2006] EWCA Civ 393.

[53] [2016] EWHC 3096 (QB).

[54] [2018] EWHC 1208 (QB).

CPA, and therefore the parties, and the judge—Hickinbottom J—referred to significant academic literature, including Jane's work.[55] After undertaking a detailed analysis of the product, including the warnings and precautions provided to consumers, the claim was rejected, and it was held that the implant was not defective.[56] Hickinbottom J, whilst not actively overruling *A v National Blood Authority*, took a very different approach to determine what was and was not a 'defect'. The approach taken moved away from the strict liability, consumer-protection approach of Burton J and instead engaged in the type of analysis Jane warned about over thirty years earlier.

In particular, in finding for the defendant, Hickinbottom J examined (and criticised) the approach and reasoning in *A v National Blood Authority*.[57] There were two elements of that decision that came in for particular criticism. First, Burton J's distinction between standard and non-standard products was rejected as being both 'unnecessary and undesirable' as a determinative criterion in its own standalone right.[58] Secondly, his Lordship disagreed with Burton's conclusion that certain factors (particularly the avoidability of the harm) ought to be excluded from defect analysis.[59] No doubt to the delight of a generation of tort law students when faced with a CPA problem question, Hickinbottom J provided a list of the relevant factors that should be considered when determining whether a product is defective. These are (i) risk-benefit of the product, (ii) the 'avoidability' of the defect, (iii) whether the product is 'standard' or 'non-standard',[60] (iv) the compliance or non-compliance with appropriate standards, (v) the compliance or non-compliance with any relevant regulatory regime, and (vi) warnings provided to consumers.[61]

Hickinbottom J recognised that 'in considering avoidability, there is a danger of unduly focusing upon the acts and omissions of the designer/producer of the product, rather than the product itself'.[62] Any such focus would of course place the analysis in a negligence-like frame, a concern articulated by Jane many years earlier. The argument raised by the claimant was not that the implant should never have fractured, but that persons are generally entitled to expect such a product to last

[55] [2016] EWHC 3096 (QB), [55] and [68].
[56] ibid, [134].
[57] See, eg, ibid, [55]–[58], [69]–[71], [80], [87], [90]–[96].
[58] ibid, [94].
[59] ibid, [81]–[89].
[60] Although, in light of His Honour's previous criticisms of this factor, he does note at [96] that: 'this appears to me to be an arid exercise, is a distraction from the exercise that the court is required to undertake, namely consideration of the appropriate level of safety taking into account all relevant circumstances. I appreciate that, where a particular specimen of a product is out of specification (or otherwise "non-standard"), then risk-benefit of an in-specification product is unlikely to have much, if any, weight: but I would not advocate a rule of law that it must have none. In assessing the safety of a product, the court should consider the relevant circumstances, in a suitably flexible way: no more and no less.'
[61] ibid, [55] and [68].
[62] ibid, [85].

ten to fifteen years, significantly longer than the three years in this case.[63] Despite the concerns of introducing a fault-based inquiry by the backdoor, Hickinbottom J went on to say that 'the ease and extent to which a risk can be eliminated or mitigated may be a circumstance that bears upon the issue of the level of safety that the public generally is entitled to expect'.[64] Hickinbottom J held that, even though the implant had lasted only three years and the damage occurred in a way that was 'rare, unpredicted and unpredictable', it was nonetheless not defective.[65] The claimant was therefore not entitled to compensation under the CPA. This decision was reached after analysis of expert evidence concerning both the implant itself and the claimant's personal factors, such as his weight and activity levels.[66]

In *Wilkes*, the finding that the implant was not defective rested on the individual circumstances of the claimant in that case. These issues therefore needed to be reconsidered on a more general basis in the following *Gee v DePuy International Ltd*.[67] This was a Group Litigation Order (GLO) claim from 312 individuals who had all experienced Adverse Reaction to Metal Debris (ARMD) with their Metal-on-Metal (MoM) hip implants. Whilst this particular implant was designed to avoid a previous complication with the traditional polyethylene-based implants, a small minority of patients with a MoM implant developed soft tissue damage because their immune systems triggered a reaction to metal debris.[68] Whilst ARMD is an inherent risk associated with these products, the claimants argued that they had developed complications far earlier than they anticipated. In a long and detailed judgment reaching 762 paragraphs,[69] Andrews J continued the general approach of *Wilkes* (which was cited with approval) and arguably continued the retreat from the consumer-focused *A v National Blood Authority*.

This pair of decisions has apparently changed the course of product liability law in the UK, taking a path that Jane warned against before the CPA was even enacted. Despite claims that the CPA set up a strict liability regime, the wording of the legislation, the defences contained in it, and its interpretation in common law have resulted in an effective return to fault-based analysis. As outlined by Andrews J:

> It is important to bear in mind that the test is not that of an absolute level of safety, nor is there an absolute liability for harm caused by a harmful characteristic. The Act does not impose a warranty of performance on a producer … All hip prostheses will eventually wear out and fail, if the patient survives long enough, and some will fail within 10 years: the natural propensity of a hip implant to fail

[63] ibid, [86].
[64] ibid, [89].
[65] ibid, [123].
[66] See discussion of these factors at ibid, [36].
[67] [2018] EWHC 1208 (QB).
[68] ibid, [117]. This was particularly problematic as the implants were specifically designed for young and/or active patients.
[69] Surprisingly, and in many ways disappointingly, the decision does not engage with Jane's work.

> therefore cannot be a 'defect', any more than the inevitable wear and tear that causes minute particles of debris to enter the patient's body. Otherwise all hip implants would be 'defective', irrespective of the materials used in the articulation.

Even though the ARMD occurred significantly earlier than anticipated, it was held that there was no identifiable defect with the implant.[70] The analysis incorporates an avoidability and risk-benefit analysis, which looks a lot like a fault-based approach to liability. The approach taken by the *DePuy* cases was subsequently endorsed by the Court of Appeal the following year in *Bailey v GlaxoSmithKline*.[71] This is the very outcome that the Council Directive was designed to avoid. It blurs the distinction between liability in common law negligence and (apparently) strict liability under the CPA regime. It shows how perspicacious Jane's analysis was, and is not good news for consumer rights. In 1986, Jane argued that strict liability was illusory, and these cases perpetuate that illusion.

6. Part 4: Product Liability and the Future

As the previous section has highlighted, the definition of 'defective' has caused a number of political, legal, and social difficulties. These challenges continued and, in fact, deepened when technological developments, particularly software and artificial intelligence (AI), came onto the consumer market. The Council Directive and CPA were drafted before these types of technologies were familiar, and in many cases before they were even envisioned. Product liability for new and emerging technologies, and the aim of ensuring that consumers can have recourse to compensation if harmed by products, has in some quarters become an important political issue. In September 2022, the European Commission introduced the draft Artificial Intelligence Liability Directive, which was then approved by the European Parliament on 12 March 2024. This Directive proposal aims to 'adapt private law to the needs of the transition to the digital economy', by which it means making it more likely that consumers will be able to recover compensation when they suffer damage as a result of AI-based products.[72] This was prompted, at least in part, by the increasing difficulty of fitting AI-based products within the established strict liability regime, as a result of scientific and technological developments.

The product liability regime for medical instruments is something that has drawn particular policy-based attention and analysis. In 2021, for example, the Medicines and Healthcare products Regulatory Agency ran a public consultation

[70] [2018] EWHC 1208 (QB), [96].

[71] [2019] EWCA Civ 1924.

[72] Proposal for a Directive of the European Parliament and of the Council on adapting non-contractual civil liability rules to artificial intelligence (AI Liability Directive) COM/2022/496.

on the future regulation of medical products, and the government's response to the results was published in June 2022.[73] After consultation with multiple stakeholders, including healthcare professionals, manufacturers of medical devices, members of the public, patients, trade associations, and businesses, the government determined that further regulation was required in relation to medical products, particularly the supply and implantation of medical devices.[74] In a similar vein, the Secretary of State for Health and Social Care announced the Equity in Medical Devices Independent Review in November 2021, chaired by Dame Margaret Whitehead.[75] The call for evidence for this review closed in October 2022 and the panel had aimed to provide their advice to the government in June 2023. At the time of writing, however, no further developments appear to have occurred.[76]

There are also larger changes in the pipeline, all of which suggest a broad political and social desire for increased focus on strict liability for consumer products. From 13 December 2023, a new key instrument—the General Product Safety Regulation (GPSR)—will be implemented in the EU to provide a complete product safety legal framework.[77] The key impetus for the GPSR is to 'increase the protection of EU consumers against dangerous non-food products sold offline or online'.[78] As a result of Brexit, UK consumers will unfortunately not benefit directly from this development. This means that whilst their European counterparts will enjoy increased consumer protection in relation to products, UK consumers will be left behind; something that will only be exacerbated by the impact of the *DePuy* cases.

There are now several reasons to see consumers in the UK as being in an increasingly unfavourable position when it comes to product liability. First is the apparent trend towards a generally more fault-based approach under the CPA, as set out above. Second is the increasing use of AI and software-based products that are probably outside of the remit of the CPA in any event. The final reason is the

[73] Medicines and Healthcare Products Regulatory Agency and The Rt Hon Sajid Javid MP, *Government response to consultation on the future regulation of medical devices in the United Kingdom* (26 June 2022).

[74] ibid. Despite the overlap and the aim of increased consumer protection, this document does not refer to the *DePuy* cases or Jane's work on the topic.

[75] Equity in Medical Devices: Independent Review; for more information, including the Terms of Reference, see https://www.gov.uk/government/groups/equity-in-medical-devices-independent-review (last accessed 6 June 2024).

[76] See Equity in Medical Devices: Independent Review Call for Evidence: (https://www.gov.uk) (last accessed 19 October 2023).

[77] Regulation (EU) 2023/988 of the European Parliament and of the Council of 10 May 2023 on general product safety, amending Regulation (EU) No 1025/2012 of the European Parliament and of the Council and Directive (EU) 2020/1828 of the European Parliament and the Council, and repealing Directive 2001/95/EC of the European Parliament and of the Council and Council Directive 87/357/EEC.

[78] For further information on these reforms, see https://commission.europa.eu/business-economy-euro/product-safety-and-requirements/product-safety/general-product-safety-regulation_en#:~:text=The%20GPSR%20requires%20that%20all,regulated%20in%20other%20EU%20legislation (last accessed 6 June 2024).

current deficit in protection afforded to consumers in the UK when compared with that provided in the EU. AI and software-based products, such as automated vehicles for example, have already reinvigorated calls for strict liability.[79] Given the essential novelty of the law having to cater for non-human decision making, and the potential for autonomous products to cause extensive harm to both users and third parties, it is unsurprising that policy arguments in favour of allocating risk to the best-placed insurer have so far won out.[80] Producers of decision-making software are, after all, in a better position than users of that software to evaluate and price the risks of its not behaving according to expectations. Those producers also have much more control over the way in which the software is designed, and in how it operates, than do its consumers.

And yet it is far from clear that such AI and software-based products would be covered by the current (ostensibly stricter than negligence) regime of the CPA in any event. The meaning of 'product' in the CPA has not been considered by a UK court. The courts have however interpreted the meaning of 'goods' under the Sale of Goods Act 1979 (SGA), the Commercial Agents (Council Directive) Regulations 1993, and at common law. The reasoning in these cases is relevant to the current discussion because 'product' is defined in the CPA by reference to 'goods'.

This body of case law has drawn a distinction between:

(1) software supplied on a tangible medium, such as a disk or USB, which is a 'good'; and
(2) 'pure' software supplied electronically, for example by being downloaded from a cloud server, which is not a 'good'.

In *St Albans CDC v International Computers Ltd*,[81] the Court of Appeal considered the supply of defective software to a local council. Sir Iain Glidewell considered whether the defendant would be liable for breach of the statutory implied terms in the SGA regarding quality and fitness for purpose. This depended on whether the software supplied by the defendant was a 'good'. The software in question was supplied by means of a disk that loaded the software onto the claimant's computer. Sir Iain Glidewell observed, obiter, that software supplied in this way was not a 'good' under the SGA.[82] In reaching this conclusion, he made the following points:

[79] See, eg, Adam Rosenberg, 'Strict Liability: Imagining a Legal Framework for Autonomous Vehicles' (2017) 20 *Tulane Journal of Technology and Intellectual Property* 205.

[80] Where an accident is caused by an automated vehicle, the insurer is directly liable for any damage caused under the Automated and Electric Vehicles Act 2018 (AEVA). However, an insurer who has settled their liability with an injured party has the right under AEVA to bring a recoupment claim against any other person responsible for the accident. This might include a claim against a producer of defective software that was installed in the automated vehicle and found to have caused the accident.

[81] [1996] 4 All ER 481.

[82] ibid, 493.

(1) The SGA defines goods as 'all personal chattels other than things in action and money'. A tangible item such as a disk falls within this definition. However, an intangible item such as a software program does not.
(2) If a software program is supplied on a disk, and the software on that disk is defective, the disk itself, containing the software, can be considered defective.
(3) In this case, the software program was not supplied on a disk or any tangible medium. The software was instead loaded from the defendant's disk onto the claimant's computer. Therefore, only the software program was supplied to the claimant, and per (1) this was not a supply of 'goods'.[83]

Sir Iain Glidewell's reasoning has been followed in a number of cases:

(1) In *London Borough of Southwark v IBM UK* (2011),[84] the claimant council sued the defendant software supplier for breach of the statutory implied terms in the SGA. Akenhead J dismissed the claim on the basis that there was no 'sale', but went on to observe, obiter, that a supply of software on a tangible medium such as a CD could in principle constitute a supply of 'goods'.
(2) In *Fern Computer Consultancy v Intergraph* (2014),[85] the claimant software company sued the defendant agent under the Commercial Agents Regulations. The defendant argued that the claim should be dismissed because, among other things, it was not a 'commercial agent' because it supplied software, which was not a 'good'. Mann J rejected this argument, holding that there was a serious question to be tried as to whether the defendant supplied 'goods' (the software in this case having been supplied on a CD).
(3) In *Your Response v Datateam Business Media* (2014),[86] the Court of Appeal held that a database stored electronically was intangible property which did not amount to 'goods', and therefore could not be the subject of a common law possessory lien. The court referred to *St Albans* as authority for the proposition that data cannot itself constitute a 'good' because it is intangible.
(4) In *Computer Associates v Software Incubator* (2018),[87] the Court of Appeal held that software which was electronically supplied (via an email containing a link to an online portal from which the software was downloaded) was not a sale of 'goods' for the purpose of the Commercial Agents Regulations. The software had not been supplied on a tangible medium and

[83] ibid, 492–493.
[84] [2011] EWHC 549.
[85] [2014] EWHC 2908 (Ch).
[86] [2014] EWCA Civ 281.
[87] [2018] 1 CLC 456.

> the Court observed that it could not 'ignore the weight of judicial authority that supports maintaining the tangible/intangible distinction'.[88]

The respondent in *Software Incubator* applied to the Supreme Court for permission to appeal. On 22 May 2019, that Court decided to refer the question of whether electronically supplied software amounts to 'goods' under the Commercial Agents Regulations to the Court of Justice of the European Union (CJEU).[89] The case is currently pending before the CJEU. It seems unlikely that the English courts will abolish the tangible/intangible distinction, based as it is on a long line of authority. If the distinction is to be abolished, this will likely require legislation.

In *Software Incubator*, Gloster LJ said:

> [D]espite the problems of principle which arise if one excludes electronically supplied software from the definition of 'goods', I am not persuaded that it is open to this court to impute what many might think was a common-sense meaning of 'goods' to the legislators ... To do so would be contrary to precedent. This court cannot simply ignore the weight of judicial authority that supports maintaining the tangible/intangible distinction.[90]

This, alongside the clear need first identified by Jane several decades ago and made more urgent by recent developments, suggests that the UK should follow the EU's lead in reviewing the product liability regime so that it is fit for purpose when applied to AI and software-based products. Perhaps if the legislature and the courts had heeded her words when the current regime was in development, the law would not require such an extensive rethink now.

7. Conclusion

Jane is an incredible person, professionally and personally. She has made a huge impact on the lives of a generation of private law scholars and has had a comparable impact on many areas of tort law. This chapter has focused on her ground-breaking work on product liability, highlighting the important role that US jurisprudence plays in this area and the risks posed by the current regime that allows movement towards a fault-based approach to liability. And, as with so many things, Jane was right on the money in her ground-breaking analysis (largely before the Act was drafted, let alone, enacted). The more recent case law in the UK has indicated a

[88] ibid, [55].

[89] *Software Incubator Ltd v Computer Associates (UK) Ltd*, Judgment of the Court (Fourth Chamber) of 16 September 2021, C-410/19.

[90] [2018] 1 CLC 456, [55].

troubling retreat from strict liability to a more fault-based approach, thereby limiting consumer rights and creating difficulties when distinguishing the CPA from common-law negligence. This has resulted in international calls for reform to move consumer protection laws back to their strict liability roots. It has also raised a number of challenges for emerging technologies and their consumers. There are moves around the world to review how these technologies are going to be governed, and it will be important for every such review to return to the clear, compelling, and rigorous analysis provided by Jane in the initial CPA debates. Her impact in this area will live on.

17
The Accident Preference, 'Unrigorous Thinking' and Injury Compensation Schemes

Genevieve Grant and Harold Luntz

1. Introduction

Professor Jane Stapleton's striking early scholarly work focused on tort law's failure to compensate people with non-traumatic injury, including disease, arising from negligence.[1] Stapleton attributed this deficit—and its neglect by commentators and reform proponents—to the preferences of tort's 'conceptual machinery' for conditions more readily attributable to specific events, times and places.[2] This is still apparent in the recent case of *Paul v Royal Wolverhampton NHS Trust*,[3] where the Supreme Court held that relatives of patients who themselves suffered injury could recover only where there was a medical 'accident'. Stapleton mapped the distorting effect of this preference, together with the data deficit around the propensity of injured people to make claims, on the broader compensation debate.[4] Stapleton also cautioned against the 'unrigorous thinking' that saw non-tort schemes designed through the limiting lens of tort: as tort replacements for a preferred subset of injured persons, rather than effecting a more just and comprehensive response to the burden of injury and disease on the basis of disability and need, rather than cause.[5]

This chapter explores the continued relevance of key aspects of Stapleton's ideas to contemporary Australian compensation schemes for personal injury. Inroads made into achieving better understanding of the underlying prevalence and distribution of injury and disease in the community and the profile of claiming activity are investigated. The chapter charts the proliferation of non-tort schemes in Australia over the last 40 years, and their replication of tort's accident preference.

[1] Jane Stapleton, 'Compensating Victims of Diseases' (1985) 5(2) Oxford Journal of Legal Studies 248; Jane Stapleton, *Disease and the Compensation Debate* (Clarendon Press, 1986).

[2] Stapleton, 'Compensating Victims' (n 1) 248–53.

[3] [2024] UKSC 1 (11 January 2024).

[4] Stapleton, 'Compensating Victims' (n 1) 248, 261–3; Stapleton, *Disease and the Compensation Debate* (n 1) 4–8, 12–15.

[5] Stapleton, 'Compensating Victims' (n 1) 259.

Genevieve Grant and Harold Luntz, *The Accident Preference, 'Unrigorous Thinking' and Injury Compensation Schemes*
In: *Torts on Three Continents.* Edited by: Kylie Burns, Jodi Gardner, Jonathan Morgan and Sandy Steel,
Oxford University Press. © Genevieve Grant and Harold Luntz 2024. DOI: 10.1093/oso/9780198889748.003.0017

It also argues that in this landscape, a new variety of unrigorous thinking has emerged: the assumption that introducing a scheme signals 'job done' in facilitating appropriate support for claimants. Legal scholars continue to neglect the performance of these large-scale legal systems that are responsible for the provision of compensation to hundreds of thousands of Australians each year. Despite this inattention, a growing and diverse range of data is now available to shed light on the adequacy and success of tort and compensation schemes, data which is fruitfully being put to work by scholars in a number of disciplines outside law. This research is answering critical questions of interest to legal scholars, practitioners and law reformers. Drawing on this transdisciplinary evidence base, the chapter calls for a new generation of injury law scholarship to enrich the policy debate around tort and compensation systems.

2. Professor Stapleton's Early Work on Non-Traumatic Injury and Disease

Professor Stapleton's early scholarship was informed by her academic background in science.[6] Her legal doctoral work, published in a short article (1985)[7] and as a monograph (1986),[8] came at the end of a busy time for reform-orientated analysis of tort and compensation schemes, much of which focused on motor crash injury. We set out some of the key developments here for context. Ison's *Forensic Lottery* was published in 1967.[9] In the US, Massachusetts was the first state to implement a form of no-fault motor crash scheme, commencing on 1 January 1971,[10] with a number of other states considering and implementing schemes of various kinds in the 1970s.[11] The first edition of Atiyah's *Accidents, Compensation and the Law* appeared in 1970 (Atiyah was later to become Stapleton's DPhil supervisor).[12] In 1974, New Zealand's comprehensive no-fault injury scheme commenced[13] and

[6] Lesley Dingle and Daniel Bates, 'A Conversation with Professor Jane Stapleton' (6 February 2012), <https://www.squire.law.cam.ac.uk/eminent-scholars-archiveprofessor-jane-stapleton/conversation-professor-jane-stapleton> accessed 14 October 2023. See also Jane Stapleton, *Three Essays on Torts* (OUP, 2021) 1.

[7] Stapleton, 'Compensating Victims' (n 1).

[8] Stapleton, *Disease and the Compensation Debate* (n 1).

[9] Terence Ison, *The Forensic Lottery: A Critique on Tort Liability as a System of Personal Injury Compensation* (Staples Press, 1967).

[10] James M Anderson, Paul Heaton and Stephen J Carroll, *The U.S. Experience with No-Fault Automobile Insurance: A Retrospective* (RAND Corporation, 2010) 36.

[11] ibid 39–44. Anderson, Heaton and Carroll chart the fall of no-fault from popularity over the ensuing 40 years, which they attribute to '(1) its unexpectedly high claim costs and (2) the political debate shifting from an overall assessment of the optimal insurance system to the impact of those high costs on consumers' (xvi).

[12] PS Atiyah, *Accidents, Compensation and the Law* (Weidenfeld & Nicolson, 1970).

[13] *Accident Compensation Act 1972*. On the history of developments in New Zealand, see Grant Duncan, 'National's Fortunate Failure to Deregulate Accident Compensation' (2019) 34(2) New Zealand Sociology 253; Simon Connell, 'Community Insurance Versus Compulsory Insurance: Competing Paradigms of No-Fault Accident Compensation in New Zealand' (2019) 39(3) Legal Studies 499.

Victoria became the first state in Australia to introduce a no-fault motor accident scheme.[14] That same year the Australian Woodhouse Inquiry recommended the establishment of a comprehensive scheme and supported the inclusion of sickness and disease;[15] however, a Senate Committee called for the Bill to be withdrawn and reconsidered and for the sickness part of the scheme to be removed.[16] In 1978 came the report of the Pearson Commission, the recommendations of which included the establishment of a no-fault scheme for road traffic injury in the United Kingdom, as well as a no-fault nuclear accident scheme.[17] In 1980 the first edition of what is now *Luntz and Hambly's Torts: Cases, Legislation and Commentary* was published in Australia,[18] with a first chapter 'seeking to place the law of torts in its social context and emphasising its actual operation', along with the operation of alternative compensation mechanisms in the form of statutory schemes.[19] In 1983, a report of the results of a study was published by the Centre for Socio-Legal Studies at the University of Oxford into compensation for illness and injury.[20] By 1986 and the publication of Stapleton's monograph (and with the benefit now of more than 35 years of hindsight), it was starting to become clear that Australia would never have a comprehensive injury compensation scheme. Still, the previous two decades had been a period where many legal scholars had engaged actively in debates about tort reform and the value and optimal design of non-tort injury compensation schemes.

In her elegant and underappreciated early work,[21] Stapleton unpacks the way tort reform debates focused on traumatic injury and neglected non-traumatic varieties of personal injury, 'where the injury cannot be unequivocally pinpointed to a certain time and place', but where the source is made by humans 'and could potentially give rise to a civil action for damages'.[22] She defines non-traumatic injuries as including a range of occupational health conditions ('dust diseases, repetitive

[14] *Motor Accidents Act 1973* (Vic).

[15] Australian National Rehabilitation and Compensation Committee of Inquiry, *Compensation and Rehabilitation in Australia: Report of the National Committee of Inquiry* (Australian Government Publishing Service, 1974).

[16] Harold Luntz, 'Looking Back at Accident Compensation: An Australian Perspective' (2003) 34 Victoria University Wellington Law Review 279, 288–9.

[17] Royal Commission on Civil Liability and Compensation for Personal Injury, *Report* (Cmnd 7054-1, 1978).

[18] See Harold Luntz et al, *Luntz & Hambly's Torts: Cases, Legislation and Commentary* (9th edn, LexisNexis Butterworths, 2021) 82–92.

[19] Harold Luntz, 'A Personal Journey through the Law of Torts' (2005) 27(3) Sydney Law Review 393, 401.

[20] Donald R Harris et al, *Compensation and Support for Illness and Injury* (Clarendon Press, 1984).

[21] Google scholar citations are a clumsy tool for measuring the reach of legal scholarship (and do not include judicial citations), but provide a very rough guide for present purposes. As at 17 June 2024, Stapleton's monograph had been cited 43 times (with 18 citations in the last 10 years) and the shorter paper 15 times. In comparison, a number of Stapleton's more recent publications have each been cited hundreds of times (eg Jane Stapleton, 'Tort, Insurance and Ideology' (1995) 58(6) Modern Law Review 820; Jane Stapleton, 'Legal Cause: Cause-in-Fact and the Scope of Liability for Consequences' (2001) 54 Vanderbilt Law Review 820.

[22] Stapleton, 'Compensating Victims' (n 1) 248.

movement injury, deafness and dermatitis'); product injuries from tobacco, pharmaceuticals and asbestos; and 'environmental injuries' such as those caused by fertilisers.[23] Stapleton explores the multiple ways the 'conceptual machinery' of tort operates to ensure that '*effective* tort liability is principally confined to cases of traumatic injury', particularly injuries occurring at work and on the road.[24] The key parts of this machinery are concerned with time, medical causation and proof of fault. She argues that the barriers survivors of non-traumatic conditions experience in recovering tort damages 'are not satisfactorily resolved' by the introduction of stricter liability or cause-based, non-tort compensation schemes. A proper and more sustainable solution to the identified limitations of tort's operation requires a needs-based scheme focused on disability (consequences) rather than cause.

Stapleton's work takes what remains a refreshingly practical approach, tackling a range of 'non-legal barriers to the initiation and success of disease claims in tort': these include difficulty accessing expert legal advisers and supporters and, in cases of latent disease, the improbability of identifying the proper defendant and getting them to court.[25] She also canvases 'psychological barriers to the disease victim bringing a common law claim', noting that while '[l]ittle work has been done on this aspect of the perception of disease', the profile, status and stigma associated with chronic disease could foreseeably diminish the extent to which potential claimants pursue compensation or even result in them completely forgoing a claim. While Stapleton's work ultimately moved away from these concerns, some 32 years later in her Clarendon Law Lectures, she reported remaining 'interested in looking at tort law as it actually is, in all its seeming messiness'.[26]

3. Assessing Injury Law in Practice: Data Deficiencies and Developments

To empirically interrogate the effectiveness of tort and injury compensation schemes, a range of information about the incidence, cause and consequences of relevant health problems is needed. At the top of this list, we might put information about the prevalence of compensable conditions, and the extent to which those entitled to make claims do so.[27] Debates about the performance of tort and injury

[23] ibid. Definitions of traumatic injury are varied, but the US Bureau of Labor Statistics defines traumatic injury in work-related contexts as 'any wound or damage to the body resulting from acute exposure to energy, such as heat or electricity; impact from a crash or fall; or from the absence of such essentials as heat or oxygen, caused by a specific event or incident within a single workday or shift' (US Bureau of Labor Statistics, *Handbook of Methods, Census of Fatal Occupational Injuries: Concepts* (8 December 2020) <https://www.bls.gov/opub/hom/cfoi/concepts.htm> accessed 14 October 2023.

[24] Stapleton, 'Compensating Victims' (n 1) 248.

[25] Stapleton, *Disease and the Compensation Debate* (n 1) 101–4.

[26] Stapleton, *Three Essays* (n 6).

[27] Peter Cane and James Goudkamp, *Atiyah's Accidents, Compensation and the Law* (9th edn, Cambridge University Press, 2018) 395.

compensation schemes have long been thwarted by the elusiveness of this kind of data. Professor Stapleton's early work engaged directly with these challenges as they applied to disease claims. Early in the monograph, Stapleton criticises the evidentiary gap around the profile of injury and disease in the community. She draws together a patchwork of sources of data on a diverse range of health conditions, highlighting the problematic absence of information about whether those conditions developed in circumstances where a responsible party could be sued in negligence.[28] This objective of understanding the prevalence of potential claims and establishing a reliable 'claiming rate' in the community remains almost insurmountably challenging in the Australian context. In the absence of information about the prevalence of potential claims in the community, we typically lack a denominator in relation to which the actual number of claims can be evaluated.

There are rare exceptions, however.[29] In the compulsory third-party transport crash scheme in New South Wales, 'propensity to claim' is calculated based on the number of claims as a proportion of road casualties, and this data is periodically made public in industry reports. In 2016 it was reported that propensity to claim in the scheme (which was then predominantly based on common law damages claims) had increased over time from 30 per cent in 2008 to 60 per cent in 2015.[30] The scheme was subsequently reformed.[31] Additionally, researchers linking cancer registries with workers' compensation claims have explored propensity to claim for these predominantly occupationally incurred diseases. Among people diagnosed with asbestos-related cancers in British Columbia (Canada) from 1970 to 2005, only 33 per cent made a claim in the provincial workers' compensation system;[32] the equivalent rate in New South Wales, Australia (1964–2002) was 60 per cent[33]

[28] Stapleton, *Disease and the Compensation Debate* (n 1) 4–11.

[29] In the United Kingdom, Annette Morris's analysis of Compensation Recovery Unit data provided a useful corrective to perceptions that the number of tort law claims was increasing rapidly in the 2000s (Annette Morris, 'Spiralling or Stabilising? The Compensation Culture and our Propensity to Claim Damages for Personal Injury' (2007) 70(3) Modern Law Review 349). That agency now provides useful data distilling the annual number of claim notifications it receives across major categories of clinical negligence, motor vehicle, employer and public liability: Department for Work & Pensions, 'Compensation Recovery Unit performance data' (1 June 2023) <https://www.gov.uk/government/publications/compensation-recovery-unit-performance-data/compensation-recovery-unit-performance-data> accessed 14 October 2023. There is unfortunately no equivalent data available in Australia, despite the requirement that insurers and claimants notify Centrelink (the social security agency) of settlements and damages awards in compensation claims (Services Australia, 'Centrelink Compensation Recovery' (2 February 2023) <https://www.servicesaustralia.gov.au/centrelink-compensation-recovery> accessed 14 October 2023).

[30] State Insurance Regulatory Authority of New South Wales, 'Review of Selected Performance Indicators of the NSW CTP Scheme 2015 (report for the State Insurance Regulatory Authority), May 2016, <https://www.sira.nsw.gov.au/resources-library/green-slip-resources/publications/archived/EY_Review-of-selected-performance-indicators-of-NSW-CTP-Scheme-2015_Final-for-web.pdf> accessed 14 October 2023.

[31] Luntz et al (n 18) 54–5.

[32] Tracy L Kirkham et al, 'Surveillance of Mesothelioma and Workers' Compensation in British Columbia, Canada' (2011) 68(1) Occupational and Environmental Medicine 30, 32.

[33] Rebecca A Hyland et al, 'Incidence Trends and Gender Differences in Malignant Mesothelioma in New South Wales' (2007) 33 Scandinavian Journal of Work and Environmental Health 286, 288.

and in Italy (2010–15) it was 66 per cent.[34] These analyses and others like them provide rare insights into the profile of claims activity.[35] Their illumination of the effectiveness of compensation schemes in reaching intended users is typically a secondary product of the commercial imperative to understand the trajectory of claims liabilities, or more transparently, entrepreneurial research activity linking population health and insurance data registries.

Another body of research has endorsed Stapleton's concerns by providing a deeper and more nuanced understanding of the inequitable impacts of tort law's 'accident preference'. This work indicates that though traumatic injury is an important cause of death and loss of quality of life, other sources have more bearing on the global burden of injury and disease. Stapleton's argument that disease has a greater impact than traumatic injury in terms of health burden in the community is supported by the Global Burden of Diseases, Injuries and Risk Factors (GBD) Study.[36] The most comprehensive assessment of human health, the GBD Study, is 'a systematic scientific effort to quantify the comparative magnitude of health loss from diseases, injuries, and risks by age, sex, and population over time'.[37] The GBD Study uses disability-adjusted life years (DALYs) to investigate trends in population health. A DALY is a metric that captures the years of healthy life lost as a result of disease and injury, and facilitates comparison between the actual health of a population and health where the population lives without disease or disability.[38] The most recent GBD Study identified that in 2019 road injuries remained the leading cause of DALYs among persons aged 10–24 years (accounting for 6.6 per cent of that group's DALYs) and 25–49 years (accounting for 5.1 per cent of that group's DALYs).[39] For those aged 50–74, road injuries ranked tenth among the leading causes and accounted for 2.1 per cent of DALYs.[40] Though disease represents a more amorphous group of conditions than traumatic injuries and is perhaps not as clearly linked to relevant negligence, it is collectively responsible for a greater share of the burden of injury in the community than road and other varieties of injury.

[34] Alessandro Marinaccio et al, 'Predictors of Filing Claims and Receiving Compensation in Malignant Mesothelioma Patients' (2020) 124(9) Health Policy 1032.

[35] Another useful example is provided by the Asbestos Injuries Compensation Fund Ltd, which was set up in 2006 to implement and administer the agreement between James Hardie Industries NV and the New South Wales Government, under which compensation is payable by certain former James Hardie entities. The fund publishes an annual actuarial report which records the actual number of claims compared with the expected number of claims: see Asbestos Injuries Compensation Fund, 'Actuarial Reports' <https://www.aicf.org.au/actuarial.php> accessed 16 October 2023.

[36] Christopher JL Murray and Alan D Lopez, 'Measuring Global Health: Motivation and Evolution of the Global Burden of Disease Study' (2017) 390 (10100) *The Lancet* 1460.

[37] ibid 1460.

[38] Australian Institute of Health and Welfare, *Australian Burden of Disease Study 2022* (Catalogue number BOD 37 AIHW, Australian Government, 2022) 2.

[39] Theo Vos et al, 'Global Burden of 369 Diseases and Injuries in 204 Countries and Territories, 1990–2019: A Systematic Analysis for the Global Burden of Disease Study 2019' (2020) *The Lancet* 396 (10258) 1204, 1211.

[40] ibid 1212.

From 2010 to 2019, the ten diseases and injuries that were the most significant drivers of increasing health burden were ischaemic heart disease, diabetes, strokes, chronic kidney disease, lung cancer, age-related hearing loss, HIV/AIDS, other musculoskeletal disorders, lower back pain and depressive disorders.[41] Together these conditions accounted for 29 per cent of DALYs. The GBD Study also identifies considerable gender differences in the distribution of injury and disease: while road injuries, cirrhosis and lung cancer were in the top ten leading causes of DALYs among males, they were replaced in the equivalent top ten for females by low back pain, gynaecological diseases and headache disorders.[42] Deaths due to drug use increased markedly over the period to 2019, with factors including prescribing practices relating to opioids reportedly playing an important role in this development.[43] The GBD Study findings lend further weight to Stapleton's concerns about the traumatic injury preference of tort and the compensation schemes created to supplement it.

To this body of evidence, we can add the now thorough understanding of the socially determined nature of health. Since the 1980s, a compelling volume of work in public health and health equity has demonstrated that some community members are more likely than others to experience poor health and non-traumatic injury and disease of various kinds.[44] It follows that the lack of effective coverage afforded to these conditions by injury law also reflects and reproduces these inequities. Health over a lifetime is materially affected by negative social and economic circumstances, and health has a social gradient: less advantaged people face a greater risk of injury and disease, and the negative health effects accumulate over time.[45] The vast research literature on the impact on health of the social gradient together with such factors as 'the conditions in which people are born, grow, live, work, and age'[46] stands in contrast to the discrete, event-based focus of tort's remedial approach. The overwhelming message of public health is that injury and disease are not randomly distributed in the community, despite the narrative of the injury circumstances in much tort case law.[47] Admittedly, although this more recent research tells us a great deal about the distribution and impact of injury and disease in the community, it tells us little about what proportion of injury and disease is the result of negligence. That lack of data about the causal origins of health conditions remains elusive at a population level for most health problems.

[41] ibid 1208.

[42] ibid 1209.

[43] ibid 1215.

[44] Michael Marmot, 'Social Determinants of Health Inequalities' (2005) 365(9464) *The Lancet* 1099; Michael G Marmot and Richard G Wilkinson, *Social Determinants of Health* (2nd edn, OUP, 2006).

[45] Commission on Social Determinants of Health, *Closing the Gap in a Generation: Health Equity Through Action on the Social Determinants of Health (Final Report of the Commission on Social Determinants of Health* (World Health Organization, 2008) 31–2.

[46] ibid 26.

[47] Samantha Hardy, 'Injury as Melodrama' in Andrew Kenyon and Peter Rush (eds), *Aesthetics of Law and Culture: Texts, Images, Screens* (Emerald Group, 2004).

Even in instances where we are able to reliably establish a baseline rate of negligently inflicted health conditions—for example, in the clearest cases of motor vehicle crash injury—a complex range of factors have a bearing on whether an injured person might take action to pursue a claim. Stapleton recognised this before the research literature had caught up, noting the possibility that people with non-traumatic conditions might be less likely than others to pursue a remedy.[48] Where we rely only on claims data, we lack insight into the group of potential claimants who do not claim and their reasons for inaction. Socio-legal understandings of the way people respond to legal problems have expanded in recent years, moving from the antiquated, linear idea of 'naming, blaming and claiming'[49] towards a more sophisticated appreciation of the diversity of responses to injurious events.[50] Importantly, evidence of how people make those choices is still overwhelmingly based on tort claims, rather than the more prevalent administrative processes associated with compensation schemes.[51] Nonetheless, across nations, legal need research has used surveys to establish the prevalence of various kinds of legal problems in the community, including injury problems, and how people experience and respond to those problems.[52] A nation-wide, representative survey of legal need in Australia in 2008 identified that over a 12-month period, 7.9 per cent of Australians aged 15 or over experienced an injury or illness problem with legal implications.[53] Approximately 3.3 per cent of Australians aged 15 or over—634,000 people—experienced a substantial illness or injury problem, meaning it had a moderate or severe impact on their everyday life.[54] More recent representative research has identified that in the two years to 2021, 3.9 per cent of Victorian survey respondents experienced a justiciable injury or illness problem.[55] Among those who had experienced an injury or illness problem, 19.2 per cent reported experiencing moderate or severe levels of mental distress and 14.2 per cent reported

[48] Stapleton, *Disease and the Compensation Debate* (n 1) 102–3.

[49] William LF Felstiner, Richard L Abel and Austin Sarat, 'The Emergence and Transformation of Disputes: Naming, Blaming, Claiming ...' (1980) 15 Law and Society Review 63.

[50] See, for example, Sachin S Pandya and Peter Siegelman, 'Underclaiming and Overclaiming' (2013) 38(4) Law & Social Inquiry 836; David M Engel, 'Perception and Decision at the Threshold of Tort Law: Explaining the Infrequency of Claims' (2012) 62 DePaul Law Review 293; Christine Coumarelos, Genevieve Grant and Zhigang Wei, 'Personal Injury Problems: New Insights from the Legal Australia-Wide Survey' (2017) 22 Justice Issues 1.

[51] Entry into compensation schemes involves an initial engagement with an administrative claims process through a claim form (online or otherwise), which is quite different from a letter of demand to a defendant with the assistance of a lawyer, as might be typical for medical negligence or public liability claims.

[52] OECD/Open Society Foundations, *Legal Needs Surveys and Access to Justice* (OECD Publishing, 2019).

[53] Coumarelos, Grant and Wei (n 50).

[54] ibid 22.

[55] Defined as 'concerning an injury caused by someone else, or an injury or illness caused by: a work accident; working conditions; negligent/wrong medical treatment' (Nigel J Balmer et al, *The Public Understanding of Law Survey (PULS) Volume 1: Everyday Problems and Legal Need* (Victoria Law Foundation, 2023) 46).

being unable to heat or cool their home.[56] People with injury and illness problems reported a range of problem resolution strategies: 3 per cent did nothing to deal with their problem, while 14.6 managed alone or with the assistance of family or friends; 54.4 per cent used independent non-legal help (such as from a doctor); and 27.9 per cent sought independent legal help.[57] Interestingly, only 20 per cent of people with a justiciable injury or illness problem regarded it as 'legal' in nature.[58] Only 4.3 per cent of people who experienced an injury or illness problem reported that a court or tribunal was involved in the resolution of the problem.[59]

From this valuable research we can see that while people with injury problems might be more likely than some others to take steps to resolve their problem, there are yet others who take no steps and do not seek legal advice. For some people, the decision not to pursue compensation is made as a result of a reasoned calculation about the extent of their loss and the costs and benefits associated with a claim.[60] Yet others encounter the kinds of barriers Stapleton described and more: concerns about the costs of taking action, a lack of understanding of the appropriate steps and variable legal capability for navigating claims processes and challenging adverse decisions.[61] The losses experienced by people affected by these kinds of barriers to claiming are ultimately allowed to lie where they fall. This result reflects on the design and operation of compensation systems, and the broader legal systems they are based in.

4. The Current Landscape of Tort Alternatives for Injury, Disease and Disability—A View from Australia

In the Australian context, it is clear that the opportunity for wholesale reform of injury compensation has long passed. A complicated network of schemes of support and compensation exists for people who have experienced a variety of kinds of injury and disease.[62] These include 11 major workers' compensation statutes and schemes (8 for the states and territories, and 3 at Commonwealth level), nine

[56] ibid 67.
[57] ibid 108.
[58] ibid 114.
[59] ibid.
[60] Coumarelos, Grant and Wei (n 50) 21.
[61] ibid; Hugh M McDonald and Julie People, 'Legal Capability and Inaction for Legal Problems: Knowledge, Stress and Cost' (2014) 41 Updating Justice 1; Pascoe Pleasence and Nigel J Balmer, 'Justice & the Capability to Function in Society' (2019) 148(1) Daedalus 140.
[62] On the equivalent in the United Kingdom, see the schemes catalogued in Geraint Howells, 'Justifications for Preferential Adoption of No-Fault Accident Compensation Schemes' (2019) 16(1) Otago Law Review 127. A comprehensive understanding requires expanding our field of view beyond tort and injury compensation schemes to social security, life insurance and other financial supports: see, eg, Alex Collie, Michael Di Donato and Ross Iles, 'Work Disability in Australia: An Overview of Prevalence, Expenditure, Support Systems and Services' (2019) 29 Journal of Occupational Rehabilitation 526.

for transport crash injury and eight for victims of crime—together, they cover a population of around 27 million.[63] These arrangements are overwhelmingly cause-based, with varying degrees of no-fault supplements to or replacements for tort, and replicate the problems Stapleton eloquently catalogued. Again and again, small-scale change is made to the major components of this network (for work and transport crash injury), often increasing scheme complexity and making it harder for claimants to understand the benefits they are entitled to and how they might go about accessing them. Beyond the work and transport crash schemes, the major development in Australia in the last ten years has been the establishment of a 'coherent national system for disability support'[64] in the form of the National Disability Insurance Scheme (NDIS) and National Injury Insurance Scheme (NIIS).[65] The NDIS began operating around Australia in 2016 as a funder of reasonable and necessary supports for people aged under 65 years with significant and permanent disabilities. It does not provide income support or lump sums of compensation.[66] Progress on the NIIS (a parallel scheme intended to cover support for survivors of catastrophic injury) has to date extended only as far as transport crash injury, with work ongoing on an equivalent set of arrangements for occupational injury. The NDIS has consistently been plagued by the familiar tension between rights and sustainable administration[67] and evidence of negative experience many participants have when engaging with the scheme.[68]

New compensation schemes and programmes emerge with increasing regularity. Their form and motivations are as varied as their subject matter.[69] Some are legislated, others are less formal and more *ex gratia*, and others still go beyond facilitating payment in also making restorative justice mechanisms available (such as conferences involving claimants and representatives of organisations responsible for the wrong).[70] A significant part of this new scheme activity is a product

[63] See Luntz et al (n 18) ch 1.

[64] Productivity Commission, *Disability Care and Support: Inquiry Report* (Report No 54, Australian Government, 2011) 10.

[65] Luntz et al (n 18) 63–6.

[66] See further Harold Luntz, 'Compensation Recovery and the National Disability Insurance Scheme' (2013) 20(3) Torts Law Journal 153.

[67] A Venning et al, 'Adjudicating Reasonable and Necessary Funded Supports in the National Disability Insurance Scheme: A Critical Review of the Values and Priorities Indicated in the Decisions of the Administrative Appeals Tribunal (2021) 80(1) Australian Journal of Public Administration 97.

[68] NDIS Review, 'What We Have Heard Report' (30 June 2023) <https://www.ndisreview.gov.au/sites/default/files/resource/download/what-we-have-heard-report.pdf> accessed 14 October 2023.

[69] Knetsch and Watts recently observed that the motivations for compensation schemes are often mixed: 'as well as public pressure to provide victims with compensation as a display of solidarity, it may also be more politically expedient to establish an *ex* post and ad hoc solution rather than a (potentially expensive and politically unpopular) ex ante solution' (Jonas Knetsch and Kim Watts, 'What is the Potential of Compensation Funds for Addressing COVID-related Personal Injury?' (2023) 48 Geneva Papers on Risk and Insurance: Issues and Practice 608, 609).

[70] For a recent survey of state-based arrangements for 'extrajudicial compensation' in Australia, see Sarah Lim, Natalie Ng and Greg Weeks, 'Government Schemes for Extrajudicial Compensation: An Assessment (2020) 100 AIAL Forum 79.

of the major examination of institutional responses to child sexual abuse by the Australian Royal Commission on these matters that was announced in 2012 and ran from early 2013 to late 2017.[71] The Royal Commission had a specific focus on exploring the role of litigation processes and procedural requirements and their impacts on the ability of survivors to recover damages.[72] Like Stapleton's analysis of disease claims, it singled out difficulties associated with limitation periods, identifying proper defendants where significant time had elapsed and the psychological barriers to claiming as key challenges facing abuse survivors seeking redress. In addition, there was the legal difficulty of suing an unincorporated association, as were many of the bodies that ran the institutions at which the abuse had occurred.[73]

The compensation schemes and programmes that have emerged in Australia over the last 20 years include:

- the National Redress Scheme for people who have experienced institutional child sexual abuse, which is accepting applications up to June 2027 and as at September 2023 had received 30,680 applications, finalised 13,532 applications and facilitated the distribution of $AUD1.13 billion in 12,762 payments;[74]
- the Defence Abuse Reparation Scheme administered by the Defence Abuse Response Taskforce, which from 2012–2016 made reparation payments totalling $67 million to 1,723 survivors of abuse in the Australian Defence Force;[75]
- redress schemes for members of the Stolen Generations, Indigenous people who were forcibly removed from their communities and families—such a programme was first established in Tasmania in 2006 and equivalents have now operated in all Australian states and territories except Western Australia and Queensland;[76]

[71] Royal Commission into Institutional Responses to Child Sexual Abuse, *Final Report: Our Inquiry (Volume 1)* (Commonwealth of Australia, 2017) 13.

[72] Royal Commission into Institutional Responses to Child Sexual Abuse, *Redress and Civil Litigation Report* (Commonwealth of Australia, 2015) pt IV.

[73] ibid 432–3.

[74] National Redress Scheme, 'National Redress Scheme—Update' (30 September 2023) <https://www.nationalredress.gov.au/about/updates/1671> accessed 14 October 2023. From 1 December 2016, the Commonwealth Ombudsman has been able to receive reports relating to serious abuse within the Australian Defence force, under the *Ombudsman Regulations 2017* (Cth). The Commonwealth Ombudsman reports that to 30 September 2023, 1,985 payments have been made by the Department of Defence on the basis of the Ombudsman's recommendations, totalling $AUD82.5 million (Commonwealth Ombudsman, 'Reporting Abuse in Defence Statistics—30 September 2023' <https://www.ombudsman.gov.au/complaints/defence-force-complaints/reporting-abuse#_statistics> accessed 14 October 2023).

[75] Defence Abuse Response Taskforce, *Final Report* (Commonwealth of Australia, March 2016) 1.

[76] Lim, Ng and Weeks (n 70) 86–7.

- schemes for occupational asbestos-related disease, separate to state and territory workers' compensation regimes (as in Tasmania,[77] the Commonwealth[78] and Western Australia);[79]
- the Parliamentary Injury Compensation Scheme, established in 2016 to cover injuries and illness experienced by federal parliamentarians in connection with their work and the Prime Minister's spouse in connection with their official activities;[80]
- the Restorative Engagement and Redress Scheme for Police Staff, which provides financial redress and restorative engagement to employees of Victoria Police who experienced sex discrimination or sexual harassment in the workplace, and was established in 2019 following the report of the Victorian Human Rights and Equal Opportunity Commission's Independent Review of Victoria Police;[81]
- the COVID-19 vaccine claims scheme, established in 2021 to provide compensation to people who have moderate to severe adverse reactions to COVID-19 vaccines, which after its first 11 months of operation in November 2022 had received 3,071 claims but determined just 263;[82] and
- the Australian Sports Commission Restorative Program for former athlete scholarship holders at the Australian Institute of Sport who experienced abuse between 1981 and 2013, which is scheduled to conclude in mid-2024.[83]

The constituent parts of this diverse compensation scheme landscape are the subject of varying levels of accountability, scrutiny and evaluation. The key public accountability mechanisms include annual reports, media scrutiny, freedom of information regimes, parliamentary oversight, independent complaints and investigation processes and research. It remains rare for the operation of schemes to be consistently scrutinised and evaluated in a way that enables their performance to be comprehensively understood in sufficiently real time.[84] Transparency about

[77] *Asbestos-Related Diseases (Occupational Exposure) Compensation Act 2011* (Tas).
[78] *Asbestos Related Claims (Management of Commonwealth Liabilities) Act 2005* (Cth).
[79] *Waterfront Workers (Compensation for Asbestos Related Diseases) Act 1986* (WA).
[80] *Parliamentary Injury Compensation Scheme Instrument 2016* (Cth).
[81] Equal Opportunity and Human Rights Commission, *Independent Review into Sex Discrimination and Sexual Harassment, Including Predatory Behaviour, in Victoria Police: Phase 1 Report—Dec 2015* (2015) <https://www.humanrights.vic.gov.au/resources/independent-review-victoria-police-phase-1/> accessed 14 October 2023; Victorian Government, 'Restorative Engagement and Redress Scheme' (22 March 2023) <https://www.vic.gov.au/redress-police-employees> accessed 14 October 2023.
[82] Mary Ward, 'Humiliating and Degrading: Patients Wait Months for Answers on Vaccine Injury Claims' (*Sydney Morning Herald*, 19 November 2022) <https://www.smh.com.au/national/humiliating-and-degrading-patients-wait-months-for-answers-on-vaccine-injury-claims-20221117-p5bz9m.html> accessed 14 October 2023.
[83] Australian Sports Commission, 'ASC Restorative Program' <https://www.ausport.gov.au/about/asc-restorative-program> accessed 14 October 2023.
[84] See Lim, Ng and Weeks (n 70) on the particular review and reform deficiencies for discretionary Commonwealth schemes.

the operation of schemes is effected primarily through annual reports they publish, often explaining the number of claims made and in some cases the extent to which various key performance indicators are met.[85] It typically takes a crisis of some kind for such reports to attract media attention. Some schemes do not provide even this basic information.[86]

In some schemes there is a degree of parliamentary oversight. The New South Wales Parliament has a statutory oversight role in relation to that state's major compensation schemes (for transport crash and work-related injury, dust diseases and lifetime care and support).[87] The periodic parliamentary reviews provide a useful opportunity for problems arising in scheme operations to be ventilated and for stakeholders to engage with the scheme and the parliamentary committee.[88] Still, this mechanism did not prevent an underpayment scandal that saw the government backpay $AUD38 million to 53,000 injured workers in November 2021.[89] Ombudsmen have also played a role in highlighting systematic failures in scheme functioning. In Victoria, the Ombudsman receives complaints about the state-run workers' compensation scheme. On the basis of those complaints, the Ombudsman initiated a series of own-motion investigations into aspects of the scheme.[90] Its 2016 report investigated the handling of complex claims and identified a range of concerning practices by insurance agents and insufficient regulatory oversight; when complaints persisted, the Ombudsman embarked on a follow-up investigation and reported in 2019 that claims managers were continuing to make unreasonable decisions. Following the Ombudsman's recommendations, the dispute resolution processes in the scheme were comprehensively reformed.[91]

[85] See, eg, National Redress Scheme (n 74) and the annual reports of compensation authorities and regulators such as the Transport Accident Commission in Victoria (Transport Accident Commission, *Annual Report 2021–22* (TAC 2022)) and WorkCover Queensland (WorkCover Queensland, *Annual Report 2021–22* (WorkCover Queensland 2022).

[86] Little information has been published about the operation of the COVID-19 vaccine injury compensation scheme, for example, with media reporting providing the only insight (see, eg, Ward (n 82)).

[87] *State Insurance and Care Governance Act 2015 (NSW)* s 27.

[88] See, eg, the discussion of legal support for claimants in New South Wales Parliament, Legislative Council Standing Committee on Law and Justice, *2022 Review of the Compulsory Third Party Insurance Scheme* (Report 82, February 2023) 23–34.

[89] icare, 'icare to Pay Out $38 million to 54,000 Injured Workers' (11 November 2021) <https://www.icare.nsw.gov.au/news-and-stories/2021/icare-to-pay-out-38-million-to-injured-workers> accessed 14 October 2023.

[90] Victorian Ombudsman, *Investigation into the Management of Complex Workers Compensation Claims and WorkSafe Oversight* (September 2016); Victorian Ombudsman, *WorkSafe 2: Follow-up Investigation into the Management of Complex Workers Compensation Claims* (December 2019); *Victorian Ombudsman, WorkSafe 3: Investigation into Victorian Self-Insurers' Claims Management and Worksafe Oversight* (June 2023).

[91] *Workplace Injury Rehabilitation and Compensation Amendment (Arbitration) Act 2021* (Vic). The reforms sought to 'provide an inexpensive, timely resolution pathway available to all workers, including those whose disputes would otherwise have remained unresolved and those whose disputes would have been taken to court, but who would prefer a less adversarial pathway' (Victorian Parliament, Legislative Assembly, *Workplace Injury Rehabilitation and Compensation Amendment (Arbitration) Bill 2021*, 3 March 2021, Second reading, Melissa Horne, 762).

Taken together, the existing transparency and accountability mechanisms are as diverse and inconsistent as the schemes themselves: they are more established and productive in the context of schemes with longevity, but even those schemes are not free from scandal. The absence of monitoring and evaluation in some schemes gives rise to the impression that the step of establishing the scheme was regarded as the solution to the problem of the financial, care and support needs of the relevant claimant group, rather than ensuring that those needs were effectively and appropriately met by the scheme in the medium and longer term. Scholars who are engaged in research on law's response to injury, disease and other harms should pay attention and hold schemes to a higher standard.

5. Why Do Compensation Schemes Deserve More Scholarly Attention?

Compensation schemes are deserving of more scholarly attention and analysis for a number of reasons. A significant number of people are directly affected by the design and operation of these schemes as claimants on them each year. In Australia, our work and transport crash schemes provide benefits to more than 300,000 new claimants annually.[92] Compensation schemes are among the major mechanisms by which redress is delivered to people who have sustained relevant categories of illness and injury, meaning that the population exposed to schemes is large. Additionally, the expenditure of public funds on the delivery of support is significant, in the absence of which the burden of support falls either on the ill or injured or the lesser safety net of the social security system (and NDIS, in cases involving injury with serious, permanent impairment), to the detriment of potential claimants and the broader community.

Where new schemes are implemented for more discrete claimant populations, there should be attention to whether the scheme has been implemented as planned and achieves the intended outcomes. While schemes periodically report on their performance, there is very limited scholarly critique and analysis. While there is, for instance, a burgeoning literature around dispute system design,[93] there is little equivalent evidence around compensation scheme design—meaning that the opportunity for new schemes to learn from those that have gone before is often limited to the transfer of knowledge between experienced personnel. There is an empirical evidence base that indicates that the design of schemes and legal processes can themselves exert an adverse health impact on claimants,[94] which is troubling in

[92] Genevieve Grant, 'Claiming Justice in Injury Law' (2015) 41(3) Monash University Law Review 618, 618–9.

[93] See, eg, Lisa Blomgren Amsler, Janet Martinez and Stephanie E Smith, *Dispute System Design: Preventing, Managing, and Resolving Conflict* (Stanford University Press 2020).

[94] See, eg, Alex Collie, *The Mental Health Impacts of Compensation Claim Assessment Processes* (Insurance Work and Health Group, School of Public Health and Preventive Medicine, Faculty of

view of the restorative objectives of both tort and compensation schemes. In some instances, claimants may have experienced significant trauma over an extended period as a direct result of their exposure to the civil justice system and in pursuit of tort claims (as highlighted by the Royal Commission into Institutional Responses to Child Sexual Abuse).[95] Given law's role in the damaging experiences such claimants may have had, there is a kind of moral imperative for legal scholarship in ensuring compensation schemes do better in respect of claimants' outcomes and experiences.

Another justification for a greater scrutiny of the way compensation schemes perform is the need for claimant-focused advocates to highlight the aspects of schemes that are unjust and inefficient, and operating in harmful ways. This role often falls to plaintiff lawyers, who may certainly see more often than most the experiences that claimants have with schemes, but may also have interests that do not entirely overlap with those of claimants. Debates about scheme development and reform often engage influential and invested stakeholders, such as insurers, employer organisations, institutions and governments.[96] With such powerful actors involved, particularly in the presence of the profit imperative of insurance entities in privately underwritten schemes or where claims management functions are outsourced to private providers, the risks to claimant interests in claims administration are ever-present. Additionally, the potential for basic access to justice problems to arise in compensation schemes is great: for example, in connection with people understanding their eligibility to make a claim, or claim particular benefits, or challenge decisions that are made by schemes or insurers about the entitlements a claimant has. It is the classic 'one shotter' claimant versus the 'repeat player' scheme, insurer or decision-maker[97]—writ large. Most fundamentally, however, greater research scrutiny of compensation schemes is needed to address the fundamentally unresolved questions about the long-term outcome schemes achieve. Do they improve community health and wellbeing in a socially and financially responsible and just manner? What is the degree of transition from schemes to other support mechanisms? What are the long-term impacts of claiming for claimants? Little is known about how tort performs in these respects, too.

Medicine Nursing and Health Sciences, Monash University 2018); Alex Collie et al, 'Does Time off Work After Injury Vary by Jurisdiction? A Comparative Study of Eight Australian Workers' Compensation Systems (2016) 6(5) BMJ Open e010910.

[95] Royal Commission into Institutional Responses to Child Sexual Abuse (n 72).

[96] See, eg, the contributors to the parliamentary committee deliberations for the *2022 Review of the Compulsory Third Party Insurance Scheme* (n 88).

[97] Marc Galanter, 'Why the Haves Come Out Ahead: Speculations on the Limits of Legal Change' (1974) 9 Law & Society Review 95, 97–104.

6. The New Landscape of Compensable Injury Research: An Opportunity and a Plea

In their preface to the 9th edition of *Atiyah's Accidents*, Cane and Goudkamp suggest that, regrettably, 'scholarly interest in the field of accident compensation has waned'.[98] This may be true, but only of legal scholarship. Despite the flurry of new scheme activity described above and the long-standing framework of work and transport crash injury compensation arrangements in Australia, few legal scholars are now working in this area. Compensation schemes are evidently unfashionable as a scholarly interest. It is rare to see undergraduate honours theses and much less Masters and PhD theses on topics to do with personal injury. This experience may be the same for tort and private law more broadly. In Australian law schools, topics in the fields of public law and human rights have long dominated the research interests of students, which has a flow-on effect to the areas of focus that are most prevalent in the community of legal scholars.

There is a notable but ultimately sparse history of empirical legal research on claiming behaviour and outcomes in connection with tort, insurance and injury compensation.[99] Much of this work is now dated and while of course still enormously valuable, it does not engage with social, legal and technological changes of the last 30 years. The impacts of major developments remain to be explored—for example, changes in the market for legal services; the nature of work;[100] and the shift to online legal information, claims processes and courts. The absence of newer generation legal research on compensation schemes is a problem. It contributes to a destructive cycle: minimal evidence of scheme performance is available to contribute to reform processes or the design of new schemes, and there is little sustained scrutiny or evaluation of the operation of reforms once they are implemented or changed. This means there is a lack of evidence available to inform subsequent reforms within and beyond jurisdictions.

Research on compensable injury remains alive and well, however: it is simply located outside law schools in universities. There is now a considerable amount of transdisciplinary research about the nature and impacts of injury claims and compensation schemes, spanning such fields as epidemiology, health services research and health economics. Prominent themes in the research include the health impacts of claims processes;[101] predicting the trajectory of claimant recovery after

[98] Cane and Goudkamp (n 27) xx.

[99] Key examples include the Pearson Commission's analysis of these matters (n 17); the report of the Centre for Socio-Legal Studies (n 20); Deborah R Hensler et al, *Compensation for Accidental Injuries in the United States* (RAND Corporation, 1991).

[100] cf Harry Glasbeek, 'The Legal Pulverisation of Social Issues: *Andar Transport Pty Ltd v Brambles Ltd*' (2005) 13 *Torts Law Journal* 217.

[101] See, eg, Christa Orchard, Nancy Carnide and Peter Smith, 'How Does Perceived Fairness in the Workers' Compensation Claims Process Affect Mental Health Following a Workplace Injury?' (2020) 30 Journal of Occupational Rehabilitation 40.

injury (to facilitate early intervention in claims management, where it could be beneficial);[102] and exploring the health and social consequences for claimants of changes in scheme benefits and design.[103] This is critical work for understanding the operation of various mechanisms of compensation for injury and disease. It has much to offer legal analysis of compensation. Consider this fundamental question: what happens to a claimant after their claim has resolved? Legal scholarship typically pays scant attention, beyond the issue of collateral benefits.[104] An emerging series of studies is linking datasets to investigate the movement of claimants in Australian workers' compensation and transport crash compensation schemes to the social security system.[105] This work enables us to better understand the degree of cost shifting between support schemes as well as the longer-term trajectories of injured claimants—issues that are central to our understanding of schemes' performance. Other novel work is employing computational social science methods to model the dynamism of compensation schemes and simulate policy changes and law reform, as a means to test their impacts and inform decisions about choices and implementation.[106]

Research that traverses disciplines in compensation settings has its risks. Inattention to the interdisciplinary context can lead to the production of flawed or misleading evidence, which can itself contribute to inappropriate reform. We have seen this over many years with the simplistic way concepts around lawyer engagement and claiming activity have been misused and misconstrued in cohort studies in injury epidemiology research, often leading to inaccurate findings about the health impacts of compensation claims processes and lawyer use.[107] High-quality collaborations between legal, health and medical and other researchers have the potential to provide evidence of the operation and effects of compensation schemes in ways that can more reliably contribute to future reforms. Through such collaborations, research could broaden the agenda of relevant evaluative criteria for the study of the compensation mechanisms.

[102] See, eg, Christopher Papic et al, 'Factors Associated with Long Term Work Incapacity Following a Non-Catastrophic Road Traffic Injury: Analysis of a Two-Year Prospective Cohort Study (2022) 22(1) BMC Public Health 1498.

[103] See, eg, Alex Collie et al, 'Impact of Legislative Reform on Benefit Access and Disability Duration in Workers' Compensation: An Interrupted Time Series Study' (2020) 77(1) Occupational and Environmental Medicine 32.

[104] Richard Lewis, 'Deducting Collateral Benefits from Damages: Principle and Policy' (1998) 18(1) Legal Studies 15; Luntz (n 66).

[105] Daniel Griffiths et al, 'Transition Between Social Protection Systems for Workers With Long Term Health Problems: A Controlled Retrospective Cohort Study' (2023) 23 SSM-Population Health 101491.

[106] Jason Thompson and Camilo Cruz-Gambardella, 'Development of a Computational Policy Model for Comparing the Effect of Compensation Scheme Policies on Recovery After Workplace Injury' (2022) 32(2) Journal of Occupational Rehabilitation 241.

[107] Clare E Scollay, 'Exploring the Dynamics of Legal Service Use in Compensation Systems' in Prue Vines and Arno Akkermans (eds), *Unexpected Consequences of Compensation Law* (Hart, 2020).

The treatment of schemes as benign administrative processes providing a better, cheaper or more restorative option than tort has meant that other implications associated with their use—expressive, equity and justice-orientated functions—are often sidelined. Evidence of these impacts will inevitably have to compete with other kinds in policy debates around compensation system performance and reform. Still, it can provide insight necessary to the development of a more complete and evidence-based approach to the operation of schemes. Greater attention from legal scholars would bring a focus on such matters as whether law is being implemented as intended, whether those who are eligible are able to claim and the nature of barriers to claiming, and where there are systematic elements of the design of schemes and claims processes that are damaging, and the ways in which these could be improved. Claimants' experiences in schemes, when understood as a matter of access to justice, have a quality that elevates them beyond a 'customer experience', in the terms preferred in many government, insurance and commercial services. A legal lens is more likely than some others to be attentive to equity considerations, including in respect of access to benefits and the role and value of legal assistance. In the contest of such competing objectives as fairness, efficiency, just compensation and financial viability, the case needs to be made for the justice dimension of the work schemes do, or it stands to be sidelined and diminished over time.

Much of the most valuable research about the operation of contemporary compensation schemes makes novel use of administrative data across insurance, health services and legal institutions,[108] linking datasets to cast new light on the nature and impacts of compensation claims and processes. Access to data has always proved a challenge for this kind of research, which often relies on the discretion of insurance organisations.[109] The growing enthusiasm for big data and its capacity to generate insight into a range of services and systems has spread to civil justice, litigation, insurance and compensation-related fields.[110] To get a better understanding of the way compensation mechanisms are performing and the experience claimants are having within and beyond them, a new generation of transdisciplinary empirical research is required and legal scholars must get involved. This call may not be well received by some private law enthusiasts. Only a few years ago a keynote speaker at an event in this scholarly community decried

[108] Administrative data is information collected and stored as part of the everyday functions of organisations, rather than for a specific research purpose (Hugh M McDonald et al, *Apples, Oranges and Lemons: The Use and Utility of Administrative Data in the Victorian Legal Assistance Sector* (Victoria Law Foundation 2020)).

[109] Tom Baker, 'Transparency Through Insurance: Mandates Dominate Discretion' in Joseph W Doherty, Robert T Reville, and Laura Zakaras, *Confidentiality, Transparency, and the US Civil Justice System* (OUP, 2012).

[110] See, for example, Natalie Byrom, *Digital Justice: HMCTS Data Strategy and Delivering Access to Justice* (The Legal Education Foundation, 2019); Khic-Houy Prang, Behrooz Hassani-Mahmooei and Alex Collie, 'Compensation Research Database: Population-Based Injury Data for Surveillance, Linkage and Mining' (2016) 9(1) BMC Research Notes 1.

what he described as the 'odd preference' among law school deans, appointment committees and research funders for expensive empirical and multidisciplinary projects, compared with the 'highly cost-effective largely solitary research' typical in private law.[111] The position of private law, it was suggested, is undermined by the growing appetite for empirical and other styles of legal scholarship.[112] On the contrary, we argue that this kind of work is essential to ensuring that high-quality, legally informed scholarship on tort and injury compensation schemes does not disappear completely.

So what might be the focus of this work? In *Atiyah's Accidents*, Cane and Goudkamp articulate the following evaluation criteria for 'judging the success of a system as a mechanism for compensating for death, injury and disability':

> First, how many of those who are entitled to compensation according to the rules of the system actually receive compensation and in the amount to which they are entitled? Secondly, how many people, by fraud or abuse of the system, receive compensation to which they are not entitled? Thirdly, what are the administrative costs of the system?[113]

A high-priority addition to this list—and an avenue for the kind of engaged interdisciplinary approach we are proposing—would be a more user-centred empirical legal scholarship on tort and compensation schemes. Such research requires us to ask: what impacts are the schemes themselves having on claimants? Focusing on the experiences claimants (and potential claimants) have in participating in schemes, and the diverse outcomes and effects of those engagements, is more than a variation on the 'administrative costs' theme—it provides a basis for evaluating compensation schemes in a more complete and informative way.[114] To do so could draw on the learnings of the patient-centred care and lived experience emphases in health and medical research, as well as more fully engaging with the insights legal needs studies have generated into the barriers and facilitators of access to justice for people who encounter legal problems related to injury, illness and disease.

[111] Andrew Burrows, 'Challenges for Private Law in the 21st Century' in Kit Barker, Karen Fairweather and Ross Grantham (eds), *Private Law in the 21st Century* (Hart, 2017) 32–3.

[112] ibid.

[113] Cane and Goudkamp (n 27) 395.

[114] See, eg, Natasha Cortis and Ilan Katz, 'Waiting for Redress: Child Sexual Abuse Survivors' Experiences of Australia's National Redress Scheme' (2022) 129 Child Abuse & Neglect 105657, 8 (noting that survivors 'encountered further harm while following the administrative processes they hoped would help realise their rights and provide justice' and suggesting that 'new Schemes should be conceptualised and designed from the outset in ways that are informed by survivors' experiences').

7. Conclusion

The reality of the performance of tort and compensation systems should be more rigorously measured, scrutinised and understood. This is a project we hope Professor Stapleton might continue to regard as worthwhile. Though she moved away from her doctoral work's exploration of the scope of coverage of injury law and scheme design, the commitment to the practical Stapleton exhibited in *Disease and the Compensation Debate* (1986) is echoed in her more recent advocacy for reflexive tort scholarship and communication with the bar and bench. In her Clarendon Law Lectures, Stapleton referred to her early scientific training as underpinning her preference for analysing tort law as it is, noting that '[r]eality always trumps the model' and 'the scientist seeks to confirm the strength of a model by its capacity to resolve real problems and to predict outcomes'.[115] A more concerted effort to apply this approach to contemporary tort and injury compensation law has the genuine potential to yield new insight and reinvigorate research in this important field.

[115] Stapleton, *Three Essays* (n 6).

Index

For the benefit of digital users, indexed terms that span two pages (e.g., 52–53) may, on occasion, appear on only one of those pages.

General and personal references to Jane Stapleton can be found under her main entry at 'Stapleton, Jane'. She is referred to elsewhere in the index as 'JS'.

Second Language Attrition and the Case of Irish

INQUIRIES IN LANGUAGE LEARNING

FORSCHUNGEN ZU PSYCHOLINGUISTIK
UND FREMDSPRACHENDIDAKTIK

Edited by/Herausgegeben von
Christiane Bongartz und Jutta Rymarczyk

VOLUME 33

Zu Qualitätssicherung und Peer Review der vorliegenden Publikation

Die Qualität der in dieser Reihe erscheinenden Arbeiten wird vor der Publikation durch eine Herausgeberin der Reihe geprüft.

Notes on the quality assurance and peer review of this publication

Prior to publication, the quality of the work published in this series is reviewed by one of the editors of the series.

Eimear Geary

Second Language Attrition and the Case of Irish

An Exploration of the Savings Paradigm with Respect to Lexical Item Knowledge

PETER LANG

Bibliographic Information published by the Deutsche Nationalbibliothek
The Deutsche Nationalbibliothek lists this publication in the Deutsche Nationalbibliografie; detailed bibliographic data is available in the internet at http://dnb.d-nb.de.

Library of Congress Cataloging-in-Publication Data
A CIP catalog record for this book has been applied for at the Library of Congress.

Zugl.: Köln, Univ., Phil. Fak., Diss., 2022

D 38
ISSN 1868-386X
ISBN 978-3-631-86905-5
E-ISBN 978-3-631-87895-8
E-ISBN 978-3-631-87896-5
DOI 10.3726/b19720

Peter Lang – Berlin · Bern · Bruxelles · New York · Oxford · Warszawa · Wien

This publication has been peer reviewed.

www.peterlang.com

In loving memory of my mother, Bridie Geary
(1942–1990)

Preface

Language Learning is a field which bridges the gap between the research conducted within Psycholinguistics and the applied research within Foreign Language Didactics. For a long time, these two fields were regarded as separate disciplines, and the emphasis lay on their differences. However, just as there has been a gradual convergence between the concepts of *language acquisition* and *language learning*, over the past few years Psycholinguistics and Foreign Language Didactics have also been moving closer together. While Psycholinguistics is taking a growing interest in the classroom context in which language learning takes place, Foreign Language Didactics have fully embraced empirical research which sheds light on the linguistic phenomena found in the interactions within the classroom.

The series *Inquiries in Language Learning (Forschungen zu Psycholinguistik und Fremdsprachendidaktik)* aims to reflect this development. Since the areas of intersection between these two research fields have a high level of interdisciplinarity, the contributions to this series are relevant in many different ways for educators and researchers who are concerned with language learning. On the one hand, good foreign language or second language teaching requires teachers whose methodological and pedagogical decisions are based on a sound knowledge of language acquisition theory. Furthermore, foreign language textbooks should have a solid empirical foundation. On the other hand, the interpretation of linguistic data requires familiarity with the types of classroom activities and rituals that shape the various learning processes. After all, psycholinguistic research design must attend to the technicalities of classroom teaching and learning in order to obtain authentic results.

In this series we hope to contribute to the cross-disciplinary efforts in our research fields, bringing together psycholinguistic principles and classroom-based developments, thus reconciling theories and methods with research and practice.

Christiane Bongartz
Jutta Rymarczyk

Vorwort

Sprachenlernen/ *Language Learning* ist das Bindeglied, das die naturwissenschaftliche Forschung der Psycholinguistik und die anwendungsorientierte Forschung der Fremdsprachendidaktik zusammenführt. Lange Zeit wurden die Disziplinen getrennt voneinander behandelt und die Betonung lag auf den disparaten Anteilen der beiden Gebiete. Vergleichbar zur Annäherung der Begriffe „Spracherwerb" und „Sprachenlernen" (*language acquisition* und *language learning*) ist jedoch seit einigen Jahren eine Annäherung der Psycholinguistik und der Fremdsprachendidaktik zu beobachten. Während die Psycholinguistik den schulischen Kontext des Spracherwerbs stärker beachtet, ist aus der Fremdsprachendidaktik die empirische Forschung nicht mehr wegzudenken, die linguistische Phänomene der Interaktion im Klassenzimmer beleuchtet.

Mit der Reihe *„Inquiries in Language Learning.* Forschungen zu Psycholinguistik und Fremdsprachendidaktik" wollen wir dieser Entwicklung Rechnung tragen. Da die Schnittstelle der beiden Forschungsgebiete, die durch die Reihe bedient wird, naturgemäß eine hohe Interdisziplinarität aufweist, strahlt ihre Relevanz in unterschiedliche Richtungen aus: Einerseits braucht guter Fremdsprachenunterricht Lehrkräfte, deren methodisch-didaktische Entscheidungen auf detaillierter Kenntnis spracherwerbstheoretischer Aspekte beruhen. Das Schreiben von Lehrbüchern für den Fremdsprachenunterricht muss auf einer soliden empirischen Basis geschehen. Andererseits bedarf die Interpretation psycholinguistischer Daten der Vertrautheit mit Unterrichtsabläufen und den Ritualen, die Vermittlungsprozesse prägen. Das Entwerfen eines psycholinguistischen Forschungsdesigns muss unterrichtstechnische Aspekte einbeziehen, um letztlich authentische Ergebnisse abbilden zu können.

Mit der Gesamtschau unserer Arbeitsbereiche hoffen wir dem Ineinandergreifen und den Verschränkungen von psycholinguistischen Grundlagen und fachdidaktischen Weiterentwicklungen, von Theorien und Methoden sowie von Forschung und Praxis gerecht werden zu können.

Christiane Bongartz
Jutta Rymarczyk

Table of contents

List of Tables

List of abbreviations

ATH	Activation Threshold Hypothesis
ATM	Attitudes and Motivation
CDST	Complex Dynamic Systems Theory
FLA	First Language Acquisition
L1	First Language
L2	Second Language
MFL	Modern Foreign Language
NTB	Neurolinguistic Theory of Bilingualism
RIF	Retrieval Induced Forgetting
SLA	Second Language Acquisition

Acknowledgements

I especially wish to express my sincere gratitude to my supervisor, Prof. Dr. Andreas Rohde, for his unfailing support and guidance throughout this project and with whom I enjoyed many interesting and stimulating conversations about the complexities and intricacies of language acquisition and attrition.

I would also very much like to thank my examiners, Prof. Dr. Christiane Bongartz and Prof. Dr. Dany Adone, for their willingness to take part in the project; I am very aware of the great amount of time and effort that that involves.

A very big thank you also to my colleague, Dr. Kim Schick, for her reassuring encouragement and sound advice and to Kerstin Chlubek and Pia Holtappels for taking time out from their busy schedules to assist in the testing procedure. I also wish to thank my colleagues at the English Department II of the University of Cologne for their support and encouragement. I feel very fortunate to be part of such a great team.

I am extremely indebted to Andrew O'Sullivan, Cara Donnellan, Éanna O'Sullivan and Eilin Geraghty for their invaluable input and assistance with this project. I am also immensely grateful to my father, Maurice Geary, and Sinéad Geary, John Oman, Tom Geary, Úna Ní Raifeartaigh, Susanne Köhler, Uwe Buchner, Alf Gündel, Katrin Geiger, Stefanie Anzinger and all my dear friends for their unwavering support and kindness.

Most of all, and from the bottom of my heart, thank you to Liam and Maeve for putting up with the long hours of writing at home and for all of the fun times in between, as well as for your remarkable wisdom and inspiring attitude to life that kept me focused and motivated throughout this project. This book could not have been written without you.

Chapter 1: Introduction

With increasing globalisation, migration and international travel, bi- and multilingualism have become normal phenomena with much linguistic research conducted accordingly on the acquisition and learning of language. When an individual acquires a second or more languages, this invariably has repercussions for the inner workings of the linguistic system and the co-existence of these languages, with increased use of one language influencing availability of the other. Increasingly in the last 30 years, researchers have begun to explore this phenomenon and how periods of disuse and inactivation can lead to attrition of linguistic knowledge.

Language attrition is considered a process whereby knowledge (partly or completely) of a particular language once known is lost or "forgotten" over time. Explorations of attrition attempt to gain a more comprehensive insight into linguistic representations in the brain and to understand what supports the consolidation of information so that language knowledge is retained, even after long periods of language disuse. It also asks if seemingly forgotten knowledge is in fact still intact and, should this be the case, what conditions need to be in place for this information to be reactivated.

The present study explores whether or not second language (L2) lexical knowledge acquired in childhood and puberty has been completely erased after a prolonged period of inactivity or whether it has merely become inaccessible, i.e., whether residual levels of knowledge of the language are still present and available and through re-exposure can be reactivated. The L2 in question is Irish, a language that is learned throughout one's entire period of school education in the Republic of Ireland, starting in the first year of primary school. It is a compulsory subject for the majority of pupils and has to be taken in the state school examination, the Leaving Certificate, at the end of the final year at secondary school.

Moreover, the present study also investigates claims put forward in language attrition research with respect to the role played by extra-linguistic factors in the attrition process and how these may account for variation in performance across participants. These include proficiency at the time of onset of attrition, age at the start of learning as well as age at the onset of attrition, context of acquiring or learning the language, and attitude and motivation (Murtagh 2003; Schmid & Mehotcheva 2012; Watson & Nic Ghiolla Phádraigh 2009).

The book also provides a sociolinguistic exploration of Irish as a minority language as this is considered an important factor in understanding learner experiences and the choices made at the individual level with respect to language use. These experiences and decisions are critical considering language activation processes and how accessibility of linguistic information in the long-term might be affected accordingly.

1.1 Research objectives

Specifically, the objectives of the study are as follows:

1. To assess levels of retention of lexical items of school-learned Irish after an incubation period of approximately 30 years.
2. To ascertain what factors might account for difference in performance across individuals.
3. To establish whether or not there is a relearning advantage with respect to seemingly forgotten lexical items in the Irish language, i.e., to test the Savings Paradigm with respect to lexical knowledge.
4. To determine whether levels of retention parallel performance in relearning.

Four adults were tested. Whilst they are currently resident in Germany and fluent speakers of German, all grew up in Ireland and completed their education there (including university). They all followed the same Irish language syllabus (Curriculum and Examinations Board 1984), with only one of the participants starting to learn Irish later, i.e., at secondary school level, after her family moved to the Republic of Ireland from Northern Ireland. The other participants began Irish lessons as soon as they started primary school, which was at the age of four or five years. The participants were in the age range 45–50 years at the time of testing and had had no official contact with Irish (i.e., in an instructional context) and little to no contact with the language in a natural setting in 25–30 years.

The project was conducted in the form of a case study with each participant being tested individually to establish current levels of knowledge of lexical items and whether or not seemingly lost knowledge is indeed still available at a sub-threshold memory level. A detailed interview was conducted to explore experiences in learning and language use in an attempt to account for variation in performance and to explore further the possible impact of extra-linguistic factors on levels of knowledge retention. An accompanying questionnaire was also distributed to collect background biographical information.

In a similar format to that of De Bot & Stoessel (2000), three levels of Irish lexical item knowledge were tested: *recall*, *recognition* and possible *savings*. When one is able to recall an item, one is able to produce the L2 translation (or an appropriate synonym) of the presented L1 item. Recognition, on the other hand, refers to the ability to understand the meaning of the presented L2 item, i.e., one can translate it into the L1 or offer an appropriate synonym in the L1. Savings, as explained in greater detail in Chapter 2, Section 2.2.4, pertains to the idea that information that is seemingly forgotten is indeed still available and through re-exposure can be reactivated and accessed again.

An important consideration in language attrition analysis is the context and circumstances under which the attrited language was acquired or learned. Therefore, Schmid & Mehotcheva (2012) highlight the importance of specifying whether the attrited language was a first language (L1), a second language (L2) acquired while immersed in a natural setting or a foreign language (FL) learned in an instructional, institutional one. In the case of Irish a strict categorization along the lines of L1, L2 and FL proves difficult. Considering its official and symbolic status in Irish society, the prolonged exposure to the language that Irish pupils have, and the varying degrees of input available across homes, the lines drawing such distinctions are blurred. Irish is, nonetheless, categorised as an L2 as opposed to an FL in this project due to its status, historical significance and omnipresence in Irish society in general, as well as the length of time pupils in Ireland have Irish as a subject at school compared to other languages. On the other hand, despite readily available input from a range of sources such as radio, TV, print media and the internet, Irish becomes for many a language that they "forget" once they leave school.

1.2 Uniqueness of the present study

The Irish language provides an interesting and unique focus for attrition analysis because unlike languages typically explored in attrition research, a monolingual Irish speaker does not exist, i.e., all Irish L1 speakers are also English L1 speakers, i.e., bilingual Irish – English speakers. This naturally has profound repercussions with respect to the practical necessity of learning the language which invariably influences learner behaviour once official instruction has ended.

Another factor that distinguishes the study of attrition of Irish from other studies on L2 attrition is that it is a not a *modern foreign language*, as is typically taught in school, such as French or German. Its status as a minority language, therefore, affects curriculum planning and design in education and the particular motivations that manifest in learning it and eventually maintaining contact with it on leaving school. Such a study can thereby add more comprehensive insight into the complex nature of motivation considering that typically motives are examined within the modern foreign language learning milieu with such observations generally being assumed to be conclusive. This study, therefore, prompts a review of such interpretations.

The case of the Irish language also presents an invaluable opportunity for the study of attrition since many speakers, especially those who leave Ireland to live and work abroad after their education, experience complete isolation from the language with little to no exposure to it over prolonged periods. This eliminates the problem highlighted by De Bot, Martens & Stoessel (2004) which is that relatively easy access to second and foreign languages in multilingual countries such as The Netherlands, where much of the research on attrition has been conducted, makes it difficult to locate clear cases of attrition. In the present study total loss of contact with the language can for the most part be guaranteed since the participants no longer live in Ireland and on leaving school had no more contact with the language apart from one participant who took it as a subject in his first year at university and had some intermittent passive input in the intervening years up to the time of testing for the present study. Unlike modern foreign languages, Irish is also not present in the international media so that unless one consciously pursues it, one will not be automatically exposed to it.

This also provides the possibility to gain further insight into, and a deeper understanding of, potential individual and social factors that may influence the attrition process, such as motivation, issues of identity and attitudes towards the language itself, as well as the role that experiences in learning the language may have.

Another advantage of this study is the long period of incubation in which the language has not been used. Only a handful of studies that have dealt with long periods of disuse have been conducted to date (e.g., Bahrick 1984a, b; De Bot & Stoessel 2000; Hansen, Umeda & McKinney 2002). The majority of L2/FL attrition studies look at incubation periods ranging from three months up to four years (see Mehotcheva & Mytara 2019 for an overview).

What is also particular about the present study is that it is an exploration of L2/FL loss starting in an L1 environment and proceeding in an L3 environment. All participants in the study are fluent German L2 speakers.

1.3 The importance of understanding attrition

Understanding language loss has important implications for the L2/FL education domain and pedagogy. If it can be established what impacts greater retention of knowledge, teaching curricula and methodologies can be revised accordingly to make them more effective with respect to long-term knowledge retention.

Mehotcheva & Mytara (2019: 2) also state that consideration of attrition processes could provide for more effective design of language maintenance courses for learners whose earlier instruction has ended and who wish to revive and build on previous knowledge.

They also argue that considering the personal effort, time and resources invested in learning, it is important to ensure that this knowledge stays intact and whilst many strive to achieve this by consuming L2 media, the efficacy of this has yet to be confirmed by linguistic research. Therefore, more research on what supports knowledge retention is vital so that previous efforts in learning are not futile.

An understanding of the process of attrition and the possibility of a reactivation of lost linguistic knowledge is also becoming increasingly important considering the surge in migrant numbers in recent years in various regions across the globe. Comprehending the linguistic demands faced by minority language communities and the possible effects these may have on the languages that they already speak, can provide invaluable guidance on how to approach language education, deepening our understanding of what appropriate and effective teaching methodologies in multilingual groups should entail and how the language knowledge which migrants bring with them can be retained and availed of to assist in the learning of new languages.

Köpke (2004: 15) also argues that understanding language decay in healthy individuals can help shed light on assessment of decay in pathological cases.

1.4 The relevance of lexical attrition research

Larson-Hall (2019: 14) states that more research on L2 lexical attrition is necessary considering that so little has been conducted to date, in particular studies with longer incubation periods and a wider range of L2s; all conditions that the present study satisfies.

Moreover, Bardovi & Stringer (2019: 14) state that difficulties that appear in the comprehension and production of syntax are at times in fact due to problems in lexical retrieval so that lexical attrition can have direct repercussions for syntactic knowledge. As the lexicon constitutes the essential foundation of syntactic structures, to understand the attrition processes of syntax requires first and foremost an examination of lexical attrition, retention levels and possible issues pertaining to accessing this information.

Weltens & Grendel (1993: 142) also point out the very pragmatic reason for looking at the lexicon. Considering how much time and effort is invested in learning vocabulary in language education, it is important to understand what is necessary for this lexical knowledge to become consolidated in memory.

Additionally, Bardovi-Harlig & Stringer (2019: 12) refer to the importance of identifying one linguistic domain for analysis since just as different domains undergo different acquisition trajectories, they also cannot be grouped together when looking at attrition.

1.5 Challenges in studying attrition

1.5.1 Establishing baseline knowledge

As Tomiyama (2000: 310) states, establishing attrition is not straightforward. To find what was lost, one must be sure that it was there in the first place so that a comparison can be made between an earlier and a later state. Pan & Berko-Gleason (1986: 198) also refer to the challenge of establishing a knowledge baseline and a reference point with respect to proficiency against which current levels of language knowledge can be compared. They argue that this is even more challenging in the case of L2 attrition than in L1 attrition since particularly in the case of languages learned in an instruction setting, native-like competence was most probably never achieved. Therefore, in such cases assumptions about what knowledge might have been attained is difficult to make. This challenge has been overcome in the present study as each of the participants learned according to same uniform, standard curriculum and all of them had studied the texts from which the vocabulary items were taken, therefore, ensuring that this vocabulary had been known previously by the participants. All of the participants were also in school within a relatively similar time-span.

Moreover, grades were also available from their performance in the Leaving Certificate examination (the final state school examination) so that retrospective self-estimates which are clearly vulnerable to unreliability did not have to be made by the participants, thereby circumventing another problem that can arise in research as pointed out by Pan & Berko-Gleason (1986: 198).

1.5.2 The interplay of extra-linguistic factors

As stated by Larson-Hall (2019: 14), problems in lexical item retrieval cannot be explained according to one simple account. Mehotcheva & Mytara (2019: 4) argue that a number of variables interrelate to varying degrees over time in determining severity of attrition. They refer to De Bot (2007) who pointed to the individual nature of the process with much variability in retention levels even across individuals who learned and attrited under similar conditions.

Mehotcheva & Mytara (2019: 4), therefore, in keeping with De Bot's proposal, suggest adopting a dynamic systems perspective, according to which the process of attrition is understood as resulting from the interplay of a number of different variables within an overall system that is dynamic and in flux with the variables mutually influencing each other. The present study also considers variables as operating in such clusters rather than as encapsulated, isolated entities, understanding also that the relative importance and influence of different variables in the system can change over time. This changing constellation and varying strength of the different factors as well as their mutual influence within a cluster, is distinctly unique to each learner and attriter, thereby complicating the exploration of attrition.

Therefore, direct comparisons are difficult given the diversity of learner and attriter profiles. Opitz (2019: 1) also argues that the dynamic nature of language and the varying interrelation of influencing factors makes research on attrition particularly challenging. Since different variables operate according to changing and varying degrees of strength in each individual, generalised predictions prove elusive since factors, other than incubation, cannot be isolated. Indeed, Scontras, Fuchs & Polinsky (2015: 1) state that the exploration of all facets of multilingualism, therefore, warrants a particularly "nuanced approach".

It is argued in the present study that understanding attrition necessitates looking at each attriter individually and examining in detail the different variables and their interplay in each case, despite the complexity and ensuing challenges of such an endeavour.

1.5.3 Practical challenges

As Mehotcheva & Mytara (2019: 15) also state, the nature of some extra-linguistic influencing variables makes the attainment of quantitative measures challenging. Factors such as motivation and attitudes are not just susceptible to variation over time but any exploration is dependent on the subjective feedback of the participant and this may not always be reliable.

In the present study, there were also practical limitations in terms of the availability of the participants for testing. It was possible, however, to test them on four different occasions spanning 13 months. Moreover, Meara (2004: 141) states that there is also a practical limitation in terms of the number of words that can be tested. This is further elaborated on in Chapter 4, Section 4.4.1.3.

1.6 The role of qualitative analysis in attrition research

Considering the heterogenous nature of attriting groups and the interaction of individual factors leading to their distinct end-states, it has been widely argued that qualitative research focused on individual profiles can potentially provide fruitful insights into understanding the process of attrition (Pan & Berko-Gleason 1986; Herdina & Jessner 2013, Yilmaz & Schmid 2018). Herdina & Jessner (2013: 753) also suggest, considering the dynamic nature of attrition and the individual path of attrition across individuals, that "qualitative research suggested by language biographies might be favoured compared to more conventional quantitative methods." Graham (2012: 151) correspondingly highlights the "different history of involvement" with the language that each individual attriter incorporates and even when a language has been acquired or learned under similar circumstances, the experience of this is distinct to each individual.

Adopting a case study approach, as is done in the present study, presents an invaluable opportunity for the necessary detailed exploration of individual attrition trajectories and the unique interplay of variables impacting this.

1.7 Organisation of the study

Chapter 2 explains the process of attrition and key theories that have been presented in the literature to explain how languages are stored and interact so that one language appears to become dormant over a period of disuse. The Savings Paradigm which is of particular relevance to this study is also explored in detail in this section. This is followed by an examination of selected studies on L2/FL attrition, in particular research that has tested the Savings Paradigm and a possible relearning advantage for previously known but seemingly forgotten knowledge. An exploration of key factors deemed to impact degrees of attrition is then presented along with relevant findings that have emerged in attrition research on their respective roles.

Chapter 3 introduces the Irish language and offers a socio-linguistic perspective of the language. This is considered necessary for the purposes of the present study because how a language is perceived both at the societal and the individual level has significant consequences with respect to language behaviour and the choices that learners make accordingly. A brief explanation of specific grammatical features of Irish that explain the justifications for the category breakdown of the lexical items in the present study is also given. Chapter 4 outlines the research design and methodologies used offering justification for why certain decisions were made with respect to the study plan and organisation. The results and accompanying analysis of the study are presented in Chapter 5. This is followed by a discussion of these in Chapter 6 in view of key extra-linguistic variables that were examined in the study. Finally, Chapter 7 offers a brief overview of the findings.

Chapter 2: Language attrition

In the following, both the phenomenon of attrition and the key theories dominating the discourse in attrition research are explained. This is followed by an overview of selected studies rendered particularly relevant for the present investigation and a synopsis of the main factors deemed influential in determining the attrition process.

2.1 Understanding attrition

Traditionally and in general terms, language attrition has been understood as a complex, non-linear process whereby knowledge (in part or complete) of a particular language once known is lost or "forgotten" over time. The term has, however, been extended to refer not only to the idea of loss but to deviations in earlier language knowledge (be it the L1 or the L2) with this divergence resulting from one of the languages being used less or not at all (accessibility) whilst the second language emerges as the more dominant one (competition) (Köpke & Keijzer 2019). Attrition, however, does not refer to any changes in linguistic knowledge or capability arising from a cognitive deterioration caused by age, illness or injury (Yilmaz & Schmid 2018: 226).

There has been much interest in this phenomenon particularly with respect to digressions in L1 knowledge due to L2 influence (Yilmaz & Schmid 2018). For example, in the acquisition of an L2, Chamorro, Sorace & Sturt (2016) argue that the L1 is affected whereby it diverges or "attrites" over time relative to its state at onset of learning the L2 as a result of this new L2 knowledge. In keeping with this, Yilmaz & Schmid (2018: 228) refer to a two-way interaction between both languages in bilingual environments whereby the lesser-used language faces competition from the dominant language. In the case of an environment where the L2 is more prevalent, they argue that all speakers in such contexts can expect their L1 to experience a degree of change accordingly. In this regard, it is argued that the L1 of bilinguals is transformed and thereby processed differently compared to monolinguals' use of that same native language.

Herdina & Jessner (2013: 753) explain that language attrition is a dynamic, persistent yet irregular phenomenon, emerging as a consequence of the load experienced by the individual speaker when managing two or more languages. In viewing the language system as dynamic, they refer to transitional bilingualism, whereby speakers transition from one dominant language to another. In the case of L1 attrition, they argue that as the L2 grows, the L1 declines accordingly. This does not, however, result in total transition but rather an imbalance in the command of each language (Herdina & Jessner 2013: 752). They refer to Opitz (2013: 702) who explains attrition as emanating from changes in the overall multilingual linguistic system, for example, as a result of the sequential acquisition of another language. This means that each language in the system is subject to change as a result. Herdina & Jessner (2013: 753) state that attrition also comprises reduction or simplification of language systems and/or the impairment of access to them, emphasizing that it is a normal, and often unavoidable, consequence of bilingual language development.

An important consideration, therefore, in language attrition analysis is the context and circumstances under which the attrited language was acquired or learned and, as delineated section 1.1 and in keeping with Schmid & Mehotcheva (2012), whether the language under investigation constitutes an L1, L2 or FL.

Schmid & Mehotcheva (2012: 2) differentiate between *language attrition* and *language shift* stating that shift, whilst also constituting a form of language loss, is intergenerational and societal in nature unlike attrition which is an individual phenomenon and not a process proceeding over generations.

Indeed Schmid & Köpke (2013: 1) state that traditionally language attrition as a focus of study was often viewed as more relevant to the subdisciplines of language contact and change, as well as language death, rather than as an area of interest that could support a greater understanding of language acquisition and bilingualism. However, developments in the study of bilingualism to incorporate more holistic perspectives, including an increased interest in observations of the simultaneous development of each language in the bilingual's repertoire (for example, examining not just the effect of the L1 on the L2 but also observing how the L1 changes due to influence from the L2), led researchers to view attrition as an individual, psycholinguistic phenomenon arising from the interaction between two systems of linguistic knowledge, with the underlying knowledge system itself however remaining unchanged (Schmid & Köpke 2013: 2).

2.1.1 Attrition: An issue of performance or competence?

Köpke & Keijzer (2019: 1–2) ask whether attrition is a phenomenon emanating from performance issues or underlying competence, i.e., does the knowledge itself decay or is attrition rather a process whereby only access to, and the processing of, that information becomes difficult. They refer to widespread agreement in the literature that it is not the case that attrition can only affect competence or performance, rather that it can refer to either and this is determined by the age at which contact with the language is discontinued or at least the age when another language becomes more dominant, i.e., in cases of bilingualism. For example, in the case of L1 attrition, if contact with the language ceases or is reduced before the onset of puberty, levels of competence might be affected. Individuals who experience reduced contact with the language after puberty, on the other hand, tend to experience problems in performance, i.e., in accessing and processing information in the attriting language; however, this knowledge is still intact. Schmid (2007: 150) refers to a "saturation threshold", a level of linguistic knowledge attained that makes it immune to complete loss. This she argues is the case for adult L1 attriters who have achieved full maturation of the L1 before contact with the language is reduced or ceases completely due to the increased or exclusive use of an L2.

Whilst De Leeuw, Mennen & Scobbie (2012: 697–698) do not completely dispute a maturation phase which they consider to be biological in nature in language development, they argue, nonetheless, that language development is not determined by it exclusively. Instead, they propose that any such maturation does not preclude change in performance after this period has ended, thereby viewing language development as evolving, dynamic and complex, continuing throughout life, and at varied rates for individual speakers.

As Yilmaz & Schmid (2018) argue, when changes are observed, this does not mean that underlying competence is necessarily affected, meaning that if one looks at maturation in terms of underlying ability, maturational constraints may indeed still be in place, even though performance suggests otherwise. De Leeuw et al. (2012: 685) argue that if permanent loss could indeed be established, then this would be more indicative of an impaired competence as opposed to performance issues.

Yilmaz & Schmid (2018: 230–231) also highlight the distinction between production and perception but argue that whilst production problems seem to be indicative of retrieval or processing problems arising in real-time, these may not be representative of deviations or loss of linguistic knowledge at a deeper level, or rather, as discussed by Sharwood (1983a, 1983b, 1989) the issue appears to pertain more to the *control* of each linguistic system but attrition does not necessarily mean that either system is, in fact, modified.

Köpke (2004: 14) argues that performance attrition arises due to online competition between the two (or more) linguistic systems with the less dominant language experiencing higher activation levels as a result. Therefore, although productive skills have been shown to be more affected by attrition compared to receptive skills, this might be due to problems in simultaneously controlling each system as the individual strives to access and process the appropriate information in real-time and might not be indicative of loss in long-term memory. Observation of problems in perception might consequently offer insight into how deeper level knowledge has changed or deteriorated and help establish attrition effects that looking at production alone might not ascertain. Essentially, shortcomings in performance (production) do not necessarily imply difficulties in the underlying competence.

With respect to L1 attrition, Yilmaz & Schmid (2018: 231–232) provide an overview of studies that suggest that deeper linguistic knowledge appears to be relatively unaffected in periods of L1 disuse. They refer, for example, to studies conducted by Herdina & Jessner (2002) and Genevska-Hanke (2017) that showed that once re-immersed in the L1 environment, L1 attrition effects were reversed.

2.1.2 Impermanency of loss and the effect of relearning

The question is if the process of attrition culminates in a state of permanent loss or change or whether any such loss or changes can be reversed. In the case of the latter and looking at L2 attrition specifically, it would be expected that attriting L2 speakers would experience an easier resurgence/restoration of previously known but seemingly forgotten L2 knowledge compared to novice L2 learners in a learning scenario. This would consequently suggest that underlying competence is still intact, even if the linguistic system had been dormant for some time, with lack of use gradually developing into problems in accessibility as opposed to a complete erasure or cancelation of previously known knowledge.

2.1.3 Consolidation of information

The question is also if certain knowledge is less vulnerable to attrition and if so what causes this. Bahrick (1984a) refers to a permastore, i.e., linguistic knowledge that endures and is retrievable even after 25 years despite disuse. Mehotcheva & Köpke (2019: 20) refer to Neisser (1984: 33) who speaks of a critical threshold of strength whereby information becomes consolidated and steadfast in memory making it immune to decay. On the other hand, Larson-Hall (2019: 13) argues with respect to L2/FL lexical attrition specifically that immunity from deterioration can never be guaranteed; a phenomenon she terms the *Constant Decay Hypothesis* as based on model simulations of lexical loss conducted by Meara (1989, 2004) who observed that lexical structures and connections remain unimpaired and sound but problems can arise in accessibility over time and it is this which constitutes attrition. i.e., it is considered a performance phenomenon. While Larson-Hall's interpretation of loss is based on issues of retrievability of information and connections between various items and Neisser looks at the very strength or "depth of entrenchment" of that information, the two are not necessarily mutually exclusive. To emphasize consolidation of information in memory is not to deny that issues of accessibility can arise over long periods of inactivation. In fact, Larson-Hall (2019: 3) agrees that a critical threshold of strength might determine the longevity of information whereby that which is best secured in memory is most resistant to loss.

A dominant question in attrition research is consequently whether specific linguistic knowledge has just become inaccessible and with the right impulses may be activated again, i.e., can knowledge of a language that had been fully acquired ever be completely "lost" or "forgotten" or in the case of deviations or changes undergone by a linguistic system under the influence of another, can these be reversed. Referring to Sharwood Smith (1989, 2007), Yilmaz & Schmid (2018: 230) argue that whilst attrition is a phenomenon whereby underlying linguistic knowledge is affected with respect to divergence or issues of retrieval and accessing, this fundamental knowledge is, nonetheless, still intact.

In the case of L2 attrition the question of whether underlying knowledge is still available after prolonged periods of disuse is equally valid and constitutes a central focus of the present study on lexical knowledge. Irish language loss as portrayed by the participants in this study represents a case of language attrition with the discontinuation of use and exposure to the language being abrupt and an individual choice.

Of interest with respect to the focus of the present study is the finding from explorations in second and foreign language attrition research that production appears to be more vulnerable to attrition than comprehension (Bahrick 1984a; Wei 2014; Weltens 1988; Hansen 2011). Wei (2014) highlights the selectivity of second language attrition based on findings that show that, for example, the lexicon appears to be more vulnerable to attrition than syntax and morphology, with phonology displaying the greatest steadfastness.

It has also been argued that a number of individual factors may influence the degree of knowledge retention or loss in each specific case of attrition (Schmid & Mehotcheva 2012). These include age and attained proficiency at the onset of attrition, motivation, attitude towards the language, time since the onset of attrition, and frequency as well as recency of use (Bardovi-Harlig & Springer 2013; Bahrick 1984b; Murtagh 2003; Paradis 2007; Schmid & Mehotcheva 2012; Watson & Nic Ghiolla Phádraigh 2009). See Section 2.3 for a further exploration of studies conducted on L2 attrition.

With respect to *rate of attrition*, Mehotcheva & Köpke (2019: 6) state, based on an overview of the literature, that the overriding view appears to be that attrition sets in immediately when contact with the language ceases but after the first three to six years of incubation it appears to level-off, remaining at a plateau thereafter during which little to no further attrition appears to take place. Mehotcheva & Köpke refer to Yoshitomi (1999) who suggested that attrition might take place in the early stages of incubation instead of the later ones because masterly bilingualism necessitates the ability to simultaneously and spontaneously regulate the various linguistic subsystems at play in the language repertoire. This appears to suggest that attrition corresponds to an inability emanating from lack of practise and experience in regulating these subsystems, hence its emergence in the early years of incubation. Mehotcheva & Köpke (2019: 12) state, however, that there can be much deviation in rate of attrition depending on which extra-linguistic factors are at play, giving the example of faster rates of attrition observed among those who are younger at age of onset of incubation.

It is not possible to examine rate of attrition in the present study as such an investigation necessitates the collection of intermittent longitudinal data starting immediately at onset of incubation. Testing in this study, however, started after an incubation period of approximately 30 years so that such a trajectory could not be observed.

2.2 Attrition theories

A number of theories have been proposed in an attempt to explain the process of attrition. These theories are not exclusive to one particular linguistic domain.

2.2.1 The Regression Hypothesis

Traditionally, one of the most prominent theories has been the Regression Hypothesis put forward by Jakobson (1968). Whilst originally based on observations of phonological loss in aphasic patients in which it was noticed that the last items learned were those which appeared to be first lost (Berko-Gleason 1982), the application of the hypothesis has been widened to explore other linguistic domains and non-pathological cases of language attrition, whereby attrition is viewed as a reverse process to that of acquisition (see Keijzer (2007) for an overview of research on the Regression Hypothesis in cases of non-pathological language loss). This view has been criticized, however, on the grounds that easily identifiable linear losses of linguistic knowledge are difficult to establish and evidence in its support is sparse (De Bot 1997). The accumulative nature of vocabulary learning which is typical for L2 and FL teaching is also difficult to assess in the case of Irish since exposure to the language was not necessarily limited to the classroom context and the complexity of the content did not, by definition, follow a linear progression. Whilst Ecke (2004: 333) refers to some studies claiming support for the hypothesis, he too admits that insufficient empirical data on actual patterns of regression restrict the scope for reliable interpretation arguing that the hypothesis appears to be “too unrefined” to properly expound language attrition processes.

Hedgcock (1991: 47) argues in his discussion of the Regression Hypothesis that the issue is not one of how late an item was learned but rather of how well it was learned. The more frequently an item is used the more entrenched this linguistic knowledge becomes. Specifically with respect to lexical items, those first learned tend to be those most frequently and commonly used in everyday life with each activation serving as a reinforcement of their linguistic representation in the brain, thereby making them more immune to attrition.

2.2.2 The Activation Threshold Hypothesis

Paradis (2004: 28) explains attrition as resulting from “long-term lack of stimulation”. Within the context of his Neurolinguistic Theory of Bilingualism (NTB), according to which Paradis (2004) views the two languages of a bilingual as existing as two independent subsystems within the general language network in the brain, he put forward the Activation Threshold Hypothesis (ATH) which suggests that each linguistic item or structure necessitates a degree of impulse for it to become activated; the level at which this occurs constituting its activation threshold. With each further activation of the item or structure, the activation threshold is lowered as are the cognitive costs associated with its stimulation, so that it may be accessed more easily and quickly. Paradis states that the activation level of any item changes constantly because the threshold rises again gradually after an activation. However, once activated anew, the threshold drops once more. For example, when a particular lexical item is selected or activated, its competitors, be they from the same language or other matching alternatives from the other language/s in the individual’s repertoire, are inhibited.

Schmid (2019: 3) states that the two languages in a bilingual do not exist in isolation and that activation of one will have repercussions for the other since the language subsystems are interconnected. According to Köpke & Keijzer (2019: 4), the ATH claims that “interaction between the languages is largely determined by the frequency of activation of each language and each item that is part of the language, and by the time elapsed since its last activation.” The ATH is, therefore, based on the idea of competition, with this arising from differences in frequency of use of equivalent structures or items across languages in bilinguals, i.e., such a rival structure or item has to exist according to Paradis's hypothesis because what determines the very rise and fall of thresholds for a particular structure or item is the existence and strength of a competitor. The degree to which this competitor is successfully inhibited dictates the threshold level of any given item in the other language. On the other hand, the threshold level itself dictates how easily a competitor can be restrained, i.e., a low threshold level implies greater strength due to regular use, thereby enabling it to inhibit the competitor whose own threshold level is thereby raised making it "weaker" in terms of competing with its rival for realisation.

Köpke & Keijzer (2019: 4) note also that this competition effect is more often seen in the case of lexical item retrieval where frequency effects are greater (considering the declarative quality of lexical knowledge) compared to grammatical knowledge that is deemed procedural in nature and generally more readily available for activation as a result. Procedural knowledge refers to that knowledge that once acquired becomes internalized and automatic like riding a bike. Declarative knowledge, on the other hand, is consciously and explicitly represented and processed in memory (Köpke & Keijzer 2019: 6). Schmid (2019: 4) also refers to the declarative/procedural nature of stored linguistic information as being decisive in terms of understanding frequency of use effects and refers to Paradis (2007) who likewise argues that activation levels of lexical items are more affected by frequency of use than grammar or phonology.

On this note, Köpke & Keijzer (2019: 4) refer to research by Köpke (2002) who found this to be the case in her study of German L1 speakers of French among German-French bilinguals that showed that the lexicon was more affected by regularity of activation than grammar since generally speaking most lexical items have an equivalent in the other language (except for lexical items that refer to culturally specific phenomena in a particular language, for example, the phenomenon of "Schultüte" in German) so that competition effects are greater for the lexicon. With respect to grammatical structures, Gürel (2004 cited in Köpke & Keijzer 2019: 4) argued, based on her findings in a study of Turkish L1 grammatical processing in a group of Turkish – English bilinguals, that competition in the case of frequency of use only emerges if there is an equivalent rival structure or item in the other language actively contending for activation. Where this is not the case, it is easier for individuals to find the sought-after structure. Therefore, these attrition effects are selective, pertaining only to information that has an equivalent in the other language, which is not always the case with respect to grammatical structures. She found for instance that only those Turkish grammatical structures that had an equivalent in English showed symptoms of attrition, structures that were unique to Turkish did not. Therefore, individuals have an easier time accessing the correct selection for items and structures that are not shared across the languages.

Yılmaz & Schmid (2013: 268) also found in their study of Turkish L1 attrition among Turkish – Dutch bilinguals in The Netherlands that the limited range of vocabulary demonstrated by the Turkish L1 attriters compared to the Turkish monolingual group is indicative of competition and higher threshold levels for certain information due to the dominance and corresponding availability of Dutch L2 variants. This L2 information needs to be suppressed whilst simultaneously the lesser used L1 variant is sought from deep within the stored linguistic inventory. Yilmaz & Schmid (2013: 268), therefore, argue that the decrease in lexical variety is not due to lack of practice per se but due to a bilingualism effect, "when there is more information to choose from, it takes longer to find it." It is, however, argued here that whilst greater choice might result in reduced diversity of lexical output, more frequent use and practice would indeed play a role in facilitating subsequent ease of access and activation of the desired information. This does not, nonetheless, contradict the bilingualism effect that Yilmaz & Schmid refer to but rather supports the idea of dynamic thresholds as put forward by Paradis; regular triggering of information lowers its threshold level.

In adopting the interpretation of attrition as a process characterised by difficulties in accessing information, according to the ATH, therefore, lack of use suffices for a language to attrite since this leads to a rise in the activation threshold, whereby more impulse and greater effort is required for an item to be accessed or reactivated. Mehotcheva (2010: 30) also states that in compliance with the ATH, total loss of linguistic knowledge does not occur; the issue is instead one of accessibility. On the other hand, she states that if learning of the L2 or FL ceases prematurely, i.e., before this knowledge has a chance to become consolidated in long-term memory, this can lead to complete loss of this knowledge (Mehotcheva 2010: 32). This highlights the importance of regular activation in the learning phase so that information can become more entrenched in memory.

In this respect, exposure to and corresponding activation of *new* information encountered during incubation is, on the other hand, a case of learning but not *relearning* which constitutes re-activation of old information. In terms of activation hindering attrition, therefore, the starting point for analysis is naturally information that had already been stored in long-term memory before contact with the language ceased.

Gharibi & Boers (2017: 55) state that since activation levels are lower with respect to receptive knowledge, it is less demanding to access this information. Mehotcheva (2010: 30) says that production is more difficult for attriters because it necessitates an impulse from within the linguistic system, whereas in the case of recognition, the presented L2 item provides an external facilitatory cue. Recalling a word in the attriting language, on the other hand, also necessitates greater effort because the language most in use provides a more readily available alternative since through regular use, its corresponding threshold is lower. The individual, therefore, has to work harder to inhibit this information and dig deeper in the memory store to select and access the correct information from the magnitude of data available.

This is also in keeping with findings from the field of psychology whereby forgetting is considered to be a case of having difficulties in retrieval due to the magnitude of information available at the storage level (see Section 2.2.4).

While some studies have established that rehearsal during incubation plays a role in aiding retention (e.g., Graham 2012; Murtagh 2003), others have failed to fully corroborate these findings (e.g., Bahrick 1984a; Mehotcheva 2010). The implications of these observations with respect to the ATH is further discussed in Chapter 6, Section 6.6.3. If attrition refers to difficulty of access as opposed to loss of language, then the positive role of rehearsal would provide some justification for the ATH. This is in keeping with Herdina & Jessner's (2002) proposed Dynamic System of Multilingualism according to which languages will grow or deteriorate depending on how much effort is invested in their maintenance. In accordance with this, activation and use of a language within the multilingual network is crucial so that proficiency in that language does not diminish.

2.2.3 Complex Dynamic Systems Theory

According to Complex Dynamic Systems Theory (CDST), attrition, just like acquisition, is considered a natural development of a language within a multilingual system (Opitz 2019: 4). Attrition is not viewed as resulting from one single factor but instead linguistic, sociolinguistic and psycholinguistic variables are seen to be in continual interaction, simultaneously influencing each other and leading to a language system that is dynamic and in a constant state of flux (De Bot, Lowie & Verspoor 2007). Patterns of language development or attrition are, therefore, considered to be non-linear. As Mehotcheva & Mytara (2019: 4) explain, this means that while one interaction between variables may not have any significant effects on the overall system, another one might and this is irrespective of how big or small this interplay is.

Mehotcheva (2010: 24) refers to Van Geert (1994: 50) who explains CDST as being based on a system in which all variables are interconnected and mutually interact, meaning that a change in one variable will have ripple effects throughout the whole system whereby each variable is impacted and altered (also the very variable itself) resulting in a restructuring of the whole system accordingly. This system is fluid and fluctuating over time with all variables co-determining each other's alterations and degree of attrition. Chan, Lowie & De Bot (2016: 188) in drawing on De Bot & Larsen-Freeman (2011), state that CDST also begins with a sensitive consideration of the "initial conditions of the individual", i.e., what is the individual profile of the attriter in terms of the distinct experience of language learning and the context of attrition.

What is noteworthy in CDST is that it highlights how unique language development is for each individual as it views all variables stemming from the linguistic system, the individual qualities, peculiarities and experiences of the learner, as well as the surrounding context, as interacting. For example, age interacts with learning experience and with proficiency which in turn interacts with motivation. Mehotcheva & Mytara (2019: 5) argue that in CDST an "interconnected perspective" is adopted as opposed to an "isolationist" one. This also permits much variability in terms of outcome across individuals, even when it appears that the same factors are at play.

It is also argued in the present study that the strength of each of the variables in a particular interaction at any given time is specific to each individual and that since these factors change in force and intensity, dominance and influence over time, this results in a path of development that is non-predictable and constantly vulnerable to change.

Within the framework of CDST, the language in a multilingual system is expected to undergo changes pertaining to quality (i.e., proficiency) due to a multitude of interconnected factors, with proficiency level dropping if the necessary input to maintain it is lacking. This drop may, however, be reversed with renewed efforts and can remain stable as long as the language remains activated (Mehotcheva 2010: 25).

Single factors, therefore, do not operate in isolation to determine a particular outcome. Language knowledge can both thrive with the right impulses or deteriorate under other circumstances and influences that are less conducive to supporting the language system.

Mehotcheva (2010: 25) refers to Briggs & Peat (1989: 148) who state that the linguistic system also consists of subsystems with a change in one also influencing the others. In a multilingual, each individual language constitutes such a subsystem which in turn comprises further subsystems; for example, syntax, morphology, phonology. Herdina & Jessner (2002: 86) state that not only do these subsystems undergo constant reconfiguration due to internal alterations of their individual components but they must also realign to external changes within the overall system that is continuously attempting to attain a state of equilibrium.

With respect to lexical representations specifically, Larson-Hall (2019: 3) refers to De Bot & Lowie (2010) who argue that these too are integrated with the constellation of this system also being dynamic, leading to, and accounting for, variation in retrieval times for individual words.

Mehotcheva (2010: 27) criticises CDST on the basis that it lacks power of precise prediction. This rebuke is disputed here, however, as contrary to what she states, CDST goes beyond the broad observation that multilinguals are susceptible to attrition due to competition effects. Instead, it aims to explain language development in the context of a living, dynamic system, highlighting the complexity and intricacy that evolves when a multitude of variables are at play at any given moment. It is this very point, i.e., that the nature of an "end state" *cannot* be predicted that deems CDST a legitimate explanation. In fact, it is not the aim of this theory to try and make predictions about end states but rather to aid in understanding why a particular state is in place at a specific point in time.

In keeping with Opitz (2019: 3) and Köpke (2007), it is also argued here that, against the backdrop of CDST, variables are not to be examined in isolation but rather as clusters of factors that influence each other in unique and individually defined ways. Opitz (2019: 3) also points out the practical advantage that this offers with respect to research proceedings in that clustering variables allows for a reduction in the number of participants in a given study.

2.2.4 Meara's model of lexical attrition

Meara (2004: 152) writes of what he terms the "Boulogne Ferry Effect"; the experience of arriving in France from the UK and French words that one believed one had forgotten seem to re-emerge from the depths of one's memory. In more formal terms it represents the process whereby words from a dormant vocabulary become activated again on re-exposure to that language.

Meara presented a model of lexical attrition in the form of a computer simulation based on a Boolean network. A Boolean Network consists of Boolean variables which are data types that have one of two possible settings; in this case activated or deactivated. Within a Boolean network, this particular state is regulated and dictated by the setting of the other variables to which it is connected. However, these variables are characterised by their dynamic nature and as they are linked, changes in their states may or may not have a spreading influence whereby the setting of other variables might too be affected. At another discrete point in time, therefore, the states of the variables might be different again.

In Meara's model, vocabulary is presented in an interrelated network in which all words are connected to two other words. It is important to note that in a real mental lexicon, items are linked to many other ones so Meara's model is, in this respect, a simplified version. According to this model if a word is linked to two activated words, it too will be activated. If it is linked to two deactivated words, then it will also be deactivated. If is connected to one activated and one deactivated word, whether or not it will be activated will depend on, what Meara calls, its activation threshold. When input falls below the threshold, the word becomes deactivated (this he equates with receptive vocabulary), when input reaches or goes beyond the threshold level, the word is activated (which he equates with productive vocabulary). He also demonstrates that the words that are activated can become deactivated at interim periods just as words that are deactivated can become activated intermittently. In setting up his model, whether or not the words were activated or deactivated was randomly set as were the links between them.

For Meara (2004: 145), lexical attrition is an attribute of the interconnection between words in the vocabulary network as opposed to being confined to words in isolation, i.e., vocabulary loss stems from underlying processes that induce changes in the structure of the network as a whole, not only the individual words of which it is composed.

According to Meara, vocabulary loss is an observable change in the number of activated words in the vocabulary pool and will always be measurable. In his simulations, vocabulary loss is always set off by attrition events but attrition events do not always necessarily lead to actual vocabulary loss.

Meara (2004: 145) explains an attrition event as a small structural change that takes place in the vocabulary network whereby it becomes more difficult to activate a particular word. This might, but not necessarily, have an influence on the number of words that can be activated overall in the system. While attrition events do not necessarily lead to vocabulary loss, they do weaken the overall structure of the network, thereby making its components more vulnerable to, even minor, variations in activation. Vocabulary loss, on the other hand, always presupposes an attrition event. Meara also found that the effect of attrition events on the overall network and the degree of vocabulary loss is essentially unpredictable.

He also shows that when words are deactivated, the effect on the overall system is limited as it springs into activation again as long as the input it receives does not drop below its activation threshold.

What is particularly interesting about Meara's model is that it predicts that even though the fundamental lexical model can be the same across individuals (when extending this model to real-life scenarios), very dissimilar patterns of attrition can emerge. Meara too shows that overall change proceeds at a slow rate and only a few attrition events actually lead to vocabulary loss, with this often being rather insignificant. This, he argues, is representative of results from research on lexical retention in real learners that have shown that vocabulary knowledge tends to be remain stable with little loss over long periods of time (e.g., De Bot & Weltens 1995). In his opinion, such episodes of stability make the challenge of establishing loss in real learners particularly challenge.

On the other hand, Meara acknowledges that his model is a rather simplistic representation of what real attrition processes might look like. Items in the mental lexicon have a variety of states; not just activated or deactivated, and are linked to more than two items with these connections being multifaceted and functioning on many different levels such as at the semantic or phonological level, for example. Besides, his model also only looks at the possible network of words within a monolingual system; such a network is unquestionably even more complicated when examples of bi-and multilingualism are considered as is the case for the participants in this study. On another point, when he writes about the input or stimulus that a word receives that alters its threshold level, he does not elaborate on what constitutes such input.

Despite it not being possible to put this model to the test in real life (as Meara himself recognises), Meara presents an interesting tool, nonetheless, with which parallels to real attrition operations can be drawn. In highlighting the importance of looking at words as part of a structured network, his simulation serves to demonstrate the randomness, variability and individuality of the process. He argues that such simulations should encourage researchers to look at attrition data anew and especially consider what various states of activation could mean in terms of the vocabulary network as a whole.

Larson-Hall (2019) states that Meara's design fits well with the idea of interconnectedness and mutual interplay that constitutes the basis of CDST, referring to a study by De Bot & Lowie (2010) that demonstrated that lexical representations are interrelated, dynamic and intermittent.

As already stated, Larson-Hall (2019: 6) proposes a hypothesis she terms the Constant Decay Hypothesis which draws on Meara's simulation of attrition processes according to which, attrition events begin once contact with the language ceases and can range from retrieval difficulties to eventual decay of information when not triggered due to lack of input. Large scale deactivation caused by constant attrition can lead to what Larson-Hall (2019: 3) calls catastrophic loss.

2.2.5 The Savings Paradigm and retrieval

Central to the Savings Paradigm is the issue of accessibility. Originally proposed as a theory to explore and understand forgetting in general, the paradigm was also later applied to language loss and the idea of a possible reactivation of what seems to be forgotten linguistic knowledge. According to the Savings Paradigm, linguistic items and structures have merely become inaccessible in real time as opposed to erased from memory (Nelson 1978). With respect to lexical attrition, for example, subconscious residual linguistic knowledge is said to still remain which should make it easier to relearn words which could neither be recalled nor recognised (but which had been known in the past) than to learn new, previously unknown ones (De Bot & Stoessel 2000). According to the paradigm, and in keeping with Paradis's Activation Threshold Hypothesis, different levels of activation (higher) are needed for recall, meaning that words in the attriting language can be readily produced and/or translated from the dominant language to the attriting language, and recognition (lower), meaning words from the attriting language can be translated into the non-attriting language or matched with words from the non-attriting language or pictures.

Weltens & Grendel (1993) refer to Retrieval-Failure Theory (Loftus & Loftus 1976) in this regard using the analogy of not being able to find an item that has been mislaid; the item has not disintegrated, rather it is just temporarily absent and with the right cue can be located again. Importantly, Weltens & Grendel draw attention to the phenomenon of varied performance over time despite the stability and unchanging nature of underlying competence, i.e, sometimes individuals can access information, other times not. The underlying information remains intact, however, so that the problems that arise are performance ones whereby difficulties arise in the retrieval and processing of this information.

Larson-Hall (2019: 1) argues that studies on the Savings Paradigm have provided evidence that lexical attrition pertains not to a loss of lexical items and a terminal dissolution of structures but a weakening and deterioration in connections (see Section 2.2.3 on Meara's model of lexical connections). This has been shown, for example, in studies by De Bot & Stoessel (2000) and Hansen et al. (2002), each of which is outlined in detail in Section 2.3.2. These demonstrated that memory traces of vocabulary do indeed exist despite long periods of disuse and claims by the participants that they believed this knowledge to have completely dissipated. Instead, it was found that there is a relearning advantage for previously known knowledge whereby it is easier to learn compared to the learning of completely new content.

Taura (2019: 6) refers to two studies that support this. Hirai (2002) tested a Japanese-English bilingual child after their return to Japan over a period of four years. The child was seven years and two months old on leaving the L2 environment and had been born there. Hirai found that during incubation, productive skills in the L2 were affected more pronouncedly and earlier than receptive skills. What is particularly interesting with respect to the Savings Paradigm was that return visits to the L2 country served to revive seemingly forgotten knowledge. Taura uses this example to demonstrate that access to seemingly forgotten knowledge can be achieved given the right stimulation. Similarly, a study by Tomiyama (1999a, b, 2000) that also looked at a Japanese-English bilingual who had lived as a young child in the US found that the problems he demonstrated in English after a period of living in Japan were more indicative of problems in accessing information than actual information depletion.

Isurin & Seidel (2015) tested the Savings Paradigm by investigating whether or not remains of a lost language could be established in an adult who had been adopted at the age of three years and who not only had had no contact with her L1 in the meantime but who could not even remember which language she had acquired prior to adoption. The linguistic background of the subject was unknown until therapy for posttraumatic stress disorder when fragments of the language she believed she had spoken before her adoption at the age of three years emerged. Since these seemed to be of a Russian or Ukrainian origin, the subject was tested on Russian/Ukrainian cognates with Isurin & Seidel finding evidence of a savings effect for words which she most probably would have known by the age of three compared to new words (words typically acquired later as a child gets older). In contrast, the comparison group revealed no effect for old words over new ones.

A key question arising from the idea of savings is why older information has an advantage in relearning over newer information. In an attempt to answer this, observations from the field of Cognitive Psychology are drawn on. Bjork (2012) argues that while the capacity of the human long-term memory is almost limitless, problems can arise in retrieval. Disuse, therefore, does not result in decay; items are still stored in long-term memory and have just become inaccessible. He distinguishes between storage strength, a measure of how interconnected or entrenched information is, and retrieval strength which is how accessible specific information is in a particular moment. Whilst retrieval strength determines whether or not something can be recalled, storage strength determines how rapidly retrieval strength is lost or how quickly it is regained. Forgetting is, according to Bjork, losing retrieval strength. If an item has high storage strength, retrieval strength can be regained relatively quickly. In other words, the more storage strength a memory has, the easier it is to re-learn this information.

Bjork argues that regardless of how well information has been learned and even when it achieves a high level of storage strength, it will become inaccessible over time if not activated. However, this information endures in memory. Storage strength, he argues, is partly determined by how connected a memory is with other elements of a person's past. He offers the analogy of one's childhood address compared to a hotel room number; one might not readily recall one's childhood address (i.e., it might have low retrieval strength) but it has high storage strength, while the room number of a hotel will most likely have a high retrieval strength but a low storage strength meaning one can easily recall this information whilst residing in the hotel but this information does not become stored in long-term memory. He argues that the reason for this is efficiency. There would be too much interference from unwanted and unnecessary competing items if all information was at the ready and online in a given moment. Instead considering the sheer amount of information that our memory holds, those items not required become "dormant".

According to Bjork, forgetting information and later relearning it can actually be beneficial in that it ensures that learners develop strategies that support retrieval with each recovery constituting a form of training and practise.

Considering the focus of the present study which is examining a language that participants, for the most part, had learned in school, it is interesting to consider the optimal conditions that Bjork & Bjork (2019: 165) outline for successful learning, i.e., the entrenchment of information in memory. These are spacing, interleaving and context change. Spacing implies enabling adequate intervals between exposure to and processing of information. This is because a degree of forgetting will take place in this time which requires the individual to work harder to access the information at a later point in time. Interleaving and context change mean inserting the information into a variety of different formats, activities and contexts to expand the variability of cues that learners can draw on to retrieve this information later. With each recovery the information becomes more entrenched so that retrieval is thereby considered a learning event.

New information might be more difficult to reproduce because it has not yet reached the necessary level of entrenchment and stabilisation and has, therefore, not yet become ingrained in the memory system. Information might even dissipate completely if it does not undergo adequate rehearsal and the necessary connections in the language network are not established or strengthened to enable it to become stored in long-term memory.

Re-exposure acts as a stimulus or trigger whereby the catalogue of stored information is searched for any associations relevant to this stimulus or indeed the very information itself. In the case of newly learned information, associations can help to guide retrieval. Once the data has become encoded, however, as is the case in older, rehearsed information, the issue is then predominantly one of relocating this information.

This is of utmost relevance for this study considering the long period of learning during which lexical items were accessed regularly and repeatedly and in a variety of contexts. It can, therefore, be anticipated that these items had attained high storage strength and that the problems in recognition and recall are, therefore, more indicative of a drastic reduction in retrieval strength but not an ultimate disintegration of this information.

2.3 Research on attrition

L1 attrition among individuals living in an L2 environment has constituted a central focus in attrition research to date. For a comprehensive overview of studies on L1 attrition, see Schmid & Köpke (2019).

Fewer studies have been conducted on attrition of an L2 acquired in a natural setting (examples, Cohen 1989; Graham 2012; Hansen, Colver, Chong, Pereira, Robinson, Sawada & Miller 2012; Olshtain 1986, 1989; Taura 2008; Tomiyama 1999a, b) and even fewer again on attrition of a language learned in an instruction setting (examples, Bahrick 1984a, b; Cohen 1986; Grendel 1993; Murtagh 2003; Weltens 1988; Weltens & Grendel 1993).

Research on savings and relearning effects of an L2 /FL specifically is scant. Hansen et al. (2002) looked at savings with respect to languages acquired through a combination of formal and incidental learning among L1 English missionaries. De Bot & Stoessel (2000) looked at relearning effects with respect to two L1 German attriters of L2 Dutch and De Bot et al. (2004) investigated possible savings in a series of studies looking at L1 English learners of German as well as L1 Dutch learners of French. These three studies will be looked at in detail below as they provide an important starting point for the orientation of the present study considering that, like these studies, it also focuses on retention and possible savings of lexical items.

Before these studies on savings are presented, however, a brief overview of selected studies on 2L/FL attrition is given. Even though these studies did not consider savings and relearning effects specifically, they, nonetheless, provide a useful portrait of preservation patterns and potential determiners of attrition (naturally keeping in mind differences and variation in exact focus, cohort characteristics and methodology used in each study) against which to interpret the individual estimates of lexical item retention levels that emerged in session 1 in the present study. In the case of those studies that looked also looked at other domains, such as syntax, only their findings with respect to lexical item attrition are presented.

2.3.1 Studies on L2/FL attrition

Firstly, studies that looked at attrition of languages learned in an instruction setting are presented. This is then followed by an overview of studies that looked at second languages acquired both naturalistically and formally (i.e., whilst immersed in the L2 environment, official instruction was also received).

2.3.1.1 Bahrick (1984a)

Bahrick (1984a) tested L2 Spanish vocabulary, grammar and idiom knowledge (recall and recognition) in a study of 587 L1 English speakers with a length of incubation ranging between one and 50 years. A questionnaire was administered to gather information on instruction and frequency as well as possible Spanish language use during incubation with participants grouped accordingly in the data analysis.

With respect to lexical knowledge specifically, participants demonstrated greater difficulty recalling vocabulary than recognizing it, with those who had more classes and instruction performing better. Those who demonstrated higher levels of proficiency at the onset of incubation retained proportionally more knowledge compared to those with a lower proficiency, although the relative amount of knowledge lost was equal across all proficiency levels. Rehearsal during incubation, such as watching TV, reading and writing, did not appear to benefit participants in terms of their performance. On the other hand, Bahrick observed that information learned later showed greater vulnerability to attrition because on average there had been less opportunity to practice or rehearse this information to ensure its consolidation in memory. This suggests that once knowledge has achieved a certain storage strength during learning, it remains intact regardless of frequency of activation during incubation. Bahrick observed that most attrition appeared to take place in the first three to six years after contact with the language had stopped, with a stabilization taking place after this. He referred to this remaining knowledge as a permastore; knowledge that has become consolidated in memory and is thereby more robust and less vulnerable to decay.

2.3.1.2 Cohen (1989)

Cohen (1989) examined the lexical knowledge of two bilingual Hebrew-English child learners of Portuguese who had lived for one year in Brazil. The ages of the siblings at the time of leaving Brazil was nine years and 13 years respectively. He tested them on three occasions after their return to Israel; one month, three months and nine months focusing on their knowledge of the meaning of words and the function of these words. In other words, he was concerned with whether or not the children could produce the intended words correctly (meaning) and in the right context (function).

This was done via a spontaneous story-telling session (the story was based on a series of pictures presented to the participants). Receptive knowledge was also tested 11 months after contact with Portuguese had ceased using an oral recognition task of the words that the participants had produced in the story-telling task at testing times 1 and 2 but did not use at testing time 3 (i.e., at nine months). These were presented to the children in a list.

Cohen observed a significant decrease in the number of words the participants were able to produce after nine months compared to what they could still recognise. He thereby concluded that the words are still in memory but increasing interference from other languages (in this case, Hebrew and English) meant access to vocabulary had become increasingly impeded from one testing session to the next (Cohen 1989: 147).

When the participants could not remember a word, he found that they used different strategies to compensate; intralanguage (i.e., not switching language but instead drawing on other resources in that same language, for example, rephrasing) and interlanguage (e.g., code-switching).

Greater attrition was observed in the younger sibling in terms of what he could still produce after nine months (with greater attrition evident in the case of nouns compared to items from other word classes).

2.3.1.3 Weltens, Van Els & Schils (1989)

Weltens, Van Els & Schils (1989) investigated the attrition of French learned in school by L1 Dutch speakers in the Netherlands. They looked at receptive skills at two different instruction levels; one group had had French as a subject for four years, the other group for six years. Testing was conducted right at the end of official instruction, two years later and four years later. With respect to lexical knowledge, participants were given a French sentence that included the word (clearly highlighted) they were requested to recognise. The Dutch translation of the sentence, but in which the translation of the target word was missing, was also given. 40 items were tested. Little attrition was established with lexical item recognition remaining relatively intact across testing times although Weltens et al. pointed out that the participants in this study were highly proficient speakers to begin with. It must also be noted that the participants were given quite a substantial amount of support by means of the testing format, i.e., not only were they given contextual support via the French sentence itself, they were also given the Dutch translation. It is argued here that even with the target word deleted from the Dutch sentence, the contextual support offered by both the original French sentence and the Dutch version could have prompted successful guessing.

2.3.1.4 Bierling (1990)

Larson-Hall (2019: 6–7) refers to a study by Bierling (1990) that looked at lexical attrition in fifteen adult learners of Spanish. Participants took part in an intensive college summer course and were tested on five occasions, the first time being after the first 20 hours of class and the final one being approximately 18 months later. In the first 20 hours of classes, they learned and were tested on 100 new content words. Some of the words were also unusual and not of high frequency. After initial exposure the teacher no longer dealt with these words in the remainder of the course. Later participants were tested on production by picture naming in which they had to write the correct Spanish word next to the picture. The receptive test consisted of matching words to pictures with this test always following the one on recall. Whilst a drop in recognition was also observed, Bierling established a drastic decline in the ability to produce Spanish words with this decrement setting in rapidly (by 3.5 weeks) and appearing to ease slightly (by seven weeks). Bierling detected this pattern of rapid loss in productive knowledge followed by a relative stabilisation even in the case of the more proficient learners. There was greater deviation however in the case of recognition. Bierling reported that four of the participants were still able to recognize more than 50 % of the words in the final testing. Bierling's findings concur with the general observation in the literature that individuals seem to more easily recognise than produce words they had previously known.

A critique of this study is that participants may not have truly "learned" the words in the first place considering that exposure to these new items was limited. On the other hand, despite the overall drop in recall capabilities, Larson-Hall points out that it is interesting to observe that some vocabulary could nonetheless still be produced after a period of time had elapsed since first contact with the items. She too argues, however, that this exposure was most likely not adequate for the knowledge to become properly entrenched in memory.

2.3.1.5 Grendel (1993)

Weltens & Grendel (1993: 151) refer to Grendel's (1993) study which similarly to Weltens et al. (1989), investigated the attrition of French lexical knowledge learned in school by L1 Dutch speakers in the Netherlands, also looking at two different instruction levels; one group had had French as a subject for four years, the other group for six years.

200 learners were tested right when they finished their official instruction, two years later and four years later. Grendel focused on recognition with respect to orthographic knowledge (familiarity with the written form of words) as well as semantic knowledge (discernment of word associations). To test orthographic knowledge, an orthographic lexical decision task was used using pseudowords (grouped according to grapheme cluster frequencies), and real words with participants having to reject items that were not real words. It was expected that pseudowords with low-frequency clusters would be rejected more quickly than those with high-frequency ones as they look more like real words, but that this effect would fade as time went on, thereby demonstrating according to Grendel, a loss in orthographic knowledge, i.e., a loss in the ability to recognize "normal" from "odd" grapheme combinations for that particular language (i.e., French). This was not found to be the case. Participants were just as good at discerning the finer differences in orthographic form in the final testing as they had been

in the earlier one two years previously. Based on this, Grendel argued that orthographic knowledge had not attrited over the four years.

To test semantic knowledge, she used a semantic priming test that used pairs of words; in some pairs, words were semantically related, in others they weren't. It was anticipated that words that were paired with semantically related ones would be recognised earlier that the words in an unrelated pair and that if attrition of semantic knowledge had taken place across testing sessions, this would manifest itself in a decreasing ability to avail of the semantic support offered via the accompanying stimulus, i.e., the semantically related word, so that little difference would be expected with respect to the participants' performance on the semantically related pairs compared to the unrelated ones. As in the case of the orthographic knowledge test, participants did not exhibit a deterioration in their semantic knowledge as they were able to correctly use the semantically related stimulus to guide them even four years after contact with the language had ceased.

Weltens at al. (1989) and Grendel (1993) used a control group to establish a baseline for proficiency in French at the time of leaving school. Murtagh & Van der Slik (2004) point out that this complicates interpretation of results since more individual factors have to be accounted for (as the control group consisted of different individuals to the ones tested for attrition).

2.3.1.6 Murtagh (2003)

Murtagh (2003) looked at retention of Irish as an L2/FL among 59 young Irish adults 18 months after they had finished school. The L1 of the participants was English and all had had Irish as a subject at school and taken the state Leaving Certificate examination at the end of their secondary school education. Three different groups were tested; one group of participants had taken ordinary level Irish at school, the second group had taken the higher level course and the third group came from an Irish immersion program. There were 239 participants from twelve different classes (across all three instructional levels/contexts) in the initial sample; 59 of these were tested after 18 months.

The participants were tested twice using a communication (listening and speaking) test and a C-Test; the first time was two to three months before they sat the Leaving Certificate Irish examination and the second time was

18 months after finishing school. Participants were also interviewed at each testing time and information was gathered on their Irish language exposure and use, their attitudes and motivation as well as on how they themselves estimated their current level of knowledge and in the case of testing at time two, how much Irish they felt they had forgotten since leaving school.

Despite participants having the feeling that they had forgotten much of the language in the 18 months since they had left school, test results showed otherwise; no attrition was detected. Indeed, a slight improvement was established in one of the sub-tests (story telling) among participants who had attended an Irish-medium school. Murtagh assumed this to be a result of the continued contact with the language that these participants reported.

The two main predictors of success in retention were initial proficiency (at onset of incubation) and regularly reading Irish texts (i.e., rehearsal during incubation). Reports of anxiety whilst being in class correlated with lower scores in testing whilst positive evaluation of the course as well as high motivation and parental encouragement correlated with higher scores. Murtagh & Van der Slik (2004: 297) also acknowledge that each of these factors is interrelated.

Participants from all three instruction backgrounds reported decreased opportunity to use the language as time progressed (however, this was less so for those who had been in the immersion program). Nonetheless, the majority continued to have some contact with the language (in the case of those who had had Irish in mainstream schools, exposure to Irish was more passive, for example, in the form of TV or radio).

2.3.1.7 Graham (2012)

Graham (2012) tested lexical knowledge of 12 adult L1 English speakers of L2 Spanish who had spent between one and a half to two years in a Spanish-speaking country. Participants were tested on four occasions over an incubation period of 15 years.

He tested participants using three instruments; storytelling, narration and a questionnaire that was split into two parts, one that was completed in the very first meeting to get information on background variables and the second part which was always conducted at the start of each subsequent testing session to gather details on language use since the previous meeting.

Graham analysed the words used by the participants in Spanish according to the categorisations of *form*, *meaning* and *use* as guided by Nation's analysis of what constitutes word knowledge (Graham 2012: 136). *Form* referred to correct pronunciation and the appropriate morphology. *Meaning* looked at the suitability of the selected words to communicate the intended meaning and *use* pertained to a consideration of whether the items were used in the right context and in a grammatically correct way.

He found that those who used Spanish least after returning from the L2 environment exhibited the greatest loss in lexical knowledge. However, even those who continued to use Spanish regularly were not completely immune to attrition with some decay still emerging. However, overall, very little lexical attrition could be observed. For the most part respondents did not have to resort to English language substitutions or use unrecognisable forms or not use any form at all. With respect to *form* specifically, Graham observed that most of the errors that emerged were in the phonetic domain suggesting, according to Graham, that while the general structure of words appeared to still be intact in memory, a decay in accuracy seems to take place with respect to phonetic representation (2012: 178).

In fact, overall, most attrition appeared to take place in terms of the use of the word with respect to collocations. As time progressed participants appeared to struggle in finding the correct words to accompany the selected word. On the other hand, little evidence of attrition could be detected generally with respect to appropriate use of grammatical structures, although there was much individual divergence across structures.

He observed distinct influence from English in form-meaning correspondence in lexical phrases and single words as well as in certain grammatical structures so that he concluded that the progression of attrition is best accounted for by looking at competition from the dominant language.

2.3.1.8 Hansen, Colver, Chong, Pereira, Robinson, Sawada & Miller (2012)

Hansen, Colver, Chong, Pereira, Robinson, Sawada & Miller (2012) assessed vocabulary retention in 432 returned missionaries (306 males and 126 females) whose L1 was English. Six L2 language groups, each consisting of 72 persons, were looked at; German, Japanese, Korean, Mandarin,

Portuguese and Spanish (i.e., three Asian languages that are typologically very different from each other and three Indo-European languages). These groups were further divided into six subsets according to length of time since leaving the L2 environment (the shortest span was three months to one year, the longest was 35–50 years). The length of time spent living in the L2 environment varied between two and three years for the males and one and a half to two years for the females.

Hansen et al. wished to examine whether or not certain factors played a role in determining success in maintaining lexical knowledge. These factors were the amount of L2 input both before and during the mission abroad, the role that typological differences between English and the L2 might play, affective factors such as motivation to maintain use of and contact with the language on returning home, and attitudes towards the L2 and its associated culture. They also looked at the possible changing influence of these variables over time and whether or not factors that had played a role in acquisition also played a role in retention.

Vocabulary testing consisted of 100 words divided into 25 concrete nouns, 25 abstract nouns, 25 verbs, and 25 adjectives, all known to the missionaries (selected from missionary texts in each respective language the missionaries had been required to memorise while they were abroad). Participants were tested over the phone on recall. The English word was read out twice and the participant was given six seconds to answer.

A survey was also completed by each participant so that information on demographics and background variables could be collected.

Hansen et al. found that overall a large amount of the vocabulary tested had been retained even in cases of an incubation period of close to 50 years. However, the main predictor of retention was length of incubation whereby those who were longest away from the L2 environment performed worse. Portuguese and Spanish items were better retained than those from Asian languages although performance of the German group was more in line with that of the L2 Chinese participants. Spanish was the best retained overall and Korean the worst. Along with the German group, the Spanish group had had more exposure to the language before and after the mission. However, Hansen et al. (2012: 130) established that the Spanish group differed to the German one in that the Spanish participants demonstrated greater motivation to maintain contact with the language and had a more positive attitude to it and its associated culture which Hansen et al. concluded positively reinforced language retention. On the other hand, they found that the Korean learners had the least exposure to the language before and after the mission and these participants were also least motivated to maintain contact with it, expressing less positive attitudes towards the language and its culture than the other groups did with respect to their respective L2s. Overall, however, Hansen et al. found that affect overall (attitudes and motivation) played a weaker role with respect to retention than observed for acquisition.

A critique of this study is the nature of the testing and the time limit of six seconds in answering. Since the testing took place over the phone, participants could not see/read the words. Additionally, not being able to answer in six seconds might just demonstrate problems with speed of retrievability but does not necessarily imply that given more time, the participants may not have come up with the correct translation.

2.3.2 Studies on relearning and savings

2.3.2.1 De Bot & Stoessel (2000)

De Bot & Stoessel (2000) investigated whether or not residual knowledge of reportedly forgotten L2 lexical items learnt in childhood was still present and could be recovered in adulthood after years of having no contact with the L2. This was done by looking at whether or not there was a learning advantage with respect to these lexical items over completely new ones.

They tested two adult L1 German siblings' knowledge of Dutch, a language they had been exposed to over a period of four years as children 30 years previously, during which time the younger sibling (3–7 years) attended kindergarten and primary school and the older just primary school (7–11 years).

All of the real words that the participants were tested on were ones that they had previously known. This was established with the aid of a list of basic vocabulary known as the *Streefwoordenlijst* which had been compiled by Dutch researchers based on feedback from primary school teachers. This is a list of words that children attending primary school in The Netherlands can be expected to know by a certain age. De Bot & Stoessel could, therefore, assume that the participants in their study would have encountered these words during their stay in The Netherlands. With respect to what constituted knowing a word, De Bot & Stoessel focused on semantic meaning.

Only nouns were tested (De Bot & Stoessel do not elaborate on why they only worked with nouns). In the case of one of the participants, three tests were conducted over a period of two months. The second participant was tested on three consecutive days (the participants were not tested together). The following table provides an overview of the tests conducted.

Table 1 Distribution of tests (De Bot & Stoessel 2000: 340)

	Recall	**Recognition**	**Sub-threshold memory**
Session 1	picture naming	L2 to L1 translation	relearning/reactivation (savings)
Session 2		L2 to L1 translation Picture-word matching	relearning/reactivation (savings) pseudo-word learning (savings)
Session 3		L2 to L1 translation	relearning/reactivation (savings)

De Bot & Stoessel (2000: 341) state that they started each session with the recognition task whereby 50 randomly selected nouns from the Streefwordenlijst were shown. They did not elaborate on why recognition was tested before recall considering that recall has been shown to be more difficult than recognition. It was also stated that there was no overlap between words used in the translation tasks and in the savings task. It is not clear what is meant here as it is assumed that the recall and recognition tests would serve to filter out words participants could remember so that the remainder could be used in the savings test, otherwise there could be no guarantee that the participants did not already know the words used in relearning. In fact, this might indeed be what De Bot & Stoessel did; their description, however, is not clear. Personal email communication with one of the authors did not lend greater clarity to this due to the long time-span between time of testing and communication.

The words were read out by the participants themselves in the first session. In subsequent sessions they were read by the L1 Dutch test administrator or a recording of a native-speaker was played. De Bot & Stoessel argued that, considering that the participants were children when they lived in the L2 environment, most of the contact they had with Dutch would have been aural-based. It is not clear, however, why the participants were requested to read the words in the first round but not in later sessions. Also considering the ages of the children when they lived in The Netherlands, both would have acquired a certain degree of literacy in Dutch (even more so in the case of the older child). It would, therefore, have been interesting to have had them read the words in each testing session to see if phonological knowledge had been affected.

Participants were tested on recall (production through picture naming and L1 to L2 translation) and on recognition (L2 to L1 translation, picture – word matching). In the subsequent learning phase, the participants were presented with 20 Dutch words they appeared to have forgotten, with their respective English translations. Twenty pseudowords, also with English translations, were mixed in with the real words. Each word with its translation was shown for 6 seconds before the next word pair appeared. Each word pair was shown in isolation to prevent learners from availing of contextual cues to guide them in their answers. The participants were then subsequently tested on these relearned/newly learned items.

Their performance was compared to a comparison group which consisted of first-time adult learners of Dutch and who had been required to learn the same words on which they were also subsequently tested.

In the case of the two participants who were the focus of the study, De Bot & Stoessel found that the words that they had previously known in their childhood were more easily learned than completely new words (in this case, the pseudowords). This led De Bot & Stoessel to conclude that that residual, sub-threshold linguistic knowledge is still present in keeping with the Saving's Paradigm. Re-exposure to the language served to reactivate previously known knowledge thereby making this information more easily retrievable than new information.

De Bot & Stoessel did observe, however, that in the comparison group some participants performed better than one of the two participants (the male sibling) after relearning, suggesting that the results could be more indicative of better learning strategies on the part of some participants as opposed to there being any relearning advantage with respect to previously known knowledge.

2.3.2.2 *De Bot, Martens & Stoessel (2004)*

In a follow-up study, De Bot, Martens & Stoessel (2004) conducted a similar series of tests to further explore whether or not residual vocabulary knowledge exists and can be restored. In these tests, a group of L1 English learners of German in the U.S. were tested as well as two separate groups of L1 Dutch learners of French in the Netherlands. This time a comparison group of first-time learners of either German or French was not included (to exclude possible effects of efficient learning strategies that emerged in De Bot & Stoessel (2000)) and a delayed post-test was carried out in the case of the two French groups which had also not been done in the previous study in 2000.

Study 1: L1 English learners of German -relearning of 30 lexical items
Twenty-one students of German at university in the USA took part in the testing. Old and new words were pretested, albeit only with respect to recall. Based on performance, 15 old and 15 news words were used in the relearning phase (the list of "new" words, i.e., words the participants had not previously learned was established based on the advice of the teacher). This time words and their English translations were presented as a list in relearning, with an overall time of three minutes allotted (average of 6 seconds per word) for this phase. After distracting activities of approximately half an hour, the participants were tested on the words they had just learned with respect to recognition. They were given the list of German words and had to fill in the English translation for each one. It is not clear why the participants were not tested on recall first.

Based on performance, a clear advantage for older items could be established. 40.9 % of the old words were recognized correctly, but only 14.6 % of the new words. However, the meaningfulness of these findings is to be approached with a degree of caution. Since the words in the pretesting phase had only been tested for recall, it could be that in the relearning phase words might have been used that the participants could possibly have recognised already in the pretesting phase, i.e., no relearning then took place as these words were in fact "known".

A list of the words was not provided and it was not specified if the investigation was focused on particular word categories.

Study 2: L1 Dutch learners of French – relearning of 30 lexical items
Thirty university students in The Netherlands who had learned French over a period of four to six years at secondary school were tested. Length of incubation was not stated. De Bot et al. used a list of typical vocabulary dealt with in French lessons in school to generate a list of old words. For the new words, a list of low-frequency words was compiled with the help of a dictionary. However, there was no guarantee that the participants did not know these words. Based on a pretesting, a list of 30 words for the relearning phase was compiled. This consisted of 15 old words and 15 new words. However, as in the aforementioned study, only recall was tested in the pretest. As stated above, this renders the following stages of the study problematic since in the (re)learning phase and following testing on recognition, words might have been included that the participants could already have recognised in the pretesting phase.

Relearning took place whereby each item and its translation were presented in isolation and participants were given six seconds to look at and learn each pair. It is not stated however how these pairs were shown. Distracting activities took place after which participants were tested again on recognition (here the question arises as to why they were not tested on recall first). A questionnaire was filled out in the distracting phase. Four months later they conducted a second savings session using different words which they tested three weeks later.

Applying the Savings Paradigm, they found that participants in the study proved more effective at relearning and retaining words they once knew but appeared to have forgotten than at learning completely new words. This was demonstrated in tests carried out both right after learning and after a delay.

Study 3: L1 Dutch learners of French – relearning of 40 lexical items
Another experiment that followed a similar format was conducted this time using a computerised set-up and with a different group of participants. Twelve students whose L1 was Dutch and who had learned French at school over a period of three to seven years were tested on two separate sessions, with an interval of three weeks between sessions.

One hundred French words were pretested. However, it is not specifically stated whether testing pertained only to recall or recognition or both.

Not having this information is problematic (see comment in study 1 and 2 just described on the issue that arises if recognition is not assessed in the pretest). The list consisted of 70 old words and 30 new ones.

For the relearning phase 20 old and 20 new French words were shown in random order (this took place after a distracting phase that included a questionnaire), each pair for eight seconds. After 10 minutes participants were tested on recognition of these words (they were presented in a random order). The words disappeared from the screen after eight seconds. Three weeks later the participants were tested on these words again which was then followed by relearning of words that they could not recall and a subsequent test of these items.

More old words than new ones could be correctly recognised across all tests. They do, however, refer to a learning effect for new items after the second relearning session, whereby participants performed better on these words compared to their performance in testing after the first learning session. De Bot et al. state that this provides evidence of residual knowledge even for words learned for a first time in session one.

Whilst in each case in the series of studies conducted by De Bot et al. just described a questionnaire was administered to participants, no further information is given on the outcome of these in terms of background variables that may possibly have influenced individual performance and with respect to the pretesting phases, demonstrated retention of knowledge. It could also be that some participants' performance was more a reflection of their motivation than actual savings. It is also stated in the case of the final study that the aim of the questionnaire was to establish whether or not the participants had comparable levels of knowledge. However, how this was established (the type of question asked) is not elaborated on.

De Bot et al. only looked at recognition in testing. Recall would have given an even greater indication of degrees of activation and a more comprehensive insight into the effects of relearning. Keeping in mind Paradis's Activation Threshold Hypothesis (see Section 2.2.1), recall is considered to require greater effort in the sense that more activation is needed than for recognition. If re-exposure to the language in the form of relearning can also be shown to affect recall and thereby these activation levels, then this speaks for an even greater effect of re-exposure.

It is also unclear why in study 3 they used a different number of lexical items for relearning or why the times allotted for each word-pair in relearning was different across studies. One could anticipate that a smaller number of items for relearning would imply less cognitive effort, even when the time allocated for each word pair was lower in that case.

2.3.2.3 Hansen, Umeda & McKinney (2002)

Hansen, Umeda & McKinney (2002) examined lexical item retention and savings effects via relearning in 304 L1 English speakers, 166 of whom had learned L2 Japanese and 138 L2 Korean, all working as missionaries in the country in question (Japan or South Korea). They had been highly proficient speakers of the language. Some had had language training prior to living in the country, others hadn't. The length of residence ranged from 18 to 24 months for the females and 18 to 36 months for the males. The average length of incubation was between one and 45 years for the Japanese group and one and 39 years for the Korean group.

In a pretest participants were tested on recognition on 160 words equally divided between concrete nouns, abstract nouns, verbs and adjectives. These consisted of items that the participants had definitely confronted in their missionary work. The participants heard the L2 word and were requested to give the English translation. Hansen et al. took the first 16 items that the participants did not know for relearning. Sixteen pseudowords were also interspersed with these in relearning to establish whether or not there is a relearning advantage for previously known but seemingly forgotten words. After a distraction, participants were tested on recognition only on the 32 items with a time limit of 6 seconds for each item.

Hansen et al. found a very large relearning advantage for old words compared to new ones (the pseudowords) for both the Japanese and the Korean cohort. Interestingly they observed that those who performed well in the pretest of the 160 original items were also those who benfitted most from the savings effect in relearning.

On the other hand, despite the mean length of incubation being shorter for the Korean group than for the Japanese group (by approximately three years), they demonstrated a faster rate of attrition. The Japanese group also appeared to retain lexical knowledge longer (as established in comparing the different age cohorts) and to benefit more from savings accordingly. Hansen et al. attribute this discrepancy in savings to the initial difference in retained vocabulary size for each cohort. They also suggest that different group experiences in each country may impact affective factors and the motivation to maintain contact with the language on returning home. In fact they further state that while availability of Japanese speaking networks is sparse in the US, there is even less opportunity to engage with Korean speakers. Such differences, according to Hansen et al. could also possibly explain the differences in results.

2.4 Influencing factors in L2 attrition

In keeping with Larson-Hall (2019) and Mehotcheva & Mytara (2019) it is also argued in the present study that extralinguistic factors have to be looked at if the attrition process is to be truly understood. For example, age at onset of L2 learning as well as age at onset of attrition/incubation, L2 learning experience, attitude as well as motivation (i.e., what is commonly referred to as ATM), with respect to L2 learning in addition to a willingness to maintain knowledge after learning has ceased, social factors, proficiency level achieved, level of education and length of incubation have all been shown to play a role in determining attrition outcomes. Mehotcheva & Mytara (2019) refer to a number of studies that found various combinations of these factors to play a role in determining degrees of attrition (e.g., Weltens & Grendel 1993; Hansen 1999, 2001; Köpke & Schmid 2004; Köpke 2007; Taura 2008; Bardovi-Harlig & Stringer 2010; Wang 2010; Xu 2010; Schmid & Mehotcheva 2012).

In the following greater detail is offered on the main factors that have emerged as playing a role in L2 attrition. Testing each factor in isolation is difficult, however, as most are interconnected and interrelated. For example, level of proficiency achieved may be linked to the factors of motivation and age at onset of learning.

The degree of attrition can vary greatly from individual to individual. Some retain vast amounts of knowledge, others less with and even within one individual, difficulty or ease of access can vary across different linguistic domains. Across individuals, one might show greater attrition in one linguistic area, another might experience it in another (Yilmaz & Schmid 2018: 233).

2.4.1 Age

Two points on the timeline of acquisition and learning warrant simultaneous attention in understanding how attrition might proceed; age at the *onset of acquisition* and age at *onset of attrition*. Contrary to Isurin (1999: 9) who argues that L2 forgetting is not a function of how early a second language is introduced but when it is no longer used, it is argued in the present study that both age of acquisition/learning and age of attrition constitute key variables and neither can be ignored if the process of attrition is to be truly understood.

Much research has been conducted to date on the benefits of an early start of second language acquisition. This so-called critical or sensitive period when the brain is said to be particularly conducive to language acquisition spans the childhood years ending approximately at the time when puberty begins (see Singleton & Ryan (2004) for a comprehensive review of studies conducted on age effects with respect to acquisition in the respective linguistic domains in Second Language Acquisition (SLA)). A critical period has also been suggested for attrition whereby if contact with the language ceases before the onset of puberty when the linguistic system is said to not yet have achieved full maturation, a significant reduction in proficiency, or in some cases a total loss of linguistic knowledge, can be detected (Köpke & Schmid 2004; Mehotcheva 2010). Bylund (2009: 708) also argues that an examination of the different degrees of susceptibility to attrition pre- and post-puberty can shed greater light on the workings of the language faculty and the plasticity of the language processing system.

In reviewing the literature on L1 attrition among individuals whose contact with the language ceased in early childhood, the focus has been predominantly on the phonological and/morpho-syntactic domain. A selection of these studies are briefly referred to in the following.

For example, studies of international adoptees who left their L1 environment at a very young age and no longer had contact with their L1 after arriving in their new homeland have shown complete attrition of the L1 with the detrimental effects resulting from deprivation setting in rapidly (e.g., Pallier, Dehaene, Poline, LeBihan, Argenti, Dupoux & Mehler 2003; Ventureyra, Pallier & Yoo 2004).

Whilst there appears to be overriding agreement that age effects have a role to play in attrition, there have, however, been studies that have shown some subthreshold traces of knowledge remaining despite cessation of contact with the language occurring in early childhood. For example, Pierce, Klein, Chen, Delcenserie & Genesee (2014) established implicit remnants of L1 phonological information in adoptees. Using fMRI neuroimaging they compared the brain activation of three groups of French speakers in Quebec in Canada during a Chinese lexical tone discrimination task. One group consisted of active Chinese-French bilingual speakers, the second of adoptees from China who came to Canada at the age of 12 months and who had had no exposure to Chinese since their arrival approximately 12 years previously (average age at testing was 13 years) and the third group comprised monolingual French speakers who did not have any knowledge of Chinese. Pierce et al. observed neural representations of Chinese in the adoptees that matched those of the bilingual group despite the adoptees no longer being able to use the language and not having had any Chinese language input since their adoption.

Bowers, Mattys & Gage (2009) argue that such implicit knowledge might only be tapped into in relearning scenarios and refer to studies that found an advantage of early exposure even if this ended before puberty in cases where individuals were re-exposed to the language. Tees & Werker (1984) also looked at phonology and examined the performance of L1 English students of Hindi as a second language at university on their perception of voiceless retroflex and dental plosives and observed that the subset that had been exposed to Hindi in the first one to two years of their lives outperformed the other members of the group. Oh, Jun, Knightly & Au (2003) also found that adult L1 speakers of English who had been exposed to Korean in early childhood outperformed first-time learners in a phoneme discrimination task in Korean. In an earlier study, Au, Knightly, Jun & Oh (2002) observed that the accent of L1 English learners of Spanish who had been exposed to Spanish in early childhood was more native-like than those who were learning it for the first time. In each of these studies, whilst contact with the language over the years had been minimal, Bowers et al., nonetheless, argue that even very small amounts of contact might suffice to ensure that knowledge can be more readily activated and thereby be more easily accessible after long periods of time.

In fact, while Bardovi-Harlig & Stringer (2019: 14) acknowledge a critical period during which the mental lexicon becomes stabilized, they also question the direct role of the critical period in the case of syntax and phonology for L1 and L2 attrition. Rather, they argue, a vulnerable lexicon (i.e., in which development has been interrupted or halted due to a severing of contact with the language before the onset of puberty) can have negative knock-on effects for syntactic processing in particular considering the centrality of the lexicon for the acquisition of syntax. With respect to L2 attrition specifically, they, nonetheless, argue that no study thus far has controlled for age to such a degree that the existence of a critical period can be proven beyond doubt (Bardovi-Harlig & Stringer 2019: 14). They argue that language knowledge is remarkably robust and even after long periods of non-exposure, much knowledge, in particular phonological, remains intact, stating that changes observed are more representative of cross-linguistic influence from the other language as opposed to attrition.

With respect to heritage languages, a number of studies have demonstrated the role of age with respect to ease of lexical retrieval, whereby heritage speakers are typically quicker in accessing words learned at an early age and before they left their L1 environment compared to words learned later (Ammerlaan 1996; Hulsen 2000; Gharibi & Boers 2017; Montrul 2009; Montrul & Foote 2014).

With respect to L2 attrition specifically, Mehotcheva & Mytara (2019: 7–8) emphasize the importance of considering age at onset of attrition in attempting to understand the attrition process even when at times research has generated conflicting results. While certain studies have found age effects in that the younger children were when they moved away from the L2 environment, the greater was the severity of attrition (Cohen 1989; Hansen-Strain 1990; Olshtain 1986; Tomiyama 2008), others have failed to establish such a relation (Kuhberg 1992; Reetz-Kurashige 1999). Whilst it is generally agreed that age and literacy are closely interrelated so that advanced age correlates positively with enhanced literacy skills which in turn aid retention (see Section 2.4.5), it is difficult to tease these two factors apart. In each of the aforementioned studies in which older children were found to have attrited less, they also had more advanced literary skills in the language. On the other hand, in Kuhberg's (1992) study of two L1 Turkish attriters of L2 German, aged seven and nine years when testing started one month after their return to Turkey, the older child demonstrated greater degrees of attrition. This illustrates the complexity of the process and shows that other factors can override the effects of older age of attrition. It also cannot be automatically assumed that the older child in this study had more advanced literacy skills. In fact, what Mehotcheva & Mytara (2019) neglect to mention is that these two girls were not siblings and grew up in completely separate environments and that Kuhberg (1992: 145) also states explicitly that the older child was under substantial pressure once back in Turkey to prioritise her Turkish language skills. Reetz-Kurashige (1999), who was also unable to isolate age as an overriding factor in her study of 18 L1 Japanese attriters of L2 English (aged between six and thirteen years) observed that it was a blend of variables that explained each individual's performance.

Mehotcheva & Mytara (2019: 7) also argue that external factors such as learning context and use, as well as attitudes and motivation, are closely interrelated with age. For example, perceptions and attitudes towards the language can be influenced by the age at which the language is learned or acquired. When children are completely immersed in an L2 environment, they often develop an intrinsic motivation to learn the language so that they can communicate with friends and progress at school. They usually acquire literacy skills in the language also in this time. Depending on their social and professional circumstances, adult learners may not necessarily need to communicate with native speakers of the language or use the language as intensely as children do. If they are taking classes, they may also have a very different learning experience, since often they are in a class with other non-native speakers so the input may not be as rich and varied as it is for a child who is immersed in the L2 environment.

Variation across individuals with respect to achievements in L2 attainment and also L2 attrition is to be expected even within a particular age-group. Pallier (2007: 4) in referring to the idea of a critical period also cautions that it is to be understood as a "statistical claim over relatively large samples of subjects." Additionally, Bardovi & Stringer (2019: 4) argue in keeping with Montrol (2008: 50) that age is a "macro-variable" in the sense that it becomes an influential factor when looked at in combination with other variables such as quality and quantity of input, for example.

With respect to L2 lexical attrition specifically, Larson-Hall (2019: 9) states that severe or total loss of lexical knowledge can be the outcome of premature cessation of contact with the L2, i.e., during childhood. She refers to studies that demonstrated substantial attrition among young learners whose contact with the language discontinued at a young age even though they had reached high levels of proficiency (Cohen 1986; Hansen-Strain 1990; Olshtain 1986, 1989; Tomiyama 2000).

In terms of lexical retrieval, Mehotcheva (2010: 39) states that high frequency words tend to be learned earlier than low frequency ones and that words that have been acquired early tend to generate quicker responses and fewer errors but that research on frequency and early word learning has predominantly been focused on L1 acquisition. It is also difficult to establish the age of acquisition of individual L2/FL words as well as how frequently they were used in an instructional setting. An examination of school text-books from early grades might be revealing to some degree but is limited in its potential as representation in a textbook might not match actual frequency of use. In the case of the present study, the rapid accumulative nature of the Irish words learned at school made the establishment of a clear linear line of progression in terms of vocabulary acquisition difficult.

Larson-Hall (2019: 13) also argues that lexical loss is random in the sense that what was learned last will not necessarily be lost first or what was learned first will not necessarily be lost last as posited by the Regression Hypothesis (Jakobson 1968) (see Section 2.2.1 for further information on the Regression Hypothesis). Instead, she argues that the path of lexical loss is arbitrary as proposed by Complex Dynamic Systems Theory (see Section 2.2.3).

Naturally age in the context of learning a language in an instruction setting cannot be looked at in isolation. Frequency of classes, quality of teaching and input and the attitudes and motivation generated accordingly, all play a key role in any endeavour to understand the level of language competencies achieved and degrees of knowledge retention even years after official instruction has ended.

On a side note; despite not constituting the focus of the present study, an interesting observation has been that age appears to clearly play an influential role in the case of aging adults where cognitive processes slow down so that their performance on linguistic tasks is impaired accordingly (Schmid & Keijzer 2009; Schmid & Dusseldorp 2010).

2.4.2 Language proficiency at onset of attrition and general level of education

2.4.2.1 Language proficiency

In attrition studies, attained proficiency refers to the proficiency level that has been achieved by the time contact with the language ceases, i.e., at the onset of incubation. Mehotcheva & Mytara (2019) present an overview of the studies and findings on the factor of attained proficiency and possible repercussions with respect to long-term retention levels. It is a particularly relevant and interesting factor in the case of L2/FL attrition, as unlike L1 attrition where generally it can be assumed that full proficiency has been reached across all persons (see Birdsong (1992) for a discussion of outcomes in First Language Acquisition (FLA) compared to SLA), attained proficiency in the case of second language learning and instruction is a distinctly individual variable. Whilst generally it is agreed that proficiency levels do impact retention, the factor does not operate in isolation. It provides however a useful starting point in combination with other interrelated variables against which individuals can be compared.

Hansen (1990: 369–370) argues that whilst studies on child attrition for the most part show that those who attain higher levels of proficiency retain more knowledge, she states that the results are more ambiguous with respect to adult attrition (referring to Weltens's (1988) observations) and also that a different pattern might arise in cases of natural acquisition through immersion in the L2 environment.

Mehotcheva & Mytara (2019: 5), however, refer to a number of studies that have in fact found support for a positive relationship between attained proficiency and retention levels (i.e., the higher the attained proficiency, the more knowledge retained) (e.g., Godsall-Myers 1981; Bahrick 1984a, b; Xu 2010; Mehotcheva 2010; Weltens et al. 1989).

2.4.2.2 Level of education

On the other hand, and as pointed out by Mehotcheva & Mytara (2019: 14), level of education overall has not been granted considerable attention thus far in L2/FL attrition research. With respect to L1 attrition, Yilmaz & Schmid (2018: 286) argue that because individuals that are more educated might read more and be able to afford trips back to their home country, this might lead to higher L1 retention levels. Köpke & Schmid (2004: 9) refer to a selection of studies that explored the role of overall level of education in L1 attrition (e.g., Jaspaert & Kroon 1989; Yağmur 1997). They observed that evidence for attrition as provided by performance on L1 proficiency tests during the attrition period was associated with level of education. As stated by Köpke & Schmid (2004: 9) and also argued here, in assessing the role of education levels it is difficult to establish if performance in evaluations is due to language forgetting or rather to shortcomings or difficulties in general with literacy and taking tests, i.e., results may not be representative of actual language loss.

With respect to L2/FL attrition, if individuals perform worse than others with a higher level of education in proficiency tests conducted during attrition, this does not mean that they experienced *more* attrition since their proficiency might have been lower in the first place (i.e., at the onset of incubation). For example, in his examination of the attrition of L2 French learned in a school setting by L1 Dutch speakers in the Netherlands, Weltens (1988) did not establish a difference in the relative *amount* of language loss across different levels of training with each group losing 10–15 % of overall language knowledge. Therefore, when level of education in general is being explored, a baseline needs to be established for each individual and a comparison made between his/her proficiency at the onset of incubation and at a later point during incubation. On the other hand, a cross-comparison between participants based only on performance in tests conducted during attrition cannot decipher the role of overall education. It is also important, as already stated, to be aware that individuals with a lower level of education may also have difficulties in test-taking in general which might obscure results.

In fact, the question of *why* and *how* level of education might affect actual retention levels is not clear and why individuals with a lower level of education might forget more. In the case of L2/FL attrition, lower education might coincide with lower proficiency levels in the L2 at the onset of incubation. However as explained, even when subsequent performance on later testing during attrition shows results indicative of deterioration, it does not mean that, relative to other attriters, *more* was lost. On the other hand, levels of proficiency at onset of incubation might be indicative of how consolidated or fortified information is in memory, with higher proficiency levels suggesting that this information has become more secure in the long-term memory store and so may be less susceptible to loss. This initial proficiency level in the L2 does not necessarily have to coincide with levels of education overall, however. At most it is argued here that if level of education is to play a role, it is a marginal one. For example, higher levels of education might be indicative of greater motivation and interest in maintaining contact with the language after official instruction has ended.

2.4.3 Length of learning/exposure and input

While it has been suggested that length of exposure might provide a more reliable indicator of long-term retention than achieved proficiency (Hansen (1999) referred to in Mehotcheva (2010: 41)), Mehotcheva & Mytara (2019: 13) citing Schmid (2006) and Bardovi-Harlig & Stringer (2010), argue that it is difficult to tease these two factors apart. It can be expected that greater exposure will support the advancement of proficiency in that language. Mehotcheva & Mytara also highlight the interplay between length of exposure and motivation whereby more motivated learners might seek out more opportunities to increase exposure to the language. It is argued here that not just length of exposure warrants greater exploration but the *frequency, type and quality of input* and *opportunities for language use* in this period are key features that have to be examined, aspects that have hitherto received only scant attention in the literature. Understanding degrees and severity of attrition warrants careful exploration of how consolidated target information is in memory in the first place, or rather how "well learned" that linguistic knowledge was as determined by length of learning and exposure but also, importantly, frequency and quality of input. In terms of language learning in an instruction context, as is the case in the present study, these considerations are particularly important, especially since they can also be decisive in determining motivation levels which contribute to attainment of proficiency and decisions to maintain contact with the language after school. In this respect, quality and frequency of input tend only to be looked at in the literature with respect to success of acquisition but not in terms of long-term knowledge retention.

There is not always a clear cut-off point, however, between exposure and a cessation of contact with the language. Weltens & Grendel (1993: 148) refer to the context of The Netherlands, for example, where, depending on the region, people are exposed to a number of different languages in their everyday lives as well as English in the media after official instruction has ceased. They state that foreign languages learned at an early age may never become truly inaccessible due to this multilingual environment that aids in the maintenance of the knowledge the learner had acquired so that once active engagement with the language takes place it may be fully reactivated again.

This is interesting in the context of Ireland where although Irish is often seen on public signs, on official documents or heard in the media, people can easily ignore it, nonetheless, since unlike the example of modern foreign languages and multilingualism in The Netherlands, Irish is rarely actively used in society. In this study, therefore, a relatively clear cut-off point at which incubation started could be established as Irish had little to no presence in the lives of the participants once they finished official instruction.

2.4.4 Length of incubation and rehearsal

As stated by Mehotcheva & Köpke (2019: 2), length of *incubation* is a term specifically used in L2/FL attrition research and refers to the period of time between the point at which active contact with the language ceases and the later time at which retention levels are tested. It is generally agreed that it constitutes the only factor that can be looked at in isolation (Mehotcheva & Mytara 2019: 14–15). Whilst all other factors are interrelated, for example, age at onset of acquisition is connected to length of learning which in turn can have repercussions with respect to attained proficiency, period of disuse is not considered to be related to other variables. Nonetheless, it has not been shown to single-handedly influence degrees of attrition. Bahrick (1984a) did observe, however, that most attrition took place in the first three to eight years after contact with the language had ceased and then levelled off. With respect to the role of rehearsal during incubation, results have been mixed (see Mehotcheva & Mytara (2019: 10–11) for a succinct overview of findings thus far on rehearsal in L2 incubation). Indeed, it is argued here that comparisons across studies should not be based on *length* of incubation alone but also in combination with the consideration of whether or not rehearsal took place in this time.

For example, in looking at lexical loss specifically, Graham (2012) did not observe significant drops in retention levels of L2 Spanish after an incubation period of 12 years in the case of L1 English missionaries that had spent 18–24 months in a Spanish speaking country. Participants were tested on four different occasions over the incubation period, whereby it was found that lexical knowledge remained for the most part intact. Rehearsal during the incubation period emerged as a strong predictor of knowledge retention. The participant that maintained the most contact with the language demonstrated higher levels of retained linguistic proficiency than other participants and those who had the least exposure to Spanish during incubation appeared to suffer the biggest losses with respect to their lexical knowledge. For example, the participant that demonstrated the lowest proficiency level had hardly any contact with the language after leaving the L2 environment and appeared to have lost all knowledge of the language. While Graham's observation that rehearsal did play a role in incubation can be related to Paradis's Activation Threshold Hypothesis whereby each time there is active engagement with an item or structure, its activation level is lowered, this interpretation is in contrast to Mehotcheva (2010: 152–153) who in her study of Dutch and German students' attrition of Spanish after a stay abroad did not find support for the positive effects of rehearsal and who thereby stated that this undermined the ATH. Rather she proposed that what is decisive is use of the language during the learning phase so that this information can become secure in memory. In keeping with Schmid's idea of a saturation threshold, she states that once this level of consolidation in memory is reached, knowledge becomes immune to loss so that rehearsal during incubation is no longer decisive. What is important to consider, however, with respect to the ATH is the *ease of access* to this information. Whilst not denying that information might still be secure in memory, the question in terms of the ATH is how much effort must be expended for it to be retrieved, or, in other words, is less effort required in subsequent attempts to access items once they have been reactivated. Rehearsal might indeed serve to make this consolidated information more easily retrievable compared to information that is never triggered.

However, on the other hand, and further complicating the issue, is the question why some items, therefore, can be more easily activated than others even when there has been *no contact* with the language, as is the case in this study, i.e., why can item *x* be activated but not item *y*, if neither item has been activated during incubation. This might be related to how consolidated this information had become in memory as proposed by Schmid's saturation threshold.

On this note, Bahrick (1984a: 11), for example, found that considerable semantic knowledge was still accessible after long periods of incubation despite not having being used. It is argued here, however, that in the case of attrition of modern foreign languages it is difficult to truly guarantee that there was no exposure to the attriting language during incubation, especially in the case of Bahrick's study which was conducted in the USA where Spanish, the L2 of interest in his study, is widely spoken.

Graham (2012) also found that even when Spanish was used regularly by some of the participants, language loss still appeared in certain areas of their Spanish knowledge. This suggests, therefore, that whilst rehearsal might be beneficial to a degree, it does not appear to guarantee total retention. On the other hand, in keeping with Graham's justifications, difficulties of access might have been due to cross-linguistic influence from English with respect to some items and structures.

On the other hand, Hansen (1999) found that much L2 Japanese knowledge had been lost after 30 years of incubation among L1 English missionaries. However, language typology might also have played a role in explaining the differing degrees of loss in each of the aforementioned studies. Japanese is typologically vastly different to English so that L1 English speakers cannot avail of their L1 linguistic knowledge in accessing Japanese items. Spanish and English, on the other hand, are not so distant typologically so that L1 knowledge might at times serve as a scaffold in processing L2 items. In this respect it could also have been that the learners of Spanish had attained higher proficiency levels considering that relative to Japanese it might have been easier to learn, in which case higher language proficiency at onset of incubation might explain the higher rates of Spanish retention relative to Japanese as opposed to length of incubation per se. Greater rehearsal among Graham's participants during incubation might also explain the difference in results. Hansen also found that those who had spent a longer time in Japan retained more knowledge of Japanese, suggesting that length of learning/residence played a positive role in supporting the attainment of higher proficiency levels.

Murtagh (2003) found in testing Irish learners after an 18-month incubation period that engagement with the language in the form of reading aided the maintenance of retention levels. This is an important finding because it highlights the importance of looking also at the *quality* of the input during rehearsal. As argued by Mehotcheva & Mytara (2019: 11), the type of engagement with the language influences the effect that it has. Input in the form of films may not be as effective in fortifying information due to its fleeting nature compared to reading, for example, which necessitates greater active engagement with the language on the part of the individual (see Section 2.4.5 on the role of literacy in language learning).

Taura (2019: 11) suggests using brain-imaging techniques to observe brain activity when each language is used. He looked at Japanese returnees up to seven years after leaving the L2 English environment. In the case of one of the participants on whom he conducted brain imaging, he found increased activation when she had to use English compared to when she used her L1 Japanese. In a later examination, however, he found less activation in the brain when English was used. It subsequently emerged that the participant had been working as a part-time teaching assistant at an international school which meant that she was completely immersed in the English language in the times when she was working. Therefore, there was less effort necessary on her part to activate English and this could be seen in the activation levels observed. Taura compares this to similar findings by Abutalebi, Tettamanti & Perani (2009), who found that it was continued use and exposure to the language after the onset of incubation that eased accessibility and thereby activation of the attriting language and not initial language proficiency levels.

Taura (2019: 2) also refers to Koike (1989) who examined L2 English retention levels among L1 Japanese children who had spent over two years in English-speaking countries before returning to Japan and who found that, despite differences in individual language proficiency, regular exposure to English on returning to Japan supported knowledge preservation.

Weltens & Grendel (1993: 153) refer to studies by Verkaik & Van der Wijst (1986) and Cohen (1989), that argue that disuse leads to problems in accessibility of information, suggesting that rehearsal functions to maintain low threshold levels of activation thereby easing retrievability.

De Bot (2004: 234) also argues that, while there is most probably no limit to memory capacity and that older languages remain in memory despite disuse, it is important to ensure that in cases where individuals know more than one language, that unused languages are, nonetheless, also activated. Otherwise, he argues, they become driven further into the background by competing, more active languages, and prove, therefore, more difficult to access.

Whilst dealing with L1 attrition it is nonetheless interesting to note Schmid's (2007) observation of higher scores of L1 lexical and fluency retention measures among speakers who spend time in official, formal settings where only the L1 is spoken because in such contexts, a one-language-only principle is usually more strictly adhered to. Since code-switching is less used in such settings, the second language is consciously suppressed, with activation levels of the L1 becoming less influenced by competition from the L2. In informal settings in bilingual contexts on the other hand, code-switching is often the norm, meaning that even in a predominantly L1 context (the home, for example), the L2 is still online with both languages having more of an equal activation than if only one language is used exclusively. Whilst Yilmaz & Schmid (2018: 234), in keeping with Schmid (2007), argue that L1 attrition may be more representative, therefore, of a failure to prevent competition and influence from the L2 as opposed to an inability to access the appropriate L1 equivalent, it is argued here that the official and formal nature of professional environments might also offer input that is qualitatively superior to that often experienced in informal settings. In other words, there appears to be both a qualitative as well as a quantitative difference in input and output in formal versus informal settings, which in turn could lead to different manifestations of L1 attrition as observed by Schmid.

2.4.5 Literacy

Considering the emphasis on reading and writing in Irish language education in the 1980s it is important to look at how literacy might impact long-term retention of lexical item knowledge. Grabe (2009: 60–64) explains how learning to read in the L2 draws on both implicit and explicit learning processes and emphasises the role of frequency of input and regular rehearsal in strengthening consolidation in memory of vocabulary knowledge acquired through reading. Implicit learning constitutes an automatisation of routines that assist effortless reading such as being able to parse syntactic structures and recognise words and is stimulated by the repeated input encountered in reading. Explicit learning, on the other hand, entails paying direct attention to the particular information in focus in that moment. As this data becomes established in working memory through rehearsal, it then becomes consolidated in long-term memory. Just as it takes time for the knowledge that emerges from implicit learning to become established and automatised, information that has been learned explicitly is very vulnerable to loss unless it is activated many times. Grabe (2009: 81), therefore, emphasises that since the implicit acquisition of knowledge is an incremental process and information that is learned through direct effort necessitates regular exposure and reproduction, learners need to engage in reading over the long-term. Li (2010: 254) refers to Wodinsky & Nation (1988) who found that in L2 learning, reading supports vocabulary acquisition.

As Li points out, learners have to have acquired a certain number of words to be able to read so that knowledge of vocabulary and reading skills reinforce each other. As learners expand their vocabulary, reading becomes easier and more advanced and conversely serves to support learners in establishing the meaning of newly encountered words as they learn to avail of the syntactic and contextual cues in the text to guide them. Hereby, Pan & Berko-Gleason (1986: 204) argue that literacy provides a "broader data base" for the learner.

Hickey (2007: 472) in exploring the acquisition of reading skills in Irish, also points out that individuals with advanced reading skills in the L2 have superior word recognition, as they interact in a form of reciprocal reinforcement, thereby enabling at the same time better text comprehension.

As well as citing Nation (2014) who emphasised the role of reading in enhancing vocabulary knowledge, Stenson & Hickey (2018:7) refer to a study by Webb (2005) that compared two tasks to foster vocabulary learning; one a language production task and the other a reading task. They observed that reading generated better results with respect to vocabulary acquisition.

By providing an additional illustration in the brain, it is therefore widely accepted that literacy strengthens language knowledge (Yilmaz & Schmid 2018: 236) which in turn aids retention. Köpke (2004: 9–10) states that an individual who has never learned to read and write in the L2 might have greater problems in accessing this language after a period of non-exposure. She argues that when the learner becomes literate in a language a second layer of representations (in addition to oral representations) is established that should "contribute significantly to the grounding of language" (p. 10).

Mehotcheva & Mytara (2019: 14) also argue that whilst the factors of age and literacy are closely confounded and are difficult to isolate in research, studies have, nonetheless, found that literacy helps to avert attrition with better retention demonstrated by participants that were literate in the L2 (Olshtain 1986, 1989; Reetz-Kurashige 1999; Hansen & Newbold 1997; Hansen & Chantril 1999). Olshtain (1989), for example, found in her analysis of L2 retention of English in L1 Hebrew speakers over a period of one year on returning to Israel, that the older age-group (8–14 years old) maintained knowledge of the correct use the irregular forms of noun plurals and verb forms in the past tense but the younger age-group (5–8 years old) appeared to have lost this knowledge. In explaining this, Olshtain argued that not only had the older group the benefits of greater cognitive maturation but they were also literate in English unlike the children in the younger group and also that they continued to read in English on their return to Israel, which Olshtain argues provided the necessary input for the preservation of this linguistic knowledge. She highlights, however, that age and the automatisation of literacy are inextricably linked. Hansen (2001: 65) agrees that it is almost impossible to isolate each of these factors as older children inevitably have better literacy skills (and often higher levels of language proficiency). For example, in a study of four American L1 English children who had lived in Japan for two and a half years, Hansen-Strain (1990) observed that the older children (aged nine and seven when they moved back to the USA) retained more of the Japanese language on returning to the USA. Hansen-Strain asserted that it was a combination of their acquisition of literacy skills and their increased exposure to Japanese since they were older and had, therefore, also attended a Japanese school, that contributed to higher levels of preservation of linguistic knowledge. Hansen-Strain (1990: 369–370) concludes, nonetheless, also in referring to other studies (Berman & Olshtain 1983; Cohen 1989; Olshtain 1989), that while variables such as age and language proficiency interact with literacy, the general tendency observed in studies on L2 attrition is that the better the literacy skills, the greater the long-term knowledge retention. This was further demonstrated by Hansen & Shewell (2002) who additionally found in their study of adult L1 English attriters of L2 Korean that literacy positively correlated with motivation to further enhance their literacy skills in the language and to use reading opportunities to maintain contact with the language after leaving the L2 environment.

The benefits of availing of L2 literacy skills after returning to the L1 environment is also highlighted by Taura (2019: 10). He refers to a study conducted by Taniguchi (2014) conducted in Japan that looked at six L1 Japanese children that had lived in an English-speaking country for approximately four years (from 4;4 to 8;3 years) with the aim of acquiring literacy. On their return they were tested on oral storytelling over a period of 24 months while they were still at primary school in Japan. It was found that those children who continued to engage in English literacy activities at home and received a lot of support from their parents as well as being involved in English-speaking groups, performed better on the retelling of stories in English. The literacy skills acquired in the L2 environment enabled continued engagement with the language after returning to the L1 environment. Taura (2019: 7–8) also refers to a study by Yukawa (1998) that compared L1 attrition in her two children after the family had moved from Japan to an L2 environment and found that her older child retained more Japanese knowledge not only because he was older when the family left Japan for the second time but because this also coincided with his attainment of literacy skills in the Japanese language unlike her daughter who had not yet acquired these skills and who then showed significantly greater lexical loss and syntactic decay in Japanese after they had left.

Yilmaz & Schmid (2018: 236) additionally argue that higher levels of education are linked to literacy (see Section 2.4.2.2 on levels of education).

2.4.6 Attitudes and motivation

Dörnyei (2000: 520) explains motivations as being "responsible for why people decide to do something, how long they are willing to sustain the activity, and how hard they are going to pursue it."

Whilst the notion of what constitutes motivation in learning has undergone a number of developments and paradigmatic shifts, the concept of motivation presented in Gardner & Smythe's (1975) socio-educational model of second language learning to explain success in achieving L2 proficiency continues to hold scholarly attention today. Within this model in which a number of individual variables such as intelligence, language aptitude and anxiety are said to interact and are dynamic, the reciprocity and interaction between the L2 learner (and his concept of self) and the L2 community to which he/she navigates is emphasised and explicated as an *integrative motivation* on the part of the learner. Instead of viewing the language learner as an isolated, encapsulated entity, Gardner & Smythe adopted a macro approach to understanding motivation, underlining the importance of understanding a learner's motivation within the greater social context as well as with respect to his attitude to his immediate learning conditions and environment (Gardner & Lysynchuk 1990). Gardner (2001: 10) speaks of an integrative orientation when learners are driven by an interest in gaining access to, and a better understanding of, the L2 speaking community. Gardner (1988: 104) however emphasises that an *integrative motive* does not exist in and of itself but is instead a practical term used to imply a number of interrelated attitudinal and motivational features that in a combined state appear to play a role in driving second-language learning. Neither does it constitute one fixed stable constellation but one that is changing and dynamic in sync with the wider context of learning.

Gardner (2001: 10) also offers reasons for learning that he categorizes under *instrumental orientation*. These motives stem from practical incentives that bring one closer to one's goals, be they academic (e.g., passing an exam) or professional (e.g., having to demonstrate knowledge of a particular language to get a job). For instance, in the Republic of Ireland one has to provide proof of a specific level of proficiency in the Irish language in order to qualify for certain professions in the public sector. Instrumental motivation thereby implies learning the language to accomplish pragmatic endeavours. These goals do not comprise the desire to become a proficient L2 speaker so that one emerges closer to the L2 speaker community although he does state that an individual might have both integrative and instrumental reasons for learning a language. Gardner (2001: 12) argues that essentially motivation to learn a second language is influenced by two connecting variables; integrativeness and attitudes toward the learning situation, while the level of language proficiency achieved, on the other hand, proceeds from motivation and language aptitude.

Two other classifications of motivation emerged in the form of Self-Determination theory (Deci, Vallerand, Pelletier & Ryan 1991; Noels, Pelletier, Clement & Vallerand 2003) which shifted from a macro view of the learner to achieve an enlarged and more micro perspective of the individual. Said to stem from within the individual without the influence of external factors, *intrinsic motivation* refers to learning for enjoyment whereby learning in and of itself constitutes the very reason for the individual's engagement in the act. The learner experiences satisfaction in making progress in learning and better understanding the language and producing it. *Extrinsic motivation*, whilst also internalised like intrinsic motivation, drives the act of learning not for the sake of enjoyment but to reap a reward or to avoid negative repercussions and is not unlike Gardner's concept of instrumental motivation. In internalising this extrinsic motivation, however, the means to achieving this external goal become driven and guided by internal forces of self-fulfilment and personal ownership.

Noels et al. (2003) argued that these expressions of motivation are more effective in stimulating the learner to act because of their very internalised and self-determined nature as shaped by one's inherent wants and needs. Noels, Clement & Pelletier (1999: 25) also argue that there is a higher likelihood that learners who are driven by intrinsic motivation will continue learning over time, thereby improving their L2 proficiency further which in turn will possibly drive motivation even more. Extrinsic motivation on the other hand, whilst it might lead to positive outcomes in the short-term, does not necessarily lead to a long-term commitment to learning since such individuals may not actually have positive attitudes towards the language itself (Noels et al. 1999: 25).

Concerned with the micro level of learning, Dörnyei (2005) proposed the idea of an L2 Motivational Self System that represents the notion that one can have various possible selves and that envisaging a particular L2 self can support the learner in encouraging their drive to learn so that they get closer to achieving the realization of this self (Flynn & Harris 2016: 373). See Chapter 6, Section 6.5 for a discussion of this in the case of minority languages.

2.4.6.1 Attitudes

Cherciov (2013: 718) refers to Oskamp's (1977: 8) explication of attitude to mean the preparation or readiness to behave or act or as Cherciov herself puts it, "the pre-disposition to respond in a favorable or unfavorable manner to an *attitude* object, which includes things, places, people, situations, or ideas" (2013: 718). She proceeds to differentiate between, on the one hand, attitudes and, on the other, opinions, values and beliefs whereby in referring to Oskamp & Schulz (2005: 18), she explains that an attitude is a "general evaluative orientation" whereas opinions, values and beliefs, whilst connected to this predilection, are more specific and constrained and governed to a greater degree by cognitive reasoning. Smyth & Darmody (2016: 1) in exploring attitudes among 13-year-olds in Ireland to Irish as a subject at school cite Cargile, Giles, Ryan & Bradac (1994: 221) who refer to the tridimensionality of the essence of language attitudes; the cognitive aspect that essentially represents "beliefs about the world", the affective dimension that refers to the (constructed) feelings about the object or phenomenon under consideration, and finally the behavioural component meaning that attitudes encourage particular behaviour.

The general attitude towards the attriting language and its status in society is also a factor that should be examined as this may influence the motivation to learn the language (Mehotcheva & Mytara 2019: 12). It is argued in the present study, however, that whilst language prestige and generally positive attitudes towards the language at a societal level might serve to stimulate an initial positive motivation in learning it, this may be nullified if teaching approaches are such as to suppress pupils' curiosity about the language over time and any ambition to acquire it.

2.4.6.2 Motivation and attitudes as dynamic and changing

It has been shown that motivation can positively influence acquisition and learning. Higher levels of motivation have been established in students that achieved higher levels of proficiency (Clement, Gardner & Smythe 1977; Gardner, Tremblay & Masgoret 1997; Gardner, Masgoret, Tennant & Mihic 2004; Tremblay & Gardner 1995; Masgoret & Gardner 2003; Schumann 1994).

Pan & Berko-Gleason (1986: 198) refer to Gardner, Lalonde & MacPherson (1985) who emphasized the importance of looking at the motivation and attitudes that dominated during the instruction phase as their influence during this time is fundamental to understanding degrees of attrition that might possibly follow in the years after learning has ended. Also considering the rather unique scenario that the learning of a minority language such as Irish constitutes, this also warrants careful reflection in the present study. Nonetheless, in keeping with Mehotcheva & Mytara (2019: 9), motivation is also considered dynamic and changing and does not exist as one isolated factor but interacts with other variables over time. Schmid & Mehotcheva (2012) refer to a number of analyses that demonstrate and discuss this variability in motivation over time (Dörnyei & Otto 1998; Nikitina & Furuoka 2005; Schmid 2006).

2.4.6.3 The interrelation of variables

In keeping with the idea of clusters of interacting factors (see Mehotcheva & Mytara 2019), it is also argued in this analysis that motivation is just one factor among a number of variables that interact to influence the attrition process. This complexity is further amplified considering that the very concept of motivation itself constitutes a multi-layered construct shaped by a number of underlying variables such as learning experience and teacher engagement. Mehotcheva & Mytara (2019: 6) highlight the interplay between attitudes and motivation, attained proficiency, and language contact and use not just in prognosticating levels of learning achievement but also in shaping the course of attrition and predicting degrees of loss. Success in mastering higher proficiency levels can lead to a greater confidence in using the language. In cases where circumstances change such that an individual is no longer in a context where they have contact with the language (be it in an instruction or natural environment), this confidence and sense of positive attainment can possibly prompt a greater motivation to actively maintain contact with the language. Mehotcheva & Mytara (2019: 9) refer to variables at the personal and the environmental level that feed into motivation. Within the sphere of personal factors, they refer to attitudes towards the language (including its speakers as well as the culture associated with it), the attitude towards the learning context and the inner motivation to learn the language.

On the other hand, the status of the language and the learning context refer to environmental factors. Here it is argued that sociopolitical and historical factors contribute to the standing or status of a particular language and these social and political associations with the language, as well as the accumulation of levels of success in acquisition not to mention the very experience of L2/FL learning, feed into each other, simultaneously shaping motivation. Personal and environmental factors are not necessarily distinct but can also interact. For example, one's subjective attitude towards a language may not just be influenced by the status and prestige of the language in a particular environment but also by how it is presented and taught in the classroom context. With respect to retention, while practical circumstances, such as time and location, can also influence whether or not contact with the language is maintained, memories from the time of learning, confidence and an assessment of whether or not it is worth the effort to actively and consciously preserve knowledge of the language can also play a role. The learning environment as well as the materials and methodologies that had been used not only have short-term consequences with respect to learning success, but the experience of learning can also leave an imprint in the memories of learners; one that might be positive and thereby motivating or one that might be negative, even traumatic and thereby demotivating.

2.4.6.4 Motivation and L1 attrition

In a review of selected studies on L1 attrition, Yilmaz and Schmid (2018: 235) observed however that attitudes, be they with respect to identification with the attriting language and its associated culture, or regarding the L2, do not always appear to play a role in determining retention outcomes.

Indeed, in attempting to draw a parallel to acquisition processes, Mehotcheva & Mytara (2019: 9) state that since attitudes and motivation do not play a role in L1 acquisition, this might explain why studies have not found a link between attitudes and L1 attrition. A couple of points should be noted here, however. Some studies, albeit fewer than those that did not establish a connection, have in fact found a link between attitudes and L1 attrition (e.g., Ben-Rafael & Schmid 2007; Bylund & Ramírez-Galan 2016; Cherciov 2013; Schmid 2002) and that acquisition and attrition are not necessarily two sides of the same coin. On the contrary, attrition constitutes a separate phenomenon susceptible to different triggers and experiences than acquisition. Whilst arguably motivation and attitudes play an interesting and unique role in the acquisition and learning of a second language, one that does not factor in first language acquisition, this does not rule out the possibility that they might, nonetheless, also play a role in explaining L1 language loss, even if this influence is significantly reduced relative to L2 attrition.

For example, Schmid's (2002) study on L1 attrition among members of the immigrant Jewish community in the UK that had been forced to flee Germany during WWII found that those individuals that had escaped from Germany later and therefore had experienced greater persecution expressed a more negative attitude towards the German language and this manifested itself in higher degrees of attrition among these persons compared to those who left earlier on in the war or just before it.

In examining the role of language aptitude with respect to L1 retention among L1 Spanish speakers of L2 Swedish who had arrived in Sweden at approximately the age of 24 years and had been living there for on average for 16 years, Bylund & Ramírez-Galan (2016) found that it was not language aptitude that predicted levels of L1 retention but attitudes and an emotional connection with the language and culture surrounding that (the majority of the participants were from Chile, with the remainder from Columbia and Peru).

Ben-Rafael & Schmid (2007) studied L1 attrition among groups of L1 Russian and L1 French speakers in Israel and also found an indirect correlation between the participants' attitudes towards the host country and tokens of L1 attrition. The French group whose emigration to Israel was propelled more by ideological reasons as elicited in the interviews, demonstrated greater L2 (Hebrew) influences in their French than the Russian cohort whose reasons for moving to Israel were more pragmatic in nature. The French, on the other hand, in more closely embracing the culture of the host country as well as the Hebrew language, demonstrated greater degrees of L1 attrition accordingly.

2.4.6.5 Motivation and L2 attrition

With respect to L2 attrition, reports on the role of attitudes and motivation are also mixed. For example, Weltens & Grendel (1993: 146) and Mehotcheva (2010: 158) claim that there is little definitive evidence that attitude and motivation can function to counteract attrition. While Taura (2019: 9) refers to a study conducted by Fujita (2002) in which a direct link between L2 maintenance and a cultural affinity to the language was not detected, Mehotcheva & Mytara (2019: 9) point to studies by Wang (2010) and Xu (2010) that found that positive attitudes towards the L2/FL served to positively influence both acquisition and retention levels. Wang (2010) found a positive correlation between high performance on language tests as well as retention of linguistic knowledge and a favourable attitude towards English as well as a motivation to actively preserve L2 knowledge in the case of adult Chinese graduates from an English-speaking college. On the other hand, Xu (2010) who examined L1 Chinese and L1 Dutch learners of English interestingly found a correlation between positive attitudes and proficiency levels in the case of the Chinese learners of English in her study but this effect did not emerge on the part of the Dutch learners of English, neither in the case of proficiency nor with respect to retention levels. Whilst Xu accounts for this discrepancy by stating that the testing instrument proved inadequate with respect to deciphering attitudes, it is unclear as to how a clear link between attitudes and proficiency could, nonetheless, be established in the case of the Chinese cohort but not for the Dutch group.

Mehotcheva (2010) also observed that positive attitudes towards the FL led to higher ratings on some of the measures used in her study of Spanish retention among L1 Dutch speakers after a stay abroad as part of an Erasmus program, although she was unable to establish distinct patterns of motivation overall.

In the case of heritage languages, Gharibi & Boers (2017: 56) refer to analyses where attitudes have been considered to be particularly influential in language maintenance at the societal level (Fishman 1991; Schmid 2002; Ben-Rafael & Schmid 2007). Indeed, Mehotcheva (2010: 159) argues that the lack of overriding evidence establishing a decisive link between attitudes and motivation and levels of retention might stem from the very testing measures themselves, suggesting that Gardner's commonly used attitude and motivation questionnaire (1985) may not be refined enough to identify finer subtleties.

It is argued in the present study that the emergence of a negative attitude might affect degrees of language retention in the sense that it might cause the learner to abandon or relinquish any desire to maintain contact with the language, thereby resulting in its deactivation within the linguistic system. In this respect, while negative attitudes and motivation do not constitute the direct cause of attrition, they do have repercussions with respect to how individuals *behave*. It is these language decisions, as prompted by motivation, that determine what actually happens to linguistic knowledge in memory and depending on degrees of entrenchment, this may or may not have significant consequences with respect to long-term retention. Whilst L1 knowledge might be more embedded and, therefore, more robust and less subject to variations in motivation beyond possible superficial indications, the L2, depending on degrees of proficiency among other aspects, may be more vulnerable to the effects of attitude and motivation for the reasons just stated.

2.4.6.6 Unique nature of motivation in minority language attrition

It is also important to consider that the specific L2 in question, its status and societal dominance, as well as what it represents and how it is presented and taught, can influence the particular manifestation of motivation that emerges from the unique interplay of these variables. Blackwood (2013: 8), for example, highlights that it has been difficult to achieve unanimous consensus among experts in the field on one definitive explanation of motivation, the different types and how these should be classified as well as the effect of positive attitudes and motivation on L2 learning. However, with reference to threatened languages in particular and the endeavours of revitalisation, she refers to Choi (2003: 82) and Baker (1988) who argue that there is a clear link between attitudes towards a language and success, or indeed lack of, in language preservation and its regeneration. Considering that Irish falls more generally into the category of a threatened language in the sense that it is not a modern foreign language actively spoken beyond a limited number of small communities, it could be surmised that positive attitudes might indeed be among the driving forces encouraging the people in these communities to speak the language. Once school has been completed in Ireland, there is no social or political compulsion to continue actively using the language. Neither does it stem from a necessity to have a common language of communication since all Irish speakers are L1 English speakers too. It would, therefore, be difficult to ignore positive attitudes as a factor driving active use of the language in such communities. Therefore, while positive attitudes and motivation may not have been found to have an overriding influence with respect to long term knowledge retention in the case of modern foreign languages, this seems to be different for minority languages which, without positive attitudes, would possibly never be spoken or pursued beyond compulsory education. In other words, there is more support *automatically* in place with respect to modern foreign languages and learners eventually continuing to build on what they have learned in school. Work, travel, relationships and media can all play a role in encouraging further use even when attitudes towards the actual learning of the language were less than positive. Minority languages, such as Irish, cannot "offer" what modern foreign languages can and so are more reliant on the positive attitudes of their users in continuing use. Flynn & Harris (2016: 381) in their analysis of 12 adult learners of Irish established motivating factors that they argue are seldom considered in L2 research; a motivation to engage with the culture and heritage of the language (see Chapter 6, Section 6.5.2).

Naturally, simply having positive attitudes does not suffice on its own to ensure preservation of language knowledge and adequate engagement with an L2. These attitudes need to be transformed into active use which in the case of minority languages is often more challenging and effortful. Considering Pintrich & Schunk's (2007: 5) reference to motivation as "the process whereby goal-directed activity is instigated and sustained" and while compulsion to learn Irish at school in Ireland ensures a degree of active involvement with the language, it is interesting to observe what happens after pupils leave school. In the case of languages such as Irish, positive attitudes are a necessary first step in stimulating motivation to actively use the language.

In looking at modern foreign languages, Mehotcheva (2010: 156) argues that FL attrition might be guided by different factors than those typically delineated for first or second language attrition, stating that, in particular, attitudes and motivation may not play the same role in the FL attrition process. She does not, however, elaborate on why attitudes and motivation might play a lesser role in the case of FL attrition apart from stating that it is difficult to establish precisely which motivations and attitudes dominate during the learning phase. Indeed, earlier in that same article (2010: 31), she proposes, in line with the Activation Threshold Hypothesis, that in the case of second language acquisition, motivation plays an important role with respect to retention because it can influence activation levels. This leads to the question, however, why motivation and attitude cannot, therefore, also influence activation levels in the case of foreign language learning and in the case of the present study, minority languages.

Indeed, as just mentioned and discussed in further detail in Chapter 6, Section 6.5, motivation and attitudes might play an even more decisive role in cases of minority language attrition or indeed, with respect to foreign language attrition. In giving the example of cases of L1 attrition where individuals connotated their L1 with negative experiences, Mehotcheva (2010: 31) demonstrates how attitudes and emotions can affect long-term language maintenance. There is no reason to suggest that this might be different with respect to L2 or FL learning in school. A negative emotional experience in learning could affect any long-term willingness to maintain contact with the language or desire to continue to engage in learning after official education has ended.

2.4.6.7 *Challenges in measuring motivation*

Considering the intricacy of the interconnection between motivation and other factors, accurately classifying a particular learner's motives as well as identifying and establishing distinct patterns of motivation can be challenging for the researcher. Jordan (2004) calls into question the robustness of such an endeavour since the concept of motivation is vulnerable to subjective interpretation. Cherciov (2013: 730) also states that the dynamic aspect of attitudes and their constant susceptibility to change means that they play an unpredictable role in attrition. Variation in the means and degrees of establishing and measuring motivation may also account for why there is much divergence in the findings on its role in language retention.

As Mehotcheva (2010: 41) points out, motivation is usually explored though self-reports which can be subjective and biased. Adding to the complexity in examining motivation with respect to retention is that such an investigation also necessitates exploring motivation during learning. In the case of very long incubation periods such as is the case in the present study, memories may no longer be accurate. Nonetheless exploration via self-reports is the most viable research option and whilst necessitating caution which respect to evaluation, if planned and executed with the necessary sensitivity to the vulnerability of memory and the dynamic aspect of attitudes and motivation in their interaction with other factors, self-reports can lead to very insightful and helpful observations. This can help to overcome the challenge presented by Mehotcheva & Mytara (2019: 9) whereby self-reports typically look at one particular point in time. See Chapter 4, Section 4.4.3 for information on how attitudes and motivation were explored in the present study considering these challenges.

Chapter 3: The Irish language

It is argued in the present study that in attempting to understand attrition it is necessary to examine the specific sociocultural peculiarities of the language in question and, its status, relevance and standing within the particular context in which it was learned as well as in the wider society, i.e., the unique "circumstances" of not just the language itself but of the very special relationship between it and its learners. Addressing matters of identification and associations is fundamental to comprehending individual justifications for discontinuing contact with the language at the end of official instruction, a scenario that manifests as inactivation of linguistic knowledge and can prove a potent instigator of attrition. Considering that the present study looks at attrition and possible savings of a language that is *not* a modern foreign language but one with a unique standing and history within Irish society, it is essential to understand its role and status (not just official) and its relevance for the narrative of Irish nationhood and interpretations of identity. Therefore, the following chapter aims to explain and explore the sociopolitical and cultural history of the Irish language, endeavours to revive it via the education system and the very distinct relationship individuals have with it as a result. Only with this insight into the special above-mentioned conditions pertaining to Irish, can one possibly understand eventual individual experiences of learning and of attrition of this particular language. In this chapter a key grammatical feature of the Irish language will also be explained, i.e., that it is a noun-centred language. This is delineated briefly to explain justifications for the item category breakdown used in testing.

3.1 Sociopolitical background of Irish

The Republic of Ireland was declared in 1948 consisting of 26 of the 32 counties of the island of Ireland, marking the culmination of a long and very complex history of struggle for independence from and resistance to British rule. One critical step towards this had been the Anglo-Irish Treaty of 1921 which led to the secession of these 26 counties from the United Kingdom and the founding of the Irish Free State in 1922. The English language had already come to dominate Irish society throughout the 16th and 17th centuries with the consolidation of British imperialism and many of those now in government in the new state and in society in general felt that reigniting the Irish language was one important step towards defining a unique identity distinct from that of Britain. Therefore, many policies of the new Irish state aimed to reestablish its prominence in society. (For a detailed exploration of the history of Ireland in the 21st century, see Ferriter 2004)

A number of ministers in government in these young years of the state had been active in the Gaelic League (*Conradh na Gaeilge*), a movement established in 1893 by Douglas Hyde (who later went on to become Ireland's first president in 1938) and which was engaged in ambitiously reviving the Irish language and culture and regenerating its active use in everyday life. Education was to constitute a key area in which this goal was to be realised. With this, Irish was first established as a compulsory subject in primary school in 1922, then in the lower part of secondary school in 1928 (up to the Intermediate Certificate, an examination taken at the end of the third year of secondary school) and finally in 1934 it was also made compulsory for the award of the Leaving Certificate (the final state school examination). It is also important to note that it was made an obligatory subject for entry to the National Universities of Ireland (NUI). This clearly constituted a driving force among learners aspiring to attend university in Ireland which was the case for the participants in the present study. It was also made an obligatory prerequisite for employment in the public service.

Under Article 8 in the revised constitution of 1937, Irish was designated the first official language and English was recognized as the second official language. Although Ireland joined the European Union (EU) in 1973, Irish was only recognised as an official language in the EU in 2005, albeit with limited status, meaning that only certain documents would be made available in Irish (although this number was increased in 2015). The Irish government has, however, actively striven to achieve full status for Irish in the EU with the Council of Europe conclusively agreeing that this be granted by 2022 (See European Commission website for more information).

3.2 Endeavours to revive Irish

Ó Riagáin (2001: 196) highlights how substantial Irish language policy has been since the founding of the state:

> Irish language policy over the 20th century includes, for all or part of the period, policies that were directed at economic development in the Irish-speaking areas, language teaching in the schools, recruitment procedures in the public sector, language standardisation, radio and television services, together with a myriad of other less important policies to promote and extend the use of Irish. Some of these policies are or were 'universal type' policies, that is, designed to affect all members of the public, while others selectively target particular geographic or social groups. (Ó Riagáin 2001: 196)

Endeavours to revive Irish have not been entirely successful however, at least with respect to making it the language of daily use, and even after the founding of the state when it could be expected that pride in Irish identity might have been particularly strong, the government faced a difficult task with respect to restoring the Irish language to a central role in the functioning of society. Outside of the Gaeltacht areas (Irish-speaking regions found for the most part along the south, west and north-west coast), English had come to be the everyday language of the Irish people. Indeed, in attempting to represent the journey of the Irish language, perhaps one should refer to Fishman (1991: 128) who in referring to the Irish case sums it up quite aptly when he asks, "what went wrong?" He proceeds with the even more emotionally laden question of whether it is even worth the effort as well as the financial cost, considering that Irishness can still be achieved (and has been) without the Irish language, with this constituting a much easier endeavour in his opinion. He argued that in the Irish context, the Reversing Language Shift movement neglected to come to terms with the real problem which was a lack of intergenerational transmission (p. 144).

According to Fishman (1991) the very need to use a language to communicate is what is considered to be a key prerequisite for language shift to be reversed. At a pragmatic level, having English, the language of international commerce, as its everyday system of communication offers the island particular advantages. Relatively isolated from the European mainland, the English language eases access to global markets and resources. Therefore, creating a *need* to speak Irish in everyday life in Irish society is challenging and to expect that parents whose L1 is English to imbue a willingness in their children to become fully competent bilingual Irish – English L1 speakers who embrace the language to such a degree that they actively use it in their daily lives can be considered utopian, especially when this is based purely on ideological, even nostalgic or romantic grounds. Languages also have utilitarian dimensions. Modern foreign languages offer a functional component that one could argue is lacking in the case of such minority languages as Irish. Whilst seemingly valued for its inherent linguistic richness and cultural heritage, the Irish language itself was not transmitted in such a way in education in the 1980s (when the participants in the present study were in school) that it was seen as practical or useful in terms of bringing its speakers closer to the benefits of a consumerist, materialistic market; gains that can only be made through interaction with an external player. The insular nature of the language in this respect made, and continues to make, revitalization challenging. Using the example of language policies in Germany and France, Garcia (2015: 45-46) demonstrates how language teaching policy differs between, on the one hand, the way that modern foreign languages are approached and, on the other, how minority languages are dealt with in education; with the utilitarian aspect clearly being embraced in the case of modern foreign language education whereas in the case of minority language education, it is the upholding of cultural identity and heritage that seems to dominate curriculum planning.

While many languages die out because they do not have the support of the government, this, as previously stated, is not the case for Irish. However, Fishman (1991: 141) refers to Irish attempts at revitalization as abounding in tokenism; indeed, he claims this to also be the case for Irish in education. He argues, nonetheless, that this is possibly an inevitable progression in scenarios where the identity of a nation is no longer dependent on having its own unique language. He refers to the model of "Xmen in Yish" in this case "Irishmen in English" where "Irishness" without the everyday use of the Irish language has become so well established that the role of the language seems predominantly symbolic.

To understand the plight of the Irish language, it is essential to look at the history of language shift in Ireland. A Gaelic language, it was the Celts who brought Irish to the island around 200 B.C. from mainland Europe. Over time it inevitably underwent change, experiencing influence from Norse, Latin and English (Hughes 2001). English and French, on the other hand, were first brought to Ireland by the Anglo-Normans in the mid 12th century. However, rather than these languages becoming consolidated as the dominant languages, these settlers instead adopted the Irish language, a trend that prevailed until the 16th and 17th centuries. Whilst undoubtedly the shift to English truly began during the plantations of this time when settlers from England and Scotland were sent to occupy land confiscated by the British crown, Bradley (2014: 538) argues that it was the acquiescence on the part of the Irish people themselves with respect to adopting English that secured the shift. The association of English with prosperity and progress coupled with the role that the famine in the mid-18th century played in eradicating colossal numbers of native-Irish speakers either through death or emigration, meant that by the end of the 19th century the number of Irish speakers had drastically diminished. He argues that it was pragmatism that led the Irish to shift to the English language; the people with political and economic power spoke English and to gain access to these and their resources, knowledge of this language was required; to the detriment of Irish. Hickey (2009: 4) also argues that the Irish language had a negative connotation among sections of the population at the time; one of poverty, exclusion and isolation. Additionally, he points to the role played by the Catholic church which controlled primary school education in Ireland and its failure to promote and defend the use of Irish. He refers to the well-documented examples of children being punished if caught speaking Irish at school and the *tallystick* (*bata scóir*) worn by children, on which a line was marked each time a child spoke Irish (Hickey 2009: 3). In fact, he contends that this shift in the 19th century also explains why neither bilingualism nor long-term diglossia emerged in Ireland.

However, at the end of the 19th century and the formation of the Gaelic League, the quest for the revival of Irish was ignited in attempts to assert distance from the British crown and reclaim a unique Irish identity.

3.3 Trends in Irish speaker numbers

While according to the census of 2011 the number of Irish speakers has tripled in the last 100 years, Bradley (2014: 539) argues that these estimates are to be treated with caution since a clear definition of what constitutes a "speaker of Irish" was not given in the census and some respondents might have overestimated their proficiency stemming from an underlying sense of nostalgia and patriotism. The issue of whether or not one could hold a fluent conversation in Irish was not probed in any detail in the census. Nonetheless, Ó Riagáin (2001) praises attempts made by the state to re-ignite the use of the Irish language:

> However, while the effort to re-establish Irish as a national language has not been successful, neither can the impact of Irish language policy be described as negligible. Not only did it slow down a long-established process of language shift, but it clearly altered the spatial and social structure of bilingualism in Ireland. After 70 years, the linguistic division of labour has changed to the extent that, while the original communities of Irish-speakers have declined in both absolute and relative terms, there has been an increase in the proportions of Irish-speakers in some, mainly middle-class, groups elsewhere. (Ó Riagáin 2001: 197)

Since the founding of the state there has been a relaxing of policies with respect to the Irish language in some domains of public life. For example, since 1973 it is no longer a requirement for employment in the public service sector. Indeed, the burden has increasingly been put on education to keep the Irish language afloat. Whilst the Official Languages Act of 2003 guarantees the right of all Irish citizens to communicate with the state through Irish should they so desire, apart from the state school system, Irish is rarely used in government itself or in any of its departments, in the civil service, the police force or legal profession (Bradley 2014: 539).

In 2010, the Irish government published the *20-Year Strategy for the Irish Language 2010–2030* with the following goal:

> The objective of Government policy in relation to Irish is to increase on an incremental basis the use and knowledge of Irish as a community language. Specifically, the Government aim is to ensure that as many citizens as possible are bilingual in both Irish and English. (Government of Ireland, 21.12.2010, p. 3)

It is also noteworthy that it is gaining greater recognition at EU level. In close cooperation with the Irish government, the EU has increased the number of Irish language speakers it employs in recent years and has built up a number of Irish language resources and already translated parts of EU legislation in Irish. This progress in terms of the recognition of the Irish language within the workings of the EU should not be overlooked in an analysis of the status of Irish and language policy. It is important that the language warrants such acknowledgement at European level as this serves to emphasize its importance and relevance, especially for those learners who may have developed an apathy towards the language and who fail to recognise that beyond its symbolic value it can have utilitarian benefits too.

3.4 Irish: A minority language

In his examination of various Reversing Language Shift (RLS) efforts around the world, Fishman (1991) dedicated a whole chapter to the Irish language; specifying the uniqueness of the Irish effort, so much so that he even coined the term "Irelandization" (p. 143) to highlight its singularity. By Irelandization he meant the "peripheralization" of the language in question in its own country of origin and in which attempts are being made to revive it, i.e., "Xish in Xland," or in this case, "Irish in Ireland". What makes the Irish scenario particularly unique according to Fishman is the sheer dedication and commitment at state level to its protection and its status as a national and official language, yet it remains a language that is, at a certain level, endangered, a situation he deems to be quite an anomaly in the field. He argues that the case of Irish is different from that of almost all other threatened languages worldwide because usually endangered languages are the "receding languages" of groups that constitute an ethnic minority in a particular society (p. 135). This is clearly not the case in Ireland.

Ó Murchú (2016: 5) writing for Mercator, the European Research Centre on Multilingualism and Language Learning, defines Irish as "an autochthonous language spoken in the Republic of Ireland and Northern Ireland."

Although Stenson & Hickey (2016: 1) refer to Irish as a "threatened minority language", Irish is not considered a regional or a minority language within EU policy. The Republic of Ireland is, therefore, not party to the European Charter for Regional or Minority Languages of the Council of Europe of 1992 and updated in April 2020. An examination of this listing shows that Irish is only considered a minority language within the context of the U.K. but not in the Republic of Ireland itself.

According to Article 1a of the charter, regional or minority languages are:

> languages traditionally used within a given territory of a state by nationals of that state who form a group numerically smaller than the rest of the state's population; they are different from the official language(s) of that state, and they

include neither dialects of the official language(s) of the state nor the languages of migrants. (European Charter for Regional or Minority Languages, 1992)

As Irish is the first official language of Ireland this renders it inapplicable to the above charter. On the other hand, The United Nations Educational, Scientific and Cultural Organisation (UNESCO) classifies it as a language that is "definitely endangered" on their 5-point scale that ranks languages according to the severity of risk of language extinction (Moseley 2010: UNESCO website):

vulnerable – definitely endangered – severely endangered – critically endangered – extinct

In fact, Li, MacSwiney Brugha & Gallagher (2017) use both the terms "endangered language" and "minority language" in reference to Irish.

Defining what constitutes a minority language or a heritage language is not straightforward and the conclusion from a thorough examination of the literature in the field of language revitalization is that there appears to be a clear lack of consensus with respect to terminology. For example, in offering an overview of the different interpretations of what constitutes a heritage language, De Bot & Gorter (2005: 1) start by referring to Cummins's explanation stemming from the North American experience according to which a heritage language refers to that language spoken by "immigrant, refugee and indigenous groups in North America". Reformulating this to fit the European context, De Bot & Gorter suggest rather the term minority language, dividing this further into immigrant minority languages (those spoken by an immigrant community in an L2 majority language environment) and regional minority languages, those originating in or "indigenous" to a particular region but that are not spoken by the majority of population within the wider society.

Ó Murchú (2001: 114) refers to Coulmas (1998: 68) who remarks that when speaking about linguistic minorities, nation-states commonly differentiate between indigenous minorities and migrants. He argues that when a person or group migrates to a particular state in which a different language to their L1 is spoken, it cannot necessarily be expected that official linguistic policy be changed in that country to accommodate the various L1s of different migrant groups. On the other hand, he argues that it is a different scenario for autochthonous groups "on which a language regime has been imposed against their will" (p. 68) and who, in his opinion, therefore, have a more legitimate claim with respect to recognition of their culture and language.

Despite Irish being the official first language of Ireland and not party to the European Charter for Regional or Minority Languages, it is, nonetheless, referred to as a minority language in the present study considering its *degree of use* in the everyday lives of the majority of the population in Ireland *relative* to English, which undeniably constitutes the majority and dominant language with respect to actual language use on the island.

3.5 Irish in education

In the Republic of Ireland, Irish is taught as an L2 in the majority of schools across the country. Considering that it is just one of a number of many different subjects that learners have to study, input is limited. It is, however, taught as an L1 in some schools in Gaeltacht areas and the number of Irish-medium schools (Gaelscoileanna) where Irish is the language of instruction has been increasing. In 2014/15 there were a reported 141 Irish-medium primary schools outside of the Gaeltacht and 50 Irish-medium secondary schools (from the Annual Statistical report 2014/15 of the Government of Ireland, Department of Education and Skills). English continues, however, to be the L1 for the majority of the pupils of these schools.

It has to be noted that a distinction is made between Irish and Modern Foreign Languages (MFLs) in education policy in Ireland. In 2020, the Department of Education and Skills published a report on the teaching of modern foreign languages in secondary school (Department of Education and Skills, April 2020). This did not include Irish. Whilst clearly the term *Modern Foreign Language* is focused on the reference to *foreign*, on page 25 of the report just mentioned, the term "modern language" is used to refer to modern foreign languages. This terminology is particularly noteworthy because even though Irish is still spoken in certain communities and there is an increasing emphasis on communication in Irish instruction, the term "modern language" does not appear to be used in an official capacity to refer to it.

The participants in this study attended school under the primary school curriculum of 1971. The *Curaclam na Bunscoile*, as it was called, incorporated content taught using a mix of audio-lingual and grammar-translation methods. This remained in place until it was revised in 1999 when new content and pedagogical approaches of a communicative and task-based nature were introduced. None of the participants attended an Irish-medium school, although one lived in a rural area where much Irish was used in the everyday life of the school, even though it was not the language of instruction in other subjects. At secondary school pupils were allocated to either a higher-level Irish class or to an ordinary-level class. Traditionally the terms "Honours" and "Pass" were used to refer to these levels with honours constituting the higher level, pass the ordinary one. The honours level was more demanding than the pass level and students were awarded more points for performance in an honours paper than in a pass paper in national examinations.

Whilst Ó Riain (2004) argues that making it compulsory at school helped to modernize it, it is, however, not entirely clear what he means by "modernization". Indeed, while one might argue that making it an obligatory subject increased awareness of it and manifested its importance, at least at an ideological level, it was not enough to fuel a need or desire among the population at large to use it in their daily lives.

That it was left to the educational system alone to revive the language, made the undertaking particularly challenging. In Joshua Fishman's (1991) renowned 8-stage model on Reversing Language Shift, a key step is number 3; ensuring that the language is passed down as a mother-tongue from one generation to the next. As long as the endangered language was not being spoken in the home, Fishman believed that a language could never be truly revived. Ó Riain (1994: 92–93) also argued that it was a tremendous folly to believe that "increased usage of Irish would automatically flow from increased learning of the language". In his exploration of Irish-medium education, Kennedy (2012: 295) reports that even though learners in Irish-medium schools use more Irish on average than fellow learners in ordinary schools, this effect does not infiltrate the wider society, in the sense that Irish-medium learners do not necessarily continue to speak Irish beyond the school context and when they do it is rather intermittent and certainly not enough to create resolute active speaker communities. He, therefore, argues that the burden of revival cannot fall on education alone but that a "multi-layered" approach is crucial in which family and the wider social environment also play an active role in supporting progress made in schools (p. 297).

Murtagh & Van der Slik (2004: 299) observed similar behaviour in the case of learners at Irish-medium schools compared to those at ordinary schools. Irish-medium schools had higher levels of success with respect to stimulating Irish language use, also outside of the school context, with learners demonstrating higher levels of motivation accordingly. Learners in such schools were also shown to receive on average more parental encouragement than their peers in ordinary schools. However, Murtagh & Van der Slik also found that this motivation and use did not extend beyond the immersion school years in a number of cases, even in the case of those who had also used Irish with each other in social settings during their school time. They thereby point out the vulnerability and precariousness of Irish as the chosen medium for communication beyond school as time progresses.

It is certainly noteworthy that even participants who had been motivated enough during their education to converse in Irish in their free time gradually stopped doing so with each other on finishing school. This raises important questions about the place of Irish in society on the island. At one level, there appears to be a lack of opportunities or contexts in which to interact in Irish. On the other hand, learners, past or present, do not appear to actively create these contexts to the extent that they impact the wider society. Indeed Darmody & Daly (2015: 88) state in agreement with Murtagh (2007) that despite the bulk of the responsibility falling on education and the time and effort invested in teaching the language, its effects are, nonetheless, rather limited when observed from a wider social perspective.

In an article in *The Irish Times* (Jan 28, 2014) on the teaching of Irish, Dr Muiris O'Laoire highlights the importance of seeking clarity on the intentions behind education policy on Irish in schools, and asks the fundamental question underlying the debate: "is it to communicate or is it for cultural reasons?" He argues that it is essential that "meaningful opportunities for use" are provided if communication is the aim. Naturally, the purpose of teaching Irish in school can encompass both cultural and social/communicative aims. If, however, learners are leaving school without either the ability or the motivation to communicate in the language, then the education system has fallen gravely short with respect to the latter.

In a comparison of the national census of 1961and 1981, Fishman (1981: 128) sees an overall increase in the numbers of teenagers and young adults learning Irish. What is notable with respect to the present study, however, is the drop in proficiency levels in the cohort that had been in school in 1961 in the subsequent 20 years after finishing official instruction in the language (i.e., in the 35–44 age-group in the 1981 census). Fishman claims that this is representative of attrition taking place after leaving school and an obvious lack of transmissibility from one generation to the next outside the school context. From the figures it appears that language transmission appears to take place exclusively between teacher and pupil in the educational context but not beyond.

Ó Riagáin (1997: 205) gives the following overview of trends in higher and ordinary level Leaving Certificate participation between the years 1980 and 1990, selected here since this is the period during which the participants in this study were in secondary school. Department of Education Statistical Reports show that in 1980 only 18.4 % of pupils did the higher-level paper in Irish, while 62.3 % did the ordinary-level paper. By 1990 the percentage of pupils who took the higher-level paper had dropped to 14.7 % with 62.1 % taking the ordinary level Leaving Certificate paper in that year.[1] The fact that the percentage is significantly lower for the higher-level paper and even drops in the ten-year period presented is particularly noteworthy and lends support to the idea that the highly demanding and academic nature of the course possibly posed a demotivating factor among learners. Three of the participants in this study took the higher-level paper for the Leaving Certificate, one of those being Cathy, who did not start learning Irish until she was ten years old.

One of the recommendations of the Irish government's *20-Year Strategy for the Irish Language* was that teachers in English-medium schools use Irish

1 On a side note, albeit an interesting one, the Department of Education reported an increase in the number of learners taking the higher-level paper between 2010 and 2015 (from 33 % to 42 %). This might have been a result of the revisions made to the Leaving Certificate assessment in 2012 in which the oral component of the exam was given a higher weighting (from 25 % to 40 %) and the amount of literature to be covered was reduced (Government of Ireland, Department of Education and Skills, *Progress Report 2015*: 4).

as the language of instruction when teaching the subject of Irish and try to use it intermittently in other classes and contexts in the school environment. However, in Ó Murchú's (2016: 17–18) report on the Irish language in education in the Republic of Ireland for Mercator (the European Research Centre on Multilingualism and Language Learning) details from the 2013–2014 progress report on this 20-year strategy are cited that show that this strategy was seldom used in primary schools. This might possibly be linked to another report published in 2011 mentioned by Ó Murchú (2016: 29) that was conducted by An Chomhairle um Oideachas Gaeltachta agus Gaelscolaíochta (COGG), (a government-funded organisation set up to support schools in Gaeltacht regions and Irish-medium schools) in which it is relayed that in ordinary schools (as opposed to Irish-medium schools) teachers themselves admitted to shortcomings in their Irish language competence and an accompanying lack of confidence in their abilities, thereby helping to partly explain why the language skills of learners at school are sometimes sub-standard.

Whilst Fishman (1991: 130) compliments the Irish education system on enhancing awareness of the Irish language and presenting its rich cultural contributions to the identity of Ireland, he states quite frankly that these endeavours, nonetheless, fail as many pupils do not achieve true mastery of the language or embrace a willingness to use it in daily life and ensure that it is transmitted thus to following generations. Indeed, with respect to revival of the Irish language, de Brún (2018: 43) regards the desire to revive past knowledge and values as debatable and asks, in quoting the founder of the Gaelic League, Douglas Hyde, if it is indeed possible or even advisable to "render the present a rational continuation of the past"?

Due to their grassroots origin and by creating competent speakers of the language, Fishman (1991: 138–139) sees great promise in community endeavours such as the creation of Irish-medium schools to accommodate children who had attended Irish-medium pre-schools (*Naíonraí*). These are often set up by parents disillusioned with teaching methods in mainstream schools. Fishman observed, however, that these schools remained restricted to middle-class, better educated groups and saw this as a limiting factor. For the language to truly flourish, these means should be available to children from all backgrounds. Fishman (1991: 143) also argues that it is rather unrealistic to expect Irish to become the language of everyday communication but it might be possible to create bilingual communities.

3.6 Attitudes to the Irish language

As argued by Smyth & Darmody (2016: 1) in referring to Baker (1992), exploring attitudes to languages can serve to explain language behaviour. This is interesting considering that Hickey (2009) found in his survey of attitudes to both Irish and English based on 1,086 questionnaires distributed between 2006 and 2008 that the cultural importance of Irish was highly valued with the majority of respondents considering it a key part of their identity even if they did not actively speak it.

Fishman (1991: 124) also refers to the overriding positive attitudes towards the Irish language not just in Gaeltacht areas but across the Republic. This is noteworthy since this was approximately the time when the participants in this study finished school. He writes that the population at large valued the symbolic significance of the language and were supportive of government measures to ensure its protection and educational policies that secured its presence in the school system. Murtagh & Harris (1999) also reported favourable attitudes towards the language among the groups of children that they interviewed. Like Fishman, they too found that this did not necessarily translate into *actual language* use however. Kennedy (2012: 296) observed similarly positive attitudes to the Irish language, but with actual language use remaining confined to the educational context.

Ó Murchú (2016) offered data gathered by the 2011 national Census on Irish language use that found that 41.4 % of the total population rated themselves as Irish speakers (a number that includes Gaeltacht residents). This result was generated based on self-reports and self-ratings and considering also that L1 Irish speakers from Gaeltacht areas were included, the result may be somewhat positively skewed and not be accurately representative of true patterns of Irish language knowledge and use. Ó Murchú also observed that these reports were not significantly different to those found in the 2006 national census.

Li et al. (2017) refer to a survey done in 2016 on attitudes of secondary school pupils to learning Irish which showed some contradictory and paradoxical results. While many pupils stated that they valued the cultural significance of the language and valued it as part of their identity, they, nonetheless, criticised the way that it was taught and the overriding focus on reading literary and historical texts and writing, with many saying that they wanted it to be taught more like a modern foreign language. Some also felt that it was not useful to them and their main motivation for learning it was simply to pass it in the Leaving Certificate examination. Li et al. see these responses as contradictory. In one respect learners appear to value its symbolic value and consider knowledge of it as important for their sense of Irish identity, on the other hand, they neglect to engage actively in it.

As state policies with respect to Irish have also changed in certain domains of public life, for example, it is no longer obligatory for acceptance for certain positions in the Civil Service, some argue that this loosening of policies trickles down to grassroot level and therefore also serves to weaken its significance in society in general. McCubbin (2010), for example, argues that such approaches on the part of the government may possibly lead to indifference at community level. It is difficult to assess whether this indeed might be a reason for a certain apathy on the part of individuals. While a certain extrinsic motivation might have been removed by no longer requiring it for certain jobs in the public sector, pressure to perform well in the Leaving Certificate examination has remained nonetheless.

3.7 Selected features of Irish grammar

Ó Glaisne's (1961) thorough explanation of Irish grammar was used as reference in explaining key aspects of Irish grammar as outlined below. Hickey's (1985) analysis of Irish syntax was also consulted. A Celtic language, the grammar and phonology of Irish is vastly different to that of English and therefore often considered rather obscure by novice learners and passive onlookers. It is closely related to Scots Gaelic and Manx (spoken on the Isle of Man), and more distantly related to Breton, Welsh and Cornish and is a member of the Indo-European family of languages. The Irish translation is *Gaeilge*, explaining why one sometimes hears it referred to as *Gaelic* in English. However, in Ireland itself, the language is referred to as *Irish*.

Despite there being regional variations in the language, there is an official standard (*Caighdeán Oifigiúil*) and the lexicon is regularly updated to accommodate new vocabulary that emerges, this being predominantly in the area of technological advances and modern media. There is an Irish radio service that operates out of the Gaeltacht, as well as a national state-funded Irish language television channel (called TG4, operating since 1996).

The focus of this section is not to delve into a detailed exploration and analysis of Irish grammar but rather to call attention to a small set of selected key features. These particular characteristics of Irish warrant brief elaboration for the purpose of this study to explain the reasoning behind the category breakdown of the selected items for testing, i.e., the predominance of nouns relative to verbs and adjectives (50 % nouns; divided equally between concrete and abstract nouns, 25 % verbs, 25 % adjectives). This momentary excursion into Irish grammar is not to undermine its richness or complexity but aims to explain one specific feature highly relevant for the purposes of the present study, that is, how nouns are also used to depict actions and adjectives and how this is made possible syntactically. This inevitably necessitates a look at the role of prepositions in the Irish language, as well as how verbs are constructed considering that in Irish they do not have an infinitive form.

3.7.1 Key characteristics

Unlike English which has a SVO (Subject-Verb-Object) word order and is considered to be verb-centred, Irish has a VSO (Verb-Subject-Object) word order and is a noun-centred language (Hickey 1985: 2; Ó Riain 2004). In this respect, it is not uncommon in Irish for actions to be conveyed through the nominalisation of a verb which in combination with a particular preposition and a support verb expresses a specific action. Whilst also possible in English, nominalisations of verbs are less frequent than in Irish. As Acquaviva (2014: 541) explains, as well as finite forms, Irish verbs additionally have two non-finite forms (i.e., they do not inflect for tense or mood); the verbal noun and the verbal adjective.

3.7.1.1 Verbal noun

In the case of a verbal noun, whilst describing an action, it is nonetheless deemed nominal in that it behaves like nouns with respect to its morphology and case-marking attributes.

The following is an adapted version of an example demonstrating a verbal noun offered by Hickey (1985: 2):

(1) Rinne sé dearmad faoi
made he forgetfulness under-it
He forgot it

Dearmad is a noun and generally means *forgetfulness* or *negligence.* In the above example, *to forget* is expressed as "to make or to do forgetfulness" with the conjugated preposition *faoi* meaning *under it* whereby *faoi* simultaneously incorporates the pronoun *it* (3rd person singular) in the masculine form. (See following section for information on prepositional pronouns in Irish)

3.7.1.2 Prepositional pronoun

Irish prepositions fuse with pronouns to create one-word forms. As Hickey (see Hickey's homepage at Düsseldorf university for more information on Irish syntax) observes, prepositions play a greater role in Irish than they do in English and it is common in Irish for prepositions to be declined. For example, Irish does not have a verb for "to have". Instead the act of possessing is indicated in the particular preposition chosen.

Example of preposition (*ag*) carrying the meaning of the verb *to have.* Literally translated *"ag"* means *"at"*in English and is declined as follows:

1st person sing.	agam (*at me*)
2nd person sing.	agut (*at you*)
3rd person sing.	aige (*masc.*) aici (*fem.*) (*at him, at her*)
1st person plural.	againn (*at us*)
2nd person plural.	agaibh (*at you, pl.*)
3rd person plural	acu (*at us*)

(2) Tá deartháir and deirfiúr amháin **agam**
is brother and sister one ***at me***
I have one brother and one sister

(3) Tá teach deas **acu**
is house nice at their having
They have a nice house

3.7.1.3 Verbal adjective

Nouns can also be used in combination with prepositions and a variation of the verb *to be* instead of using an adjectival form:

(4) Tá ocras orm
Is hunger on me
I'm hungry

(5) Bhí ocras orthu
Was hunger on them
They were hungry

(6) Tá áthas orainn
is happiness on us
We are happy

(7) Tá dóchas orainn
Is hope on us
We're hopeful

Here the preposition "*ar*" that literally means "on" is used and is conjugated as follows:

1st person sing.	orm (on me)
2nd person sing.	ort (on you)
3rd person sing.	air (*masc.*) uirthi (*fem.*) (on her, on him)
1st person plural.	orainn (on us)
2nd person plural	oraibh (on you, pl)
3rd person plural	orthu (on them)

One preposition can carry a number of different verb meanings and this is contingent on the specific noun used in the sentence. It was quite normal for the declension of each preposition to be repeatedly recited out loud across classrooms in Ireland.

3.7.1.4 Verb roots

What is particularly interesting about Irish verbs is that they do not have an infinitive. The root of the verb as represented in the 2nd person singular imperative in Irish is that which offers the basis for further conjugation in the majority of cases. Therefore the imperative form was always given, i.e., the verb root, in the listings of the verbs that were presented to the participants in the present study, but the equivalent translation in English was given as the infinitive in each case to ensure that the participants knew that a verb was being referred to.

3.7.2 Irish orthography

The orthographical system of Irish deviates greatly from that of English with Hickey (2007: 473) referring to Ó Murchú (1977) who in his analysis of Irish orthography described it as "bewildering" for new learners. Stenson & Hickey (2016) provide a thorough explanation of key characteristics of Irish orthography in which they point out, based on a comparison of frequent words in Irish and English, that while Irish orthography is more regular and consistent than that of English, this should not be confused with transparency. Therefore, although not as opaque as English, Irish orthography is, nonetheless, complex and significantly different to that of English, thereby warranting explicit attention when teaching reading in Irish to learners (2016: 3). For further information on Irish orthography, Stenson & Hickey (2018) have created a comprehensive explanation of Irish language spelling patterns and phonological features in a handbook for Irish language educators. Irish orthography was not dealt with explicitly at school in the 1980s when the participants in the present study were in education and vocabulary lists were category-based, not phonics-based.

3.8 Peig

The primary literature used to provide the lexical items used in this study was the autobiography, *Peig*, obligatory reading for all pupils of Irish in Ireland in the 1980s as it constituted the primary literature on the Leaving Certificate syllabus. Narrated by Peig Sayers, an Irish language "seanchaí" (storyteller) who lived between 1873 and 1958, *Peig* was published in 1936. In her book, Sayers offers an autobiographical account of a difficult and dismal life on a small island off the south-west coast of Ireland called "The Great Blasket Island" to which she moved at the age of 19 years from the Dingle Peninsula on the mainland after having married a native of the island. Whilst Peig Sayers herself was illiterate, her stories and accounts were recorded by her son, Micheál. The themes of the novel were based on poverty, emigration and death and the melancholic, depressing and tragic tone of the novel did not gain popularity among pupils, all in their teenage years and emotionally removed from the experiences they were forced to read about. This is not, however, to underestimate the cultural significance of Peig and her stories. She undoubtedly offered a wealth of insight into life at that time on Great Blasket and enriched the folkloric repertoire available to scholars (see Bo Almqvist's (2011) writings on Peig's scholarly legacy and his overview of those scholars who travelled to the Blaskets and recorded Peig's stories). It seemed though that this cultural significance was not transmitted to the pupils at school who had to study Peig.

It must also be noted that the items chosen for testing were not exclusive to this particular literature but were words generally used across texts and in many cases learned already in the early years of instruction. Nonetheless, *Peig* constituted a secure starting point of reference as all participants had definitely studied it.

Chapter 4: Research design

The following offers a detailed overview of each testing session followed by an elaboration on selected criteria and relevant issues and justifications with respect to planning the testing format. To understand the chosen methodology, the aims of the study are outlined here once again:

1. To assess levels of retention of lexical items of school-learned Irish after an incubation period of approximately 30 years.
2. To ascertain what factors might account for difference in performance across individuals.
3. To establish whether or not there is a relearning advantage with respect to seemingly forgotten lexical items in the Irish language, i.e., to test the Savings Paradigm with respect to lexical knowledge.
4. To determine whether levels of retention parallel performance in relearning.

It is argued here that a "one fits all" approach should not be enforced with respect to investigations of attrition. Whilst Schmid (2004) remarks on the lack of uniformity across studies on attrition in terms of testing procedures, arguing that if there was such consistency it would enable more solid comparisons, the very individual nature and experience of attrition makes such invariability difficult and indeed in enforcing such, might result in missing key elements pertaining to the particular group in question. What we should expect to glean from studies in attrition are broad patterns but which allow for individual variation. It is also argued here that the actual *experience* of language learning itself is a distinctly individual phenomenon which must be seen as separate from the institutionalised and official education in that language. Therefore, the long-term outcomes of this personal experience as part of the greater institutionalised whole are unique and particular to each individual. In looking at attrition this must be accounted for and the study planned accordingly. Other factors such as length of exposure to the language, type of exposure, proficiency achieved and length of incubation and whether or not there was rehearsal during this time have to be taken into consideration and are unique for each individual under study. It is deemed advantageous therefore to look at the individal as a case study as is done in this research (see Section 4.4.4).

4.1 Background information

Four individuals from the Republic of Ireland who had been living in Germany for a number of years (ranging from 13 years to 29 years) and were still residing there at the time of testing, took part in the study. The participants were recruited from a group of personal acquaintances of the author. English was the L1 of all of the participants and all of them spoke German as an L2. Each of the participants had acquired Irish in a similar context which was at school, starting immediately at the start of primary education (except for one who started learning it at the age of ten years after her family moved from Northern Ireland to the Republic of Ireland). They all continued with Irish throughout their whole school education until the final national state examination (Leaving Certificate), taken at the end of the last year at school. The participants had very little to no contact with the Irish language since leaving school or university. The education level of all participants was relatively equal with all of them having completed third level university education and all of them having acquired a Master's degree or equivalent.

More detailed background information on each candidate is given in Chapter 5 along with an analysis of their test results and their individual input in the questionnaire and interview.

Four sessions of testing took place over a period of 13.5 months and were conducted at the University of Cologne. Each participant was tested individually in every session. The same two research assistants were present with the main test administrator at each of the testing sessions. These research assistants had been thoroughly briefed in advance and were rehearsed in the testing procedure. They did not know the participants personally and did not have any knowledge of the Irish language. This was not necessary in conducting the tests. The main test administrator did have knowledge of Irish having learned it in school in Ireland and was also present to assess whether or not the participants also pronounced the Irish lexical items correctly. To ensure that she knew the correct pronunciation of each word, the main researcher went through each item with a highly proficient L2 speaker of Irish in Ireland prior to testing.

In each testing session, items were presented in lowercase 54-point black calibri font against a white background. One of the research assistants was responsible for manually clicking to the next word when the participant was ready. In the recognition tests, the participants were requested to also read the Irish word out loud so that it could be established if there was any phonological deviation from the correct standard form.

There was no variation in the testing procedure or how items were presented across participants. The language of testing was English. Clear instructions were given at the start of each testing phase. No feedback was given to participants on their performance. Picture-testing was not used as some items were considered too abstract to be captured in an unambiguous visual. Considering also that the participants had learned Irish predominantly through texts, they were familiar with the written word.

The advantage of the testing conducted in this study was that it was done in person and so could be better monitored which would have been difficult had it been done online or long-distance. Not doing it online meant that many of the challenges associated with such studies as outlined by Mickan (2021: 201) could be avoided, such as techinical issues and participants not taking the testing seriously.

4.2 Breakdown of sessions

In keeping with Schmid (2004) who proposes gathering multi-modal data, in this study introspection and reflection, as well as formal elicitation tasks with the linguistic focus on lexical knowledge, were conducted. The following table provides an overview of the breakdown of each testing session.

Table 2 Breakdown of testing sessions

Session 1 Pretest of 200 items	Session 2 (Two weeks after session 1)	Session 3 (Four weeks after session 2)	Session 4 (One year after session 3)
Recall	Relearning	Recall	Relearning
Questionnaire and Interview	Distraction activity	Recognition	Distraction activity
Recognition	Recall		Recall
	Recognition		Recognition

4.2.1 Session 1

To get a general estimation of amount of retention of lexical items, participants had to take an initial productive and receptive vocaubulary test in session one. This session also served to filter out words still known to the participants, thereby eliminating them for the savings/relearning testing sessions. A questionnaire was also completed and interview questions answered orally in this session.

Two hundred words were selected for the test. A large number of words was used to ensure that enough words would be left over for the relearning phase of testing as it was not known in advance what could be expected from the participants with respect to their lexical knowledge. The 200 words selected consisted of 50 concrete nouns, 50 abstract nouns, 50 verbs and 50 adjectives. This breakdown was guided by Hansen et al. (2012) (see Chapter 2, Section 2.3.1, for more information on Hansen et al.'s study). As outlined in Chapter 3, Section 3.7, since Irish is a noun-centred language, it was decided to have nouns constituting 50 % of the words (split evenly between concrete and abstract nouns). A time restraint on responses was not set to give participants time to deal with possible competition from other languages. See Section 4.4.1.6 below for elaboration on issues of timing.

Participants were first tested on recall (production). Each of the 200 items was shown individually in English and the participants had to give the Irish language equivalent or an appropriate synonym in Irish. After this testing phase, the items that participants could recall (i.e., they were able to give the Irish translation for the English item) were then removed from the list by the main test administrator in another room. During this time, participants completed the questionnaire and the interview was conducted under the guidance of the two research assitants.

On completion of the questionnaire and interview, the participants were then tested on recognition. In this phase, the participants were shown the Irish translations of the remaining words from the original list of 200 words (i.e., which excluded those items that they had successfully recalled in the first phase of testing). Once again, each item was shown individually and the participants had to both read the word out loud and offer the English translation or an appropriate synonym in English.

The items that participants could successfully recognise were then also removed from the list, leaving only those items that participants claimed to not know, i.e., they could neither be recalled (the Irish translation for the English word could not be given) nor recognised (the English translation for the Irish word could not be given).

These remaining unknown items were set aside for the subsequent testing sessions on relearning and savings.

4.2.2 Session 2

Session two took place two weeks later. In the relearning phase, Irish lexical items from session one that the participants did not know were shown with their English translation to the participants. Each word-pair was shown in isolation (Irish – English), with each pair being shown for eight seconds (in keeping with De Bot et al. 2004, see Chapter 2, Section 2.3.2, for greater elaboration on their study). Twenty-six pseudowords each with an English translation were mixed in with the real word pairs. Pseudowords were used instead of real words to ensure that words were used that the participants could not have known so that a difference between reactivation of old knowledge and first-time learning of new information could be established. It is important to note that the participants did not know that pseudowords were used. See Section 4.4.2 for elaboration on the creation of the pseudowords.

After a distractor activity of approximately one hour, participants were tested on the words they had just learned; first on recall, then on recognition for the remaining words that they had not been able to recall. Since recall is considered more difficult than recognition, it was deemed necessary to firstly eliminate those words that could be recalled, as it is assumed that words that can be recalled will also be recognised. Participants were shown each word separately and asked for the translation in Irish. In the recognition phase of testing, participants were shown the Irish word and were asked for the English translation. See Section 4.4.1.5 for further elaboration on the procedure.

The number of words available for relearning could not be determined in advance as it depended on how participants performed in session 1 when words were filtered accordingly. In cases where a large number of words were unknown, it was decided that no more than 50 of these unknown words would be reused for relearning. In fact De Bot & Stoessel (2000) and Hansen et al. (2002) can be criticised for using a very small number of words for relearning as this might have provided participants with a greater opportunity for "learning" (in freeing up cognitive resources) making it difficult to establish whether or not performance was due to a reactivation of previously known items due to triggering via re-exposure or in fact pure, first-time learning (see Chapter 2, Section 2.3.1, for an outline of these studies).

It is argued in the present study that an advantage for previously known words is better demonstrated and a triggering effect more apparent when more items are tested. On the other hand, it was important not to overtax the participants by using too many items for relearning. See Section 4.4.2.2 in which a detailed explanation is given on the criteria for establishing the number of real and pseudowords.

4.2.3 Session 3

Session 3 took place one month later. In this round, all of the words relearned in session two (including the pseudowords) were tested; first for recall followed by a recognition test for those remaining words they had not been able to recall. No relearning took place in this session. The testing procedure for recall and recognition was the same as in previous sessions.

4.2.4 Session 4

As in De Bot at al.'s study (2004), a second relearning phase also took place in the present study to see if there was an advantage of re-exposure, even for new words (in this case, the pseudowords). This was conducted in session four which was one year after session 3. The subjects had not been informed that there would be another testing session one year later and were asked at short notice to take part. This session consisted of a relearning phase of the same words used in session 2. After a one-hour distraction activity (the interval between relearning and testing was the same in sessions 2 and 4), participants were once again tested on recall and recognition according to the same format in rounds two and three. The order of test items was randomised in each new round of testing, however, to avoid a test effect.

4.3 General comments on study design

Whilst impossible to completely ensure that there is complete homogeneity across participants in a research project, in the present study there was a relative degree of consistency across certain variables, enabling a closer scrutiny of those variables that might explain variation in performance. Schmid (2011a) also advises controlling for certain variables as far as is possible to constrain their explicatory potential.

4.3.1 Establishing baseline knowledge and proficiency

With respect to L1 attrition research, Opitz (2019: 3) highlights the challenge of disinguishing between incomplete acquisition and attrition. In this study on L2 attrition, this issue is eliminated as it can be assured that all of the participants had acquired the lexical items tested. As Gharibi & Boers (2017: 52) state (in reference to heritage speakers) a gap in vocabulary knowledge can only be considered to be the result of attrition if it can be established that the speaker had this knowledge previously. Therefore, to explore this phenomenon they argue that two measures have to be ascertained; one the level of knowledge at an earlier stage (in this study, at the time when contact with the language ceased) and second, the level of knowledge at a later stage (in this case at the time of testing which was 25–30 years later). Mehotcheva (2010: 16) too argues that it has to be certain that the target group had a command of the linguistic feature under scrutiny at a certain time in the past. She refers to the challenge FL attrition research in particular presents with respect to establishing a baseline. Comparing FL attriters to a monolingual group could lead to false interpretations of results as the attriting group may never have reached that level of linguistic competence in the first place which could lead to an overestimation of the degree of attrition they have actually undergone. In her dissertation, Mehotcheva compared her participants' performance to the language level of another group of FL learners which still had contact with the language and was engaged in learning, i.e., contact with the language had not yet ceased. Beyond this, the groups shared a number of characteristics thereby making such a comparison practicable. It should be noted, however, that the group she studied had stopped learning the language between 12 and 72 months previously. One could, therefore, assume that teaching methodologies and the content of the curriculum would have remained relatively constant in that time, thereby rendering such a comparison of these two groups a viable opportunity to assess degrees of attrition. With respect to the present study, a similar comparison is not possible as many factors would most certainly have changed in the previous 25–30 years with respect to educaton in general in Ireland, and the teaching of Irish specifically – the methodologies used, the content of the curriculm, testing formats and the standards expected at the end of school education.

In fact, in referring to Weltens (1989) and Grendel (1993), who also worked with a control group to establish a baseline proficiency against which to measure their participants' performance, Murtagh & Van der Slik (2004: 283) argue that such comparisons are problematic since more individual factors need to be taken into consideration, thereby complicating the analysis and undermining its robustness. In the present study, this problem has been avoided as the focus is on the same four individuals with no comparisons to other individuals being made; only the exclusive evolution of each individual's own language knowledge and experience is analysed.

Even when not including a comparison group, Weltens & Grendel (1993: 145) state that it can be difficult to establish a homogeneous baseline across participants in a study. For example, one semester of study can be very different across universities in terms of proficiency levels expected and achieved. Since all of the participants in this study learned Irish according to the same curriculum and took the standardised final school Leaving Certificate examination in Irish, it could be determined what content was covered and what levels of language knowledge would have been expected of them on finishing school. The author herself had attended school while this curriculum was in place and still had access to the majority of the materials and books used during the time that the participants were in school. Most of the words were taken from the compulsory reading, *Peig*, with the remaining items taken from various other obligatory texts. Whilst the participants took the final state examination in different years, the standard was quite consistent overall year to year.

4.3.2 Limitations in study design

As Mehotcheva & Mytara (2019: 16) state, tracking learners over long periods of time in a longitudinal study format can be beneficial with respect to analysing correlations between various factors. They recognise, however, the challenges in recruiting suitable participants who are willing to commit to such a long-term endeavour as well as the very effect of retesting on the overall outcome. It would no doubt have been fascinating to track the participants of this study in the 25–30 years since they had left school. Nonetheless, discovering the group of individuals for the present study who both fit the criteria and were available for testing over a period of 13.5 months was deemed extremely fortunate.

4.4 Considerations in study design

Background information is now given on specific aspects and criteria relevant to the planning of the testing; selecting the testing format, choosing the words, constructing pseudowords, planning and creating the questionnaire and interview.

4.4.1 Lexical item testing

In the following, reasons for focusing on lexical items as well as justifications for the item and category selection will be elaborated on.

4.4.1.1 Why only look at the lexicon?

Up until the Education Act of 1998, the focus of Irish lessons in the 1980s in Ireland was predominantly on the written word. There was a strong focus on literature; its analysis and interpretation. While this was then combined with increased oral practise as the final Leaving Certificate examination in Irish approached, (a section of which constituted an oral exam) teaching for the most part followed a mix of the audio-lingual and grammar-translation method with much rote-learning of content taking place. This constitutes one of the reasons why this study focuses on the lexicon; Irish was traditionally not taught like a modern foreign language as we know it today; there was little focus on communication and interaction and more on reading and writing. Weltens & Grendel (1993: 142) also argue that focussing on vocabulary is important and justified considering the vast amount of time dedicated to the learning of vocabulary in instruction contexts.

There was also a practical reason behind the decision to focus only on vocabulary. Larson-Hall (2019: 10) argues that inability to actively produce the language apart from a few formulaic expressions is representative of severe linguistic diminution and decay. She states that in the case of recognition or receptive ability with respect to vocabulary, performance below 50 % can be deemed catastrophic loss and inability to produce more than 25 % output can be regarded as indicative of extreme loss. Prior to official testing, it was ascertained that the participants in this study had experienced extreme loss of Irish in line with Larson-Hall's proposed demarcations. They were unable to converse in Irish or understand written or spoken texts apart from random words.

Their performance in session 1 is also in line with these suggested cut-off points with respect to recognition and production. Only one of the participants (Enda) achieved results beyond these levels in session 1 but when tested in a second round on less frequent words, with the aim of sifting out words he could remember, he performed below the 50 % level for recognition and just reached 25 % on recall (see Chapter 5, Section 5.1.1.2).

In view of this pronounced and blatant loss, looking at lexical items as the building blocks to greater linguistic knowledge was deemed the most reasonable and practical approach in this study considering also that one of the aims was to see if a degree of retention could, nonetheless, be established, despite the severity of diminution.

Another reason for focusing on the lexicon is put forward by Gharibi & Boers (2017: 52–53). In referring to Polinsky (1997, 2007) and Montrul (2009), they argue that exploring attrition by looking at vocabularly specifically might offer insights into the state of the language system overall because of the substantial correlation between levels of grammar understanding and proficiency and vocabulary knowledge, with outcomes on vocabulary tests having been shown to be rather reliable indicators of language proficiency overall, although considering that the participants had clearly lost vast amounts of knowledge and they were no longer exposed to the language, this appears not to apply in this study. Whilst this latter point is to be approached with caution in the case of the participants in this study who have been far removed from the language both physically and emotionally for many years, it could be that, if a relearning advantage is found for vocabulary, this would extend to other linguistic domains given enough time and input.

Another key deliberation is the nature of the knowledge under investigation. Whilst grammatical knowledge is considered for the most part to be procedural and thereby easier to activate, lexical knowledge is declarative (see Chapter 2, Section 2.1). Its activation can, therefore, be more more revealing as the risk of automated responses that are not necessarily indicative of knowledge per se is reduced.

One might then possibly conclude that through more intense re-exposure, individuals would more easily relearn the language thus in keeping with the idea that forgetting a language, despite its severity, does not inevitably imply irreversible loss.

As Bardovi-Harlig & Stringer (2019: 13) also point out, if one was to look at syntax, finding cases of true attrition is challenging because in cases where individuals appear to struggle with respect to syntactic complexity, it is sometimes more of an issue of problems in lexical repesentation and retrieval. In such analyses it can, therefore, be difficult to tease such difficulties apart.

Additionally, with respect to forgetting specifically, Mickan, McQueen & Lemhöfer (2020: 15) state that being unable to retrieve a word or having difficulty in doing so is usually what is understood by forgetting. Therefore, they argue that naming accuracy is a clear and instinctive way to determine this and indeed in Retrieval Induced Forgetting (RIF) studies naming accuracy is considered to be only the only reliable measure of interference-induced forgetting.

4.4.1.2 Knowing a word

Paribakht & Wesche (1997) refer to five categories of word knowledge elicited from self-reports using a Vocabulary Knowledge Scale (VKS).

I I don't remember having seen this word before.

II I have seen this word before, but I don't know what it means.

III I have seen this word before, and I think it means .... (synonym or translation)

IV I know this word. It means .... (synonym or translation)

V I can use this word in a sentence: .... (Write a sentence)

(Paribakht & Wesche, 1997: 180)

However, the context of Paribakht & Wesche's study was very different to the present one; in their study, the participants were students of the L2 in question at university at the time of testing. Moreover, their focus was on new words, i.e., items just acquired and using the VKS, Paribakht & Wesche aimed to gain insight into an individual's depth, as opposed to breadth, of vocabulary knowledge.

Paribakht & Wesche explored how vocabulary learning via reading could be fostered with the aid of supplementary exercises and which type best benefitted learning outcomes. Participants were, therefore, presented with the VKS directly after having been exposed to the vocabulary items of interest in the form of reading texts; in one case with participants having been given supplementary vocabulary exercises to aid learning of specific vocabulary as well as comprehension exercises on the text and in the other, they were given additional texts containing the target vocabulary as well as further comprehension exercises based on these. It must also be noted that in Paribakht & Wesche's study the test item was presented in the L2 and participants were not tested to see if they could produce the L2 form if only confronted with the L1.

The different focus as well as the long period of incubation (participants had not been exposed to the language immediately prior to testing as had been the case in Paribakht & Wesche's study) and the reported language loss reported by participants rendered categories I and V obsolete for the purposes of the present study. On the other hand, categories II, III and IV were subsumed under the recognition phase of testing when participants were given the opportunity to offer an L1 translation of the L2 item. In particular with respect to the purposes of the present study establishing *whether* and *what* specifically one knows as ascertained via responses in the recall and recognition phases of testing was deemed more relevant, regardless of whether this was prompted by categories III or IV in the VKS.

Instead, in this study, *form* and *meaning* constituted the focus with respect to word knowledge based on Nation (2001) and Graham (2012). Whilst Nation also included a third category, *use* (see also Chapter 2, Section 2.3.1 for an explanation of *form*, *meaning* and *use* according to Nation's categories), that refers to grammatical functions and patterns of implementation, on analysis of the severity of language loss experienced by the participants in the present study it was deemed appropriate and relevant to concentrate on *form* (specifically pronunciation) and, in particular, on *meaning* (i.e., what meaning the item as presented in the L2 represents/ refers to and, alternatively, what form represents the meaning of an L1 item in the L2). This is also in keeping with De Bot & Stoessel (2000).

4.4.1.3 Justifications for testing procedure

It was also deemed necessary to exclude any opportunities for the participants to avail of avoidance strategies or contextual cues. The elicitation of their vocabulary knowledge was, therefore, distinctly targeted and thereby constrained to exclude any such cues or faciliatory circumstances. De Bot & Stoessel (2000) also argued against the use of contextual cues in eliciting vocabulary knowledge arguing that testing words in isolation eliminated the possibilty that participants would rely on syntactic context to guide them in retrieving the meaning of the word. Graham (2012: 177), for example, criticises some aspects of attrition research typically conducted in which rather loose testing formats (narratives or open-ended questions) are used that enable participants to avoid or work around particular items or structures. Clear-cut targeted elicitation also reduces the cognitive load placed on participants so that they can focus fully on the item or structure of interest (Graham 2012: 148). Considering the length of incubation in this study, this was an important consideration in planning the focus and method of elicitation. One of the central aims was to establish if specific items had been retained and with respect to what had been lost, if that could be easily retrieved. An open task format would have made such a direct comparison difficult.

Addditionally, key studies on relearning that were conducted on vocabulary knowledge specifically looked at lexical items in isolation (De Bot & Stoessel 2000; De Bot et al. 2004; Hansen et al. 2012). (See Chapter 2, Section 2.3, for more information on these studies) It was also, therefore, explicitly decided to follow up on these studies to see if correlating findings would emerge and if not, to explore why.

Therefore, in keeping with the aforementioned studies, words were presented in isolation in the present study also. If words had been presented in context, it would have made it more difficult to assess to what degree success depends on word knowledge or effective use of the context or an interaction between the two. De Bot & Stoessel argue that whilst testing words in isolation might lead to an underestimation of participants' knowledge, offering context might lead to an overestimation.

For example, the following from Weltens et al.'s (1989) testing on L1 Dutch speakers' receptive lexical knowledge of L2 French, cited in Weltens & Grendel (1993: 142), demonstrates how syntax and overall sentence meaning can aid in the recognition of items. With respect to lexical knowledge, participants were given a French sentence that included the word (clearly highlighted) they were requested to recognise. The Dutch translation of the sentence, but in which the translation of the target word was missing, was also given:

French-Dutch recognition j'ai nettoyé l'évier
Dutch translation Ik heb de.......... schoongemaakt.
(*the target word "gootsteen" was deleted*)
(Example from Weltens et al. (1989), cited in Weltens & Grendel (1993: 142))

In such an example, the participant is given much contextual support which might lead to performance being guided by correct guessing as opposed to being truly indicative of knowing the word.

Also, traditionally at school in the 1980s in Ireland the norm was for vocabulary to be learned in the form of rote learning and word lists. Whilst vocabulary then reappeared in texts and this context would have served to reinforce knowledge, the main approach to learning was one of habitual repetition. Much explicit instruction and learning of isolated vocabulary items took place in the Irish classroom to aid learners in reading and writing, the focus of much of Irish learning at the time. Li (2010), for example, highlights the importance of vocabulary acquisition for reading in the L2 since words provide the building blocks of language learning, especially with respect to the mastery of reading comprehension, a focus that was also central to learning Irish in classrooms in the 1980s where content was predominantly text-based. Li refers to studies that have found that not having acquired the necessary amount of vocabulary hinders reading comprehension (Stahl & Nagy 2006; Stanovich 1986). Teachers in Ireland correspondingly put great emphasis on vocabulary learning to ensure that learners would be able to handle the many texts on the curriculum at the time.

Stoessel & de Bot also found that even though presented in isolation, compound nouns were able to trigger the retrieval of the meaning of their distinct components. They gave the example of *fishing boat* which successfully guided participants in accessing the item *fish*. This conforms to Meara's (2004) idea of the interconnectedness of lexical items (see Chapter 2, Section 2.2.3). Therefore, despite the situation that the words were shown separately and out of context, and that for practical reasons, only a specific number of items could be selected for testing, one can assume that the knowledge of other associated words might well have been triggered through the retrieval of the items tested in the present study.

Whilst Meara (2004: 1) argues that vocabularies need to be viewed as interlocking networks as represented in his computer simulations, he states that in real-life research this is very difficult to capture with few alternatives to the testing methods traditionally used in attrition research available.

It must also be stated once again that the aim of this study was not just to assess whether or not, and to what degree, vocabulary was lost but also whether or not there is a relearning advantage for previously known words. Therefore, it was practical and useful to work with individual lexical items.

4.4.1.4 Number of words and breakdown

Even though there are different dialects in Irish, it was ensured that the words chosen were not specific to any particular dialect but were items from the literature of the national curriculum. Although two of the participants were from the south-west of the country (Munster) and the other two from the east (Leinster) so that certain variations in dialect were to be expected, the items selected for testing were standard varieties.

In the first round of testing, 200 words were tested. The aim of the first test was twofold: (1) to get a general estimate of degrees of retention of vocabulary knowledge after many years of not having any contact with the language and (2) to gather test items for the subsequent relearning and savings sessions. Since it was not known in advance how participants would perform, it was decided to test a large number of words to increase the chance of being able to filter enough words for relearning. It was also calculated that performance on a larger number of words would give a better idea overall of degrees of retention since the point when contact with the language had ceased. Participants were not informed of the aim of the testing.

Meara (2004: 151) also argues that whether or not examples of items that have been attrited can be found will depend on the size of the initial test instrument. If the number of words tested is too small, examples of attrition may not be established and the amount of actual attrition that has taken place may be underestimated, potentially leading to the presumption that vocabulary is stable, which it might in fact not be. Therefore, he argues for using an instrument that uses as large a number of lexical items as is practicable in a particular testing scenario. It was decided, however, not to go above 200 words in this study so as not to overburden participants.

In keeping with Hansen et al. (2002: 661) and (2012: 120), whose test items were divided equally between concrete nouns, abstract nouns, verbs and adjectives, in this study, the 200 test items were also broken down accordingly, i.e., 50 concrete nouns, 50 abstract nouns, 50 verbs and 50 adjectives were tested. As previously stated, considering that Irish is a noun-centred language, it felt justified to dedicate half of the items overall to nouns. De Bot & Stoessel (2000) only used nouns in their testing; however, they did not elaborate on why they chose to do so. In fact, whilst it would have been interesting had they looked at other word categories to see if there was any variation of significance in performance across categories and why this might be the case, the main aim, however, as is the case in the present study, was to create a list of unknown words so that the Savings Paradigm could be tested. Interestingly, in a review of studies on lexical retrieval, it emerged that there was an overriding focus on nouns. For example, in testing the role of context on lexical retrieval, Van den Broek, Takashima, Segers & Verhoeven (2018: 553) also only used nouns in their study of L1 Dutch speakers that were required to learn words from a language that they had no prior knowledge of, in this case, Swahili. Mickan (2021: 165) also only looked at nouns (using 137 pictures of everyday objects and animals) which the L1 German speakers in her study had to name in Spanish.

On a side note considering that the nouns in the present study are divided according to concrete nouns and abstract nouns, it is noteworthy to observe and as highlighted by Scorolli, Binkofski, Buccino, Nicoletti, Riggio & Borghi (2011), that there is not always agreement in the literature on what constitutes an abstract noun compared to a concrete one. In fact Scorolli et al. argue that it is more useful to view concrete and abstract nouns as existing on a continuum rather than in a concrete/abstract dichotomy and that interpretations can also even be culture-dependent. Generally speaking, concrete nouns can be detected using our physical senses while abstract nouns are intangible in a physical sense such as feelings, qualities, conditions and theories. In their indepth guide to grammar, Fergussen & Manser (2011) explain how abstract nouns can be created by adding various suffixes. For example, suffixes such as *-age, -al, -ance, -ation, -ence, -ery* can be added to verbs, *-ism, -ity, -ness, -th* to adjectives and *cy, -dom –hood, -icide, -ism, -ship* to other nouns.

Based on the participants' performance in session 1, the number of words remaining for relearning (i.e., words that could neither be recalled nor recognised) varied from individual to individual. However, a *maximum* of 76 items was presented for relearning (this included 26 pseudowords). In other words, only 50 items from session 1 were carried over for relearning when the number of unknown words in this filtering session exceeded this limit. This was the case for two of the participants.

See Section 4.4.2.2 for elaboration on ratio of pseudowords to real words used in the present study.

As stated in Section 4.2.2, it is also important to have more rather than fewer words for relearning since what is essentially being examined is whether a retriggering of information takes place. If too few words are used in relearning, this gives more time for actual learning (also with respect to the new words) by allowing more cognitive space to engage in *learning strategies* as opposed to *triggering*, which is what essentially studies on savings aim to establish. Also, if it is the case of triggering previous knowledge whereby old words are easier to learn, then such a number should not overtax the participants. Therefore, in using only a total of only 30 items and 32 items respectively, De Bot & Stoessel (2000) and Hansen et al. (2002) may not have reliably been able to distinguish between good learning strategies and actual triggering of previous knowledge. For example, as stated in Chapter 2, Section 2.3.2, in a comparison with a group of first-time learners of Dutch, De Bot & Stoessel (2000) observed that some of the participants in this group performed better than one of the two siblings tested in the experiment group. Whilst De Bot proposed not using such a comparison group in further studies of attrition to avoid the issue of effective learning strategies as opposed to reactivating old knowledge, it is argued here that increasing the number of words to be relearned as was done in the present study, could also possibly eliminate this issue. For example, Bahrick & Phelps (1987) had 50 lexical items for relearning in their study of L1 English speakers' retention of Spanish.

It was anticipated that working with more items (as opposed to fewer ones) would offer results with enough scope for analysis from which general conclusions could be drawn, whilst not overburdening participants at the same time. These observations would have been more restricted in the case of a more limited number of test items. It should be noted, nonetheless, that in the case of the strongest performer in session 1, only 25 words could be filtered (after a second filtering phase) for relearning. His relearning, therefore, consisted of a total of 51 items (25 real words, 26 pseudowords). This is taken into account in the analysis of results.

The majority of the items used in the pretest were taken from the novel, *Peig*, an autobiography written in Irish which was on the Leaving Certificate syllabus at the time when the participants were in school and which was compulsory reading (see Chapter 3, Section 3.8). Much class-time was dedicated to analysing this work and a lot of homework dedicated to writing essays in Irish accordingly. It could, therefore, be certain that all of the participants would have been familiar with this vocabulary considering that a large chunk of the final two years of Irish at school was dedicated to reading, analysing and writing about it. The remaining vocabulary was taken from various other obligatory reading from the syllabus at the time. This was necessary as the vocabulary in *Peig* is somewhat repetitive and taking words from other literature allowed for greater choice.

All participants were tested on the same 200 words in the pretest (i.e., in session 1). The real words used in relearning and subsequent tests were different, however, as these comprised the respective items that each individual didn't know in the pretest.

4.4.1.5 Frequency of the items

Mickan and colleagues (2021: 199) observed that word frequency only played a negligible role in the retention of L3 Spanish nouns among L1 German – L2 English students in the six months subsequent to returning to Germany after a six-month stay in Spain.

Lexical items in the present study were of varying levels of frequency. However, since it was not known in advance what could be expected with respect to lexical knowledge of the participants, the leaning overall was towards relatively frequent words with very obscure options being eliminated. This was generally ascertained through an analysis of samples of the language used in *Peig* and based on the advice of a highly proficient L2 speaker of Irish in Ireland who was consulted on the word list prior to testing. It must be noted, however, that a steadfast scale of frequent to less frequent words could not be established; instead a general estimation was made. This is in keeping with Gharibi & Boers (2017: 58) who, in a study of L1 vocabulary attrition of Persian heritage speakers in New Zealand, created a general categorisation of frequency of words in Persian according to low, medium and high. Since a validated word frequency list was not available for Persian, they grouped their words according to feedback from postgraduate Persian L1 speakers in New Zealand. This was based on the vocabulary that they thought young native Persian speakers were using in their native country. Again, this was a general estimation as was the case in this study.

Considering also that the language used in *Peig* was not complex and that most of this language had for the most part been acquired over the years prior to the actual reading of the book, as well as the fact that the participants had been learning Irish since they were children and had, therefore, accumulated much vocabulary over time, it was very difficult to assess in which order specifically the items tested had been acquired.

4.4.1.6 Why a productive and a receptive test?

As in Gharibi & Boers (2017: 60), receptive and productive vocabulary were tested separately in this study. Traditionally in attrition research, examinations of each have led to different observations of how recall and recognition are affected by periods of disuse with recall needing greater effort than recognition in terms of retrieval (see Chapter 2, Section 2.1). Weltens & Grendel (1993: 154) also stressed the necessity to explore production as well as recognition since only looking at reception could give rise to a false impression of degrees or quality of attrition, i.e., whereby it appears that less attrition has taken place than is, in reality, the case.

Therefore, in each round of testing, participants were always tested on recall first (they were shown the English word and had to give the Irish translation). It was decided to always begin with recall since this necessitates that participants exert greater effort in activating this knowledge. Therefore, it is assumed that if words could be recalled, then they would also be readily recognised. Words that could be recalled were then removed and the remaining words were tested for recognition, i.e., in this case, the Irish word was shown and participants were asked to give the English translation.

Whilst there are other tests of vocabulary knowledge available (see Schmid 2019: 7), this one was selected because of its focused and controlled nature whereby it enabled a direct comparison between what was known and what knowledge is still available, whilst at the same time eliminating the availability of syntactic cues, thereby allowing for the collection of raw data untainted by other factors. Gharibi & Boers (2017: 58), for example, used the Verbal Fluency Test (VFT) whereby participants are either required to name as many words as they can pertaining to a particular semantic field, alternatively they are requested to generate as many words as possible that start with a certain letter, in each case within a fixed time-frame. While they state (in referring to Schmid 2007; Schmid & Keijzer 2009; Schmid & Jarvis 2014) that the VFT is commonly availed of in attrition research since difficulties in lexical access are regarded as initial signs of attrition, it was deemed unsuitable for this study because whilst it would reveal to a degree what items participants could still access, it would not be informative as to what they could *no longer* retrieve. The VFT does not allow for the linguistic progression within one individual to be assessed, i.e., an ascertainment of how large the gap between previous and present knowledge is; so although it can offer a comparison of performance across participants, it does not capture the linguistic development and change in the individual over time.

The VFT would possibly have been useful had the performance of the participants been compared to a group of current school-leavers (as explained in Section 4.3.1, such a comparison was not made because not only has the syllabus changed significantly since the time the participants of this study were in school, but also the circumstances of learning and tools available to school-goers are much different and more varied today).

Since the vocabulary that the participants had known in the past was available for this study, a constrained testing of this was considered the most effective way of establishing degrees of loss of lexical items.

Gharibi & Boers (2017: 65) also question the reliability of the VFT with respect to capturing accurate representations of vocabulary knowledge since in their study comparing bilingual to monolingual lexical attrition among heritage speakers of Persian, some of the monolinguals generated very low results in the test despite being fully proficient in their L1.

The participants in Bahrick & Phelp's (1987: 2) study were also tested on recall and recognition. However, they were sent the list of 50 English words. They were then requested to write down the Spanish equivalent in each case (doing so at their own pace) in one session and to send the test back immediately. Once the recalled words were eliminated, the participants were then sent the remainder of the words but this time in Spanish, which they then had to translate into English and then send back to the researchers. Whilst practical limitations possibly led to the decision to conduct the testing in this way, it was decided in the present study to have each participant attend each session in person with testing conducted in a controlled environment with two research assistants and the main test administrator present in each case. There is the danger in Bahrick & Phelp's study that some participants may have looked up the Spanish translations for the words that were on the list that they were sent or they may have consulted someone.

4.4.1.7 Issue of timing

There was no time limit set for responses. The participants in this study are bilingual speakers of English and German, switching regularly between both languages on a daily basis depending on the particular social or work context they are in. Mehotcheva (2010: 161), whose tests were also untimed, refers to studies which have shown that multilingual speakers are slower than monolingual speakers in picture naming tasks due to competition between languages (Gollan, Montoya, Fennema-Notestine & Morris 2005; Ivanova & Costa 2008). The participants in this study not only had to deal with competition from one other language but two. Also had a time limit been set, it might have given a false impression of degrees of attrition; participants might fail to retrieve an item quickly but that does not mean that they would not eventually succeed in retrieving the word. Keeping in mind the competition effects just mentioned as well as the very long incubation period, it was expected that participants would need more time in accessing previous knowledge. The focus was on *if* they could retrieve it, not on how *quickly*. Also considering that one of the aims of the study was to test for a relearning advantage, it was important to establish a list of words that appeared to be truly unknown. Had a time-limit been set, it could not have been certain that these words used for relearning were in actual fact unknown since the issue might simply have been that the participants did not have enough time to retrieve the words and that they would have been successful in this had they been given more time. Such words would then have skewed the results of the testing after relearning.

Weltens & Grendel (1993: 150) quote Weltens (1989) in whose study of L1 Dutch attriters of French, a time-limit was also not set on responses. As Weltens (1989: 93) points out: "[the] subjects had ample time to squeeze out of their memories anything that was still there, however vaguely and remotely".

Comparing the findings of a pilot test (lexical decision experiment) conducted by Verkaik & Van der Wijst (1986) in which a time-limit was set, to Welten's (1989) test in which a time-limit was not imposed, Weltens & Grendel (1993: 151) observe that more vocabulary "loss" emerged in the test that had time-constraints. The question then became not one of whether the knowledge was still there but how quickly it could be accessed, i.e., speed might have suffered but not necessarily accuracy (Weltens 1989: 93).

In fact Mickan et al. (2020: 15) refer to naming accuracy versus naming latencies, whereby naming accuracy refers to whether or not the word can be retrieved correctly and naming latencies refer to naming speed, i.e., how quickly the item can be accessed. They argue that whilst timing offers insight into difficulties or ease of retrieval, which they state is the first step before actual forgetting, it does not depict whether or not the word was forgotten. Therefore, setting a time limit on responses might give a false impression of the number of words actually forgotten. For instance, in the studies described in Chapter 2, Section 2.3.2 (De Bot & Stoessel 2000; De Bot et al. 2004; Hansen et al. 2002) a limit had been set on responses and this might have skewed results accordingly; perhaps the participants did indeed know the words they apparently could neither recall nor recognise in the initial round of testing but simply did not have enough time to access them. In the relearning and testing phase, such words would have been more readily available for retrieval, thereby emerging as examples of easily relearned words.

Time constraints in the present study were only set for the length of time that word pairs appeared in the relearning phase with each pair remaining on the screen for eight seconds.

4.4.2 Pseudowords

Pseudowords were created for relearning and the testing of savings to avoid possible recognition which could occur in cases where real new items were used. Arndt, Lee & Flora (2008: 4) explain pseudowords as items that abide by the orthographic norms of the language in question but the "pseudoword orthography does not correspond to any meaningful referent in the language".

4.4.2.1 Creating pseudowords

Apart from ensuring that the pseudowords obey the phonotactic and morphological rules of the language, Beeckmans, Eyckmans, Janssens, Dufranne & Van de Velde (2001: 239) state that there are no fixed guidelines on how they should be created and in how far they can deviate from existing words. They refer to the most common approach which is the changing of at least two letters in a real word. They caution against the danger here, however, that in making such changes one unwittingly creates a pseudoword that is very similar to another already existing real word. In the creation of the pseudowords for this study, Ó Glaisne (1961) and Stenson & Hickey (2018, 2016) who provide comprehensive information on Irish spelling were followed very closely. Consisting of high-frequency clusters, the pseudowords' accordance with Irish orthograhic and phonological norms was carefully checked by consulting dictionaries and confirmed by two highly proficient Irish speakers in Ireland.

As also done by Hansen et. al (2012) in the case of the 16 pseudowords that they used, the pseudowords in the present study were also divided according to conrete nouns, abstract nouns, adjectives and verbs in accordance with the relative split of the real words.

In understanding high-frequency clusters and why they were used in the creation of the pseudowords in this study, it is helpful to look at Weltens & Grendel (1993: 151) who give the example from Grendel's (1993) study of a high-frequency pseudoword for *pepper* in French, a low-frequency one (i.e., contains a less common, but nonetheless acceptable grapheme combination) and a non-word (i.e., it contains a disallowed grapheme combination).

High-frequency pseudoword: *poible*
Low-frequency pseudoword: *poifle*
Non-word: *poizve*

The focus of the present study was different, however (Grendel was concerned with orthographic attrition) and it was important that all pseudowords looked as real and were as convincing as possible. Had participants been confronted with any unusual clusters (albeit acceptable in terms of obeying the rules for grapheme combinations in Irish), an alarm might have been raised and the participants might have become distracted by this, thereby deflecting from the main aim of the study.

Care was also taken that the pseudowords did not represent frequently used words for which participants might know the correct translation. On the other hand, it was important that highly unusual meanings were also avoided so as not to prompt suspicion on the part of the participants.

4.4.2.2 Ratio of real words to pseudowords

With respect to the ratio of pseudowords to real words, In'nami (2013: 76) in referring to Nation & Webb (2011) states that there is no clear consensus on how many pseudowords should be used relative to real words in any given test. He cites Schmitt (2010: 2000) who states that pseudowords usually account for approximately 25–33 % of the total number of test items.

Beeckmans et al. (2001: 239) refer to typical ratios used in yes/no tests. These are tests where participants are told that there are pseudowords mixed in with real words and they have to demonstrate that they can decipher which words are real and which not. Whilst the function of the pseudowords in the present study was different, it was useful, nonetheless, to observe the general ratio used in the studies referred to by Beeckmans et al., such as the 60 % real words to 40 % pseudowords ratio used in the studies by Meara & Buxton (1987) and Abels (1994) and which they also used in their own testing, as well as the 66.6 % real words to 33.3 % pseudowords ratio used by Meara (1992). Beeckmans et al. (2001: 239) refer to personal communication with Meara in which he recommends (and also observed in further work he conducted), a ratio of 66.6 % to 33.3 %

for real to pseudowords. Whilst the overall number of words was greater in Meara's work and the focus different to that of the present study, the approximate ratio of 66.6 % to 33.3 % (2:1) for real to pseudowords was deemed, nonetheless, a useful orientation.

However in this study, the exact ratio of pseudowords to real words could not be fixed in advance as it depended on how many words emerged as "unknown" or forgotten in the pretesting phase. Since it was decided that a limit of 50 real words would be taken from the unknown words left over in the pretest (should more than 50 words have been neither recognised nor recalled), a ratio of approximately 66.6 % real words to 33.3 % pseudowords was set as constituting the baseline ratio of pseudowords to real words (i.e. 26 pseudowords were to be used). This was indeed the case for two of the participants: Cathy and Brendan. For Ellen, this was also the approximate ratio, although she had slightly fewer real words relative to pseudowords in comparison to Cathy and Brendan, i.e., 62 % (real): 38 % (pseudowords). For Enda, the ratio was 50 %:50 %. Hansen et al.'s (2012) ratio was also 50 %:50 %. However, they tested their participants on fewer words in total; 16 pseudowords and 16 real words, i.e., a total of 32 words.

4.4.3 Questionnaire and interview

There was no time limitation in this part of testing. Participants could take as long as they wished in answering all questions. Section 1 was conducted in the form of a questionnaire and consisted of questions pertaining to background information which the participants completed in written form on site. They were presented with the questions for Section 2 in printed form which they answered orally in the presence of the two research assistants. The interview was recorded and responses subsequently transcribed.

The main researcher was absent in this phase. The reason for this was twofold. Firstly, she prepared the items for the recognition testing phase in this time. Secondly, since the participants knew the main researcher and she too had been educated in the Irish school system, it was decided that it would be better if she wasn't present so that the participants wouldn't unwittingly feel a subconscious pressure to give answers they might have deemed desirable but that did not represent the truth. The research assistants had no knowledge of Irish or the Irish educational system and were, therefore, neutral observers. The advantage of answering the questions in Section 2 orally was that participants were not limited as they would have been had they had to write their answers. In such a case where answers had to be written, it was expected that they might opt for brevity and "cut corners" in their answers. It was hoped that the relative effortlessness of speaking in comparison would reap more elaborate and comprehensive responses. The interviewees were informed in advance that they were to give as much information as possible and to elaborate as much as they felt was necessary and that they were also free to backtrack, rephrase and clarify.

The two research assistants to whom answers were addressed were not active interviewers in the sense that apart from nodding or the odd filler in recognition of comprehensibility they did not contribute to the conversation or ask follow-on questions. This was deemed important so that the interviewees would not be influenced in any way in their responses.

Also in keeping with the idea of clusters of factors that are intricately interconnected, where attempts to examine them in isolation have failed to arrive at unanimous consensus (Opitz 2019), it was considered more helpful for the purposes of this study to approach the factors as interlinked and dynamic as opposed to each one constituting a distinct quantifiable unit that could be statistically measured. Such empirical methods can be problematic when it comes to understanding the complexity and variability of these interconnections. Opitz also argues that looking at such clusters reduces the need to study large numbers of participants, although it is stated here that a large number of participants and the ability to study even more individual variations would offer even more insight into how these variables interact and influence each other. However for practical reasons, it was not possible to study more than four individuals in this project.

Being of a case study nature, dealing with a small number of participants had the advantage that more in-depth interviews could be engaged in. In this scenario, working exclusively with questionnaires would also not have yielded reliable averages that could possibly be generalised to the greater society. The section consisting of a *questionnaire* in this study dealt with background information such as age at time of testing, age at start of exposure to Irish, years of learning it in an educational context, age at onset of attrition, level of proficiency achieved at onset of attrition, other languages learned, languages currrently used and length of residence in Germany.

The status of Irish as a minority language and the *context* of learning were considered very relevant aspects in attempting to understand attrition in the case of the participants in this study. The goal of the *interview* was to explore these and probe other factors highlighted in the literature as relevant for promoting retention or inducing attrition, with questions attempting to gain insight into the context of learning and use of Irish in education, at home and in the wider community and to explore attitudes to the language as well as its relevance for one's sense of identity. More specifically, participants were asked about their attitudes to the general focus/content of the syllabus, the learning situation, teaching methodologies and their feelings about the lessons, as well as how this compared to how they perceived modern foreign language classes. The legacy of their individual experience of learning Irish was probed via an exploration of their motivations in learning and reasons for a discontinuation or continued contact with the language. Their families' attitude was also investigated and to what degree there was support for the learning and use of Irish in the wider community, as well as how they feel about their own diminished proficiency and to what degree they are aware of the language on return trips to Ireland. The questions were formulated in such a way that the answers would offer more of an autobiographical narrative considering the case study approach of the research with participants being invited to offer as much information as possible.

4.4.3.1 Development of questionnaire and interview

The focus of the questionnaire and interview questions was partly guided by R.C. Gardner's (1985) Attitude/Motivation Test Battery that focuses on attitudes and motivation in second language learning and Schmid's sociolinguistic questionnaire from her comprehensive Language Attrition Test Battery (LATB; Schmid, 2005; cf. https://languageattrition.org/), a set of instruments and tests, essentially concentrating on L1 attrition. Nonetheless, the sociolinguistic questionnaire from this battery provided a useful orientation for the question formation in the present study of L2 attrition. Schmid also argues that when a questionnaire is used, it should be followed by an interview format so that more detailed information can be obtained. A number of existing questionnaires from various studies conducted within the sphere of acquisition and attrition research as well as on attitudes to Irish specifically were also consulted (e.g., Harris, Forde, Archer, Nic Fhearaile & O'Gorman 2006; Harris & Murtagh 1999; Mehotcheva 2010; Murtagh 2003; Schmid 2002).

4.4.3.2 Issue of memory

It must be noted, however, that both Gardner's and Schmid's questionnaires focus on the current attitudes respondents have at the time of testing towards the target language and its speakers. In the present study the context and experience of learning constitutes a key area of interest. Considering, however, the long incubation period and that the learning situation had taken place many years previously, a blurring of memory in the intervening years was expected and that participants would not be able to offer specific detail retrospectively. It was deemed nonetheless important, despite such limitations, to at least derive an overall *impression* of how participants had felt about learning Irish at the time and to offer them the opportunity to explore and contemplate various aspects pertaining to the learning experience in hindsight. Despite the vulnerability of memory, this "lasting impression" was, nonetheless, considered meaningful and significant.

4.4.3.3 Limitations of questionnaires

Apart from the short questionnaire that constituted section 1 that was used to gather background information on the participants, an interview was chosen over the format of a questionnaire similar to Gardner's or Schmid's mentioned above in which participants are presented with an item using a pre-assigned alternative response format, the most fitting of which they have to select. This was considered too limiting in terms of the insight that could be gained on the individual nature of motivation and attitude and understanding the context, as well as the experience, of learning. It was also deemed impractical und unreasonable to apply Gardner's questions and format directly in attempting to capture attitudes and emotions concerning a learning experience that had taken place many years previously as it was doubtful that participants would remember the particularities pertaining to the learning scenario as covered in his questionnaire.

Considering also that this is a case study with each individual being looked at in his own right, it is argued that a true understanding of this particular experience cannot be gained by generalised quantitative measures. The small number of participants presented the advantage of being able to probe the issues of interest in a more detailed manner without having to constrain answers in line with a such a response format. Ticking boxes, whilst advantageous with respect to getting general impressions and when dealing with large populations, may evade the fundamental issues or might be too constrained to not allow participants to express their true feelings or opinions because the options fail to capture the essence of their beliefs. It is also difficult to judge what categories such as "slightly disagree" or "slightly agree" essentially represent. Knowing that someone "slightly disagrees" with something does not offer tangible information in terms of why someone feels this way or how each individual participant interprets that specific categorisation.

Additionally, it is argued in the present study that the emotional and individual nature of the learning experience is beyond statistical quantification, or at least in attempting to do so, would severely restrict the potential to gain a comprehensive insight into the personal experience, which is what this study is interested in. Therefore, in the present study, attempts to quantify experiences, attitudes and motivations were not made, as they are deemed subjective as well as dynamic, and phenomena existing beyond the limitations of numerical measures. At most the motivation of individuals may be compared in terms of seemingly "more or less" but even this is open to diverse interpretation. Indeed, this emphasises the risk that the patterns that emerge from statistical analyses of data elicited in such a constrained manner as is typical of questionnaires, might be deemed to be indisputably valid and unquestionably representative of the essence of the phenomenon under scrutiny when the focus is on inter-individual perceptions. This highlights the need and benefits of conducting qualitative analysis so as not to miss the unique individual experience. Chen, Warden & Chang (2005: 615) in referring to Dörnyei (2001) also indicate the challenge of creating surveys on motivation that are quantitative measures but that aim to understand qualitative variables. Cherciov (2013) highlights the importance of conducting qualitative analyses on an individual basis and in reviewing studies on language attitudes and levels of L1 attrition, she observed that interviews as opposed to questionnaires proved more useful and constructive in establishing links between attitudes and language retention (Ben-Rafael & Schmid 2007; Schmid 2002). This is most likely due to the nature of the interview format which is more flexible and malleable than a pre-assigned, fixed questionnaire, and, therefore, provides greater opportunity to probe possible predictors of retention in more detail. Likewise, it is argued here that an over-dependence on empirical data but without seemingly questioning their validity, representation and power of statement, can lead to skewed interpretations of those phenomena that are in fact highly subjective and decidedly individual. While Mehotcheva (2010: 43) states that explorations of language use and attitudes based on individuals' self-reports are prone to subjectivity, it is this very reason that renders interviews more conducive to revealing what fueled or prompted that unique personal perspective.

Naturally, and as argued by Lowie, van Dijk, Chan & Verspoor (2017: 145), the individual variables that are related to second language acquisition are not stable, meaning that what is captured at one point in time has to be treated with caution. They argue that it is important to look at the individual path of development as this can be just as revealing as looking at global generalisations emanating from group analyses.

With respect to the focus of this study, attitudes towards Irish were intricately interconnected with emotion. The Irish language played a very disinctive role in the history of the island and for many individuals it is strongly associated with the identity of the Irish people. This constituted another reason why the interview format was deemed most appropriate in an attempt to potentially capture any emotional underpinnings.

It is also argued here that in such cases where participants are simply required to tick a box, that they might just do so randomly or without much reflection. An interview-style with no predetermined responses is considered more thought-provoking and stimulating and allows for the individual's own unique input.

In fact, it was anticipated that having to complete the questionnaire and interview in the presence of the researchers meant that the participants would take it seriously and give it their full attention. It also had the advantage that anything that was unclear could be addressed. Some researchers send participants the questionnaire (e.g., Bahrick 1987; Gardner 1985).

This is sometimes unavoidable for practical reasons. However, reliability may be compromised in such cases as the questionnaire might assume an anonymous, random nature with respondents not giving it the necessary earnest attention as a result.

4.4.3.4 Distinction between attitudes and motivation

As stated in Chapter 2, Section 2.4.7, and in keeping with Dörnyei (2000: 520), what separates motivation from a wish or desire is the willingness to act and, it is argued here, the very *engagement in an act*, i.e., motivation constitutes the driving force behind why actions are carried out. While questionnaires often focus on the respondents' attitudes to the language and its speakers, these responses are often automatically translated into depictions of motivation. However, such responses should be treated with caution and as explained in Chapter 6, Section 6.5, positive attitudes towards the language, its speakers and associated culture may not result in *actively* learning or using that language. Positive attitudes certainly feed into motivation but a positive attitude is not in itself synonymous with motivation. Therefore, it was considered important that not only the *attitudes* that might lead to certain motivations were explored, but also the motivations themselves in explaining what propelled the very act of learning.

4.4.3.5 Number of questions

Considering that the participants in this interview were given open-ended questions and so answers were to be more comprehensive and elaborate than allowed in an item-based, Likert scale response system, 34 questions was considered a realistic number that would provide the opportunity to offer ample insight but at the same time not overburden the participants. This is also based on a consultation of questionnaires as stated above and in keeping with Dostert (2009: 98) who in creating a questionnaire as part of her analysis of L1 English attriters in Germany, also deemed a number of the items typically used in sociolinguistic questionnaires to be repetitive and needlessly long and who, therefore, recommends fine-tuning questions accordingly.

4.4.3.6 *Sensitivity of the data*

As Gardner (1985: 5) also argues, sensitivity and discretion are required with respect to the handling of the data collected considering that questionnaires and interviews aim to gather information on attitudes towards a language and, therefore, inevitably attitudes towards the community that speaks that language, the status of the language and what it represents, as well as the context in which learning took place. The participants in this study were assured that the videos that were filmed would not be made public or passed on to third parties but were to be used exclusively by the main researcher for the purposes of this study alone.

4.4.4 Case study research

In attempting to understand attrition, much valuable insight can be gathered from observing the personal experience due to the unique manifestation of the interconnection of potentially influential factors in each individual.

It is difficult to establish the nuances and intricacies of attrition on a mass scale and to isolate the impact and particular weighting of individual predictors in such studies. Mass studies can miss the essential detail. Hereby a bottom-up approach as opposed to a top-down approach is useful. The advantage of looking at a small number of participants but with certain common characteristics allows one to focus on specific variables in detail. Large-scale quantitative measures have no doubt a role to play in attrition research, however, the non-linear and dynamic nature of attrition shows that focusing on the individual as opposed to remaining at a more removed level of group analysis is essential to gaining a true understanding of the fluctuating and irregular quality of attrition operations.

As defined by Duff (2018: 306), in applied linguistic research a case typically refers to a language user or speaker and in case study research the emphasis is on the individual as a "multi-faceted" entity whose experience functioning in and navigating various different social groups and constellations constitutes an essential component of understanding the trajectory of language acquisition and in this exploration, attrition. Duff (2018: 307) states that, for example, in exploring the role of motivation in L2 learning, or in observing L2 linguistic development, researchers are increasingly availing of case studies as a mean to attain a more holistic understanding of the "individuals' dynamic socio-effective and cognitive learning processes in ways that larger quantitative studies cannot do as well." Lowie et al. (2017: 13) also argue in favour of case study research as opposed to relying exclusively on group studies and generating statistical generalisations from inter-individual generalisation since these do not consider the individual's development over time. Gharibi & Boers (2017: 53) refer to Montrul (2008) who maintains that the level of thorough analysis that is characteristic of case study research serves to ensure that what is being explored is indeed attrition and not incomplete acquisition, a distinction that a large-scale empirical study might fail to pick up on.

Herdina & Jessner (2013: 752–756) too argue that the very non-linearity of the attrition process that emerges in bilingualism and multilingualism poses a number of methodological challenges. Considering that bilinguals and multilinguals do not consitute homogenous, uniform cohorts, they call for availing of qualitative research procedures such as case studies, in exploring language loss phenomena and indeed argue that these possibly be preferred over typical quantitative methods when deciding on the appropriate research path to follow in examining the attrition of language knowledge.

With respect to the selection of the cases for this study, participants who had learned the Irish language at school, who had little to no contact with that language since finishing their formal education and for whom the incubation period spanned a long period of time and were living abroad (i.e., not in Ireland) were the focus of this study. Four individuals who satisfied these requirements were available in the environment in which the study was conducted (in Cologne, Germany). Practical issues such as availability for testing also played a role in selection.

While Duff (2018: 308) states that the focus in case study research is not so much on the outcome as it is on the process, it is argued here that in case study research both the process and the outcome are equally important. Looking at the outcome as this study does in the sense that it attempted to establish approximate degrees of attrition and possible savings of seemingly forgotten knowledge, it is insightful with the aid of analysis of each case to look at the path taken by each individual that might explain these outcomes.

Duff (2008: 2) states that contemporary case studies in applied linguistics focus less on the structural features of language learning, practice and loss and more on poststructuralist elements, and rely to a greater degree on interpretative and critical analyses. Case studies enable an indepth exploration of the unique experiences of the language users in question, their participation in the langauge community, their individual attitudes and motivations within that group and, how they identify with and position themselves accordingly within that culture. In keeping with Duff who lists trends in case study research that work progressively with sociocultural, anthropological, and phenomenological approaches, this investigation also explores extralinguistic variables and the participants' individual narratives. A case study approach enables one to accommodate the multidimensional nature of the language speaker and consider all of these aspects combined with the uniqueness of their language learning and maintenance profile.

This is an exploratory case study combined with empirical data based on performance in the lexical learning tasks to assess whether or not the Savings Paradigm can be supported. In looking at each individual's personal profile with respect to their experience of both learning the language and whether or not they made efforts to maintain proficiency in the language, enables an exploration of the possible role played by extra-linguistic factors and how a Complex Dynamic Systems Theory may explain the varied outcomes observed. It is argued here that a key component of case study research is reflection, not just on the part of the researcher in analysis of the data but just as importantly, on the part of the participants who, in being confronted with in-depth and open-ended questions, are compelled to reflect on their language learning trajectory and possible reasons, be they emotional or pragmatic, why their relationship with the language evolved in a particular way.

Such an in-depth exploratory analysis is challenging if working with large groups of study participants; whilst analysis of data stemming from large groups can be interesting in establishing patterns and trends with respect to factors that may or may not affect the attrition process, in such studies the individual experience fades as it becomes subsumed under the totality of group bias. Therefore, the danger in conducting mass data analysis of large numbers of participants is that the individual profile which can reveal much about acquisition and attrition might get lost, thereby depriving the field of what could be very insightful data.

Duff (2008: 95) refers to some such case studies in applied linguistics research that have reaped interesting results in the field of language teaching and learning. One of those mentioned was Borg (1998) who, over a two-week period, observed and interviewed an EFL teacher in a language school on the island of Malta, and in doing so was able to generate useful insights into the psychological aspects of the teaching of grammar. Duff (2008: 93) also makes reference to a study conducted by Kanno (2003) who looked at four L1 Japanese children after their return to Japan from Canada where they had been immersed in English. Kanno attempted to use the individual narratives of each child to understand the unique L2 learning experience and their continued relationship with the language after having returned to their L1 environment.

While Taura (2019: 6) criticizes the case study approach adopted by Yoshitomi in which she studied the L2 English attrition in four Japanese children after their return to Japan, arguing that working with such a limited number of participants does not allow for generalisation, it is this very point which makes a case study approach with respect to language attrition so valuable. Whilst general trends have been established in language attrition research in terms of what may or may not slow down the attrition process, scholars in the field are progressively calling for more detailed observations of the individual experience because increasingly it is being found that the very nature of attrition is one that is unique to each individual and that cannot necessarily be generalised. Therefore, caution is expressed in attempting to establish a fixed definition of the process that attempts to present it as one that is unform across all individuals and experienced by everyone alike.

Graham (2012: 151) also argues that while it is beneficial to look at overall results from mass data, analysing particular participants more closely adds valuable insight into the attrition process. In his study of the L2 attrition of returnee missionaries whose L1 was English, he therefore selected four individuals who took part in his study to explore their language learning experience in greater detail and the trajectory that this L2 knowledge took on returning to the L1 environment (for more details on this study, see Chapter 2, Section 2.3.1). Graham highlights the uniqueness of each individual history and path with respect to the experience of learning the language and the willingness or lack thereof, to maintain contact with the language on returning to the L1 environment. What is particularly interesting and, also observed in the present study, are the differing outcomes in terms of language proficiency and long-term maintenance among these individuals despite the similarity and uniformity of conditions in learning. This, he claims, can most aptly be explained through an exploration of each individual experience, as such insight might possibly not emerge in large quantitative studies.

It is not argued in this book that case studies should replace empirical and quantitative approaches to applied linguistic research, rather that they have the potential to complement such approaches with insightful and perceptive details that may possibly go unnoticed otherwise. This comforms with the arguments put forward by Dewaele & Pavelescu (2021: 66) who in exploring the role that emotions play in the language learning experience, highlight the benefits of adopting a case study approach, contending, in keeping with the position of this study, that emotions cannot be "essentialised". This idea is extended here in which it is argued that neither can the attitudes and motivations that possibly lead to these emotions necessarily be extracted and categorised according to some predetermined classification as would be the procedure in a purely quantitative analysis. Emotions, attitides and motivations which are crucial to understanding the acquisition process are apt to change and fluctuate over time. Such flux is difficult to establish when looking exclusively at group trends that overshadow the individual trajectory.

The very reasons and experiences of the individual are of utmost importance in attempting to understand attrition and why one individual's path takes such a different course to another despite uniformity in the external institutional framework of learning as is the case in the present study. It is this unique combination and interplay of internal and external variables which is particular to each individual that necessitates such detailed exploration and renders case study research an exceptionally useful accompaniment to empirical research.

Indeed Duff (2018: 306) argues that case study research plays a bigger role than ever before in knowledge generation despite developments in technology enabling the generation of "big data". As she so aptly puts it, "they provide an in-depth, contextualised, concrete, multi-dimensional exemplification of a subject that can help tell an important story but at the same time vividly encapsulate abstract principles as well" (Duff 2018: 306).

Chapter 5: Results and analysis

In the following the individual participant profiles are presented, including insight gained in the interviews. In each case scores for their performance in each test are given which is followed by an overview and analysis of all results combined.

5.1 Individual participant profiles and test results

5.1.1 Enda

Table 3 Participant profile – Enda

Enda	
Age at start of learning	4 years and 2 months old
Age at onset of attrition	19 years old
Length of learning	15 years
Proficiency level at onset of incubation	Leaving Certificate: Grade **B** in **higher level paper** plus one year at university (exact grade unavailable)
Exposure and learning context	Predominantly in education context, passive input in immediate environment
Length of period of disuse	29 years
Rehearsal during incubation	Intermittent (Receptive: listening to Irish-language radio)
Motivation	*
Other language/s at school and grade in Leaving Certificate	French: Awarded a grade A in higher level paper
University Education	B.A. Modern languages (French and German) M.A. German studies
Age on leaving Ireland	27 years old
Country of residence at time of testing	Germany
Length of residence in Germany at first session of testing	21 years
Occupation at time of testing	Secondary school (German Gymnasium) teacher of French and English
Languages actively used other than English	German: native-like, CEFR: C2 French: CEFR: C1 (teaches French at Gymnasium)

*Unquantifiable but described in detail in Chapter 6, Section 6.5

5.1.1.1 Enda's learning profile

In the following an overview is given of Enda's background with respect to the factors typically put forward in the literature as affecting degrees of knowledge retention.

5.1.1.1.1 Experiences in learning Irish

Enda had been exposed to some Irish before starting school as his parents spoke a certain amount of Irish at home. He grew up in an area on the south-west coast of the island where financial incentives for speaking Irish were offered by the government. He added that it was, therefore, important for his parents to give the impression of being Irish speakers, especially when inspectors visited. However, he claims that this input or, more specifically, his uptake was passive as he wasn't addressed directly and neither was the language used in a functional way.

In his primary school, however, there was a heavy focus on Irish with it being, for the most part, the working language of the school. In the Irish lessons, which he had on a daily basis, the language of teaching was Irish and not English and Irish was also sometimes spoken in other subjects. He states that the school principal was very passionate about the Irish language and this trickled down to the pupils, all of whom, Enda claims, tried to always speak Irish, even during the school breaks. On the other hand, English was, for the most part, the classroom language in his Irish lessons at secondary school.

Whilst he enjoyed it in primary school, Enda claims to not have derived as much satisfaction from the experience of learning Irish in the first years of secondary school when he felt that there was too much focus on grammar and drills and that it was taught like Latin, i.e., a written language as opposed to a spoken one. He also remembers it as a pressurised learning situation with an ever-present discernible underlying fear that one might not know the answer to the teacher's question when asked. Whilst he states that he tended to have a more positive attitude to the language than most of his peers, this tension in the classroom, mixed with boredom, led to a negativity towards the language. He argues that motivation wasn't nurtured in the case of the Irish language, stating, "there wasn't any kind of group learning or role-plays or anything like that. It was based on the written word and the accuracy of the written word."

He also adds that the Irish language had a lot of "emotional baggage" attached to it, giving the example of the 1.5 million people that were forced to emigrate to North America during the famine in the mid 19th century. With their departure and the death of many who remained on the island from hunger and malnourishment, the Irish language lost many of its speakers. The trauma emanating from such tragic phases in the history of the island and which also came to be associated with the language itself constituted an underlying element in the content of the syllabus with much of the poetry and literature dealing with this sorrowful past and the poverty, sufferings and struggles of the nation. He claims that there was, accordingly, a sense of obligation associated with the language.

Enda states that his motivation also depended on the phase in which he was learning Irish. Towards the end of his secondary school education, for example, he developed a renewed interest in it, even deciding to study it at university. In his final year at school in preparation for the oral section of the Leaving Certificate examination, he claims that he very much attempted to immerse himself in the language, listening to Irish-language radio and engaging as much as he could with the content at school. In establishing this goal for himself, i.e., to prepare for the oral examination, he discovered an enjoyment in learning that he had not experienced in the earlier years at secondary school. The oral examination was a very positive experience for him. He felt a sense of achievement and positively rewarded in the engagement and feedback he got from his examiner.

Despite his reservations about the way the language was taught and the content of the syllabus, Enda left school with a positive attitude towards the Irish language and embarked on a study of languages, choosing Irish, German and French in his first year at university. Since only two languages could be studied after the first year, he had to give up Irish as he achieved higher marks in French and German. He states that the standards and demands in the Irish department were much higher than in the French and German departments and that it was difficult to achieve high grades, even though he recalls investing more time and effort in his Irish studies. He felt that this tended to push willing learners away.

However, he enjoyed his experience of studying Irish at university. He recalls a meeting room there where all of the students of Irish could get together and chat in Irish. He states that it was taken for granted that everyone would speak in Irish, so it created a very pleasant situation. He says that it was the only language to have that facility at university, so that made it quite special for him.

5.1.1.1.2 Experience in modern foreign language education

He started learning French at secondary school and German at university and claims that the fact that he now speaks French and German better than Irish is a reflection of his greater motivation at the time to learn these languages. At school French was more appealing because it also had an "international aspect" for him. He further states, "[…] you learn French and that's somehow associated with modernity. It's got a certain appeal, a certain attractiveness." He also enjoyed French class because there was more engagement in communication than in Irish class and generally the content of the French syllabus was more interesting.

He states that he was very motivated in learning German at university too, not only due to the teaching methodologies that were used which put a greater emphasis on communication, but also due to the interesting content.

5.1.1.1.3 Attitudes, identity and Irish language use

In the past, Enda stated that he tried to listen to Irish language radio again but his efforts, whilst conscious, were not systematic and over time he became less engaged. He feels that he has lost a lot of his Irish language knowledge when he compares it to the level he had after his first year at university.

He states that because Irish was always connected with Irish history, one never truly managed to transcend this and bring Irish into the present, immediate environment. He was bothered by what he terms the "insularity" of the language. He states:

> […] if you listen to Irish radio in Ireland, a lot of the programmes are talking about people who speak Irish. So, it's very self-obsessed if you would, so, you have programmes where people do interviews about why so many people are not speaking Irish, you know it's a microcosm and that can be very claustrophobic

> sometimes I find. Or they are dealing with the few areas of Ireland where Irish is spoken. So, there's a heavy focus on its talking about itself all the time.

Nonetheless, for him Irish is still vibrant and he continues to view it not just as an important part of his own identity, but of the Irish identity in general and as a very positive element of the cultural heritage of Ireland. However, he claims to no longer be "obsessed" about his attitude to the Irish language. This is an interesting comment as it suggests that Enda had reflected deeply on his attitude to and feelings towards the language and highlights how inherent it was to his identification with the island, its history and his own narrative growing up there. He refers to its presence in his environment:

> It was sourced everywhere. [...] People in an authority when you grow up in Ireland; priests, teachers, they all had this kind of indoctrination that they wanted to speak Irish. So, [...] it was just a daily thing that you had contact with.

Within his family in Ireland, he says that his youngest sister has a very positive attitude towards the Irish language and sends both of her children to an Irish-medium school. His brother also likes to speak Irish and Enda expects that his family has generally a more positive attitude to the language than the average person in Ireland. He believes that this is because he is from a region where traditionally much Irish had been spoken. He also states that he himself is keenly aware of the language when he visits Ireland, for example, on public notices or in community efforts to blend Irish into daily life. He, nonetheless, believes that one can be truly Irish without having any knowledge of the language. He sees the identity of the country as naturally evolving and as the island becomes more multi-cultural and diverse the definitions and interpretations of what constitutes "Irishness" inevitably change.

5.1.1.2 Test results – Enda

5.1.1.2.1 Session 1

Table 4 Results of session 1 – Enda

Session 1 (Enda)				
Pretest	Recalled	Recognised	Recalled or recognised	Neither (forgotten)
200 words (All previously known)	145 (72.5%)	49 (24.5%)	194 (97%)	6 (3%)

Enda performed extremely well in session 1 (testing of words from *Peig* and other texts that were on the Leaving Certificate curriculum when the participants attended school). He was able to recall 72.5 % of the words previously known and recognised 24.5 % of the words. Overall, he was able to recall or recognise 97 % of the 200 words presented to him. This left only six words for a relearning session. As this number was too low, it was decided to have a second filtering phase to accumulate new words for relearning.

5.1.1.2.2 Session 2a (Second pretest)

Two weeks later, a second (filtering) pretest took place. This time he was presented with a total of 95 words which included 26 pseudowords so as to not trigger any suspicion in the relearning phase when these words appeared again. The difficulty level of the words taken for this test was adjusted in attempts to increase the chances that some words would become available for relearning. On careful analysis of the literature and in consultation with a highly proficient L2 speaker of Irish in Ireland, 69 words that were deemed to be less frequent were selected for the filtering phase.

Table 5 Results of session 2a – Enda

Session 2a (Enda)				
Pretest	Recalled	Recognised	Recalled or recognised	Neither (forgotten)
69 real, previously known words	17 (25%)	27 (39%)	44 (64%)	25 (36%)
26 pseudo-word	0	0	0	26 (100%)

From this group of words, he was able to recall 25 % of the 69 real words and recognise 39 % of them. Overall, he could recall or recognise 64 % of the words. This left 25 words for relearning and 26 pseudowords. Suggestions he made for these pseudowords were ignored (in the sense that they were considered incorrect) in the filtering phase. These 69 items were divided accordingly: 18 concrete nouns, 18 abstract nouns, 16 verbs, 17 adjectives (see Chapter 4, Section 4.4.1.3, for explanation of category breakdown).

The category breakdown of the 44 real items that he could recall or recognise was as follows:

Table 6 Session 2a, Category breakdown of known items – Enda

Enda (44 items)	Concrete nouns	Abstract nouns	Verbs	Adjectives
Recalled	7	2	6	2
Recognised	2	7	8	10
Total (recalled or recognised)	9	9	14	12
% of total number of items (44)	20.5%	20.5%	32%	27%

The arbitrary (as determined by pretest in session 2a) breakdown of the 25 remaining items for relearning was as follows:

Table 7 Category breakdown of real items for relearning – Enda

Concrete nouns	Abstract nouns	Verbs	Adjectives
9	9	2	5

Considering the arbitrary and uneven breakdown of the word categories of the 25 items left over for relearning, further comparative analysis of category breakdown in subsequent testing was not conducted in Enda's case.

5.1.1.2.3 Tests on savings: Results

Table 8 Session 2b (after relearning) – Enda

Session 2b (Enda)				
Relearning (51 items)	Recall	Recognition	Recalled or recognised	Neither
25 real, previously known words	17 (68%)	8 (32%)	25 (100%)	0
26 pseudo-word	4 (15%)	5 (19%)	9 (35%)	17 (65%)

Table 9 Session 3 (no relearning) – Enda

Session 3 (Enda)				
No relearning	Recall	Recognition	Recalled or recognised	Neither
25 real, previously known words	4 (16%)	12 (48%)	16 (64%)	9 (36%)
26 pseudo-word	0	2 (8%)	2 (8%)	24 (92%)

Table 10 Session 4 (after relearning) – Enda

Session 4 (Enda)				
Relearning (51 items)	Recall	Recognition	Recalled or recognised	Neither
25 real, previously known words	18 (72%)	5 (20%)	23 (92%)	2 (8%)
26 pseudo-word	5 (19%)	8 (31%)	13 (50%)	13 (50%)

Session 2b

Due to time constraints on Enda's part with respect to availability for testing sessions, the relearning session took place on the same day of the second filtering phase. He took a one hour break after the pretest (2a) before relearning and another one hour break after relearning before testing (2b) on relearned items.

In the testing after relearning, Enda was able to recall 68 % of the real words and could recognise 32 % of them. Overall, he was able to recall or recognise 100 % of the real words. Considering that he had fewer items to relearn compared to the other participants, the issue arises if he had an advantage in the relearning phase since he had fewer words to learn and, therefore, a smaller cognitive load.

In the case of the pseudowords he was able to recall 15 % of them (4 words) and recognised 19 % of them (5 words). This is an interesting observation. Whilst there was clearly a relearning advantage for words previously known, Enda could still recall or recognise 35 % (9 words) of the pseudowords. This suggests that he may have good learning strategies. (See Chapter 4, Section 4.4.2.1 for information on creating the pseudowords).

Session 3

Enda was able to recall 16 % of the 25 real words and recognised 48 % of them. Overall, he was able to recall or recognise 64 % of the real words (compared to 100 % a month earlier directly after relearning).

A drop in performance is evident across the four-week period between sessions 2 and 3, suggesting a raising of threshold levels for items in this interval of disuse in keeping with the ATH.

He was unable to recall any of the pseudowords but was able to recognise 8 % of them (compared to 35 % for combined recall and recognition a month earlier directly after relearning)

Session 4

In this round, Enda was able to recall 72 % of the previously known words and could recognise 20 % of them. Overall directly after relearning he was able to recall or recognise 92 % of the words. Thirteen months earlier directly after relearning he had been able to recall or recognise 100 % of the words. After the second relearning, however, he was able to recall more words (18 words compared to 17 words) but could recognise fewer words (5 words compared to 8 words).

In this final round, he performed better on the pseudowords. He was able to recall 19 % of them and recognise 31 %. Overall, he was able to recall or recognise 50 % of the pseudowords (compared to 35 %) 13 months previously directly after relearning. This seems to support the Savings Paradigm and a relearning advantage even in the case of words to which exposure had been brief.

It is also important to note that at no time did Enda appear to suspect that the pseudowords were not real. This was the case for all of the participants.

5.1.2 Ellen

Table 11 Participant profile – Ellen

Ellen	
Age at start of learning	4 years and 1 month
Age at onset of attrition	18 years old
Length of learning	14 years
Proficiency level at onset of incubation	Leaving Certificate: Grade **B** in **higher level paper**
Exposure and learning context	School
Length of period of disuse	30 years
Rehearsal during incubation	No
Motivation	*
Other language/s at school and grade in Leaving Certificate	French: Awarded a grade A in the higher level paper
University Education	B.A. in English, Sociology & Political Science M.A. in Community Development
Age on leaving Ireland	34 years old
Country of residence at time of testing	Germany
Length of residence in Germany at first session of testing	13 years
Occupation at time of testing	Project Manager in International Communications
Languages actively used other than English	German: very good (B2)

*Unquantifiable but described in detail in Chapter 6, Section 6.5

5.1.2.1 Ellen's learning profile

Ellen appeared to have been a studious pupil at school based on her responses in the interview and her result in the final Leaving Certificate Irish exam.

5.1.2.1.1 Experiences in learning Irish

Ellen did not have any contact with Irish prior to starting official instruction and claims to have been very motivated in primary school where she found Irish classes engaging and enjoyable. During this time, she felt that Irish was a part of her identity.

However, this motivation seemed to ebb in lower and middle secondary school, increasing once again in the final year when the preparation for the oral section of the Leaving Certificate examination provided a motivating impetus for her. Up until this time, communication activities were rare and pupils were not given the necessary encouragement to converse in Irish, with classes typically being conducted through English. With respect to the scholarly nature of the focus of the Irish language syllabus, Ellen states that it was assumed that "you already had a mastery of academic language and that you were in the position then to apply that and the people [...] got somewhat lost in that process [...] in secondary school."

She expressed disappointment with respect to the content of the syllabus and how the language was taught stating that it was outdated and antiquated with too much emphasis on a literary canon that was unappealing to the youth of the day. Instead of Irish adding to her academic interests in secondary school, she found that her enjoyment of other subjects substituted her desire to engage with the Irish language.

Apart from a sense of obligation, the reasons for attaining a high proficiency in Irish were more instrumental and short-term oriented; the better one's grade, the more points one would get, thereby increasing one's chances of getting into the university course of one's choice.

What is also noteworthy is that even though Ellen's father was a fluent Irish speaker (he had grown up in a Gaeltacht area and was, therefore, bilingual), he did not speak Irish with his children. On the other hand, Ellen recounted being sent to Paris in the summer holidays as a teenager to improve her French but her parents did not avail of the Irish language knowledge that was at their disposal in their own home to support her in becoming fluent in that language. This is not necessarily to suggest a lack of interest in the Irish language but instead an explanation may be found in the status that the language had at the time in Irish society, where despite its symbolic role in enhancing national identity, it lacked "modernity". Indeed, Ellen states that her father would probably have spoken Irish at home had she and her siblings been receptive to it but due to the reasons just mentioned, the language presented no such appeal at that time.

Even though her father did not speak Irish to his children (Ellen has four siblings), she states that the attitude of her family to the Irish language was very positive. Her mother was always very open about acquiring languages being a French language teacher herself at a secondary school.

5.1.2.1.2 Experience in modern foreign language education

At secondary school Ellen also had French which she found much more interesting than Irish and was more motivated to learn. The French language was regarded as up-to-date and modern and it was "brought to life" through communication in class. Indeed, she argues that her rationality in learning was not induced by some long-term economic goal, in terms of job opportunities, for example, but arose from what she terms the "contemporary" nature of the content, with the focus and exercises representing everyday life. It was interesting for her to learn about the experiences of a typical teenager in another country and culture at that time.

In terms of the wider social surroundings and attitudes to the Irish language when she was growing up, she states that whilst there was no negativity towards the language itself, there was a degree of bitterness stemming from what was almost seen to be state coercion to take it as a subject at school as well as the nature of the content and the teaching approaches used. The issue of obligation is often raised in Ireland with respect to the Irish language but this is rarely discussed in the case of other compulsory subjects, suggesting that the benefits of Irish compared to these other subjects are not evident to learners. Leading on from possible considerations of long-term practicality or useability, it seems that the content of the Irish syllabus was not stimulating enough to compensate for what some might have viewed as shortcomings with respect to achieving greater career or economic aspirations. This seems to suggest that learning Irish based purely on the reason that it is a national language of notable historical and cultural significance is not enough to entice pupils who instead feel that they have to "suffer" for it, at least as was seemingly the case with respect to the syllabus that had been in place in the 1980s.

5.1.2.1.3 Attitudes, identity and Irish language use

Whilst Ellen has positive feelings towards the Irish language itself, she has "very negative" feelings towards how it was taught in the school. In terms of how her attitude has changed over the years, she states:

> To some degree I wouldn't say that I started with a negative attitude towards the Irish language. It was always positive. It always felt like a part of my identity even though it's gone. [...] but if anything it has got more positive in that I have a lot of regrets that I can't speak Irish, or use Irish anymore.

On leaving school Ellen did not have any more contact with the Irish language and did not make a conscious effort to retain knowledge of the language. With respect to the question of whether on return visits to Ireland, she is aware of any changes with respect to the frequency and quantity of Irish language use in her immediate social and familial environment and in Irish society in general compared to when she was growing up, she states that whilst she cannot judge whether Irish is actually used more often in the public sphere, she herself is certainly more aware of the language, referring, for example, to the news in Irish (on the radio or TV) even though she claims that she does not understand it. She thinks that the language has become "more attractive" and that people outside of the Gaeltacht regions who speak it are very proud of this fact.

While her brother completed a diploma in Irish as an adult, she has observed, on the other hand, that none of her nieces and nephews in Ireland appear to enjoying learning Irish and it seems to be their least favourite subject at school. This suggests that the situation appears not to have changed greatly since the participants in the present study were in school.

Overall, she feels regret that she has lost her knowledge of the language, stating, "I always had this notion in my head that it will be something that is just there and it's passive and it could be revived. I doubt that now, I think it's shame."

For her, the connection between her identity and the Irish language is somewhat complex. While she argues that knowledge of the language should not be used to separate communities, she, nonetheless, considers the language to be a significant part of her identity and even though she cannot speak it, she asserts that it characterises what is for her one of the essential and unique qualities of the Irish nation and how she personally has been shaped by that particular history. She does, however, believe that one can still be "Irish" without ever having learnt the language.

5.1.2.2 Test results – Ellen

5.1.2.2.1 Session 1

Table 12 Results of session 1 – Ellen

Session 1 (Ellen)				
Pretest	Recalled	Recognised	Recalled or recognised	Neither (forgotten)
200 words (All previously known)	67 (33.5%)	90 (45%)	157 (78.5%)	43 (21.5%)

Ellen was able to recall or recognise 78.5 % of the words, recognising more items than she could recall. Greater ease of recognition compared to recall after a period of disuse has also been found in other studies on retention (see Chapter 2, Section 2.1) These results are also noteworthy because in the interview she claims to have forgotten 95 %, if not all, of her knowledge. She performed well and despite this being a filtering phase, it is noteworthy how many lexical entries she could access after 30 years of disuse.

The category breakdown of the 157 items that she could recall or recognise was as follows:

Table 13 Session 1, Category breakdown of known items – Ellen

Ellen (157 items)	Concrete nouns	Abstract nouns	Verbs	Adjectives
Recalled (67)	21	16	11	19
Recognised (90)	23	22	25	20
Total (recalled or recognised)	44	38	36	39
% of total number of items (157)	28%	24%	23%	25%
% of total words tested in that category (50 respectively)	88%	76%	72%	78%

In the above breakdown, one sees a relatively even spread across categories, although like Cathy and Brendan, she performed slightly better on concrete nouns overall. She demonstrated the most trouble *recalling* the verbs but this was compensated for in recognition, with her performance being better on recognition across all word categories (albeit it very only very slightly in the case of concrete nouns and adjectives).

For the relearning session, the remaining 43 items that she had neither been able to recall nor recognise in the filtering phase were used along with 26 pseudowords.

The breakdown of these 43 items was as follows:

Table 14 Category breakdown of real items for relearning – Ellen

Concrete nouns	Abstract nouns	Verbs	Adjectives
6	12	14	11

5.1.2.2.2 Tests on saving: Results

Table 15 Session 2 (after relearning) – Ellen

Session 2 (Ellen)				
Relearning (69 items)	Recall	Recognition	Recalled or recognised	Neither
43 real, previously known words	21 (49%)	5 (12%)	26 (61%)	17 (39%)
26 pseudo-words	1 (4%)	0	1 (4 %)	25 (96%)

Table 16 Session 3 (no relearning) – Ellen

Session 3 (Ellen)				
No relearning	Recall	Recognition	Recalled or recognised	Neither
43 real, previously known words	19 (44%)	9 (21%)	28 (65%)	15 (35%)
26 pseudo-words	1 (4%)	1 (4%)	2 (8%)	24 (92%)

Table 17 Session 4 (after relearning) – Ellen

Session 4 (Ellen)				
Relearning (69 items)	Recall	Recognition	Recalled or recognised	Neither
43 real, previously known words	28 (65%)	5 (12%)	33 (77%)	10 (23%)
26 pseudo-words	1 (4%)	1 (4%)	2 (8%)	24 (92%)

Session 2

There was no second pretest for Ellen, Brendan and Cathy since enough words were left over after testing in session 1 for relearning.

Ellen could recall 49 % of the 43 words previously known and recognise 12 % of them. It is noteworthy that unlike in the pretest of session 1 in which she performed better on recognition, in session 2 after relearning she was able to directly recall more items, suggesting that re-exposure in the form of relearning had significantly lowered threshold levels considering that these were items that she did not know in the pretest.

In total she could recall or recognise 61 % of the previously known words. In contrast she could only recall one of the pseudowords, which constituted only 4 % of the total number of pseudowords. This suggests a strong relearning advantage for previously known words.

Session 3

Ellen could recall 44 % of the words and recognise 21 %, overall recalling or recognising 65 % of the words. On a closer examination of the 19 words that she could recall in session 3, in which no relearning took place, 16 overlapped with recalled words in session 2. Of the remaining three words that she could recall in round 3, she had been able to recognise two of them in session 2 and had neither been able to recall nor recognise the remaining one. This is interesting because it suggests a drop in threshold levels for these items despite there not having been further exposure in the interval between testing sessions.

On the other hand, whilst she recalled slightly less compared to session 2, suggesting that overall threshold levels had risen again, she could recognise more in session 3, resulting in a slight increase overall (i.e., recall and recognition combined) in session 3 compared to session 2. This is noteworthy considering that no relearning took place in this session so a drop in performance relative to session 2 would have been expected. On a closer examination of the nine items recognised in session 3, three of these were items she had also recognised in session 2. Of the remaining six items, she had been able to recall five of these in session 2 (suggesting an increase in threshold levels in the interlude between session 2 and 3 for these items) and neither recall nor recognise the remaining item (suggesting a drop in the threshold level for this item).

With respect to the pseudowords, as well as recalling one, she was also able to recognise one, which she had not been able to do in session 2 after relearning.

Session 4

One year later another relearning took place of the same words. Ellen had had no contact with the Irish language in the year between testing sessions 3 and 4.

There was once again a relearning advantage for words previously known with an improvement on performance 13 months earlier directly after relearning, suggesting that another exposure, even 12 months later reinforces knowledge previously known.

Ellen was able to recall 65 % of the words compared to 49 % after relearning in session 2. The fact that she was able to recall more words suggests that the threshold level had been lowered due to a triggering or reactivation of these items. Of these 28 words, 20 words overlapped with words she could recall after relearning in session 2. Of the remaining eight words, she had been able to recognise three of them in session 2, and the remaining five were words she had neither be able to recall nor recognize after relearning in session 2. Once again it appears that the activation threshold of these items dropped through re-exposure.

On the other hand, the remaining recalled word from session 2 that did not overlap with the words that she could recall in round 4 after relearning was the verb *siúil (to walk),* although she did recognise it in round 4, suggesting a rise in the threshold level for this item. She was able to recognise 12 % of the words. This was on par with recognition in round 2. Overall, she was able to recall or recognise 77 % of the real words compared to 61 % a year earlier directly after relearning. This was due to an increase in recall suggesting that threshold levels had dropped due to re-exposure making items more readily accessible.

With respect to the pseudowords, performance was once again much weaker, reinforcing the observation that there appears to be a clear relearning advantage for words she had previously known. She was able to recall one of the pseudowords and recognise one of them.

5.1.2.2.3 Category performance

Table 18 Session 2, Category performance – Ellen

Ellen (Session 2)	Concrete nouns	Abstract nouns	Verbs	Adjectives
Recalled	4	5	7	5
Recognised	1	2	2	0
Total (Recognised or Recalled)	5	7	9	5
Neither	1	5	5	6

Table 19 Session 3, Category performance – Ellen

Ellen (Session 3)	Concrete nouns	Abstract nouns	Verbs	Adjectives
Recalled	4	5	7	3
Recognised	1	4	2	2
Total (Recognised or Recalled)	5	9	9	5
Neither	1	3	5	6

Table 20 Session 4, Category performance – Ellen

Ellen (Session 4)	Concrete nouns	Abstract nouns	Verbs	Adjectives
Recalled	5	6	10	7
Recognised	1	2	2	0
Total (Recognised or Recalled)	6	8	12	7
Neither	0	4	2	4

Table 21 Overview of word categories, Sessions 2–4 – Ellen

Ellen (Sessions 2-4)	Total (items recalled or recognised)	Concrete nouns	Abstract nouns	Verbs	Adjectives
Session 2 (after relearning)	26	5	7	9	5
Session 3	28	5	9	9	5
Session 4 (after relearning)	33	6	8	12	7

It is important to point out that a direct comparison across the categories cannot be made considering that there was an unequal number of items per category. Instead, this illustration is purely to show development across sessions *within each category* itself to see if any idiosyncrasies emerge. No noteworthy patterns emerge across sessions with respect to each individual category. It is interesting to note, however, that in comparing sessions 2 and 4, improvement spans all categories.

Table 22 Participant profile – Brendan

Brendan	
Age at start of learning	5 years old
Age at onset of attrition	18 years old
Length of learning	13 years
Proficiency level at onset of incubation	Leaving Certificate: Grade **B** in **lower level paper**
Exposure and learning context	School
Length of period of disuse	32 years
Rehearsal during incubation	No
Motivation	*
Other language/s at school and grade in Leaving Certificate	French: Awarded a grade B in the higher level paper Latin: Awarded a grade C in the higher level paper
University Education	B.A. in German and Linguistics Diploma in Business Studies
Age on leaving Ireland	21 years old
Country of residence at time of testing	Germany
Length of residence in Germany at first session of testing	29 years
Occupation at time of testing	IT consultant
Languages actively used other than English	German: native-like, CEFR: C2

*Unquantifiable but described in detail in Chapter 6, Section 6.5

5.1.3 Brendan

5.1.3.1 Brendan's learning profile

School was Brendan's primary source of Irish language input since none of his family spoke Irish at home. His mother was from England so she did not have any knowledge of the language and his father who was Irish did not speak any Irish with his children, although he did sometimes listen to the news in Irish. However, Brendan did not know if his father could actually speak the language apart from using the odd greeting. His family had also not been particularly interested in the Irish language but it was not that they didn't care about it either. For instance, one of his sisters went on to become a primary school teacher for which knowledge of the Irish language is required. When he was growing up, he did not know anybody who spoke Irish or showed any interest in speaking it in his wider environment.

5.1.3.1.1 Experiences in learning Irish

While he enjoyed Irish more in primary school where he had lessons more frequently than in secondary education, he regrets that not enough emphasis was put on speaking and communication. This was the only time, however, when he felt that it was close to being a second language for him although he states that he never got to that level where he used it outside of the classroom. In secondary school it became "just another subject". He does not recall Irish being used beyond the lessons and in the Irish lesson itself both English and Irish were used, although the ratio depended very much on the teacher.

At secondary school he found Irish both extremely difficult and boring. He stresses how tedious and demanding he perceived the syllabus; very "academic" with a heavy emphasis on complex literary texts and accompanying analysis, material that he described as alienating with no relevance for his teenage life, and leaving no room for enjoyment or developing a love for the language. Neither was the opportunity to actively use the language made available to him and his peers at school. For him, the experience consisted mainly of memorising interpretations of literature and the accompanying vocabulary just so that this could be repeated in exams.

He is of the opinion that the opportunity to revive Irish through the education system was wasted and feels frustrated that neither he nor his friends developed any cultural identification with it. He believes that had people enjoyed learning Irish at school, they may also have spoken it outside and beyond education. When asked if he took more from the Irish classes than just language, he states that what he learned was how a language should not be taught, "in everything that is bad in life you learn that there is a way not to do it and that's what I took away from my experience learning Irish."

He also believed the standards that were expected in Irish were very high and maintains that it was difficult to achieve a grade A in the higher-level paper in Irish in the Leaving Certificate. He sees this as discouraging and despite hard work and diligence he never felt duly rewarded for his efforts. Because of these challenges the majority of pupils in his school opted to take the lower-level course so that they could invest more time in other subjects and gain more points there. This, he argued, was a strategic move and not necessarily indicative of weaker academic ability.

Brendan argues that it was the combination of alienating content and almost unachievable academic expectations that led to pupils at the time becoming indifferent with respect to the language, learning it only for short-term gains. His motivation was based on compulsion and knowing that he would not use Irish once official instruction ended, learning it was purely "a means to an end".

5.1.3.1.2 Experience in modern foreign language education

Brendan states that he enjoyed learning all other languages more than Irish and even went on to study languages at university. Whilst he found the syllabi of these other languages a lot more interesting, he also considered them easier, referring in particular to the complexity of Irish grammar. There was also more communication and interaction in these classes compared to Irish. He was additionally very curious to learn about the societies associated with modern foreign languages. For example, he states that he closely identified with France and its culture, which he experienced as fresh, exciting and modern, offering him "something to look forward to". His French teacher was clearly passionate about the language and very interested in French culture, instilling a curiosity and willingness to learn among her pupils.

In comparison he felt that the Irish language teachers were not entirely motivated in their teaching and there seemed to be an aura of apathy and disinterest compared to what he experienced among his French teachers. Naturally this is a subjective observation and one that may not be generalised. It could have been that the Irish teachers themselves were frustrated with the syllabus and system and potentially may have been just as motivated as their colleagues who taught foreign languages had they felt more inspired by the content they had to teach.

5.1.3.1.3 Attitudes, identity and Irish language use

Whilst he was relieved to have finished Irish once he left school, his attitude to the language has changed somewhat over the years and he now feels a certain regret that he cannot communicate in it. On the other hand, whilst he is proud of his Irish heritage and identifies very much with many aspects of Irish history and culture, the Irish language does not constitute a key part of this for him. For example, at one point he made a very brief effort to refresh his knowledge with the help of a CD but abandoned the endeavour very quickly on realising his interest was not great enough to match the necessary effort.

Apart from his sister and brother, no other family members or friends have actively striven to maintain knowledge of the language and indeed apart from the bilingual public signs in Ireland, he is not aware of the presence of the language or of any change in the quantity or frequency of Irish language use in his immediate social environment when he visits Ireland.

5.1.3.2 Test results – Brendan

5.1.3.2.1 Session 1

Table 23 Results of session 1 – Brendan

Session 1 (Brendan)				
Pretest	Recalled	Recognised	Recalled or recognised	Neither (forgotten)
200 words (All previously known)	43 (21.5%)	42 (21%)	85 (42.5%)	115 (57.5%)

Of the 200 previously known words, he performed just as well in recall as in recognition. Overall, he could recall or recognise 42.5 % of the previously known words, with the remaining 57.5 % (115 words) being neither recognised nor recalled.

The category breakdown of the 85 items that he could recall or recognise was as follows:

Table 24 Session 1, Category breakdown of known items – Brendan

Brendan (85 items)	Concrete nouns	Abstract nouns	Verbs	Adjectives
Recalled (43)	22	4	6	11
Recognised (42)	9	19	5	9
Total (recalled or recognised)	31	23	11	20
% of total number of items (85)	36.5%	27%	13%	23.5%
% of total words tested in that category (50 respectively)	62%	46%	22%	40%

In the above table one sees that concrete nouns followed by abstract nouns constituted the highest percentage and verbs the lowest. As discussed in Section 5.2.5.1, this may stem from the fact that Irish is a noun-centred language.

Since Brendan did not know 115 items in testing in the pretesting phase (session 1), 12 concrete nouns, 12 abstract nouns, 13 verbs and 13 adjectives were chosen from this pool of leftover items giving a total of 50 words for relearning. The division of 50 items meant that the split could not be completely even across categories; since there was an excess of nouns overall (i.e., in combining concrete and abstract nouns), the adjective and the verb categories were each allocated an extra item.

5.1.3.2.2 Tests on savings: Results

Table 25 Session 2 (after relearning) – Brendan

Session 2 (Brendan)				
Relearning (76 items)	Recall	Recognition	Recalled or recognised	Neither
50 real, previously known words	30 (60%)	9 (18%)	39 (78%)	11 (22%)
26 pseudo-words	0	0	0	26 (100%)

Table 26 Session 3 (no relearning) – Brendan

Session 3 (Brendan)				
No relearning	Recall	Recognition	Recalled or recognised	Neither
50 real, previously known words	9 (18%)	16 (32%)	25 (50%)	25 (50%)
26 pseudo-words	0	0	0	26 (100%)

Table 27 Session 4 (after relearning) – Brendan

Session 4 (Brendan)				
Relearning (76 items)	Recall	Recognition	Recalled or recognised	Neither
50 real, previously known words	29 (58%)	8 (16%)	37 (74%)	13 (26%)
26 pseudo-words	1 (4%)	1 (4%)	2 (8%)	24 (92%)

Session 2

After relearning in round 2, Brendan was able to recall 60 % and recognise 18 % of the previously known words. Overall, he was able to recall or recognise 78 % of the real words but none of the pseudowords. A clear relearning advantage for previously known words is apparent in keeping with the Savings Paradigm.

Session 3

One month later (without any further relearning), Brendan was able to recall 18 % of the previously known words and recognise 32 % of them. Overall, he was able to recall or recognise 50 % of the real words (compared to 78 % a month earlier directly after relearning).

It is interesting to note that whilst his performance in production dropped significantly (from 30 words to 9 words), of the 21 remaining words that he could no longer recall in session 3, he could still recognise almost half of them (i.e., 10 words). This leaves 11 words that were no longer accessible four weeks later. Threshold levels for these items had clearly risen in the four weeks of disuse.

Contrary to findings in session 1 where he performed equally well on recall and recognition, these findings support observations made in other studies that recognition proves easier than recall after a period of disuse (see Chapter 2, Section 2.1).

As in session 2, he was not able to recall or recognise any of the pseudowords offering further support for the Savings Paradigm.

Session 4

One year later another round of relearning followed by testing took place.

Interestingly, there is very little difference in session 4 compared to his performance directly after relearning in session 2, which had taken place 13 months previously. Brendan was able to recall 58 % of the words that he had previously known compared to 60 % in session 2 after relearning (i.e., one word less). In session 4 he was able to recognise 16 % of the items compared to 18 % in round 2 after relearning (i.e., one word less). Overall, in session 4 he was able to recall or recognise 74 % of words that he had previously known compared to an overall score of 78 % in session 2 after relearning. Interestingly in session 4 after relearning there was a slight improvement visible with respect to the pseudowords, whereby he could recall one of them as well as recognise another one.

On comparing the words that he could recall and recognise after relearning sessions 2 and 4, 23 of the 29 words that he could recall in session 4 were the same words that he could recall in session 2. Of the remaining six words that he could recall in session 4, four of these were words he had recognised after relearning in session 2 and the remaining two were words he had neither been able to recall nor recognise, i.e, threshold levels fell with re-exposure in the case of these six remaining items.

On the other hand, of the seven remaining words from the 30 that he could recall after relearning in session 2, he could recognise four of these after relearning in session 4, leaving three items that he could neither recall nor recognise. In this case threshold levels rose for these seven items.

Comparing sessions 2 and 4, the threshold level fell for six items due to re-exposure (making them more easily retrievable as demonstrated in their recall in session 4). However, the threshold level rose for seven items (items that were recalled in session 2 were no longer available for recall in session 4). It must be noted, however, that there was an interval of 12 months between session 4 and the previous exposure in session 3 so that it can be expected that threshold levels might rise for some items; the degree might be determined by how consolidated information pertaining to a particular item had been in long-term memory, which would explain variation across individual words. Nonetheless, despite this variation in activation patterns for individual items, overall, one sees a rise in threshold levels in the four weeks between sessions 2 and 3 but a further drop again on re-exposure in session 4 suggesting that re-exposure re-activates items and in doing so threshold levels tend to drop.

5.1.3.2.3 Category performance

Since Brendan did not know 115 items in testing in the pretesting phase (session 1), 12 concrete nouns, 12 abstract nouns, 13 verbs and 13 adjectives were chosen from this pool of leftover items giving a total of 50 words for relearning. The division of 50 items meant that the split could not be completely even across categories; since there was an excess of nouns overall (i.e., in combining concrete and abstract nouns), the adjective and the verb categories were each allocated an extra item.

Table 28 Category breakdown of real items for relearning – Brendan

Concrete nouns	Abstract nouns	Verbs	Adjectives
12	12	13	13

Table 29 Session 2, Category performance – Brendan

Brendan (Session 2)	Concrete nouns	Abstract nouns	Verbs	Adjectives
Recalled	6	9	10	5
Recognised	4	1	0	4
Total (Recognised or Recalled)	10	10	10	9
Neither	2	2	3	4

Table 30 Session 3, Category performance – Brendan

Brendan (Session 3)	Concrete nouns	Abstract nouns	Verbs	Adjectives
Recalled	0	3	4	2
Recognised	5	1	5	5
Total (Recognised or Recalled)	5	4	9	7
Neither	7	8	4	6

Table 31 Session 4, Category performance – Brendan

Brendan (Session 4)	Concrete nouns	Abstract nouns	Verbs	Adjectives
Recalled	7	8	6	8
Recognised	1	1	3	3
Total (Recognised or Recalled)	8	9	9	11
Neither	4	3	4	2

Table 32 Overview of word categories, Sessions 2–4 – Brendan

Brendan (Sessions 2-4)	Total (items recalled or recognised)	Concrete nouns	Abstract nouns	Verbs	Adjectives
Session 2 (after relearning)	39	10	10	10	9
Session 3	25	5	4	9	7
Session 4 (after relearning)	37	8	9	9	11

No noteworthy patterns emerge across sessions with respect to each individual category, apart from the observation that verbs and adjectives emerged as somewhat more robust on relearning compared to nouns, i.e., performance on nouns drops more between sessions 2 and 3 than it does for verbs and adjectives.

5.1.4 Cathy

Table 33 Participant profile – Cathy

Cathy	
Age at start of learning	10 years and 2 months
Age at onset of attrition	18 years old
Length of learning	8 years
Proficiency level at onset of incubation	Leaving Certificate: Grade **D** in **higher level paper**
Exposure and learning context	School
Length of period of disuse	28 years
Rehearsal during incubation	No
Motivation	*
Other language/s at school and grade in Leaving Certificate	German: Awarded a grade A in the higher level paper. French: Awarded a C in the higher level paper
University Education	B.A. in German and Greek and Roman Civilisation M.A. in Film Studies
Age on leaving Ireland	22 years old but moved back on two occasions for a period of 2 and 4 years respectively.
Country of residence at time of testing	Germany
Length of residence in Germany at first session of testing	17 years
Occupation at time of testing	Works in film production
Languages actively used other than English	German: native-like, CEFR: C2

*Unquantifiable but described in detail in Chapter 6, Section 6.5

5.1.4.1 Cathy's language learning profile

Cathy did not start learning Irish until the age of ten when her family moved to the Republic of Ireland from Northern Ireland.

5.1.4.1.1 Experiences in learning Irish

Like her siblings, Cathy's main contact with Irish on moving to the Republic of Ireland was within the school context. Although her father had also learned Irish at school, he did not use it with his children. Cathy did not know whether this was because he no longer had any knowledge of it or he lacked the motivation to speak it. Her mother grew up in Northern Ireland and had, therefore, never learned Irish. Consequently, Cathy did not receive any Irish language input at home. There was, nonetheless, strong support and encouragement with respect to learning in her home environment which might also be a decisive factor in explaining Cathy's decision to take the high-level paper in the final state examination at the end of school, which she successfully completed. On the other hand, the fact that she had missed the first six years of Irish classes at school might explain why she did not perform as well in Irish as she did in her other subjects considering that she appears to have been a studious and conscientious pupil.

Beyond the actual Irish class itself, Cathy recalls Irish being used at school only in connection with religion whereby religious songs and prayers were recited verbatim with little understanding of the texts themselves.

She had, along with her peers, Irish classes more frequently in primary school compared to secondary school and whether or not instruction took place through Irish was teacher-dependent. On starting to learn Irish she states for instance that her teacher mainly spoke English to her and in secondary school she recalls only one Irish teacher in particular that resolutely spoke exclusively in Irish.

Cathy did not enjoy learning Irish at school and felt very intimidated when she started Irish class being the only one among her peers who had no knowledge of the language. She argues that this was not helped by some of the approaches, whilst well-intentioned, that were used in teaching. She recollects one game which served only to demotivate instead of encourage her.

> Our teacher often did things where we would stand – people would stand around the class at the wall and then if you answered a question correctly, you would sit down and I was usually the last person standing and in the end she would – the teacher would always help me a little bit and give me a very big clue to help me get the right answer, so that I could sit down but I was always the worst in class and always the last one standing at the wall.

However, as time progressed and she made more advances in the language, she realised that she was not necessarily weaker than other pupils in Irish arguing that there was an overall disinterest and apathy with respect to the language among her peers who tended to view it as something they had to endure. She argues that the language was presented in such an uninteresting way that it dampened any enthusiasm pupils had with respect to a readiness and willingness to learn and use it. In primary school, she recalls the content as being rather childish and simple and in secondary school "it was just very boring, very old fashioned, nothing that seemed relevant to a teenager or to a modern city life [...] very ancient texts or old texts [...] things that did not seem interesting to me at the time." She claims that the experience of learning Irish in school was demotivating as well as uninspiring. This also appears to have had long-term repercussions in the sense that she didn't have any desire to consciously preserve the knowledge of the language that she had acquired on leaving school. Instead, she states that she was very happy that she didn't have to be concerned about Irish or "worry about it". Her motivation in learning Irish was purely instrumental as she states, "well, I had to learn to it, it was compulsory and not my choice."

5.1.4.1.2 Experience in modern foreign language education

This is in contrast to how she feels about other subjects. She enjoyed learning in general, viewing other subjects favourably, and was very positive about school overall. While the Irish syllabus was predominantly focused on literature and not on communication, German, for example, was conveyed as a "living language" with the focus being on interpersonal interaction and verbal expression. She highlights how interesting and stimulating both the content of the syllabus and the teaching approaches were in contrast to Irish arguing that had Irish been viewed as a modern foreign language and not as a language that was supposed to constitute some form of an L1, perhaps more appropriate content and teaching methodologies might have been availed of. With respect to German, on the other hand, she considered the content "exciting" and "modern", whereby she and her peers felt that they could relate to it claiming that it was "much more in tune with the lives of young people."

She also very much liked her German teacher, whom she had throughout secondary school and who was well-structured, supportive and inspirational, transmitting her own love of German to her pupils whilst simultaneously engaging them in actively using the language with confidence. Cathy also states that while the syllabus in all subjects will inevitably comprise tedious elements, a committed and inspiring teacher can compensate for such components. The fact too that the initial decision to learn German was a matter of choice and not obligation instilled a certain intrinsic motivation in her.

5.1.4.1.3 Attitudes, identity and Irish language use

Cathy's attitude towards Irish, however, has changed over the years, albeit more with respect to a certain nostalgia that she has come to develop and associate with it. She continues to feel that the Irish language is an important part of her identity at a deep, emotional level and regrets that she is unable to actively use it. When she visits Ireland, she spends most of her time in areas where the language is not spoken and states that the Irish language plays no part in her social life nor is it used by any of her friends. Neither is she aware of any changes with respect to the frequency and quantity of Irish language used in Irish society in general compared to when she was growing up. Whilst the children of her cousins learn Irish in school, there are no other people in her environment who are actively using the language and even these children do not appear to use it outside of class.

With respect to Irish identity, she states that she cannot strictly define what it means to be "Irish" but argues, nonetheless, that one can be truly Irish without having ever learned the language.

She now claims to have forgotten all of the knowledge that she had acquired of the language and argues that had perhaps the language been taught differently at school she might have retained active knowledge of it.

5.1.4.2 Test results

5.1.4.2.1 Session 1

Table 34 Results of session 1 – Cathy

Session 1 (Cathy)				
Pretest	Recall	Recognition	Recalled or recognised	Neither (forgotten)
200 words (All previously known)	17 (8.5%)	36 (18%)	53 (26.5%)	147 (73.5%)

Of the 200 words tested for filtering for the relearning phase, Cathy could recall 8.5 % of the words and recognise 18 % of the words. Greater ease of recognition compared to recall after a period of disuse has also been found in other studies on retention (see Chapter 2, Section 2.1). Overall, she was able to recall or recognise just over one quarter (26.5 %) of the words.

The category breakdown of the 53 items that she could recall or recognise was as follows:

Table 35 Session 1, Category breakdown of known items – Cathy

Cathy (53 items)	Concrete nouns	Abstract nouns	Verbs	Adjectives
Recalled (17)	10	3	1	3
Recognised (36)	12	9	8	7
Total (recalled or recognised)	22	12	9	10
% of total number of items (53)	41%	23%	17%	19%
% of total words tested in that category (50 respectively)	44%	24%	18%	20%

In the above breakdown one sees that Cathy performed best on concrete nouns; of the 53 words that she was able to recall or recognise, 22 of them were concrete nouns. Whilst her weakest performance was on verbs, the difference was slight compared to adjectives. She performed better on recognition across all four categories: this was especially evident in the case of abstract nouns, verbs and adjectives.

Since Cathy did not know 147 items in the pretesting phase (session 1), 12 concrete nouns, 12 abstract nouns, 13 verbs and 13 adjectives were chosen from this pool of leftover items giving a total of 50 words for relearning.

As in Brendan's case, the division of 50 items meant that the split could not be completely even across categories; since there was an excess of nouns overall (i.e., in combining concrete and abstract nouns), the adjective and the verb categories were each allocated an extra item.

5.1.4.2.2 Tests on savings: Results

Table 36 Session 2 (after relearning) – Cathy

Session 2 (Cathy)				
Relearning (76 items)	Recall	Recognition	Recalled or recognised	Neither
50 real, previously known words	27 (54%)	6 (12%)	33 (66%)	17 (34%)
26 pseudo-words	0	2 (8%)	2 (8%)	24 (92%)

Table 37 Session 3 (no relearning) – Cathy

Session 3 (Cathy)				
No relearning	Recall	Recognition	Recalled or recognised	Neither
50 real, previously known words	5 (10%)	17 (34%)	22 (44%)	28 (56%)
26 pseudo-words	0	2 (8%)	2(8%)	24 (92%)

Table 38 Session 4 (after relearning) – Cathy

Session 4 (Cathy)				
Relearning (76 items)	Recall	Recognition	Recalled or recognised	Neither
50 real, previously known words	27 (54%)	10 (20%)	37 (74%)	13 (26%)
26 pseudo-words	0	2 (8%)	2(8%)	24 (92%)

Session 2

After relearning, Cathy could recall 54 % of the words she had previously known. She was able to recognise 12 % of the words, overall being able to recall or recognise 66 % of the words she had previously known but had apparently forgotten. She could not recall any of the pseudowords but could recognise two of them. Overall, a clear relearning advantage is apparent in the case of words once known but seemingly forgotten compared to words never known.

Session 3

A considerable drop in recall is evident in this session compared to session 2 a month earlier directly after relearning. In this third session where no relearning took place, Cathy was able to recall 10 % (compared to 54 % in session 2) of the words previously known years earlier as a pupil at school.

She performed better in recognition in this session, recognising 34 % of the words compared to 12 % in round 2. Overall, she was able to recall or recognise 44 % of the words (compared to 66 % in round 2).

Of the five words that she could recall in session 3, four of them she had also been able to recall in session 2. She had recognised the remaining word in session 2.

On the other hand, looking at the 27 items she had been able to recall in session 2, threshold levels rose for 23 of these items in the one-month interval between sessions 2 and 3. In session 3, she was, nonetheless, able to recognise 11 of these items with the remaining 12 items being neither recalled nor recognised.

She was not able to recall any of the pseudowords but could recognise two of them, once more showing an advantage for words previously known compared to words she had never learned at school.

Of the 17 words that she could recognise in session 3, four of these overlapped with recognised words in session 2, 11 of these were items that she had been able to recall in session 2, and two were items she had neither been able to recognise nor recall. Comparing session 2 to 3, threshold levels remained the same for four of the items, they rose for 11 of the items (from recall to recognition) but dropped for two of them (from "neither" to recognition).

In looking at the individual data for items across sessions 2 and 3, threshold levels rose overall in this interval. This rise in threshold levels is expected in a period of disuse in keeping with the ATH (see Chapter 2, Section 2.2.1).

Session 4

One sees that even though 13 months had passed since initial relearning in session 2, another episode of exposure appears to reinforce learning. Cathy improved from 66 % in session 2 to 74 % overall in session 4 in terms of recalling or recognising words. More regular exposure in this time might possibly have reaped even better results. This suggests that a first relearning activated knowledge and a second exposure lowered the threshold, even though there was a relatively long spacing in between. Cathy did not have any contact with the Irish language in the 12-month period between testing in session 3 and 4. Looking at the results in round 4, recognition improved whilst recall accumulatively stayed the same as that after relearning in session 2.

She was unable to recall any of the pseudowords but could recognise two of them.

With respect to the 27 real words that she was able to recall in session 4, 24 were the same as those she could recall after relearning in round 2.

Of the remaining three words that she could recall in session 4, one of these she had been able to recognise in session 2 and the remaining two she had not known in session 2 after relearning, suggesting therefore that threshold levels for these three items had dropped after a second relearning.

On the other hand, on analysing the 27 recalled items in session 2, one of these she recognised in session 4, and the other two she did not know at all. In these three cases the threshold levels rose in session 4 despite relearning.

Of the 10 items that she was able to recognise in session 4, five overlapped with recognised items in session 2. Of the remaining five words, she had been able to recall one of them and the other four she did not know after relearning in session 2, i.e., the threshold level had risen for one item (recall to recognition) but fallen for four ("neither" to recognition).

Whilst overall threshold levels dropped so that more items were known after relearning in session 4, there was, nonetheless, some individual variation among items, with items not necessarily overlapping in terms of recall or recognition across sessions demonstrating, in keeping the ATH, that threshold levels are dynamic. Although for the most part relearning promoted a reactivation to the degree that threshold levels dropped, this was not always the case for each individual item. This may be due to how well information is consolidated in memory (see Chapter 2, Section 2.2.4).

5.1.4.2.3 Category performance

Table 39 Category breakdown of real items for relearning – Cathy

Concrete nouns	Abstract nouns	Verbs	Adjectives
12	12	13	13

Table 40 Session 2, Category performance – Cathy

Cathy (Session 2)	Concrete nouns	Abstract nouns	Verbs	Adjectives
Recalled	10	8	5	4
Recognised	2	1	1	2
Total (Recognised or Recalled)	12	9	6	6
Neither	0	3	7	7

Table 41 Session 3, Category performance – Cathy

Cathy (Session 3)	Concrete nouns	Abstract nouns	Verbs	Adjectives
Recalled	0	4	1	0
Recognised	8	2	2	5
Total (Recognised or Recalled)	8	6	3	5
Neither	4	6	10	8

Table 42 Session 4, Category performance – Cathy

Cathy (Session 4)	Concrete nouns	Abstract nouns	Verbs	Adjectives
Recalled	11	8	5	3
Recognised	1	3	1	5
Total (Recognised or Recalled)	12	11	6	8
Neither	0	1	7	5

Table 43 Overview of word categories, Sessions 2–4 – Cathy

Cathy (Sessions 2-4)	Total (items recalled or recognised)	Concrete nouns	Abstract nouns	Verbs	Adjectives
Session 2 (after relearning)	33	12	9	6	6
Session 3	22	8	6	3	5
Session 4 (after relearning)	37	12	11	6	8

No noteworthy patterns emerge across sessions with respect to each individual category. The rise in threshold levels expected in the interval between session 2 and session 3 spans all categories as seen in the figures for session 3. These fall again due to re-exposure in the form of relearning in session 4 with greater ease of retrieval evident in all categories in this session.

5. 2 Overview and analysis of overall results

5.2.1 Session 1

Table 44 Comparison of participant performance – Session 1

Session 1 - Enda				
Pretest	Recalled	Recognised	Recalled or recognised	Neither (forgotten)
200 words (All previously known)	145 (72.5%)	49 (24.5%)	194 (97%)	6 (3%)
Session 1 - Ellen				
Pretest	Recalled	Recognised	Recalled or recognised	Neither (forgotten)
200 words (All previously known)	67 (33.5%)	90 (45%)	157 (78.5%)	43 (21.5%)
Session 1 - Brendan				
Pretest	Recalled	Recognised	Recalled or recognised	Neither (forgotten)
200 words (All previously known)	43 (21.5%)	42 (21%)	85 (42.5%)	115 (57.5%)
Session 1 - Cathy				
Pretest	Recall	Recognition	Recalled or recognised	Neither (forgotten)
200 words (All previously known)	17 (8.5%)	36 (18%)	53 (26.5%)	147 (73.5%)

The participants have had no contact with the language since leaving school, except for Enda who studied it in his first year at university and made intermittent attempts to retain knowledge in the initial few years after onset of attrition.

Despite an apparent frustration on the part of the participants with respect to the content of the curriculum, as well as the teaching styles and approaches, knowledge of the language has not been completely erased.

Contrary to observations made in other studies, *an obvious pattern of candidates performing better on recognition over recall is not readily apparent here.* Only two of the participants (Ellen and Cathy) performed better on recognition but only by approximately 10 %. Brendan performed almost equally well in both with a very slight increase in recall. Enda, on the other hand, could recall the majority of the items. However, in the second filtering test (Session 2a) that Enda took where less frequent words were used, he performed better on recognition (of the 69 real words that he was tested on, he could recall 25 % of them and recognise 39 %). What is interesting also is that phonological knowledge appeared to be completely intact. None of the participants had problems pronouncing either the real words or expressing the intended pronunciation of the pseudowords. This was observed by the main researcher who was present during testing. This is in keeping with findings referred to Bardovi-Harlig & Stringer (2019: 12–13) where it was found that phonology proved rather robust despite periods of disuse.

The *nature of the exposure* to Irish experienced by the participants in school might explain the results in session 1 in which recall did not necessarily lag behind recognition. Reasons for the relative strength of recall will be discussed in Chapter 6, Section 6.4, in terms of the role of literacy in knowledge consolidation.

The observations were conducted as case studies so each result is looked at in its own right. However, the difference in performance across the participants in session one warrants closer examination. For example, why would one candidate, i.e., Enda, perform so strongly, recalling or recognising almost all of the words and why would another, i.e., Cathy, only accomplish success in just one quarter of the items, i.e., *why does one individual seemingly retain a high amount of knowledge and another much less?* Considering that many different factors play a role and their influence is constantly changing and in flux as proposed by Complex Dynamic Systems Theory, striving to isolate one decisive factor as an encapsulated entity might lead to the overlooking of some important key elements. There are certain differences among the profiles of the participants which might possibly play a role in explaining differences in levels of retention after such a long period of incubation. *These differences are discussed in detail in Chapter 6 against the backdrop of possible explanatory factors typically put forward in the literature with respect to language attrition and levels of retention.*

5.2.2 Session 2 (after relearning)

Table 45 Comparison of participant performance after relearning – Session 2

Session 2b - Enda				
Relearning (51 items)	Recall	Recognition	Recalled or recognised	Neither
25 real, previously known words	17 (68%)	8 (32%)	25 (100%)	0
26 pseudo-word	4 (15%)	5 (19%)	9 (35%)	17 (65%)
Session 2 - Ellen				
Relearning (69 items)	Recall	Recognition	Recalled or recognised	Neither
43 real, previously known words	21 (49%)	5 (12%)	26 (61%)	17 (39%)
26 pseudo-words	1 (4%)	0	1 (4 %)	25 (96%)
Session 2 - Brendan				
Relearning (76 items)	Recall	Recognition	Recalled or recognised	Neither
50 real, previously known words	30 (60%)	9 (18%)	39 (78%)	11 (22%)
26 pseudo-words	0	0	0	26 (100%)
Session 2 - Cathy				
Relearning (76 items)	Recall	Recognition	Recalled or recognised	Neither
50 real, previously known words	27 (54%)	6 (12%)	33 (66%)	17 (34%)
26 pseudo-words	0	2 (8%)	2 (8%)	24 (92%)

With respect to the real, previously known words, considering that each of the participants not only had different *items* in the relearning session but also, apart from Brendan and Cathy, a different *number* of words, the candidates cannot be compared in terms of relative performance since it is anticipated that those with more items to relearn would have had a greater cognitive load to process. For example, whilst it might appear at first glance that Enda performed strongest, he only had 25 real items to relearn compared to 43 for Ellen and 50 for Brendan and Cathy respectively, as illustrated here:

Table 46 Total number of items per participant

Participant	Total items	Real items	Pseudowords
Enda	51	25	26
Ellen	69	43	26
Brendan	76	50	26
Cathy	76	50	26

Despite this divergence, some general observations can, nonetheless, be made.

While Enda was able to recall or recognise all of the real words, one could argue that he had a certain advantage compared to the other participants in that he had fewer words to relearn and, thus, less of a cognitive load. In session 1, he had, however, demonstrated that he had retained quite a lot of knowledge despite such a long incubation period, suggesting that the threshold level with respect to access was still quite low. This might also possibly pertain to those items he claimed to have forgotten. Whilst the threshold level overall was higher for these forgotten items than for the items he could successfully recall or recognise in session 1, it was, nonetheless, low enough to make those items more easily accessible on re-exposure and relearning. The question is also whether Enda had effective learning strategies enabling him to successfully relearn the items or whether these items were still more readily available to him and simply needed a prompt in the form of re-exposure to be reactivated. This is an important point considering that he also managed to recall or recognise approximately one third of the pseudowords which supports the idea that he, despite having more ready access to previously known items due to lower threshold levels, may avail of certain strategies to promote successful learning. His strong performance in session 1 leads to the question of whether he had "learned better" when he was at school whereby this information became more consolidated in memory or if, rather than being carried over from his school-days, he developed these learning strategies throughout his adult life. The impact of learning context is discussed further in detail in section Chapter 6, Section 6.3.

While Ellen had 7 fewer real items than Brendan and Cathy to relearn, it is interesting to look at performance across these three participants, also keeping in mind their background profiles. Interestingly, Brendan, who had recalled or recognised 42 % of previously known items in the pretest compared to 78.5 % for Ellen, he was able to recall or recognise 78 % after relearning in round 2, compared to 61 % for Ellen even though he had more items to relearn. Cathy also performed stronger than Ellen even though she had only been able to recall or recognise 26 % of previously known items in the pretest. After relearning she was able to recall or recognise 66 % of the items, and like Brendan, she too had more items than Ellen to relearn. However, considering that they each had different words for relearning, the question of whether Brendan and Cathy had more effective learning strategies needs to be treated with caution. It could be that the words that they had to relearn were "easier" in the sense that they had a lower threshold in their mental lexicon and so were more readily accessible on relearning compared to those that Ellen had to relearn.

While Cathy succeeded in recognising two of the pseudowords and Ellen was able to recall one of them, there was a clear relearning advantage in all cases for previously known words compared to pseudowords lending support to the Savings Paradigm.

In the case of all participants there was *greater success in recall of previously known items than for recognition directly after relearning.* Interestingly the difference in performance between recall and recognition was relatively uniform across participants:

Enda: recalled 36 % more items than recognised
Ellen: recalled 37 % more items than recognised
Brendan: recalled 42 % more items than recognised
Cathy: recalled 42 % more items than recognized

5.2.3 Session 3 (no relearning)

Table 47 Comparison of participant performance – Session 3

Session 3 - Enda				
No relearning	Recall	Recognition	Recalled or recognised	Neither
25 real, previously known words	4 (16%)	12 (48%)	16 (64%)	9 (36%)
26 pseudo-word	0	2 (8%)	2 (8%)	24 (92%)
Session 3 - Ellen				
No relearning	Recall	Recognition	Recalled or recognised	Neither
43 real, previously known words	19 (44%)	9 (21%)	28 (65%)	15 (35%)
26 pseudo-words	1 (4%)	1 (4%)	2 (8%)	24 (92%)
Session 3 - Brendan				
No relearning	Recall	Recognition	Recalled or recognised	Neither
50 real, previously known words	9 (18%)	16 (32%)	(25) 50%	25 (50%)
26 pseudo-words	0	0	0	26 (All) 100%
Session 3 - Cathy				
No relearning	Recall	Recognition	Recalled or recognised	Neither
50 real, previously known words	5 (10%)	17 (34%)	22 (44%)	28 (56%)
26 pseudo-words	0	2 (8%)	2(8%)	24 (92%)

Once again with respect to the real, previously known words, considering that each of the participants were tested on different items in the relearning session and in subsequent testing sessions, and had a different number of words, apart from Brendan and Cathy, a direct comparison between each individual cannot be made. However, some observations warrant commentary, nonetheless, in comparing performance on real items to session 2. See following Section 5.2.3.1.

5.2.3.1 Comparison to session 2

The following serves only to demonstrate trends across sessions for each *individual*. It looks at real items only since not enough pseudowords were known to render such a breakdown useful. The participants are not compared to each other due to the different items and partly different numbers of items.

Table 48 Comparison of sessions 2 and 3: Total items known (all participants)

Total (items recalled or recognised)				
	Enda	Ellen	Brendan	Cathy
Session 2 (after relearning)	25	26	39	33
Session 3	16	28	25	22
	36% decrease	4% increase	28% decrease	22% decrease

Table 49 Comparison of sessions 2 and 3: Items recalled (all participants)

Items **recalled** only				
	Enda	Ellen	Brendan	Cathy
Session 2 (after relearning)	17	21	30	27
Session 3	4	19	9	5
	52% decrease	5% decrease	42% decrease	44% decrease

Table 50 Comparisons of sessions 2 and 3: Items recognised (all participants)

Items **recognised** only				
	Enda	Ellen	Brendan	Cathy
Session 2 (after relearning)	8	5	9	6
Session 3	12	9	16	17
	16% increase	9% increase	14% increase	22% increase

With respect to the real items, apart from Ellen (whose performance improved slightly), the participants demonstrated a weaker performance in session 3 when no relearning took place.

These same three participants also performed better on recognition whereas Ellen was able to recall more than double the number of items she could recognise (compare Table 49 to Table 50), although her performance on recall dropped overall and her performance on recognition increased. Apart from one item (the abstract noun: *dlí; law*) the increase in recognition was on items that she had been able to recall in round 2.

On comparing the percentage change overall for each individual participant, Enda who had had fewer items in relearning, appeared, nonetheless, to have more trouble re-activating knowledge he had successfully retrieved directly after relearning in session 2 compared to the others, who had had more items for relearning and thereby a greater cognitive load. His performance dropped by 36 %. One would expect that focussing on fewer items at a time in relearning might lead to a greater entrenchment of this knowledge in memory. However, learning success is not necessarily based on quantity alone as demonstrated in this result.

This is also evident in the case of Cathy who did not know as many items after relearning in session 2 as Brendan who was tested on an equal number of words. However, she proved slightly more successful in retaining knowledge than he did. She lost 22 % of knowledge of the individual items she had successfully relearned in round 2 compared to 28 % for Brendan.

Indeed, retention is not only interesting in this study with respect to the incubation period between cessation of contact with Irish and re-exposure approximately 30 years later but also in terms of what is retained after relearning.

In keeping with the idea of the *better learned, the better retained* this leads to the question of what it means to have learned something, a question also put forward by Bjork & Bjork (2019) who argue that a distinction needs to be made between learning and performance. In keeping with Bjork & Bjork, it is argued here that learning manifests in the emergence of knowledge that can sustain the elements of time and other noise and interference whereas performance emerges in the guise of short-term materialisation but does not necessarily attest to long-term entrenched knowledge. Whilst Enda may have been more successful initially in learning in the sense of performance as he had a lower cognitive load to process, he may not have necessarily *learned* this better, i.e., greater *immediate* success in performance does not inevitably imply greater levels of retention. While Enda and Brendan might have been more effective in availing of certain learning strategies that offered immediate rewards, Cathy, and in particular Ellen, might have engaged other tools and devices that enable greater retention overall. This is also particularly interesting because despite the difference in content and number of items, each participant relearned and was tested under exactly the same environmental conditions.

5.2.3.2 Activation levels

On the other hand, *relearning* has taken place here and not first-time learning. Therefore, success rates might be more an issue of threshold levels if one, in keeping with the ATH, argues that these differ according to item. This is a possibility considering that not only did the participants have different items, but each distinct item inevitably has a different level of activation in each individual. It could be that the specific items that Brendan and Cathy had had to relearn had different threshold levels in each of their unique linguistic systems.

The participants' performance in session 2 after relearning suggested a lowering of the threshold level for those items they had been able to successfully recall or recognise. Further re-exposure in session 3 would, therefore, predict that participants would at least be partly successful in recognising words relearned one month earlier even if they struggled with recall in session 3 and indeed all of the participants were able to recognise more items in this session than they had been able to in session 2. Contrary to results in the pretest where a mixed pattern emerged in terms of ratio of success in terms of recall to recognition, here findings are in line with other studies where individuals tend to perform better on recognition after a period of having no contact with the language.

The fact that, apart from Ellen, performance had, nonetheless, dropped overall suggests that more re-exposure is necessary for the threshold level of certain items to be lowered and appears to provide support for the Activation Threshold Hypothesis, whereby once an item receives input its threshold is lowered only to rise again in the absence of a trigger in the form of exposure and the accompanying activation. The immediate exposure in session 2 in the form of relearning as opposed to session 3 where, apart from the recognition phase, no re-exposure had taken place, served to provide a trigger in lowering activation levels which rose again in the interlude of one month between testing in session 2 and session 3. This would also explain why the threshold level for recall appears to have increased more for all participants than the threshold level for recognition. This drop in recall was considerable in the cases of Enda, Brendan and Cathy while in the case of Ellen it was only slight. It is also worth noting that, apart from Ellen's case, there was less of an increase in recognition than there was a drop in recall, therefore, depicting an increase in activation levels overall after a period of one month in which there was no exposure to the language.

5.2.4 Session 4 (after relearning)

Table 51 Comparison of participant performance after relearning – Session 4

Session 4 - Enda				
Relearning (51 items)	Recall	Recognition	Recalled or recognised	Neither
25 real, previously known words	18 (72%)	5 (20%)	23 (92%)	2 (8%)
26 pseudo-word	5 (19%)	8 (31%)	13 (50%)	13 (50%)
Session 4 - Ellen				
Relearning (69 items)	Recall	Recognition	Recalled or recognised	Neither
43 real, previously known words	28 (65%)	5 (12%)	33 (77%)	10 (23%)
26 pseudo-words	1 (4%)	1 (4%)	2 (8%)	24 (92%)
Session 4 - Brendan				
Relearning (76 items)	Recall	Recognition	Recalled or recognised	Neither
50 real, previously known words	29 (58%)	8 (16%)	37 (74%)	13 (26%)
26 pseudo-words	1 (4%)	1 (4%)	2 (8%)	24 (92%)
Session 4 - Cathy				
Relearning (76 items)	Recall	Recognition	Recalled or recognised	Neither
50 real, previously known words	27 (54%)	10 (20%)	37 (74%)	13 (26%)
26 pseudo-words	0	2 (8%)	2(8%)	24 (92%)

5.2.4.1 Comparison with results from session 2

Table 52 Comparison of sessions 2 and 4: Total items known (all participants)

Total (items recalled or recognised)				
	Enda	Ellen	Brendan	Cathy
Session 2 (after relearning)	25	26	39	33
Session 3	16	28	25	22
Session 4 (after relearning)	23	33	37	37
Difference between sessions **2** and **4**	8% decrease	16% increase	4% decrease	8% increase

Table 53 Comparison of sessions 2 and 4: Items recalled (all participants)

Items **recalled** only				
	Enda	Ellen	Brendan	Cathy
Session 2 (after relearning)	17	21	30	27
Session 3	4	19	9	5
Session 4 (after relearning)	18	28	29	27
Difference between sessions **2** and **4**	4% increase	16% increase	2% decrease	No change

Table 54 Comparison of sessions 2 and 4: Items recognised (all participants)

Items **recognised** only				
	Enda	Ellen	Brendan	Cathy
Session 2 (after relearning)	8	5	9	6
Session 3	12	9	16	17
Session 4 (after relearning)	5	5	8	10
Difference between sessions **2** and **4**	12% decrease	No change	2% decrease	8% increase

Considering that relearning had taken place both in session 2 and session 4 (with an interlude of one year between session 3 and session 4), it is interesting to compare the participants' individual performance across these two sessions.

There was a slight decline in performance overall in the case of Enda and Brendan, whereas both Ellen and Cathy improved. This improvement appears to have been due to an improvement in recall in the case of Ellen (no change in recognition levels was evident in her case) and due to an increase in recognition in the case of Cathy (there was no change in recall evident in her case). While Enda improved in recall overall compared to session 2, his performance in recognition decreased, whilst Brendan's slight decrease overall was due to an equal decrease (albeit it very slight; only 2 %) in both recall and recognition.

5.2.4.2 Activation levels

Overall, however, the differences in performance between session 2 and 4 are very small. Where there is a decrease, it is too slight to be deemed indicative of any considerable changes in overall activation. What the pattern does seem to suggest in looking at performance across sessions 2, 3 and 4, is that in keeping with Paradis's ATH (see Chapter 2, Section 2.2.2) that overall threshold levels rise again when there is no activation (as depicted in results for session 3) and drop on re-exposure (as depicted in results for session 4). Despite cases where there is a decrease, all participants, nonetheless, succeed in recalling or recognising considerably more words than is the case for pseudowords demonstrating that re-exposure stimulates ease of retrievability. Words that appeared to have become forgotten over time with the right trigger can be reactivated again.

5.2.5 Overview of performance in individual word categories

5.2.5.1 Category breakdown of words known in session 1

The following provides a breakdown of items that the participants successfully recalled or recognised in the filtering test with respect to word category (session 1 in the case of Ellen, Brendan and Cathy, session 2a in the case of Enda).

Table 55 Category performance across participants (Session 1 / Session 2a (Enda))

Enda (44 items)	Concrete nouns	Abstract nouns	Verbs	Adjectives
Recalled (17)	7	2	6	2
Recognised (27)	2	7	8	10
Total: Recalled or Recognised (44)	9	9	14	12
% of total items (44)	20.5%	20.5%	32%	27%
Ellen (157 items)	Concrete nouns	Abstract nouns	Verbs	Adjectives
Recalled (67)	21	16	11	19
Recognised (90)	23	22	25	20
Total: recalled or recognised (157)	44	38	36	39
% of total items (157)	28%	24%	23%	25%
Brendan (85 items)	Concrete nouns	Abstract nouns	Verbs	Adjectives
Recalled (43)	22	4	6	11
Recognised (42)	9	19	5	9
Total: recalled or recognised (85)	31	23	11	20
% of total items (85)	36.5%	27%	13%	23.5%
Cathy (53 items)	Concrete nouns	Abstract nouns	Verbs	Adjectives
Recalled (17)	10	3	1	3
Recognised (36)	12	9	8	7
Total: recalled or recognised (53)	22	12	9	10
% of total items (53)	41%	23%	17%	19%

Performance across categories is relatively consistent for the two strongest candidates, Enda and Ellen. Considering, however, that Enda had performed so well across all categories in session 1, he was given somewhat less frequent items in session 2a. Therefore, a direct comparison between his performance and that of the other participants cannot be made.

In the case of the two weakest candidates, verbs emerged as more vulnerable, in particular relative to concrete nouns. In fact, except for Enda, concrete nouns constituted the highest percentage of known words followed by abstract nouns with this being especially apparent in the case of the weaker candidates. This suggests that nouns had become more entrenched in memory which may be due to the fact that Irish is a noun-centred language so that in terms of frequency, the verbs might not have been activated as much as the nouns in the years of learning so that their storage strength might be smaller, thereby making them more difficult to activate compared to nouns (see Chapter 3, Section 3.7).

On the other hand, in her study of four L1 English children's loss of Japanese as an L2, Hansen-Strain (1990: 375) found that L2 nouns were most vulnerable to loss. She also refers to Cohen (1989) who reported similar findings in the case of Portuguese L2 forgetting. It must be noted, however, that in Hansen-Strain's study, the children only spent two-and-a-half-years in Japan and two of them were very young (pre-school) during this time so that they spent most of their time with their English-speaking mother. The two older children, aged nine years and seven years on their return to the U.S., did, however, attend school in Japan. Tests were also only conducted, however, in the first year after their return to the U.S. Cohen (1989) also looked at children (two bilingual Hebrew-English speakers aged nine years and 13 years on leaving the L2 environment) and they had spent even less time in the L2 environment (Brazil), only living there for one year. In Cohen's study, tests were also conducted in the first year after onset of attrition (see Chapter 2, Section 2.3, for more information on this study).

These very different conditions compared to the present study make a direct comparison difficult also considering the extended and intense instruction as well as the very long incubation period of the present project.

5.2.5.2 Category breakdown after relearning

Since the participants had different items as well as numbers of items resulting from filtering in session 1, each individual is looked at separately and only a brief overview of word category patterns across sessions 2 and 4 after relearning is presented. This is not done for Enda because he had a very unequal category breakdown (see Section 5.1.1.2.2).

(The following tables have been copied from previous individual participant sections)

Ellen

As stated, the breakdown of the 43 real items left over from session one was as follows: 6 concrete nouns, 12 abstract nouns, 14 verbs and 11 adjectives. As already illustrated in Section 5.1.2.2.3 on category performance across sessions 2, 3 and 4, out of a total of 43 real items for relearning, Ellen performed as follows:

Table 21 Overview of word categories, Sessions 2–4 (Ellen) (the following is a copy of Table 21 from page 178)

Ellen (Sessions 2-4)	Total (items recalled or recognised)	Concrete nouns	Abstract nouns	Verbs	Adjectives
Session 2 (after relearning)	26	5	7	9	5
Session 3	28	5	9	9	5
Session 4 (after relearning)	33	6	8	12	7

No noteworthy pattern is evident here. Ellen showed improvement across all categories. This is, however, interesting in its own right as it shows that no one category appears to be more or less affected by relearning.

Brendan

As stated, the category breakdown of the 50 real words for relearning was as follows: 12 concrete nouns, 12 abstract nouns, 13 verbs and 13 adjectives. As already illustrated in Section 5.1.3.2.3 on category performance across sessions 2, 3 and 4, out of a total of 50 real items for relearning, Brendan performed as follows:

Table 32 Overview of word categories, Sessions 2–4 (Brendan) (the following is a copy of Table 32 from page 188)

Brendan (Sessions 2-4)	Total (items recalled or recognised)	Concrete nouns	Abstract nouns	Verbs	Adjectives
Session 2 (after relearning)	39	10	10	10	9
Session 3	25	5	4	9	7
Session 4 (after relearning)	37	8	9	9	11

Once again, no striking relearning effect for a specific category is evident here with a relatively consistent pattern emerging across each one.

Cathy

Like Brendan, the category breakdown of the 50 real words for relearning was as follows: 12 concrete nouns, 12 abstract nouns, 13 verbs and 13 adjectives. As already illustrated in Section 5.1.4.2.3 on category performance across sessions 2, 3 and 4, out of a total of 50 real items for relearning, Cathy performed as follows:

Table 43 Overview of word categories, Sessions 2–4 (Cathy) (the following is a copy of Table 43 from page 199)

Cathy (Sessions 2-4)	Total (items recalled or recognised)	Concrete nouns	Abstract nouns	Verbs	Adjectives
Session 2 (after relearning)	33	12	9	6	6
Session 3	22	8	6	3	5
Session 4 (after relearning)	37	12	11	6	8

The one noteworthy pattern in Cathy's case is that she performed better in each of the noun categories and worse in the case of the verbs. This may be due, as stated previously, to the fact that Irish is a noun-centred language so storage strength might be greater for nouns than for verbs.

Larson-Hall (2019: 12) states that perceptible patterns have not been established with respect to whether or not certain words are more or less vulnerable to attrition and suggests in keeping with Meara's (2004) observations based on simulated lexical activation models (see Chapter 2, Section 2.2.4) that lexical loss may in fact not be determined by how regularly words had been activated in the past but is instead random in nature. Whilst neither Larson-Hall nor Meara were referring specifically to word categories, it is, nonetheless, argued here that contrary to their conclusions, that past activation prevalence and recurrence might in fact play a role in determining which items are more readily accessible or not. Meara himself states that his model is a simulation that makes "a few very minimal assumptions about the way words might be linked in a vocabulary" (2004: 138) and as such is limited in its power of prediction. Rather in keeping with the idea of storage strength as discussed by Bjork (see Chapter 2, Section 2.2.4) and Paradis's ATH (see Chapter 2, Section 2.2.1), it is argued here that there may indeed be threshold levels for activation as suggested by general performance on nouns in both session 1 and subsequent relearning, considering the noun-based nature of Irish. In fact, the observation that this pattern was more pronounced in the case of the weaker candidates reinforces the idea of activation thresholds. Those with a higher level of language proficiency, Enda and Ellen, would have repeatedly actively engaged all words as is necessary in achieving such levels, explaining their performance across categories. Weaker, possibly less engaged learners might have activated less but due to the prominence of nouns would have inevitably activated these more thereby leading to their greater entrenchment in memory, hence, the patterns that have emerged.

5.2.6 Summary of observations

Based on the above results and analysis, the following patterns emerged:

1. Evidence of retention of lexical items after an average incubation period of 30 years was established but this varied greatly across participants.

2. Subsequent relearning and testing provide support for the Savings Paradigm.
3. Re-exposure in the form of relearning appears to cause threshold levels for activation to drop overall, with these rising again in periods of disuse.
4. No one word category emerged as particularly vulnerable to attrition. However, in scores on retention levels in session 1, verbs emerged as less robust, in particular relative to concrete and abstract nouns, in the case of the two weakest candidates. This suggests that nouns had become more entrenched in memory which, as already stated, may be because Irish is a noun-centred language (see Chapter 3, Section 3.7).

Chapter 6: Discussion

In the following an examination of the factors that may explain differences in performance is conducted. Finally, the findings are presented within the framework of Complex Dynamic Systems Theory and the Savings Paradigm.

6.1 Overview of participant profiles

Table 56 Overview of participant profiles

	Enda	**Ellen**	**Brendan**	**Cathy**
Percentage of items retained (Session 1)	97%	78.5%	42%	26%
Language Distance	Irish-English	Irish-English	Irish-English	Irish-English
Age at start of learning	4 years and 2 months	4 years and 1 month	5 years and 2 months	10 years and 2 months
Age at onset of attrition	19 years old	18 years old	18 years old	18 years old
Length of learning	15 years	14 years	13 years	8 years
Proficiency level at onset of incubation	Leaving Certificate: Grade **B** in **higher level paper** plus one year at university (exact grade unavailable)	Leaving Certificate: Grade **B** in **higher level paper**	Leaving Certificate: Grade **B** in **lower level Irish paper**	Leaving Certificate: Grade **D** in **higher level paper**
Exposure and learning context	Predominantly school, intermittent input in immediate environment	School	School	School
Length of period of disuse	30 years	31 years	33 years	28 years
Rehearsal during incubation	Intermittent (Receptive: listening to Irish-language radio)	No	No	No
Motivation	*	*	*	*
Other language/s at school and grade in Leaving Certificate	French: Awarded a grade A in the higher level paper	French: Awarded a grade A in the higher level paper	French: Awarded a grade B in the higher level paper. Latin: Awarded a grade C in the higher level paper	German: Awarded a grade A in the higher level paper. French: Awarded a C in the higher level paper
Current languages other than English	German: fluent (CEFR: C2) French: very good (teaches French at Gymnasium, CEFR: C1)	German: very good (B2)	German: fluent (CEFR: C2)	German: fluent (CEFR: C2)

*Unquantifiable but described in detail in Section 6.5

6.2 Age at start of learning, length of learning and level of proficiency

At an initial glance on comparing *proficiency levels at onset of incubation* with retention levels, it appears that, in keeping with studies referred to in Chapter 2, Section 2.4.2.1, *the more you know, the less you lose*. If knowing denotes the entrenchment of knowledge in the memory store then this leads to the matter of what causes the fortifying of this knowledge. Exposure alone may not suffice in achieving this. If this was the case, one would expect less variation in results. Instead, activation and reactivation are said to support information becoming more embedded in memory (Bjork 2012), in keeping with Paradis's Activation Threshold Hypothesis. Greater language awareness might prompt a closer learner engagement and thereby more regular activation of linguistic knowledge.

In the present study, both candidates who performed strongest in session 1 (pretest) with respect to retention levels had a high level of proficiency at age of onset of incubation. Enda continued to study Irish at university, taking it as a subject in his first year studying languages. Like Ellen, he was awarded a grade B in the higher-level Leaving Certificate paper. Considering the demanding and high standard of the Irish higher-level course and paper in the national state examination, this can be considered as constituting an advanced level of proficiency. Although Brendan, who took the lower-level paper in the Leaving Certificate, did demonstrate some vocabulary knowledge retention (42 %), his lower proficiency level at onset of attrition might, nonetheless, explain why he recalled and recognised less overall than Ellen or Enda. It is also interesting to note that his length of incubation was the longest (albeit only by two years compared to Ellen). Considering that studies have shown a levelling-off of attrition effects after the first five years of non-exposure (Bahrick 1984a), it is also not expected that there would be a drop in retention levels between 30 years (Ellen) and 32 years (Brendan) of non-exposure to a language. This difference is, therefore, not considered striking enough to explain his performance. One factor that did emerge as significant in the interview with him was his *lack of motivation and interest in the language*. This might have fed into his *level of proficiency* at onset of incubation.

With respect to Cathy who demonstrated the lowest levels of retention, two factors stand out in which she differs to the other participants; *age at the start of learning* and correspondingly *length of learning*. This might explain her *level of proficiency* at onset of attrition. The age at onset of acquisition effect is a much-discussed factor with respect to language learning attainment and achievement levels, especially with respect to language acquisition in which the language learner is immersed in the L2 environment (see Chapter 2, Section 2.4.1). However, Cathy's results in session 1 prompt not only the question of what role *age at the start of learning* and correspondingly the *length of learning* play with respect to level of proficiency achieved, but it also gives rise to the question if this also has repercussions in terms of long-term *retention* of knowledge, as her results indeed suggest. Additionally, with respect to her shorter period of learning it is important to note that in missing the first six years of Irish in primary school she was deprived of a fundamental foundation in language knowledge. The first years of Irish learning at primary school are traditionally very intense with much time and effort dedicated to ensuring that the children achieve a solid grammar and vocabulary base. This is repeatedly reinforced throughout the subsequent years in secondary school so that quite possibly this is why the other three participants performed better in session 1. A longer and intense period of learning might have led to information achieving a higher storage strength in memory (see Chapter 2, Section 2.2.4).

On the other hand, the fact that she decided to take the higher-level Irish course which was renowned for its challenging and demanding content suggests a high *motivation* and drive on her part. This highlights in keeping with the idea of clusters that factors do not operate in isolation.

For example, looking at Enda's profile compared to the other three participants in terms of *length of learning, exposure and learning context* and *rehearsal during incubation* certain differences emerge. Enda, who achieved the highest score in terms of levels of lexical item retention, had had Irish for one year more than Ellen, having it at university where he claimed Irish was more interesting and interactive and during which time, he felt more motivated in using it. This then gives rise to the question whether it was not so much the *longer period of learning* per se that enhanced knowledge entrenchment but the *content* and *nature of learning* in this time that stimulated such positive results and high levels of retention.

With respect to *exposure and learning context* he states that beyond the school context he was also exposed to some input in his home and immediate social environment growing up in a very remote part of the Dingle Peninsula on the south-west coast where Irish is still partly used (at least Irish vocabulary) in daily interactions. In terms of *rehearsal during incubation* he took time to listen to Irish language radio on occasions during some of his visits to Ireland, although this was irregular. Looking at these three factors overall, the common underlying component across all three factors is *increased input* compared to the other three participants which it appears, drawing on his responses in the interview, was based on a genuine interest in the language. What is also important to note is that there seemed to be an *attentive active uptake* of this input on Enda's part whereby he was not simply an onlooker but attentively listened, read and produced the language both orally and in written form during his years of active learning. Rehearsal over the years of incubation, albeit sporadic and intermittent, might possibly have served to reinforce this knowledge in the memory store (see Section 6.6 where this is discussed further). The fact that he also chose to study Linguistics and Languages at university demonstrates his curiosity and enthusiasm for the workings of language.

Ellen was a studious pupil with a high level of proficiency at onset of incubation. Whilst she did not have the retention levels that Enda had, she differed also to him in that she did not carry on to study it at university and had no input in the language during incubation. However, she appears to have had a more positive attitude towards Irish in school than either Cathy or Brendan.

6.2.1 Establishing levels of proficiency

Mickan (2021) and colleagues tested a group of L1 German – L2 English learners of L3 Spanish on 137 nouns. Participants were tested on these same items, before they left for a six month stay in a Spanish-speaking country, at the end of their stay (which Mickan terms the pre-attrition baseline) and six months after their return to Germany. Scores on the three picture-naming vocabulary tests were compared across the testing sessions. Participants came from diverse backgrounds in terms of Spanish knowledge. Some had quite a lot of prior knowledge, others had only been learning Spanish for a short time. Proficiency levels were determined using self-ratings.

Mickan's conclusion that retention levels as revealed in the vocabulary test at time 3 (i.e., six months after arrival back in Germany) do not strongly correlate with initial proficiency is somewhat problematic as proficiency levels based on self-reports can be unreliable since they are not standardised and are based on subjectivity. Many individuals also simply do not know how to objectively assess their own language skills and even when information is obtained on exposure or length of learning this may not provide accurate indications of true levels of proficiency.

It is argued here that had a more comprehensive language test been distributed greater reliability on proficiency levels might have been achieved and possibly even greater variety on scores established. Instead, Mickan determined Spanish proficiency in speaking, listening, writing and reading based on self-ratings using a 7-point Likert scale (1 = very poor, 7 = near native-like). The amount of exposure (in years) was established but was not explored further. Considering that a number of the participants were studying Spanish or Latin American Studies at the time at university meant that they would have most likely been exposed to a richer and more academic language quality compared to individuals who might possibly only have been exposed to the language more passively or just used it in informal situations (Yilmaz & Schmid (2018: 234) refer to Schmid (2007) who argues that context influences the quality of language used). Whilst the length of exposure might be the same across some individuals, it could be expected that the very difference in quality of input would result in varied levels of retention.

Therefore, although the procedure in Mickan's study might have enabled a general impression with respect to the possible influence of initial proficiency, her findings should, nonetheless, be observed with a degree of caution.

In the present study, on the other hand, all four participants had learned according to the same curriculum in the same instruction context and were examined in a highly comprehensive state national examination at the end of their schooling so that grades achieved were consistent with a standardised assessment scheme according to which levels attained were measured.

6.2.2 Proficiency as established in the Leaving Certificate examination

Performance in an examination can be dictated by a number of factors; the particular questions asked, the correction scheme, levels of difficulty and corresponding expectations on the part of teachers and in this case considering the undeniable importance of the Leaving Certificate examination, affective factors such as nervousness and the ability to focus in stressful situations. Therefore, it is argued here that when it comes to second language learning in a school context, gaining access to an accumulated grade based on a series of tests in which all skills are accessed over time as well as the teacher's overall impression might offer an even more accurate representation of proficiency levels. Access to such information was not possible, however, in the present study considering the very long incubation period.

However, considering the long period of learning Irish that spanned the participants' entire school career in the Republic of Ireland and the importance and seriousness of the Leaving Certificate examination whereby it could be expected that participants would have given their best in completing it and would have been at optimal levels in terms of preparation, it can be stated with a high degree of confidence that these grades represent reliable indicators of proficiency levels at onset of incubation.

Indeed, level of proficiency as established based on Leaving Certificate results does appear to correlate with scores on retention levels in session one.

6.2.3 Level of proficiency and relearning effects

It is proposed here that over the period of incubation, the participants will have evolved cognitively and possibly developed processing strategies that can serve to aid them in later re-activation endeavours as seems to have been the case in the present study when looking at the participants' scores after relearning. With respect to attrition itself, it appears that levels of proficiency at onset of incubation might play a role in terms of how much one remembers initially before re-exposure to the language (although, as stated, the factor of proficiency level does not exist in isolation). Nonetheless, it has been observed in this study that performance *after* relearning, on the other hand, in the form of re-accessing old knowledge, does not necessarily correlate directly with performance in initial testing on re-exposure. This is contrary to the findings of Hansen et al. (2002) who observed that those who performed better in relearing had also higher scores in the pretest (see Chapter 2, Section 2.3.2).

Although, as stated, a direct comparison between each candidate is not possible due to the difference in both the test items themselves as well as the number of items used in relearning for each individual participant, it was, nonetheless, detected that those participants who had lower results in initial testing of retention levels (session 1) compensated for this deficit in subsequent relearning, demonstrating considerable improvement in accessibility after re-exposure relative to their initial performance, suggesting, therefore, that success in relearning appears to be *independent* of initial retention levels assessed after a period of disuse (in this case the average incubation period was 30 years).

6.3 Context of learning: Input and learning experience

Bjork & Bjork (2019: 166) state:

> Individual differences do matter, and matter greatly, because all new learning builds on, and depends on, old learning. In addition, personal, family, and cultural histories can affect such things as our motivation to learn, the degree to which learning is valued, and the aspirations and expectations we have with respect to our acquiring skills and knowledge.

Pan & Berko-Gleason (1986: 193) highlight the importance of examining not just language loss itself but also critically inspecting the pedagogical approaches that had been employed in the teaching of that language previously to see if particular approaches hasten and induce more loss than others.

Mehotcheva & Mytara (2019: 11) argue, nonetheless, that the *context of learning* is an area largely underexplored in attrition research and that qualitative analysis has the potential to yield interesting insights into these factors. Context of learning constitutes a central focus in the present study and in keeping with Mehotcheva & Mytara (p. 15), a qualitative analysis approach is adopted, hence, the incorporation of an interview in striving to gather knowledge on not just the context of learning in terms of environment but also, and very importantly, on the very *experience* of learning in this context.

In relation to this, it is worth noting that language environment was among the influential factors isolated by Weltens & Grendel (1993: 148) in accounting for the variation in results on lexical attrition. Murtagh (2003) also observed that learning environment played a decisive role not just in proficiency outcome but in levels of knowledge retention of Irish exhibited in testing 18 months after official instruction had ended.

6.3.1 Context of learning Irish

With the founding of the independent Irish state in 1922, education became a key instrument in accentuating and firmly establishing the value of the unique and rich cultural heritage of Ireland and its language (see Chapter 3, Section 3.1).

Keeping in mind that official education policy in Ireland distinguishes between Irish and Modern Foreign Languages (see Chapter 3, Section 3.5), Li et al. (2017: 128) argue that Irish pedagogy needs to be looked at differently to the modern foreign language teaching context. Irish is not a language mandatory for transaction in any facet of the contemporary world, be that economic, political or social, but it is of significant and inherent cultural and biographical relevance for the historical identity and trajectory of the Irish nation.

Language environment and input was relatively similar for each of the participants in this study, especially in terms of the curriculum, although interpersonal relations and experiences with teachers as well as teacher engagement, dedication and encouragement not to mention instruction style and manner would inevitably have varied, not just across participants but also for each participant individually throughout the course of their own education.

While Irish lessons in the 1980s were teacher-centred with less focus placed on interaction and communication, at least relative to present-day language classrooms, this is not, however, to suggest that oral competencies were not important. Indeed Harris et al. (2006: 3) in reviewing national surveys conducted throughout the 1970s and 1980s state that, on the contrary, expectations with respect to spoken Irish were very high with only approximately one third of primary school learners achieving the standards stipulated in the curriculum. They argue that this was not due to poor teaching; rather the high standard that was set was established according to the higher levels of proficiency in Irish observed in settings where children had more access to the language, be it in the family environment or in attending Irish-medium schools. Children in ordinary schools were, nonetheless, expected to match these Gaeltacht or immersion learners in their speaking proficiency. In fact, Harris et al. (2006: 3) refer to the aims of the Irish curriculum as being "rather ambitious" with some educators arguing that attaining such levels was a particularly challenging endeavour and not necessarily realistic (Harris et al. 2006: 167). All of the participants in the present study still recalled, despite the length of time that had passed since they were in school, how demanding and "academic" the Irish syllabus was, particularly at post-primary level.

Mehotcheva (2010: 14) in referring to studies conducted by Grendel (1993), Bahrick (1984a, b) and Weltens (1988) on L2 learning, also highlights that formal language instruction in the 1980s was different to the approaches typical of contemporary language teaching in which there is a greater focus on oral skills and interaction, input is more varied and, of course, learners have greater access to the language beyond the classroom, ranging from online learning platforms to opportunities to interact in the language by way of social media.

On the other hand, the great deal of time that was dedicated to reading and writing in Irish learning, whilst not always undertaken with fervor in that moment, might have contributed to a more resilient trace of that knowledge in long-term memory, compared to the evanescent nature of oral communication, in which attempts to articulate one's thoughts in a fleeting interaction may not transpire into entrenched linguistic knowledge (see Section 6.4 on the role of literacy in consolidating linguistic information). For example, Hansen et al. (2012: 197) state that words learned explicitly and with regular exposure achieve a greater degree of entrenchment in memory. In the present study, evidence of lexical knowledge retention was still found despite a long incubation period.

Naturally this is not meant to devalue the importance of oral communication and interaction in language instruction. Indeed, had approaches been different, learners might have maintained contact with the language and the fact that none of the participants can actively use the language today testifies to shortcomings in teaching approaches at the time.

Indeed, Harris & Murtagh (1999) as reported in Harris et al. (2006: 8) found that higher achievements in Irish emerged in classes in which there had been a greater emphasis on communication. However, it was observed that speaking in the sense of "real communication" or the negotiation of meaning rarely took place in the Irish lessons and stronger pupils spoke more than those who were less proficient (Harris et al. 2006: 8). Weaker performance was evident in classes dominated by the more traditional approaches of rote-learning, drills and routines such as reading aloud which additionally led to less positive attitudes to learning and difficulties in paying attention in class.

In a comparison of data that was generated in a survey in 2002 with data from a survey conducted in 1985 with respect to Irish listening and speaking proficiency among 6th grade primary school learners (when pupils are approximately 11–12 years old) as well as external variables such as teacher and parental attitudes and teaching materials, Harris et al. (2006) observed a significant downward trend in achievement levels in listening and speaking in the 17-year interval among pupils who attend ordinary schools (i.e., not Irish-medium schools or ones located in Gaeltacht areas) with Harris et al. (2006: 162) even referring to the change in performance as "dramatic".

The results of the above comparison conducted by Harris et al. (2006) are particularly noteworthy within the context of the present study since the four participants were still, or had just left, primary school when the first survey was conducted in 1985. In this regard it is interesting to note that two of the participants, Ellen and Enda, recalled enjoying Irish more in primary school than in secondary school where they felt it had more of a second language character because not only did the instruction take place in Irish in the core Irish lessons, it was also used by teachers in regular interaction beyond Irish lessons and the classes themselves were livelier and more communicative. This might have contributed to their higher levels of achievement overall in Irish and the higher rates of long-term retention that they exhibited in the pretest in this study. It is, therefore, particularly interesting to observe that despite the criticisms expressed among the participants, the learning situation in the 1980s was, nonetheless, apparently more conducive to attaining higher levels of proficiency than in later years.[2]

Considering, as stated in Chapter 2, Section 2.4.2.1, that higher proficiency levels at onset of incubation appear to affect retention levels, the factors that may have aided consolidation of knowledge in the case of the participants in this study was the frequency of input during learning, in particular in primary school, as well as the overall length of learning which spanned 13–14 years. Bahrick, Hall & Baker (2013) also emphasise the importance of learning taking place over a long period of time to support consolidation of knowledge in memory. They give the example of content just dealt with in one semester as not being conducive to long-term retention.

6.3.2 The role of home and social environment

Darmody & Daly (2015: 85) found that the main factor influencing attitudes to Irish was the home environment, not just in terms of whether the language was actually used at home but also with respect to whether learners felt that their parents supported their learning of Irish at school and considered this an important part of their education. Murtagh & Van der Slik (2004: 297) also report on significant positive correlations between parental encouragement and spoken Irish proficiency achievement. Children of parents who have, on average, higher Irish language competencies and engage in school content, also tend to achieve higher levels of performance in tests (Harris et al. 2006: 172).

None of the participants in this study reported actively using Irish at home, although all of them stated that they felt that their parents valued

2 Harris et al. (2006) explain the substantial decreases in achievement between 1985 and 2002 as evolving from a range of factors; a reduction in the core time dedicated to Irish in the intervening years, a growth in disillusionment among teachers, and the widening discrepancy between the nature of materials and methods used in the Irish classroom compared to developments and advances in pedagogy in the wider educational context and changes in society at large with teachers complaining that the content was "dated and unsuitable" (Harris et al. 2006: 167).

education, even if they did not actively participate in their children's schooling, for example, helping with homework. Neither did any of the participants sense any indication of negativity on their parents' part with respect to the teaching of Irish specifically.

With regard to the role of the social environment in supporting language learning, Chan et al. (2016: 184) argue, using the example of EFL learning, that authentic engagement with the language outside the classroom and natural input via various media channels and communication with other speakers of that language not only plays a pivotal role in supporting vocabulary learning but also in the preservation of this knowledge. This was lacking, however, in the learning experience of the participants in the present study. Apart from Enda, they had very little exposure to Irish beyond the classroom.

Place & Hoff (2011) state that findings which demonstrate that the level of the child's skills in each language is related to the amount and frequency of exposure to each language is consistent with theories that describe language acquisition as a process of obtaining information from experience, i.e., more language experience results in more rapid language development. This, in keeping with arguments thus far outlined, was, therefore, problematic in the Irish learning context when experience was for the most part limited to the classroom context. Pearson (2007: 401) (referred to in Gharibi & Boers (2017: 55)), for example, wrote of the input-proficiency-use cycle, whereby input feeds proficiency which in turn stimulates use. In the case of Irish, however, the input did not necessarily promote vast leaps forward in proficiency and for the most part it did not stimulate greater use of the language.

6.3.3 Context of teaching Irish in the 1980s

Language teaching resources are also closely dependent on greater market forces and it has to be noted that the period spanning the 1980s when the participants in this study were in school, was a time of severe economic hardship in Ireland. This led to a distinct deceleration in developments in education and significant spending cuts with Colleges of Education being forced to reduce student numbers (Department of Education (1984: 34) as referred to in O'Donovan (2013: 29)). Referring to the comment made by Brendan in this study (Chapter 5, Section 5.1.3.1) that the teachers themselves seemed apathetic could possibly be explained by these austere measures in which teachers might have felt that they were "under-resourced and over-worked" and this might have been reflected in their teaching approaches. On the other hand, in a comparison of reports on teacher attitudes from 1985 and 2002, teachers in the 1980s reported substantially higher rates of job satisfaction and positive attitudes towards the teaching of Irish than in 2002 (Harris et al. 2006).

6.4 Literacy

Murtagh (2003: 37) states that literary skills that are well developed help learners reach a critical threshold of knowledge that once attained makes it more resilient to attrition effects and quotes Hansen (1999: 16) who stated that literacy "anchors linguistic knowledge", thereby making it less vulnerable to loss. Given the emphasis on reading and writing in the Irish curriculum in the 1980s, its relevance for possible long-term knowledge retention cannot be overlooked in the context of this study. Stenson & Hickey (2018: 6) state that reading is highly effective not just in promoting vocabulary acquisition but also in aiding retention of that information.

Murtagh (2003) in her study on school-leavers' retention of Irish 18 months after finishing school found that it was not only initial proficiency that emerged as a predictive factor of linguistic knowledge retention but also that those individuals who continued to read in Irish after contact with the language had officially stopped, showed higher levels of overall proficiency in the testing conducted as part of the study.

6.4.1 Focus on literacy in Irish learning

The nature of the exposure to Irish experienced by the participants in school might explain the results in session 1 in which recall did not necessarily lag behind recognition. The findings of other studies which found higher rates for recognition than recall (see Schmid & Mehotcheva 2012) are based on the learning and retention of modern second/foreign languages in which there is generally a greater emphasis on speaking, unlike a typical Irish lesson in schools in Ireland in the 1980s although it is accepted, as argued by Mehotcheva 2010, that the participants in Weltens's (1988) study (L1

Dutch learners of French) would also have experienced more traditional methods considering the period when he conducted his study. It is argued here, however, that there would still have been more communication in these lessons than there was in a typical Irish class where teaching methods were more in keeping with pedagogical approaches used in the teaching of Latin. Also compared to other studies that looked at second or foreign languages learned in an instruction setting, the age at the start of learning Irish was much younger than in these studies where the school children usually start learning a foreign language in secondary school. Therefore, learners of Irish were exposed to the written form of the language earlier.

The nature of this (distinct when compared to the teaching of modern foreign languages) exposure combined with age may explain why the findings in session 1 tend to deviate somewhat from that reported in other studies. With respect to Irish learners, most of the input was in the written form and much of the output was also written, for example, in the form of essays and literary analysis.

6.4.2 Reading

Guthrie (2004: 3–4) refers to a number of different aspects that literacy engagement encompasses. One of them is *time on task* which implies paying attention to the text and maintaining cognitive effort in establishing meaning. Another aspect refers to *affect* which centres on enthusiasm and enjoyment in reading. A third, *cognitive qualities of the reader*, highlights the depth of cognitive processing engaged in while reading and the corresponding promotion of conceptual learning that takes place through the comprehension monitoring that reading frequently necessitates. The fourth component is *activity-based* which focuses on the variety and amount of literacy activities sought out by learners.

With respect to the above delineation of literacy engagement, two aspects appear to have been satisfied by schools in Ireland in the 1980s in the case of teaching Irish and literacy; *time on task* and *cognitive qualities of the reader*. While the particular texts prescribed by the Irish language curriculum served to regrettably dampen learner enthusiasm for reading Irish texts in many learners and dissuade them from seeking out more possibilities to read, on the other hand, the sustained and sometimes even exclusive engagement with printed text, especially in secondary school, would, in keeping with Guthrie, have enhanced the depth of cognitive processing of the language and reinforced vocabulary knowledge. The texts dealt with on the Irish curriculum were frequently reread and closely analysed thereby necessitating greater attention and absorption than fleeting spoken language.

Attentional processes play a key role in registering and learning new information. Grabe refers to Schmidt's (2001) explanation of attention as a set of processes that "allow us to notice input and focus consciously on that input. In this way, attention supports explicit learning" (Grabe 2009: 67). This gives rise to the important matter of providing learners with engaging and animating texts, an undertaking in which the Department of Education in Ireland floundered with the texts stipulated by the Irish curriculum falling short content-wise in terms of stimulating learner reading ambition, as already stated.

However, even in the case of learners who were not always attentive; encountering words repeatedly in print form, nonetheless, appeared to help consolidate this information in long-term memory. Cummins (2011: 142) also points out that academic language is primarily found in texts and this aids greatly in the expansion of vocabulary knowledge.

Grabe (2009: 102) highlights that word recognition plays a central role in reading and considering that vast amounts of time were dedicated to reading and writing in learning Irish, meant that this was constantly being reinforced. He also argues that to be able to attain enough vocabulary to master academic texts in the L2, repeated encounters with words in the form of reading is vital with rates of understanding new words as well as retention benefitting substantially when reading is accompanied by supportive instruction and repeated exposure, especially when attention to vocabulary is conscious (p. 273).

He also states that learners need to be engaged in reading over a long period of time for knowledge acquired implicitly to be become automatised and information that is learned explicitly to become consolidated in memory (2009: 81).

This directly addresses Murtagh's (2003: 37) question if the reason why reading aids proficiency is due to the very practical reason that through this practice one spends even more time engaging with the language. It is argued here that while any extra time invested in the L2 is positive, it is not this extra time per se but the act of reading itself that supports consolidation of acquired knowledge in memory. Through working repeatedly with selected texts, the same vocabulary was encountered frequently, thereby supporting the transformation of input into intake and automatic implicit knowledge, which, according to R. Ellis (1994) (referred to in Sanatullova-Allison 2014: 5), occurs if the learner practices adequate noticing and attentive examination, which naturally would have occurred to varying degrees with respect to the participants in this study. Nonetheless, the strong emphasis on reading in the Irish curriculum in the 1980s might help to explain why even after such a long incubation period, participants still demonstrated lexical knowledge retention.

6.4.3 Writing

Accompanying extensive reading in the Irish learning context was also much written output in the form of essays and literary analysis. Written composition necessitates a heightened discernment of the workings of the language, as in actively producing a written form, individuals have to draw on more linguistic information (e.g., phoneme to grapheme correspondence) than in producing the word orally. In writing, greater attention needs to be paid to accuracy, precision and specificity. Years of writing and the necessary deliberation this entails might have led to a greater reinforcement of the language.

Nation (2014: 24) emphasises the importance and benefits of output for learning to be effective as does Sanatullova-Allison (2014: 7) who speaks of higher involvement load and argues that the greater this is, the greater the retention of words will be. The role of the reading texts in Irish served not just as input but also as a basis for very focussed output in the form of written texts. Whilst reading supports extensive receptive vocabulary development and fosters immediate word recognition, repeated recall will strengthen this information in long-term memory (Grabe 2009: 60–61).

Chan et al. (2016: 187) emphasise the important role played by intentional learning which they refer to as "explicitly drawing the learners' attention to the lexical items themselves" in the consolidation and ensuing retention of vocabulary knowledge. Incidental learning further supports this through repetition and expanding the contextual scope in which items appear. They refer in particular to a study by Hulstijn & Laufer (2001) in which it was found that words learned incidentally in reading became more reinforced and consolidated in memory when reused in writing composition compared to when exposure was purely receptive or aided by context.

On the other hand, the emphasis on writing in the Irish instruction context might have been to the detriment of fluency in oral communication. Considering that the participants were not able to hold even a rudimentary conversation in Irish suggests that the long-term effects of such approaches are selective. In future research it would be useful to compare results to performance in a modern foreign language that had been learned at school. However, considering that the learning of modern second languages begins in secondary school, such a comparison would have been challenging as age at onset of learning/acquisition would have been significantly different. Also, the very nature of such languages, as modern actively used languages, means that contact with these languages during the period of incubation cannot necessarily be ruled out. Nonetheless, this would prove an interesting observation. Indeed, should results on the modern language reveal better performance *despite* a later age of onset of acquisition, this would force a more in-depth look at teaching approaches for Irish.

6.4.4 BICS and CALP

Cummins (1979) distinguished between Basic Interpersonal Communicative Skills (BICS) and Cognitive Academic Language Proficiency (CALP) stating that many heritage language speakers (i.e., speakers of a language at home that is not the majority societal language), for example, whilst having a solid mastery of BICS in the majority language, fall short in terms of their academic language proficiency. He argued that mastering CALP was an important step in advancing one's chances to excel academically and later professionally. It is proposed here that the concept of BICS and CALP can also be applied to the Irish language learning setting but in this scenario the reverse seems to be the case with respect to mastery of the language. In the 1980s, by the time they did their Leaving Certificate, learners had achieved a high level of CALP as shaped by the content of the curriculum, not necessarily to the detriment of, but rather in spite of, not having highly developed BICS. This might, however, also explain why, even after approximately thirty years of incubation, after triggering activation of the language through re-exposure in the form of relearning, all participants demonstrated, albeit varying, degrees of retention.

6.4.5 Reading in the maintenance of language knowledge

Köpke (2004: 9) additionally states that not only does literacy help to reinforce language representations in the brain but it might also equate the language with a certain prestige and in doing so may encourage a greater motivation among individuals to both learn the language and strive to retain knowledge of it later. Although the choice of literature on the Irish curriculum failed to evoke a relish for pursuing literary pursuits in Irish beyond the school context among three of the participants in this study, Irish literature was still perceived as a valuable component of the Irish language and identity. In Li et al.'s (2017) examination of learner attitudes to Irish, they also found that even though learners were discontented with the fact that there was a very high focus on literature in the Irish syllabus, they, nevertheless, expressed a strong appreciation of the very existence of such a rich repertoire of Irish heritage in written form. Unfortunately, however, the majority of young adults do not continue reading in Irish after they leave school. In referring to a selection of national surveys conducted on language use between 1973 and 1993, approximately 93 % of respondents claimed that they never read Irish language publications (Murtagh 2003: 10).

6.5 Attitudes and motivation

Mickan (2021: 156) criticises previous attrition studies for only looking at motivation at the point when attrition starts (although she does not specifically state which studies she is referring to), arguing that motivation during learning is just as important. In this study motivation during learning is also regarded as an extremely important consideration and one that should not be overlooked in attempting to understand attrition of languages learned in an instruction setting. While Weltens & Grendel (1993: 146) state that there is little evidence that attitude and motivation play a role in counteracting L2 attrition, it must also be noted that they were looking at L2 retention in the case of modern foreign languages and not minority languages (see Chapter 3, Section 3.5, for the distinction made between

Irish and MFLs in education policy in Ireland). Mehotcheva (2010: 159) suggests, for example, that attitudinal factors may not play such a decisive role in the case of L2 attrition because due to the dynamic and changing nature of attitudes and motivation it is difficult to capture the essence of the motivation that might have dominated during the learning period. Whilst accepting that this is a challenging task, the difficulty of assessing a phenomenon does not supersede the possibility that it may, nonetheless, be relevant.

Most of the literature on motivation, however, is dominated by the teaching of English and other modern foreign languages. Motivations in learning Irish are unique to its status as a minority language with no exclusively monolingual L1 speakers of Irish and only a relatively small number of Irish speaking communities on the island.

In discussing motivation among pupils in the Republic of Ireland, two key points have to be highlighted. The first is that learning Irish is compulsory at school. The second is that one has to take the Irish examination in the state Leaving Certificate and a pass grade is necessary for admittance to most of the country's national universities.

6.5.1 Instrumental motivation in learning Irish

It appears, based on the responses of the participants in the present study that the overriding motivation during learning was instrumental. This is in line with findings from other surveys and studies conducted on attitudes and motivations in learning Irish. For example, Darmody & Daly (2015: 88) refer to Murtagh (2007) who found that the learners of Irish in her study were driven more by instrumental goals such as possible career opportunities as opposed to intrinsic motivations. These observations were confirmed by Darmody & Daly who found that passing exams constituted the primary driving force among learners. Li et al. (2017) also established in a survey of 495 young adults that the prevailing motivation that emerged was instrumental with 50.21 % of respondents stating that they "simply wanted to pass Irish in their exams" and 40.42 % claiming that they learned Irish "to help get high points in the Leaving Certificate" (Li et al. 2017: 121). These findings also highlight that the main motivation in learning Irish has not changed since the participants in the present study were in school.

As stated in Chapter 2, Section 2.4.7, intrinsic motivation might prompt a closer, more regular engagement with the material during learning that results in information becoming more securely entrenched in long-term memory. On the other hand, purely instrumental reasons for learning, such as achieving immediate positive results in the guise of passing exams or getting a job, may lead to more superficial, short-term learning (see Chapter 2, Section 2.2.4, on memory consolidation).

This might also explain why in the present study less lexical knowledge was maintained over time in the case of those two participants who seemed most driven by instrumental motivation when they learned Irish at school. Ellen, and in particular Enda, on the other hand, exhibited a stronger identification with the language which may have prompted more constant, and attentive learning thereby explaining their higher proficiency levels as demonstrated in the Leaving Certificate examination and their better scores in testing on retention in session 1.

6.5.2 Cultural identity and heritage as a motivator

While one overwhelming statement cannot unequivocally be made with respect to motivation and long-term retention of Irish considering this is determined by an aggregate of closely interrelated factors and as stated in Chapter 4, Section 4.4.3, obtaining truly reliable indicators of motivation during learning after such a long incubation period is particularly challenging, it appears, nonetheless, that motivation driven by an identification with the language and an appreciation of its cultural significance might be a predictor of long-term retention. This identification may be critical when it comes to learning a minority language because in internalizing the cultural value of the language, learners might commit more to the learning process and in doing so be prompted to engage more regularly with the language, thereby not only leading to higher proficiency levels in the short-term but also fortifying its storage in long-term memory.

In their study of 12 adults learning Irish, Flynn & Harris (2016) also found that engaging with cultural identity and heritage emerged as a key motivator among these learners who, it must be noted, had undertaken the language course by choice. Flynn & Harris refer to this as a socio-cultural motivating factor which is based on "the learner's cultural background, linguistic heritage and personal identity", a factor they argue warrants greater attention and one that is hardly reported on in the literature on L2 learning motivation (2016: 381).

While all participants in the present study had positive attitudes towards the cultural significance of Irish, the question arises why participants differed with respect to the degree to which they internalised this identification, considering that all followed the same Irish curriculum and they all appeared interested in language learning in general. In examining the profiles of the participants, the geographical location where they grew up and went to school might be a decisive factor. While Cathy and Brendan grew up on the east coast, Enda and Ellen lived in the south-west region where traditionally more Irish was spoken and present in everyday life. This may have led to a more heightened sense of the importance of the language and possibly a more positive attitude toward the language in general despite the lack of enthusiasm experienced at times in the school context. Harris et al. (2006: 5) also refer to research that found a relation between region/location where school was attended and proficiency attained in Irish. It was found that pupils from the south-west of the island reached the highest levels of Irish language proficiency overall with learners in Dublin and surrounding areas demonstrating lower proficiency levels (where Cathy and Brendan attended school).

6.5.3 Irish language learning and the future self

Mehotcheva & Mytara (2019: 12) state that the status of the language can play a role in predicting retention levels during the incubation period. They argue that how a particular language is regarded can be context-dependent, so that in societies where a particular language enjoys prestige, attriters tend to lose less knowledge of that language than in contexts where associations with the language are less positive. In the present study it is also argued that not only societal status can influence how speakers perceive a language but also the personal associations that the individual has with it, and in the present study, particularly those stemming from experiences learning Irish in the classroom.

In exploring the issue of identity and Dörnyei's (2005) concept of the L2 motivational Self System which represents the notion that one can have various possible selves and that envisaging an ideal L2 self can support the individual in learning, it is relevant to ask what this self might have constituted in the case of Irish and in doing so look at the L2 persona that was *presented* to learners. As stated in Chapter 5, Section 5.1, the associations with the language as presented in school in the 1980s served more to distance learners rather than engage them. Considering also that part of the attraction of learning another language might be its exoticism, the insularity and the very familiarity of Irish culture might have dampened learner enthusiasm rather than encouraged the formation of an L2 self.

In learning a modern foreign language, on the other hand, there is an element of *anticipation* with respect to future possibilities. As the people of what was, back then, a relatively isolated island, many young people tended to look outward, beyond the island. Whilst Hourihane & Keane (2004) argue that for a long time Ireland was insular and slow to change, this might be more indicative of state level. The responses of the participants in the present study suggest that it was this very insularity that propelled them in learning modern foreign languages, especially in comparison to their experience in the Irish classroom in which the "future" did not entail any of the exciting opportunity and prospects that the modern classroom suggested.

Blackwood (2013:8) also emphasises the concept of future in understanding individual actions when it comes to language choice and states that the decision to use a particular language is a pragmatic one based on whether or not a language is perceived as important for one's future.

6.5.3.1 Motivation and minority languages

Masgoret & Gardner (2003: 172) also proposed that learners who wish to identify with an L2, its speakers and culture, will be more motivated than learners who are more indifferent with respect to the target language group. Gardner (2001: 11), for example, argues that:

> If the student learns the language just to get a good grade in the course, then he or she has little incentive to go beyond the class requirements. If he or she views it as a means of establishing real bonds of communication with another people, then she or he is truly learning the language.

However, as already stated, a challenge that is very specific to Irish and other such minority languages is that a monolingual group of speakers does not exist, therefore, rendering superfluous any real *need* to learn the language for integrative purposes. Flynn & Harris (2016: 382) refer to Singleton (1987), Duran (1995) and Ó Laoire (2000) who argue that communicative approaches to teaching Irish with the aim that learners may integrate into a target speaking community of Irish speakers need, therefore, to be reassessed.

On the other hand, Murtagh & Van der Slik (2004: 297) found that not only did learners who attended Irish-medium schools continue to use more Irish with their school-peers in social contexts but they were also more motivated overall in learning Irish compared to learners attending ordinary schools with this correlating with better performance in the final Leaving Certificate examination, and higher retention levels 18 months after instruction had ended.

6.5.3.2 Attitudes to the learning scenario

In the present study, negative attitudes with respect to how the language was taught as opposed to any unfavourable attitudes towards the language itself appeared to feed into motivation and its role in possibly impacting long-term knowledge retention.

In a survey published by Li et al. (2017) in which they explored attitudes to learning Irish in the current-day school system, the overriding opinion did not deviate greatly from that reported by the participants in this study. Current-day learners still lament the time invested in reading and literary engagement but paradoxically emphasise the importance of the language with respect to the identity of the nation. The participants in Harris & Murtagh (1999) also experienced Irish lessons as "boring, old-fashioned and repetitious" expressing lower motivation in terms of "commitment to learning Irish" despite recognising the symbolic value of the language (cited in Harris et al. 2006: 7).

As also stated in the interviews, standards and expectations were very high in Irish and this might also have led to demotivation among learners. This is in keeping with Dornyei (1994a: 276) who stated that in an education context, students' intrinsic interest can be overridden by extrinsic demands whereby the initial engagement with learning and the enjoyment experienced by the learner is transformed due to external pressures.

6.5.4 Motivation to maintain knowledge of the language

Practical circumstances can also influence whether or not contact with a language is maintained, such as time and the opportunity to find other interlocutors. In this respect inaction must not necessarily be viewed as indicative of negative attitudes. Pan & Berko-Gleason (1986: 198) refer to Gardner, Lalonde & MacPherson (1985) who also found that use of the language after official learning ended was not related to attitudes. Nonetheless, reasons for discontinued use can also have an emotional basis; memories from the time and experience of learning might encourage renewed engagement or when these are negative, block any desire to continue learning.

As already stated, the dynamic aspect of motivation renders examination particularly challenging, especially after very long incubation periods, (see Chapter 4, Section 4.4.3). However, despite the vulnerability of memory, it is interesting and noteworthy to observe and explore which, as well as why, particular overriding memories emerge to dominate the individual's narrative. In this study, the memory of the experience of learning Irish was primarily laced with negativity despite any nostalgia that might have emerged in the intervening years.

6.6 Incubation: Length, input and rehearsal

6.6.1 Length of incubation

Length of period of disuse was one of the factors isolated by Weltens & Grendel (1993: 144) that was observed as affecting attrition levels. In the present study, the length of incubation is relatively consistent across participants. Therefore, whether differing lengths of period of disuse lead to diverse retention levels could not be assessed. On the other hand, it was interesting to observe that there still was variation in performance among the participants *despite* similar lengths of incubation suggesting that other factors may be more decisive in influencing attrition.

6.6.2 Input and rehearsal during incubation

In fact, what might be more critical is not so much length of time since contact with the language ceased as *rehearsal* during incubation.[3] As stated in Chapter 2, Section 2.4.4., some studies have indeed established a positive role with respect to language use in this time, for example, Murtagh (2003) and Taura (2008), although as stated by Mehotcheva & Mytara (2019: 10), results overall on the role of rehearsal are mixed. More recently, Mickan and her colleagues (2021: 202) observed, however, that rehearsal did indeed emerge as a beneficial and affirming factor with respect to supporting knowledge retention. In an examination of L3 Spanish attrition among L1 German – L2 English university students in the six months after their study abroad in Spain, they established that variation in retention levels among their participants was strongly accounted for by degrees, as well as quality, of language use during this period. Those who maintained contact with native Spanish speakers performed better on tests after six months. It must be noted, however, that this was a very short incubation period relative to the present investigation.

In the present study, the participants, apart from Enda, had virtually no contact with the language since they finished school. Enda's input was mainly in the form of listening to radio or engaging in small talk with Irish speakers he randomly encountered when holidaying in Ireland. His retention levels were higher than the other three participants but it is surmised here that this is perhaps not due to rehearsal necessarily, or certainly not exclusively, since his language use and exposure during incubation was intermittent and passive, but rather it is due to a combination of factors; motivation combined with longer learning (one year at university, during which time it can be assumed that input was qualitatively rich) and higher proficiency levels at onset of incubation.

With respect to rehearsal during incubation, it is important to probe in greater detail the issue of frequency and quality of input in this time, aspects

3 The term *incubation* is used in L2/FL attrition studies to refer to the period between the point at which active contact with the language ceases and the later point in time when retention levels are tested (see Chapter 2, Section 2.4.4).

hitherto not always necessarily distinguished and comprehensively investigated within the attrition research domain. When they were asked if they were aware of the language on their visits to Ireland, two of the participants (Brendan and Cathy) claimed that they weren't, despite its pervasive but unobtrusive presence on public signs, in state ceremonial announcements and intermittent use in state-run media, for example. This suggests that the participants themselves did not actively engage with this input and so its effect in terms of activating the language could be expected to be minimal.

Considering, therefore, that in the present study active rehearsal did not truly take place during incubation, it cannot be answered here whether or not, or to what degree, it plays a role in averting attrition.

Seeking out opportunities to use Irish in Ireland, on the other hand, requires more effort and the practical limitations the participants in this study might have found themselves confronted with on their visits to Ireland such as family obligations and time limitations may not have made this a feasible option.

O'Grady & Lee (2011) argue that input is relevant not only for language acquisition, but also for maintaining the language. They refer to research which has shown that languages which are not used with sufficient frequency in a wide range of situations are subject to attrition (Cook 2003). They put forward the view that language maintenance is, in essence, a process of activation and strengthening whereby acquisition takes place when particular form-meaning mappings are activated in response to unambivalent instantiations in the input. Subsequent maintenance of the mapping depends on the extent to which it is strengthened through repeated opportunities for activation, presumably in both production and comprehension (O'Grady & Lee 2011:12).

Whilst necessary to explore in further studies, this prompts the tentative suggestion that language use might indeed play a role in supporting retention (in accordance with the findings of Mickan 2021) in that, had participants continued to engage to some degree with the language since finishing it at school, high levels of loss might have been avoided.

However, cessation of engagement with Irish after school seems to be a common occurrence, even among those who remain in Ireland and not just in the case of learners that had attended ordinary schools. Murtagh (2003), referred to in Murtagh & Van der Slik (2004: 299), found that only half the participants in her study that had attended an Irish-medium school and who in that time had also regularly interacted with their peers in Irish outside of the school context, continued to do so 18 months after leaving school. It is certainly noteworthy that even participants who had been motivated enough during school time to converse in Irish gradually stopped doing so with each other after school. Murtagh states that this serves to highlight the vulnerability of the Irish language, even considering that a social context in which the language could be used was already in place for the participants in her study.

Estimating the amount of rehearsal based on feedback in questionnaires as appears to be the typical approach, might lead to skewed assessments as respondents may not be keenly aware of how frequently and how much of the target language they actually use. (See Chapter 4, Section 4.4.3.3 on the challenges presented by questionnaires).

6.6.3 Re-exposure and the Activation Threshold Hypothesis

As stated, and further discussed in Chapter 2, Section 2.4.4, effects of rehearsal during incubation have either been put forward as corroboration of the ATH (Graham 2012) or in cases where rehearsal could not be isolated as a determining factor, as a negation of its explanatory power (Mehotcheva 2010). It is argued here that in the case of lexical items, *knowledge loss despite use* can only be said to defy the ATH if it can be ascertained that those specific items rehearsed during incubation can no longer be assessed at a later testing time. What was tested may not be that which was rehearsed and what was rehearsed cannot be reliably assessed in a questionnaire.

According to the ATH, rehearsal leads to reactivation which lowers threshold levels making subsequent access easier. As discussed in Chapter 2, Section 2.4.4, it could be expected that the retrieval of items used in rehearsal would be less effortful than in the case of those not triggered. This cannot be established in one test or even in a comparison of findings from an initial test conducted at onset of attrition and one at a later point in time, unless it can be assured that the tested items had actually been used in rehearsal. Instead, it is argued here, this can be more reliably assessed via intermittent testing during incubation in which reactivated items are tested again.

In this regard, it was indeed observed in the present study that items reactivated via relearning were more easily accessible in later testing than items newly learned (the pseudowords), suggesting, therefore, that the Activation Threshold Hypothesis cannot be categorically refuted (see Chapter 5, Section 5.2).

6.7 Complex Dynamic Systems Theory and the attrition of Irish

Observations in this study are in keeping with Complex Dynamic Systems Theory. As argued by Herdina & Jessner (2013: 2), language attrition is propelled by a number of interacting factors that are dynamic in nature with each language in the bi- or multilingual system being vulnerable to change accordingly. As emphasized by Herdina & Jessner and corroborated in this study, no one factor can be isolated and made accountable for attrition. Instead, it is driven by the interplay of interrelated psycholinguistic and sociolinguistic variables that is distinct to each individual. Attrition can, therefore, best be understood by observing how different factors interact and influence each other within each individual separately, so that generalisations about the trajectory of attrition can only be made with caution.

In this study, despite coherence across certain variables, especially in the case of three of the participants, performance was, nonetheless, individual and distinct. In the case of each individual, the factors typically put forward in explaining attrition interact to differing degrees. Cathy, who started learning Irish later than the other three participants, had, therefore, a shorter length of learning and she also achieved a lower grade than the other two participants that took the higher-level paper. On the other hand, the fact that she only started learning in 5th grade in primary school (age 11) but proceeded, nonetheless, to study the higher-level curriculum suggests high motivation on her part. Claiming that this was instrumental in nature, it was strong enough to propel her forward in pursuing the academically more demanding level.

In exploring the individual profile of Ellen and Brendan, despite similar age of acquisition, onset of attrition, length of incubation and content of learning, they differed in that Brendan took the lower-level paper and Ellen the higher one. Considering their results in other modern foreign languages were relatively on par without any significant deviation suggests that motivation might have been the factor that distinguished them. Brendan's lower motivation might have stemmed from a number of factors, his subjective experience of learning as dictated by his feelings towards the content of the curriculum and relationship with teachers. Ellen's could have been driven be the very environment where she grew up, a greater identification with the language and the teaching styles that she experienced. On the other hand, Enda's interest in the language was great enough to make him decide to study it at university.

Despite the dynamic interplay of factors in each individual giving rise to different scores in the pretest, all of the participants demonstrated reactivation of previously known knowledge. See Chapter 5 for a detailed overview of each individual's performance.

6.8 Support for the Savings Paradigm

In keeping with findings in Hansen et al. (2002), De Bot & Stoessel (2000) and De Bot et al. (2004) and as delineated in Chapter 5, support for the Savings Paradigm has been established in the present study. This is also in accordance with what Meara (2004: 152) observed in his simulations of vocabulary attrition, i.e., that lexical structures remain complete and unimpaired and that this knowledge can be reactivated even when it appears that it has been lost. Despite the long period of disuse in the present study, in which it seems lexical items had become dormant rather than depleted, on re-exposure, participants successfully accessed and retrieved some of this knowledge. In this regard it demonstrates a relearning advantage for previously known but seemingly forgotten items as demonstrated in the higher scores for old words compared to items learned for the first time (pseudowords) across all tests conducted after relearning.

In view of Meara's simulation of attrition events (see Chapter 2, Section 2.2.3), and the observation of the positive effects of relearning in the form of re-exposure in the present study, it might be expected that further exposure would also have knock-on effects for other items in the system not directly singled out for selection.

The degree of consolidation of knowledge in memory appears to be a critical aspect in understanding attrition. This was evident in the testing in the present study whereby performance was significantly better on previously known items compared to newly learned ones due seemingly to a deeper entrenchment of information that previously had experienced regular access and re-activation unlike information encountered for the first time.

6.8.1 Cross-linguistic influence

Failure to successfully recognise or recall the pseudowords in this study was seen as first and foremost to result from their *novelty*, meaning that they had not been accessed enough to leave a trace in memory. It was not considered to be due primarily to cross-linguistic effects stemming from the co-existence of English and German in the linguistic system; otherwise, it would be expected that more difficulties would also have emerged in accessing the previously known words too. What made a difference was the strength of information in memory, i.e., how well it was stored and consolidated. With each activation in the past, this trace was reinforced.

Whilst not denying that cross-linguistic influence can occur when more than one language is present within an individual's linguistic system (see Chapter 2, Section 2.1), cross-linguistic effects, it is argued here, are, nonetheless, not always indicative of attrition, or rather, one can only speak of attrition if the information had been truly learned or known in the first place.

For example, Mickan (2021: 151) reported evidence of cross-linguistic influence in a laboratory set-up whereby picture naming in L1 German and L2 English appeared to interfere in subsequent recall of recently learned L3 Spanish words (learned one day previously). Whilst undoubtedly cross-linguistic influence was at play, it is questionable if the problems in recall of these Spanish items can be deemed to be representative of attrition as Mickan surmises. Considering that the acquisition and learning of linguistic knowledge constitutes a *process,* as does attrition, and since in Mickan's study, participants only had one brief exposure to the target items, it is questionable whether these new Spanish words had in fact been truly *learned* in the first place.

Bjork & Bjork (2019: 166) also highlight the difference between learning and performance and argue that "learning can only be accessed at a delay" whereas performance refers to the "speed of producing the required information during instruction." They, therefore, caution against confusing performance with actual learning (see Chapter 2, Section 2.2.4, for more information on the role of storage strength in subsequent retrieval).

This leads therefore, to the question if one can truly talk about attrition of linguistic knowledge in such cases as in Mickan's study, considering also that there was no follow-up exposure to the items to ensure that the knowledge could become embedded in memory.

In the present study, on the other hand, the process of attrition is understood to refer to difficulties in accessing linguistic information that had become *fully consolidated* in the linguistic system in the past through regular attentive engagement with that language over time. It is argued that only in establishing difficulties in access (e.g., due to cross-linguistic influence – see Graham (2012) and Bardovi & Stringer (2010) for further discussion of cross-linguistic effects) to *such information* can one refer to attrition. However, as already stated, not all cross-linguistic influence necessarily implies attrition (e.g., everyday tip of the tongue phenomena, code-switching in bilinguals or cases, as in Mickan's study, where information has not become consolidated in the linguistic system in the first place). This is not to undermine the role played by cross-linguistic effects associated with the attrition process (see Chapter 2, Section 2.1). Cessation of use of a language inevitably results in greater quantitative use of another language or languages leading to a decrease in their threshold levels and a raising of the activation levels of the dormant language in keeping with the Activation Threshold Hypothesis.

For example, whilst it was not the main focus of present study, it was interesting to observe that when participants struggled to recall an Irish word when presented with the English translation, they sometimes offered the German translation first. This is not surprising considering that German was evidently also online (testing was conducted in the L2 environment of the participants (Germany)) so that competition was both expected and clearly at play as evidenced in the German translations that were intermittently offered when participants struggled to recall the Irish equivalent.

In this regard, it is also pertinent to note that all of the participants have long since moved away from Ireland and have acquired another language, German, to native-like levels (in the case of three of the participants). It is suggested that Irish knowledge long-term was, therefore, affected not only by disuse but also by competition from an already existing dominant language (English) followed subsequently by the extra cognitive effort dedicated to leaning a second language (German) later.

On this note, it is interesting to look at a different study (to the one just mentioned) of L3 Spanish attrition among L1 German – L2 English speakers after a six-month stay abroad in a Spanish-speaking country, in which Mickan (2021: 192–193) did not find conclusive evidence in terms of the role of L1 versus L2 effects in the case of L3 attrition, observing that regardless of how much of the L1 was used in the incubation period compared to the L2, L3 attrition rates were comparable across participants.

6.8.2 Ease of retrievability

Also posed in Chapter 2, Section 2.1.1, is the question why some items can be recalled or recognised in initial testing and others seem to have been forgotten years after contact with the language has ceased during which time there has been no exposure to it. Meara observed that activation was random in his simulation models of lexical attrition (see Chapter 2, Section 2.2.3) and based on findings in session 1 in the present study, this might indeed be the case.

On the other hand, it could possibly be that some items have become more entrenched in memory due to greater frequency of activation during learning in keeping with Schmid's (2007: 150) suggestion of a "saturation threshold", whereby linguistic knowledge becomes more robust through repeated activation during acquisition or learning that renders it less vulnerable to attrition (see Chapter 2, Section 2.1.1).

On the other hand, Larsen-Hall (2019: 13) argues that items that had been "better learned" will not necessarily be more immune to attrition than items that had been learned later, stating that support for such a threshold of knowledge has not yet been found. She refers to Graham (2012) who interpreted the L2 Spanish lexical loss in his study as being random since some basic words could also not be remembered. It must be noted, however, that the subjects in his study had only spent between one and a half to two years in the Spanish-speaking country and considering the nature of their work as missionaries, would most likely have predominantly used the language orally. It is, therefore, possible that such a threshold as posited by Schmid had not been reached. Larson-Hall also refers to the seemingly arbitrary pattern of L2 Spanish lexical loss in Bierling's (1990) investigation of lexical attrition among L1 English speakers to support her proposition that loss is random. However, Larson-Hall, admits that a failing in Bierling's study was that the items tested may not have been truly learned in the first place (see Chapter 2, Section 2.3.1 for a review of the abovementioned studies).

Due to rapid accumulation of vocabulary in learning Irish in Ireland and engagement with texts from an early age, it is difficult to reliably assess frequency of individual lexical item use in the case of Irish (see Chapter 4, Section 4.4.1.4). Nonetheless, on a closer examination of words known in session 1, basic everyday vocabulary such as *teach* (house), *uisce* (water), *doras* (door), *buachaill* (boy), *lá* (day), *bainne* (milk), and scéal (story) was still known, even by Cathy whose performance was significantly weaker than the other participants, suggesting that regular exposure to items during learning strengthens their consolidation in memory and this might indeed result in easier retrieval years later.

Chapter 7: Overview and closing remarks

This study has served as an important confirmation of the Savings Paradigm on which, apart from a small number of studies, little research has been conducted. Whilst vast amounts of Irish linguistic knowledge had attrited in the 27–32 years of incubation for the participants in this study during which they, apart from one, had no active engagement or contact with the language, evidence of lexical information that was still intact was found on re-exposure to selected items in the form of relearning. On subsequent testing, participants scored significantly better in recall and recognition tests of words previously known but forgotten during incubation compared to newly learned words in the guise of pseudowords which they encountered for the first time in the testing sessions. It was also explored whether scores on the pretest could predict scores on subsequent tests after relearning, although this was only a tentative observation since each participant was tested on his/own individual set and number of words as dictated by performance in the pretest. It was, nonetheless, observed that those two participants who scored lowest in the pretest gained ground in relearning, clearly benefitting from the effects of re-exposure (see analysis of results, Chapter 5).

Participants scored very differently in the pretest. Words not recalled or recognised in this pretest were deemed forgotten even though they had featured in the literature that the participants had dealt with intensely during their school-time in preparation for their final state school examination. Whilst the main aim of this pretest was to filter words to be used in relearning, the divergent range of scores that emerged across participants despite certain similarities in their learning histories and length of incubation is notable but not surprising in keeping with Complex Dynamic Systems Theory. As also found by Weltens & Grendel (1993: 144), a number of interrelated factors were found to drive attrition. The potency of each factor and the degree to which it influenced, and in turn was simultaneously impacted itself by these other factors, was unique to each individual. The factors explored in this study were age of acquisition, length of learning and learning environment/context of learning, the content of the curriculum and teaching methodologies, motivation in learning, attitudes towards Irish, proficiency level attained, age at onset of incubation, rehearsal during incubation and length of incubation.

Whilst there was a certain coherence across participants with respect to some of the variables, such as age at onset of incubation and context of learning and school curriculum as well as length of incubation, there were differences with respect to level of proficiency attained, attitudes and motivation as well as rehearsal during incubation. While three of the participants starting learning Irish once they began primary school which was at the age of four or five, Cathy did not start learning until she was 11 years old.

7.1 Key findings

A detailed analysis of the responses of the participants and their performance in the tests shows:

(i) Great amounts of language loss after no contact with the language over a period of 25–30 years, despite instruction over a period of 13–14 years.
(ii) Evidence, nonetheless, for language traces in the form of savings.
(iii) Relearning prompts reactivation and leads to more rapid learning of previously known information than the learning of completely new items.
(iv) It appears that proficiency at onset of attrition seems to be a significant indicator of attrition levels. Interestingly, higher proficiency is also not completely independent of motivation and whilst motivation is dynamic and fluid, it emerged in the interviews that the two strongest cases in terms of initial proficiency also showed a greater interest in learning Irish at school due to a stronger identification with the cultural and historical value of the language.
(v) Language rehearsal during incubation might serve to deter extreme levels of attrition. Considering the proficiency levels that the participants had attained in a demanding examination at the end of secondary school education and the vast drop in retention levels in the intervening years up until re-exposure in this study, it is suggested that rehearsal during incubation might indeed have served to keep the language online with each exposure to an item lowering its activation level. That rehearsal might aid ease of access

is also suggested by the participants' performance on the tests post-relearning where through activation in the form of relearning, information that participants claimed to have forgotten had become more easily accessible and could be retrieved again. Considering that this emerged after just a few exposures and after such a long incubation period, it is highly conceivable that had this happened throughout incubation, activation levels would have been lower and information correspondingly more easily accessible.

(vi) Sociolinguistic factors appear to play a decisive role in understanding attrition of minority languages that have been learned in an instruction setting. Variables such as status of the language, content of curriculum, approaches to teaching, learner attitudes and motivation and context and frequency of language use are interrelated and aid in predicting degrees of long-term engagement with the language, which can have significant psycholinguistic repercussions with respect to storage of that language in the linguistic system and corresponding activation thresholds.

(vii) No one factor can be singled out as predicting attrition but instead factors interact in unique and dynamic ways that are distinct for each individual.

(viii) While it appeared initially that great loss had taken place, evidence of traces of lexical knowledge was still found, thereby suggesting that the issue is not one of loss per se but of difficulty in accessing information that is still intact. This consolidation of lexical knowledge in long-term memory might have been aided by the strong emphasis on reading and writing in the Irish curriculum at the time when the participants in this study attended school. This is in keeping with findings on the importance of literacy in language acquisition (see Chapter 6, Section 6.4).

(ix) Difficulties in initial remembering do not inevitably imply lower levels of success in subsequent learning. The participants with the two lowest scores in the pretest gained significant ground in relearning, despite having more words to learn in total. Due to a different number and set of items, the participants cannot be directly compared with each other on this aspect.

Whilst words were tested in isolation in this study (see Chapter 4, Section 4.4.1 for justifications), a model similar to Meara's would suggest that even more reactivation could be anticipated given the interconnections between individual items (see Chapter 2, Section 2.2.3).

With respect to what was remembered in the initial testing, even though two of the participants performed below 50 % on the lexical item recall and recognition test, this serves as an invaluable snapshot of retention levels. How one views this performance draws on the age-old adage of seeing the glass as half full or half-empty. This book adopts the former view, embracing the evidence that linguistic knowledge can survive years of inactivation and be reactivated when the right cues become available, thereby speeding up any ensuing relearning endeavours.

The advantage of this study was that it enabled a narrowing down of the factors that could impact attrition levels since the participants had very similar profiles in a number of respects. Weltens & Grendel (1993) refer to other studies on second language attrition, which, when compared to each other, reach quite different outcomes, demonstrating that "one fits all", generalized statements cannot be made as to what factors specifically affect attrition since not only the influencing variables themselves vary across studies but also their potency in each individual case and how they interrelate accordingly. In the present study, it was insightful to see how different the results were. These insights were achieved by personal engagement with the participants in the form of a case study, a sensibility that might not have been attainable had the numbers of participants been bigger and the interview taken more the form of a survey in which participants only had the chance to select from a limited pre-specified choice of responses by ticking a box.

The nature of the interviews was such to encourage as much input as possible from the participants. Whilst guided by research on attrition, it was important that the questions were open enough to allow for participants to contribute their own input, allowing them to perhaps refer to an issue or factors that may have missed in the preparation of the interview or had not yet been dealt with in the literature.

7.2 Considerations in supporting long-term retention of linguistic information

A number of interpretations of attrition commonly proposed in the literature were explored in Chapter 2, Section 2.1 of this book. It is argued in this study that if linguistic information has the chance to become properly consolidated in memory, then whilst difficulties in access arise due to long periods of disuse and inactivation of this knowledge, whereby simultaneously competing languages become more readily available for selection, this knowledge remains available nonetheless and through re-exposure and the right cues can become accessible again. Therefore, an important consideration in language learning in the first place is ensuring that linguistic information reaches this level of entrenchment in memory, highlighting, therefore, the necessity to examine learning processes and educational approaches to teaching languages, and in particular, in the case of this study, minority languages. In exploring possibilities to stave off attrition, the learning context and the social status of the individual language consequently warrant careful exploration as was done in this study.

With respect to learning, Bjork & Bjork (2019: 166) highlight the effort that is necessary on the part of the learner and emphasise the importance of "training, practice, and experience." It is important, therefore, in education that efforts are made to motivate pupils and awaken in them a positive connection to the language to increase chances of a continued interest in maintaining and even developing knowledge of the language after official instruction has ceased. More immediately this might lead to higher proficiency levels, which have been shown to positively correlate with better retention over time.

7.3 Qualifications and future research

A cross-case analysis was conducted in this study whereby each case is presented individually followed by a comparison against common variables to see if certain issues or aspects emerge as salient and insightful with respect to explaining experiences of attrition and possible savings.

Inevitably, and as also argued by Duff (2018: 322), the issue of generalisability needs to be addressed in terms of interpretation of observations emanating from case study research. Duff refers to scholars who argue that in case study research a degree of generalisation of the observations and conclusions drawn is possible, even when a small number of cases has been examined. Naturally, the greater the number of cases, the increase in the potential for generalisation. Nonetheless, working with a small number of cases may facilitate a more intense examination and focus on the part of the researcher, increasing the potential to make novel observations. Whilst the aim is not to generalise these findings to speak for larger populations, their significance can be twofold; one is to provide further support for, or, on the other hand, to refute and call into question, findings and common verdicts in the literature. The second is that the tool used, i.e., the case study, to generate these findings offers a unique possibility to discover factors or elements that could be relevant to the field as a whole and possibly indicative of larger cohorts but which had hitherto been overlooked in the vastness of larger empirical endeavours. One could possibly view the case study as a first step in the exploration of these factors, followed by the testing of larger groups using a variety of tools and approaches to explore these phenomena further. As Duff argues, should generalisation take place, it should be for analytic purposes, i.e., the generation or supporting of a theory or model, but not to make statements that aim to unambiguously portray the overall experience of the language learner or, in this case, the language attriter, as one single, uniform or standard phenomenon.

In this study, and perhaps somewhat contrary to Duff, the observations made, whilst not deemed unequivocally representative of larger populations, are, nonetheless, availed of to inspire further conversations about the nature of the language learning experience in general, the focus and content that ministries of education enforce in curriculum planning and the role that language status can play in encouraging individuals to learn the language or not. Most importantly with respect to the focus of this study, it also sheds light on possible reasons why some individuals attrite more than others and the question of whether the observations found in this study are transferable to other language landscapes. Interestingly too, and as already stated, it is the very individual nature of the attrition process that has prompted a greater call among some scholars for more qualitative research in this area (see Chapter 4, Section 4.4). At most it is argued here, general trends in the attrition process can be established but there is much individual variation and deviation from this path with case studies enabling a closer exploration of possible reasons for this.

Considering that this was a group of bilingual English-German speakers, it would have been interesting to compare their performance to an equivalent group, also living abroad and who had learned Irish at school in the same period but who are monolingual English (e.g., individuals now living in the U.K. or North America) to see if being bilingual and having two languages in the linguistic system affects access to the target language.

Indeed, one general critique pertaining to some of the available literature and studies on attrition is a lack of clarity and unity across research on a definition of attrition. Without such a general consensus and the determination of a clear starting point for analysis (i.e., how entrenched must knowledge be to label problems in accessing it *attrition* and not difficulties arising from incomplete acquisition or learning), results might be misleading and interpretations inappropriately extended to account for the general phenomenon and the wider population.

Despite limitations with respect to number of participants, evidence in support of savings and in keeping with Hansen (2002) and De Bot & Stoessel (2000) was established. Naturally however, attrition of lexical items is a phenomenon that is complex, random and distinctly individual. Therefore, more research in this area can benefit findings hitherto detected.

In this respect, a quote by Gardner (1985: 5) is considered particularly fitting, whereby he states that "there is no substitute for replication. One study, no matter how carefully conducted, cannot be taken as conclusive. It is only with repeated investigation that the complexities of an area can be truly appreciated and comprehended."

References

Abels, M. (1994), *Ken ik dit woord?* Unpublished doctoral dissertation, Nijmegen: Catholic University of Nijmegen.

Abutalebi, Jubin; Tettamanti, Marco & Perani, Daniela (2009), The bilingual brain: Linguistic and non- linguistic skills. *Brain and Language* 109: 2, 51–54.

Acquaviva, Paolo (2014), The categories of Modern Irish verbal inflection. *Journal of Linguistics* 50, 537–586.

Almqvist, Bo (2011), Kenneth Jackson and Peig Sayers: The perfect memory. *Béaloideas; The Journal of the Folklore of Ireland Society* 79, 22–43.

Ammerlaan, Ton (1996), *'You get a bit wobbly...' – Exploring bilingual lexical retrieval processes in the context of first language attrition.* Unpublished Doctoral Dissertation, Nijmegen: Katholieke Universiteit.

Arndt, Jason; Lee, Karen & Flora, David B. (2008), Recognition without identification for words, pseudowords and nonwords. *Journal of Memory and Language* 59: 3, 346–360.

Au, Terry Kit-Fong; Knightly, Leah M.; Jun, Sun-Ah & Oh, Janet S. (2002), Overhearing a language during childhood. *Psychological Science* 13, 238–243.

Bahrick, Harry P. (1984a), Fifty Years of Second Language Attrition: Implications for Programmatic Research. *The Modern Language Journal* 68: 2, 105–118.

Bahrick, Harry P. (1984b), Semantic memory content in permastore: Fifty years of memory for Spanish learned in school. *Journal of Experimental Psychology: General* 113: 3, 1–29.

Bahrick, Harry P.; Hall, Lynda K. & Baker, Melinda, K. (2013), *Life-Span Maintenance of Knowledge*. New York: Psychology Press.

Bahrick, Harry P. & Phelphs, Elizabeth (1987), Retention of Spanish vocabulary over 8 years. *Journal of Experimental Psychology Learning Memory and Cognition* 13: 2, 344–349.

Baker, Colin (1988), *Key Issues in Bilingualism and Bilingual Education.* Clevedon: Multilingual Matters Ltd.

Baker, Colin (1992), *Attitudes and Language*. Clevedon: Multilingual Matters Ltd.

Bardovi-Harlig, Kathleen & Stringer, David (2010), Variables in second language attrition: Advancing the state of the art. *Studies in Second Language Acquisition* 32, 1–45.

Bardovi-Harlig, Kathleen & Stringer, David (2013), The lexicon in second language attrition: What happens when the cat's got your tongue? In: Altarriba, Jeanette & Isurin, Ludmila (eds.), *Memory, Language and Bilingualism: Theoretical and Applied Approaches*. Cambridge: Cambridge University Press, 291–318.

Bardovi-Harlig, Kathleen & Stringer, David (2019), Syntax and Phonology in L2 Attrition: Modularity and resilience. In: Schmid, Monika S. & Köpke, Barbara (eds.), *The Oxford Handbook of Language Attrition*. UK: Oxford University Press, 1–15.

Beeckmans, Renaud; Eyckmans, June; Janssens, Vera; Dufranne, Michel & Van de Velde, Hans (2001), Examining the Yes/No vocabulary test: Some methodological issues in theory and practice. *Language Testing* 18:3, 235–274.

Ben-Rafael, Miriam & Schmid, Monika S. (2007), Language attrition and ideology: Two groups of immigrants in Israel. In: Köpke, Barbara; Schmid, Monika S.; Keijzer, Merel & Dostert, Susan (eds.), *Language Attrition: Theoretical Perspectives*. Amsterdam: John Benjamins, 205–226.

Berko-Gleason, Jean (1982), Insights from child language acquisition for second language loss. In: Lambert, Richard D. & Freed, Barbara F. (eds.), *The Loss of Language Skills*. Rowley, MA: Newbury House, 13–23.

Berman, Ruth & Olshtain, Elite (1983), Features of first language transfer in second language attrition. *Applied Linguistics* 4, 222–234.

Bierling, Marilyn R. (1990), *Foreign language vocabulary attrition: A study of college-level Spanish students over an 18-month period*. PhD thesis, Michigan: Michigan State University.

Birdsong, David (1992), Ultimate attainment in second language acquisition. *Language* 68: 4, 706–755.

Bjork, Robert (2012a), "Storage Strength versus Retrieval Strength." GoCognitive: Education Tools for Cognitive Neuroscience. https://www.youtube.com/watch?v=1FQoGUCgb5w (accessed March 29, 2021).

Bjork, Robert (2012b), "The Theory of Disuse and the Role of Forgetting in Human Memory." GoCognitive: Education Tools for Cognitive Neuroscience. https://www.youtube.com/watch?v=Hv6Vye1JCjo (accessed March 29, 2021).

Bjork, Robert A. & Bjork, Elizabeth L. (2019), Forgetting as learning. *Advances in Physiology Education* 43, 164–167.

Blackwood, Janet (2013), Language choice motivations in a Bribri community in Costa Rica. *Faculty Publications* 2, 6–20.

Borg, Simon (1998), Teachers' pedagogical systems and grammar teaching: A qualitative study. *TESOL Quarterly* 32, 9–38.

Bowers, Jeffrey S.; Mattys, Sven L. & Gage, Suzanne H. (2009), Preserved implicit knowledge of a forgotten childhood language. *Psychological Science* 20: 9, 1064–1069.

Bradley, Michael (2014), Is it possible to revitalize a dying language? An examination of attempts to halt the decline of Irish. *Open Journal of Modern Linguistics* 4, 537–543.

Briggs, John & Peat, David (1989), *Turbulent mirror. An Illustrated Guide to Chaos Theory and the Science of Wholeness*. New York: Harper & Row Publishers.

Bylund, Emanuel (2009), Maturational constraints and first language attrition. *Language Learning* 59: 3, 687–715.

Bylund, Emanuel & Ramírez-Galan, Pedro (2016), Language aptitude in first language attrition: A study on late Spanish-Swedish bilinguals. *Applied Linguistics* 37: 5, 621–638.

Cargile, Aaron C.; Giles, Howard; Ryan, Ellen & Bradac, James (1994), Language attitudes as a social process: A conceptual model and new directions. *Language and Communication* 14: 3, 211–236.

Chamorro, Gloria; Sorace, Antonella & Sturt, Patrick (2016), What is the source of L1 attrition? The effect of recent L1 re-exposure on Spanish speakers under L1 attrition. *Bilingualism: Language and Cognition* 19, 520–532.

Chan, HuiPing; Lowie, Wander & De Bot, Kees (2016), Input outside the classroom and vocabulary development: A dynamic perspective. In: Reiff, Monika & Robinson, Justyna (eds.), *Cognitive perspectives on bilingualism* (Trends in Applied Linguistics). Boston, Berlin: De Gruyter Mouton, 183–203.

Chen, Judy F.; Warden, Clyde A. & Chang, Huo-Tsan (2005), Motivators that do not motivate: The case of Chinese EFL learners and the influence of culture on motivation. *TESOL Quarterly* 39: 4, 609–633.

Cherciov, Mirela (2013), Investigating the impact of attitude on first language attrition and second language acquisition from a Dynamic Systems Theory perspective. *International Journal of Bilingualism* 17: 6, 716–733.

Choi, Jinny K. (2003), Language attitudes and the future of bilingualism: The case of Paraguay. *International Journal of Bilingual Education and Bilingualism* 6, 81–94.

Clément, Richard; Gardner, Robert C. & Smythe, Padric C. (1977), Motivational variables in SLA: A study of Francophones learning English. *Canadian Journal of Behavioural Science* 9: 2, 123–133.

Cohen, Andrew D. (1986), Forgetting foreign language vocabulary. In: Weltens, Bert; De Bot, Kees & Van Els, Theo (eds.), *Language Attrition in Progress*. Dordrecht: Foris, 143–158.

Cohen, Andrew D. (1989), Attrition in the productive lexicon of two Portuguese third language speakers. *Studies in Second Language Acquisition* 11: 2, 135–149.

Cook, Vivian (2003), *Effects of the Second Language on the First*. Clevedon: Multilingual Matters.

Coulmas, Florian (1998), Language rights: Interests of state, language groups and the individual. *Language Sciences* 20: 1, 63–72.

Council of Europe (1992), *European Charter for Regional or Minority Languages*. European Treaty Series – No. 148. https://www.coe.int/en/web/european-charter-regional-or-minority-languages (accessed March 29, 2021).

Cummins, Jim (1979), Cognitive/academic language proficiency, linguistic interdependence, the optimum age question and some other matters. *Working papers on bilingualism* 19, 197–205.

Cummins, Jim (2011), Literacy engagement: Fuelling academic growth for English learners. *The Reading Teacher* 65, 142–146.

Darmody, Merike & Daly, Tania (2015), *Attitudes towards the Irish Language on the Island of Ireland*. ESRI, Dublin: Foras na Gaeilge.

De Bot, Kees (1997), Language attrition. In: Tucker, G. Richard & Corson, David (eds.), *Encyclopedia of Language and Education. Vol. 1.* Dordrecht: Kluwer Academic Publishers, 51–62.

De Bot, Kees (2007), Dynamic systems theory, lifespan development and language attrition. In: Köpke, Barbara; Schmid, Monika S.; Keijzer, Merel & Dostert, Susan (eds.), *Language Attrition: Theoretical Perspectives.* Amsterdam: John Benjamins, 53–68.

De Bot, Kees & Gorter, Durk (2005), A European perspective on heritage languages. *The Modern Language Journal* 89: 4, 612–616.

De Bot, Kees & Larsen-Freeman, Diane (2011), A dynamic approach to second language development: Methods and techniques. In: Verspoor, Marjolijn; Lowie, Wander & De Bot, Kees (eds.), *Researching Second Language Development from a Dynamic Systems Theory Perspective.* Amsterdam, Philadelphia: John Benjamins, 5–24.

De Bot, Kees & Lowie, Wander (2010), On the stability of representations in the multilingual lexicon. In: Sicola, Laura & Pütz, Martin (eds.), *Cognitive Processing in Second Language Acquisition: Inside the Learner's Mind.* Amsterdam: John Benjamins, 117–133.

De Bot, Kees & Stoessel, Saskia (2000), In search of yesterday's words: Reactivating a long forgotten language. *Applied Linguistics* 21: 3, 364–388.

De Bot, Kees & Weltens, Bert (1995), Foreign language attrition. *Annual Review of Applied Linguistics* 15, 151–164.

De Bot, Kees; Martens, Vanessa & Stoessel, Saskia (2004), Finding residual lexical knowledge: The "savings" approach to testing vocabulary. *International Journal of Bilingualism* 8, 373–382.

De Bot, Kees; Lowie, Wander & Verspoor, Marjolijn (2007), A Dynamic Systems Theory approach to second language acquisition. *Bilingualism: Language and Cognition* 10: 1, 7–21.

De Brún, Fionntán (2018), Escaping the 'Shower of Folly' – the Irish language, revivalism and the history of ideas. *Maynooth Philosophical Papers* 9, 43–58.

Deci, Edward L.; Vallerand, Robert J.; Pelletier, Luc G. & Ryan, Richard M. (1991), Motivation in education: The self-determination perspective. *The Educational Psychologist* 26, 325–246.

De Leeuw, Esther; Mennen, Ineke & Scobbie, James (2012), Dynamic systems, maturational constraints and L1 phonetic attrition. *International Journal of Bilingualism* 17: 6, 683–700.

Dewaele, Jean-Marc & Pavelescu, Liana (2021), The relationship between incommensurable emotions and willingness to communicate in English as a foreign language: A multiple case study. *Innovation in Language Learning and Teaching* 15: 1, 66–80.

Dörnyei, Zoltan (1994), Motivation and motivating in the foreign language classroom. *The Modern Language Journal* 78: 3, 273–284.

Dörnyei, Zoltan (2000), Motivation in action: Towards a process-orientated conceptualization of student motivation. *British Journal of Educational Psychology* 70: 4, 519–538.

Dörnyei, Zoltan (2001), *Teaching and Researching Motivation.* Essex, England: Pearson.

Dörnyei, Zoltan (2005), *The Psychology of the Language Learner: Individual Differences in Second Language Acquisition*. London: Lawrence Erlbaum.

Dörnyei, Zoltan & Otto, Istvan (1998), Motivation in action: A process model of L2 motivation. *Working Papers in Applied Linguistics* 4, 43–69.

Dostert, Susan (2009), *Multilingualism, L1 Attrition and the Concept of 'native speaker'*. PhD thesis, Düsseldorf: Heinrich-Heine Universität.

Duff, Patricia (2008), *Case Study Research in Applied Linguistics.* New York: Routledge.

Duff, Patricia (2018), Case study research in applied linguistics. In: Litosseliti, Lia (ed.), *Research methods in linguistics (2nd ed.)*. London: Bloomsbury, 305–330.

Duran, James (1995), Dialects, speech communities, and Applied Linguistics: A realistic approach to the teaching of Irish in non-Irish speaking areas. *Journal of Celtic Language Learning* 1, 21–37.

Ecke, Peter (2004), Language attrition and theories of forgetting: A cross-disciplinary review. *International Journal of Bilingualism* 8: 3, 321–354.

Ellis, Rod (1994), A theory of instructed second language acquisition. In: Ellis, Nick C. (ed.), *Implicit and Explicit Learning of Languages.* London, UK: Academic Press, 81–113.

Fergussen, Rosalind & Manser, Martin H. (2011), *The Complete Guide to Grammar*. UK: Arcturus Publishing Limited.

Ferriter, Diarmaid (2004), *The Transformation of Ireland 1900–2000.* UK: Profile Books Ltd.

Fishman, Joshua (1991), *Reversing Language Shift.* UK: Multilingual Matters.

Flynn, Colin & Harris, John (2016), Motivational diversity among adult minority language learners: Are current theoretical constructs adequate? *Journal of Multilingual and Multicultural Development* 37, 1–14.

Fujita, Mariko (2002), *Second Language English Attrition of Japanese Bilingual Children.* PhD thesis, Tokyo: Temple University.

Garcia, Nuria (2015), Tensions between cultural and utilitarian dimensions of language: A comparative analysis of 'multilingual' education policies in France and Germany. *Current Issues in Language Planning* 16, 43–59.

Gardner, Robert C. (1985a), The Attitude/Motivation Test Battery: Technical Report. https://publish.uwo.ca/~gardner/docs/AMTBmanual.pdf (Accessed March 29, 2021).

Gardner, Robert C. (1985b), *Social Psychology and Second Language Learning: The Role of Attitudes and Motivation.* London: Edward Arnold.

Gardner, Robert C. (1988), The socio-educational model of second-language learning: Assumptions, findings, and issues. *Language Learning* 38: 1, 101–126.

Gardner, Robert C. (2001), Integrative motivation and second language acquisition. *Motivation and Second Language Acquisition* 23, 1–19.

Gardner, Robert C. & Lysynchuk, Linda M. (1990), The role of aptitude, attitudes, motivation and language use on second-language acquisition and retention. *Canadian Journal of Behavioural Science* 22: 3, 254–270.

Gardner, Robert C. & Smythe, Padric C. (1975), Motivation and second language acquisition. *The Canadian Modern Language Review* 31, 218–230.

Gardner, Robert C.; Lalonde, Richard N. & MacPherson, Janet (1985), Social factors in second language attrition. *Language Learning* 35, 519–540.

Gardner, Robert C.; Tremblay, Paul F. & Masgoret, Anne-Marie (1997), Towards a full model of second language learning: An empirical investigation. *The Modern Language Journal* 81: 3, 344–362.

Gardner, Robert C.; Masgoret, Anne-Marie; Tennant, Jeff & Mihic, Ljiljana (2004), Integrative motivation: Changes during a year-long intermediate-level language course. *Language Learning* 54: 1, 1–34.

Genevska-Hanke, Dobrinka (2017), Interpersonal variation in late L1 attrition and its implications for the competence/performance debate. In: Levkovych, Nataliya & Urdze, Aina (eds.), *Linguistik im Nordwesten: Beiträge zum 8. Nordwestdeutschen Linguistischen Kolloquium, Bremen, 13.-14.11.2015.* Bochum: Universitätsverlag Brockmeyer (Diversitas Linguarum 42), 1–31.

Gharibi, Khadijeh & Boers, Frank (2017), Influential factors in incomplete acquisition and attrition of young heritage speakers' vocabulary knowledge. *Language Acquisition* 24: 1, 52–69.

Godsall-Myers, Jean (1981), *The Attrition of Language Skills in German Classroom Bilinguals – A Case Study.* Unpublished PhD Dissertation, Pennsylvania: Bryn Mawr College.

Gollan, Tanar H.; Montoya, Rosa; Fennema-Notestine, Christine & Morris, Shaunna (2005), Bilingualism affects picture naming but not picture classification. *Memory and Cognition* 33: 7, 1220–1234.

Government of Ireland: Department of Education and Skills (1984), *The Report on the In-Service Education of Teachers.* Dublin: The Stationery Office.

Government of Ireland: Department of Education and Skills (2010). *20-Year Strategy for the Irish Language 2010–2030.* https://www.gov.ie/en/publication/f67134-20-year-strategy-for-the-irish-language-2010-2030/ (accessed March 29, 2021).

Government of Ireland: Department of Education and Skills (2015). *20-Year Strategy for the Irish Language 2010–2030. Progress Report: 2010 to 2015.* https://www.gov.ie/en/publication/f67134-20-year-strategy-for-the-irish-language-2010-2030/ (accessed March 29, 2021).

Government of Ireland: Department of Education and Skills (2015). *Annual Statistical Report 2014/15.* https://www.education.ie/en/publications/statistics/statistical-reports/ (accessed March 29, 2021).

Government of Ireland: Department of Education and Skills (2020). *Modern Foreign Languages: A Report on the Quality of Practice in Post-Primary Schools.* https://www.education.ie/en/publications/inspection-reports-publications/evaluation-reports-guidelines/?pageNumber=1 (accessed March 29, 2021).

Grabe, William (2009), *Reading in a Second Language: Moving from Theory to Practice*. New York: University Press.

Graham, C. Ray (2012), Vocabulary attrition in adult speakers of Spanish as an L2. In: Hansen, Lynne (ed.), *Second Language Acquisition Abroad: The LDS Missionary Experience*. Amsterdam: John Benjamins, 135–184.

Grendel, Marjon (1993), *Verlies en herstel van lexicale kennis [The loss and regaining of lexical knowledge]*. Unpublished Ph.D. Dissertation, Nijmegen: University of Nijmegen.

Gürel, Ayşe (2004), Selectivity in L2-incuded L1 attrition: A psycholinguistic account. *Journal of Neurolinguistics* 17: 1, 53–78.

Guthrie, John T. (2004), Teaching for literacy engagement. *Journal of Literacy Research* 36: 1, 1–30.

Hansen, Lynne (1990), The attrition of Japanese by English-speaking children: An interim report. *Language Sciences* 12: 4, 367–377.

Hansen, Lynne (1999), Not a total loss: The attrition of Japanese negation over three decades. In: Hansen, Lynne (ed.), *Second Language Attrition in Japanese Contexts*. New York: Oxford University Press, 142–153.

Hansen, Lynne (2001), Language attrition: The fate of the start. *Annual Review of Applied Linguistics* 21, 60–73.

Hansen, Lynne (2011), The acquisition, attrition, and relearning of mission vocabulary. In: Schmid, Monika S. & Lowie, Wander (eds.), *Modeling Bilingualism: From Structure to Chaos*. Amsterdam, Philadelphia: John Benjamins, 115– 134.

Hansen, Lynne & Chantril, Ching-Fen (1999), Literacy as a second language anchor: Evidence from L2 Japanese and L2 Chinese. In: Robinson, Peter (ed.), *Representation and Process: Proceedings of the 3rd Pacific Second Language Research Forum 1*. Tokyo: PacSLRF 99, Aoyama University, 279–286.

Hansen, Lynne & Newbold, Jesse (1997), Literacy as an anchor of the spoken language: Evidence from the acquisition and attrition of Japanese. Paper presented at the AAAL Annual Conference, Orlando, FL.

Hansen, Lynne & Shewell, Justin (2002), Keeping a second language: The influence of literacy and motivation in the attrition of Korean, Japanese, and Chinese. *Journal of the Applied Linguistics Association of Korea* 18: 2, 61–82.

Hansen, Lynne; Umeda, Yukako & McKinney, Melanie (2002), Savings in the relearning of second language vocabulary: The effects of time and proficiency. *Language Learning* 52: 4, 653–678.

Hansen, Lynne; Colver, Andrew; Chong, Wonhye; Pereira, Helama; Robinson, Jeremy; Sawada, Akihiro & Miller, Ronald M. (2012), The lost word: Vocabulary attrition in six mission languages. In: Hansen, Lynne (ed.), *Second Language Acquisition Abroad: The LDS Missionary Experience*. Amsterdam: John Benjamins, 111–134.

Harris, John & Murtagh, Lelia (1999), *Teaching and Learning Irish in Primary Schools*. Dublin: Institiúid Teangeolaíochta Éireann.

Harris, John; Forde, Patrick; Archer, Peter; Nic Fhearaile, Siobhán & O'Gorman, Mary (2006), *Irish in Primary Schools: Long-Term National Trends in Achievement*. Dublin: An Roinn Oideachais agus Eolaíochta (Department of Education and Science).

Hedgcock, John (1991), Foreign language retention and attrition: A study of regression models. *Foreign Language Annals* 24: 1, 43–55.

Herdina, Philip & Jessner, Ulrike (2002), *A Dynamic Model of Multilingualism. Perspectives of Change in Psycholinguistics*. Clevedon: Multilingual Matters.

Herdina, Philip & Jessner, Ulrike (2013), The implications of language attrition for dynamic systems theory: Next steps and consequences. *International Journal of Bilingualism* 17: 6, 752–756.

Hickey, Raymond (1985), Salient features of Irish syntax. *Lingua Posnaniensis* 26, 15–26.

Hickey, Raymond (2009), Language use and attitudes in Ireland. A preliminary evaluation of survey results. In: Ó Catháin, Brian (ed.), *Sochtheangeolaíocht na Gaeilge*. Léachtaí Cholm Cille 39, 62–89.

Hickey, Tina (2007), Fluency in reading Irish as L1 or L2: Promoting high-frequency word recognition in emergent readers. *International Journal of Bilingual Education and Bilingualism* 10: 4, 471–493.

Hirai, Seiko (2002), Longitudinal L2 attrition: A case study of a Japanese-English bilingual child. *Japan Journal of Multilingualism and Multiculturalism* 8, 26–57.

Hourihane, Jim & Keane, Margaret (2004), Geography in Ireland. *Belgeo* 1, 95–121.

Hughes, Art J. (2001), Advancing the language: Irish in the twenty-first century. *New Hibernia Review* 5: 1, 101–126.

Hulsen, Madeleine (2000), *Language Loss and Language Processing: Three Generations of Dutch Migrants in New Zealand.* PhD dissertation, Nijmegen, The Netherlands: Katholieke Universiteit Nijmegen.

Hulstijn, Jan & Laufer, Batia (2001), Some empirical evidence for the involvement load hypothesis in vocabulary acquisition. *Language Learning* 51: 3, 539–558.

In'nami, Yo (2013), Second-language vocabulary assessment research: Issues and challenges. *Vocabulary Learning and Instruction* 2: 1, 74–81.

Isurin, Ludmila (1999), *Mechanisms of First Language Forgetting*. PhD dissertation, Louisiana: Louisiana State University.

Isurin, Ludmila & Seidel, Christy (2015), Traces of memory for a lost childhood language: The Savings Paradigm expanded. *Language Learning* 65: 4, 761–790.

Ivanova, Iva & Costa, Albert (2008), Does bilingualism hamper lexical access in speech production? *Acta Psychologica* 127, 277–288.

Jakobson, Roman (1968), *Child Language, Aphasia and Phonological Universals, Vol. 72*. Berlin: Walter de Gruyter GmbH & Co KG.

Jaspaert, Koen & Kroon, Sjaak (1989), Social determinants of language loss. *Review of Applied Linguistics (I.T.L.)* 83/84, 75–98.

Jordan, Geoffrey (2004), *Theory Construction in Second Language Acquisition*. Philadelphia: John Benjamins.

Kanno, Yasuko (2003), *Negotiating Bilingual and Bicultural Identities: Japanese Returnees Betwixt Two Worlds*. Mahwah, NJ: Lawrence Erlbaum.

Keijzer, Merel (2007), *Last in First Out? An Investigation of the Regression Hypothesis in Dutch Emigrants in Anglophone Canada*. PhD thesis, Amsterdam: Vrije Universiteit.

Kennedy, Ivan (2012), *Irish Medium Education: Cognitive Skills, Linguistic Skills, and Attitudes towards Irish*. Unpublished PhD thesis, Bangor: College of Education and Lifelong Learning, Bangor University.

Koike, I. (1989), *Descriptive Analysis of Loss of English as a Second Language.* Paper presented at the 28th JACET Conference, Fukuoka, Japan.

Köpke, Barbara (2002), Activation thresholds and non-pathological first language attrition. In Fabbro, Franco (ed.), *Advances in the Neurolinguistics of Bilingualism.* Udine: Forum, 119–142.

Köpke, Barbara (2004), Neurolinguistic aspects of attrition. *Journal of Neurolinguistics* 17: 1, 3–30.

Köpke, Barbara (2007), Language attrition at the crossroads of brain, mind, and society. In: Köpke, Barbara; Schmid, Monika S.; Keijzer, Merel & Dostert Susan (eds.), *Language Attrition: Theoretical Perspectives.* Amsterdam: John Benjamins, 9–38.

Köpke, Barbara & Keijzer, Merel (2019), Introduction to psycholinguistic and neurolinguistic approaches to language attrition. In: Schmid, Monika S. & Köpke, Barbara (eds.), *The Oxford Handbook of Language Attrition.* Oxford: Oxford University Press, 63–72.

Köpke, Barbara & Schmid, Monika S. (2004), Language attrition: The next phase. In: Schmid, Monika S.; Köpke, Barbara; Keijzer, Merel & Weilemar Lina (eds.), *First Language Attrition: Interdisciplinary Perspectives on Methodological Issues.* Amsterdam: John Benjamins, 1–43.

Kuhberg, Heinz (1992), Longitudinal L2-attrition versus L2-acquisition in three Turkish children: Empirical findings. *Second Language Research* 8: 2, 138–154.

Larson-Hall, Jenifer (2019), L2 lexical attrition. In: Schmid, Monika S. & Köpke, Barbara (eds.), *The Oxford Handbook of Language Attrition.* UK: Oxford University Press, 1–15.

Li, Jia (2010), Learning vocabulary via computer-assisted scaffolding for text processing. *Computer Assisted Language Learning* 23: 3, 253–275.

Li, Lan; MacSwiney Brugha, Cathal & Gallagher, Mary (2017), Protecting endangered languages: The case of Irish. *Studies in Arts and Humanities* 3: 2, 109–130.

Loftus, Geoffrey & Loftus, Elizabeth (1976), *Human Memory: The Processing of Information.* Hillsdale, NJ: Erlbaum.

Lowie, Wander; Van Dijk, Marijn; Chan, Huiping & Verspoor, Marjolijn (2017), Finding the key to successful L2 learning in groups and individuals. *Studies in Second Language Learning and Teaching* 7: 1, 127–148.

Masgoret, Anne-Marie & Gardner, Robert C. (2003), Attitudes, motivation, and second language learning: A meta-analysis of studies conducted by Gardner and associates. *Language Learning* 53: 1, 123–163.

McCubbin, Justin (2010), Irish-language policy in a multi-ethnic state: Competing discourses on ethno-cultural membership and language ownership. *Journal of Multilingual and Multicultural Development* 31: 5, 457–478.

Meara, Paul (1989), Matrix models of vocabulary acquisition. *AILA Review*, 6, 66–74. http://www.aila.info/download/publications/review/AILA06.pdf (accessed March 29, 2021).

Meara, Paul (1992), *EFL Vocabulary Test*. Swansea, UK: Centre for Applied Language Studies.

Meara, Paul (2004), Modelling vocabulary loss. *Applied Linguistics* 25, 137–155.

Meara, Paul & Buxton, Barbara (1987), An alternative to multiple choice vocabulary tests. *Language Testing* 4, 142–51.

Mehotcheva, Teodora H. (2010), *After the Fiesta Is Over: Foreign Language Attrition of Spanish in Dutch and German Erasmus Students*. PhD dissertation, The Netherlands: University of Groningen.

Mehotcheva, Teodora H. & Mytara, Kleopatra (2019), Exploring the impact of extralinguistic factors on L2/FL attrition. In: Schmid, Monika S. & Köpke, Barbara (eds.), *The Oxford Handbook of Language Attrition*. UK: Oxford University Press, 1–17.

Mickan, Anne (2021), *What Was That Spanish Word Again? Investigations into the Cognitive Mechanisms Underlying Foreign Language Attrition*. PhD dissertation, Nijmegen, The Netherlands: Max Planck Institute for Psycholinguistics.

Mickan, Anne; McQueen, James M. & Lemhöfer, Kristin (2020), Between-language competition as a driving force in foreign language attrition. *Cognition* 198, 1– 23.

Montrul, Silvina (2008), *Incomplete Acquisition in Bilingualism: Re-examining the Age Factor*. Amsterdam: Benjamins.

Montrul, Silvina (2009), "Lexical knowledge and access in Spanish heritage speakers: Theoretical and pedagogical issues." Paper presented at the *Third Heritage Language Summer Institute*. University of Illinois, Urbana-Champaign.

Montrul, Silvina & Foote, Rebecca (2014), Age of acquisition interactions in bilingual lexical access: A study of the weaker language of L2 learners and heritage speakers. *International Journal of Bilingualism* 18: 3, 274–303.

Moseley, Christopher (ed.) (2010), *Atlas of the World's Languages in Danger*, 3rd ed. Paris: UNESCO Publishing. Online version: http://www.unesco.org/culture/en/endangeredlanguages/atlas (accessed March 29, 2021).

Murtagh, Lelia (2003), *Retention and Attrition of Irish as a Second Language. A Longitudinal Study of General and Communicative Proficiency in Irish among Second Level School Leavers and the Influence of Instructional Background, Language Use and Attitude/Motivation Variables.* Unpublished Ph.D. Dissertation, Groningen, The Netherlands: Rijksuniversiteit Groningen.

Murtagh, Lelia (2007), Out-of-school use of Irish, motivation and proficiency in immersion and subject- only post-primary programmes. *The International Journal of Bilingual Education and Bilingualism* 10: 4, 428–453.

Murtagh, Lelia & Van der Slik, Frans (2004), Retention of Irish skills: A longitudinal study of a school-acquired second language. *International Journal of Bilingualism* 8: 3, 279–302.

Nation, Paul (2001), *Learning Vocabulary in Another Language.* Cambridge: CUP.

Nation, Paul (2014), *What you need to know to learn a foreign language.* http://www.wgtn.ac.nz/lals/about/staff/publications/paul-nation/foreign-language_1125.pdf (Accessed March 29, 2021).

Nation, Paul & Webb, Stuart (2011), *Researching and Analyzing Vocabulary*. Boston, MA: Heinle, Cengage Learning.

Neisser, Ulric (1984), Interpreting Harry Bahrick's discovery: What confers immunity against forgetting? *Journal of Experimental Psychology: General* 113: 1, 32–35.

Nelson, Thomas (1978), Detecting small amounts of information in memory: Savings for non- recognized items. *Journal of Experimental Psychology: Human Learning and Memory* 4: 5, 453–68.

Nikitina, Larisa & Furuoka, Fumitaka (2005), Integrative motivation in a foreign language classroom: A study on the nature of motivation of the

Russian language learners in Universiti Malaysia Sabah. *Jurnal Kinabalu, Jurnal Perniagaan & Sains Sosial* 11, 23–34.

Noels, Kimberly A.; Clément, Richard & Pelletier, Luc (1999), Perceptions of teachers' communicative style and students' intrinsic and extrinsic motivation. *The Modern Language Journal* 83: 1, 23–34.

Noels, Kimberly A.; Pelletier, Luc; Clément, Richard & Vallerand, Robert (2003), Why are you learning a second language? Motivational orientations and self-determination theory. *Language Learning* 53: 1, 33–63.

O'Donovan, Bernadette (2013), *Primary School Teachers' Understanding of Themselves as Professionals*. Unpublished Doctor of Education dissertation, Dublin: Dublin City University.

Ó Glaisne, Risteard (1961), *Bun-Ghaeilge: A Concise Guide to Irish*. Baile Átha Cliath: Cara.

O'Grady, William & Lee, On-Soon (2011), Practical and theoretical issues in the study of heritage language acquisition. *Heritage Language Journal* 8, 23–40.

Oh, Janet; Jun, Sun-Ah; Knightly, Leah & Au, Terry (2003), Holding on to childhood language memory. *Cognition* 86: 3, 53–64.

Ó Laoire, Muiris (2000), Learning Irish for participation in the Irish language speech community outside the Gaeltacht. *Journal of Celtic Language Learning* 5, 20–33.

Ó Laoire, Muiris (2014, January 28), Education: How did we get the teaching of Irish so wrong? *The Irish Times*. https://www.irishtimes.com/news/education/education-how-did-we-get-the-teaching-of-irish-so-wrong-1.1667065 (accessed March 29, 2021).

Olshtain, Elite (1986), The attrition of English as a second language with speakers of Hebrew. In: Weltens, Bert; De Bot, Kees & Van Els, Theo (eds.), *Language Attrition in Progress*. Dordrecht: Foris, 187–204.

Olshtain, Elite (1989), Is second language attrition the reversal of second language acquisition? *Studies in Second Language Acquisition* 11, 151–165.

Ó Murchú, Helen (2001), The European charter for regional or minority languages: Ceart domsa ceart duitse? *Teanga* 20: 2, 110–143.

Ó Murchú, Helen (2016), *The Irish Language in Education in the Republic of Ireland. Mercator European Research Centre on Multilingualism and Language Learning.*

O Murchú, Máirtín (1977), Successes and failures in the modernization of Irish spelling. In: Fishman, Joshua (ed.), *Advances in the Creation and Revision of Writing Systems*. The Hague: Mouton, 267–289.

Opitz, Conny (2013), A dynamic perspective on late bilinguals' linguistic development in an L2 environment. *International Journal of Bilingualism* 17: 6, 701–715.

Opitz, Conny (2019), A complex dynamic systems perspective on personal background variables in L1 attrition. In: Schmid, Monika S. & Köpke, Barbara (eds.), *The Oxford Handbook of Language Attrition.* UK: Oxford University Press, 1–14.

Ó Riagáin, Pádraig (1997), *Language Policy and Social Reproduction: Ireland 1893–1993* (Oxford Studies in Language Contact). UK: Oxford University Press.

Ó Riagáin, Pádraig (2001), Irish production and reproduction 1981–1996. In: Fishman, Joshua (ed.), *Can Threatened Languages Be Saved? Reversing Language Shift Revisted: A 21st Century Perspective.* Clevedon: Multilingual Matters, 195–215.

Ó Riain, Seán (1994), *Pleanáil Teanga in Éirinn (Language Planning in Ireland) 1919–1985*. Baile Átha Cliath (Dublin): Carbad/Bord na Gaeilge.

Ó Riain, Seán (2004), "The Irish Language." Talk given in Maribor on behalf of Embassy of Ireland, Berlin. http://www.seanoriain.eu/Media/The%20Irish%20Language.pdf (accessed March 29, 2021).

Oskamp, Stuart (1977), *Attitudes and Opinions*. Englewood Cliffs, NJ: Prentice-Hall.

Oskamp, Stuart & Schultz, P. Wesley (2005), *Attitudes and Opinions.* Mahwah, NJ: L. Erlbaum Associates.

Pallier, Christophe (2007), Critical periods in language acquisition and language attrition. In: Köpke, Barbara; Schmid, Monika S.; Keijzer, Merel & Dostert, Susan (eds.), *Language Attrition: Theoretical Perspectives.* Amsterdam/Philadelphia: John Benjamins, 155–168.

Pallier, Christophe; Dehaene, Stanislas; Poline, Jean-Baptiste; Le Bihan, Denis; Argenti, Anne-Marie; Dupoux, Emmanuel & Mehler, Jacques

(2003), Brain imaging of language plasticity in adopted adults: Can a second language replace the first? *Cerebral Cortex* 13, 155–161.

Pan, B. Alexander & Berko-Gleason, Jean (1986), The study of language loss: Models and hypotheses for an emerging discipline. *Applied Psycholinguistics* 7, 193–206.

Paradis, Michel (2004), *A Neurolinguistic Theory of Bilingualism*. Amsterdam: John Benjamins.

Paradis, Michel (2007), L1 attrition features predicted by a neurolinguistic theory of bilingualism. In: Köpke, Barbara; Schmid, Monika S.; Keijzer, Merel & Dostert, Susan (eds.), *Language Attrition: Theoretical Perspectives*. Amsterdam/Philadelphia: John Benjamins, 121–135.

Paribakht, T. Sima & Wesche, Marjorie (1997), Vocabulary enhancement activities and reading for meaning in second language vocabulary acquisition. In: Coady, James & Huckin, Thomas (eds.), *Second Language Vocabulary Acquisition: A Rationale for Pedagogy*. Cambridge: Cambridge University Press, 174–200.

Pearson, Barbara (2007), Social factors in childhood bilingualism in the United States. *Applied Psycholinguistics* 28: 3, 399–410.

Pierce, Lara J.; Klein, Denise; Chen, Jen-Kai; Delcenserie, Audrey & Genesee, Fred (2014), Mapping the unconscious maintenance of a lost first language. *Proceedings of the National Academy of Sciences* 111: 48, 17314–17319.

Pintrich, Paul R. & Schunk, Dale H. (2007), *Motivation in Education: Theory, Research and Applications* (3rd ed.). Upper Saddle River, NJ: Merrill Prentice Hall.

Place, Silvia & Hoff, Erika (2011), Properties of dual language exposure that influence 2-year-olds' bilingual proficiency. *Child Development* 82, 1834–1849.

Polinsky, Maria (1997), American Russian: Language loss meets language acquisition. In: Browne, Wayles; Dornische, Ewa; Kondrashova, Natasha & Zec, Draga (eds.), *Annual Workshop on Formal Approaches to Slavic Linguistics. The Cornell Meeting, 1995*. Ann Arbor, MI: Michigan Slavic Publications, 370–406.

Polinsky, Maria (2007), Reaching the end point and stopping midway: Different scenarios in the acquisition of Russian. *Russian Linguistics* 31: 2, 157–199.

Reetz-Kurashige, Anita (1999), Japanese returnees' retention of English-speaking skills: Changes in verb usage over time. In: Hansen, Lynne (ed.), *Second Language Attrition in Japanese Contexts*. Oxford: Oxford University Press, 21–58.

Sanatullova-Allison, Elvira (2014), Memory retention in second language acquisition and instruction: Insights from literature and research. *The IAFOR Journal of Language Learning* 1: 1, 1–13.

Schmid, Monika S. (2002), *First Language Attrition, Use and Maintenance: The Case of German Jews in Anglophone Countries*. Amsterdam, The Netherlands: John Benjamins Publishing Company.

Schmid, Monika S. (2004), Language attrition research: An annotated bibliography. In: Schmid, Monika S.; Köpke, Barbara; Keijzer, Merel & Weilemar, Lina (eds.), *First Language Attrition: Interdisciplinary Perspectives on Methodological Issues*. Amsterdam: John Benjamins, 317–348.

Schmid, Monika S. (2006), Second language attrition. In: Brown, Keith (ed.), *The Encyclopedia of Language and Linguistics, Vol. 11*. Amsterdam: Elsevier, 74–81.

Schmid, Monika S. (2007), The role of L1 use for L1 attrition. In: Köpke, Barbara; Schmid, Monika S.; Keijzer, Merel & Dostert, Susan (eds.), *Language Attrition: Theoretical Perspectives*. Amsterdam/Philadelphia: John Benjamins, 135–154.

Schmid, Monika S. (2019), The impact of frequency of use and length of residence on L1 attrition. In: Schmid, Monika S. & Köpke, Barbara (eds.), *The Oxford Handbook of Language Attrition*. UK: Oxford University Press, 1–42.

Schmid, Monika S. (n.y.), Resources for researchers: Sociolinguistic questionnaires. https://languageattrition.org/resources-for-researchers/experiment-materials/ (accessed March 29, 2021).

Schmid, Monika S. & Dusseldorp, Elise (2010), Quantitative analyses in a multivariate study of language attrition. *Second Language Research* 26: 1, 125–160.

Schmid, Monika S. & Jarvis, Scott (2014), Lexical access and lexical diversity in first language attrition. *Bilingualism: Language and Cognition* 17: 4, 729–748.

Schmid, Monika S. & Keijzer, Merel (2009), First language attrition and reversion among older migrants. *International Journal of the Sociology of Language* 200, 83–101.

Schmid, Monika S. & Köpke, Barbara (2013), Second language acquisition and attrition. In: Schmid, Monika S. & Köpke, Barbara (eds.), *First Language Attrition*. Amsterdam: John Benjamins, 1–5.

Schmid, Monika S. & Mehotcheva, Teodora (2012), Foreign language attrition. *Dutch Journal of Applied Linguistics (DuJAL)* 1: 1, 102–124.

Schmidt, Richard (2001), Attention. In: Robinson, Peter (ed.), *Cognition and Second Language Instruction*. New York: Cambridge University Press, 3–32.

Schmitt, Norbert (2010), *Researching Vocabulary: A Vocabulary Research Manual*. Hampshire, UK: Palgrave Macmillan.

Schumann, John H. (1994), Where is cognition? Emotion and cognition in second language acquisition. *Studies in Second Language Acquisition* 16, 231–242.

Scontras, Gregory; Fuchs, Zuzanna & Polinsky, Maria (2015), Heritage language and linguistic theory. *Frontiers in Psychology* 6, 291–352.

Scorolli, Claudia; Binkofski, Ferdinand; Buccino, Giovanni; Nicoletti, Roberto; Riggio, Lucia & Borghi, Anna Maria (2011), Abstract and concrete sentences, embodiment, and languages. *Frontiers in Psychology* 2, 1–11.

Sharwood Smith, Michael A. (1983a), On first language loss in the second language acquirer: Problems of transfer. In: Gass, Susan & Selinker, Larry (eds.), *Language Transfer in Language Learning*. Rowley, MA: Newbury House, 222–231.

Sharwood Smith, Michael A. (1983b), On explaining language loss. In: Felix, Sascha W. & Wode, Henning (eds.), *Language Development at the Crossroads*. Tübingen: Gunter Narr Verlag, 49–59.

Sharwood Smith, Michael A. (1989), Crosslinguistic influence in language loss. In: Hyltenstam, Kenneth & Obler, Loraine (eds.), *Bilingualism Across the Lifespan*. Cambridge: Cambridge University Press, 185–201.

Sharwood Smith, Michael A. (2007), Understanding attrition within a MOGUL framework. In Köpke, Barbara; Schmid, Monika S.; Keijzer, Merel & Dostert, Susan (eds.), *Language Attrition: Theoretical Perspectives*. Amsterdam/Philadelphia: John Benjamins, 39–52.

Singleton, David (1987), Communicative needs: The case of Irish. In: Valdman, Albert (ed.), *Proceedings from the Symposium on the Evaluation of Foreign Language Proficiency*. Bloomington, Indiana: Indiana University, 79–83.

Singleton, David & Ryan, Lisa (2004), *Language Acquisition: The Age Factor*. Clevedon: Multilingual Matters Ltd.

Smyth, Emer & Darmody, Merike (2016), Attitudes to Irish as a school subject among 13-year-olds. The Economic and Social Research Institute (ESRI) working paper No. 525. https://www.econstor.eu/bitstream/10419/174258/1/WP525.pdf (accessed March 29, 2021).

Stahl, Steven A. & Nagy, William E. (2006), *Teaching Word Meanings*. Mahwah, NJ: Lawrence Erlbaum Associates.

Stanovich, Keith E. (1986), Matthew effects in reading: Some consequences of individual differences in the acquisition of literacy. *Reading Research Quarterly* 21, 360–407.

Stenson, Nancy & Hickey, Tina M. (2016), When regular is not easy: Cracking the code of Irish orthography. *Writing Systems Research* 8: 2, 187–217.

Stenson, Nancy & Hickey, Tina M. (2018), Understanding Irish Spelling: A Handbook for Teachers and Learners. An Chomhairle um Oideachas Gaeltachta agus Gaelscolaíochta. https://www.cogg.ie/wp-content/uploads/Understanding-Irish-Spelling-A-Handbook-for-Teachers-and-Learners-by-Dr.-Nancy-Stenson-and-Dr.-Tina-Hickey.pdf (accessed March 29, 2021).

Taniguchi, Joy (2014), Literacy retention and development in Japanese returnee children. *The Japanese Journal of Language in Society* 17: 1, 77–84.

Taura, Hideyuki (2008), *Language Attrition and Retention in Japanese Returnee Students*. Tokyo: Akashi Shoten.

Taura, Hideyuki (2019), Attrition studies on Japanese returnees. In: Schmid, Monika S. & Köpke, Barbara (eds.), *The Oxford Handbook of Language Attrition*. UK: Oxford University Press, 1–13.

Tees, Richard C. & Werker, Janet F. (1984), Perceptual flexibility: Maintenance or recovery of the ability to discriminate non-native speech sounds. *Canadian Journal of Experimental Psychology* 38, 579–590.

Tomiyama, Machiko (1999a), The first stage of second language attrition: A case study of a Japanese returnee. In: Hansen, Lynne (ed.), *Second Language Attrition in Japanese Contexts*. Oxford: Oxford University Press, 59–79.

Tomiyama, Machiko (1999b), The later stages of natural L2 attrition. In: Robinson, Peter (ed.), *Representation and Processes: Proceedings of the 3rd Pacific Second Language Forum, Vol. 1*. Tokyo: PacSLRF, 309–318.

Tomiyama, Machiko (2000), Child second language attrition: A longitudinal case study. *Applied Linguistics* 21: 3, 304–332.

Tomiyama, Machiko (2008), Age and proficiency in L2 attrition: Data from two siblings. *Applied Linguistics* 29: 1, 1–23.

Tremblay, Paul F. & Gardner, Robert C. (1995), Expanding the motivation construct in language learning. *The Modern Language Journal* 79: 4, 505–518.

Van den Broek, Gesa; Takashima, Atsuko; Segers, Eliane & Verhoeven, Ludo (2018), Contextual richness and word learning: Context enhances comprehension but retrieval enhances retention. *Language Learning* 68: 2, 546–585.

Van Geert, Paul (1994), A dynamic system model of cognitive and language growth. *Psychological Review* 98, 3–53.

Ventureyra, Valérie A. G.; Pallier, Christophe & Yoo, Hi-Yon (2004), The loss of first language phonetic perception in adopted Koreans. *Journal of Neurolinguistics* 17, 79–91.

Verkaik, P. & Van der Wijst, P. (1986), *Taalverlies en Woordherkenning in het Frans ls Vreemde Taal. [Language Attrition and Word Recognition in French as a Foreign Language.]* Master's thesis, Nijmegen, The Netherlands: University of Nijmegen.

Wang, Xiangyun (2010), *Patterns and Causes of Attrition of English as a Foreign Language*. PhD thesis, China: Shandong University.

Watson, Iarfhlaith & Nic Ghiolla Phádraigh, Máire (2009), Is there an educational advantage to speaking Irish? An investigation of the relationship between education and ability to speak Irish. *International Journal of the Sociology of Language* 199, 143–156.

Webb, Stuart (2005), Receptive and productive vocabulary learning – The effects of reading and writing on word knowledge. *Studies in Second Language Acquisition* 27: 1, 33–52.

Wei, Junyan (2014), Selectivity of second language attrition. *Theory and Practice in Language Studies* 4: 8, 1603–1608.

Weltens, Bert (1988), *The Attrition of French as a Foreign Language*. Dordrecht/Providence: Foris Publications.

Weltens, Bert & Grendel, Marjon (1993), Attrition of vocabulary knowledge. In: Schreuder, Robert & Weltens, Bert (eds.), *The Bilingual Lexicon*. Amsterdam: John Benjamins, 135–157.

Weltens, Bert; Van Els, Theo & Schils, Erik (1989), The long-term retention of French by Dutch students. *Studies in Second Language Acquisition* 11: 2, 205–216.

Wodinsky, Marilyn & Nation, Paul (1988), Learning from graded readers. *Reading in a Foreign Language* 5:1, 155–161.

Xu, Xiaoyan (2010), *English Language Attrition and Retention in Chinese and Dutch University Students*. PhD thesis, Groningen, The Netherlands: University of Groningen.

Yağmur, Kutlay (1997), *First Language Attrition Among Turkish Speakers in Sydney*. Tilburg: Tilburg University Press.

Yilmaz, Gülsen & Schmid, Monika S. (2018), First language attrition and bilingualism. In: Miller, David; Bayram, Fatih; Rothman, Jason & Serratrice, Ludovica (eds.), *Bilingual Cognition and Language: The State of the Science across Its Subfields*. Amsterdam/Philadelphia: John Benjamins, 225–250.

Yoshitomi, Asako (1999), On the loss of English as a L2 by Japanese returnee children. In: Hansen, Lynne (ed.), *Second Language Attrition in Japanese Contexts*. Oxford: Oxford University Press, 80–113.

Yukawa, Emiko (1998), *L1 Japanese Attrition and Regaining: Three Case Studies of Two Early Bilingual Children*. Tokyo: Kurosio Publishing.

Appendices

List of appendices
Appendix A: Test items: Session 1 (200 items)
Appendix B: Sociolinguistic questionnaire and interview schedule

All results, interview transcripts and original sociolinguistic questionnaire and interview format are available on request. Please contact the author.

Appendix A: Test items: Session 1 (200 items)

Concrete nouns		
1	house	teach
2	water	uisce
3	door	doras
4	boy	buachaill
5	way/path	slí / caoi
6	bread	arán
7	woman	Mná , bean
8	book	leabhar
9	sea/ocean	farraige
10	money	airgead
11	parent	tuismithóir
12	donkey	asal
13	people	daoine
14	brother	deartháir
15	child	leanbh / páiste
16	teacher	múinteoir
17	river	abhainn
18	star	réalta
19	wall	balla
20	turf/peat	móin
21	bike	rothar
22	island	oileán
23	hand	lámh
24	food	bia
25	moon	gealach
26	mouth	béal
27	light (noun)	solas
28	milk	bainne

Abstract nouns		
1	care/attention	aire
2	knowledge	eolas
3	family	clann/ teaghlach
4	year	bliain
5	authority	údarás
6	government	oireachtas
7	side (noun)	taobh
8	week	seachtain
9	riches/wealth	saibhreas
10	weather	aimsir
11	England	Sasana
12	Wednesday	Dé Céadaoin
13	Friday	Dé hAoine
14	minute	nóiméad
15	fool	amadán
16	shame (noun)	náire
17	(social) company	comhluadar
18	schoolbreak	sos
19	wind	gaoth
20	language	teanga
21	day	lá
22	night	oíche
23	country	tír
24	laughter	gáire
25	war	cogadh
26	noise	fuaim
27	trip/tour (noun)	turas
28	hunger	ocras

Concrete nouns		
29	potato	práta
30	heart	croí
31	neighbour	comharsa
32	will/testament	uacht
33	meat	feoil
34	composition / essay	aiste
35	priest	sagart
36	sun	grian
37	tree	crann
38	city	cathair
39	exam	scrúdú
40	daughter	iníon
41	floor	urlár
42	apple	úll
43	school	scoil
44	husband	fear céile
45	umbrella	scáth fearthainne
46	nurse	banaltra
47	present (noun)	bronntanas
48	stick	bata
49	fish	iasc
50	leg	cos

Abstract nouns		
29	surprise	ionadh
30	story	scéal
31	environment	timpeallacht
32	reference/allusion	tagairt
33	lifetime	saol
34	devil	diabhal
35	crowd	slua
36	law	dlí
37	colour	dath
38	happiness/pleasure	sóchas / sásamh
39	news	nuacht
40	hope (noun)	dóchas
41	heaven	neamh
42	spring	earrach
43	autumn	fómhar
44	fear/worry	eagla
45	power	smacht
46	winter	geimhreadh
47	fright	scanradh
48	relief	faoiseamh
49	question (noun)	ceist
50	blame / fault	locht

Verbs		
1	to be able to	féad
2	to see	feic
3	to open	oscail
4	to defend/protect	cosain
5	to get up/to rise	éirigh
6	to buy	ceannaigh
7	to appear	dealraigh
8	to lie (flat/down)	luigh
9	to cry	caoin
10	to call/cry out	glaoigh
11	to choose /select	roghnaigh
12	to think	smaoinigh
13	to hurry	brostaigh
14	to travel	taistil
15	to play (children)	súgradh
16	to put	cuir
17	to do	déan
18	to walk	siúil
19	to get/receive	faigh
20	to grow	fás
21	to eat	ith
22	to drink	ól
23	to hit	buail
24	to speak	labhair / can
25	to hear	clois / cluin
26	to sleep	codail
27	to go	téigh
28	to come	tar

Adjectives		
1	dry	tirim
2	strong	láidir
3	sad	brónach
4	naughty	dána
5	easy	éasca
6	finished (adj)	críochnaithe
7	horrible/terrible	uafásach
8	hot	te
9	wise/prudent	críonna
10	cloudy	scamallach
11	alive	beo
12	guilty	ciontach
13	generous/lavish	flaithiúil
14	clean	glan
15	dirty	salach
16	short	gearr
17	free/unrestained	saor
18	difficult	deacair
19	weak	lag
20	quiet	ciúin
21	long	fada
22	fast/quickly	luath
23	friendly	cairdiúil
24	empty	folamh
25	tired	tuirseach
26	brave	cróga
27	slow	mall
28	great	íontach

Verbs		
29	to rain	báistigh
30	to show	taispeáin
31	to move/to stir	corraigh
32	to start	tosaigh
33	to dance	damhsaigh
34	to run	rith
35	to bless	beannaigh
36	to ensure	cinntigh
37	to leave/go away	imigh
38	to read	léigh
39	to break	bris
40	to ask	iarr
41	to wash	nigh
42	to answer	freagair
43	to take	gabh
44	to wait	fan
45	to listen to	éist
46	to work	oibrigh
47	to follow	lean
48	to dwell/reside	cónaigh
49	to write	scríobh
50	to give/grant	tabhair

Adjectives		
29	cold	fuar
30	full	iomlán
31	poor	bocht
32	boring	leadránach
33	funny/amusing	greannmhar
34	correct	ceart
35	soft	bog
36	tall/high	ard
37	deep	domhain
38	wet/soaked	báite
39	dark	dorcha
40	controversial	conspóideach
41	early	moch
42	true	fíor
43	young	óg
44	tormented	cráite
45	wild	fiáin
46	heavy	trom
47	old	sean
48	last (adj)	déanach
49	ready	ullamh
50	national	náisiúnta

Appendix B: Questionnaire and Interview

*Sociolinguistic Questionnaire and Interview Schedule

Section I
Sociolinguistic Questionnaire

Section II
Interview

*For original presentation and format, as well as interview transcripts, please contact the author

Sociolinguistic Questionnaire and Interview Schedule

Section I: Sociolinguistic Questionnaire Schedule

Please complete the following:

A. Personal details

1. Name:
2. Place and Date of birth:
3. Where did you grow up?

B. Education - general

4. Primary School name and location/address:
5. Secondary School name and location/address:
6. In what year did you do your Leaving Certificate examination?
7. Where applicable: Name/s of third level education institution/s:
8. Third level education: Years in attendance (from - to):
9. Subject/s studied in third level education:
10. Qualification/s awarded:

C. Irish language education and exposure

11. What age were you and in which class did you start learning Irish at school?
12. What age were you when you finished official Irish instruction (please specify whether this was at school or at university and school year/level, example, Leaving Certificate)?
13. What grade did you get in Irish in the Irish Leaving Certificate examination? Please state whether it was the honours or ordinary (pass) course/paper.
14. Did you still have contact with Irish when you finished school (or university should you have studied Irish in third level education)? In what capacity? Elaborate where applicable.
15. When did you leave/move away from Ireland (year/age)?
16. After you moved away from Ireland, did you have any contact with Irish on return visits or whilst living abroad? In what capacity? Elaborate/give details.
17. Is this still the case or has this contact ceased? If so, *when* and *why*? Elaborate/give details.

D. Other language education

18. What other language/s did you learn at school/university?
19. In what period did you have these lessons/courses (school years/ age)?
20. If you studied any of these for the Leaving Certificate, what grade did you get? Please state whether it was the honours or ordinary (pass) course/paper.

Section II: Interview schedule

Please give your answers orally to the following questions and elaborate as much as possible.

You may take some time beforehand to take notes and gather your thoughts. There is room for notes under each question. Extra paper is also provided.

A. Home and Social Environment

1. Were you exposed to Irish before starting school?
   If yes, give details (where/context, approximate frequency, quantity and quality (type) of input)
2. Did other members of your family/friends speak Irish? Which ones? In what capacity?
3. Did they speak Irish to you directly/address you in Irish? Elaborate.

B. School/Education

4. Do you remember how often you had Irish lessons in primary school?
5. Do you remember how often you had Irish lessons in secondary school?
6. Do you remember how often you had Irish lessons at university? (*where applicable*)
7. Was Irish used at school beyond Irish lessons? Elaborate on the contexts in which Irish was used.
8. Can you remember if Irish was taught through the medium of Irish or English or were both languages used in class? Elaborate.
9. How would you divide up Irish / English time (approximate percentage ratio)?

10. Did you enjoy learning Irish at school? Why/why not? Elaborate on your experiences.
11. Do you remember how you felt about the content of the syllabus? Elaborate.
12. What do you remember about the learning situation and can you recall how you felt in class? What made you feel this way? Elaborate.
13. When you look back at your school experience of learning Irish, what do you feel overall?
14. What prompts this feeling?
15. Is this feeling different compared to what you feel when you think of other subjects (not just other language classes)? Elaborate.
16. Do you think that you took more from the Irish course than just language? What were/are your feelings towards the language itself?
17. What was your main motivation in learning Irish? Did you enjoy the learning process? Did you feel under pressure to achieve good grades? Did you see long-term benefits in knowing Irish?

C. Experiences in other language education

18. With respect to the other language/s that you learned at school/university, did you enjoy learning these languages more, the same or less than Irish? Why? Elaborate on your experiences.
19. What was your motivation in learning these languages? Were you more, the same or less motivated in learning them compared to learning Irish? Why? Elaborate.
20. Did you feel that the syllabi of these other language courses were more, the same, or less interesting than that of Irish? Elaborate.
21. Were the teaching methodologies and approaches different for these other languages? In what way? Elaborate.

D. Attitudes, Awareness and Identity

22. On having left school (or where applicable, university, should you have studied Irish) what were your feelings with respect to the Irish language?
23. Has your attitude to the Irish language changed over the years? Elaborate.

24. Do you feel that the Irish language is an important part of your identity? Has this changed over the years since you finished school and since you have moved away from Ireland?
25. Do you think that you can be truly Irish without ever having learnt the language? Elaborate.
26. Growing up in Ireland, did you see Irish as one of your first languages or did you view it as a second language?
27. How would you describe the attitude of your family to the Irish language?
28. How would you describe the attitude of your social surroundings to the Irish language (on your return visits to Ireland)?
29. On your return visits to Ireland, are you *aware* of the language? Do you pay attention to it when it is used in the media/social environment in general? Have you noticed any changes with respect to the frequency and quantity of Irish language use in your immediate familial and social environment and in Irish society in general compared to when you were growing up?
30. Do you think that you have forgotten a lot of your Irish or even all of it? In approximate terms, how much knowledge do you think you have lost? All, a lot, some, little, none?
31. Where applicable, how do you feel about losing knowledge of Irish?
32. On having finished Irish classes/studies, did you ever make a conscious effort to maintain your knowledge of Irish? Elaborate.
33. Was there a fear at home or in your immediate environment that English language knowledge might suffer due to focusing on Irish?
34. Is there anything that you would like to add pertaining to the above focus on your Irish language knowledge/learning experience/use/attitudes?

Thank you for your participation

Inquiries in Language Learning
Forschungen zu Psycholinguistik und Fremdsprachendidaktik

Edited by / herausgegeben von Christiane Bongartz and / und Jutta Rymarczyk

Vol. / Bd. 1 Bärbel Diehr /Jutta Rymarczyk (eds. / Hrsg.): Researching Literacy in a Foreign Language among Primary School Learners. Forschung zum Schrifterwerb in der Fremdsprache bei Grundschülern. 2010.

Vol. / Bd. 2 Alexander Heimes: *Psycholinguistic Thought Meets Sociocultural Theory*. Die integrativen Zusammenhänge von Fachmethodik und Fremdsprachenlernen im bilingualen (Geschichts-)Unterricht. 2011.

Vol. / Bd. 3 Markus Kötter/Jutta Rymarczyk (Hrsg.): Fremdsprachenunterricht in der Grundschule. Forschungsergebnisse und Vorschläge zu seiner weiteren Entwicklung. 2011.

Vol. / Bd. 4 Nicole Ogasa: Gefühle und Lernen im Fremdsprachenunterricht. Der Einfluss von Gefühlen auf das Lernen. 2011.

Vol. / Bd. 5 Henning Rossa: Mentale Prozesse beim Hörverstehen in der Fremdsprache. Eine Studie zur Validität der Messung sprachlicher Kompetenzen. 2012.

Vol. / Bd. 6 Udo Ohm / Christiane Bongartz (Hrsg.): Soziokulturelle und psycholinguistische Untersuchungen zum Zweitspracherwerb. Ansätze zur Verbindung zweier Forschungsparadigmen. 2012.

Vol. / Bd. 7 Bärbel Diehr / Lars Schmelter (Hrsg.): Bilingualen Unterricht weiterdenken. Programme, Positionen, Perspektiven. 2012.

Vol. / Bd. 8 Jutta Rymarczyk (ed.): Foreign Language Learning Outside School. Places to See, Learn and Enjoy. 2013.

Vol. / Bd. 9 Sabrina Bechler: Bilinguale Module in der Grundschule. Integriertes Inhalts- und Sprachlernen im Fächerverbund Mensch, Natur und Kultur. 2014.

Vol. / Bd. 10 Michał B. Paradowski (ed.): Teaching Languages off the Beaten Track. 2014.

Vol. / Bd. 11 Christiane Schöneberger: Grammar Growth in Child Second Language German. Investigating DP Development in an Immersion Setting. 2015.

Vol. / Bd. 12 Stéfanie Witzigmann / Jutta Rymarczyk (Hrsg.): Mehrsprachigkeit als Chance. Herausforderungen und Potentiale individueller und gesellschaftlicher Mehrsprachigkeit. 2015.

Vol. / Bd. 13 Martina Weitz: Die Rolle des L2-Inputs in bilingualen Kindergärten. 2015.

Vol. / Bd. 14 Markus Kötter / Jutta Rymarczyk (Hrsg.): Englischunterricht auf der Primarstufe. Neue Forschungen – weitere Entwicklungen. 2015.

Vol. / Bd. 15 Sylvie Méron-Minuth / Senem Özkul (Hrsg.): Fremde Sprachen lehren und lernen. Aktuelle Fragen und Forschungsaufgaben. 2015.

Vol. / Bd. 16 Sviatlana Karpava: Vulnerable Domains for Cross-Linguistic Influence in L2 Acquisition of Greek. 2015.

Vol. / Bd. 17 Christiane M. Bongartz / Andreas Rohde (Hrsg.): Inklusion im Englischunterricht. 2015.

Vol. / Bd. 18 Bärbel Diehr / Angelika Preisfeld / Lars Schmelter (Hrsg.): Bilingualen Unterricht weiterentwickeln und erforschen. 2016.

Vol. / Bd. 19 Jan Springob: Inklusiver Englischunterricht am Gymnasium. Evidenz aus der Schulpraxis im Spiegel von Spracherwerbstheorie und Fremdsprachendidaktik. 2017.

Vol. / Bd. 20 Alexander Kautzsch: The Attainment of an English Accent. British and American Features in Advanced German Learners. 2017.

Vol. / Bd. 21 Jaewon Nielbock-Yoon: A Case Study on Trilingual Siblings' Code Switching. Focus on Minority Language Development. 2017.

Vol. / Bd. 22 Stefanie Rottschäfer: Accent and Identity in Learner Varieties of English. A Study with German and French University Students in an English as a Lingua Franca Setting. 2018.

Vol. / Bd. 23 Almut Ketzer-Nöltge: Satzverarbeitung in der Erst- und Zweitsprache Deutsch. Gedächtnisleistung und kognitive Belastung. 2019.

Vol. / Bd. 24 Eva M. Knopp: From Bilingual to Biliterate: Secondary Discourse Abilities in Bilingual Children's Story Telling. Evidence from Greek Heritage Language Speakers in Germany and the United States. 2019.

Vol. / Bd. 25 Katharina Hirt: Genustransfer durch «Thinking for Speaking» Kognitive Muster und ihre Bedeutung für den DaF-Unterricht. 2019.

Vol. / Bd. 26 Holger Limberg / Karen Glaser (Hrsg.): Pragmatische Kompetenzen im schulischen Fremdsprachenunterricht. 2020.

Vol. / Bd. 27 Christiane Bongartz / Jacopo Torregrossa (eds.): What's in a Narrative? Variation in Storytelling at the Interface Between Language and Literacy. 2020.

Vol. / Bd. 28 Maria Andreou: Language and Cognitive Aspects of Child Bilingualism. Research Observations and Classroom Applications. 2021.

Vol. / Bd. 29 Sandra Bellet: Gesamtsprachliche Bildung und früher Englischunterricht. Professionalisierung von Lehramtsstudierenden im Kontext von Mehrsprachigkeit. 2022.

Vol. / Bd. 30 Stefanie Frisch / Jutta Rymarczyk (eds.): Current Research into Young Foreign Language Learners' Literacy Skills. 2021.

Vol. / Bd. 31 Kim-Sarah Schick / Andreas Rohde (Hrsg.): Von integrativem zu inklusivem Englischunterricht. Weiterentwicklung sprachdidaktischer Prinzipien vor dem Hintergrund sonderpädagogischer Förderung. 2022.

Vol. / Bd. 32 Nadine Kolb: Child Second Language Development in Immersion Education. A Study on Generic Determiner Phrases in L2 German and L2 French. 2022.

Vol ./ Bd. 33 Eimear Geary: Second Language Attrition and the Case of Irish. An Exploration of the Savings Paradigm with Respect to Lexical Item Knowledge. 2022.